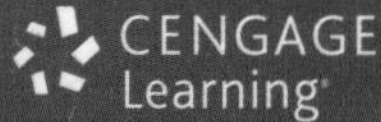
CENGAGE
Learning

Options That **SAVE** Your Students Money

MEDIA AND TECHNOLOGY

UP TO 60% OFF

Increased Engagement.
Improved Outcomes.
Superior Service.

Larson's proven pedagogy and exercise sets become even more effective in Enhanced WebAssign®, the exclusive Cengage Learning online homework solution that allows students to develop a deeper conceptual understanding of mathematics.

Learning resources include:

- Cengage YouBook, an interactive and customizable eBook
- Personalized Study Plan
- Show My Work option that allows instructors to see students' step by step solutions
- Math Evaluator functionality that accepts equivalent answers

Multi Term Access ISBN: 978-0-538-73807-1
Single Term Access ISBN: 978-0-538-73808-8

ACCESS, RENT, SAVE, AND ENGAGE

UP TO 60% OFF

CENGAGE brain .com
Buy direct and save!

Students can save up to 60% on their course materials from purchasing directly from CengageBrain.com.

Purchase print textbooks, eTextbooks, or individual eChapters and supplemental materials all for substantial savings over average retail prices.

CengageBrain.com also includes single sign-on access to Cengage Learning's broad range of online homework and study tools, including a selection of free or sample content to preview before purchasing.

Additional formats available on CengageBrain.com include:

- eTextbooks
- eChapters
- Textbook rental

Visit **CengageBrain.com** to save more!

CUSTOM SOLUTIONS

FIT EVERY NEED

Available books in the series include:

- Elementary Algebra
- Intermediate Algebra
- Elementary and Intermediate Algebra

Contact your Cengage Learning representative to learn more about what custom solutions are available to meet your course needs.

Custom options can include:

- Adapting existing Cengage Learning content by adding or removing content
- Adapting media and technology resources to better match the course
- Incorporate unique instructor materials
- Add student solutions to the text

CourseSmart

Cengage Learning and CourseSmart: Real Value for You and Your Students
Evaluate eTextbooks at **www.coursesmart.com**.

Source Code: 14M-MA0028

CourseCare provides exceptional service and training support to integrate media and technology tools into your course. Our dedicated team of digital solutions experts will ensure you, your colleagues and your students have a successful and engaging course experience. Learn more at cengage.com/coursecare

TRANSFORMING LEARNING TRANSFORMING LIVES

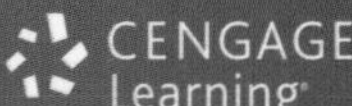

Annotated Instructor's Edition

INTERMEDIATE
Algebra Within Reach

SIXTH EDITION

Annotated Instructor's Edition

INTERMEDIATE Algebra Within Reach

SIXTH EDITION

Ron Larson
The Pennsylvania State University
The Behrend College

With the assistance of Kimberly Nolting
Hillsborough Community College

BROOKS/COLE
CENGAGE Learning™

Australia • Brazil • Japan • Korea • Mexico • Singapore • Spain • United Kingdom • United States

Intermediate Algebra Within Reach, Sixth Edition

Ron Larson

Senior Publisher: Charlie Van Wagner
Senior Acquiring Sponsoring Editor: Marc Bove
Associate Development Editor: Stefanie Beeck
Assistant Editor: Lauren Crosby
Senior Editorial Assistant: Jennifer Cordoba
Media Editor: Bryon Spencer
Senior Brand Manager: Gordon Lee
Senior Market Development Manager: Danae April
Content Project Manager: Jill Quinn
Manufacturing Planner: Doug Bertke
Rights Acquisition Specialist: Shalice Shah-Caldwell
Text and Cover Designer: Larson Texts, Inc.
Cover Image: front cover, ©iStockphoto.com/Christopher Futcher; back cover, ©iStockphoto.com/4x6
Compositor: Larson Texts, Inc.

Library of Congress Control Number: 2012947671

Student Edition:
ISBN-13: 978-1-285-08741-2
ISBN-10: 1-285-08741-0

Annotated Instructor's Edition:
ISBN-13: 978-1-285-16028-3
ISBN-10: 1-285-16028-2

Brooks/Cole
20 Channel Center Street
Boston, MA 02210
USA

Cengage Learning is a leading provider of customized learning solutions with office locations around the globe, including Singapore, the United Kingdom, Australia, Mexico, Brazil and Japan. Locate your local office at **International.cengage.com/region**

Cengage Learning products are represented in Canada by Nelson Education, Ltd.

For your course and learning solutions, visit **www.cengage.com**.

Purchase any of our products at your local college store or at our preferred online store **www.cengagebrain.com**.

Instructors: Please visit **login.cengage.com** and log in to access instructor-specific resources.

Printed in the United States of America

1 2 3 4 5 6 7 16 15 14 13 12

Contents

9 ▶ EXPONENTIAL AND LOGARITHMIC FUNCTIONS 471

10 ▶ CONICS 531

11 ▶ SEQUENCES, SERIES, AND THE BINOMIAL THEOREM 575

APPENDICES

*Available at the text-specific website *www.cengagebrain.com*

Preface

Welcome to *Intermediate Algebra Within Reach*, Sixth Edition. I am proud to present to you this new edition. As with all editions, I have been able to incorporate many useful comments from you, our user. And, while much has changed with this revision, you will still find what you expect—a pedagogically sound, mathematically precise, and comprehensive textbook.

I'm very excited about this edition. As I was writing, I kept one thought in mind—provide students what they need to learn algebra *within reach*. As you study from this book, you should notice right away that something is different. I've structured the book so that examples and exercises are on the same page—*within reach*. I am also offering something brand new with this edition: a companion website at **AlgebraWithinReach.com**. This site offers many resources that will help you as you study algebra. All of these resources are just a click away—*within reach*.

My goal for every edition of this textbook is to provide students with the tools that they need to master algebra. I hope that you find the changes in this edition, together with AlgebraWithinReach.com, will accomplish just that.

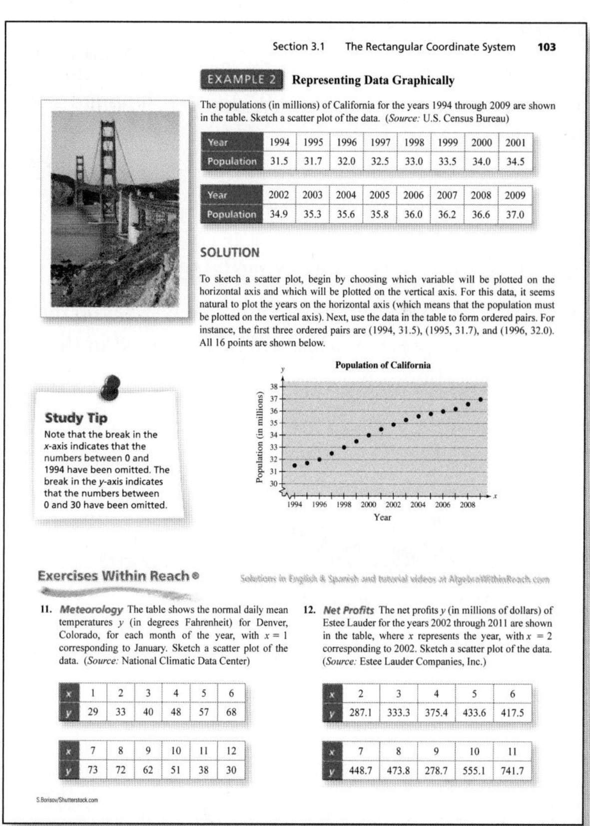

Section 3.1 The Rectangular Coordinate System 103

EXAMPLE 2 **Representing Data Graphically**

The populations (in millions) of California for the years 1994 through 2009 are shown in the table. Sketch a scatter plot of the data. (*Source:* U.S. Census Bureau)

Year	1994	1995	1996	1997	1998	1999	2000	2001
Population	31.5	31.7	32.0	32.5	33.0	33.5	34.0	34.5

Year	2002	2003	2004	2005	2006	2007	2008	2009
Population	34.9	35.3	35.6	35.8	36.0	36.2	36.6	37.0

SOLUTION

To sketch a scatter plot, begin by choosing which variable will be plotted on the horizontal axis and which will be plotted on the vertical axis. For this data, it seems natural to plot the years on the horizontal axis (which means that the population must be plotted on the vertical axis). Next, use the data in the table to form ordered pairs. For instance, the first three ordered pairs are (1994, 31.5), (1995, 31.7), and (1996, 32.0). All 16 points are shown below.

Study Tip

Note that the break in the x-axis indicates that the numbers between 0 and 1994 have been omitted. The break in the y-axis indicates that the numbers between 0 and 30 have been omitted.

Exercises Within Reach® Solutions in English & Spanish and tutorial videos at AlgebraWithinReach.com

11. ***Meteorology*** The table shows the normal daily mean temperatures y (in degrees Fahrenheit) for Denver, Colorado, for each month of the year, with $x = 1$ corresponding to January. Sketch a scatter plot of the data. (*Source:* National Climatic Data Center)

x	1	2	3	4	5	6
y	29	33	40	48	57	68

x	7	8	9	10	11	12
y	73	72	62	51	38	30

12. ***Net Profits*** The net profits y (in millions of dollars) of Estee Lauder for the years 2002 through 2011 are shown in the table, where x represents the year, with $x = 2$ corresponding to 2002. Sketch a scatter plot of the data. (*Source:* Estee Lauder Companies, Inc.)

x	2	3	4	5	6
y	287.1	333.3	375.4	433.6	417.5

x	7	8	9	10	11
y	448.7	473.8	278.7	555.1	741.7

S.Borisov/Shutterstock.com

New To This Edition

REVISED Exercises Within Reach

The exercise sets have been carefully and extensively reviewed to ensure they are relevant and cover all topics suggested by our users. Additionally, the exercises have been completely restructured. Exercises now appear on the *same* page and immediately follow a corresponding example. There is no need to flip back and forth from example to exercise. The end-of-section exercises focus on mastery of conceptual understanding. View and listen to worked-out solutions at AlgebraWithinReach.com.

NEW Data Spreadsheets

Download editable spreadsheets from AlgebraWithinReach.com, and use this data to solve exercises.

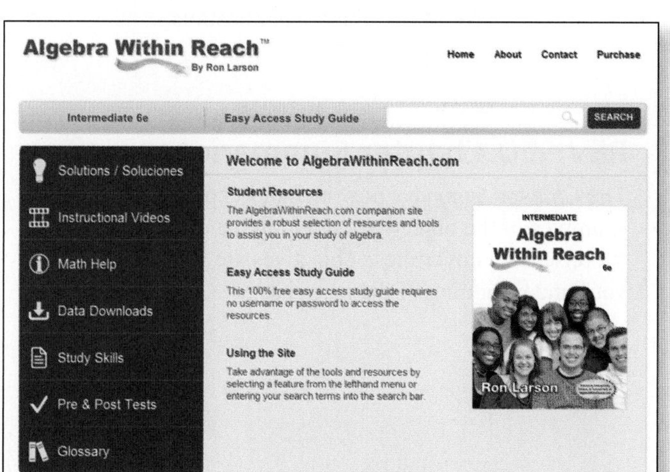

NEW AlgebraWithinReach.com

This companion website offers multiple tools and resources to supplement your learning. Access to these features is free. View and listen to worked-out solutions of thousands of exercises in English or Spanish, download data sets, take diagnostic tests, watch lesson videos and much more.

NEW Concept Summary

This simple review of important concepts appears at the end of every section. Each Concept Summary reviews *What*, *How*, and *Why*—what concepts you studied, how to apply the concepts, and why the concepts are important. The Concept Summary includes four exercises to check your understanding.

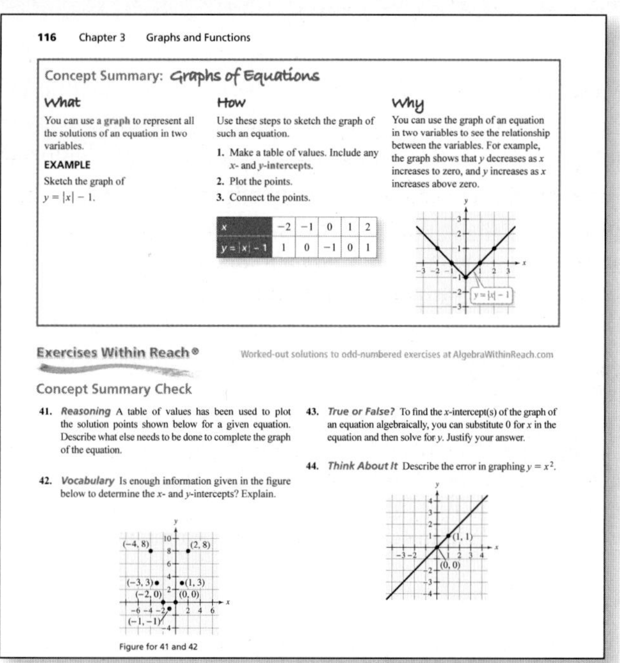

116 Chapter 3 Graphs and Functions

Concept Summary: Graphs of Equations

What

You can use a **graph** to represent all the solutions of an equation in two variables.

EXAMPLE

Sketch the graph of $y = |x| - 1$.

How

Use these steps to sketch the graph of such an equation.

1. Make a table of values. Include any x- and y-**intercepts**.
2. Plot the points.
3. Connect the points.

x	-2	-1	0	1	2
$y = \|x\| - 1$	1	0	-1	0	1

Why

You can use the graph of an equation in two variables to see the relationship between the variables. For example, the graph shows that y decreases as x increases to zero, and y increases as x increases above zero.

Exercises Within Reach® Worked-out solutions to odd-numbered exercises at AlgebraWithinReach.com

Concept Summary Check

41. ***Reasoning*** A table of values has been used to plot the solution points shown below for a given equation. Describe what else needs to be done to complete the graph of the equation.

42. ***Vocabulary*** Is enough information given in the figure below to determine the x- and y-intercepts? Explain.

Figure for 41 and 42

43. ***True or False?*** To find the x-intercept(s) of the graph of an equation algebraically, you can substitute 0 for x in the equation and then solve for y. Justify your answer.

44. ***Think About It*** Describe the error in graphing $y = x^2$.

NEW Math Helps

Additional instruction is available for every example and many exercises at AlgebraWithinReach.com. Just click on *Math Help*.

REVISED Section Objectives

A bulleted list of learning objectives provides you the opportunity to preview what will be presented in the upcoming section.

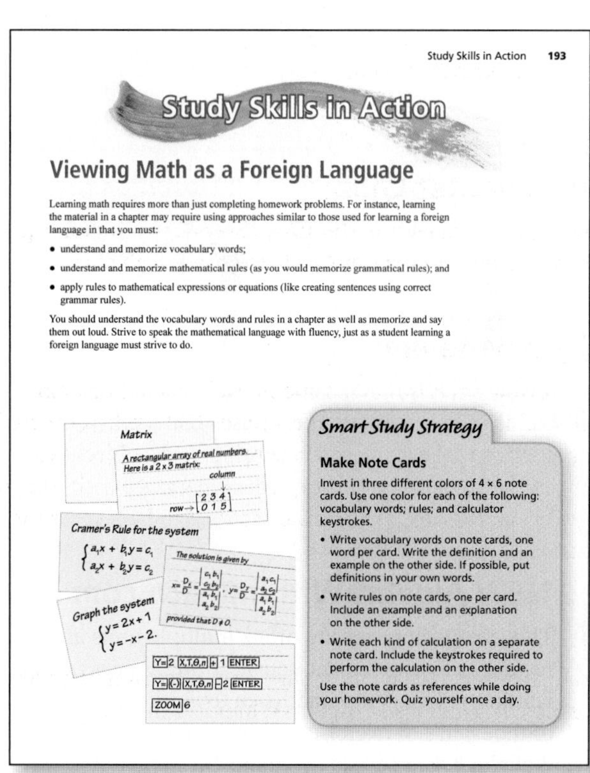

Study Skills in Action 193

Study Skills in Action

Viewing Math as a Foreign Language

Learning math requires more than just completing homework problems. For instance, learning the material in a chapter may require using approaches similar to those used for learning a foreign language in that you must:

- understand and memorize vocabulary words;
- understand and memorize mathematical rules (as you would memorize grammatical rules); and
- apply rules to mathematical expressions or equations (like creating sentences using correct grammar rules).

You should understand the vocabulary words and rules in a chapter as well as memorize and say them out loud. Strive to speak the mathematical language with fluency, just as a student learning a foreign language must strive to do.

Smart Study Strategy

Make Note Cards

Invest in three different colors of 4 × 6 note cards. Use one color for each of the following: vocabulary words; rules; and calculator keystrokes.

- Write vocabulary words on note cards, one word per card. Write the definition and an example on the other side. If possible, put definitions in your own words.
- Write rules on note cards, one per card. Include an example and an explanation on the other side.
- Write each kind of calculation on a separate note card. Include the keystrokes required to perform the calculation on the other side.

Use the note cards as references while doing your homework. Quiz yourself once a day.

REVISED Study Skills in Action

Each chapter presents a study skill essential to success in mathematics. Read and apply these throughout the course. Print them out at AlgebraWithinReach.com to keep them as reminders to develop strong study skills.

REVISED Applications

A wide variety of real-life applications are integrated throughout the text in examples and exercises. These applications demonstrate the relevance of algebra in the real world. Many of these applications use current, real data.

REVISED Chapter Summaries

The *Chapter Summary* now includes explanations and examples of the objectives taught in the chapter. Review exercises that cover these objectives are listed to check your understanding of the material.

Trusted Features

Examples

Each example has been carefully chosen to illustrate a particular mathematical concept or problem-solving technique. The examples cover a wide variety of problems and are titled for easy reference. Many examples include detailed, step-by-step solutions with side comments, which explain the key steps of the solution process.

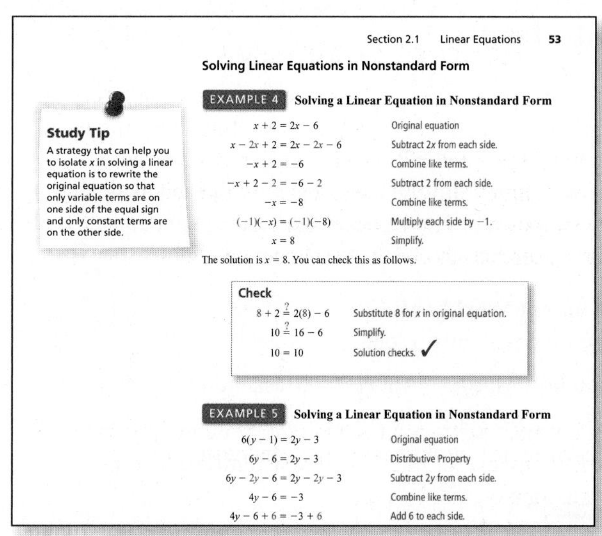

Section 2.1 Linear Equations 53

Solving Linear Equations in Nonstandard Form

Study Tip

A strategy that can help you to isolate x in solving a linear equation is to rewrite the original equation so that only variable terms are on one side of the equal sign and only constant terms are on the other side.

EXAMPLE 4 **Solving a Linear Equation in Nonstandard Form**

$x + 2 = 2x - 6$	Original equation
$x - 2x + 2 = 2x - 2x - 6$	Subtract $2x$ from each side.
$-x + 2 = -6$	Combine like terms.
$-x + 2 - 2 = -6 - 2$	Subtract 2 from each side.
$-x = -8$	Combine like terms.
$(-1)(-x) = (-1)(-8)$	Multiply each side by -1.
$x = 8$	Simplify.

The solution is $x = 8$. You can check this as follows.

Check

$8 + 2 \stackrel{?}{=} 2(8) - 6$	Substitute 8 for x in original equation.
$10 \stackrel{?}{=} 16 - 6$	Simplify.
$10 = 10$	Solution checks. ✓

EXAMPLE 5 **Solving a Linear Equation in Nonstandard Form**

$6(y - 1) = 2y - 3$	Original equation
$6y - 6 = 2y - 3$	Distributive Property
$6y - 2y - 6 = 2y - 2y - 3$	Subtract $2y$ from each side.
$4y - 6 = -3$	Combine like terms.
$4y - 6 + 6 = -3 + 6$	Add 6 to each side.

Study Tips

Study Tips offer students specific point-of-use suggestions for studying algebra, as well as pointing out common errors and discussing alternative solution methods. They appear in the margins.

Technology Tips

Point-of-use instructions for using graphing calculators or software appear in the margins as *Technology Tips*. These features encourage the use of graphing technology as a tool for visualization of mathematical concepts, for verification of other solution methods, and for help with computations.

Cumulative Review

Each exercise set (except in Chapter 1) is followed by *Cumulative Review* exercises that cover concepts from previous sections. This serves as a review and also a way to connect old concepts with new concepts.

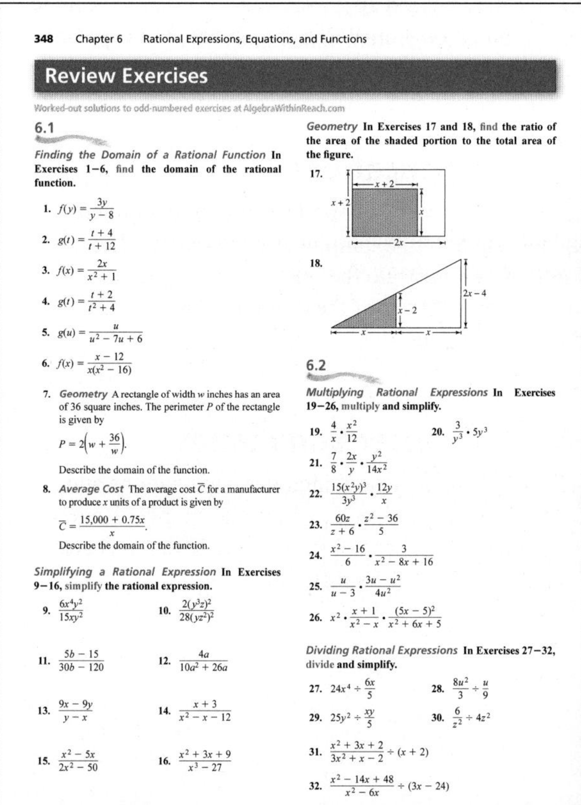

348 Chapter 6 Rational Expressions, Equations, and Functions

Review Exercises

Worked-out solutions to odd-numbered exercises at AlgebraWithinReach.com

6.1

Finding the Domain of a Rational Function **In Exercises 1–6, find the domain of the rational function.**

1. $f(y) = \frac{3y}{y - 8}$
2. $g(t) = \frac{t + 4}{t + 12}$
3. $f(x) = \frac{2x}{x^2 + 1}$
4. $g(t) = \frac{t + 2}{t^2 + 4}$
5. $g(u) = \frac{u}{u^2 - 7u + 6}$
6. $f(x) = \frac{x - 12}{x(x^2 - 16)}$
7. *Geometry* A rectangle of width w inches has an area of 36 square inches. The perimeter P of the rectangle is given by $P = 2\left(w + \frac{36}{w}\right)$. Describe the domain of the function.
8. *Average Cost* The average cost $\overline{C}$ for a manufacturer to produce x units of a product is given by $\overline{C} = \frac{15,000 + 0.75x}{x}$. Describe the domain of the function.

Simplifying a Rational Expression **In Exercises 9–16, simplify the rational expression.**

9. $\frac{6x^4y^2}{15xy^2}$
10. $\frac{2(y^3z)^2}{28(yz^2)^2}$
11. $\frac{5b - 15}{30b - 120}$
12. $\frac{4a}{10a^2 + 26a}$
13. $\frac{9x - 9y}{y - x}$
14. $\frac{x + 3}{x^2 - x - 12}$
15. $\frac{x^2 - 5x}{2x^2 - 50}$
16. $\frac{x^2 + 3x + 9}{x^3 - 27}$

Geometry **In Exercises 17 and 18, find the ratio of the area of the shaded portion to the total area of the figure.**

17\.

18\.

6.2

Multiplying Rational Expressions **In Exercises 19–26, multiply and simplify.**

19. $\frac{4}{x} \cdot \frac{x^2}{12}$
20. $\frac{3}{y^3} \cdot 5y^3$
21. $\frac{7}{8} \cdot \frac{2x}{y} \cdot \frac{y^2}{14x^2}$
22. $\frac{15(x^2y)^3}{3y^3} \cdot \frac{12y}{x}$
23. $\frac{60z}{z + 6} \cdot \frac{z^2 - 36}{5}$
24. $\frac{x^2 - 16}{6} \cdot \frac{3}{x^2 - 8x + 16}$
25. $\frac{u}{u - 3} \cdot \frac{3u - u^2}{4u^2}$
26. $x^2 \cdot \frac{x + 1}{x^2 - x} \cdot \frac{(5x - 5)^2}{x^2 + 6x + 5}$

Dividing Rational Expressions **In Exercises 27–32, divide and simplify.**

27. $24x^4 \div \frac{6x}{5}$
28. $\frac{8u^2}{3} \div \frac{u}{9}$
29. $25y^2 \div \frac{xy}{5}$
30. $\frac{6}{z^2} \div 4z^2$
31. $\frac{x^2 + 3x + 2}{3x^2 + x - 2} \div (x + 2)$
32. $\frac{x^2 - 14x + 48}{x^2 - 6x} \div (3x - 24)$

Mid-Chapter Quiz

Each chapter contains a *Mid-Chapter Quiz*. View and listen to worked-out solutions at AlgebraWithinReach.com.

Chapter Review

The *Review Exercises* at the end of each chapter contain skill-building and application exercises that are first ordered by section, and then grouped according to the objectives stated at the start of the section. This organization allows you to easily identify the appropriate sections and concepts for study and review.

Chapter Test

Each chapter ends with a *Chapter Test*. View and listen to worked-out solutions at AlgebraWithinReach.com.

Cumulative Test

The *Cumulative Tests* that follow Chapters 4, 7, and 10 provide a comprehensive self-assessment tool that helps you check your mastery of previously covered material. View and listen to worked-out solutions at AlgebraWithinReach.com.

Supplements

Student

Student Solutions Manual

ISBN 978-1-285-41985-5

Author: Ron Larson

The Student Solutions Manual provides detailed, step-by-step solutions to all odd-numbered problems in both the section exercise sets and review exercises. It also contains detailed, step-by-step solutions to all Mid-Chapter Quiz, Chapter Test, and Cumulative Test questions.

Student Workbook

ISBN 978-1-285-41994-7

Author: Maria H. Andersen, Muskegon Community College

Get a head start! The Student Workbook contains assessments, activities and worksheets for classroom discussions, in-class activities, and group work.

ENHANCED WebAssign

Printed Access Card: 978-0-538-73810-1

Online Access Code: 978-1-285-18181-3

Enhanced WebAssign (assigned by the instructor) provides you with instant feedback on homework assignments. This online homework system is easy to use and includes helpful links to textbook sections, video examples, and problem-specific tutorials.

Instructor

Complete Solutions Manual

ISBN 978-1-285-41992-3

Author: Ron Larson

The Complete Solutions Manual provides detailed step-by-step solutions to all problems in the text. It contains Chapter and Final Exam test forms with answer keys as well as individual test items and answers for Chapters 1–11.

Instructor's Resource Binder

ISBN 978-0-538-73675-6

Author: Maria H. Andersen, Muskegon Community College

The Instructor's Resource Binder contains uniquely designed Teaching Guides, which include instruction tips, examples, activities, worksheets, overheads, and assessments with answers to accompany them.

Printed Access Card: 978-0-538-73810-1

Online Access Code: 978-1-285-18181-3

Exclusively from Cengage Learning, Enhanced WebAssign combines the exceptional mathematics content that you know and love with the most powerful online homework solution, WebAssign. Enhanced WebAssign engages students with immediate feedback, rich tutorial content, and interactive, fully customizable eBooks (YouBook), helping students to develop a deeper conceptual understanding of their subject matter. Online assignments can be built by selecting from thousands of text-specific problems or supplemented with problems from any Cengage Learning textbook.

PowerLecture with Examview

ISBN: 978-1-285-41996-1

Author: Ron Larson

This supplement provides the instructor with dynamic media tools for teaching. Create, deliver and customize tests (both print and online) in minutes with *Examview® Computerized Testing Featuring Algorithmic Equations*. Easily build solution sets for homework or exams using *Solution Builder*'s online solution manual. Microsoft® Powerpoint® lecture slides including *all* examples from the text, figures from the book, and easy-to-use PDF testbanks, in electronic format, are also included on this DVD-ROM.

Solution Builder

This online instructor database offers complete worked-out solutions to all exercises in the text, allowing you to create customized, secure solutions printouts (in PDF format) matched exactly to the problems you assign in class. For more information, visit www.cengage.com/solutionbuilder.

Acknowledgements

I would like to thank the many people who have helped me revise the various editions of this text. Their encouragement, criticisms, and suggestions have been invaluable.

Reviewers

Tom Anthony, *Central Piedmont Community College*
Tina Cannon, *Chattanooga State Technical Community College*
LeAnne Conaway, *Harrisburg Area Community College and Penn State University*
Mary Deas, *Johnson County Community College*
Jeremiah Gilbert, *San Bernadino Valley College*
Jason Pallett, *Metropolitan Community College-Longview*
Laurence Small, *L.A. Pierce College*
Dr. Azar Raiszadeh, *Chattanooga State Technical Community College*
Patrick Ward, *Illinois Central College*

My thanks to Robert Hostetler, The Behrend College, The Pennsylvania State University, David Heyd, The Behrend College, The Pennsylvania State University, and Patrick Kelly, Mercyhurst University, for their significant contributions to previous editions of this text.

I would also like to thank the staff of Larson Texts, Inc., who assisted in preparing the manuscript, rendering the art package, and typesetting and proofreading the pages and the supplements.

On a personal level, I am grateful to my spouse, Deanna Gilbert Larson, for her love, patience, and support. Also, a special thanks goes to R. Scott O'Neil.

If you have suggestions for improving this text, please feel free to write to me. Over the past two decades I have received many useful comments from both instructors and students, and I value these comments very much.

Ron Larson
Professor of Mathematics
Penn State University
www.RonLarson.com

Fundamentals of Algebra

MASTERY IS WITHIN REACH!

"I get distracted very easily. If I study at home, other things call out to me. My instructor suggested studying on campus before going home or to work. I didn't like the idea at first, but tried it anyway. After a few times, I realized that it was the best thing for me—I got things done and it took less time. I also did better on my next test."

Cathy
Music

See page 27 for suggestions about keeping a positive attitude.

1.1 The Real Number System

- Understand the set of real numbers and the subsets of real numbers.
- Use the real number line to order real numbers.
- Find the distance between two real numbers.
- Find the absolute value of a real number.

Real Numbers

The numbers you use in everyday life are called **real numbers**. They are classified into different categories, as shown at the right.

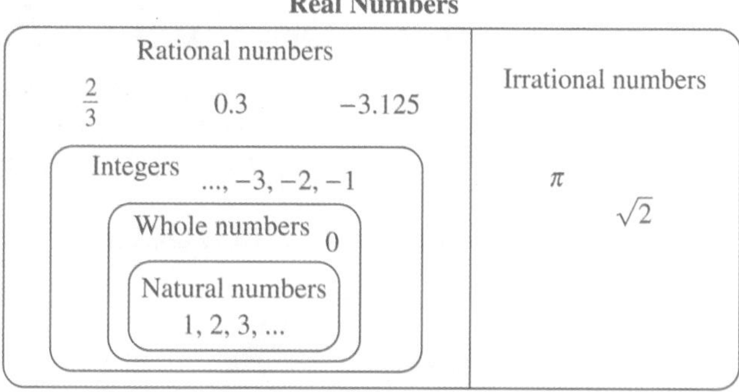

There are other classifications that are not shown in the diagram above. For instance, the set of **integers** can be divided into 3 categories: **negative integers**, zero, and **positive integers**.

Whole numbers

$\{\ldots, -3, -2, -1, 0, 1, 2, 3, \ldots\}$ The set of integers

Negative integers Positive integers

EXAMPLE 1 **Classifying Real Numbers**

Which of the numbers in the set $\left\{-7, -\sqrt{3}, -1, -\frac{1}{5}, 0, \frac{3}{4}, \sqrt{2}, \pi, 5\right\}$ are (a) natural numbers, (b) integers, (c) rational numbers, and (d) irrational numbers?

SOLUTION

a. Natural numbers: $\{5\}$

b. Integers: $\{-7, -1, 0, 5\}$

c. Rational numbers: $\left\{-7, -1, -\frac{1}{5}, 0, \frac{3}{4}, 5\right\}$

d. Irrational numbers: $\left\{-\sqrt{3}, \sqrt{2}, \pi\right\}$

Exercises Within Reach®

Solutions in English & Spanish and tutorial videos at AlgebraWithinReach.com

Classifying Real Numbers **In Exercises 1–4, determine which of the numbers in the set are (a) natural numbers, (b) integers, (c) rational numbers, and (d) irrational numbers.** See Additional Answers.

1. $\left\{-6, -\sqrt{6}, -\frac{4}{3}, 0, \frac{5}{8}, 1, \sqrt{2}, 2, \pi, 6\right\}$

2. $\left\{-\frac{10}{3}, -\pi, -\sqrt{3}, -1, 0, \frac{2}{5}, \sqrt{3}, \frac{5}{2}, 5, 101\right\}$

3. $\left\{-4.2, \sqrt{4}, -\frac{1}{9}, 0, \frac{3}{11}, \sqrt{11}, 5.\overline{5}, 5.543\right\}$

4. $\left\{-\sqrt{25}, -\sqrt{6}, -0.\overline{1}, -\frac{5}{3}, 0, 0.85, 3, 110\right\}$

The Real Number Line and Order

The picture that represents the real numbers is called the **real number line**. It consists of a horizontal line with a point (the **origin**) labeled 0. Numbers to the left of zero are **negative** and numbers to the right of zero are **positive**.

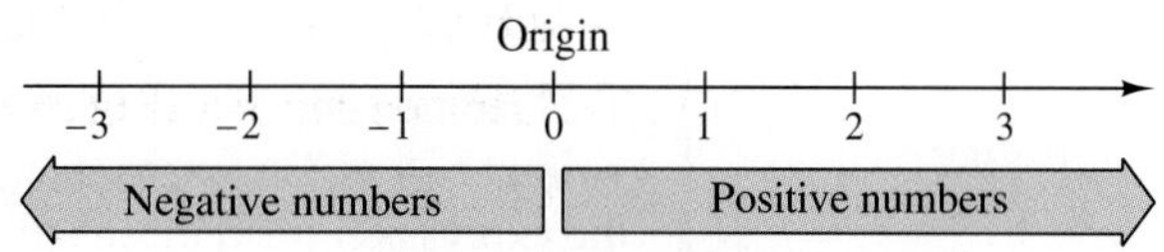

The Real Number Line

Zero is neither positive nor negative. So, to describe a real number that might be either positive or zero, you can use the term **nonnegative real number**.

Each point on the real number line corresponds to exactly one real number, and each real number corresponds to exactly one point on the real number line. When you draw the point (on the real number line) that corresponds to a real number, you are **plotting** the real number.

EXAMPLE 2 Plotting Real Numbers

Plot each number on the real number line.

a. $-\dfrac{5}{3}$ **b.** 2.3

c. $\dfrac{9}{4}$ **d.** -0.3

SOLUTION

All four points are shown in the figure.

a. The point representing the real number $-\frac{5}{3} = -1.666\ldots$ lies between -2 and -1, but closer to -2, on the real number line.

b. The point representing the real number 2.3 lies between 2 and 3, but closer to 2, on the real number line.

c. The point representing the real number $\frac{9}{4} = 2.25$ lies between 2 and 3, but closer to 2, on the real number line. Note that the point representing $\frac{9}{4}$ lies slightly to the left of the point representing 2.3.

d. The point representing the real number -0.3 lies between -1 and 0, but closer to 0, on the real number line.

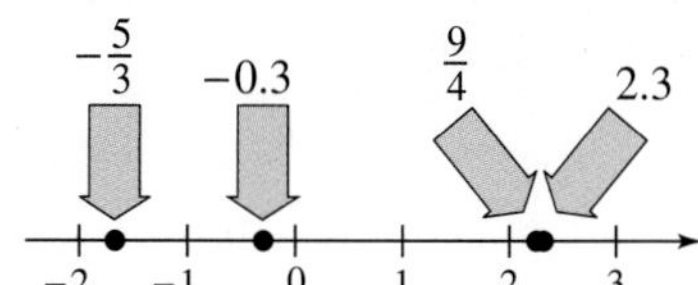

Exercises Within Reach ®

Solutions in English & Spanish and tutorial videos at AlgebraWithinReach.com

Plotting Real Numbers **In Exercises 5 and 6, plot the numbers on the real number line.**

5. (a) 3 (b) $\frac{5}{2}$ (c) $-\frac{7}{2}$ (d) -5.2

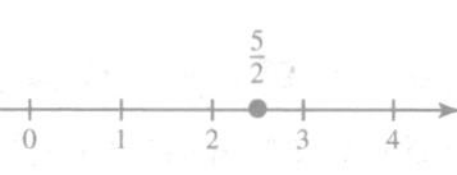

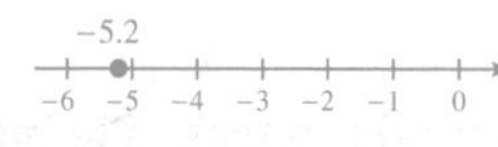

6. (a) 8 (b) $\frac{4}{3}$ (c) -6.75 (d) $-\frac{9}{2}$

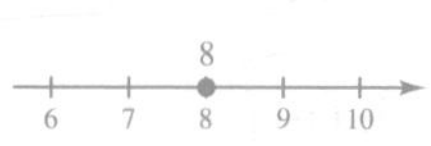

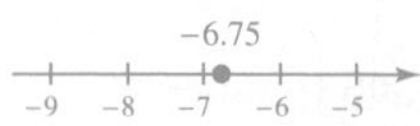

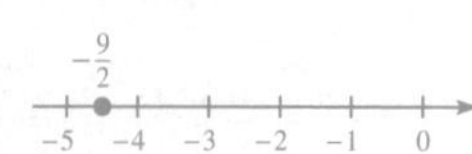

Order on the Real Number Line

If the real number a lies to the left of the real number b on the real number line, then a is **less than** b, which is written as

$$a < b.$$

This relationship can also be described by saying that b is **greater than** a and writing $b > a$. The expression $a \le b$ means that a is **less than or equal to** b, and the expression $b \ge a$ means that b is **greater than or equal to** a. The symbols $<, >, \le,$ and $\ge$ are called **inequality symbols**.

EXAMPLE 3 Ordering Real Numbers

Place the correct inequality symbol (< or >) between the real numbers.

a. $-4 \;\square\; 0$ **b.** $-3 \;\square\; -5$ **c.** $\frac{1}{5} \;\square\; \frac{1}{3}$ **d.** $-\frac{1}{4} \;\square\; -\frac{1}{2}$

SOLUTION

a. Because -4 lies to the left of 0 on the real number line, you can say that -4 is less than 0, and write $-4 < 0$.

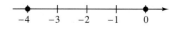

b. Because -3 lies to the right of -5 on the real number line, you can say that -3 is greater than -5, and write $-3 > -5$.

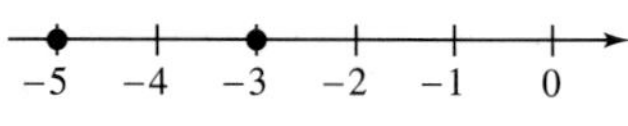

c. Because $\frac{1}{5}$ lies to the left of $\frac{1}{3}$ on the real number line, you can say that $\frac{1}{5}$ is less than $\frac{1}{3}$, and write $\frac{1}{5} < \frac{1}{3}$.

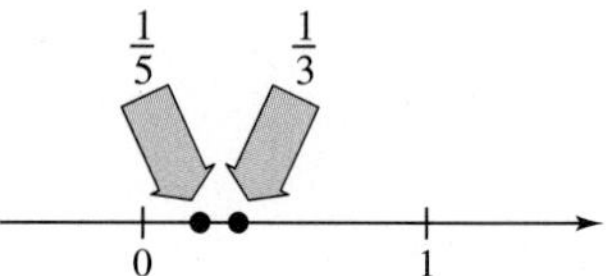

d. Because $-\frac{1}{4}$ lies to the right of $-\frac{1}{2}$ on the real number line, you can say that $-\frac{1}{4}$ is greater than $-\frac{1}{2}$ and write $-\frac{1}{4} > -\frac{1}{2}$.

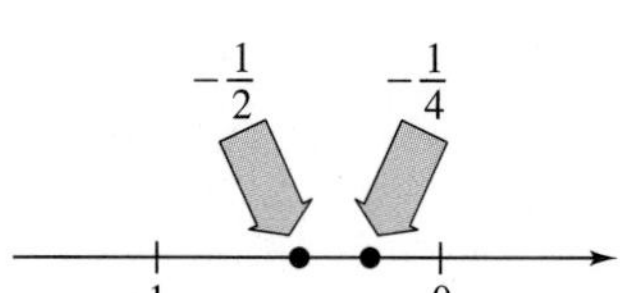

Exercises Within Reach®

Solutions in English & Spanish and tutorial videos at AlgebraWithinReach.com

Ordering Real Numbers **In Exercises 7–16, place the correct inequality symbol (< or >) between the real numbers.**

7. $\frac{4}{5} < 1$

8. $2 > \frac{5}{3}$

9. $-5 < 2$

10. $9 > -1$

11. $-5 < -2$

12. $-8 < -3$

13. $\frac{5}{8} > \frac{1}{2}$

14. $\frac{3}{2} < \frac{5}{2}$

15. $-\frac{2}{3} > -\frac{10}{3}$

16. $-\frac{5}{3} < -\frac{3}{2}$

Distance on the Real Number Line

Distance Between Two Real Numbers

If a and b are two real numbers such that $a \le b$, then the **distance between *a* and *b*** is given by $b - a$.

Note from this definition that if $a = b$, the distance between a and b is zero. If $a \neq b$, then the distance between a and b is positive.

Study Tip

Recall that when you subtract a negative number, as in Example 4(a), you add the opposite of the second number to the first. Because the opposite of -2 is 2, you add 2 to 3.

EXAMPLE 4 Finding the Distance Between Two Numbers

Find the distance between the real numbers.

a. -2 and 3 **b.** 0 and 4

c. -4 and 0 **d.** 1 and $-\frac{1}{2}$

SOLUTION

a. Because $-2 \le 3$, the distance between -2 and 3 is

$$3 - (-2) = 3 + 2 = 5.$$

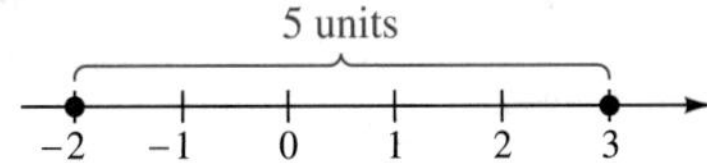

b. Because $0 \le 4$, the distance between 0 and 4 is

$$4 - 0 = 4.$$

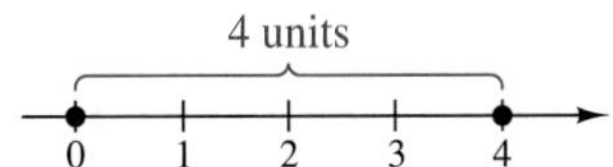

c. Because $-4 \le 0$, the distance between -4 and 0 is

$$0 - (-4) = 0 + 4 = 4.$$

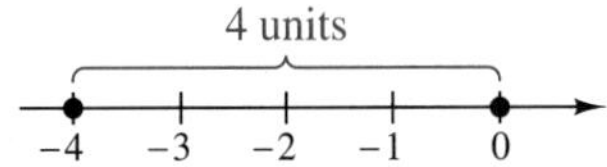

d. Because $-\frac{1}{2} \le 1$, let $a = -\frac{1}{2}$ and $b = 1$. So, the distance between 1 and $-\frac{1}{2}$ is

$$1 - \left(-\frac{1}{2}\right) = 1 + \frac{1}{2} = 1\frac{1}{2}.$$

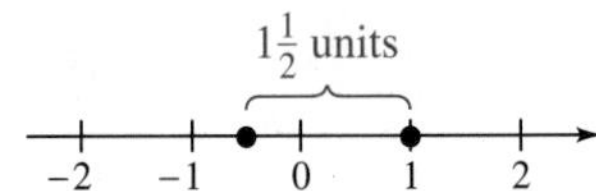

Exercises Within Reach®

Solutions in English & Spanish and tutorial videos at AlgebraWithinReach.com

Finding the Distance Between Two Numbers **In Exercises 17–28, find the distance between the real numbers.**

17. 4 and 10 6

18. 75 and 20 55

19. -12 and 7 19

20. -54 and 32 86

21. 18 and -32 50

22. 14 and -6 20

23. -8 and 0 8

24. 0 and 125 125

25. 0 and 35 35

26. -35 and 0 35

27. -6 and -9 3

28. -12 and -7 5

Finding Absolute Value

Two real numbers are called **opposites** of each other if they lie the same distance from, but on opposite sides of, 0 on the real number line.

Study Tip

Because *opposite* numbers lie the same distance from 0 on the real number line, they have the same absolute value. So, $|5| = 5$ and $|-5| = 5$.

Opposites and Additive Inverses

Let a be a real number.

1. $-a$ is the opposite of a.
2. $-(-a) = a$ Double negative
3. $a + (-a) = 0$ Additive inverse

Definition of Absolute Value

If a is a real number, then the **absolute value** of a is

$$|a| = \begin{cases} a, & \text{if } a \geq 0 \\ -a, & \text{if } a < 0 \end{cases}.$$

Commuting to and from work can be a good example for illustrating distance and absolute value. Assuming that the same route is used, the distance traveled to work is the same as the distance traveled back home, and the real number value is never negative.

EXAMPLE 5 Evaluating Absolute Values

a. $|-10| = 10$ The absolute value of -10 is 10.

b. $\left|\frac{3}{4}\right| = \frac{3}{4}$ The absolute value of $\frac{3}{4}$ is $\frac{3}{4}$.

c. $-|-6| = -(6) = -6$ The opposite of $|-6|$ is -6.

EXAMPLE 6 Comparing Real Numbers

Place the correct symbol (<, >, or =) between the real numbers.

a. $|-2| \quad \square \quad 1$ **b.** $|-4| \quad \square \quad |4|$ **c.** $2 \quad \square \quad -|-2|$

SOLUTION

a. $|-2| > 1$, because $|-2| = 2$ and 2 is greater than 1.

b. $|-4| = |4|$, because $|-4| = 4$ and $|4| = 4$.

c. $2 > -|-2|$, because $-|-2| = -2$ and 2 is greater than -2.

Exercises Within Reach®

Solutions in English & Spanish and tutorial videos at AlgebraWithinReach.com

Evaluating an Absolute Value **In Exercises 29–34, evaluate the expression.**

29. $|10|$ 10 **30.** $|62|$ 62 **31.** $|-225|$ 225

32. $|-14|$ 14 **33.** $-\left|-\frac{3}{4}\right|$ $-\frac{3}{4}$ **34.** $-\left|\frac{3}{8}\right|$ $-\frac{3}{8}$

Comparing Real Numbers **In Exercises 35–38, place the correct symbol (<, >, or =) between the real numbers.**

35. $|-6| \; > \; |2|$ **36.** $|-2| \; = \; |2|$ **37.** $|47| \; > \; |-27|$ **38.** $|150| \; < \; |-310|$

Application EXAMPLE 7 Translating Words into Symbols

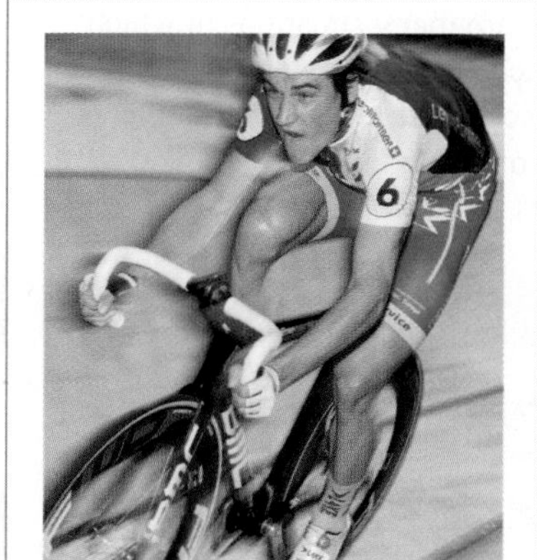

Write each statement using inequality symbols.

a. A bicycle racer's speed s is at least 16 miles per hour and at most 28 miles per hour.

b. The tire pressure p is at least 30 pounds per square inch and no more than 35 pounds per square inch.

c. The price p is less than \$225.

d. The average a will exceed 5000.

SOLUTION

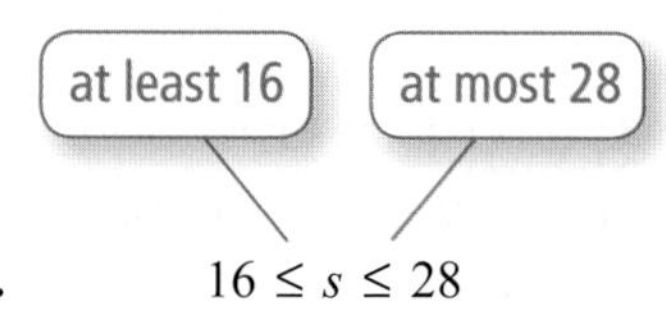

a. $16 \leq s \leq 28$

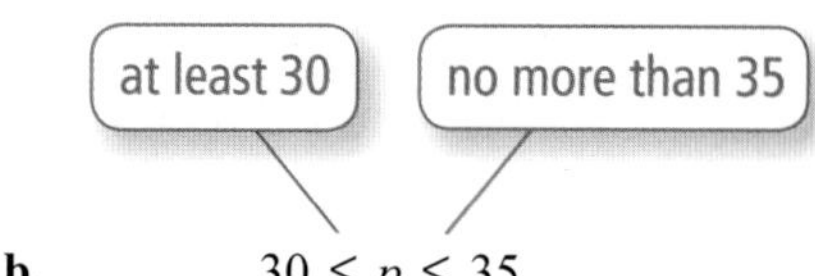

b. $30 \leq p \leq 35$

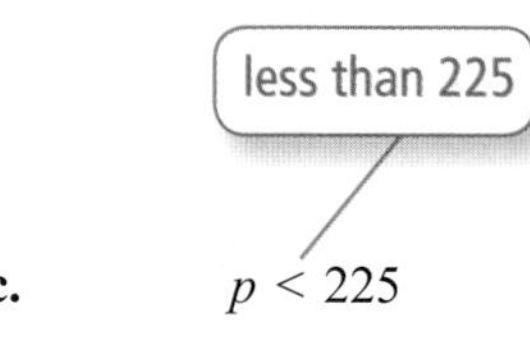

c. $p < 225$

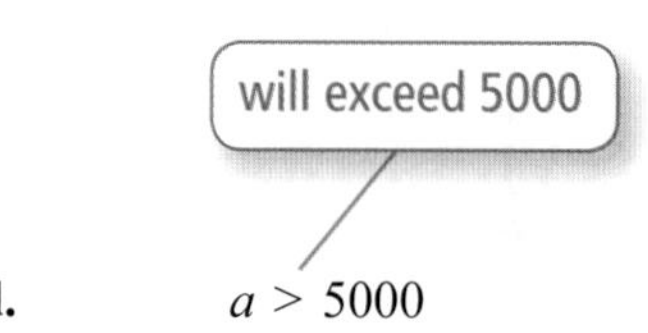

d. $a > 5000$

Exercises Within Reach®

Solutions in English & Spanish and tutorial videos at AlgebraWithinReach.com

Translating Words into Symbols **In Exercises 39–44, write the statement using inequality symbols.**

39. The weight on the elevator cannot exceed 2500 pounds.
$x \leq 2500$

40. You can drive at most 65 miles per hour on the interstate.
$x \leq 65$

41. The contestant's weight must be more than 200 pounds.
$x > 200$

42. You can save no more than \$2 with the coupon.
$x \leq 2$

43. A person must be 52 inches tall or taller to ride the roller coaster.
$x \geq 52$

44. It takes your friend at least 8 minutes and at most 10 minutes to run a mile.
$8 \leq x \leq 10$

45. ***Checking Account*** During the past month, the balance of your checking account did not exceed \$700 and did not drop below \$200. Write this statement using inequality symbols.
$200 \leq x \leq 700$

46. ***Reading*** Last night you read more than 40 pages and less than 70 pages of a new book. Write this statement using inequality symbols.
$40 < x < 70$

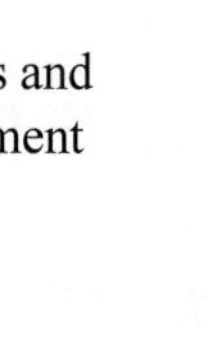

Martin Lehmann/Shutterstock.com; Utemov Alexey/Shutterstock.com

Concept Summary: Ordering Real Numbers

What

When you are asked to order two **real numbers**, the goal is to determine which of the two numbers is greater.

EXAMPLE

Order $-\frac{6}{3}$ and $-\frac{5}{2}$.

How

You can use the **real number line** to order two real numbers. For example, to order two fractions, rewrite them with the same denominator, or rewrite them as decimals. Then **plot** each number on a number line.

EXAMPLE

$-\frac{6}{3} = -2, \quad -\frac{5}{2} = -2.5$

−2.5

−3 −2 −1 0

$-\frac{6}{3} > -\frac{5}{2}$

Why

There are many situations in which you need to order real numbers. For instance, to determine the standings at a golf tournament, you order the scores of the golfers.

Exercises Within Reach ®

Worked-out solutions to odd-numbered exercises at AlgebraWithinReach.com

Concept Summary Check

47. ***Using a Number Line*** Explain how the number line above shows that $-2.5 < -2$. -2.5 is to the left of -2.

48. ***Using a Number Line*** Two real numbers are plotted on the real number line. How can you tell which number is greater? The number on the right is greater than the number on the left.

49. ***Ordering Methods*** Which method for ordering fractions is shown in the solution above? See Additional Answers.

50. ***Rewriting Fractions*** Describe another way to rewrite and order $-\frac{6}{3}$ and $-\frac{5}{2}$. See Additional Answers.

Extra Practice

Identifying Numbers **In Exercises 51–54, list all members of the set.**

51. The integers between -5.8 and 3.2
$-5, -4, -3, -2, -1, 0, 1, 2, 3$

52. The even integers between -2.1 and 10.5
$-2, 0, 2, 4, 6, 8, 10$

53. The odd integers between π and 10
5, 7, 9

54. The prime numbers between 4 and 25
5, 7, 11, 13, 17, 19, 23

Approximating and Ordering Numbers **In Exercises 55–58, approximate the two numbers and order them.**

55.
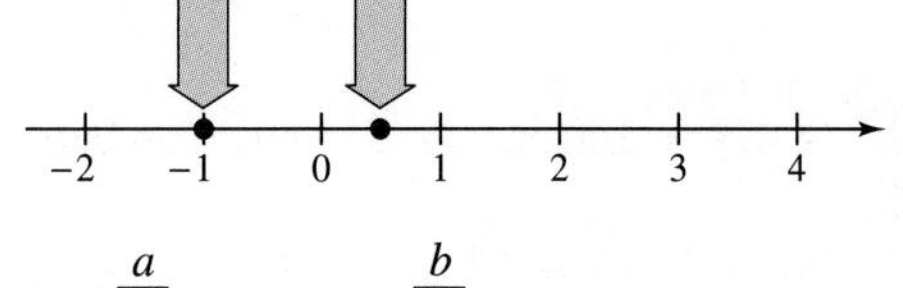

$-1 < \frac{1}{2}$

56.
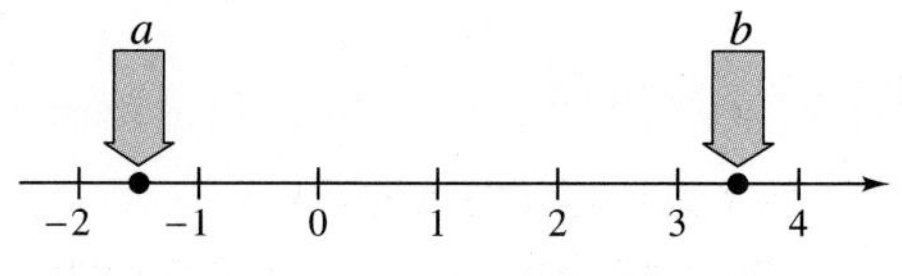

$-\frac{3}{2} < \frac{7}{2}$

57.
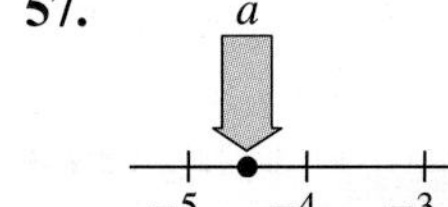
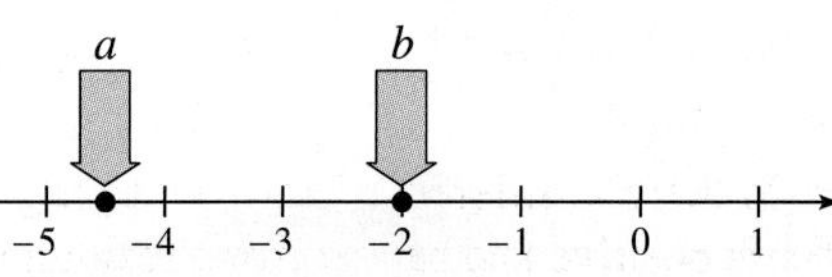

$-\frac{9}{2} < -2$

58.
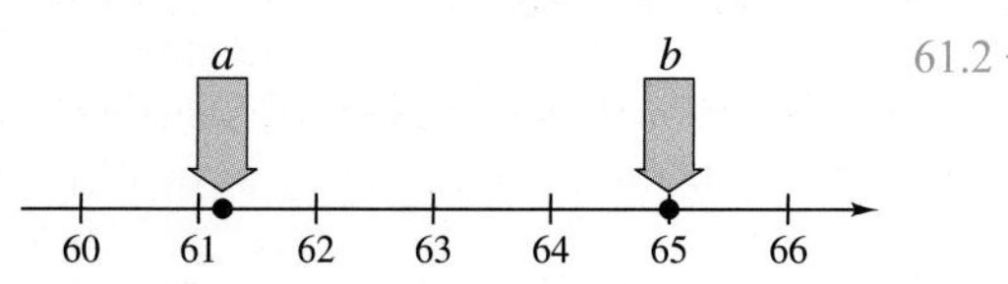

$61.2 < 65$

Evaluating an Absolute Value **In Exercises 59–64, evaluate the expression.**

59. $-|-85|$ -85

60. $-|-36.5|$ -36.5

61. $-|3.5|$ -3.5

62. $|-1.4|$ 1.4

63. $|-\pi|$ π

64. $-|\pi|$ $-\pi$

Plotting Numbers **In Exercise 65–74, plot the number and its opposite on the real number line. Determine the distance of each from 0.** See Additional Answers.

65. -7 7

66. -4 4

67. 5 5

68. 6 6

69. $-\frac{3}{5}$ $\frac{3}{5}$

70. $\frac{7}{4}$ $\frac{7}{4}$

71. $\frac{5}{3}$ $\frac{5}{3}$

72. $-\frac{3}{4}$ $\frac{3}{4}$

73. -4.25 4.25

74. 3.5 3.5

Translating Words into Symbols **In Exercise 75–78, write the statement using inequality notation.**

75. x is negative.
$x < 0$

76. y is more than 25.
$y > 25$

77. u is at least 16.
$u \geq 16$

78. x is nonnegative.
$x \geq 0$

79. ***Coin Collection*** Write a statement represented by $30 < x < 50$, where x is the number of coins in a jar.
You have more than 30 coins and fewer than 50 coins in a jar.

80. ***Basketball*** Write a statement represented by $280 \leq x \leq 310$, where x is the number points a basketball player scored this season. A basketball player scored no more than 310 points and no less than 280 points this season.

Think About It **In Exercise 81–84, find two possible values of *a*.**

81. $|a| = 4$
$-4, 4$

82. $-|a| = -7$
$7, -7$

83. The distance between a and 3 is 5.
$-2, 8$

84. The distance between a and -1 is 6.
$-7, 5$

Identifying Real Numbers **In Exercise 85–92, give three examples of numbers that satisfy the given conditions.**

85. A real number that is a negative integer
Sample answer: $-3, -100, -\frac{4}{1}$

86. A real number that is a whole number
Sample answer: $0, 2, \frac{9}{3}$

87. A real number that is a not a rational number
Sample answer: $\sqrt{2}, \pi, -\sqrt{3}$

88. A real number that is not an irrational number
Sample answer: $-\frac{4}{5}, -5, \sqrt{4}$

89. A rational number that is not an integer
Sample answer: $\frac{1}{2}, -\frac{3}{2}, -\frac{5}{6}$

90. A rational number that is not a negative number
Sample answer: $\frac{1}{2}, 10, 20\frac{1}{5}$

91. A real number that is not a positive rational number
Sample answer: $-\sqrt{7}, 0, -\frac{1}{3}$

92. An integer that is not a whole number
Sample answer: $-7, -\frac{4}{2}, -\frac{5}{1}$

Explaining Concepts

True or False? **In Exercises 93 and 94, decide whether the statement is true or false. Justify your answer.**

93. Every real number is either rational or irrational.
True. If a number can be written as the ratio of two integers, it is rational. If not, the number is irrational.

94. The distance between a number b and its opposite is equal to the distance between 0 and twice the number b.
True. The distance between zero and the number b is the same as the distance between zero and the opposite of b.

95. ***Number Sense*** Compare the rational numbers 0.15 and $0.\overline{15}$.
0.15 is a terminating rational number and $0.\overline{15}$ is a repeating rational number.

96. ***Precision*** Is there a difference between saying that a real number is positive and saying that a real number is nonnegative? Explain your answer.
Yes. The nonnegative real numbers include zero.

1.2 Operations with Real Numbers

- Add and subtract real numbers.
- Multiply and divide real numbers.
- Evaluate exponential expressions.
- Use the order of operations to evaluate expressions.

Adding and Subtracting Real Numbers

Addition of Two Real Numbers

1. To add two real numbers with *like signs*, add their absolute values and attach the common sign to the result.
2. To add two real numbers with *unlike signs*, subtract the lesser absolute value from the greater absolute value and attach the sign of the number with the greater absolute value.

EXAMPLE 1 Adding Real Numbers

a. $-84 + 14 = -(84 - 14)$
$= -70$

b. $-138 + (-62) = -(138 + 62)$
$= -200$

c. $3.2 + (-0.4) = +(3.2 - 0.4)$
$= 2.8$

d. $-26.41 + (-0.53) = -(26.41 + 0.53)$
$= -26.94$

Subtraction of Two Real Numbers

To subtract the real number b from the real number a, add the opposite of b to a. That is, $a - b = a + (-b)$.

EXAMPLE 2 Subtracting Real Numbers

a. $9 - 21 = 9 + (-21)$
$= -(21 - 9)$
$= -12$

b. $-2.5 - (-2.7) = -2.5 + 2.7$
$= +(2.7 - 2.5)$
$= 0.2$

Exercises Within Reach ®

Solutions in English & Spanish and tutorial videos at AlgebraWithinReach.com

Adding and Subtracting Real Numbers **In Exercises 1–12, evaluate the expression.**

1. $-8 + 12$ 4
2. $-5 + 9$ 4
3. $13 + (-6)$ 7
4. $12 + (-10)$ 2
5. $-17 + (-6)$ −23
6. $-6.4 + (-3.7)$ −10.1
7. $-8 - 12$ −20
8. $-3 - 17$ −20
9. $13 - (-9)$ 22
10. $4 - (-11)$ 15
11. $-15 - (-18)$ 3
12. $-21.5 - (-6.3)$ −15.2

Addition and Subtraction of Fractions

1. *Like Denominators:* The sum and difference of two fractions with like denominators ($c \neq 0$) are:

$$\frac{a}{c} + \frac{b}{c} = \frac{a+b}{c} \qquad \frac{a}{c} - \frac{b}{c} = \frac{a-b}{c}$$

2. *Unlike Denominators:* To add or subtract two fractions with unlike denominators, first rewrite the fractions so that they have the same denominator and then apply the first rule.

EXAMPLE 3 Adding and Subtracting Fractions

a. $\frac{5}{17} + \frac{9}{17} = \frac{5+9}{17}$ Add numerators.

$= \frac{14}{17}$ Simplify.

b. $\frac{3}{8} - \frac{5}{12} = \frac{3(3)}{8(3)} - \frac{5(2)}{12(2)}$ Least common denominator is 24.

$= \frac{9}{24} - \frac{10}{24}$ Simplify.

$= \frac{9-10}{24}$ Subtract numerators.

$= -\frac{1}{24}$ Simplify.

c. $1\frac{4}{5} + \frac{11}{7} = \frac{9}{5} + \frac{11}{7}$ Write $1\frac{4}{5}$ as $\frac{9}{5}$.

$= \frac{9(7)}{5(7)} + \frac{11(5)}{7(5)}$ Least common denominator is 35.

$= \frac{63}{35} + \frac{55}{35}$ Simplify.

$= \frac{63+55}{35}$ Add numerators.

$= \frac{118}{35}$ Simplify.

Study Tip

A quick way to convert the mixed number $1\frac{4}{5}$ into the fraction $\frac{9}{5}$ is to multiply the whole number by the denominator of the fraction and add the result to the numerator, as follows.

$$1\frac{4}{5} = \frac{1(5)+4}{5} = \frac{9}{5}$$

Exercises Within Reach ®

Solutions in English & Spanish and tutorial videos at AlgebraWithinReach.com

Adding and Subtracting Fractions **In Exercises 13–24, evaluate the expression. Write the result in simplest form.**

13. $\frac{3}{8} + \frac{7}{8}$ $\frac{5}{4}$

14. $\frac{5}{6} + \frac{7}{6}$ 2

15. $\frac{3}{4} - \frac{1}{4}$ $\frac{1}{2}$

16. $\frac{5}{9} - \frac{1}{9}$ $\frac{4}{9}$

17. $\frac{3}{5} + \left(-\frac{1}{2}\right)$ $\frac{1}{10}$

18. $\frac{6}{7} + \left(-\frac{3}{7}\right)$ $\frac{3}{7}$

19. $\frac{5}{8} - \frac{1}{8}$ $\frac{1}{2}$

20. $\frac{3}{10} - \frac{5}{2}$ $-\frac{11}{5}$

21. $3\frac{1}{2} + 4\frac{3}{8}$ $\frac{63}{8}$

22. $5\frac{3}{4} + 7\frac{3}{8}$ $\frac{105}{8}$

23. $10\frac{5}{8} - 6\frac{1}{4}$ $\frac{35}{8}$

24. $8\frac{1}{2} - 4\frac{2}{3}$ $\frac{23}{6}$

Multiplying and Dividing Real Numbers

Initiate a discussion about why a negative number times a negative number is a positive number. A pattern approach can be used to illustrate the product. Alternatively, assign the question as an essay.

Multiplication of Two Real Numbers

1. To multiply two real numbers with *like signs*, find the product of their absolute values. The product is *positive*.
2. To multiply two real numbers with *unlike signs*, find the product of their absolute values, and attach a minus sign. The product is *negative*.
3. The product of zero and any other real number is zero.

EXAMPLE 4 Multiplying Real Numbers

a. $-6 \cdot 9 = -54$ (Unlike signs)

b. $(-5)(-7) = 35$ (Like signs)

c. $5(-3)(-4)(7) = 420$ (Like signs; Like signs)

Study Tip

When operating with fractions, you should check to see whether your answers can be simplified by dividing out factors that are common to the numerator and denominator. For instance, the fraction $\frac{4}{6}$ can be written in simplified form as

$$\frac{4}{6} = \frac{\overset{1}{\cancel{2}} \cdot 2}{\underset{1}{\cancel{2}} \cdot 3} = \frac{2}{3}.$$

Note that dividing out a common factor is the division of a number by itself, and what remains is a factor of 1.

Multiplication of Two Fractions

The product of the two fractions a/b and c/d is given by

$$\frac{a}{b} \cdot \frac{c}{d} = \frac{ac}{bd}, \quad b \neq 0, \quad d \neq 0.$$

EXAMPLE 5 Multiplying Fractions

$$-\frac{3}{8}\left(\frac{11}{6}\right) = -\frac{3(11)}{8(6)}$$ Multiply numerators and denominators.

$$= -\frac{\cancel{3}(11)}{8(2)(\cancel{3})}$$ Factor and divide out common factor.

$$= -\frac{11}{16}$$ Simplify.

Exercises Within Reach®

Solutions in English & Spanish and tutorial videos at AlgebraWithinReach.com

Multiplying Real Numbers **In Exercises 25–38, evaluate the expression. Write all fractions in simplest form.**

25. $5(-6)$ -30

26. $3(-9)$ -27

27. $(-8)(-6)$ 48

28. $(-4)(-7)$ 28

29. $2(4)(-5)$ -40

30. $3(-7)(10)$ -210

31. $(-1)(12)(-3)$ 36

32. $(-2)(-6)(4)$ 48

33. $\frac{1}{2}\left(\frac{1}{6}\right)$ $\frac{1}{12}$

34. $\frac{1}{3}\left(\frac{2}{3}\right)$ $\frac{2}{9}$

35. $-\frac{3}{2}\left(\frac{8}{5}\right)$ $-\frac{12}{5}$

36. $\frac{10}{13}\left(-\frac{3}{5}\right)$ $-\frac{6}{13}$

37. $-\frac{5}{8}\left(-\frac{4}{5}\right)$ $\frac{1}{2}$

38. $-\frac{4}{7}\left(-\frac{4}{5}\right)$ $\frac{16}{35}$

The **reciprocal** of a nonzero real number a is defined as the number by which a must be multiplied to obtain 1. For instance, the reciprocal of 3 is $\frac{1}{3}$ because

$$3\left(\frac{1}{3}\right) = 1.$$

Similarly, the reciprocal of $-\frac{4}{5}$ is $-\frac{5}{4}$ because

$$-\frac{4}{5}\left(-\frac{5}{4}\right) = 1.$$

Study Tip

Division by 0 is not defined because 0 has no reciprocal.

Division of Two Real Numbers

To divide the real number a by the nonzero real number b, multiply a by the reciprocal of b. That is,

$$a \div b = a \cdot \frac{1}{b}, \quad b \neq 0.$$

The result of dividing two real numbers is the **quotient** of the numbers. The number a is the **dividend** and the number b is the **divisor**. When the division is expressed as a/b or $\frac{a}{b}$, a is the **numerator** and b is the **denominator**.

Additional Examples

Perform the indicated operations.

a. $6 + (-13) + 10$

b. $\left(\frac{2}{3}\right)\left(\frac{4}{5}\right)$

c. $(10)(-1.5)(6)(2.4)$

d. $-\frac{9}{14} \div -\frac{1}{3}$

Answers:

a. 3

b. $\frac{8}{15}$

c. -216

d. $\frac{27}{14}$

EXAMPLE 6 Dividing Real Numbers

a. $-30 \div 5 = -30 \cdot \frac{1}{5}$

$= -\frac{30}{5}$

$= -\frac{6 \cdot \cancel{5}}{\cancel{5}}$

$= -6$

b. $\frac{5}{16} \div 2\frac{3}{4} = \frac{5}{16} \div \frac{11}{4}$

$= \frac{5}{16} \cdot \frac{4}{11}$

$= \frac{5(4)}{16(11)}$

$= \frac{5(\cancel{4})}{\cancel{4}(4)(11)}$

$= \frac{5}{44}$

Exercises Within Reach®

Solutions in English & Spanish and tutorial videos at AlgebraWithinReach.com

Dividing Real Numbers **In Exercises 39–52, evaluate the expression. If it is not possible, state the reason. Write all fractions in simplest form.**

39. $\frac{-18}{-3}$ 6

40. $-\frac{30}{-15}$ 2

41. $-48 \div 16$ -3

42. $63 \div (-7)$ -9

43. $-10 \div 0$ Undefined

44. $-125 \div 0$ Undefined

45. $-\frac{4}{5} \div \frac{8}{25}$ $-\frac{5}{2}$

46. $-\frac{11}{12} \div \frac{5}{24}$ $-\frac{22}{5}$

47. $\left(-\frac{1}{3}\right) \div \left(-\frac{5}{6}\right)$ $\frac{2}{5}$

48. $\left(-\frac{3}{8}\right) \div \left(-\frac{4}{3}\right)$ $\frac{9}{32}$

49. $4\frac{1}{8} \div 4\frac{1}{2}$ $\frac{11}{12}$

50. $26\frac{2}{3} \div 10\frac{5}{6}$ $\frac{32}{13}$

51. $-4\frac{1}{4} \div -5\frac{5}{8}$ $\frac{34}{45}$

52. $-3\frac{5}{6} \div -2\frac{2}{3}$ $\frac{23}{16}$

Exponential Expressions

Exponential Notation

Let n be a positive integer and let a be a real number. Then the product of n factors of a is given by

$$a^n = \underbrace{a \cdot a \cdot a \cdots a}_{n \text{ factors}}.$$

In the exponential form a^n, a is the **base** and n is the **exponent**. Writing the exponential form a^n is called "raising a to the nth power."

When a number, for instance 5, is raised to the *first* power, you would usually write 5 rather than 5^1. Raising a number to the *second* power is called **squaring** the number. Raising a number to the *third* power is called **cubing** the number.

EXAMPLE 7 Evaluating Exponential Expressions

a. $(-3)^4 = (-3)(-3)(-3)(-3)$ Negative sign is part of the base.

$= 81$

b. $-3^4 = -(3)(3)(3)(3)$ Negative sign is not part of the base.

$= -81$

c. $\left(\frac{2}{5}\right)^3 = \left(\frac{2}{5}\right)\left(\frac{2}{5}\right)\left(\frac{2}{5}\right)$

$= \frac{8}{125}$

d. $(-5)^3 = (-5)(-5)(-5)$ Negative raised to odd power

$= -125$

e. $(-5)^4 = (-5)(-5)(-5)(-5)$ Negative raised to even power

$= 625$

Exercises Within Reach®

Solutions in English & Spanish and tutorial videos at AlgebraWithinReach.com

Using Exponential Notation **In Exercises 53–58, write the expression using exponential notation.**

53. $(-7) \cdot (-7) \cdot (-7)$ $(-7)^3$

54. $(-4)(-4)(-4)(-4)(-4)(-4)$ $(-4)^6$

55. $\left(\frac{1}{4}\right) \cdot \left(\frac{1}{4}\right) \cdot \left(\frac{1}{4}\right) \cdot \left(\frac{1}{4}\right)$ $\left(\frac{1}{4}\right)^4$

56. $\left(\frac{5}{8}\right) \cdot \left(\frac{5}{8}\right) \cdot \left(\frac{5}{8}\right) \cdot \left(\frac{5}{8}\right)$ $\left(\frac{5}{8}\right)^4$

57. $-(7 \cdot 7 \cdot 7)$ -7^3

58. $-(5 \cdot 5 \cdot 5 \cdot 5 \cdot 5 \cdot 5)$ -5^6

Evaluating an Exponential Expression **In Exercises 59–74, evaluate the exponential expression.**

59. 2^5 32

60. 5^3 125

61. $(-2)^4$ 16

62. $(-3)^3$ -27

63. -4^3 -64

64. -6^4 -1296

65. $\left(\frac{4}{5}\right)^3$ $\frac{64}{125}$

66. $\left(\frac{2}{3}\right)^4$ $\frac{16}{81}$

67. $\left(-\frac{1}{2}\right)^2$ $\frac{1}{4}$

68. $\left(-\frac{3}{4}\right)^3$ $-\frac{27}{64}$

69. $-\left(-\frac{1}{2}\right)^5$ $\frac{1}{32}$

70. $-\left(-\frac{1}{4}\right)^3$ $\frac{1}{64}$

71. $(0.3)^3$ 0.027

72. $(0.2)^4$ 0.0016

73. $5(-0.4)^3$ -0.32

74. $-3(0.8)^2$ -1.92

Order of Operations

Order of Operations

To evaluate an expression involving more than one operation, use the following order.

1. First do operations that occur within symbols of grouping.
2. Then evaluate powers.
3. Then do multiplications and divisions from left to right.
4. Finally, do additions and subtractions from left to right.

EXAMPLE 8 Order of Operations Without Grouping Symbols

a. $20 - 2 \cdot 3^2 = 20 - 2 \cdot 9$ Evaluate power.

$= 20 - 18 = 2$ Multiply, then subtract.

b. $5 - 6 - 2 = (5 - 6) - 2$ Left-to-Right Rule

$= -1 - 2 = -3$ Subtract.

c. $8 \div 2 \cdot 2 = (8 \div 2) \cdot 2$ Left-to-Right Rule

$= 4 \cdot 2 = 8$ Divide, then multiply.

EXAMPLE 9 Order of Operations With Grouping Symbols

This is a good time to demonstrate informally the use of the calculator. Have students do each of the following problems by hand and then check their answers on their calculators.

$3 + 8 - 2 - 1 = 8$

$5 - 2^2 \cdot 3 + 7 = 0$

a. $7 - 3(4 - 2) = 7 - 3(2)$ Subtract within symbols of grouping.

$= 7 - 6 = 1$ Multiply, then subtract.

b. $4 - 3(2)^3 = 4 - 3(8)$ Evaluate power.

$= 4 - 24 = -20$ Multiply, then subtract.

c. $1 - [4 - (5 - 3)] = 1 - (4 - 2)$ Subtract within symbols of grouping.

$= 1 - 2 = -1$ Subtract within symbols of grouping, then subtract.

d. $\frac{2 \cdot 5^2 - 10}{3^2 - 4} = (2 \cdot 5^2 - 10) \div (3^2 - 4)$ Rewrite using parentheses.

$= (50 - 10) \div (9 - 4)$ Evaluate powers and multiply within symbols of grouping.

$= 40 \div 5 = 8$ Subtract within symbols of grouping, then divide.

Exercises Within Reach®

Solutions in English & Spanish and tutorial videos at AlgebraWithinReach.com

Using the Order of Operations **In Exercises 75–90, evaluate the expression.**

75. $16 - 6 - 10$ 0

76. $18 - 12 + 4$ 10

77. $24 - 5 \cdot 2^2$ 4

78. $18 + 3^2 - 12$ 15

79. $28 \div 4 + 3 \cdot 5$ 22

80. $6 \cdot 7 - 6^2 \div 4$ 33

81. $14 - 2(8 - 4)$ 6

82. $21 - 5(7 - 5)$ 11

83. $17 - 5(16 \div 4^2)$ 12

84. $72 - 8(6^2 \div 9)$ 40

85. $5^2 - 2[9 - (18 - 8)]$ 27

86. $8 \cdot 3^2 - 4(12 + 3)$ 12

87. $5^3 + |-14 + 4|$ 135

88. $|(-2)^5| - (25 + 7)$ 0

89. $\frac{6 + 8(3)}{7 - 12}$ −6

90. $\frac{9 + 6(2)}{3 + 4}$ 3

Concept Summary: Using the Order of Operations

What

To evaluate expressions consisting of grouping symbols, **exponents**, or more than one operation correctly, you need to use the established **order of operations**.

EXAMPLE

Evaluate

$(7 + 3)2^4 + 5.$

How

You can apply the order of operations as follows.

1. **P**arentheses
2. **E**xponents
3. **M**ultiplication and **D**ivision
4. **A**ddition and **S**ubtraction

EXAMPLE

$$\begin{aligned}(7 + 3)2^4 + 5 &= 10 \cdot 2^4 + 5\\ &= 10 \cdot 16 + 5\\ &= 160 + 5\\ &= 165\end{aligned}$$

Why

When you use the order of operations, along with the properties of real numbers, you will be able to evaluate expressions correctly and efficiently.

Exercises Within Reach®

Worked-out solutions to odd-numbered exercises at AlgebraWithinReach.com

Concept Summary Check

91. ***Writing*** In your own words, describe how to use the order of operations to evaluate an expression.
Apply the order of operations as follows: Parentheses, Exponents, Multiplication and Division, Addition and Subtraction.

92. ***Using the Order of Operations*** What is the first step in evaluating the expression $6^2 - 8(3 + 4)$?
Evaluate 3 + 4.

93. ***Subtracting Real Numbers*** Explain how to subtract one real number from another.
To subtract one real number from another, add the first number to the opposite of the second number.

94. ***Using Exponents*** If $a > 0$, state the values of n such that $(-a)^n = -a^n$.
n would be an odd integer.

Extra Practice

Evaluating an Expression **In Exercises 95–106, evaluate the expression.**

95. $85 - |-25|$ 60

96. $-36 + |-8|$ −28

97. $-(-11.325) + |34.625|$ 45.95

98. $|-16.25| - 54.78$ −38.53

99. $-\left|-6\frac{7}{8}\right| - 8\frac{1}{4}$ $-\frac{121}{8}$

100. $-\left|-15\frac{2}{3}\right| - 12\frac{1}{3}$ −28

101. $\dfrac{4^2 - 5}{11} - 7$ −6

102. $\dfrac{5^3 - 50}{-15} + 27$ 22

103. $\dfrac{6 \cdot 2^2 - 12}{3^2 + 3}$ 1

104. $\dfrac{7^2 - 2(11)}{5^2 + 8(-2)}$ 3

105. $\dfrac{3 + \frac{3}{4}}{\frac{1}{8}}$ 30

106. $\dfrac{6 - \frac{2}{3}}{\frac{4}{9}}$ 12

Circle Graphs **In Exercises 107 and 108, find the unknown fractional part of the circle graph.**

107. $\frac{17}{180}$

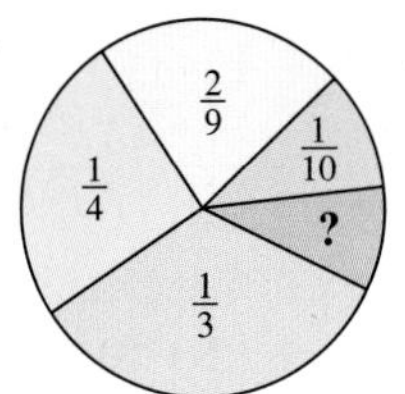

108. $\frac{11}{70}$

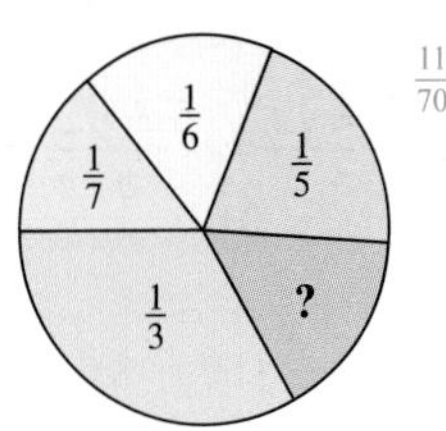

109. ***Account Balance*** During one month, you make the following transactions in your non-interest-bearing checking account. Find the balance at the end of the month. $2533.56

NUMBER OR CODE	DATE	TRANSACTION DESCRIPTION	PAYMENT AMOUNT	✓	FEE	DEPOSIT AMOUNT	BALANCE $2618.68
	3/1	Pay				$1236 45	
2154	3/3	Magazine	$ 25 62				
2155	3/6	Insurance	$455 00				
	3/12	Withdrawal	$ 125 00				
2156	3/15	Mortgage	$ 715 95				

Geometry **In Exercises 110–113, find the area of the figure. (The area A of a rectangle is given by $A = \text{length} \cdot \text{width}$, and the area A of a triangle is given by $A = \frac{1}{2} \cdot \text{base} \cdot \text{height}$.)**

110.

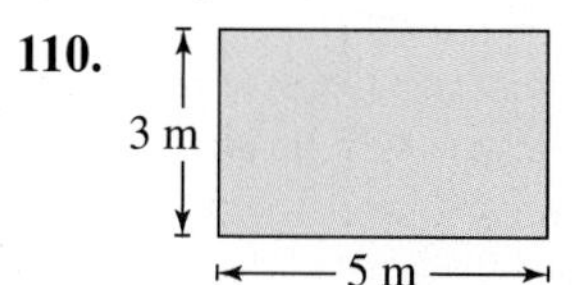

15 square meters

111.

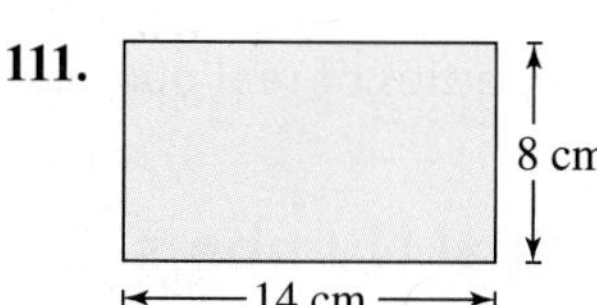

112 square centimeters

112.

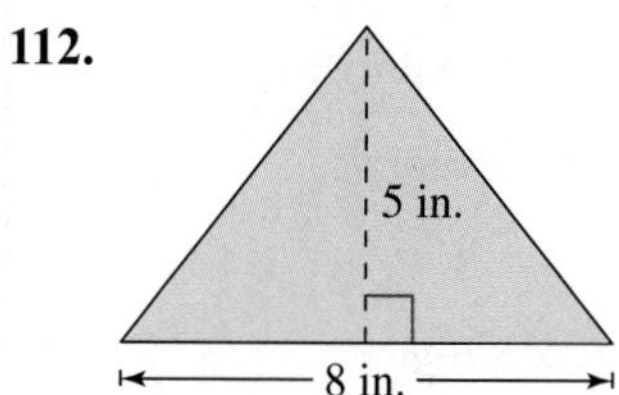

20 square inches

113.

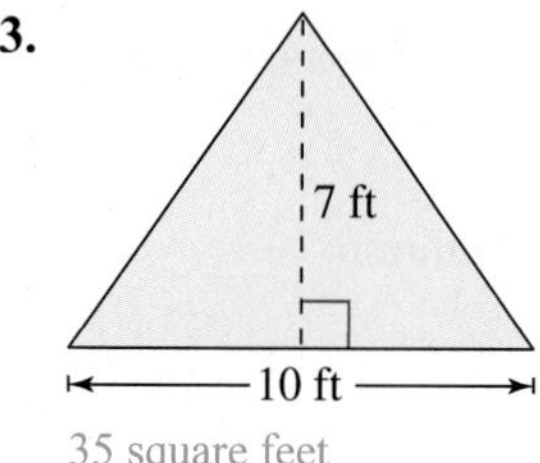

35 square feet

114. ***Savings Plan*** You plan to deposit \$50 per month in a savings account for 18 years.

(a) How much money will you deposit during the 18 years?
\$10,800

(b) The savings account earns 4% interest compounded monthly. The total amount in the account after 18 years will be

$$50\left[\left(1+\frac{0.04}{12}\right)^{216}-1\right]\left(1+\frac{12}{0.04}\right).$$

Use a calculator to determine this amount.
\$15,832.22

(c) How much of the amount in part (b) is earnings from interest?
\$5032.22

Explaining Concepts

True or False? **In Exercises 115–119, determine whether the statement is true or false. Justify your answer.**

115. The reciprocal of every nonzero rational number is a rational number.
True. A nonzero rational number is of the form $\frac{a}{b}$, where a and b are integers and $a \neq 0$, $b \neq 0$. The reciprocal will be $\frac{b}{a}$, which is also rational.

116. The product of two fractions is the product of the numerators over the LCD.
False. The product of two fractions is the product of the numerators over the product of the denominators.

117. When a negative real number is raised to the 12th power, the result is positive.
True. When a negative number is raised to an even power, the result is positive.

118. When a negative real number is raised to the 11th power, the result is positive. False. When a negative number is raised to an odd power, the result is negative.

119. $a \div b = b \div a$ False. $6 \div 3 = 2 \neq \frac{1}{2} = 3 \div 6$

120. ***Number Sense*** Are the expressions $(2^2)^3$ and $2^{(2^3)}$ equal? Explain.
$(2^2)^3 = 4^3 = 64$ and $2^{(2^3)} = 2^8 = 256$.
Because $64 \neq 256$, $(2^2)^3 \neq 2^{(2^3)}$.

121. ***Writing*** In your own words, describe the rules for determining the sign of the product or the quotient of two real numbers.
If the numbers have like signs, the product or quotient is positive. If the numbers have unlike signs, the product or quotient is negative.

122. ***Reasoning*** Decide which expressions are equal to 27 when you follow the standard order of operations. For the expressions that are not equal to 27, insert symbols of grouping to make the expression equal to 27. Discuss the value of symbols of grouping in mathematical communication.

(a) $40 - 10 + 3$
$40 - (10 + 3) = 27$

(b) $5^2 + \frac{1}{2} \cdot 4$
$5^2 + \frac{1}{2} \cdot 4 = 27$

(c) $8 \cdot 3 + 30 \div 2$
$(8 \cdot 3 + 30) \div 2 = 27$

(d) $75 \div 2 + 1 + 2$
$75 \div (2 + 1) + 2 = 27$

Error Analysis **In Exercises 123 and 124, describe and correct the error.**

123. $\frac{2}{3} + \frac{3}{2} = \frac{2+3}{3+2} = 1$ (crossed out)

To add fractions with unlike denominators, first find the least common denominator.
$\frac{2}{3}+\frac{3}{2}=\frac{2(2)}{3(2)}+\frac{3(3)}{2(3)}=\frac{4}{6}+\frac{9}{6}=\frac{13}{6}$

124. $\frac{5+12}{5} = \frac{\cancel{5}+12}{\cancel{5}} = 12$ (crossed out)

Only common factors (not terms) of the numerator and denominator can be divided out.
$\frac{5+12}{5}=\frac{17}{5}$

1.3 Properties of Real Numbers

- Identify and use the properties of real numbers.
- Develop additional properties of real numbers.

Basic Properties of Real Numbers

Properties of Real Numbers

Let a, b, and c represent real numbers, variables, or algebraic expressions.

Property	*Example*
Commutative Property of Addition:	
$a + b = b + a$	$3 + 5 = 5 + 3$
Commutative Property of Multiplication:	
$ab = ba$	$2 \cdot 7 = 7 \cdot 2$
Associative Property of Addition:	
$(a + b) + c = a + (b + c)$	$(4 + 2) + 3 = 4 + (2 + 3)$
Associative Property of Multiplication:	
$(ab)c = a(bc)$	$(2 \cdot 5) \cdot 7 = 2 \cdot (5 \cdot 7)$
Distributive Property:	
$a(b + c) = ab + ac$	$4(7 + 3) = 4 \cdot 7 + 4 \cdot 3$
$(a + b)c = ac + bc$	$(2 + 5)3 = 2 \cdot 3 + 5 \cdot 3$
$a(b - c) = ab - ac$	$6(5 - 3) = 6 \cdot 5 - 6 \cdot 3$
$(a - b)c = ac - bc$	$(7 - 2)4 = 7 \cdot 4 - 2 \cdot 4$
Additive Identity Property:	
$a + 0 = 0 + a = a$	$9 + 0 = 0 + 9 = 9$
Multiplicative Identity Property:	
$a \cdot 1 = 1 \cdot a = a$	$-5 \cdot 1 = 1 \cdot (-5) = -5$
Additive Inverse Property:	
$a + (-a) = 0$	$3 + (-3) = 0$
Multiplicative Inverse Property:	
$a \cdot \frac{1}{a} = 1, \quad a \neq 0$	$8 \cdot \frac{1}{8} = 1$

Study Tip

The operations of subtraction and division are not listed at the right because they do not have many of the properties of real numbers. For instance, subtraction and division are not commutative or associative. To see this, consider the following.

$4 - 3 \neq 3 - 4$

$15 \div 5 \neq 5 \div 15$

$8 - (6 - 2) \neq (8 - 6) - 2$

$20 \div (4 \div 2) \neq (20 \div 4) \div 2$

This is a good opportunity for a writing assignment. Ask students to give informal examples in life that illustrate some of these properties of real numbers.

EXAMPLE 1 Identifying Properties of Real Numbers

Identify the property of real numbers illustrated by each statement.

a. $4(a + 3) = 4 \cdot a + 4 \cdot 3$ **b.** $6 \cdot \frac{1}{6} = 1$

c. $-3 + (2 + b) = (-3 + 2) + b$ **d.** $(b + 8) + 0 = b + 8$

SOLUTION

a. This statement illustrates the Distributive Property.

b. This statement illustrates the Multiplicative Inverse Property.

c. This statement illustrates the Associative Property of Addition.

d. This statement illustrates the Additive Identity Property, where $(b + 8)$ is an algebraic expression.

Set of Numbers

Operations with the Numbers

Properties of the Operations

The properties of real numbers make up the third component of what is called a **mathematical system**. These three components are a *set of numbers* (Section 1.1), *operations* with the set of numbers (Section 1.2), and *properties* of the operations with the numbers (Section 1.3).

EXAMPLE 2 Using Properties of Real Numbers

Complete each statement using the specified property of real numbers.

a. Multiplicative Identity Property: $(4a)1 = \square$

b. Associative Property of Addition: $(b + 8) + 3 = \square$

c. Additive Inverse Property: $0 = 5c + \square$

SOLUTION

a. By the Multiplicative Identity Property, $(4a)1 = 4a$.

b. By the Associative Property of Addition, $(b + 8) + 3 = b + (8 + 3)$.

c. By the Additive Inverse Property, $0 = 5c + (-5c)$.

Exercises Within Reach ®

Solutions in English & Spanish and tutorial videos at AlgebraWithinReach.com

Identifying a Property of Real Numbers **In Exercises 1–6, identify the property of real numbers illustrated by the statement.**

1. $18 - 18 = 0$ Additive Inverse Property

2. $5 + 0 = 5$ Additive Identity Property

3. $\frac{1}{12} \cdot 12 = 1$ Multiplicative Inverse Property

4. $52 \cdot 1 = 52$ Multiplicative Identity Property

5. $(8 - 5)(10) = 8 \cdot 10 - 5 \cdot 10$ Distributive Property

6. $7(9 + 15) = 7 \cdot 9 + 7 \cdot 15$ Distributive Property

Using a Property of Real Numbers **In Exercises 7–10, complete the statement using the specified property of real numbers.**

7. Commutative Property of Multiplication:

$15(-3) = -3(15)$

8. Associative Property of Addition:

$6 + (5 + y) = (6 + 5) + y$

9. Distributive Property:

$5(6 + z) = 5 \cdot 6 + 5 \cdot z$

10. Distributive Property:

$(8 - y)(4) = 8 \cdot 4 - y \cdot 4$

Additional Properties of Real Numbers

Once you have determined the basic properties (or *axioms*) of a mathematical system, you can go on to develop other properties. These additional properties are **theorems**, and the formal arguments that justify the theorems are **proofs**.

Additional Properties of Real Numbers

Let a, b, and c represent real numbers, variables, or algebraic expressions.

Properties of Equality

Addition Property of Equality:	If $a = b$, then $a + c = b + c$.
Multiplication Property of Equality:	If $a = b$, then $ac = bc$.
Cancellation Property of Addition:	If $a + c = b + c$, then $a = b$.
Cancellation Property of Multiplication:	If $ac = bc$ and $c \neq 0$, then $a = b$.

Properties of Zero

Multiplication Property of Zero:	$0 \cdot a = 0$
Division Property of Zero:	$\frac{0}{a} = 0, a \neq 0$
Division by Zero Is Undefined:	$\frac{a}{0}$ is undefined.

Properties of Negation

Multiplication by -1:	$(-1)(a) = -a, (-1)(-a) = a$
Placement of Negative Signs:	$(-a)(b) = -(ab) = (a)(-b)$
Product of Two Opposites:	$(-a)(-b) = ab$

Each of the additional properties in the above list can be proved by using the basic properties of real numbers.

EXAMPLE 3 Proof of the Cancellation Property of Addition

Prove that if $a + c = b + c$, then $a = b$. (Use the Addition Property of Equality.)

SOLUTION

Notice how each step is justified from the preceding step by means of a property of real numbers.

$a + c = b + c$	Write original equation.
$(a + c) + (-c) = (b + c) + (-c)$	Addition Property of Equality
$a + [c + (-c)] = b + [c + (-c)]$	Associative Property of Addition
$a + 0 = b + 0$	Additive Inverse Property
$a = b$	Additive Identity Property

EXAMPLE 4 **Proof of a Property of Negation**

Prove that $(-1)a = -a$. (You may use any of the properties of equality and properties of zero.)

SOLUTION

At first glance, it is a little difficult to see what you are being asked to prove. However, a good way to start is to consider carefully the definitions of the three numbers in the equation.

$a =$ given real number

$-1 =$ the additive inverse of 1

$-a =$ the additive inverse of a

Now, by showing that $(-1)a$ has the same properties as the additive inverse of a, you will be showing that $(-1)a$ must be the additive inverse of a.

$(-1)a + a = (-1)a + (1)(a)$	Multiplicative Identity Property
$= (-1 + 1)a$	Distributive Property
$= (0)a$	Additive Inverse Property
$= 0$	Multiplication Property of Zero

Because $(-1)a + a = 0$, you can use the fact that $-a + a = 0$ to conclude that $(-1)a + a = -a + a$. From this, you can complete the proof as follows.

$(-1)a + a = -a + a$	Shown in first part of proof
$(-1)a = -a$	Cancellation Property of Addition

Exercises Within Reach®

Solutions in English & Spanish and tutorial videos at AlgebraWithinReach.com

Identifying a Property of Real Numbers **In Exercises 11 and 12, identify the property of real numbers illustrated by the statement.**

11. $x + 4 = 5$
$(x + 4) - 4 = 5 - 4$ Addition Property of Equality

12. $7x = 14$
$\frac{1}{7}(7x) = \frac{1}{7}(14)$ Multiplication Property of Equality

Using the Distributive Property **In Exercises 13–18, rewrite the expression using the Distributive Property.**

13. $20(2 + 5)$ $20 \cdot 2 + 20 \cdot 5$

14. $-3(4 - 8)$ $-3 \cdot 4 + (-3)(-8)$

15. $(x + 6)(-2)$ $x(-2) + 6(-2)$ or $-2x - 12$

16. $(z - 10)(12)$ $z \cdot 12 + (-10)12$ or $12z - 120$

17. $-6(2y - 5)$ $-6(2y) + (-6)(-5)$ or $-12y + 30$

18. $-4(10 - b)$ $-4 \cdot 10 + (-4)(-b)$ or $-40 + 4b$

Using the Distributive Property **In Exercises 19–22, use the Distributive Property to simplify the expression.**

19. $7x + 2x$ $(7 + 2)x = 9x$

20. $8x - 6x$ $(8 - 6)x = 2x$

21. $\frac{7x}{8} - \frac{5x}{8}$ $\frac{x}{8}(7 - 5) = \frac{x}{4}$

22. $\frac{3x}{5} + \frac{x}{5}$ $\frac{x}{5}(3 + 1) = \frac{4x}{5}$

Proving a Statement **In Exercises 23–26, use the basic properties of real numbers to prove the statement.** See Additional Answers.

23. If $ac = bc$ and $c \neq 0$, then $a = b$.

24. $a \cdot 1 = 1 \cdot a$

25. $a = (a + b) + (-b)$

26. $a + (-a) = 0$

The list of additional properties of real numbers on page 20 forms a very important part of algebra. Knowing the names of the properties is useful, but knowing how to use each property is extremely important. The next two examples show how several of the properties can be used to solve equations. (You will study these techniques in detail in Section 2.1.)

EXAMPLE 5 Identifying Properties of Real Numbers

In the solution of the equation

$b + 2 = 6$

identify the property of real numbers that justifies each step.

SOLUTION

$b + 2 = 6$ Write original equation.

Solution Step	*Property*
$(b + 2) + (-2) = 6 + (-2)$	Addition Property of Equality
$b + [2 + (-2)] = 6 - 2$	Associative Property of Addition
$b + 0 = 4$	Additive Inverse Property
$b = 4$	Additive Identity Property

Exercises Within Reach®

Solutions in English & Spanish and tutorial videos at AlgebraWithinReach.com

Identifying a Property of Real Numbers **In Exercises 27–40, identify the property of real numbers illustrated by the statement.**

27. $13 + 12 = 12 + 13$
Commutative Property of Addition

28. $(5 + 10)(8) = 8(5 + 10)$
Commutative Property of Multiplication

29. $(-4 \cdot 10) \cdot 8 = -4(10 \cdot 8)$
Associative Property of Multiplication

30. $3 + (12 - 9) = (3 + 12) - 9$
Associative Property of Addition

31. $10(2x) = (10 \cdot 2)x$
Associative Property of Multiplication

32. $1 \cdot 9k = 9k$
Multiplicative Identity Property

33. $10x \cdot \frac{1}{10x} = 1$ Multiplicative Inverse Property

34. $0 + 4x = 4x$ Additive Identity Property

35. $2x - 2x = 0$ Additive Inverse Property

36. $4 + (3 - x) = (4 + 3) - x$ Associative Property of Addition

37. $3(2 + x) = 3 \cdot 2 + 3x$ Distributive Property

38. $3(6 + b) = 3 \cdot 6 + 3 \cdot b$ Distributive Property

39. $(x + 1) - (x + 1) = 0$ Additive Inverse Property

40. $6(x + 3) = 6 \cdot x + 6 \cdot 3$ Distributive Property

Identifying Properties of Real Numbers **In Exercises 41 and 42, identify the property of real numbers that justifies each step.** See Additional Answers.

41.
$$x + 5 = 3$$
$$(x + 5) + (-5) = 3 + (-5)$$
$$x + [5 + (-5)] = -2$$
$$x + 0 = -2$$
$$x = -2$$

42.
$$x - 8 = 20$$
$$(x - 8) + 8 = 20 + 8$$
$$x + (-8 + 8) = 28$$
$$x + 0 = 28$$
$$x = 28$$

EXAMPLE 6 Identifying Properties of Real Numbers

In the solution of the equation

$$3x = 15$$

identify the property of real numbers that justifies each step.

SOLUTION

$3x = 15$	Write original equation.
Solution Step	*Property*
$\frac{1}{3}(3x) = \frac{1}{3}(15)$	Multiplication Property of Equality
$\left(\frac{1}{3} \cdot 3\right)x = 5$	Associative Property of Multiplication
$(1)(x) = 5$	Multiplicative Inverse Property
$x = 5$	Multiplicative Identity Property

Exercises Within Reach®

Solutions in English & Spanish and tutorial videos at AlgebraWithinReach.com

Identifying Properties of Real Numbers **In Exercises 43–46, identify the property of real numbers that justifies each step.** See Additional Answers.

43.
$$2x - 5 = 6$$
$$(2x - 5) + 5 = 6 + 5$$
$$2x + (-5 + 5) = 11$$
$$2x + 0 = 11$$
$$2x = 11$$
$$\tfrac{1}{2}(2x) = \tfrac{1}{2}(11)$$
$$\left(\tfrac{1}{2} \cdot 2\right)x = \tfrac{11}{2}$$
$$1 \cdot x = \tfrac{11}{2}$$
$$x = \tfrac{11}{2}$$

44.
$$3x + 4 = 10$$
$$(3x + 4) + (-4) = 10 + (-4)$$
$$3x + [4 + (-4)] = 6$$
$$3x + 0 = 6$$
$$3x = 6$$
$$\tfrac{1}{3}(3x) = \tfrac{1}{3}(6)$$
$$\left(\tfrac{1}{3} \cdot 3\right)x = 2$$
$$1 \cdot x = 2$$
$$x = 2$$

45.
$$-4x - 4 = 0$$
$$(-4x - 4) + 4 = 0 + 4$$
$$-4x + (-4 + 4) = 4$$
$$-4x + 0 = 4$$
$$-4x = 4$$
$$-\tfrac{1}{4}(-4x) = -\tfrac{1}{4}(4)$$
$$\left[-\tfrac{1}{4} \cdot (-4)\right]x = -1$$
$$1 \cdot x = -1$$
$$x = -1$$

46.
$$-5x + 25 = 5$$
$$(-5x + 25) + (-25) = 5 + (-25)$$
$$-5x + [25 + (-25)] = -20$$
$$-5x + 0 = -20$$
$$-5x = -20$$
$$-\tfrac{1}{5}(-5x) = -\tfrac{1}{5}(-20)$$
$$\left[-\tfrac{1}{5} \cdot (-5)\right]x = 4$$
$$1 \cdot x = 4$$
$$x = 4$$

Concept Summary: Using Properties of Real Numbers

What

You can use the properties of real numbers to rewrite and simplify expressions and solve equations.

EXAMPLE

Identify the property of real numbers illustrated by the statement.

$(2 + 3) + 5 = 2 + (3 + 5)$

How

To help you understand each property of real numbers, try stating each property in your own words. For instance, the Associative Property of Addition can be stated as follows: *When three real numbers are added, it makes no difference which two are added first.*

EXAMPLE

$(2 + 3) + 5 = 2 + (3 + 5)$

Associative Property of Addition

Why

Knowing the names of the properties is useful, but knowing how to use each property is extremely important. You need to know how to use these properties to successfully rewrite and simplify expressions as well as solve equations.

Exercises Within Reach®

Worked-out solutions to odd-numbered exercises at AlgebraWithinReach.com

Concept Summary Check

47–50. See Additional Answers.

47. ***Additive Inverse*** Does every real number have an additive inverse? Explain.

48. ***Multiplicative Inverse*** Does every real number have a multiplicative inverse? Explain.

49. ***Reasoning*** Are subtraction and division commutative? If not, show a counterexample.

50. ***Reasoning*** Are subtraction and division associative? If not, show a counterexample.

Extra Practice

Using Associative Properties **In Exercises 51–54, rewrite the expression using the Associative Property of Addition or the Associative Property of Multiplication.**

51. $32 + (4 + y)$
$(32 + 4) + y$

52. $15 + (3 - x)$
$(15 + 3) - x$

53. $9(6m)$
$(9 \cdot 6)m$

54. $11(4n)$
$(11 \cdot 4)n$

Using Properties of Real Numbers **In Exercises 55–58, the right side of the statement does not equal the left side. Change the right side so that it *does* equal the left side.**

55. $3(x + 5) \neq 3x + 5$
$3x + 15$

56. $4(x + 2) \neq 4x + 2$
$4x + 8$

57. $-2(x + 8) \neq -2x + 16$
$-2x - 16$

58. $-9(x + 4) \neq -9x + 36$
$-9x - 36$

Mental Math **In Exercises 59–64, use the Distributive Property to perform the arithmetic mentally. For example, you work in an industry in which the wage is \$14 per hour with time-and-a-half for overtime. So, your hourly wage for overtime is**

$$14(1.5) = 14\left(1 + \frac{1}{2}\right) = 14 + 7 = \$21.$$

59. $16(1.75) = 16\left(2 - \frac{1}{4}\right)$ 28

60. $15\left(1\frac{2}{3}\right) = 15\left(2 - \frac{1}{3}\right)$ 25

61. $7(62) = 7(60 + 2)$ 434

62. $5(51) = 5(50 + 1)$ 255

63. $9(6.98) = 9(7 - 0.02)$ 62.82

64. $12(19.95) = 12(20 - 0.05)$ 239.4

65. ***Geometry*** The figure shows two adjoining rectangles. Demonstrate the Distributive Property by filling in the blanks to write the total area of the two rectangles in two ways.

$a(b + c) = ab + ac$

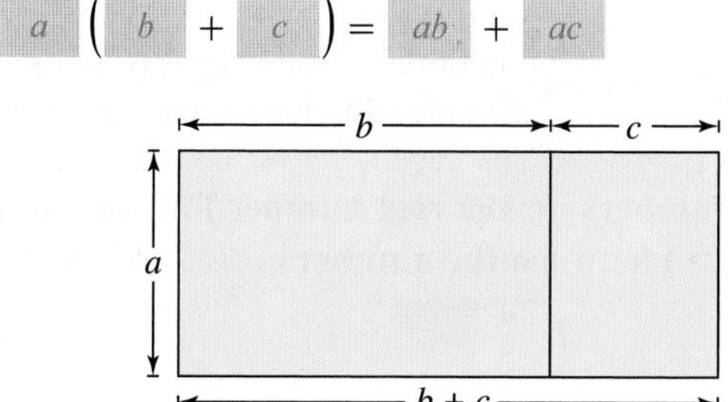

66. ***Geometry*** The figure shows two adjoining rectangles. Demonstrate the "subtraction version" of the Distributive Property by filling in the blanks to write the area of the left rectangle in two ways.

$a(b - c) = ab - ac$

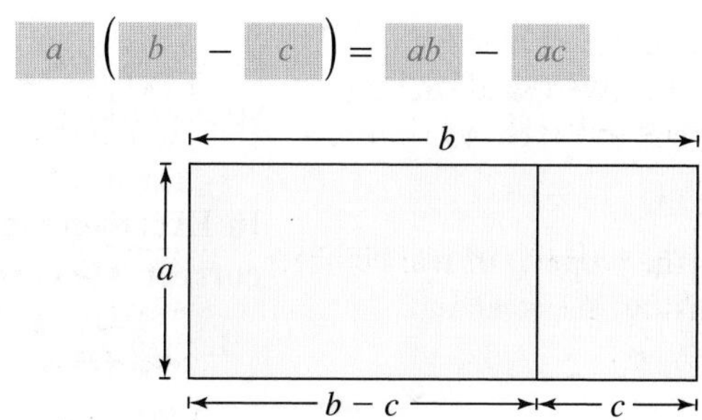

Geometry **In Exercises 67 and 68, write the expression for the perimeter of the triangle shown in the figure. Use the properties of real numbers to simplify the expression.**

67.

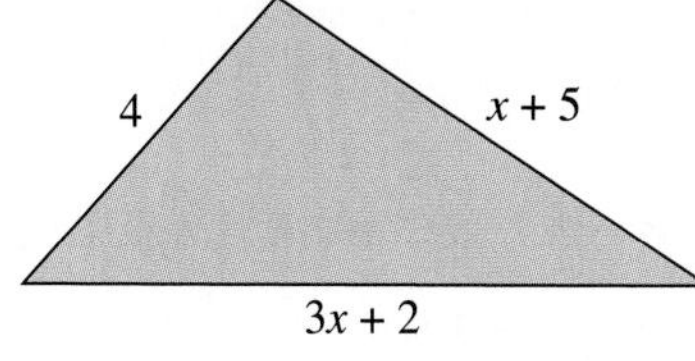

$4 + (x + 5) + (3x + 2); 4x + 11$

68.

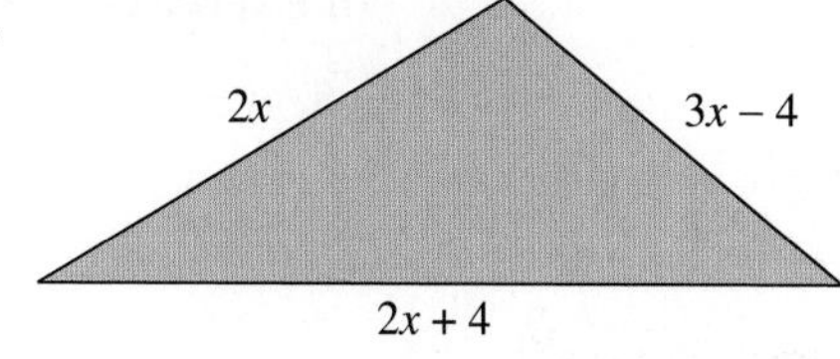

$(2x) + (3x - 4) + (2x + 4); 7x$

Geometry **In Exercises 69 and 70, write and simplify the expression for (a) the perimeter and (b) the area of the rectangle.**

69.

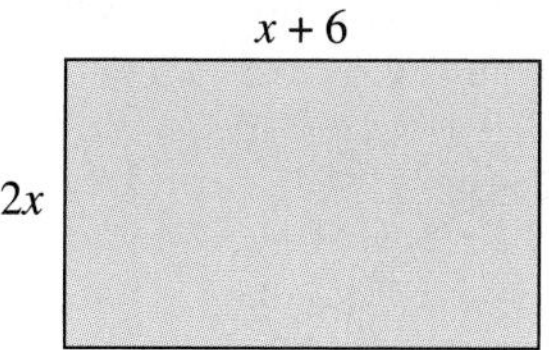

(a) $2(x + 6) + 2(2x); 6x + 12$
(b) $(x + 6)(2x); 2x^2 + 12x$

70. 5x; 2x − 1

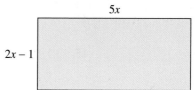

(a) $2(5x) + 2(2x - 1); 14x - 2$
(b) $5x(2x - 1); 10x^2 - 5x$

Explaining Concepts

71. ***Additive Inverse*** What is the additive inverse of a real number? Give an example of the Additive Inverse Property.

The additive inverse of a real number is its opposite. The sum of a number and its additive inverse is the additive identity zero. For example, $-3.2 + 3.2 = 0$.

72. ***Multiplicative Inverse*** What is the multiplicative inverse of a real number? Given an example of the Multiplicative Inverse Property.

The multiplicative inverse of a real number a, where $a \neq 0$, is the number $1/a$. The product of a number and its multiplicative inverse is the multiplicative identity 1. For example, $8 \cdot \frac{1}{8} = 1$.

73. ***Writing*** In your own words, give a verbal description of the Commutative Property of Addition.

Given two real numbers a and b, the sum a plus b is the same as the sum b plus a.

74. ***Writing*** Explain how the Addition Property of Equality can be used to allow you to subtract the same number from each side of an equation.

To subtract a number a from each side of the equation, use the Addition Property of Equality to add $-a$ to each side.

75. ***Reasoning*** You define a new mathematical operation using the symbol $\odot$. This operation is defined as $a \odot b = 2 \cdot a + b$. Give examples to show that this operation is neither commutative nor associative.

Sample answer: $4 \odot 7 = 15 \neq 18 = 7 \odot 4$;
$3 \odot (4 \odot 7) = 21 \neq 27 = (3 \odot 4) \odot 7$

76. ***Reasoning*** You define a new mathematical operation using the symbol ‡. This operation is defined as $a ‡ b = a - (b + 1)$. Give examples to show that this operation is neither commutative nor associative.

Sample answer: $9 ‡ 6 = 2 \neq -4 = 6 ‡ 9$;
$(10 ‡ 2) ‡ 7 = -1 \neq 15 = 10 ‡ (2 ‡ 7)$

Mid-Chapter Quiz: Sections 1.1–1.3

Solutions in English & Spanish and tutorial videos at AlgebraWithinReach.com

Take this quiz as you would take a quiz in class. After you are done, check your work against the answers in the back of the book.

In Exercises 1 and 2, plot the numbers on the real number line and place the correct inequality symbol (< or >) between the numbers. See Additional Answers.

1. -4.5 $>$ -6 **2.** $\frac{3}{4}$ $<$ $\frac{3}{2}$

In Exercises 3 and 4, find the distance between the real numbers.

3. -15 and 7 22 **4.** -8.75 and -2.25 6.5

In Exercises 5 and 6, evaluate the expression.

5. $|-7.6|$ 7.6 **6.** $-|9.8|$ -9.8

In Exercises 7–16, evaluate the expression. Write fractions in simplest form.

7. $32 + (-18)$ 14 **8.** $-12 - (-17)$ 5

9. $\frac{3}{4} + \frac{7}{4}$ $\frac{5}{2}$ **10.** $\frac{2}{3} - \frac{1}{6}$ $\frac{1}{2}$

11. $(-3)(2)(-10)$ 60 **12.** $\left(-\frac{4}{5}\right)\left(\frac{15}{32}\right)$ $-\frac{3}{8}$

13. $\frac{7}{12} \div \frac{5}{6}$ $\frac{7}{10}$ **14.** $\left(-\frac{3}{2}\right)^3$ $-\frac{27}{8}$

15. $3 - 2^2 + 25 \div 5$ 4 **16.** $\dfrac{18 - 2(3 + 4)}{6^2 - (12 \cdot 2 + 10)}$ 2

17. (a) Distributive Property
(b) Additive Inverse Property

18. (a) Associative Property of Addition
(b) Multiplicative Identity Property

In Exercises 17 and 18, identify the property of real numbers illustrated by each statement.

17. (a) $8(u - 5) = 8 \cdot u - 8 \cdot 5$ (b) $10x - 10x = 0$

18. (a) $(7 + y) - z = 7 + (y - z)$ (b) $2x \cdot 1 = 2x$

Applications

19. During one month, you made the following transactions in your non-interest-bearing checking account. Find the balance at the end of the month. $1068.20

NUMBER OR CODE	DATE	TRANSACTION DESCRIPTION	PAYMENT AMOUNT	✓	FEE	DEPOSIT AMOUNT	BALANCE
							$1406.98
2103	1/5	Car Payment	$375 03				
2104	1/7	Phone	$ 59 20				
	1/8	Withdrawal	$225 00				
	1/12	Deposit				$320 45	

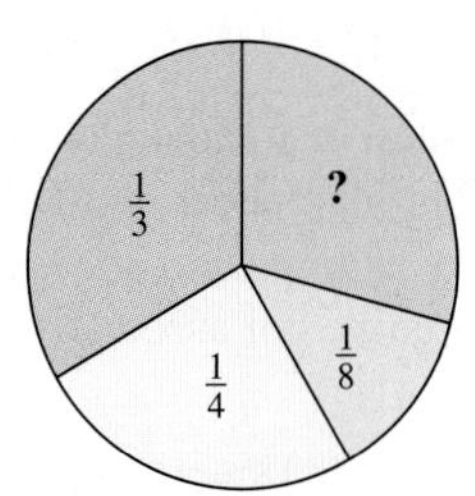

20. You deposit $45 in a retirement account twice each month. How much will you deposit in the account in 8 years? $8640

21. Find the unknown fractional part of the circle graph at the left. Explain how you were able to make this determination.
$\frac{7}{24}$; The sum of the parts of the circle graph is equal to 1.

Study Skills in Action

Keeping a Positive Attitude

A student's experiences during the first three weeks in a math course often determine whether the student sticks with it or not. You can get yourself off to a good start by immediately acquiring a positive attitude and the study behaviors to support it.

Using Study Strategies

In each *Study Skills in Action* feature, you will learn a new study strategy that will help you progress through the course. Each strategy will help you do the following.

- Set up good study habits
- Organize information into smaller pieces
- Create review tools
- Memorize important definitions and rules
- Learn the math at hand

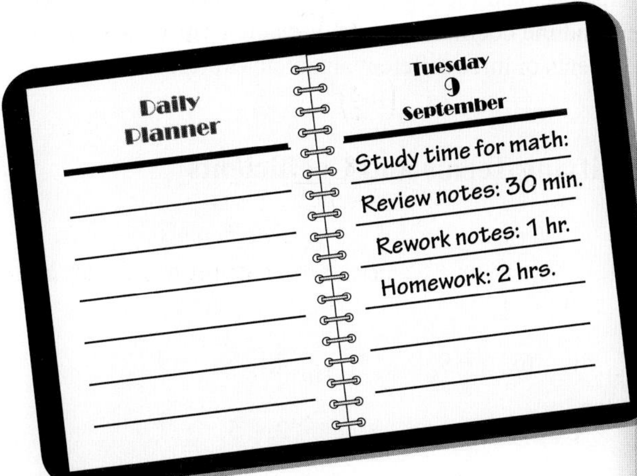

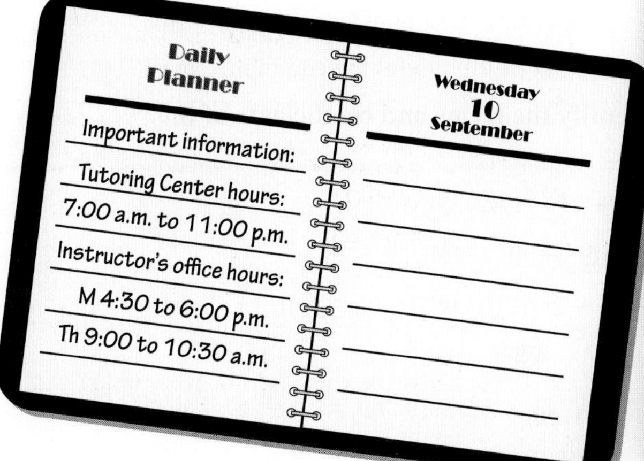

Smart Study Strategy

Create a Positive Study Environment

1. After the first math class, set aside time for reviewing your notes and the textbook, reworking your notes, and completing homework.
2. Find a productive study environment on campus. Most colleges have a tutoring center where students can study and receive assistance as needed.
3. Set up a place for studying at home that is comfortable, but not too comfortable. It needs to be away from all potential distractions.
4. Make at least two other *collegial friends* in class. Collegial friends are students who study well together, help each other out when someone gets sick, and keep each other's attitudes positive.
5. Meet with your instructor at least once during the first two weeks. Ask the instructor what he or she advises for study strategies in the class. This will help you and let the instructor know that you really want to do well.

1.4 Algebraic Expressions

- Identify the terms and coefficients of algebraic expressions.
- Simplify algebraic expressions by combining like terms.
- Evaluate algebraic expressions by substituting values for the variables.

Algebraic Expressions

Algebraic Expression

A collection of letters (called **variables**) and real numbers (called **constants**) combined using the operations of addition, subtraction, multiplication, or division is called an **algebraic expression**.

Study Tip

It is important to understand the difference between a *term* and a *factor*. Terms are separated by addition, whereas factors are separated by multiplication. For instance, the expression $4x(x + 2)$ has three factors: 4, x, and $(x + 2)$.

The **terms** of an algebraic expression are those parts that are separated by *addition*. For example, the algebraic expression $x^2 - 3x + 6$ has three terms: x^2, $-3x$, and 6. Note that $-3x$ is a term, rather than $3x$, because

$$x^2 - 3x + 6 = x^2 + (-3x) + 6.$$ Think of subtraction as a form of addition.

The terms x^2 and $-3x$ are the **variable terms** of the expression, and 6 is the **constant term**. The numerical factor of a term is called the **coefficient**. For instance, the coefficient of the variable term $-3x$ is -3, and the coefficient of the variable term x^2 is 1. Example 1 identifies the terms and coefficients of three different algebraic expressions.

Additional Examples

Identify the terms and coefficients in each algebraic expression.

a. $-3y^2 - 5xy + 7$

b. $3.4a^2 + 5b^2 - 6ab + 1$

Answers:

	Terms	Coefficients
a.	$-3y^2, -5xy, 7$	$-3, -5, 7$
b.	$3.4a^2, 5b^2, -6ab, 1$	$3.4, 5, -6, 1$

EXAMPLE 1 Identifying Terms and Coefficients

	Algebraic Expression	*Terms*	*Coefficients*
a.	$5x - \frac{1}{3}$	$5x, \quad -\frac{1}{3}$	$5, \quad -\frac{1}{3}$
b.	$4y + 6x - 9$	$4y, \quad 6x, \quad -9$	$4, \quad 6, \quad -9$
c.	$\frac{2}{x} + 5x^4 - y$	$\frac{2}{x}, \quad 5x^4, \quad -y$	$2, \quad 5, \quad -1$

Exercises Within Reach®

Solutions in English & Spanish and tutorial videos at AlgebraWithinReach.com

Identifying Terms and Coefficients **In Exercises 1–14, identify the terms and coefficients of the algebraic expression.**

1. $10x + 5$ $10x, 5; 10, 5$

2. $4 + 17y$ $17y, 4; 17, 4$

3. $12 - 6x^2$ $-6x^2, 12; -6, 12$

4. $-16t^2 + 48$ $-16t^2, 48; -16, 48$

5. $-3y^2 + 2y - 8$ $-3y^2, 2y, -8; -3, 2, -8$

6. $9t^2 + 2t + 10$ $9t^2, 2t, 10; 9, 2, 10$

7. $1.2a - 4a^3$ $-4a^3, 1.2a; -4, 1.2$

8. $25z^3 - 4.8z^2$ $25z^3, -4.8z^2; 25, -4.8$

9. $4x^2 - 3y^2 - 5x + 21$ $4x^2, -3y^2, -5x, 21; 4, -3, -5, 21$

10. $7a^2 + 4a - b^2 + 19$ $7a^2, 4a, -b^2, 19; 7, 4, -1, 19$

11. $xy - 5x^2y + 2y^2$ $-5x^2y, 2y^2, xy; -5, 2, 1$

12. $14u^2 + 25uv - 3v^2$ $14u^2, 25uv, -3v^2; 14, 25, -3$

13. $\frac{1}{4}x^2 - \frac{3}{8}x + 5$ $\frac{1}{4}x^2, -\frac{3}{8}x, 5; \frac{1}{4}, -\frac{3}{8}, 5$

14. $\frac{2}{3}y + 8z + \frac{5}{6}$ $\frac{2}{3}y, 8z, \frac{5}{6}; \frac{2}{3}, 8, \frac{5}{6}$

Simplifying Algebraic Expressions

In an algebraic expression, two terms are said to be **like terms** if they are both constant terms or if they have the same variable factor. For example, $2x^2y$, $-x^2y$, and $\frac{1}{2}(x^2y)$ are like terms because they have the same variable factor x^2y. Note that $4x^2y$ and $-x^2y^2$ are not like terms because their variable factors x^2y and x^2y^2 are different.

One way to **simplify an algebraic expression** is to combine like terms.

Study Tip

As you gain experience with the rules of algebra, you may want to combine some of the steps in your work. For instance, you might feel comfortable listing only the following steps to solve Example 2(c).

$$5x + 3y - 4x = 3y + (5x - 4x)$$
$$= 3y + x$$

EXAMPLE 2 Combining Like Terms

a. $2x + 3x - 4 = (2 + 3)x - 4$ Distributive Property

$= 5x - 4$ Simplest form

b. $-3 + 5 + 2y - 7y = (-3 + 5) + (2 - 7)y$ Distributive Property

$= 2 - 5y$ Simplest form

c. $5x + 3y - 4x = 3y + 5x - 4x$ Commutative Property

$= 3y + (5x - 4x)$ Associative Property

$= 3y + (5 - 4)x$ Distributive Property

$= 3y + x$ Simplest form

EXAMPLE 3 Combining Like Terms

a. $7x + 7y - 4x - y = (7x - 4x) + (7y - y)$ Group like terms.

$= 3x + 6y$ Combine like terms.

b. $2x^2 + 3x - 5x^2 - x = (2x^2 - 5x^2) + (3x - x)$ Group like terms.

$= -3x^2 + 2x$ Combine like terms.

c. $3xy^2 - 4x^2y^2 + 2xy^2 + x^2y^2$

$= (3xy^2 + 2xy^2) + (-4x^2y^2 + x^2y^2)$ Group like terms.

$= 5xy^2 - 3x^2y^2$ Combine like terms.

Exercises Within Reach®

Solutions in English & Spanish and tutorial videos at AlgebraWithinReach.com

Simplifying an Algebraic Expression **In Exercises 15–30, simplify the expression by combining like terms.**

15. $3x + 4x$ $7x$

16. $18z + 14z$ $32z$

17. $-2x^2 + 4x^2$ $2x^2$

18. $20a^2 - 5a^2$ $15a^2$

19. $7x - 11x$ $-4x$

20. $-23t + 11t$ $-12t$

21. $9y - 5y + 4y$ $8y$

22. $8y + 7y - y$ $14y$

23. $3x - 2y + 5x + 20y$ $8x + 18y$

24. $-2a + 4b - 7a - b$ $-9a + 3b$

25. $7x^2 - 2x - x^2$ $6x^2 - 2x$

26. $9y + y^2 - 6y$ $y^2 + 3y$

27. $-3z^4 + 6z - z + 8 + z^4 - 4z^2$ $-2z^4 - 4z^2 + 5z + 8$

28. $-5y^3 + 3y - 6y^2 + 8y^3 + y - 4$ $3y^3 - 6y^2 + 4y - 4$

29. $x^2 + 2xy - 2x^2 + xy + y$ $-x^2 + 3xy + y$

30. $3a - 5ab + 9a^2 + 4ab - a$ $9a^2 - ab + 2a$

EXAMPLE 4 **Removing Symbols of Grouping**

Simplify each expression.

a. $3(x - 5) - (2x - 7)$ **b.** $-4(x^2 + 4) + x^2(x + 4)$

Study Tip

The exponential notation described in Section 1.2 can also be used when the base is a variable or an algebraic expression. For instance, in Example 4(b), $x^2 \cdot x$ can be written as

$x^2 \cdot x = x \cdot x \cdot x$
$= x^3.$ 3 factors of x

SOLUTION

a. $3(x - 5) - (2x - 7) = 3x - 15 - 2x + 7$ Distributive Property
$= (3x - 2x) + (-15 + 7)$ Group like terms.
$= x - 8$ Combine like terms.

b. $-4(x^2 + 4) + x^2(x + 4) = -4x^2 - 16 + x^2 \cdot x + 4x^2$ Distributive Property
$= -4x^2 - 16 + x^3 + 4x^2$ Exponential form
$= x^3 + (4x^2 - 4x^2) - 16$ Group like terms.
$= x^3 + 0 - 16$ Combine like terms.
$= x^3 - 16$ Additive Identity Property

EXAMPLE 5 **Removing Symbols of Grouping**

Additional Examples
Simplify each expression.
a. $7x + 4 - 3x + 8$
b. $2x^2 - 5x + 3x^2 - 4$
c. $6y(4 + 8y) + 2y$
Answers:
a. $4x + 12$
b. $5x^2 - 5x - 4$
c. $48y^2 + 26y$

a. $5x - 2x[3 + 2(x - 7)] = 5x - 2x(3 + 2x - 14)$ Distributive Property
$= 5x - 2x(2x - 11)$ Combine like terms.
$= 5x - 4x^2 + 22x$ Distributive Property
$= -4x^2 + 27x$ Combine like terms.

b. $-3x(5x^4) + 2x^5 = -15x \cdot x^4 + 2x^5$ Multiply.
$= -15x^5 + 2x^5$ Exponential form
$= -13x^5$ Combine like terms.

Exercises Within Reach®

Solutions in English & Spanish and tutorial videos at AlgebraWithinReach.com

Simplifying an Algebraic Expression **In Exercises 31–50, simplify the expression.**

31. $10(x - 3) + 2x - 5$ $12x - 35$
32. $3(x + 1) + x - 6$ $4x - 3$
33. $x - (5x + 9)$ $-4x - 9$
34. $y - (3y - 1)$ $-2y + 1$
35. $5a - (4a - 3)$ $a + 3$
36. $7x - (2x + 5)$ $5x - 5$
37. $-3(3y - 1) + 2(y - 5)$ $-7y - 7$
38. $5(a + 6) - 4(2a - 1)$ $-3a + 34$
39. $-3(y^2 - 2) + y^2(y + 3)$ $y^3 + 6$
40. $x(x^2 - 5) - 4(4 - x)$ $x^3 - x - 16$
41. $x(x^2 + 3) - 3(x + 4)$ $x^3 - 12$
42. $5(x + 1) - x(2x + 6)$ $-2x^2 - x + 5$
43. $9a - [7 - 5(7a - 3)]$ $44a - 22$
44. $12b - [9 - 7(5b - 6)]$ $47b - 51$
45. $3[2x - 4(x - 8)]$ $-6x + 96$
46. $4[5 - 3(x^2 + 10)]$ $-12x^2 - 100$
47. $8x + 3x[10 - 4(3 - x)]$ $12x^2 + 2x$
48. $5y - y[9 + 6(y - 2)]$ $-6y^2 + 8y$
49. $2[3(b - 5) - (b^2 + b + 3)]$ $-2b^2 + 4b - 36$
50. $5[3(z + 2) - (z^2 + z - 2)]$ $-5z^2 + 10z + 40$

Evaluating Algebraic Expressions

EXAMPLE 6 Evaluating Algebraic Expressions

Evaluate each algebraic expression when $x = -2$.

a. $5 + x^2$ **b.** $5 - x^2$

SOLUTION

a. When $x = -2$, the expression $5 + x^2$ has a value of

$$5 + (-2)^2 = 5 + 4 = 9.$$

b. When $x = -2$, the expression $5 - x^2$ has a value of

$$5 - (-2)^2 = 5 - 4 = 1.$$

A common error is to confuse the operation of subtraction with the substitution of a negative number. Point out how parentheses are used when a numerical value is substituted for a variable in an expression.

EXAMPLE 7 Evaluating Algebraic Expressions

Evaluate each algebraic expression when $x = 2$ and $y = -1$.

a. $|y - x|$ **b.** $x^2 - 2xy + y^2$

SOLUTION

a. When $x = 2$ and $y = -1$, the expression $|y - x|$ has a value of

$$|(-1) - (2)| = |-3| = 3.$$

b. When $x = 2$ and $y = -1$, the expression $x^2 - 2xy + y^2$ has a value of

$$2^2 - 2(2)(-1) + (-1)^2 = 4 + 4 + 1 = 9.$$

Exercises Within Reach ®

Solutions in English & Spanish and tutorial videos at AlgebraWithinReach.com

Evaluating an Algebraic Expression **In Exercises 51–62, evaluate the expression for the specified values of the variable(s). If not possible, state the reason.**

	Expression	*Values*		*Expression*	*Values*
51.	$5 - 3x$	(a) $x = \frac{2}{3}$ 3	**52.**	$\frac{3}{2}x - 2$	(a) $x = 6$ 7
		(b) $x = 5$ -10			(b) $x = -3$ $-\frac{13}{2}$
53.	$10 - 4x^2$	(a) $x = -1$ 6	**54.**	$3y^2 + 10$	(a) $y = -2$ 22
		(b) $x = \frac{1}{2}$ 9			(b) $y = \frac{1}{2}$ $\frac{43}{4}$
55.	$y^2 - y + 5$	(a) $y = 2$ 7	**56.**	$2x^2 + 5x - 3$	(a) $x = 2$ 15
		(b) $y = -2$ 11			(b) $x = -3$ 0
57.	$\frac{1}{x^2} + 3$	(a) $x = 0$ Not possible; undefined	**58.**	$5 - \frac{3}{x}$	(a) $x = 0$ Not possible; undefined
		(b) $x = 3$ $\frac{28}{9}$			(b) $x = -6$ $\frac{11}{2}$
59.	$3x + 2y$	(a) $x = 1,\ y = 5$ 13	**60.**	$6x - 5y$	(a) $x = -2,\ y = -3$ 3
		(b) $x = -6\ y = -9$ -3			(b) $x = 1,\ y = 1$ 1
61.	$x^2 - xy + y^2$	(a) $x = 2,\ y = -1$ 7	**62.**	$y^2 + xy - x^2$	(a) $x = 5,\ y = 2$ -11
		(b) $x = -3\ y = -2$ 7			(b) $x = -3,\ y = 3$ -9

EXAMPLE 8 Evaluating an Algebraic Expression

Evaluate

$$\frac{2xy}{x+1}$$

when $x = -4$ and $y = -3$.

Additional Examples

Evaluate each expression.

a. $2y - 3x$ when $x = -5$ and $y = 6$

b. $y^2 + 3x - 7$ when $x = 1$ and $y = -1$

Answers:

a. 27

b. −3

SOLUTION

When $x = -4$ and $y = -3$, the expression $\frac{2xy}{x+1}$ has a value of

$$\frac{2xy}{x+1} = \frac{2(-4)(-3)}{-4+1}$$ Substitute −4 for x and −3 for y.

$$= \frac{24}{-3}$$ Simplify.

$$= -8.$$ Divide.

Application

EXAMPLE 9 Geometry: Finding the Area of a Roof

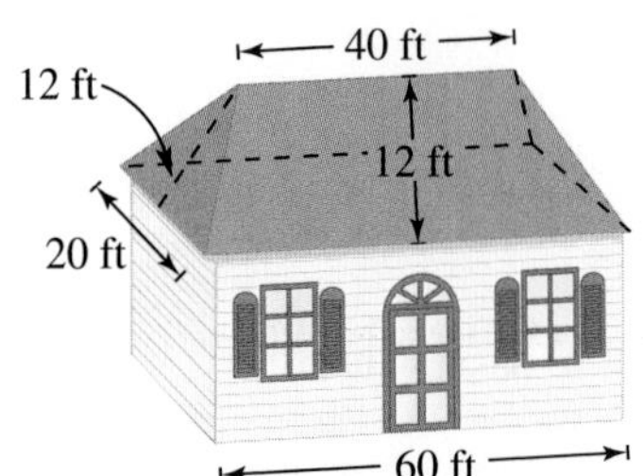

The roof shown at the left is made up of two trapezoids and two triangles. Find the total area of the roof. (*Note:* For a trapezoid,

$$\text{Area} = \frac{1}{2}h(b_1 + b_2)$$

where b_1 and b_2 are the lengths of the bases and h is the height.)

SOLUTION

$$\text{Total area} = 2 \cdot \frac{1}{2}h(b_1 + b_2) + 2 \cdot \frac{1}{2}bh$$

$$= 2\left(\frac{1}{2}\right)(12)(60 + 40) + 2\left(\frac{1}{2}\right)(20)(12)$$

$$= 1200 + 240$$

$$= 1440$$

The total area of the roof is 1440 square feet.

Exercises Within Reach®

Solutions in English & Spanish and tutorial videos at AlgebraWithinReach.com

Evaluating an Algebraic Expression **In Exercises 63–68, evaluate the expression for the specified values of the variable(s). If not possible, state the reason.**

	Expression	*Values*		*Expression*	*Values*
63.	$\frac{x}{y^2 - x}$ (a) Not possible; undefined (b) $\frac{1}{2}$	(a) $x = 4,\ y = 2$ (b) $x = 3,\ y = 3$	**64.**	$\frac{x}{x - y}$ (a) 0 (b) Not possible; undefined	(a) $x = 0, y = 10$ (b) $x = 4, y = 4$
65.	$\lvert y - x \rvert$ (a) 3 (b) 0	(a) $x = 2,\ y = 5$ (b) $x = -2,\ y = -2$	**66.**	$\lvert x^2 - y \rvert$ (a) 2 (b) 24	(a) $x = 0, y = -2$ (b) $x = 3, y = -15$
67.	Distance traveled: rt (a) 210 (b) 140	(a) $r = 40,\ t = 5\frac{1}{4}$ (b) $r = 35,\ t = 4$	**68.**	Simple interest: Prt (a) 4550 (b) 2646	(a) $P = 7000, r = 0.065, t = 10$ (b) $P = 4200,\ r = 0.07,\ t = 9$

Application

EXAMPLE 10 Using a Mathematical Model

The yearly revenues (in billions of dollars) for gambling industries in the United States for the years 2002 through 2010 can be modeled by

$$\text{Revenue} = -0.169t^2 + 2.76t + 14.5, \quad 2 \le t \le 10$$

where t represents the year, with $t = 2$ corresponding to 2002. Create a table that shows the revenue for each of these years. (*Source:* 2010 Service Annual Survey)

SOLUTION

To create a table of values that shows the revenues (in billions of dollars) for the gambling industries for the years 2002 through 2010, evaluate the expression $-0.169t^2 + 2.76t + 14.5$ for each integer value of t from $t = 2$ to $t = 10$.

Year	2002	2003	2004	2005	2006	2007	2008	2009	2010
t	2	3	4	5	6	7	8	9	10
Revenue	19.3	21.3	22.8	24.1	25.0	25.5	25.8	25.7	25.2

Exercises Within Reach ®

Solutions in English & Spanish and tutorial videos at AlgebraWithinReach.com

Using a Model **The annual sales (in billions of dollars) of hunting equipment in the United States for the years 2003 through 2009 can be modeled by**

$$\textbf{Sales} = 0.37t + 1.6, \quad 3 \le t \le 9$$

where t represents the year, with $t = 3$ corresponding to 2003. In Exercises 69 and 70, use the model. (*Source:* National Sporting Goods Association)

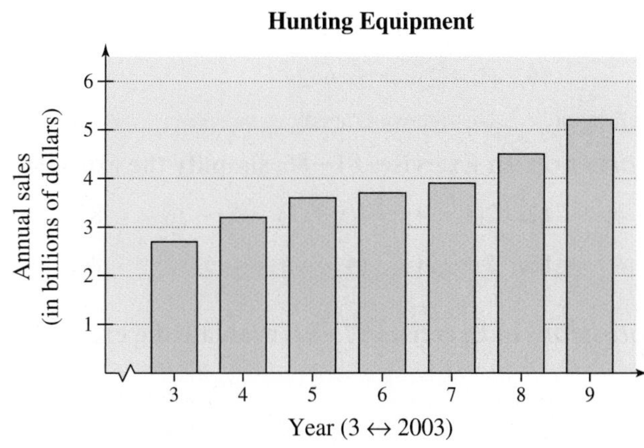

69. Graphically approximate the sales of hunting equipment in 2005. Then use the model to confirm your estimate algebraically. \$3.5 billion; \$3.45 billion

70. Use the model to complete the table showing the sales for the years 2003 through 2009.

Year	2003	2004	2005	2006	2007	2008	2009
Sales	2.71	3.08	3.45	3.82	4.19	4.56	4.93

Concept Summary: Simplifying Algebraic Expressions

What

You can use properties of algebra to **simplify algebraic expressions**. To simplify these types of expressions usually means to remove symbols of grouping and combine **like terms**.

EXAMPLE

Simplify the expression $3x + 2(4 + x)$.

How

The main tool for removing symbols of grouping and combining like terms is the Distributive Property.

EXAMPLE

$3x + 2(4 + x)$

$= 3x + 8 + 2x$ Distributive Prop.

$= 3x + 2x + 8$ Comm. Prop.

$= x(3 + 2) + 8$ Distributive Prop.

$= 5x + 8$ Simplify.

Why

Simplifying an algebraic expression into a more usable form is one of the most frequently used skills in algebra.

Exercises Within Reach ®

Worked-out solutions to odd-numbered exercises at AlgebraWithinReach.com

Concept Summary Check

71. ***Precision*** Explain the difference between terms and factors in an algebraic expression. In an algebraic expression, terms are separated by addition whereas factors are separated by multiplication.

72. ***Distributive Property*** Explain how you can use the Distributive Property to simplify the expression $5x + 3x$. See Additional Answers.

73. ***Combining Like Terms*** Explain how to combine like terms in an algebraic expression. Give an example. See Additional Answers.

74. ***Writing*** Explain the difference between simplifying an algebraic expression and evaluating an algebraic expression. See Additional Answers.

Extra Practice

Identifying a Property of Algebra **In Exercises 75–80, identify the property of algebra illustrated by the statement.**

75. $4 - 3x = -3x + 4$ Commutative Property of Addition

76. $(10 + x) - y = 10 + (x - y)$ Associative Property of Addition

77. $-5(2x) = (-5 \cdot 2)x$ Associative Property of Multiplication

78. $(x - 2)(3) = 3(x - 2)$ Commutative Property of Multiplication

79. $(5 - 2)x = 5x - 2x$ Distributive Property

80. $7y + 2y = (7 + 2)y$ Distributive Property

Simplifying an Algebraic Expression **In Exercises 81–86, simplify the expression.**

81. $3(x + 2) - 5(x - 7)$ $-2x + 41$

82. $2(x - 4) - 4(x + 2)$ $-2x - 16$

83. $2[x + 2(x + 7)]$ $6x + 28$

84. $-2[3x + 2(4 - x)]$ $-2x - 16$

85. $2x - 3[x - (4 - x)]$ $-4x + 12$

86. $-7x + 3[x - (3 - 2x)]$ $2x - 9$

Evaluating an Algebraic Expression **In Exercises 87–92, evaluate the expression for the specified values of the variable(s).**

87. $-3 + 4x$
(a) $x = 3$ 9
(b) $x = -2$ -11

88. $12 - 2x^2$
(a) $x = 2$ 4
(b) $x = -4$ -20

89. $b^2 - 4ab$
(a) $a = 2,\ b = -3$ 33
(b) $a = 6,\ b = -4$ 112

90. $a^2 + 2ab$
(a) $a = -2,\ b = 3$ -8
(b) $a = 4,\ b = -2$ 0

91. $\dfrac{-y}{x^2 + y^2}$
(a) $x = 0,\ y = 5$ $-\frac{1}{5}$
(b) $x = 1,\ y = -3$ $\frac{3}{10}$

92. $\dfrac{2x - y}{y^2 + 1}$
(a) $x = 1\ y = 2$ 0
(b) $x = 1,\ y = 3$ $-\frac{1}{10}$

Geometry **In Exercises 93–96, find the volume of the rectangular solid by evaluating the expression *lwh* for the dimensions given in the figure.**

93.

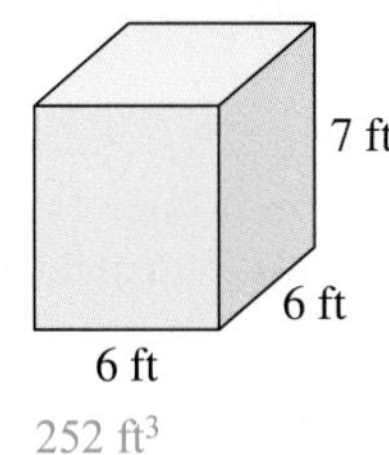

252 ft^3

94.

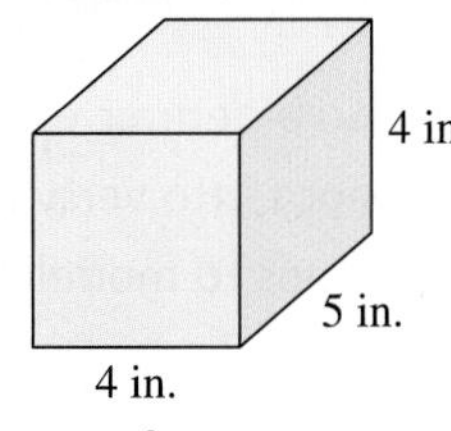

80 in.^3

95.

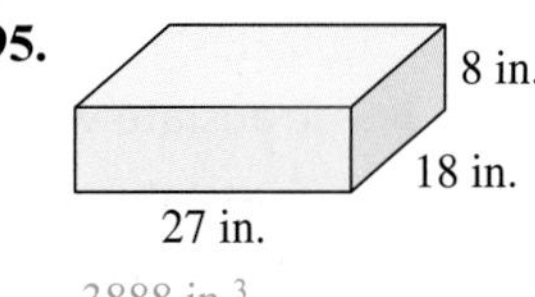

3888 in.^3

96.

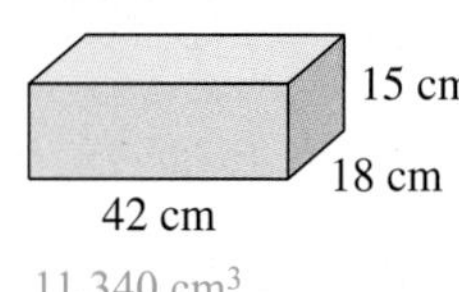

$11{,}340 \text{ cm}^3$

Evaluating an Expression **In Exercises 97–100, evaluate the expression $0.01p + 0.05n + 0.10d + 0.25q$ to find the value of the given number of pennies p, nickels n, dimes d, and quarters q.**

97. 11 pennies, 7 nickels, 3 quarters $1.21

98. 8 pennies, 13 nickels, 6 dimes $1.33

99. 43 pennies, 27 nickels, 17 dimes, 15 quarters $7.23

100. 111 pennies, 22 nickels, 2 dimes, 42 quarters $12.91

Geometry **In Exercises 101 and 102, write and simplify an expression for the area of the figure. Then evaluate the expression for the given value of the variable.**

101. $b = 15$ $\frac{1}{2}b^2 - \frac{3}{2}b$; 90

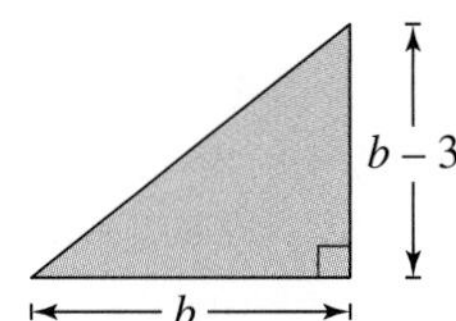

102. $h = 12$ $\frac{5}{4}h^2 + 10h$; 300

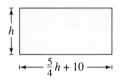

h

$\frac{5}{4}h + 10$

Explaining Concepts

103. (a) A convex polygon with n sides has

$$\frac{n(n-3)}{2}, \quad n \geq 4$$

diagonals. Verify the formula for a square, a pentagon, and a hexagon. See Additional Answers.

(b) Explain why the formula in part (a) will always yield a natural number. See Additional Answers.

104. ***Reasoning*** Is it possible to evaluate the expression

$$\frac{x+2}{y-3}$$

when $x = 5$ and $y = 3$? Explain.

No. When $y = 3$, the expression is undefined.

105. ***Reasoning*** Is it possible to evaluate the expression $3x + 5y - 18z$ when $x = 10$ and $y = 8$? Explain.

No. To evaluate the expression is to find the value of the expression for given values of the variables x, y, and z. Because z does not have a value, the expression cannot be evaluated.

106. ***Writing*** State the procedure for simplifying an algebraic expression by removing a set of parentheses preceded by a minus sign, such as the parentheses in $a - (b + c)$. Then given an example.

To remove a set of parentheses preceded by a minus sign, distribute -1 to each term inside the parentheses. For example: $13 - (-10 + 5) = 13 + 10 - 5 = 18$.

107. ***Writing*** How can a factor be part of a term in an algebraic expression? Explain and give an example.

Factors in a term are separated by multiplication. Because the coefficient and the variable(s) in a term are multiplied together, the coefficient and the variable(s) are all factors of the term. For example, in the algebraic expression $3x + 2$, the term $3x$ has the factors 3 and x.

108. ***Writing*** How can an algebraic term be part of a factor in an algebraic expression? Give an example.

A factor can consist of a sum of terms. For example, the sum $x + y$ is a factor of $(x + y) \cdot z$.

109. ***Number Sense*** You know that the expression $180 - 10x$ has a value of 100. Is it possible to determine the value of x with this information? Explain and find the value, if possible.

Yes. To determine the value of x, set the expression equal to 100. Because $180 - 10x = 100$, the value of x is 8.

110. ***Number Sense*** You know that the expression $8y - 5x$ has a value of 14. Is it possible to determine the values of x and y with this information? Explain and find the values, if possible.

No. There are an infinite number of values of x and y that would satisfy $8y - 5x = 14$. For example, $x = 10$ and $y = 8$ would be a solution, and so would $x = 2$ and $y = 3$.

1.5 Constructing Algebraic Expressions

- Translate verbal phrases into algebraic expressions.
- Translate algebraic expressions into verbal phrases.
- Construct algebraic expressions to model real-life problems.

Translating Verbal Phrases into Algebraic Expressions

Translating Key Words and Phrases

Key Words and Phrases	*Verbal Description*	*Algebraic Expression*
Addition:		
Sum, plus, greater than, increased by, more than,	The sum of 5 and x	$5 + x$
exceeds, total of	Seven more than y	$y + 7$
Subtraction:		
Difference, minus, less than, decreased by,	b is subtracted from 4.	$4 - b$
subtracted from, reduced by, the remainder	Three less than z	$z - 3$
Multiplication:		
Product, multiplied by, twice, times, percent of	Two times x	$2x$
Division:		
Quotient, divided by, ratio, per	The ratio of x and 8	$\frac{x}{8}$

EXAMPLE 1 Translating Verbal Phrases

	Verbal Description	*Algebraic Expression*
a.	Seven more than three times x	$3x + 7$
b.	Four less than the product of 6 and n	$6n - 4$
c.	The quotient of x and 3, decreased by 6	$\frac{x}{3} - 6$

Exercises Within Reach®

Solutions in English & Spanish and tutorial videos at AlgebraWithinReach.com

Translating a Verbal Phrase **In Exercises 1–6, translate the verbal phrase into an algebraic expression.**

1. The sum of 23 and a number n $23 + n$
2. Twelve more than a number n $n + 12$
3. The sum of 12 and twice a number n $12 + 2n$
4. The total of 25 and three times a number n $25 + 3n$
5. Six less than a number n $n - 6$
6. Fifteen decreased by three times a number n $15 - 3n$

EXAMPLE 2 Translating Verbal Phrases

	Verbal Description	*Algebraic Expression*
a.	Eight added to the product of 2 and n	$2n + 8$
b.	Four times the sum of y and 9	$4(y + 9)$
c.	The difference of a and 7, all divided by 9	$\dfrac{a - 7}{9}$

In Examples 1 and 2, the verbal description specified the name of the variable. In most real-life situations, however, the variables are not specified and it is your task to assign variables to the *appropriate* quantities.

EXAMPLE 3 Translating Verbal Phrases

	Verbal Description	*Label*	*Algebraic Expression*
a.	The sum of 7 and a number	The number $= x$	$7 + x$
b.	Four decreased by the product of 2 and a number	The number $= n$	$4 - 2n$
c.	Seven less than twice the sum of a number and 5	The number $= y$	$2(y + 5) - 7$

Exercises Within Reach ®

Solutions in English & Spanish and tutorial videos at AlgebraWithinReach.com

Translating a Verbal Phrase **In Exercises 7–24, translate the verbal phrase into an algebraic expression.**

7. Four times a number n minus 10
$4n - 10$

8. The product of a number y and 10 is decreased by 35.
$10y - 35$

9. Half of a number n $(1/2)n$

10. Seven-fifths of a number n $(7/5)n$

11. The quotient of a number x and 6 $x/6$

12. The ratio of y and 3 $y/3$

13. Eight times the ratio of N and 5 $8 \cdot (N/5)$

14. Fifteen times the ratio of x and 32 $15 \cdot (x/32)$

15. The number c is quadrupled and the product is increased by 10. $4c + 10$

16. The number u is tripled and the product is increased by 250. $3u + 250$

17. Thirty percent of the original price L $0.30L$

18. Twenty-five percent of the bill B $0.25B$

19. The sum of a number and 5, all divided by 10
$(n + 5)/10$

20. The sum of 7 and twice a number, all divided by 8
$(7 + 2x)/8$

21. The absolute value of the difference between a number and 8 $|n - 8|$

22. The absolute value of the quotient of a number and 4
$\left|\dfrac{x}{4}\right|$

23. The product of 3 and the square of a number is decreased by 4
$3x^2 - 4$

24. The sum of 10 and one-fourth the square of a number
$10 + \frac{1}{4}x^2$

Study Tip

When you write a verbal model or construct an algebraic expression, watch out for statements that may be interpreted in more than one way. For instance, the statement "The sum of x and 1 divided by 5" is ambiguous because it could mean

$$\frac{x+1}{5} \text{ or } x+\frac{1}{5}.$$

Notice in Example 4(b) that the verbal description for

$$\frac{3+x}{4}$$

contains the phrase "all divided by 4."

Translating Algebraic Expressions into Verbal Phrases

A good way to learn algebra is to do it forward and backward. For instance, the next example translates algebraic expressions into verbal form. Keep in mind that other key words could be used to describe the operations in each expression.

EXAMPLE 4 Translating Algebraic Expressions

Without using a variable, write a verbal description for each expression.

a. $5x - 10$

b. $\dfrac{3+x}{4}$

c. $2(3x + 4)$

d. $\dfrac{4}{x-2}$

SOLUTION

a. 10 less than the product of 5 and a number

b. The sum of 3 and a number, all divided by 4

c. Twice the sum of 3 times a number and 4

d. Four divided by a number reduced by 2

Exercises Within Reach ®

Solutions in English & Spanish and tutorial videos at AlgebraWithinReach.com

Translating an Algebraic Expression **In Exercises 25–40, write a verbal description of the algebraic expression without using the variable.**

25. $t - 2$
A number decreased by 2

26. $5 - x$
The difference of 5 and a number

27. $y + 50$
A number increased by 50

28. $2y + 3$
Three more than the product of a number and 2

29. $2 - 3x$
Two decreased by three times a number

30. $7y - 4$
Four less than seven times a number

31. $\dfrac{z}{2}$
The ratio of a number and 2

32. $\dfrac{y}{8}$
The ratio of a number and 8

33. $\frac{4}{5}x$
Four-fifths of a number

34. $\frac{2}{3}t$
Two-thirds of a number

35. $8(x - 5)$
Eight times the difference of a number and 5

36. $(y + 6)4$
The sum of a number and 6, multiplied by 4

37. $\dfrac{x+10}{3}$
The sum of a number and 10, divided by 3

38. $\dfrac{3-n}{9}$
The difference of 3 and a number, divided by 9

39. $y^2 - 3$
The square of a number, decreased by 3

40. $x^2 + 2$
Two more than the square of a number

Constructing Mathematical Models

Translating a verbal phrase into a mathematical model is critical in problem solving. The next four examples will demonstrate three steps for creating a mathematical model.

1. Construct a verbal model that represents the problem situation.
2. Assign labels to all quantities in the verbal model.
3. Construct a mathematical model (algebraic expression).

Application

EXAMPLE 5 Constructing a Mathematical Model

A cash register contains x quarters. Write an algebraic expression for this amount of money in dollars.

SOLUTION

Verbal Model: [Value of coin] • [Number of coins]

Labels: Value of coin = 0.25 (dollars per quarter)
Number of coins = x (quarters)

Expression: $0.25x$ (dollars)

Application

EXAMPLE 6 Constructing a Mathematical Model

A cash register contains n nickels and d dimes. Write an algebraic expression for this amount of money in cents.

SOLUTION

Verbal Model: [Value of nickel] • [Number of nickels] + [Value of dime] • [Number of dimes]

Labels: Value of nickel = 5 (cents per nickel)
Number of nickels = n (nickels)
Value of dime = 10 (cents per dime)
Number of dimes = d (dimes)

Expression: $5n + 10d$ (cents)

Exercises Within Reach®

Solutions in English & Spanish and tutorial videos at AlgebraWithinReach.com

Constructing a Mathematical Model **In Exercises 41–46, write an algebraic expression that represents the specified quantity.**

41. The amount of money (in dollars) represented by n quarters $0.25n$

42. The amount of money (in dollars) represented by x nickels $0.05x$

43. The amount of money (in dollars) represented by m dimes $0.10m$

44. The amount of money (in dollars) represented by y pennies $0.01y$

45. The amount of money (in cents) represented by m nickels and n dimes $5m + 10n$

46. The amount of money (in cents) represented by m dimes and n quarters $10m + 25n$

Application **EXAMPLE 7** **Constructing a Mathematical Model**

A person riding a bicycle travels at a constant rate of 12 miles per hour. Write an algebraic expression showing how far the person can ride in t hours.

Study Tip

Using unit analysis, you can see that the expression in Example 7 has miles as its units of measure.

$$12\,\frac{\text{miles}}{\cancel{\text{hour}}} \cdot t\,\cancel{\text{hours}}$$

SOLUTION

For this problem, use the formula Distance = (Rate)(Time).

Verbal Model: Rate • Time

Labels: Rate = 12 (miles per hour)
Time = t (hours)

Expression: $12t$ (miles)

When translating verbal phrases involving percents, be sure you write the percent in *decimal form*.

Percent	*Decimal Form*	*Percent*	*Decimal Form*
4%	0.04	140%	1.40
62%	0.62	25%	0.25

Remember that when you find a percent of a number, you multiply. For instance, 25% of 78 is given by

$0.25(78) = 19.5.$ 25% of 78

Application **EXAMPLE 8** **Constructing a Mathematical Model**

A person adds k liters of fluid containing 55% antifreeze to a car radiator. Write an algebraic expression that indicates how much antifreeze was added.

Study Tip

In Example 8, note that the algebraic expression uses the decimal form of 55%. That is, you compute with 0.55 rather than 55%.

SOLUTION

Verbal Model: Percent antifreeze • Number of liters

Labels: Percent of antifreeze = 0.55 (in decimal form)
Number of liters = k (liters)

Expression: $0.55k$ (liters)

Exercises Within Reach®

Solutions in English & Spanish and tutorial videos at AlgebraWithinReach.com

Constructing a Mathematical Model **In Exercises 47–52, write an algebraic expression that represents the specified quantity.**

47. The distance traveled in t hours at an average speed of 55 miles per hour $55t$

48. The distance traveled in 5 hours at an average speed of r miles per hour $5r$

49. The time required to travel 320 miles at an average speed of r miles per hour
$\frac{320}{r}$

50. The average rate of speed when traveling 320 miles in t hours
$\frac{320}{t}$

51. The amount of antifreeze in a cooling system containing y gallons of coolant that is 45% antifreeze $0.45y$

52. The amount of water in q quarts of a food product that is 65% water $0.65q$

Application **EXAMPLE 9** **Geometry: Constructing Mathematical Models**

Write expressions for the perimeter and area of the rectangle show at the left.

2w in.

(w + 12) in.

SOLUTION

For the perimeter of the rectangle, use the formula

Perimeter $= 2(\text{Length}) + 2(\text{Width})$

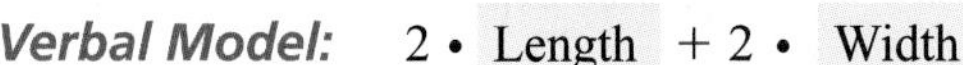

Verbal Model: $2 \cdot$ Length $+ 2 \cdot$ Width

Labels: Length $= w + 12$ (inches)
Width $= 2w$ (inches)

Expression: $2(w + 12) + 2(2w) = 2w + 24 + 4w$
$= 6w + 24$ (inches)

For the area of the rectangle, use the formula

Area $= (\text{Length})(\text{Width})$

Verbal Model: Length $\cdot$ Width

Labels: Length $= w + 12$ (inches)
Width $= 2w$ (inches)

Expression: $(w + 12)(2w) = 2w^2 + 24w$ (square inches)

Application **EXAMPLE 10** **Consecutive Integer Problem**

Write an expression for the following phrase.

"The sum of two consecutive integers, the first of which is n"

SOLUTION

The first integer is n. The next consecutive integer is $n + 1$. So the sum of two consecutive integers is $n + (n + 1) = 2n + 1$.

Exercises Within Reach®

Solutions in English & Spanish and tutorial videos at AlgebraWithinReach.com

Geometry **In Exercises 53–56, write expressions for the perimeter and area of the region. Simplify the expressions.** See Additional Answers.

53. 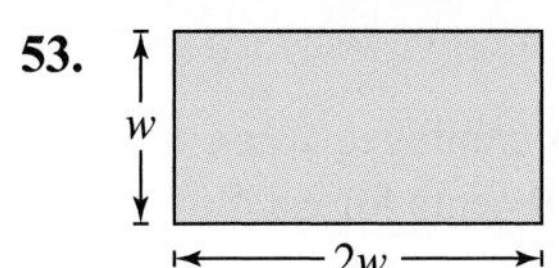

54. 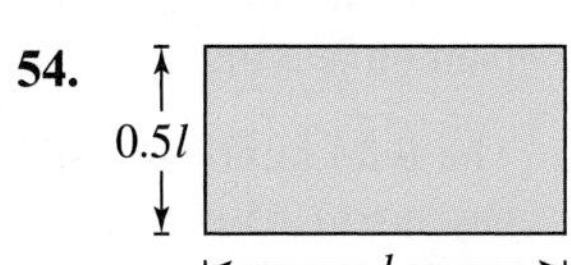

55. 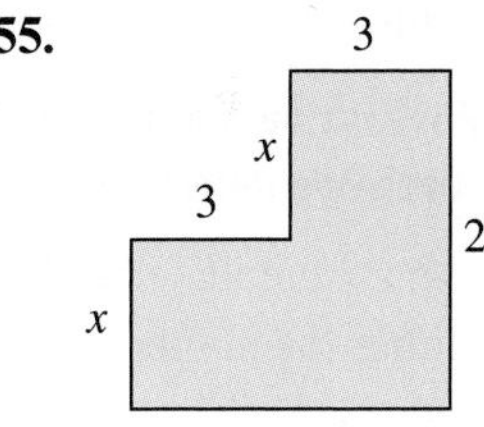

56. 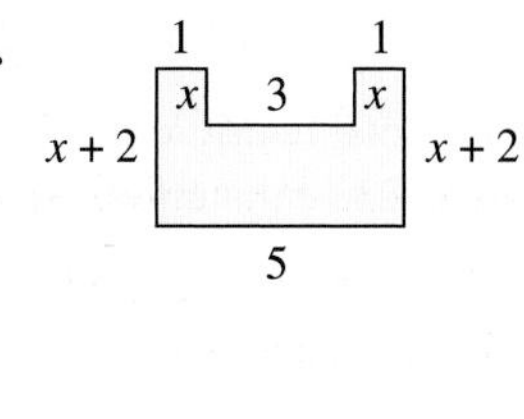

Constructing a Mathematical Model **In Exercises 57 and 58, write an algebraic expression that represents the specified quantity, and simplify, if possible.**

57. The sum of three consecutive integers, the first of which is n $(n) + (n + 1) + (n + 2) = 3n + 3$

58. The sum of four consecutive integers, the first of which is n $(n) + (n + 1) + (n + 2) + (n + 3) = 4n + 6$

auremar/Shutterstock.com

Concept Summary: Using Verbal Models to Write Algebraic Expressions

What

You can construct algebraic expressions from written statements by constructing verbal mathematical models.

EXAMPLE

A gallon of milk costs \$3.29. Write an algebraic expression that represents the total cost of buying g gallons of milk.

How

Determine any known and unknown quantities from the written statement. Then construct a verbal model. Use the model to write an algebraic expression.

EXAMPLE

Cost per gallon • Number of gallons =

\$3.29 • g gallons =

$$3.29 \cdot g = 3.29g$$

Note that nowhere in the written statement does it say to multiply 3.29 by g. It is *implied* in the statement.

Why

Algebra is a problem-solving language that you can use to solve real-life problems. Some of these problems can consist of several known and unknown quantities. Using a verbal mathematical model can help you organize your thoughts about the known and unknown quantities, as well as the overall solution to the real-life problem.

Exercises Within Reach®

Worked-out solutions to odd-numbered exercises at AlgebraWithinReach.com

Concept Summary Check

59. ***Identifying Operations*** The phrase *reduced by* implies what operation? Subtraction

60. ***Identifying Operations*** The word *ratio* indicates what operation? Division

61. ***Known and Unknown Quantities*** What are the known and unknown quantities in the example above?
Known: cost per gallon; Unknown: number of gallons

62. ***Assigning Variables*** Two unknown quantities in a verbal model are "Number of cherries" and "Number of strawberries." What variables would you use to represent these quantities? Explain.
Sample answer: Let c be the number of cherries and s be the number of strawberries. Using the first letter of each word helps you to remember what each variable represents.

Extra Practice

Constructing a Mathematical Model **In Exercises 63–74, write an algebraic expression that represents the specified quantity, and simplify, if possible.**

63. The amount of wage tax due on a taxable income of I dollars when the tax rate is 1.25% $0.0125I$

64. The amount of sales tax on a purchase valued at L dollars when the tax rate is 5.5% $0.055L$

65. The sale price of a coat that has an original price of L dollars when it is a "20% off" sale $L - 0.20L = 0.80L$

66. The total bill for a meal that cost C dollars when you plan on leaving a 15% tip $C + 0.15C = 1.15C$

67. The total hourly wage for an employee when the base pay is \$8.25 per hour plus 60 cents for each of q units produced per hour $8.25 + 0.60q$

68. The total hourly wage for an employee when the base pay is \$11.65 per hour plus 80 cents for each of q units produced per hour $11.65 + 0.80q$

69. The sum of a number n and five times the number
$n + 5n = 6n$

70. The sum of a number n and eight times the number
$n + 8n = 9n$

71. The sum of three consecutive odd integers, the first of which is $2n + 1$ $(2n + 1) + (2n + 3) + (2n + 5) = 6n + 9$

72. The sum of three consecutive even integers, the first of which is $2n$ $(2n) + (2n + 2) + (2n + 4) = 6n + 6$

73. The product of two consecutive even integers, divided by 4
$\frac{2n(2n + 2)}{4} = n^2 + n$

74. The absolute value of the difference of two consecutive integers, divided by 2
$\frac{|(x) - (x + 1)|}{2} = \frac{|-1|}{2} = \frac{1}{2}$

Geometry **In Exercises 75–78, write an expression for the area of the figure. Simplify the expression.**

75.

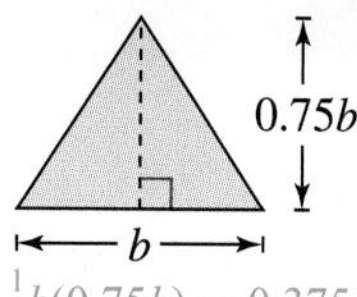

$\frac{1}{2}b(0.75b) = 0.375b^2$

76.

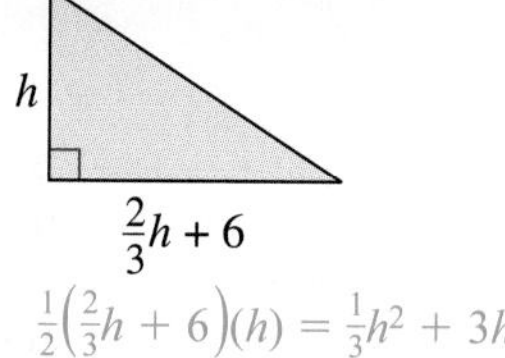

$\frac{1}{2}\left(\frac{2}{3}h + 6\right)(h) = \frac{1}{3}h^2 + 3h$

77.

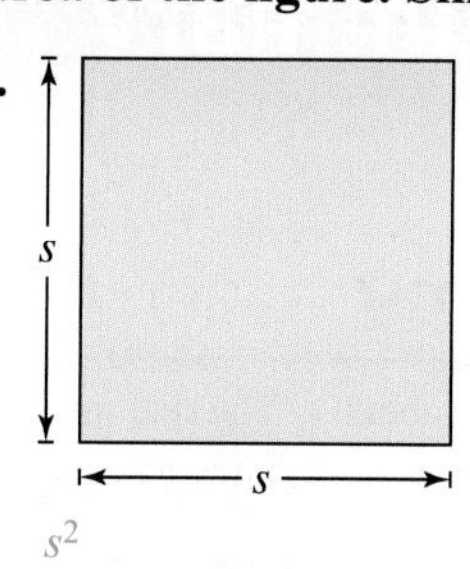

s^2

78.

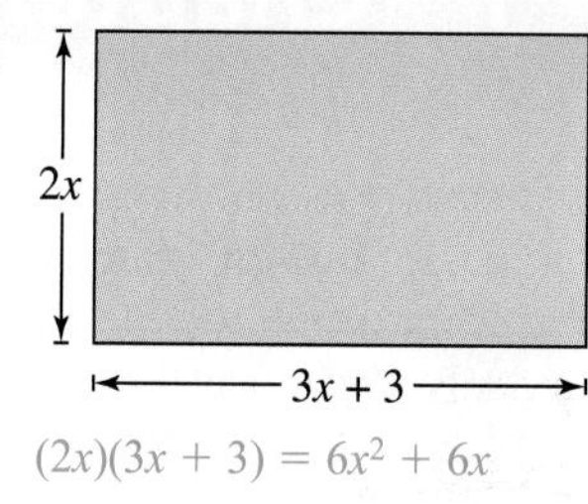

$(2x)(3x + 3) = 6x^2 + 6x$

79. ***Geometry*** Write an expression for the area of the soccer field shown in the figure. What is the unit of measure for the area? $b(b - 50) = b^2 - 50b$; square meters

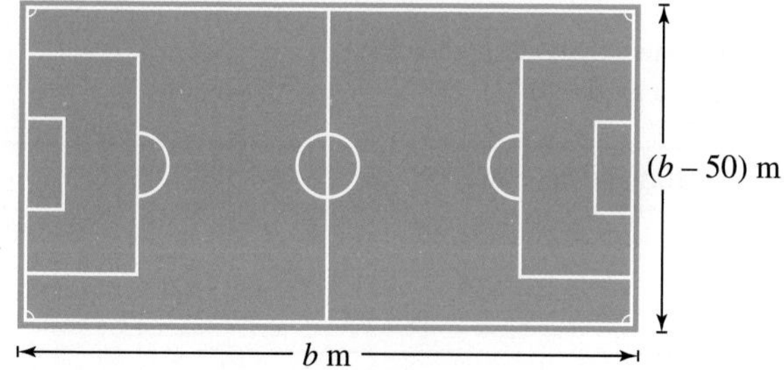

80. ***Geometry*** A plane is flying an advertising banner. The length of the banner is 6 times its width w (in feet). Write an expression for the area of the banner. What is the unit of measure for the area? $6w^2$; square feet

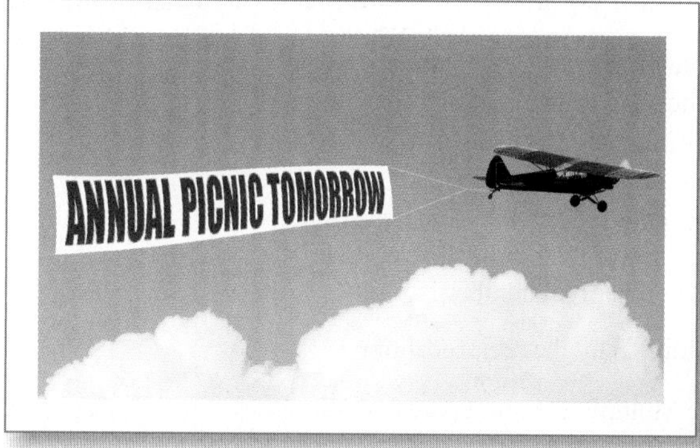

81. ***Finding a Pattern*** Complete the table. The third row contains the differences between consecutive entries of the second row. Describe the pattern of the third row.
The differences are constant.

n	0	1	2	3	4	5
$5n - 3$	−3	2	7	12	17	22
Differences	5	5	5	5	5	

82. ***Finding a Pattern*** Complete the table. The third row contains the differences between consecutive entries of the second row. Describe the pattern of the third row.
The differences are constant.

n	0	1	2	3	4	5
$3n + 1$	1	4	7	10	13	16
Differences	3	3	3	3	3	

83. ***Finding a Pattern*** Use the results of Exercises 81 and 82 to determine the common third-row difference when the algebraic expression is $an + b$. a

84. ***Think About It*** Find a and b such that the expression $an + b$ yields the following table. $4n + 3$

n	0	1	2	3	4	5
$an + b$	3	7	11	15	19	23

Explaining Concepts

85. ***Think About It*** Which are equivalent to $4x$? a and c

(a) x multiplied by 4

(b) x increased by 4

(c) The product of x and 4

(d) The ratio of 4 and x

86. ***Number Sense*** When n is an integer, how are the integers $2n - 1$ and $2n + 1$ related? Explain.
The two integers $2n - 1$ and $2n + 1$ are consecutive odd integers. Any integer multiplied by 2 will be even, so $2n$ will be even. Therefore, adding and subtracting 1 will result in consecutive odd integers.

87. ***Writing*** When translating a phrase into an algebraic expression, explain why it may be helpful to use a specific case before writing the expression.
Using a specific case may make it easier to see the form of the expression for the general case.

88. ***Reasoning*** When each phrase is translated into an algebraic expression, is order important? Explain.

(a) y multiplied by 5 No. Multiplication is commutative.

(b) 5 deceased by y Yes. Subtraction is not commutative.

(c) y divided by 5 Yes. Division is not commutative.

(d) the sum of 5 and y No. Addition is commutative.

1 Chapter Summary

	What did you learn?	Explanation and Examples	Review Exercises
1.1	Understand the set of real numbers and the subsets of real numbers *(p. 2)*.	The real numbers include rational numbers, irrational numbers, integers, whole numbers, and natural numbers.	1–4
	Use the real number line to order real numbers *(p. 3)*.	If the real number a lies to the left of the real number b on the real number line, then a is less than b, which is written as $a < b$. This relationship can also be described by saying that b is greater than a and writing $b > a$.	5–10
	Find the distance between two real numbers *(p. 5)*.	If a and b are two real numbers such that $a \le b$, then the distance between a and b is given by $b - a$.	11–14
	Find the absolute value of a real number *(p. 6)*.	If a is a real number, then the absolute value of a is $\lvert a \rvert = \begin{cases} a, & \text{if } a \ge 0 \\ -a, & \text{if } a < 0 \end{cases}.$	15–18
1.2	Add and subtract real numbers *(p. 10)*.	**1.** *To add with like signs:* Add the absolute values and attach the common sign. **2.** *To add with unlike signs:* Subtract the lesser absolute value from the greater absolute value and attach the sign of the number with the greater absolute value. **3.** *To subtract b from a:* Add $-b$ to a. **4.** *To add fractions:* Write the fractions so that they have the same denominator. Then add the numerators over the common denominator.	19–28
	Multiply and divide real numbers *(p. 12)*.	**1.** *To multiply two real numbers*: With like signs, the product is positive. With unlike signs, the product is negative. The product of zero and any other real number is zero. **2.** *To divide a by b:* Multiply a by the reciprocal of b. **3.** *To multiply fractions:* Multiply the numerators and the denominators.	29–38
	Evaluate exponential expressions *(p. 14)*.	Let n be a positive integer and let a be a real number. Then the product of n factors of a is given by $a^n = \underbrace{a \cdot a \cdot a \cdots a}_{n \text{ factors}}.$	39–42
	Use the order of operations to evaluate expressions *(p. 15)*.	**1.** First do operations within symbols of grouping. **2.** Then evaluate powers. **3.** Then do multiplications and divisions from left to right. **4.** Finally, do additions and subtractions from left to right.	43–48

	What did you learn?	Explanation and Examples	Review Exercises
1.3	Identify and use the properties of real numbers *(p. 18)*.	Let a, b, and c represent real numbers, variables, or algebraic expressions. *Commutative Properties:* $a + b = b + a$ $ab = ba$ *Associative Properties:* $(a + b) + c = a + (b + c)$ $(ab)c = a(bc)$ *Distributive Properties:* $a(b + c) = ab + ac$ $a(b - c) = ab - ac$ $(a + b)c = ac + bc$ $(a - b)c = ac - bc$ *Identity Properties:* $a + 0 = 0 + a = a$ $a \cdot 1 = 1 \cdot a = a$ *Inverse Properties:* $a + (-a) = 0$ $a \cdot \frac{1}{a} = 1, \quad a \neq 0$	49–56
	Develop additional properties of real numbers *(p. 20)*.	See page 20 for additional properties of real numbers.	57, 58
1.4	Identify the terms and coefficients of algebraic expressions *(p. 28)*.	The *terms* of an algebraic expression are those parts that are separated by addition. The *coefficient* of a term is its numerical factor. In the expression $4x + 3y$, $4x$ and $3y$ are the terms of the expression, and 4 and 3 are the coefficients.	59–62
	Simplify algebraic expressions by combining like terms *(p. 29)*.	To simplify an algebraic expression, remove the symbols of grouping and combine like terms.	63–70
	Evaluate algebraic expressions by substituting values for the variables *(p. 31)*.	To evaluate an algebraic expression, substitute numerical values for each of the variables in the expression and simplify.	71–74
1.5	Translate verbal phrases into algebraic expressions *(p. 36)*.	*Addition:* sum, plus, greater than, increased by, more than, exceeds, total of *Subtraction:* difference, minus, less than, decreased by, subtracted from, reduced by, the remainder *Multiplication:* product, multiplied by, twice, times, percent of *Division:* quotient, divided by, ratio, per	75–78
	Translate algebraic expressions into verbal phrases *(p. 38)*.	*Verbal Description* — *Algebraic Expression* Two added to the product of 3 and x — $3x + 2$	79–82
	Construct algebraic expressions to model real-life problems *(p. 39)*.	**1.** Construct a verbal model that represents the problem situation. **2.** Assign labels to all quantities in the verbal model. **3.** Construct a mathematical model (algebraic expression).	83–86

Review Exercises

Worked-out solutions to odd-numbered exercises at AlgebraWithinReach.com

1.1

Classifying Real Numbers **In Exercises 1 and 2, determine which of the numbers in the set are (a) natural numbers, (b) integers, (c) rational numbers, and (d) irrational numbers.**

1. $\left\{\frac{3}{5}, -4, 0, \sqrt{2}, 52, -\frac{1}{8}, \sqrt{9}\right\}$

(a) $\{52, \sqrt{9}\}$ (b) $\{-4, 0, 52, \sqrt{9}\}$

(c) $\left\{\frac{3}{5}, -4, 0, 52, -\frac{1}{8}, \sqrt{9}\right\}$ (d) $\{\sqrt{2}\}$

2. $\left\{98, -141, -\frac{7}{8}, 3.99, -\sqrt{12}, -\frac{54}{11}\right\}$

(a) $\{98\}$ (b) $\{98, -141\}$

(c) $\left\{98, -141, -\frac{7}{8}, 3.99, -\frac{54}{11}\right\}$ (d) $\{-\sqrt{12}\}$

Identifying Numbers **In Exercises 3 and 4, list all members of the set.**

3. The natural numbers between -2.3 and 6.1

$\{1, 2, 3, 4, 5, 6\}$

4. The even integers between -5.5 and 2.5

$\{-4, -2, 0, 2\}$

Plotting Real Numbers **In Exercises 5 and 6, plot the numbers on the real number line.**

See Additional Answers.

5. (a) 4 (b) -3 (c) $\frac{3}{4}$ (d) -2.4

6. (a) -9 (b) 7 (c) $-\frac{3}{2}$ (d) 5.25

Ordering Real Numbers **In Exercises 7–10, place the correct inequality symbol (< or >) between the real numbers.**

7. $-5 \; < \; 3$ **8.** $-2 \; > \; -8$

9. $-\frac{8}{5} \; < \; -\frac{2}{5}$ **10.** $8.4 \; > \; -3.2$

Finding the Distance Between Two Numbers **In Exercises 11–14, find the distance between the real numbers.**

11. 11 and -3 14

12. 4 and -13 17

13. -13.5 and -6.2 7.3

14. -8.4 and -0.3 8.1

Evaluating an Absolute Value **In Exercises 15–18, evaluate the expression.**

15. $|-5|$ 5 **16.** $|6|$ 6

17. $-|-7.2|$ -7.2 **18.** $|-3.6|$ 3.6

1.2

Evaluating an Expression **In Exercises 19–38, evaluate the expression. If it is not possible, state the reason. Write all fractions in simplest form.**

19. $15 + (-4)$ 11 **20.** $-12 + 3$ -9

21. $-63.5 + 21.7$ -41.8 **22.** $14.35 - 10.3$ 4.05

23. $\frac{4}{21} + \frac{7}{21}$ $\frac{11}{21}$ **24.** $\frac{21}{16} - \frac{13}{16}$ $\frac{1}{2}$

25. $-\frac{5}{6} + 1$ $\frac{1}{6}$ **26.** $3 + \frac{4}{9}$ $\frac{31}{9}$

27. $8\frac{3}{4} - 6\frac{5}{8}$ $\frac{17}{8}$ **28.** $-2\frac{9}{10} + 5\frac{3}{20}$ $\frac{9}{4}$

29. $-7 \cdot 4$ -28 **30.** $9 \cdot (-5)$ -45

31. $120(-5)(7)$ -4200 **32.** $(-16)(-15)(-4)$ -960

33. $\frac{3}{8} \cdot \left(-\frac{2}{15}\right)$ $-\frac{1}{20}$ **34.** $\frac{5}{21} \cdot \frac{21}{5}$ 1

35. $\frac{-56}{-4}$ 14 **36.** $\frac{85}{0}$ Undefined

37. $-\frac{7}{15} \div \left(-\frac{7}{30}\right)$ 2 **38.** $-\frac{2}{3} \div \frac{4}{15}$ $-\frac{5}{2}$

Evaluating an Exponential Expression **In Exercises 39–42, evaluate the exponential expression.**

39. $(-6)^4$ 1296

40. $-(-3)^4$ -81

41. -4^2 -16

42. 2^5 32

Using the Order of Operations **In Exercises 43–46, evaluate the expression.**

43. $120 - (5^2 \cdot 4)$ 20

44. $45 - 45 \div 3^2$ 40

45. $8 + 3[6^2 - 2(7 - 4)]$ 98

46. $2^4 - [10 + 6(1 - 3)^2]$ -18

47. ***Total Charge*** You purchased an entertainment system and made a down payment of \$395 plus nine monthly payments of \$45 each. What is the total amount you paid for the system? \$800

48. ***Savings Plan*** You deposit \$80 per month in a savings account for 10 years. The account earns 2% interest compounded monthly. The total amount in the account after 10 years will be

$$80\left[\left(1 + \frac{0.02}{12}\right)^{120} - 1\right]\left(1 + \frac{12}{0.02}\right).$$

Use a calculator to determine this amount. \$10,635.27

1.3

Identifying a Property of Real Numbers **In Exercises 49–56, identify the property of real numbers illustrated by the statement.**

49. $13 - 13 = 0$ Additive Inverse Property

50. $7\left(\frac{1}{7}\right) = 1$ Multiplicative Inverse Property

51. $7(9 + 3) = 7 \cdot 9 + 7 \cdot 3$ Distributive Property

52. $15(4) = 4(15)$ Commutative Property of Multiplication

53. $5 + (4 - y) = (5 + 4) - y$ Associative Property of Addition

54. $6(4z) = (6 \cdot 4)z$ Associative Property of Multiplication

55. $xy \cdot 1 = xy$ Multiplicative Identity Property

56. $xz - yz = (x - y)z$ Distributive Property

Identifying Properties of Real Numbers **In Exercises 57 and 58, identify the property of real numbers that justifies each step.** See Additional Answers.

57.

$$x + 2 = 4$$
$$(x + 2) + (-2) = 4 + (-2)$$
$$x + [2 + (-2)] = 2$$
$$x + 0 = 2$$
$$x = 2$$

58.

$$3x - 8 = 1$$
$$(3x - 8) + 8 = 1 + 8$$
$$3x + (-8 + 8) = 9$$
$$3x + 0 = 9$$
$$3x = 9$$
$$\tfrac{1}{3}(3x) = \tfrac{1}{3}(9)$$
$$\left(\tfrac{1}{3} \cdot 3\right)x = 3$$
$$1 \cdot x = 3$$
$$x = 3$$

1.4

Identifying Terms and Coefficients **In Exercises 59–62, identify the terms and coefficients of the algebraic expression.**

59. $4y^3 - y^2 + \dfrac{17}{2}y$ $4y^3, -y^2, \frac{17}{2}y$; $4, -1, \frac{17}{2}$

60. $\dfrac{x}{3} + 2xy^2 + \dfrac{1}{5}$ $2xy^2, \frac{x}{3}, \frac{1}{5}$; $2, \frac{1}{3}, \frac{1}{5}$

61. $52 - 1.2x^3 + \dfrac{1}{x}$ $-1.2x^3, \frac{1}{x}, 52$; $-1.2, 1, 52$

62. $2ab^2 + a^2b^2 - \dfrac{1}{a}$ $a^2b^2, 2ab^2, -\frac{1}{a}$; $1, 2, -1$

Simplifying an Algebraic Expression **In Exercises 63–70, simplify the expression.**

63. $6x + 3x$ $9x$

64. $10y - 7y$ $3y$

65. $3u - 2v + 7v - 3u$ $5v$

66. $9m - 4n + m - 3n$ $10m - 7n$

67. $5(x - 4) + 10$ $5x - 10$

68. $3x - (y - 2x)$ $5x - y$

69. $3[b + 5(b - a)]$ $-15a + 18b$

70. $-2t[8 - (6 - t)] + 5t$ $-2t^2 + t$

Evaluating an Algebraic Expression **In Exercises 71–74, evaluate the algebraic expression for the specified values of the variable(s). If not possible, state the reason.**

71. $x^2 - 2x - 3$
(a) $x = 3$ 0
(b) $x = 0$ -3

72. $\dfrac{x}{y + 2}$
(a) $x = 0, y = 3$ 0
(b) $x = 5, y = -2$ Not possible, undefined

73. $y^2 - 2y + 4x$
(a) $x = 4, y = -1$ 19
(b) $x = -2, y = 2$ -8

74. $|2x - x^2| - 2y$
(a) $x = -6, y = 3$ 42
(b) $x = 4, y = -5$ 18

1.5

Translating a Verbal Phrase **In Exercises 75–78, translate the verbal phrase into an algebraic expression.**

75. Twelve decreased by twice the number n $12 - 2n$

76. One hundred increased by the product of 15 and a number x $100 + 15x$

77. The sum of the square of a number y and 49 $y^2 + 49$

78. Three times the absolute value of the difference of a number n and 3, all divided by 5 $\dfrac{3|n - 3|}{5}$

Translating an Algebraic Expression **In Exercises 79–82, write a verbal description of the algebraic expression without using the variable.**

79. $2y + 7$ The sum of two times a number and 7

80. $5u - 3$ Three less than the product of 5 and a number

81. $\dfrac{x - 5}{4}$ The difference of a number and 5, divided by 4

82. $4(a - 1)$ Four times the difference of a number and 1

Constructing a Mathematical Model **In Exercises 83–86, write an algebraic expression that represents the specified quantity, and simplify, if possible.**

83. The amount of income tax due on a taxable income of I dollars when the tax rate is 18% $0.18I$

84. The distance traveled in 8 hours at an average speed of r miles per hour $8r$

85. The area of a rectangle whose length is l units and whose width is 5 units less than the length $l(l - 5) = l^2 - 5l$

86. The sum of two consecutive odd integers, the first of which is $2n + 1$ $(2n + 1) + (2n + 3) = 4n + 4$

Chapter Test

Solutions in English & Spanish and tutorial videos at AlgebraWithinReach.com

Take this test as you would take a test in class. After you are done, check your work against the answers in the back of the book.

1. Place the correct inequality symbol (< or >) between the real numbers.

(a) $\frac{5}{2}$ < $|-3|$ (b) $-\frac{2}{3}$ > $-\frac{3}{2}$

2. Find the distance between -4.4 and 6.9. 11.3

In Exercises 3–10, evaluate the expression. Write all fractions in simplest form.

3. $-14 + 9 - 15$ -20

4. $\frac{2}{3} + \left(-\frac{7}{6}\right)$ $-\frac{1}{2}$

5. $-2(225 - 150)$ -150

6. $(-3)(4)(-5)$ 60

7. $\left(-\frac{7}{16}\right)\left(-\frac{8}{21}\right)$ $\frac{1}{6}$

8. $\frac{5}{18} \div \frac{15}{8}$ $\frac{4}{27}$

9. $\left(-\frac{3}{5}\right)^3$ $-\frac{27}{125}$

10. $\frac{4^2 - 6}{5} + 13$ 15

11. Identify the property of real numbers illustrated by each statement.

(a) $(-3 \cdot 5) \cdot 6 = -3(5 \cdot 6)$ Associative Property of Multiplication

(b) $3y \cdot \frac{1}{3y} = 1$ Multiplicative Inverse Property

12. Rewrite the expression $-6(2x - 1)$ using the Distributive Property. $-12x + 6$

In Exercises 13–16, simplify the expression.

13. $3x^2 - 2x - 5x^2 + 7x - 1$ $-2x^2 + 5x - 1$

14. $x(x + 2) - 2(x^2 + x - 13)$ $-x^2 + 26$

15. $a(5a - 4) - 2(2a^2 - 2a)$ a^2

16. $4t - [3t - (10t + 7)]$ $11t + 7$

17. Explain the meaning of "evaluating an algebraic expression." Evaluate the expression $7 + (x - 3)^2$ for each value of x.

(a) $x = -1$ 23 (b) $x = 3$ 7

17. Evaluating an expression is solving the expression when values are provided for its variables.

18. An electrician divides 102 inches of wire into 17 pieces with equal lengths. How long is each piece? 6 inches

19. A *cord* of wood is a pile 4 feet high, 4 feet wide, and 8 feet long. The volume of a rectangular solid is its length times its width times its height. Find the number of cubic feet in 5 cords of wood. 640 cubic feet

20. Translate the phrase into an algebraic expression.

"The product of a number n and 5, decreased by 8" $5n - 8$

21. Write an algebraic expression for the sum of two consecutive even integers, the first of which is $2n$. $2n + (2n + 2) = 4n + 2$

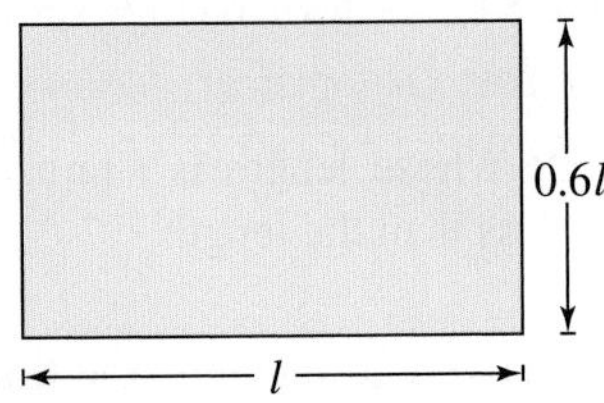

22. Write expressions for the perimeter and area of the rectangle shown at the left. Simplify the expressions and evaluate them when $l = 45$.

Perimeter: $2l + 2(0.6l) = 3.2l$, 144; Area: $l(0.6l) = 0.6l^2$, 1215

2 Linear Equations and Inequalities

MASTERY IS WITHIN REACH!

"I have to work and I have two kids. The only time I sit still is in class and it is hard to stay awake. I went to talk to a counselor and she had lots of suggestions about organization. I found more time to study my math and made sure that I was really learning while I did my homework. Things made more sense in class and it was easier to understand and remember everything."

Tamika
Nursing

See page 77 for suggestions about improving your memory.

2.1 Linear Equations

- Check solutions of equations.
- Solve linear equations in standard form.
- Solve linear equations in nonstandard form.

Introduction

An **equation** is a statement that equates two algebraic expressions. A **linear equation in one variable** x is an equation that can be written in the standard form $ax + b = 0$, where a and b are real numbers with $a \neq 0$. **Solving** an equation involving a variable means finding all values of the variable for which the equation is true. Such values are **solutions** and are said to **satisfy** the equation.

The **solution set** of an equation is the set of all solutions of the equation. A linear equation that has the set of all real numbers as its solution set is called an **identity**. A linear equation that is true for some real numbers is called a **conditional equation**. A linear equation that is false for all real numbers is called a **contradiction**.

Study Tip

When checking a solution, you should write a question mark over the equal sign to indicate that you are uncertain whether the "equation" is true for a given value of the variable.

EXAMPLE 1 Checking a Solution of an Equation

Determine whether $x = -3$ is a solution of $-3x - 5 = 4x + 16$.

SOLUTION

$-3x - 5 = 4x + 16$	Write original equation.
$-3(-3) - 5 \stackrel{?}{=} 4(-3) + 16$	Substitute -3 for x.
$9 - 5 \stackrel{?}{=} -12 + 16$	Simplify.
$4 = 4$	Solution checks. ✓

Because each side turns out to be the same number, you can conclude that $x = -3$ *is* a solution of the original equation. Try checking to see whether $x = -2$ is a solution.

Exercises Within Reach®

Solutions in English & Spanish and tutorial videos at AlgebraWithinReach.com

Checking Solutions of an Equation **In Exercises 1–6, determine whether each value of the variable is a solution of the equation.**

	Equation	*Values*		*Equation*	*Values*
1.	$3x - 7 = 2$ (a) Not a solution (b) Solution	(a) $x = 0$ (b) $x = 3$	**2.**	$5x + 9 = 4$ (a) Solution (b) Not a solution	(a) $x = -1$ (b) $x = 2$
3.	$x + 8 = 3x$ (a) Solution (b) Not a solution	(a) $x = 4$ (b) $x = -4$	**4.**	$10x - 3 = 7x$ (a) Not a solution (b) Not a solution	(a) $x = 0$ (b) $x = -1$
5.	$\frac{1}{4}x = 3$ (a) Not a solution (b) Solution	(a) $x = -4$ (b) $x = 12$	**6.**	$3(y + 2) = y - 5$ (a) Not a solution (b) Solution	(a) $y = -\frac{3}{2}$ (b) $y = -5.5$

Solving Linear Equations in Standard Form

Point out to students that they should not multiply or divide each side of an equation by a variable, because the value of that variable could be zero.

Forming Equivalent Equations: Properties of Equality

An equation can be transformed into an **equivalent equation** using one or more of the following procedures.

	Original Equation	*Equivalent Equation*
1. *Simplify Either Side:* Remove symbols of grouping, combine like terms, or simplify fractions on one or both sides of the equation.	$4x - x = 8$	$3x = 8$
2. *Apply the Addition Property of Equality:* Add (or subtract) the same quantity to (from) *each* side of the equation.	$x - 3 = 5$	$x = 8$
3. *Apply the Multiplication Property of Equality:* Multiply (or divide) *each* side of the equation by the same *nonzero* quantity.	$3x = 12$	$x = 4$
4. *Interchange Sides:* Interchange the two sides of the equation.	$7 = x$	$x = 7$

Study Tip

Be sure you see that solving an equation such as the one in Example 2 has two important parts. The first part is *finding* the solution(s). The second part is *checking* that each solution you find actually satisfies the original equation. You can improve your accuracy in algebra by developing the habit of checking each solution.

EXAMPLE 2 **Solving a Linear Equation in Standard Form**

$4x - 12 = 0$ Original equation

$4x - 12 + 12 = 0 + 12$ Add 12 to each side.

$4x = 12$ Combine like terms.

$\frac{4x}{4} = \frac{12}{4}$ Divide each side by 4.

$x = 3$ Simplify.

The solution is $x = 3$. Check this solution in the original equation.

Exercises Within Reach®

Solutions in English & Spanish and tutorial videos at AlgebraWithinReach.com

Justifying Steps **In Exercises 7 and 8, justify each step of the solution.**

7. $3x + 15 = 0$ Original equation

$3x + 15 - 15 = 0 - 15$ Subtract 15 from each side.

$3x = -15$ Combine like terms.

$\frac{3x}{3} = \frac{-15}{3}$ Divide each side by 3.

$x = -5$ Simplify.

8. $7x - 21 = 0$ Original equation

$7x - 21 + 21 = 0 + 21$ Add 21 to each side.

$7x = 21$ Combine like terms.

$\frac{7x}{7} = \frac{21}{7}$ Divide each side by 7.

$x = 3$ Simplify.

EXAMPLE 3 Solving a Linear Equation in Standard Form

Solve $2x + 2 = 0$. Then check the solution.

SOLUTION

$2x + 2 = 0$	Write original equation.
$2x + 2 - 2 = 0 - 2$	Subtract 2 from each side.
$2x = -2$	Combine like terms.
$\frac{2x}{2} = \frac{-2}{2}$	Divide each side by 2.
$x = -1$	Simplify.

Additional Examples
Solve each equation.
a. $3x + 9 = 0$
b. $-5x + 45 = 0$

Answers:
a. $x = -3$
b. $x = 9$

The solution is $x = -1$. You can check this as follows.

Check

$2(-1) + 2 \stackrel{?}{=} 0$	Substitute -1 for x in original equation.
$-2 + 2 \stackrel{?}{=} 0$	Simplify.
$0 = 0$	Solution checks. ✓

As you gain experience involving linear equations, you will probably be able to perform some of the solution steps in your head. For instance, you might solve the equation given in Example 3 by performing two of the steps mentally and writing only three steps, as follows.

$2x + 2 = 0$	Write original equation.
$2x = -2$	Subtract 2 from each side.
$x = -1$	Divide each side by 2.

Exercises Within Reach®

Solutions in English & Spanish and tutorial videos at AlgebraWithinReach.com

Identifying Equivalent Equations **In Exercises 9–16, determine whether the equations are equivalent.**

9. $3x = 10, 4x = x + 10$ Equivalent
10. $5x = 22, 4x = 22 - x$ Equivalent
11. $x + 5 = 12, 2x + 15 = 24$ Not equivalent
12. $x - 3 = 8, 3x - 6 = 24$ Not equivalent
13. $3(4 - 2t) = 5, 12 - 6t = 5$ Equivalent
14. $(3 + 2^2)z = 16, 7z = 16$ Equivalent
15. $2x - 7 = 3, x = 3$ Not equivalent
16. $6 - 5x = -4, x = -4$ Not equivalent

Solving an Equation **In Exercises 17–22, solve the equation. Check your solution.**

17. $x - 3 = 0$ 3
18. $x + 8 = 0$ -8
19. $3x - 12 = 0$ 4
20. $-14x - 28 = 0$ -2
21. $6x + 4 = 0$ $-\frac{2}{3}$
22. $8z - 10 = 0$ $\frac{5}{4}$

Solving Linear Equations in Nonstandard Form

Study Tip

A strategy that can help you to isolate x in solving a linear equation is to rewrite the original equation so that only variable terms are on one side of the equal sign and only constant terms are on the other side.

EXAMPLE 4 Solving a Linear Equation in Nonstandard Form

$x + 2 = 2x - 6$	Original equation
$x - 2x + 2 = 2x - 2x - 6$	Subtract $2x$ from each side.
$-x + 2 = -6$	Combine like terms.
$-x + 2 - 2 = -6 - 2$	Subtract 2 from each side.
$-x = -8$	Combine like terms.
$(-1)(-x) = (-1)(-8)$	Multiply each side by -1.
$x = 8$	Simplify.

The solution is $x = 8$. You can check this as follows.

Check

$8 + 2 \stackrel{?}{=} 2(8) - 6$	Substitute 8 for x in original equation.
$10 \stackrel{?}{=} 16 - 6$	Simplify.
$10 = 10$	Solution checks. ✓

EXAMPLE 5 Solving a Linear Equation in Nonstandard Form

$6(y - 1) = 2y - 3$	Original equation
$6y - 6 = 2y - 3$	Distributive Property
$6y - 2y - 6 = 2y - 2y - 3$	Subtract $2y$ from each side.
$4y - 6 = -3$	Combine like terms.
$4y - 6 + 6 = -3 + 6$	Add 6 to each side.
$4y = 3$	Combine like terms.
$\frac{4y}{4} = \frac{3}{4}$	Divide each side by 4.
$y = \frac{3}{4}$	Simplify.

The solution is $y = \frac{3}{4}$. Check this in the original equation.

Exercises Within Reach®

Solutions in English & Spanish and tutorial videos at AlgebraWithinReach.com

Solving an Equation **In Exercises 23–32, solve the equation. Check your solution.**

23. $3t + 8 = -2$ $-\frac{10}{3}$

24. $10 - 6x = -5$ $\frac{5}{2}$

25. $7 - 8x = 13x$ $\frac{1}{3}$

26. $2s - 16 = 34s$ $-\frac{1}{2}$

27. $3x - 1 = 2x + 14$ 15

28. $9y + 4 = 12y - 2$ 2

29. $8(x - 8) = 24$ 11

30. $6(x + 2) = 30$ 3

31. $3(x - 4) = 7x + 6$ $-\frac{9}{2}$

32. $-2(t + 3) = 9 - 5t$ 5

EXAMPLE 6 Solving a Linear Equation Containing Fractions

$\frac{x}{18} + \frac{3x}{4} = 2$	Original equation
$36\left(\frac{x}{18} + \frac{3x}{4}\right) = 36(2)$	Multiply each side by LCD 36.
$36 \cdot \frac{x}{18} + 36 \cdot \frac{3x}{4} = 36(2)$	Distributive Property
$2x + 27x = 72$	Simplify.
$29x = 72$	Combine like terms.
$\frac{29x}{29} = \frac{72}{29}$	Divide each side by 29.
$x = \frac{72}{29}$	Simplify.

The solution is $x = \frac{72}{29}$. Check this in the original equation.

Additional Examples

Solve each equation.

a. $2x - 4 = x - 3$

b. $\frac{2x}{16} + \frac{x}{24} = 5$

c. $0.29x + 0.04(200 - x) = 450$

Answers:

a. $x = 1$

b. $x = 30$

c. $x = 1768$

Study Tip

A different approach to Example 7 is to begin by multiplying each side of the equation by 100. This clears the equation of decimals to produce

$12x + 9(5000 - x) = 51{,}300.$

Try solving this equation to see that you obtain the same solution.

EXAMPLE 7 Solving a Linear Equation Containing Decimals

$0.12x + 0.09(5000 - x) = 513$	Original equation
$0.12x + 450 - 0.09x = 513$	Distributive Property
$0.03x + 450 = 513$	Combine like terms.
$0.03x + 450 - 450 = 513 - 450$	Subtract 450 from each side.
$0.03x = 63$	Combine like terms.
$\frac{0.03x}{0.03} = \frac{63}{0.03}$	Divide each side by 0.03.
$x = 2100$	Simplify.

The solution is $x = 2100$. Check this in the original equation.

Exercises Within Reach®

Solutions in English & Spanish and tutorial videos at AlgebraWithinReach.com

Solving an Equation **In Exercises 33–44, solve the equation. Check your solution.**

33. $t - \frac{2}{5} = \frac{3}{2}$ $\frac{19}{10}$

34. $z + \frac{1}{15} = -\frac{3}{10}$ $-\frac{11}{30}$

35. $\frac{t}{5} - \frac{t}{2} = 1$ $-\frac{10}{3}$

36. $\frac{t}{6} + \frac{t}{8} = 1$ $\frac{24}{7}$

37. $\frac{8x}{5} - \frac{x}{4} = -3$ $-\frac{20}{9}$

38. $\frac{11x}{6} + \frac{1}{3} = 2x$ 2

39. $0.3x + 1.5 = 8.4$ 23

40. $16.3 - 0.2x = 7.1$ 46

41. $1.2(x - 3) = 10.8$ 12

42. $6.5(1 - 2x) = 13$ $-\frac{1}{2}$

43. $\frac{2}{3}(2x - 4) = \frac{1}{2}(x + 3) - 4$ $\frac{1}{5}$

44. $\frac{3}{4}(6 - x) = \frac{1}{3}(4x + 5) + 2$ $\frac{2}{5}$

Some equations in nonstandard form have *no solution* or *infinitely many solutions*. Two such cases are illustrated in Example 8.

EXAMPLE 8 Solving Equations: Special Cases

No Solution

a. $2x - 4 = 2(x - 3)$

$2x - 4 = 2x - 6$

$-4 \neq -6$

Because -4 does not equal -6, you can conclude that the original equation has no solution.

Infinitely Many Solutions

b. $3x + 2 + 2(x - 6) = 5(x - 2)$

$3x + 2 + 2x - 12 = 5x - 10$

$5x - 10 = 5x - 10$

$-10 = -10$

Because the last equation is true for any value of x, you can conclude that the original equation has infinitely many solutions.

Application

EXAMPLE 9 Using a Model

The bar graph at the left shows the numbers y of bridges (in thousands) in the United States from 2005 through 2010. An equation that models the data is

$$y = 2.01t + 585.0, \quad 5 \le t \le 10$$

where t represents the year, with $t = 5$ corresponding to 2005. Determine when the number of bridges reached 600,000. (*Source:* U.S. Federal Highway Administration)

Bridges in U.S.

Bridges (in thousands): 575, 580, 585, 590, 595, 600, 605, 610

Year (5 ↔ 2005): 5, 6, 7, 8, 9, 10

SOLUTION

Let $y = 600$ and solve the resulting equation for t.

$y = 2.01t + 585.0$	Write original equation.
$600 = 2.01t + 585.0$	Substitute 600 for y.
$15 = 2.01t$	Subtract 585.0 from each side.
$7 \approx t$	Divide each side by 2.01.

Because $t = 5$ corresponds to 2005, it follows that $t = 7$ corresponds to 2007. So, the number of bridges reached 600,000 in 2007.

Exercises Within Reach®

Solutions in English & Spanish and tutorial videos at AlgebraWithinReach.com

Solving an Equation **In Exercises 45–48, solve the equation. Justify your answer.**

45. $4y - 3 = 4y$ No solution, because $-3 \neq 0$.

46. $24 - 2x = -2x$ No solution, because $24 \neq 0$.

47. $4(2x - 3) = 8x - 12$ Infinitely many, because $-12 = -12$.

48. $9x + 6 = 3(3x + 2)$ Infinitely many, because $6 = 6$.

49. ***Height*** Consider the fountain shown. The initial velocity of the stream of the water is 48 feet per second. The velocity v of the water at any time t (in seconds) is given by $v = 48 - 32t$. Find the time for a drop of water to travel from the base to the maximum height of the fountain. (*Hint:* The maximum height is reached when $v = 0$.) 1.5 seconds

50. ***Height*** The velocity v of an object projected vertically upward with an initial velocity of 64 feet per second is given by $v = 64 - 32t$, where t is time in seconds. When does the object reach its maximum height? 2 seconds

Concept Summary: Solving Linear Equations

What

When given a **linear equation in one variable**, the goal is usually to **solve** the equation by **isolating the variable** on one side of the equation.

EXAMPLE

Solve the equation

$3x + 2 = 11$.

How

To isolate the variable, you "get rid" of terms and factors by using **inverse operations**.

EXAMPLE

$3x + 2 = 11$

$3x + 2 - 2 = 11 - 2$ Subtract.

$3x = 9$

$\frac{3x}{3} = \frac{9}{3}$ Divide.

$x = 3$

The solution is $x = 3$. ✓

Why

An equation is like a scale. To keep the equation balanced, you must do the same thing to each side of the equation. The resulting equation is said to be **equivalent** to the original equation.

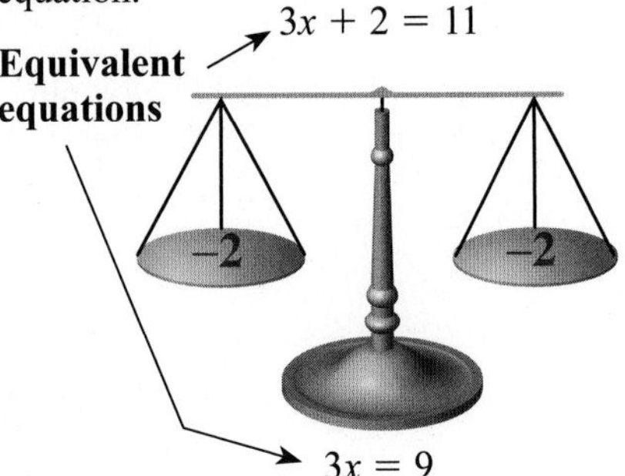

Exercises Within Reach®

Worked-out solutions to odd-numbered exercises at AlgebraWithinReach.com

Concept Summary Check

51. ***Vocabulary*** In your own words, explain what is meant by the term *equivalent equations*. Equivalent equations are equations that have the same set of solutions.

52. ***Writing*** Describe the steps that can be used to transform an equation into an equivalent equation. Simplify either side, apply the Addition Property of Equality, apply the Multiplication Property of Equality, interchange sides

53. ***Number Sense*** When dividing each side of an equation by the same quantity, why must the quantity be nonzero? Dividing by zero is undefined.

54. ***Checking a Solution*** Explain how the process of evaluating an algebraic expression can be used to check a solution of an equation. A solution of an equation can be checked by substituting in the value(s) of the variable(s). The algebraic expressions on each side of the equation can be simplified.

Extra Practice

Classifying an Equation **In Exercises 55–58, classify the equation as an identity, a conditional equation, or a contradiction.**

55. $6(x + 3) = 6x + 3$ Contradiction

56. $3x + 11 - 6x = 11 - 3x$ Identity

57. $\frac{2}{3}x + 4 = \frac{1}{3}x + 12$ Conditional

58. $\frac{4}{3}(2x - 6) = \frac{8}{3}x - 8$ Identity

Solving an Equation **In Exercises 59–64, solve the equation. If there is exactly one solution, check your solution. If not, justify your answer.**

59. $12(x + 3) = 7(x + 3)$ -3

60. $-5(x - 10) = 6(x - 10)$ 10

61. $-9y - 4 = -9y$ No solution, because $-4 \neq 0$.

62. $6a + 2 = 6a$ No solution, because $2 \neq 0$.

63. $7(x + 6) = 3(2x + 14) + x$ Infinitely many, because $42 = 42$.

64. $5(x + 8) = 4(2x + 10) - 3x$ Infinitely many, because $40 = 40$.

65. ***Car Repair*** The bill for the repair of your car was \$257. The cost for parts was \$162. The cost for labor was \$38 per hour. How many hours did the repair work take? 2.5 hours

66. ***Appliance Repair*** The bill for the repair of your refrigerator was \$187. The cost for parts was \$74. The cost for the service call and the first half hour of service was \$50. The additional cost for labor was \$21 per half hour. How many hours did the repair work take? 2 hours

67. ***Geometry*** The length of a rectangle is t times its width (see figure). So, the perimeter P is given by $P = 2w + 2(tw)$, where w is the width of the rectangle. The perimeter of the rectangle is 1000 meters.

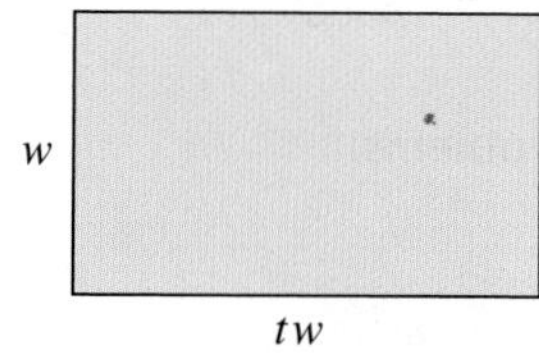

(a) Complete the table of widths, lengths, and areas of the rectangle for the specified values of t.

t	1	1.5	2	3	4	5
Width	250	200	166.7	125	100	83.3
Length	250	300	333.4	375	400	416.5
Area	62,500	60,000	55,577.8	46,875	40,000	34,694.5

(b) Use the table to write a short paragraph describing the relationship among the width, length, and area of a rectangle that has a *fixed* perimeter. Because the length is t times the width and the perimeter is fixed, as t increases, the length increases and the width and area decrease. The maximum area occurs when the length and width are equal.

68. ***Geometry*** Repeat parts (a) and (b) of Exercise 67 for a rectangle with a fixed perimeter of 60 inches and a length that is t inches greater than the width (see figure). Use the values $t = 0, 1, 2, 3, 5$, and 10 to create a table given that $P = 2w + 2(t + w)$. (Spreadsheet at AlgebraWithinReach.com) See Additional Answers.

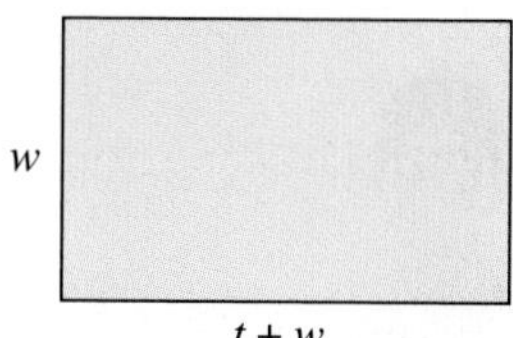

71. No. To write an identity in standard form, $ax + b = 0$, the values of a and b must both be zero. Because the equation is no longer in standard form, the identity cannot be written as a linear equation in standard form.

Explaining Concepts

69. ***True or False?*** Multiplying each side of an equation by zero yields an equivalent equation. Justify your answer. False. This does not follow the Multiplication Property of Equality.

70. ***True or False?*** Subtracting zero from each side of an equation yields an equivalent equation. Justify your answer. True. This follows the Addition Property of Equality.

71. ***Reasoning*** Can an identity be written as a linear equation in standard form? Explain.

Classifying an Equation **In Exercises 72–75, classify the equation as an identity, a conditional equation, or a contradiction. Discuss real-life situations that could be represented by the equation, or could be used to show that the equation is a contradiction.**

72. $x + 0.20x = 50.16$ Conditional equation; *Sample answer:* 20% tip on a restaurant bill

73. $3x = 3x + 1$ Contradiction; *Sample answer:* Let $x = 1$ pound.

74. $0.25(40 + x) = 10 + 0.25x$ Identity; *Sample answer:* value of 1 roll of quarters plus x additional quarters

75. $5w + 3 = 28$ Conditional equation; *Sample answer:* 28 days are worked, where w represents the number of full (5-day) work weeks.

Cumulative Review

In Exercises 76–79, evaluate the expression.

76. $\frac{2}{5} + \frac{4}{5}$ $\frac{6}{5}$

77. $\frac{5}{6} - \frac{2}{3}$ $\frac{1}{6}$

78. $-5 - (-3)$ -2

79. $-12 - (6 - 5)$ -13

In Exercises 80–83, evaluate the expression for the specified values of the variable.

80. $8 + 7x; x = 2, x = 3$ 22; 29

81. $\frac{2x}{x + 1}; x = 1, x = 5$ $1; \frac{5}{3}$

82. $x^2 - 1; x = -4, x = 3$ 15; 8

83. $|3x - 7|; x = -1, x = 1$ 10; 4

In Exercises 84–87, translate the verbal phrase into an algebraic expression.

84. Eight less than a number n $n - 8$

85. The ratio of n and four $\frac{n}{4}$

86. Twice the sum of n and three $2(n + 3)$

87. Five less than half of a number n $\frac{1}{2}n - 5$

2.2 Linear Equations and Problem Solving

- Write algebraic equations representing real-life situations.
- Solve percent problems.
- Solve ratio, rate, and proportion problems.

Mathematical Modeling

In this section, you will see how algebra can be used to solve problems that occur in real-life situations. This process is called **mathematical modeling**, and its basic steps are as follows.

Verbal description ⇨ Verbal model ⇨ Assign labels ⇨ Algebraic equation

Application

EXAMPLE 1 Mathematical Modeling

Write an algebraic equation that represents the following problem. Then solve the equation and answer the question.

You have accepted a job at an annual salary of $48,500. This salary includes a year-end bonus of $500. You are paid twice a month. What will your gross pay be for each paycheck?

SOLUTION

Because there are 12 months in a year and you will be paid twice a month, it follows that you will receive 24 paychecks during the year. Construct an algebraic equation for this problem as follows. Begin by constructing a verbal model, then assign labels, and finally form an algebraic equation.

Verbal Model: Income for year $= 24 \times$ Amount of each paycheck $+$ Bonus

Labels: Income for year $= \$48{,}500$ (dollars)
Amount of each paycheck $= x$ (dollars)
Bonus $= 500$ (dollars)

Equation:

$48{,}500 = 24x + 500$ Original equation

$48{,}000 = 24x$ Subtract 500 from each side.

$\dfrac{48{,}000}{24} = \dfrac{24x}{24}$ Divide each side by 24.

$2000 = x$ Simplify.

Each paycheck will be $2000. Check this in the original statement of the problem.

Exercises Within Reach ®

Solutions in English & Spanish and tutorial videos at AlgebraWithinReach.com

Mathematical Modeling **In Exercises 1 and 2, construct a verbal model and write an algebraic equation that represents the problem. Solve the equation.** See Additional Answers.

1. You have accepted a job offer at an annual salary of $37,120. This salary includes a year-end bonus of $2800. You are paid every 2 weeks. What will your gross pay be for each paycheck?

2. You have a job on an assembly line for which you earn $10.00 per hour plus $0.75 per unit assembled. Find the number of units produced in an eight-hour day if your earnings for the day are $146.

wavebreakmedia ltd/Shutterstock.com

Study Tip

Recall that a *rate* is a fraction that compares two quantities measured in different units.

Percent Problems

Rates that describe increases, decreases, and discounts are often given as **percents**. Percent means *per hundred*, so 40% means 40 per hundred or, equivalently, $\frac{40}{100}$.

Percent	10%	$12\frac{1}{2}\%$	20%	25%	$33\frac{1}{3}\%$	50%	$66\frac{2}{3}\%$	75%
Decimal	0.1	0.125	0.2	0.25	$0.\overline{3}$	0.5	$0.\overline{6}$	0.75
Fraction	$\frac{1}{10}$	$\frac{1}{8}$	$\frac{1}{5}$	$\frac{1}{4}$	$\frac{1}{3}$	$\frac{1}{2}$	$\frac{2}{3}$	$\frac{3}{4}$

EXAMPLE 2 Solving a Percent Problem

The number 15.6 is 26% of what number?

SOLUTION

Verbal Model: Compared number = Percent (decimal form) • Base number

Labels: Compared number = 15.6
Percent = 0.26 (decimal form)
Base number = b

Equation:
$15.6 = 0.26b$ Original equation
$60 = b$ Divide each side by 0.26.

Check that 15.6 is 26% of 60 by multiplying 60 by 0.26 to obtain 15.6.

Exercises Within Reach ®

Solutions in English & Spanish and tutorial videos at AlgebraWithinReach.com

Finding Equivalent Forms of a Percent **In Exercises 3–10, complete the table showing the equivalent forms of the percent.**

	Percent	Parts out of 100	Decimal	Fraction		Percent	Parts out of 100	Decimal	Fraction
3.	30%	30	0.30	$\frac{3}{10}$	**4.**	75%	75	0.75	$\frac{3}{4}$
5.	7.5%	7.5	0.075	$\frac{3}{40}$	**6.**	8%	8	0.8	$\frac{2}{25}$
7.	$66\frac{2}{3}\%$	$66\frac{2}{3}$	$0.\overline{6}$	$\frac{2}{3}$	**8.**	12.5%	12.5	0.125	$\frac{1}{8}$
9.	100%	100	1.00	1	**10.**	42%	42	0.42	$\frac{21}{50}$

Solving a Percent Problem **In Exercises 11–26, solve using a percent equation.**

11. What is 35% of 250? 87.5

12. What is 65% of 800? 520

13. What is 42.5% of 816? 346.8

14. What is 70.2% of 980? 687.96

15. What is $12\frac{1}{2}\%$ of 1024? 128

16. What is $33\frac{1}{3}\%$ of 816? 272

17. What is 0.4% of 150,000? 600

18. What is 0.1% of 8925? 8.925

19. What is 250% of 32? 80

20. What is 300% of 16? 48

21. 84 is 24% of what number? 350

22. 416 is 65% of what number? 640

23. 42 is 120% of what number? 35

24. 168 is 350% of what number? 48

25. 22 is 0.8% of what number? 2750

26. 18 is 2.4% of what number? 750

EXAMPLE 3 Solving a Percent Problem

The number 28 is what percent of 80?

SOLUTION

Verbal Model: [Compared number] = [Percent (decimal form)] • [Base number]

Labels: Compared number = 28
Percent = p (decimal form)
Base number = 80

Equation:

$28 = p(80)$ Original equation

$\frac{28}{80} = p$ Divide each side by 80.

$0.35 = p$ Simplify.

So, 28 is 35% of 80. Check this solution by multiplying 80 by 0.35 to obtain 28.

Application

EXAMPLE 4 Solving a Percent Application

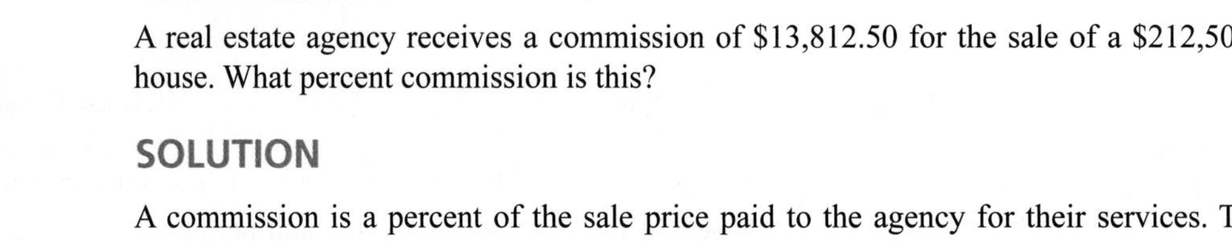

A real estate agency receives a commission of $13,812.50 for the sale of a $212,500 house. What percent commission is this?

SOLUTION

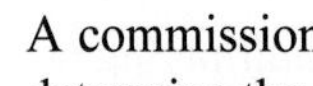

A commission is a percent of the sale price paid to the agency for their services. To determine the percent commission, start with a verbal model.

Verbal Model: [Commission] = [Percent (decimal form)] • [Sale price]

Labels: Commission = 13,812.50 (dollars)
Percent = p (decimal form)
Sale price = 212,500 (dollars)

Equation:

$13{,}812.50 = p(212{,}500)$ Original equation

$\frac{13{,}812.50}{212{,}500} = p$ Divide each side by 212,500.

$0.065 = p$ Simplify.

So, the real estate agency receives a commission of 6.5%. Check this solution by multiplying $212,500 by 0.065 to obtain $13,812.50.

Exercises Within Reach ®

Solutions in English & Spanish and tutorial videos at AlgebraWithinReach.com

Solving a Percent Problem **In Exercises 27–32, solve using a percent equation.**

27. 496 is what percent of 800? 62%

28. 1650 is what percent of 5000? 33%

29. 2.4 is what percent of 480? 0.5%

30. 3.3 is what percent of 220? 1.5%

31. 2100 is what percent of 1200? 175%

32. 900 is what percent of 500? 180%

33. ***Real Estate Commission*** A real estate agency receives a commission of $12,250 for the sale of a $175,000 house. What percent commission is this? 7%

34. ***Real Estate Commission*** A real estate agency receives a commission of $24,225 for the sale of a $285,000 house. What percent commission is this? 8.5%

Ratios, Rates, and Proportions

You know that a percent compares a number with 100. A **ratio** is a more generic rate form that compares one number with another. If a and b represent two quantities, then a/b is called the ratio of a to b. Note the *order* implied by a ratio. The ratio of a to b means a/b, whereas the ratio of b to a means b/a.

EXAMPLE 5 Using a Ratio

Find the ratio of 4 feet to 8 inches using the same units.

SOLUTION

Because the units of feet and inches are not the same, you must first convert 4 feet into its equivalent in inches or convert 8 inches into its equivalent in feet. You can convert 4 feet to 48 inches (by multiplying 4 by 12) to obtain

$$\frac{4 \text{ feet}}{8 \text{ inches}} = \frac{48 \text{ inches}}{8 \text{ inches}} = \frac{48}{8} = \frac{6}{1}.$$

Or, you can convert 8 inches to $\frac{8}{12}$ foot (by dividing 8 by 12) to obtain

$$\frac{4 \text{ feet}}{8 \text{ inches}} = \frac{4 \text{ feet}}{\frac{8}{12} \text{ foot}} = 4 \div \frac{8}{12} = 4 \cdot \frac{12}{8} = \frac{6}{1}.$$

Study Tip

The **unit price** of an item is the quotient of the total price divided by the total units.

$$\text{Unit price} = \frac{\text{Total price}}{\text{Total units}}$$

To state unit prices, use the word "per." For instance, the unit price for a brand of coffee might be $7.59 *per* pound.

EXAMPLE 6 Comparing Unit Prices

Which is the better buy, a 12-ounce box of breakfast cereal for $2.79 or a 16-ounce box of the same cereal for $3.59?

SOLUTION

The unit price for the 12-ounce box is

$$\text{Unit price} = \frac{\text{Total price}}{\text{Total units}} = \frac{\$2.79}{12 \text{ ounces}} = \$0.23 \text{ per ounce}.$$

The unit price for the 16-ounce box is

$$\text{Unit price} = \frac{\text{Total price}}{\text{Total units}} = \frac{\$3.59}{16 \text{ ounces}} = \$0.22 \text{ per ounce}.$$

The 16-ounce box has a slightly lower unit price, and so it is the better buy.

Exercises Within Reach ®

Solutions in English & Spanish and tutorial videos at AlgebraWithinReach.com

Using a Ratio **In Exercises 35–38, write the verbal expression as a ratio. Use the same units in both the numerator and denominator, and simplify.**

35. 120 meters to 180 meters $\frac{2}{3}$

36. 12 ounces to 20 ounces $\frac{3}{5}$

37. 40 milliliters to 1 liter $\frac{1}{25}$

38. 125 centimeters to 2 meters $\frac{5}{8}$

Consumer Awareness **In Exercises 39–42, use unit prices to determine the better buy.**

39. (a) A $14\frac{1}{2}$-ounce bag of chips for $2.32
(b) A $5\frac{1}{2}$-ounce bag of chips for $0.99
$14\frac{1}{2}$-ounce bag

40. (a) A $10\frac{1}{2}$-ounce package of cookies for $1.79
(b) A 16-ounce package of cookies for $2.39
16-ounce package

41. (a) A four-ounce tube of toothpaste for $1.69
(b) A six-ounce tube of toothpaste for $2.39
six-ounce tube

42. (a) A two-pound package of hamburger for $4.79
(b) A three-pound package of hamburger for $6.99
three-pound package

A **proportion** is a statement that equates two ratios. For example, if the ratio of a to b is the same as the ratio of c to d, you can write the proportion as

$$\frac{a}{b} = \frac{c}{d}.$$

In typical problems, you know three of the values and need to find the fourth. The quantities a and d are called the **extremes** of the proportion, and the quantities b and c are called the **means** of the proportion. In a proportion, the product of the extremes is equal to the product of the means. This is done by **cross-multiplying**. That is, if

$$\frac{a}{b} = \frac{c}{d}$$

then $ad = bc$.

Application EXAMPLE 7 Geometry: Solving a Proportion

The triangles shown below are similar triangles. Use this fact to find the length of the unknown side x of the larger triangle.

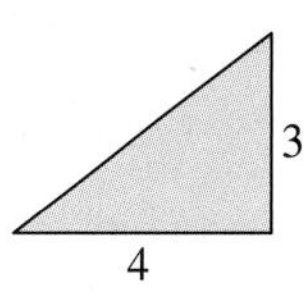

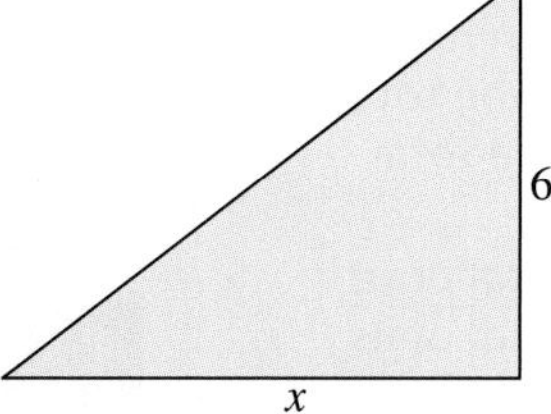

Study Tip

The proportion in Example 7 could also be written as

$$\frac{4}{3} = \frac{x}{6}.$$

After cross-multiplying, you would obtain the same equation

$$24 = 3x.$$

SOLUTION

$\frac{4}{x} = \frac{3}{6}$	Set up proportion.
$4 \cdot 6 = x \cdot 3$	Cross-multiply.
$24 = 3x$	Simplify.
$8 = x$	Divide each side by 3.

So, the length of the unknown side of the larger triangle is 8 units. Check this in the original statement of the problem.

Students can review the concept of similarity in Appendix B, located on our website *www.cengage.com/math/larson/algebra*.

Exercises Within Reach®

Solutions in English & Spanish and tutorial videos at AlgebraWithinReach.com

Solving a Proportion **In Exercises 43–46, solve the proportion.**

43. $\frac{x}{6} = \frac{2}{3}$ 4

44. $\frac{t}{4} = \frac{3}{2}$ 6

45. $\frac{y}{36} = \frac{6}{7}$ $\frac{216}{7}$

46. $\frac{5}{16} = \frac{x}{4}$ $\frac{5}{4}$

Geometry **In Exercises 47–50, the triangles are similar. Solve for the length x by using the fact that corresponding sides of similar triangles are proportional.**

47. $5.\overline{09}$

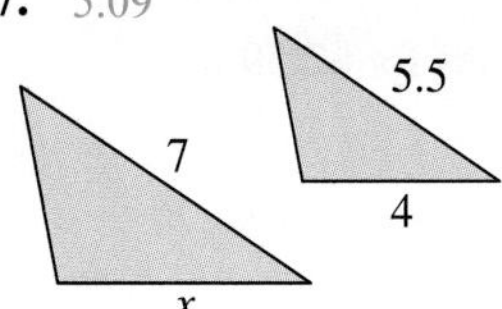

48. $\frac{15}{2}$

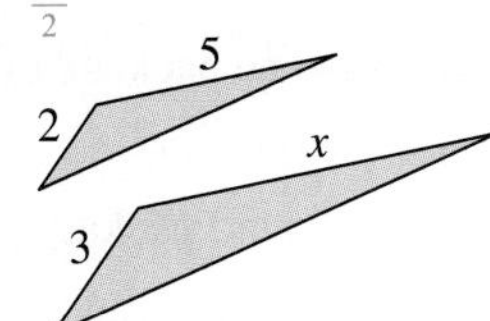

49. 3

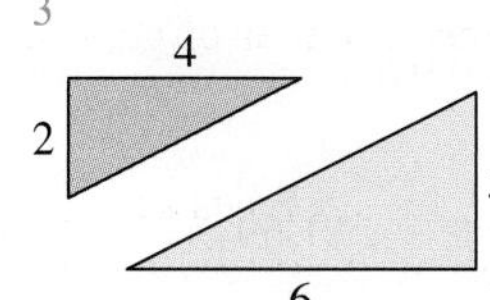

50. $\frac{45}{4}$

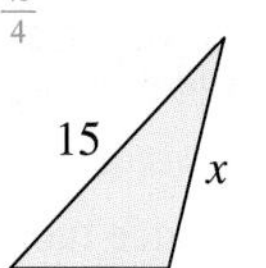

EXAMPLE 8 Finding Gasoline Cost

You are driving from New York City to Phoenix, a trip of 2450 miles. You begin the trip with a full tank of gas. After traveling 424 miles, you refill the tank for \$52. Assuming gas prices will be the same for the duration of your trip, how much should you plan to spend on gasoline for the entire trip?

Study Tip

You can write a proportion in several ways. Just be sure to put like quantities in similar positions on each side of the proportion.

SOLUTION

Verbal Model: $\dfrac{\text{Cost for entire trip}}{\text{Cost for one tank}} = \dfrac{\text{Miles for entire trip}}{\text{Miles for one tank}}$

Labels:
Cost of gas for entire trip $= x$ (dollars)
Cost of gas for one tank $= 52$ (dollars)
Miles for entire trip $= 2450$ (miles)
Miles for one tank $= 424$ (miles)

Equation:

$\dfrac{x}{52} = \dfrac{2450}{424}$ Original proportion

$x \cdot 424 = 52 \cdot 2450$ Cross-multiply.

$424x = 127{,}400$ Simplify.

$x \approx 300.47$ Divide each side by 424.

You should plan to spend approximately \$300 for gasoline on the trip. Check this in the original statement of the problem.

Strategy for Solving Word Problems

1. Ask yourself what you need to know to solve the problem. Then *write a verbal model* that includes arithmetic operations to describe the problem.
2. *Assign labels* to each part of the verbal model—numbers to the known quantities and letters (or expressions) to the variable quantities.
3. Use the labels to *write an algebraic model* based on the verbal model.
4. *Solve* the resulting algebraic equation.
5. *Answer* the original question and check that your answer satisfies the original problem as stated.

Exercises Within Reach®

Solutions in English & Spanish and tutorial videos at AlgebraWithinReach.com

51. ***Property Tax*** The tax on a property with an assessed value of \$110,000 is \$1650. Find the tax on a property with an assessed value of \$160,000. \$2400

52. ***Recipe*** Three cups of flour are required to make one batch of cookies. How many cups are required to make $3\frac{1}{2}$ batches? $10\frac{1}{2}$ cups

53. ***Fuel Usage*** A tractor uses 5 gallons of diesel fuel to plow for 105 minutes. Determine the number of gallons of fuel used in 6 hours. 17.1 gallons

54. ***Spring Length*** A force of 32 pounds stretches a spring 6 inches. Determine the number of pounds of force required to stretch it 1.25 feet. 80 pounds

Concept Summary: Solving Proportions

What

A **proportion** is a statement that equates two **ratios**. When solving a proportion, you usually know three of the values and are asked to find the fourth.

EXAMPLE

Solve $\frac{6}{18} = \frac{x}{6}$.

How

One way to solve a proportion is to use **cross-multiplication**.

EXAMPLE

$\frac{6}{18} = \frac{x}{6}$ Write proportion.

$6(6) = 18x$ Cross-multiply.

$\frac{36}{18} = x$ Divide each side by 18.

$2 = x$ Simplify.

Why

There are many real-life applications involving ratios and proportions. For example, knowing how to use ratios will help you identify **unit prices**.

Exercises Within Reach®

Worked-out solutions to odd-numbered exercises at AlgebraWithinReach.com

Concept Summary Check

55. ***Vocabulary*** Define the term *ratio*. Give an example of a ratio. A ratio is a generic rate form that compares one number with another, such as 5 teachers to 46 students.

56. ***Vocabulary*** Define the term *proportion*. Give an example of a proportion.

Equates two ratios; $\frac{1}{2} = \frac{2}{4}$

57. ***Number Sense*** If $\frac{a}{b} = \frac{c}{d}$, does this mean that $a = c$ and $b = d$?

Not necessarily. In general, you need to cross-multiply to solve for an unknown quantity.

58. ***Solving a Proportion*** Is is possible to solve the proportion $\frac{4}{b} = \frac{6}{d}$? Explain.

No. A proportion can only have one unknown quantity.

Extra Practice

Solving a Percent Problem **In Exercises 59–64, solve using a percent equation.**

59. What is 20% of 225? 45

60. What is 0.3% of 3000? 9

61. 13 is 26% of what number? 50

62. 54 is 2.5% of what number? 2160

63. 66 is what percent of 220? 30%

64. 132 is what percent of 1000? 13.2%

Unit Price **In Exercises 65–68, find the unit price (in dollars per ounce) of the product.**

65. A 20-ounce can of pineapple for \$1.10 \$0.06

66. A 64-ounce bottle of juice for \$1.89 \$0.03

67. A one-pound, four-ounce loaf of bread for \$2.29 \$0.11

68. A one-pound, six-ounce box of cereal for \$5.19 \$0.24

Solving a Proportion **In Exercises 69–72, solve the proportion.**

69. $\frac{y}{6} = \frac{y-2}{4}$ 6

70. $\frac{a}{5} = \frac{a+4}{8}$ $\frac{20}{3}$

71. $\frac{z-3}{3} = \frac{z+8}{12}$ $\frac{20}{3}$

72. $\frac{y+1}{10} = \frac{y-1}{6}$ 4

73. ***Company Layoff*** Because of decreasing sales, a small company laid off 25 of its 160 employees. What percent of the work force was laid off?

15.625%

74. ***Monthly Rent*** You spend \$748 of your monthly income of \$3400 for rent. What percent of your monthly income is your monthly rent payment?

22%

75. ***Gratuity*** You want to leave a 15% tip for a meal that costs \$32.60. How much should you leave? \$4.89

76. ***Gratuity*** You want to leave a 20% tip for a meal that costs \$49.24. How much should you leave? \$9.85

77. ***Quality Control*** A quality control engineer reported that 1.5% of a sample of parts were defective. The engineer found three defective parts. How large was the sample? 200 parts

78. ***Price Inflation*** A new car costs \$29,750, which is approximately 115% of what a comparable car cost 3 years ago. What did the car cost 3 years ago? About \$25,870

79. ***Income Tax*** You have \$12.50 of state tax withheld from your paycheck per week when your gross pay is \$625. Find the ratio of tax to gross pay. $\frac{1}{50}$

80. ***Price-Earnings Ratio*** The **price-earnings ratio** is the ratio of the price of a stock to its earnings. Find the price-earnings ratio of a stock that sells for \$56.25 per share and earns \$6.25 per share. $\frac{9}{1}$

81. ***Quality Control*** A quality control engineer finds one defective unit in a sample of 75. At this rate, what is the expected number of defective units in a shipment of 200,000? 2667 units

82. ***Quality Control*** A quality control engineer finds 3 defective units in a sample of 120. At this rate, what is the expected number of defective units in a shipment of 5000? 125 units

83. ***Geometry*** A man who is 6 feet tall walks directly toward the tip of the shadow of a tree. When the man is 75 feet from the tree, he starts forming his own shadow beyond the shadow of the tree (see figure). The length of the shadow of the tree beyond this point is 11 feet. Find the height h of the tree. $\frac{516}{11} \approx 46.9$ feet

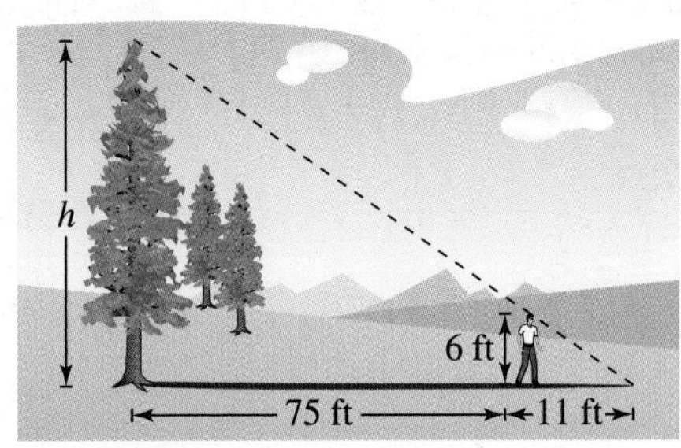

84. ***Geometry*** Find the length l of the shadow of a man who is 6 feet tall and is standing 15 feet from a streetlight that is 20 feet high. (see figure). $\frac{45}{7} \approx 6.43$ feet

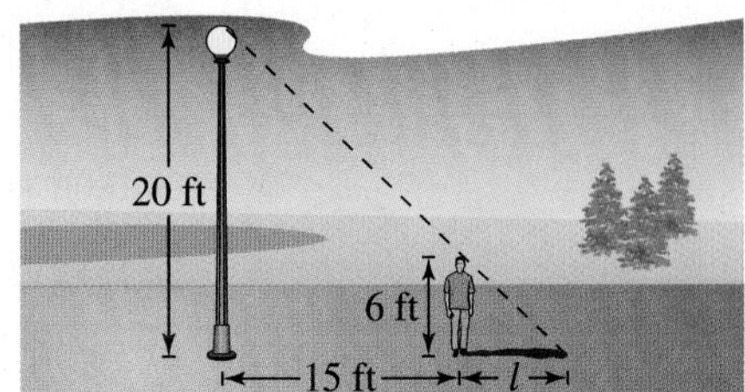

Explaining Concepts

85. ***Writing*** Explain how to change percents to decimals and decimals to percents. Give examples.
See Additional Answers.

86. ***Precision*** Is it true that $\frac{1}{2}\% = 50\%$? Explain.
No. $\frac{1}{2}\% = 0.5\% \neq 50\%$

87. ***Writing*** In your own words, describe the meaning of *mathematical modeling*. Give an example.
See Additional Answers.

88. ***Reasoning*** During a year of financial difficulties, your company reduces your salary by 7%. What percent increase in this reduced salary is required to raise your salary to the amount it was prior to the reduction? Why isn't the percent increase the same as the percent of the reduction? See Additional Answers.

Cumulative Review

In Exercises 89–92, evaluate the expression.

89. $-\frac{4}{15} \cdot \frac{15}{16}$ $-\frac{1}{4}$

90. $\frac{3}{8} \div \frac{5}{16}$ $\frac{6}{5}$

91. $(12 - 15)^3$ -27

92. $\left(-\frac{5}{8}\right)^2$ $\frac{25}{64}$

In Exercises 93–96, identify the property of real numbers illustrated by the statement.

93. $5 + x = x + 5$ Commutative Property of Addition

94. $3x \cdot \frac{1}{3x} = 1$ Multiplicative Inverse Property

95. $6(x - 2) = 6x - 6 \cdot 2$ Distributive Property

96. $3 + (4 + x) = (3 + 4) + x$
Associative Property of Addition

In Exercises 97–102, solve the equation.

97. $2x - 5 = x + 9$ 14

98. $6x + 8 = 8 - 2x$ 0

99. $2x + \frac{3}{2} = \frac{3}{2}$ 0

100. $-\frac{x}{10} = 1000$ $-10{,}000$

101. $-0.35x = 70$ -200

102. $0.60x = 24$ 40

2.3 Business and Scientific Problems

- Use mathematical models to solve business-related problems.
- Use mathematical models to solve mixture problems and rate problems.
- Use formulas to solve application problems.

Rates in Business Problems

The **markup** on a consumer item is the difference between the cost a retailer pays for an item and the price at which the retailer sells the item. A verbal model for this relationship is as follows.

Selling price = Cost + Markup

The markup is the product of the **markup rate** and the cost.

Markup = Markup rate • Cost

Application

EXAMPLE 1 Finding a Markup Rate

Jeans were invented in 1873 by Levi Strauss and Jacob Davis. Billions of pairs of jeans are sold each year throughout the world.

A clothing store sells a pair of jeans for \$42. The cost of the jeans is \$16.80. What is the markup rate?

SOLUTION

Verbal Model: Selling price = Cost + Markup

Labels: Selling price = 42 (dollars)
Cost = 16.80 (dollars)
Markup rate = p (percent in decimal form)
Markup = $p(16.80)$ (dollars)

Equation:

$42 = 16.80 + p(16.80)$ Original equation

$25.2 = p(16.80)$ Subtract 16.80 from each side.

$\frac{25.2}{16.80} = p$ Divide each side by 16.80.

$1.5 = p$ Simplify.

Because $p = 1.5$, it follows that the markup rate is 150%. Check this in the original statement of the problem.

Exercises Within Reach ®

Solutions in English & Spanish and tutorial videos at AlgebraWithinReach.com

Solving a Markup Problem **In Exercises 1–8, find the missing quantities.**

	Cost	Selling Price	Markup	Markup Rate		Cost	Selling Price	Markup	Markup Rate
1.	\$45.97	\$64.33	\$18.36	40%	2.	\$62.40	\$96.72	\$34.32	55%
3.	\$152.00	\$250.80	\$98.80	65%	4.	\$402.40	\$623.72	\$221.32	55%
5.	\$22,250.00	\$26,922.50	\$4672.50	21%	6.	\$12,550.00	\$16,440.50	\$3890.50	31%
7.	\$225.00	\$416.70	\$191.70	85.2%	8.	\$732.00	\$976.00	\$244.00	$33\frac{1}{3}\%$

The model for a **discount** is similar to that for a markup.

Sale price = Original price − Discount

The discount is the product of the **discount rate** and the original price.

Discount = Discount rate • Original price

Application **EXAMPLE 2** **Finding a Discount and a Discount Rate**

A DVD player is marked down from its original price of $90 to a sale price of $63. What is the discount rate?

Study Tip

Although markup and discount are similar, it is important to remember that markup is based on cost and discount is based on original price.

SOLUTION

Verbal Model: Discount = Discount rate • Original price

Labels: Discount $= 90 - 63 = 27$ (dollars)
Original price $= 90$ (dollars)
Discount rate $= p$ (percent in decimal form)

Equation:

$27 = p(90)$ Original equation

$\frac{27}{90} = p$ Divide each side by 90.

$0.30 = p$ Simplify.

The discount rate is 30%. Check this in the original statement of the problem.

Exercises Within Reach ®

Solutions in English & Spanish and tutorial videos at AlgebraWithinReach.com

Solving a Discount Problem **In Exercises 9–16, find the missing quantities.**

	Original Price	*Sale Price*	*Discount*	*Discount Rate*		*Original Price*	*Sale Price*	*Discount*	*Discount Rate*
9.	$49.95	$25.74	$24.21	48.5%	**10.**	$119.00	$79.73	$39.27	33%
11.	$300.00	$111.00	$189.00	63%	**12.**	$345.00	$210.45	$134.55	39%
13.	$45.00	$27.00	$18.00	40%	**14.**	$24.88	$19.90	$4.98	20%
15.	$1155.50	$831.96	$323.54	28%	**16.**	$459.50	$257.32	$202.18	44%

17. ***Discount*** A shoe store sells a pair of athletic shoes for $75. The shoes go on sale for $45. What is the discount? $30

18. ***Discount*** A bakery sells a dozen rolls for $2.25. You can buy a dozen day-old rolls for $0.75. What is the discount? $1.50

19. ***Discount Rate*** An auto store sells a pair of car mats for $20. On sale, the car mats sell for $16. What is the discount rate? 20%

20. ***Discount Rate*** A department store sells a beach towel for $32. On sale, the beach towel sells for $24. What is the discount rate? 25%

Igor Grochev/Shutterstock.com

Mixture Problems and Rate Problems

Many real-life problems involve combinations of two or more quantities that make up a new or different quantity. Such problems are called **mixture problems**.

$$\boxed{\text{First rate}} \cdot \boxed{\text{Amount}} + \boxed{\text{Second rate}} \cdot \boxed{\text{Amount}} = \boxed{\text{Final rate}} \cdot \boxed{\text{Final amount}}$$

Application

EXAMPLE 3 Solving a Mixture Problem

Study Tip

When you set up a verbal model, be sure to check that you are working with *the same type of units* in each part of the model. For instance, in Example 3 note that each of the three parts of the verbal model measures cost. (If two parts measured cost and the other part measured pounds, you would know that the model was incorrect.)

A nursery wants to mix two types of lawn seed. Type A sells for \$10 per pound and type B sells for \$15 per pound. To obtain 20 pounds of a mixture at \$12 per pound, how many pounds of each type of seed are needed?

SOLUTION

The rates are the unit prices for each type of seed.

Verbal Model: $\boxed{\text{Total cost of \$10 seed}} + \boxed{\text{Total cost of \$15 seed}} = \boxed{\text{Total cost of \$12 seed}}$

Labels:
Unit price of type A $= 10$ (dollars per pound)
Pounds of \$10 seed $= x$ (pounds)
Unit price of type B $= 15$ (dollars per pound)
Pounds of \$15 seed $= 20 - x$ (pounds)
Unit price of mixture $= 12$ (dollars per pound)
Pounds of \$12 seed $= 20$ (pounds)

Equation:
$10x + 15(20 - x) = 12(20)$ Original equation
$10x + 300 - 15x = 240$ Distributive Property
$300 - 5x = 240$ Combine like terms.
$-5x = -60$ Subtract 300 from each side.
$x = 12$ Divide each side by -5.

The mixture should contain 12 pounds of the \$10 seed and $20 - x = 20 - 12 = 8$ pounds of the \$15 seed.

Exercises Within Reach ®

Solutions in English & Spanish and tutorial videos at AlgebraWithinReach.com

Mixture Problem **In Exercises 21–24, determine the numbers of units of solutions 1 and 2 needed to obtain a final solution of the specified amount and concentration.**

	Concentration of Solution 1	*Concentration of Solution 2*	*Concentration of Final Solution*	*Amount of Final Solution*
21.	20%	60%	40%	100 gal
	50 gallons of solution 1, 50 gallons of solution 2			
22.	50%	75%	60%	10 L
	6 liters of solution 1, 4 liters of solution 2			
23.	15%	60%	45%	24 qt
	8 quarts of solution 1, 16 quarts of solution 2			
24.	45%	85%	70%	600 ml
	225 milliliters of solution 1, 375 milliliters of solution 2			

25. ***Seed Mixture*** A nursery wants to mix two types of lawn seed. Type 1 sells for \$12 per pound, and type 2 sells for \$20 per pound. To obtain 100 pounds of a mixture at \$14 per pound, how many pounds of each type of seed are needed? 75 pounds of type 1, 25 pounds of type 2

26. ***Nut Mixture*** A grocer mixes two kinds of nuts costing \$7.88 per pound and \$8.88 per pound to make 100 pounds of a mixture costing \$8.28 per pound. How many pounds of each kind of nut are in the mixture? 60 pounds at \$7.88, 40 pounds at \$8.88

Time-dependent problems such as distance traveled at a given speed and work done at a specified rate are classic types of **rate problems**. The distance-rate-time problem fits the verbal model

Distance = Rate • Time .

Application

EXAMPLE 4 Solving a Distance-Rate Problem

Students are traveling in two cars to a football game 150 miles away. The first car leaves on time and travels at an average speed of 48 miles per hour. The second car starts $\frac{1}{2}$ hour later and travels at an average speed of 58 miles per hour. At these speeds, how long will it take the second car to catch up to the first car?

SOLUTION

Verbal Model: Distance of first car = Distance of second car

Labels:
Time for first car $= t$ (hours)
Distance of first car $= 48t$ (miles)
Time for second car $= t - \frac{1}{2}$ (hours)
Distance of second car $= 58\left(t - \frac{1}{2}\right)$ (miles)

Equation:

$48t = 58\left(t - \frac{1}{2}\right)$ Original equation

$48t = 58t - 29$ Distributive Property

$48t - 58t = 58t - 58t - 29$ Subtract $58t$ from each side.

$-10t = -29$ Combine like terms.

$\frac{-10t}{-10} = \frac{-29}{-10}$ Divide each side by -10.

$t = 2.9$ Simplify.

After the first car travels for 2.9 hours, the second car catches up to it. So, it takes the second car $t - 0.5 = 2.9 - 0.5 = 2.4$ hours to catch up to the first car.

Additional Examples

1. How long will it take you to ride your bike 30 kilometers at 18 kilometers per hour?
   Answer: $1\frac{2}{3}$ hours
2. A train can travel a distance of 252 kilometers in 2 hours and 15 minutes without making any stops. What is the average speed of the train?
   Answer: 112 kilometers per hour

Exercises Within Reach®

Solutions in English & Spanish and tutorial videos at AlgebraWithinReach.com

27. ***Distance*** Two planes leave Chicago's O'Hare International Airport at approximately the same time and fly in opposite directions. How far apart are the planes after $1\frac{1}{3}$ hours if their average speeds are 480 miles per hour and 600 miles per hour?
1440 miles

28. ***Distance*** Two trucks leave a depot at approximately the same time and travel the same route. How far apart are the trucks after $4\frac{1}{2}$ hours if their average speeds are 52 miles per hour and 56 miles per hour? 18 miles

29. ***Travel Time*** On the first part of a 317-mile trip, a sales representative averaged 58 miles per hour. The sales representative averaged only 52 miles per hour on the remainder of the trip because of an increased volume of traffic (see figure). The total time of the trip was 5 hours and 45 minutes. Find the amount of driving time at each speed.
3 hours at 58 miles per hour, $2\frac{3}{4}$ hours at 52 miles per hour

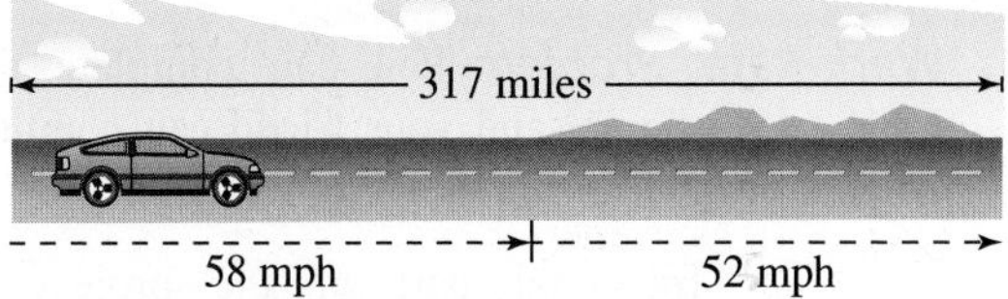

30. ***Travel Time*** Two cars start at the same location and travel in the same direction at average speeds of 30 miles per hour and 45 miles per hour. How much time must elapse before the two cars are 5 miles apart? $\frac{1}{3}$ hour = 20 minutes

In work-rate problems, the **rate of work** is the *reciprocal* of the time needed to do the entire job. For instance, if it takes 5 hours to complete a job, then the per hour work rate is $\frac{1}{5}$ job per hour. In general,

$$\text{Per hour work rate} = \frac{1}{\text{Total hours to complete a job}}.$$

Application **EXAMPLE 5** **Solving a Work-Rate Problem**

Consider two machines in a paper manufacturing plant. Machine 1 can complete one job in 4 hours. Machine 2 is newer and can complete one job in $2\frac{1}{2}$ hours. How long will it take the two machines working together to complete one job?

Study Tip

Notice the hidden product, rate • time, in the portion of work done by each machine in Example 5. Watch for such products in the exercise set.

SOLUTION

Verbal Model: Work done = Portion done by machine 1 + Portion done by machine 2

Labels: Work done by both machines = 1 (job)

Time for each machine = t (hours)

Per hour work rate for machine 1 = $\frac{1}{4}$ (job per hour)

Per hour work rate for machine 2 = $\frac{2}{5}$ (job per hour)

Equation:

$1 = \left(\frac{1}{4}\right)(t) + \left(\frac{2}{5}\right)(t)$ Rate • time + rate • time

$1 = \left(\frac{1}{4} + \frac{2}{5}\right)(t)$ Distributive Property

$1 = \left(\frac{5}{20} + \frac{8}{20}\right)(t)$ Least common denominator is 20.

$1 = \left(\frac{13}{20}\right)(t)$ Simplify.

$1 \div \frac{13}{20} = t$ Divide each side by $\frac{13}{20}$.

$\frac{20}{13} = t$ Simplify.

The equation in Example 5 can also be written as

$$\frac{1}{t} = \frac{1}{4} + \frac{2}{5}.$$

In this case, $1/t$ = the portion of the work done together in 1 hour, $\frac{1}{4}$ = the portion of the work done by machine 1 in 1 hour, and $\frac{2}{5}$ = the portion of work done by machine 2 in 1 hour.

It will take $\frac{20}{13}$ hours (or about 1.54 hours) for both machines to complete the job. Check this solution in the original statement of the problem.

Exercises Within Reach®

Solutions in English & Spanish and tutorial videos at AlgebraWithinReach.com

31. ***Work-Rate Problem*** You can complete a typing project in 5 hours, and your friend can complete it in 8 hours.

(a) What fractional part of the project can be accomplished by each person in 1 hour?

$\frac{1}{5}, \frac{1}{8}$

(b) How long will it take both of you to complete the project working together?

$\frac{40}{13} = 3\frac{1}{13}$ hours

32. ***Work-Rate Problem*** You can mow a lawn in 3 hours, and your friend can mow it in 4 hours.

(a) What fractional part of the lawn can each of you mow in 1 hour?

$\frac{1}{3}, \frac{1}{4}$

(b) How long will it take both of you to mow the lawn working together?

$\frac{12}{7} = 1\frac{5}{7}$ hours

Using Formulas to Solve Problems

Miscellaneous Common Formulas

Temperature: F = degrees Fahrenheit, C = degrees Celsius

$$F = \frac{9}{5}C + 32$$

Simple Interest: I = interest, P = principal, r = interest rate (decimal form), t = time (years)

$$I = Prt$$

Distance: d = distance traveled, r = rate, t = time

$$d = rt$$

Application

EXAMPLE 6 Finding an Interest Rate

A deposit of $8000 earned $300 in interest in 6 months. What was the annual interest rate for this account?

SOLUTION

Verbal Model: Interest = Principal • Rate • Time

Labels:
Interest = 300 (dollars)
Principal = 8000 (dollars)
Annual interest rate = r (percent in decimal form)
Time = $\frac{1}{2}$ (year)

Equation:

$$300 = 8000(r)\left(\frac{1}{2}\right)$$ Original equation

$$\frac{300}{4000} = r \quad \Rightarrow \quad 0.075 = r$$ Divide each side by 4000.

The annual interest rate is $r = 0.075$ or 7.5%.

Exercises Within Reach®

Solutions in English & Spanish and tutorial videos at AlgebraWithinReach.com

33. ***Simple Interest*** Find the interest on a $5000 bond that pays an annual percentage rate of $6\frac{1}{2}\%$ for 6 years. $1950

34. ***Simple Interest*** Find the annual interest rate on a certificate of deposit that accumulated $400 interest in 2 years on a principal of $2500. 8%

35. ***Simple Interest*** The interest on a savings account is 7%. Find the principal required to earn $500 in interest in 2 years. $3571.43

36. ***Simple Interest*** The interest on a bond is $5\frac{1}{2}\%$. Find the principal required to earn $900 in interest in 3 years. $5454.55

Smit/Shutterstock.com; Kristijan Zontar/Shutterstock.com

Common Formulas for Area, Perimeter or Circumference, and Volume

Square

$A = s^2$

$P = 4s$

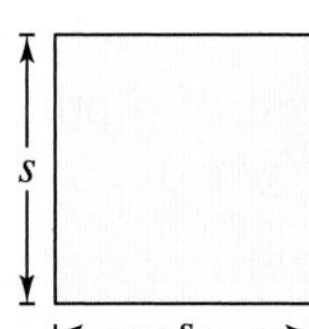

Rectangle

$A = lw$

$P = 2l + 2w$

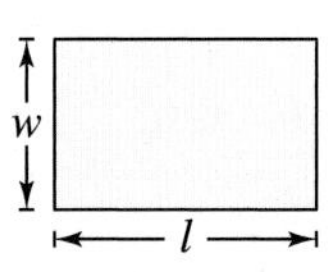

Circle

$A = \pi r^2$

$C = 2\pi r$

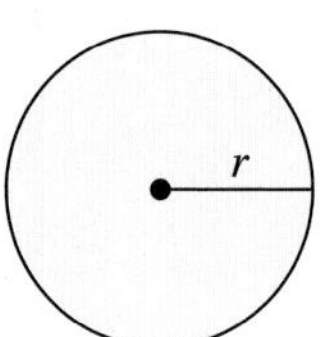

Triangle

$A = \frac{1}{2}bh$

$P = a + b + c$

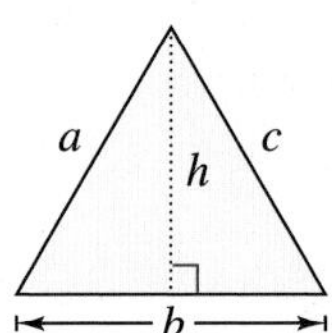

Cube

$V = s^3$

Rectangular Solid

$V = lwh$

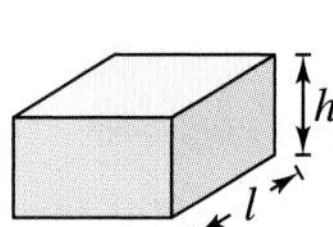

Circular Cylinder

$V = \pi r^2 h$

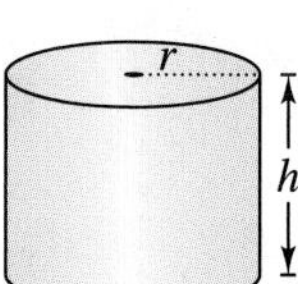

Sphere

$V = \frac{4}{3}\pi r^3$

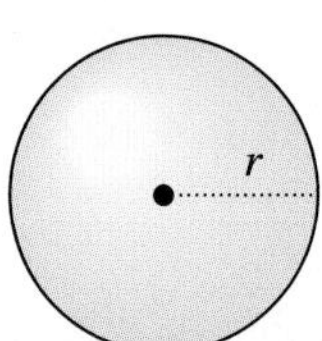

Application **EXAMPLE 7** Geometry: **Rewriting a Formula**

In the perimeter formula $P = 2l + 2w$, solve for w.

SOLUTION

$P = 2l + 2w$	Original formula
$P - 2l = 2w$	Subtract $2l$ from each side.
$\frac{P - 2l}{2} = w$	Divide each side by 2.

Exercises Within Reach®

Solutions in English & Spanish and tutorial videos at AlgebraWithinReach.com

Rewriting a Formula **In Exercises 37–44, solve for the specified variable.**

37. Solve for R.

Ohm's Law: $E = IR$ $R = \frac{E}{I}$

38. Solve for r.

Simple Interest: $A = P + Prt$ $r = \frac{A - P}{Pt}$

39. Solve for L.

Discount: $S = L - rL$ $L = \frac{S}{1 - r}$

40. Solve for C.

Markup: $S = C + rC$ $C = \frac{S}{1 + r}$

41. Solve for a.

Free-Falling Body: $h = 36t + \frac{1}{2}at^2 + 50$

$a = \frac{2h - 72t - 100}{t^2}$

42. Solve for a.

Free-Falling Body: $h = -15t + \frac{1}{2}at^2 + 9$

$a = \frac{2h + 30t - 19}{t^2}$

43. Solve for h.

Surface Area of a Circular Cylinder:

$S = 2\pi r^2 + 2\pi rh$ $h = \frac{S}{2\pi r} - r$

44. Solve for b.

Area of a Trapezoid: $A = \frac{1}{2}(a + b)h$

$b = \frac{2A}{h} - a$

Application **EXAMPLE 8** **Using a Geometric Formula**

Study Tip

When solving problems such as the one in Example 8, you may find it helpful to draw and label a diagram.

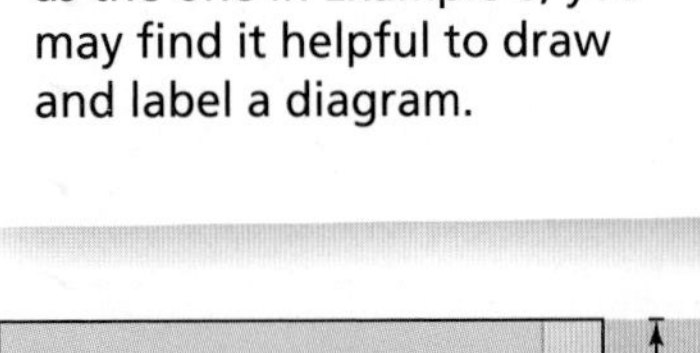

The city plans to put sidewalks along the two streets that border your corner lot, which is 250 feet long on one side and has an area of 30,000 square feet. Each lot owner is to pay \$1.50 per foot of sidewalk bordering his or her lot.

a. Find the width of your lot.

b. How much will you have to pay for the sidewalks put on your lot?

SOLUTION

The figure shows a labeled diagram of your lot.

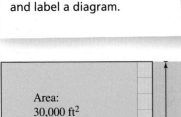

a. *Verbal Model:* Area = Length • Width

Labels:
Area of lot = 30,000 (square feet)
Length of lot = 250 (feet)
Width of lot = w (feet)

Equation:

$30{,}000 = 250 \cdot w$ — Original equation

$\dfrac{30{,}000}{250} = w$ — Divide each side by 250.

$120 = w$ — Simplify.

Your lot is 120 feet wide.

b. *Verbal Model:* Cost = Rate per foot • Length of sidewalk

Labels:
Cost of sidewalks = C (dollars)
Rate per foot = 1.50 (dollars per foot)
Total length of sidewalk = 120 + 250 (feet)

Equation:

$C = 1.50(120 + 250)$ — Original equation

$C = 1.50 \cdot 370$ — Add within parentheses.

$C = 555$ — Multiply.

You will have to pay \$555 for the sidewalks put on your lot.

Exercises Within Reach®

Solutions in English & Spanish and tutorial videos at AlgebraWithinReach.com

45. *Geometry* A rectangular picture frame has a perimeter of 40 inches. The width of the frame is 4 inches less than its height. Find the height of the frame.
12 inches

46. *Geometry* A rectangular stained glass window has a perimeter of 18 feet. The height of the window is 1.25 times its width. Find the width of the window.
4 feet

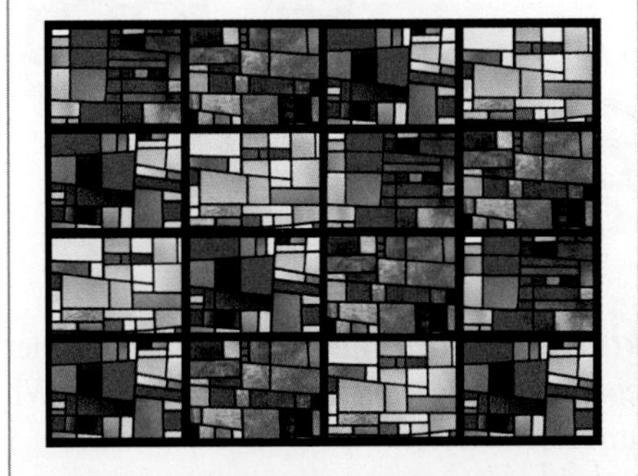

Philip Meyer/Shutterstock.com

Concept Summary: Using Formulas to Solve Real-Life Applications

What

Many real-life problems involve geometric applications such as perimeter, area, and volume. Other problems might involve distance, temperature, or interest.

EXAMPLE

You jog at an average rate of 8 kilometers per hour. How long will it take you to jog 14 kilometers?

How

To solve such problems, you can use **formulas**. For example, to solve a problem involving distance, rate, and time, you can use the distance formula.

EXAMPLE

Distance = Rate • Time

$d = rt$ Distance formula

$14 = 8(t)$ Substitute for d and r.

$\frac{14}{8} = t$ Divide each side by 8.

$1.75 = t$ Simplify.

It will take you 1.75 hours.

Why

You can use formulas to solve many real-life applications. Some formulas occur so frequently that it is to your benefit to memorize them.

Exercises Within Reach®

Worked-out solutions to odd-numbered exercises at AlgebraWithinReach.com

Concept Summary Check

47. ***Formula*** What is the formula for the volume of a cube? $V = s^3$

48. ***Create an Example*** Give an example in which you need to find the area of a real-life object. Installing carpet in a room

49. ***Writing*** In your own words, describe the units of measure used for perimeter, area, and volume. Give examples of each. See Additional Answers.

50. ***Structure*** Rewrite the formula for simple interest by solving for P. $\frac{I}{rt} = P$

Extra Practice

51. ***Markup*** A department store sells a jacket for \$157.14. The cost of the jacket to the store is \$130.95. What is the markup? \$26.19

52. ***Markup*** A shoe store sells a pair of shoes for \$89.95. The cost of the shoes to the store is \$46.50. What is the markup? \$43.45

53. ***Markup Rate*** A jewelry store sells a bracelet for \$84. The cost of the bracelet to the store is \$46.67. What is the markup rate? 80%

54. ***Markup Rate*** A department store sells a sweater for \$60. The cost of the sweater to the store is \$35. What is the markup rate? 71.4%

55. ***International Calling Card Rate*** The weekday rate for a telephone call is \$0.02 per minute. Determine the length of a call that costs \$0.70. What would have been the cost of the call if it had been made during the weekend, when there is a 20% discount? 35 minutes; \$0.56

56. ***International Calling Card Rate*** The weekday rate for a telephone call is \$0.04 per minute. Determine the length of a call that costs \$1.20. What would have been the cost of the call if it had been made during the weekend, when there is a 15% discount? 30 minutes; \$1.02

57. ***Cost*** An auto store gives the original price of a tire as \$79.42. During a promotional sale, the store is selling four tires for the price of three. The store needs a markup on cost of 10% during the sale. What is the cost to the store of each tire? \$54.15

58. ***Price*** The produce manager of a supermarket pays \$22.60 for a 100-pound box of bananas. The manager estimates that 10% of the bananas will spoil before they are sold. At what price per pound should the bananas be sold to give the supermarket an average markup rate on cost of 30%. \$0.33

59. ***Ticket Sales*** Ticket sales for a play total \$2200. There are three times as many adult tickets sold as children's tickets. The prices of the tickets for adults and children are \$6 and \$4, respectively. Find the number of children's tickets sold. 100 children's tickets

60. ***Ticket Sales*** Ticket sales for a spaghetti dinner total \$1350. There are four times as many adult tickets sold as children's tickets. The prices of the tickets for adults and children are \$6 and \$3, respectively. Find the number of children's tickets sold. 50 children's tickets

61. ***Antifreeze Mixture*** The cooling system on a truck contains 5 gallons of coolant that is 40% antifreeze. How much must be withdrawn and replaced with 100% antifreeze to bring the coolant in the system to 50% antifreeze. $\frac{5}{6}$ gallon

62. ***Fuel Mixture*** You mix gasoline and oil to obtain $2\frac{1}{2}$ gallons of mixture for an engine. The mixture is 40 parts gasoline and 1 part two-cycle oil. How much gasoline must be added to bring the mixture to 50 parts gasoline and 1 part oil? 0.61 gallon

63. ***Travel Time*** Determine the time required for a space shuttle to travel a distance of 5000 miles in orbit when its average speed is 17,500 miles per hour. 17.14 minutes

64. ***Speed of Light*** The distance between the Sun and Earth is 93,000,000 miles, and the speed of light is 186,282.397 miles per second. Determine the time required for light to travel from the Sun to Earth.

499.24 seconds $\approx$ 8.32 minutes

65. ***Work-Rate Problem*** It takes 30 minutes for a pump to empty a water tank. A larger pump can empty the tank in half the time. How long would it take to empty the tank with both pumps operating? 10 minutes

66. ***Work-Rate Problem*** It takes 90 minutes to mow a lawn with a push mower. It takes half the time using a riding mower. How long would it take to mow the lawn using both mowers? 30 minutes

67. ***Geometry*** Find the volume of the circular cylinder shown in the figure. $147\pi \approx 461.8$ cubic centimeters

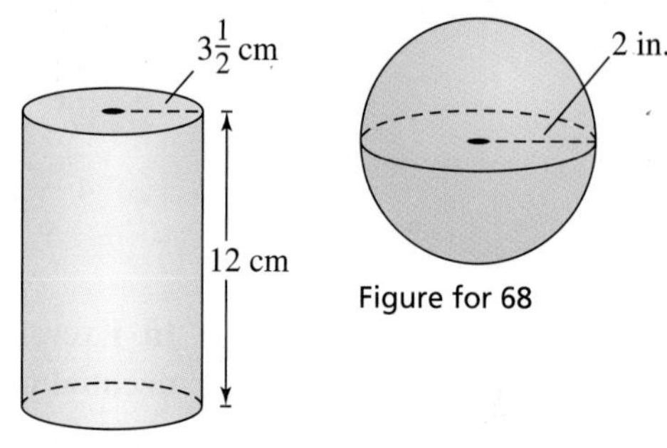

Figure for 67

Figure for 68

68. ***Geometry*** Find the volume of the sphere shown in the figure.

$\frac{32\pi}{3} \approx 33.5$ cubic inches

Explaining Concepts

69. ***Writing*** Explain how to find the sale price of an item when you are given the original price and the discount rate. The sale price is the original price times the quantity 1 minus the discount rate.

70. ***Precision*** When the sides of a square are doubled, does the perimeter double? Explain. Yes. The perimeter of a square with side s is $4s$. If the length of each side is $2s$, the perimeter is $4(2s) = 8s$.

71. ***Precision*** When the sides of a square are doubled, does the area double? Explain. No, it quadruples. The area of a square with side s is s^2. If the length of the side is $2s$, the area is $(2s)^2 = 4s^2$.

72. ***Writing*** If you forget the formula for the volume of a right circular cylinder, how can you derive it? The area of the base, which would be the area of a circle, multiplied by the height.

Cumulative Review

In Exercises 73–76, give (a) the additive inverse and (b) the multiplicative inverse of the quantity.

73. 21
(a) -21 (b) $\frac{1}{21}$

74. -34
(a) 34 (b) $-\frac{1}{34}$

75. $-5x$
(a) $5x$ (b) $-\frac{1}{5x}$

76. $8m$
(a) $-8m$ (b) $\frac{1}{8m}$

In Exercises 77–80, simplify the expression.

77. $2x(x - 4) + 3$ $2x^2 - 8x + 3$

78. $4x - 5(1 - x)$ $9x - 5$

79. $x^2(x - 4) - 2x^2$ $x^3 - 6x^2$

80. $x^2(2x + 1) + 2x(3 - x)$ $2x^3 - x^2 + 6x$

In Exercises 81–84, solve using a percent equation.

81. 52 is 40% of what number? 130

82. 72 is 48% of what number? 150

83. 117 is what percent of 900? 13%

84. 287 is what percent of 350? 82%

Mid-Chapter Quiz: Sections 2.1–2.3

Solutions in English & Spanish and tutorial videos at AlgebraWithinReach.com

Take this quiz as you would take a quiz in class. After you are done, check your work against the answers in the back of the book.

In Exercises 1–8, solve the equation and check your solution. (If it is not possible, state the reason.)

1. $4x - 8 = 0$ 2
2. $-3(z - 2) = 0$ 2
3. $2(y + 3) = 18 - 4y$ 2
4. $5t + 7 = 7(t + 1) - 2t$ Infinitely many, because $7 = 7$.
5. $\frac{1}{4}x + 6 = \frac{3}{2}x - 1$ $\frac{28}{5}$
6. $\frac{2b}{5} + \frac{b}{2} = 3$ $\frac{10}{3}$
7. $\frac{4 - x}{5} + 5 = \frac{5}{2}$ $\frac{33}{2}$
8. $3x + \frac{11}{12} = \frac{5}{16}$ $-\frac{29}{144}$

In Exercises 9 and 10, solve the equation and round your answer to two decimal places.

9. $0.25x + 6.2 = 4.45x + 3.9$ 0.55
10. $0.42x + 6 = 5.25x - 0.80$ 1.41
11. Write the decimal 0.45 as a fraction and as a percent. $\frac{9}{20}$, 45%
12. 500 is 250% of what number? 200

Applications

13. Find the unit price (in dollars per ounce) of a 12-ounce box of cereal that sells for $4.85. $0.40 per ounce
14. A quality control engineer finds one defective unit in a sample of 150. At this rate, what is the expected number of defective units in a shipment of 750,000? 5000 defective units
15. A store is offering a discount of 25% on a computer with an original price of $1080. A mail-order catalog has the same computer for $799 plus $14.95 for shipping. Which is the better buy? The computer from the store
16. Last week you earned $616. Your regular hourly wage is $12.25 for the first 40 hours, and your overtime hourly wage is $18. How many hours of overtime did you work? 7 hours
17. Fifty gallons of a 30% acid solution is obtained by combining solutions that are 25% acid and 50% acid. How much of each solution is required? 40 gallons of 25% solution, 10 gallons of 50% solution
18. On the first part of a 300-mile trip, a sales representative averaged 62 miles per hour. The sales representative averaged 46 miles per hour on the remainder of the trip because of an increased volume of traffic. The total time of the trip was 6 hours. Find the amount of driving time at each speed. 1.5 hours at 62 miles per hour, 4.5 hours at 46 miles per hour
19. You can paint a room in 3 hours, and your friend can paint it in 5 hours. How long will it take both of you to paint the room working together? $\frac{15}{8} = 1.875$ hours

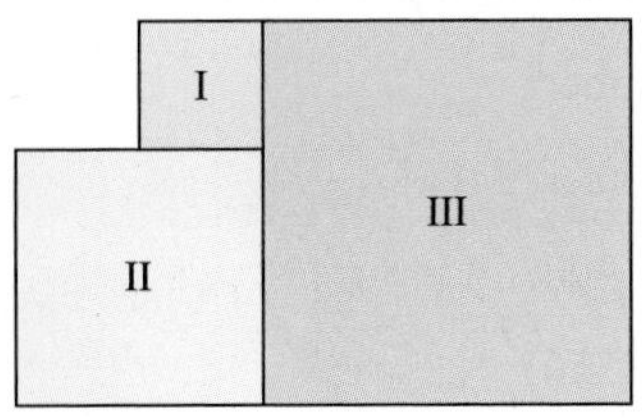

20. The figure at the left shows three squares. The perimeters of squares I and II are 20 inches and 32 inches, respectively. Find the area of square III. 169 square inches

Study Skills in Action

Improving Your Memory

Have you ever driven on a highway for ten minutes when all of a sudden you kind of woke up and wondered where the last ten miles had gone? It was like the car was on autopilot. The same thing happens to many college students as they sit through back-to-back classes. The longer students sit through classes on "autopilot," the more likely they will "crash" when it comes to studying outside of class on their own.

While on autopilot, you do not process and retain new information effectively. You can improve your memory by learning how to focus during class and while studying on your own.

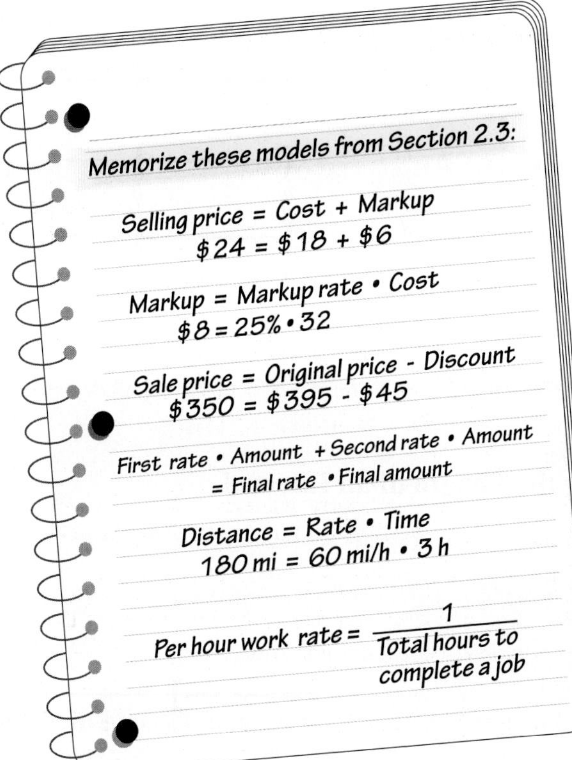

Memorize these models from Section 2.3:
Selling price = Cost + Markup
$24 = $18 + $6
Markup = Markup rate • Cost
$8 = 25% • 32
Sale price = Original price - Discount
$350 = $395 - $45
First rate • Amount + Second rate • Amount
= Final rate • Final amount
Distance = Rate • Time
180 mi = 60 mi/h • 3 h
Per hour work rate = 1 / Total hours to complete a job

Smart Study Strategy

Keep Your Mind Focused

During class

- When you sit down at your desk, get all other issues out of your mind by reviewing your notes from the last class and focusing just on math.
- Repeat in your mind what you are writing in your notes.
- When the math is particularly difficult, ask your instructor for another example.

While completing homework

- Before doing homework, review the concept boxes and examples. Talk through the examples out loud.
- Complete homework as though you were also preparing for a quiz. Memorize the different types of problems, formulas, rules, and so on.

Between classes

- Review the concept boxes, the Concept Summaries, and the Concept Summary Check exercises.

Preparing for a test

- Review all your notes that pertain to the upcoming test. Review examples of each type of problem that could appear on the test.

2.4 Linear Inequalities

- Sketch the graphs of inequalities.
- Solve linear inequalities.
- Solve application problems involving inequalities.

Intervals on the Real Number Line

As with an equation, you **solve an inequality** in the variable x by finding all values of x for which the inequality is true. Such values are called **solutions** and are said to satisfy the inequality. The set of all solutions of the inequality is the **solution set** of the inequality. The **graph** of an inequality is obtained by plotting its solution set on the real number line. Often, these graphs are intervals—either bounded or unbounded.

Study Tip

The **length** of an interval is the distance between its endpoints.

Bounded Intervals on the Real Number Line

Let a and b be real numbers such that $a < b$. The following intervals on the real number line are called **bounded intervals**. The numbers a and b are the **endpoints** of each interval. A bracket indicates that the endpoint is included in the interval, and a parenthesis indicates that the endpoint is excluded.

Notation	*Interval Type*	*Inequality*	*Graph*
$[a, b]$	Closed	$a \le x \le b$	
(a, b)	Open	$a < x < b$	
$[a, b)$		$a \le x < b$	
$(a, b]$		$a < x \le b$	

EXAMPLE 1 Finding the Length of an Interval

Find the length of the interval $[-1, 1]$.

SOLUTION

The length of the interval $[-1, 1]$ is

$$|1 - (-1)| = 2.$$

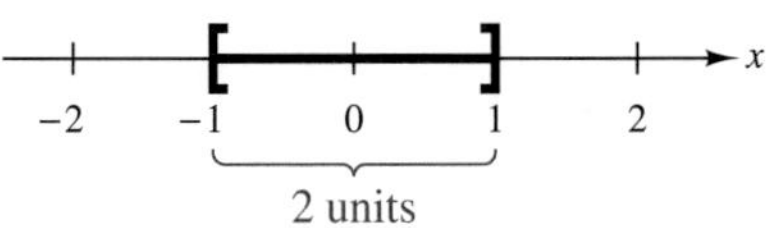

Exercises Within Reach®

Solutions in English & Spanish and tutorial videos at AlgebraWithinReach.com

Finding the Length of an Interval **In Exercises 1–6, find the length of the interval.**

1. $[-3, 5]$ 8 **2.** $[4, 10]$ 6 **3.** $(-9, 2]$ 11

4. $[5, 13)$ 8 **5.** $(-3, 0)$ 3 **6.** $(0, 7)$ 7

Study Tip

The symbols ∞ **(positive infinity)** and $-\infty$ **(negative infinity)** do not represent real numbers. They are simply convenient symbols used to describe the unboundedness of an interval such as $(5, \infty)$.

Unbounded Intervals on the Real Number Line

Let a and b be real numbers. The following intervals on the real number line are called **unbounded intervals**.

Notation	*Interval Type*	*Inequality*	*Graph*
$[a, \infty)$		$x \geq a$	
(a, ∞)	Open	$x > a$	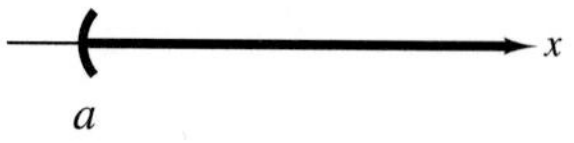
$(-\infty, b]$		$x \leq b$	
$(-\infty, b)$	Open	$x < b$	
$(-\infty, \infty)$	Entire real line		

EXAMPLE 2 Graphing Inequalities

Sketch the graph of each inequality.

a. $-3 < x \leq 1$

b. $0 < x < 2$

c. $-3 < x$

d. $x \leq 2$

SOLUTION

a. The graph of $-3 < x \leq 1$ is a bounded interval.

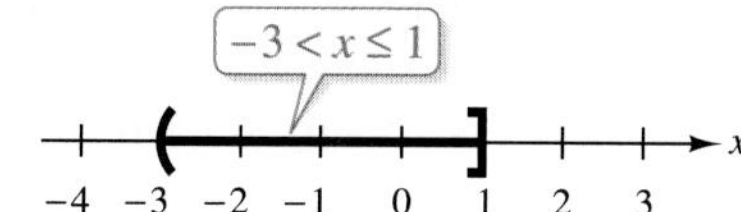

b. The graph of $0 < x < 2$ is a bounded interval.

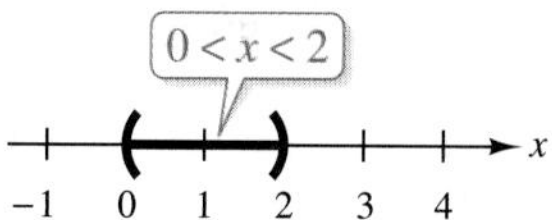

c. The graph of $-3 < x$ is an unbounded interval.

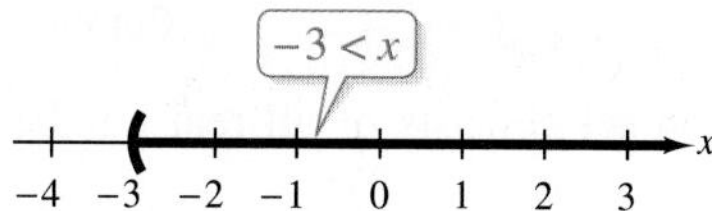

d. The graph of $x \leq 2$ is an unbounded interval.

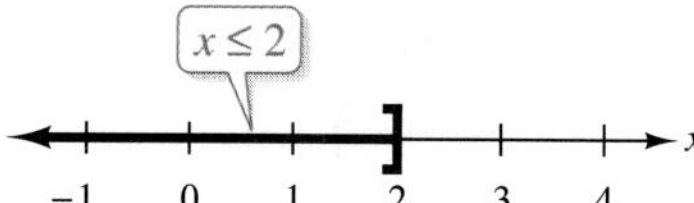

Exercises Within Reach®

Solutions in English & Spanish and tutorial videos at AlgebraWithinReach.com

Graphing an Inequality **In Exercises 7–20, sketch the graph of the inequality.** See Additional Answers.

7. $x \leq 4$

8. $x > -6$

9. $x > 3.5$

10. $x \leq -2.5$

11. $x \geq \frac{1}{2}$

12. $x < \frac{1}{4}$

13. $-5 < x \leq 3$

14. $-1 < x \leq 5$

15. $4 > x \geq 1$

16. $9 \geq x \geq 3$

17. $\frac{3}{2} \geq x > 0$

18. $-\frac{15}{4} < x < -\frac{5}{2}$

19. $3.5 < x \leq 4.5$

20. $6.5 \geq x > -2.5$

Solving Linear Inequalities

Properties of Inequalities

1. *Addition and Subtraction Properties*

Adding the same quantity to, or subtracting the same quantity from, each side of an inequality produces an equivalent inequality.

If $a < b$, then $a + c < b + c$.

If $a < b$, then $a - c < b - c$.

2. *Multiplication and Division Properties: Positive Quantities*

Multiplying or dividing each side of an inequality by a positive quantity produces an equivalent inequality.

If $a < b$ and c is positive, then $ac < bc$.

If $a < b$ and c is positive, then $\frac{a}{c} < \frac{b}{c}$.

3. *Multiplication and Division Properties: Negative Quantities*

Multiplying or dividing each side of an inequality by a negative quantity produces an equivalent inequality in which the inequality symbol is reversed.

If $a < b$ and c is negative, then $ac > bc$. Reverse inequality.

If $a < b$ and c is negative, then $\frac{a}{c} > \frac{b}{c}$. Reverse inequality.

4. *Transitive Property*

Consider three quantities for which the first quantity is less than the second, and the second is less than the third. It follows that the first quantity must be less than the third quantity.

If $a < b$ and $b < c$, then $a < c$.

EXAMPLE 3 Solving a Linear Inequality

$x + 6 < 9$ Original inequality

$x + 6 - 6 < 9 - 6$ Subtract 6 from each side.

$x < 3$ Combine like terms.

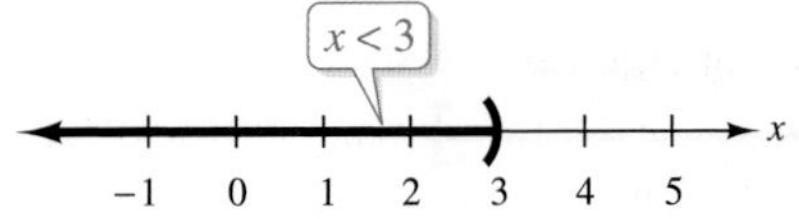

The solution set consists of all real numbers that are less than 3. The solution set in interval notation is $(-\infty, 3)$.

Exercises Within Reach®

Solutions in English & Spanish and tutorial videos at AlgebraWithinReach.com

Identifying Equivalent Inequalities **In Exercises 21–24, determine whether the inequalities are equivalent.**

21. $3x - 2 < 12, \quad 3x < 10$ Not equivalent

22. $6x + 7 \geq 11, \quad 6x \geq 18$ Not equivalent

23. $7x - 6 \leq 3x + 12, \quad 4x \leq 18$ Equivalent

24. $11 - 3x \geq 7x + 1, \quad 10 \geq 10x$ Equivalent

Solving an Inequality **In Exercises 25–30, solve the inequality and sketch the solution on the real number line.** See Additional Answers.

25. $x - 4 \geq 0$ $x \geq 4$

26. $x + 1 < 0$ $x < -1$

27. $x + 7 \leq 9$ $x \leq 2$

28. $z - 5 > 0$ $z > 5$

29. $2x < 8$ $x < 4$

30. $3x \geq 12$ $x \geq 4$

EXAMPLE 4 Solving a Linear Inequality

$8 - 3x \le 20$	Original inequality
$8 - 8 - 3x \le 20 - 8$	Subtract 8 from each side.
$-3x \le 12$	Combine like terms.
$\dfrac{-3x}{-3} \ge \dfrac{12}{-3}$	Divide each side by -3 and reverse the inequality symbol.
$x \ge -4$	Simplify.

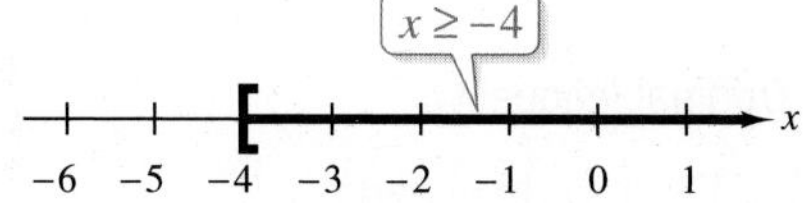

The solution set consists of all real numbers that are greater than or equal to -4. The solution set in interval notation is $[-4, \infty)$.

Additional Example
Solve the inequality.
$3(2x - 6) < 10x + 2$
Answer:
$x > -5$

EXAMPLE 5 Solving a Linear Inequality

$7x - 3 > 3(x + 1)$	Original inequality
$7x - 3 > 3x + 3$	Distributive Property
$7x - 3x - 3 > 3x - 3x + 3$	Subtract $3x$ from each side.
$4x - 3 > 3$	Combine like terms.
$4x - 3 + 3 > 3 + 3$	Add 3 to each side.
$4x > 6$	Combine like terms.
$\dfrac{4x}{4} > \dfrac{6}{4}$	Divide each side by 4.
$x > \dfrac{3}{2}$	Simplify.

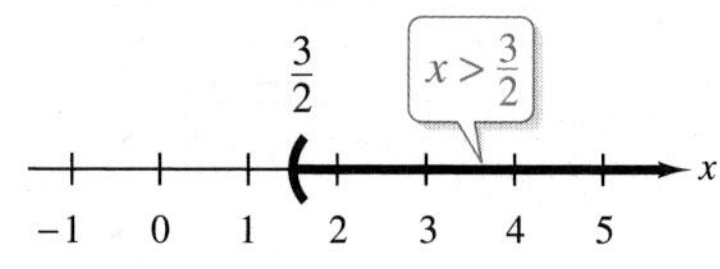

The solution set consists of all real numbers that are greater than $\frac{3}{2}$. The solution set in interval notation is $\left(\frac{3}{2}, \infty\right)$.

Exercises Within Reach®

Solutions in English & Spanish and tutorial videos at AlgebraWithinReach.com

Solving an Inequality **In Exercises 31–46, solve the inequality and sketch the solution on the real number line.** See Additional Answers.

31. $-9x \ge 36$ $x \le -4$

32. $-6x \le 24$ $x \ge -4$

33. $-\frac{3}{4}x < -6$ $x > 8$

34. $-\frac{1}{5}x > -2$ $x < 10$

35. $5 - x \le -2$ $x \ge 7$

36. $1 - y \ge -5$ $y \le 6$

37. $2x - 5.3 > 9.8$ $x > 7.55$

38. $1.6x + 4 \le 12.4$ $x \le 5.25$

39. $5 - 3x < 7$ $x > -\frac{2}{3}$

40. $12 - 5x > 5$ $x < \frac{7}{5}$

41. $3x - 11 > -x + 7$ $x > \frac{9}{2}$

42. $21x - 11 \le 6x + 19$ $x \le 2$

43. $-3x + 7 < 8x - 13$ $x > \frac{20}{11}$

44. $6x - 1 > 3x - 11$ $x > -\frac{10}{3}$

45. $-3(y + 10) \ge 4(y + 10)$ $y \le -10$

46. $2(4 - z) \ge 8(1 + z)$ $z \le 0$

Two inequalities joined by the word *and* or the word *or* constitute a **compound inequality**. When two inequalities are joined by the word *and*, the solution set consists of all real numbers that satisfy *both* inequalities. The solution set for the compound inequality $-4 \le 5x - 2$ *and* $5x - 2 < 7$ can be written more simply as the **double inequality**

$$-4 \le 5x - 2 < 7.$$

Additional Examples

Solve each inequality.

a. $-1 \le 5 - 2x < 7$

b. $x + 3 < -7$ or $x + 3 > 14$

Answers:

a. $-1 < x \le 3$

b. $x < -10$ or $x > 11$

EXAMPLE 6 Solving a Double Inequality

$-7 \le 5x - 2 < 8$	Original inequality
$-7 + 2 \le 5x - 2 + 2 < 8 + 2$	Add 2 to all three parts.
$-5 \le 5x < 10$	Combine like terms.
$\frac{-5}{5} \le \frac{5x}{5} < \frac{10}{5}$	Divide each part by 5.
$-1 \le x < 2$	Simplify.

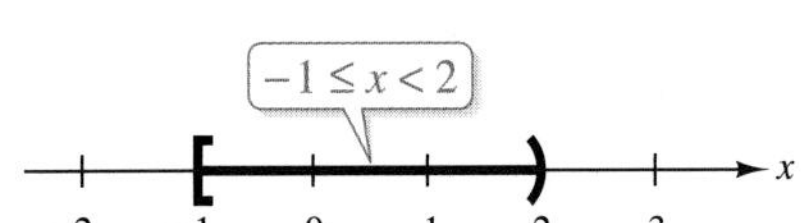

The solution set consists of all real numbers that are greater than or equal to -1 and less than 2. The solution set in interval notation is $[-1, 2)$.

EXAMPLE 7 Solving a Compound Inequality

$-3x + 6 \le 2$	or	$-3x + 6 \ge 7$	Original inequality
$-3x + 6 - 6 \le 2 - 6$		$-3x + 6 - 6 \ge 7 - 6$	Subtract 6 from all parts.
$-3x \le -4$		$-3x \ge 1$	Combine like terms.
$\frac{-3x}{-3} \ge \frac{-4}{-3}$		$\frac{-3x}{-3} \le \frac{1}{-3}$	Divide all parts by -3 and reverse both inequality symbols.
$x \ge \frac{4}{3}$		$x \le -\frac{1}{3}$	Solution set

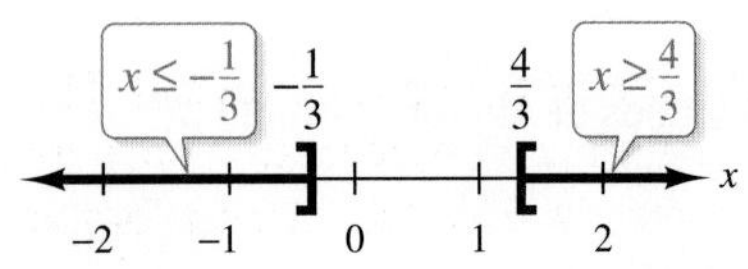

The solution set consists of all real numbers that are less than or equal to $-\frac{1}{3}$ or greater than or equal to $\frac{4}{3}$.

Exercises Within Reach®

Solutions in English & Spanish and tutorial videos at AlgebraWithinReach.com

Solving an Inequality **In Exercises 47–60, solve the inequality and sketch the solution on the real number line. (Some inequalities have no solution.)** See Additional Answers.

47. $0 < 2x - 5 < 9$ $\frac{5}{2} < x < 7$

48. $-6 \le 3x - 9 < 0$ $1 \le x < 3$

49. $8 < 6 - 2x \le 12$ $-3 \le x < -1$

50. $-10 \le 4 - 7x < 10$ $-\frac{6}{7} < x \le 2$

51. $-1 < -0.2x < 1$ $-5 < x < 5$

52. $-2 < -0.5s \le 0$ $0 \le s < 4$

53. $2x - 4 \le 4$ and $2x + 8 > 6$ $-1 < x \le 4$

54. $7 + 4x < -5 + x$ and $2x + 10 \le -2$ $x \le -6$

55. $8 - 3x > 5$ and $x - 5 \ge 10$ No solution

56. $9 - x \le 3 + 2x$ and $3x - 7 \le -22$ No solution

57. $3x + 11 < 3$ or $4x - 1 \ge 9$ $x < -\frac{8}{3}$ or $x \ge \frac{5}{2}$

58. $4x + 10 \le -6$ or $-2x + 5 < -4$ $x \le -4$ or $x > \frac{9}{2}$

59. $7.2 - 1.1x > 1$ or $1.2x - 4 > 2.7$ $-\infty < x < \infty$

60. $0.4x - 3 \le 8.1$ or $4.2 - 1.6x \le 3$ $-\infty < x < \infty$

Application

Application

EXAMPLE 8 Finding the Maximun Width of a Package

An overnight delivery service will not accept any package with a combined length and girth (perimeter of a cross section perpendicular to the length) exceeding 132 inches. Consider a rectangular box that is 68 inches long and has square cross sections. What is the maximum acceptable width of such a box?

SOLUTION

First make a sketch as shown below. The length of the box is 68 inches, and because a cross section is square, the width and height are each x inches.

Verbal Model: Length + Girth $\leq$ 132 inches

Labels:
Width of a side $= x$ (inches)
Length $= 68$ (inches)
Girth $= 4x$ (inches)

Inequality:
$$68 + 4x \leq 132$$
$$4x \leq 64$$
$$x \leq 16$$

The width of the box can be at most 16 inches.

Exercises Within Reach®

Solutions in English & Spanish and tutorial videos at AlgebraWithinReach.com

61. ***Budget*** A student group has \$4500 budgeted for a field trip. The cost of transportation for the trip is \$1900. To stay within the budget, all other costs C must be no more than what amount? \$2600

62. ***Budget*** You have budgeted \$1800 per month for your total expenses. The cost of rent per month is \$600 and the cost of food is \$350. To stay within your budget, all other costs C must be no more than what amount? \$850

Monkey Business Images/Shutterstock.com

63. ***Geometry*** The width of a rectangle is 22 meters. The perimeter of the rectangle must be at least 90 meters and not more than 120 meters. Find the interval for the length x. $23 \leq x \leq 38$

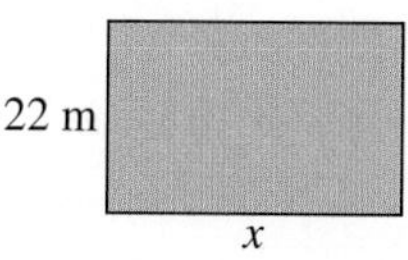

64. ***Geometry*** The length of a rectangle is 12 centimeters. The perimeter of the rectangle must be at least 30 centimeters and not more than 42 centimeters. Find the interval for the width x. $3 \leq x \leq 9$

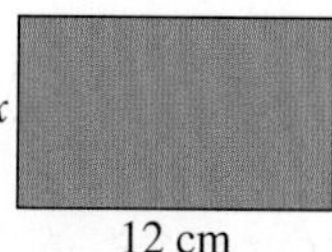

Concept Summary: Solving Linear Inequalities

What

When you **solve an inequality**, you find all the values for which the inequality is true. The procedures for solving inequalities are similar to those for solving equations.

EXAMPLE

Solve the inequality $-2x + 4 > 8$.

How

To solve an inequality, isolate the variable by using the properties of inequalities.

EXAMPLE

$-2x + 4 > 8$ Write inequality.

$-2x + 4 - 4 > 8 - 4$ Subtract 4.

$-2x > 4$ Combine terms.

$\frac{-2x}{-2} < \frac{4}{-2}$ Divide by -2 and reverse inequality.

$x < -2$ Solution set

Why

Many real-life applications involve phrases like "at least" or "no more than." Using inequalities, you will be able to solve such problems.

Exercises Within Reach®

Worked-out solutions to odd-numbered exercises at AlgebraWithinReach.com

Concept Summary Check

65. ***Writing*** Does the graph of $x < -2$ contain a parenthesis or a square bracket? Explain. Parenthesis; A parenthesis is used when the endpoint is excluded.

66. ***Structure*** State whether each inequality is equivalent to $x > 3$. Explain your reasoning in each case. See Additional Answers.

(a) $x < 3$ (b) $3 < x$

(c) $-x < -3$ (d) $-3 < x$

67. ***Reasoning*** Is dividing each side of an inequality by 5 the same as multiplying each side by $\frac{1}{5}$? Explain. Yes. By definition, dividing by a number is the same as multiplying by its reciprocal.

68. ***Writing*** Describe two types of situations in which you must reverse the inequality symbol of an inequality. The inequality symbol is reversed when both sides of an inequality are multiplied or divided by a negative number.

Extra Practice

Matching In Exercises 69–74, **match** the inequality with its graph.

(a) 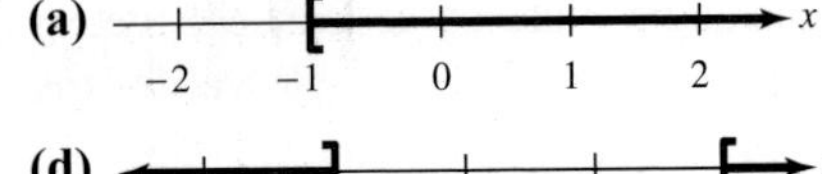

(b) 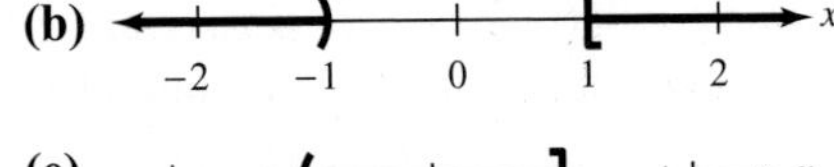

(c) 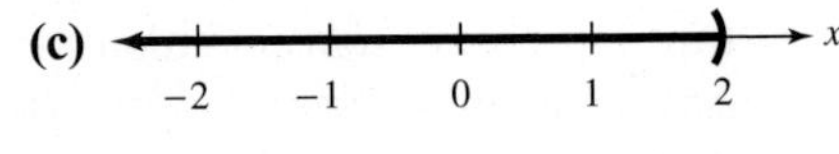

(d) 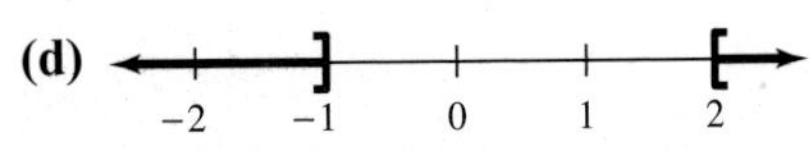

(e) 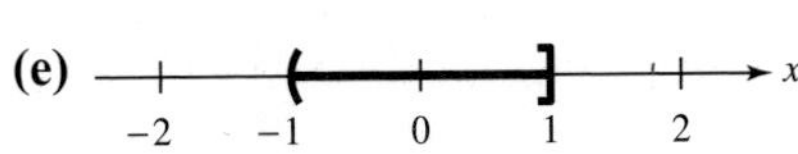

(f) 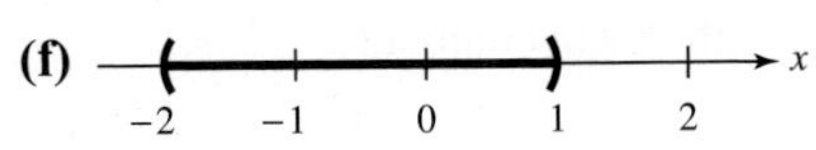

69. $x \geq -1$ a

70. $-1 < x \leq 1$ e

71. $x \leq -1$ or $x \geq 2$ d

72. $x < -1$ or $x \geq 1$ b

73. $-2 < x < 1$ f

74. $x < 2$ c

Solving an Inequality In Exercises 75–86, **solve** the inequality and **sketch** the solution on the real number line. See Additional Answers.

75. $\frac{x}{4} > 2 - \frac{x}{2}$ $x > \frac{8}{3}$

76. $\frac{x}{6} - 1 \leq \frac{x}{4}$ $x \geq -12$

77. $\frac{x-4}{3} + 3 \leq \frac{x}{8}$ $x \leq -8$

78. $\frac{x+3}{6} + \frac{x}{8} \geq 1$ $x \geq \frac{12}{7}$

79. $\frac{3x}{5} - 4 < \frac{2x}{3} - 3$ $x > -15$

80. $\frac{4x}{7} + 1 > \frac{x}{2} + \frac{5}{7}$ $x > -4$

81. $-4 \leq 2 - 3(x + 2) < 11$ $-5 < x \leq 0$

82. $16 < 4(y + 2) - 5(2 - y) \leq 24$ $2 < y \leq \frac{26}{9}$

83. $-3 < \frac{2x-3}{2} < 3$ $-\frac{3}{2} < x < \frac{9}{2}$

84. $0 \leq \frac{x-5}{2} < 4$ $5 \leq x < 13$

85. $1 > \frac{x-4}{-3} > -2$ $1 < x < 10$

86. $-\frac{2}{3} < \frac{x-4}{-6} \leq \frac{1}{3}$ $2 \leq x < 8$

87. ***Meteorology*** Miami's average temperature is greater than the average temperature in Washington, D.C., and the average temperature in Washington, D.C., is greater than the average temperature in New York City. How does the average temperature in Miami compare with the average temperature in New York City?
The average temperature in Miami is greater than the average temperature in New York City.

88. ***Elevation*** The elevation (above sea level) of San Francisco is less than the elevation of Dallas, and the elevation of Dallas is less than the elevation of Denver. How does the elevation of San Francisco compare with the elevation of Denver?
The elevation of San Francisco is less than the elevation of Denver.

89. ***Operating Costs*** A utility company has a fleet of vans. The annual operating cost per van is $C = 0.35m + 2900$, where m is the number of miles traveled by a van in a year. What is the maximum number of miles that will yield an annual operating cost that is no more than \$12,000?
26,000 miles

90. ***Operating Costs*** A fuel company has a fleet of trucks. The annual operating cost per truck is $C = 0.58m + 7800$, where m is the number of miles traveled by a truck in a year. What is the maximum number of miles that will yield an annual operating cost that is less than \$25,000? 29,655 miles

91. ***Hourly Wage*** Your company requires you to select one of two payment plans. One plan pays a straight \$12.50 per hour. The second plan pays \$8.00 per hour plus \$0.75 per unit produced per hour. Write an inequality for the number of units that must be produced per hour so that the second option yields the greater hourly wage. Solve the inequality.
$12.50 < 8 + 0.75n$; $n > 6$

92. ***Monthly Wage*** Your company requires you to select one of two payment plans. One plan pays a straight \$3000 per month. The second plan pays \$1000 per month plus a commission of 4% of your gross sales. Write an inequality for the gross sales per month for which the second option yields the greater monthly wage. Solve the inequality. $3000 < 1000 + 0.04s$; $s > \$50,000$

Explaining Concepts

93. ***Writing*** Describe any differences between properties of equalities and properties of inequalities.
See Additional Answers.

94. ***Precision*** If $-3 \le x \le 10$, then $-x$ must be in what interval? Explain. See Additional Answers.

95. ***Logic*** Discuss whether the solution set of a linear inequality is a *bounded* interval or an *unbounded* interval. See Additional Answers.

96. ***Logic*** Two linear inequalities are joined by the word *or* to form a compound inequality. Discuss whether the solution set is a bounded interval.
See Additional Answers.

Writing a Compound Inequality **In Exercises 97–100, let a and b be real numbers such that $a < b$. Use a and b to write a compound algebraic inequality in x with the given type of solution. Explain your reasoning.**

97. A bounded interval
$a < x < b$; A double inequality is always bounded.

98. Two unbounded intervals See Additional Answers.

99. The set of all real numbers See Additional Answers.

100. No solution
$x < a$ and $x > b$; Because $a < b$, there are no values of x that would be less than a *and* greater than b.

Cumulative Review

In Exercises 101–104, place the correct symbol ($<$, $>$, or $=$) between the real numbers.

101. $|4| \; < \; |-5|$

102. $|-4| \; < \; |-6|$

103. $|-7| \; = \; |7|$

104. $-|5| \; = \; -(5)$

In Exercises 105–108, determine whether each value of the variable is a solution of the equation.

105. $3x = 27$; $x = 6, x = 9$ No; Yes

106. $x - 14 = 8$; $x = 6, x = 22$ No; Yes

107. $7x - 5 = 7 + x$; $x = 2, x = 6$ Yes; No

108. $2 + 5x = 8x - 13$; $x = 3, x = 5$ No; Yes

In Exercises 109–112, solve the equation.

109. $2x - 17 = 0$ $\frac{17}{2}$

110. $x - 17 = 4$ 21

111. $32x = -8$ $-\frac{1}{4}$

112. $14x + 5 = 2 - x$ $-\frac{1}{5}$

2.5 Absolute Value Equations and Inequalities

- Solve absolute value equations.
- Solve absolute value inequalities.
- Solve real-life problems involving absolute value.

Solving Equations Involving Absolute Value

Solving an Absolute Value Equation

Let x be a variable or an algebraic expression and let a be a real number such that $a \geq 0$. The solutions of the equation $|x| = a$ are given by $x = -a$ and $x = a$. That is,

$$|x| = a \implies x = -a \quad \text{or} \quad x = a.$$

Study Tip

The strategy for solving an absolute value equation is to *rewrite* the equation in *equivalent forms* that can be solved by previously learned methods. This is a common strategy in mathematics. That is, when you encounter a new type of problem, you try to rewrite the problem so that it can be solved by techniques you already know.

EXAMPLE 1 Solving Absolute Value Equations

Solve each absolute value equation.

a. $|x| = 10$ **b.** $|x| = 0$ **c.** $|y| = -1$

SOLUTION

a. This equation is equivalent to the two linear equations

$x = -10$ and $x = 10$. Equivalent linear equations

So, the absolute value equation has two solutions: $x = -10$ and $x = 10$.

b. This equation is equivalent to the two linear equations

$x = -(0) = 0$ and $x = 0$. Equivalent linear equations

Because both equations are the same, you can conclude that the absolute value equation has only one solution: $x = 0$.

c. This absolute value equation has *no solution* because it is not possible for the absolute value of a real number to be negative.

Exercises Within Reach ®

Solutions in English & Spanish and tutorial videos at AlgebraWithinReach.com

Checking a Solution **In Exercises 1–4, determine whether the value of the variable is a solution of the equation.**

	Equation	*Value*		*Equation*	*Value*
1.	$\|4x + 5\| = 10$	$x = -3$ Not a solution	2.	$\|2x - 16\| = 10$	$x = 3$ Solution
3.	$\|6 - 2w\| = 2$	$w = 4$ Solution	4.	$\left\|\frac{1}{2}t + 4\right\| = 8$	$t = 6$ Not a solution

Solving an Equation **In Exercises 5–10, solve the equation. (Some equations have no solution.)**

5. $|x| = 4$ 4, −4
6. $|x| = 3$ 3, −3
7. $|t| = -45$ No solution
8. $|s| = 16$ 16, −16
9. $|h| = 0$ 0
10. $|x| = -82$ No solution

EXAMPLE 2 **Solving an Absolute Value Equation**

Solve $|3x + 4| = 10$.

SOLUTION

$|3x + 4| = 10$			Write original equation.
$3x + 4 = -10$	or	$3x + 4 = 10$	Equivalent equations
$3x + 4 - 4 = -10 - 4$		$3x + 4 - 4 = 10 - 4$	Subtract 4 from each side.
$3x = -14$		$3x = 6$	Combine like terms.
$x = -\frac{14}{3}$		$x = 2$	Divide each side by 3.

Check

$$|3x + 4| = 10$$
$$\left|3\left(-\tfrac{14}{3}\right) + 4\right| \stackrel{?}{=} 10$$
$$|-14 + 4| \stackrel{?}{=} 10$$
$$|-10| = 10 \checkmark$$

$$|3x + 4| = 10$$
$$|3(2) + 4| \stackrel{?}{=} 10$$
$$|6 + 4| \stackrel{?}{=} 10$$
$$|10| = 10 \checkmark$$

EXAMPLE 3 **Solving an Absolute Value Equation**

Additional Example
Solve $-2|4x + 5| = -6$.
Answer:
$x = -2, x = -\frac{1}{2}$

Solve $|2x - 1| + 3 = 8$.

SOLUTION

$|2x - 1| + 3 = 8$			Write original equation.
$|2x - 1| = 5$			Write in standard form.
$2x - 1 = -5$	or	$2x - 1 = 5$	Equivalent equations
$2x = -4$		$2x = 6$	Add 1 to each side.
$x = -2$		$x = 3$	Divide each side by 2.

The solutions are $x = -2$ and $x = 3$. Check these in the original equation.

Exercises Within Reach®

Solutions in English & Spanish and tutorial videos at AlgebraWithinReach.com

Solving an Equation **In Exercises 11–24, solve the equation. (Some equations have no solution.)**

11. $|5x| = 15$ 3, −3

12. $\left|\frac{1}{3}x\right| = 2$ 6, −6

13. $|x + 1| = 5$ 4, −6

14. $|x + 5| = 7$ 2, −12

15. $|4 - 3x| = 0$ $\frac{4}{3}$

16. $|3x - 2| = -5$ No solution

17. $\left|\frac{2x + 3}{5}\right| = 5$ 11, −14

18. $\left|\frac{7a + 6}{4}\right| = 2$ $\frac{2}{7}, -2$

19. $|5 - 2x| + 10 = 6$ No solution

20. $|5x - 3| + 8 = 22$ $\frac{17}{5}, -\frac{11}{5}$

21. $\left|\frac{x - 2}{3}\right| + 6 = 6$ 2

22. $\left|\frac{x - 2}{5}\right| + 4 = 4$ 2

23. $3|2x - 5| + 4 = 7$ 2, 3

24. $2|4 - 3x| - 6 = -2$ $2, \frac{2}{3}$

EXAMPLE 4 **Solving an Equation Involving Two Absolute Values**

Solve $|3x - 4| = |7x - 16|$.

SOLUTION

$|3x - 4| = |7x - 16|$		Write original equation.
$3x - 4 = 7x - 16$	or $3x - 4 = -(7x - 16)$	Equivalent equations
$-4x - 4 = -16$	$3x - 4 = -7x + 16$	
$-4x = -12$	$10x = 20$	
$x = 3$	$x = 2$	Solutions

The solutions are $x = 3$ and $x = 2$. Check these in the original equation.

EXAMPLE 5 **Solving an Equation Involving Two Absolute Values**

Solve $|x + 5| = |x + 11|$.

SOLUTION

By equating the expression $(x + 5)$ to the opposite of $(x + 11)$, you obtain

$x + 5 = -(x + 11)$	Equivalent equation
$x + 5 = -x - 11$	Distributive Property
$2x + 5 = -11$	Add x to each side.
$2x = -16$	Subtract 5 from each side.
$x = -8.$	Divide each side by 2.

However, by setting the two expressions equal to each other, you obtain

$x + 5 = x + 11$	Equivalent equation
$x = x + 6$	Subtract 5 from each side.
$0 = 6$	Subtract x from each side.

which is a false statement. So, the original equation has only one solution: $x = -8$. Check this solution in the original equation.

Study Tip

When solving an equation of the form

$$|ax + b| = |cx + d|$$

it is possible that one of the resulting equations will not have a solution. Note this occurrence in Example 5.

Exercises Within Reach®

Solutions in English & Spanish and tutorial videos at AlgebraWithinReach.com

Solving an Equation **In Exercises 25–32, solve the equation.**

25. $|2x + 1| = |x - 4|$ $-5, 1$

26. $|2x - 5| = |x + 10|$ $-\frac{5}{3}, 15$

27. $|x + 8| = |2x + 1|$ $7, -3$

28. $|10 - 3x| = |x + 7|$ $\frac{3}{4}, \frac{17}{2}$

29. $|3x + 1| = |3x - 3|$ $\frac{1}{3}$

30. $|2x + 7| = |2x + 9|$ -4

31. $|4x - 10| = 2|2x + 3|$ $\frac{1}{2}$

32. $3|2 - 3x| = |9x + 21|$ $-\frac{5}{6}$

Solving Inequalities Involving Absolute Value

Solving an Absolute Value Inequality

Let x be a variable or an algebraic expression and let a be a real number such that $a > 0$.

1. The solutions of $|x| < a$ are all values of x that lie between $-a$ and a. That is,

$$|x| < a \quad \text{if and only if} \quad -a < x < a.$$

2. The solutions of $|x| > a$ are all values of x that are less than $-a$ or greater than a. That is,

$$|x| > a \quad \text{if and only if} \quad x < -a \text{ or } x > a.$$

These rules are also valid if $<$ is replaced by $\leq$ and $>$ is replaced by $\geq$.

EXAMPLE 6 **Solving an Absolute Value Inequality**

Solve $|x - 5| < 2$.

SOLUTION

$\|x - 5\| < 2$	Write original inequality.
$-2 < x - 5 < 2$	Equivalent double inequality
$-2 + 5 < x - 5 + 5 < 2 + 5$	Add 5 to all three parts.
$3 < x < 7$	Combine like terms.

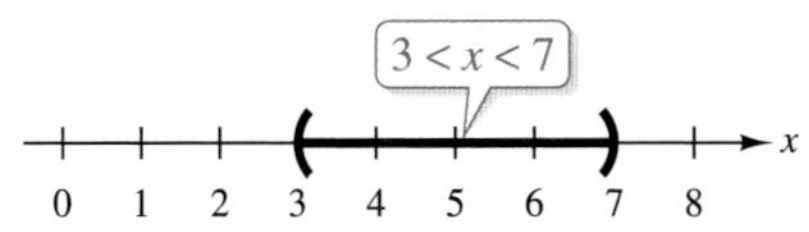

The solution set consists of all real numbers that are greater than 3 and less than 7. The solution set in interval notation is (3, 7).

Exercises Within Reach®

Solutions in English & Spanish and tutorial videos at AlgebraWithinReach.com

Checking a Solution **In Exercises 33–36, determine whether the x-value is a solution of the inequality.**

	Inequality	*Value*		*Inequality*	*Value*
33.	$\|x\| < 3$	$x = 2$ Solution	**34.**	$\|x\| \leq 5$	$x = -7$ Not a solution
35.	$\|x - 7\| \geq 3$	$x = 9$ Not a solution	**36.**	$\|x - 3\| > 5$	$x = 16$ Solution

Solving an Inequality **In Exercises 37–46, solve the inequality.**

37. $|y| < 4$ $\quad -4 < y < 4$

38. $|x| < 6$ $\quad -6 < x < 6$

39. $|x| \geq 6$ $\quad x \leq -6$ or $x \geq 6$

40. $|y| \geq 4$ $\quad y \leq -4$ or $y \geq 4$

41. $|x + 6| > 10$ $\quad x < -16$ or $x > 4$

42. $|y - 2| \leq 4$ $\quad -2 \leq y \leq 6$

43. $|2x| < 14$ $\quad -7 < x < 7$

44. $|4z| \leq 9$ $\quad -\frac{9}{4} \leq z \leq \frac{9}{4}$

45. $\left|\frac{y}{3}\right| \leq \frac{1}{3}$ $\quad -1 \leq y \leq 1$

46. $\left|\frac{t}{5}\right| < \frac{3}{5}$ $\quad -3 < t < 3$

EXAMPLE 7 Solving an Absolute Value Inequality

$\lvert 3x - 4 \rvert \geq 5$		Original inequality
$3x - 4 \leq -5$ or	$3x - 4 \geq 5$	Equivalent inequalities
$3x - 4 + 4 \leq -5 + 4$	$3x - 4 + 4 \geq 5 + 4$	Add 4 to all parts.
$3x \leq -1$	$3x \geq 9$	Combine like terms.
$\frac{3x}{3} \leq \frac{-1}{3}$	$\frac{3x}{3} \geq \frac{9}{3}$	Divide each side by 3.
$x \leq -\frac{1}{3}$	$x \geq 3$	Simplify.

The solution set consists of all real numbers that are less than or equal to $-\frac{1}{3}$ or greater than or equal to 3.

Additional Examples

Solve each inequality.

a. $\lvert 8x + 1 \rvert \geq 5$

b. $\lvert 7 - x \rvert < 2$

Answers:

a. $x \leq -\frac{3}{4}$ or $x \geq \frac{1}{2}$

b. $5 < x < 9$

EXAMPLE 8 Solving an Absolute Value Inequality

$\left\lvert 2 - \frac{x}{3} \right\rvert \leq 0.01$	Original inequality
$-0.01 \leq 2 - \frac{x}{3} \leq 0.01$	Equivalent double inequality
$-0.01 - 2 \leq 2 - 2 - \frac{x}{3} \leq 0.01 - 2$	Subtract 2 from all three parts.
$-2.01 \leq -\frac{x}{3} \leq -1.99$	Combine like terms.
$-2.01(-3) \geq -\frac{x}{3}(-3) \geq -1.99(-3)$	Multiply all three parts by -3 and reverse both inequality symbols.
$6.03 \geq x \geq 5.97$	Simplify.
$5.97 \leq x \leq 6.03$	Solution set in standard form

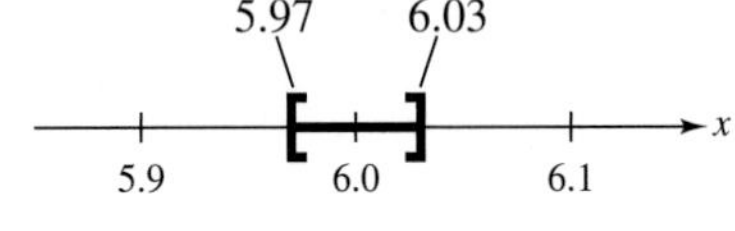

The solution set consists of all real numbers that are greater than or equal to 5.97 and less than or equal to 6.03. The solution set in interval notation is [5.97, 6.03].

Exercises Within Reach®

Solutions in English & Spanish and tutorial videos at AlgebraWithinReach.com

Solving an Inequality **In Exercises 47–56, solve the inequality. (Some inequalities have no solution.)**

47. $\lvert 2x + 3 \rvert > 9$

$x < -6$ or $x > 3$

48. $\lvert 7r - 3 \rvert > 11$

$r < -\frac{8}{7}$ or $r > 2$

49. $\lvert 2x - 1 \rvert \leq 7$

$-3 \leq x \leq 4$

50. $\lvert 6t + 15 \rvert \geq 30$

$t \leq -\frac{15}{2}$ or $t \geq \frac{5}{2}$

51. $\lvert 3x + 10 \rvert < -1$

No solution

52. $\lvert 4x - 5 \rvert > -3$

$-\infty < x < \infty$

53. $\dfrac{\lvert a + 6 \rvert}{2} \geq 16$

$a \leq -38$ or $a \geq 26$

54. $\dfrac{\lvert y - 16 \rvert}{4} < 30$

$-104 < y < 136$

55. $\lvert 0.2x - 3 \rvert < 4$

$-5 < x < 35$

56. $\lvert 1.5t - 8 \rvert \leq 16$

$-5.\overline{3} \leq t \leq 16$

Applications

Application

EXAMPLE 9 Oil Production

The estimated daily production at an oil refinery is given by the absolute value inequality $|x - 200{,}000| \le 25{,}000$, where x is measured in barrels of oil. Solve the inequality to determine the maximum and minimum production levels.

SOLUTION

$|x - 200{,}000| \le 25{,}000$ Write original inequality.

$-25{,}000 \le x - 200{,}000 \le 25{,}000$ Equivalent double inequality

$175{,}000 \le x \le 225{,}000$ Add 200,000 to all three parts.

So, the oil refinery produces a maximum of 225,000 barrels of oil a minimum of 175,000 barrels of oil per day.

Application

EXAMPLE 10 Creating a Model

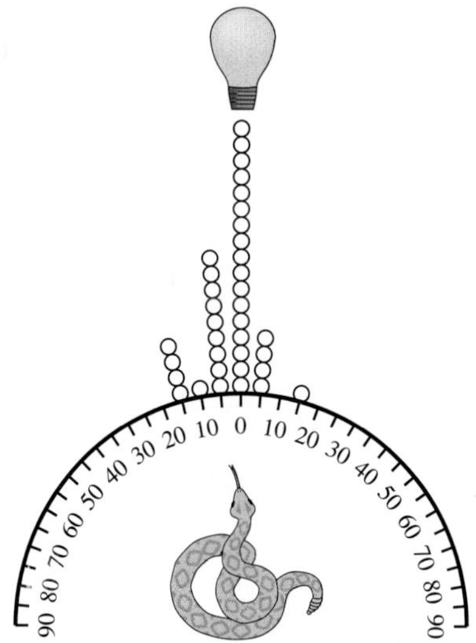

To test the accuracy of a rattlesnake's "pit-organ sensory system," a biologist blindfolds a rattlesnake and presents the snake with a warm "target." Of 36 strikes, the snake is on target 17 times. Let A represent the number of degrees by which the snake is off target. Then $A = 0$ represents a strike that is aimed directly at the target. Positive values of A represent strikes to the right of the target, and negative values of A represent strikes to the left of the target. Use the diagram shown to write an absolute value inequality that describes the interval in which the 36 strikes occurred.

SOLUTION

From the diagram, you can see that in the 36 strikes, the snake is never off by more than 15 degrees in either direction. As a compound inequality, this can be represented by $-15 \le A \le 15$. As an absolute value inequality, this interval can be represented by $|A| \le 15$.

Exercises Within Reach®

Solutions in English & Spanish and tutorial videos at AlgebraWithinReach.com

57. ***Speed Skating*** Each skater in a 500-meter short track speed skating final had a time that satisfied the inequality $|t - 42.238| \le 0.412$, where t is the time in seconds. Sketch the graph of the solution of the inequality. What were the fastest and slowest possible times? See Additional Answers.

58. ***Time Study*** A time study was conducted to determine the length of time required to perform a task in a manufacturing process. The times required by approximately two-thirds of the workers in the study satisfied the inequality

$$\left|\frac{t - 15.6}{1.9}\right| \le 1$$

where t is time in minutes. Sketch the graph of the solution of the inequality. What were the maximum and minimum possible times? See Additional Answers.

Concept Summary: Solving Absolute Value Equations

What

The **absolute value equation** $|x| = a$, $a \geq 0$, has exactly two solutions: $x = -a$ and $x = a$.

EXAMPLE

Solve $|x + 1| = 12$.

How

Use these steps to solve an absolute value equation.

1. Write the equation in **standard form**, if necessary.
2. Rewrite the equation in equivalent forms and solve using previously learned methods.

EXAMPLE

$|x + 1| = 12$

$x + 1 = 12$ or $x + 1 = -12$

$x = 11$ or $x = -13$

Why

Understanding how to solve absolute value equations will help you when you learn how to solve absolute value inequalities.

Exercises Within Reach®

Worked-out solutions to odd-numbered exercises at AlgebraWithinReach.com

Concept Summary Check

59. ***Solving an Equation*** In your own words, explain how to solve an absolute value equation. Illustrate your explanation with an example. See Additional Answers.

60. ***Number of Solutions*** In the equation $|x| = b$, b is a positive real number. How many solutions does this equation have? Explain. There are two solutions to the equation $|x| = b$ if b is a positive real number: $x = b$ and $x = -b$.

61. ***Writing an Equation*** Write an absolute value equation that represents the verbal statement.

"The distance between x and zero is a."

$|x| = a$

62. ***True or False?*** The solutions of $|x| = -2$ are $x = -2$ and $x = 2$. Justify your answer. False. It is not possible for the absolute value of a real number to be negative.

Extra Practice

Solving an Equation **In Exercises 63 and 64, solve the equation.**

63. $|4x + 1| = \frac{1}{2}$ $-\frac{1}{8}, -\frac{3}{8}$

64. $\frac{1}{4}|3x + 1| = 4$ $5, -\frac{17}{3}$

Think About It **In Exercises 65 and 66, write an absolute value equation that represents the verbal statement.**

65. The distance between x and 4 is 9. $|x - 4| = 9$

66. The distance between -3 and t is 5. $|-3 - t| = 5$

Writing an Inequality **In Exercises 67–72, write an absolute value inequality that represents the interval.**

67. 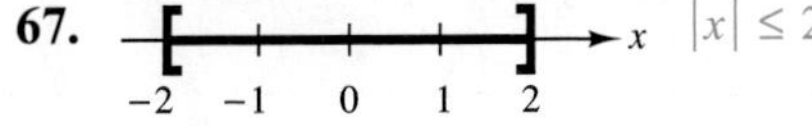 $|x| \leq 2$

68. 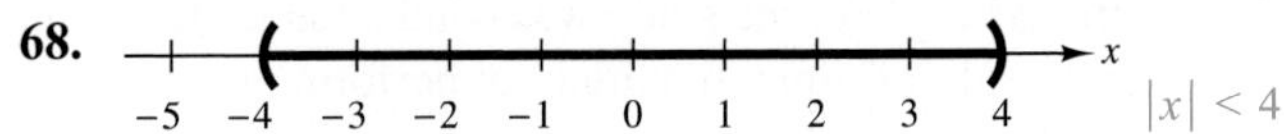 $|x| < 4$

69. 15 16 17 18 19 20 21 22 23 x $|x - 19| > 2$

70. 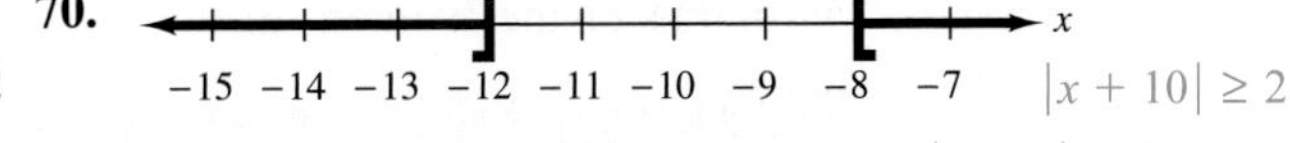 $|x + 10| \geq 2$

71. 0 1 2 3 4 5 6 7 8 9 x $|x - 4| \geq 2$

72. 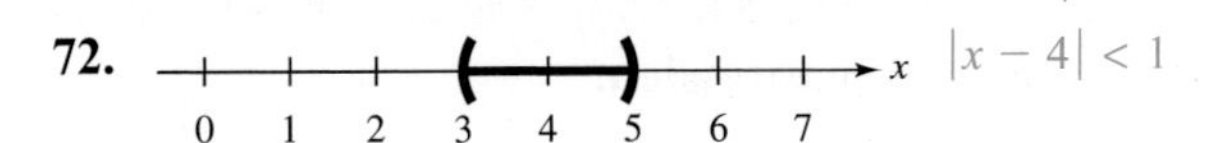 $|x - 4| < 1$

Solving an Inequality **In Exercises 73 and 74, solve the inequality.**

73. $\left|\frac{3x - 2}{4}\right| + 5 \geq 5$ $-\infty < x < \infty$

74. $\left|\frac{2x - 4}{5}\right| - 9 \leq 3$ $-28 \leq x \leq 32$

Writing an Inequality **In Exercises 75–78, write an absolute value inequality that represents the verbal statement.**

75. The set of all real numbers x whose distance from 0 is less than 3. $|x| < 3$

76. The set of all real numbers x whose distance from 0 is more than 2. $|x| > 2$

77. The set of all real numbers x for which the distance from 0 to 3 less than twice x is more than 5. $|2x - 3| > 5$

78. The set of all real numbers x for which the distance from 0 to 5 more than half of x is less than 13. $\left|\frac{1}{2}x + 5\right| < 13$

79. ***Accuracy of Measurements*** In woodshop class, you must cut several pieces of wood to within $\frac{3}{16}$ inch of the teacher's specifications. Let $(s - x)$ represent the difference between the specification s and the measured length x of a cut piece.

(a) Write an absolute value inequality that describes the values of x that are within specifications.
$|s - x| \le \frac{3}{16}$

(b) The length of one piece of wood is specified to be $5\frac{1}{8}$ inches. Describe the acceptable lengths for this piece. $4\frac{15}{16} \le x \le 5\frac{5}{16}$

80. ***Body Temperature*** Physicians generally consider an adult's body temperature x to be normal if it is within 1°F of the temperature 98.6°F.

(a) Write an absolute value inequality that describes the values of x that are considered normal.
$|x - 98.6| \le 1$

(b) Describe the range of body temperatures that are considered normal.
$97.6 \le x \le 99.6$

81. ***Height*** The heights h of two-thirds of the members of a population satisfy the inequality

$$\left|\frac{h - 68.5}{2.7}\right| \le 1$$

where h is measured in inches. Determine the interval on the real number line in which these heights lie.
65.8 inches $\le h \le$ 71.2 inches

82. ***Geometry*** The side of a square is measured as 10.4 inches with a possible error of $\frac{1}{16}$ inch. Using these measurements, determine the interval containing the possible areas of the square.
106.864 in.^{2} $\le$ area $\le$ 109.464 in.^{2}

Explaining Concepts

83. ***Reasoning*** The graph of the inequality $|x - 3| < 2$ can be described as *all real numbers that are within two units of 3*. Give a similar description of $|x - 4| < 1$.
All real numbers that are within one unit of 4

84. ***Precision*** Write an absolute value inequality to represent all the real numbers that are more than $|a|$ units from b. Then write an example showing the solution of the inequality for sample values of a and b.
See Additional Answers.

85. ***Reasoning*** Complete $|2x - 6| \le$ 6 so that the solution is $0 \le x \le 6$.

86. ***Writing*** Describe and correct the error. Explain how you can recognize that the solution is wrong without solving the inequality. See Additional Answers.

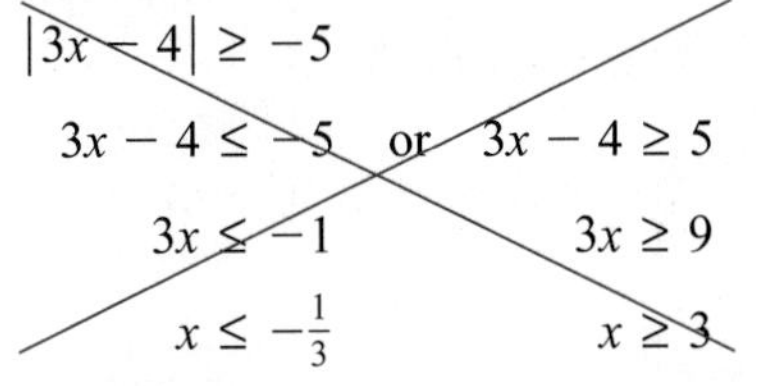

Cumulative Review

In Exercises 87 and 88, translate the verbal phrase into an algebraic expression.

87. Four times the sum of a number n and 3 $4(n + 3)$

88. Eight less than two times a number n $2n - 8$

In Exercises 89–92, solve the inequality.

89. $x - 7 > 13$ $x > 20$

90. $x + 7 \le 13$ $x \le 6$

91. $4x + 11 \ge 27$
$x \ge 4$

92. $-4 < x + 2 < 12$
$-6 < x < 10$

Stanislav Komogorov/Shutterstock.com

2 Chapter Summary

	What did you learn?	Explanation and Examples	Review Exercises
2.1	Check solutions of equations *(p. 50).*	Substitute the solution into the original equation and simplify each side. If each side turns out to be the same number, you can conclude that you found the solution of the original equation.	1–4
	Solve linear equations in standard form *(p. 51).*	To solve a linear equation in the standard form $ax + b = 0$, where a and b are real numbers with $a \neq 0$, isolate x by rewriting the standard equation in the form $x =$ a number .	5–8
	Solve linear equations in nonstandard form *(p. 53).*	To solve a linear equation in nonstandard form, rewrite the equation in the form $x =$ a number .	9–30
2.2	Write algebraic equations representing real-life situations *(p. 58).*	**1.** Ask yourself what you need to know to solve the problem. Then *write a verbal model* that includes arithmetic operations to describe the problem. **2.** *Assign labels* to each part of the verbal model. **3.** Use the labels to *write an algebraic model.* **4.** *Solve* the resulting algebraic equation. **5.** *Answer* the original question and check that your answer satisfies the original problem as stated.	31, 32
	Solve percent problems *(p. 59).*	The primary use of percents is to compare two numbers: 3 is 50% of 6. (3 is compared to 6.) Many percent problems can be solved using the *percent equation:* Compared number = Percent (decimal form) • Base number	33–46
	Solve ratio, rate, and proportion problems *(p. 61).*	The proportion $\frac{a}{b} = \frac{c}{d}$ is equivalent to $ad = bc$.	47–60
2.3	Use mathematical models to solve business-related problems *(p. 66).*	Selling price = Cost + Markup Markup = Markup rate • Cost Sale price = Original price − Discount Discount = Discount rate • Original price	61–66
	Use mathematical models to solve mixture problems and rate problems *(p. 68).*	First rate • Amount + Second rate • Amount = Final rate • Final amount Distance = Rate • Time	67–74

	What did you learn?	Explanation and Examples	Review Exercises
2.3	Use formulas to solve application problems *(p. 71)*.	Temperature: $F = \frac{9}{5}C + 32$ Simple Interest: $I = Prt$ Distance: $d = rt$ *Square* $A = s^2$ $P = 4s$ 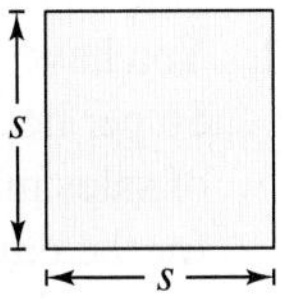 *Rectangle* $A = lw$ $P = 2l + 2w$ 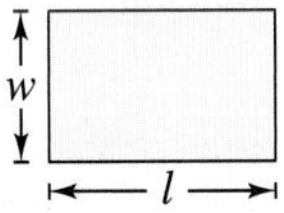 *Circle* $A = \pi r^2$ $C = 2\pi r$ 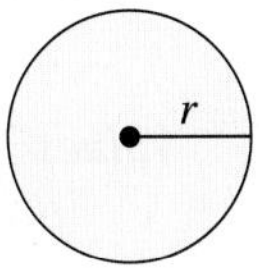 *Triangle* $A = \frac{1}{2}bh$ $P = a + b + c$ 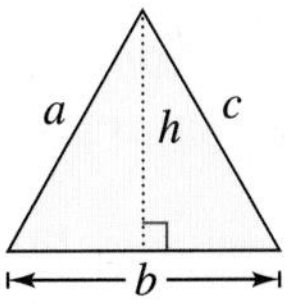 *Cube* $V = s^3$ 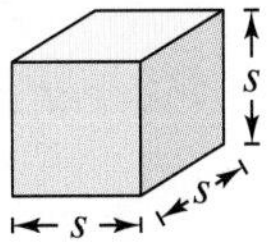 *Rectangular Solid* $V = lwh$ 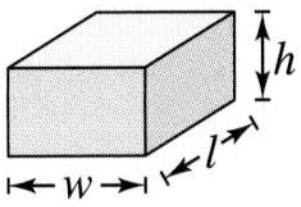 *Circular Cylinder* $V = \pi r^2 h$ 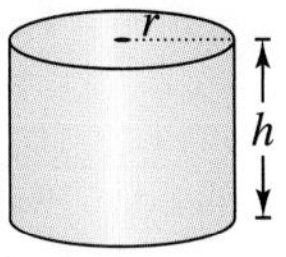 *Sphere* $V = \frac{4}{3}\pi r^3$ 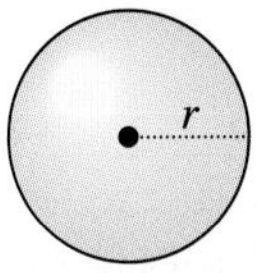	75–82
2.4	Sketch the graphs of inequalities *(p. 78)*.	See examples of bounded and unbounded intervals on pages 78 and 79.	83–86
	Solve linear inequalities *(p. 80)*.	**1.** Addition and subtraction: If $a < b$, then $a + c < b + c$. If $a < b$, then $a - c < b - c$. **2.** Multiplication and division (c is positive): If $a < b$, then $ac < bc$. If $a < b$, then $\frac{a}{c} < \frac{b}{c}$. **3.** Multiplication and division (c is negative): If $a < b$, then $ac > bc$. If $a < b$, then $\frac{a}{c} > \frac{b}{c}$. **4.** Transitive property: If $a < b$ and $b < c$, then $a < c$.	87–104
	Solve application problems involving inequalities *(p. 83)*.	Use a verbal model to write an inequality and then solve the inequality.	105, 106
2.5	Solve absolute value equations *(p. 86)*.	$\|x\| = a \Rightarrow x = -a$ or $x = a$.	107–114
	Solve absolute value inequalities *(p. 89)*.	**1.** $\|x\| < a$ if and only if $-a < x < a$. **2.** $\|x\| > a$ if and only if $x < -a$ or $x > a$.	115–124
	Solve real-life problems involving absolute value *(p. 91)*.	Use the information given in the problem to write an absolute value equation or an absolute value inequality. Then solve the equation or inequality.	125, 126

Review Exercises

Worked-out solutions to odd-numbered exercises at AlgebraWithinReach.com

2.1

Checking Solutions of an Equation **In Exercises 1–4, determine whether each value of the variable is a solution of the equation.**

	Equation	*Values*	
1.	$45 - 7x = 3$	(a) $x = 3$ Not a solution	(b) $x = 6$ Solution
2.	$3(3 + 4x) = 15x$	(a) $x = 3$ Solution	(b) $x = -2$ Not a solution
3.	$\frac{x}{7} + \frac{x}{5} = 12$	(a) $x = 28$ Not a solution	(b) $x = 35$ Solution
4.	$\frac{t+2}{6} = \frac{7}{2}$	(a) $t = -12$ Not a solution	(b) $t = 19$ Solution

Solving an Equation **In Exercises 5–30, solve the equation and check your solution. (Some equations have no solution.)**

5. $3x + 21 = 0$ -7
6. $-4x + 64 = 0$ 16
7. $5x - 120 = 0$ 24
8. $7x - 49 = 0$ 7
9. $x + 4 = 9$ 5
10. $x - 7 = 3$ 10
11. $-3x = 36$ -12
12. $11x = 44$ 4
13. $-\frac{1}{8}x = 3$ -24
14. $\frac{1}{10}x = 5$ 50
15. $5x + 4 = 19$ 3
16. $3 - 2x = 9$ -3
17. $17 - 7x = 3$ 2
18. $3 + 6x = 51$ 8
19. $7x - 5 = 3x + 11$ 4
20. $9 - 2x = 4x - 7$ $\frac{8}{3}$
21. $3(2y - 1) = 9 + 3y$ 4
22. $-2(x + 4) = 2x - 7$ $-\frac{1}{4}$
23. $4y - 4(y - 2) = 8$ Infinitely many
24. $7x + 2(7 - x) = 5x + 14$ Infinitely many
25. $4(3x - 5) = 6(2x + 3)$ No solution
26. $8(x - 2) = 4(2x + 3)$ No solution
27. $\frac{4}{5}x - \frac{1}{10} = \frac{3}{2}$ 2
28. $\frac{1}{4}s + \frac{3}{8} = \frac{5}{2}$ $\frac{17}{2}$
29. $1.4t + 2.1 = 0.9t$ -4.2
30. $2.5x - 6.2 = 3.7x - 5.8$ $-0.\overline{3}$

2.2

Mathematical Modeling **In Exercises 31 and 32, construct a verbal model and write an algebraic equation that represents the problem. Solve the equation.**

31. An internship pays \$320 per week plus an additional \$75 for a training session. The total pay for the internship and training is \$2635. How many weeks long is the internship? See Additional Answers.

32. You have a job as a sales person for which you are paid \$6 per hour plus \$1.25 per sale made. Find the number of sales made in an eight-hour day when your earnings for the day are \$88. See Additional Answers.

Finding Equivalent Forms of a Percent **In Exercises 33–36, complete the table showing the equivalent forms of the percent.**

	Percent	Parts out of 100	Decimal	Fraction
33.	68%	68	0.68	$\frac{68}{100}$
34.	35%	35	0.35	$\frac{7}{20}$
35.	60%	60	0.6	$\frac{3}{5}$
36.	$16\frac{2}{3}\%$	$16\frac{2}{3}$	$0.1\overline{6}$	$\frac{1}{6}$

Solving a Percent Problem **In Exercises 37–42, solve using a percent equation.**

37. What is 130% of 50? 65
38. What is 0.8% of 2450? 19.6
39. 645 is $21\frac{1}{2}\%$ of what number? 3000
40. 498 is 83% of what number? 600
41. 250 is what percent of 200? 125%
42. 162.5 is what percent of 6500? 2.5%
43. ***Real Estate Commission*** A real estate agency receives a commission of \$9000 for the sale of a \$150,000 house. What percent commission is this? 6%
44. ***Pension Fund*** Your employer withholds \$216 of your gross income each month for your retirement. Determine the percent of your total monthly gross income of \$3200 that is withheld for retirement. 6.75%
45. ***Quality Control*** A quality control engineer reported that 1.6% of a sample of parts were defective. The engineer found six defective parts. How large was the sample? 375 parts
46. ***Sales Tax*** The state sales tax on an item you purchase is $6\frac{1}{4}\%$. How much sales tax will you pay on an item that costs \$44? \$2.75

Consumer Awareness **In Exercises 47 and 48, use unit prices to determine the better buy.**

47. (a) A 39-ounce can of coffee for $9.79

(b) An 8-ounce can of coffee for $1.79

8-ounce can

48. (a) A 3.5-pound bag of sugar for $3.08

(b) A 2.2-pound bag of sugar for $1.87

2.2-pound bag

49. ***Income Tax*** You have $9.90 of state tax withheld from your paycheck per week when your gross pay is $396. Find the ratio of tax to gross pay. $\frac{1}{40}$

50. ***Price-Earnings Ratio*** The ratio of the price of a stock to its earnings is called the price-earnings ratio. Find the price-earnings ratio of a stock that sells for $46.75 per share and earns $5.50 per share. $\frac{17}{2}$

Solving a Proportion **In Exercises 51–54, solve the proportion.**

51. $\frac{7}{8} = \frac{y}{4}$ $\frac{7}{2}$

52. $\frac{x}{16} = \frac{5}{12}$ $\frac{20}{3}$

53. $\frac{b}{6} = \frac{5 + b}{15}$ $\frac{10}{3}$

54. $\frac{x + 1}{3} = \frac{x - 1}{2}$ 5

Geometry **In Exercises 55 and 56, the triangles are similar. Solve for the length x by using the fact that corresponding sides of similar triangles are proportional.**

55.

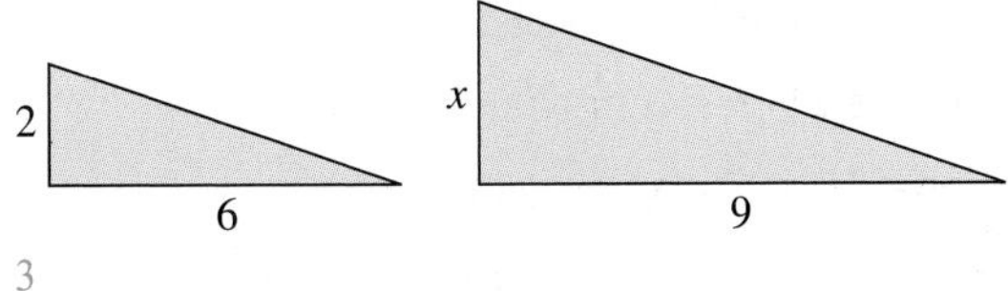

3

56.

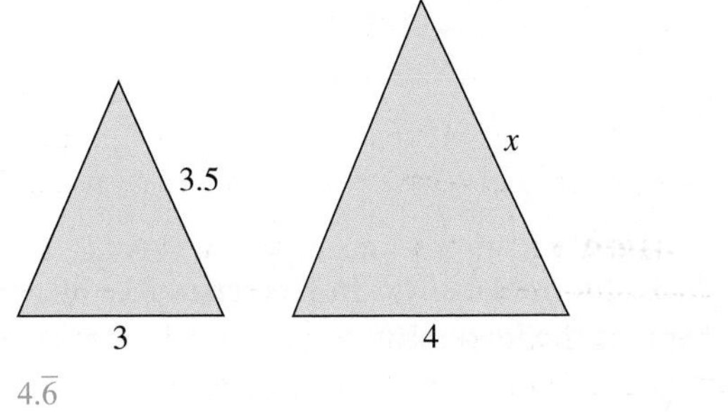

$4.\overline{6}$

57. ***Property Tax*** The tax on a property with an assessed value of $105,000 is $1680. Find the tax on a property with an assessed value of $125,000. $2000

58. ***Masonry*** You use $1\frac{1}{2}$ bags of mortar mix to lay 42 bricks. How many bags will you use to lay 336 bricks? 12 bags

59. ***Geometry*** You want to measure the height of a flagpole. To do this, you measure the flagpole's shadow and find that it is 30 feet long. You also measure the height of a five-foot lamp post and find its shadow to be 3 feet long (see figure). Find the height h of the flagpole. 50 feet

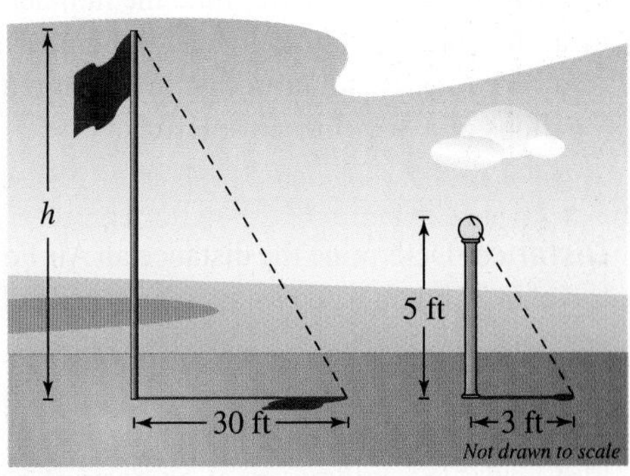

60. ***Geometry*** You want to measure the height of a silo. To do this, you measure the silo's shadow and find that it is 20 feet long. You are 6 feet tall and your shadow is $1\frac{1}{2}$ feet long. Find the height of the silo. 80 feet

2.3

Solving a Markup Problem **In Exercises 61 and 62, find the missing quantities.**

	Cost	Selling Price	Markup	Markup Rate
61.	$99.95	$149.93	$49.98	50%
62.	$23.50	$31.33	$7.83	33.3%

Solving a Discount Problem **In Exercises 63 and 64, find the missing quantities.**

	Original Price	Sale Price	Discount	Discount Rate
63.	$71.95	$53.96	$17.99	25%
64.	$559.95	$279.98	$279.97	50%

65. ***Amount Financed*** You buy a motorcycle for $2795 plus 6% sales tax. Find the amount of sales tax and the total bill. You make a down payment of $800. Find the amount financed.

Sales tax: $167.70; Total bill: $2962.70;
Amount financed: $2162.70

66. ***Labor*** An automobile repair bill of $366.44 lists $208.80 for parts, a 5% tax on parts, and the rest for labor. The labor rate is $32 per hour. How many hours did it take to repair the automobile?

4.6 hours

67. ***Mixture Problem*** Determine the number of liters of a 30% saline solution and the number of liters of a 60% saline solution that are required to obtain 10 liters of a 50% saline solution.
$3\frac{1}{3}$ liters of 30% solution; $6\frac{2}{3}$ liters of 60% solution

68. ***Mixture Problem*** Determine the number of gallons of a 20% bleach solution and the number of gallons of a 50% bleach solution that are required to obtain 8 gallons of a 40% bleach solution.
$2\frac{2}{3}$ gallons of 20% solution; $5\frac{1}{3}$ gallons of 50% solution

69. ***Distance*** Determine the distance an Air Force jet can travel in $2\frac{1}{3}$ hours when its average speed is 1500 miles per hour. 3500 miles

70. ***Travel Time*** Determine the time for a migrating bird to fly 185 kilometers at an average speed of 66 kilometers per hour. $\frac{185}{66} \approx 2.8$ hours

71. ***Speed*** A truck driver traveled at an average speed of 48 miles per hour on a 100-mile trip to pick up a load of freight. On the return trip with the truck fully loaded, the average speed was 40 miles per hour. Find the average speed for the round trip.
43.6 miles per hour

72. ***Speed*** For 2 hours of a 400-mile trip, your average speed is 40 miles per hour. Determine the average speed that must be maintained for the remainder of the trip if you want the average speed for the entire trip to be 50 miles per hour. $53\frac{1}{3}$ miles per hour

73. ***Work-Rate Problem*** Find the time for two people working together to complete a task when it takes them 4.5 hours and 6 hours working individually.
$\frac{18}{7} \approx 2.57$ hours

74. ***Work-Rate Problem*** Find the time for two people working together to complete a task when it takes them 8 hours and 10 hours to complete the entire task working individually. $\frac{20}{9} = 2.\overline{2}$ hours

75. ***Simple Interest*** Find the interest on a $1000 corporate bond that matures in 4 years and has an 8.5% interest rate. $340

76. ***Simple Interest*** Find the annual interest rate on a certificate of deposit that pays $37.50 per year in interest on a principal of $500. 7.5%

77. ***Simple Interest*** The interest on a savings account is 9.5%. Find the principal required to earn $20,000 in interest in 4 years. $52,631.58

78. ***Simple Interest*** A corporation borrows $3.25 million for 2 years at an annual interest rate of 12% to modernize one of its manufacturing facilities. What is the total principal and interest that must be repaid?
$4.03 million = $4,030,000

79. ***Simple Interest*** An inheritance of $50,000 is divided into two investments earning 8.5% and 10% simple interest. (The 10% investment has a greater risk.) Your objective is to obtain a total annual interest income of $4700 from the investments. What is the smallest amount you can invest at 10% in order to meet your objective? $30,000

80. ***Simple Interest*** You invest $1000 in a certificate of deposit that has an annual interest rate of 7%. After 6 months, the interest is computed and added to the principal. During the second 6 months, the interest is computed using the original investment plus the interest earned during the first 6 months. What is the total interest earned during the first year of the investment? $71.23

81. ***Geometry*** The perimeter of the rectangle shown in the figure is 64 feet. Find its dimensions.
20 feet × 12 feet

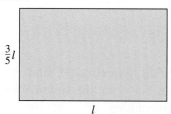

82. ***Geometry*** The area of the triangle shown in the figure is 60 square meters. Solve for x. 4 meters

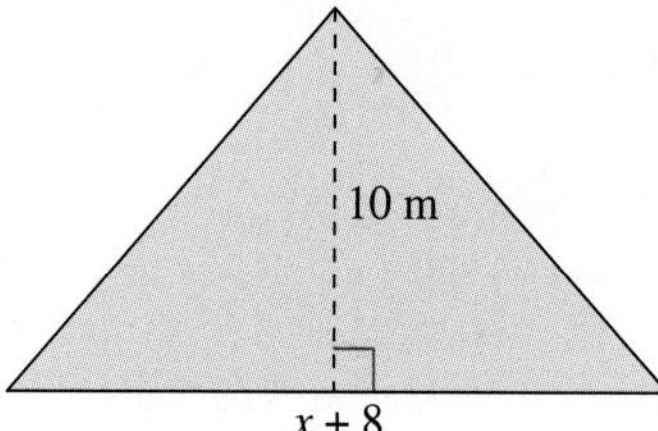

2.4

Graphing an Inequality **In Exercises 83–86, sketch the graph of the inequality.**

83. $-3 \le x < 1$

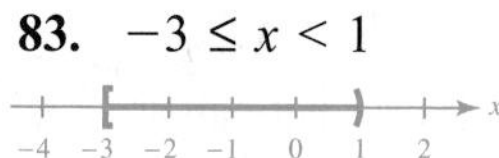

84. $-2.5 \le x < 4$

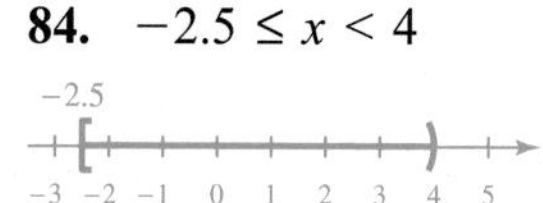

85. $-7 < x$

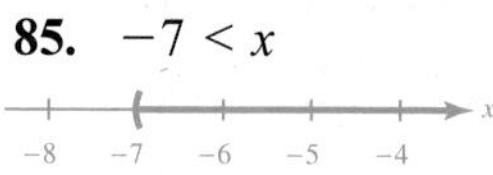

86. $x \ge -2$

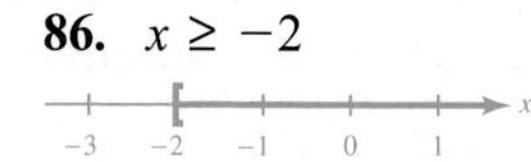

Solving an Inequality **In Exercises 87–104, solve the inequality and sketch the solution on the real number line.** See Additional Answers.

87. $x - 5 \leq -1$ $x \leq 4$

88. $x + 8 > 5$ $x > -3$

89. $-6x < -24$ $x > 4$

90. $-16x \geq -48$ $x \leq 3$

91. $5x + 3 > 18$ $x > 3$

92. $3x - 11 \leq 7$ $x \leq 6$

93. $8x + 1 \geq 10x - 11$ $x \leq 6$

94. $12 - 3x < 4x - 2$ $x > 2$

95. $\frac{1}{3} - \frac{1}{2}y < 12$ $y > -\frac{70}{3}$

96. $\frac{x}{4} - 2 < \frac{3x}{8} + 5$ $x > -56$

97. $-4(3 - 2x) \leq 3(2x - 6)$ $x \leq -3$

98. $3(2 - y) \geq 2(1 + y)$ $y \leq \frac{4}{5}$

99. $-6 \leq 2x + 8 < 4$ $-7 \leq x < -2$

100. $-13 \leq 3 - 4x < 13$ $-\frac{5}{2} < x \leq 4$

101. $5 > \frac{x + 1}{-3} > 0$ $-16 < x < -1$

102. $12 \geq \frac{x - 3}{2} > 1$ $5 < x \leq 27$

103. $5x - 4 < 6$ and $3x + 1 > -8$ $-3 < x < 2$

104. $6 - 2x \leq 1$ or $10 - 4x > -6$ $-\infty < x < \infty$

105. ***International Calling Card Rate*** The cost of an international telephone call is \$0.10 per minute. Your prepaid calling card has \$12.50 left to pay for a call. How many minutes can you talk? 125 minutes

106. ***Earnings*** A country club waiter earns \$6 per hour plus tips of at least 15% of the restaurant tab from each table served. What total amount of restaurant tabs assures the waiter of making at least \$150 in a five-hour shift? At least \$800

2.5

Solving an Equation **In Exercises 107–114, solve the equation. (Some equations have no solution.)**

107. $|x| = 6$ $6, -6$

108. $|x| = -4$ No solution

109. $|4 - 3x| = 8$ $4, -\frac{4}{3}$

110. $|2x + 3| = 7$ $2, -5$

111. $|5x + 4| - 10 = -6$ $0, -\frac{8}{5}$

112. $|x - 2| - 2 = 4$ $8, -4$

113. $|3x - 4| = |x + 2|$ $\frac{1}{2}, 3$

114. $|5x + 6| = |2x - 1|$ $-\frac{7}{3}, -\frac{5}{7}$

Solving an Inequality **In Exercises 115–122, solve the inequality. (Some inequalities have no solution.)**

115. $|x - 4| > 3$ $x < 1$ or $x > 7$

116. $|t + 3| > 2$ $t < -5$ or $t > -1$

117. $|3x| < 12$ $-4 < x < 4$

118. $\left|\frac{t}{3}\right| < 1$ $-3 < t < 3$

119. $|2x - 7| < 15$ $-4 < x < 11$

120. $|4x - 1| > 7$ $x < -\frac{3}{2}$ or $x > 2$

121. $|b + 2| - 6 > 1$ $b < -9$ or $b > 5$

122. $|2y - 1| + 4 < -1$ No solution

Writing an Inequality **In Exercises 123 and 124, write an absolute value inequality that represents the interval.**

123.
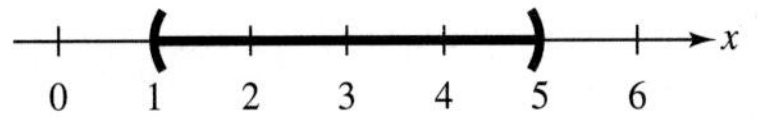

$|x - 3| < 2$

124.
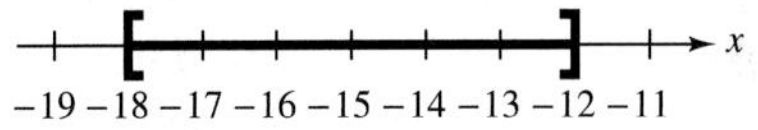

$|x + 15| \leq 3$

125. ***Temperature*** The storage temperature of a computer must satisfy the inequality

$|t - 78.3| \leq 38.3$

where t is the temperature in degrees Fahrenheit. Sketch the graph of the solution of the inequality. What are the maximum and minimum temperatures?
See Additional Answers.
Maximum: 116.6 degrees Fahrenheit
Minimum: 40 degrees Fahrenheit

126. ***Temperature*** The operating temperature of a computer must satisfy the inequality

$|t - 77| \leq 27$

where t is the temperature in degrees Fahrenheit. Sketch the graph of the solution of the inequality. What are the maximum and minimum temperatures?
See Additional Answers.
Maximum: 104 degrees Fahrenheit
Minimum: 50 degrees Fahrenheit

Chapter Test

Solutions in English & Spanish and tutorial videos at AlgebraWithinReach.com

Take this test as you would take a test in class. After you are done, check your work against the answers in the back of the book.

In Exercises 1–4, solve the equation.

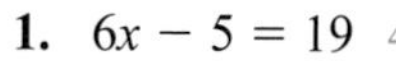

1. $6x - 5 = 19$ 4

2. $5x - 6 = 7x - 12$ 3

3. $15 - 7(1 - x) = 3(x + 8)$ 4

4. $\frac{2x}{3} = \frac{x}{2} + 4$ 24

5. What is 125% of 3200? 4000

6. 32 is what percent of 8000? 0.4%

7. A store is offering a 20% discount on all items in its inventory. Find the original price of a tractor that has a sale price of \$8900. \$11,125

8. Which of the products at the left is a better buy? Explain your reasoning. 15-ounce package; It has a lower unit price.

9. The tax on a property with an assessed value of \$110,000 is \$1650. What is the tax on a property with an assessed value of \$145,000? \$2175

10. The bill for the repair of a home appliance was \$165. The cost for parts was \$85. The labor rate was \$16 per half hour. How many hours did the repair work take? $2\frac{1}{2}$ hours

11. A pet store owner mixes two types of dog food costing \$2.60 per pound and \$3.80 per pound to make 40 pounds of a mixture costing \$3.35 per pound. How many pounds of each kind of dog food are in the mixture? 15 pounds at \$2.60 per pound; 25 pounds at \$3.80 per pound

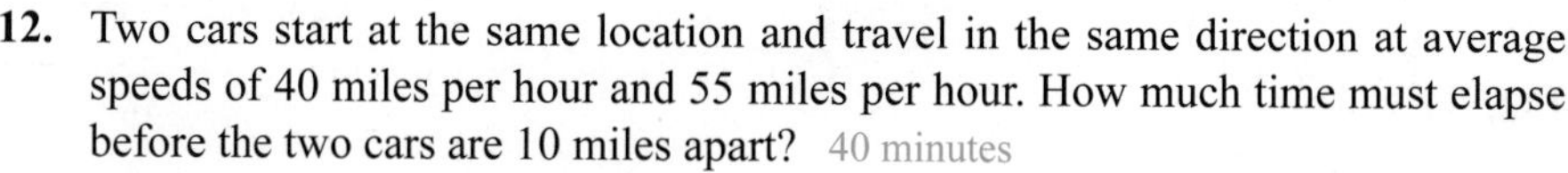

12. Two cars start at the same location and travel in the same direction at average speeds of 40 miles per hour and 55 miles per hour. How much time must elapse before the two cars are 10 miles apart? 40 minutes

13. The interest on a savings account is 7.5%. Find the principal required to earn \$300 in interest in 2 years. \$2000

14. Solve each equation.

(a) $|3x - 6| = 9$ 5, −1

(b) $|3x - 5| = |6x - 1|$ $\frac{2}{3}, -\frac{4}{3}$

(c) $|9 - 4x| + 4 = 1$ No solution

15. Solve each inequality and sketch the solution on the real number line.

(a) $3x + 12 \geq -6$

(b) $9 - 5x < 5 - 3x$

(c) $0 \leq \frac{1 - x}{4} < 2$

(d) $-7 < 4(2 - 3x) \leq 20$

15. (a) $x \geq -6$

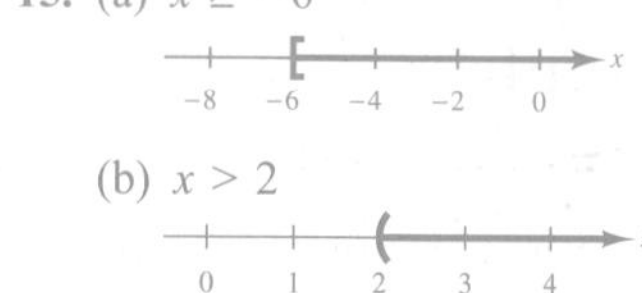

(b) $x > 2$

(c) $-7 < x \leq 1$

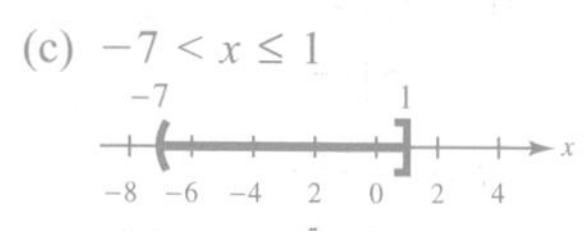

(d) $-1 \leq x < \frac{5}{4}$

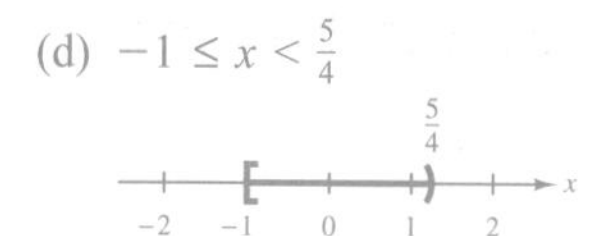

16. Rewrite the statement "t is at least 8" using inequality notation. $t \geq 8$

17. Solve each inequality.

(a) $|x - 3| \leq 2$ $1 \leq x \leq 5$

(b) $|5x - 3| > 12$ $x < -\frac{9}{5}$ or $x > 3$

(c) $\left|\frac{x}{4} + 2\right| < 0.2$ $-8.8 < x < -7.2$

18. A utility company has a fleet of vans. The annual operating cost per van is

$$C = 0.37m + 2700$$

where m is the number of miles traveled by a van in a year. What is the maximum number of miles that will yield an annual operating cost that is less than or equal to \$11,950? 25,000 miles

Graphs and Functions

MASTERY IS WITHIN REACH!

"I used to be really afraid of math, so reading the textbook was difficult. I have learned that it just takes different strategies to read the textbook. It's my resource book when I do homework. I take it to class because it helps me follow along. I'm not afraid of math anymore because I know how to study it—finally."

James
Criminal Justice

See page 135 for suggestions about reading your textbook like a manual.

3.1 The Rectangular Coordinate System

- Plot points on a rectangular coordinate system.
- Determine whether ordered pairs are solutions of equations.
- Use the Distance Formula and the Midpoint Formula.

The Rectangular Coordinate System

Point out the signs of x and y in each of the four quadrants. Discuss questions such as: If $(-a, b)$ is in Quadrant IV, in which quadrant would you find $(a, -b)$? Students tend to think that $(-a, b)$ must be in Quadrant II. This discussion will help them realize that "$-a$" does not necessarily represent a negative value.

A **rectangular coordinate system** is formed by two real number lines intersecting at right angles. The horizontal number line is the ***x*-axis** and the vertical number line is the ***y*-axis**. (The plural of axis is *axes*.)

The point of intersection of the two axes is called the **origin**, and the axes separate the plane into four regions called **quadrants**. Each point in the plane corresponds to an **ordered pair** (x, y) of real numbers x and y called the **coordinates** of the point.

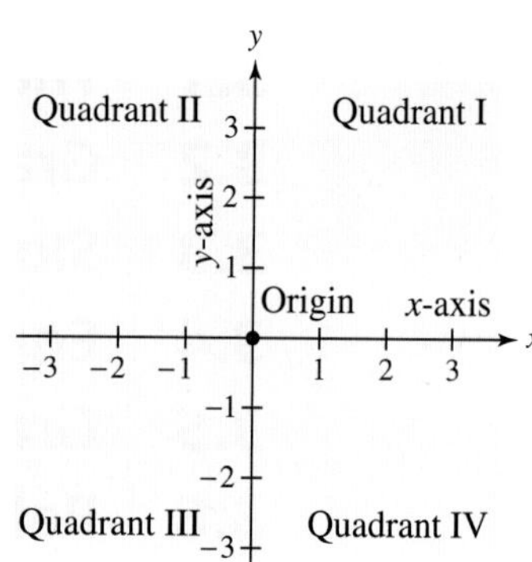

EXAMPLE 1 Plotting Points on a Rectangular Coordinate System

Plot the points $(-2, 1)$, $(4, 0)$, $(3, -1)$, $(4, 3)$, $(0, 0)$, and $(-1, -3)$ on a rectangular coordinate system.

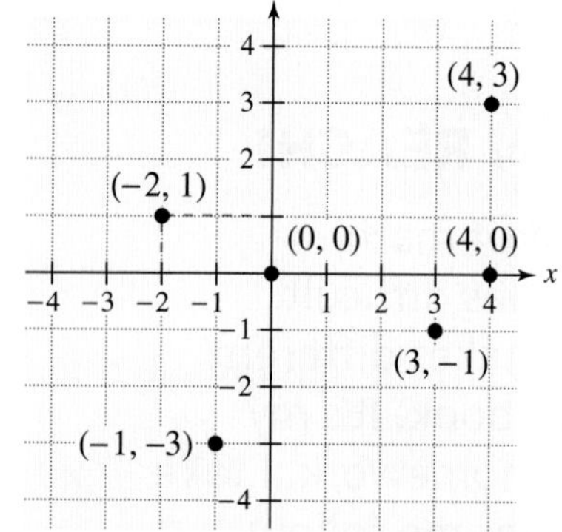

SOLUTION

The point $(-2, 1)$ is two units to the *left* of the vertical axis and one unit *above* the horizontal axis.

Two units to the left of the vertical axis

One unit above the horizontal axis

$(-2, 1)$

Similarly, the point $(4, 0)$ is four units to the *right* of the vertical axis and *on* the horizontal axis. (It is on the horizontal axis because its y-coordinate is 0.) The other four points can be plotted in a similar way.

Exercises Within Reach®

Solutions in English & Spanish and tutorial videos at AlgebraWithinReach.com

Plotting Points **In Exercises 1–6, plot the points on a rectangular coordinate system.** See Additional Answers.

1. $(4, 3), (-5, 3), (3, -5)$
2. $(-2, 5), (-2, -5), (3, 5)$
3. $(-8, -2), (6, -2), (5, 0)$
4. $(0, 4), (0, 0), (-7, 0)$
5. $\left(\frac{5}{2}, -2\right), \left(-2, \frac{1}{4}\right), \left(\frac{3}{2}, -\frac{7}{2}\right)$
6. $\left(-\frac{2}{3}, 3\right), \left(\frac{1}{4}, -\frac{5}{4}\right), \left(-5, -\frac{7}{4}\right)$

Finding Coordinates **In Exercises 7–10, determine the coordinates of the points.** See Additional Answers.

7.

8.

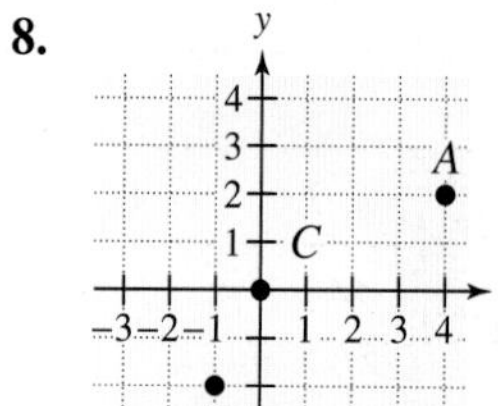

9.

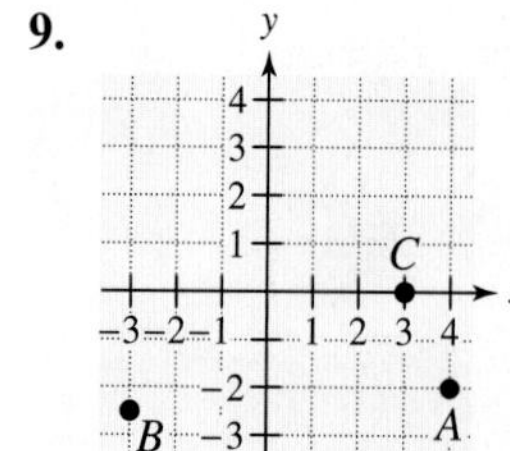

10.

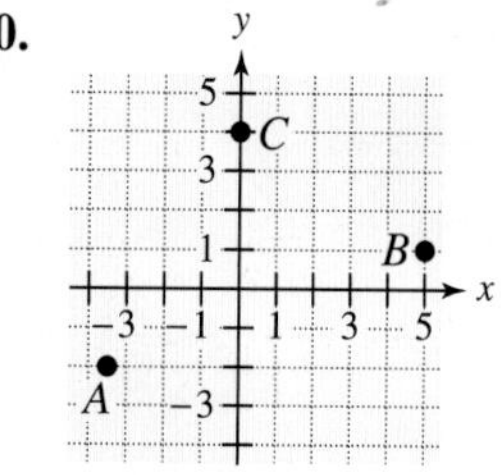

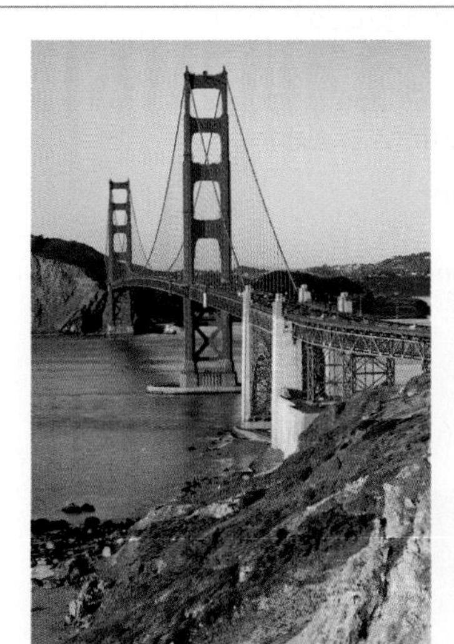

EXAMPLE 2 Representing Data Graphically

The populations (in millions) of California for the years 1994 through 2009 are shown in the table. Sketch a scatter plot of the data. (*Source:* U.S. Census Bureau)

Year	1994	1995	1996	1997	1998	1999	2000	2001
Population	31.5	31.7	32.0	32.5	33.0	33.5	34.0	34.5

Year	2002	2003	2004	2005	2006	2007	2008	2009
Population	34.9	35.3	35.6	35.8	36.0	36.2	36.6	37.0

SOLUTION

To sketch a scatter plot, begin by choosing which variable will be plotted on the horizontal axis and which will be plotted on the vertical axis. For this data, it seems natural to plot the years on the horizontal axis (which means that the population must be plotted on the vertical axis). Next, use the data in the table to form ordered pairs. For instance, the first three ordered pairs are (1994, 31.5), (1995, 31.7), and (1996, 32.0). All 16 points are shown below.

Study Tip

Note that the break in the x-axis indicates that the numbers between 0 and 1994 have been omitted. The break in the y-axis indicates that the numbers between 0 and 30 have been omitted.

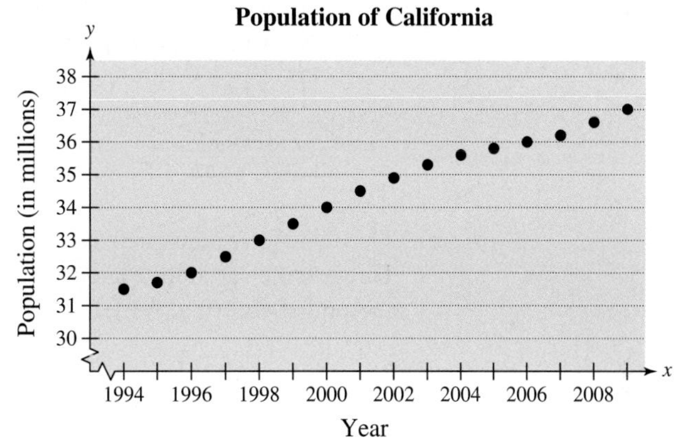

Exercises Within Reach®

Solutions in English & Spanish and tutorial videos at AlgebraWithinReach.com

11. ***Meteorology*** The table shows the normal daily mean temperatures y (in degrees Fahrenheit) for Denver, Colorado, for each month of the year, with $x = 1$ corresponding to January. Sketch a scatter plot of the data. (*Source:* National Climatic Data Center)

See Additional Answers.

x	1	2	3	4	5	6
y	29	33	40	48	57	68

x	7	8	9	10	11	12
y	73	72	62	51	38	30

12. ***Net Profits*** The net profits y (in millions of dollars) of Estee Lauder for the years 2002 through 2011 are shown in the table, where x represents the year, with $x = 2$ corresponding to 2002. Sketch a scatter plot of the data. (*Source:* Estee Lauder Companies, Inc.)

See Additional Answers.

x	2	3	4	5	6
y	287.1	333.3	375.4	433.6	417.5

x	7	8	9	10	11
y	448.7	473.8	278.7	555.1	741.7

S.Borisov/Shutterstock.com

Ordered Pairs as Solutions

EXAMPLE 3 Constructing a Table of Values

Construct a table of values for $y = 3x + 2$. Then plot the solution points on a rectangular coordinate system. Choose x-values of $-3, -2, -1, 0, 1, 2$, and 3.

SOLUTION

For each x-value, you must calculate the corresponding y-value. For example, if you choose $x = 1$, then the y-value is

$$y = 3(1) + 2 = 5.$$

The ordered pair $(x, y) = (1, 5)$ is a **solution point** (or **solution**) of the equation.

Choose x	Calculate y from $y = 3x + 2$	Solution point
$x = -3$	$y = 3(-3) + 2 = -7$	$(-3, -7)$
$x = -2$	$y = 3(-2) + 2 = -4$	$(-2, -4)$
$x = -1$	$y = 3(-1) + 2 = -1$	$(-1, -1)$
$x = 0$	$y = 3(0) + 2 = 2$	$(0, 2)$
$x = 1$	$y = 3(1) + 2 = 5$	$(1, 5)$
$x = 2$	$y = 3(2) + 2 = 8$	$(2, 8)$
$x = 3$	$y = 3(3) + 2 = 11$	$(3, 11)$

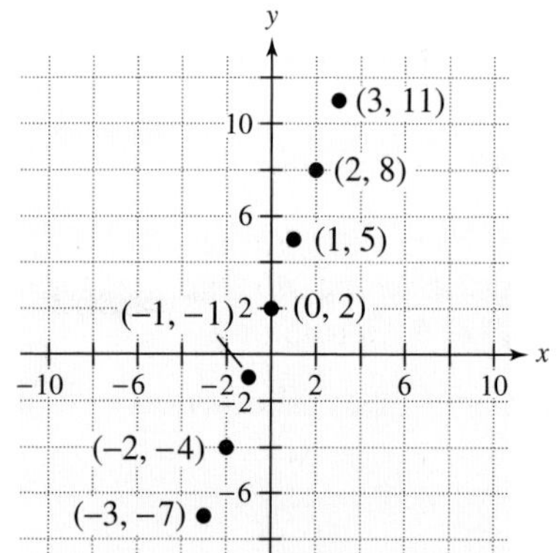

Once you have constructed a table of values, you can get a visual idea of the relationship between the variables x and y by plotting the solution points on a rectangular coordinate system, as shown at the left.

Exercises Within Reach®

Solutions in English & Spanish and tutorial videos at AlgebraWithinReach.com

Constructing a Table of Values **In Exercises 13–16, complete the table of values. Then plot the solution points on a rectangular coordinate system.** See Additional Answers.

13.

x	-2	0	2	4	6
$y = 5x + 3$	-7	3	13	23	33

14.

x	-3	0	3	6	9
$y = 6x - 7$	-25	-7	11	29	47

15.

x	-4	0	3	5	10
$y = \|2x - 7\| + 2$	17	9	3	5	15

16.

x	-5	-1	0	3	8
$y = \|-3x + 1\| - 5$	11	-1	-4	3	18

Guidelines for Verifying Solutions

To verify that an ordered pair (x, y) is a solution of an equation with variables x and y, use the steps below.

1. Substitute the values of x and y into the equation.
2. Simplify each side of the equation.
3. If each side simplifies to the same number, then the ordered pair is a solution. If the two sides yield different numbers, then the ordered pair is not a solution.

EXAMPLE 4 Verifying Solutions of an Equation

Which of the ordered pairs are solutions of $x^2 - 2y = 6$?

a. $(2, 1)$ **b.** $(0, -3)$ **c.** $\left(1, -\frac{5}{2}\right)$

SOLUTION

a. For the ordered pair $(2, 1)$, substitute $x = 2$ and $y = 1$ into the equation.

$(2)^2 - 2(1) \stackrel{?}{=} 6$ Substitute 2 for x and 1 for y.

$2 \neq 6$ Not a solution 

Because the substitution does not satisfy the original equation, you can conclude that the ordered pair $(2, 1)$ *is not* a solution of the original equation.

b. For the ordered pair $(0, -3)$, substitute $x = 0$ and $y = -3$ into the equation.

$(0)^2 - 2(-3) \stackrel{?}{=} 6$ Substitute 0 for x and -3 for y.

$6 = 6$ Solution ✓

Because the substitution satisfies the original equation, you can conclude that the ordered pair $(0, -3)$ *is* a solution of the original equation.

c. For the ordered pair $\left(1, -\frac{5}{2}\right)$, substitute $x = 1$ and $y = -\frac{5}{2}$ into the equation.

$(1)^2 - 2\left(-\frac{5}{2}\right) \stackrel{?}{=} 6$ Substitute 1 for x and $-\frac{5}{2}$ for y.

$6 = 6$ Solution ✓

Because the substitution satisfies the original equation, you can conclude that the ordered pair $\left(1, -\frac{5}{2}\right)$ *is* a solution of the original equation.

Study Tip

Approaching a problem in different ways can help you understand the problem better.

1. **Algebraic Approach** Use algebra to find several solutions.
2. **Numerical Approach** Construct a table that shows several solutions.
3. **Graphical Approach** Draw a graph that shows several solutions.

Exercises Within Reach®

Solutions in English & Spanish and tutorial videos at AlgebraWithinReach.com

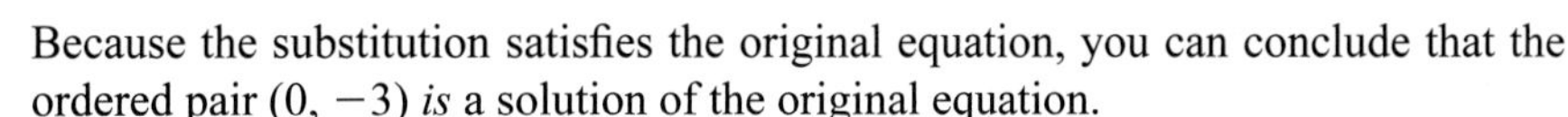

Verifying Solutions of an Equation **In Exercises 17–20, determine whether each ordered pair is a solution of the equation.**

17. $4y - 2x + 1 = 0$
(a) $(0, 0)$ Not a solution
(b) $\left(\frac{1}{2}, 0\right)$ Solution
(c) $\left(-3, -\frac{7}{4}\right)$ Solution
(d) $\left(1, -\frac{3}{4}\right)$ Not a solution

18. $5x - 2y + 50 = 0$
(a) $(-10, 0)$ Solution
(b) $\left(\frac{4}{5}, -27\right)$ Not a solution
(c) $\left(-9, \frac{5}{2}\right)$ Solution
(d) $(20, -2)$ Not a solution

19. $x^2 + 3y = -5$
(a) $(3, -2)$ Not a solution
(b) $(-2, -3)$ Solution
(c) $(3, -5)$ Not a solution
(d) $(4, -7)$ Solution

20. $y^2 - 4x = 8$
(a) $(0, 6)$ Not a solution
(b) $(-4, 2)$ Not a solution
(c) $(-1, 3)$ Not a solution
(d) $(7, 6)$ Solution

The Distance Formula and the Midpoint Formula

Additional Example
Find the distance between the points (6, −3) and (6, 5).
Answer: 8

The Distance Formula

The distance d between two points (x_1, y_1) and (x_2, y_2) is

$$d = \sqrt{(x_2 - x_1)^2 + (y_2 - y_1)^2}.$$

EXAMPLE 5 Finding the Distance Between Two Points

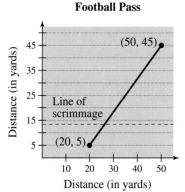

Find the distance between the points $(-1, 2)$ and $(2, 4)$, as shown at the left.

SOLUTION

Let $(x_1, y_1) = (-1, 2)$ and let $(x_2, y_2) = (2, 4)$. Then apply the Distance Formula.

$d = \sqrt{(x_2 - x_1)^2 + (y_2 - y_1)^2}$ Distance Formula

$= \sqrt{[2 - (-1)]^2 + (4 - 2)^2}$ Substitute coordinates of points.

$= \sqrt{3^2 + 2^2}$ Simplify.

$= \sqrt{13} \approx 3.61$ Simplify and use a calculator.

Application

EXAMPLE 6 Finding the Length of a Football Pass

Football Pass

Line of scrimmage
(50, 45)
(20, 5)
Distance (in yards)

A quarterback throws a pass from the five-yard line, 20 yards from the sideline. A wide receiver catches the pass on the 45-yard line, 50 yards from the same sideline, as shown at the left. How long is the pass?

SOLUTION

The length of the pass is the distance between the points (20, 5) and (50, 45).

$d = \sqrt{(50 - 20)^2 + (45 - 5)^2}$ Substitute coordinates of points into the Distance Formula.

$= \sqrt{900 + 1600}$ Simplify.

$= \sqrt{2500}$ Add.

$= 50$ Simplify.

So, the pass is 50 yards long.

Exercises Within Reach®

Solutions in English & Spanish and tutorial videos at AlgebraWithinReach.com

Using the Distance Formula **In Exercises 21–24, find the distance between the points.**

21. (1, 3), (5, 6) 5

22. (3, 10), (15, 5) 13

23. (3, 7), (4, 5) $\sqrt{5}$

24. (5, 2), (8, 3) $\sqrt{10}$

25. ***Football Pass*** A quarterback throws a pass from the 10-yard line, 10 yards from the sideline. A wide receiver catches the pass on the 40-yard line, 35 yards from the same sideline. How long is the pass?
$\sqrt{1525} \approx 39.05$ yards

26. ***Soccer Pass*** A soccer player passes the ball from a point that is 18 yards from the endline and 12 yards from the sideline. A teammate receives the pass 42 yards from the same endline and 50 yards from the same sideline. How long is the pass? $\sqrt{2020} \approx 44.94$ yards

The **midpoint** of a line segment that joins two points is the point that divides the segment into two equal parts. To find the midpoint of the line segment that joins two points in a coordinate plane, you can simply find the average values of the respective coordinates of the two endpoints.

Additional Example

Find the midpoint of the line segment joining the points $(-1, -4)$ and $(3, -2)$.

Answer: $(1, -3)$

The Midpoint Formula

The midpoint of the line segment joining the points (x_1, y_1) and (x_2, y_2) is given by

$$\text{Midpoint} = \left(\frac{x_1 + x_2}{2}, \frac{y_1 + y_2}{2}\right).$$

EXAMPLE 7 Finding the Midpoint of a Line Segment

Find the midpoint of the line segment joining the points $(-5, -3)$ and $(9, 3)$, as shown at the left.

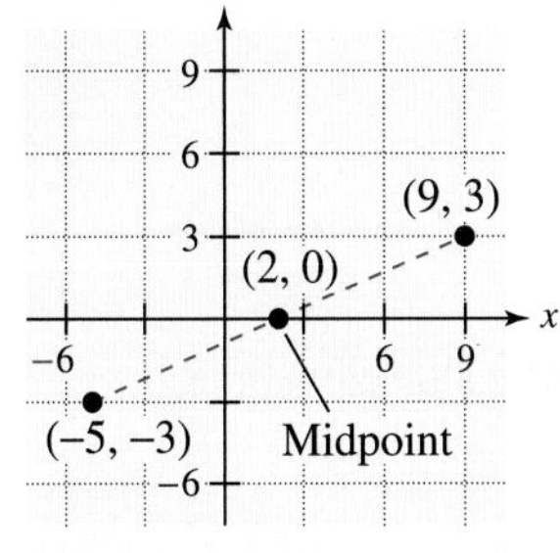

SOLUTION

Let $(x_1, y_1) = (-5, -3)$ and let $(x_2, y_2) = (9, 3)$.

$$\text{Midpoint} = \left(\frac{x_1 + x_2}{2}, \frac{y_1 + y_2}{2}\right) \qquad \text{Midpoint Formula}$$

$$= \left(\frac{-5 + 9}{2}, \frac{-3 + 3}{2}\right) \qquad \text{Substitute for } x_1, y_1, x_2, \text{ and } y_2.$$

$$= (2, 0) \qquad \text{Simplify.}$$

Exercises Within Reach®

Solutions in English & Spanish and tutorial videos at AlgebraWithinReach.com

Using the Midpoint Formula **In Exercises 27–30, find the midpoint of the line segment.**

27.

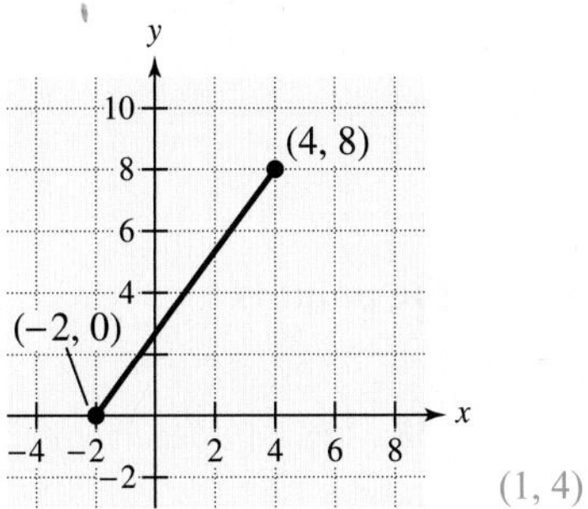

$(1, 4)$

28.

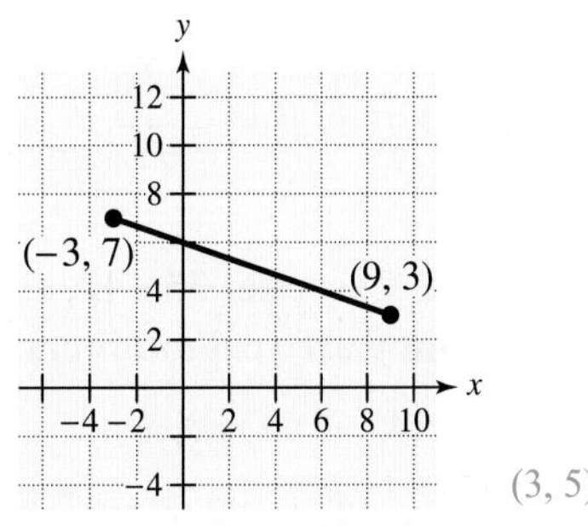

$(3, 5)$

29.

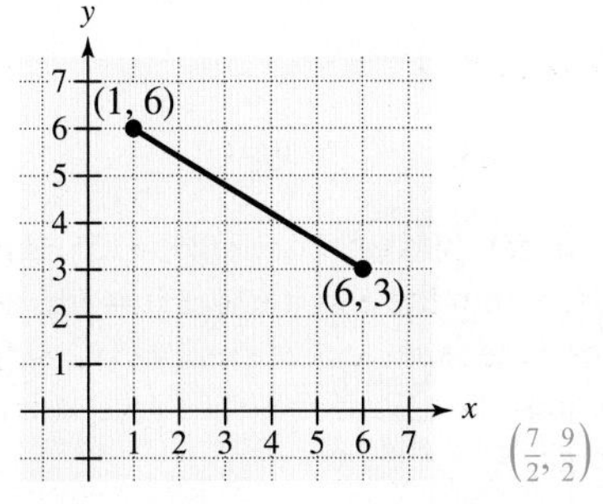

$\left(\frac{7}{2}, \frac{9}{2}\right)$

30.

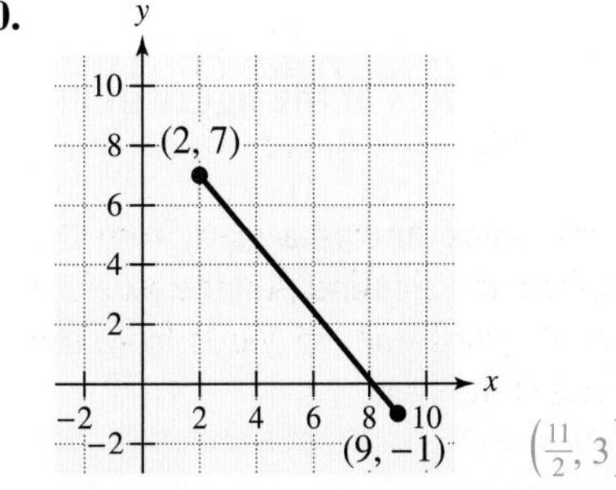

$\left(\frac{11}{2}, 3\right)$

Concept Summary: Using the Rectangular Coordinate System

What

You can represent **ordered pairs** of real numbers graphically by plotting points on a **rectangular coordinate system.**

You can use the **coordinates** of two points to find

- the distance between the points, or
- the **midpoint** of the segment that joins the points.

EXAMPLE

a. Plot the points $(-2, 1)$ and $(1, 5)$.

b. Find the distance between the points $(-2, 1)$ and $(1, 5)$.

How

EXAMPLE

a. The point $(-2, 1)$ is 2 units to left of the vertical axis and 1 unit above the horizontal axis. Plot the point $(1, 5)$ in a similar way.

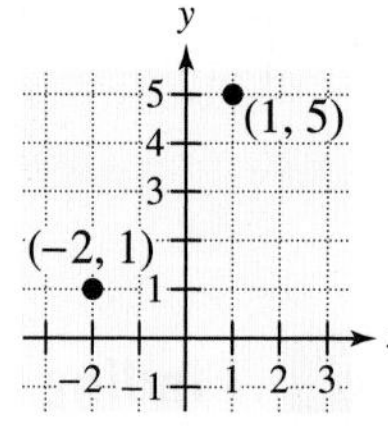

b. Use the **Distance Formula** with $(x_1, y_1) = (-2, 1)$, $(x_2, y_2) = (1, 5)$.

$$d = \sqrt{(x_2 - x_1)^2 + (y_2 - y_1)^2}$$

$$= \sqrt{[1 - (-2)]^2 + (5 - 1)^2}$$

$$= \sqrt{9 + 16} = 5$$

Why

As you study algebra, you will realize that knowing how to plot points on a rectangular coordinate system is an invaluable skill. For example, by plotting points, you can create scatter plots to represent relationships between two variables.

Additionally, you now have a third approach to problem solving.

1. Algebraic Approach
2. Numerical Approach
3. Graphical Approach

Exercises Within Reach®

Worked-out solutions to odd-numbered exercises at AlgebraWithinReach.com

Concept Summary Check

31. ***The Graph of an Ordered Pair*** Describe the graph of an ordered pair of real numbers. The graph of an ordered pair of real numbers is a point on a rectangular coordinate system.

32. ***The x-Coordinate*** What is the x-coordinate of the point $(-2, 1)$? -2 What does it represent?
The x-coordinate -2 represents the horizontal position of the point, two units to the left of the vertical axis.

33. ***The Distance Formula*** In the example above, does the Distance Formula give the same result when $(x_1, y_1) = (1, 5)$ and $(x_2, y_2) = (-2, 1)$? Yes

34. ***The Midpoint Formula*** State the Midpoint Formula.
The midpoint of the line segment joining the points (x_1, y_1) and (x_2, y_2) is given by the Midpoint Formula

$$\text{Midpoint} = \left(\frac{x_1 + x_2}{2}, \frac{y_1 + y_2}{2}\right)$$

Extra Practice

Determining the Quadrant **In Exercises 35–42, determine the quadrant in which the point is located without plotting it. (x and y are real numbers.)**

35. $(-3, -5)$ Quadrant III

36. $(4, -2)$ Quadrant IV

37. $\left(-\frac{8}{9}, \frac{3}{4}\right)$ Quadrant II

38. $\left(\frac{5}{11}, \frac{3}{8}\right)$ Quadrant I

39. (x, y), $x > 0, y < 0$
Quadrant IV

40. (x, y), $x > 0, y > 0$
Quadrant I

41. (x, y), $xy > 0$
Quadrant I or III

42. (x, y), $xy < 0$
Quadrant II or IV

Shifting a Graph **In Exercises 43 and 44, the figure is shifted to a new location in the plane. Find the coordinates of the vertices of the figure in its new location.**

43.

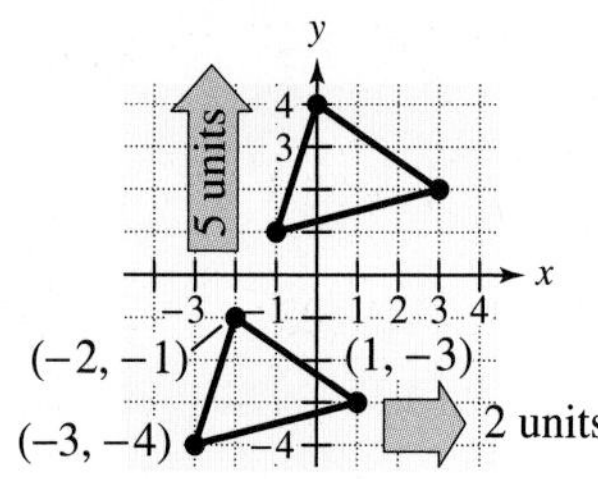

$(-3, -4) \Rightarrow (-1, 1)$
$(1, -3) \Rightarrow (3, 2)$
$(-2, -1) \Rightarrow (0, 4)$

44.

$(-3, 5) \Rightarrow (3, 2)$
$(-1, 2) \Rightarrow (5, -1)$
$(-3, -1) \Rightarrow (3, -4)$
$(-5, 2) \Rightarrow (1, -1)$

Horizontal and Vertical Distance **In Exercises 45–48, plot the points and find the distance between them. State whether the points lie on a horizontal or a vertical line.** See Additional Answers.

45. $(3, -2), (3, 5)$ 7; Vertical line

46. $(-2, 8), (-2, 1)$ 7; Vertical line

47. $(3, 2), (10, 2)$ 7; Horizontal line

48. $(-12, -2), (13, -2)$ 25; Horizontal line

Geometry **In Exercises 49 and 50, find the perimeter of the triangle.**

49.

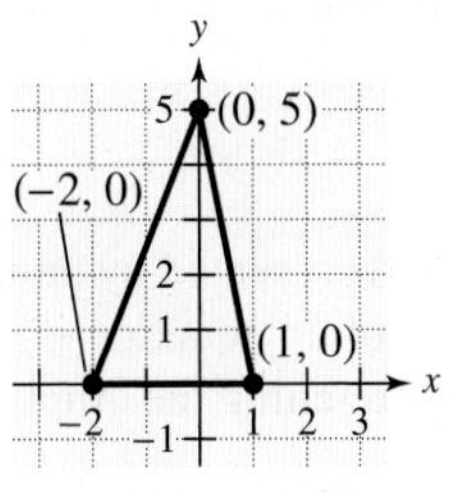

$3 + \sqrt{26} + \sqrt{29} \approx 13.48$

50.

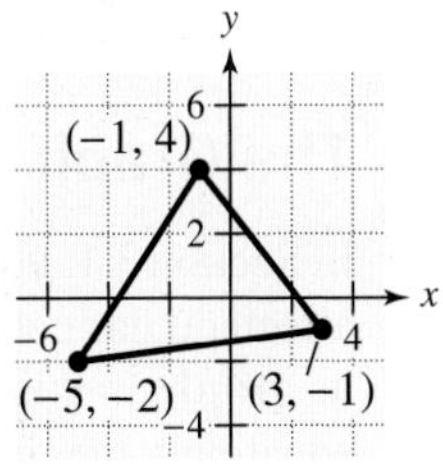

$\sqrt{52} + \sqrt{41} + \sqrt{65} \approx 21.68$

51. ***Net Sales*** The net sales of Apple grew from \$36.5 billion in 2009 to \$108.2 billion in 2011. Use the Midpoint formula to estimate the net sales in 2010. (*Source:* Apple Inc.) \$72.35 billion

52. ***Revenue*** The revenues of DIRECTV grew from \$21.6 billion in 2009 to \$27.2 billion in 2011. Use the Midpoint formula to estimate the revenues in 2010. (*Source:* DIRECTV Group Inc.) \$24.4 billion

Explaining Concepts

53. ***Number Sense*** Point C is in Quadrant I and point D is in Quadrant III. Is it possible for the distance from C to D to be -5? Explain. No. The distance between two points is always positive.

54. ***Reasoning*** Describe the coordinates of a point on the x-axis. The x-coordinate is any real number and the y-coordinate is 0.

55. ***Writing*** Explain why the ordered pair $(-3, 4)$ is not a solution point of the equation $y = 4x + 15$.

$4 \neq 4(-3) + 15$

56. ***Precision*** When points are plotted on a rectangular coordinate system, do the scales on the x- and y-axes need to be the same? Explain. No. The scales on the x- and y-axes are determined by the magnitudes of the quantities being measured by x and y.

57. ***Conjecture*** Plot the points $(2, 1)$, $(-3, 5)$, and $(7, -3)$ on a rectangular coordinate system. Then change the sign of the y-coordinate of each point and plot the three new points on the same rectangular coordinate system. What conjecture can you make about the location of a point when the sign of the y-coordinate is changed?
See Additional Answers; Reflection in the x-axis

58. ***Conjecture*** Plot the points $(2, 1)$, $(-3, 5)$, and $(7, -3)$ on a rectangular coordinate system. Then change the sign of the x-coordinate of each point and plot the three new points on the same rectangular coordinate system. What conjecture can you make about the location of a point when the sign of the x-coordinate is changed?
See Additional Answers; Reflection in the y-axis

Cumulative Review

In Exercises 59–64, solve the proportion.

59. $\frac{3}{4} = \frac{x}{28}$ 21

60. $\frac{5}{6} = \frac{y}{36}$ 30

61. $\frac{a}{27} = \frac{4}{9}$ 12

62. $\frac{m}{49} = \frac{5}{7}$ 35

63. $\frac{z+1}{10} = \frac{z}{9}$ 9

64. $\frac{n}{16} = \frac{n-3}{8}$ 6

In Exercises 65–68, find the missing quantities.

	Original Price	*Sale Price*	*Discount*	*Discount Rate*
65.	\$80.00	\$52.00	\$28.00	35%
66.	\$55.00	\$35.75	\$19.25	35%
67.	\$112.50	\$81.00	\$31.50	28%
68.	\$258.50	\$134.42	\$124.08	48%

3.2 Graphs of Equations

- Sketch graphs of equations using the point-plotting method.
- Find and use x- and y-intercepts as aids to sketching graphs.
- Solve application problems involving graphs of equations.

The Graph of an Equation

In Section 3.1, you saw that the solutions of an equation can be represented by points on a rectangular coordinate system. The set of all solution points of an equation is called its **graph**. In this section, you will study a basic technique for sketching the graph of an equation—the **point-plotting method**.

EXAMPLE 1 Sketching the Graph of a Linear Equation

To sketch the graph of $3x - y = 2$, first solve the equation for y.

$3x - y = 2$ Write original equation.

$-y = -3x + 2$ Subtract $3x$ from each side.

$y = 3x - 2$ Divide each side by -1.

Next, create a table of values. The choice of x-values to use in the table is somewhat arbitrary. However, the more x-values you choose, the easier it will be to recognize a pattern.

x	-2	-1	0	1	2	3
$y = 3x - 2$	-8	-5	-2	1	4	7
Solution point	$(-2, -8)$	$(-1, -5)$	$(0, -2)$	$(1, 1)$	$(2, 4)$	$(3, 7)$

Now, plot the solution points. It appears that all six points lie on a line, so complete the sketch by drawing a line through the points, as shown below.

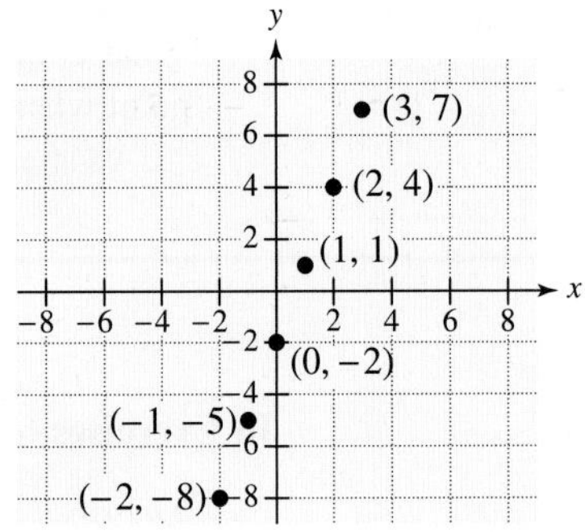

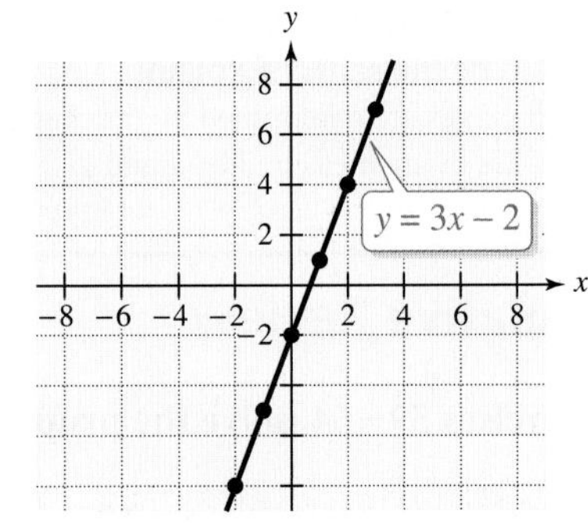

Exercises Within Reach® Solutions in English & Spanish and tutorial videos at AlgebraWithinReach.com

Sketching the Graph of a Linear Equation **In Exercises 1–8, sketch the graph of the equation.**

See Additional Answers.

1. $y = 3x$ **2.** $y = -2x$ **3.** $y = 4 - x$ **4.** $y = x - 7$

5. $2x - y = 3$ **6.** $3x - y = -2$ **7.** $3x + 2y = 2$ **8.** $2y + 5x = 6$

The Point-Plotting Method of Sketching a Graph

1. If possible, rewrite the equation by isolating one of the variables.
2. Make a table of values showing several solution points.
3. Plot these points on a rectangular coordinate system.
4. Connect the points with a smooth curve or line.

EXAMPLE 2 Sketching the Graph of a Nonlinear Equation

Sketch the graph of $-x^2 + 2x + y = 0$.

SOLUTION

Begin by solving the equation for y.

$-x^2 + 2x + y = 0$ Write original equation.

$2x + y = x^2$ Add x^2 to each side.

$y = x^2 - 2x$ Subtract $2x$ from each side.

Next, create a table of values.

x	-2	-1	0	1	2	3	4
$y = x^2 - 2x$	8	3	0	-1	0	3	8
Solution point	$(-2, 8)$	$(-1, 3)$	$(0, 0)$	$(1, -1)$	$(2, 0)$	$(3, 3)$	$(4, 8)$

Now, plot the seven solution points. Finally, connect the points with a smooth curve, as shown below.

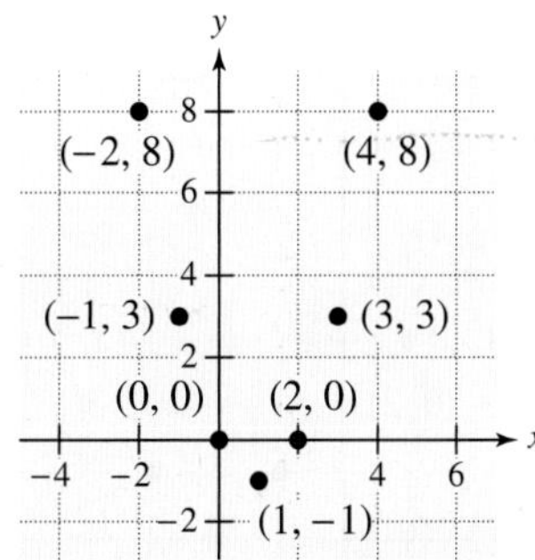

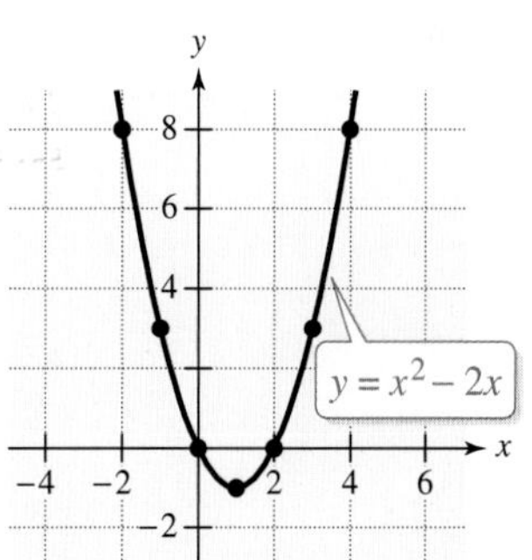

Exercises Within Reach®

Solutions in English & Spanish and tutorial videos at AlgebraWithinReach.com

Sketching the Graph of a Nonlinear Equation **In Exercises 9–16, sketch the graph of the equation.**

See Additional Answers.

9. $y = -x^2$

10. $y = x^2$

11. $y = x^2 - 3$

12. $y = 4 - x^2$

13. $-x^2 - 3x + y = 0$

14. $-x^2 + x + y = 0$

15. $x^2 - 2x - y = 1$

16. $x^2 + 3x - y = 4$

EXAMPLE 3 The Graph of an Absolute Value Equation

Sketch the graph of $y = |x - 2|$.

SOLUTION

This equation is already written in a form with y isolated on the left. So, begin by creating a table of values. Be sure that you understand how the absolute value is evaluated. For instance, when $x = -2$, the value of y is

$y = |x - 2|$ Write original equation.

$= |-2 - 2|$ Substitute -2 for x.

$= |-4|$ Simplify.

$= 4$ Simplify.

and when $x = 3$, the value of y is

$y = |x - 2|$ Write original equation.

$= |3 - 2|$ Substitute 3 for x.

$= |1|$ Simplify.

$= 1.$ Simplify.

x	-2	-1	0	1	2	3	4	5
$y = \|x - 2\|$	4	3	2	1	0	1	2	3
Solution point	$(-2, 4)$	$(-1, 3)$	$(0, 2)$	$(1, 1)$	$(2, 0)$	$(3, 1)$	$(4, 2)$	$(5, 3)$

Next, plot the solution points. It appears that the points lie in a "V-shaped" pattern, with the point (2, 0) at the bottom of the "V." Connect the points to form the graph shown below.

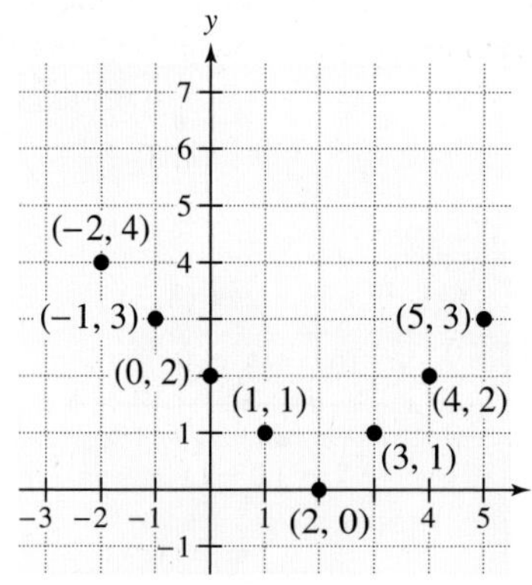

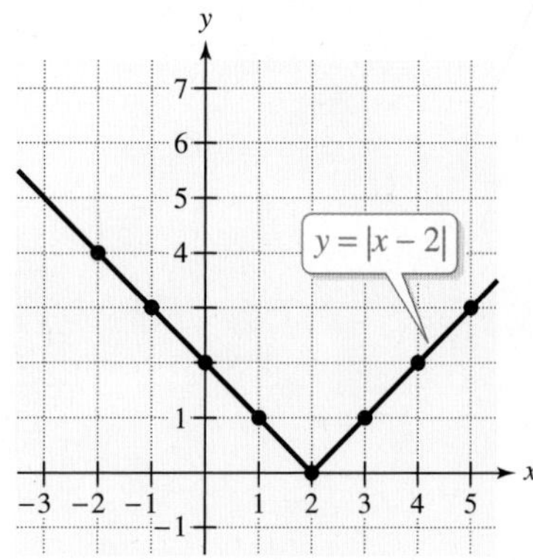

Exercises Within Reach®

Solutions in English & Spanish and tutorial videos at AlgebraWithinReach.com

Sketching the Graph of an Absolute Value Equation **In Exercises 17–22, sketch the graph of the equation.** See Additional Answers.

17. $y = |x|$

18. $y = -|x|$

19. $y = |x| + 3$

20. $y = |x| - 1$

21. $y = |x + 3|$

22. $y = |x - 1|$

Intercepts: Aids to Sketching Graphs

Study Tip

When you create a table of values for a graph, include any intercepts you have found. You should also include points to the left and right of the intercepts of the graph. This helps to give a more complete view of the graph.

Definition of Intercepts

The point $(a, 0)$ is called an ***x*-intercept** of the graph of an equation if it is a solution point of the equation. To find the x-intercept(s), let $y = 0$ and solve the equation for x.

The point $(0, b)$ is called a ***y*-intercept** of the graph of an equation if it is a solution point of the equation. To find the y-intercept(s), let $x = 0$ and solve the equation for y.

EXAMPLE 4 Finding the Intercepts of a Graph

Find the intercepts and sketch the graph of $y = 2x - 3$.

SOLUTION

Find the x-intercept by letting $y = 0$ and solving for x.

$y = 2x - 3$ Write original equation.

$0 = 2x - 3$ Substitute 0 for y.

$3 = 2x$ Add 3 to each side.

$\frac{3}{2} = x$ Solve for x.

Find the y-intercept by letting $x = 0$ and solving for y.

$y = 2x - 3$ Write original equation.

$y = 2(0) - 3$ Substitute 0 for x.

$y = -3$ Solve for y.

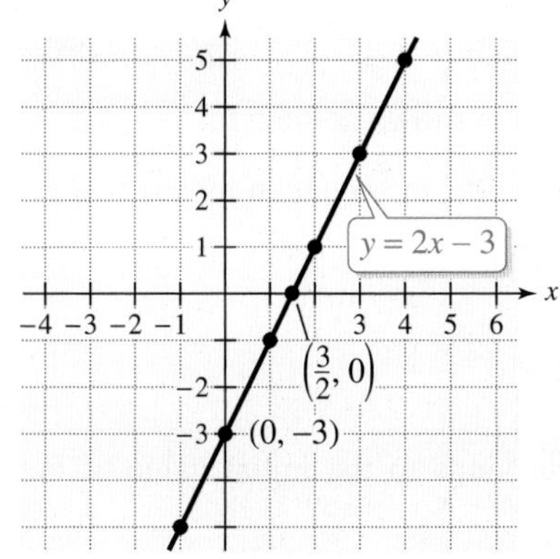

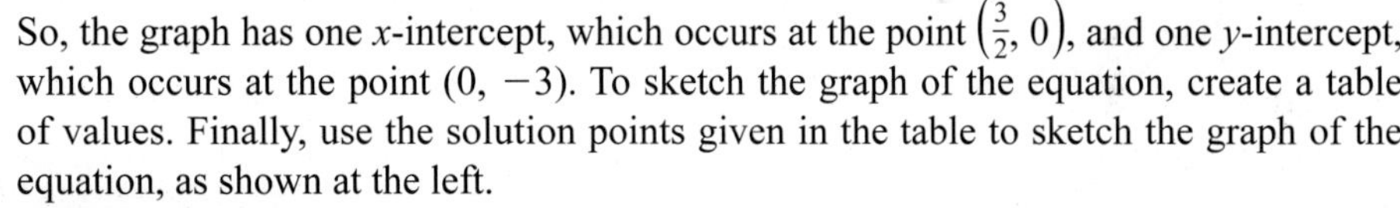

So, the graph has one x-intercept, which occurs at the point $\left(\frac{3}{2}, 0\right)$, and one y-intercept, which occurs at the point $(0, -3)$. To sketch the graph of the equation, create a table of values. Finally, use the solution points given in the table to sketch the graph of the equation, as shown at the left.

x	-1	0	1	$\frac{3}{2}$	2	3	4
$y = 2x - 3$	-5	-3	-1	0	1	3	5
Solution point	$(-1, -5)$	$(0, -3)$	$(1, -1)$	$\left(\frac{3}{2}, 0\right)$	$(2, 1)$	$(3, 3)$	$(4, 5)$

Exercises Within Reach®

Solutions in English & Spanish and tutorial videos at AlgebraWithinReach.com

Finding the Intercepts of a Graph **In Exercises 23–36, find the intercepts and sketch the graph of the equation.** See Additional Answers.

23. $y = 3 - x$

24. $y = x - 3$

25. $y = 2x - 3$

26. $y = -4x + 8$

27. $4x + y = 3$

28. $y - 2x = -4$

29. $2x - 3y = 6$

30. $3x - 2y = 8$

31. $3x + 4y = 12$

32. $4x + 5y = 10$

33. $x + 5y = 10$

34. $x + 3y = 15$

35. $5x - y = 10$

36. $7x - y = -21$

Applications of Graphs

Application

EXAMPLE 5 **Straight-Line Depreciation: Finding the Pattern**

Your small business buys a new printing press for \$65,000. For income tax purposes, you decide to depreciate the printing press over a 10-year period. At the end of 10 years, the value of the printing press is expected to be \$5000.

a. Find an equation that relates the value of the printing press to the number of years since its purchase.

b. Sketch the graph of the equation.

c. What is the y-intercept of the graph and what does it represent?

SOLUTION

a. The total depreciation over the 10-year period is $65{,}000 - 5000 = \$60{,}000$. Because the same amount is depreciated each year, it follows that the annual depreciation is

$$\frac{60{,}000}{10} = \$6000.$$

So, after 1 year, the value of the printing press is

$$\text{Value after 1 year} = 65{,}000 - (1)6000 = \$59{,}000.$$

By similar reasoning, you can see that the values after 2, 3, and 4 years are

$$\text{Value after 2 years} = 65{,}000 - (2)6000 = \$53{,}000$$
$$\text{Value after 3 years} = 65{,}000 - (3)6000 = \$47{,}000$$
$$\text{Value after 4 years} = 65{,}000 - (4)6000 = \$41{,}000.$$

Let y represent the value of the printing press after t years and follow the pattern determined for the first 4 years to obtain

$$y = 65{,}000 - 6000t.$$

Printing Press

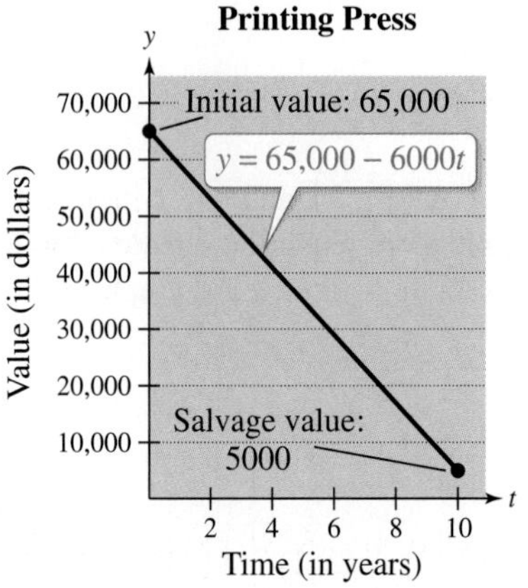

b. A sketch of the graph of the depreciation equation is shown at the left.

c. To find the y-intercept of the graph, let $t = 0$ and solve the equation for y.

$y = 65{,}000 - 6000t$ — Write original equation.

$y = 65{,}000 - 6000(0)$ — Substitute 0 for t.

$y = 65{,}000$ — Simplify.

So, the y-intercept is $(0, 65{,}000)$, which corresponds to the initial value of the printing press.

Exercises Within Reach®

Solutions in English & Spanish and tutorial videos at AlgebraWithinReach.com

37. ***Straight-Line Depreciation*** Your company purchases a new delivery van for \$40,000. For tax purposes, the van will be depreciated over a seven-year period. At the end of 7 years, the value of the van is expected to be \$5000. See Additional Answers.

(a) Find an equation that relates the value of the van to the number of years since its purchase.

(b) Sketch the graph of the equation.

(c) What is the y-intercept of the graph and what does it represent?

38. ***Straight-Line Depreciation*** Your company purchases a new limousine for \$65,000. For tax purposes, the limousine will be depreciated over a 10-year period. At the end of 10 years, the value of the limousine is expected to be \$10,000. See Additional Answers.

(a) Find an equation that relates the value of the limousine to the number of years since its purchase.

(b) Sketch the graph of the equation.

(c) What is the y-intercept of the graph and what does it represent?

Application

EXAMPLE 6 Renewable Biomass Energy

The table shows the yearly production P of renewable biomass energy (in quadrillion Btu) in the United States from 2003 through 2010. (*Source:* U.S. Energy Information Administration)

Year	2003	2004	2005	2006	2007	2008	2009	2010
P	2.8	3.0	3.1	3.2	3.5	3.9	3.9	4.3

A model for this data is $P = 0.21t + 2.1$, where t represents the year, with $t = 3$ corresponding to 2003.

a. Plot the data and graph the model on the same coordinate system.

b. How well does the model represent the data? Explain your reasoning.

c. Use the model to predict the total production of renewable biomass energy in 2015.

Production of Renewable Biomass Energy

P; 1, 2, 3, 4, 5, 6; Renewable biomass energy use (in quadrillion Btu)

t; 2, 4, 6, 8, 10; Year (3 ↔ 2003)

SOLUTION

a. Plot the points from the table. Then sketch the graph of $P = 0.21t + 2.1$ on the same coordinate system, as shown at the left.

b. The model represents the data quite well. All 8 points lie close to the line given by the model.

c. To predict the total production of renewable biomass energy in 2015, substitute 15 for t in the model and solve for P.

$P = 0.21t + 2.1$ Write original equation.

$= 0.21(15) + 2.1$ Substitute 15 for t.

$= 3.15 + 2.1$ Multiply.

$= 5.25$ Add.

The total production of renewable biomass energy will be about 5.25 quadrillion Btu in 2015.

Exercises Within Reach®

Solutions in English & Spanish and tutorial videos at AlgebraWithinReach.com

39. ***Hooke's Law*** The force F (in pounds) required to stretch a spring x inches from its natural length is given by

$$F = \frac{4}{3}x, \quad 0 \le x \le 12.$$

(a) Use the model to complete the table.

x	0	3	6	9	12
F	0	4	8	12	16

(b) Plot the data and graph the model on the same coordinate system. See Additional Answers.

40. ***Using a Model*** Use the model in Exercise 39 to determine how F changes when x doubles. F doubles

Concept Summary: Graphs of Equations

What

You can use a **graph** to represent all the solutions of an equation in two variables.

EXAMPLE

Sketch the graph of

$y = |x| - 1$.

How

Use these steps to sketch the graph of such an equation.

1. Make a table of values. Include any x- and y-**intercepts**.
2. Plot the points.
3. Connect the points.

x	-2	-1	0	1	2
$y = \|x\| - 1$	1	0	-1	0	1

Why

You can use the graph of an equation in two variables to see the relationship between the variables. For example, the graph shows that y decreases as x increases to zero, and y increases as x increases above zero.

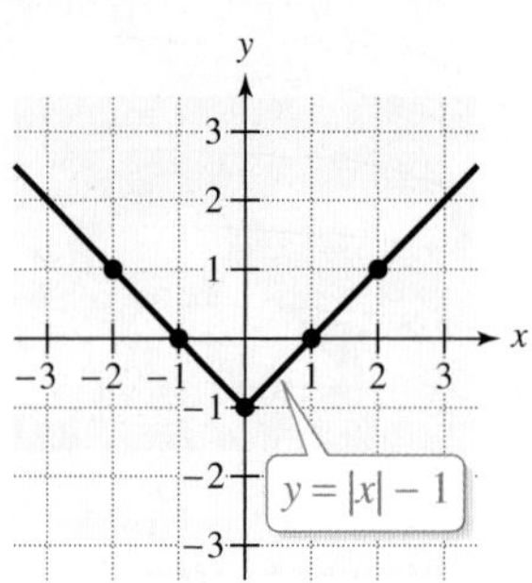

Exercises Within Reach®

Worked-out solutions to odd-numbered exercises at AlgebraWithinReach.com

Concept Summary Check

41. ***Reasoning*** A table of values has been used to plot the solution points shown below for a given equation. Describe what else needs to be done to complete the graph of the equation. To complete the graph of the equation, connect the points with a smooth curve.

42. ***Vocabulary*** Is enough information given in the figure below to determine the x- and y-intercepts? Explain. See Additional Answers.

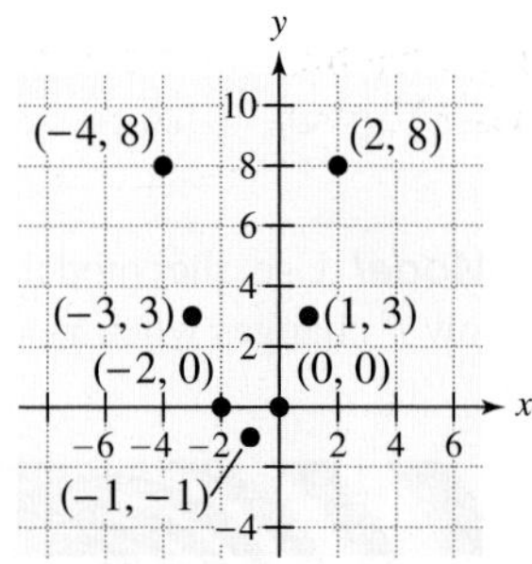

Figure for 41 and 42

43. ***True or False?*** To find the x-intercept(s) of the graph of an equation algebraically, you can substitute 0 for x in the equation and then solve for y. Justify your answer. See Additional Answers.

44. ***Think About It*** Describe the error in graphing $y = x^2$.

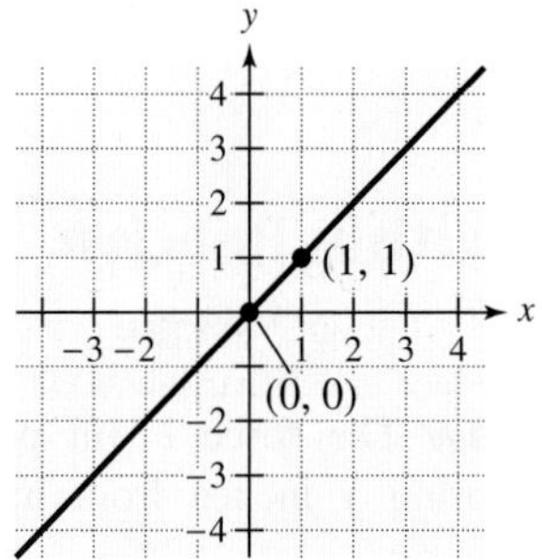

More points need to be used in order to see the general shape of the graph.

Extra Practice

Sketching the Graph of an Equation **In Exercises 45–56, sketch the graph of the equation and show the coordinates of three solution points (including x- and y-intercepts).** See Additional Answers.

45. $y = x^2 - 9$
46. $y = x^2 - 16$
47. $y = 9 - x^2$
48. $y = 16 - x^2$
49. $y = x(x - 2)$
50. $y = x(x + 2)$
51. $y = -x(x + 4)$
52. $y = -x(x - 6)$
53. $y = |x + 2|$
54. $y = |x + 4|$
55. $y = |x - 3|$
56. $y = |x - 4|$

Explaining Concepts

57. ***Misleading Graphs*** Graphs can help you visualize relationships between two variables, but they can also be misused to imply results that are not correct. The two graphs below represent the same data points. Why do the graphs appear different? Identify ways in which the graphs could be misleading.

(a) **Company Sales**

s
Sales (in thousands)
750
600
450
300
150
'10 '12
Year

(b)

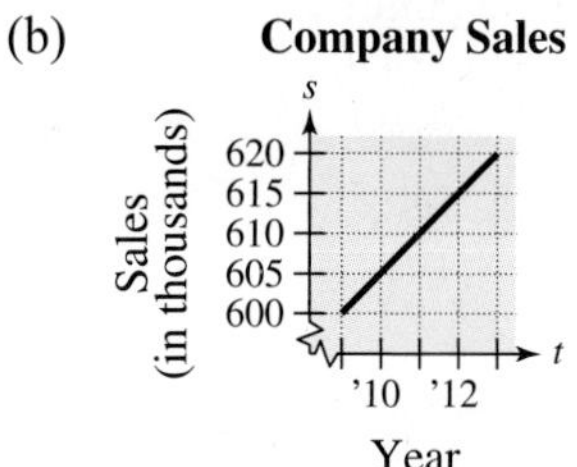

The scales on the y-axes are different. From graph (a) it appears that sales have not increased. From graph (b) it appears that sales have increased dramatically.

58. ***Exploration*** Graph the equations $y = x^2 + 1$ and $y = -(x^2 + 1)$ on the same set of coordinate axes. Explain how the graph of an equation changes when the expression for y is multiplied by -1. Justify your answer by giving additional examples.

See Additional Answers.

The graph is reflected in the x-axis when the expression for y is multiplied by -1.

59. ***Modeling*** A company's profits decrease rapidly for a time, but then begin decreasing at a lower rate. Sketch an example of a graph representing such a situation, showing the profit y in terms of the time t.

See Additional Answers.

60. ***Writing*** Discuss the possible numbers of x- and y-intercepts of the graph of

$$y = ax + 5$$

where a is a positive integer.

The x-intercept is $(-5/a, 0)$, where a is a positive integer. The y-intercept is $(0, 5)$. So, infinitely many x-intercepts are possible and only one y-intercept is possible.

61. ***Writing*** Discuss the possible numbers of x- and y-intercepts of a horizontal line on a rectangular coordinate system.

A horizontal line has no x-intercept. The y-intercept of a horizontal line is $(0, b)$, where b is any real number. So, a horizontal line has one y-intercept.

62. ***Think About It*** The graph shown represents the distance d in miles that a person drives during a 10-minute trip from home to work.

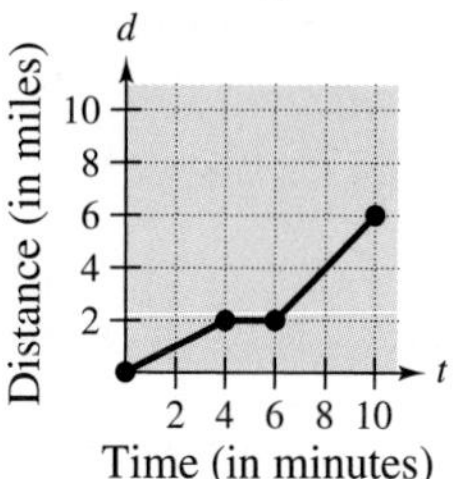

(a) How far is the person's home from the person's place of work? Explain.
6 miles; The graph shows that the total distance traveled during the 10-minute trip is 6 miles.

(b) Describe the trip for time $4 < t < 6$. Explain.
The person is stopped, because the graph remains constant on the interval $4 < t < 6$.

(c) During what time interval is the person's speed greatest? Explain. The speed is greatest on the interval $6 < t < 10$, because the graph is steepest on that interval.

Cumulative Review

In Exercises 63–66, find the reciprocal.

63. 7 $\frac{1}{7}$

64. $\frac{1}{7}$ 7

65. $\frac{4}{5}$ $\frac{5}{4}$

66. $\frac{8}{5}$ $\frac{5}{8}$

In Exercises 67–70, solve the equation and check your solution.

67. $x - 8 = 0$ 8

68. $x + 11 = 24$ 13

69. $4x + 15 = 23$ 2

70. $13 - 6x = -5$ 3

In Exercises 71 and 72, plot the points and connect them with line segments to form a triangle.

See Additional Answers.

71. $(0, 5), (3, 5), (3, 7)$

72. $(2, 4), (6, 2), (2, 2)$

3.3 Slope and Graphs of Linear Equations

- Determine the slope of a line through two points.
- Write linear equations in slope-intercept form and graph the equations.
- Use slopes to determine whether two lines are parallel, perpendicular, or neither.
- Use slopes to describe rates of change in real-life problems.

The Slope of a Line

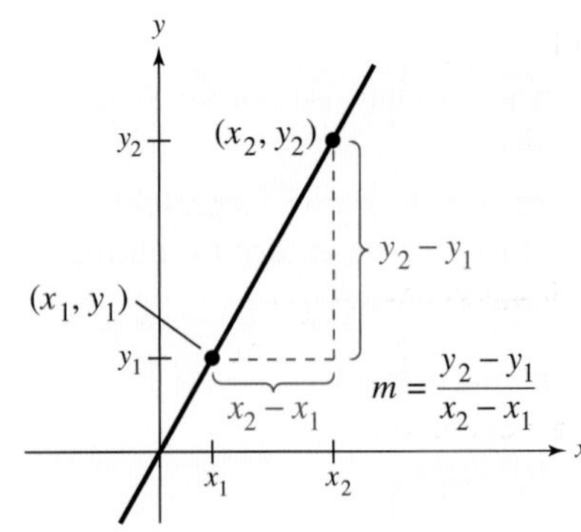

Definition of the Slope of a Line

The **slope** m of a nonvertical line that passes through the points (x_1, y_1) and (x_2, y_2) is

$$m = \frac{y_2 - y_1}{x_2 - x_1} = \frac{\text{Change in } y}{\text{Change in } x} = \frac{\text{Rise}}{\text{Run}}$$

where $x_1 \neq x_2$.

EXAMPLE 1 **Finding the Slope of a Line Through Two Points**

a. The slope of the line through (1, 2) and (4, 5) is

$$m = \frac{y_2 - y_1}{x_2 - x_1}$$

$$= \frac{5 - 2}{4 - 1}$$

$$= 1.$$

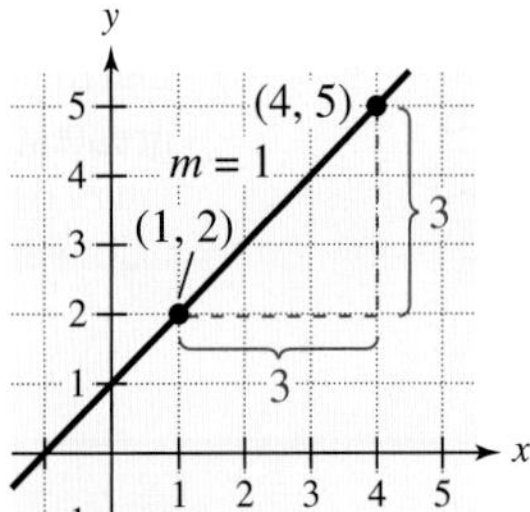

Positive slope

b. The slope of the line through (−1, 4) and (2, 1) is

$$m = \frac{1 - 4}{2 - (-1)}$$

$$= \frac{-3}{3}$$

$$= -1.$$

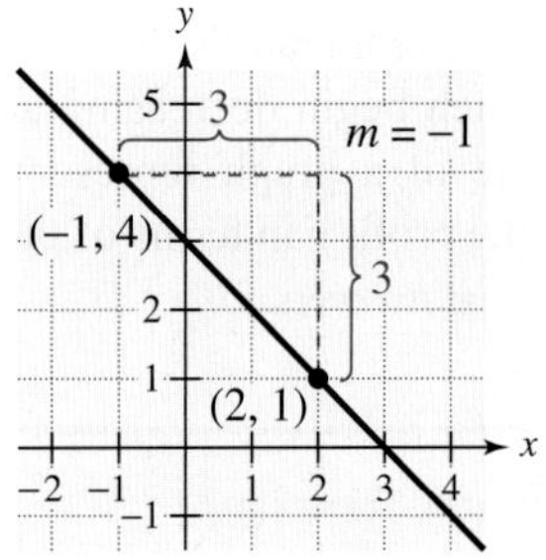

Negative slope

Exercises Within Reach®

Solutions in English & Spanish and tutorial videos at AlgebraWithinReach.com

Finding the Slope of a Line **In Exercises 1–8, plot the points and find the slope of the line that passes through them.** See Additional Answers.

1. (0, 0), (4, 8) $m = 2$
2. (0, 0), (−3, −12) $m = 4$
3. (−4, 3), (−2, 5) $m = 1$
4. (7, 1), (4, −5) $m = 2$
5. (3, 7), (−2, 10) $m = -\frac{3}{5}$
6. (−6, 7), (−3, −1) $m = -\frac{8}{3}$
7. $\left(\frac{3}{4}, 2\right), \left(5, -\frac{5}{2}\right)$ $m = -\frac{18}{17}$
8. $\left(\frac{1}{2}, -1\right), \left(3, \frac{2}{3}\right)$ $m = \frac{2}{3}$

EXAMPLE 2 Finding the Slope of a Line Through Two Points

a. The slope of the line through (1, 4) and (3, 4) is

$$m = \frac{y_2 - y_1}{x_2 - x_1}$$

$$= \frac{4 - 4}{3 - 1}$$

$$= \frac{0}{2} = 0.$$

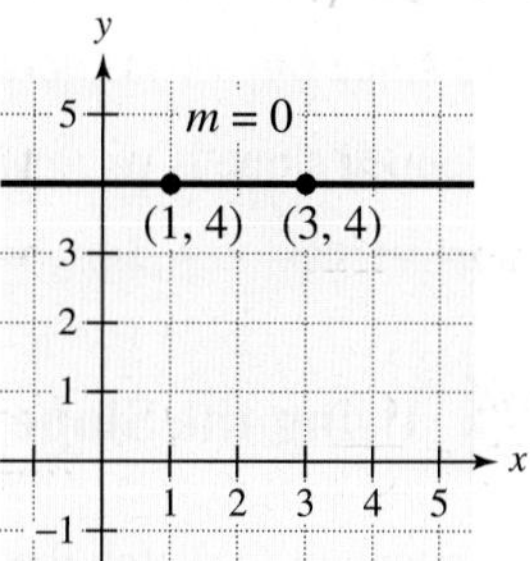

Zero slope

b. The slope of the line through (3, 1) and (3, 3) is undefined.

$$\frac{3 - 1}{3 - 3} = \frac{2}{0}$$

Because division by zero is not defined, the slope of a vertical line is not defined.

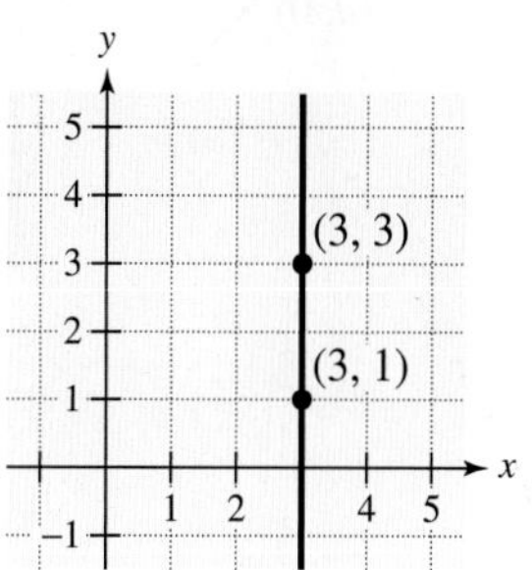

Slope is undefined.

Exercises Within Reach®

Solutions in English & Spanish and tutorial videos at AlgebraWithinReach.com

Finding the Slope of a Line **In Exercises 9–14, find the slope of the line.**

9. Undefined

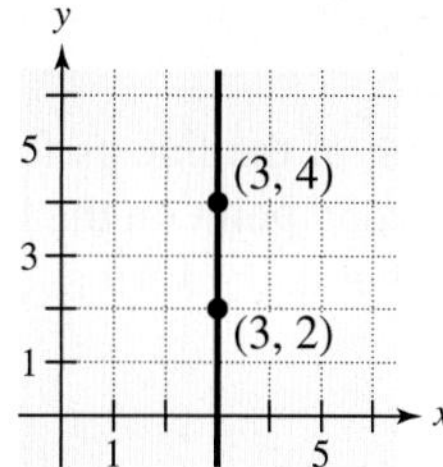

10. 0

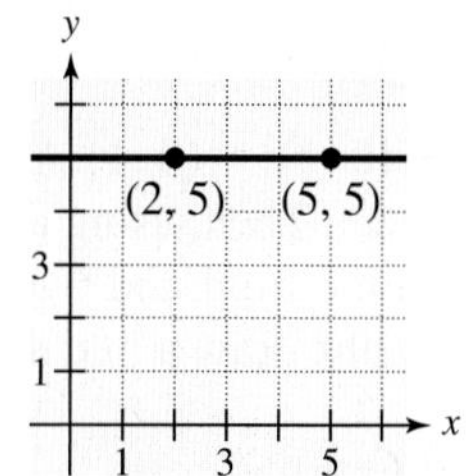

11. $\frac{2}{5}$

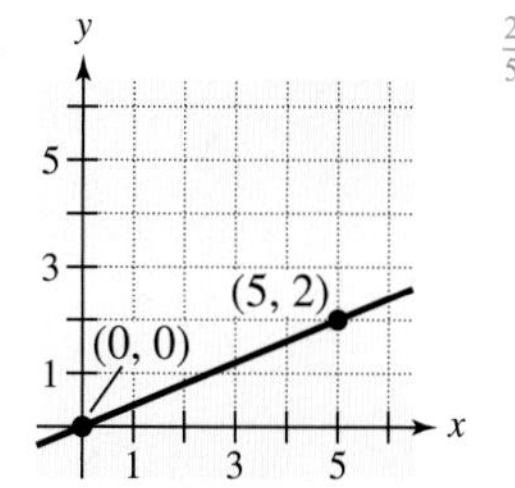

12. $-\frac{3}{2}$

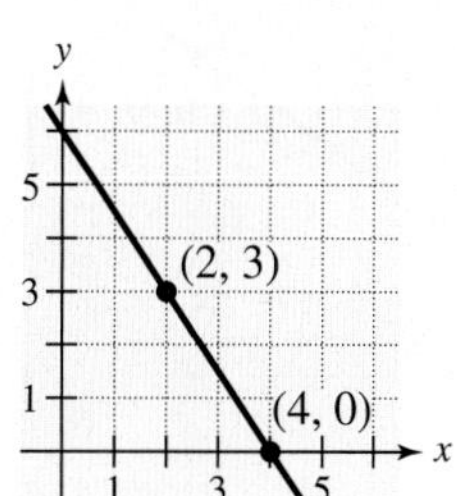

13. -2

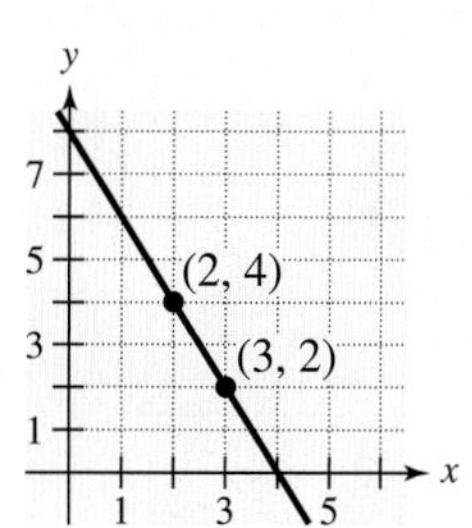

14. 4

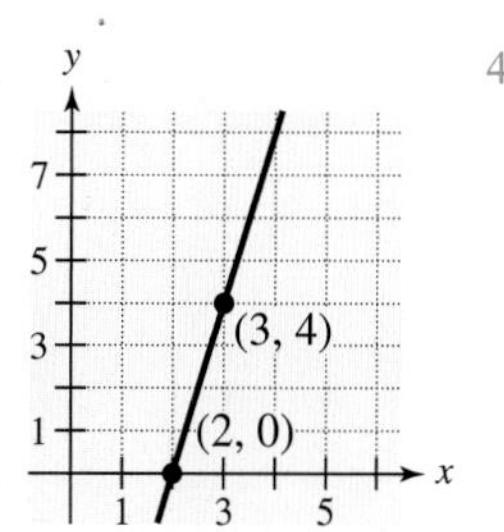

Using Slope **In Exercises 15 and 16, identify the line that has each slope *m*.**

15. 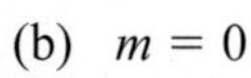 (a) $m = \frac{3}{4}$

(b) $m = 0$

(c) $m = -3$

(a) L_3
(b) L_2
(c) L_1

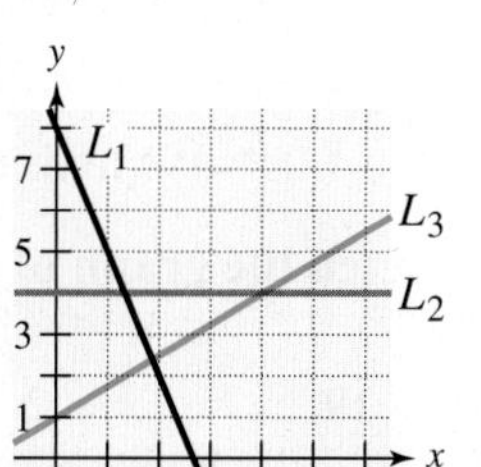

16. 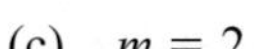 (a) $m = -\frac{5}{2}$

(b) m is undefined.

(c) $m = 2$

(a) L_2
(b) L_3
(c) L_1

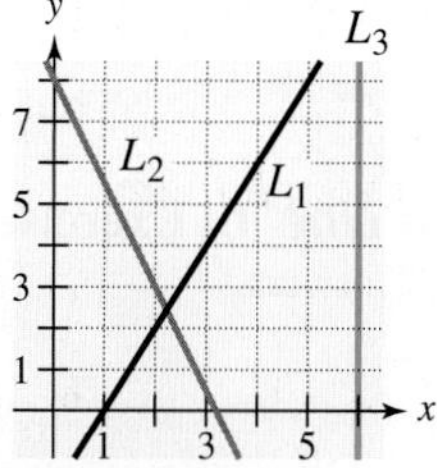

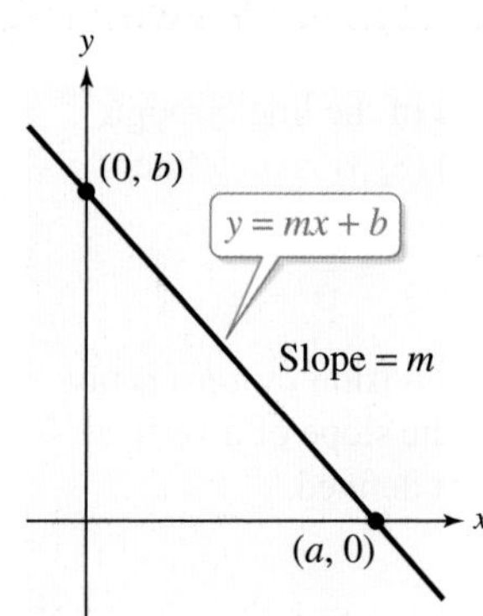

Slope as a Graphing Aid

Slope-Intercept Form of the Equation of a Line

The graph of the equation

$$y = mx + b$$

is a line whose slope is m and whose y-intercept is $(0, b)$.

1. A line with positive slope $(m > 0)$ *rises* from left to right.
2. A line with negative slope $(m < 0)$ *falls* from left to right.
3. A line with zero slope $(m = 0)$ is *horizontal*.
4. A line with undefined slope is *vertical*.

EXAMPLE 3 Using the Slope and y-Intercept to Sketch a Line

Use the slope and y-intercept to sketch the graph of $y = \frac{2}{3}x + 1$.

SOLUTION

The equation is already in slope-intercept form.

$$y = \frac{2}{3}x + 1 \qquad \text{Slope-intercept form}$$

So, the slope of the line is

$$m = \frac{2}{3} = \frac{\text{Change in } y}{\text{Change in } x}$$

and the y-intercept is $(0, b) = (0, 1)$. You can sketch the graph of the line as follows. First, plot the y-intercept. Then, using a slope of $\frac{2}{3}$, locate a second point on the line by moving three units to the right and two units upward (or two units upward and three units to the right). Finally, draw a line through the two points.

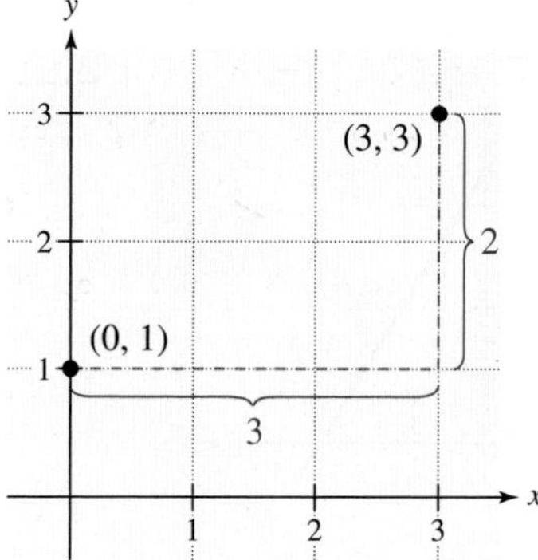

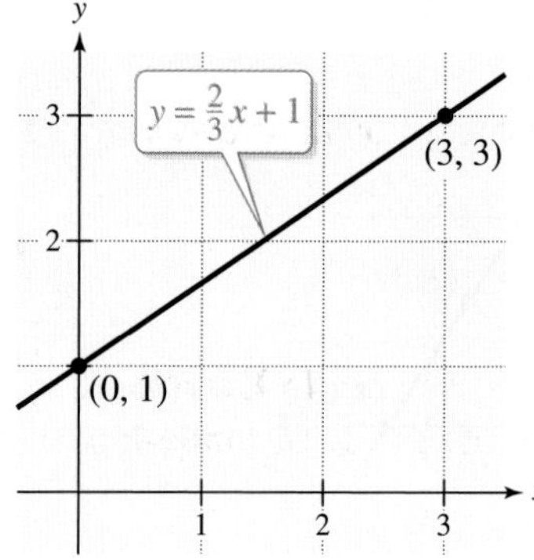

Exercises Within Reach®

Solutions in English & Spanish and tutorial videos at AlgebraWithinReach.com

Sketching a Line **In Exercises 17–22, use the slope and y-intercept to sketch the graph of the equation.**
See Additional Answers.

17. $y = 2x - 1$ **18.** $y = 3x + 2$ **19.** $y = -x - 2$

20. $y = -3x + 6$ **21.** $y = -\frac{1}{2}x + 4$ **22.** $y = \frac{3}{4}x - 5$

EXAMPLE 4 Using the Slope and *y*-Intercept to Sketch a Line

Use the slope and *y*-intercept to sketch the graph of $12x + 3y = 6$.

SOLUTION

Begin by writing the equation in slope-intercept form.

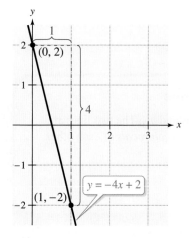

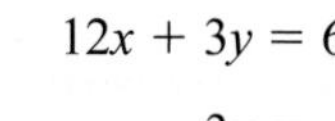

$12x + 3y = 6$ Write original equation.

$3y = -12x + 6$ Subtract $12x$ from each side.

$y = \dfrac{-12x + 6}{3}$ Divide each side by 3.

$y = -4x + 2$ Slope-intercept form

So, the slope of the line is $m = -4$ and the *y*-intercept is $(0, b) = (0, 2)$. You can sketch the graph of the line as follows. First, plot the *y*-intercept. Then, using a slope of -4, locate a second point on the line by moving one unit to the right and four units downward (or four units downward and one unit to the right). Finally, draw a line through the two points.

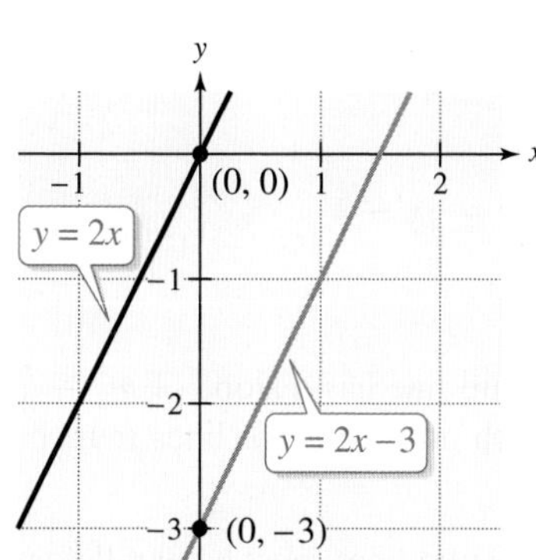

EXAMPLE 5 Using Slopes and *y*-Intercepts to Sketch Lines

On the same set of coordinate axes, sketch the lines given by

$y = 2x$ and $y = 2x - 3$.

SOLUTION

For the line given by $y = 2x$, the slope is $m = 2$ and the *y*-intercept is $(0, 0)$. For the line given by $y = 2x - 3$, the slope is also $m = 2$ and the *y*-intercept is $(0, -3)$. The graphs of these two lines are shown at the left.

Exercises Within Reach®

Solutions in English & Spanish and tutorial videos at AlgebraWithinReach.com

Finding the Slope and y-Intercept **In Exercises 23–26, write the equation of the line in slope-intercept form, if necessary. Then find the slope and *y*-intercept of the line.**

23. $y = 3x - 2$ $m = 3; (0, -2)$

24. $y = 4 - 2x$ $y = -2x + 4; m = -2; (0, 4)$

25. $4x - 6y = 24$ $y = \frac{2}{3}x - 4; m = \frac{2}{3}; (0, -4)$

26. $4x + 8y = -1$ $y = -\frac{1}{2}x - \frac{1}{8}; m = -\frac{1}{2}; \left(0, -\frac{1}{8}\right)$

Sketching a Line **In Exercises 27–32, write the equation of the line in slope-intercept form, and then use the slope and *y*-intercept to sketch the line.** See Additional Answers.

27. $x + y = 0$ $y = -x$

28. $x - y = 0$ $y = x$

29. $3x - y - 2 = 0$ $y = 3x - 2$

30. $2x - y - 3 = 0$ $y = 2x - 3$

31. $3x + 2y - 2 = 0$ $y = -\frac{3}{2}x + 1$

32. $x - 2y - 2 = 0$ $y = \frac{1}{2}x - 1$

Sketching Lines **In Exercises 33 and 34, sketch the lines given by the equations on the same set of coordinate axes.** See Additional Answers.

33. L_1: $y = \frac{1}{2}x$
L_2: $y = \frac{1}{2}x + 2$

34. L_1: $y = -3x + 1$
L_2: $y = -3x - 2$

Parallel and Perpendicular Lines

Parallel Lines

Two distinct nonvertical lines are parallel if and only if they have the same slope.

Perpendicular Lines

Consider two nonvertical lines whose slopes are m_1 and m_2. The two lines are perpendicular if and only if their slopes are *negative reciprocals* of each other. That is,

$$m_1 = -\frac{1}{m_2}, \quad \text{or equivalently,} \quad m_1 \cdot m_2 = -1.$$

EXAMPLE 6 **Parallel and Perpendicular Lines**

Are the pairs of lines parallel, perpendicular, or neither?

a. $y = -2x + 4, y = \frac{1}{2}x + 1$ **b.** $y = \frac{1}{3}x + 2, y = \frac{1}{3}x - 3$

SOLUTION

a. The first line has a slope of $m_1 = -2$, and the second line has a slope of $m_2 = \frac{1}{2}$. Because these slopes are negative reciprocals of each other, the two lines must be perpendicular, as shown below on the left.

b. Each of these two lines has a slope of $m = \frac{1}{3}$. So, the two lines must be parallel, as shown below on the right.

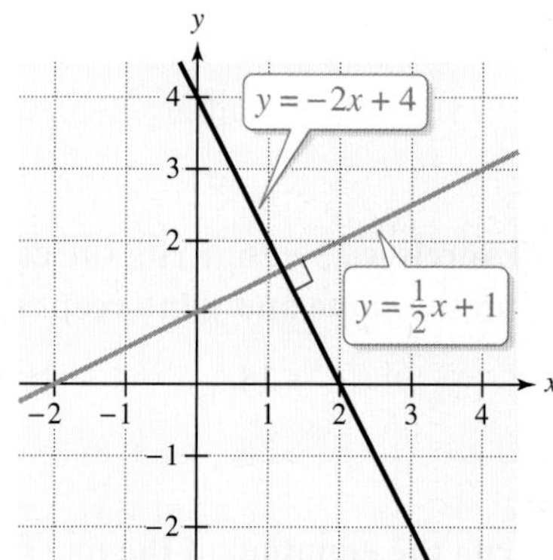

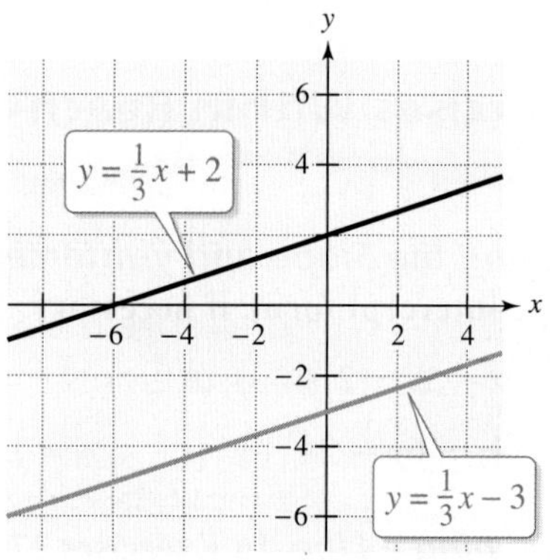

Exercises Within Reach®

Solutions in English & Spanish and tutorial videos at AlgebraWithinReach.com

Parallel and Perpendicular Lines **In Exercises 35–40, determine whether the lines are parallel, perpendicular, or neither.**

35. L_1: $y = \frac{1}{2}x - 2$ Parallel
L_2: $y = \frac{1}{2}x + 3$

36. L_1: $y = -\frac{2}{3}x - 5$ Perpendicular
L_2: $y = \frac{3}{2}x + 1$

37. L_1: $y = \frac{3}{4}x - 3$ Perpendicular
L_2: $y = -\frac{4}{3}x + 1$

38. L_1: $y = 3x - 2$ Parallel
L_2: $y = 3x + 1$

39. L_1: $y = -4x + 6$ Neither
L_2: $y = -\frac{1}{4}x - 1$

40. L_1: $y = \frac{3}{2}x + 7$ Neither
L_2: $y = \frac{2}{3}x - 4$

Slope as a Rate of Change

In real-life problems, slope can describe a **constant rate of change** or an **average rate of change**. In such cases, units of measure are used, such as miles per hour.

Application

EXAMPLE 7 Slope as a Rate of Change

In 2004, the average seating capacity of a regional aircraft was 47. By 2010, the average seating capacity had risen to 56. Find the average rate of change in seating capacity of a regional aircraft from 2004 to 2010. (*Source:* OAG)

SOLUTION

Let c represent the average seating capacity and let t represent the year. The two given data points are represented by (t_1, c_1) and (t_2, c_2).

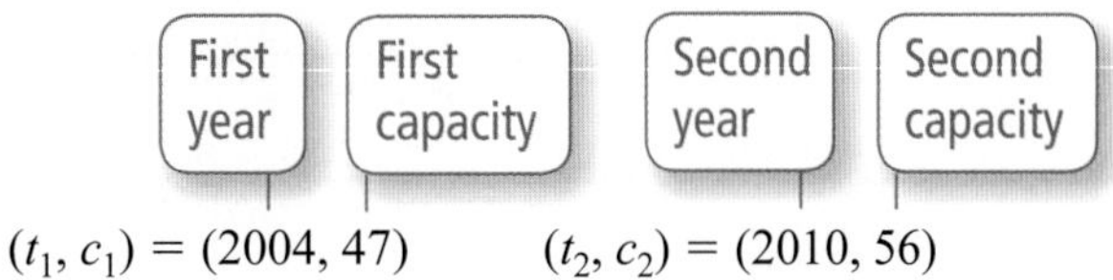

$(t_1, c_1) = (2004, 47)$ $\quad$ $(t_2, c_2) = (2010, 56)$

U.S. Regional Aircraft

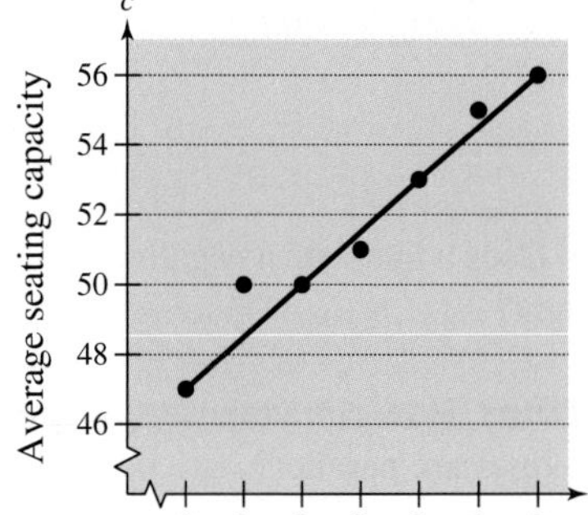

Now use the formula for slope to find the average rate of change.

$$\text{Rate of change} = \frac{c_2 - c_1}{t_2 - t_1} \qquad \text{Slope formula}$$

$$= \frac{56 - 47}{2010 - 2004} \qquad \text{Substitute values.}$$

$$= \frac{9}{6} = 1.5 \qquad \text{Simplify.}$$

From 2004 through 2010, the average rate of change in the seating capacity of a regional aircraft was 1.5 seats per year. The exact change in seating capacity varied from one year to the next, as shown in the scatter plot at the left.

Exercises Within Reach®

Solutions in English & Spanish and tutorial videos at AlgebraWithinReach.com

41. ***Hourly Wages*** In 2005, the average hourly wage of health services employees in the United States was \$17.05. By 2010, the average hourly wage had risen to \$20.43. Find the average rate of change in the hourly wage from 2005 to 2010. (*Source:* U.S. Bureau of Labor Statistics) \$0.68

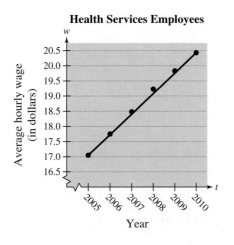

Health Services Employees

w; Average hourly wage (in dollars): 16.5, 17.0, 17.5, 18.0, 18.5, 19.0, 19.5, 20.0, 20.5; t; Year: 2005, 2006, 2007, 2008, 2009, 2010

42. ***Consumer Awareness*** In 2004, the average retail price of 16 ounces of potato chips was \$3.35. By 2010, the average price had risen to \$4.74. Find the average rate of change in the retail price from 2004 to 2010. (*Source:* U.S. Bureau of Labor Statistics) \$0.23

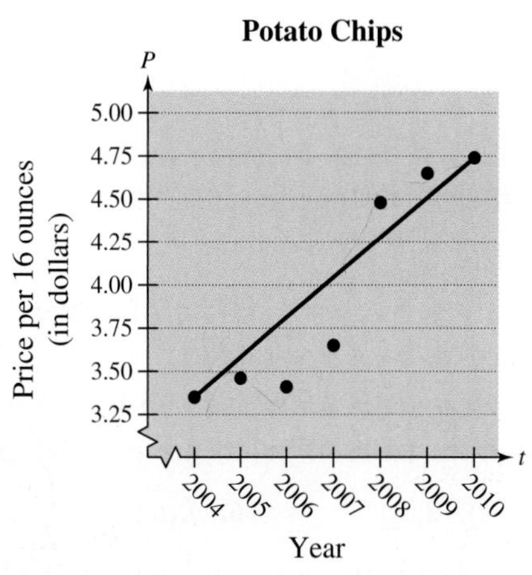

Concept Summary: Finding the Slope of a Line

What

The **slope** of a line is a measure of the "steepness" of the line. The slope also indicates whether the line *rises* or *falls* from left to right, is *horizontal*, or is *vertical*.

EXAMPLE

Find the slope of the line that passes through the points (2, 3) and (5, 1).

How

You can use a formula to find the slope of a nonvertical line that passes through any two points (x_1, y_1) and (x_2, y_2).

Slope: $m = \frac{y_2 - y_1}{x_2 - x_1} = \frac{\text{Rise}}{\text{Run}}$

EXAMPLE

Let $(x_1, y_1) = (2, 3)$ and let $(x_2, y_2) = (5, 1)$.

$$m = \frac{y_2 - y_1}{x_2 - x_1} = \frac{1 - 3}{5 - 2} = -\frac{2}{3}$$

Why

You can use slope to

- help sketch the graphs of linear equations.
- determine whether lines are parallel, perpendicular, or neither.

Exercises Within Reach®

Worked-out solutions to odd-numbered exercises at AlgebraWithinReach.com

Concept Summary Check

43. *Rise and Run* In the solution above, what is the rise from (2, 3) to (5, 1)? What is the run? −2; 3

44. *Steepness* Which slope is steeper, $-\frac{2}{3}$ or $\frac{1}{3}$? Explain.
$-\frac{2}{3}$; For each 3 horizontal units of change, the vertical change is 2 units for a slope of $-\frac{2}{3}$, but only 1 unit for a slope of $\frac{1}{3}$.

45. *The Sign of a Slope* Does a line with a negative slope rise or fall from left to right? It falls from left to right.

46. *Slope and Parallel Lines* How can you use slope to determine whether two lines are parallel? Two lines with the same slope are parallel.

Extra Practice

Using Slope to Describe a Line **In Exercises 47–50, find the slope of the line that passes through the points. State whether the line rises, falls, is horizontal, or is vertical.**

47. $\left(\frac{3}{4}, \frac{1}{4}\right), \left(-\frac{3}{2}, \frac{1}{8}\right)$ $m = \frac{1}{18}$; rises

48. $\left(-\frac{3}{2}, -\frac{1}{2}\right), \left(\frac{4}{3}, -\frac{3}{2}\right)$ $m = -\frac{6}{17}$; falls

49. $(4.2, -1), (-4.2, 6)$ $m = -\frac{5}{6}$; falls

50. $(3.4, 0), (3.4, 1)$ m is undefined; vertical

Finding a Coordinate **In Exercises 51 and 52, solve for x so that the line that passes through the points has the given slope.**

51. $(4, 5), (x, 7)$
$m = -\frac{2}{3}$ $x = 1$

52. $(x, 3), (6, 6)$
$m = \frac{3}{4}$ $x = 2$

Finding a Coordinate **In Exercises 53 and 54, solve for y so that the line that passes through the points has the given slope.**

53. $(-3, y), (9, 3)$
$m = \frac{3}{2}$ $y = -15$

54. $(-3, 20), (2, y)$
$m = -6$ $y = -10$

Parallel or Perpendicular Lines **In Exercises 55–58, determine whether the lines L_1 and L_2 that pass through the pairs of points are parallel, perpendicular, or neither.**

55. L_1: $(0, 4), (2, 8)$
L_2: $(0, -1), (3, 5)$
Parallel

56. L_1: $(3, 4), (-2, 3)$
L_2: $(0, -3), (2, -1)$
Neither

57. L_1: $(0, 2), (6, -2)$
L_2: $(2, 0), (8, 4)$
Neither

58. L_1: $(3, 2), (-1, -2)$
L_2: $(2, 0), (3, -1)$
Perpendicular

Using Intercepts **In Exercises 59–62, plot the x- and y-intercepts and sketch the line.** See Additional Answers.

59. $3x - 5y - 15 = 0$ **60.** $3x + 5y + 15 = 0$ **61.** $-4x - 2y + 16 = 0$ **62.** $-5x + 2y - 20 = 0$

63. ***Road Grade*** When driving down a mountain road, you notice warning signs indicating an "8% grade." This means that the slope of the road is $-\frac{8}{100}$. Over a stretch of the road, your elevation drops by 2000 feet (see figure). What is the horizontal change in your position? 25,000 feet

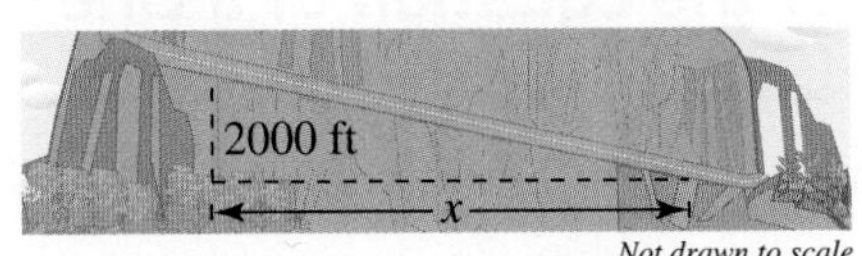

64. ***Ramp*** A loading dock ramp rises 4 feet above the ground. The ramp has a slope of $\frac{1}{10}$. What is the horizontal length of the ramp? 40 feet

65. ***Roof Pitch*** The slope, or pitch, of a roof (see figure) is such that it rises (or falls) 3 feet for every 4 feet of horizontal distance. Determine the maximum height of the attic of the house. $\frac{45}{4} = 11.25$ feet

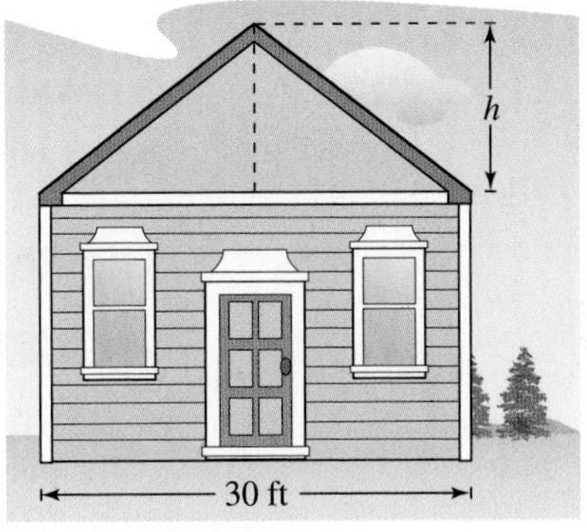

Figure for 65 and 66

66. ***Roof Pitch*** The slope, or pitch, of a roof (see figure) is such that it rises (or falls) 4 feet for every 5 feet of horizontal distance. Determine the maximum height of the attic of the house. 12 feet

Explaining Concepts

67. ***Writing*** Can any pair of points on a nonvertical line be used to calculate the slope of the line? Explain. Yes. When different pairs of points are selected, the change in y and the change in x are the lengths of the sides of similar triangles. Corresponding sides of similar triangles are proportional.

68. ***Structure*** Explain how to use the form of the equation $y = 3x + 4$ to sketch its graph. The equation is in slope-intercept form with $m = 3$ and $b = 4$. Plot the y-intercept $(0, 4)$ and use the slope of 3 to plot another point. Draw a line through the points.

69. ***Reasoning*** Two lines have undefined slopes. Are the lines parallel? Yes; Because the slopes are undefined, the lines are vertical, so they are parallel.

70. ***Reasoning*** The line through $(1, 2)$ and $(2, 4)$ has a slope of 2. The line through $(2, 4)$ and (a, b) also has a slope of 2. Are the lines parallel? Explain. No; Because the lines have the same slope and share a common point, they must be the same line.

71. ***Reasoning*** Can two lines with positive slopes be perpendicular to each other? Explain. No. The slopes of perpendicular lines are negative reciprocals of each other.

72. ***Reasoning*** On a line with a slope of $\frac{3}{4}$, how much does y increase when x increases by 8 units? Explain. 6 units; Because the slope is $\frac{3}{4}$, for each increase of 4 units in x, y increases by 3 units. So, when x increases by 8 units, y increases by 6 units.

Cumulative Review

In Exercises 73–76, rewrite the statement using inequality notation.

73. x is negative. $x < 0$

74. m is at least -3. $m \geq -3$

75. z is at least 85, but no more than 100. $85 \leq z \leq 100$

76. n is less than 20, but no less than 16. $16 \leq n < 20$

In Exercises 77–84, solve the equation.

77. $|x| = 8$ 8, −8 **78.** $|g| = -4$ No solution

79. $|4h| = 24$ 6, −6 **80.** $\left|\frac{1}{5}m\right| = 2$ −10, 10

81. $|x + 4| = 5$ 1, −9

82. $|2t - 3| = 11$ 7, −4

83. $|6b + 8| = -2b$ −2, −1

84. $|n - 3| = |2n + 9|$ −12, −2

3.4 Equations of Lines

- Write equations of lines using point-slope form.
- Write equations of horizontal, vertical, parallel, and perpendicular lines.
- Use linear models to solve application problems.

The Point-Slope Form of the Equation of a Line

Point-Slope Form of the Equation of a Line

The **point-slope form** of the equation of the line that passes through the point (x_1, y_1) and has a slope of m is

$$y - y_1 = m(x - x_1).$$

Additional Example

Write an equation of the line that passes through the point $(-4, 1)$ and has slope $m = \frac{1}{2}$.

Answer:

$y = \frac{1}{2}x + 3$

EXAMPLE 1 Using Point-Slope Form

Write an equation of the line that passes through the point $(2, -3)$ and has slope $m = -2$.

SOLUTION

Use the point-slope form with $(x_1, y_1) = (2, -3)$ and $m = -2$.

$y - y_1 = m(x - x_1)$	Point-slope form
$y - (-3) = -2(x - 2)$	Substitute -3 for y_1, 2 for x_1, and -2 for m.
$y + 3 = -2x + 4$	Simplify.
$y = -2x + 1$	Subtract 3 from each side.

So, an equation of the line is $y = -2x + 1$. Note that this is the slope-intercept form of the equation. The graph of the line is shown at the left.

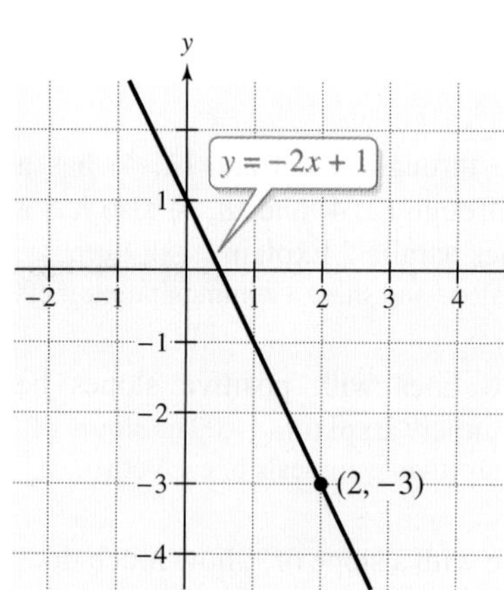

Exercises Within Reach ®

Solutions in English & Spanish and tutorial videos at AlgebraWithinReach.com

Using Point-Slope Form **In Exercises 1–10, use the point-slope form of the equation of a line to write an equation of the line that passes through the point and has the specified slope. Write the equation in slope-intercept form.**

1. $(0, 0)$
$m = -\frac{1}{2}$ $y = -\frac{1}{2}x$

2. $(0, 0)$
$m = \frac{1}{5}$ $y = \frac{1}{5}x$

3. $(0, -4)$
$m = 3$ $y = 3x - 4$

4. $(0, 5)$
$m = -3$ $y = -3x + 5$

5. $(0, 6)$
$m = -\frac{3}{4}$ $y = -\frac{3}{4}x + 6$

6. $(0, -8)$
$m = \frac{2}{3}$ $y = \frac{2}{3}x - 8$

7. $(-2, 8)$
$m = -2$ $y = -2x + 4$

8. $(4, -1)$
$m = 3$ $y = 3x - 13$

9. $(-4, -7)$
$m = \frac{5}{4}$ $y = \frac{5}{4}x - 2$

10. $(6, -8)$
$m = -\frac{2}{3}$ $y = -\frac{2}{3}x - 4$

EXAMPLE 2 An Equation of a Line Through Two Points

Write the general form of the equation of the line that passes through the points $(4, 2)$ and $(-2, 3)$.

SOLUTION

Let $(x_1, y_1) = (4, 2)$ and $(x_2, y_2) = (-2, 3)$. Then apply the formula for the slope of a line that passes through two points, as follows.

$$m = \frac{y_2 - y_1}{x_2 - x_1}$$

$$= \frac{3 - 2}{-2 - 4}$$

$$= -\frac{1}{6}$$

Now, using the point-slope form, you can find the equation of the line.

$$y - y_1 = m(x - x_1) \qquad \text{Point-slope form}$$

$$y - 2 = -\frac{1}{6}(x - 4) \qquad \text{Substitute 2 for } y_1, 4 \text{ for } x_1, \text{ and } -\tfrac{1}{6} \text{ for } m.$$

$$6(y - 2) = -(x - 4) \qquad \text{Multiply each side by 6.}$$

$$6y - 12 = -x + 4 \qquad \text{Distributive Property}$$

$$x + 6y - 12 = 4 \qquad \text{Add } x \text{ to each side.}$$

$$x + 6y - 16 = 0 \qquad \text{Subtract 4 from each side.}$$

The general form of the equation of the line is $x + 6y - 16 = 0$. The graph of the line is shown at the left.

Exercises Within Reach ®

Solutions in English & Spanish and tutorial videos at AlgebraWithinReach.com

Writing an Equation of a Line **In Exercises 11–16, write the general form of the equation of the line.**

11.

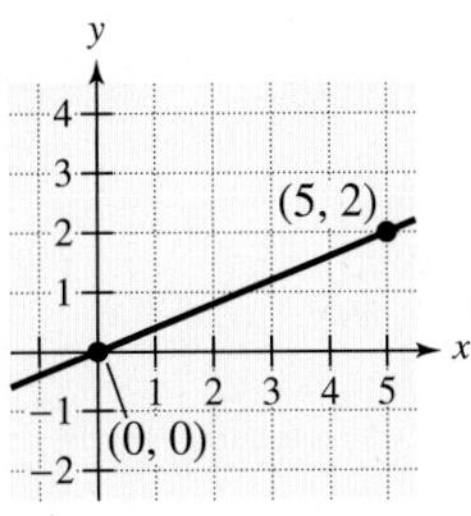

$2x - 5y = 0$

12.

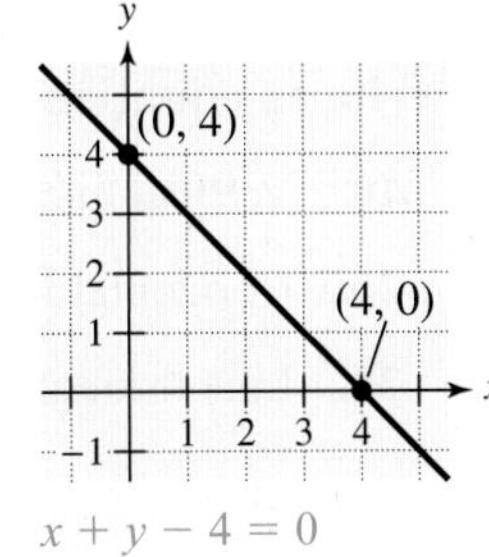

$x + y - 4 = 0$

13.

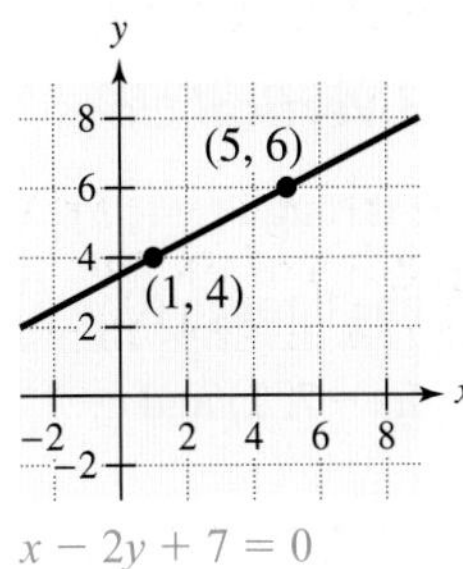

$x - 2y + 7 = 0$

14.

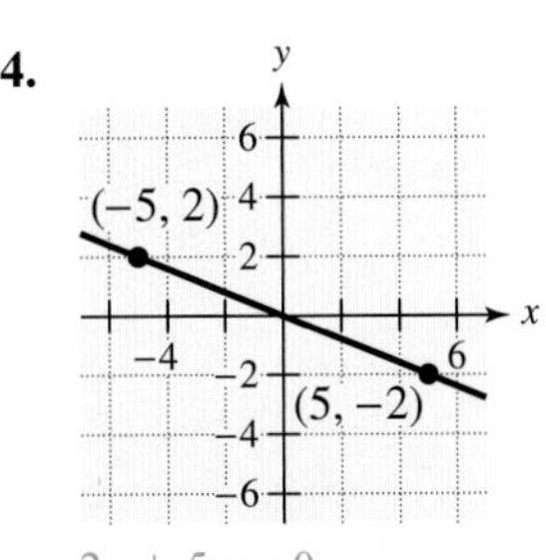

$2x + 5y = 0$

15.

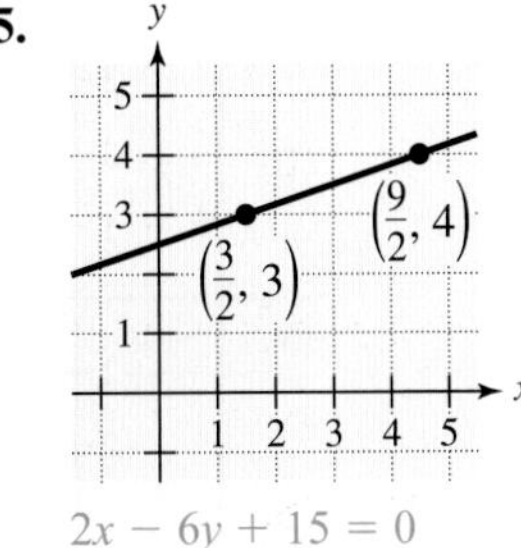

$2x - 6y + 15 = 0$

16.

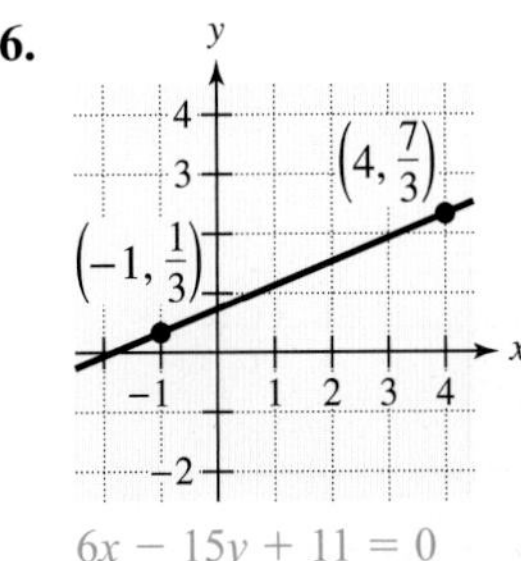

$6x - 15y + 11 = 0$

Other Equations of Lines

Horizontal lines have a slope of 0 and are of the form $y = b$.

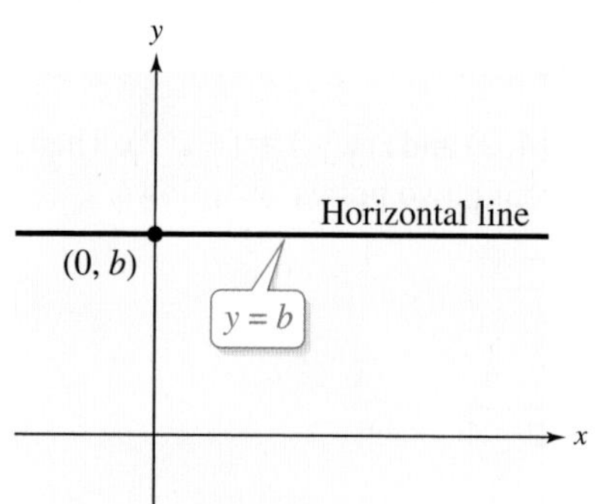

Vertical lines have an undefined slope and are of the form $x = a$.

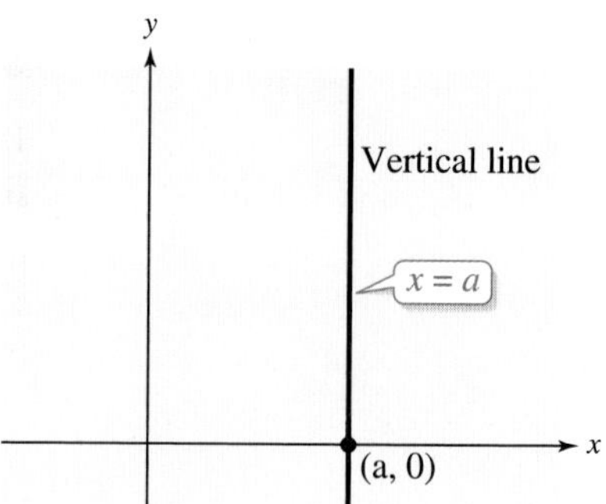

EXAMPLE 3 **Horizontal and Vertical Lines**

Write an equation for each line.

a. Vertical line through $(-2, 4)$

b. Horizontal line through $(0, 6)$

c. Line through $(-2, 3)$ and $(3, 3)$

d. Line through $(-1, 2)$ and $(-1, 3)$

SOLUTION

a. Because the line is vertical and passes through the point $(-2, 4)$, you know that every point on the line has an x-coordinate of -2. So, an equation of the line is $x = -2$.

b. Because the line is horizontal and passes through the point $(0, 6)$, you know that every point on the line has a y-coordinate of 6. So, an equation of the line is $y = 6$.

c. Because both points have the same y-coordinate, the line through $(-2, 3)$ and $(3, 3)$ is horizontal. So, its equation is $y = 3$.

d. Because both points have the same x-coordinate, the line through $(-1, 2)$ and $(-1, 3)$ is vertical. So, its equation is $x = -1$.

Exercises Within Reach ®

Solutions in English & Spanish and tutorial videos at AlgebraWithinReach.com

Horizontal and Vertical Lines **In Exercises 17–22, write an equation of the line.**

17. Vertical line through $(-1, 5)$ $x = -1$

18. Vertical line through $(2, -3)$ $x = 2$

19. Horizontal line through $(0, -5)$ $y = -5$

20. Horizontal line through $(-4, 6)$ $y = 6$

21. Line through $(-7, 2)$ and $(-7, -1)$ $x = -7$

22. Line through $(6, 4)$ and $(-9, 4)$ $y = 4$

Writing an Equation of a Line **In Exercises 23 and 24, write an equation of the line.**

23. $y = -2$

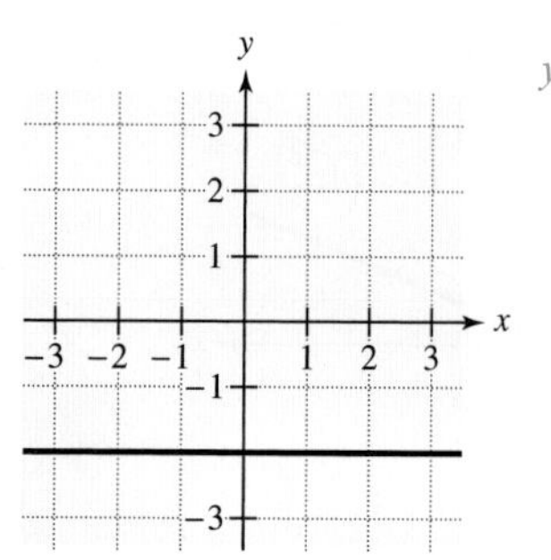

24. $x = -2$

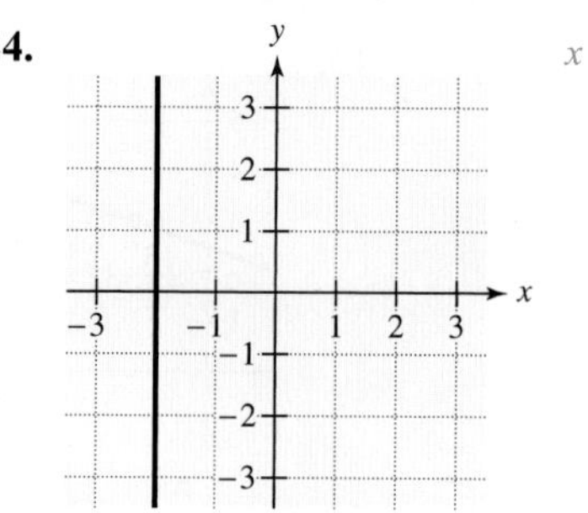

EXAMPLE 4 Parallel and Perpendicular Lines

Write equations of the lines that pass through the point $(3, -2)$ and are (a) parallel and (b) perpendicular to the line $x - 4y = 6$, as shown at the left.

SOLUTION

By writing the given line in slope-intercept form, $y = \frac{1}{4}x - \frac{3}{2}$, you can see that it has a slope of $\frac{1}{4}$. So, a line parallel to it must also have a slope of $\frac{1}{4}$ and a line perpendicular to it must have a slope of -4.

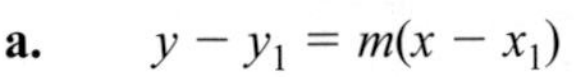

a. $y - y_1 = m(x - x_1)$ — Point-slope form

$y - (-2) = \frac{1}{4}(x - 3)$ — Substitute -2 for y_1, 3 for x_1, and $\frac{1}{4}$ for m.

$y = \frac{1}{4}x - \frac{11}{4}$ — Equation of parallel line

b. $y - y_1 = m(x - x_1)$ — Point-slope form

$y - (-2) = -4(x - 3)$ — Substitute -2 for y_1, 3 for x_1, and -4 for m.

$y = -4x + 10$ — Equation of perpendicular line

Study Tip

The slope-intercept form of the equation of a line is better suited for *sketching a line*. On the other hand, the point-slope form of the equation of a line is better suited for *creating the equation of a line*, given its slope and a point on the line.

Summary of Equations of Lines

1. Slope of a line through (x_1, y_1) and (x_2, y_2): $m = \dfrac{y_2 - y_1}{x_2 - x_1}$
2. General form of equation of line: $ax + by + c = 0$
3. Equation of vertical line: $x = a$
4. Equation of horizontal line: $y = b$
5. Slope-intercept form of equation of line: $y = mx + b$
6. Point-slope form of equation of line: $y - y_1 = m(x - x_1)$
7. Parallel lines (equal slopes): $m_1 = m_2$
8. Perpendicular lines (negative reciprocal slopes): $m_1 = -\dfrac{1}{m_2}$

Exercises Within Reach ®

Solutions in English & Spanish and tutorial videos at AlgebraWithinReach.com

Writing Equations of Lines **In Exercises 25–30, write equations of the lines that pass through the point and are (a) parallel and (b) perpendicular to the given line.**

25. $(2, 1)$
$6x - 2y = 3$ (a) $y = 3x - 5$ (b) $y = -\frac{1}{3}x + \frac{5}{3}$

26. $(-3, 4)$
$x + 6y = 12$ (a) $y = -\frac{1}{6}x + \frac{7}{2}$ (b) $y = 6x + 22$

27. $(-5, 4)$
$5x + 4y = 24$ (a) $y = -\frac{5}{4}x - \frac{9}{4}$ (b) $y = \frac{4}{5}x + 8$

28. $(6, -4)$
$3x + 10y = 24$ (a) $y = -\frac{3}{10}x - \frac{11}{5}$ (b) $y = \frac{10}{3}x - 24$

29. $(5, -3)$
$4x - y - 3 = 0$ (a) $y = 4x - 23$ (b) $y = -\frac{1}{4}x - \frac{7}{4}$

30. $(-5, -10)$
$2x + 5y - 12 = 0$ (a) $y = -\frac{2}{5}x - 12$ (b) $y = \frac{5}{2}x + \frac{5}{2}$

Applications

When you are given two or more points that determine a line and you use the line to estimate a point beyond the given points, you are **extrapolating**. When you use the line to estimate a point between the given points, you are **interpolating**.

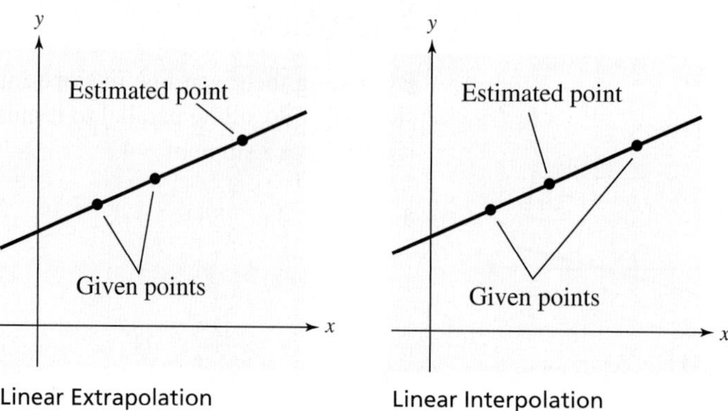

Linear Extrapolation   Linear Interpolation

Application **EXAMPLE 5** **Using Linear Extrapolation and Interpolation**

When soft drinks sold for \$0.80 per cup at football games, approximately 6000 cups were sold. When the price was raised to \$1.00 per cup, the demand dropped to 4000 cups. Assume that the relationship between the price p and demand d is linear.

a. Write an equation of the line giving the demand d in terms of the price p.

b. Use the equation in part (a) to predict the number of cups of soft drinks sold when the price is raised to \$1.10.

c. Use the equation in part (a) to estimate the number of cups of soft drinks sold when the price is \$0.90.

SOLUTION

a. Using the point-slope form of the equation of a line, you can write

$$d - d_1 = \frac{d_2 - d_1}{p_2 - p_1}(p - p_1)$$

$$d - 4000 = \frac{6000 - 4000}{0.8 - 1}(p - 1)$$

$$d = -10{,}000p + 14{,}000.$$

b. When $p = \$1.10$, the demand is $-10{,}000(1.10) + 14{,}000 = 3000$ cups.

c. When $p = \$0.90$, the demand is $-10{,}000(0.90) + 14{,}000 = 5000$ cups.

Exercises Within Reach ®

Solutions in English & Spanish and tutorial videos at AlgebraWithinReach.com

31. ***Education*** A small college had an enrollment of 1500 students in 2005. During the next 5 years, the enrollment increased by approximately 60 students per year. Write an equation of the line giving the enrollment N in terms of the year t. (Let $t = 5$ correspond to 2005.)
$N = 60t + 1200$

32. ***Linear Extrapolation and Interpolation*** Use the equation in Exercise 31 to (a) predict the enrollment in 2015 and (b) estimate the enrollment in 2008.
(a) 2100 students (b) 1680 students

Application **EXAMPLE 6** **Using Linear Extrapolation**

Texas Instruments had total sales of \$12.5 billion in 2008 and \$14.0 billion in 2010. (*Source:* Texas Instruments, Inc.)

a. Using only this information, write a linear equation that models the sales (in billions of dollars) in terms of the year.

b. Interpret the meaning of the slope in the context of the problem.

c. Predict the sales for 2015.

SOLUTION

a. Let $t = 8$ represent 2008. Then the two given values are represented by the data points (8, 12.5) and (10, 14.0). The slope of the line through these points is

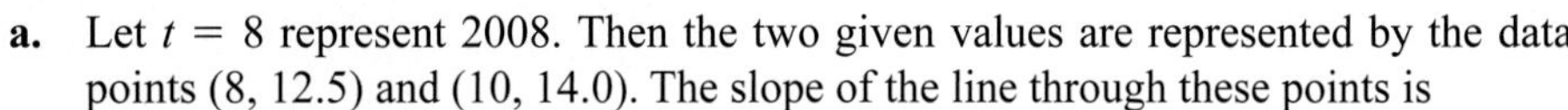

$$m = \frac{y_2 - y_1}{t_2 - t_1}$$
$$= \frac{14.0 - 12.5}{10 - 8}$$
$$= 0.75$$

Using the point-slope form, you can find an equation that relates the sales y and the year t to be

$$y - y_1 = m(t - t_1) \qquad \text{Point-slope form}$$
$$y - 12.5 = 0.75(t - 8) \qquad \text{Substitute for } y_1, m, \text{ and } t_1.$$
$$y - 12.5 = 0.75t - 6 \qquad \text{Distributive Property}$$
$$y = 0.75t + 6.5. \qquad \text{Write in slope-intercept form.}$$

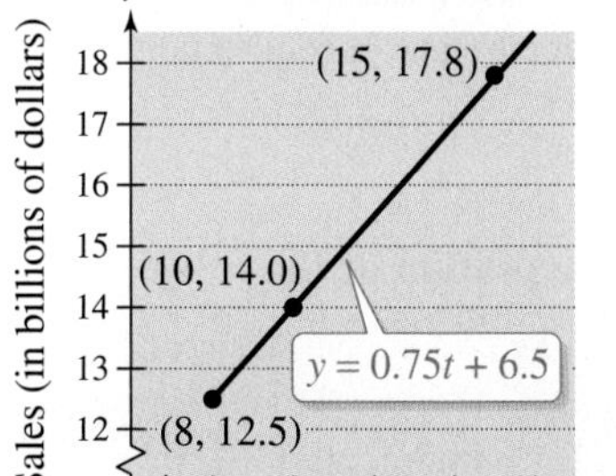

b. The slope of the equation in part (a) indicates that the total sales for Texas Instruments increased by \$0.75 billion each year.

c. Using the equation in part (a), you can predict the sales for 2015 ($t = 15$) to be

$$y = 0.75t + 6.5 \qquad \text{Equation in part (a)}$$
$$y = 0.75(15) + 6.5 \qquad \text{Substitute 15 for } t.$$
$$y = 17.75 \qquad \text{Simplify.}$$

So, the predicted sales for 2015 are \$17.8 billion. The graph of this equation is shown at the left.

Exercises Within Reach®

Solutions in English & Spanish and tutorial videos at AlgebraWithinReach.com

33. ***Cost*** The cost C (in dollars) of producing x units of a product is shown in the table. Write a linear equation that models the data. Predict the cost of producing 400 units. $C = 20x + 5000$; \$13,000

x	0	50	100	150	200
C	5000	6000	7000	8000	9000

34. ***Temperature Conversion*** The relationship between the Fahrenheit F and Celsius C temperature scales is shown in the table. Write a linear equation that models the data. Estimate the Fahrenheit temperature when the Celsius temperature is 18 degrees. $F = \frac{9}{5}C + 32$; $64.4°F$

C	0	5	10	15	20
F	32	41	50	59	68

Concept Summary: Writing Equations of Lines

What

You can use the **point-slope form** to write an equation of a line passing through any two points.

EXAMPLE

Write an equation of the line passing through the points (2, 3) and (3, 5).

How

Here are the steps to write an equation of the line.

1. Find the slope of the line.

$$m = \frac{y_2 - y_1}{x_2 - x_1} = \frac{5 - 3}{3 - 2} = 2$$

2. Use the point-slope form.

$$y - y_1 = m(x - x_1)$$
$$y - 3 = 2(x - 2)$$
$$y - 3 = 2x - 4$$
$$y = 2x - 1$$

Why

When you are given

- the slope of a line and a point on the line, or
- any two points on a line,

you can use point-slope form to write an equation of the line.

Exercises Within Reach®

Worked-out solutions to odd-numbered exercises at AlgebraWithinReach.com

Concept Summary Check

35. ***Reasoning*** The equation $y - 1 = 4(x - 1)$ is written in point-slope form. What is the slope? By observation, what point does the equation of the line pass through? The slope is 4 and a point is (1, 1).

36. ***Vocabulary*** Is the general form of an equation of a line equivalent to the point-slope form? Explain. Yes. You can rewrite one form to produce the other.

37. ***Logic*** A horizontal line passes through the point (−4, 5). Do you have enough information to write an equation of the line? If so, what is the equation of the line? Yes. $y = 5$

38. ***Writing*** A line passes through the point (3, 2). Do you have enough information to write an equation of the line? Explain. No. To write the equation of the line through (3, 2), you would need either the slope or another point through which the line passes.

Extra Practice

Writing an Equation of a Line **In Exercises 39–42, write the intercept form of the equation of the line with intercepts $(a, 0)$ and $(0, b)$. The equation is given by $\frac{x}{a} + \frac{y}{b} = 1, a \neq 0, b \neq 0$.** See Additional Answers.

39. x-intercept: (3, 0)
y-intercept: (0, 2)

40. x-intercept: (−6, 0)
y-intercept: (0, 2)

41. x-intercept: $\left(-\frac{5}{6}, 0\right)$
y-intercept: $\left(0, -\frac{7}{3}\right)$

42. x-intercept: $\left(-\frac{8}{3}, 0\right)$
y-intercept: (0, −4)

43. ***Sales*** The total sales for a new camera equipment store were $200,000 for the second year and $500,000 for the fifth year. Write a linear equation that models the data. Predict the total sales for the sixth year.

$S = 100{,}000t$; $600,000

44. ***Sales*** The total sales for a new sportswear store were $150,000 for the third year and $250,000 for the fifth year. Write a linear equation that models the data. Predict the total sales for the sixth year.

$S = 50{,}000t$; $300,000

45. ***Straight-Line Depreciation*** A small business purchases a photocopier for $7400. After 4 years, its depreciated value will be $1500.

(a) Assuming straight-line depreciation, write an equation of the line giving the value V of the copier in terms of time t in years. $V = 7400 - 1475t$

(b) Use the equation in part (a) to find the value of the copier after 2 years. $4450

46. ***Straight-Line Depreciation*** A business purchases a van for $27,500. After 5 years, its depreciated value will be $12,000.

(a) Assuming straight-line depreciation, write an equation of the line giving the value V of the van in terms of time t in years. $V = 27{,}500 - 3100t$

(b) Use the equation in part (a) to find the value of the van after 2 years. $21,300

47. ***Depth Markers*** A swimming pool is 40 feet long, 20 feet wide, 4 feet deep at the shallow end, and 9 feet deep at the deep end. Position the side of the pool on a rectangular coordinate system as shown in the figure and write an equation of the line representing the edge of the inclined bottom of the pool. Use this equation to determine the distances from the deep end at which markers must be placed to indicate each one-foot change in the depth of the pool. $x - 8y = 0$
See Additional Answers.

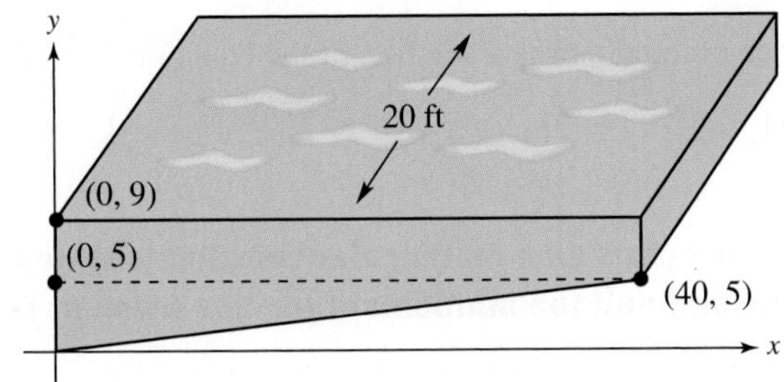

48. ***Carpentry*** A carpenter uses a wedge-shaped block of wood to support heavy objects. The block of wood is 12 inches long, 2 inches wide, 6 inches high at the tall end, and 2 inches high at the short end. Position the side of the block on a rectangular coordinate system as shown in the figure and write an equation of the line representing the edge of the slanted top of the block. Use this equation to determine the distances from the tall end at which marks must be made to indicate each one-inch change in the height of the block. $x + 3y - 18 = 0$
See Additional Answers.

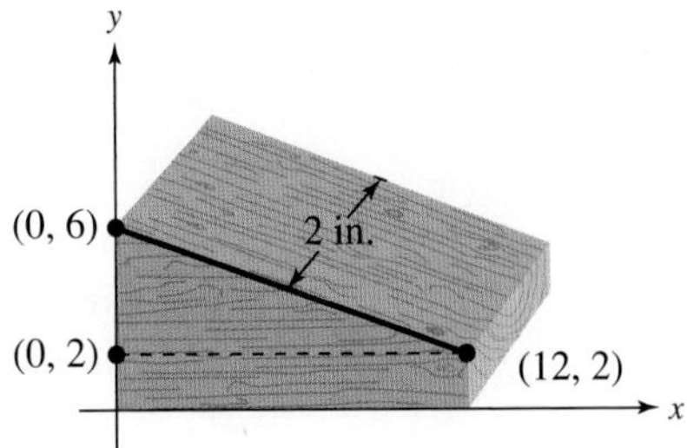

Explaining Concepts

49. ***Vocabulary*** Write, from memory, the point-slope form, the slope-intercept form, and the general form of an equation of a line.

Point-slope form: $y - y_1 = m(x - x_1)$
Slope-intercept form: $y = mx + b$
General form: $ax + by + c = 0$

50. ***Reasoning*** In the equation

$y = 3x + 5$

what does the 3 represent? What does the 5 represent?
The slope is 3 and the y-coordinate of the y-intercept is 5.

51. ***Think About It*** In the equation of a vertical line, the variable y is missing. Explain why.
Any point on a vertical line is independent of y.

52. ***Writing*** Can any pair of points on a line be used to determine the equation of the line? Explain.
Yes. When different pairs of points are selected, the change in y and the change in x are the lengths of the sides of similar triangles. Corresponding sides of similar triangles are proportional. After finding the slope, you can find the equation of the line by using the point-slope form of the equation of a line.

Cumulative Review

In Exercises 53–56, sketch the graph of the equation.
See Additional Answers.

53. $y = 4x$

54. $y = 2x - 3$

55. $y = x^2 - 1$

56. $y = -|x + 1|$

In Exercises 57–64, solve for a so that the line through the points has the given slope.

57. $(1, 2), (a, 4)$ 2
$m = 2$

58. $(0, 1), (2, a)$ 7
$m = 3$

59. $(-4, a), (-2, 3)$
$m = \frac{1}{2}$
2

60. $(a, 3), (6, 3)$
$m = 0$
All real numbers

61. $(5, 0), (0, a)$ 3
$m = -\frac{3}{5}$

62. $(0, a), (4, -6)$ $-\frac{10}{3}$
$m = -\frac{2}{3}$

63. $(-7, -2), (-1, a)$ -8
$m = -1$

64. $(9, -a), (-3, 4)$ $\frac{28}{5}$
$m = -\frac{4}{5}$

Mid-Chapter Quiz: Sections 3.1–3.4

Solutions in English & Spanish and tutorial videos at AlgebraWithinReach.com

Take this quiz as you would take a quiz in class. After you are done, check your work against the answers in the back of the book.

1. Determine the quadrant(s) in which the point $(x, 4)$ is located if x is a real number. Explain your reasoning. Quadrant I or II

2. Determine whether each ordered pair is a solution of the equation $4x - 3y = 10$.

   (a) $(2, 1)$ Not a solution (b) $(1, -2)$ Solution (c) $(2.5, 0)$ Solution (d) $\left(2, -\frac{2}{3}\right)$ Solution

In Exercises 3 and 4, plot the points on a rectangular coordinate system, find the distance between them, and find the midpoint of the line segment joining the two points. See Additional Answers.

3. $(-1, 5), (3, 2)$ Distance: 5 Midpoint: $\left(1, \frac{7}{2}\right)$

4. $(-4, 3), (6, -7)$ Distance: $10\sqrt{2}$ Midpoint: $(1, -2)$

In Exercises 5–7, sketch the graph of the equation and show the coordinates of three solution points (including x- and y-intercepts). See Additional Answers.

5. $3x + y = 6$
6. $y = 6x - x^2$
7. $y = |x - 2| - 3$

In Exercises 8–10, find the slope of the line that passes through the points. State whether the line rises, falls, is horizontal, or is vertical.

8. $(-3, 8), (7, 8)$ $m = 0$; Horizontal
9. $(3, 0), (6, 5)$ $m = \frac{5}{3}$; Rises
10. $(-2, 7), (4, -1)$ $m = -\frac{4}{3}$; Falls

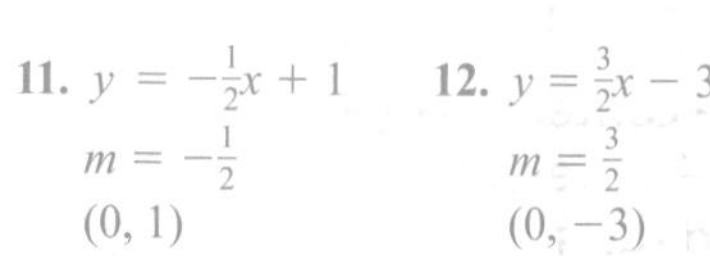

In Exercises 11 and 12, write the equation of the line in slope-intercept form, and then use the slope and y-intercept to sketch the graph of the equation. See Additional Answers.

11. $3x + 6y = 6$
12. $6x - 4y = 12$

In Exercises 13 and 14, determine whether the lines are parallel, perpendicular, or neither.

13. $y = 3x + 2,\ y = -\frac{1}{3}x - 4$ Perpendicular

14. L_1: $(4, 3), (-2, -9)$; L_2: $(0, -5), (5, 5)$ Parallel

15. Write the general form of the equation of the line that passes through the point $(6, -1)$ and has a slope of $\frac{1}{2}$. $x - 2y - 8 = 0$

Application

16. Your company purchases a new printing press for \$124,000. For tax purposes, the printing press will be depreciated over a 10-year period. At the end of 10 years, the value of the printing press is expected to be \$4000. Find an equation that relates the value of the printing press to the number of years since its purchase. Then sketch the graph of the equation. $V = 124{,}000 - 12{,}000t,\ 0 \le t \le 10$
See Additional Answers.

Study Skills in Action

Reading Your Textbook Like a Manual

Many students avoid opening their textbooks for the same reason many people avoid opening their checkbooks—anxiety and frustration. The truth? Not opening your math textbook will cause more anxiety and frustration! Your textbook is a manual designed to help you master skills and understand and remember concepts. It contains many features and resources that can help you be successful in your course.

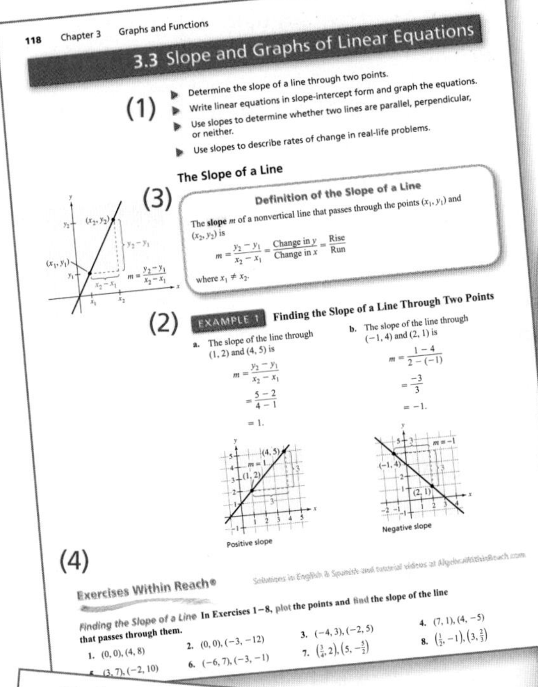

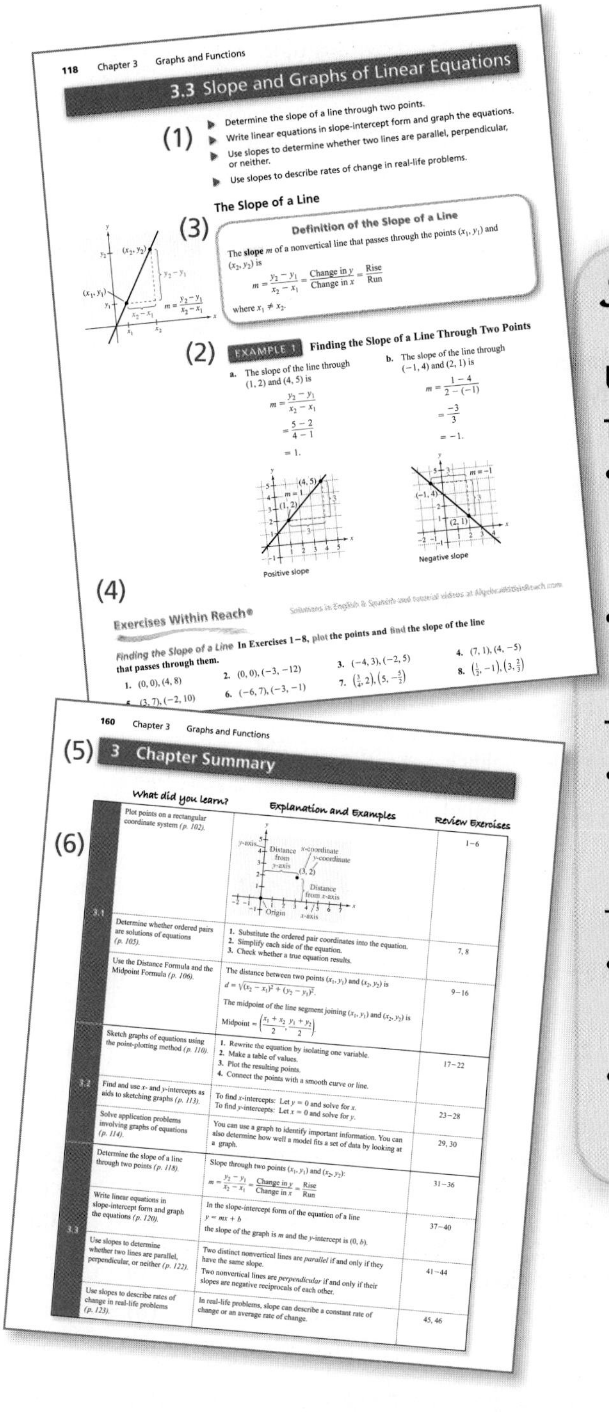

118 Chapter 3 Graphs and Functions

3.3 Slope and Graphs of Linear Equations

(1)
- Determine the slope of a line through two points.
- Write linear equations in slope-intercept form and graph the equations.
- Use slopes to determine whether two lines are parallel, perpendicular, or neither.
- Use slopes to describe rates of change in real-life problems.

The Slope of a Line

(3) **Definition of the Slope of a Line**

The **slope** m of a nonvertical line that passes through the points (x_1, y_1) and (x_2, y_2) is

$$m = \frac{y_2 - y_1}{x_2 - x_1} = \frac{\text{Change in } y}{\text{Change in } x} = \frac{\text{Rise}}{\text{Run}}$$

where $x_1 \neq x_2$.

(2) EXAMPLE 1 **Finding the Slope of a Line Through Two Points**

a. The slope of the line through (1, 2) and (4, 5) is

$$m = \frac{y_2 - y_1}{x_2 - x_1} = \frac{5 - 2}{4 - 1} = 1.$$

b. The slope of the line through (−1, 4) and (2, 1) is

$$m = \frac{1 - 4}{2 - (-1)} = \frac{-3}{3} = -1.$$

Positive slope

Negative slope

(4)

Exercises Within Reach®

Finding the Slope of a Line **In Exercises 1–8, plot the points and find the slope of the line that passes through them.**

1. (0, 0), (4, 8)
2. (0, 0), (−3, −12)
3. (−4, 3), (−2, 5)
4. (7, 1), (4, −5)
5. (3, 7), (−2, 10)
6. (−6, 7), (−3, −1)
7. $(\frac{3}{4}, 2), (5, -\frac{5}{2})$
8. $(\frac{1}{2}, -1), (3, \frac{2}{3})$

160 Chapter 3 Graphs and Functions

(5) **3 Chapter Summary**

(6)

	What did you learn?	Explanation and Examples	Review Exercises
3.1	Plot points on a rectangular coordinate system *(p. 102)*.	[graph: y-axis, Distance from y-axis, x-coordinate, y-coordinate, (3, 2), Distance from x-axis, Origin, x-axis]	1–6
	Determine whether ordered pairs are solutions of equations *(p. 105)*.	1. Substitute the ordered pair coordinates into the equation. 2. Simplify each side of the equation. 3. Check whether a true equation results.	7, 8
	Use the Distance Formula and the Midpoint Formula *(p. 106)*.	The distance between two points (x_1, y_1) and (x_2, y_2) is $d = \sqrt{(x_2 - x_1)^2 + (y_2 - y_1)^2}$. The midpoint of the line segment joining (x_1, y_1) and (x_2, y_2) is Midpoint $= \left(\frac{x_1 + x_2}{2}, \frac{y_1 + y_2}{2}\right)$.	9–16
3.2	Sketch graphs of equations using the point-plotting method *(p. 110)*.	1. Rewrite the equation by isolating one variable. 2. Make a table of values. 3. Plot the resulting points. 4. Connect the points with a smooth curve or line.	17–22
	Find and use x- and y-intercepts as aids to sketching graphs *(p. 113)*.	To find x-intercepts: Let $y = 0$ and solve for x. To find y-intercepts: Let $x = 0$ and solve for y.	23–28
	Solve application problems involving graphs of equations *(p. 114)*.	You can use a graph to identify important information. You can also determine how well a model fits a set of data by looking at a graph.	29, 30
3.3	Determine the slope of a line through two points *(p. 118)*.	Slope through two points (x_1, y_1) and (x_2, y_2): $m = \frac{y_2 - y_1}{x_2 - x_1} = \frac{\text{Change in } y}{\text{Change in } x} = \frac{\text{Rise}}{\text{Run}}$	31–36
	Write linear equations in slope-intercept form and graph the equations *(p. 120)*.	In the slope-intercept form of the equation of a line $y = mx + b$ the slope of the graph is m and the y-intercept is $(0, b)$.	37–40
	Use slopes to determine whether two lines are parallel, perpendicular, or neither *(p. 122)*.	Two distinct nonvertical lines are *parallel* if and only if they have the same slope. Two nonvertical lines are *perpendicular* if and only if their slopes are negative reciprocals of each other.	41–44
	Use slopes to describe rates of change in real-life problems *(p. 123)*.	In real-life problems, slope can describe a constant rate of change or an average rate of change.	45, 46

Smart Study Strategy

Use the Features of Your Textbook

To review what you learned in a previous class:

- Read the list of skills you should learn (1) at the beginning of the section. If you cannot remember how to perform a skill, review the appropriate example (2) in the section.
- Read and understand the contents of all tinted concept boxes (3)—these contain important definitions and rules.

To prepare for homework:

- Complete a few of the exercises (4) following each example. If you have difficulty with any of these, reread the example or seek help from a peer or instructor.

To review for quizzes and tests:

- Make use of the Chapter Summary (5). Check off the concepts (6) you know, and review those you do not know.
- Complete the Review Exercises. Then take the Mid-Chapter Quiz, Chapter Test, or Cumulative Test, as appropriate.

3.5 Graphs of Linear Inequalities

- Verify solutions of linear inequalities in two variables.
- Sketch graphs of linear inequalities in two variables.
- Use linear inequalities to model and solve real-life problems.

Linear Inequalities in Two Variables

A **linear inequality in two variables,** x and y, is an inequality that can be written in one of the forms below (where a and b are not both zero).

$$ax + by < c, \qquad ax + by > c, \qquad ax + by \le c, \qquad ax + by \ge c$$

An ordered pair (x_1, y_1) is a **solution** of a linear inequality in x and y if the inequality is true when x_1 and y_1 are substituted for x and y, respectively.

EXAMPLE 1 Verifying Solutions of a Linear Inequality

Determine whether each ordered pair is a solution of $2x - 3y \ge -2$.

a. $(0, 0)$ **b.** $(0, 1)$

SOLUTION

To determine whether an ordered pair (x_1, y_1) is a solution of the inequality, substitute x_1 and y_1 into the inequality.

a. $2x - 3y \ge -2$ Write original inequality.

$2(0) - 3(0) \overset{?}{\ge} -2$ Substitute 0 for x and 0 for y.

$0 \ge -2$ Inequality is satisfied. ✓

Because the inequality is satisfied, the point $(0, 0)$ *is* a solution.

b. $2x - 3y \ge -2$ Write original inequality.

$2(0) - 3(1) \overset{?}{\ge} -2$ Substitute 0 for x and 1 for y.

$-3 \not\ge -2$ Inequality is not satisfied. ✗

Because the inequality is not satisfied, the point $(0, 1)$ *is not* a solution.

Exercises Within Reach ®

Solutions in English & Spanish and tutorial videos at AlgebraWithinReach.com

Verifying Solutions of a Linear Inequality **In Exercises 1–6, determine whether each ordered pair is a solution of the inequality.**

1. $x - 2y < 4$ (a) $(0, 0)$ (b) $(2, -1)$ (c) $(3, 4)$ (d) $(5, 1)$

(a) Solution (b) Not a solution (c) Solution (d) Solution

2. $x + y < 3$ (a) $(0, 6)$ (b) $(4, 0)$ (c) $(0, -2)$ (d) $(1, 1)$

(a) Not a solution (b) Not a solution (c) Solution (d) Solution

3. $3x + y \ge 10$ (a) $(1, 3)$ (b) $(-3, 1)$ (c) $(3, 1)$ (d) $(2, 15)$

(a) Not a solution (b) Not a solution (c) Solution (d) Solution

4. $-3x + 5y \ge 6$ (a) $(2, 8)$ (b) $(-10, -3)$ (c) $(0, 0)$ (d) $(3, 3)$

(a) Solution (b) Solution (c) Not a solution (d) Solution

5. $y > 0.2x - 1$ (a) $(0, 2)$ (b) $(6, 0)$ (c) $(4, -1)$ (d) $(-2, 7)$

(a) Solution (b) Not a solution (c) Not a solution (d) Solution

6. $y < -3.5x + 7$ (a) $(1, 5)$ (b) $(5, -1)$ (c) $(-1, 4)$ (d) $\left(0, \frac{4}{3}\right)$

(a) Not a solution (b) Not a solution (c) Solution (d) Solution

Study Tip

The graph of the corresponding equation is a dashed line for an inequality involving $<$ or $>$ because the points on the line are *not* solutions of the inequality. It is a solid line for an inequality involving $\leq$ or $\geq$ because the points on the line *are* solutions of the inequality. In either case, to test a half-plane you should choose a test point that is *not* on the line of the corresponding equation.

The Graph of a Linear Inequality in Two Variables

The **graph of a linear inequality** is the collection of all solution points of the inequality.

Sketching the Graph of a Linear Inequality in Two Variables

1. Replace the inequality symbol by an equal sign, and sketch the graph of the resulting equation. (Use a dashed line for $<$ or $>$, and a solid line for $\leq$ or $\geq$.)
2. Test one point in each of the half-planes formed by the graph in Step 1.
   a. If the point satisfies the inequality, shade the entire half-plane to denote that every point in the region satisfies the inequality.
   b. If the point does not satisfy the inequality, then shade the other half-plane.

EXAMPLE 2 Sketching the Graphs of Linear Inequalities

Sketch the graph of each linear inequality.

a. $x \geq -3$

b. $y < 4$

SOLUTION

a. The graph of the corresponding equation $x = -3$ is a vertical line. The line is solid because the inequality symbol is $\geq$. By testing the point $(0, 0)$, you can see that the points satisfying the inequality are those in the half-plane to the right of the line $x = -3$. The graph is shown below on the left.

b. The graph of the corresponding equation $y = 4$ is a horizontal line. The line is dashed because the inequality symbol is $<$. By testing the point $(0, 0)$, you can see that the points satisfying the inequality are those in the half-plane below the line $y = 4$. The graph is shown below on the right.

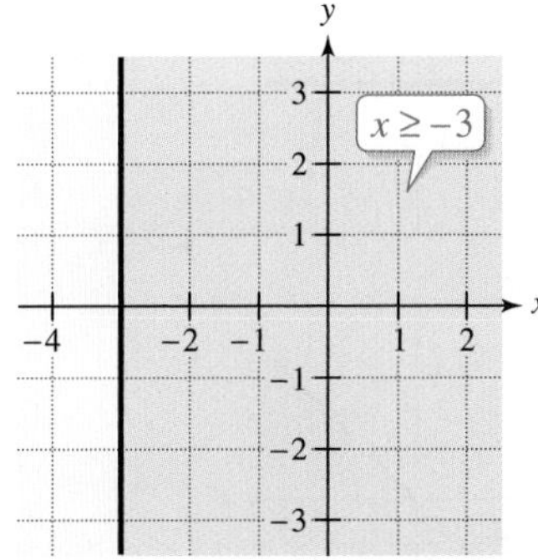

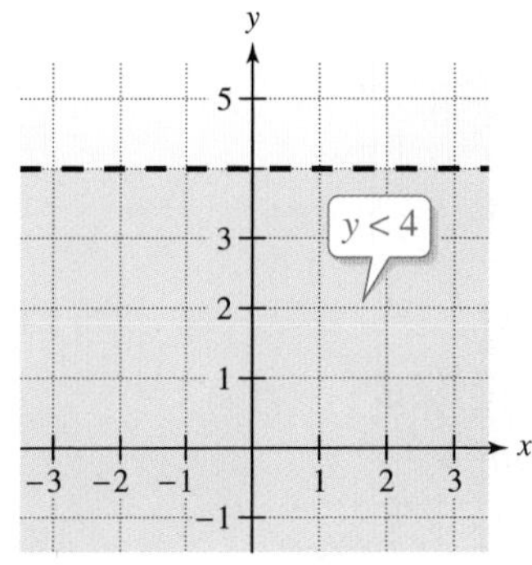

Exercises Within Reach®

Solutions in English & Spanish and tutorial videos at AlgebraWithinReach.com

Sketching the Graph of a Linear Inequality **In Exercises 7–12, sketch the graph of the linear inequality.** See Additional Answers.

7. $x \geq 6$

8. $x < -3$

9. $y < 5$

10. $y > 2$

11. $y > \frac{1}{2}x$

12. $y \leq 2x$

Study Tip

A convenient test point for determining which half-plane contains solutions to an inequality is the origin (0, 0). In Example 3, when you substitute 0 for x and 0 for y, you can easily see that (0, 0) does not satisfy the inequality.

$$x + y > 3$$
$$0 + 0 \not> 3$$

Remember that the origin cannot be used as a test point when it lies on the graph of the corresponding equation.

EXAMPLE 3 Sketching the Graph of a Linear Inequality

Sketch the graph of the linear inequality $x + y > 3$.

SOLUTION

The graph of the corresponding equation $x + y = 3$ is a line. To begin, find the x-intercept by letting $y = 0$ and solving for x.

$x + 0 = 3 \Rightarrow x = 3$ Substitute 0 for y and solve for x.

Find the y-intercept by letting $x = 0$ and solving for y.

$0 + y = 3 \Rightarrow y = 3$ Substitute 0 for x and solve for y.

So, the graph has an x-intercept at the point (3, 0) and a y-intercept at the point (0, 3). Plot these points and connect them with a dashed line. Because the origin (0, 0) does not satisfy the inequality, the graph consists of the half-plane lying above the line, as shown.

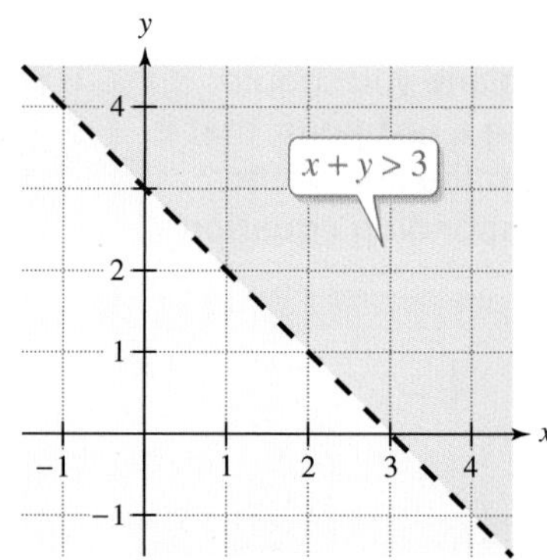

Exercises Within Reach ®

Solutions in English & Spanish and tutorial videos at AlgebraWithinReach.com

Matching **In Exercises 13–18, match the inequality with its graph.**

(a)

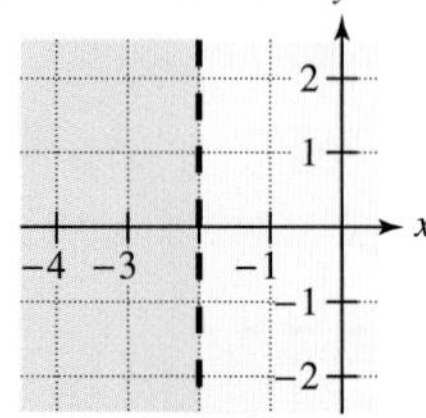

(b)

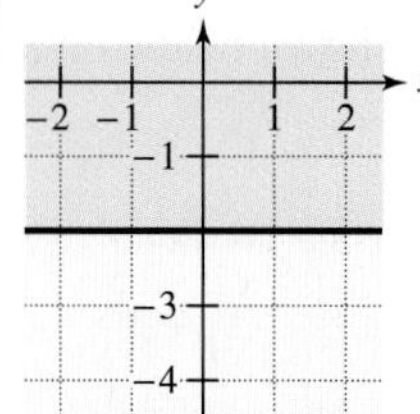

(c)

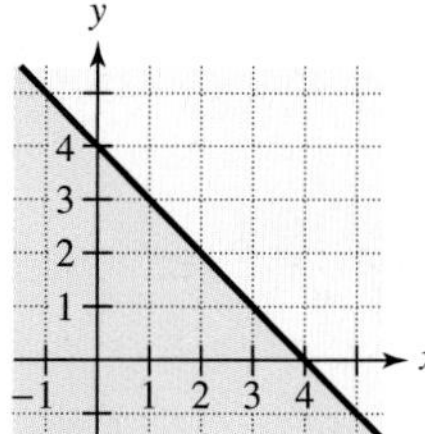

(d)

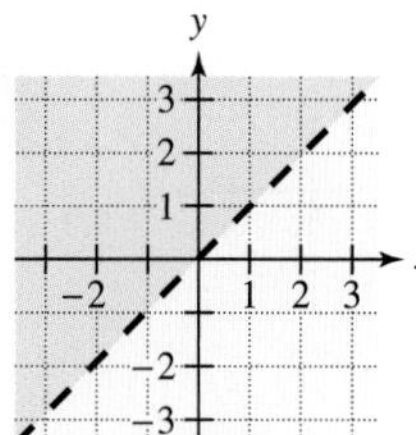

(e)

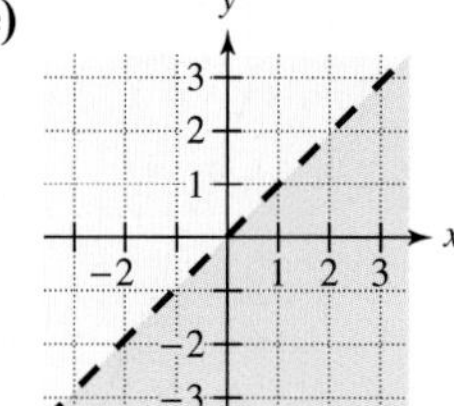

(f)

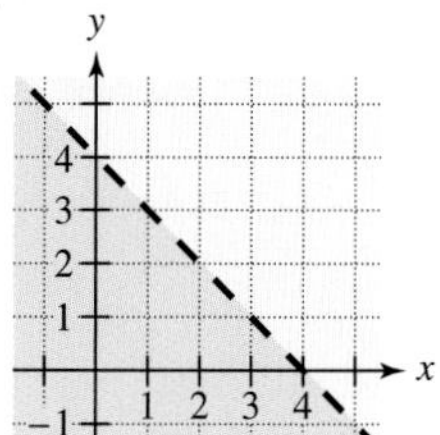

13. $y \geq -2$ b

14. $x < -2$ a

15. $x - y < 0$ d

16. $x - y > 0$ e

17. $x + y < 4$ f

18. $x + y \leq 4$ c

Sketching the Graph of a Linear Inequality **In Exercises 19–24, sketch the graph of the linear inequality.** See Additional Answers.

19. $y \geq 3 - x$

20. $y > x + 6$

21. $y \leq x + 2$

22. $y \leq 1 - x$

23. $x + y \geq 4$

24. $x + y \leq 5$

EXAMPLE 4 Sketching the Graph of a Linear Inequality

Sketch the graph of the linear inequality $2x + y \leq 2$.

SOLUTION

The graph of the corresponding equation $2x + y = 2$ is a line. To begin, find the x-intercept by letting $y = 0$ and solving for x.

$2x + 0 = 2 \Rightarrow x = 1$ Substitute 0 for y and solve for x.

Find the y-intercept by letting $x = 0$ and solving for y.

$2(0) + y = 2 \Rightarrow y = 2$ Substitute 0 for x and solve for y.

So, the graph has an x-intercept at the point $(1, 0)$ and a y-intercept at the point $(0, 2)$. Plot these points and connect them with a solid line. Because the origin $(0, 0)$ satisfies the inequality, the graph consists of the half-plane lying on or below the line, as shown below.

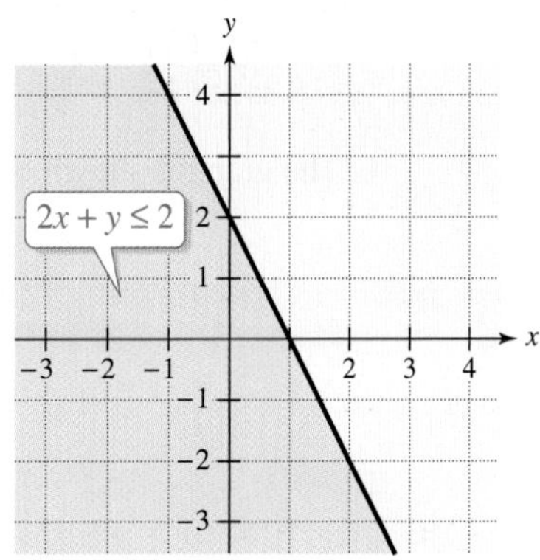

Exercises Within Reach®

Solutions in English & Spanish and tutorial videos at AlgebraWithinReach.com

Shading a Half-Plane **In Exercises 25 and 26, complete the graph of the inequality by shading the correct half-plane.** See Additional Answers.

25. $3x - y > 2$

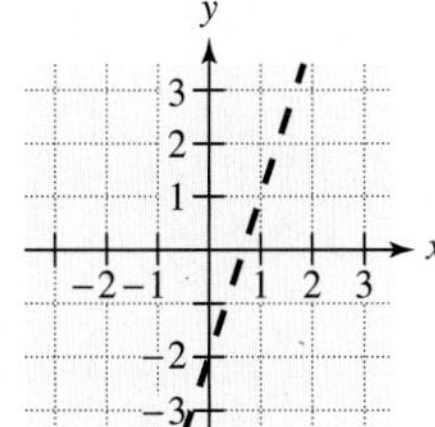

26. $2x + y \leq 4$

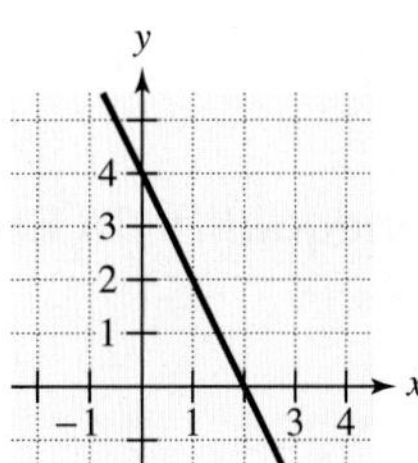

Sketching the Graph of a Linear Inequality **In Exercises 27–34, sketch the graph of the linear inequality.** See Additional Answers.

27. $2x + y \geq 4$

28. $3x + y \leq 9$

29. $-x + 2y < 4$

30. $x + 3y < 9$

31. $\frac{1}{2}x + y > 6$

32. $\frac{2}{3}x + y > 3$

33. $x - 4y \leq 8$

34. $x - 2y \geq 6$

Study Tip

For a linear inequality in two variables, you can sometimes simplify the graphing procedure by writing the inequality in slope-intercept form, as shown in Example 5.

EXAMPLE 5 Sketching the Graph of a Linear Inequality

Use the slope-intercept form of a linear equation as an aid in sketching the graph of the inequality

$$2x - 3y \leq 15.$$

SOLUTION

To begin, rewrite the inequality in slope-intercept form.

$2x - 3y \leq 15$	Write original inequality.
$-3y \leq -2x + 15$	Subtract $2x$ from each side.
$y \geq \frac{2}{3}x - 5$	Divide each side by -3 and reverse the inequality symbol.

From this form, you can conclude that the solution is the half-plane lying *on or above* the line

$$y = \frac{2}{3}x - 5.$$

The graph is shown below. To verify the solution, test any point in the shaded region.

Additional Example

Sketch the graph of the linear inequality.

$x - 3y \leq 6$

Answer:

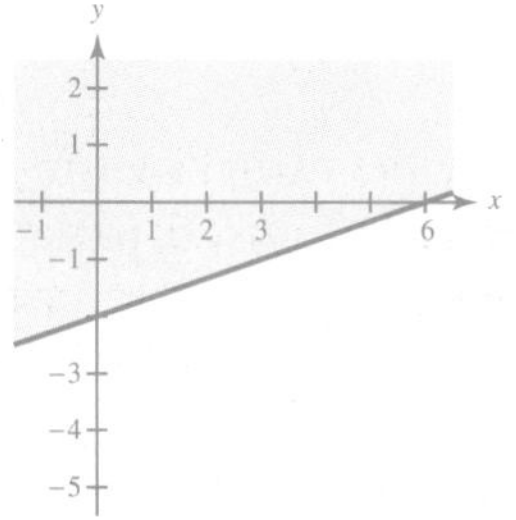

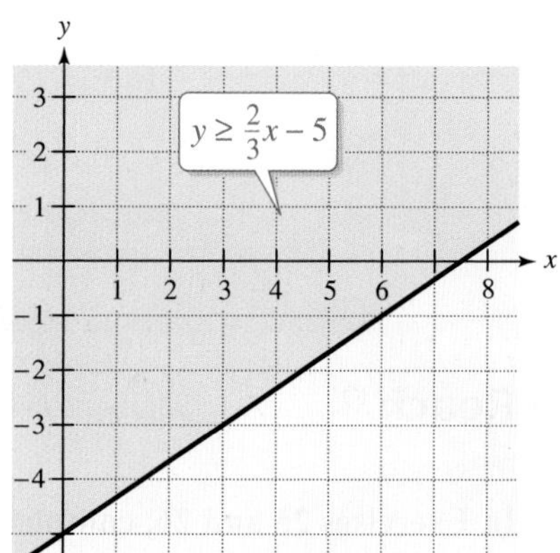

Exercises Within Reach ®

Solutions in English & Spanish and tutorial videos at AlgebraWithinReach.com

Sketching the Graph of a Linear Inequality **In Exercises 35–48, sketch the graph of the linear inequality.** See Additional Answers.

35. $3x + 2y \geq 2$

36. $3x + 5y \leq 15$

37. $5x + 4y < 20$

38. $4x - 7y > 21$

39. $x - 3y - 9 < 0$

40. $x + 4y + 12 > 0$

41. $3x - 2 \leq 5x + y$

42. $2x - 2y \geq 8 + 2y$

43. $0.2x + 0.3y < 2$

44. $0.25x - 0.75y > 6$

45. $y - 1 > -\frac{1}{2}(x - 2)$

46. $y - 2 < -\frac{2}{3}(x - 3)$

47. $\frac{x}{3} + \frac{y}{4} \leq 1$

48. $\frac{x}{2} + \frac{y}{6} \geq 1$

Application

Application

EXAMPLE 6 Working to Meet a Budget

Your budget requires you to earn *at least* \$160 per week. You work two part-time jobs. One is tutoring, which pays \$10 per hour, and the other is at a fast-food restaurant, which pays \$8 per hour. Let x represent the number of hours tutoring and let y represent the number of hours worked at the fast-food restaurant.

a. Write a linear inequality that represents the different numbers of hours you can work at each job in order to meet your budget requirements.

b. Graph the inequality and identify at least two ordered pairs (x, y) that represent numbers of hours you can work at each job in order to meet your budget requirements.

SOLUTION

a. *Verbal Model:* 10 • [Number of hours tutoring] + 8 • [Number of hours at fast-food restaurant] ≥ 160

Labels: Number of hours tutoring $= x$ (hours)

Number of hours at fast-food restaurant $= y$ (hours)

Inequality: $10x + 8y \geq 160$

b. Rewrite the inequality in slope-intercept form.

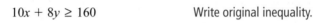

$10x + 8y \geq 160$	Write original inequality.
$8y \geq -10x + 160$	Subtract $10x$ from each side.
$y \geq -1.25x + 20$	Divide each side by 8.

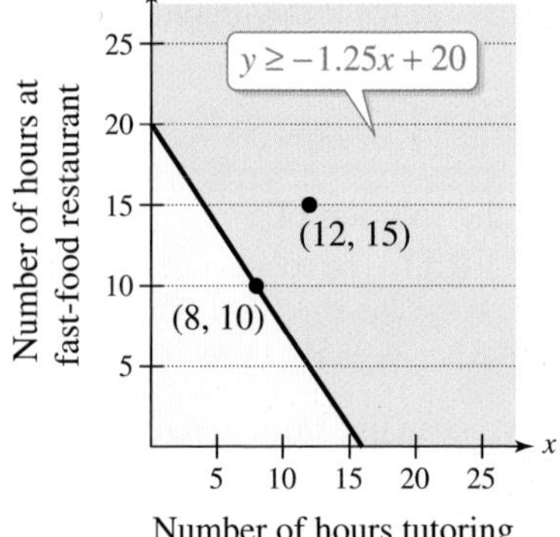

Graph the corresponding equation $y = -1.25x + 20$ and shade the half-plane lying above the line, as shown at the left. From the graph, you can see that two solutions that yield the desired weekly earnings of at least \$160 are (8, 10) and (12, 15). (There are many other solutions.)

Exercises Within Reach®

Solutions in English & Spanish and tutorial videos at AlgebraWithinReach.com

49. ***Weekly Pay*** You have two part-time jobs. One is at a grocery store, which pays \$11 per hour, and the other is mowing lawns, which pays \$9 per hour. Between the two jobs, you want to earn at least \$240 per week. See Additional Answers.

(a) Write a linear inequality that represents the different numbers of hours you can work at each job.

(b) Graph the inequality and identify three ordered pairs that are solutions of the inequality.

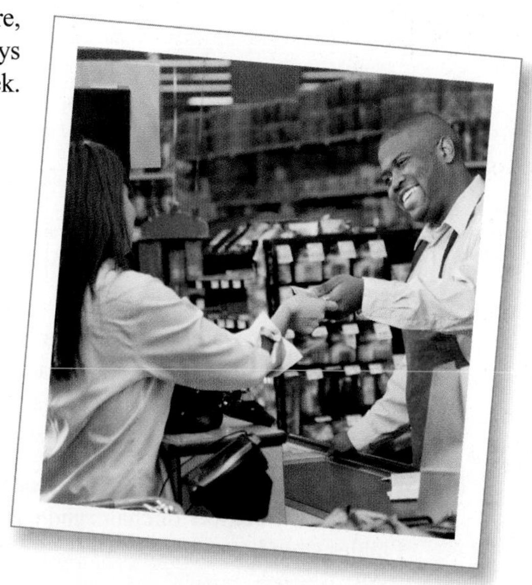

50. ***Weekly Pay*** You have two part-time jobs. One is at a candy store, which pays \$8 per hour, and the other is providing childcare, which pays \$10 per hour. Between the two jobs, you want to earn at least \$160 per week. See Additional Answers.

(a) Write a linear inequality that represents the different numbers of hours you can work at each job.

(b) Graph the inequality and identify three ordered pairs that are solutions of the inequality.

Steve Debenport/iStockphoto.com

Concept Summary: Graphing Linear Inequalities

What

You can use a graph to represent all the **solutions** of a **linear inequality in two variables**.

EXAMPLE

Sketch the **graph** of the linear inequality $y < x + 1$.

How

Use these steps to sketch the graph of such an inequality.

1. Replace the inequality sign by an equal sign.
2. Sketch the graph of the resulting equation.
   - Use a dashed line for $<$ and $>$.
   - Use a solid line for $\le$ and $\ge$.
3. Shade the **half-plane** that contains points that satisfy the inequality.

Why

You can use linear inequalities to model and solve many real-life problems. The graphs of inequalities help you see all the possible solutions. For example, one solution of the inequality $y < x + 1$ is (1, 1).

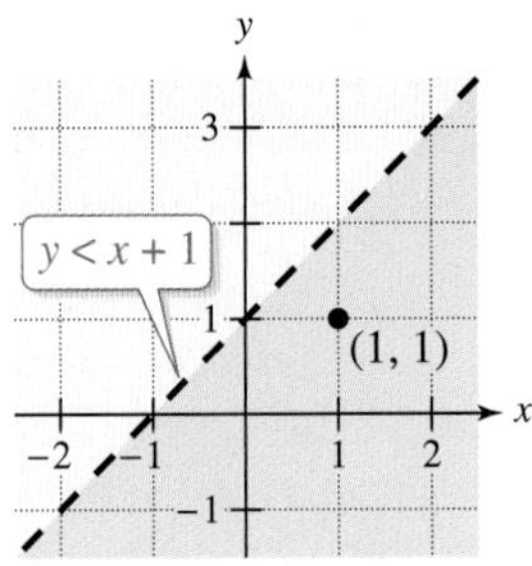

Exercises Within Reach®

Worked-out solutions to odd-numbered exercises at AlgebraWithinReach.com

Concept Summary Check

51. ***True or False?*** To determine whether the point (2, 4) is a solution of the inequality $2x + 3y > 12$, you must graph the inequality. Justify your answer. See Additional Answers.

52. ***True or False?*** Any point in the Cartesian plane can be used as a test point to determine the half-plane that represents the solution of a linear inequality. Justify your answer. See Additional Answers.

53. ***Reasoning*** How does the solution of $x - y > 1$ differ from the solution of $x - y \ge 1$? The solution of $x - y > 1$ does not include the points on the line $x - y = 1$. The solution of $x - y \ge 1$ includes all the points on the line $x - y = 1$.

54. ***Writing*** Explain how you can use *slope-intercept form* to graph a linear inequality without using a test point. To graph a linear inequality in slope-intercept form, graph above the line if the inequality symbol is >. Graph below the line if the inequality symbol is <.

Extra Practice

Writing an Inequality **In Exercises 55–58, write an inequality for the shaded region shown in the figure.**

55.

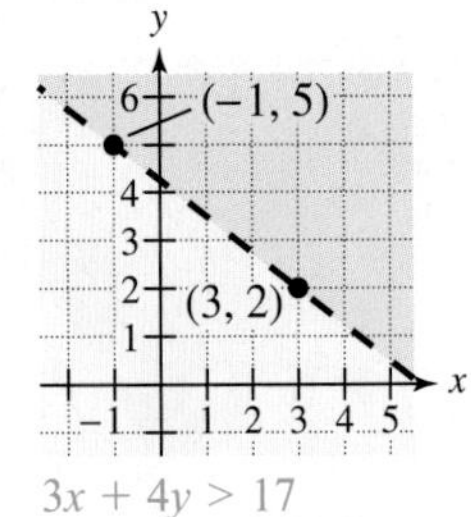

$3x + 4y > 17$

56.

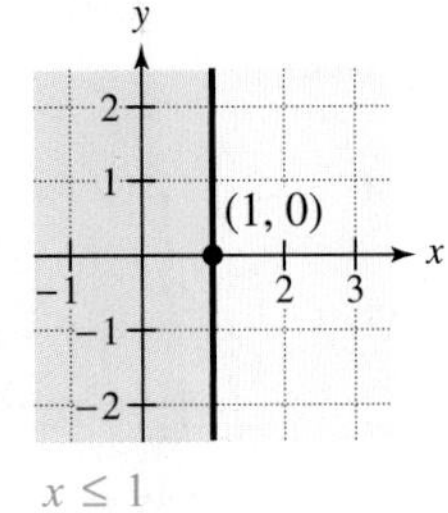

$x \le 1$

57.

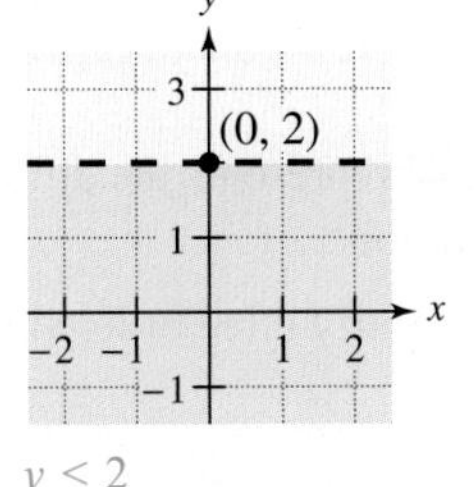

$y < 2$

58.

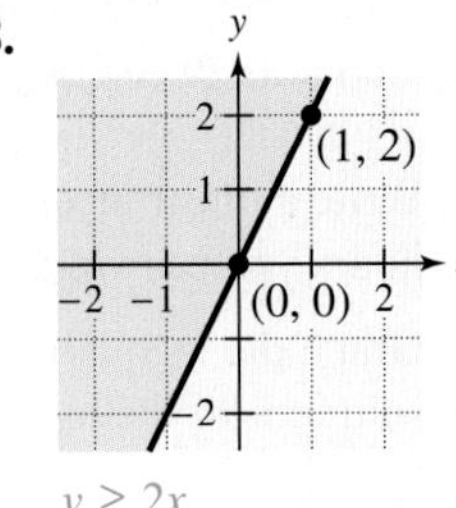

$y \ge 2x$

59. ***Storage Space*** A warehouse for storing chairs and tables has 1000 square feet of floor space. Each chair requires 10 square feet of floor space and each table requires 15 square feet. See Additional Answers.

    (a) Write a linear inequality for this space constraint, where x is the number of chairs and y is the number of tables stored.

    (b) Sketch the graph of the inequality.

60. ***Storage Space*** A warehouse for storing desks and filing cabinets has 2000 square feet of floor space. Each desk requires 15 square feet of floor space and each filing cabinet requires 6 square feet. See Additional Answers.

    (a) Write a linear inequality for this space constraint, where x is the number of desks and y is the number of filing cabinets stored.

    (b) Sketch the graph of the inequality.

61. ***Consumerism*** You and some friends go out for pizza. Together you have \$32. You want to order two large pizzas with cheese at \$10 each. Each additional topping costs \$0.60 and each small soft drink costs \$1.00.

(a) Write an inequality that represents the different numbers of toppings x and drinks y that your group can afford. (Assume there is no sales tax.)

$0.60x + 1.00y \le 12;\ x \ge 0,\ y \ge 0$

(b) Sketch the graph of the inequality.

See Additional Answers.

(c) What are the coordinates for an order of six soft drinks and two large pizzas with cheese, each with three additional toppings? Is this a solution of the inequality? (6, 6); yes

62. ***Nutrition*** A dietitian is asked to design a special diet supplement using two foods. Each ounce of food X contains 30 units of calcium and each ounce of food Y contains 20 units of calcium. The minimum daily requirement in the diet is 300 units of calcium.

(a) Write an inequality that represents the different amounts of food X and food Y required.

$30x + 20y \ge 300;\ x \ge 0,\ y \ge 0$

(b) Sketch the graph of the inequality and identify three ordered pairs that are solutions of the inequality.

See Additional Answers.

Sample answers: (x, y): (2, 12), (4, 9), (8, 3)

63. ***Sports*** Your hockey team needs at least 70 points for the season in order to advance to the playoffs. Your team finishes with w wins, each worth 2 points, and t ties, each worth 1 point.

(a) Write a linear inequality that represents the different numbers of wins and ties your team must record in order to advance to the playoffs.

$2w + t \ge 70;\ w \ge 0,\ t \ge 0$

(b) Sketch the graph of the inequality and identify three ordered pairs that are solutions of the inequality.

See Additional Answers.

Sample answer: (w, t): (10, 50), (20, 30), (30, 10)

Explaining Concepts

64. ***Logic*** What is meant by saying that (x_1, y_1) is a solution of a linear inequality in x and y? The inequality is true when x_1 and y_1 are substituted for x and y, respectively.

65. ***Precision*** Explain the difference between graphing the solution of the inequality $x \le 3$ on the real number line and graphing it on a rectangular coordinate system.

On the number line, the solution of $x \le 3$ is an unbounded interval. On a rectangular coordinate system, the solution is a half-plane.

66. ***Reasoning*** The graph of a particular inequality consists of the half-plane above a dashed line. Explain how to modify the inequality to represent all of the points in the plane *other* than the half-plane represented by the graph.

To represent the other points in the plane, change the inequality symbol so that it is $\le$ rather than $>$.

67. ***Think About It*** Discuss how you could write a double inequality whose solution set is the graph of a line in the plane. Give an example. See Additional Answers.

68. ***Think About It*** The origin *cannot* be used as the test point to determine which half-plane to shade in graphing a linear inequality. What is the y-intercept of the graph of the corresponding non-vertical linear equation? Explain.

(0, 0); If the point (0, 0) cannot be used as a test point, the point (0, 0) must lie on the boundary line. Therefore, the line passes through the origin, and the y-intercept is (0, 0).

69. ***Writing*** Every point that is in the solution set of the linear inequality $y \le ax + b$ is a solution of the linear inequality $y < cx + d$. What can you say about the values of a, b, c, and d? See Additional Answers.

Cumulative Review

In Exercises 70–73, transform the absolute value equation into two linear equations.

70. $|x| = 6$ $x = 6, x = -6$

71. $|x + 2| = 3$ $x + 2 = 3, x + 2 = -3$

72. $|2x + 3| = 9$ $2x + 3 = 9, 2x + 3 = -9$

73. $|8 - 3x| = 10$ $8 - 3x = 10, 8 - 3x = -10$

In Exercises 74–77, plot the points on a rectangular coordinate system. See Additional Answers.

74. $(0, 0), (-3, 4), (-4, -2)$

75. $(5, -1), (-7, 4), (3, 3)$

76. $(1, 1), \left(\frac{3}{4}, \frac{3}{4}\right), \left(\frac{5}{4}, \frac{5}{4}\right)$

77. $\left(\frac{1}{2}, \frac{3}{2}\right), \left(2, \frac{1}{2}\right), \left(\frac{1}{2}, \frac{7}{2}\right)$

In Exercises 78 and 79, write equations of the lines that pass through the point and are (a) parallel and (b) perpendicular to the given line.

78. $(3, 0);\ y = -2x + 1$

(a) $y = -2x + 6$ (b) $y = \frac{1}{2}x - \frac{3}{2}$

79. $(5, 7);\ 2x - 3y = -6$

(a) $y = \frac{2}{3}x + \frac{11}{3}$ (b) $y = -\frac{3}{2}x + \frac{29}{2}$

3.6 Relations and Functions

- Identify the domains and ranges of relations.
- Determine if relations are functions by inspection.
- Use function notation and evaluate functions.
- Identify the domains and ranges of functions.

Relations

Many everyday occurrences involve two quantities that are paired or matched with each other by some rule of correspondence. The mathematical term for such a correspondence is *relation*.

Definition of Relation

A **relation** is any set of ordered pairs. The set of first components in the ordered pairs is the **domain** of the relation. The set of second components is the **range** of the relation.

EXAMPLE 1 Analyzing a Relation

Find the domain and range of the relation $\{(0, 1), (1, 3), (2, 5), (3, 5), (0, 3)\}$.

SOLUTION

The domain is the set of all first components of the relation, and the range is the set of all second components.

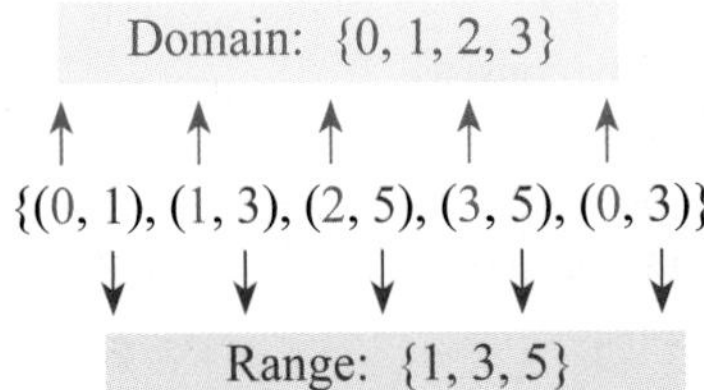

A graphical representation of this relation is shown below.

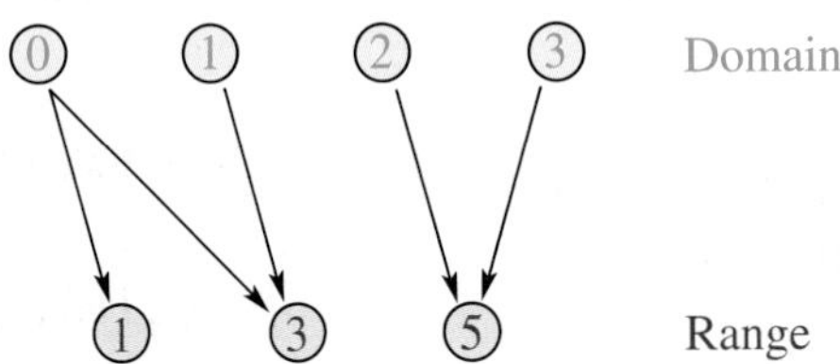

Study Tip

When you write the domain or range of a relation, it is not necessary to list repeated components more than once.

Exercises Within Reach®

Solutions in English & Spanish and tutorial videos at AlgebraWithinReach.com

Analyzing a Relation **In Exercises 1–4, find the domain and the range of the relation. Then draw a graphical representation of the relation.** See Additional Answers.

1. $\{(-2, 0), (0, 1), (1, 4), (0, -1)\}$
   Domain: $\{-2, 0, 1\}$; Range: $\{-1, 0, 1, 4\}$
2. $\{(3, 10), (4, 5), (6, -2), (8, 3)\}$
   Domain: $\{3, 4, 6, 8\}$; Range: $\{-2, 3, 5, 10\}$
3. $\{(0, 0), (4, -3), (2, 8), (5, 5), (6, 5)\}$
   Domain: $\{0, 2, 4, 5, 6\}$; Range: $\{-3, 0, 5, 8\}$
4. $\{(-3, 6), (-3, 2), (-3, 5)\}$
   Domain: $\{-3\}$; Range: $\{2, 5, 6\}$

Functions

Definition of a Function

A **function** f from a set A to a set B is a rule of correspondence that assigns to each element x in the set A exactly one element y in the set B. The set A is called the **domain** (or set of inputs) of the function f, and the set B contains the **range** (or set of outputs) of the function.

Study Tip

In Example 2, the set in part (a) has only a finite number of ordered pairs, whereas the set in part (b) has an infinite number of ordered pairs.

EXAMPLE 2 Input-Output Ordered Pairs for Functions

Write a set of ordered pairs that represents the rule of correspondence.

a. Winners of the Super Bowl in 2009, 2010, 2011, and 2012

b. The squares of all real numbers

SOLUTION

a. For the function that pairs the year from 2009 to 2012 with the winner of the Super Bowl, each ordered pair is of the form (year, winner).

b. For the function that pairs each real number with its square, each ordered pair is of the form (x, x^2).

$\{$All points (x, x^2), where x is a real number$\}$

Functions are commonly represented in four ways.

1. *Verbally* by a sentence that describes how the input variable is related to the output variable.
2. *Numerically* by a table or a list of ordered pairs that matches input values with output values.
3. *Graphically* by points of the form (input value, output value) on a rectangular coordinate system.
4. *Algebraically* by an equation in two variables.

Exercises Within Reach®

Solutions in English & Spanish and tutorial videos at AlgebraWithinReach.com

Writing Ordered Pairs **In Exercises 5–10, write a set of ordered pairs that represents the rule of correspondence.**

5. The cubes of all positive integers less than 8
(1, 1), (2, 8), (3, 27), (4, 64), (5, 125), (6, 216), (7, 343)

6. The cubes of all integers greater than -2 and less than 5
$(-1, -1)$, (0, 0), (1, 1), (2, 8), (3, 27), (4, 64)

7. The winners of the World Series from 2008 to 2011
(2008, Philadelphia Phillies), (2009, New York Yankees), (2010, San Francisco Giants), (2011, St. Louis Cardinals)

8. The men inaugurated as president of the United States in 1981, 1989, 1993, 2001, 2009 (1981, R. Reagan), (1989, G. Bush), (1993, W. Clinton), (2001, G.W. Bush), (2009, B. Obama)

9. The fuel used by a vehicle on a trip is a function of the driving time in hours. Fuel is used at a rate of 3 gallons per hour on trips of 3 hours, 1 hour, 2 hours, 8 hours, and 7 hours. (3, 9), (1, 3), (2, 6), (8, 24), (7, 21)

10. The time it takes a court stenographer to transcribe a testimony is a function of the number of words. Working at a rate of 120 words per minute, the stenographer transcribes testimonies of 360 words, 600 words, 1200 words, and 2040 words.
(360, 3), (600, 5), (1200, 10), (2040, 17)

Characteristics of a Function

1. Each element in the domain A must be matched with an element in the range, which is contained in the set B.
2. Some elements in set B may not be matched with any element in the domain A.
3. Two or more elements of the domain may be matched with the same element in the range.
4. No element of the domain is matched with two different elements in the range.

EXAMPLE 3 Testing for Functions

Decide whether the description represents a function.

a. The input value x is any of the 50 states in the United States and the output value y is the number of governors of that state.

b. $\{(1, 1), (2, 4), (3, 9), (4, 16), (5, 25), (6, 36)\}$

c.

Input, x	Output, y
-1	7
3	2
4	0
3	4

d. Let $A = \{a, b, c\}$ and let $B = \{1, 2, 3, 4, 5\}$.

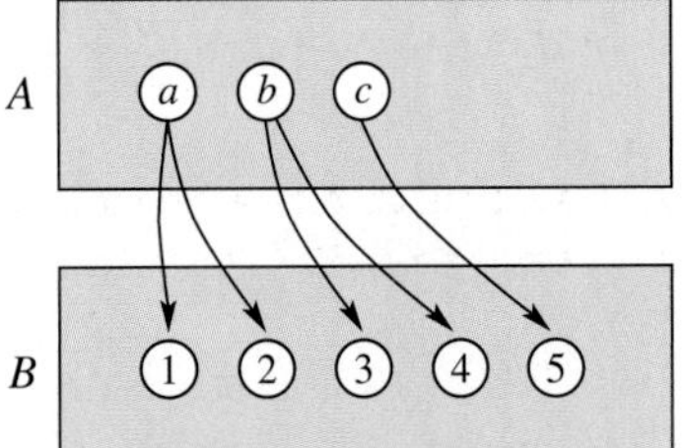

SOLUTION

a. This set of ordered pairs *does* represent a function. Regardless of the input value x, the output value is always 1.

b. This set of ordered pairs *does* represent a function. No input value is matched with *two* output values.

c. This table *does not* represent a function. The input value 3 is matched with *two* different output values, 2 and 4.

d. This diagram *does not* represent a function. The element a in set A is matched with *two* elements in set B. This is also true of element b.

Exercises Within Reach ®

Solutions in English & Spanish and tutorial videos at AlgebraWithinReach.com

Testing for a Function **In Exercises 11 and 12, determine whether each set of ordered pairs represents a function from A to B.**

11. $A = \{0, 1, 2, 3\}$ and $B = \{-2, -1, 0, 1, 2\}$

(a) $\{(0, 1), (1, -2), (2, 0), (3, 2)\}$
Function from A to B

(b) $\{(0, -1), (2, 2), (1, -2), (3, 0), (1, 1)\}$
Not a function from A to B

(c) $\{(0, 0), (1, 0), (2, 0), (3, 0)\}$
Function from A to B

(d) $\{(0, 2), (3, 0), (1, 1)\}$ Not a function from A to B

12. $A = \{1, 2, 3\}$ and $B = \{9, 10, 11, 12\}$

(a) $\{(1, 10), (3, 11), (3, 12), (2, 12)\}$
Not a function from A to B

(b) $\{(1, 10), (2, 11), (3, 12)\}$
Function from A to B

(c) $\{(1, 10), (1, 9), (3, 11), (2, 12)\}$
Not a function from A to B

(d) $\{(3, 9), (2, 9), (1, 12)\}$ Function from A to B

The function $y = x^2$ represents the variable y as a function of the variable x. The variable x is the **independent variable** and the variable y is the **dependent variable**. In this context, the domain of the function is the set of all *allowable* real values of the independent variable x, and the range of the function is the *resulting* set of all values taken on by the dependent variable y.

EXAMPLE 4 Testing for Functions Represented by Equations

Which of the equations represent y as a function of x?

a. $y = x^2 + 1$ **b.** $x - y^2 = 2$ **c.** $-2x + 3y = 4$

SOLUTION

a. For the equation

$$y = x^2 + 1$$

just one value of y corresponds to each value of x. For instance, when $x = 1$, the value of y is

$$y = 1^2 + 1 = 2.$$

So, y *is* a function of x.

b. By writing the equation $x - y^2 = 2$ in the form

$$y^2 = x - 2$$

you can see that two values of y correspond to some values of x. For instance, when $x = 3$,

$$y^2 = 3 - 2$$

$$y^2 = 1$$

$$y = 1 \quad \text{or} \quad y = -1.$$

So, the solution points $(3, 1)$ and $(3, -1)$ show that y *is not* a function of x.

c. By writing the equation $-2x + 3y = 4$ in the form

$$y = \frac{2}{3}x + \frac{4}{3}$$

you can see that just one value of y corresponds to each value of x. For instance, when $x = 2$, the value of y is $\frac{4}{3} + \frac{4}{3} = \frac{8}{3}$. So, y *is* a function of x.

Exercises Within Reach®

Solutions in English & Spanish and tutorial videos at AlgebraWithinReach.com

Solutions of an Equation **In Exercises 13–16, show that both ordered pairs are solutions of the equation, and explain why this implies that y is not a function of x.**

13. $x^2 + y^2 = 25$; $(0, 5)$, $(0, -5)$
There are two values of y associated with one value of x.

14. $x^2 + 4y^2 = 16$; $(0, 2)$, $(0, -2)$
There are two values of y associated with one value of x.

15. $|y| = x + 2$; $(1, 3)$, $(1, -3)$
There are two values of y associated with one value of x.

16. $|y - 2| = x$; $(2, 4)$, $(2, 0)$
There are two values of y associated with one value of x.

Testing for a Function **In Exercises 17–20, determine whether the equation represents y as a function of x.**

17. $y^2 = x$ Not a function **18.** $y = x^2$ Function **19.** $y = |x|$ Function **20.** $|y| = x$ Not a function

Function Notation

Function Notation

In the notation $f(x)$, f is the **name** of the function, x is the **domain** (or input) value, and $f(x)$ is the **range** (or output) value y for a given x. The symbol $f(x)$ is read as *the value of f at x* or simply *f of x*.

Study Tip

Note that

$$g(x + 1) \neq g(x) + g(1).$$

In general, $g(a + b)$ is not equal to $g(a) + g(b)$.

Point out that the absolute value function $f(x) = |x|$ can be piecewise-defined as follows.

$$f(x) = \begin{cases} x, & \text{if } x \geq 0 \\ -x, & \text{if } x < 0 \end{cases}$$

Additional Example:

Evaluate the function for $x = -3$, $x = 0$, and $x = 5$.

$$f(x) = \begin{cases} 2x^2, & x < 0 \\ 4, & 0 \leq x \leq 3 \\ 3x - 1, & x > 3 \end{cases}$$

Solution:

For $x = -3$, use $f(x) = 2x^2$ to obtain $f(-3) = 18$.

For $x = 0$, use $f(x) = 4$ to obtain $f(0) = 4$.

For $x = 5$, use $f(x) = 3x - 1$ to obtain $f(5) = 14$.

EXAMPLE 5 Evaluating a Function

Let $g(x) = 3x - 4$. Find each value of the function.

a. $g(1)$ **b.** $g(-2)$ **c.** $g(y)$ **d.** $g(x + 1)$ **e.** $g(x) + g(1)$

SOLUTION

a. Replacing x with 1 produces $g(1) = 3(1) - 4 = 3 - 4 = -1$.

b. Replacing x with -2 produces $g(-2) = 3(-2) - 4 = -6 - 4 = -10$.

c. Replacing x with y produces $g(y) = 3(y) - 4 = 3y - 4$.

d. Replacing x with $(x + 1)$ produces

$$g(x + 1) = 3(x + 1) - 4 = 3x + 3 - 4 = 3x - 1.$$

e. Using the result of part (a) for $g(1)$, you have

$$g(x) + g(1) = (3x - 4) + (-1) = 3x - 4 - 1 = 3x - 5.$$

EXAMPLE 6 Evaluating a Piecewise Defined Function

Let $f(x) = \begin{cases} x^2 + 1, & \text{if } x < 0 \\ x - 2, & \text{if } x \geq 0 \end{cases}$. Find each value of the function.

a. $f(-1)$ **b.** $f(0)$

SOLUTION

a. Because $x = -1 < 0$, use $f(x) = x^2 + 1$ to obtain

$$f(-1) = (-1)^2 + 1 = 1 + 1 = 2.$$

b. Because $x = 0 \geq 0$, use $f(x) = x - 2$ to obtain $f(0) = 0 - 2 = -2$.

Exercises Within Reach®

Solutions in English & Spanish and tutorial videos at AlgebraWithinReach.com

Evaluating a Function **In Exercises 21–24, evaluate the function as indicated, and simplify.**

21. $f(x) = 12x - 7$

(a) $f(3)$ 29 (b) $f\left(\frac{3}{2}\right)$ 11

(c) $f(a) + f(1)$ $12a - 2$ (d) $f(a + 1)$ $12a + 5$

22. $f(x) = \sqrt{x + 5}$

(a) $f(-1)$ 2 (b) $f(4)$ 3

(c) $f(z - 5)$ $\sqrt{z}$ (d) $f(5z)$ $\sqrt{5z + 5}$

23. $f(x) = \dfrac{3x}{x - 5}$

(a) $f(0)$ 0 (b) $f\left(\frac{5}{3}\right)$ $-\frac{3}{2}$

(c) $f(2) - f(-1)$ $-\frac{5}{2}$ (d) $f(x + 4)$ $\dfrac{3x + 12}{x - 1}$

24. $f(x) = \begin{cases} x + 8, & \text{if } x < 0 \\ 10 - 2x, & \text{if } x \geq 0 \end{cases}$

(a) $f(4)$ 2 (b) $f(-10)$ -2

(c) $f(0)$ 10 (d) $f(6) - f(-2)$ -8

Finding the Domain and Range of a Function

The domain of a function may be explicitly described along with the function, or it may be *implied* by the expression used to define the function. The **implied domain** is the set of all real numbers (inputs) that yield real number values for the function. Here are two examples.

The function given by

$$f(x) = \frac{1}{x-3} \qquad \text{Domain: all } x \neq 3$$

has an implied domain that consists of all real values of x other than $x = 3$. The value $x = 3$ is excluded from the domain because division by zero is undefined.

The function given by

$$f(x) = \sqrt{x} \qquad \text{Domain: all } x \geq 0$$

has an implied domain that consists of all real values of x greater than or equal to 0. Negative values are excluded because the square root of a negative number is not a real number.

Application **EXAMPLE 7** **Finding the Domain and Range of a Function**

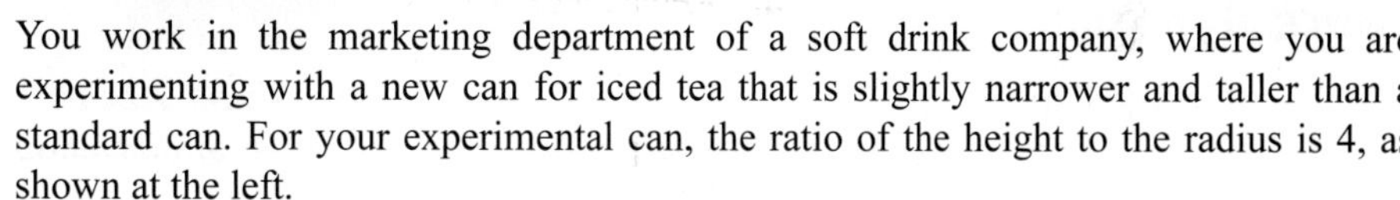

You work in the marketing department of a soft drink company, where you are experimenting with a new can for iced tea that is slightly narrower and taller than a standard can. For your experimental can, the ratio of the height to the radius is 4, as shown at the left.

$\frac{h}{r} = 4$

a. Write the volume of the can as a function of the radius r.

b. Find the domain and range of the function.

SOLUTION

a. The volume of a circular cylinder is given by the formula $V = \pi r^2 h$. Because the ratio of the height of the can to the radius is 4, $h = 4r$. Substitute this value of h into the formula to obtain

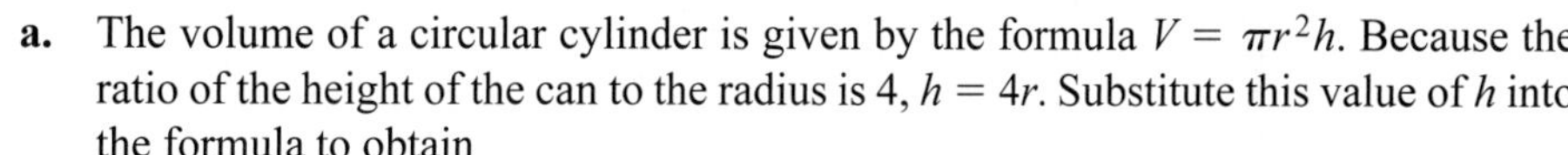

$$V(r) = \pi r^2 h = \pi r^2(4r) = 4\pi r^3. \qquad \text{Write } V \text{ as a function of } r.$$

b. The radius r must be a positive number. So, the domain is the set of all real numbers r such that $r > 0$.

The volume V must be a positive number. So, the range is the set of all real numbers V such that $V > 0$.

Exercises Within Reach®

Solutions in English & Spanish and tutorial videos at AlgebraWithinReach.com

Finding the Domain of a Function **In Exercises 25–30, find the domain of the function.**

25. $f(x) = x^2 + x - 2$ All real numbers x

26. $h(x) = 3x^2 - x$ All real numbers x

27. $f(t) = \frac{t+3}{t(t+2)}$ All real numbers t such that $t \neq 0, -2$

28. $g(s) = \frac{s-2}{(s-6)(s-10)}$ All real numbers s such that $s \neq 6, 10$

29. $g(x) = \sqrt{x+4}$ All real numbers x such that $x \geq -4$

30. $f(x) = \sqrt{2-x}$ All real numbers x such that $x \leq 2$

Finding the Domain and Range of a Function **In Exercises 31–34, find the domain and range of the function.** See Additional Answers.

31. Circumference of a circle: $C = 2\pi r$

32. Area of a square with side s: $A = s^2$

33. Area of a circle with radius r: $A = \pi r^2$

34. Volume of a sphere with radius r: $V = \frac{4}{3}\pi r^3$

Concept Summary: *Finding the Domain and Range of a Function*

What

When using a **function** that represents a real-life situation, it is important to know the **domain** and **range** of the function.

EXAMPLE

The total cost C of producing x video games is approximated by the function $C(x) = 1.95x + 8000$. Find the domain and range of the function.

How

Find the domain by identifying the set of all *allowable* real values of x. Find the range by identifying the *resulting* set of all values taken on by y.

EXAMPLE

Domain: $x \geq 0$; the number of video games sold must be greater than or equal to zero.

Range: $C(x) \geq 8000$; the cost of producing any number of games costs more than $8000.

Why

Many real-life situations are represented by functions. These real-life situations have physical limitations. Finding and stating the domain and range of a function is as important as the function itself.

Exercises Within Reach ®

Worked-out solutions to odd-numbered exercises at AlgebraWithinReach.com

Concept Summary Check

35. ***Vocabulary*** Explain the meanings of the terms *domain* and *range* in the context of a function.
See Additional Answers.

36. ***Number Sense*** Describe the meaning of the notation $C(100)$. Then use the function in the example above to find $C(100)$. $C(100)$ means "the value of C at 100"; $8195

37. ***Reasoning*** Explain why negative values are excluded from the domain of the function in the example above.
You cannot produce a negative number of video games.

38. ***Think About It*** Use the following terms to make a list of domain-related terms and a list of range-related terms.

y, second component, x, input, independent variable, dependent variable, output, first component, $f(x)$

Domain: x, input, independent variable, first component

Range: y, $f(x)$, output, dependent variable, second component

Extra Practice

Testing for a Function **In Exercises 39–42, determine whether the relation is a function.**

39. 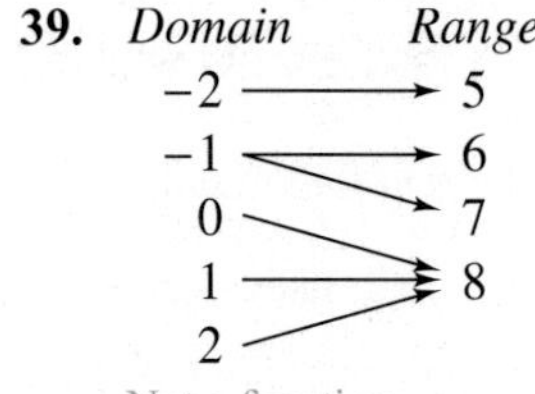

Not a function

40. 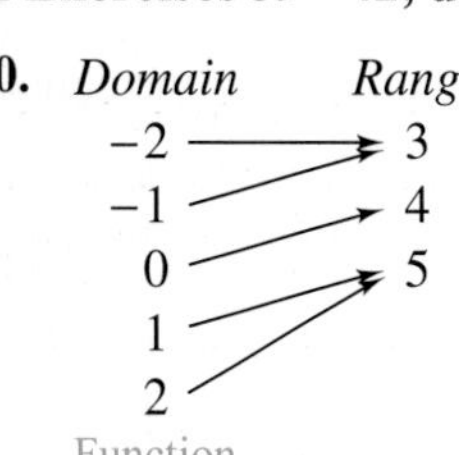
Function

41. Domain: 0, 2, 4, 6, 8 — Range: 15, 20, 25
Function

42. 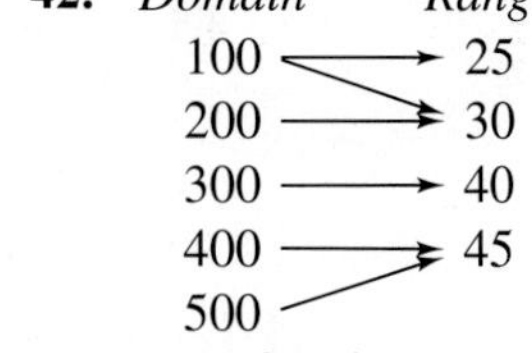
Not a function

Evaluating a Function **In Exercises 43 and 44, evaluate the function as indicated, and simplify.**

43. $g(x) = 2 - 4x + x^2$

(a) $g(4)$ 2 (b) $g(0)$ 2

(c) $g(2y)$ $4y^2 - 8y + 2$ (d) $g(4) + g(6)$ 16

44. $g(x) = 8 - |x - 4|$

(a) $g(0)$ 4 (b) $g(8)$ 4

(c) $g(16) - g(-1)$ -7 (d) $g(x - 2)$ $8 - |x - 6|$

Finding the Domain of a Function **In Exercises 45–48, find the domain of the function.** See Additional Answers.

45. $f(x) = \sqrt{2x - 1}$

46. $G(x) = \sqrt{8 - 3x}$

47. $f(t) = |t - 4|$

48. $f(x) = |x + 3|$

49. ***Cost*** The cost of producing a software program is $3.25 per unit with fixed costs of $495. Write the total cost C as a function of x, the number of units produced.
$C(x) = 3.25x + 495, x > 0$

50. ***Distance*** A train travels at a speed of 65 miles per hour. Write the distance d traveled by the train as a function of the time t in hours. $d = 65t$

51. ***Geometry*** An open box is to be made from a square piece of material 24 inches on a side by cutting equal squares from the corners and turning up the sides (see figure). Write the volume V of the box as a function of x.

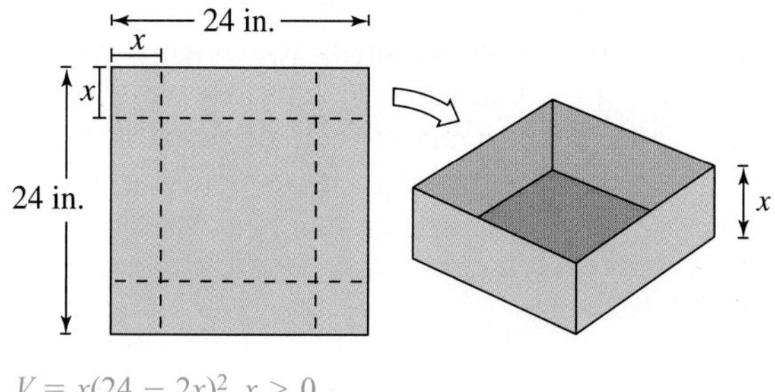

$V = x(24 - 2x)^2, x > 0$

52. ***Profit*** The marketing department of a business has determined that the profit from selling x units of a product is approximated by the model

$P(x) = 50\sqrt{x} - 0.5x - 500.$

Find (a) $P(1600)$ and (b) $P(2500)$. (a) \$700 (b) \$750

53. ***Safe Load*** A solid rectangular beam has a height of 6 inches and a width of 4 inches. The safe load S of the beam with the load at the center is a function of its length L and is approximated by the model

$$S(L) = \frac{128{,}160}{L}$$

where S is measured in pounds and L is measured in feet. Find (a) $S(12)$ and (b) $S(16)$. (a) 10,680 pounds (b) 8010 pounds

Wages **In Exercises 54 and 55, use the following information. A wage earner is paid \$12 per hour for regular time and time-and-a-half for overtime. The weekly wage function is**

$$W(h) = \begin{cases} 12h, & 0 \le h \le 40 \\ 18(h - 40) + 480, & h > 40 \end{cases}$$

where h represents the number of hours worked in a week. See Additional Answers.

54. (a) Evaluate $W(30)$, $W(40)$, $W(45)$, and $W(50)$.
    (b) Could you use values of h for which $h < 0$ in this model? Why or why not?

55. (a) Evaluate $W(20)$, $W(25)$, $W(35)$, and $W(55)$.
    (b) Describe the domain implied by the situation.

Explaining Concepts

Precision **In Exercises 56 and 57, determine whether the statement uses the word *function* in a way that is mathematically correct. Explain.**

56. The sales tax on clothes is 5.5% of the selling price, so the sales tax on clothes is a function of the price. Correct

57. A shoe store sells shoes at prices from \$19.99 to \$189.99 per pair, so the daily revenue of the store is a function of the number of pairs of shoes sold. See Additional Answers.

58. ***Think About It*** Set A is a relation but not a function. Is it possible that a subset of set A is a function? Explain. See Additional Answers.

59. ***Think About It*** Set B is a function. Is it possible that a subset of set B is a relation but not a function? Explain. See Additional Answers.

60. ***Writing*** Describe a real-life situation that is a relation. Identify the domain and the range and discuss whether the relation is also a function. See Additional Answers.

61. ***Writing*** Describe a real-life situation that is a function. Identify the domain and the range and discuss whether the function represents a finite or an infinite number of ordered pairs. See Additional Answers.

Cumulative Review

In Exercises 62–65, identify the property of real numbers illustrated by the statement.

62. $(6x) \cdot 1 = 6x$ Multiplicative Identity Property
63. $0(3x + 17) + 12 = 12$ Multiplication Property of Zero
64. $2a \cdot \dfrac{1}{2a} = 1$ Multiplicative Inverse Property
65. $8 - 2(x + 7) = 8 - 2x - 2(7)$ Distributive Property

In Exercises 66–69, sketch the graph of the equation. See Additional Answers.

66. $y = -x$
67. $y = \frac{1}{2}x + 3$
68. $x - 4y = 16$
69. $5x - 3y = 0$

In Exercises 70–73, sketch the graph of the inequality. See Additional Answers.

70. $x \le -2$
71. $y > 3$
72. $y > \frac{3}{4}x - 1$
73. $2x + 3y \le 6$

3.7 Graphs of Functions

- Sketch graphs of functions on rectangular coordinate systems.
- Use the Vertical Line Test to determine if graphs represent functions.
- Use vertical and horizontal shifts and reflections to sketch graphs of functions.

The Graph of a Function

Study Tip

In Example 1, the (implied) domain of the function is the set of all real numbers. When writing the equation of a function, you may choose to restrict its domain by writing a condition to the right of the equation. For instance, the domain of the function

$$f(x) = 4x + 5, \ x \geq 0$$

is the set of all nonnegative real numbers (all $x \geq 0$).

EXAMPLE 1 Sketching the Graph of a Function

Sketch the graph of $f(x) = 2x - 1$.

SOLUTION

One way to sketch the graph is to begin by making a table of values.

x	-1	0	1	2	3	4
$f(x)$	-3	-1	1	3	5	7

Next, plot the six points shown in the table. Finally, connect the points with a line, as shown.

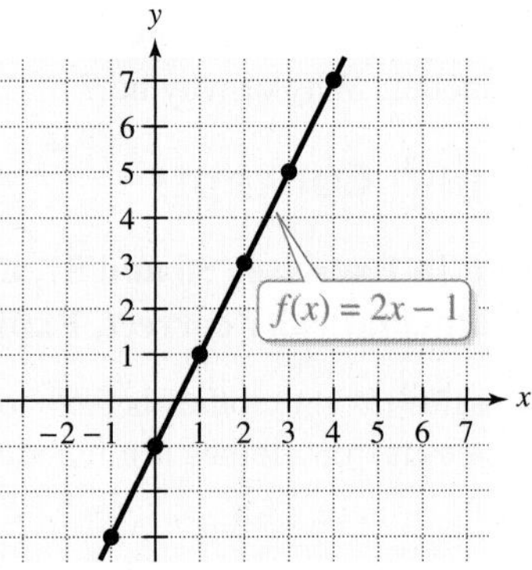

Exercises Within Reach ®

Solutions in English & Spanish and tutorial videos at AlgebraWithinReach.com

Sketching the Graph of a Function **In Exercises 1–4, complete the table and use the results to sketch the graph of the function.** See Additional Answers.

1. $f(x) = 2x - 7$

x	0	1	2	3	4	5
$f(x)$	-7	-5	-3	-1	1	3

2. $f(x) = 1 - 3x$

x	-2	-1	0	1	2
$f(x)$	7	4	1	-2	-5

3. $h(x) = -(x - 1)^2$

x	-1	0	1	2	3
$h(x)$	-4	-1	0	-1	-4

4. $g(x) = (x + 2)^2 + 3$

x	-4	-3	-2	-1	0
$g(x)$	7	4	3	4	7

Sketching the Graph of a Function **In Exercises 5 and 6, sketch the graph of the function.** See Additional Answers.

5. $f(x) = |x + 3|$

6. $g(x) = |x - 1|$

EXAMPLE 2 Sketching the Graph of a Piecewise-Defined Function

Sketch the graph of

$$f(x) = \begin{cases} x + 3, & x < 0 \\ -x + 4, & x \geq 0 \end{cases}.$$

Then determine its domain and range.

SOLUTION

Begin by graphing $f(x) = x + 3$ for $x < 0$, as shown below. You will recognize that this is the graph of the line $y = x + 3$ with the restriction that the x-values are negative. Because $x = 0$ is not in the domain, the right endpoint of the line is an open dot. Next, graph $f(x) = -x + 4$ for $x \geq 0$ on the same set of coordinate axes, as shown below. This is the graph of the line $y = -x + 4$ with the restriction that the x-values are nonnegative. Because $x = 0$ is in the domain, the left endpoint of the line is a solid dot. From the graph, you can see that the domain is $-\infty \leq x \leq \infty$ and the range is $y \leq 4$.

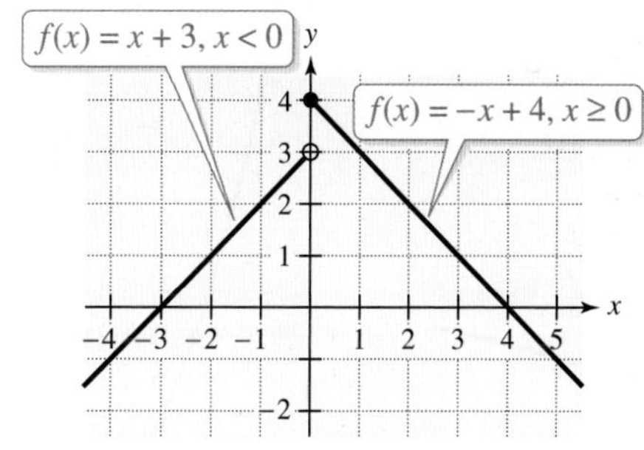

Exercises Within Reach®

Solutions in English & Spanish and tutorial videos at AlgebraWithinReach.com

Sketching the Graph of a Piecewise-Defined Function **In Exercises 7–12, sketch the graph of the function. Then determine its domain and range.** See Additional Answers.

7. $h(x) = \begin{cases} 2x + 3, & x < 0 \\ 3 - x, & x \geq 0 \end{cases}$ Domain: $-\infty < x < \infty$; Range: $-\infty < y \leq 3$

8. $f(x) = \begin{cases} x + 6, & x < 0 \\ 6 - 2x, & x \geq 0 \end{cases}$ Domain: $-\infty < x < \infty$; Range: $-\infty < y \leq 6$

9. $f(x) = \begin{cases} 3 - x, & x < -3 \\ x^2 + x, & x \geq -3 \end{cases}$ Domain: $-\infty < x < \infty$; Range: $-\frac{1}{4} \leq y < \infty$

10. $h(x) = \begin{cases} 4 - x^2, & x \leq 2 \\ x - 2, & x > 2 \end{cases}$ Domain: $-\infty < x < \infty$; Range: $-\infty < y < \infty$

11. $g(x) = \begin{cases} 1, & x \leq 1 \\ x, & x > 2 \end{cases}$ Domain: $-\infty < x \leq 1$ and $2 < x < \infty$; Range: $y = 1$ and $2 < y < \infty$

12. $g(x) = \begin{cases} -x, & x < 0 \\ 2x - 2, & x > 1 \end{cases}$ Domain: $-\infty < x < 0$ and $1 < x < \infty$; Range: $0 < y < \infty$

To become good at sketching the graphs of functions, it helps to be familiar with the graphs of some basic functions. The functions shown below, and variations of them, occur frequently in applications.

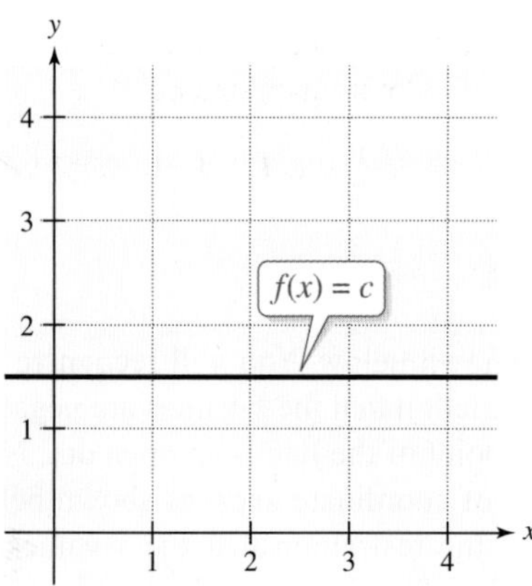

Constant function

Identity function

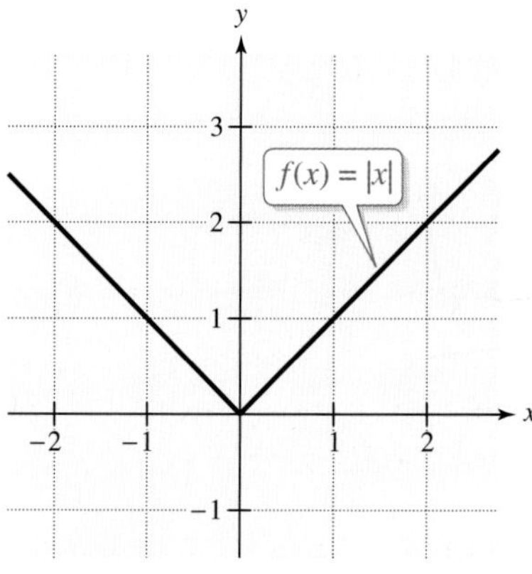

Absolute value function

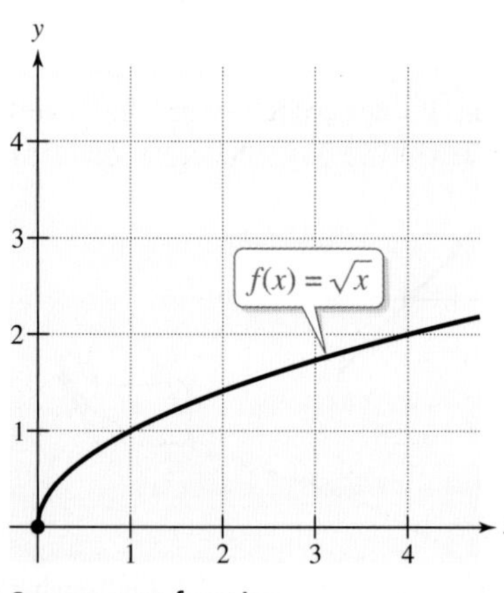

Square root function

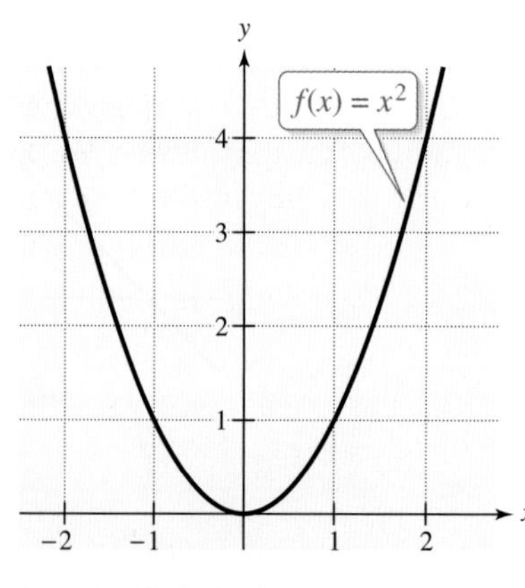

Squaring function

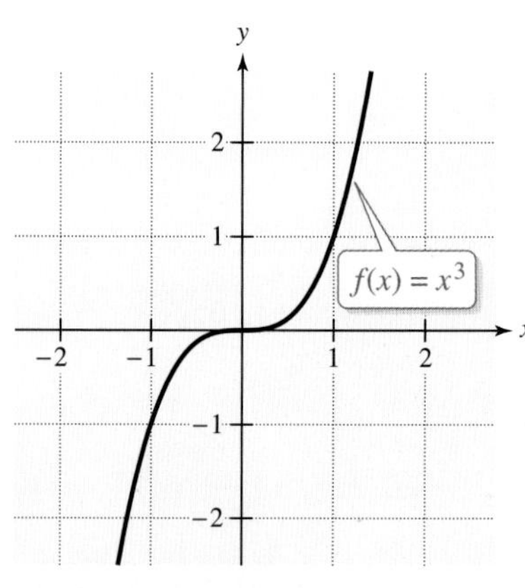

Cubing function

Exercises Within Reach®

Solutions in English & Spanish and tutorial videos at AlgebraWithinReach.com

Variations of Graphs of Basic Functions **In Exercises 13–18, identify the basic function related to the graph. Explain how the graph is different from the graph of the basic function.**

13.

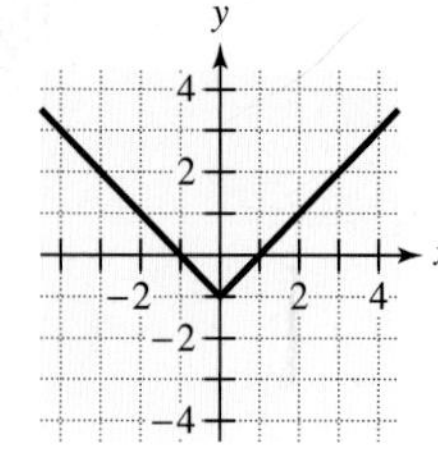

$f(x) = |x|$; The graph is shifted 1 unit downward.

14.

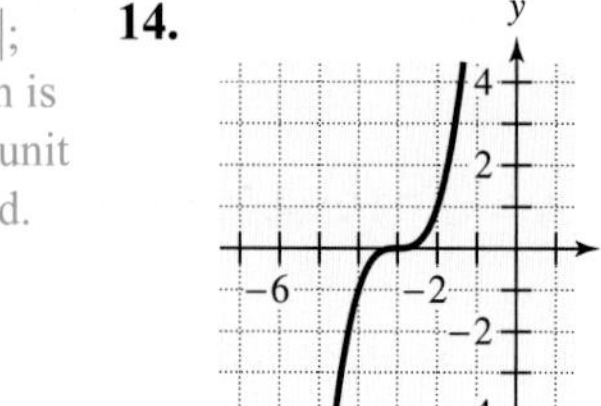

$f(x) = x^3$; The graph is shifted 3 units to the left.

15.

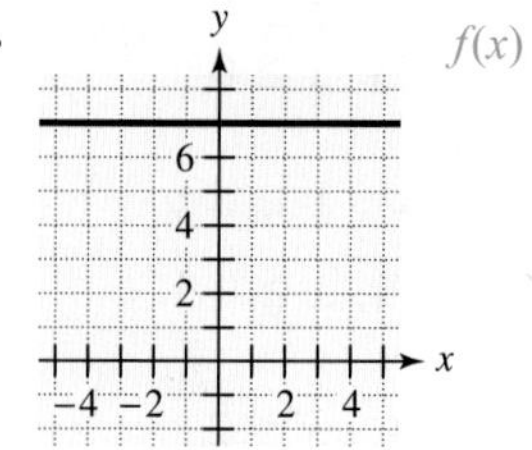

$f(x) = c$; $c = 7$

16.

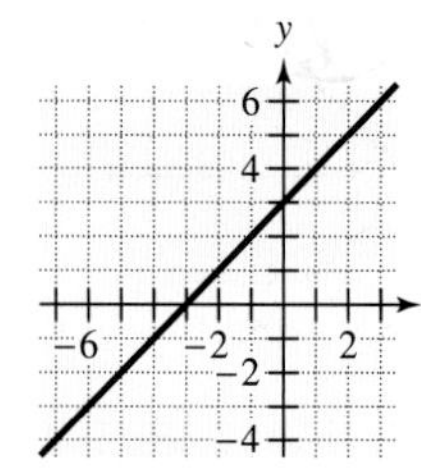

$f(x) = x$; The graph is shifted 3 units to the left or 3 units upward.

17.

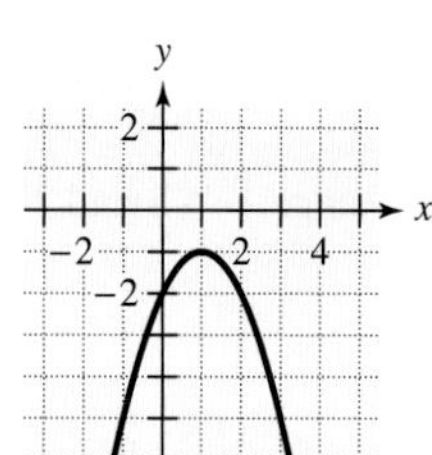

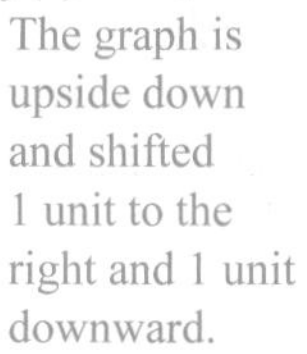

$f(x) = x^2$; The graph is upside down and shifted 1 unit to the right and 1 unit downward.

18.

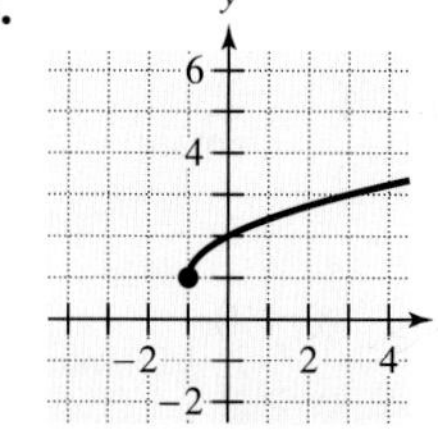

$f(x) = \sqrt{x}$; The graph is shifted 1 unit to the left and 1 unit upward.

The Vertical Line Test

Vertical Line Test for Functions

A set of points on a rectangular coordinate system is the graph of y as a function of x if and only if no vertical line intersects the graph at more than one point.

EXAMPLE 3 Using the Vertical Line Test

Determine whether each equation represents y as a function of x.

a. $y = x^2 - 3x + \frac{1}{4}$

b. $x = y^2 - 1$

SOLUTION

a. From the graph of the equation shown below on the left, you can see that every vertical line intersects the graph at most once. So, by the Vertical Line Test, the equation *does* represent y as a function of x.

b. From the graph of the equation shown below on the right, you can see that a vertical line intersects the graph twice. So, by the Vertical Line Test, the equation *does not* represent y as a function of x.

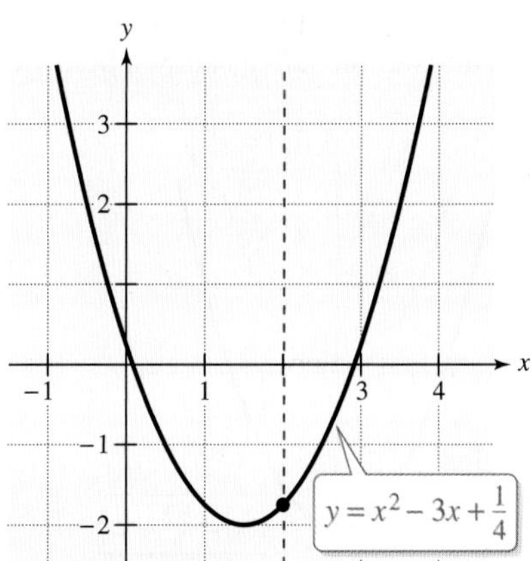

Graph of a function of x
Vertical line intersects once.

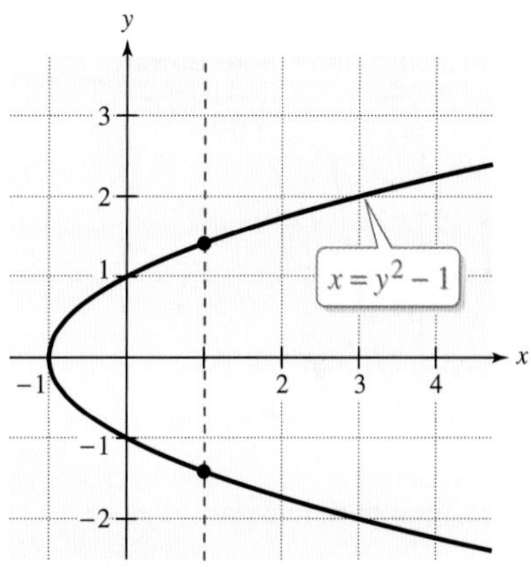

Not a graph of a function of x
Vertical line intersects twice.

Exercises Within Reach®

Solutions in English & Spanish and tutorial videos at AlgebraWithinReach.com

Using the Vertical Line Test **In Exercises 19–22, use the Vertical Line Test to determine whether y is a function of x.**

19. $y = \frac{1}{3}x^3$

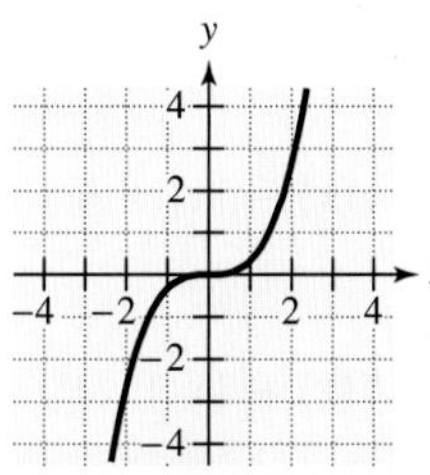

Function

20. $y = x^2 - 2x$

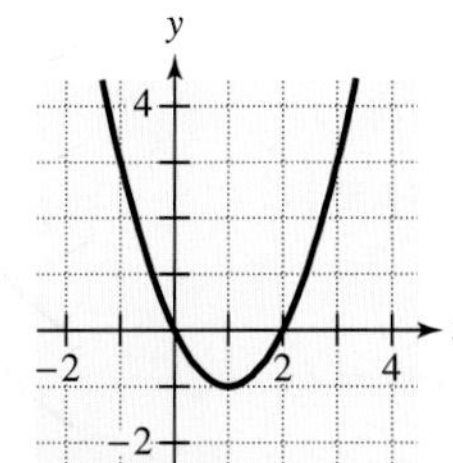

Function

21. $-2x + y^2 = 6$

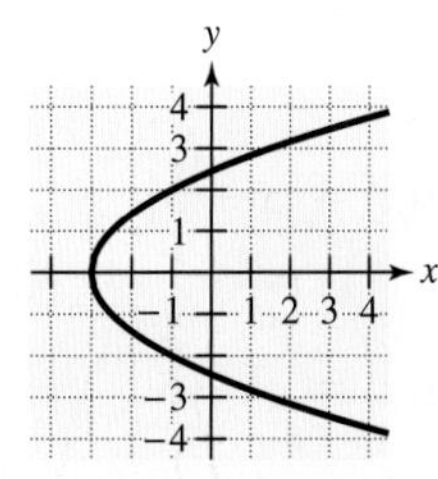

Not a function

22. $|y| = x$

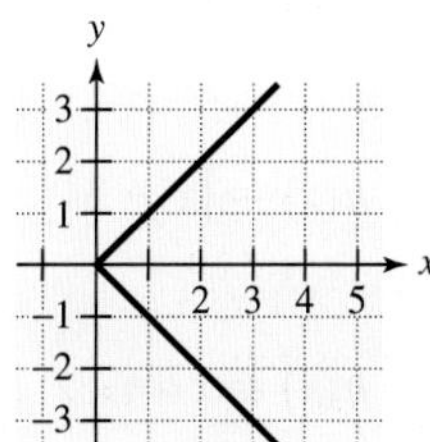

Not a function

Transformations of Graphs of Functions

Vertical and Horizontal Shifts

Let c be a positive real number. **Vertical** and **horizontal shifts** of the graph of the function $y = f(x)$ are represented as follows.

1. Vertical shift c units *upward*: $h(x) = f(x) + c$
2. Vertical shift c units *downward*: $h(x) = f(x) - c$
3. Horizontal shift c units to the *right*: $h(x) = f(x - c)$
4. Horizontal shift c units to the *left*: $h(x) = f(x + c)$

Additional Examples

Use the graph of $g(x) = x^2$ to sketch the graph of each function.

a. $f(x) = x^2 + 3$

b. $h(x) = (x - 2)^2$

Answers:

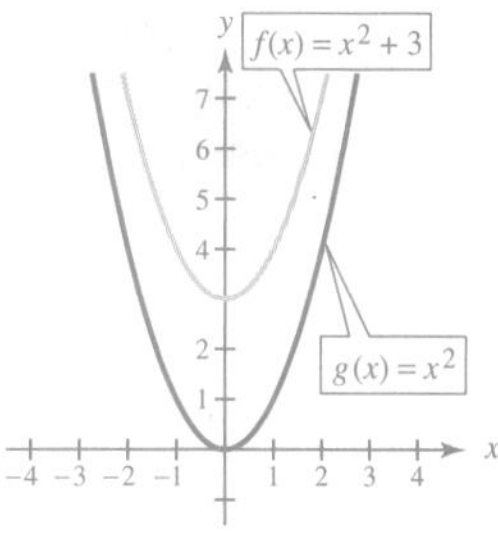

a. Vertical shift: three units upward

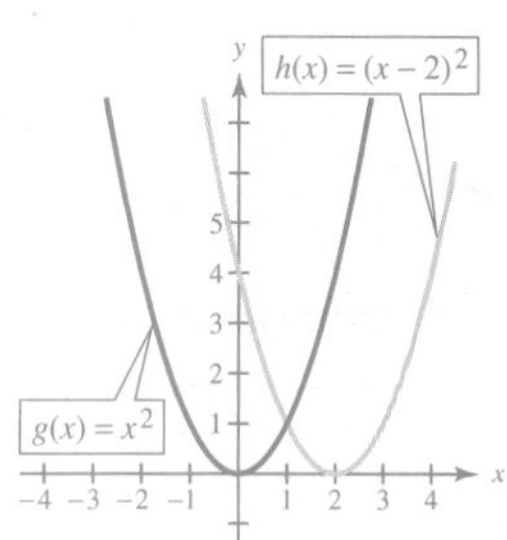

b. Horizontal shift: two units right

EXAMPLE 4 Shifts of the Graphs of Functions

Use the graph of $f(x) = x^2$ to sketch the graph of each function.

a. $g(x) = x^2 - 2$ **b.** $h(x) = (x + 3)^2$

SOLUTION

a. Relative to the graph of $f(x) = x^2$, the graph of $g(x) = x^2 - 2$ represents a shift of two units *downward*, as shown below on the left.

b. Relative to the graph of $f(x) = x^2$, the graph of $h(x) = (x + 3)^2$ represents a shift of three units to the *left*, as shown below on the right.

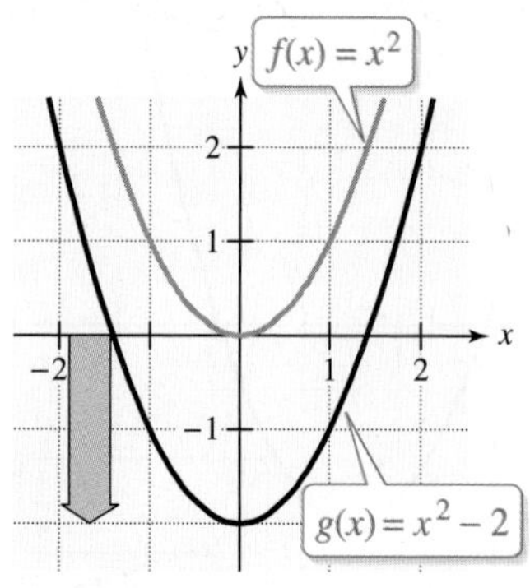

Vertical shift: Two units downward

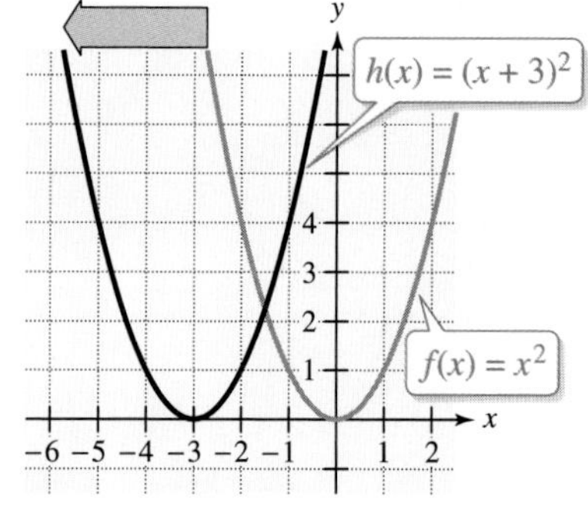

Horizontal shift: Three units left

Exercises Within Reach®

Solutions in English & Spanish and tutorial videos at AlgebraWithinReach.com

Shifts of the Graphs of Functions **In Exercises 23–26, identify the transformation of f, and sketch the graph of the function h.** See Additional Answers.

23. $f(x) = x^2$

(a) $h(x) = x^2 + 2$ (b) $h(x) = x^2 - 4$ (c) $h(x) = (x + 2)^2$ (d) $h(x) = (x - 4)^2$

24. $f(x) = |x|$

(a) $h(x) = |x - 3|$ (b) $h(x) = |x + 3|$ (c) $h(x) = |x| - 4$ (d) $h(x) = |x| + 6$

25. $f(x) = x^3$

(a) $h(x) = x^3 + 3$ (b) $h(x) = x^3 - 5$ (c) $h(x) = (x - 3)^3$ (d) $h(x) = (x + 2)^3$

26. $f(x) = \sqrt{x}$

(a) $h(x) = \sqrt{x - 2}$ (b) $h(x) = \sqrt{x} + 2$ (c) $h(x) = 4 + \sqrt{x}$ (d) $h(x) = \sqrt{4 + x}$

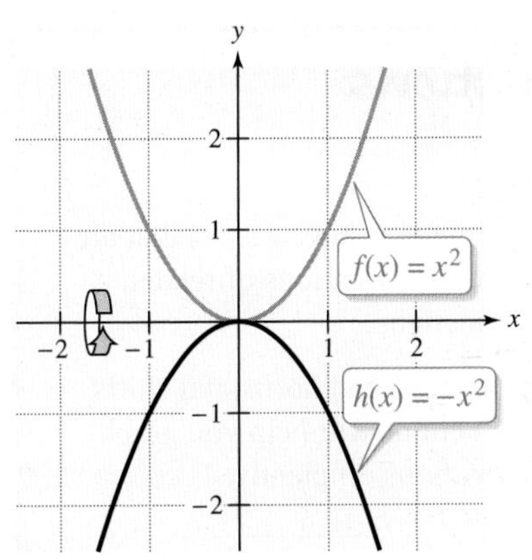

The second basic type of transformation is a reflection. For instance, if you imagine that the x-axis represents a mirror, then the graph of

$$h(x) = -x^2$$

is the mirror image (or reflection) of the graph of $f(x) = x^2$.

Reflections in the Coordinate Axes

Reflections of the graph of $y = f(x)$ are represented as follows.

1. Reflection in the x-axis: $h(x) = -f(x)$
2. Reflection in the y-axis: $h(x) = f(-x)$

EXAMPLE 5 Reflections of the Graphs of Functions

Use the graph of $f(x) = \sqrt{x}$ to sketch the graph of each function.

a. $g(x) = -\sqrt{x}$ **b.** $h(x) = \sqrt{-x}$

SOLUTION

a. Relative to the graph of $f(x) = \sqrt{x}$, the graph of $g(x) = -\sqrt{x} = -f(x)$ represents a *reflection in the x-axis*, as shown below on the left.

b. Relative to the graph of $f(x) = \sqrt{x}$, the graph of $h(x) = \sqrt{-x} = f(-x)$ represents a *reflection in the y-axis*, as shown below on the right.

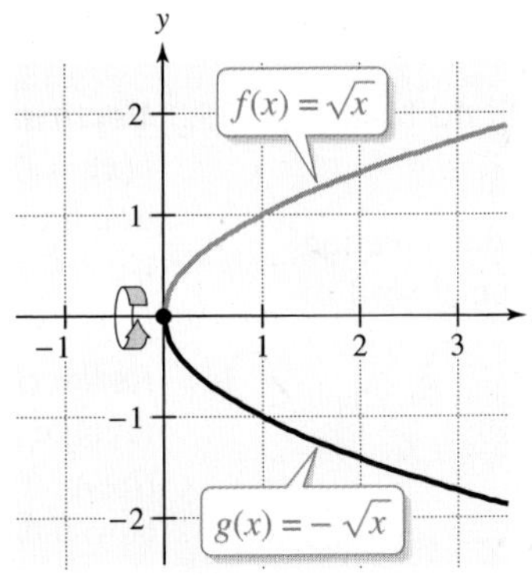

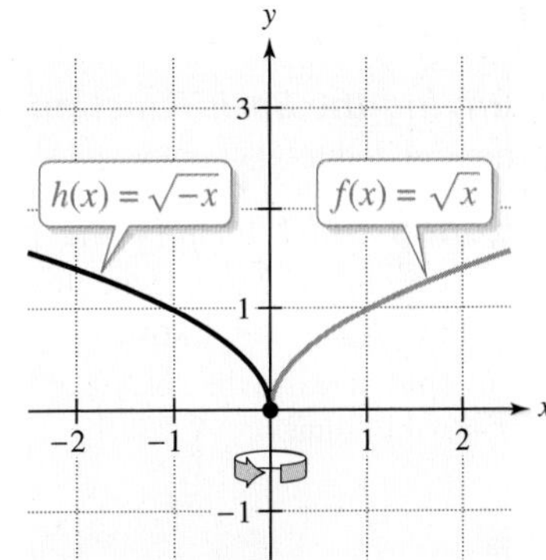

Exercises Within Reach®

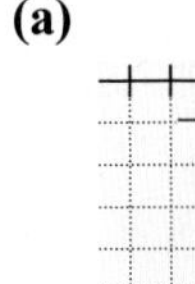

Matching **In Exercises 27–32, match the function with its graph.**

(a) 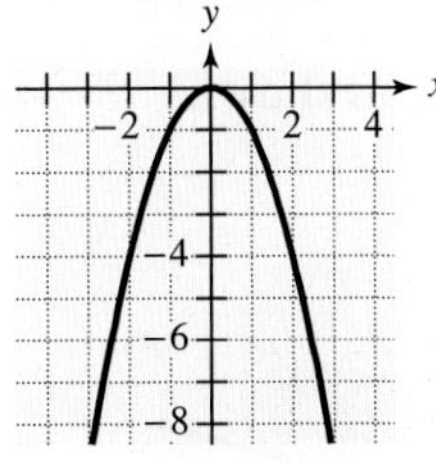

(b) 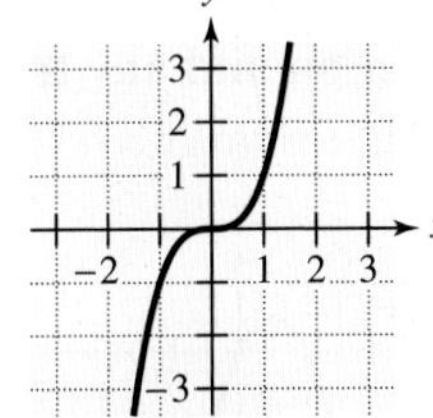

(c) 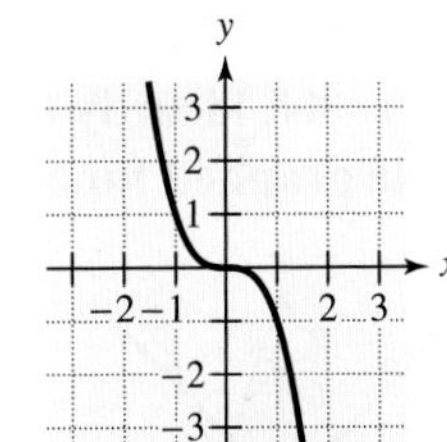

(d) 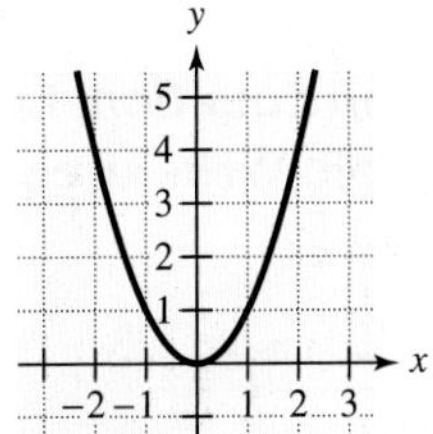

27. $y = x^2$ d

28. $y = -x^2$ a

29. $y = (-x)^2$ d

30. $y = x^3$ b

31. $y = -x^3$ c

32. $y = (-x)^3$ c

Concept Summary: Transformations of Graphs of Functions

What

Once you can identify the **graphs** of basic functions, you can use **vertical** and **horizontal shifts** and **reflections** to sketch the graphs of many functions.

EXAMPLE

Sketch the graph of

$h(x) = (x - 1)^2 + 2.$

How

Shifts (c is a positive real number.)

Up: $h(x) = f(x) + c$
Down: $h(x) = f(x) - c$
Right: $h(x) = f(x - c)$
Left: $h(x) = f(x + c)$

Reflections

In the x-axis: $h(x) = -f(x)$
In the y-axis: $h(x) = f(-x)$

EXAMPLE

The basic function is $f(x) = x^2$. To graph $h(x) = (x - 1)^2 + 2$, shift the graph of the basic function 1 unit to the right and 2 units up.

Why

Many functions have graphs that are simple transformations of related basic functions.

Using vertical and horizontal shifts and reflections will help you graph such functions efficiently.

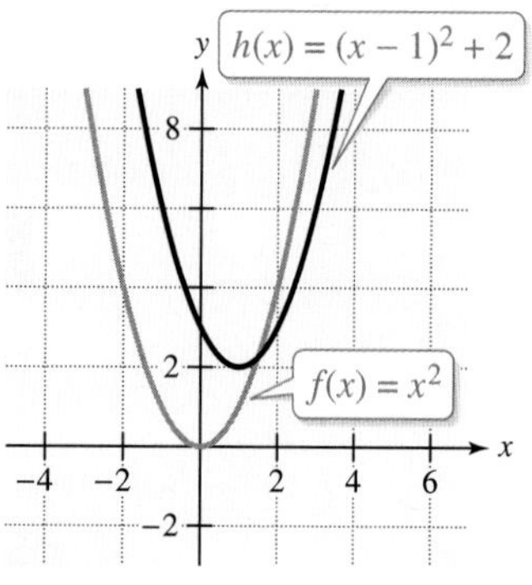

Exercises Within Reach ®

Worked-out solutions to odd-numbered exercises at AlgebraWithinReach.com

Concept Summary Check

33. ***Identifying the Basic Function*** Identify the basic function related to $h(x) = (x - 1)^2 + 2$. $f(x) = x^2$

34. ***Horizontal Shift*** Identify the part of the function $h(x) = (x - 1)^2 + 2$ that indicates a horizontal shift of the graph of $f(x) = x^2$. Explain.
$(x - 1)^2$; This increases the x-value of each point of the graph of $f(x) = x^2$ by 1 unit.

35. ***Vertical Shift*** Identify the part of the function $h(x) = (x - 1)^2 + 2$ that indicates a vertical shift of the graph of $f(x) = x^2$. Explain.
$+2$; This increases the y-value of each point of the graph of $f(x) = x^2$ by 2 units.

36. ***Reflection*** Does the graph of $h(x) = (x - 1)^2 + 2$ involve a reflection of the graph of $f(x) = x^2$? Explain.
No; A reflection would be indicated by a negative sign with x or with the squared expression $(x - 1)^2$.

Extra Practice

Using the Vertical Line Test **In Exercises 37–40, sketch the graph of the equation. Use the Vertical Line Test to determine whether y is a function of x.** See Additional Answers.

37. $-2x + 3y = 12$
y is a function of x.

38. $y = x^2 + 2$
y is a function of x.

39. $y^2 = x + 1$
y is not a function of x.

40. $x = y^4$
y is not a function of x.

Writing an Equation **In Exercises 41–44, identify the basic function and any transformation shown in the graph. Write an equation for the graphed function.** See Additional Answers.

41. 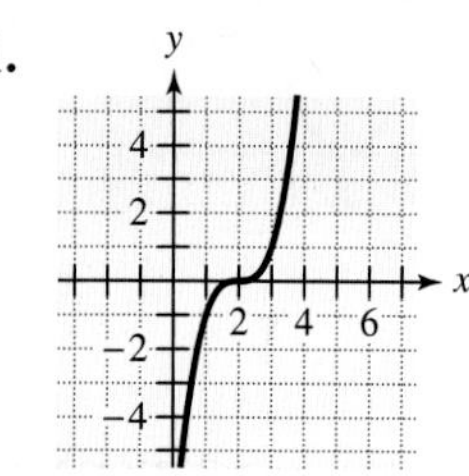

42. 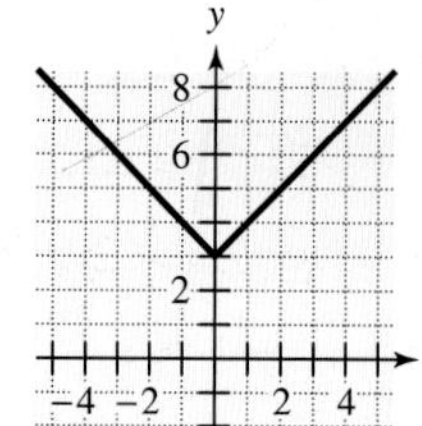

43. 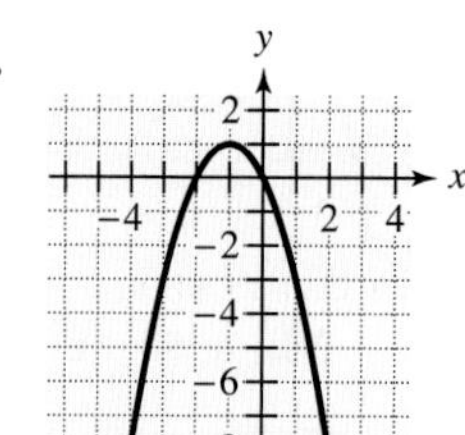

44. 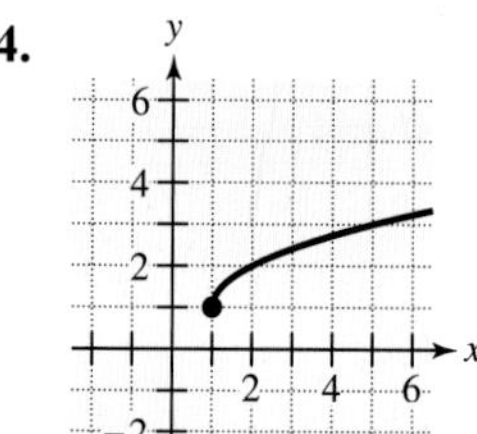

45. ***Sketching Transformations*** Use the graph of f to sketch each graph. See Additional Answers.

(a) $y = f(x) + 2$
(b) $y = -f(x)$
(c) $y = f(x - 2)$
(d) $y = f(x + 2)$
(e) $y = f(x) - 1$
(f) $y = f(-x)$

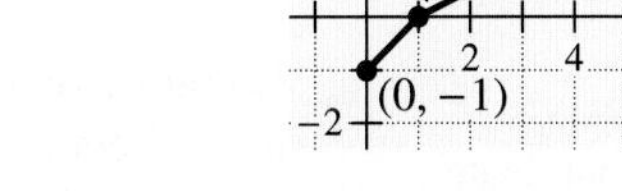

46. ***Sketching Transformations*** Use the graph of f to sketch each graph. See Additional Answers.

(a) $y = f(x) - 1$
(b) $y = f(x + 1)$
(c) $y = f(x - 1)$
(d) $y = -f(x)$
(e) $y = f(-x)$
(f) $y = f(x) + 2$

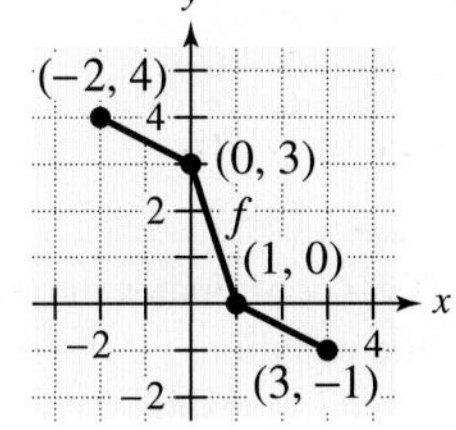

47. ***Geometry*** The perimeter of a rectangle is 200 meters. See Additional Answers.

(a) Show algebraically that the area of the rectangle is given by $A = l(100 - l)$, where l is its length.

(b) Sketch the graph of the area function.

(c) Use the graph to determine the value of l that yields the greatest value of A. Interpret the result.

48. ***Graphical Reasoning*** A thermostat in a home is programmed to lower the temperature during the night. The graph shows the temperature T (in degrees Fahrenheit) in terms of t, the time on a 24-hour clock.

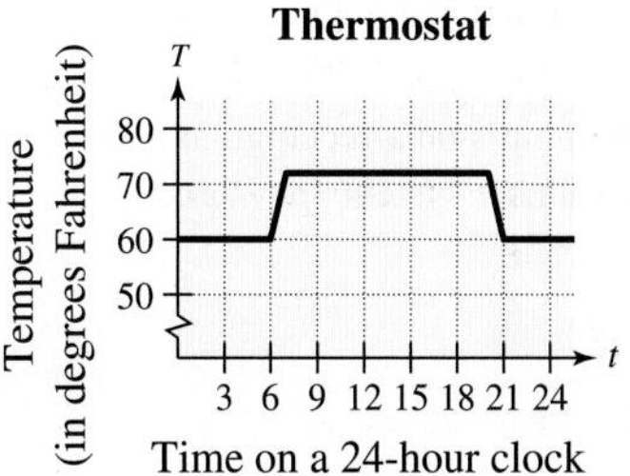

(a) Explain why T is a function of t. For each time t, there corresponds one and only one temperature T.

(b) Find $T(4)$ and $T(15)$. 60°F, 72°F

(c) The thermostat is reprogrammed to produce a temperature H, where $H(t) = T(t - 1)$. Explain how this changes the temperature in the house. All of the temperature changes occur 1 hour later.

(d) The thermostat is reprogrammed to produce a temperature H, where $H(t) = T(t) - 1$. Explain how this changes the temperature in the house. All of the temperatures decrease by 1 degree.

Explaining Concepts

49. ***Writing*** In your own words, explain how to use the Vertical Line Test. If the graph of an equation has the property that no vertical line intersects the graph at two (or more) points, the equation represents y as a function of x.

50. ***Types of Shifts*** Describe the four types of shifts of the graph of a function. Vertical shift upward, vertical shift downward, horizontal shift to the left, horizontal shift to the right

51. ***Describing a Transformation*** Describe the relationship between the graphs of $f(x)$ and $g(x) = f(-x)$. The graph of $g(x)$ is a reflection in the y-axis of the graph of $f(x)$.

52. ***Describing a Transformation*** Describe the relationship between the graphs of $f(x)$ and $g(x) = f(x - 2)$. The graph of $g(x)$ is a horizontal shift two units to the right of the graph of $f(x)$.

Cumulative Review

In Exercises 53–56, write a verbal description of the algebraic expression.

53. $4x + 1$ The sum of four times a number and 1

54. $x(x - 2)$ A number multiplied by the difference of the number and 2

55. $\dfrac{2n}{3}$ The ratio of two times a number and 3

56. $x^2 + 6$ A number squared increased by 6

In Exercises 57–60, solve for y in terms of x.

57. $2x + y = 4$ $y = -2x + 4$

58. $3x - 6y = 12$ $y = \frac{1}{2}x - 2$

59. $-4x + 3y + 3 = 0$ $y = \frac{4}{3}x - 1$

60. $3x + 4y - 5 = 0$ $y = -\frac{3}{4}x + \frac{5}{4}$

In Exercises 61–64, determine whether the point is a solution of the inequality.

61. $y < 2x + 1$, $(0, 1)$ Not a solution

62. $y \le 4 - |2x|$, $(-2, -1)$ Solution

63. $2x - 3y > 2y$, $(6, 2)$ Solution

64. $4x + 1 > x - y$, $(-1, 3)$ Solution

3 Chapter Summary

	What did you learn?	Explanation and Examples	Review Exercises
3.1	Plot points on a rectangular coordinate system *(p. 102).*	y-axis; Distance from y-axis; x-coordinate; y-coordinate; (3, 2); Distance from x-axis; Origin; x-axis	1–6
	Determine whether ordered pairs are solutions of equations *(p. 105).*	**1.** Substitute the ordered pair coordinates into the equation. **2.** Simplify each side of the equation. **3.** Check whether a true equation results.	7, 8
	Use the Distance Formula and the Midpoint Formula *(p. 106).*	The distance between two points (x_1, y_1) and (x_2, y_2) is $d = \sqrt{(x_2 - x_1)^2 + (y_2 - y_1)^2}$. The midpoint of the line segment joining (x_1, y_1) and (x_2, y_2) is $\text{Midpoint} = \left(\frac{x_1 + x_2}{2}, \frac{y_1 + y_2}{2}\right)$.	9–16
3.2	Sketch graphs of equations using the point-plotting method *(p. 110).*	**1.** Rewrite the equation by isolating one variable. **2.** Make a table of values. **3.** Plot the resulting points. **4.** Connect the points with a smooth curve or line.	17–22
	Find and use x- and y-intercepts as aids to sketching graphs *(p. 113).*	To find x-intercepts: Let $y = 0$ and solve for x. To find y-intercepts: Let $x = 0$ and solve for y.	23–28
	Solve application problems involving graphs of equations *(p. 114).*	You can use a graph to identify important information. You can also determine how well a model fits a set of data by looking at a graph.	29, 30
3.3	Determine the slope of a line through two points *(p. 118).*	Slope through two points (x_1, y_1) and (x_2, y_2): $m = \frac{y_2 - y_1}{x_2 - x_1} = \frac{\text{Change in } y}{\text{Change in } x} = \frac{\text{Rise}}{\text{Run}}$	31–36
	Write linear equations in slope-intercept form and graph the equations *(p. 120).*	In the slope-intercept form of the equation of a line $y = mx + b$ the slope of the graph is m and the y-intercept is $(0, b)$.	37–40
	Use slopes to determine whether two lines are parallel, perpendicular, or neither *(p. 122).*	Two distinct nonvertical lines are *parallel* if and only if they have the same slope. Two nonvertical lines are *perpendicular* if and only if their slopes are negative reciprocals of each other.	41–44
	Use slopes to describe rates of change in real-life problems *(p. 123).*	In real-life problems, slope can describe a constant rate of change or an average rate of change.	45, 46

	What did you learn?	Explanation and Examples	Review Exercises
3.4	Write equations of lines using point-slope form *(p. 126).*	The point-slope form of the equation of the line that passes through the point (x_1, y_1) and has a slope of m is $y - y_1 = m(x - x_1)$.	47–54
	Write equations of horizontal, vertical, parallel, and perpendicular lines *(p. 128).*	The equation of a *horizontal line* has the form $y = b$. The equation of a *vertical line* has the form $x = a$.	55–60
	Use linear models to solve application problems *(p. 130).*	You can use a linear model to extrapolate and interpolate.	61, 62
3.5	Verify solutions of linear inequalities in two variables *(p. 136).*	An ordered pair (x_1, y_1) is a solution of a linear inequality in x and y if the inequality is true when x_1 and y_1 are substituted for x and y, respectively.	63, 64
	Sketch graphs of linear inequalities in two variables *(p. 137).*	**1.** Form the corresponding linear equation. Graph the equation using a dashed line for $<$ or $>$, or a solid line for $\le$ or $\ge$. **2.** Test a point in one of the half-planes. **a.** If the point satisfies the inequality, shade its half-plane. **b.** If not, shade the other half-plane.	65–72
	Use linear inequalities to model and solve real-life problems *(p. 141).*	Construct a verbal model and assign labels to write a linear inequality for a real-life problem. Graph the inequality to find the solutions of the problem.	73, 74
3.6	Identify the domains and ranges of relations *(p. 144).*	A relation is any set of ordered pairs. The set of first components in the ordered pairs is the domain of the relation. The set of second components is the range of the relation.	75, 76
	Determine if relations are functions by inspection *(p. 146).*	A function is a relation in which no two ordered pairs have the same first component and different second components.	77–80
	Use function notation and evaluate functions *(p. 148).*	$f(x)$ is read as *the value of f at x* or simply *f of x*. *Independent variable, input,* and x are terms associated with a function's domain. *Dependent variable, output, y,* and $f(x)$ are terms associated with a function's range.	81–84
	Identify the domains and ranges of functions *(p. 149).*	A function's *implied domain* is the set of all real numbers that yield real number values for the function. The *range* consists of the function values that result from all of the values of the domain.	85–90
3.7	Sketch graphs of functions on rectangular coordinate systems *(p. 152).*	Make a table of values for the function. Then plot the points and connect them with a line or smooth curve.	91–100
	Use the Vertical Line Test to determine if graphs represent functions *(p. 155).*	A vertical line can intersect the graph of a function at no more than one point.	101, 102
	Use vertical and horizontal shifts and reflections to sketch graphs of functions *(p. 156).*	Vertical shift of c units: $h(x) = f(x) \pm c$ Horizontal shift of c units: $h(x) = f(x \pm c)$ Reflection in x-axis: $h(x) = -f(x)$ Reflection in y-axis: $h(x) = f(-x)$	103–106

Review Exercises

Worked-out solutions to odd-numbered exercises at AlgebraWithinReach.com

3.1

Plotting Points **In Exercises 1 and 2, plot the points on a rectangular coordinate system.**
See Additional Answers.

1. $(0, 2), \left(-4, \frac{1}{2}\right), (2, -3)$

2. $\left(1, -\frac{3}{2}\right), \left(-2, 2\frac{3}{4}\right), (5, 10)$

Determining the Quadrant **In Exercises 3–6, determine the quadrant in which the point is located without plotting it. (x and y are real numbers.)**

3. $(6, 4)$
Quadrant I

4. $(-4.8, -2)$
Quadrant III

5. $(4, y),\ y \neq 0$
Quadrant I or IV

6. $(x, y),\ xy > 0$
Quadrant I or III

Verifying Solutions of an Equation **In Exercises 7 and 8, determine whether each ordered pair is a solution of the equation.**

7. $y = 4 - \frac{1}{2}x$

(a) $(2, 3)$
Solution

(b) $(-1, 5)$
Not a solution

(c) $(-6, 1)$
Not a solution

(d) $(8, 0)$
Solution

8. $3x - 2y + 18 = 0$

(a) $(3, 10)$
Not a solution

(b) $(0, 9)$
Solution

(c) $(-4, 3)$
Solution

(d) $(-8, 0)$
Not a solution

Using the Distance Formula **In Exercises 9–12, find the distance between the points.**

9. $(4, 3), (4, 8)$ 5

10. $(2, -5), (6, -5)$ 4

11. $(-5, -1), (1, 2)$ $3\sqrt{5}$

12. $(-3, 3), (6, -1)$ $\sqrt{97}$

Using the Midpoint Formula **In Exercises 13–16, find the midpoint of the line segment joining the points.**

13. $(1, 4), (7, 2)$ $(4, 3)$

14. $(-1, 3), (5, 5)$ $(2, 4)$

15. $(5, -2), (-3, 5)$ $\left(1, \frac{3}{2}\right)$

16. $(1, 6), (6, 1)$ $\left(\frac{7}{2}, \frac{7}{2}\right)$

3.2

Sketching the Graph of an Equation **In Exercises 17–22, sketch the graph of the equation.**
See Additional Answers.

17. $3y - 2x - 3 = 0$

18. $3x + 4y + 12 = 0$

19. $y = x^2 - 1$

20. $y = (x + 3)^2$

21. $y = |x| - 2$

22. $y = |x - 3|$

Finding the Intercepts of a Graph **In Exercises 23–28, sketch the graph of the equation and show the coordinates of three solution points (including x- and y-intercepts).** See Additional Answers.

23. $8x - 2y = -4$
$\left(-\frac{1}{2}, 0\right), (0, 2)$

24. $5x + 4y = 10$
$(2, 0), \left(0, \frac{5}{2}\right)$

25. $y = 5 - |x|$
$(-5, 0), (5, 0), (0, 5)$

26. $y = |x| + 4$
$(0, 4)$

27. $y = |2x + 1| - 5$
$(-3, 0), (2, 0), (0, -4)$

28. $y = |3 - 6x| - 15$
$(-2, 0), (3, 0), (0, -12)$

29. ***Straight-Line Depreciation*** Your family purchases a new SUV for \$35,000. For financing purposes, the SUV will be depreciated over a five-year period. At the end of 5 years, the value of the SUV is expected to be \$15,000.

(a) Find an equation that relates the value of the SUV to the number of years since its purchase.
$y = 35{,}000 - 4000t,\ 0 \le t \le 5$

(b) Sketch the graph of the equation.
See Additional Answers.

(c) What is the y-intercept of the graph and what does it represent? $(0, 35{,}000)$; Initial purchase price

30. ***Straight-Line Depreciation*** A company purchases a new computer system for \$20,000. For tax purposes, the computer system will be depreciated over a six-year period. At the end of the 6 years, the value of the system is expected to be \$2000.

(a) Find an equation that relates the value of the computer system to the number of years since its purchase.
$y = 20{,}000 - 3000t,\ 0 \le t \le 6$

(b) Sketch the graph of the equation.
See Additional Answers.

(c) What is the y-intercept of the graph and what does it represent? $(0, 20{,}000)$; Initial purchase price

3.3

Finding the Slope of a Line **In Exercises 31–36, find the slope of the line that passes through the points.**

31. $(4, 2), (-3, -1)$ $\frac{3}{7}$

32. $(-2, 5), (3, -8)$ $-\frac{13}{5}$

33. $\left(-6, \frac{3}{4}\right), \left(4, \frac{3}{4}\right)$ 0

34. $(7, 2), (7, 8)$ Undefined

35. $(0, 6), (8, 0)$ $-\frac{3}{4}$

36. $(0, 0), \left(\frac{7}{2}, 6\right)$ $\frac{12}{7}$

Sketching a Line **In Exercises 37–40, write the equation of the line in slope-intercept form, and then use the slope and *y*-intercept to sketch the line.**

See Additional Answers.

37. $5x - 2y - 4 = 0$
$y = \frac{5}{2}x - 2$

38. $x - 3y - 6 = 0$
$y = \frac{1}{3}x - 2$

39. $6x + 2y + 5 = 0$
$y = -3x - \frac{5}{2}$

40. $y - 6 = 0$
$y = 6$

Parallel and Perpendicular Lines **In Exercises 41–44, determine whether the lines are parallel, perpendicular, or neither.**

41. L_1: $y = \frac{3}{2}x + 1$
L_2: $y = \frac{2}{3}x - 1$
Neither

42. L_1: $y = 2x - 5$
L_2: $y = 2x + 3$
Parallel

43. L_1: $y = \frac{3}{2}x - 2$
L_2: $y = -\frac{2}{3}x + 1$
Perpendicular

44. L_1: $y = -0.3x - 2$
L_2: $y = 0.3x + 1$
Neither

45. ***Consumer Awareness*** In 2000, the average price of a gallon of whole milk was \$2.79. By 2010, the average price had risen to \$3.32. Find the average rate of change in the price of a gallon of whole milk from 2000 to 2010. (*Source:* U.S. Bureau of Labor Statistics) \$0.053

46. ***Consumer Awareness*** In 2004, the average price of a pound of ground roast coffee was \$2.78. By 2010, the average price had risen to \$4.15. Find the average rate of change in the price of a pound of ground roast coffee from 2004 to 2010. (*Source:* U.S. Bureau of Labor Statistics) \$0.228

3.4

Using Point-Slope Form **In Exercises 47–50, use the point-slope form of the equation of a line to write an equation of the line that passes through the point and has the specified slope. Write the equation in slope-intercept form.**

47. $(1, -4)$
$m = -4$ $y = -4x$

48. $(-5, -5)$
$m = 3$ $y = 3x + 10$

49. $(-6, 5)$
$m = \frac{1}{4}$ $y = \frac{1}{4}x + \frac{13}{2}$

50. $(3, 7)$
$m = -\frac{3}{5}$ $y = -\frac{3}{5}x + \frac{44}{5}$

Writing an Equation of a Line **In Exercises 51–54, write the general form of the equation of the line that passes through the points.**

51. $(-6, 0), (0, -3)$ $x + 2y + 6 = 0$

52. $(0, 10), (6, 8)$ $x + 3y - 30 = 0$

53. $(-2, -3), (4, 6)$ $3x - 2y = 0$

54. $(-10, 2), (4, -7)$ $9x + 14y + 62 = 0$

Horizontal and Vertical Lines **In Exercises 55–58, write an equation of the line.**

55. Horizontal line through $(5, -9)$ $y = -9$

56. Vertical line through $(-4, -6)$ $x = -4$

57. Line through $(-5, -2)$ and $(-5, 3)$ $x = -5$

58. Line through $(1, 8)$ and $(15, 8)$ $y = 8$

Writing Equations of Lines **In Exercises 59 and 60, write equations of the lines that pass through the point and are (a) parallel and (b) perpendicular to the given line.** See Additional Answers.

59. $\left(\frac{3}{5}, -\frac{4}{5}\right)$
$3x + y = 2$

60. $(-1, 5)$
$2x + 4y = 1$

61. ***Annual Salary*** Your annual salary in 2007 was \$32,000. During the next 6 years, your annual salary increased by approximately \$1050 per year.

(a) Write an equation of the line giving the annual salary S in terms of the year t. (Let $t = 7$ correspond to the year 2007.) $S = 1050t + 24{,}650$

(b) Use the equation in part (a) to predict your annual salary in the year 2015. \$40,400

(c) Use the equation in part (a) to estimate what your annual salary was in 2010. \$35,150

62. ***Rent*** The rent for an apartment was \$525 per month in 2008. During the next 5 years, the rent increased by approximately \$45 per year.

(a) Write an equation of the line giving the rent R in terms of the year t. (Let $t = 8$ correspond to the year 2008.) $R = 45t + 165$

(b) Use the equation in part (a) to predict the rent in the year 2016. \$885

(c) Use the equation in part (a) to estimate the rent in 2012. \$705

3.5

Verifying Solutions of a Linear Inequality **In Exercises 63 and 64, determine whether each ordered pair is a solution of the inequality.**

63. $5x - 8y \geq 12$

(a) $(-1, 2)$ Not a solution (b) $(3, -1)$ Solution

(c) $(4, 0)$ Solution (d) $(0, 3)$ Not a solution

64. $2x + 4y - 14 < 0$

(a) $(0, 0)$ Solution (b) $(3, 2)$ Not a solution

(c) $(1, 3)$ Not a solution (d) $(-4, 1)$ Solution

Sketching the Graph of a Linear Inequality **In Exercises 65–72, sketch the graph of the linear inequality.** See Additional Answers.

65. $y > -2$ 66. $x \leq 5$

67. $x - 2 \geq 0$ 68. $y - 4 < 0$

69. $2x + y < 1$ 70. $3x - 4y > 2$

71. $-(x - 1) \leq 4y - 2$ 72. $(y - 3) \geq 2(x - 5)$

73. ***Geometry*** The perimeter of a rectangle of length x and width y cannot exceed 800 feet.

(a) Write a linear inequality for this constraint. $x + y \leq 400; x \geq 0, y \geq 0$

(b) Sketch the graph of the inequality. See Additional Answers.

74. ***Weekly Pay*** You have two part-time jobs. One is at a grocery store, which pays \$8.50 per hour, and the other is mowing lawns, which pays \$10 per hour. Between the two jobs, you want to earn at least \$250 per week.

(a) Write a linear inequality that represents the different numbers of hours you can work at each job. $8.5x + 10y \geq 250; x \geq 0, y \geq 0$

(b) Graph the inequality and identify three ordered pairs that are solutions of the inequality. See Additional Answers.

3.6

Analyzing a Relation **In Exercises 75 and 76, find the domain and the range of the relation. Then draw a graphical representation of the relation.**
See Additional Answers.

75. $\{(-3, 4), (-1, 0), (0, 1), (1, 4), (-3, 5)\}$
Domain: $\{-3, -1, 0, 1\}$; Range: $\{0, 1, 4, 5\}$

76. $\{(-2, 4), (-1, 1), (0, 0), (1, 1), (2, 4)\}$
Domain: $\{-2, -1, 0, 1, 2\}$; Range: $\{0, 1, 4\}$

Testing for a Function **In Exercises 77–80, determine whether the relation is a function.**

77. 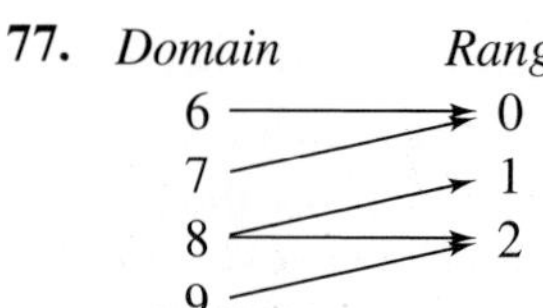

Not a function

78. 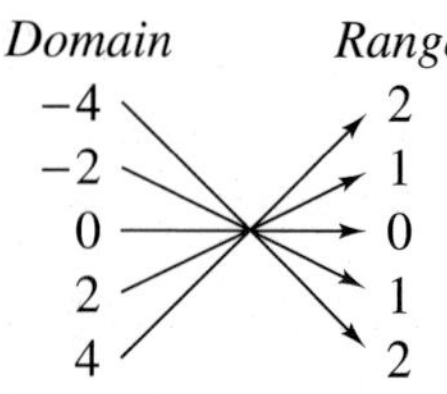

Function

79.

Input, x	Output, y
1	10
2	10
3	10
4	10
5	10

Function

80.

Input, x	Output, y
3	0
6	3
9	8
6	12
0	2

Not a function

Evaluating a Function **In Exercises 81–84, evaluate the function as indicated, and simplify.**

81. $f(t) = \sqrt{5 - t}$

(a) $f(-4)$ 3 (b) $f(5)$ 0

(c) $f(3)$ $\sqrt{2}$ (d) $f(5z)$ $\sqrt{5 - 5z}$

82. $g(x) = |x + 4|$

(a) $g(0)$ 4 (b) $g(-8)$ 4

(c) $g(2) - g(-5)$ 5 (d) $g(x - 2)$ $|x + 2|$

83. $f(x) = \begin{cases} -3x, & x \leq 0 \\ 1 - x^2, & x > 0 \end{cases}$

(a) $f(3)$ -8 (b) $f\left(-\frac{2}{3}\right)$ 2

(c) $f(0)$ 0 (d) $f(4) - f(3)$ -7

84. $h(x) = \begin{cases} x^3, & x \leq 1 \\ (x - 1)^2 + 1, & x > 1 \end{cases}$

(a) $h(1)$ 1 (b) $h\left(-\frac{1}{2}\right)$ $-\frac{1}{8}$

(c) $h(-2)$ -8 (d) $h(4) - h(3)$ 5

Finding the Domain of a Function **In Exercises 85 and 86, find the domain of the function.**

85. $h(x) = 4x^2 - 7$
All real numbers x

86. $g(s) = \dfrac{s + 1}{(s - 1)(s + 5)}$
All real numbers s such that $s \neq 1, -5$

Finding the Domain and Range of a Function **In Exercises 87 and 88, find the domain and range of the function.**

87. $f(x) = \sqrt{3x + 10}$
Domain: All real numbers x such that $x \geq -\frac{10}{3}$
Range: All real numbers such that $f(x) \geq 0$

88. $f(x) = |x - 6| + 10$
Domain: All real numbers x
Range: All real numbers such that $f(x) \geq 10$

89. ***Geometry*** A wire 150 inches long is cut into four pieces to form a rectangle whose shortest side has a length of x. Write the area A of the rectangle as a function of x. What is the domain of the function?
$A = x(75 - x);\ 0 < x < \frac{75}{2}$

90. ***Geometry*** A wire 240 inches long is cut into four pieces to form a rectangle whose shortest side has a length of x. Write the area A of the rectangle as a function of x. What is the domain of the function?
$A = x(120 - x);\ 0 \leq x < 60$

3.7

Sketching the Graph of a Function **In Exercises 91–94, complete the table and use the results to sketch the graph of the function.** See Additional Answers.

91. $y = 4 - (x - 3)^2$

x	0	1	2	3	4	5
$f(x)$	−5	0	3	4	3	0

92. $h(x) = 9 - (x - 2)^2$

x	−1	0	1	2	3	4
$h(x)$	0	5	8	9	8	5

93. $g(x) = 6 - 3x,\ -2 \leq x < 4$

x	−2	−1	0	1	2	4
$g(x)$	12	9	6	3	0	−6

94. $h(x) = x(4 - x),\ 0 \leq x \leq 4$

x	0	1	2	3	4
$h(x)$	0	3	4	3	0

A Piecewise-Defined Function **In Exercises 95 and 96, sketch the graph of the function. Then determine its domain and range.** See Additional Answers.

95. $f(x) = \begin{cases} 2 - (x - 1)^2, & x < 1 \\ 2 + (x - 1)^2, & x \geq 1 \end{cases}$
Domain: $-\infty < x < \infty$
Range: $-\infty < y < \infty$

96. $f(x) = \begin{cases} 2x, & x \leq 0 \\ x^2 + 1, & x > 0 \end{cases}$
Domain: $-\infty < x < \infty$
Range: $-\infty < y \leq 0$ and $1 < y < \infty$

Matching **In Exercises 97–100, match the function with its graph.**

(a)

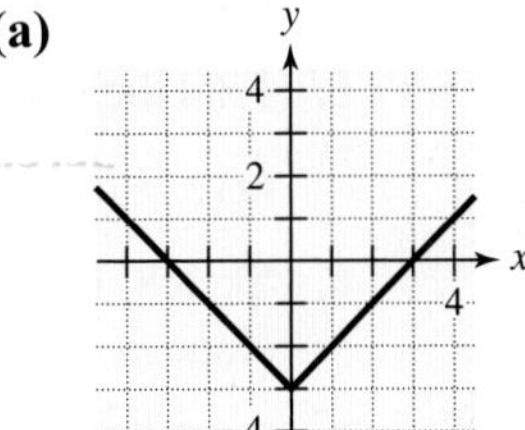

(b)

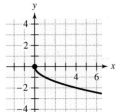

(c)

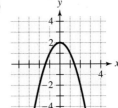

(d)

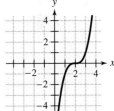

97. $f(x) = -x^2 + 2$ c

98. $f(x) = |x| - 3$ a

99. $f(x) = -\sqrt{x}$ b

100. $f(x) = (x - 2)^3$ d

Using the Vertical Line Test **In Exercises 101 and 102, use the Vertical Line Test to determine whether y is a function of x.**

101. $9y^2 = 4x^3$

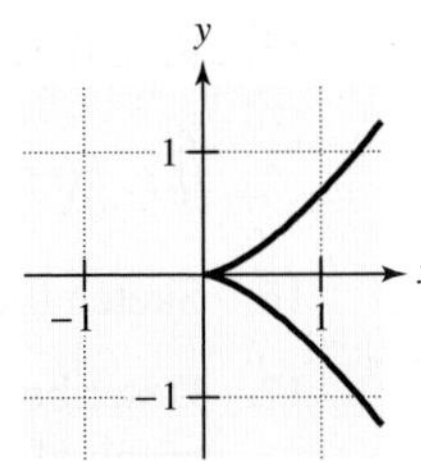

Not a function

102. $y = x^3 - 3x^2$

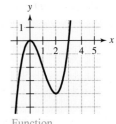

Function

Transformations of Graphs of Functions **In Exercises 103–106, identify the transformation of the graph of $f(x) = \sqrt{x}$, and sketch the graph of h.**
See Additional Answers.

103. $h(x) = -\sqrt{x}$ Reflection in the x-axis

104. $h(x) = \sqrt{x} + 3$ Vertical shift 3 units upward

105. $h(x) = \sqrt{x + 3}$ Horizontal shift 3 units to the left

106. $h(x) = \sqrt{x - 2} - 1$ Horizontal shift 2 units to the right and 1 unit downward

Chapter Test

Solutions in English & Spanish and tutorial videos at AlgebraWithinReach.com

Take this test as you would take a test in class. After you are done, check your work against the answers in the back of the book.

1. Quadrant IV
2. See Additional Answers. Distance: 5; Midpoint: $\left(5, -\frac{1}{2}\right)$
3. $(-1, 0), (0, -3)$
8. $3x - 2y - 18 = 0$
10. (a) $y = \frac{3}{5}x + \frac{21}{5}$
    (b) $y = -\frac{5}{3}x - \frac{1}{3}$
12. No. The graph does not pass the Vertical Line Test.
13. (a) Function; For each value of x there is only one value of y.
    (b) Not a function; Zero is matched with two different elements in the range.
15. (a) All real values of t such that $t \le 9$
    (b) All real values of x such that $x \neq 4$
17. Horizontal shift 2 units to the right, vertical shift 1 unit upward
18. $V = 26{,}000 - 4000t$; 2.5 years since the car was purchased
19. (a) $y = |x - 2|$
    (b) $y = |x| - 2$
    (c) $y = 2 - |x|$

---

1. Determine the quadrant in which the point (x, y) lies when $x > 0$ and $y < 0$.
2. Plot the points $(7, -2)$ and $(3, 1)$. Then find the distance between them and the coordinates of the midpoint of the line segment joining the two points.
3. Find the x- and y-intercepts of the graph of the equation $y = -3(x + 1)$.
4. Sketch the graph of the equation $y = |x - 2|$. See Additional Answers.
5. Find the slope of the line that passes through each pair of points.
   (a) $(-4, 7), (2, 3)$ $-\frac{2}{3}$ (b) $(3, -2), (3, 6)$ Undefined
6. Sketch the graph of the line that passes through the point $(-3, 5)$ with slope $m = \frac{4}{3}$. See Additional Answers.
7. Find the x- and y-intercepts of the graph of $2x + 5y - 10 = 0$. Use the results to sketch the graph. See Additional Answers.
8. Write an equation of the line through the points $(2, -6)$ and $(8, 3)$.
9. Write an equation of the vertical line through the point $(-2, 4)$. $x + 2 = 0$
10. Write equations of the lines that pass through the point $(-2, 3)$ and are (a) parallel and (b) perpendicular to the line $3x - 5y = 4$.
11. Sketch the graph of the inequality $x + 4y \le 8$. See Additional Answers.
12. The graph of $y^2(4 - x) = x^3$ is shown at the left. Does the graph represent y as a function of x? Explain your reasoning.
13. Determine whether each relation represents a function. Explain.
    (a) $\{(2, 4), (-6, 3), (3, 3), (1, -2)\}$ (b) $\{(0, 0), (1, 5), (-2, 1), (0, -4)\}$
14. Evaluate $g(x) = x/(x - 3)$ as indicated, and simplify.
    (a) $g(2)$ -2 (b) $g\left(\frac{7}{2}\right)$ 7 (c) $g(x + 2)$ $\frac{x+2}{x-1}$
15. Find the domain of each function.
    (a) $h(t) = \sqrt{9 - t}$ (b) $f(x) = \dfrac{x + 1}{x - 4}$
16. Sketch the graph of the function $g(x) = \sqrt{2 - x}$. See Additional Answers.
17. Describe the transformation of the graph of $f(x) = x^2$ that would produce the graph of $g(x) = (x - 2)^2 + 1$.
18. After 4 years, the value of a \$26,000 car will have depreciated to \$10,000. Write the value V of the car as a linear function of t, the number of years since the car was purchased. When will the car be worth \$16,000?
19. Use the graph of $f(x) = |x|$ to write a function that represents each graph.

(a)

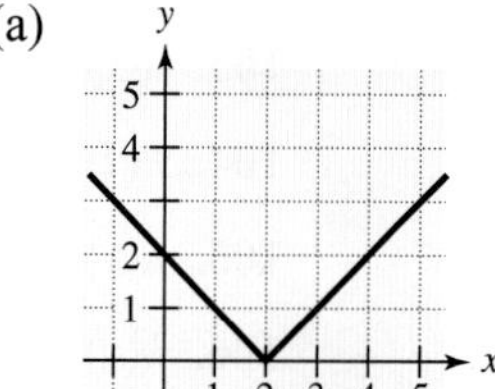

(b)

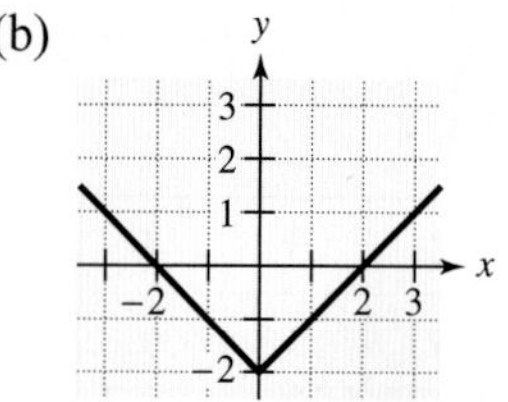

(c)

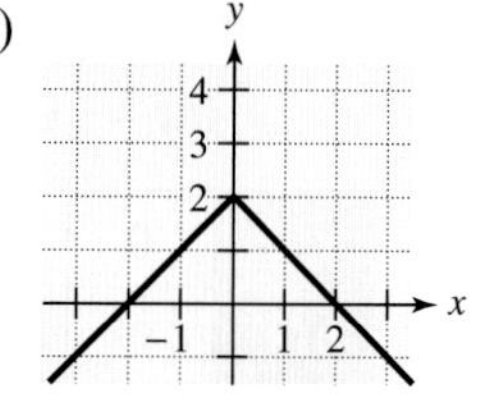

4 Systems of Equations and Inequalities

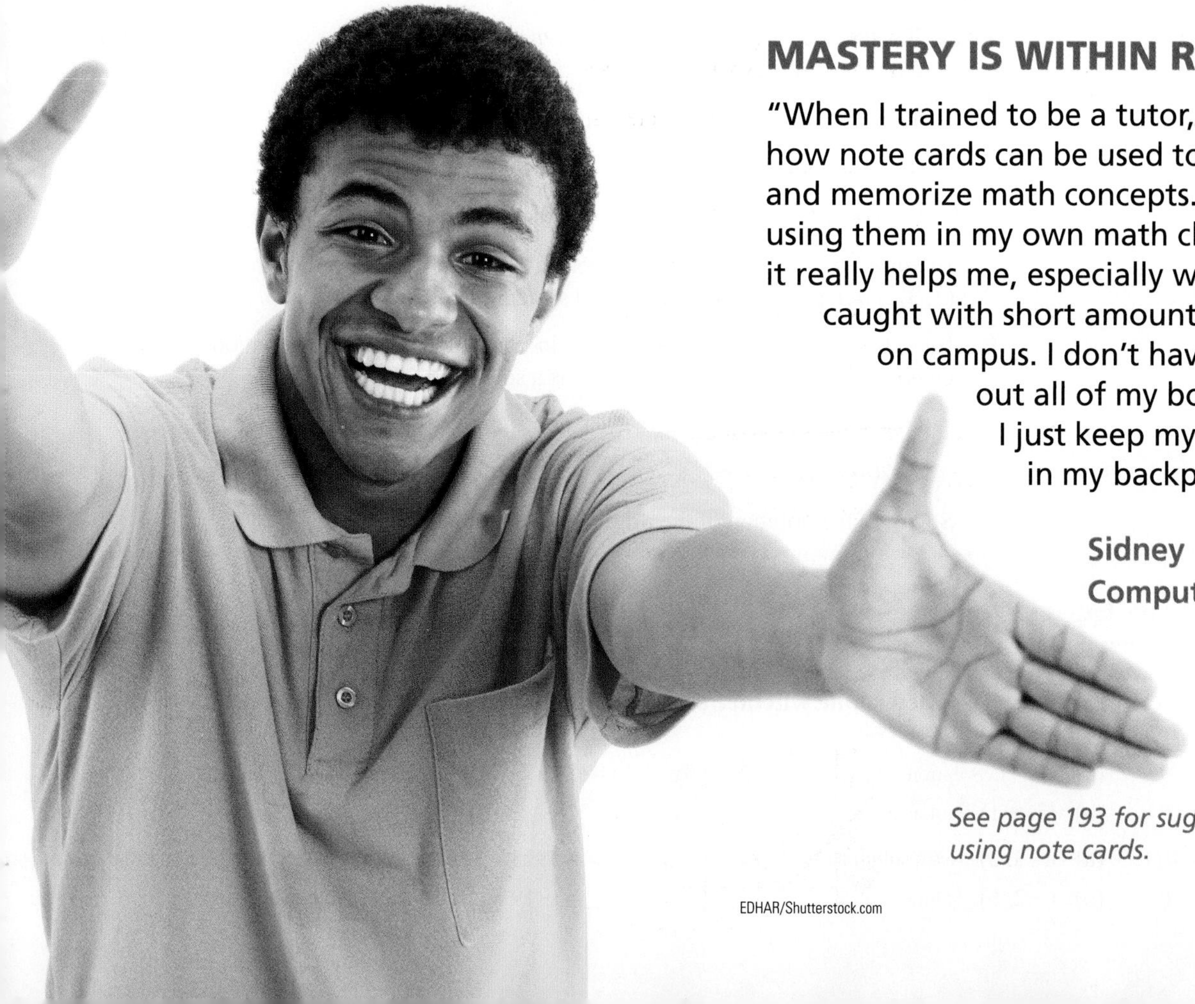

MASTERY IS WITHIN REACH!

"When I trained to be a tutor, I learned how note cards can be used to help review and memorize math concepts. I started using them in my own math class, and it really helps me, especially when I am caught with short amounts of time on campus. I don't have to pull out all of my books to study. I just keep my note cards in my backpack."

Sidney
Computer Science

See page 193 for suggestions about using note cards.

4.1 Systems of Equations

- Determine whether ordered pairs are solutions of systems of equations.
- Solve systems of equations graphically and algebraically.
- Use systems of equations to model and solve real-life problems.

Systems of Equations

Many problems in business and science involve **systems of equations**. These systems consist of two or more equations involving two or more variables.

$$\begin{cases} ax + by = c & \text{Equation 1} \\ dx + ey = f & \text{Equation 2} \end{cases}$$

A **solution** of such a system is an ordered pair (x, y) of real numbers that satisfies *each* equation in the system. When you find the set of all solutions of the system of equations, you are **solving the system of equations**.

EXAMPLE 1 Checking Solutions of a System of Equations

Determine whether each ordered pair is a solution of the system of equations.

$$\begin{cases} x + y = 6 & \text{Equation 1} \\ 2x - 5y = -2 & \text{Equation 2} \end{cases}$$

a. $(3, 3)$ **b.** $(4, 2)$

SOLUTION

a. To determine whether the ordered pair $(3, 3)$ is a solution of the system of equations, substitute 3 for x and 3 for y in *each* of the equations.

$3 + 3 = 6$ ✓ Substitute 3 for x and 3 for y in Equation 1.

$2(3) - 5(3) \neq -2$ ✗ Substitute 3 for x and 3 for y in Equation 2.

Because the check fails in Equation 2, you can conclude that the ordered pair $(3, 3)$ *is not* a solution of the original system of equations.

b. By substituting 4 for x and 2 for y in each of the original equations, you can determine that the ordered pair $(4, 2)$ is a solution of *both* equations.

$4 + 2 = 6$ ✓ Substitute 4 for x and 2 for y in Equation 1.

$2(4) - 5(2) = -2$ ✓ Substitute 4 for x and 2 for y in Equation 2.

So, $(4, 2)$ *is* a solution of the original system of equations.

Exercises Within Reach®

Solutions in English & Spanish and tutorial videos at AlgebraWithinReach.com

Checking Solutions **In Exercises 1–4, determine whether each ordered pair is a solution of the system of equations.**

1. $\begin{cases} x + 2y = 9 \\ -2x + 3y = 10 \end{cases}$ (a) $(1, 4)$ Solution (b) $(3, -1)$ Not a solution

2. $\begin{cases} 5x - 4y = 34 \\ x - 2y = 8 \end{cases}$ (a) $(0, 3)$ Not a solution (b) $(6, -1)$ Solution

3. $\begin{cases} -2x + 7y = 46 \\ 3x + y = 0 \end{cases}$ (a) $(-3, 2)$ Not a solution (b) $(-2, 6)$ Solution

4. $\begin{cases} -5x - 2y = 23 \\ x + 4y = -19 \end{cases}$ (a) $(-3, -4)$ Solution (b) $(3, 7)$ Not a solution

Solving Systems of Equations

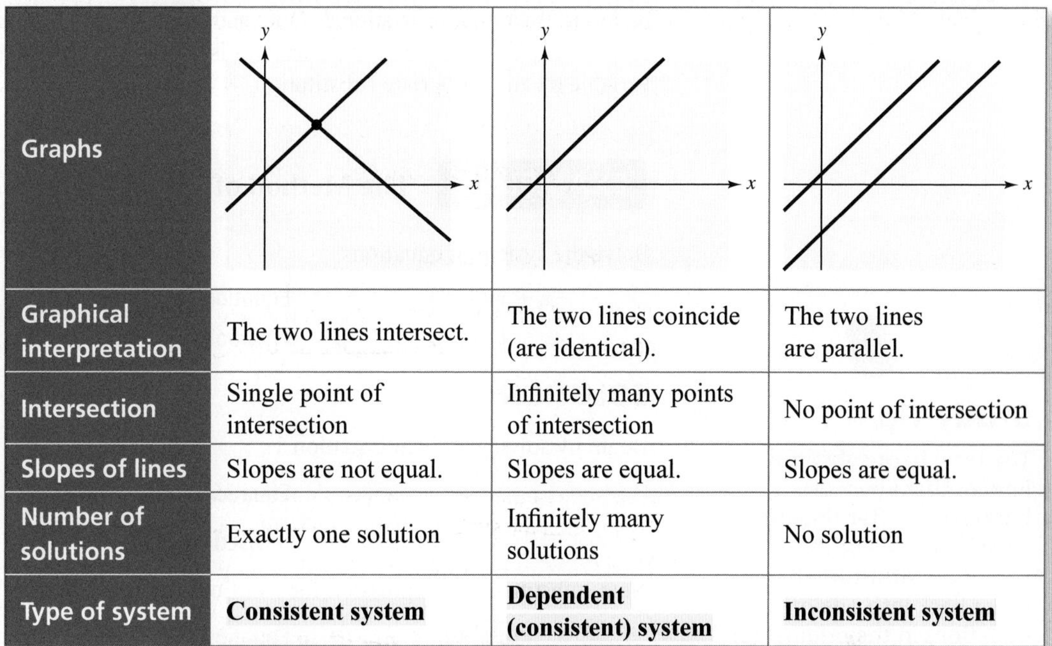

Graphs	(intersecting lines)	(coinciding lines)	(parallel lines)
Graphical interpretation	The two lines intersect.	The two lines coincide (are identical).	The two lines are parallel.
Intersection	Single point of intersection	Infinitely many points of intersection	No point of intersection
Slopes of lines	Slopes are not equal.	Slopes are equal.	Slopes are equal.
Number of solutions	Exactly one solution	Infinitely many solutions	No solution
Type of system	**Consistent system**	**Dependent (consistent) system**	**Inconsistent system**

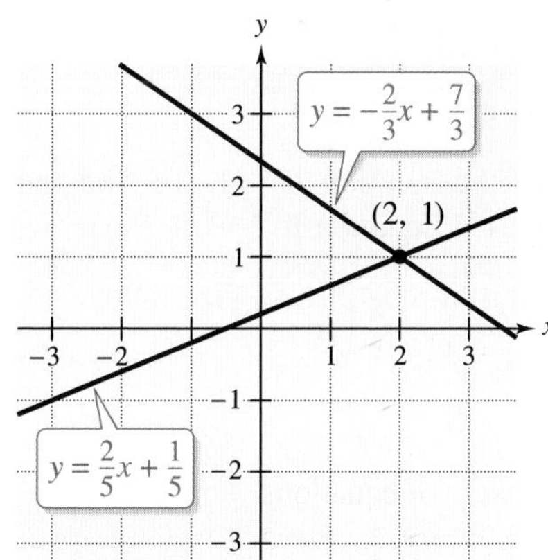

EXAMPLE 2 Solving a System of Equations

Solve the system of equations by graphing.

$$\begin{cases} y = -\frac{2}{3}x + \frac{7}{3} & \text{Equation 1} \\ y = \frac{2}{5}x + \frac{1}{5} & \text{Equation 2} \end{cases}$$

SOLUTION

Because their slopes are not equal, you can conclude that the graphs intersect at a single point. The lines corresponding to these two equations are shown at the left. From this graph, it appears that the two lines intersect at the point (2, 1). Check this in the original system of equations.

Exercises Within Reach®

Solutions in English & Spanish and tutorial videos at AlgebraWithinReach.com

Solving a System of Equations **In Exercises 5–12, solve the system of equations by graphing.**

5. $\begin{cases} y = -x + 3 \\ y = x + 1 \end{cases}$ (1, 2)

6. $\begin{cases} y = 2x - 4 \\ y = -\frac{1}{2}x + 1 \end{cases}$ (2, 0)

7. $\begin{cases} x - y = 2 \\ x + y = 2 \end{cases}$ (2, 0)

8. $\begin{cases} x - y = 0 \\ x + y = 4 \end{cases}$ (2, 2)

9. $\begin{cases} 3x - 4y = 5 \\ x = 3 \end{cases}$ (3, 1)

10. $\begin{cases} 5x + 2y = 24 \\ y = 2 \end{cases}$ (4, 2)

11. $\begin{cases} 4x + 5y = 20 \\ \frac{4}{5}x + y = 4 \end{cases}$
Infinitely many solutions

12. $\begin{cases} -x + 3y = 7 \\ 2x - 6y = 6 \end{cases}$
No solution

Solving a system of equations graphically is limited by the ability to sketch an accurate graph. An accurate solution is difficult to obtain if one or both coordinates of a solution point are fractional or irrational. One analytic way to determine an exact solution of a system of two equations in two variables is to convert the system to *one* equation in *one* variable by an appropriate substitution.

EXAMPLE 3 The Method of Substitution: One-Solution Case

Solve the system of equations.

$$\begin{cases} -x + y = 3 & \text{Equation 1} \\ 3x + y = -1 & \text{Equation 2} \end{cases}$$

Study Tip

The term **back-substitute** implies that you work backwards. After finding a value for one of the variables, substitute that value back into one of the equations in the original (or revised) system to find the value of the other variable.

SOLUTION

Begin by solving for y in Equation 1.

$-x + y = 3$ Equation 1

$y = x + 3$ Revised Equation 1

Next, substitute this expression for y in Equation 2.

$3x + y = -1$ Equation 2

$3x + (x + 3) = -1$ Substitute $x + 3$ for y.

$4x + 3 = -1$ Combine like terms.

$4x = -4$ Subtract 3 from each side.

$x = -1$ Divide each side by 4.

At this point, you know that the x-coordinate of the solution is -1. To find the y-coordinate, *back-substitute* the x-value into the revised Equation 1.

$y = x + 3$ Revised Equation 1

$y = -1 + 3$ Substitute -1 for x.

$y = 2$ Simplify.

The solution is $(-1, 2)$. Check this in the original system of equations.

Exercises Within Reach®

Solutions in English & Spanish and tutorial videos at AlgebraWithinReach.com

Solving a System of Equations **In Exercises 13–24, solve the system of equations by the method of substitution.**

13. $\begin{cases} y = 2 \\ x - 6y = -6 \end{cases}$ $(6, 2)$

14. $\begin{cases} x = 4 \\ x - 2y = -2 \end{cases}$ $(4, 3)$

15. $\begin{cases} x - 2y = 0 \\ 3x + 2y = 8 \end{cases}$ $(2, 1)$

16. $\begin{cases} x - y = 0 \\ 5x - 2y = 6 \end{cases}$ $(2, 2)$

17. $\begin{cases} x + y = 3 \\ 2x - y = 0 \end{cases}$ $(1, 2)$

18. $\begin{cases} -x + y = 5 \\ x - 4y = 0 \end{cases}$ $\left(-\frac{20}{3}, -\frac{5}{3}\right)$

19. $\begin{cases} x + y = 2 \\ x - 4y = 12 \end{cases}$ $(4, -2)$

20. $\begin{cases} x - 2y = -1 \\ x - 5y = 2 \end{cases}$ $(-3, -1)$

21. $\begin{cases} x + 6y = 19 \\ x - 7y = -7 \end{cases}$ $(7, 2)$

22. $\begin{cases} x - 5y = -6 \\ 4x - 3y = 10 \end{cases}$ $(4, 2)$

23. $\begin{cases} 8x + 5y = 100 \\ 9x - 10y = 50 \end{cases}$ $(10, 4)$

24. $\begin{cases} x + 4y = 300 \\ x - 2y = 0 \end{cases}$ $(100, 50)$

EXAMPLE 4 The Method of Substitution: No-Solution Case

To solve the system of equations

$$\begin{cases} 2x - 2y = 0 & \text{Equation 1} \\ x - y = 1 & \text{Equation 2} \end{cases}$$

begin by solving for y in Equation 2.

$y = x - 1$ Revised Equation 2

Next, substitute this expression for y in Equation 1.

$2x - 2(x - 1) = 0$ Substitute $x - 1$ for y in Equation 1.

$2x - 2x + 2 = 0$ Distributive Property

$2 = 0$ False statement

Because the substitution process produces a false statement ($2 = 0$), you can conclude that the original system of equations is inconsistent and has no solution. You can check your solution graphically, as shown below.

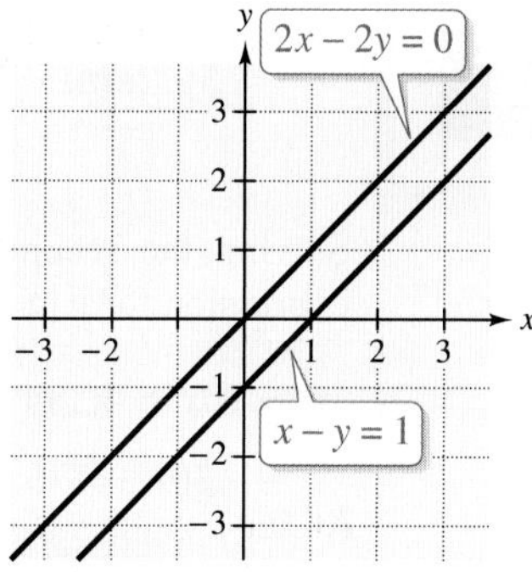

Study Tip

The two lines in Example 4 are parallel, so the system has no solution. You can also see that the system has no solution by writing each equation in slope-intercept form, as follows.

$y = x$

$y = x - 1$

Because the slopes are equal and the y-intercepts are different, the system has no solution.

Additional Example

Solve the system of equations.

$$\begin{cases} x + 3y = -4 \\ -2x - 6y = 8 \end{cases}$$

Answer:

Infinitely many solutions

The steps for using the method of substitution to solve a system of two equations involving two variables are summarized as follows.

The Method of Substitution

1. Solve one of the equations for one variable in terms of the other.
2. Substitute the expression obtained in Step 1 into the other equation to obtain an equation in one variable.
3. Solve the equation obtained in Step 2.
4. Back-substitute the solution from Step 3 into the expression obtained in Step 1 to find the value of the other variable.
5. Check the solution to see that it satisfies *both* of the original equations.

Exercises Within Reach®

Solutions in English & Spanish and tutorial videos at AlgebraWithinReach.com

Solving a System of Equations **In Exercises 25–30, solve the system of equations by the method of substitution.**

25. $\begin{cases} 3x + y = 8 \\ 3x + y = 6 \end{cases}$ No solution

26. $\begin{cases} x - 3y = 12 \\ -2x + 6y = -18 \end{cases}$ No solution

27. $\begin{cases} 8x + 6y = 6 \\ 12x + 9y = 6 \end{cases}$ No solution

28. $\begin{cases} 12x - 14y = 15 \\ 18x - 21y = 10 \end{cases}$ No solution

29. $\begin{cases} -x + 4y = 7 \\ 3x - 12y = -21 \end{cases}$ Infinitely many solutions

30. $\begin{cases} 2x - 4y = 9 \\ x - 2y = 4.5 \end{cases}$ Infinitely many solutions

Applications

Application

EXAMPLE 5 A Mixture Problem

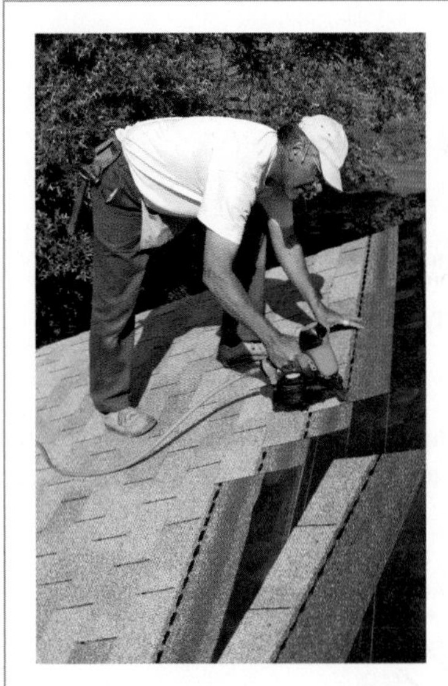

A roofing contractor buys 30 bundles of shingles and 4 rolls of roofing paper for \$732. In a second purchase (at the same prices), the contractor pays \$194 for 8 bundles of shingles and 1 roll of roofing paper. Find the price per bundle of shingles and the price per roll of roofing paper.

SOLUTION

Verbal Model: $30\begin{pmatrix}\text{Price of}\\\text{a bundle}\end{pmatrix} + 4\begin{pmatrix}\text{Price of}\\\text{a roll}\end{pmatrix} = 732$

$8\begin{pmatrix}\text{Price of}\\\text{a bundle}\end{pmatrix} + 1\begin{pmatrix}\text{Price of}\\\text{a roll}\end{pmatrix} = 194$

Labels: Price of a bundle of shingles $= x$ (dollars)

Price of a roll of roofing paper $= y$ (dollars)

System: $\begin{cases} 30x + 4y = 732 & \text{Equation 1} \\ 8x + y = 194 & \text{Equation 2} \end{cases}$

Solving the second equation for y produces $y = 194 - 8x$, and substituting this expression for y in the first equation produces the following.

$30x + 4(194 - 8x) = 732$ Substitute $194 - 8x$ for y.

$30x + 776 - 32x = 732$ Distributive Property

$-2x = -44$ Simplify.

$x = 22$ Divide each side by -2.

Back-substituting 22 for x in revised Equation 2 produces

$y = 194 - 8(22) = 18.$

So, you can conclude that the contractor pays \$22 per bundle of shingles and \$18 per roll of roofing paper. Check this in the original statement of the problem.

Exercises Within Reach ®

Solutions in English & Spanish and tutorial videos at AlgebraWithinReach.com

31. ***Hay Mixture*** A farmer wants to mix two types of hay. The first type sells for \$125 per ton and the second type sells for \$75 per ton. The farmer wants a total of 100 tons of hay at a cost of \$90 per ton. How many tons of each type of hay should be used in the mixture?
30 tons at \$125 per ton, 70 tons at \$75 per ton

32. ***Seed Mixture*** Ten pounds of mixed birdseed sells for \$6.97 per pound. The mixture is obtained from two kinds of birdseed, with one variety priced at \$5.65 per pound and the other at \$8.95 per pound. How many pounds of each variety of birdseed are used in the mixture?
6 pounds at \$5.65 per pound, 4 pounds at \$8.95 per pound

Christina Richards/Shutterstock.com; Marina Jay/Shutterstock.com

Application EXAMPLE 6 Break-Even Analysis

A small business invests \$14,000 to produce a new energy bar. Each bar costs \$0.80 to produce and sells for \$1.50. How many bars must be sold before the business breaks even?

SOLUTION

Verbal Model:

Total cost	=	Cost per bar	•	Number of bars	+	Initial cost
Total revenue	=	Price per bar	•	Number of bars		

Labels:

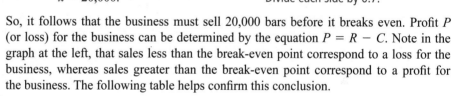

Total cost $= C$	(dollars)
Cost per bar $= 0.80$	(dollars per bar)
Number of bars $= x$	(bars)
Initial cost $= 14{,}000$	(dollars)
Total revenue $= R$	(dollars)
Price per bar $= 1.50$	(dollars per bar)

System:

$$\begin{cases} C = 0.80x + 14{,}000 & \text{Equation 1} \\ R = 1.50x & \text{Equation 2} \end{cases}$$

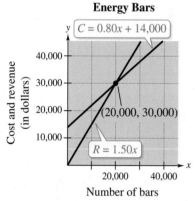

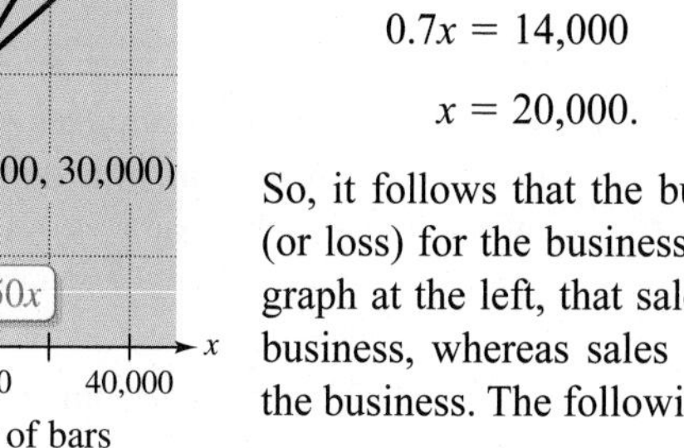

Because the break-even point occurs when $R = C$, you have

$1.50x = 0.80x + 14{,}000$ $R = C$

$0.7x = 14{,}000$ Subtract $0.80x$ from each side.

$x = 20{,}000.$ Divide each side by 0.7.

So, it follows that the business must sell 20,000 bars before it breaks even. Profit P (or loss) for the business can be determined by the equation $P = R - C$. Note in the graph at the left, that sales less than the break-even point correspond to a loss for the business, whereas sales greater than the break-even point correspond to a profit for the business. The following table helps confirm this conclusion.

Units, x	0	5000	10,000	15,000	20,000	25,000
Revenue, R	\$0	\$7500	\$15,000	\$22,500	\$30,000	\$37,500
Cost, C	\$14,000	\$18,000	\$22,000	\$26,000	\$30,000	\$34,000
Profit, P	−\$14,000	−\$10,500	−\$7000	−\$3500	\$0	\$3500

Exercises Within Reach®

Solutions in English & Spanish and tutorial videos at AlgebraWithinReach.com

33. ***Break-Even Analysis*** A small business invests \$8000 to produce a new candy bar. Each bar costs \$1.20 to produce and sells for \$2.00. How many candy bars must be sold before the business breaks even?
10,000 candy bars

34. ***Break-Even Analysis*** A business spends \$10,000 to produce a new board game. Each game costs \$1.50 to produce and is sold for \$9.99. How many games must be sold before the business breaks even? 1178 games

Concept Summary: Solving Systems of Linear Equations by Substitution

What

You can **solve a system of equations** using an algebraic method called the method of substitution.

EXAMPLE

Solve the system of linear equations.

$$\begin{cases} y = x + 2 & \text{Equation 1} \\ x + y = 6 & \text{Equation 2} \end{cases}$$

How

The goal is to reduce the system of two linear equations in two variables to a single equation in one variable. Then solve the equation.

EXAMPLE

Substitute $x + 2$ for y in Equation 2. Then solve for x.

$$x + (x + 2) = 6$$
$$x = 2$$

Now substitute 2 for x in Equation 1 to find the value of y.

$$y = 2 + 2$$
$$y = 4$$

Why

Solving a system of equations graphically is limited by your ability to sketch an accurate graph, especially when the coordinates of a solution point are fractional or irrational.

One way to obtain an exact solution is to solve the system by substitution.

Exercises Within Reach®

Worked-out solutions to odd-numbered exercises at AlgebraWithinReach.com

Concept Summary Check

35. ***Vocabulary*** What does it mean to *back-substitute* when solving a system of equations? Back-substitution is the method used after the value is found for one of the variables. The value is substituted back into one of the original equations.

36. ***Reasoning*** When solving a system of equations by the method of substitution, how do you recognize that it has no solution? When you obtain a false result such as $-4 = 2$, then the system of equations has no solution.

37. ***Reasoning*** When solving a system of equations by the method of substitution, how do you recognize that it has infinitely many solutions? When you obtain a true result such as $15 = 15$, then the system of equations has infinitely many solutions.

38. ***Writing*** Describe the geometric properties of the graph of each possible type of solution set of a system of linear equations in two variables. If the system has exactly one solution, the lines intersect at one point. If the system has infinitely many solutions, the lines coincide, and if the system has no solution, the lines are parallel.

Extra Practice

Solving a System of Equations **In Exercises 39–44, solve the system of equations by the method of substitution.**

39. $\begin{cases} 4x - 14y = -15 \\ 18x - 12y = 9 \end{cases}$ $\left(\frac{3}{2}, \frac{3}{2}\right)$

40. $\begin{cases} 5x - 24y = -12 \\ 17x - 24y = 36 \end{cases}$ $\left(4, \frac{4}{3}\right)$

41. $\begin{cases} \frac{1}{5}x + \frac{1}{2}y = 8 \\ x + y = 20 \end{cases}$ $\left(\frac{20}{3}, \frac{40}{3}\right)$

42. $\begin{cases} \frac{1}{2}x + \frac{3}{4}y = 10 \\ \frac{3}{2}x - y = 4 \end{cases}$ $(8, 8)$

43. $\begin{cases} \frac{1}{8}x + \frac{1}{2}y = 1 \\ \frac{3}{5}x + y = \frac{3}{5} \end{cases}$ $(-4, 3)$

44. $\begin{cases} \frac{1}{8}x - \frac{1}{4}y = \frac{3}{4} \\ -\frac{1}{4}x + \frac{3}{4}y = -1 \end{cases}$ $(10, 2)$

Think About It **In Exercises 45–48, write a system of equations having the given solution. (There are many correct answers.)**

45. $(4, 5)$ $\begin{cases} 2x - 3y = -7 \\ x + y = 9 \end{cases}$

46. $(-1, -2)$ $\begin{cases} 7x + y = -9 \\ -x + 3y = -5 \end{cases}$

47. No solution $\begin{cases} 2x + y = -1 \\ 4x + 2y = 7 \end{cases}$

48. Infinitely many solutions, including $\left(4, -\frac{1}{2}\right)$ and $(-1, 1)$ $\begin{cases} 3x + 10y = 7 \\ 6x + 20y = 14 \end{cases}$

Think About It **In Exercises 49–52, determine the value of a and the value of b for which the system of equations is dependent.**

49. $\begin{cases} y = \frac{2}{3}x + 1 \\ 3y = ax + b \end{cases}$ $a = 2, b = 3$

50. $\begin{cases} 4y = 5x \\ 2y = ax + b \end{cases}$ $a = \frac{5}{2}, b = 0$

51. $\begin{cases} 6x - 8y - 48 = 0 \\ \frac{3}{4}x - ay + b = 0 \end{cases}$ $a = 1, b = -6$

52. $\begin{cases} \frac{1}{2}y = -\frac{3}{4}x + \frac{9}{4} \\ ax = -\frac{5}{6}y + b \end{cases}$ $a = \frac{5}{4}, b = \frac{15}{4}$

Number Problem **In Exercises 53–56, find two positive integers that satisfy the given requirements.**

53. The sum of the greater number and twice the lesser number is 61, and their difference is 7. 18, 25

54. The sum of the two numbers is 52, and the greater number is 8 less than twice the lesser number. 20, 32

55. The difference of twice the lesser number and the greater number is 13, and the sum of the lesser number and twice the greater number is 114. 28, 43

56. The difference of the numbers is 86, and the greater number is three times the lesser number. 43, 129

Geometry **In Exercises 57 and 58, find the dimensions of the rectangle meeting the specified conditions.**

	Perimeter	*Condition*
57.	68 yards	The width is $\frac{7}{10}$ of the length. 14 yards × 20 yards
58.	90 meters	The length is $1\frac{1}{2}$ times the width. 18 meters × 27 meters

59. ***Investment*** A total of $12,000 is invested in two bonds that pay 8.5% and 10% simple interest. The annual interest is $1140. How much is invested in each bond? $4000 at 8.5%, $8000 at 10%

60. ***Investment*** A total of $25,000 is invested in two funds paying 8% and 8.5% simple interest. The annual interest is $2060. How much is invested in each fund? $13,000 at 8%, $12,000 at 8.5%

Explaining Concepts

61. ***Precision*** Describe any advantages of the method of substitution over the graphical method of solving a system of equations. One advantage of the substitution method is that it generates exact answers, whereas the graphical method usually yields approximate solutions.

62. ***Writing*** Is it possible for a consistent system of linear equations to have exactly two solutions? Explain. No. The lines represented by a consistent system of linear equations intersect at exactly one point (one solution) or the lines coincide (infinitely many solutions).

63. ***Reasoning*** In a consistent system of three linear equations in two variables, exactly two of the equations are dependent. How many solutions does the system have? Explain. The system will have exactly one solution because the system is consistent.

64. ***Reasoning*** Is it possible for a system of three linear equations in two variables to be inconsistent if two of the equations are dependent? Explain. Yes. The system will be inconsistent if the third equation is parallel to the other two equations.

65. ***Logic*** You want to create several systems of equations with relatively simple solutions that students can use for practice. Discuss how to create a system of equations that has a given solution. Illustrate your method by creating a system of linear equations that has one of the following solutions: (1, 4), (−2, 5), (−3, 1), or (4, 2). See Additional Answers.

Cumulative Review

In Exercises 66–69, solve the equation and check your solution.

66. $x - 4 = 1$ 5

67. $6x - 2 = -7$ $-\frac{5}{6}$

68. $3x - 12 = x + 2$ 7

69. $4x + 21 = 4(x + 5)$ No solution

In Exercises 70–73, sketch the graph of the function. See Additional Answers.

70. $f(x) = x^2 - 6$

71. $g(x) = 5$

72. $h(t) = 3t - 1$

73. $s(t) = |t - 2|$

In Exercises 74 and 75, verify that both ordered pairs are solutions of the equation, and explain why this implies that y is not a function of x.

74. $-x + y^2 = 0$; $(4, 2), (4, -2)$
There are two values of y associated with one value of x.

75. $|y| = x + 4$; $(3, 7), (3, -7)$
There are two values of y associated with one value of x.

4.2 Linear Systems in Two Variables

- Solve systems of linear equations algebraically using the method of elimination.
- Use systems of linear equations to model and solve real-life problems.

The Method of Elimination

EXAMPLE 1 The Method of Elimination

Solve the system of linear equations.

$$\begin{cases} 3x + 2y = 4 & \text{Equation 1} \\ 5x - 2y = 8 & \text{Equation 2} \end{cases}$$

SOLUTION

Begin by noting that the coefficients of y are opposites. By adding the two equations, you can eliminate y.

$$\begin{array}{ll} 3x + 2y = \ \ 4 & \text{Equation 1} \\ \underline{5x - 2y = \ \ 8} & \text{Equation 2} \\ 8x \qquad = 12 & \text{Add equations.} \end{array}$$

So, $x = \frac{3}{2}$. Back-substitute this value into Equation 1 and solve for y.

$$\begin{array}{ll} 3x + 2y = 4 & \text{Equation 1} \\ 3\left(\frac{3}{2}\right) + 2y = \ \ 4 & \text{Substitute } \frac{3}{2} \text{ for } x. \\ 2y = -\frac{1}{2} & \text{Subtract } \frac{9}{2} \text{ from each side.} \\ y = -\frac{1}{4} & \text{Divide each side by 2.} \end{array}$$

The solution is $\left(\frac{3}{2}, -\frac{1}{4}\right)$. Check this in the original system of equations.

Exercises Within Reach®

Solutions in English & Spanish and tutorial videos at AlgebraWithinReach.com

Solving a System of Linear Equations **In Exercises 1–4, solve the system of linear equations by the method of elimination.**

1. $\begin{cases} 2x + y = 4 \\ x - y = 2 \end{cases}$

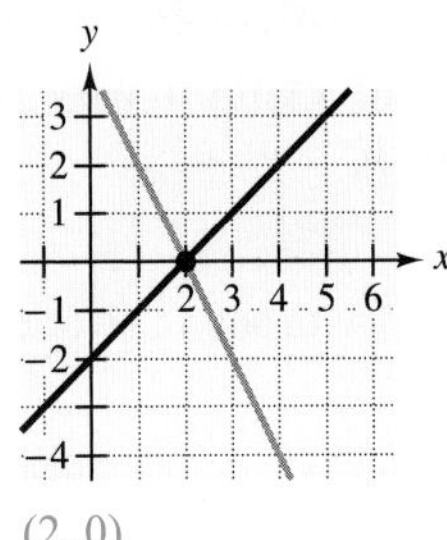

(2, 0)

2. $\begin{cases} x + 3y = 2 \\ -x + 2y = 3 \end{cases}$

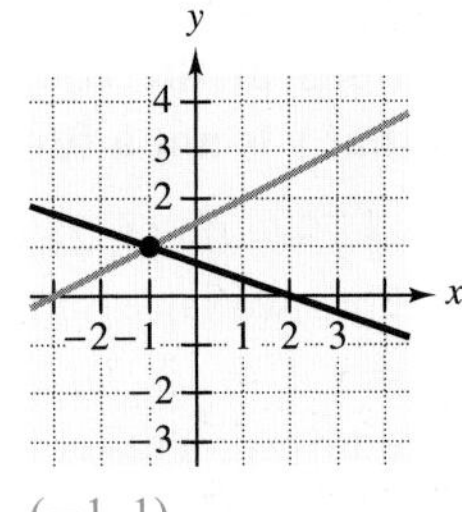

(−1, 1)

3. $\begin{cases} -x + 2y = 1 \\ x - \ \ y = 2 \end{cases}$

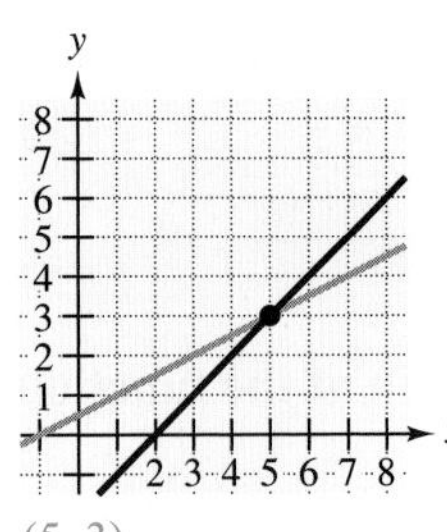

(5, 3)

4. $\begin{cases} 3x + y = 3 \\ 2x - y = 7 \end{cases}$

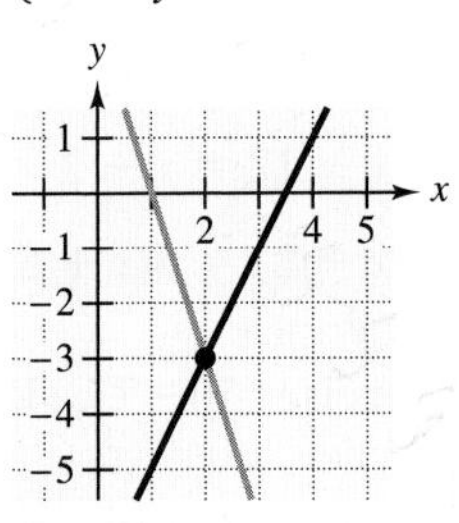

(2, −3)

The Method of Elimination

1. Obtain coefficients for x (or y) that are opposites by multiplying all terms of one or both equations by suitable constants.
2. Add the equations to eliminate one variable and solve the resulting equation.
3. Back-substitute the value obtained in Step 2 into either of the original equations and solve for the other variable.
4. Check your solution in *both* of the original equations.

EXAMPLE 2 The Method of Elimination

Solve the system of linear equations.

$$\begin{cases} 4x - 5y = 13 & \text{Equation 1} \\ 3x - y = 7 & \text{Equation 2} \end{cases}$$

SOLUTION

To obtain coefficients of y that are opposites, multiply Equation 2 by -5.

$$\begin{cases} 4x - 5y = 13 \\ 3x - y = 7 \end{cases} \Rightarrow \begin{array}{rl} 4x - 5y = & 13 \\ -15x + 5y = & -35 \\ \hline -11x \quad = & -22 \end{array} \quad \begin{array}{l} \text{Equation 1} \\ \text{Multiply Equation 2 by } -5. \\ \text{Add equations.} \end{array}$$

So, $x = 2$. Back-substitute this value into Equation 2 and solve for y.

$3x - y = 7$ Equation 2

$3(2) - y = 7$ Substitute 2 for x.

$y = -1$ Solve for y.

The solution is $(2, -1)$. Check this in the original system of equations.

Additional Example

Solve the system of linear equations.

$$\begin{cases} 5x + 3y = 8 \\ 2x - 4y = 11 \end{cases}$$

Answer: $\left(\frac{5}{2}, -\frac{3}{2}\right)$

Exercises Within Reach®

Solutions in English & Spanish and tutorial videos at AlgebraWithinReach.com

Solving a System of Linear Equations **In Exercises 5–14, solve the system of linear equations by the method of elimination.**

5. $\begin{cases} 6x - 6y = 25 \\ 3y = 11 \end{cases}$ $\left(\frac{47}{6}, \frac{11}{3}\right)$

6. $\begin{cases} x + 3y = 4 \\ 2x = 2 \end{cases}$ $(1, 1)$

7. $\begin{cases} x + 7y = -6 \\ x - 5y = 18 \end{cases}$ $(8, -2)$

8. $\begin{cases} 4x + y = -2 \\ -6x + y = 18 \end{cases}$ $(-2, 6)$

9. $\begin{cases} 5x + 2y = 7 \\ 3x - y = 13 \end{cases}$ $(3, -4)$

10. $\begin{cases} 4x + 3y = 8 \\ x - 2y = 13 \end{cases}$ $(5, -4)$

11. $\begin{cases} x - 3y = 2 \\ 3x - 7y = 4 \end{cases}$ $(-1, -1)$

12. $\begin{cases} 2s - t = 9 \\ 3s + 4t = -14 \end{cases}$ $(2, -5)$

13. $\begin{cases} 4x + 3y = -10 \\ 3x - y = -14 \end{cases}$ $(-4, 2)$

14. $\begin{cases} 7r - s = -25 \\ 2r + 5s = 14 \end{cases}$ $(-3, 4)$

EXAMPLE 3 The Method of Elimination: No-Solution Case

To solve the system of linear equations

$$\begin{cases} 3x + 9y = 8 & \text{Equation 1} \\ 2x + 6y = 7 & \text{Equation 2} \end{cases}$$

obtain coefficients of x that are opposites by multiplying Equation 1 by 2 and Equation 2 by -3.

$$\begin{cases} 3x + 9y = 8 \\ 2x + 6y = 7 \end{cases} \Rightarrow \begin{array}{rl} 6x + 18y = & 16 \\ -6x - 18y = & -21 \\ \hline 0 = & -5 \end{array} \quad \begin{array}{l} \text{Multiply Equation 1 by 2.} \\ \text{Multiply Equation 2 by } -3. \\ \text{Add equations.} \end{array}$$

Because $0 = -5$ is a false statement, you can conclude that the system is inconsistent and has no solution. You can confirm this by graphing the two equations as shown at the left. Because the lines are parallel, the system is inconsistent.

EXAMPLE 4 The Method of Elimination: Many-Solutions Case

To solve the system of linear equations

$$\begin{cases} -2x + 6y = 3 & \text{Equation 1} \\ 4x - 12y = -6 & \text{Equation 2} \end{cases}$$

obtain coefficients of x that are opposites by multiplying Equation 1 by 2.

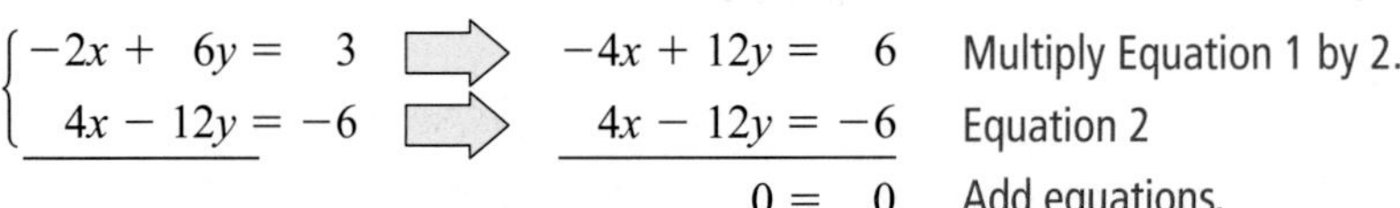

$$\begin{cases} -2x + 6y = 3 \\ 4x - 12y = -6 \end{cases} \Rightarrow \begin{array}{rl} -4x + 12y = & 6 \\ 4x - 12y = & -6 \\ \hline 0 = & 0 \end{array} \quad \begin{array}{l} \text{Multiply Equation 1 by 2.} \\ \text{Equation 2} \\ \text{Add equations.} \end{array}$$

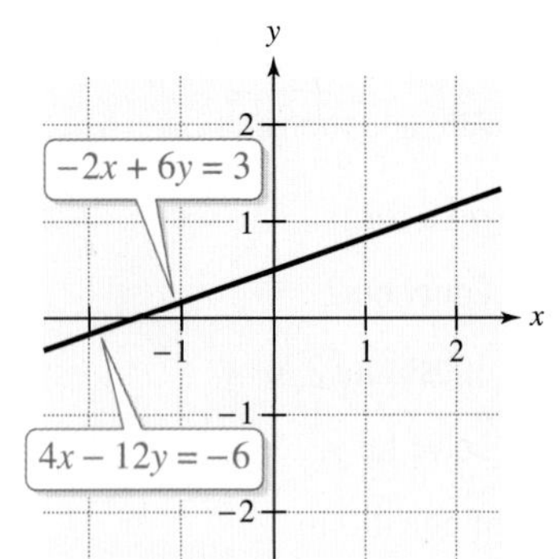

Because $0 = 0$ is a true statement, you can conclude that the system is dependent and has infinitely many solutions. You can confirm this by graphing the two equations, as shown at the left. So, the solution set consists of all points (x, y) lying on the line $-2x + 6y = 3$.

Exercises Within Reach®

Solutions in English & Spanish and tutorial videos at AlgebraWithinReach.com

Solving a System of Linear Equations **In Exercises 15–22, solve the system of linear equations by the method of elimination.**

15. $\begin{cases} x + y = 0 \\ -3x - 3y = 0 \end{cases}$

Infinitely many solutions

16. $\begin{cases} x - 3y = 5 \\ -2x + 6y = -10 \end{cases}$

Infinitely many solutions

17. $\begin{cases} 12x - 5y = 2 \\ -24x + 10y = 6 \end{cases}$

No solution

18. $\begin{cases} 4x - 5y = 3 \\ -8x + 10y = 14 \end{cases}$

No solution

19. $\begin{cases} -2x + 3y = 9 \\ 6x - 9y = -27 \end{cases}$

Infinitely many solutions

20. $\begin{cases} 5x + 7y = 25 \\ x + 1.4y = 5 \end{cases}$

Infinitely many solutions

21. $\begin{cases} 4x - 8y = 36 \\ 3x - 6y = 15 \end{cases}$

No solution

22. $\begin{cases} 5x + 15y = 20 \\ 2x + 6y = 12 \end{cases}$

No solution

Applications

Application

EXAMPLE 5 **Solving a Mixture Problem**

A company with two stores buys six large delivery vans and five small delivery vans. The first store receives 4 of the large vans and 2 of the small vans for a total cost of \$200,000. The second store receives 2 of the large vans and 3 of the small vans for a total cost of \$160,000. What is the cost of each type of van?

SOLUTION

The two unknowns in this problem are the costs of the two types of vans.

Verbal Model:

$$4\begin{pmatrix}\text{Cost of}\\ \text{large van}\end{pmatrix} + 2\begin{pmatrix}\text{Cost of}\\ \text{small van}\end{pmatrix} = \$200{,}000$$

$$2\begin{pmatrix}\text{Cost of}\\ \text{large van}\end{pmatrix} + 3\begin{pmatrix}\text{Cost of}\\ \text{small van}\end{pmatrix} = \$160{,}000$$

Labels: Cost of large van $= x$ (dollars)
Cost of small van $= y$ (dollars)

System:

$$\begin{cases} 4x + 2y = 200{,}000 & \text{Equation 1} \\ 2x + 3y = 160{,}000 & \text{Equation 2} \end{cases}$$

To solve this system of linear equations, use the method of elimination. To obtain coefficients of x that are opposites, multiply Equation 2 by -2.

$$\begin{cases} 4x + 2y = 200{,}000 \\ 2x + 3y = 160{,}000 \end{cases} \Rightarrow \begin{aligned} 4x + 2y &= 200{,}000 && \text{Equation 1} \\ -4x - 6y &= -320{,}000 && \text{Multiply Equation 2 by } -2. \end{aligned}$$

$$\begin{aligned} -4y &= -120{,}000 && \text{Add equations.} \\ y &= 30{,}000 && \text{Divide each side by } -4. \end{aligned}$$

So, the cost of each small van is $y = \$30{,}000$. Back-substitute this value into Equation 1 to find the cost of each large van.

$$\begin{aligned} 4x + 2y &= 200{,}000 && \text{Equation 1} \\ 4x + 2(30{,}000) &= 200{,}000 && \text{Substitute 30,000 for } y. \\ 4x &= 140{,}000 && \text{Simplify.} \\ x &= 35{,}000 && \text{Divide each side by 4.} \end{aligned}$$

The cost of each large van is $x = \$35{,}000$. Check this solution in the original statement of the problem.

Exercises Within Reach®

Solutions in English & Spanish and tutorial videos at AlgebraWithinReach.com

23. ***Comparing Costs*** A band charges \$500 to play for 4 hours plus \$50 for each additional hour. A DJ charges \$300 to play for 4 hours plus \$75 for each additional hour. After how many hours will the cost of the DJ exceed the cost of the band? 12 hours

24. ***Comparing Costs*** An SUV costs \$26,445 and costs an average of \$0.18 per mile to maintain. A hybrid model of the SUV costs \$31,910 and costs an average of \$0.13 per mile to maintain. After how many miles will the cost of the gas-only SUV exceed the cost of the hybrid? 109,300 miles

Application **EXAMPLE 6** **An Application Involving Two Speeds**

You take a motorboat trip on a river (18 miles upstream and 18 miles downstream). You run the motor at the same speed going up and down the river, but because of the river's current, the trip upstream takes $1\frac{1}{2}$ hours and the trip downstream takes only 1 hour. Determine the speed of the current.

SOLUTION

Verbal Model: Boat speed − Current speed = Upstream speed

Boat speed + Current speed = Downstream speed

Labels: Boat speed (in still water) $= x$ (miles per hour)

Current speed $= y$ (miles per hour)

Upstream speed $= \dfrac{18}{1.5} = 12$ (miles per hour)

Downstream speed $= \dfrac{18}{1} = 18$ (miles per hour)

System: $\begin{cases} x - y = 12 & \text{Equation 1} \\ x + y = 18 & \text{Equation 2} \end{cases}$

To solve this system of linear equations, use the method of elimination.

$$x - y = 12 \qquad \text{Equation 1}$$
$$\underline{x + y = 18} \qquad \text{Equation 2}$$
$$2x \quad = 30 \qquad \text{Add equations.}$$

So, the speed of the boat in still water is 15 miles per hour. Back-substitution yields $y = 3$. So, the speed of the current is 3 miles per hour.

Exercises Within Reach ®

Solutions in English & Spanish and tutorial videos at AlgebraWithinReach.com

25. ***Air Speed*** An airplane flying into a headwind travels 1800 miles in 3 hours and 36 minutes. On the return flight, the same distance is traveled in 3 hours. Find the speed of the plane in still air and the speed of the wind, assuming that both remain constant throughout the round trip. See Additional Answers.

26. ***Air Speed*** An airplane flying into a headwind travels 3000 miles in 6 hours and 15 minutes. On the return flight, the same distance is traveled in 5 hours. Find the speed of the plane in still air and the speed of the wind, assuming that both remain constant throughout the round trip. See Additional Answers.

27. ***Average Speed*** A van travels for 2 hours at an average speed of 40 miles per hour. How much longer must the van travel at an average speed of 55 miles per hour so that the average speed for the total trip will be 50 miles per hour? 4 hours

28. ***Average Speed*** A truck travels for 4 hours at an average speed of 42 miles per hour. How much longer must the truck travel at an average speed of 55 miles per hour so that the average speed for the total trip will be 50 miles per hour? 6.4 hours

Application

EXAMPLE 7 Data Analysis: Best-Fitting Line

The slope and y-intercept of the line

$$y = mx + b$$

that best fits the three noncollinear points (1, 0), (2, 1), and (3, 4) are given by the solution of the following system of linear equations.

$$\begin{cases} 3b + 6m = 5 & \text{Equation 1} \\ 6b + 14m = 14 & \text{Equation 2} \end{cases}$$

Solve this system. Then find the equation of the best-fitting line.

SOLUTION

To solve this system of linear equations, use the method of elimination.

$$\begin{cases} 3b + 6m = 5 \\ 6b + 14m = 14 \end{cases} \Rightarrow \begin{array}{rl} -6b - 12m = -10 & \text{Multiply Equation 1 by } -2. \\ 6b + 14m = 14 & \text{Equation 2} \\ \hline 2m = 4 & \text{Add equations.} \end{array}$$

So, $m = 2$. Back-substitution yields $b = -\frac{7}{3}$. So, the equation of the best-fitting line is $y = 2x - \frac{7}{3}$. The figure below shows the points and the best-fitting line.

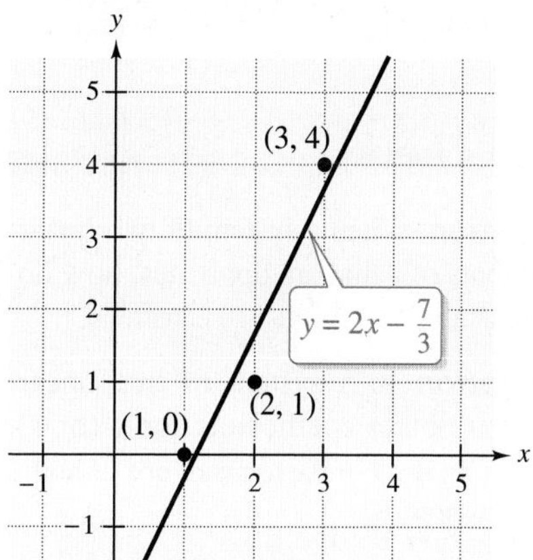

Exercises Within Reach®

Solutions in English & Spanish and tutorial videos at AlgebraWithinReach.com

29. ***Best-Fitting Line*** The slope and y-intercept of the line

$$y = mx + b$$

that best fits the three noncollinear points (0, 0), (1, 1), and (2, 3) are given by the solution of the following system of linear equations.

$$\begin{cases} 5m + 3b = 7 \\ 3m + 3b = 4 \end{cases}$$

(a) Solve the system and find the equation of the best-fitting line.

$y = -\frac{3}{2}x + \frac{23}{6}$

(b) Plot the three points and sketch the best-fitting line.

See Additional Answers.

30. ***Best-Fitting Line*** The slope and y-intercept of the line

$$y = mx + b$$

that best fits the three noncollinear points (0, 4), (1, 2), and (2, 1) are given by the solution of the following system of linear equations.

$$\begin{cases} 3b + 3m = 7 \\ 3b + 5m = 4 \end{cases}$$

(a) Solve the system and find the equation of the best-fitting line.

$y = \frac{3}{2}x - \frac{1}{6}$

(b) Plot the three points and sketch the best-fitting line.

See Additional Answers.

Concept Summary: Solving Systems of Linear Equations by Elimination

What

You can solve a system of linear equations using an algebraic method called the **method of elimination**.

EXAMPLE

Solve the system of linear equations.

$$\begin{cases} 3x + 5y = 7 & \text{Equation 1} \\ -3x - 2y = -1 & \text{Equation 2} \end{cases}$$

How

The key step in this method is to obtain opposite coefficients for one of the variables so that adding the two equations eliminates this variable.

EXAMPLE

$$\begin{array}{rll} 3x + 5y = & 7 & \text{Equation 1} \\ -3x - 2y = & -1 & \text{Equation 2} \\ \hline 3y = & 6 & \text{Add equations.} \end{array}$$

After eliminating the variable, solve for the other variable. Then use back-substitution to find the value of the eliminated variable.

Why

When solving a system of linear equations, choose the method that is most efficient.

1. Use graphing to approximate the solution.
2. To find exact solutions, use substitution or elimination.
   - When one of the variables has a coefficient of 1, use substitution.
   - When the coefficients of one of the variables are opposites, use elimination.
3. When you are not sure, use elimination. It is usually more efficient.

Exercises Within Reach®

Worked-out solutions to odd-numbered exercises at AlgebraWithinReach.com

Concept Summary Check

31. ***Solving by Elimination*** When a system of two linear equations has coefficients of y that are opposites, how do you eliminate the y variable? Add the two equations.

32. ***Solving by Elimination*** In a system of two linear equations, how do you obtain coefficients for x (or y) that are opposites? Multiply all of the terms of one or both equations by suitable constants.

33. ***Writing*** Explain what is meant by an *inconsistent* system of linear equations. The system has no solution and its graph consists of parallel lines.

34. ***Reasoning*** Both (1, 3) and (5, 5) are solutions of a system of two linear equations in two variables. How many solutions does the system have? Explain. Infinitely many; The points (1, 3) and (5, 5) are solutions of both equations in the system, so they must lie on the line of both equations. So, the lines coincide, and there are infinitely many solutions.

Extra Practice

Solving a System of Linear Equations **In Exercises 35–42, solve the system of linear equations by any convenient method.**

35. $\begin{cases} 2x - y = 20 \\ -x + y = -5 \end{cases}$ (15, 10)

36. $\begin{cases} 3x - 2y = -20 \\ 5x + 6y = 32 \end{cases}$ (−2, 7)

37. $\begin{cases} y = 5x - 3 \\ y = -2x + 11 \end{cases}$ (2, 7)

38. $\begin{cases} 3y = 2x + 21 \\ x = 50 - 4y \end{cases}$ (6, 11)

39. $\begin{cases} \frac{3}{2}x + 2y = 12 \\ \frac{1}{4}x + y = 4 \end{cases}$ (4, 3)

40. $\begin{cases} x + 2y = 4 \\ \frac{1}{2}x + \frac{1}{3}y = 1 \end{cases}$ $\left(1, \frac{3}{2}\right)$

41. $\begin{cases} 2u + 3v = 8 \\ 3u + 4v = 13 \end{cases}$ (7, −2)

42. $\begin{cases} 4x - 3y = 25 \\ -3x + 8y = 10 \end{cases}$ (10, 5)

Describing a System **In Exercises 43–46, determine whether the system is consistent or inconsistent.**

43. $\begin{cases} 4x - 5y = 3 \\ -8x + 10y = -6 \end{cases}$ Consistent

44. $\begin{cases} -10x + 15y = 25 \\ 2x - 3y = -24 \end{cases}$ Inconsistent

45. $\begin{cases} -2x + 5y = 3 \\ 5x + 2y = 8 \end{cases}$ Consistent

46. $\begin{cases} x + 10y = 12 \\ -2x + 5y = 2 \end{cases}$ Consistent

47. ***Gasoline Mixture*** Twelve gallons of regular unleaded gasoline plus 8 gallons of premium unleaded gasoline cost \$76.48. Premium unleaded gasoline costs \$0.11 more per gallon than regular unleaded. Find the price per gallon for each grade of gasoline. \$3.78 per gallon of regular unleaded, \$3.89 per gallon of premium unleaded

48. ***Alcohol Mixture*** How many liters of a 40% alcohol solution must be mixed with a 65% solution to obtain 20 liters of a 50% alcohol solution?
12 liters of 40% solution, 8 liters of 65% solution

49. ***Acid Mixture*** Thirty liters of a 46% acid solution is obtained by mixing a 40% solution with a 70% solution. How many liters of each solution must be used to obtain the desired mixture?
24 liters of 40% solution, 6 liters of 70% solution

50. ***Nut Mixture*** Ten pounds of mixed nuts sells for \$6.87 per pound. The mixture is obtained from two kinds of nuts, peanuts priced at \$5.70 per pound and cashews at \$8.70 per pound. How many pounds of each variety of nut are used in the mixture? 6.1 pounds of peanuts, 3.9 pounds of cashews

Explaining Concepts

51. ***Precision*** When solving a system by elimination, how do you recognize that it has infinitely many solutions?
When a nonzero multiple of one equation is added to another equation to eliminate a variable and the result is $0 = 0$, the system has infinitely many solutions.

52. ***Precision*** When solving a system by elimination, how do you recognize that a system of linear equations has no solution? See Additional Answers.

53. ***Writing*** In your own words, explain how to solve a system of linear equations by elimination.
See Additional Answers.

54. ***Think About It*** Under what conditions might substitution be better than elimination for solving a system of linear equations? Substitution might be better than elimination when it is easy to solve for one of the variables in one of the equations in the system. For instance, if a variable has a coefficient of 1, that equation can be written easily in terms of that variable.

Cumulative Review

In Exercises 55–58, write the general form of the equation of the line that passes through the points.

55. (0, 0), (4, 2)
$x - 2y = 0$

56. (1, 2), (6, 3)
$x - 5y + 9 = 0$

57. (−1, 2), (5, 2)
$y - 2 = 0$

58. (−3, 3), (8, −6)
$9x + 11y - 6 = 0$

In Exercises 59–62, determine whether the set of ordered pairs represents a function.

59. {(0, 0), (2, 1), (4, 2),(6, 3)}
Function

60. {(0, 2), (1, 4), (4, 1), (0, 4)}
Not a function

61. {(−4, 5), (−1, 0), (3, −2), (3, −4)}
Not a function

62. {(−3, 1), (−1, 3), (1, 3), (3, 1)}
Function

In Exercises 63 and 64, determine whether each ordered pair is a solution of the system of equations.

63. $\begin{cases} 3x - 4y = 10 \\ 2x + 6y = -2 \end{cases}$

(a) (2, −1)
Solution

(b) (−1, 0)
Not a solution

64. $\begin{cases} -4x - y = 2 \\ 2x + 7y = 38 \end{cases}$

(a) (2, −10)
Not a solution

(b) (−2, 6)
Solution

3Digitalworks/Shutterstock.com; wavebreakmedia ltd/Shutterstock.com

4.3 Linear Systems in Three Variables

- Solve systems of linear equations in row-echelon form using back-substitution.
- Solve systems of linear equations using the method of Gaussian elimination.
- Solve application problems using the method of Gaussian elimination.

Row-Echelon Form

When the method of elimination is used to solve a system of linear equations, the goal is to rewrite the system in a form to which back-substitution can be applied. This method can be applied to a system of linear equations in more than two variables, as shown below. The system on the right is in **row-echelon form**, which means that it has a "stair-step" pattern with leading coefficients of 1.

$$\begin{cases} x - 2y + 2z = 9 \\ -x + 3y = -4 \\ 2x - 5y + z = 10 \end{cases} \quad \Longrightarrow \quad \begin{cases} x - 2y + 2z = 9 \\ y + 2z = 5 \\ z = 3 \end{cases}$$

EXAMPLE 1 Using Back-Substitution

In the following system of linear equations, you know the value of z from Equation 3.

$$\begin{cases} x - 2y + 2z = 9 & \text{Equation 1} \\ y + 2z = 5 & \text{Equation 2} \\ z = 3 & \text{Equation 3} \end{cases}$$

To solve for y, substitute $z = 3$ into Equation 2 to obtain

$$y + 2(3) = 5 \quad \Longrightarrow \quad y = -1 \qquad \text{Substitute 3 for } z.$$

Finally, substitute $y = -1$ and $z = 3$ into Equation 1 to obtain

$$x - 2(-1) + 2(3) = 9 \quad \Longrightarrow \quad x = 1. \qquad \text{Substitute } -1 \text{ for } y \text{ and 3 for } z.$$

The solution is $x = 1$, $y = -1$, and $z = 3$, which can also be written as the **ordered triple** $(1, -1, 3)$. Check this in the original system of equations.

Study Tip

When checking a solution, remember that the solution must satisfy each equation in the original system.

Exercises Within Reach ®

Solutions in English & Spanish and tutorial videos at AlgebraWithinReach.com

Using Back-Substitution **In Exercises 1–4, use back-substitution to solve the system of linear equations.**

1. $\begin{cases} x - 2y + 4z = 4 \\ 3y - z = 2 \\ z = -5 \end{cases}$

   $(22, -1, -5)$

2. $\begin{cases} 5x + 4y - z = 0 \\ 10y - 3z = 11 \\ z = 3 \end{cases}$

   $(-1, 2, 3)$

3. $\begin{cases} x - 2y + 4z = 4 \\ y = 3 \\ y + z = 2 \end{cases}$

   $(14, 3, -1)$

4. $\begin{cases} x = 10 \\ 3x + 2y = 2 \\ x + y + 2z = 0 \end{cases}$

   $(10, -14, 2)$

The Method of Gaussian Elimination

Study Tip

Two systems of equations are **equivalent systems** when they have the same solution set. Rewriting a system of linear equations in row-echelon form usually involves a chain of equivalent systems, each of which is obtained by using one of the three basic row operations. This process is called **Gaussian elimination**.

Operations That Produce Equivalent Systems

Each of the following **row operations** on a system of linear equations produces an *equivalent* system of linear equations.

1. Interchange two equations.
2. Multiply one of the equations by a nonzero constant.
3. Add a multiple of one of the equations to another equation to replace the latter equation.

EXAMPLE 2 Using Gaussian Elimination to Solve a System

Solve the system of linear equations.

$$\begin{cases} x - 2y + 2z = 9 & \text{Equation 1} \\ -x + 3y = -4 & \text{Equation 2} \\ 2x - 5y + z = 10 & \text{Equation 3} \end{cases}$$

SOLUTION

Equation 1 has a leading coefficient of 1, so leave it alone to keep the x in the upper left position. Begin by eliminating the other x terms from the first column, as follows.

$$\begin{cases} x - 2y + 2z = 9 \\ y + 2z = 5 \\ 2x - 5y + z = 10 \end{cases}$$

Adding the first equation to the second equation produces a new second equation.

$$\begin{cases} x - 2y + 2z = 9 \\ y + 2z = 5 \\ -y - 3z = -8 \end{cases}$$

Adding -2 times the first equation to the third equation produces a new third equation.

Now work on the second column. (You need to eliminate y from the third equation.)

$$\begin{cases} x - 2y + 2z = 9 \\ y + 2z = 5 \\ -z = -3 \end{cases}$$

Adding the second equation to the third equation produces a new third equation.

Finally, you need a coefficient of 1 for z in the third equation.

$$\begin{cases} x - 2y + 2z = 9 \\ y + 2z = 5 \\ z = 3 \end{cases}$$

Multiplying the third equation by -1 produces a new third equation.

This is the same system that was solved in Example 1, and, as in that example, you can conclude by back-substitution that the solution is

$x = 1, \quad y = -1, \quad \text{and} \quad z = 3.$ The solution is $(1, -1, 3)$.

Additional Examples

Solve each system of linear equations.

a. $\begin{cases} 3x - y + z = 0 \\ 4x - 2y = 1 \\ -x + 6y + 4z = 2 \end{cases}$

b. $\begin{cases} 5x - 2y + 8z = -4 \\ x + 3y - 6z = 9 \\ 2x + y + 4z = 6 \end{cases}$

Answers:

a. $\left(1, \frac{3}{2}, -\frac{3}{2}\right)$

b. $\left(0, 4, \frac{1}{2}\right)$

Exercises Within Reach®

Solutions in English & Spanish and tutorial videos at AlgebraWithinReach.com

Using Gaussian Elimination **In Exercises 5–8, solve the system of linear equations.**

5. $\begin{cases} x + z = 4 \\ y = 2 \\ 4x + z = 7 \end{cases}$ (1, 2, 3)

6. $\begin{cases} x + y = 6 \\ 3x - y = 2 \\ z = 3 \end{cases}$ (2, 4, 3)

7. $\begin{cases} x + y + z = 6 \\ 2x - y + z = 3 \\ 3x - z = 0 \end{cases}$ (1, 2, 3)

8. $\begin{cases} x + y + z = 2 \\ -x + 3y + 2z = 8 \\ 4x + y = 4 \end{cases}$ (0, 4, −2)

EXAMPLE 3 **Using Gaussian Elimination to Solve a System**

Solve the system of linear equations.

$$\begin{cases} 4x + y - 3z = 11 & \text{Equation 1} \\ 2x - 3y + 2z = 9 & \text{Equation 2} \\ x + y + z = -3 & \text{Equation 3} \end{cases}$$

SOLUTION

$$\begin{cases} x + y + z = -3 \\ 2x - 3y + 2z = 9 \\ 4x + y - 3z = 11 \end{cases}$$

Interchange the first and third equations.

$$\begin{cases} x + y + z = -3 \\ -5y \quad 15 \\ 4x + y - 3z = 11 \end{cases}$$

Adding -2 times the first equation to the second equation produces a new second equation.

$$\begin{cases} x + y + z = -3 \\ -5y = 15 \\ -3y - 7z = 23 \end{cases}$$

Adding -4 times the first equation to the third equation produces a new third equation.

$$\begin{cases} x + y + z = -3 \\ y = -3 \\ -3y - 7z = 23 \end{cases}$$

Multiplying the second equation by $-\frac{1}{5}$ produces a new second equation.

$$\begin{cases} x + y + z = -3 \\ y = -3 \\ -7z = 14 \end{cases}$$

Adding 3 times the second equation to the third equation produces a new third equation.

$$\begin{cases} x + y + z = -3 \\ y = -3 \\ z = -2 \end{cases}$$

Multiplying the third equation by $-\frac{1}{7}$ produces a new third equation.

Now you can back-substitute $z = -2$ and $y = -3$ into Equation 1 to find that $x = 2$. So,

$x = 2, \quad y = -3, \quad \text{and} \quad z = -2.$ The solution is $(2, -3, -2)$.

Exercises Within Reach®

Solutions in English & Spanish and tutorial videos at AlgebraWithinReach.com

Using Gaussian Elimination **In Exercises 9–16, solve the system of linear equations.**

9. $\begin{cases} x + y + z = -3 \\ 4x + y - 3z = 11 \\ 2x - 3y + 2z = 9 \end{cases}$ $(2, -3, -2)$

10. $\begin{cases} x - y + 2z = -4 \\ 3x + y - 4z = -6 \\ 2x + 3y - 4z = 4 \end{cases}$ $(-2, 4, 1)$

11. $\begin{cases} x + 6y + 2z = 9 \\ 3x - 2y + 3z = -1 \\ 5x - 5y + 2z = 7 \end{cases}$ $(5, 2, -4)$

12. $\begin{cases} 2x + 2z = 2 \\ 5x + 3y = 4 \\ 3y - 4z = 4 \end{cases}$ $(-4, 8, 5)$

13. $\begin{cases} 6y + 4z = -12 \\ 3x + 3y = 9 \\ 2x - 3z = 10 \end{cases}$ $(5, -2, 0)$

14. $\begin{cases} 2x - 4y + z = 0 \\ 3x + 2z = -1 \\ -6x + 3y + 2z = -10 \end{cases}$ $(1, 0, -2)$

15. $\begin{cases} 2x + y + 3z = 1 \\ 2x + 6y + 8z = 3 \\ 6x + 8y + 18z = 5 \end{cases}$ $\left(\frac{3}{10}, \frac{2}{5}, 0\right)$

16. $\begin{cases} 3x - y - 2z = 5 \\ 2x + y + 3z = 6 \\ 6x - y - 4z = 9 \end{cases}$ $(2, -1, 1)$

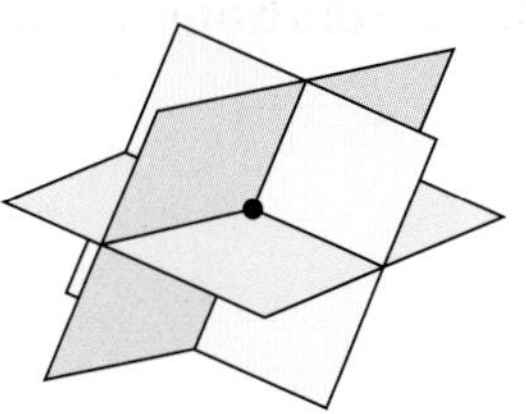
Solution: one point

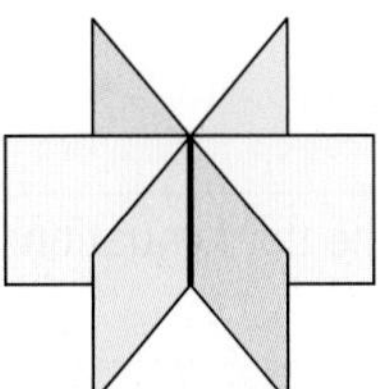
Solution: one line

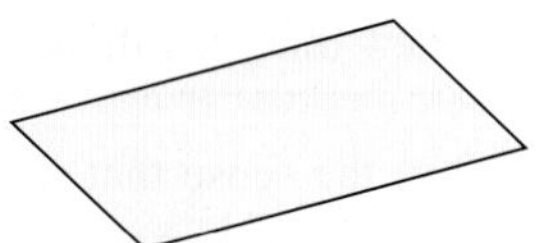
Solution: one plane

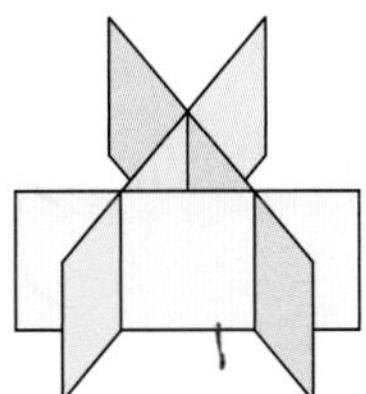
Solution: none

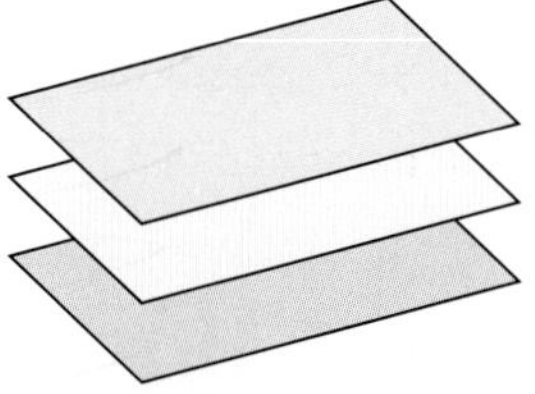
Solution: none

EXAMPLE 4 An Inconsistent System

Solve the system of linear equations.

$$\begin{cases} x - 3y + z = 1 & \text{Equation 1} \\ 2x - y - 2z = 2 & \text{Equation 2} \\ x + 2y - 3z = -1 & \text{Equation 3} \end{cases}$$

SOLUTION

$$\begin{cases} x - 3y + z = 1 \\ 5y - 4z = 0 \\ x + 2y - 3z = -1 \end{cases}$$

Adding -2 times the first equation to the second equation produces a new second equation.

$$\begin{cases} x - 3y + z = 1 \\ 5y - 4z = 0 \\ 5y - 4z = -2 \end{cases}$$

Adding -1 times the first equation to the third equation produces a new third equation.

$$\begin{cases} x - 3y + z = 1 \\ 5y - 4z = 0 \\ 0 = -2 \end{cases}$$

Adding -1 times the second equation to the third equation produces a new third equation.

Because the third "equation" is a false statement, you can conclude that this system is inconsistent and therefore has no solution. Moreover, because this system is equivalent to the original system, you can conclude that the original system also has no solution.

The Number of Solutions of a Linear System

For a system of linear equations, exactly one of the following is true.

1. There is exactly one solution.
2. There are infinitely many solutions.
3. There is no solution.

The graph of a system of three linear equations in three variables consists of *three planes*. When these planes intersect in a single point, the system has exactly one solution (see figure). When the three planes intersect in a line or a plane, the system has infinitely many solutions (see figure). When the three planes have no point in common, the system has no solution (see figure).

Exercises Within Reach®

Solutions in English & Spanish and tutorial videos at AlgebraWithinReach.com

Using Gaussian Elimination **In Exercises 17–20, solve the system of linear equations.**

17. $\begin{cases} x + 2y + 6z = 5 \\ -x + y - 2z = 3 \\ x - 4y - 2z = 1 \end{cases}$

No solution

18. $\begin{cases} x + y + 8z = 3 \\ 2x + y + 11z = 4 \\ x + 3z = 0 \end{cases}$

No solution

19. $\begin{cases} y + z = 5 \\ 2x + 4z = 4 \\ 2x - 3y = -14 \end{cases}$

$(-4, 2, 3)$

20. $\begin{cases} 5x + 2y = -8 \\ z = 5 \\ 3x - y + z = 9 \end{cases}$

$(0, -4, 5)$

EXAMPLE 5 A System with Infinitely Many Solutions

Solve the system of linear equations.

$$\begin{cases} x + y - 3z = -1 & \text{Equation 1} \\ y - z = 0 & \text{Equation 2} \\ -x + 2y = 1 & \text{Equation 3} \end{cases}$$

SOLUTION

Begin by rewriting the system in row-echelon form.

$$\begin{cases} x + y - 3z = -1 \\ y - z = 0 \\ 3y - 3z = 0 \end{cases}$$

Adding the first equation to the third equation produces a new third equation.

$$\begin{cases} x + y - 3z = -1 \\ y - z = 0 \\ 0 = 0 \end{cases}$$

Adding -3 times the second equation to the third equation produces a new third equation.

This means that Equation 3 depends on Equations 1 and 2 in the sense that it gives no additional information about the variables. So, the original system is equivalent to the system

$$\begin{cases} x + y - 3z = -1 \\ y - z = 0 \end{cases}.$$

In the last equation, solve for y in terms of z to obtain $y = z$. Back-substituting for y in the previous equation produces

$$x = 2z - 1.$$

Finally, letting $z = a$, where a is any real number, you can see that there are an infinite number of solutions to the original system, all of the form

$$x = 2a - 1, y = a, \text{ and } z = a.$$

So, every ordered triple of the form

$(2a - 1, a, a)$, a is a real number.

is a solution of the system.

Exercises Within Reach®

Solutions in English & Spanish and tutorial videos at AlgebraWithinReach.com

Using Gaussian Elimination **In Exercises 21–26, solve the system of linear equations.**

21. $\begin{cases} 2x + z = 1 \\ 5y - 3z = 2 \\ 6x + 20y - 9z = 11 \end{cases}$ $\left(-\frac{1}{2}a + \frac{1}{2}, \frac{3}{5}a + \frac{2}{5}, a\right)$

22. $\begin{cases} 3x + y + z = 2 \\ 4x + 2z = 1 \\ 5x - y + 3z = 0 \end{cases}$ $\left(-\frac{1}{2}a + \frac{1}{4}, \frac{1}{2}a + \frac{5}{4}, a\right)$

23. $\begin{cases} x + 4y - 2z = 2 \\ -3x + y + z = -2 \\ 5x + 7y - 5z = 6 \end{cases}$ $\left(\frac{6}{13}a + \frac{10}{13}, \frac{5}{13}a + \frac{4}{13}, a\right)$

24. $\begin{cases} x - 2y - z = 3 \\ 2x + y - 3z = 1 \\ x + 8y - 3z = -7 \end{cases}$ $\left(\frac{7}{5}a + 1, \frac{1}{5}a - 1, a\right)$

25. $\begin{cases} x + 2y - 7z = -4 \\ 2x + y + z = 13 \\ 3x + 9y - 36z = -33 \end{cases}$ $(-3a + 10, 5a - 7, a)$

26. $\begin{cases} 2x + y - 3z = 4 \\ 4x + 2z = 10 \\ -2x + 3y - 13z = -8 \end{cases}$ $\left(-\frac{1}{2}a + \frac{5}{2}, 4a - 1, a\right)$

Application

Application **EXAMPLE 6 Vertical Motion**

The height at time t of an object that is moving in a (vertical) line with constant acceleration a is given by the **position equation**

$$s = \frac{1}{2}at^2 + v_0t + s_0.$$

The height s is measured in feet, the acceleration a is measured in feet per second squared, the time t is measured in seconds, v_0 is the initial velocity (at time $t = 0$), and s_0 is the initial height. Find the values of a, v_0, and s_0 for a projected object given that $s = 164$ feet at 1 second, $s = 180$ feet at 2 seconds, and $s = 164$ feet at 3 seconds.

SOLUTION

By substituting the three values of t and s into the position equation, you obtain three linear equations in a, v_0, and s_0.

When $t = 1, s = 164$: $\frac{1}{2}a(1)^2 + v_0(1) + s_0 = 164$

When $t = 2, s = 180$: $\frac{1}{2}a(2)^2 + v_0(2) + s_0 = 180$

When $t = 3, s = 164$: $\frac{1}{2}a(3)^2 + v_0(3) + s_0 = 164$

By multiplying the first and third equations by 2, this system can be rewritten as

$$\begin{cases} a + 2v_0 + 2s_0 = 328 & \text{Equation 1} \\ 2a + 2v_0 + s_0 = 180 & \text{Equation 2} \\ 9a + 6v_0 + 2s_0 = 328 & \text{Equation 3} \end{cases}$$

and you can apply Gaussian elimination to obtain

$$\begin{cases} a + 2v_0 + 2s_0 = 328 & \text{Equation 1} \\ -2v_0 - 3s_0 = -476. & \text{Equation 2} \\ 2s_0 = 232 & \text{Equation 3} \end{cases}$$

From the third equation, $s_0 = 116$, so back-substitution into Equation 2 yields

$-2v_0 - 3(116) = -476$ Substitute 116 for s_0.

$-2v_0 = -128$ Simplify.

$v_0 = 64.$ Divide each side by -2.

Finally, back-substituting $v_0 = 64$ and $s_0 = 116$ into Equation 1 yields

$a + 2(64) + 2(116) = 328$ Substitute 64 for v_0 and 116 for s_0.

$a = -32.$ Simplify.

So, the position equation for this object is $s = -16t^2 + 64t + 116$.

Exercises Within Reach®

Solutions in English & Spanish and tutorial videos at AlgebraWithinReach.com

Vertical Motion **In Exercises 27 and 28, find the position equation $s = \frac{1}{2}at^2 + v_0t + s_0$ for an object that has the indicated heights at the specified times.**

27. $s = 128$ feet at $t = 1$ second
$s = 80$ feet at $t = 2$ seconds
$s = 0$ feet at $t = 3$ seconds $s = -16t^2 + 144$

28. $s = 48$ feet at $t = 1$ second
$s = 64$ feet at $t = 2$ seconds
$s = 48$ feet at $t = 3$ seconds $s = -16t^2 + 64t$

Concept Summary: Solving Systems of Linear Equations by Gaussian Elimination

What

You can solve a system of linear equations in three variables by rewriting the system in **row-echelon form**. This usually involves a chain of **equivalent systems**, each of which is obtained by using one of three basic **row operations**. This process is called **Gaussian elimination**.

How

Each of the following row operations on a system of linear equations produces an equivalent system of linear equations.

1. Interchange two equations.
2. Multiply one of the equations by a nonzero constant.
3. Add a multiple of one of the equations to another equation to replace the latter equation.

Why

When solving an equivalent system, you can conclude that the solution is also the solution of the original system.

Exercises Within Reach®

Worked-out solutions to odd-numbered exercises at AlgebraWithinReach.com

Concept Summary Check

29. ***Gaussian Elimination*** How can the process of Gaussian elimination help you to solve a system of equations? In general, after applying Gaussian elimination in a system of equations, what are the next steps you take to find the solution of the system?
See Additional Answers.

30. ***Logic*** Describe the three row operations that you can use to produce an equivalent system of equations while applying Gaussian elimination. See Additional Answers.

31. ***Vocabulary*** Give an example of a system of three linear equations in three variables that is in row-echelon form.

32. ***Writing*** Show how to use back-substitution to solve the system of equations you wrote in Exercise 31.

31. $\begin{cases} x + 2y - 3z = 13 \\ y + 4z = -11 \\ z = -3 \end{cases}$

32. Substitute $z = -3$ in the second equation to obtain $y = 1$. Substitute $y = 1$ and $z = -3$ in the first equation to obtain $x = 2$. The solution of the system is $(2, 1, -3)$.

Extra Practice

Finding an Equivalent System **In Exercises 33 and 34, determine whether the two systems of linear equations are equivalent. Justify your answer.** See Additional Answers.

33. $\begin{cases} x + 3y - z = 6 \\ 2x - y + 2z = 1 \\ 3x + 2y - z = 2 \end{cases}$ $\begin{cases} x + 3y - z = 6 \\ -7y + 4z = -11 \\ -7y + 2z = -16 \end{cases}$

34. $\begin{cases} x - 2y + 3z = 9 \\ -x + 3y = -4 \\ 2x - 5y + 5z = 17 \end{cases}$ $\begin{cases} x - 2y + 3z = 9 \\ y + 3z = 5 \\ -y - z = -1 \end{cases}$

Finding a System in Three Variables **In Exercises 35 and 36, find a system of linear equations in three variables with integer coefficients that has the given point as a solution. (There are many correct answers.)**

35. $(4, -3, 2)$ See Additional Answers.

36. $(5, 7, -10)$ See Additional Answers.

37. ***Geometry*** The sum of the measures of two angles of a triangle is twice the measure of the third angle. The measure of the second angle is 28° less than the measure of the third angle. Find the measures of the three angles.
88°, 32°, 60°

38. ***Geometry*** The measure of one angle of a triangle is two-thirds the measure of a second angle, and the measure of the second angle is 12° greater than the measure of the third angle. Find the measures of the three angles. 48°, 72°, 60°

39. ***Coffee*** A coffee manufacturer sells a 10-pound package that consists of three flavors of coffee. Vanilla coffee costs \$6 per pound, hazelnut coffee costs \$6.50 per pound, and French roast coffee costs \$7 per pound. The package contains the same amount of hazelnut coffee as French roast coffee. The cost of the 10-pound package is \$66. How many pounds of each type of coffee are in the package?
2 pounds of vanilla, 4 pounds of hazelnut, 4 pounds of French roast

40. ***Hot Dogs*** A vendor sells three sizes of hot dogs at prices of \$1.50, \$2.50, and \$3.25. On a day when the vendor had a total revenue of \$289.25 from sales of 143 hot dogs, 4 times as many \$1.50 hot dogs were sold as \$3.25 hot dogs. How many hot dogs were sold at each price? 84 hot dogs at \$1.50, 38 hot dogs at \$2.50, 21 hot dogs at \$3.25

41. ***School Orchestra*** The table shows the percents of each section of the North High School orchestra that were chosen to participate in the city orchestra, the county orchestra, and the state orchestra. Thirty members of the city orchestra, 17 members of the county orchestra, and 10 members of the state orchestra are from North High. How many members are in each section of North High's orchestra? 50 students in strings, 20 students in winds, 8 students in percussion

Orchestra	String	Wind	Percussion
City orchestra	40%	30%	50%
County orchestra	20%	25%	25%
State orchestra	10%	15%	25%

42. ***Sports*** The table shows the percents of each unit of the North High School football team that were chosen for academic honors, as city all-stars, and as county all-stars. Of all the players on the football team, 5 were awarded with academic honors, 13 were named city all-stars, and 4 were named county all-stars. How many members of each unit are there on the football team? 20 players on defense, 30 players on offense, 10 players on special teams

	Defense	Offense	Special teams
Academic honors	0%	10%	20%
City all-stars	10%	20%	50%
County all-stars	10%	0%	20%

Explaining Concepts

43. ***Logic*** You apply Gaussian elimination to a system of three equations in the variables x, y, and z. From the row-echelon form, the solution $(1, -3, 4)$ is apparent *without* applying back-substitution or any other calculations. Explain why. See Additional Answers.

44. ***Reasoning*** A system of three linear equations in three variables has an infinite number of solutions. Is it possible that the graphs of two of the three equations are parallel planes? Explain. No. If two planes are parallel, the system has no solution.

45. ***Precision*** Two ways that a system of three linear equations in three variables can have no solution are shown on page 187. Describe the graph for a third type of situation that results in no solution. Three planes have no point in common when two of the planes are parallel and the third plane intersects the other two planes.

46. ***Think About It*** Describe the graphs and numbers of solutions possible for a system of three linear equations in three variables in which at least two of the equations are dependent. See Additional Answers.

47. ***Think About It*** Describe the graphs and numbers of solutions possible for a system of three linear equations in three variables when each pair of equations is consistent and *not* dependent. See Additional Answers.

48. ***Writing*** Write a system of four linear equations in four unknowns, and use Gaussian elimination with back-substitution to solve it. Answers will vary.

Cumulative Review

In Exercises 49–52, identify the terms and coefficients of the algebraic expression.

49. $3x + 2$ $3x, 2; 3, 2$

50. $4x^2 + 5x - 4$ $4x^2, 5x, -4; 4, 5, -4$

51. $14t^5 - t + 25$ $14t^5, -t, 25; 14, -1, 25$

52. $5s^2 + 3st + 2t^2$ $5s^2, 3st, 2t^2; 5, 3, 2$

In Exercises 53–56, solve the system of linear equations by the method of elimination.

53. $\begin{cases} 2x + 3y = 17 \\ 4y = 12 \end{cases}$ $(4, 3)$

54. $\begin{cases} x - 2y = 11 \\ 3x + 3y = 6 \end{cases}$ $(5, -3)$

55. $\begin{cases} 3x - 4y = -30 \\ 5x + 4y = 14 \end{cases}$ $(-2, 6)$

56. $\begin{cases} 3x + 5y = 1 \\ 4x + 15y = 5 \end{cases}$ $\left(-\frac{2}{5}, \frac{11}{25}\right)$

Mid-Chapter Quiz: Sections 4.1–4.3

Solutions in English & Spanish and tutorial videos at AlgebraWithinReach.com

Take this quiz as you would take a quiz in class. After you are done, check your work against the answers in the back of the book.

1. Determine whether each ordered pair is a solution of the system of linear equations: (a) $(1, -2)$ (b) $(10, 4)$

$$\begin{cases} 5x - 12y = 2 \\ 2x + 1.5y = 26 \end{cases}$$

In Exercises 2–4, graph the equations in the system. Use the graphs to determine the number of solutions of the system.

2. $\begin{cases} -6x + 9y = 9 \\ 2x - 3y = 6 \end{cases}$

3. $\begin{cases} x - 2y = -4 \\ 3x - 2y = 4 \end{cases}$

4. $\begin{cases} 0.5x - 1.5y = 7 \\ -2x + 6y = -28 \end{cases}$

In Exercises 5–7, solve the system of equations by graphing.

5. $\begin{cases} x - y = 0 \\ 2x = 8 \end{cases}$

6. $\begin{cases} 2x + 7y = 16 \\ 3x + 2y = 24 \end{cases}$

7. $\begin{cases} 4x - y = 9 \\ x - 3y = 16 \end{cases}$

In Exercises 8–10, solve the system of equations by the method of substitution.

8. $\begin{cases} 2x - 3y = 4 \\ y = 2 \end{cases}$

9. $\begin{cases} 5x - y = 32 \\ 6x - 9y = 18 \end{cases}$

10. $\begin{cases} 6x - 2y = 2 \\ 9x - 3y = 1 \end{cases}$

In Exercises 11–14, use elimination or Gaussian elimination to solve the linear system.

11. $\begin{cases} x + 10y = 18 \\ 5x + 2y = 42 \end{cases}$

12. $\begin{cases} x - 3y = 6 \\ 3x + y = 8 \end{cases}$

13. $\begin{cases} a + b + c = 1 \\ 4a + 2b + c = 2 \\ 9a + 3b + c = 4 \end{cases}$

14. $\begin{cases} x + 4z = 17 \\ -3x + 2y - z = -20 \\ x - 5y + 3z = 19 \end{cases}$

In Exercises 15 and 16, write a system of linear equations having the given solution. (There are many correct answers.)

15. $(10, -12)$

16. $(2, -5, 10)$

Applications

17. Twenty gallons of a 30% brine solution is obtained by mixing a 20% solution with a 50% solution. How many gallons of each solution are required?

18. In a triangle, the measure of one angle is 14° less than twice the measure of a second angle. The measure of the third angle is 30° greater than the measure of the second angle. Find the measures of the three angles.

Study Skills in Action

Viewing Math as a Foreign Language

Learning math requires more than just completing homework problems. For instance, learning the material in a chapter may require using approaches similar to those used for learning a foreign language in that you must:

- understand and memorize vocabulary words;
- understand and memorize mathematical rules (as you would memorize grammatical rules); and
- apply rules to mathematical expressions or equations (like creating sentences using correct grammar rules).

You should understand the vocabulary words and rules in a chapter as well as memorize and say them out loud. Strive to speak the mathematical language with fluency, just as a student learning a foreign language must strive to do.

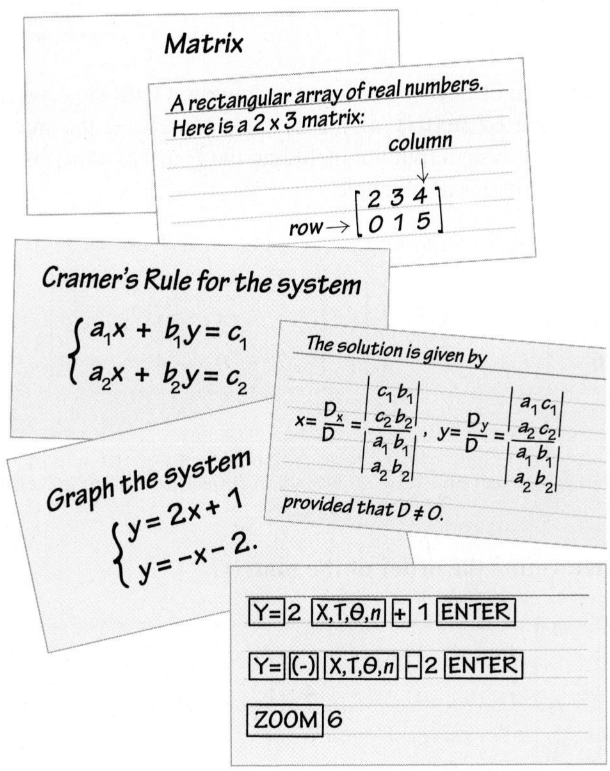

Smart Study Strategy

Make Note Cards

Invest in three different colors of 4 × 6 note cards. Use one color for each of the following: vocabulary words; rules; and calculator keystrokes.

- Write vocabulary words on note cards, one word per card. Write the definition and an example on the other side. If possible, put definitions in your own words.
- Write rules on note cards, one per card. Include an example and an explanation on the other side.
- Write each kind of calculation on a separate note card. Include the keystrokes required to perform the calculation on the other side.

Use the note cards as references while doing your homework. Quiz yourself once a day.

4.4 Matrices and Linear Systems

- Form augmented matrices and form linear systems from augmented matrices.
- Perform elementary row operations to solve systems of linear equations.
- Use matrices and Gaussian elimination to solve systems of linear equations.

Augmented and Coefficient Matrices

A **matrix** is a rectangular array of numbers. The plural of matrix is *matrices*.

EXAMPLE 1 Determining the Order of a Matrix

Determine the order of each matrix.

a. $\begin{bmatrix} 1 & -2 & 4 \\ 0 & 1 & -2 \end{bmatrix}$ **b.** $\begin{bmatrix} 0 & 0 \\ 0 & 0 \end{bmatrix}$ **c.** $\begin{bmatrix} 1 & -3 \\ -2 & 0 \\ 4 & -2 \end{bmatrix}$

SOLUTION

a. This matrix has two rows and three columns, so the order is 2×3.

b. This matrix has two rows and two columns, so the order is 2×2.

c. This matrix has three rows and two columns, so the order is 3×2.

Study Tip

The **order** of a matrix is always given as *row by column*. A matrix with the same number of rows as columns is called a **square matrix**. For instance, the 2×2 matrix in Example 1(b) is square.

A matrix derived from a system of linear equations (each written with the constant term on the right) is the **augmented matrix** of the system. Moreover, the matrix derived from the coefficients of the system (but not including the constant terms) is the **coefficient matrix** of the system. Here is an example.

System *Coefficient Matrix* *Augmented Matrix*

$$\begin{cases} x - 4y + 3z = 5 \\ -x + 3y - z = -3 \\ 2x - 4z = 6 \end{cases} \quad \begin{bmatrix} 1 & -4 & 3 \\ -1 & 3 & -1 \\ 2 & 0 & -4 \end{bmatrix} \quad \left[\begin{array}{ccc:c} 1 & -4 & 3 & 5 \\ -1 & 3 & -1 & -3 \\ 2 & 0 & -4 & 6 \end{array}\right]$$

Exercises Within Reach®

Solutions in English & Spanish and tutorial videos at AlgebraWithinReach.com

Determining the Order of a Matrix **In Exercises 1–10, determine the order of the matrix.**

1. $\begin{bmatrix} 3 & -2 \\ -4 & 0 \\ 2 & -7 \\ -1 & -3 \end{bmatrix}$ 4×2

2. $\begin{bmatrix} 3 & 4 \\ 2 & -1 \\ 8 & 10 \\ -6 & -6 \\ 12 & 50 \end{bmatrix}$ 5×2

3. $\begin{bmatrix} 4 \\ -2 \\ 0 \\ 1 \end{bmatrix}$ 4×1

4. $\begin{bmatrix} 5 & -8 & 32 \\ 7 & 15 & 28 \end{bmatrix}$ 2×3

5. $\begin{bmatrix} -2 & 5 \\ 0 & -1 \end{bmatrix}$ 2×2

6. $\begin{bmatrix} 4 & 0 & -5 \\ -1 & 8 & 9 \\ 0 & -3 & 4 \end{bmatrix}$ 3×3

7. $[5]$ 1×1

8. $\begin{bmatrix} 6 \\ -13 \\ 22 \end{bmatrix}$ 3×1

9. $[13 \quad 12 \quad -9 \quad 0]$ 1×4

10. $[1 \quad -1 \quad 2 \quad 3]$ 1×4

EXAMPLE 2 Forming Coefficient and Augmented Matrices

Study Tip

In Example 2(b), note the use of 0 for the missing *y*-variable in the second equation and also for the missing *z*-variable in the third equation.

Form the coefficient matrix and the augmented matrix for each system.

a. $\begin{cases} -x + 5y = 2 \\ 7x - 2y = -6 \end{cases}$ **b.** $\begin{cases} 3x + 2y - z = 1 \\ x + 2z = -3 \\ -2x - y = 4 \end{cases}$

SOLUTION

	System	*Coefficient Matrix*	*Augmented Matrix*
a.	$\begin{cases} -x + 5y = 2 \\ 7x - 2y = -6 \end{cases}$	$\begin{bmatrix} -1 & 5 \\ 7 & -2 \end{bmatrix}$	$\left[\begin{array}{cc:c} -1 & 5 & 2 \\ 7 & -2 & -6 \end{array}\right]$
b.	$\begin{cases} 3x + 2y - z = 1 \\ x + 2z = -3 \\ -2x - y = 4 \end{cases}$	$\begin{bmatrix} 3 & 2 & -1 \\ 1 & 0 & 2 \\ -2 & -1 & 0 \end{bmatrix}$	$\left[\begin{array}{ccc:c} 3 & 2 & -1 & 1 \\ 1 & 0 & 2 & -3 \\ -2 & -1 & 0 & 4 \end{array}\right]$

EXAMPLE 3 Forming Linear Systems from Their Matrices

Write the system of linear equations represented by each augmented matrix.

a. $\left[\begin{array}{cc:c} 3 & -5 & 4 \\ -1 & 2 & 0 \end{array}\right]$ **b.** $\left[\begin{array}{cc:c} 1 & 3 & 2 \\ 0 & 1 & -3 \end{array}\right]$ **c.** $\left[\begin{array}{ccc:c} 2 & 0 & -8 & 1 \\ -1 & 1 & 1 & 2 \\ 5 & -1 & 7 & 3 \end{array}\right]$

SOLUTION

a. $\begin{cases} 3x - 5y = 4 \\ -x + 2y = 0 \end{cases}$ **b.** $\begin{cases} x + 3y = 2 \\ y = -3 \end{cases}$ **c.** $\begin{cases} 2x - 8z = 1 \\ -x + y + z = 2 \\ 5x - y + 7z = 3 \end{cases}$

Exercises Within Reach®

Solutions in English & Spanish and tutorial videos at AlgebraWithinReach.com

Forming Coefficient and Augmented Matrices **In Exercises 11–14, form (a) the coefficient matrix and (b) the augmented matrix for the system of linear equations.**

11. $\begin{cases} 4x - 5y = -2 \\ -x + 8y = 10 \end{cases}$ (a) $\begin{bmatrix} 4 & -5 \\ -1 & 8 \end{bmatrix}$ (b) $\left[\begin{array}{cc:c} 4 & -5 & -2 \\ -1 & 8 & 10 \end{array}\right]$

12. $\begin{cases} 8x + 3y = 25 \\ 3x - 9y = 12 \end{cases}$ (a) $\begin{bmatrix} 8 & 3 \\ 3 & -9 \end{bmatrix}$ (b) $\left[\begin{array}{cc:c} 8 & 3 & 25 \\ 3 & -9 & 12 \end{array}\right]$

13. $\begin{cases} x + y = 0 \\ 5x - 2y - 2z = 12 \\ 2x + 4y + z = 5 \end{cases}$ See Additional Answers.

14. $\begin{cases} 9x - 3y + z = 13 \\ 12x - 8z = 5 \\ 3x + 4y - z = 6 \end{cases}$ See Additional Answers.

Forming a Linear System **In Exercises 15–18, write the system of linear equations represented by the augmented matrix. (Use variables *x*, *y*, and *z*.)**

15. $\left[\begin{array}{cc:c} 4 & 3 & 8 \\ 1 & -2 & 3 \end{array}\right]$ $\begin{cases} 4x + 3y = 8 \\ x - 2y = 3 \end{cases}$

16. $\left[\begin{array}{cc:c} 9 & -4 & 0 \\ 6 & 1 & -4 \end{array}\right]$ $\begin{cases} 9x - 4y = 0 \\ 6x + y = -4 \end{cases}$

17. $\left[\begin{array}{ccc:c} 1 & 0 & 2 & -10 \\ 0 & 3 & -1 & 5 \\ 4 & 2 & 0 & 3 \end{array}\right]$ $\begin{cases} x + 2z = -10 \\ 3y - z = 5 \\ 4x + 2y = 3 \end{cases}$

18. $\left[\begin{array}{ccc:c} 4 & -1 & 3 & 5 \\ 2 & 0 & -2 & -1 \\ -1 & 6 & 0 & 3 \end{array}\right]$ $\begin{cases} 4x - y + 3z = 5 \\ 2x - 2z = -1 \\ -x + 6y = 3 \end{cases}$

Study Tip

Although elementary row operations are simple to perform, they involve a lot of arithmetic. So that you can check your work, you should get in the habit of noting the elementary row operations performed in each step. People use different schemes to do this. The scheme that is used in this text is to write an abbreviated version of the row operation at the left of the row that has been changed, as shown in Example 4.

Elementary Row Operations

Elementary Row Operations

Any of the following **elementary row operations** performed on an augmented matrix will produce a matrix that is row-equivalent to the original matrix. Two matrices are **row-equivalent** if one can be obtained from the other by a sequence of elementary row operations.

1. Interchange two rows.
2. Multiply a row by a nonzero constant.
3. Add a multiple of a row to another row.

EXAMPLE 4 Performing Elementary Row Operations

a. Interchange the first and second rows.

Original Matrix $\begin{bmatrix} 0 & 1 & 3 & 4 \\ -1 & 2 & 0 & 3 \\ 2 & -3 & 4 & 1 \end{bmatrix}$ *New Row-Equivalent Matrix* $\begin{matrix} R_2 \\ R_1 \\ \\ \end{matrix}\begin{bmatrix} -1 & 2 & 0 & 3 \\ 0 & 1 & 3 & 4 \\ 2 & -3 & 4 & 1 \end{bmatrix}$

b. Multiply the first row by $\frac{1}{2}$.

Original Matrix $\begin{bmatrix} 2 & -4 & 6 & -2 \\ 1 & 3 & -3 & 0 \\ 5 & -2 & 1 & 2 \end{bmatrix}$ *New Row-Equivalent Matrix* $\begin{matrix} \frac{1}{2}R_1 \rightarrow \\ \\ \\ \end{matrix}\begin{bmatrix} 1 & -2 & 3 & -1 \\ 1 & 3 & -3 & 0 \\ 5 & -2 & 1 & 2 \end{bmatrix}$

c. Add -2 times the first row to the third row.

Original Matrix $\begin{bmatrix} 1 & 2 & -4 & 3 \\ 0 & 3 & -2 & -1 \\ 2 & 1 & 5 & -2 \end{bmatrix}$ *New Row-Equivalent Matrix* $\begin{matrix} \\ \\ -2R_1 + R_3 \rightarrow \end{matrix}\begin{bmatrix} 1 & 2 & -4 & 3 \\ 0 & 3 & -2 & -1 \\ 0 & -3 & 13 & -8 \end{bmatrix}$

d. Add 6 times the first row to the second row.

Original Matrix $\begin{bmatrix} 1 & 2 & 2 & -4 \\ -6 & -11 & 3 & 18 \\ 0 & 0 & 4 & 7 \end{bmatrix}$ *New Row-Equivalent Matrix* $\begin{matrix} \\ 6R_1 + R_2 \rightarrow \\ \\ \end{matrix}\begin{bmatrix} 1 & 2 & 2 & -4 \\ 0 & 1 & 15 & -6 \\ 0 & 0 & 4 & 7 \end{bmatrix}$

Exercises Within Reach®

Solutions in English & Spanish and tutorial videos at AlgebraWithinReach.com

Performing an Elementary Row Operation **In Exercises 19–24, fill in the entries of the row-equivalent matrix formed by performing the indicated elementary row operation.**

19. $\begin{bmatrix} 1 & 1 & -4 & 2 \\ 0 & 0 & 8 & 3 \\ 0 & 4 & 5 & 5 \end{bmatrix} \begin{matrix} \\ R_3 \\ R_2 \end{matrix} \begin{bmatrix} 1 & 1 & -4 & 2 \\ 0 & 4 & 5 & 5 \\ 0 & 0 & 8 & 3 \end{bmatrix}$

20. $\begin{bmatrix} 0 & 0 & -5 & 2 \\ 0 & -7 & -3 & 3 \\ 1 & 4 & 5 & 4 \end{bmatrix} \begin{matrix} R_3 \\ \\ R_1 \end{matrix} \begin{bmatrix} 1 & 4 & 5 & 4 \\ 0 & -7 & -3 & 3 \\ 0 & 0 & -5 & 2 \end{bmatrix}$

21. $\begin{bmatrix} 9 & -18 & 27 \\ 3 & 4 & 5 \end{bmatrix} \begin{matrix} \frac{1}{9}R_1 \rightarrow \\ \\ \end{matrix} \begin{bmatrix} 1 & -2 & 3 \\ 3 & 4 & 5 \end{bmatrix}$

22. $\begin{bmatrix} 1 & 21 & 7 \\ 0 & -7 & 14 \end{bmatrix} \begin{matrix} \\ -\frac{1}{7}R_2 \rightarrow \end{matrix} \begin{bmatrix} 1 & 21 & 7 \\ 0 & 1 & -2 \end{bmatrix}$

23. $\begin{bmatrix} 1 & 4 & 3 \\ 2 & 8 & 6 \end{bmatrix} \begin{matrix} \\ -2R_1 + R_2 \rightarrow \end{matrix} \begin{bmatrix} 1 & 4 & 3 \\ 0 & 0 & 0 \end{bmatrix}$

24. $\begin{bmatrix} 1 & 4 & 5 \\ 4 & -7 & 3 \end{bmatrix} \begin{matrix} \\ -4R_1 + R_2 \rightarrow \end{matrix} \begin{bmatrix} 1 & 4 & 5 \\ 0 & -23 & -17 \end{bmatrix}$

EXAMPLE 5 Solving a System of Linear Equations

Linear System | *Associated Augmented Matrix*

$$\begin{cases} x - 2y + 2z = 9 \\ -x + 3y = -4 \\ 2x - 5y + z = 10 \end{cases} \qquad \begin{bmatrix} 1 & -2 & 2 & \vdots & 9 \\ -1 & 3 & 0 & \vdots & -4 \\ 2 & -5 & 1 & \vdots & 10 \end{bmatrix}$$

Add the first equation to the second equation. | Add the first row to the second row.

$$\begin{cases} x - 2y + 2z = 9 \\ y + 2z = 5 \\ 2x - 5y + z = 10 \end{cases} \qquad R_1 + R_2 \rightarrow \begin{bmatrix} 1 & -2 & 2 & \vdots & 9 \\ 0 & 1 & 2 & \vdots & 5 \\ 2 & -5 & 1 & \vdots & 10 \end{bmatrix}$$

Add -2 times the first equation to the third equation. | Add -2 times the first row to the third row.

$$\begin{cases} x - 2y + 2z = 9 \\ y + 2z = 5 \\ -y - 3z = -8 \end{cases} \qquad -2R_1 + R_3 \rightarrow \begin{bmatrix} 1 & -2 & 2 & \vdots & 9 \\ 0 & 1 & 2 & \vdots & 5 \\ 0 & -1 & -3 & \vdots & -8 \end{bmatrix}$$

Add the second equation to the third equation. | Add the second row to the third row.

$$\begin{cases} x - 2y + 2z = 9 \\ y + 2z = 5 \\ -z = -3 \end{cases} \qquad R_2 + R_3 \rightarrow \begin{bmatrix} 1 & -2 & 2 & \vdots & 9 \\ 0 & 1 & 2 & \vdots & 5 \\ 0 & 0 & -1 & \vdots & -3 \end{bmatrix}$$

Multiply the third equation by -1. | Multiply the third row by -1.

$$\begin{cases} x - 2y + 2z = 9 \\ y + 2z = 5 \\ z = 3 \end{cases} \qquad -R_3 \rightarrow \begin{bmatrix} 1 & -2 & 2 & \vdots & 9 \\ 0 & 1 & 2 & \vdots & 5 \\ 0 & 0 & 1 & \vdots & 3 \end{bmatrix}$$

At this point, you can use back-substitution to find that the solution is $x = 1, y = -1$, and $z = 3$. The solution can be written as the ordered triple $(1, -1, 3)$.

Study Tip

The last matrix in Example 5 is in row-echelon form. The term *echelon* refers to the stair-step pattern formed by the nonzero elements of the matrix.

Definition of Row-Echelon Form of a Matrix

A matrix in **row-echelon form** has the following properties.

1. All rows consisting entirely of zeros occur at the bottom of the matrix.
2. For each row that does not consist entirely of zeros, the first nonzero entry is 1 (called a **leading 1**).
3. For two successive (nonzero) rows, the leading 1 in the higher row is farther to the left than the leading 1 in the lower row.

Exercises Within Reach®

Solutions in English & Spanish and tutorial videos at AlgebraWithinReach.com

Solving a System of Linear Equations **In Exercises 25 and 26, use matrices to solve the system of linear equations.**

25. $\begin{cases} x - 2y - z = 6 \\ y + 4z = 5 \\ 4x + 2y + 3z = 8 \end{cases}$ (2, −3, 2)

26. $\begin{cases} x - 3z = -2 \\ 3x + y - 2z = 5 \\ 2x + 2y + z = 4 \end{cases}$ (4, −3, 2)

Solving a System of Equations

Gaussian Elimination with Back-Substitution

To use matrices and Gaussian elimination to solve a system of linear equations, use the following steps.

1. Write the augmented matrix of the system of linear equations.
2. Use elementary row operations to rewrite the augmented matrix in row-echelon form.
3. Write the system of linear equations corresponding to the matrix in row-echelon form, and use back-substitution to find the solution.

EXAMPLE 6 Gaussian Elimination with Back-Substitution

Solve the system of linear equations.

$$\begin{cases} 2x - 3y = -2 \\ x + 2y = 13 \end{cases}$$

SOLUTION

$$\left[\begin{array}{rr:r} 2 & -3 & -2 \\ 1 & 2 & 13 \end{array}\right]$$ Augmented matrix for system of linear equations

$$\begin{array}{l} R_2 \\ R_1 \end{array}\left[\begin{array}{rr:r} 1 & 2 & 13 \\ 2 & -3 & -2 \end{array}\right]$$ First column has leading 1 in upper left corner.

$$-2R_1 + R_2 \rightarrow \left[\begin{array}{rr:r} 1 & 2 & 13 \\ 0 & -7 & -28 \end{array}\right]$$ First column has a zero under its leading 1.

$$-\tfrac{1}{7}R_2 \rightarrow \left[\begin{array}{rr:r} 1 & 2 & 13 \\ 0 & 1 & 4 \end{array}\right]$$ Second column has leading 1 in second row.

The system of linear equations that corresponds to the (row-echelon) matrix is

$$\begin{cases} x + 2y = 13 \\ y = 4 \end{cases}.$$

Using back-substitution, you can find that the solution of the system is $x = 5$ and $y = 4$, which can be written as the ordered pair (5, 4). Check this solution in the original system, as follows.

Check

Equation 1: $2(5) - 3(4) = -2$ ✓

Equation 2: $5 + 2(4) = 13$ ✓

Exercises Within Reach®

Solutions in English & Spanish and tutorial videos at AlgebraWithinReach.com

Solving a System of Linear Equations **In Exercises 27–30, use matrices to solve the system of linear equations.**

27. $\begin{cases} 6x - 4y = 2 \\ 5x + 2y = 7 \end{cases}$ (1, 1)

28. $\begin{cases} 2x + 6y = 16 \\ 2x + 3y = 7 \end{cases}$ (−1, 3)

29. $\begin{cases} 12x + 10y = -14 \\ 4x - 3y = -11 \end{cases}$ (−2, 1)

30. $\begin{cases} -x - 5y = -10 \\ 2x - 3y = 7 \end{cases}$ (5, 1)

EXAMPLE 7 Gaussian Elimination with Back-Substitution

Solve the system of linear equations.

$$\begin{cases} 3x + 3y = 9 \\ 2x - 3z = 10 \\ 6y + 4z = -12 \end{cases}$$

SOLUTION

$$\begin{bmatrix} 3 & 3 & 0 & \vdots & 9 \\ 2 & 0 & -3 & \vdots & 10 \\ 0 & 6 & 4 & \vdots & -12 \end{bmatrix}$$ Augmented matrix for system of linear equations

$$\tfrac{1}{3}R_1 \rightarrow \begin{bmatrix} 1 & 1 & 0 & \vdots & 3 \\ 2 & 0 & -3 & \vdots & 10 \\ 0 & 6 & 4 & \vdots & -12 \end{bmatrix}$$ First column has leading 1 in upper left corner.

$$-2R_1 + R_2 \rightarrow \begin{bmatrix} 1 & 1 & 0 & \vdots & 3 \\ 0 & -2 & -3 & \vdots & 4 \\ 0 & 6 & 4 & \vdots & -12 \end{bmatrix}$$ First column has zeros under its leading 1.

$$-\tfrac{1}{2}R_2 \rightarrow \begin{bmatrix} 1 & 1 & 0 & \vdots & 3 \\ 0 & 1 & \frac{3}{2} & \vdots & -2 \\ 0 & 6 & 4 & \vdots & -12 \end{bmatrix}$$ Second column has leading 1 in second row.

$$-6R_2 + R_3 \rightarrow \begin{bmatrix} 1 & 1 & 0 & \vdots & 3 \\ 0 & 1 & \frac{3}{2} & \vdots & -2 \\ 0 & 0 & -5 & \vdots & 0 \end{bmatrix}$$ Second column has zero under its leading 1.

$$-\tfrac{1}{5}R_3 \rightarrow \begin{bmatrix} 1 & 1 & 0 & \vdots & 3 \\ 0 & 1 & \frac{3}{2} & \vdots & -2 \\ 0 & 0 & 1 & \vdots & 0 \end{bmatrix}$$ Third column has leading 1 in third row.

The system of linear equations that corresponds to the (row-echelon) matrix is

$$\begin{cases} x + y \phantom{+ \frac{3}{2}z} = 3 \\ y + \frac{3}{2}z = -2. \\ z = 0 \end{cases}$$

Using back-substitution, you can find that the solution is

$x = 5$ and $y = -2$, and $z = 0$

which can be written as the ordered triple $(5, -2, 0)$.

Exercises Within Reach®

Solutions in English & Spanish and tutorial videos at AlgebraWithinReach.com

Solving a System of Linear Equations **In Exercises 31–34, use matrices to solve the system of linear equations.**

31. $\begin{cases} 2x + 4y = 10 \\ 2x + 2y + 3z = 3 \\ -3x + y + 2z = -3 \end{cases}$ $(1, 2, -1)$

32. $\begin{cases} 2x - y + 3z = 24 \\ 2y - z = 14 \\ 7x - 5y = 6 \end{cases}$ $(8, 10, 6)$

33. $\begin{cases} -2x - 2y - 15z = 0 \\ x + 2y + 2z = 18 \\ 3x + 3y + 22z = 2 \end{cases}$ $(34, -4, -4)$

34. $\begin{cases} 2x + 4y + 5z = 5 \\ x + 3y + 3z = 2 \\ 2x + 4y + 4z = 2 \end{cases}$ $(-1, -2, 3)$

EXAMPLE 8 **A System with No Solution**

Linear System

$$\begin{cases} 6x - 10y = -4 \\ 9x - 15y = 5 \end{cases}$$

Associated Augmented Matrix

$$\begin{bmatrix} 6 & -10 & \vdots & -4 \\ 9 & -15 & \vdots & 5 \end{bmatrix}$$

$$\tfrac{1}{6}R_1 \rightarrow \begin{bmatrix} 1 & -\frac{5}{3} & \vdots & -\frac{2}{3} \\ 9 & -15 & \vdots & 5 \end{bmatrix}$$

$$-9R_1 + R_2 \rightarrow \begin{bmatrix} 1 & -\frac{5}{3} & \vdots & -\frac{2}{3} \\ 0 & 0 & \vdots & 11 \end{bmatrix}$$

The "equation" that corresponds to the second row of this matrix is $0 = 11$. Because this is a false statement, the system of equations has no solution.

EXAMPLE 9 **A System with Infinitely Many Solutions**

Linear System

$$\begin{cases} 12x - 6y = -3 \\ -8x + 4y = 2 \end{cases}$$

Associated Augmented Matrix

$$\begin{bmatrix} 12 & -6 & \vdots & -3 \\ -8 & 4 & \vdots & 2 \end{bmatrix}$$

$$\tfrac{1}{12}R_1 \rightarrow \begin{bmatrix} 1 & -\frac{1}{2} & \vdots & -\frac{1}{4} \\ -8 & 4 & \vdots & 2 \end{bmatrix}$$

$$8R_1 + R_2 \rightarrow \begin{bmatrix} 1 & -\frac{1}{2} & \vdots & -\frac{1}{4} \\ 0 & 0 & \vdots & 0 \end{bmatrix}$$

Point out the difference between the solution to Example 8 and the solution to Example 9. Emphasize that $0 = 11$ is never true (no solution), whereas $0 = 0$ is always true (infinitely many solutions).

Because the second row of the matrix is all zeros, the system of equations has an infinite number of solutions, represented by all points (x, y) on the line

$$x - \frac{1}{2}y = -\frac{1}{4}.$$

Because this line can be written as

$$x = \frac{1}{2}y - \frac{1}{4}$$

you can write the solution set as

$$\left(\frac{1}{2}a - \frac{1}{4}, a\right), \text{ where } a \text{ is any real number.}$$

Exercises Within Reach®

Solutions in English & Spanish and tutorial videos at AlgebraWithinReach.com

Solving a System of Linear Equations **In Exercises 35–38, use matrices to solve the system of linear equations.**

35. $\begin{cases} x + y - 5z = 3 \\ x - 2z = 1 \\ 2x - y - z = 0 \end{cases}$

$(2a + 1, 3a + 2, a)$

36. $\begin{cases} 2x + 3z = 3 \\ 4x - 3y + 7z = 5 \\ 8x - 9y + 15z = 9 \end{cases}$

$\left(-\frac{3}{2}a + \frac{3}{2}, \frac{1}{3}a + \frac{1}{3}, a\right)$

37. $\begin{cases} 2x + 4z = 1 \\ x + y + 3z = 0 \\ x + 3y + 5z = 0 \end{cases}$

No solution

38. $\begin{cases} 3x + y - 2z = 2 \\ 6x + 2y - 4z = 1 \\ -3x - y + 2z = 1 \end{cases}$

No solution

Application **EXAMPLE 10** **Investment Portfolio**

You have $219,000 to invest in municipal bonds, blue-chip stocks, and growth stocks. The municipal bonds pay 6% annually. Over the investment period, you expect blue-chip stocks to return 10% annually and growth stocks to return 15% annually. You want a combined annual return of 8%, and you also want to have only one-fourth of the portfolio invested in stocks. How much should be allocated to each type of investment?

SOLUTION

Let M, B, and G represent the amounts invested in municipal bonds, blue-chip stocks, and growth stocks, respectively. This situation is represented by the following system.

$$\begin{cases} M + B + G = 219{,}000 & \text{Equation 1: Total investment is \$219,000.} \\ 0.06M + 0.10B + 0.15G = 17{,}520 & \text{Equation 2: Combined annual return is 8\%.} \\ B + G = 54{,}750 & \text{Equation 3: } \tfrac{1}{4} \text{ of investment is in stocks.} \end{cases}$$

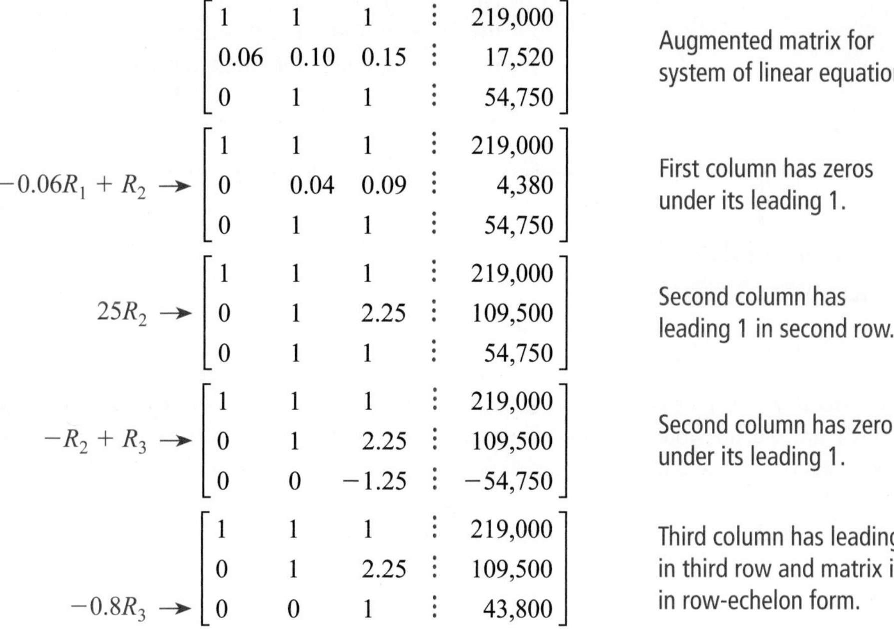

$$\begin{bmatrix} 1 & 1 & 1 & \vdots & 219{,}000 \\ 0.06 & 0.10 & 0.15 & \vdots & 17{,}520 \\ 0 & 1 & 1 & \vdots & 54{,}750 \end{bmatrix}$$ Augmented matrix for system of linear equations

$$-0.06R_1 + R_2 \rightarrow \begin{bmatrix} 1 & 1 & 1 & \vdots & 219{,}000 \\ 0 & 0.04 & 0.09 & \vdots & 4{,}380 \\ 0 & 1 & 1 & \vdots & 54{,}750 \end{bmatrix}$$ First column has zeros under its leading 1.

$$25R_2 \rightarrow \begin{bmatrix} 1 & 1 & 1 & \vdots & 219{,}000 \\ 0 & 1 & 2.25 & \vdots & 109{,}500 \\ 0 & 1 & 1 & \vdots & 54{,}750 \end{bmatrix}$$ Second column has leading 1 in second row.

$$-R_2 + R_3 \rightarrow \begin{bmatrix} 1 & 1 & 1 & \vdots & 219{,}000 \\ 0 & 1 & 2.25 & \vdots & 109{,}500 \\ 0 & 0 & -1.25 & \vdots & -54{,}750 \end{bmatrix}$$ Second column has zero under its leading 1.

$$-0.8R_3 \rightarrow \begin{bmatrix} 1 & 1 & 1 & \vdots & 219{,}000 \\ 0 & 1 & 2.25 & \vdots & 109{,}500 \\ 0 & 0 & 1 & \vdots & 43{,}800 \end{bmatrix}$$ Third column has leading 1 in third row and matrix is in row-echelon form.

From the row-echelon form, you can see that $G = 43{,}800$. By back-substituting G into the revised second equation, you can determine the value of B.

$$B + 2.25(43{,}800) = 109{,}500 \quad \Rightarrow \quad B = 10{,}950$$

By back-substituting B and G into Equation 1, you can solve for M.

$$M + 10{,}950 + 43{,}800 = 219{,}000 \quad \Rightarrow \quad M = 164{,}250$$

So, you should invest $164,250 in municipal bonds, $10,950 in blue-chip stocks. and $43,800 in growth or speculative stocks.

Exercises Within Reach®

Solutions in English & Spanish and tutorial videos at AlgebraWithinReach.com

39. ***Investment*** A corporation borrows $1,500,000 to expand its line of clothing. Some of the money is borrowed at 8%, some at 9%, and the remainder at 12%. The annual interest payment to the lenders is $133,000. The amount borrowed at 8% is four times the amount borrowed at 12%. How much is borrowed at each rate?
$800,000 at 8%, $500,000 at 9%, $200,000 at 12%

40. ***Nut Mixture*** A grocer wants to mix three kinds of nuts to obtain 50 pounds of a mixture priced at $4.10 per pound. Peanuts cost $3.00 per pound, pecans cost $4.00 per pound, and cashews cost $6.00 per pound. Three-quarters of the mixture is composed of peanuts and pecans. How many pounds of each variety should the grocer use? 20 pounds of peanuts, 17.5 pounds of pecans, 12.5 pounds of cashews

Concept Summary: Using Matrices to Solve Systems of Linear Equations

What

You can use **matrices** to solve a system of linear equations.

EXAMPLE

Use matrices to solve the system of linear equations.

$$\begin{cases} x - 2y + 3z = 9 \\ -x + 3y = -4 \\ 2x - 5y + 5z = 17 \end{cases}$$

How

Form an **augmented matrix** by using the coefficients and constants of the system.

Augmented Matrix

$$\begin{bmatrix} 1 & -2 & 3 & \vdots & 9 \\ -1 & 3 & 0 & \vdots & -4 \\ 2 & -5 & 5 & \vdots & 17 \end{bmatrix}$$

Then use Gaussian elimination and **elementary row operations** to write the matrix in **row-echelon form**.

Once the matrix is written in row-echelon form, use the matrix to write a new system of equations and then use back-substitution to find the solution.

Why

Using matrices makes solving a system less complex because you do not need to keep writing the variables.

Exercises Within Reach ®

Worked-out solutions to odd-numbered exercises at AlgebraWithinReach.com

Concept Summary Check

41. ***Vocabulary*** A matrix contains exactly four entries. What are the possible orders of the matrix? State the numbers of rows and columns in each possible order. See Additional Answers.

42. ***Logic*** For a given system of equations, which has more entries, the coefficient matrix or the augmented matrix? Explain. The augmented matrix has more entries because it includes the constant terms, whereas the coefficient matrix does not.

43. ***Vocabulary*** What is the primary difference between performing row operations on a system of equations and performing elementary row operations? See Additional Answers.

44. ***Writing*** After using matrices to perform Gaussian elimination, what steps are generally needed to find the solution of the original system of equations? Write the system of linear equations corresponding to the matrix in row-echelon form, and use back-substitution to find the solution.

Extra Practice

Solving a System of Linear Equations **In Exercises 45–48, use matrices to solve the system of linear equations.**

45. $\begin{cases} 4x + 3y = 10 \\ 2x - y = 10 \\ -2x + z = -9 \end{cases}$ $(4, -2, -1)$

46. $\begin{cases} 4x - y + z = 4 \\ -6x + 3y - 2z = -5 \\ 2x + 5y - z = 7 \end{cases}$ $\left(\frac{1}{2}, 2, 4\right)$

47. $\begin{cases} 2x + y - 2z = 4 \\ 3x - 2y + 4z = 6 \\ -4x + y + 6z = 12 \end{cases}$ $\left(2, 5, \frac{5}{2}\right)$

48. $\begin{cases} 3x + 3y + z = 4 \\ 2x + 6y + z = 5 \\ -x - 3y + 2z = -5 \end{cases}$ $\left(1, \frac{2}{3}, -1\right)$

49. ***Ticket Sales*** A theater owner wants to sell 1500 total tickets at his three theaters for a total revenue of \$10,050. Tickets cost \$1.50 at theater A, \$7.50 at theater B, and \$8.50 at theater C. Theaters B and C each have twice as many seats as theater A. How many tickets must be sold at each theater to reach the owner's goal? Theater A: 300 tickets, theater B: 600 tickets, theater C: 600 tickets

50. ***Investment*** An inheritance of \$25,000 is divided among three investments yielding a total of \$1890 in simple interest per year. The interest rates for the three investments are 5%, 7%, and 10%. The 5% and 7% investments are \$2000 and \$3000 less than the 10% investment, respectively. Find the amount placed in each investment. \$8000 at 5%, \$7000 at 7%, \$10,000 at 10%

51. ***Number Problem*** The sum of three positive numbers is 33. The second number is 3 greater than the first, and the third is four times the first. Find the three numbers. 5, 8, 20

52. ***Number Problem*** The sum of three positive numbers is 24. The second number is 4 greater than the first, and the third is three times the first. Find the three numbers. 4, 8, 12

53. ***Production*** A company produces computer chips, resistors, and transistors. Each computer chip requires 2 units of copper, 2 units of zinc, and 1 unit of glass. Each resistor requires 1 unit of copper, 3 units of zinc, and 2 units of glass. Each transistor requires 3 units of copper, 2 units of zinc, and 2 units of glass. There are 70 units of copper, 80 units of zinc, and 55 units of glass available for use. Find the numbers of computer chips, resistors, and transistors the company can produce. 15 computer chips, 10 resistors, 10 transistors

54. ***Production*** A gourmet baked goods company specializes in chocolate muffins, chocolate cookies, and chocolate brownies. Each muffin requires 2 units of chocolate, 3 units of flour, and 2 units of sugar. Each cookie requires 1 unit of chocolate, 1 unit of flour, and 1 unit of sugar. Each brownie requires 2 units of chocolate, 1 unit of flour, and 1.5 units of sugar. There are 550 units of chocolate, 525 units of flour, and 500 units of sugar available for use. Find the numbers of chocolate muffins, chocolate cookies, and chocolate brownies the company can produce.
75 chocolate muffins, 200 chocolate cookies, 100 chocolate brownies

Explaining Concepts

55. ***Reasoning*** The entries in a matrix consist of the whole numbers from 1 to 15. The matrix has more than one row and there are more columns than rows. What is the order of the matrix? Explain. 3×5. There are 15 entries in the matrix, so the order is 3×5, 5×3, or 15×1. Because there are more columns than rows, the second number in the order must be larger than the first.

56. ***Vocabulary*** Give an example of a matrix in *row-echelon form.* (There are many correct answers.)

$$\begin{bmatrix} 1 & -2 & 6 \\ 0 & 1 & 5 \\ 0 & 0 & 1 \end{bmatrix}$$

57. ***Writing*** Describe the row-echelon form of an augmented matrix that corresponds to a system of linear equations that is inconsistent. There will be a row in the matrix with all zero entries except in the last column.

58. ***Writing*** Describe the row-echelon form of an augmented matrix that corresponds to a system of linear equations that has an infinite number of solutions. The number of rows with nonzero entries in the row-echelon form is less than the number of variables in the system.

59. ***Logic*** An augmented matrix in row-echelon form represents a system of three variables in three equations that has exactly one solution. The matrix has six nonzero entries, and three of them are in the last column. Discuss the possible entries in the first three columns of this matrix. See Additional Answers.

60. ***Precision*** An augmented matrix in row-echelon form represents a system of three variables in three equations with exactly one solution. What is the smallest number of nonzero entries that this matrix can have? Explain. 3; If the solution is (0, 0, 0), the matrix entries consist of ones in the main diagonal and zeros everywhere else.

Cumulative Review

In Exercises 61–64, evaluate the expression.

61. $6(-7)$ -42

62. $45 \div (-5)$ -9

63. $5(4) - 3(-2)$ 26

64. $\dfrac{(-45) - (-20)}{-5}$ 5

In Exercises 65 and 66, solve the system of linear equations.

65. $\begin{cases} x = 4 \\ 3y + 2z = -4 \\ x + y + z = 3 \end{cases}$ $(4, -2, 1)$

66. $\begin{cases} x - 2y - 3z = 4 \\ 2x + 2y + z = -4 \\ -2x + z = 0 \end{cases}$ $(0, -2, 0)$

4.5 Determinants and Linear Systems

- Find determinants of 2×2 matrices and 3×3 matrices.
- Use determinants and Cramer's Rule to solve systems of linear equations.
- Use determinants to find areas of regions, to test for collinear points, and to find equations of lines.

Study Tip

Note that det(A) and $|A|$ are used interchangeably to represent the determinant of A. Although vertical bars are also used to denote the absolute value of a real number, the context will show which use is intended.

The Determinant of a Matrix

Definition of the Determinant of a 2 × 2 Matrix

$$\det(A) = |A| = \begin{vmatrix} a_1 & b_1 \\ a_2 & b_2 \end{vmatrix} = a_1b_2 - a_2b_1$$

A convenient method for remembering the formula for the determinant of a 2×2 matrix is shown in the diagram below.

$$\det(A) = |A| = \begin{vmatrix} a_1 & b_1 \\ a_2 & b_2 \end{vmatrix} = a_1b_2 - a_2b_1$$

Note that the determinant is given by the difference of the products of the two diagonals of the matrix.

EXAMPLE 1 Finding the Determinant of a 2 × 2 Matrix

Find the determinant of each matrix.

a. $A = \begin{bmatrix} 2 & -3 \\ 1 & 4 \end{bmatrix}$ **b.** $B = \begin{bmatrix} -1 & 2 \\ 2 & -4 \end{bmatrix}$ **c.** $C = \begin{bmatrix} 1 & 3 \\ 2 & 5 \end{bmatrix}$

SOLUTION

a. $\det(A) = \begin{vmatrix} 2 & -3 \\ 1 & 4 \end{vmatrix} = 2(4) - 1(-3) = 8 + 3 = 11$

b. $\det(B) = \begin{vmatrix} -1 & 2 \\ 2 & -4 \end{vmatrix} = (-1)(-4) - 2(2) = 4 - 4 = 0$

c. $\det(C) = \begin{vmatrix} 1 & 3 \\ 2 & 5 \end{vmatrix} = 1(5) - 2(3) = 5 - 6 = -1$

Exercises Within Reach®

Solutions in English & Spanish and tutorial videos at AlgebraWithinReach.com

Finding the Determinant **In Exercises 1–12, find the determinant of the matrix.**

1. $\begin{bmatrix} 2 & 1 \\ 3 & 4 \end{bmatrix}$ 5

2. $\begin{bmatrix} -3 & 1 \\ 5 & 2 \end{bmatrix}$ −11

3. $\begin{bmatrix} 5 & 2 \\ -6 & 3 \end{bmatrix}$ 27

4. $\begin{bmatrix} 2 & -2 \\ 4 & 3 \end{bmatrix}$ 14

5. $\begin{bmatrix} -4 & 0 \\ 9 & 0 \end{bmatrix}$ 0

6. $\begin{bmatrix} 4 & -3 \\ 0 & 0 \end{bmatrix}$ 0

7. $\begin{bmatrix} 3 & -3 \\ -6 & 6 \end{bmatrix}$ 0

8. $\begin{bmatrix} -2 & 3 \\ 6 & -9 \end{bmatrix}$ 0

9. $\begin{bmatrix} -7 & 6 \\ \frac{1}{2} & 3 \end{bmatrix}$ −24

10. $\begin{bmatrix} \frac{2}{3} & \frac{5}{6} \\ 14 & -2 \end{bmatrix}$ −13

11. $\begin{bmatrix} 0.4 & 0.7 \\ 0.7 & 0.4 \end{bmatrix}$ −0.33

12. $\begin{bmatrix} -1.2 & 4.5 \\ 0.4 & -0.9 \end{bmatrix}$ −0.72

Study Tip

The *signs* of the terms used in expanding by minors follow the alternating pattern shown below.

$$\begin{bmatrix} + & - & + \\ - & + & - \\ + & - & + \end{bmatrix}$$

Expanding by Minors

$$\det(A) = \begin{vmatrix} a_1 & b_1 & c_1 \\ a_2 & b_2 & c_2 \\ a_3 & b_3 & c_3 \end{vmatrix}$$

$$= a_1(\text{minor of } a_1) - b_1(\text{minor of } b_1) + c_1(\text{minor of } c_1)$$

$$= a_1\begin{vmatrix} b_2 & c_2 \\ b_3 & c_3 \end{vmatrix} - b_1\begin{vmatrix} a_2 & c_2 \\ a_3 & c_3 \end{vmatrix} + c_1\begin{vmatrix} a_2 & b_2 \\ a_3 & b_3 \end{vmatrix}$$

This pattern is called **expanding by minors** along the first row. A similar pattern can be used to expand by minors along any row or column.

Additional Examples

Find the determinant of each matrix.

a. $A = \begin{bmatrix} 1 & -2 & 2 \\ -1 & 3 & 0 \\ 2 & -5 & 1 \end{bmatrix}$

b. $B = \begin{bmatrix} -4 & -2 & 1 \\ 0 & 3 & 5 \\ 3 & 2 & 0 \end{bmatrix}$

Answers:

a. −1

b. 1

A zero entry in a matrix will always yield a zero term when expanding by minors. So, when students are finding the determinant of a matrix, they should choose to expand along the row or column that has the most zero entries.

EXAMPLE 2 Finding the Determinant of a 3 × 3 Matrix

Find the determinant of **a.** $A = \begin{bmatrix} -1 & 1 & 2 \\ 0 & 2 & 3 \\ 3 & 4 & 2 \end{bmatrix}$ and **b.** $B = \begin{bmatrix} 1 & 2 & 1 \\ 3 & 0 & 2 \\ 4 & 0 & -1 \end{bmatrix}$.

SOLUTION

a. By expanding by minors along the *first column*, you obtain

$$\det(A) = \begin{vmatrix} -1 & 1 & 2 \\ 0 & 2 & 3 \\ 3 & 4 & 2 \end{vmatrix} = (-1)\begin{vmatrix} 2 & 3 \\ 4 & 2 \end{vmatrix} - (0)\begin{vmatrix} 1 & 2 \\ 4 & 2 \end{vmatrix} + (3)\begin{vmatrix} 1 & 2 \\ 2 & 3 \end{vmatrix}$$

$$= (-1)(4 - 12) - (0)(2 - 8) + (3)(3 - 4)$$

$$= 8 - 0 - 3$$

$$= 5.$$

b. By expanding by minors along the *second column*, you obtain

$$\det(B) = \begin{vmatrix} 1 & 2 & 1 \\ 3 & 0 & 2 \\ 4 & 0 & -1 \end{vmatrix} = -(2)\begin{vmatrix} 3 & 2 \\ 4 & -1 \end{vmatrix} + (0)\begin{vmatrix} 1 & 1 \\ 4 & -1 \end{vmatrix} - (0)\begin{vmatrix} 1 & 1 \\ 3 & 2 \end{vmatrix}$$

$$= -(2)(-3 - 8) + 0 - 0$$

$$= 22.$$

Exercises Within Reach®

Solutions in English & Spanish and tutorial videos at AlgebraWithinReach.com

Finding the Determinant **In Exercises 13–20, find the determinant of the matrix. Expand by minors along the row or column that appears to make the computation easiest.**

13. $\begin{bmatrix} 2 & 3 & -1 \\ 6 & 0 & 0 \\ 4 & 1 & 1 \end{bmatrix}$ −24

14. $\begin{bmatrix} 10 & 2 & -4 \\ 8 & 0 & -2 \\ 4 & 0 & 2 \end{bmatrix}$ −48

15. $\begin{bmatrix} 1 & 1 & 2 \\ 3 & 1 & 0 \\ -2 & 0 & 3 \end{bmatrix}$ −2

16. $\begin{bmatrix} 2 & 1 & 3 \\ 1 & 4 & 4 \\ 1 & 0 & 2 \end{bmatrix}$ 6

17. $\begin{bmatrix} 2 & 4 & 6 \\ 0 & 3 & 1 \\ 0 & 0 & -5 \end{bmatrix}$ −30

18. $\begin{bmatrix} 2 & 3 & 1 \\ 0 & 5 & -2 \\ 0 & 0 & -2 \end{bmatrix}$ −20

19. $\begin{bmatrix} -2 & 2 & 3 \\ 1 & -1 & 0 \\ 0 & 1 & 4 \end{bmatrix}$ 3

20. $\begin{bmatrix} -2 & 3 & 0 \\ 3 & 1 & -4 \\ 0 & 4 & 2 \end{bmatrix}$ −54

Cramer's Rule

Cramer's Rule

1. For the system of linear equations

$$\begin{cases} a_1x + b_1y = c_1 \\ a_2x + b_2y = c_2 \end{cases}$$

the solution is given by $x = \dfrac{D_x}{D} = \dfrac{\begin{vmatrix} c_1 & b_1 \\ c_2 & b_2 \end{vmatrix}}{\begin{vmatrix} a_1 & b_1 \\ a_2 & b_2 \end{vmatrix}}$, $y = \dfrac{D_y}{D} = \dfrac{\begin{vmatrix} a_1 & c_1 \\ a_2 & c_2 \end{vmatrix}}{\begin{vmatrix} a_1 & b_1 \\ a_2 & b_2 \end{vmatrix}}$, $D \neq 0$.

2. For the system of linear equations

$$\begin{cases} a_1x + b_1y + c_1z = d_1 \\ a_2x + b_2y + c_2z = d_2 \\ a_3x + b_3y + c_3z = d_3 \end{cases}$$

the solution is given by

$$x = \frac{D_x}{D} = \frac{\begin{vmatrix} d_1 & b_1 & c_1 \\ d_2 & b_2 & c_2 \\ d_3 & b_3 & c_3 \end{vmatrix}}{\begin{vmatrix} a_1 & b_1 & c_1 \\ a_2 & b_2 & c_2 \\ a_3 & b_3 & c_3 \end{vmatrix}}, \quad y = \frac{D_y}{D} = \frac{\begin{vmatrix} a_1 & d_1 & c_1 \\ a_2 & d_2 & c_2 \\ a_3 & d_3 & c_3 \end{vmatrix}}{\begin{vmatrix} a_1 & b_1 & c_1 \\ a_2 & b_2 & c_2 \\ a_3 & b_3 & c_3 \end{vmatrix}}, \quad z = \frac{D_z}{D} = \frac{\begin{vmatrix} a_1 & b_1 & d_1 \\ a_2 & b_2 & d_2 \\ a_3 & b_3 & d_3 \end{vmatrix}}{\begin{vmatrix} a_1 & b_1 & c_1 \\ a_2 & b_2 & c_2 \\ a_3 & b_3 & c_3 \end{vmatrix}}, \quad D \neq 0.$$

EXAMPLE 3 Using Cramer's Rule for a 2 × 2 System

Use Cramer's Rule to solve the system of linear equations.

$$\begin{cases} 4x - 2y = 10 \\ 3x - 5y = 11 \end{cases}$$

SOLUTION

The determinant of the coefficient matrix is

$$D = \begin{vmatrix} 4 & -2 \\ 3 & -5 \end{vmatrix} = -20 - (-6) = -14$$

$$x = \frac{D_x}{D} = \frac{\begin{vmatrix} 10 & -2 \\ 11 & -5 \end{vmatrix}}{-14} = \frac{-50 - (-22)}{-14} = \frac{-28}{-14} = 2$$

$$y = \frac{D_y}{D} = \frac{\begin{vmatrix} 4 & 10 \\ 3 & 11 \end{vmatrix}}{-14} = \frac{44 - 30}{-14} = \frac{14}{-14} = -1$$

The solution is $(2, -1)$. Check this in the original system of equations.

Exercises Within Reach®

Solutions in English & Spanish and tutorial videos at AlgebraWithinReach.com

Using Cramer's Rule **In Exercises 21–26, use Cramer's Rule to solve the system of linear equations. (If not possible, state the reason.)**

21. $\begin{cases} x + 2y = 5 \\ -x + y = 1 \end{cases}$ (1, 2)

22. $\begin{cases} 2x - y = -10 \\ 3x + 2y = -1 \end{cases}$ (−3, 4)

23. $\begin{cases} 3x + 4y = -2 \\ 5x + 3y = 4 \end{cases}$ (2, −2)

24. $\begin{cases} 3x + 2y = -3 \\ 4x + 5y = -11 \end{cases}$ (1, −3)

25. $\begin{cases} 13x - 6y = 17 \\ 26x - 12y = 8 \end{cases}$ Not possible, $D = 0$

26. $\begin{cases} -0.4x + 0.8y = 1.6 \\ 2x - 4y = 5 \end{cases}$ Not possible, $D = 0$

EXAMPLE 4 Using Cramer's Rule for a 3 × 3 System

Study Tip

When using Cramer's Rule, remember that the method *does not* apply when the determinant of the coefficient matrix is zero.

Use Cramer's Rule to solve the system of linear equations.

$$\begin{cases} -x + 2y - 3z = 1 \\ 2x \qquad + z = 0 \\ 3x - 4y + 4z = 2 \end{cases}$$

SOLUTION

The determinant of the coefficient matrix is $D = 10$.

$$x = \frac{D_x}{D} = \frac{\begin{vmatrix} 1 & 2 & -3 \\ 0 & 0 & 1 \\ 2 & -4 & 4 \end{vmatrix}}{10} = \frac{8}{10} = \frac{4}{5}$$

$$y = \frac{D_y}{D} = \frac{\begin{vmatrix} -1 & 1 & -3 \\ 2 & 0 & 1 \\ 3 & 2 & 4 \end{vmatrix}}{10} = \frac{-15}{10} = -\frac{3}{2}$$

$$z = \frac{D_z}{D} = \frac{\begin{vmatrix} -1 & 2 & 1 \\ 2 & 0 & 0 \\ 3 & -4 & 2 \end{vmatrix}}{10} = \frac{-16}{10} = -\frac{8}{5}$$

The solution is $\left(\frac{4}{5}, -\frac{3}{2}, -\frac{8}{5}\right)$. Check this in the original system of equations.

Exercises Within Reach®

Solutions in English & Spanish and tutorial videos at AlgebraWithinReach.com

Using Cramer's Rule **In Exercises 27–34, use Cramer's Rule to solve the system of linear equations. (If not possible, state the reason.)**

27. $\begin{cases} 4x - y + z = -5 \\ 2x + 2y + 3z = 10 \\ 5x - 2y + 6z = 1 \end{cases}$ $(-1, 3, 2)$

28. $\begin{cases} 4x - 2y + 3z = -2 \\ 2x + 2y + 5z = 16 \\ 8x - 5y - 2z = 4 \end{cases}$ $(5, 8, -2)$

29. $\begin{cases} 4x + 3y + 4z = 1 \\ 4x - 6y + 8z = 8 \\ -x + 9y - 2z = -7 \end{cases}$ $\left(\frac{1}{2}, -\frac{2}{3}, \frac{1}{4}\right)$

30. $\begin{cases} 5x + 4y - 6z = -10 \\ -4x + 2y + 3z = -1 \\ 8x + 4y + 12z = 2 \end{cases}$ $\left(0, -\frac{3}{2}, \frac{2}{3}\right)$

31. $\begin{cases} 2x + 3y + 5z = 4 \\ 3x + 5y + 9z = 7 \\ 5x + 9y + 17z = 13 \end{cases}$ Not possible, $D = 0$

32. $\begin{cases} 5x - 3y + 2z = 2 \\ 2x + 2y - 3z = 3 \\ x - 7y + 8z = -4 \end{cases}$ Not possible, $D = 0$

33. $\begin{cases} 3x - 2y + 3z = 8 \\ x + 3y + 6z = -3 \\ x + 2y + 9z = -5 \end{cases}$ $\left(\frac{51}{16}, -\frac{7}{16}, -\frac{13}{16}\right)$

34. $\begin{cases} 6x + 4y - 8z = -22 \\ -2x + 2y + 3z = 13 \\ -2x + 2y - z = 5 \end{cases}$ $\left(-2, \frac{3}{2}, 2\right)$

Applications

In addition to Cramer's Rule, determinants have many other practical applications. For instance, you can use a determinant to find the area of a triangle whose vertices are given by three points on a rectangular coordinate system.

Area of a Triangle

The area of a triangle with vertices (x_1, y_1), (x_2, y_2), and (x_3, y_3) is

$$\text{Area} = \pm\frac{1}{2}\begin{vmatrix} x_1 & y_1 & 1 \\ x_2 & y_2 & 1 \\ x_3 & y_3 & 1 \end{vmatrix}$$

where the symbol $(\pm)$ indicates that the appropriate sign should be chosen to yield a positive area.

Application

EXAMPLE 5 Geometry: **Finding the Area of a Triangle**

Find the area of the triangle whose vertices are (2, 0), (1, 3), and (3, 2), as shown at the left.

y
(1, 3)
(3, 2)
(2, 0)
x
1 2 3

SOLUTION

Let $(x_1, y_1) = (2, 0)$, $(x_2, y_2) = (1, 3)$, and $(x_3, y_3) = (3, 2)$. To find the area of the triangle, evaluate the determinant by expanding by minors along the first row.

$$\begin{vmatrix} x_1 & y_1 & 1 \\ x_2 & y_2 & 1 \\ x_3 & y_3 & 1 \end{vmatrix} = \begin{vmatrix} 2 & 0 & 1 \\ 1 & 3 & 1 \\ 3 & 2 & 1 \end{vmatrix}$$

$$= 2\begin{vmatrix} 3 & 1 \\ 2 & 1 \end{vmatrix} - 0\begin{vmatrix} 1 & 1 \\ 3 & 1 \end{vmatrix} + 1\begin{vmatrix} 1 & 3 \\ 3 & 2 \end{vmatrix}$$

$$= 2(1) - 0 + 1(-7)$$

$$= -5$$

Using this value, you can conclude that the area of the triangle is

$$\text{Area} = -\frac{1}{2}\begin{vmatrix} 2 & 0 & 1 \\ 1 & 3 & 1 \\ 3 & 2 & 1 \end{vmatrix}$$

$$= -\frac{1}{2}(-5) = \frac{5}{2}.$$

Additional Example

Find the area of the triangle whose vertices are (1, 0), (2, 2), and (4, 3).

Answer: $\frac{3}{2}$

Exercises Within Reach®

Solutions in English & Spanish and tutorial videos at AlgebraWithinReach.com

Finding the Area of a Triangle **In Exercises 35–42, use a determinant to find the area of the triangle with the given vertices.**

35. (0, 3), (4, 0), (8, 5) 16

36. (2, 0), (0, 5), (6, 3) 13

37. (−3, 4), (1, −2), (6, 1) 21

38. (−2, −3), (2, −3), (0, 4) 14

39. (−2, 1), (3, −1), (1, 6) $\frac{31}{2}$

40. (−1, 4), (−4, 0), (1, 3) $\frac{11}{2}$

41. $\left(0, \frac{1}{2}\right), \left(\frac{5}{2}, 0\right)$ (4, 3) $\frac{33}{8}$

42. $\left(\frac{1}{4}, 0\right), \left(0, \frac{3}{4}\right)$, (8, −2) $\frac{85}{32}$

Application **EXAMPLE 6** **Finding the Area of a Region**

Find the area of the shaded region of the figure.

SOLUTION

Let $(x_1, y_1) = (-3, -1)$, $(x_2, y_2) = (2, -2)$, and $(x_3, y_3) = (1, 2)$.

To find the area of the triangle, evaluate the determinant by expanding by minors along the first column.

$$\begin{vmatrix} x_1 & y_1 & 1 \\ x_2 & y_2 & 1 \\ x_3 & y_3 & 1 \end{vmatrix} = \begin{vmatrix} -3 & -1 & 1 \\ 2 & -2 & 1 \\ 1 & 2 & 1 \end{vmatrix}$$

$$= -3\begin{vmatrix} -2 & 1 \\ 2 & 1 \end{vmatrix} - 2\begin{vmatrix} -1 & 1 \\ 2 & 1 \end{vmatrix} + 1\begin{vmatrix} -1 & 1 \\ -2 & 1 \end{vmatrix}$$

$$= -3(-4) - 2(-3) + 1(1)$$

$$= 19$$

Using this value, you can conclude that the area of the triangle is

$$\text{Area} = \frac{1}{2}(19)$$

$$= 9.5 \text{ square units.}$$

Now find the area of the shaded region.

Verbal Model: $\boxed{\text{Length of rectangle}} \cdot \boxed{\text{Height of rectangle}} - \boxed{\text{Area of triangle}}$

Labels: Length of rectangle $= 6 - (-3) = 9$ (units)
Height of rectangle $= 2 - (-2) = 4$ (units)
Area of triangle $= 9.5$ (square units)

Expression: $A = (9)(4) - 9.5 = 36 - 9.5 = 26.5$

The area of the shaded region is 26.5 square units.

Exercises Within Reach®

Solutions in English & Spanish and tutorial videos at AlgebraWithinReach.com

Finding the Area of a Region **In Exercises 43–46, find the area of the shaded region of the figure.**

43. 42

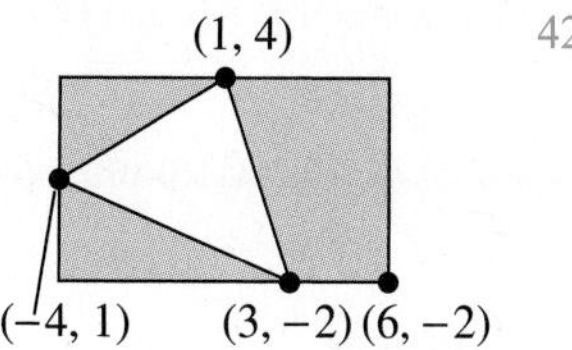

44. 10

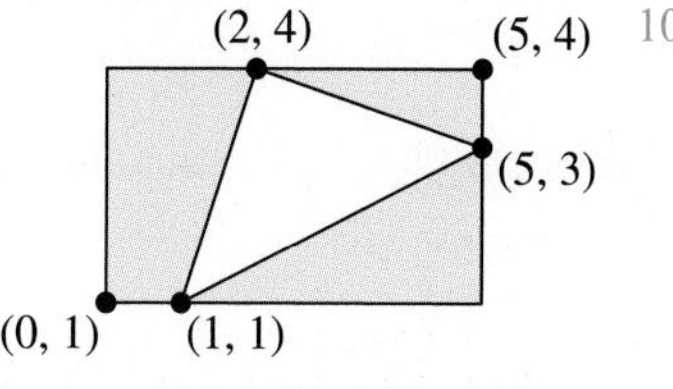

45. 16

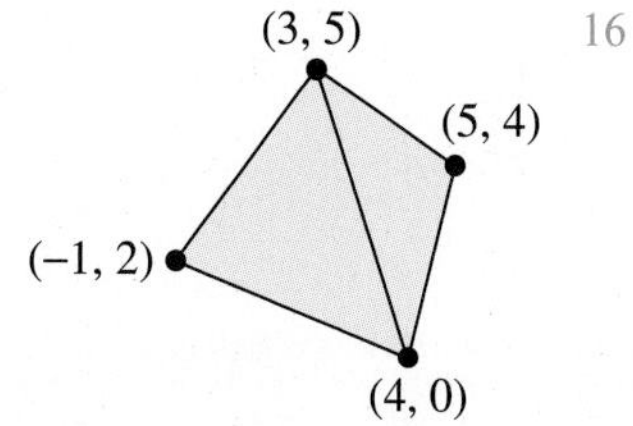

46. 15

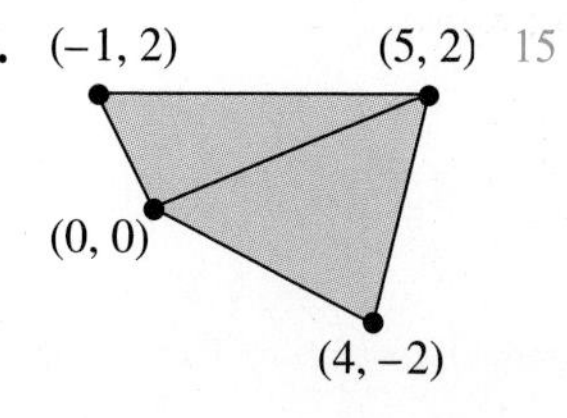

Test for Collinear Points

Three points (x_1, y_1), (x_2, y_2), and (x_3, y_3) are collinear (lie on the same line) if and only if

$$\begin{vmatrix} x_1 & y_1 & 1 \\ x_2 & y_2 & 1 \\ x_3 & y_3 & 1 \end{vmatrix} = 0.$$

Application **EXAMPLE 7** **Testing for Collinear Points**

Determine whether the points $(-2, -2)$, $(1, 1)$, and $(7, 5)$ are collinear.

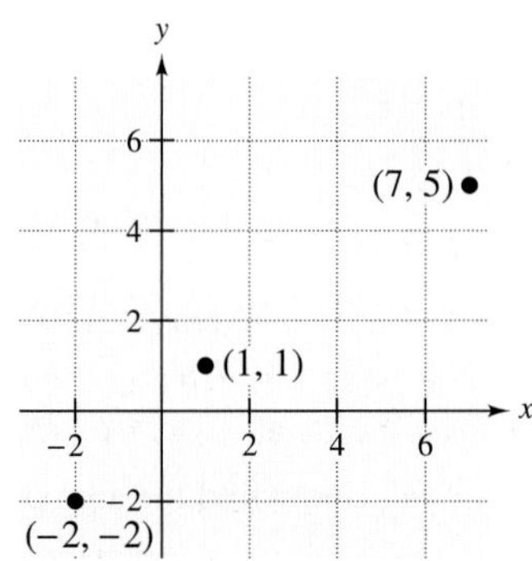

SOLUTION

Letting $(x_1, y_1) = (-2, -2)$, $(x_2, y_2) = (1, 1)$, and $(x_3, y_3) = (7, 5)$, you have

$$\begin{vmatrix} x_1 & y_1 & 1 \\ x_2 & y_2 & 1 \\ x_3 & y_3 & 1 \end{vmatrix} = \begin{vmatrix} -2 & -2 & 1 \\ 1 & 1 & 1 \\ 7 & 5 & 1 \end{vmatrix}$$

$$= -2\begin{vmatrix} 1 & 1 \\ 5 & 1 \end{vmatrix} - (-2)\begin{vmatrix} 1 & 1 \\ 7 & 1 \end{vmatrix} + 1\begin{vmatrix} 1 & 1 \\ 7 & 5 \end{vmatrix}$$

$$= -2(-4) - (-2)(-6) + 1(-2)$$

$$= -6.$$

Because the value of this determinant *is not* zero, you can conclude that the three points *are not* collinear.

Exercises Within Reach ®

Solutions in English & Spanish and tutorial videos at AlgebraWithinReach.com

Testing for Collinear Points **In Exercises 47–52, determine whether the points are collinear.**

47. $(-1, 11), (0, 8), (2, 2)$
Collinear

48. $(-1, -1), (1, 9), (2, 13)$
Not collinear

49. $(2, -4), (5, 2), (10, 10)$
Not collinear

50. $(1, 8), (3, 2), (6, -7)$
Collinear

51. $\left(-2, \frac{1}{3}\right), (2, 1), \left(3, \frac{1}{5}\right)$
Not collinear

52. $\left(0, \frac{1}{2}\right), \left(1, \frac{7}{6}\right), \left(9, \frac{13}{2}\right)$
Collinear

Two-Point Form of the Equation of a Line

An equation of the line passing through the distinct points (x_1, y_1) and (x_2, y_2) is given by

$$\begin{vmatrix} x & y & 1 \\ x_1 & y_1 & 1 \\ x_2 & y_2 & 1 \end{vmatrix} = 0.$$

Application

EXAMPLE 8 Finding an Equation of a Line

Find an equation of the line passing through $(-2, 1)$, and $(3, -2)$.

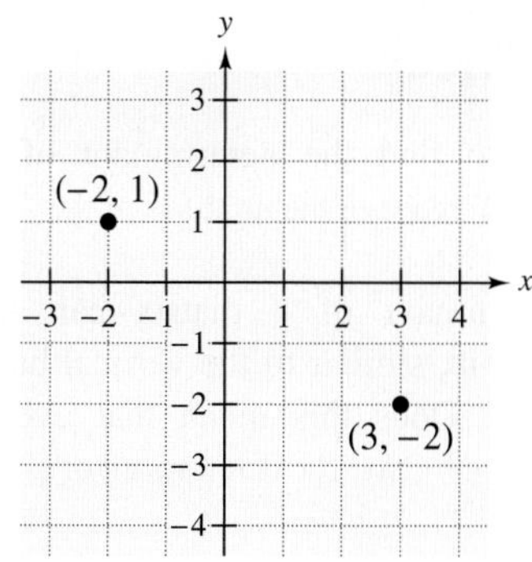

SOLUTION

Applying the determinant formula for the equation of a line produces

$$\begin{vmatrix} x & y & 1 \\ -2 & 1 & 1 \\ 3 & -2 & 1 \end{vmatrix} = 0.$$

To evaluate this determinant, you can expand by minors along the first row to obtain the following.

$$x\begin{vmatrix} 1 & 1 \\ -2 & 1 \end{vmatrix} - y\begin{vmatrix} -2 & 1 \\ 3 & 1 \end{vmatrix} + 1\begin{vmatrix} -2 & 1 \\ 3 & -2 \end{vmatrix} = 0$$

$$3x + 5y + 1 = 0$$

So, an equation of the line is $3x + 5y + 1 = 0$.

Exercises Within Reach®

Solutions in English & Spanish and tutorial videos at AlgebraWithinReach.com

Finding an Equation of a Line **In Exercises 53–60, use a determinant to find an equation of the line passing through the points.**

53. $(-2, -1), (4, 2)$ $x - 2y = 0$

54. $(-1, 3), (2, -6)$ $3x + y = 0$

55. $(10, 7), (-2, -7)$ $7x - 6y - 28 = 0$

56. $(-8, 3), (4, 6)$ $x - 4y + 20 = 0$

57. $\left(-2, \frac{3}{2}\right), (3, -3)$ $9x + 10y + 3 = 0$

58. $\left(-\frac{1}{2}, 3\right), \left(\frac{5}{2}, 1\right)$ $2x + 3y - 8 = 0$

59. $(2, 3.6), (8, 10)$ $16x - 15y + 22 = 0$

60. $(3, 1.6), (5, -2.2)$ $19x + 10y - 73 = 0$

Concept Summary: Finding the Determinant of a Matrix

What

Associated with each square matrix is a real number called a **determinant**.

EXAMPLE

Find the determinant of the matrix.

$$A = \begin{vmatrix} 2 & 4 \\ 1 & 6 \end{vmatrix}$$

How

To find the determinant of a square matrix, find the difference of the products of the two diagonals of the matrix.

$$\det(A) = \begin{vmatrix} 2 & 4 \\ 1 & 6 \end{vmatrix} = 2(6) - 1(4) = 12 - 4 = 8$$

Why

You can use determinants to do the following.

1. Solve systems of linear equations.
2. Find areas of triangles.
3. Test for collinear points.
4. Find equations of lines.

Exercises Within Reach®

Worked-out solutions to odd-numbered exercises at AlgebraWithinReach.com

Concept Summary Check

61. ***Writing*** Explain how to find the determinant of a 2×2 matrix. Find the difference of the products of the two diagonals of the matrix.

62. ***Vocabulary*** The determinant of a matrix can be represented by vertical bars, similar to the vertical bars used for absolute value. Does this mean that every determinant is nonnegative? No. The determinant of a matrix can be positive, negative, or zero.

63. ***Logic*** Is it possible to find the determinant of a 2×3 matrix? Explain. No. Only square matrices have determinants.

64. ***Reasoning*** When one column of a 3×3 matrix is all zeros, what is the determinant of the matrix? Explain. 0; In expanding by minors, the zero entries will cause each product to be zero.

Extra Practice

Finding the Determinant **In Exercises 65–70, find the determinant of the matrix. Expand by minors along the row or column that appears to make the computation easiest.**

65. $\begin{bmatrix} 5 & -3 & 2 \\ 7 & 5 & -7 \\ 0 & 6 & -1 \end{bmatrix}$ 248

66. $\begin{bmatrix} 3 & -1 & 2 \\ 1 & -1 & 2 \\ -2 & 3 & 10 \end{bmatrix}$ -32

67. $\begin{bmatrix} -\frac{1}{2} & -1 & 6 \\ 8 & -\frac{1}{4} & -4 \\ 1 & 2 & 1 \end{bmatrix}$ 105.625

68. $\begin{bmatrix} \frac{1}{2} & \frac{3}{2} & \frac{1}{2} \\ 4 & 8 & 10 \\ -2 & -6 & 12 \end{bmatrix}$ -28

69. $\begin{bmatrix} 0.6 & 0.4 & -0.6 \\ 0.1 & 0.5 & -0.3 \\ 8 & -2 & 12 \end{bmatrix}$ 4.32

70. $\begin{bmatrix} 0.4 & 0.3 & 0.3 \\ -0.2 & 0.6 & 0.6 \\ 3 & 1 & 1 \end{bmatrix}$ 0

71. ***Area of a Region*** A large region of forest has been infested with gypsy moths. The region is roughly triangular, as show in the figure. Find the area of this region. (*Note:* The measurements in the figure are in miles.) 250 square miles

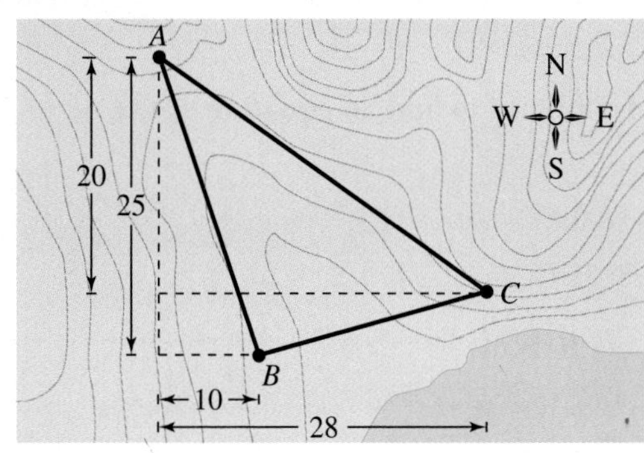

72. ***Area of a Region*** You have purchased a triangular tract of land, as shown in the figure. What is the area of this tract of land? (*Note:* The measurements in the figure are in feet.) 3100 square feet

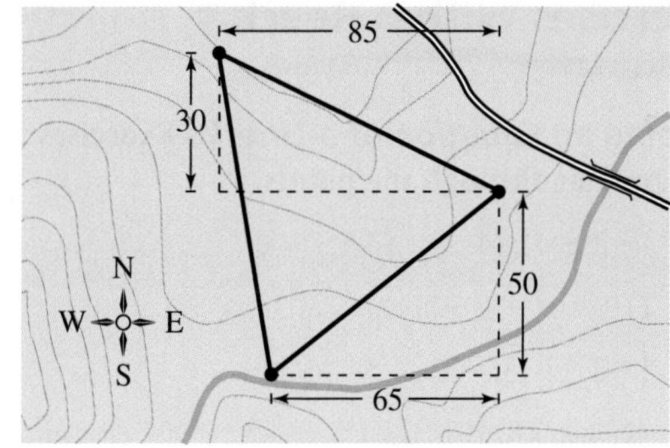

73. ***Electrical Networks*** When Kirchhoff's Laws are applied to the electrical network shown in the figure, the currents I_1, I_2, and I_3 are the solution of the system

$$\begin{cases} I_1 + I_2 - I_3 = 0 \\ I_1 \qquad + 2I_3 = 12. \\ I_1 - 2I_2 \qquad = -4 \end{cases}$$

Find the currents. $I_1 = 2, I_2 = 3, I_3 = 5$

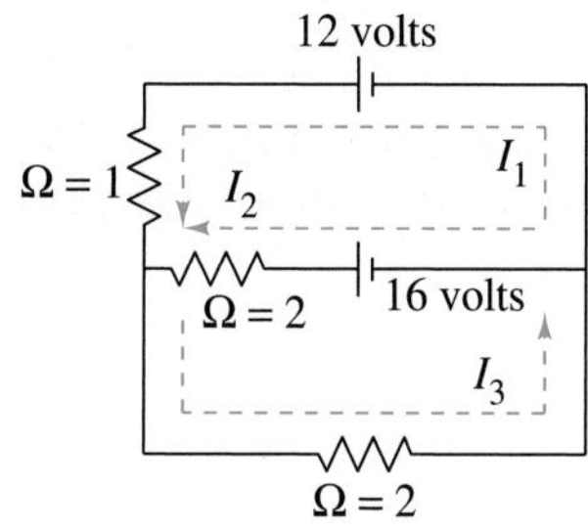

74. ***Electrical Networks*** When Kirchhoff's Laws are applied to the electrical network shown in the figure, the currents I_1, I_2, and I_3 are the solution of the system

$$\begin{cases} I_1 - I_2 + I_3 = 0 \\ \qquad I_2 + 4I_3 = 8. \\ 4I_1 + I_2 \qquad = 16 \end{cases}$$

Find the currents. $I_1 = 3, I_2 = 4, I_3 = 1$

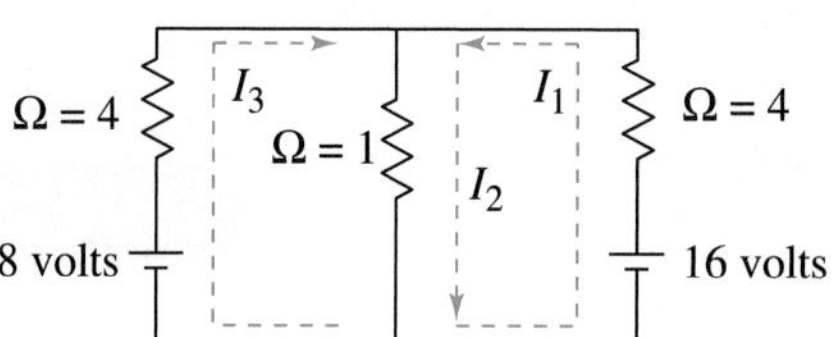

75. (a) Use Cramer's Rule to solve the system of linear equations.

$$\begin{cases} kx + 3ky = 2 \\ (2 + k)x + ky = 5 \end{cases} \quad \left(\frac{13}{2k + 6}, \frac{3k - 4}{-2k^2 - 6k}\right)$$

(b) State the values of k for which Cramer's Rule does not apply. $0, -3$

76. (a) Use Cramer's Rule to solve the system of linear equations.

$$\begin{cases} kx + (1 - k)y = 1 \\ (1 - k)x + \qquad ky = 3 \end{cases} \quad \left(\frac{4k - 3}{2k - 1}, \frac{4k - 1}{2k^2 - 1}\right)$$

(b) For what value(s) of k will the system be inconsistent? $\frac{1}{2}$

Explaining Concepts

77. ***Vocabulary*** Explain the difference between a square matrix and its determinant. A square matrix is a square array of numbers. The determinant of a square matrix is a real number.

78. ***Writing*** What is meant by the minor of an entry of a square matrix? The minor of an entry of a matrix is the determinant of the matrix that remains after deletion of the row and column in which the entry occurs.

79. ***Writing*** When two rows of a 3×3 matrix have identical entries, what is the value of the determinant? Explain. The determinant is zero. Because two rows are identical, each term is zero when expanding by minors along the other row. Therefore, the sum is zero.

80. ***Precision*** What conditions must be met in order to use Cramer's Rule to solve a system of linear equations? The coefficient matrix of the system must be square, its determinant must be a nonzero real number, and the system must have exactly one solution.

Cumulative Review

In Exercises 81–84, sketch the graph of the linear inequality. See Additional Answers.

81. $4x - 2y < 0$

82. $2x + 8y \geq 0$

83. $-x + 3y > 12$

84. $-3x - y \leq 2$

85. Given a function $f(x)$, describe how the graph of $h(x) = f(x) + c$ compares with the graph of $f(x)$ for a positive real number c. The graph of $h(x)$ has a vertical shift c units upward.

86. Given a function $f(x)$, describe how the graph of $h(x) = f(x + c)$ compares with the graph of $f(x)$ for a positive real number c. The graph of $h(x)$ has a horizontal shift c units to the left.

In Exercises 87–90, use the graph of f to sketch the graph. See Additional Answers.

87. $f(x) - 2$

88. $f(x - 2)$

89. $f(-x)$

90. $f(x - 1) + 3$

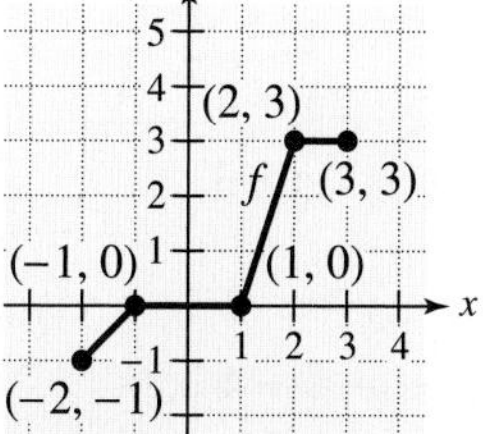

4.6 Systems of Linear Inequalities

- Solve systems of linear inequalities in two variables.
- Use systems of linear inequalities to model and solve real-life problems.

Systems of Linear Inequalities in Two Variables

EXAMPLE 1 Graphing a System of Linear Inequalities

Sketch the graph of the system of linear inequalities.

$$\begin{cases} 2x - y \le 5 \\ x + 2y \ge 2 \end{cases}$$

SOLUTION

Begin by rewriting each inequality in slope-intercept form.

$$\begin{cases} 2x - y \le 5 \\ x + 2y \ge 2 \end{cases} \quad \Rightarrow \quad \begin{array}{l} y \ge 2x - 5 \\ y \ge -\frac{1}{2}x + 1 \end{array}$$

Then sketch the line for the corresponding equation of each inequality.

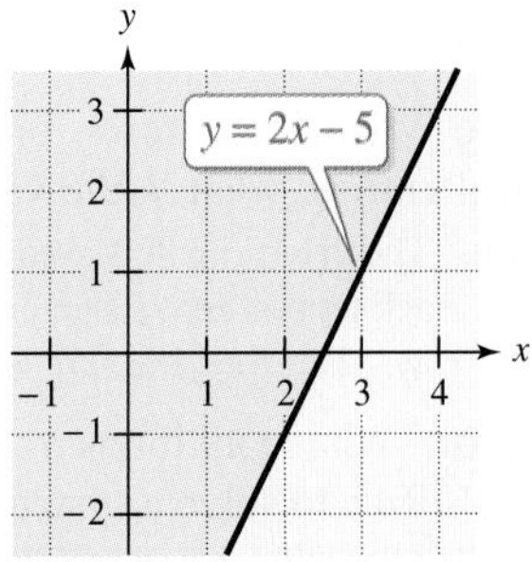

Graph of $2x - y \le 5$ is all points on and above $y = 2x - 5$.

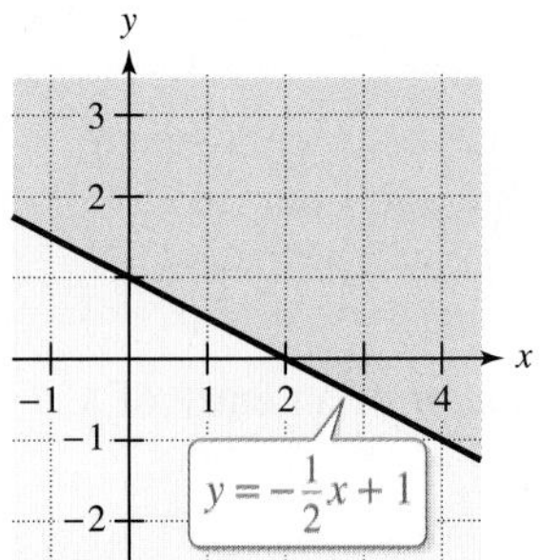

Graph of $x + 2y \ge 2$ is all points on and above $y = -\frac{1}{2}x + 1$.

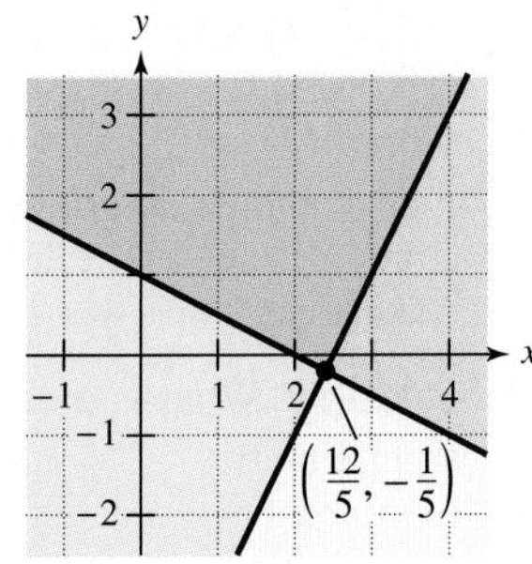

Graph of system is the purple wedge-shaped region.

Exercises Within Reach®

Solutions in English & Spanish and tutorial videos at AlgebraWithinReach.com

Graphing a System of Linear Inequalities **In Exercises 1–12, sketch the graph of the system of linear inequalities.** See Additional Answers.

1. $\begin{cases} x + y \le 3 \\ x - y \le 1 \end{cases}$

2. $\begin{cases} x + y \ge 2 \\ x - y \le 2 \end{cases}$

3. $\begin{cases} 2x - 4y \le 6 \\ x + y \ge 2 \end{cases}$

4. $\begin{cases} 4x + 10y \le 5 \\ x - y \le 4 \end{cases}$

5. $\begin{cases} x + 2y \le 6 \\ x - 2y \le 0 \end{cases}$

6. $\begin{cases} 2x + y \le 0 \\ x - y \le 8 \end{cases}$

7. $\begin{cases} x - 2y > 4 \\ 2x + y > 6 \end{cases}$

8. $\begin{cases} 3x + y < 6 \\ x + 2y > 2 \end{cases}$

9. $\begin{cases} x + y > -1 \\ x + y < 3 \end{cases}$

10. $\begin{cases} x - y > 2 \\ x - y < -4 \end{cases}$

11. $\begin{cases} y \ge \frac{4}{3}x + 1 \\ y \le 5x - 2 \end{cases}$

12. $\begin{cases} y \ge \frac{1}{2}x + \frac{1}{2} \\ y \le 4x - \frac{1}{2} \end{cases}$

Graphing a System of Linear Inequalities

1. Sketch the line that corresponds to each inequality. (Use dashed lines for inequalities with $<$ or $>$ and solid lines for inequalities with $\leq$ or $\geq$.)
2. Lightly shade the half-plane that is the graph of each linear inequality. (Colored pencils may help distinguish different half-planes.)
3. The graph of the system is the intersection of the half-planes. (If you use colored pencils, it is the region that is shaded with *every* color.)

EXAMPLE 2 Graphing a System of Linear Inequalities

Sketch the graph of the system of linear inequalities: $\begin{cases} y < 4 \\ y > 1 \end{cases}$.

SOLUTION

The graph of the first inequality is the half-plane below the horizontal line

$y = 4.$ Upper boundary

The graph of the second inequality is the half-plane above the horizontal line

$y = 1.$ Lower boundary

The graph of the system is the horizontal band that lies *between* the two horizontal lines (where $y < 4$ *and* $y > 1$), as shown below.

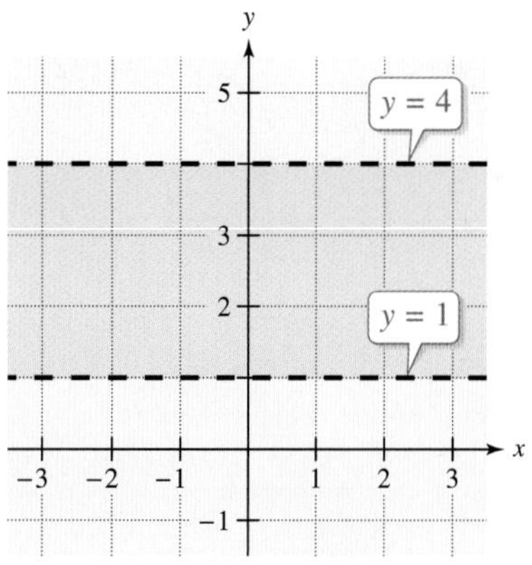

Exercises Within Reach®

Solutions in English & Spanish and tutorial videos at AlgebraWithinReach.com

Graphing a System of Linear Inequalities **In Exercises 13–20, sketch the graph of the system of linear inequalities.** See Additional Answers.

13. $\begin{cases} x > -4 \\ x \leq 2 \end{cases}$

14. $\begin{cases} y \leq 4 \\ y > -2 \end{cases}$

15. $\begin{cases} x < 3 \\ x > -2 \end{cases}$

16. $\begin{cases} y > -1 \\ y \leq 2 \end{cases}$

17. $\begin{cases} x \leq 5 \\ x > -6 \end{cases}$

18. $\begin{cases} y > -7 \\ y < 6 \end{cases}$

19. $\begin{cases} x < 3 \\ x > -3 \end{cases}$

20. $\begin{cases} y \leq 5 \\ y \geq -5 \end{cases}$

EXAMPLE 3 **Graphing a System of Linear Inequalities**

Sketch the graph of the system of linear inequalities, and label the vertices.

$$\begin{cases} x - y < \ \ 2 \\ x \qquad > -2 \\ \qquad y \leq \ \ 3 \end{cases}$$

SOLUTION

Begin by sketching the half-planes represented by the three linear inequalities. The graph of

$$x - y < 2$$

is the half-plane lying above the line $y = x - 2$, the graph of

$$x > -2$$

is the half-plane lying to the right of the line $x = -2$, and the graph of

$$y \leq 3$$

is the half-plane lying on and below the line $y = 3$. As shown below, the region that is common to all three of these half-planes is a triangle. The vertices of the triangle are found as follows.

Vertex A: $(-2, -4)$	*Vertex B:* $(5, 3)$	*Vertex C:* $(-2, 3)$
Solution of the system	Solution of the system	Solution of the system
$\begin{cases} x - y = \ \ 2 \\ x \qquad = -2 \end{cases}$	$\begin{cases} x - y = 2 \\ \qquad y = 3 \end{cases}$	$\begin{cases} x = -2 \\ y = \ \ 3 \end{cases}$

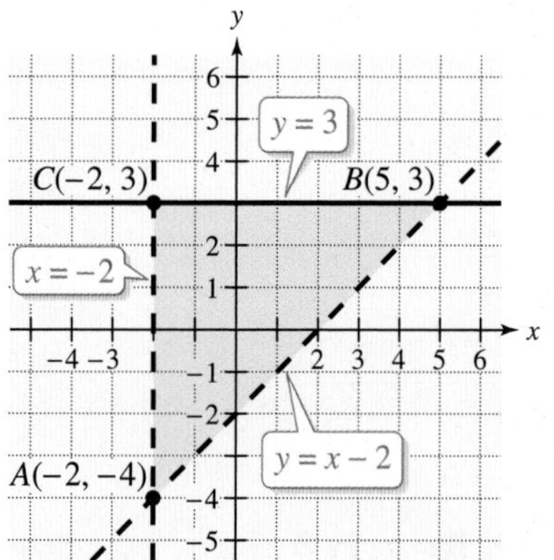

Exercises Within Reach® Solutions in English & Spanish and tutorial videos at AlgebraWithinReach.com

Graphing a System of Linear Inequalities **In Exercises 21–26, sketch the graph of the system of linear inequalities, and label the vertices.** See Additional Answers.

21. $\begin{cases} x + y \leq 4 \\ x \qquad \geq 0 \\ \qquad y \geq 0 \end{cases}$

22. $\begin{cases} 2x + y \leq 6 \\ x \qquad \geq 0 \\ \qquad y \geq 0 \end{cases}$

23. $\begin{cases} 4x - 2y > 8 \\ x \qquad \geq 0 \\ \qquad y \leq 0 \end{cases}$

24. $\begin{cases} 2x - 6y > 6 \\ x \qquad \leq 0 \\ \qquad y \leq 0 \end{cases}$

25. $\begin{cases} y > -5 \\ x \leq \ \ 2 \\ y \leq \ \ x + 2 \end{cases}$

26. $\begin{cases} y \geq -1 \\ x < \ \ 3 \\ y \geq \ \ x - 1 \end{cases}$

EXAMPLE 4 Graphing a System of Linear Inequalities

Sketch the graph of the system of linear inequalities, and label the vertices.

$$\begin{cases} x + y \le 5 \\ 3x + 2y \le 12 \\ x \ge 0 \\ y \ge 0 \end{cases}$$

SOLUTION

Begin by sketching the half-planes represented by the four linear inequalities. The graph of $x + y \le 5$ is the half-plane lying on and below the line $y = -x + 5$. The graph of $3x + 2y \le 12$ is the half-plane lying on and below the line $y = -\frac{3}{2}x + 6$. The graph of $x \ge 0$ is the half-plane lying on and to the right of the y-axis, and the graph of $y \ge 0$ is the half-plane lying on and above the x-axis. As shown below, the region that is common to all four of these half-planes is a four-sided polygon. The vertices of the region are found as follows.

Vertex A: (0, 5)	*Vertex B:* (2, 3)	*Vertex C:* (4, 0)	*Vertex D:* (0, 0)
Solution of the system	Solution of the system	Solution of the system	Solution of the system
$\begin{cases} x + y = 5 \\ x = 0 \end{cases}$	$\begin{cases} x + y = 5 \\ 3x + 2y = 12 \end{cases}$	$\begin{cases} 3x + 2y = 12 \\ y = 0 \end{cases}$	$\begin{cases} x = 0 \\ y = 0 \end{cases}$

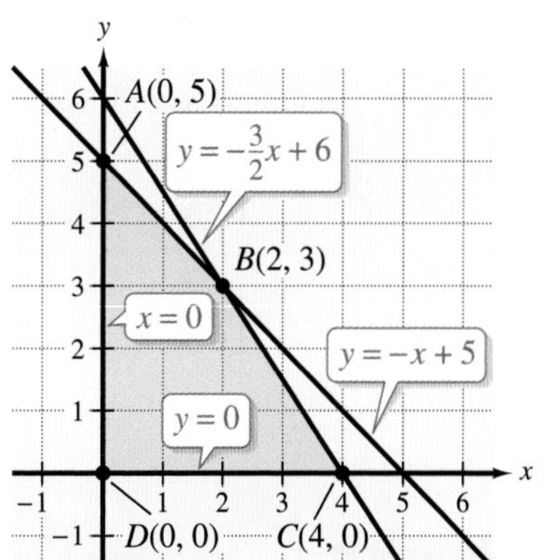

Exercises Within Reach®

Solutions in English & Spanish and tutorial videos at AlgebraWithinReach.com

Graphing a System of Linear Inequalities **In Exercises 27–30, sketch the graph of the system of linear inequalities, and label the vertices.** See Additional Answers.

27. $\begin{cases} x \ge 1 \\ x - 2y \le 3 \\ 3x + 2y \ge 9 \\ x + y \le 6 \end{cases}$

28. $\begin{cases} x + y \le 4 \\ x + y \ge -1 \\ x - y \ge -2 \\ x - y \le 2 \end{cases}$

29. $\begin{cases} x - y \le 8 \\ 2x + 5y \le 25 \\ x \ge 0 \\ y \ge 0 \end{cases}$

30. $\begin{cases} 4x - y \le 13 \\ -x + 2y \le 22 \\ x \ge 0 \\ y \ge 0 \end{cases}$

EXAMPLE 5 Finding the Boundaries of a Region

Write a system of inequalities that describes the region shown below.

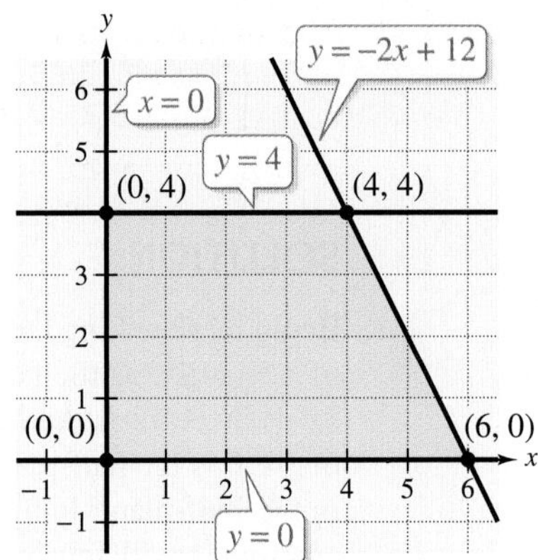

SOLUTION

Three of the boundaries of the region are horizontal or vertical—they are easy to find. To find the diagonal boundary line, you can use the techniques of Section 3.4 to find the equation of the line passing through the points (4, 4) and (6, 0). Use the formula for slope to find $m = -2$, and then use the point-slope form with point (6, 0) and $m = -2$ to obtain

$$y - 0 = -2(x - 6).$$

So, the equation is $y = -2x + 12$. The system of linear inequalities that describes the region is as follows.

$$\begin{cases} y \le 4 & \text{Region lies on and below line } y = 4. \\ y \ge 0 & \text{Region lies on and above } x\text{-axis.} \\ x \ge 0 & \text{Region lies on and to the right of } y\text{-axis.} \\ y \le -2x + 12 & \text{Region lies on and below line } y = -2x + 12. \end{cases}$$

Exercises Within Reach®

Finding the Boundaries of a Region **In Exercises 31–34, write a system of linear inequalities that describes the shaded region.**

31.

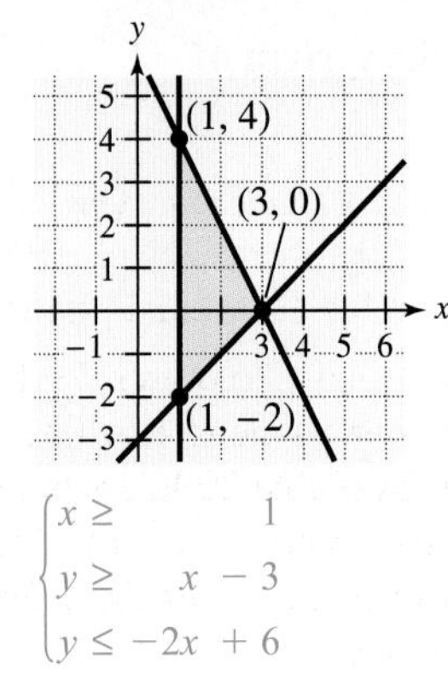

$$\begin{cases} x \ge 1 \\ y \ge x - 3 \\ y \le -2x + 6 \end{cases}$$

32.

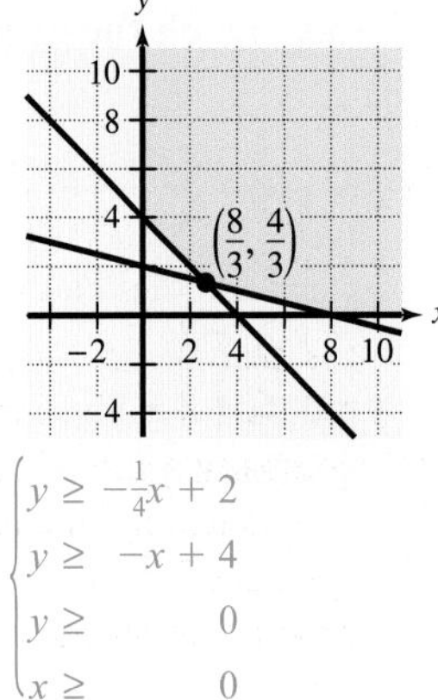

$$\begin{cases} y \ge -\frac{1}{4}x + 2 \\ y \ge -x + 4 \\ y \ge 0 \\ x \ge 0 \end{cases}$$

33.

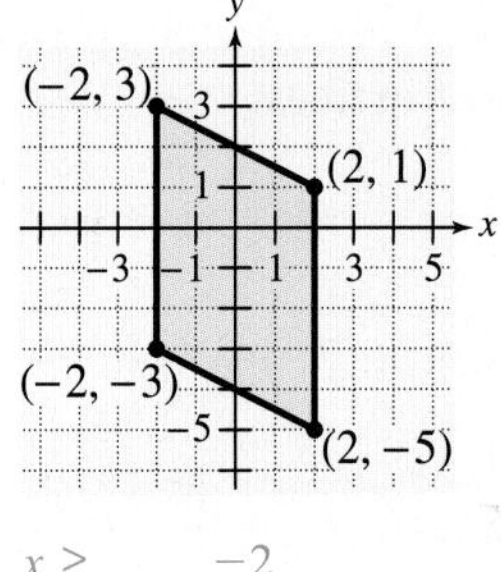

$$x \ge -2$$
$$x \le 2$$
$$y \ge -\tfrac{1}{2}x - 4$$
$$y \le -\tfrac{1}{2}x + 2$$

34.

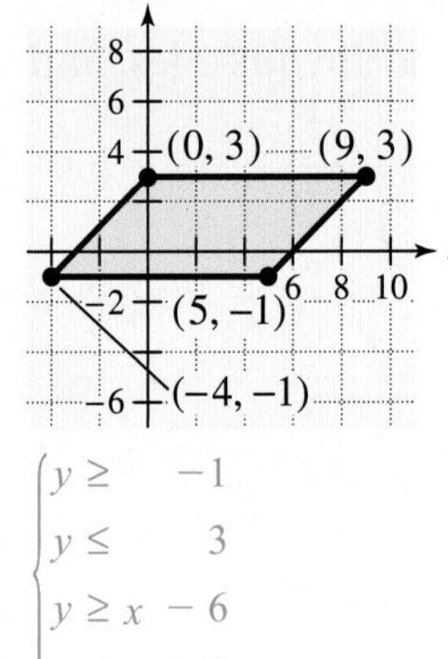

$$\begin{cases} y \ge -1 \\ y \le 3 \\ y \ge x - 6 \\ y \le x + 3 \end{cases}$$

Application

Application **EXAMPLE 6** **Nutrition**

The minimum daily requirements for the liquid portion of a diet are 300 calories, 36 units of vitamin A, and 90 units of vitamin C. A cup of dietary drink X provides 60 calories, 12 units of vitamin A, and 10 units of vitamin C. A cup of dietary drink Y provides 60 calories, 6 units of vitamin A, and 30 units of vitamin C. Write a system of linear inequalities that describes how many cups of each drink should be consumed each day to meet the minimum daily requirements for calories and vitamins.

SOLUTION

Begin by letting x and y represent the following.

$x =$ number of cups of dietary drink X

$y =$ number of cups of dietary drink Y

To meet the minimum daily requirements, the following inequalities must be satisfied.

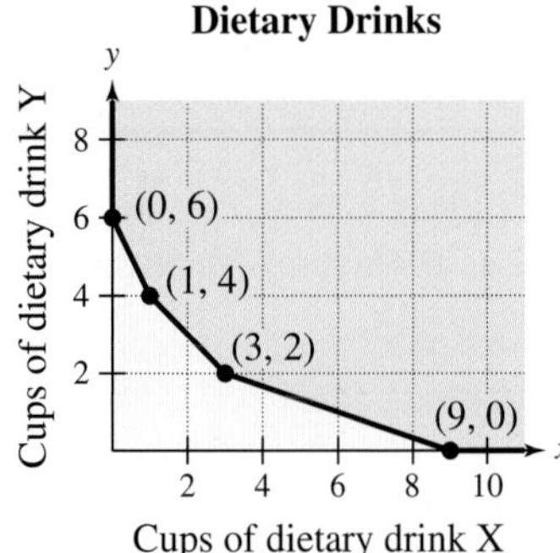

$$\begin{cases} 60x + 60y \geq 300 & \text{Calories} \\ 12x + 6y \geq 36 & \text{Vitamin A} \\ 10x + 30y \geq 90 & \text{Vitamin C} \\ x \geq 0 & \\ y \geq 0 & \end{cases}$$

The last two inequalities are included because x and y cannot be negative. The graph of this system of inequalities is shown at the left.

Exercises Within Reach ®

Solutions in English & Spanish and tutorial videos at AlgebraWithinReach.com

35. ***Production*** A furniture company can sell all the tables and chairs it produces. Each table requires 1 hour in the assembly center and $1\frac{1}{3}$ hours in the finishing center. Each chair requires $1\frac{1}{2}$ hours in the assembly center and $\frac{3}{4}$ hour in the finishing center. The company's assembly center is available 12 hours per day, and its finishing center is available 16 hours per day. Write a system of linear inequalities that describes the different production levels. Graph the system.

See Additional Answers.

36. ***Production*** An electronics company can sell all the HD TVs and DVD players it produces. Each HD TV requires 3 hours on the assembly line and $1\frac{1}{4}$ hours on the testing line. Each DVD player requires $2\frac{1}{2}$ hours on the assembly line and 1 hour on the testing line. The company's assembly line is available 20 hours per day, and its testing line is available 16 hours per day. Write a system of linear inequalities that describes the different production levels. Graph the system.

See Additional Answers.

Concept Summary: Graphing Systems of Linear Inequalities

What

The graph of a **system of linear inequalities** shows *all* of the **solutions** of the system.

EXAMPLE

Sketch the graph of the system of linear inequalities.

$$\begin{cases} y < 2x + 1 \\ y \geq -3x - 2 \end{cases}$$

How

Sketch the line that corresponds to each inequality. Be sure to use dashed lines and solid lines appropriately.

Then shade the half-plane that is the graph of each linear inequality.

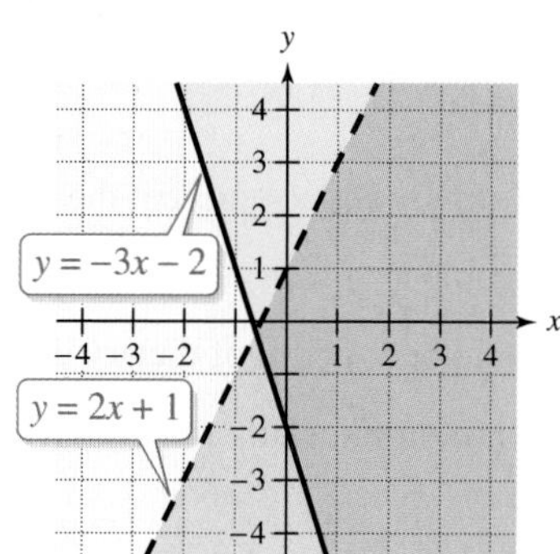

Why

Many practical problems in business, science, and engineering involve multiple constraints. These problems can be solved with systems of inequalities.

Exercises Within Reach®

Worked-out solutions to odd-numbered exercises at AlgebraWithinReach.com

Concept Summary Check

37. ***Vocabulary*** What is a system of linear inequalities in two variables? See Additional Answers.

38. ***Reasoning*** Explain when you should use dashed lines and when you should use solid lines in sketching a system of linear inequalities. See Additional Answers.

39. ***Precision*** Does the point of intersection of each pair of boundary lines correspond to a vertex? Explain. See Additional Answers.

40. ***Logic*** Is it possible for a system of linear inequalities to have no solution? Explain. See Additional Answers.

Extra Practice

Graphing a System of Linear Inequalities **In Exercises 41–46, sketch the graph of the system of linear inequalities.** See Additional Answers.

41. $\begin{cases} x + y \leq 1 \\ -x + y \leq 1 \\ y \geq 0 \end{cases}$

42. $\begin{cases} 3x + 2y < 6 \\ x - 3y \geq 1 \\ y \geq 0 \end{cases}$

43. $\begin{cases} x + y \leq 5 \\ x - 2y \geq 2 \\ y \geq 3 \end{cases}$

44. $\begin{cases} 2x + y \geq 2 \\ x - 3y \leq 2 \\ y \leq 1 \end{cases}$

45. $\begin{cases} -3x + 2y < 6 \\ x - 4y > -2 \\ 2x + y < 3 \end{cases}$

46. $\begin{cases} x + 2y > 14 \\ -2x + 3y > 15 \\ x + 3y < 3 \end{cases}$

47. ***Investment*** A person plans to invest up to $25,000 in two different interest-bearing accounts, account X and account Y. Account Y is to contain at least $4000. Moreover, account X should have at least three times the amount in account Y. Write a system of linear inequalities that describes the various amounts that can be deposited in each account. Graph the system. See Additional Answers.

48. ***Nutrition*** A veterinarian is asked to design a special canine dietary supplement using two different dog foods. Each ounce of food X contains 12 units of calcium, 8 units of iron, and 6 units of protein. Each ounce of food Y contains 10 units of calcium, 10 units of iron, and 8 units of protein. The minimum daily requirements of the diet are 200 units of calcium, 100 units of iron, and 120 units of protein. Write a system of linear inequalities that describes the different amounts of dog food X and dog food Y that can be used. See Additional Answers.

49. ***Ticket Sales*** For a concert event, there are \$30 reserved seat tickets and \$20 general admission tickets. There are 2000 reserved seats available, and fire regulations limit the number of paid ticket holders to 3000. The promoter must take in at least \$75,000 in ticket sales. Write a system of linear inequalities that describes the different numbers of tickets that can be sold.

$$\begin{cases} 30x + 20y \geq 75{,}000 \\ x + y \leq 3{,}000 \\ x \leq 2{,}000 \end{cases}$$

50. ***Geometry*** The figure shows a cross section of a roped-off swimming area at a beach. Write a system of linear inequalities that describes the cross section. (Each unit in the coordinate system represents 1 foot.)

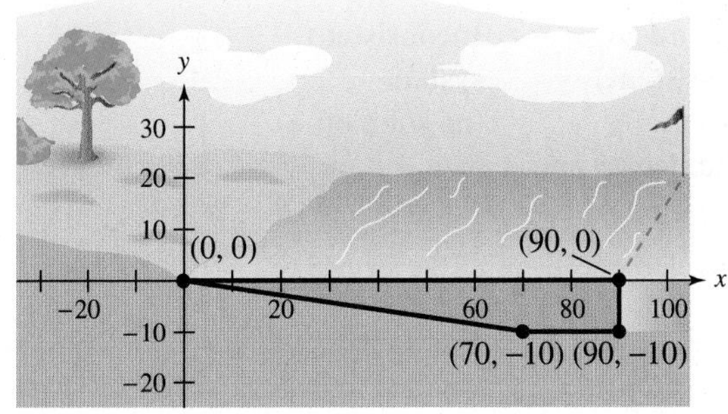

51. ***Geometry*** The figure shows the chorus platform on a stage. Write a system of linear inequalities that describes the part of the audience that can see the full chorus. (Each unit in the coordinate system represents 1 meter.)

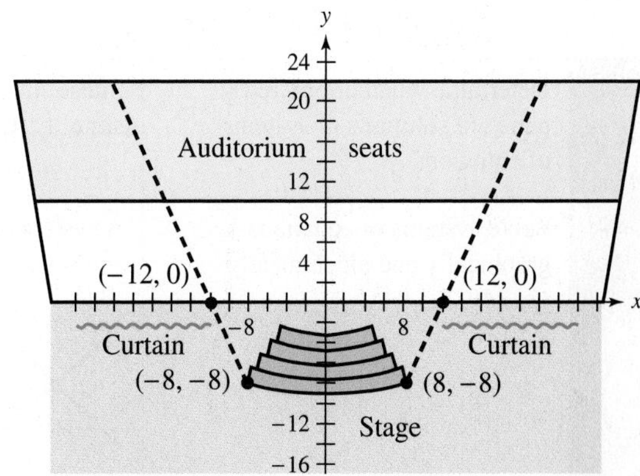

50. $\begin{cases} x \leq 90 \\ y \leq 0 \\ y \geq -10 \\ y \geq -\frac{1}{7}x \end{cases}$ 51. $\begin{cases} y \leq 22 \\ y \geq 10 \\ y \geq 2x - 24 \\ y \geq -2x - 24 \end{cases}$

Explaining Concepts

52. ***Writing*** Explain the meaning of the term *half-plane*. Give an example of an inequality whose graph is a half-plane. The graph of a linear equation splits the xy-plane into two parts, each of which is a half-plane. The graph of $y < 5$ is a half-plane.

53. ***Precision*** Explain how you can check any single point (x_1, y_1) to determine whether the point is a solution of a system of linear inequalities. The point (x_1, y_1) is a solution if it satisfies each inequality in the system.

54. ***Reasoning*** Explain how to determine the vertices of the solution region for a system of linear inequalities. To determine the vertices of the region, find all intersections between the lines corresponding to the inequalities. The vertices are the intersection points that satisfy each inequality.

55. ***Think About It*** Describe the difference between the solution set of a system of linear equations and the solution set of a system of linear inequalities. See Additional Answers.

Cumulative Review

In Exercises 56–61, find the x- and y-intercepts (if any) of the graph of the equation.

56. $y = 4x + 2$ $\left(-\frac{1}{2}, 0\right), (0, 2)$

57. $y = 8 - 3x$ $\left(\frac{8}{3}, 0\right), (0, 8)$

58. $-x + 3y = -3$ $(3, 0), (0, -1)$

59. $3x - 6y = 12$ $(4, 0), (0, -2)$

60. $y = |x + 2|$ $(-2, 0), (0, 2)$

61. $y = |x - 1| - 2$ $(-1, 0), (3, 0), (0, -1)$

In Exercises 62–65, evaluate the function as indicated, and simplify.

62. $f(x) = 3x - 7$
(a) $f(-1)$ -10
(b) $f\left(\frac{2}{3}\right)$ -5

63. $f(x) = x^2 + x$
(a) $f(3)$ 12
(b) $f(-2)$ 2

64. $f(x) = 3x - x^2$
(a) $f(0)$ 0
(b) $f(2m)$ $6m - 4m^2$

65. $f(x) = \dfrac{x + 3}{x - 1}$
(a) $f(8)$ $\dfrac{11}{7}$
(b) $f(k - 2)$ $\dfrac{k + 1}{k - 3}$

4 Chapter Summary

Section	What did you learn?	Explanation and Examples	Review Exercises
4.1	Determine whether ordered pairs are solutions of systems of equations *(p. 168)*.	Substitute the values of x and y into each of the equations and simplify. If the check does *not* fail, the ordered pair is a solution.	1, 2
	Solve systems of equations graphically and algebraically *(p. 169)*.	A system of equations can have one solution, infinitely many solutions, or no solution. 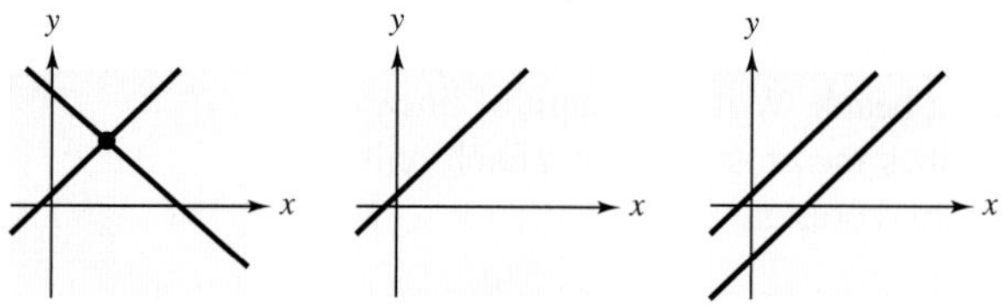 **Consistent system** (one solution); **Dependent (consistent) system** (infinitely many solutions); **Inconsistent system** (no solution) Solve systems of equations algebraically using the method of substitution. **1.** Solve one equation for one variable in terms of the other variable. **2.** Substitute the expression obtained in Step 1 into the other equation to obtain an equation in one variable. **3.** Solve the equation obtained in Step 2. **4.** Back-substitute the Step 3 solution into the expression found in Step 1 to find the value of the other variable. **5.** Check the solution to see that it satisfies both of the original equations.	3–14
	Use systems of equations to model and solve real-life problems *(p. 172)*.	You can use verbal models to write a system of linear equations that represents a real-life problem.	15, 16
4.2	Solve systems of linear equations algebraically using the method of elimination *(p. 176)*.	**1.** Obtain coefficients for x (or y) that are opposites by multiplying one or both equations by suitable constants. **2.** Add the equations to eliminate one variable. Solve the resulting equation. **3.** Back-substitute the value from Step 2 into either of the original equations and solve for the other variable. **4.** Check the solution in both of the original equations.	17–22
	Use systems of linear equations to model and solve real-life problems *(p. 179)*.	**1.** Does the problem involve more than one unknown quantity? **2.** Are there two (or more) equations or conditions to be satisfied? If one or both of these conditions occur, the appropriate mathematical model for the problem may be a system of linear equations.	23–26

	What did you learn?	Explanation and Examples	Review Exercises
4.3	Solve systems of linear equations in row-echelon form using back-substitution *(p. 184)*.	A system of equations in *row-echelon form* has a stair-step pattern with leading coefficients of 1. You can use back-substitution to solve a system in row-echelon form.	27–30
	Solve systems of linear equations using the method of Gaussian elimination *(p. 185)*.	*Gaussian elimination* is the process of forming a chain of equivalent systems by performing one row operation at a time to obtain an equivalent system in row-echelon form.	31–34
	Solve application problems using the method of Gaussian elimination *(p. 189)*.	You can use Gaussian elimination to find the solution of application problems involving vertical motion and the position equation.	35, 36
4.4	Form augmented matrices and form linear systems from augmented matrices *(p. 194)*.	A matrix derived from a system of linear equations (each written in standard form with the constant term on the right) is the augmented matrix of the system.	37–40
	Perform elementary row operations to solve systems of linear equations *(p. 196)*.	**1.** Interchange two rows. **2.** Multiply a row by a nonzero constant. **3.** Add a multiple of a row to another row.	41–48
	Use matrices and Gaussian elimination to solve systems of linear equations *(p. 198)*.	**1.** Write the augmented matrix of the system of equations. **2.** Use elementary row operations to rewrite the augmented matrix in row-echelon form. **3.** Write the system of equations corresponding to the matrix in row-echelon form. Then use back-substitution to find the solution.	41–48
4.5	Find determinants of 2×2 matrices and 3×3 matrices *(p. 204)*.	$\begin{vmatrix} a_1 & b_1 \\ a_2 & b_2 \end{vmatrix} = a_1b_2 - a_2b_1$ $\begin{vmatrix} a_1 & b_1 & c_1 \\ a_2 & b_2 & c_2 \\ a_3 & b_3 & c_3 \end{vmatrix} = a_1\begin{vmatrix} b_2 & c_2 \\ b_3 & c_3 \end{vmatrix} - b_1\begin{vmatrix} a_2 & c_2 \\ a_3 & c_3 \end{vmatrix} + c_1\begin{vmatrix} a_2 & b_2 \\ a_3 & b_3 \end{vmatrix}$	49–54
	Use determinants and Cramer's Rule to solve systems of linear equations *(p. 206)*.	Cramer's Rule is not as general as the elimination method because Cramer's Rule requires that the coefficient matrix of the system be square *and* that the system have exactly one solution.	55–58
	Use determinants to find areas of regions, to test for collinear points, and to find equations of lines *(p. 208)*.	These are examples of practical applications of determinants.	59–66
4.6	Solve systems of linear inequalities in two variables *(p. 214)*.	**1.** Sketch a dashed or solid line corresponding to each inequality. **2.** Shade the half-plane for each inequality. **3.** The intersection of all half-planes represents the system.	67–70
	Use systems of linear inequalities to model and solve real-life problems *(p. 219)*.	A system of linear inequalities can be used to find a number of acceptable solutions to real-life problems.	71, 72

Review Exercises

Worked-out solutions to odd-numbered exercises at AlgebraWithinReach.com

4.1

Checking Solutions **In Exercises 1 and 2, determine whether each ordered pair is a solution of the system of equations.**

1. $\begin{cases} 3x + 7y = 2 \\ 5x + 6y = 9 \end{cases}$

(a) $(3, 4)$ Not a solution

(b) $(3, -1)$ Solution

2. $\begin{cases} -2x + 5y = 21 \\ 9x - y = 13 \end{cases}$

(a) $(2, 5)$ Solution

(b) $(-2, 4)$ Not a solution

Solving a System of Equations **In Exercises 3–8, solve the system of equations by graphing.**
See Additional Answers.

3. $\begin{cases} x + y = 2 \\ x - y = 2 \end{cases}$ $(2, 0)$

4. $\begin{cases} 2x - 3y = -3 \\ y = x \end{cases}$ $(3, 3)$

5. $\begin{cases} x + y = -1 \\ 3x + 2y = 0 \end{cases}$ $(2, -3)$

6. $\begin{cases} 2x - y = 0 \\ -x + y = 4 \end{cases}$ $(4, 8)$

7. $\begin{cases} 2x + y = 4 \\ -4x - 2y = -8 \end{cases}$
Infinitely many solutions

8. $\begin{cases} 3x - 2y = 6 \\ -6x + 4y = 12 \end{cases}$
No solution

Solving a System of Equations **In Exercises 9–14, solve the system of equations by the method of substitution.**

9. $\begin{cases} 2x - 3y = -1 \\ x + 4y = 16 \end{cases}$ $(4, 3)$

10. $\begin{cases} 3x - 7y = 10 \\ -2x + y = -14 \end{cases}$ $(8, 2)$

11. $\begin{cases} -x + y = 6 \\ 15x + y = -10 \end{cases}$
$(-1, 5)$

12. $\begin{cases} -3x + y = 6 \\ 2x + 3y = -3 \end{cases}$
$\left(-\frac{21}{11}, \frac{3}{11}\right)$

13. $\begin{cases} -3x - 3y = 3 \\ x + y = -1 \end{cases}$
Infinitely many solutions

14. $\begin{cases} x + y = 9 \\ x + y = 0 \end{cases}$
No solution

15. ***Break-Even Analysis*** A small business invests \$25,000 to produce a one-time-use camera. Each camera costs \$4.45 to produce and sells for \$8.95. How many one-time-use cameras must be sold before the business breaks even? 5556 cameras

16. ***Seed Mixture*** Fifteen pounds of mixed birdseed sells for \$8.85 per pound. The mixture is obtained from two kinds of birdseed, with one variety priced at \$7.05 per pound and the other at \$9.30 per pound. How many pounds of each variety of birdseed are used in the mixture? 3 pounds at \$7.05 per pound, 12 pounds at \$9.30 per pound

4.2

Solving a System of Linear Equations **In Exercises 17–22, solve the system of linear equations by the method of elimination.**

17. $\begin{cases} x + y = 0 \\ 2x + y = 0 \end{cases}$ $(0, 0)$

18. $\begin{cases} 4x + y = 1 \\ x - y = 4 \end{cases}$ $(1, -3)$

19. $\begin{cases} 2x - y = 2 \\ 6x + 8y = 39 \end{cases}$ $\left(\frac{5}{2}, 3\right)$

20. $\begin{cases} 3x + 2y = 11 \\ x - 3y = -11 \end{cases}$ $(1, 4)$

21. $\begin{cases} 4x + y = -3 \\ -4x + 3y = 23 \end{cases}$
$(-2, 5)$

22. $\begin{cases} -3x + 5y = -23 \\ 2x - 5y = 22 \end{cases}$
$(1, -4)$

23. ***Ticket Sales*** Five hundred tickets were sold for a fundraising dinner. The receipts totaled \$3400.00. Adult tickets were \$7.50 each and children's tickets were \$4.00 dollars each. How many tickets of each type were sold? 400 adult tickets, 100 children's tickets

24. ***Ticket Sales*** A fundraising dinner, was held on two consecutive nights. On the first night, 100 adult tickets and 175 children's tickets were sold, for a total of \$937.50. On the second night, 200 adult tickets and 316 children's tickets were sold, for a total of \$1790.00. Find the price of each type of ticket.
\$5 for an adult ticket, \$2.50 for a children's ticket

25. ***Alcohol Mixture*** Fifty gallons of a 90% alcohol solution is obtained by mixing a 100% solution with a 75% solution. How many gallons of each solution must be used to obtain the desired mixture?
30 gallons of 100% solution, 20 gallons of 75% solution

26. ***Acid Mixture*** Forty gallons of a 60% acid solution is obtained by mixing a 75% solution with a 50% solution. How many gallons of each solution must be used to obtain the desired mixture?
16 gallons of 75% solution, 24 gallons of 50% solution

Africa Studio/Shutterstock.com

4.3

Using Back-Substitution **In Exercises 27–30, use back-substitution to solve the system of linear equations.**

27. $\begin{cases} x = 3 \\ x + 2y = 7 \\ -3x - y + 4z = 9 \end{cases}$ (3, 2, 5)

28. $\begin{cases} 2x + 3y = 9 \\ 4x - 6z = 12 \\ y = 5 \end{cases}$ (−3, 5, −4)

29. $\begin{cases} x + 2y = 6 \\ 3y = 9 \\ x + 2z = 12 \end{cases}$ (0, 3, 6)

30. $\begin{cases} 3x - 2y + 5z = -10 \\ 3y = 18 \\ 6x - 4y = -6 \end{cases}$ $\left(3, 6, -\frac{7}{5}\right)$

Using Gaussian Elimination **In Exercises 31–34, solve the system of linear equations.**

31. $\begin{cases} -x + y + 2z = 1 \\ 2x + 3y + z = -2 \\ 5x + 4y + 2z = 4 \end{cases}$ (2, −3, 3)

32. $\begin{cases} 2x + 3y + z = 10 \\ 2x - 3y - 3z = 22 \\ 4x - 2y + 3z = -2 \end{cases}$ (5, 2, −6)

33. $\begin{cases} x - y - z = 1 \\ -2x + y + 3z = -5 \\ 3x + 4y - z = 6 \end{cases}$ (0, 1, −2)

34. $\begin{cases} -3x + y + 2z = -13 \\ -x - y + z = 0 \\ 2x + 2y - 3z = -1 \end{cases}$ (4, −3, 1)

35. ***Investment*** An inheritance of \$20,000 is divided among three investments yielding a total of \$1780 in interest per year. The interest rates for the three investments are 7%, 9%, and 11%. The amounts invested at 9% and 11% are \$3000 and \$1000 less than the amount invested at 7%, respectively. Find the amount invested at each rate.
\$8000 at 7%, \$5000 at 9%, \$7000 at 11%

36. ***Vertical Motion*** Find the position equation

$$s = \frac{1}{2}at^2 + v_0t + s_0$$

for an object that has the indicated heights at the specified times.

$s = 192$ feet at $t = 1$ second
$s = 152$ feet at $t = 2$ seconds
$s = 80$ feet at $t = 3$ seconds $s = -16t^2 + 8t + 200$

photolinc/Shutterstock.com

4.4

Forming Coefficient and Augmented Matrices **In Exercises 37 and 38, form (a) the coefficient matrix and (b) the augmented matrix for the system of linear equations.**

37. $\begin{cases} 7x - 5y = 11 \\ x - y = -5 \end{cases}$

(a) $\begin{bmatrix} 7 & -5 \\ 1 & -1 \end{bmatrix}$ (b) $\left[\begin{array}{cc:c} 7 & -5 & 11 \\ 1 & -1 & -5 \end{array}\right]$

38. $\begin{cases} x + 2y + z = 4 \\ 3x - z = 2 \\ -x + 5y - 2z = -6 \end{cases}$

See Additional Answers.

Forming a Linear System **In Exercises 39 and 40, write the system of linear equations represented by the matrix. (Use variables x, y, and z.)**

39. $\left[\begin{array}{ccc:c} 4 & -1 & 0 & 2 \\ 6 & 3 & 2 & 1 \\ 0 & 1 & 4 & 0 \end{array}\right]$ $\begin{cases} 4x - y = 2 \\ 6x + 3y + 2z = 1 \\ y + 4z = 0 \end{cases}$

40. $\left[\begin{array}{cc:c} 7 & 8 & -26 \\ 4 & -9 & -12 \end{array}\right]$ $\begin{cases} 7x + 8y = -26 \\ 4x - 9y = -12 \end{cases}$

Solving a System of Linear Equations **In Exercises 41–48, use matrices to solve the system.**

41. $\begin{cases} 5x + 4y = 2 \\ -x + y = -22 \end{cases}$ (10, −12)

42. $\begin{cases} 2x - 5y = 2 \\ 3x - 7y = 1 \end{cases}$ (−9, −4)

43. $\begin{cases} 0.2x - 0.1y = 0.07 \\ 0.4x - 0.5y = -0.01 \end{cases}$ (0.6, 0.5)

44. $\begin{cases} 2x + y = 0.3 \\ 3x - y = -1.3 \end{cases}$ (−0.2, 0.7)

45. $\begin{cases} x + 4y + 4z = 7 \\ -3x + 2y + 3z = 0 \\ 4x - 2z = -2 \end{cases}$ (3, −6, 7)

46. $\begin{cases} -x + 3y - z = -4 \\ 2x + 6z = 14 \\ -3x - y + z = 10 \end{cases}$ (−2, −1, 3)

47. $\begin{cases} 2x + 3y + 3z = 3 \\ 6x + 6y + 12z = 13 \\ 12x + 9y - z = 2 \end{cases}$ $\left(\frac{1}{2}, -\frac{1}{3}, 1\right)$

48. $\begin{cases} -x + 2y + 3z = 4 \\ 2x - 4y - z = -13 \\ 3x + 2y - 4z = -1 \end{cases}$ (−3, 2, −1)

4.5

Finding the Determinant **In Exercises 49–54, find the determinant of the matrix.**

49. $\begin{bmatrix} 9 & 8 \\ 10 & 10 \end{bmatrix}$ 10

50. $\begin{bmatrix} -3.4 & 1.2 \\ -5 & 2.5 \end{bmatrix}$ -2.5

51. $\begin{bmatrix} 8 & 6 & 3 \\ 6 & 3 & 0 \\ 3 & 0 & 2 \end{bmatrix}$ -51

52. $\begin{bmatrix} 7 & -1 & 10 \\ -3 & 0 & -2 \\ 12 & 1 & 1 \end{bmatrix}$ 5

53. $\begin{bmatrix} 8 & 3 & 2 \\ 1 & -2 & 4 \\ 6 & 0 & 5 \end{bmatrix}$ 1

54. $\begin{bmatrix} 4 & 0 & 10 \\ 0 & 10 & 0 \\ 10 & 0 & 34 \end{bmatrix}$ 360

Using Cramer's Rule **In Exercises 55–58, use Cramer's Rule to solve the system of linear equations. (If not possible, state the reason.)**

55. $\begin{cases} 7x + 12y = 63 \\ 2x + 3y = 15 \end{cases}$ $(-3, 7)$

56. $\begin{cases} 12x + 42y = -17 \\ 30x - 18y = 19 \end{cases}$ $\left(\frac{1}{3}, -\frac{1}{2}\right)$

57. $\begin{cases} -x + y + 2z = 1 \\ 2x + 3y + z = -2 \\ 5x + 4y + 2z = 4 \end{cases}$ $(2, -3, 3)$

58. $\begin{cases} 2x + y + 2z = 4 \\ 2x + 2y = 5 \\ 2x - y + 6z = 2 \end{cases}$ Not possible, $D = 0$

Finding the Area of a Triangle **In Exercises 59–62, use a determinant to find the area of the triangle with the given vertices.**

59. $(1, 0), (5, 0), (5, 8)$ 16

60. $(-6, 0), (6, 0), (0, 5)$ 30

61. $(1, 2), (4, -5), (3, 2)$ 7

62. $\left(\frac{3}{2}, 1\right), \left(4, -\frac{1}{2}\right), (4, 2)$ $\frac{25}{8}$

Testing for Collinear Points **In Exercises 63 and 64, determine whether the points are collinear.**

63. $(1, 2), (5, 0), (10, -2)$ Not collinear

64. $(-4, 3), (1, 1), (6, -1)$ Collinear

Finding an Equation of a Line **In Exercises 65 and 66, use a determinant to find an equation of the line passing through the points.**

65. $(-4, 0), (4, 4)$ $x - 2y + 4 = 0$

66. $\left(-\frac{5}{2}, 3\right), \left(\frac{7}{2}, 1\right)$ $2x + 6y - 13 = 0$

4.6

Graphing a System of Linear Inequalities **In Exercises 67–70, sketch the graph of the system of linear inequalities.** See Additional Answers.

67. $\begin{cases} x + y < 5 \\ x > 2 \\ y \geq 0 \end{cases}$

68. $\begin{cases} \frac{1}{2}x + y > 4 \\ x < 6 \\ y < 3 \end{cases}$

69. $\begin{cases} x + 2y \leq 160 \\ 3x + y \leq 180 \\ x \geq 0 \\ y \geq 0 \end{cases}$

70. $\begin{cases} 2x + 3y \leq 24 \\ 2x + y \leq 16 \\ x \geq 0 \\ y \geq 0 \end{cases}$

71. ***Soup Distribution*** A charitable organization can purchase up to 500 cartons of soup to be divided between a soup kitchen and a homeless shelter. These two organizations need at least 150 cartons and 220 cartons, respectively. Write a system of linear inequalities that describes the various numbers of cartons that can go to each organization. Graph the system.

$\begin{cases} x + y \leq 500 \\ x \geq 150 \\ y \geq 220 \end{cases}$ See Additional Answers.

72. ***Inventory Costs*** A warehouse operator has up to 24,000 square feet of floor space in which to store two products. Each unit of product X requires 20 square feet of floor space and costs \$12 per day to store. Each unit of product Y requires 30 square feet of floor space and costs \$8 per day to store. The total storage cost per day cannot exceed \$12,400. Write a system of linear inequalities that describes the various ways the two products can be stored. Graph the system.

$\begin{cases} 20x + 30y \leq 24{,}000 \\ 12x + 8y \leq 12{,}400 \\ x \geq 0 \\ y \geq 0 \end{cases}$ See Additional Answers.

Chapter Test

Solutions in English & Spanish and tutorial videos at AlgebraWithinReach.com

Take this test as you would take a test in class. After you are done, check your work against the answers in the back of the book.

$$\begin{cases} 2x - 2y = 2 \\ -x + 2y = 0 \end{cases}$$

1. Determine whether each ordered pair is a solution of the system at the left.

(a) (2, 1) Solution (b) (4, 3) Not a solution

In Exercises 2–11, use the indicated method to solve the system.

2. *Graphical:* $\begin{cases} x - 2y = -1 \\ 2x + 3y = 12 \end{cases}$

3. *Substitution:* $\begin{cases} 4x - y = 1 \\ 4x - 3y = -5 \end{cases}$

4. *Substitution:* $\begin{cases} 2x - 2y = -2 \\ 3x + y = 9 \end{cases}$

5. *Elimination:* $\begin{cases} 3x - 4y = -14 \\ -3x + y = 8 \end{cases}$

6. *Elimination:* $\begin{cases} x + 2y - 4z = 0 \\ 3x + y - 2z = 5 \\ 3x - y + 2z = 7 \end{cases}$

7. *Matrices:* $\begin{cases} x - 3z = -10 \\ -2y + 2z = 0 \\ x - 2y = -7 \end{cases}$

8. *Matrices:* $\begin{cases} x - 3y + z = -3 \\ 3x + 2y - 5z = 18 \\ y + z = -1 \end{cases}$

9. *Cramer's Rule:* $\begin{cases} 2x - 7y = 7 \\ 3x + 7y = 13 \end{cases}$

2. (3, 2) 3. (1, 3)
4. (2, 3) 5. (−2, 2)
6. $(2, 2a - 1, a)$ 7. (−1, 3, 3)
8. (2, 1, −2) 9. $\left(4, \frac{1}{7}\right)$

10. *Any Method:* $\begin{cases} 3x - 2y + z = 12 \\ x - 3y = 2 \\ -3x - 9z = -6 \end{cases}$ (5, 1, −1)

11. *Any Method:* $\begin{cases} 4x + y + 2z = -4 \\ 3y + z = 8 \\ -3x + y - 3z = 5 \end{cases}$ $\left(-\frac{11}{5}, \frac{56}{25}, \frac{32}{25}\right)$

$$\begin{bmatrix} 2 & -2 & 0 \\ -1 & 3 & 1 \\ 2 & 8 & 1 \end{bmatrix}$$

12. Find the determinant of the matrix shown at the left. −16

13. Use a determinant to find the area of the triangle with vertices (0, 0), (5, 4), and (6, 0). 12

14. Graph the system of linear inequalities.

$$\begin{cases} x - 2y > -3 \\ 2x + 3y \le 22 \\ y \ge 0 \end{cases}$$ See Additional Answers.

15. The perimeter of a rectangle is 68 feet, and its width is $\frac{8}{9}$ times its length. Find the dimensions of the rectangle. 16 feet × 18 feet

16. An inheritance of $25,000 is divided among three investments yielding a total of $1275 in interest per year. The interest rates for the three investments are 4.5%, 5%, and 8%. The amounts invested at 5% and 8% are $4000 and $10,000 less than the amount invested at 4.5%, respectively. Find the amount invested at each rate. $13,000 at 4.5%, $9000 at 5%, $3000 at 8%

17. Two types of tickets are sold for a concert. Reserved seat tickets cost $30 per ticket and floor seat tickets cost $40 per ticket. The promoter of the concert can sell at most 9000 reserved seat tickets and 4000 floor seat tickets. Gross receipts must total at least $300,000 in order for the concert to be held. Write a system of linear inequalities that describes the different numbers of tickets that can be sold. Graph the system. See Additional Answers.

17. $\begin{cases} 30x + 40y \ge 300{,}000 \\ x \le 9{,}000 \\ y \le 4{,}000 \end{cases}$

where x is the number of reserved seat tickets and y is the number of floor seat tickets.

Cumulative Test: Chapters 1–4

Solutions in English & Spanish and tutorial videos at AlgebraWithinReach.com

Take this test as you would take a test in class. After you are done, check your work against the answers in the back of the book.

1. Place the correct symbol (<, >, or =) between the real numbers.

   (a) -2 [>] -4 (b) $\frac{2}{3}$ [>] $\frac{1}{2}$ (c) -4.5 [=] $-|-4.5|$

2. Write an algebraic expression for the statement, "The number n is tripled and the product is decreased by 8." $3n - 8$

In Exercises 3 and 4, simplify the expression.

3. $t(3t - 1) - 2(t + 4)$ $3t^2 - 3t - 8$
4. $4x(x + x^2) - 6(x^2 + 4)$ $4x^3 - 2x^2 - 24$

In Exercises 5–8, solve the equation or inequality.

5. $12 - 5(3 - x) = x + 3$ $\frac{3}{2}$
6. $1 - \frac{x + 2}{4} = \frac{7}{8}$ $-\frac{3}{2}$
7. $|x - 2| \ge 3$ $x \le -1$ or $x \ge 5$
8. $-12 \le 4x - 6 < 10$ $-\frac{3}{2} \le x < 4$

9 | 4.5

13 | x

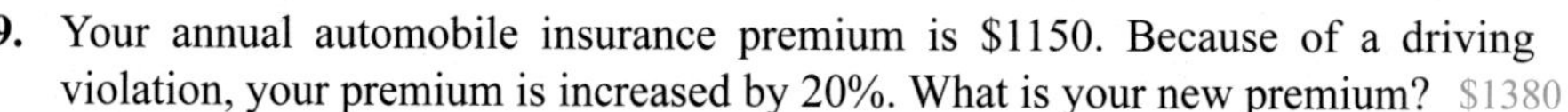

9. Your annual automobile insurance premium is \$1150. Because of a driving violation, your premium is increased by 20%. What is your new premium? \$1380
10. The triangles at the left are similar. Solve for the length x by using the fact that corresponding sides of similar triangles are proportional. 6.5
11. Company A rents a subcompact car for \$240 per week with no extra charge for mileage. Company B rents a similar car for \$100 per week plus an additional 25 cents for each mile driven. How many miles must you drive in a week so that the rental fee of company B is more than that of company A? More than 560 miles
12. Does the equation $x - y^2 = 0$ represent y as a function of x? No
13. Find the domain of the function $f(x) = \sqrt{x - 2}$. $2 \le x < \infty$
14. Given $f(x) = x^2 - 2x$, find (a) $f(3)$ and (b) $f(-3c)$. (a) 3 (b) $9c^2 + 6c$
15. Find the slope of the line that passes through $(-4, 0)$ and $(4, 6)$. Then find the distance between the points and the midpoint of the line segment joining the points. $m = \frac{3}{4}$; Distance: 10; Midpoint: $(0, 3)$
16. Write equations of the lines that pass through the point $(-2, 1)$ and are (a) parallel to $2x - y = 1$ and (b) perpendicular to $3x + 2y = 5$.

16. (a) $y = 2x + 5$
    (b) $y = \frac{2}{3}x + \frac{7}{3}$

In Exercises 17 and 18, graph the equation. See Additional Answers.

17. $4x + 3y - 12 = 0$
18. $y = 2 - (x - 3)^2$

In Exercises 19–21, use the indicated method to solve the system.

19. *Substitution:*

$$\begin{cases} x + y = 6 \\ 2x - y = 3 \end{cases}$$

(3, 3)

20. *Elimination:*

$$\begin{cases} 2x + y = 6 \\ 3x - 2y = 16 \end{cases}$$

(4, −2)

21. *Matrices:*

$$\begin{cases} 2x + y - 2z = 1 \\ x - z = 1 \\ 3x + 3y + z = 12 \end{cases}$$

(4, −1, 3)

5 Polynomials and Factoring

MASTERY IS WITHIN REACH!

"I understand math but never did as well as I wanted to until this semester. My instructor pointed out that I set problems up correctly and know all the steps, but I mess up in the details. I took a learning modality inventory and I am a visual learner, but I discovered that if I talk out loud and move my finger over each little step, I catch my mistakes."

Katie
Major undecided

See page 255 for suggestions about using different approaches to learning.

5.1 Integer Exponents and Scientific Notation

- Use the rules of exponents to simplify expressions.
- Rewrite exponential expressions involving negative and zero exponents.
- Write very large and very small numbers in scientific notation.

Rules of Exponents

Rules of Exponents

Let m and n be positive integers, and let a and b represent real numbers, variables, or algebraic expressions.

Rule	*Example*
1. Product: $a^m \cdot a^n = a^{m+n}$	$x^5(x^4) = x^{5+4} = x^9$
2. Product-to-Power: $(ab)^m = a^m \cdot b^m$	$(2x)^3 = 2^3(x^3) = 8x^3$
3. Power-to-Power: $(a^m)^n = a^{mn}$	$(x^2)^3 = x^{2 \cdot 3} = x^6$
4. Quotient: $\frac{a^m}{a^n} = a^{m-n},\ m > n,\ a \neq 0$	$\frac{x^5}{x^3} = x^{5-3} = x^2,\ x \neq 0$
5. Quotient-to-Power: $\left(\frac{a}{b}\right)^m = \frac{a^m}{b^m},\ b \neq 0$	$\left(\frac{x}{4}\right)^2 = \frac{x^2}{4^2} = \frac{x^2}{16}$

Additional Examples

Use the rules of exponents to simplify each expression.

a. $(5xy^4)(3x^2)$

b. $-3(xy^2)^2$

c. $(-3xy^2)^2$

d. $\frac{a^m b^{2m}}{a^3 b^3}$

e. $\left(\frac{y^{2n}}{3x}\right)^2$

Answers:

a. $15x^3y^4$

b. $-3x^2y^4$

c. $9x^2y^4$

d. $a^{m-3}b^{2m-3}$

e. $\frac{y^{4n}}{9x^2}$

EXAMPLE 1 Using Rules of Exponents

a. $(x^2y^4)(3x) = 3(x^2 \cdot x)(y^4) = 3(x^{2+1})(y^4) = 3x^3y^4$

b. $-2(y^2)^3 = (-2)(y^{2 \cdot 3}) = -2y^6$

c. $(-2y^2)^3 = (-2)^3(y^2)^3 = -8(y^{2 \cdot 3}) = -8y^6$

d. $(3x^2)(-5x)^3 = 3(-5)^3(x^2 \cdot x^3) = 3(-125)(x^{2+3}) = -375x^5$

e. $\frac{14a^5b^3}{7a^2b^2} = 2(a^{5-2})(b^{3-2}) = 2a^3b$

f. $\left(\frac{x^2}{2y}\right)^3 = \frac{(x^2)^3}{(2y)^3} = \frac{x^{2 \cdot 3}}{2^3y^3} = \frac{x^6}{8y^3}$

g. $\frac{x^ny^{3n}}{x^2y^4} = x^{n-2}y^{3n-4}$

Exercises Within Reach®

Solutions in English & Spanish and tutorial videos at AlgebraWithinReach.com

Using Rules of Exponents **In Exercises 1–16, use the rules of exponents to simplify the expression.**

1. $(u^3v)(2v^2)$ $2u^3v^3$

2. $(x^5y^3)(2y^3)$ $2x^5y^6$

3. $-3(x^3)^2$ $-3x^6$

4. $-5(y^4)^3$ $-5y^{12}$

5. $(-5z^2)^3$ $-125z^6$

6. $(-5z^4)^2$ $25z^8$

7. $(2u)^4(4u)$ $64u^5$

8. $(3y)^3(2y^2)$ $54y^5$

9. $\frac{27m^5n^6}{9mn^3}$ $3m^4n^3$

10. $\frac{-18m^3n^6}{-6mn^3}$ $3m^2n^3$

11. $-\left(\frac{2a}{3y}\right)^2$ $-\frac{4a^2}{9y^2}$

12. $\left(\frac{5u}{3v}\right)^3$ $\frac{125u^3}{27v^3}$

13. $\frac{x^ny^{2n}}{x^3y^2}$ $x^{n-3}y^{2n-2}$

14. $\frac{x^{2n}y^n}{x^ny}$ x^ny^{n-1}

15. $\frac{(-2x^2y)^3}{9x^2y^2}$ $-\frac{8x^4y}{9}$

16. $\frac{(-2xy^3)^2}{6y^2}$ $\frac{2x^2y^4}{3}$

Integer Exponents

Study Tip

Notice that by definition, $a^0 = 1$ for all real *nonzero* values of a. Zero cannot have a zero exponent, because the expression 0^0 is undefined.

Definitions of Zero Exponents and Negative Exponents

Let a and b be real numbers such that $a \neq 0$ and $b \neq 0$, and let m be an integer.

1. $a^0 = 1$ 2. $a^{-m} = \dfrac{1}{a^m}$ 3. $\left(\dfrac{a}{b}\right)^{-m} = \left(\dfrac{b}{a}\right)^m$

EXAMPLE 2 Using Rules of Exponents

a. $3^0 = 1$ — Definition of zero exponents

b. $3^{-2} = \dfrac{1}{3^2} = \dfrac{1}{9}$ — Definition of negative exponents

c. $\left(\dfrac{3}{4}\right)^{-1} = \left(\dfrac{4}{3}\right)^1 = \dfrac{4}{3}$ — Definition of negative exponents

Summary of Rules of Exponents

Let m and n be integers, and let a and b represent real numbers, variables, or algebraic expressions. (All denominators and bases are nonzero.)

Product and Quotient Rules	*Example*
1. $a^m \cdot a^n = a^{m+n}$	$x^4(x^3) = x^{4+3} = x^7$
2. $\dfrac{a^m}{a^n} = a^{m-n}$	$\dfrac{x^3}{x} = x^{3-1} = x^2$
Power Rules	
3. $(ab)^m = a^m \cdot b^m$	$(3x)^2 = 3^2(x^2) = 9x^2$
4. $(a^m)^n = a^{mn}$	$(x^3)^3 = x^{3 \cdot 3} = x^9$
5. $\left(\dfrac{a}{b}\right)^m = \dfrac{a^m}{b^m}$	$\left(\dfrac{x}{3}\right)^2 = \dfrac{x^2}{3^2} = \dfrac{x^2}{9}$
Zero and Negative Exponent Rules	
6. $a^0 = 1$	$(x^2 + 1)^0 = 1$
7. $a^{-m} = \dfrac{1}{a^m}$	$x^{-2} = \dfrac{1}{x^2}$
8. $\left(\dfrac{a}{b}\right)^{-m} = \left(\dfrac{b}{a}\right)^m$	$\left(\dfrac{x}{3}\right)^{-2} = \left(\dfrac{3}{x}\right)^2 = \dfrac{3^2}{x^2} = \dfrac{9}{x^2}$

Exercises Within Reach®

Using Rules of Exponents **In Exercises 17–24, use the rules of exponents to evaluate the expression.**

17. $(-3)^0$ 1

18. 25^0 1

19. 5^{-2} $\frac{1}{25}$

20. 2^{-4} $\frac{1}{16}$

21. $\left(\frac{2}{3}\right)^{-1}$ $\frac{3}{2}$

22. $\left(\frac{4}{5}\right)^{-3}$ $\frac{125}{64}$

23. -10^{-3} $-\frac{1}{1000}$

24. -20^{-2} $-\frac{1}{400}$

Study Tip

As you become accustomed to working with negative exponents, you will probably not write as many steps as shown in Example 4. For instance, to rewrite a fraction involving exponents, you might use the following simplified rule. *To move a factor from the numerator to the denominator or vice versa, change the sign of its exponent.* You can apply this rule to the expression in Example 4(a) by "moving" the factor x^{-2} to the numerator and changing the exponent to 2. That is,

$$\frac{3}{x^{-2}} = 3x^2.$$

Remember, you can move only *factors* in this manner, not terms.

EXAMPLE 3 Using Rules of Exponents

a. $2x^{-1} = 2(x^{-1}) = 2\left(\frac{1}{x}\right) = \frac{2}{x}$ — Use negative exponent rule and simplify.

b. $(2x)^{-1} = \frac{1}{(2x)^1} = \frac{1}{2x}$ — Use negative exponent rule and simplify.

EXAMPLE 4 Using Rules of Exponents

Rewrite each expression using only positive exponents. (Assume that $x \neq 0$.)

a. $\frac{3}{x^{-2}} = \frac{3}{\left(\frac{1}{x^2}\right)}$ — Use negative exponent rule.

$= 3\left(\frac{x^2}{1}\right) = 3x^2$ — Invert divisor and multiply.

b. $\frac{1}{(3x)^{-2}} = \frac{1}{\left[\frac{1}{(3x)^2}\right]}$ — Use negative exponent rule.

$= \frac{1}{\left(\frac{1}{9x^2}\right)}$ — Use product-to-power rule and simplify.

$= (1)\left(\frac{9x^2}{1}\right) = 9x^2$ — Invert divisor and multiply.

Exercises Within Reach®

Solutions in English & Spanish and tutorial videos at AlgebraWithinReach.com

Using Rules of Exponents **In Exercises 25–42, rewrite the expression using only positive exponents, and simplify. (Assume that any variables in the expression are nonzero.)**

25. $7x^{-4}$ $\frac{7}{x^4}$
26. $3y^{-3}$ $\frac{3}{y^3}$
27. $(8x)^{-1}$ $\frac{1}{8x}$
28. $(11x)^{-1}$ $\frac{1}{11x}$
29. $(4x)^{-3}$ $\frac{1}{64x^3}$
30. $(5u)^{-2}$ $\frac{1}{25u^2}$
31. $y^4 \cdot y^{-2}$ y^2
32. $z^5 \cdot z^{-3}$ z^2
33. $\frac{1}{x^{-6}}$ x^6
34. $\frac{4}{y^{-1}}$ $4y$
35. $\frac{(4t)^0}{t^{-2}}$ t^2
36. $\frac{(x^2)^0}{x^{-3}}$ x^3
37. $\frac{1}{(5y)^{-2}}$ $25y^2$
38. $\frac{3}{(3a)^{-3}}$ $81a^3$
39. $\frac{(5u)^{-4}}{(5u)^0}$ $\frac{1}{625u^4}$
40. $\frac{(9n)^{-2}}{(9n^2)^0}$ $\frac{1}{81n^2}$
41. $\left(\frac{x}{10}\right)^{-1}$ $\frac{10}{x}$
42. $\left(\frac{4}{z}\right)^{-2}$ $\frac{z^2}{16}$

EXAMPLE 5 Using Rules of Exponents

Rewrite each expression using only positive exponents. (Assume that $x \neq 0$ and $y \neq 0$.)

a. $(-5x^{-3})^2 = (-5)^2(x^{-3})^2$ — Product-to-power rule

$= 25x^{-6}$ — Power-to-power rule

$= \dfrac{25}{x^6}$ — Negative exponent rule

b. $-\left(\dfrac{7x}{y^2}\right)^{-2} = -\left(\dfrac{y^2}{7x}\right)^2$ — Negative exponent rule

$= -\dfrac{(y^2)^2}{(7x)^2}$ — Quotient-to-power rule

$= -\dfrac{y^4}{49x^2}$ — Power-to-power and product-to-power rules

c. $\dfrac{12x^2y^{-4}}{6x^{-1}y^2} = 2(x^{2-(-1)})(y^{-4-2})$ — Quotient rule

$= 2x^3y^{-6}$ — Simplify.

$= \dfrac{2x^3}{y^6}$ — Negative exponent rule

EXAMPLE 6 Using Rules of Exponents

Rewrite each expression using only positive exponents. (Assume that $x \neq 0$ and $y \neq 0$.)

a. $\left(\dfrac{8x^{-1}y^4}{4x^3y^2}\right)^{-3} = \left(\dfrac{2y^2}{x^4}\right)^{-3}$ — Simplify.

$= \left(\dfrac{x^4}{2y^2}\right)^3$ — Negative exponent rule

$= \dfrac{x^{12}}{2^3y^6} = \dfrac{x^{12}}{8y^6}$ — Quotient-to-power rule

b. $\dfrac{3xy^0}{x^2(5y)^0} = \dfrac{3x(1)}{x^2(1)} = \dfrac{3}{x}$ — Zero exponent rule

Exercises Within Reach®

Solutions in English & Spanish and tutorial videos at AlgebraWithinReach.com

Using Rules of Exponents **In Exercises 43–52, rewrite the expression using only positive exponents, and simplify. (Assume that any variables in the expression are nonzero.)**

43. $(2x^2)^{-2}$ $\dfrac{1}{4x^4}$

44. $(4a^{-2})^{-3}$ $\dfrac{a^6}{64}$

45. $-\left(\dfrac{5x}{y^3}\right)^{-3}$ $-\dfrac{y^9}{125x^3}$

46. $-\left(\dfrac{4n^2}{m^4}\right)^{-2}$ $-\dfrac{m^8}{16n^4}$

47. $\dfrac{6x^3y^{-3}}{12x^{-2}y}$ $\dfrac{x^5}{2y^4}$

48. $\dfrac{2y^{-1}z^{-3}}{4yz^{-3}}$ $\dfrac{1}{2y^2}$

49. $\left(\dfrac{3u^2v^{-1}}{3^3u^{-1}v^3}\right)^{-2}$ $\dfrac{81v^8}{u^6}$

50. $\left(\dfrac{5^2x^3y^{-3}}{125xy}\right)^{-1}$ $\dfrac{5y^4}{x^2}$

51. $\dfrac{5x^0y}{(6x)^0y^3}$ $\dfrac{5}{y^2}$

52. $\dfrac{(4x)^0y^4}{(3y)^2x^0}$ $\dfrac{y^2}{9}$

Scientific Notation

Exponents provide an efficient way of writing and computing with very large and very small numbers. For instance, a drop of water contains more than 33 billion billion molecules—that is, 33 followed by 18 zeros. It is convenient to write such numbers in **scientific notation**. This notation has the form $c \times 10^n$, where $1 \le c < 10$ and n is an integer. So, the number of molecules in a drop of water can be written in scientific notation as follows.

$$33{,}000{,}000{,}000{,}000{,}000{,}000 = 3.3 \times 10^{19}$$

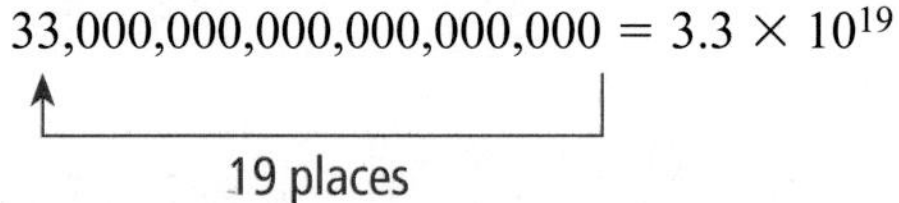

EXAMPLE 7 Writing in Scientific Notation

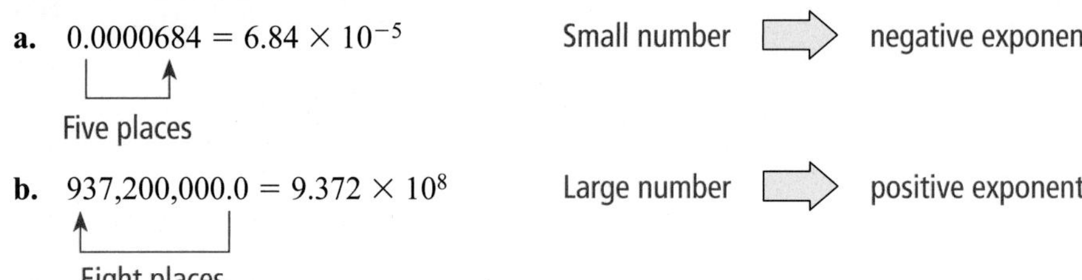

Additional Examples
Write each number in scientific notation.
a. 15,700 b. 0.0026
Answers:
a. 1.57×10^4 b. 2.6×10^{-3}

EXAMPLE 8 Writing in Decimal Notation

a. $2.486 \times 10^2 = 248.6$ Positive exponent ⇒ large number
Two places

b. $1.81 \times 10^{-6} = 0.00000181$ Negative exponent ⇒ small number
Six places

Additional Examples
Write each number in decimal notation.
a. 6.28×10^5 b. 3.05×10^{-4}
Answers:
a. 628,000 b. 0.000305

Exercises Within Reach®

Solutions in English & Spanish and tutorial videos at AlgebraWithinReach.com

Writing in Scientific Notation **In Exercises 53–58, write the number in scientific notation.**

53. 0.00031 3.1×10^{-4}

54. 0.0000000000692 6.92×10^{-11}

55. 3,600,000 3.6×10^6

56. 841,000,000,000 8.41×10^{11}

57. *Light Year:* 9,460,800,000,000 kilometers 9.4608×10^{12} kilometers

58. *Thickness of a Soap Bubble:* 0.0000001 meter 1×10^{-7} meter

Writing in Decimal Notation **In Exercises 59–64, write the number in decimal notation.**

59. 7.2×10^8 720,000,000

60. 7.413×10^{11} 741,300,000,000

61. 1.359×10^{-7} 0.0000001359

62. 8.6×10^{-9} 0.0000000086

63. *Interior Temperature of the Sun:* 1.5×10^7 degrees Celsius 15,000,000°C

64. *Width of an Air Molecule:* 9.0×10^{-9} meter 0.000000009 m

EXAMPLE 9 **Using Scientific Notation**

$$\frac{(2{,}400{,}000{,}000)(0.0000045)}{(0.00003)(1500)} = \frac{(2.4 \times 10^9)(4.5 \times 10^{-6})}{(3.0 \times 10^{-5})(1.5 \times 10^3)}$$

$$= \frac{(2.4)(4.5)(10^3)}{(4.5)(10^{-2})}$$

$$= (2.4)(10^5) = 2.4 \times 10^5$$

EXAMPLE 10 **Using Scientific Notation with a Calculator**

Use a calculator to evaluate each expression.

a. $65{,}000 \times 3{,}400{,}000{,}000$ **b.** $0.000000348 \div 870$

SOLUTION

a. 6.5 [EXP] 4 [×] 3.4 [EXP] 9 [=] Scientific calculator

6.5 [EE] 4 [×] 3.4 [EE] 9 [ENTER] Graphing calculator

The calculator display should read [2.21E 14], which implies that

$$(6.5 \times 10^4)(3.4 \times 10^9) = 2.21 \times 10^{14} = 221{,}000{,}000{,}000{,}000.$$

b. 3.48 [EXP] 7 [±] [÷] 8.7 [EXP] 2 [=] Scientific calculator

3.48 [EE] [(−)] 7 [÷] 8.7 [EE] 2 [ENTER] Graphing calculator

The calculator display should read [4E −10], which implies that

$$\frac{3.48 \times 10^{-7}}{8.7 \times 10^2} = 4.0 \times 10^{-10} = 0.0000000004.$$

Exercises Within Reach®

Solutions in English & Spanish and tutorial videos at AlgebraWithinReach.com

Using Scientific Notation **In Exercises 65–68, evaluate the expression by first writing each number in scientific notation.**

65. $(4{,}500{,}000)(2{,}000{,}000{,}000)$ 9×10^{15}

66. $(62{,}000{,}000)(0.0002)$ 1.24×10^4

67. $\dfrac{64{,}000{,}000}{0.00004}$ 1.6×10^{12}

68. $\dfrac{72{,}000{,}000{,}000}{0.00012}$ 6×10^{14}

Using Scientific Notation with a Calculator **In Exercises 69–72, use a calculator to evaluate the expression. Write the answer in scientific notation, $c \times 10^n$, with c rounded to two decimal places.**

69. $7900 \times 5{,}700{,}000{,}000$ 4.50×10^{13}

70. $0.0000000452 \div 767$ 5.89×10^{-11}

71. $\dfrac{(0.0000565)(2{,}850{,}000{,}000{,}000)}{0.00465}$ 3.46×10^{10}

72. $\dfrac{(3{,}450{,}000{,}000)(0.000125)}{(52{,}000{,}000)(0.000003)}$ 2.76×10^3

73. ***Masses of Earth and the Sun*** The masses of Earth and the Sun are approximately 5.98×10^{24} kilograms and 1.99×10^{30} kilograms, respectively. The mass of the Sun is approximately how many times that of Earth? 3.33×10^5

74. ***Light Year*** One light year (the distance light can travel in 1 year) is approximately 9.46×10^{15} meters. Approximate the time to the nearest minute for light to travel from the Sun to Earth if that distance is approximately 1.50×10^{11} meters. 1.59×10^{-5} year $\approx$ 8 minutes

Concept Summary: Using Scientific Notation

What

You can use exponents and **scientific notation** to write very large or very small numbers.

EXAMPLE

Write each number in scientific notation.

a. 240,000,000,000

b. 0.000000012

How

A number written in scientific notation has the form $c \times 10^n$, where $1 \le c < 10$ and n is an integer.

A positive exponent indicates a large number.

a. $240{,}000{,}000{,}000 = 2.4 \times 10^{11}$

11 places

A negative exponent indicates a small number.

b. $0.000000012 = 1.2 \times 10^{-8}$

8 places

Why

When multiplying or dividing very large or very small numbers, you can use scientific notation and the rules of exponents to find the products or quotients more efficiently.

Exercises Within Reach®

Worked-out solutions to odd-numbered exercises at AlgebraWithinReach.com

Concept Summary Check

75. *Scientific Notation* In the solution above, 240,000,000,000 is rewritten in the form $c \times 10^n$. What is the value of c? 2.4

76. *Scientific Notation* In the solution above, 0.000000012 is rewritten in the form $c \times 10^n$. What is the value of n? -8

77. *Writing* Explain how to write a small number in scientific notation. Write the number in the form $c \times 10^{-n}$, where $1 \le c < 10$, by moving the decimal point n places to the right.

78. *Writing* Explain how to write a large number in scientific notation. Write the number in the form $c \times 10^n$, where $1 \le c < 10$, by moving the decimal point n places to the left.

Extra Practice

Using Rules of Exponents **In Exercises 79–82, use the rules of exponents to simplify the expression.**

79. $\left[\dfrac{(-5u^3v)^2}{10u^2v}\right]^2$ $\dfrac{25u^8v^2}{4}$

80. $\left[\dfrac{-5(u^3v)^2}{10u^2v}\right]^2$ $\dfrac{u^8v^2}{4}$

81. $\dfrac{x^{2n+4}y^{4n}}{x^5y^{2n+1}}$ $x^{2n-1}y^{2n-1}$

82. $\dfrac{x^{6n}y^{n-7}}{x^{4n+2}y^5}$ $x^{2n-2}y^{n-12}$

Using Rules of Exponents **In Exercises 83–90, rewrite the expression using only positive exponents, and simplify. (Assume that any variables in the expression are nonzero.)**

83. $(2x^3y^{-1})^{-3}(4xy^{-6})$ $\dfrac{1}{2x^8y^3}$

84. $(ab)^{-2}(a^2b^2)^{-1}$ $\dfrac{1}{a^4b^4}$

85. $u^4(6u^{-3}v^0)(7v)^0$ $6u$

86. $x^5(3x^0y^4)(7y)^0$ $3x^5y^4$

87. $[(x^{-4}y^{-6})^{-1}]^2$ x^8y^{12}

88. $[(2x^{-3}y^{-2})^2]^{-2}$ $\dfrac{x^{12}y^8}{16}$

89. $\dfrac{(2a^{-2}b^4)^3b}{(10a^3b)^2}$ $\dfrac{2b^{11}}{25a^{12}}$

90. $\dfrac{(5x^2y^{-5})^{-1}}{2x^{-5}y^4}$ $\dfrac{x^3y}{10}$

Using Scientific Notation **In Exercises 91–96, evaluate the expression without using a calculator.**

91. $(2 \times 10^9)(3.4 \times 10^{-4})$ 6.8×10^5

92. $(6.5 \times 10^6)(2 \times 10^4)$ 1.3×10^{11}

93. $(5 \times 10^4)^2$ 2.5×10^9

94. $(4 \times 10^6)^3$ 6.4×10^{19}

95. $\dfrac{3.6 \times 10^{12}}{6 \times 10^5}$ 6×10^6

96. $\dfrac{2.5 \times 10^{-3}}{5 \times 10^2}$ 5×10^{-6}

97. ***Stars*** A study by Australian astronomers estimated the number of stars within range of modern telescopes to be 70,000,000,000,000,000,000,000. Write this number in scientific notation. (*Source:* The Australian National University) 7×10^{22} stars

98. ***Fishery Products*** In 2010, the total supply of edible fishery products in the United States weighed 1.2389×10^{10} pounds. Write this weight in decimal notation. (*Source:* NOAA Fisheries)
12,389,000,000 pounds

99. ***Federal Debt*** In 2010, the resident population of the United States was about 309 million people, and it would have cost each resident about \$43,783 to pay off the federal debt. Use these two numbers to approximate the federal debt in 2010. (*Source:* U.S. Census Bureau and U.S. Office of Management and Budget)
$\$1.35 \times 10^{13} = \$13{,}500{,}000{,}000{,}000$

100. ***Metal Expansion*** When the temperature of a 200-foot-long iron steam pipe is increased by 75°C, the length of the pipe will increase by $75(200)(1.1 \times 10^{-5})$ foot. Find this amount and write the answer in decimal notation. 0.165 foot

Explaining Concepts

101. ***Think About It*** Discuss whether you feel that using scientific notation to multiply or divide very large or very small numbers makes the process *easier* or *more difficult*. Support your position with an example.
Scientific notation makes it easier to multiply or divide very large or very small numbers because the properties of exponents make it more efficient.

102. ***Think About It*** You multiply an expression by a^5. The product is a^{12}. What was the original expression? Explain how you found your answer. The original expression was a^7. To find the expression, divide a^{12} by a^5, which yields a^7.

True or False? **In Exercises 103 and 104, determine whether the statement is true or false. Justify your answer.**

103. The value of $\frac{1}{3^{-3}}$ is less than 1.

False. $\frac{1}{3^{-3}} = 27$, which is greater than 1.

104. The expression 0.142×10^{10} is in scientific notation.
See Additional Answers.

Using Rules of Exponents **In Exercises 105–108, use the rules of exponents to explain why the statement is *false*.**

105. $a^m \cdot b^n = ab^{m+n}$ ✗
The product rule can be applied only to exponential expressions with the same base.

106. $(ab)^m = a^m + b^m$ ✗
The product-to-power rule applied to this expression is the *product* of each base raised to the same power.

107. $(a^m)^n = a^{m+n}$ ✗
The power-to-power rule applied to this expression raises the base to the *product* of the exponents.

108. $\frac{a^m}{a^n} = a^m - a^n$ ✗
The quotient rule applied to this expression is the base raised to the difference of the *exponents*.

Cumulative Review

In Exercises 109–112, simplify the expression by combining like terms.

109. $3x + 4x - x$ $6x$

110. $y - 3x + 4y - 2$ $-3x + 5y - 2$

111. $a^2 + 2ab - b^2 + ab + 4b^2$ $a^2 + 3ab + 3b^2$

112. $x^2 + 5x^2y - 3x^2y + 4x^2$ $5x^2 + 2x^2y$

In Exercises 113 and 114, sketch the graph of the system of linear inequalities.

113. $\begin{cases} x > 2 \\ x - y \le 0 \\ y < 0 \end{cases}$
No solution

114. $\begin{cases} x - 2y \le 6 \\ x + y \le 0 \\ y > 0 \end{cases}$
See Additional Answers.

5.2 Adding and Subtracting Polynomials

- Identify the degrees and leading coefficients of polynomials.
- Add and subtract polynomials using a horizontal format and a vertical format.
- Use polynomials to model and solve real-life problems.

Polynomials

Definition of a Polynomial in x

Let $a_n, a_{n-1}, \ldots, a_2, a_1, a_0$ be real numbers and let n be a *nonnegative integer*. A **polynomial in x** is an expression of the form

$$a_nx^n + a_{n-1}x^{n-1} + \cdots + a_2x^2 + a_1x + a_0$$

where $a_n \neq 0$. The polynomial is of **degree n**, and the number a_n is called the **leading coefficient**. The number a_0 is called the **constant term**.

A polynomial that is written in order of descending powers of the variable is said to be in **standard form**. A polynomial with only one term is a **monomial**. Polynomials with two unlike terms are called **binomials**, and those with three unlike terms are called **trinomials**.

EXAMPLE 1 Determining Degrees and Leading Coefficients

	Polynomial	*Standard Form*	*Degree*	*Leading Coefficient*
a.	$5x^2 - 2x^7 + 4 - 2x$	$-2x^7 + 5x^2 - 2x + 4$	7	-2
b.	$16 + x^3$	$x^3 + 16$	3	1
c.	10	10	0	10

Exercises Within Reach®

Solutions in English & Spanish and tutorial videos at AlgebraWithinReach.com

Determining the Degree and Leading Coefficient **In Exercises 1–10, write the polynomial in standard form, and determine its degree and leading coefficient.**

1. $4y + 16$ $4y + 16; 1; 4$
2. $50 - x$ $-x + 50; 1; -1$
3. $2x + x^2 - 6$ $x^2 + 2x - 6; 2; 1$
4. $12 + 4y - y^2$ $-y^2 + 4y + 12; 2; -1$
5. $3x - 10x^2 + 5 - 42x^3$ $-42x^3 - 10x^2 + 3x + 5; 3; -42$
6. $9x^3 - 2x^2 + 5x - 7$ $9x^3 - 2x^2 + 5x - 7; 3; 9$
7. $4 - 14t^4 + t^5 - 20t$ $t^5 - 14t^4 - 20t + 4; 5; 1$
8. $6t + 4t^5 - t^2 + 3$ $4t^5 - t^2 + 6t + 3; 5; 4$
9. -4 $-4; 0; -4$
10. 28 $28; 0; 28$

Adding and Subtracting Polynomials

EXAMPLE 2 Adding Polynomials Horizontally

Use a horizontal format to find each sum.

a. $(2x^3 + x^2 - 5) + (x^2 + x + 6)$ Original polynomials

$= (2x^3) + (x^2 + x^2) + (x) + (-5 + 6)$ Group like terms.

$= 2x^3 + 2x^2 + x + 1$ Combine like terms.

b. $(3x^2 + 2x + 4) + (3x^2 - 6x + 3) + (-x^2 + 2x - 4)$

$= (3x^2 + 3x^2 - x^2) + (2x - 6x + 2x) + (4 + 3 - 4)$

$= 5x^2 - 2x + 3$

EXAMPLE 3 Adding Polynomials Vertically

Use a vertical format to find each sum.

$$(5x^3 + 2x^2 - x + 7) + (3x^2 - 4x + 7) + (-x^3 + 4x^2 - 8)$$

SOLUTION

To use a vertical format, align the terms of the polynomials by their degrees.

$$\begin{array}{rrrr} 5x^3 & + 2x^2 & - x & + 7 \\ & 3x^2 & - 4x & + 7 \\ -x^3 & + 4x^2 & & - 8 \\ \hline 4x^3 & + 9x^2 & - 5x & + 6 \end{array}$$

Exercises Within Reach®

Solutions in English & Spanish and tutorial videos at AlgebraWithinReach.com

Adding Polynomials **In Exercises 11–20, use a horizontal format to find the sum.**

11. $5 + (2 + 3x)$ $3x + 7$

12. $(6 - 2x) + 4x$ $2x + 6$

13. $(2x^2 - 3) + (5x^2 + 6)$ $7x^2 + 3$

14. $(3x^2 + 2) + (4x^2 - 8)$ $7x^2 - 6$

15. $(5y + 6) + (4y^2 - 6y - 3)$ $4y^2 - y + 3$

16. $(3x^3 - 2x + 8) + (3x - 5)$ $3x^3 + x + 3$

17. $(2 - 8y) + (-2y^4 + 3y + 2)$ $-2y^4 - 5y + 4$

18. $(z^3 + 6z - 2) + (3z^2 - 6z)$ $z^3 + 3z^2 - 2$

19. $(x^3 + 9) + (2x^2 + 5) + (x^3 - 14)$ $2x^3 + 2x^2$

20. $(y^5 - 4y) + (3y - y^5) + (y^5 - 5)$ $y^5 - y - 5$

Adding Polynomials **In Exercises 21–28, use a vertical format to find the sum.**

21.
$$\begin{array}{rrr} 5x^2 & - 3x & + 4 \\ -3x^2 & & - 4 \\ \hline \end{array}$$
$2x^2 - 3x$

22.
$$\begin{array}{rrr} 3x^4 & - 2x^2 & - 9 \\ -5x^4 & + x^2 & \\ \hline \end{array}$$
$-2x^4 - x^2 - 9$

23. $(4x^3 - 2x^2 + 8x) + (4x^2 + x - 6)$

$4x^3 + 2x^2 + 9x - 6$

24. $(4x^3 + 8x^2 - 5x + 3) + (x^3 - 3x^2 - 7)$

$5x^3 + 5x^2 - 5x - 4$

25. $(5p^2 - 4p + 2) + (-3p^2 + 2p - 7)$

$2p^2 - 2p - 5$

26. $(16 - 32t) + (64 + 48t - 16t^2)$

$-16t^2 + 16t + 80$

27. $(2.5b - 3.6b^2) + (7.1 - 3.1b - 2.4b^2) + 6.6b^2$

$0.6b^2 - 0.6b + 7.1$

28. $(2.9n^3 - 6.1n) + 1.6n + (12.2 + 3.1n - 5.3n^3)$

$-2.4n^3 - 1.4n + 12.2$

To subtract one polynomial from another, *add the opposite*. You can do this by changing the sign of each term of the polynomial that is being subtracted and then adding the resulting like terms.

Additional Examples
Use a horizontal format to find the sum or difference.

a. $(2y^4 - 3y^2 + y - 6) + (-y^3 + 6y^2 + 8) + (5y^2 + 8y + 1)$

b. $(x^3 - 6x + 5) - (2x^3 - 3x^2 + 7x - 1)$

Answers:

a. $2y^4 - y^3 + 8y^2 + 9y + 3$

b. $-x^3 + 3x^2 - 13x + 6$

EXAMPLE 4 Subtracting Polynomials Horizontally

Use a horizontal format to subtract $x^3 + 2x^2 - x - 4$ from $3x^3 - 5x^2 + 3$.

SOLUTION

$(3x^3 - 5x^2 + 3) - (x^3 + 2x^2 - x - 4)$ Write original polynomials.

$= (3x^3 - 5x^2 + 3) + (-x^3 - 2x^2 + x + 4)$ Add the opposite.

$= (3x^3 - x^3) + (-5x^2 - 2x^2) + (x) + (3 + 4)$ Group like terms.

$= 2x^3 - 7x^2 + x + 7$ Combine like terms.

Be especially careful to use the correct signs when subtracting one polynomial from another. One of the most common mistakes in algebra is forgetting to change signs correctly when subtracting one expression from another. Here is an example.

Wrong sign

$(x^2 - 2x + 3) - (x^2 + 2x - 2) \neq x^2 - 2x + 3 - x^2 + 2x - 2$ Common error

Wrong sign

EXAMPLE 5 Subtracting Polynomials Vertically

Use a vertical format to find the difference.

$$(4x^4 - 2x^3 + 5x^2 - x + 8) - (3x^4 - 2x^3 + 3x - 4)$$

SOLUTION

$(4x^4 - 2x^3 + 5x^2 - x + 8)$
$-(3x^4 - 2x^3 \quad + 3x - 4)$

$$\begin{array}{r} 4x^4 - 2x^3 + 5x^2 - x + 8 \\ -3x^4 + 2x^3 \qquad\quad - 3x + 4 \\ \hline x^4 \qquad\quad + 5x^2 - 4x + 12 \end{array}$$

Exercises Within Reach ®

Solutions in English & Spanish and tutorial videos at AlgebraWithinReach.com

Subtracting Polynomials **In Exercises 29–34, use a horizontal format to find the difference.**

29. $(4 - y^3) - (4 + y^3)$ $-2y^3$

30. $(5y^4 - 2) - (3y^4 + 2)$ $2y^4 - 4$

31. $(3x^2 - 2x + 1) - (2x^2 + x - 1)$ $x^2 - 3x + 2$

32. $(5q^2 - 3q + 5) - (4q^2 - 3q - 10)$ $q^2 + 15$

33. $(6t^3 - 12) - (-t^3 + t - 2)$ $7t^3 - t - 10$

34. $(-2v^3 + v^2 - 4) - (-5v^3 - 10)$ $3v^3 + v^2 + 6$

Subtracting Polynomials **In Exercises 35–40, use a vertical format to find the difference.**

35. $(x^2 - x + 3) - (x - 2)$ $x^2 - 2x + 5$

36. $(3t^4 - 5t^2) - (-t^4 + 2t^2 - 14)$ $4t^4 - 7t^2 + 14$

37. $(2x^2 - 4x + 5) - (4x^2 + 5x - 6)$ $-2x^2 - 9x + 11$

38. $(4x^2 + 5x - 6) - (2x^2 - 4x + 5)$ $2x^2 + 9x - 11$

39. $(6x^4 - 3x^7 + 4) - (8x^7 + 10x^5 - 2x^4 - 12)$
$-11x^7 - 10x^5 + 8x^4 + 16$

40. $(13x^3 - 9x^2 + 4x - 5) - (5x^3 + 7x + 3)$
$8x^3 - 9x^2 - 3x - 8$

EXAMPLE 6 Combining Polynomials

Use a horizontal format to perform the indicated operations and simplify.

a. $(2x^2 - 7x + 2) - (4x^2 + 5x - 1) + (-x^2 + 4x + 4)$

b. $(-x^2 + 4x - 3) - [(4x^2 - 3x + 8) - (-x^2 + x + 7)]$

SOLUTION

a. $(2x^2 - 7x + 2) - (4x^2 + 5x - 1) + (-x^2 + 4x + 4)$

$= 2x^2 - 7x + 2 - 4x^2 - 5x + 1 - x^2 + 4x + 4$

$= (2x^2 - 4x^2 - x^2) + (-7x - 5x + 4x) + (2 + 1 + 4)$

$= -3x^2 - 8x + 7$

b. $(-x^2 + 4x - 3) - [(4x^2 - 3x + 8) - (-x^2 + x + 7)]$

$= (-x^2 + 4x - 3) - (4x^2 - 3x + 8 + x^2 - x - 7)$

$= (-x^2 + 4x - 3) - [(4x^2 + x^2) + (-3x - x) + (8 - 7)]$

$= (-x^2 + 4x - 3) - (5x^2 - 4x + 1)$

$= -x^2 + 4x - 3 - 5x^2 + 4x - 1$

$= (-x^2 - 5x^2) + (4x + 4x) + (-3 - 1)$

$= -6x^2 + 8x - 4$

Exercises Within Reach®

Solutions in English & Spanish and tutorial videos at AlgebraWithinReach.com

Combining Polynomials **In Exercises 41–56, perform the indicated operations and simplify.**

41. $-(2x^3 - 3) + (4x^3 - 2x)$ $2x^3 - 2x + 3$

42. $(2x^2 + 1) - (x^2 - 2x + 1)$ $x^2 + 2x$

43. $(4x^5 - 10x^3 + 6x) - (8x^5 - 3x^3 + 11) + (4x^5 + 5x^3 - x^2)$ $-2x^3 - x^2 + 6x - 11$

44. $(15 - 2y + y^2) + (3y^2 - 6y + 1) - (4y^2 - 8y + 16)$ 0

45. $(5n^2 + 6) + [(2n - 3n^2) - (2n^2 + 2n + 6)]$ 0

46. $(p^3 + 4) - [(p^2 + 4) + (3p - 9)]$ $p^3 - p^2 - 3p + 9$

47. $(8x^3 - 4x^2 + 3x) - [(x^3 - 4x^2 + 5) + (x - 5)]$ $7x^3 + 2x$

48. $(5x^4 - 3x^2 + 9) - [(2x^4 + x^3 - 7x^2) - (x^2 + 6)]$ $3x^4 - x^3 + 5x^2 + 15$

49. $3(4x^2 - 1) + (3x^3 - 7x^2 + 5)$ $3x^3 + 5x^2 + 2$

50. $(x^3 - 2x^2 - x) - 5(2x^3 + x^2 - 4x)$ $-9x^3 - 7x^2 + 19x$

51. $2(t^2 + 12) - 5(t^2 + 5) + 6(t^2 + 5)$ $3t^2 + 29$

52. $-10(v + 2) + 8(v - 1) - 3(v - 9)$ $-5v - 1$

53. $15v - 3(3v - v^2) + 9(8v + 3)$ $3v^2 + 78v + 27$

54. $9(7x^2 - 3x + 3) - 4(15x + 2) - (3x^2 - 7x)$ $60x^2 - 80x + 19$

55. $5s - [6s - (30s + 8)]$ $29s + 8$

56. $3x^2 - 2[3x + (9 - x^2)]$ $5x^2 - 6x - 18$

Applications

One commonly used second-degree polynomial is called a **position function**. This polynomial is a function of time and has the form

$$h(t) = -16t^2 + v_0t + s_0 \qquad \text{Position function}$$

where the height h is measured in feet and the time t is measured in seconds. This position function gives the height (above ground) of a free-falling object. The coefficient of t, v_0, is the initial velocity of the object, and the constant term, s_0, is the initial height of the object. The initial velocity is positive for an object projected upward, and the initial velocity is negative for an object projected downward.

Application

EXAMPLE 7 Free-Falling Object

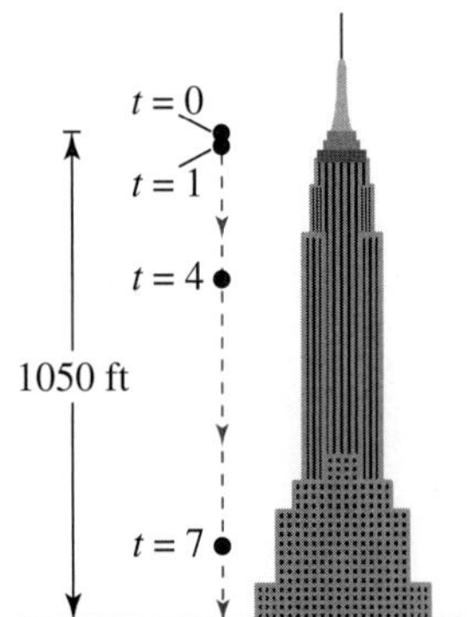

An object is thrown downward from the 86th floor observatory at the Empire State Building, which is 1050 feet high. The initial velocity is −15 feet per second. Use the position function

$$h(t) = -16t^2 - 15t + 1050$$

to find the height of the object when $t = 1$, $t = 4$, and $t = 7$. (See figure.)

SOLUTION

When $t = 1$, the height of the object is

$$h(1) = -16(1)^2 - 15(1) + 1050 \qquad \text{Substitute 1 for } t.$$

$$= -16 - 15 + 1050 \qquad \text{Simplify.}$$

$$= 1019 \text{ feet.}$$

When $t = 4$, the height of the object is

$$h(4) = -16(4)^2 - 15(4) + 1050 \qquad \text{Substitute 4 for } t.$$

$$= -256 - 60 + 1050 \qquad \text{Simplify.}$$

$$= 734 \text{ feet.}$$

When $t = 7$, the height of the object is

$$h(7) = -16(7)^2 - 15(7) + 1050 \qquad \text{Substitute 7 for } t.$$

$$= -784 - 105 + 1050 \qquad \text{Simplify.}$$

$$= 161 \text{ feet.}$$

Exercises Within Reach®

Solutions in English & Spanish and tutorial videos at AlgebraWithinReach.com

57. ***Free-Falling Object*** An object is thrown upward from a hot-air balloon that is 500 feet above the ground. The initial velocity is 40 feet per second. Use the position function

$$h(t) = -16t^2 + 40t + 500$$

to find the heights of the object when $t = 1$, $t = 5$, and $t = 6$.
524 feet; 300 feet; 164 feet

58. ***Free-Falling Object*** An object is dropped from a hot-air balloon that is 200 feet above the ground. Use the position function

$$h(t) = -16t^2 + 200$$

to find the heights of the object when $t = 1$, $t = 2$, and $t = 3$.
184 feet; 136 feet; 56 feet

Mike Blanchard/Shutterstock.com

Application **EXAMPLE 8** **Geometry: Modeling Perimeter**

Write an expression for the perimeter of the trapezoid shown below. Then find the perimeter when $y = 5$.

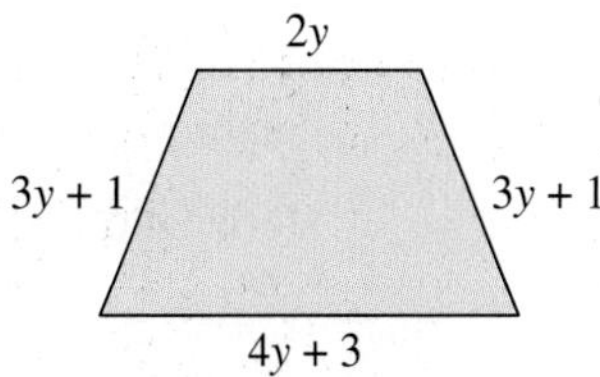

SOLUTION

To write an expression for the perimeter P of the trapezoid, find the sum of the lengths of the sides.

$$\begin{aligned} P &= (2y) + (3y + 1) + (3y + 1) + (4y + 3) \\ &= (2y + 3y + 3y + 4y) + (1 + 1 + 3) \\ &= 12y + 5 \end{aligned}$$

To find the perimeter when $y = 5$, substitute 5 for y in the expression for the perimeter.

$$\begin{aligned} P &= 12y + 5 \\ &= 12(5) + 5 \\ &= 60 + 5 \\ &= 65 \text{ units} \end{aligned}$$

Exercises Within Reach®

Solutions in English & Spanish and tutorial videos at AlgebraWithinReach.com

Geometry **In Exercises 59–62, write and simplify an expression for the perimeter of the figure. Then find the perimeter of the figure for the given value of x.**

59. $x = 6$

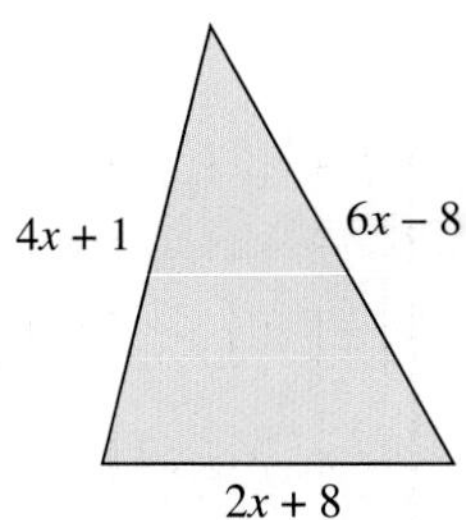

$12x + 1$; 73 units

60. $x = 8$

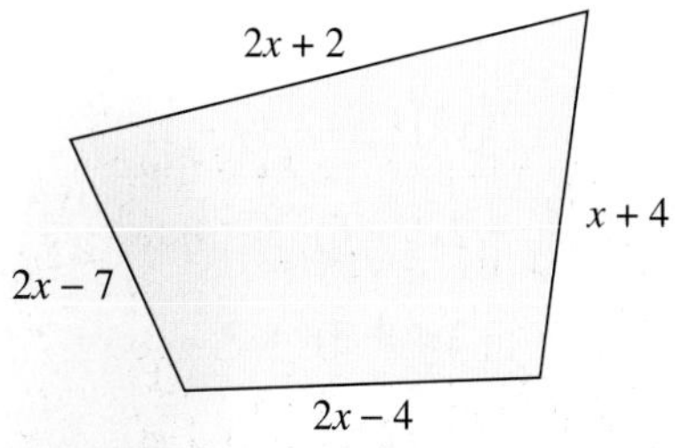

$7x - 5$; 51 units

61. $x = 4$

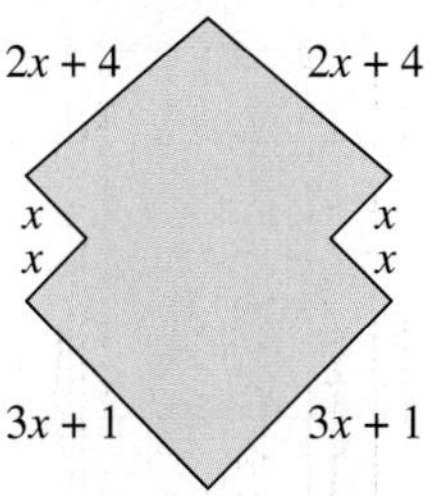

$14x + 10$; 66 units

62. $x = 5$

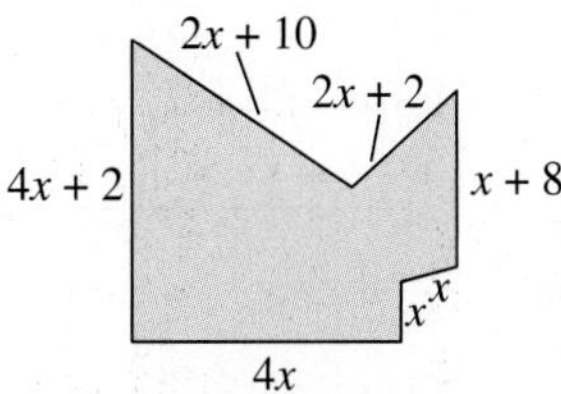

$15x + 22$; 97 units

Concept Summary: Adding and Subtracting Polynomials

What

The key to adding or subtracting **polynomials** is to recognize like terms. Like terms have the same **degree**.

EXAMPLE

Find the sum

$(3x^2 + 2x) + (x^2 - 4x)$.

How

To add two polynomials, use a horizontal or a vertical format to combine like terms.

Horizontal format:

$$\begin{aligned}&(3x^2 + 2x) + (x^2 - 4x)\\ &= (3x^2 + x^2) + (2x - 4x)\\ &= 4x^2 - 2x\end{aligned}$$

Vertical format:

$$\begin{array}{r} 3x^2 + 2x \\ x^2 - 4x \\ \hline 4x^2 - 2x \end{array}$$

Why

When you know how to add any two polynomials, you can subtract any two polynomials by *adding the opposite*. Be sure to change the sign of each term in the polynomial that is being subtracted.

Exercises Within Reach®

Worked-out solutions to odd-numbered exercises at AlgebraWithinReach.com

Concept Summary Check

63. ***Identifying Like Terms*** What are the like terms in the expression $(3x^2 + 2x) + (x^2 - 4x)$?
$3x^2$ and x^2, $2x$ and $-4x$

64. ***Combining Like Terms*** Explain how to find the sum $3x^2 + x^2$.
Find the coefficient of x^2 by adding the coefficients: $3x^2 + x^2 = (3 + 1)x^2 = 4x^2$.

65. ***Adding Polynomials*** What step in adding polynomials vertically corresponds to grouping like terms when adding polynomials horizontally? Aligning the terms of the polynomials by their degrees

66. ***Writing*** What does *adding the opposite* mean?
Subtracting

Extra Practice

Combining Polynomials **In Exercises 67–72, perform the indicated operation and simplify.**

67. $\left(\frac{2}{3}x^3 - 4x + 1\right) + \left(-\frac{3}{5} + 7x - \frac{1}{2}x^3\right)$ $\frac{1}{6}x^3 + 3x + \frac{2}{5}$

68. $\left(2 - \frac{1}{4}y^2 + y^4\right) + \left(\frac{1}{3}y^4 - \frac{3}{2}y^2 - 3\right)$ $\frac{4}{3}y^4 - \frac{7}{4}y^2 - 1$

69. $\left(\frac{1}{4}y^2 - 5y\right) - \left(12 + 4y - \frac{3}{2}y^2\right)$ $\frac{7}{4}y^2 - 9y - 12$

70. $\left(12 - \frac{2}{3}x + \frac{1}{2}x^2\right) - \left(x^3 + 3x^2 - \frac{1}{6}x\right)$
$-x^3 - \frac{5}{2}x^2 - \frac{1}{2}x + 12$

71. $(2x^{2r} - 6x^r - 3) + (3x^{2r} - 2x^r + 6)$ $5x^{2r} - 8x^r + 3$

72. $(6x^{2r} - 5x^r + 4) + (2x^{2r} + 2x^r + 3)$ $8x^{2r} - 3x^r + 7$

Geometry **In Exercises 73–76, find an expression for the area of the shaded region.**

73.

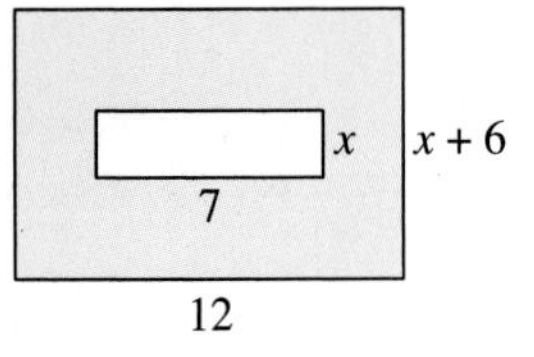

$5x + 72$

74.

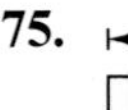

4x + 7
4
5
2x

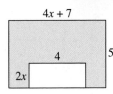

$12x + 35$

75.

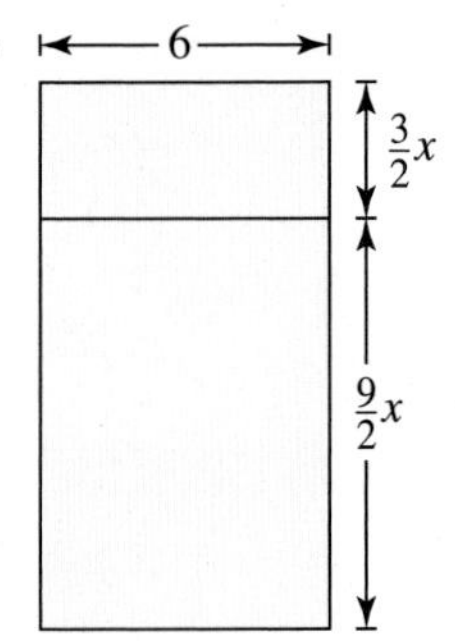

$36x$

76.

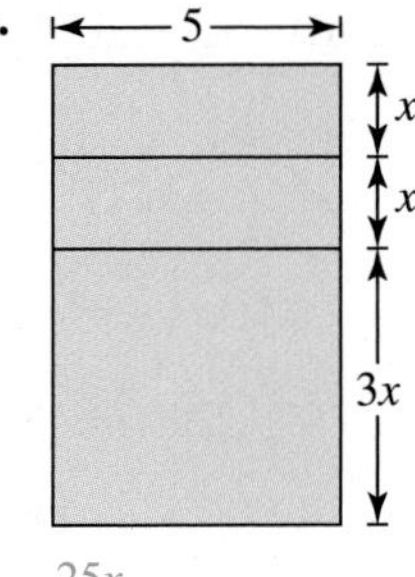

$25x$

Free-Falling Object **In Exercises 77–80, use the position function to determine whether the object was dropped, thrown upward, or thrown downward. Also determine the height (in feet) of the object at time $t = 0$.**

77. $h(t) = -16t^2 + 100$ Dropped; 100 feet

78. $h(t) = -16t^2 + 50t$ Thrown upward; 0 feet

79. $h(t) = -16t^2 + 40t + 12$ Thrown upward; 12 feet

80. $h(t) = -16t^2 - 28t + 150$ Thrown downward; 150 feet

81. ***Cost, Revenue, and Profit*** A manufacturer can produce and sell x radios per week. The total cost C (in dollars) of producing the radios is given by

$C = 8x + 15{,}000$

and the total revenue R is given by $R = 14x$. Find the profit P obtained by selling 5000 radios per week. (*Note:* $P = R - C$) \$15,000

82. ***Stopping Distance*** The total stopping distance of an automobile is the distance R (in feet) traveled during the driver's reaction time plus the distance B (in feet) traveled after the brakes are applied. In an experiment, these distances were related to the speed x of the automobile (in miles per hour) by the functions $R = 1.1x$ and $B = 0.0475x^2 - 0.001x$.

(a) Determine the polynomial that represents the total stopping distance T. $T = 0.0475x^2 + 1.099x$

(b) Estimate the total stopping distances when $x = 25$ and $x = 50$ miles per hour. 57.2 feet; 173.7 feet

Explaining Concepts

83. ***Writing*** Explain the difference between the degree of a term of a polynomial in x and the degree of a polynomial.
The degree of the term ax^k is k. The degree of a polynomial is the degree of its highest-degree term.

84. ***Structure*** What algebraic operation separates the terms of a polynomial? What operation separates the factors of a term? Addition (or subtraction) separates terms. Multiplication separates factors.

85. ***Reasoning*** Determine whether each statement is always, sometimes, or never true. Explain.

(a) A polynomial is a trinomial. Sometimes true; A polynomial is a trinomial when it has three terms.

(b) A trinomial is a polynomial.
Always true; Every trinomial is a polynomial.

86. ***Reasoning*** Can two third-degree polynomials be added to produce a second-degree polynomial? If so, give an example.
Yes. $(x^3 - 2x^2 + 4) + (-x^3 + x^2 + 3x) = -x^2 + 3x + 4$

87. ***Structure*** Is every trinomial a second-degree polynomial? If not, give an example of a trinomial that is not a second-degree polynomial. No. $x^3 + 2x + 3$

88. ***Writing*** Describe the method for subtracting polynomials. To subtract one polynomial from another, add the opposite. This is done by changing the sign of each term of the polynomial that is being subtracted and then adding like terms.

Cumulative Review

In Exercises 89–98, consider the following matrices.

$$A = \begin{bmatrix} 3 & 2 \\ 1 & 4 \end{bmatrix},\quad B = \begin{bmatrix} 6 & -3 & \vdots & 0 \\ 2 & 1 & \vdots & 4 \end{bmatrix},$$

$$C = \begin{bmatrix} 2 & 0 & -2 \\ 4 & -1 & 1 \\ 0 & 4 & -6 \end{bmatrix}$$

89. What is the order of B? 2×3

90. What is the order of C? 3×3

91. Is A a square matrix? Yes

92. Is B a square matrix? No

93. Write B in row-echelon form. $\begin{bmatrix} 1 & -\frac{1}{2} & 0 \\ 0 & 1 & 2 \end{bmatrix}$

94. Write C in row-echelon form. $\begin{bmatrix} 1 & 0 & -1 \\ 0 & 1 & -5 \\ 0 & 0 & 1 \end{bmatrix}$

95. Write the system of linear equations represented by B.
$\begin{cases} 6x - 3y = 0 \\ 2x + y = 4 \end{cases}$

96. Solve the system of linear equations obtained in Exercise 95. (1, 2)

97. Find $|A|$.
10

98. Find $|C|$.
-28

5.3 Multiplying Polynomials

- Use the Distributive Property and the FOIL Method to multiply polynomials.
- Use special product formulas to multiply two binomials.
- Use multiplication of polynomials in application problems.

Multiplying Polynomials

EXAMPLE 1 Finding Products with Monomial Multipliers

Multiply the polynomial by the monomial.

a. $(2x - 7)(3x)$ **b.** $4x^2(3x - 2x^3 + 1)$

SOLUTION

a. $(2x - 7)(3x) = 2x(3x) - 7(3x)$ Distributive Property

$= 6x^2 - 21x$ Rules of exponents

b. $4x^2(3x - 2x^3 + 1)$

$= 4x^2(3x) - 4x^2(2x^3) + 4x^2(1)$ Distributive Property

$= 12x^3 - 8x^5 + 4x^2$ Rules of exponents

$= -8x^5 + 12x^3 + 4x^2$ Standard form

EXAMPLE 2 Finding Products with Negative Monomial Multipliers

a. $(-x)(5x^2 - x) = (-x)(5x^2) - (-x)(x)$ Distributive Property

$= -5x^3 + x^2$ Rules of exponents

b. $(-2x^2)(x^2 - 8x + 4)$

$= (-2x^2)(x^2) - (-2x^2)(8x) + (-2x^2)(4)$ Distributive Property

$= -2x^4 + 16x^3 - 8x^2$ Rules of exponents

Exercises Within Reach®

Finding a Product with a Monomial Multiplier **In Exercises 1–12, multiply the polynomial by the monomial.**

1. $(3x - 4)(6x)$ $18x^2 - 24x$

2. $(5x + 3)(4x)$ $20x^2 + 12x$

3. $2y(5 - y)$ $-2y^2 + 10y$

4. $4z(6 - 3z)$ $-12z^2 + 24z$

5. $4x(2x^2 - 3x + 5)$ $8x^3 - 12x^2 + 20x$

6. $3y(-3y^2 + 7y - 3)$ $-9y^3 + 21y^2 - 9y$

7. $(-3x)(2 + 5x)$ $-15x^2 - 6x$

8. $(-7t)(6 - 3t)$ $21t^2 - 42t$

9. $-2m^2(7 - 4m + 2m^2)$ $-4m^4 + 8m^3 - 14m^2$

10. $-3a^2(8 - 2a - a^2)$ $3a^4 + 6a^3 - 24a^2$

11. $-x^3(x^4 - 2x^3 + 5x - 6)$ $-x^7 + 2x^6 - 5x^4 + 6x^3$

12. $-y^4(7y^3 - 4y^2 + y - 4)$ $-7y^7 + 4y^6 - y^5 + 4y^4$

To multiply two *binomials*, you can use both (left and right) forms of the Distributive Property. For example, if you treat the binomial $(2x + 7)$ as a single quantity, you can multiply $(3x - 2)$ by $(2x + 7)$ as follows.

$$(3x - 2)(2x + 7) = 3x(2x + 7) - 2(2x + 7)$$
$$= (3x)(2x) + (3x)(7) - (2)(2x) - (2)(7)$$
$$= 6x^2 + 21x - 4x - 14$$

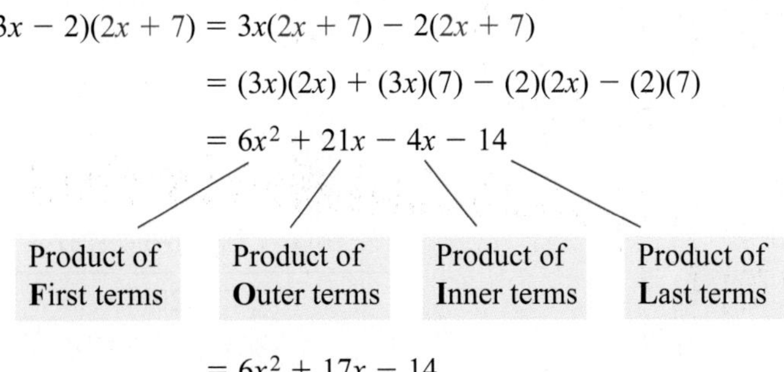

$$= 6x^2 + 17x - 14$$

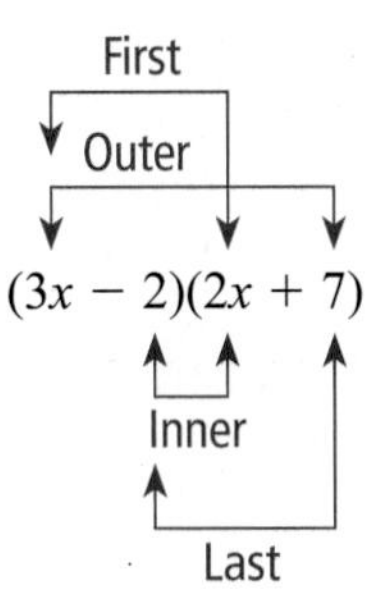

The FOIL Method

With practice, you should be able to multiply two binomials without writing out all of the steps shown above. In fact, the four products in the boxes above suggest that you can write the product of two binomials in just one step. This is called the **FOIL Method**. Note that the words *First*, *Outer*, *Inner*, and *Last* refer to the positions of the terms in the original product, as shown at the left.

EXAMPLE 3 Using the FOIL Method

Use the FOIL Method to find the product.

a. $(x - 3)(x + 3)$

b. $(3x + 4)(2x + 1)$

SOLUTION

a. $(x - 3)(x + 3) = \overbrace{x^2}^{F} + \overbrace{3x}^{O} - \overbrace{3x}^{I} - \overbrace{9}^{L}$

$= x^2 - 9$ Combine like terms.

b. $(3x + 4)(2x + 1) = \overbrace{6x^2}^{F} + \overbrace{3x}^{O} + \overbrace{8x}^{I} + \overbrace{4}^{L}$

$= 6x^2 + 11x + 4$ Combine like terms.

Additional Examples
Use the FOIL Method to find each product.

a. $(2y + 3)(5y - 1)$

b. $(6a - 1)(2a - 3)$

Answers:

a. $10y^2 + 13y - 3$

b. $12a^2 - 20a + 3$

Exercises Within Reach®

Solutions in English & Spanish and tutorial videos at AlgebraWithinReach.com

Using the FOIL Method **In Exercises 13–22, use the FOIL Method to find the product.**

13. $(x + 2)(x + 4)$ $x^2 + 6x + 8$

14. $(x - 5)(x - 3)$ $x^2 - 8x + 15$

15. $(x - 3)(x - 3)$ $x^2 - 6x + 9$

16. $(x + 7)(x - 1)$ $x^2 + 6x - 7$

17. $(2x - 3)(x + 5)$ $2x^2 + 7x - 15$

18. $(3x + 1)(x - 4)$ $3x^2 - 11x - 4$

19. $(5x - 2)(2x - 6)$ $10x^2 - 34x + 12$

20. $(4x + 7)(3x + 7)$ $12x^2 + 49x + 49$

21. $(2x^2 - 1)(x + 2)$
$2x^3 + 4x^2 - x - 2$

22. $(4x - 5)(-3x^2 + 2)$
$-12x^3 + 15x^2 + 8x - 10$

EXAMPLE 4 Multiplying Polynomials (Vertical Format)

Write the polynomials in standard form and use a vertical format to find the product.

$(4x^2 + x - 2)(5 + 3x - x^2)$

SOLUTION

$$\begin{array}{rrrrrl}
 & & 4x^2 & + \; x & - \; 2 & \text{Standard form} \\
\times & & -x^2 & + \; 3x & + \; 5 & \text{Standard form} \\
\hline
 & & 20x^2 & + \; 5x & - \; 10 & \Leftarrow \; 5(4x^2 + x - 2) \\
 & 12x^3 & + \; 3x^2 & - \; 6x & & \Leftarrow \; 3x(4x^2 + x - 2) \\
-4x^4 & - \; x^3 & + \; 2x^2 & & & \Leftarrow \; -x^2(4x^2 + x - 2) \\
\hline
-4x^4 & + \; 11x^3 & + \; 25x^2 & - \; x & - \; 10 & \text{Combine like terms.}
\end{array}$$

EXAMPLE 5 An Area Model for Multiplying Polynomials

Show that $(2x + 1)(x + 2) = 2x^2 + 5x + 2$.

SOLUTION

An appropriate area model for demonstrating the multiplication of two binomials would be $A = lw$, the area formula for a rectangle. Think of a rectangle whose sides are $x + 2$ and $2x + 1$. The area of this rectangle is

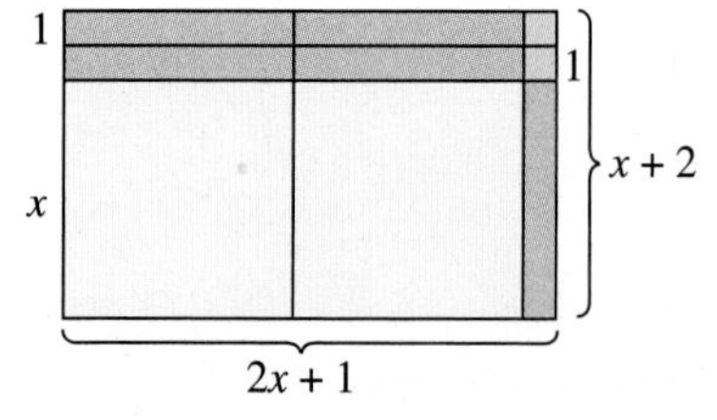

$(2x + 1)(x + 2)$. Area = (length)(width)

Another way to find the area is to add the areas of the rectangular parts, as shown at the left. There are two squares whose sides are x, five rectangles whose sides are x and 1, and two squares whose sides are 1. The total area of these nine rectangles is

$2x^2 + 5x + 2$. Area = sum of rectangular areas

Because each method must produce the same area, you can conclude that

$(2x + 1)(x + 2) = 2x^2 + 5x + 2$.

Exercises Within Reach®

Solutions in English & Spanish and tutorial videos at AlgebraWithinReach.com

Multiplying Polynomials **In Exercises 23–28, write the polynomials in standard form and use a vertical format to find the product.**

23. $(2x + 1)(9 - 14x + 7x^2)$ $14x^3 - 21x^2 + 4x + 9$

24. $(2x^2 + 3)(9 - 6x^2 + 4x^4)$ $8x^6 + 27$

25. $(2x - x^2 - 1)(1 + 2x)$ $-2x^3 + 3x^2 - 1$

26. $(2s^2 + 6 - 5s)(3s - 4)$ $6s^3 - 23s^2 + 38s - 24$

27. $(t^2 + t - 2)(t^2 + 2 - t)$ $t^4 - t^2 + 4t - 4$

28. $(y^2 + 3y + 5)(2y^2 - 1 - 3y)$ $2y^4 + 3y^3 - 18y - 5$

An Area Model for Multiplying Polynomials **In Exercises 29–32, draw an area model that represents the polynomial product.** See Additional Answers.

29. $(x + 2)(x + 3) = x^2 + 5x + 6$

30. $(2x + 2)(x + 1) = 2x^2 + 4x + 2$

31. $(x + 3)(2x + 1) = 2x^2 + 7x + 3$

32. $(x + 1)(3x + 2) = 3x^2 + 5x + 2$

Special Products

Special Products

Let u and v be real numbers, variables, or algebraic expressions. Then the following formulas are true.

Special Product	*Example*
Sum and Difference of Two Terms	
$(u+v)(u-v)=u^2-v^2$	$(3x-4)(3x+4)=9x^2-16$
Square of a Binomial	
$(u+v)^2=u^2+2uv+v^2$	$(2x+5)^2=4x^2+2(2x)(5)+25$
	$=4x^2+20x+25$
$(u-v)^2=u^2-2uv+v^2$	$(x-6)^2=x^2-2(x)(6)+36$
	$=x^2-12x+36$

Study Tip

When squaring a binomial, note that the resulting middle term, $\pm 2uv$, is always twice the product of the two terms.

EXAMPLE 6 Finding Special Products

a. $(3x-2)(3x+2)=(3x)^2-2^2$ — $(u+v)(u-v)=u^2-v^2$

$=9x^2-4$ — Simplify.

b. $(6+5x)(6-5x)=6^2-(5x)^2=36-25x^2$

c. $(2x-7)^2=(2x)^2-2(2x)(7)+7^2$ — Square of a binomial

$=4x^2-28x+49$ — Simplify.

d. $(x+4)^3=(x+4)^2(x+4)$ — Rules of exponents

$=(x^2+8x+16)(x+4)$ — Square of a binomial

$=x^2(x+4)+8x(x+4)+16(x+4)$ — Distributive Property

$=x^3+4x^2+8x^2+32x+16x+64$ — Distributive Property

$=x^3+12x^2+48x+64$ — Combine like terms.

Exercises Within Reach®

Solutions in English & Spanish and tutorial videos at AlgebraWithinReach.com

Finding a Special Product **In Exercises 33–48, use a special product formula to find the product.**

33. $(x-8)(x+8)$ x^2-64

34. $(x-5)(x+5)$ x^2-25

35. $(2+7y)(2-7y)$ $4-49y^2$

36. $(4+3z)(4-3z)$ $16-9z^2$

37. $(6x-9y)(6x+9y)$ $36x^2-81y^2$

38. $(8x-5y)(8x+5y)$ $64x^2-25y^2$

39. $(x+5)^2$ $x^2+10x+25$

40. $(x+2)^2$ x^2+4x+4

41. $(x-10)^2$ $x^2-20x+100$

42. $(u-7)^2$ $u^2-14u+49$

43. $(2x+5)^2$ $4x^2+20x+25$

44. $(3x+8)^2$ $9x^2+48x+64$

45. $(k+5)^3$ $k^3+15k^2+75k+125$

46. $(y-2)^3$ $y^3-6y^2+12y-8$

47. $(u+v)^3$ $u^3+3u^2v+3uv^2+v^3$

48. $(u-v)^3$ $u^3-3u^2v+3uv^2-v^3$

Applications

Application EXAMPLE 7 **Geometry: Area and Volume**

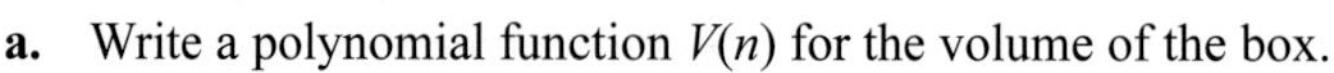

The closed rectangular box shown at the left has sides whose lengths (in inches) are consecutive integers.

a. Write a polynomial function $V(n)$ for the volume of the box.

b. What is the volume when the length of the shortest side is 4 inches?

c. Write a polynomial function $A(n)$ for the area of the base of the box.

d. Write a polynomial function for the area of the base when n is increased by 3. That is, find $A(n + 3)$.

SOLUTION

a. $V(n) = n(n + 1)(n + 2)$

$= n(n^2 + 3n + 2)$

$= n^3 + 3n^2 + 2n$

b. When $n = 4$, the volume of the box is

$V(4) = (4)^3 + 3(4)^2 + 2(4)$

$= 64 + 48 + 8$

$= 120$ cubic inches.

c. $A(n) = \text{(Length)(Width)} = n(n + 1) = n^2 + n$

d. $A(n + 3) = (n + 3)^2 + (n + 3)$

$= n^2 + 6n + 9 + n + 3$

$= n^2 + 7n + 12$

Exercises Within Reach®

Solutions in English & Spanish and tutorial videos at AlgebraWithinReach.com

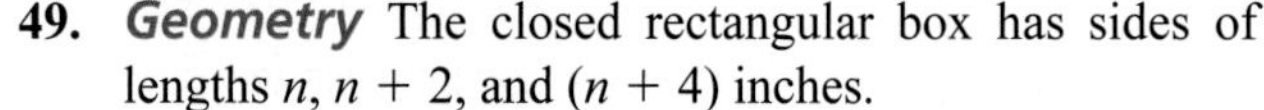

49. ***Geometry*** The closed rectangular box has sides of lengths n, $n + 2$, and $(n + 4)$ inches.

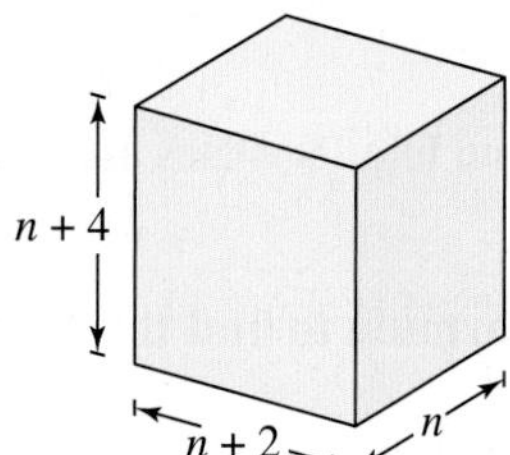

(a) Write a polynomial function $V(n)$ for the volume of the box. $V(n) = n^3 + 6n^2 + 8n$

(b) What is the volume when the length of the shortest side is 3 inches? 105 cubic inches

(c) Write a polynomial function $A(n)$ for the area of the base of the box. $A(n) = n^2 + 2n$

(d) Write a polynomial function for the area of the base when n is increased by 5. That is, find $A(n + 5)$.
$A(n + 5) = n^2 + 12n + 35$

50. ***Geometry*** The closed rectangular box has sides of lengths $2n - 2$, $2n + 2$, and $2n$ inches.

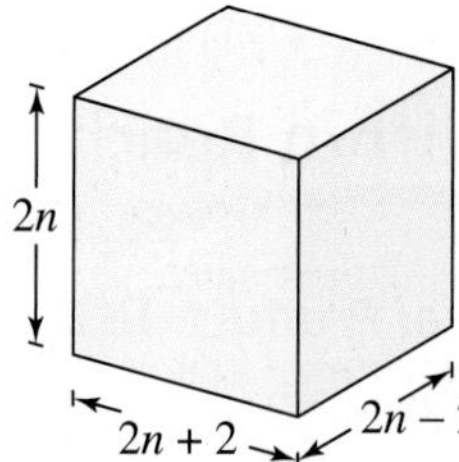

(a) Write a polynomial function $V(n)$ for the volume of the box. $V(n) = 8n^3 - 8n$

(b) What is the volume when the length of the shortest side is 4 inches? 192 cubic inches

(c) Write a polynomial function $A(n)$ for the area of the base of the box. $A(n) = 4n^2 - 4$

(d) Write a polynomial function for the area of the base when n is increased by 4. That is, find $A(n + 4)$.
$A(n + 4) = 4n^2 + 32n + 60$

Application **EXAMPLE 8** **Revenue**

A software manufacturer has determined that the demand for its new video game is given by the equation

$$p = 50 - 0.001x$$

where p is the price of the game (in dollars) and x is the number of units sold. The total revenue R from selling x units of a product is given by the equation $R = xp$.

a. Find the revenue equation for the video game.

b. Find the revenue when 3000 units are sold.

SOLUTION

a.

$R = xp$	Revenue equation
$= x(50 - 0.001x)$	Substitute for p.
$= 50x - 0.001x^2$	Distributive Property

So, the revenue equation for the video game is $R = 50x - 0.001x^2$.

b. To find the revenue when 3000 units are sold, substitute 3000 for x in the revenue equation

$R = 50x - 0.001x^2$	Revenue equation
$= 50(3000) - 0.001(3000)^2$	Substitute 3000 for x.
$= 141{,}000$	Simplify.

So, the revenue when 3000 units are sold is \$141,000.

Exercises Within Reach®

Solutions in English & Spanish and tutorial videos at AlgebraWithinReach.com

51. ***Revenue*** A shop owner has determined that the demand for his daily newspapers is given by the equation

$$p = 175 - 0.02x$$

where p is the price of the newspaper (in cents) and x is the number of papers sold. The total revenue R from selling x units of a product is given by the equation $R = xp$.

(a) Find the revenue equation for the shop owner's daily newspaper sales. $R = -0.02x^2 + 175x$

(b) Find the revenue when 3000 newspapers are sold. \$3450

52. ***Revenue*** A supermarket has determined that the demand for its apple pies is given by the equation

$$p = 20 - 0.015x$$

where p is the price of the apple pie (in dollars) and x is the number of pies sold. The total revenue R from selling x units of a product is given by the equation $R = xp$.

(a) Find the revenue equation for the supermarket's apple pie sales. $R = -0.015x^2 + 20x$

(b) Find the revenue when 50 apple pies are sold. \$962.50

Concept Summary: Multiplying Two Binomials

What

Here are two methods you can use to multiply binomials.

1. The Distributive Property
2. The **FOIL Method**

Each method applies many of the rules for simplifying algebraic expressions.

EXAMPLE

Find the product $(x - y)(x + 2)$.

How

To use the FOIL Method, add the products of the following pairs of terms.

1. **F**irst terms
2. **O**uter terms
3. **I**nner terms
4. **L**ast terms

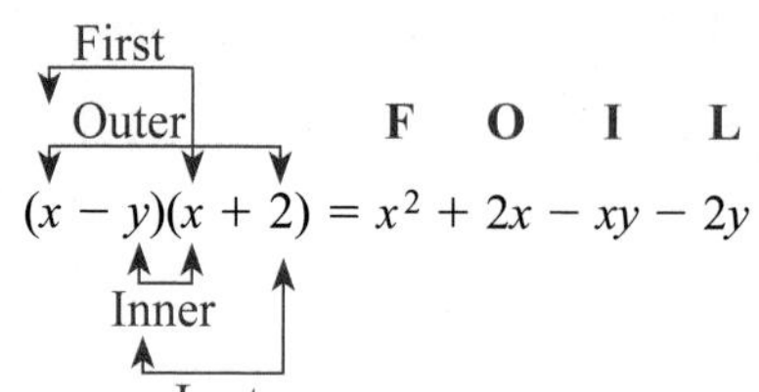

Why

When multiplying two binomials, you can use the FOIL Method. It is a useful tool to help you remember all the products.

To multiply any two polynomials, use a vertical format or a horizontal format to apply the Distributive Property.

Exercises Within Reach®

Worked-out solutions to odd-numbered exercises at AlgebraWithinReach.com

Concept Summary Check

53. ***Identifying Binomials*** Identify the binomials in the example above. $x - y$ and $x + 2$

54. ***Identifying Terms*** Identify the first terms, outer terms, inner terms, and last terms in the example above. First: x, x; Outer: $2, x$; Inner: $x, -y$; Last: $2, -y$

55. ***A Special Product*** Describe a way to multiply $(x + y)$ and $(x - y)$ without using the FOIL Method. Use the special product formula for the sum and difference of two terms.

56. ***Alternate Methods*** Describe a way to multiply any two polynomials. *Sample answers:* Use a vertical format; Use a horizontal format to apply the Distributive Property.

Extra Practice

Using a Horizontal Format **In Exercises 57–62, use a horizontal format to find the product.**

57. $(x - 1)(x^2 - 4x + 6)$ $x^3 - 5x^2 + 10x - 6$

58. $(z + 2)(z^2 - 4z + 4)$ $z^3 - 2z^2 - 4z + 8$

59. $(3a + 2)(a^2 + 3a + 1)$ $3a^3 + 11a^2 + 9a + 2$

60. $(2t + 3)(t^2 - 5t + 1)$ $2t^3 - 7t^2 - 13t + 3$

61. $(2u^2 + 3u - 4)(4u + 5)$ $8u^3 + 22u^2 - u - 20$

62. $(2x^2 - 5x + 1)(3x - 4)$ $6x^3 - 23x^2 + 23x - 4$

Multiplying Polynomials **In Exercises 63–70, perform the indicated operation(s).**

63. $u^2v(3u^4 - 5u^2v + 6uv^3)$ $3u^6v - 5u^4v^2 + 6u^3v^4$

64. $ab^3(2a - 9a^2b + 3b)$ $2a^2b^3 - 9a^3b^4 + 3ab^4$

65. $(2t - 1)(t + 1) + (2t - 5)(t - 1)$ $4t^2 - 6t + 4$

66. $(s - 3t)(s + t) - (s - 3t)(s - t)$ $2st - 6t^2$

67. $[u - (v - 3)][u + (v - 3)]$ $u^2 - v^2 + 6v - 9$

68. $[z + (y + 1)][z - (y + 1)]$ $z^2 - y^2 - 2y - 1$

69. $(6x^m - 5)(2x^{2m} - 3)$ $12x^{3m} - 10x^{2m} - 18x^m + 15$

70. $(x^{3m} - x^{2m})(x^{2m} + 2x^{4m})$ $2x^{7m} - 2x^{6m} + x^{5m} - x^{4m}$

Geometry **In Exercises 71–74, write an expression for the area of the shaded region of the figure. Then simplify the expression.**

71. 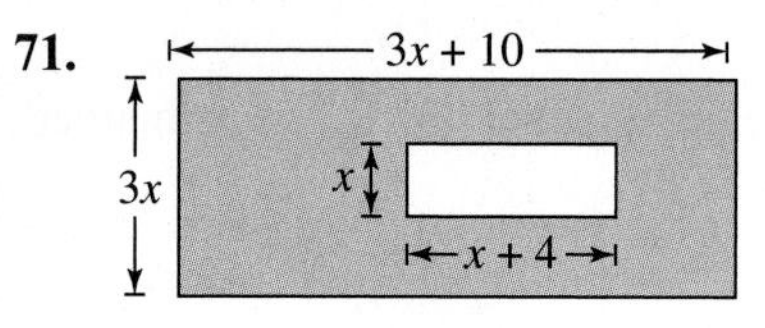

$3x(3x + 10) - x(x + 4)$; $8x^2 + 26x$

72. 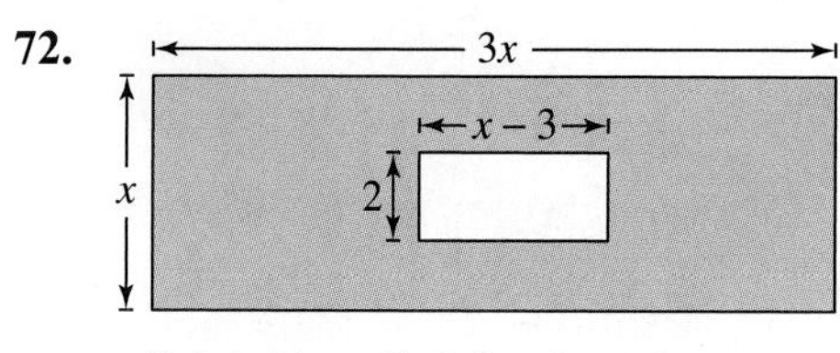

$x(3x) - 2(x - 3)$; $3x^2 - 2x + 6$

73.

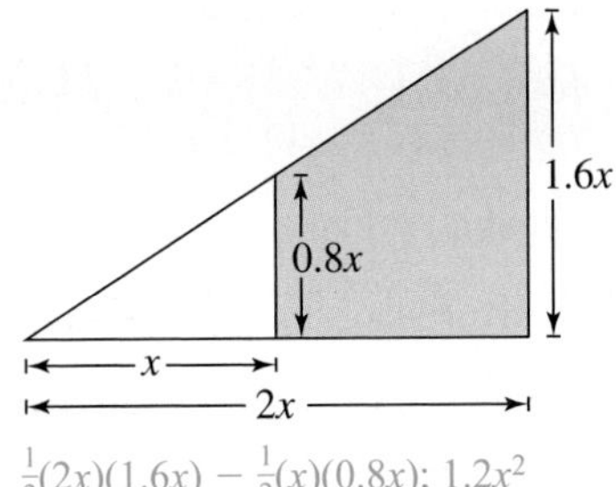

$\frac{1}{2}(2x)(1.6x) - \frac{1}{2}(x)(0.8x)$; $1.2x^2$

74.

x + 5
0.8x
2x

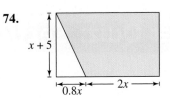

$(2x + 0.8x)(x + 5) - \frac{1}{2}(0.8x)(x + 5)$; $2.4x^2 + 12x$

75. ***Geometric Modeling*** Write the polynomial product represented by the area model. What is the name of this special product?
$(x + a)^2 = x^2 + 2ax + a^2$; square of a binomial

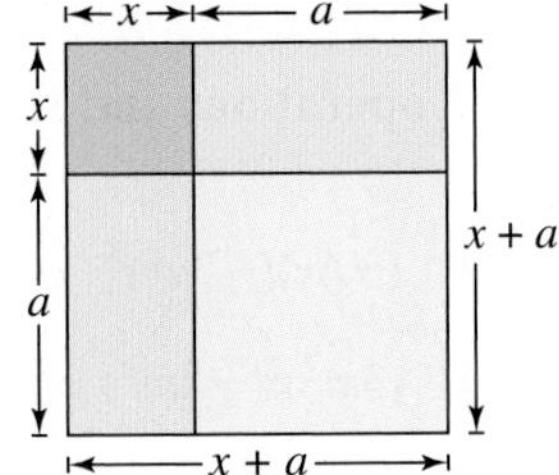

76. ***Geometric Modeling*** Draw an area model that represents the FOIL Method for the product $(x + a)(x + b)$.
See Additional Answers.

77. ***Geometry*** The length of a rectangle is $\frac{5}{2}$ times its width, and its width is $2x$. Find expressions for (a) the perimeter and (b) the area of the rectangle.
(a) $14x$ (b) $10x^2$

78. ***Geometry*** The base of a triangle is $3x$ and its height is $x + 8$. Find an expression for the area A of the triangle.
$A = \frac{3}{2}x^2 + 12x$

79. ***Compound Interest*** After 2 years, an investment of \$5000 compounded annually at interest rate r (in decimal form) will yield the amount $5000(1 + r)^2$. Find this product. $5000r^2 + 10{,}000r + 5000$

80. ***Compound Interest*** After 2 years, an investment of \$1000 compounded annually at an interest rate of 5.5% will yield the amount $1000(1 + 0.055)^2$. Find this product.
\$1113.03

81. ***Repeated Reasoning*** Find each product.

(a) $(x - 1)(x + 1)$ $x^2 - 1$

(b) $(x - 1)(x^2 + x + 1)$ $x^3 - 1$

(c) $(x - 1)(x^3 + x^2 + x + 1)$ $x^4 - 1$

From the pattern formed by these products, can you predict the result of $(x - 1)(x^4 + x^3 + x^2 + x + 1)$?
Yes, $x^5 - 1$

82. ***Verifying Special Product Formulas*** Use the FOIL Method to verify each of the following.

(a) $(x + y)^2 = x^2 + 2xy + y^2$ Proof

(b) $(x - y)^2 = x^2 - 2xy + y^2$ Proof

(c) $(x - y)(x + y) = x^2 - y^2$ Proof

Explaining Concepts

83. ***Writing*** Explain why an understanding of the Distributive Property is essential in multiplying polynomials. See Additional Answers.

84. ***Think About It*** Explain why the square of a binomial has three terms, whereas the product of two binomials found using the FOIL Method has four terms.
See Additional Answers.

85. ***Reasoning*** What is the degree of the product of two polynomials in x of degrees m and n? $m + n$

86. ***True or False?*** Determine whether the statement is true or false. Justify your answer.

(a) The product of two monomials is a monomial.
True. $(a^b)(c^d) = a^bc^d$, which is a monomial.

(b) The product of any two binomials is a binomial.
False. $(x + 2)(x - 3) = x^2 - x - 6$

Cumulative Review

In Exercises 87–92, graph the equation. Use the Vertical Line Test to determine whether y is a function of x. See Additional Answers.

87. $y = 5 - \frac{1}{2}x$

88. $y = \frac{3}{2}x - 2$

89. $y - 4x + 1 = 0$

90. $5x + 3y - 9 = 0$

91. $|y| + 2x = 0$

92. $|y| = 3 - x$

In Exercises 93–98, evaluate the expression.

93. 2^{-5} $\frac{1}{32}$

94. $\frac{1}{5^{-2}}$ 25

95. $\frac{4^2}{4^{-1}}$ 64

96. $\frac{3^{-5}}{3^{-6}}$ 3

97. $(6^3 + 3^{-6})^0$ 1

98. $2^{-4} + 16^{-1}$ 0

Mid-Chapter Quiz: Sections 5.1–5.3

Solutions in English & Spanish and tutorial videos at AlgebraWithinReach.com

Take this quiz as you would take a quiz in class. After you are done, check your work against the answers in the back of the book.

1. Write the polynomial $3 - 2x + 4x^3 - 2x^4$ in standard form and determine its degree and leading coefficient. $-2x^4 + 4x^3 - 2x + 3$; 4; -2

2. Write each number in scientific notation.

   (a) 0.00000054 5.4×10^{-7} (b) 664,000,000 6.64×10^8

In Exercises 3–22, perform the indicated operations and simplify (use only positive exponents).

3. $(5y^2)(-y^4)(2y^3)$ $-10y^9$
4. $(-6x)(-3x^2)^2$ $-54x^5$
5. $(-5n^2)(-2n^3)$ $10n^5$
6. $(3m^3)^2(-2m^4)$ $-18m^{10}$
7. $\dfrac{6x^{-7}}{(-2x^2)^{-3}}$ $-\dfrac{48}{x}$
8. $\left(\dfrac{4y^2}{5x}\right)^{-2}$ $\dfrac{25x^2}{16y^4}$
9. $\left(\dfrac{3a^{-2}b^5}{9a^{-4}b^0}\right)^{-2}$ $\dfrac{9}{a^4b^{10}}$
10. $\left(\dfrac{5x^0y^{-7}}{2x^{-2}y^4}\right)^{-3}$ $\dfrac{8y^{33}}{125x^6}$

11. $3t^3 + 3t^2 + 7$
12. $7y^2 - 5y$
13. $9x^3 - 4x^2 + 1$
14. $2u^2 - u + 1$
17. $24x^2 - 26xy + 5y^2$

11. $(2t^3 + 3t^2 - 2) + (t^3 + 9)$
12. $(3 - 7y) + (7y^2 + 2y - 3)$
13. $(7x^3 - 3x^2 + 1) - (x^2 - 2x^3)$
14. $(5 - u) + 2[3 - (u^2 + 1)]$
15. $7y(4 - 3y)$ $28y - 21y^2$
16. $(k + 8)(k + 5)$ $k^2 + 13k + 40$
17. $(4x - y)(6x - 5y)$
18. $2z(z + 5) - 7(z + 5)$ $2z^2 + 3z - 35$
19. $(6r + 5)(6r - 5)$ $36r^2 - 25$
20. $(2x - 3)^2$ $4x^2 - 12x + 9$
21. $(x + 1)(x^2 - x + 1)$ $x^3 + 1$
22. $(x^2 - 3x + 2)(x^2 + 5x - 10)$ $x^4 + 2x^3 - 23x^2 + 40x - 20$

Applications

23. Write an expression for the area of the shaded region of the figure. Then simplify the expression.
$\frac{1}{2}(x + 2)^2 - \frac{1}{2}x^2$; $2x + 2$

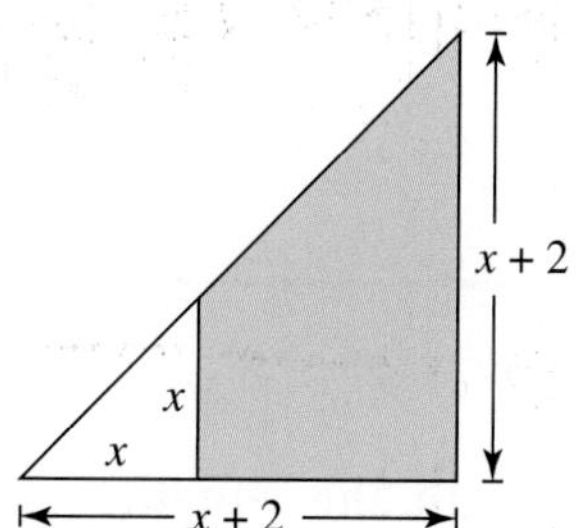

24. An object is thrown upward from the top of a 100-foot building with an initial velocity of 32 feet per second. Use the position function $h(t) = -16t^2 + 32t + 100$ to find the height of the object when $t = \frac{3}{2}$ and $t = 3$. 112 feet; 52 feet

25. A manufacturer can produce and sell x T-shirts per week. The total cost C (in dollars) of producing the T-shirts is given by $C = 5x + 2250$, and the total revenue R is given by $R = 24x$. Find the profit P obtained by selling 1500 T-shirts per week. (*Hint:* $P = R - C$.) \$26,250

Study Skills in Action

Using Different Approaches

In math, one small detail such as a misplaced parenthesis or exponent can lead to a wrong answer, even if you set up the problem correctly. One way you can minimize detail mistakes is to be aware of learning modalities. A learning modality is a preferred way of taking in information that is then transferred into the brain for processing. The three modalities are *visual*, *auditory*, and *kinesthetic*. The following are brief descriptions of these modalities.

- **Visual** You take in information more productively if you can see the information.
- **Auditory** You take in information more productively when you listen to an explanation and talk about it.
- **Kinesthetic** You take in information more productively if you can experience it or use physical activity in studying.

You may find that one approach, or even a combination of approaches, works best for you.

$$4x^2(3x - 2x^3 + 1)$$
$$= 4x^2(3x) - 4x^2(2x^3) + 4x^2(1)$$
$$= 12x^3 \quad - 8x^5 \quad + 4x^2$$
$$= -8x^5 + 12x^3 + 4x^2$$

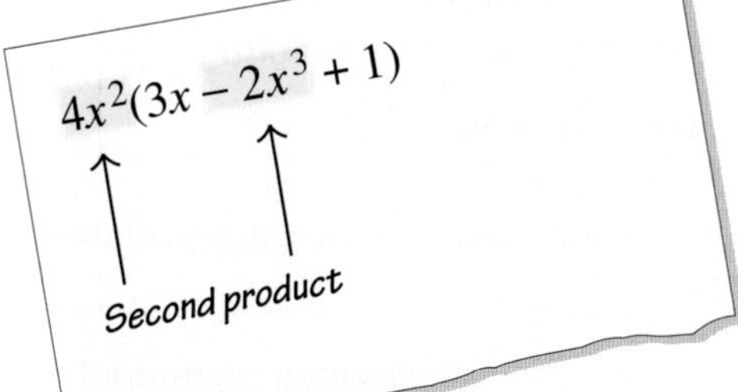

Smart Study Strategy

Use All Three Modalities

Design a way to solve a problem using all three learning modalities.

Visual Highlight or use a different color of pencil to indicate when you have simplified something in a problem. This is done in the example at the left.

Auditory Say each calculation step out loud.

"I need to multiply four *x* to the second power and two *x* to the third power. Because two *x* to the third power is negative, I must include the negative sign when I multiply by four *x* to the second power."

Kinesthetic When completing and checking your calculations, use your fingers to point to each step. Double-check signs, parentheses, exponents, and so on. For the given example, you can use your fingers to point to the products formed by the monomial and each term of the trinomial.

5.4 Factoring by Grouping and Special Forms

- Factor greatest common monomial factors from polynomials.
- Factor polynomials by grouping terms.
- Factor special products and factor completely.

Common Monomial Factors

Greatest Common Monomial Factor

When a polynomial in x with integer coefficients has a greatest common monomial factor of the form ax^n, the following statements must be true.

1. The coefficient a must be the greatest integer that *divides* each of the coefficients in the polynomial.
2. The variable factor x^n is the highest-powered variable factor that *is common* to all terms of the polynomial.

Additional Examples

Factor out the greatest common monomial factor from each polynomial.

a. $12x^2y + 28xy^2$

b. $-8ab - 6ab^2 + 4b$

Answers:

a. $4xy(3x + 7y)$

b. $2b(-4a - 3ab + 2)$ or $-2b(4a + 3ab - 2)$

EXAMPLE 1 **Factoring Out a Greatest Common Monomial Factor**

$$24x^3 - 32x^2 = (8x^2)(3x) - (8x^2)(4) = 8x^2(3x - 4)$$

EXAMPLE 2 **A Negative Common Monomial Factor**

Factor the polynomial $-3x^2 + 12x - 18$ in two ways.

a. Factor out a 3. **b.** Factor out a -3.

SOLUTION

a. By factoring out the common monomial factor of 3, you obtain

$$\begin{aligned} -3x^2 + 12x - 18 &= 3(-x^2) + 3(4x) + 3(-6) \\ &= 3(-x^2 + 4x - 6). \end{aligned}$$

b. By factoring out the common monomial factor of -3, you obtain

$$\begin{aligned} -3x^2 + 12x - 18 &= -3(x^2) + (-3)(-4x) + (-3)(6) \\ &= -3(x^2 - 4x + 6). \end{aligned}$$

Study Tip

The greatest common monomial factor of the terms of a polynomial is usually considered to have a positive coefficient. However, sometimes it is convenient to factor a negative number out of a polynomial, as shown in Example 2.

Exercises Within Reach ®

Solutions in English & Spanish and tutorial videos at AlgebraWithinReach.com

Greatest Common Monomial Factor **In Exercises 1–4, factor out the greatest common monomial factor from the polynomial.**

1. $8t^2 + 8t$
$8t(t + 1)$

2. $12x^2 + 6x$
$6x(2x + 1)$

3. $28x^2 + 16x - 8$
$4(7x^2 + 4x - 2)$

4. $9 - 27y - 15y^2$
$3(3 - 9y - 5y^2)$

A Negative Common Monomial Factor **In Exercises 5–8, factor out a negative real number from the polynomial and then write the polynomial factor in standard form.**

5. $7 - 14x$
$-7(2x - 1)$

6. $15 - 5x$
$-5(x - 3)$

7. $2y - 2 - 6y^2$
$-2(3y^2 - y + 1)$

8. $9x - 9x^2 - 24$
$-3(3x^2 - 3x + 8)$

Application EXAMPLE 3 Geometry: Dimensions of a Rectangle

The area of a rectangle of width $4x$ is given by the polynomial

$$12x^2 + 32x$$

as shown below.

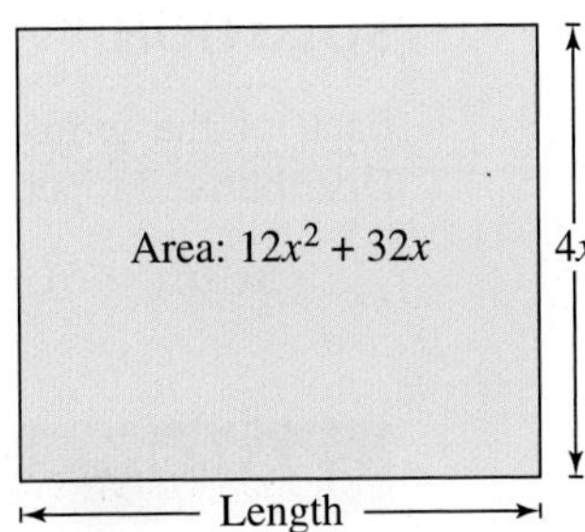

a. Factor the expression for the area to determine an expression for the length of the rectangle.

b. Find the dimensions of the rectangle when $x = 4$ inches.

SOLUTION

a. The polynomial for the area of the rectangle factors as follows.

$$12x^2 + 32x = 4x(3x + 8) \qquad \text{Factor out common monomial factor.}$$

So, the length of the rectangle is $3x + 8$.

b. When $x = 4$ inches, the dimensions of the rectangle are

$$\text{Width} = 4(4) = 16 \text{ inches}$$

$$\text{Length} = 3(4) + 8 = 20 \text{ inches.}$$

Exercises Within Reach ®

Solutions in English & Spanish and tutorial videos at AlgebraWithinReach.com

9. ***Geometry*** The area of a rectangle is given by the polynomial $45l - l^2$, where l is the length (in feet) of the rectangle.

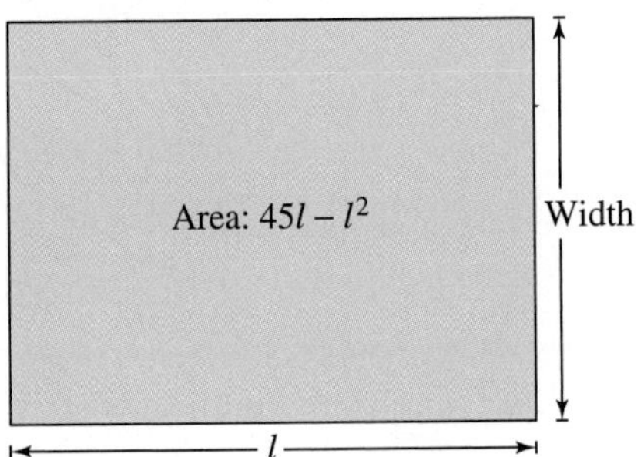

(a) Factor the expression for the area to determine an expression for the width of the rectangle.
$45l - l^2 = l(45 - l)$; Width $= 45 - l$

(b) Find the width of the rectangle when $l = 26$ feet.
19 feet

10. ***Farming*** A farmer has enough fencing to construct a rectangular pig pen that encloses an area given by the polynomial $35x^2 + 28x$, where $7x$ is the length (in feet) of the pen.

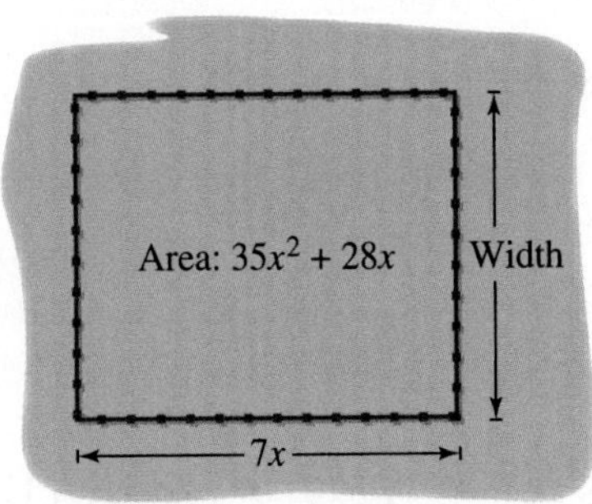

(a) Factor the expression for the area to determine an expression for the width of the pig pen.
$35x^2 + 28x = 7x(5x + 4)$; Width $= 5x + 4$

(b) Find the dimensions of the pig pen when $x = 4$ feet.
24 feet by 28 feet

Additional Examples

Factor each expression.

a. $(7a + 2)a + (7a + 2)4$

b. $2x^3 + 4x^2 - x - 2$

Answers:

a. $(7a + 2)(a + 4)$

b. $(2x^2 - 1)(x + 2)$

Study Tip

It often helps to write a polynomial in standard form before trying to factor by grouping. Then group and remove a common monomial factor from the first two terms and the last two terms. Finally, if possible, factor out the common binomial factor.

Factoring by Grouping

EXAMPLE 4 A Common Binomial Factor

Factor the expression $5x^2(6x - 5) - 2(6x - 5)$.

SOLUTION

Each of the terms of this expression has a binomial factor of $(6x - 5)$. Use the Distributive Property to factor out this common binomial factor.

$$5x^2(6x - 5) - 2(6x - 5) = (6x - 5)(5x^2 - 2)$$

EXAMPLE 5 Factoring by Grouping

Factor each polynomial by grouping.

a. $x^3 - 5x^2 + x - 5$ **b.** $4x^3 + 3x - 8x^2 - 6$

SOLUTION

a. $x^3 - 5x^2 + x - 5 = (x^3 - 5x^2) + (x - 5)$ Group terms.

$= x^2(x - 5) + 1(x - 5)$ Factor grouped terms.

$= (x - 5)(x^2 + 1)$ Common binomial factor

b. $4x^3 + 3x - 8x^2 - 6 = 4x^3 - 8x^2 + 3x - 6$ Write in standard form.

$= (4x^3 - 8x^2) + (3x - 6)$ Group terms.

$= 4x^2(x - 2) + 3(x - 2)$ Factor grouped terms.

$= (x - 2)(4x^2 + 3)$ Common binomial factor

Exercises Within Reach®

Solutions in English & Spanish and tutorial videos at AlgebraWithinReach.com

A Common Binomial Factor **In Exercises 11–16, factor the expression by factoring out the common binomial factor.**

11. $2y(y - 4) + 5(y - 4)$ $(y - 4)(2y + 5)$

12. $7t(s + 9) + 6(s + 9)$ $(s + 9)(7t - 6)$

13. $5x(3x + 2) - 3(3x + 2)$ $(3x + 2)(5x - 3)$

14. $6(4t - 3) - 5t(4t - 3)$ $(4t - 3)(6 - 5t)$

15. $2(7a + 6) - 3a^2(7a + 6)$ $(7a + 6)(2 - 3a^2)$

16. $4(5y - 12) + 3y^2(5y - 12)$ $(5y - 12)(4 + 3y^2)$

Factoring by Grouping **In Exercises 17–24, factor the polynomial by grouping.**

17. $x^2 + 25x + x + 25$ $(x + 25)(x + 1)$

18. $x^2 - 9x + x - 9$ $(x - 9)(x + 1)$

19. $y^2 - 6y + 2y - 12$ $(y - 6)(y + 2)$

20. $y^2 + 3y + 4y + 12$ $(y + 3)(y + 4)$

21. $x^3 + 2x^2 + x + 2$ $(x + 2)(x^2 + 1)$

22. $t^3 - 11t^2 + t - 11$ $(t - 11)(t^2 + 1)$

23. $3a^3 - 12a^2 - 2a + 8$ $(a - 4)(3a^2 - 2)$

24. $3s^3 + 6s^2 + 5s + 10$ $(s + 2)(3s^2 + 5)$

Factoring Special Products and Factoring Completely

Difference of Two Squares

Let u and v be real numbers, variables, or algebraic expressions. Then the expression $u^2 - v^2$ can be factored as follows.

$$u^2 - v^2 = (u + v)(u - v)$$

Difference (points to the minus sign in $u^2 - v^2$); Opposite signs (point to the signs in the two factors)

Additional Examples
Factor each polynomial.
a. $9x^2 - 49$
b. $4x^2 - 25y^2$
Answers:
a. $(3x + 7)(3x - 7)$
b. $(2x + 5y)(2x - 5y)$

EXAMPLE 6 Factoring the Difference of Two Squares

Factor each polynomial.

a. $x^2 - 64$ **b.** $49x^2 - 81y^2$

SOLUTION

a. $x^2 - 64 = x^2 - 8^2$ Write as difference of two squares.

$= (x + 8)(x - 8)$ Factored form

b. $49x^2 - 81y^2 = (7x)^2 - (9y)^2$ Write as difference of two squares.

$= (7x + 9y)(7x - 9y)$ Factored form

EXAMPLE 7 Factoring the Difference of Two Squares

Factor the expression $(x + 2)^2 - 9$.

SOLUTION

$(x + 2)^2 - 9 = (x + 2)^2 - 3^2$ Write as difference of two squares.

$= [(x + 2) + 3][(x + 2) - 3]$ Factored form

$= (x + 5)(x - 1)$ Simplify.

To check this result, write the original polynomial in standard form. Then multiply the factored form to see that you obtain the same standard form.

Exercises Within Reach®

Solutions in English & Spanish and tutorial videos at AlgebraWithinReach.com

Factoring the Difference of Two Squares **In Exercises 25–38, factor the difference of two squares.**

25. $x^2 - 9$ $(x + 3)(x - 3)$

26. $y^2 - 4$ $(y + 2)(y - 2)$

27. $1 - a^2$ $(1 + a)(1 - a)$

28. $16 - b^2$ $(4 + b)(4 - b)$

29. $4z^2 - y^2$ $(2z + y)(2z - y)$

30. $9u^2 - v^2$ $(3u + v)(3u - v)$

31. $36x^2 - 25y^2$ $(6x + 5y)(6x - 5y)$

32. $100a^2 - 49b^2$ $(10a + 7b)(10a - 7b)$

33. $(x - 1)^2 - 16$ $(x + 3)(x - 5)$

34. $(x - 3)^2 - 4$ $(x - 5)(x - 1)$

35. $81 - (z + 5)^2$ $(14 + z)(4 - z)$

36. $36 - (y - 6)^2$ $y(-y + 12)$

37. $(2x + 5)^2 - (x - 4)^2$ $(x + 9)(3x + 1)$

38. $(3y - 1)^2 - (x + 6)^2$ $(x + 3y + 5)(-x + 3y - 7)$

Sum or Difference of Two Cubes

Let u and v be real numbers, variables, or algebraic expressions. Then the expressions $u^3 + v^3$ and $u^3 - v^3$ can be factored as follows.

Like signs

1. $u^3 + v^3 = (u + v)(u^2 - uv + v^2)$

Unlike signs

Like signs

2. $u^3 - v^3 = (u - v)(u^2 + uv + v^2)$

Unlike signs

EXAMPLE 8 Factoring the Sum or Difference of Two Cubes

Factor each polynomial.

a. $x^3 - 125$ **b.** $8y^3 + 1$ **c.** $y^3 - 27x^3$

SOLUTION

a. This polynomial is the difference of two cubes because x^3 is the cube of x and 125 is the cube of 5.

$x^3 - 125 = x^3 - 5^3$ Write as difference of two cubes.

$= (x - 5)(x^2 + 5x + 5^2)$ Factored form

$= (x - 5)(x^2 + 5x + 25)$ Simplify.

b. This polynomial is the sum of two cubes because $8y^3$ is the cube of $2y$ and 1 is the cube of 1.

$8y^3 + 1 = (2y)^3 + 1^3$ Write as sum of two cubes.

$= (2y + 1)[(2y)^2 - (2y)(1) + 1^2]$ Factored form

$= (2y + 1)(4y^2 - 2y + 1)$ Simplify.

c. $y^3 - 27x^3 = y^3 - (3x)^3$ Write as difference of two cubes.

$= (y - 3x)[y^2 + 3xy + (3x)^2]$ Factored form

$= (y - 3x)(y^2 + 3xy + 9x^2)$ Simplify.

Exercises Within Reach®

Solutions in English & Spanish and tutorial videos at AlgebraWithinReach.com

Factoring the Sum or Difference of Two Cubes **In Exercises 39–48, factor the sum or difference of two cubes.**

39. $x^3 - 8$ $(x - 2)(x^2 + 2x + 4)$

40. $t^3 - 1$ $(t - 1)(t^2 + t + 1)$

41. $y^3 + 64$ $(y + 4)(y^2 - 4y + 16)$

42. $z^3 + 125$ $(z + 5)(z^2 - 5z + 25)$

43. $8t^3 - 27$ $(2t - 3)(4t^2 + 6t + 9)$

44. $27s^3 + 64$ $(3s + 4)(9s^2 - 12s + 16)$

45. $27u^3 + 1$ $(3u + 1)(9u^2 - 3u + 1)$

46. $64v^3 - 125$ $(4v - 5)(16v^2 + 20v + 25)$

47. $x^3 + 27y^3$ $(x + 3y)(x^2 - 3xy + 9y^2)$

48. $u^3 + 125v^3$ $(u + 5v)(u^2 - 5uv + 25v^2)$

Additional Examples

Factor each polynomial completely.

a. $3y^4 - 24y$

b. $2x^4y + 128xy$

Answers:

a. $3y(y - 2)(y^2 + 2y + 4)$

b. $2xy(x + 4)(x^2 - 4x + 16)$

EXAMPLE 9 Factoring Completely

Factor the polynomial $125x^2 - 80$ completely.

SOLUTION

Because both terms have a common factor of 5, begin by factoring 5 out of the expression.

$125x^2 - 80 = 5(25x^2 - 16)$ — Factor out common monomial factor.

$= 5[(5x)^2 - 4^2]$ — Write as difference of two squares.

$= 5(5x + 4)(5x - 4)$ — Factored form

EXAMPLE 10 Factoring Completely

Factor each polynomial completely.

a. $x^4 - y^4$

b. $81m^4 - 1$

SOLUTION

a. $x^4 - y^4 = (x^2)^2 - (y^2)^2$ — Write as difference of two squares.

$= (x^2 + y^2)(x^2 - y^2)$ — Factor as difference of two squares.

$= (x^2 + y^2)(x + y)(x - y)$ — Factor second difference of two squares.

b. $81m^4 - 1 = (9m^2)^2 - 1^2$ — Write as difference of two squares.

$= (9m^2 + 1)(9m^2 - 1)$ — Factor as difference of two squares.

$= (9m^2 + 1)(3m + 1)(3m - 1)$ — Factor second difference of two squares.

Study Tip

The sum of two squares, such as $9m^2 + 1$ in Example 10(b), cannot be factored further using integer coefficients. Such polynomials are called **prime** with respect to the integers. Some other prime polynomials are $x^2 + 4$ and $4x + 9$.

Exercises Within Reach®

Solutions in English & Spanish and tutorial videos at AlgebraWithinReach.com

Factoring Completely **In Exercises 49–58, factor the polynomial completely.**

49. $8 - 50x^2$ $2(2 + 5x)(2 - 5x)$

50. $8y^2 - 18$ $2(2y + 3)(2y - 3)$

51. $8x^3 + 64$ $8(x + 2)(x^2 - 2x + 4)$

52. $a^3 - 16a$ $a(a + 4)(a - 4)$

53. $y^4 - 81$ $(y - 3)(y + 3)(y^2 + 9)$

54. $u^4 - 16$ $(u^2 + 4)(u + 2)(u - 2)$

55. $3x^4 - 300x^2$ $3x^2(x + 10)(x - 10)$

56. $6x^5 + 30x^3$ $6x^3(x^2 + 5)$

57. $6x^6 - 48y^6$ $6(x^2 - 2y^2)(x^4 + 2x^2y^2 + 4y^4)$

58. $2u^6 + 54v^6$ $2(u^2 + 3v^2)(u^4 - 3u^2v^2 + 9v^4)$

Concept Summary: Factoring Polynomials

What

Factoring polynomials is the reverse process of multiplying polynomials. There are many ways to factor a polynomial.

EXAMPLE

Factor $2x^3 - 18x$ completely.

How

- Factor out the **greatest common monomial factor**.
- **Factor by grouping** terms.
- Use the formulas for special polynomial forms.
- **Factor completely** by using a combination of methods.

EXAMPLE

$$2x^3 - 18x = 2x(x^2 - 9)$$
$$= 2x(x + 3)(x - 3)$$

Why

Factoring polynomials is a useful algebraic technique.

For example, as you continue your study of algebra, you will factor polynomials to rewrite expressions and to help solve equations.

Exercises Within Reach ®

Worked-out solutions to odd-numbered exercises at AlgebraWithinReach.com

Concept Summary Check

59. ***Greatest Common Monomial Factor*** What is the greatest common monomial factor of $2x^3 - 18x$? $2x$

60. ***Difference of Two Squares*** What expression in the solution above is a difference of two squares? $x^2 - 9$

61. ***Completely Factored Form*** What is the completely factored form of $2x^3 - 18x$? $2x(x + 3)(x - 3)$

62. ***Factoring Completely*** Describe the two factoring methods used in the solution above. Factoring out the greatest common monomial factor and factoring a difference of two squares.

Extra Practice

Factoring an Expression **In Exercises 63–68, factor the expression. (Assume that all exponents represent positive integers.)**

63. $4x^{2n} - 25$ $(2x^n + 5)(2x^n - 5)$

64. $3x^{n+1} + 6x^n - 15x^{n+2}$ $-3x^n(5x^2 - x - 2)$

65. $81 - 16y^{4n}$ $(9 + 4y^{2n})(3 + 2y^n)(3 - 2y^n)$

66. $4y^{m+n} + 7y^{2m+n} - y^{m+2n}$ $y^{m+n}(7y^m - y^n + 4)$

67. $2x^{3r} + 8x^r + 4x^{2r}$ $2x^r(x^{2r} + 2x^r + 4)$

68. $x^{2r+s} - 5x^{r+3s} + 10x^{2r+2s}$ $x^{r+s}(10x^{r+s} - 5x^{2s} + x^r)$

Think About It **In Exercises 69 and 70, show all the different groupings that can be used to factor the polynomial completely. Carry out the various factorizations to show that they yield the same result.**
See Additional Answers.

69. $3x^3 + 4x^2 - 3x - 4$

70. $6x^3 - 8x^2 + 9x - 12$

71. ***Geometry*** The surface area of a right circular cylinder is given by $S = 2\pi r^2 + 2\pi rh$ (see figure). Factor this expression. $S = 2\pi r(r + h)$

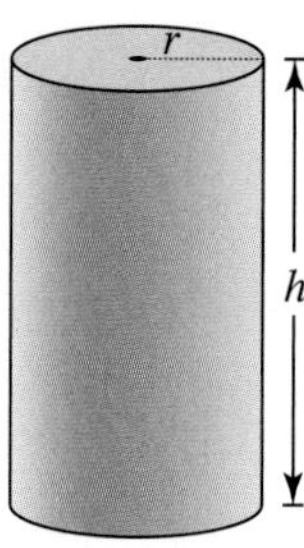

72. ***Product Design*** A washer on the drive train of a car has an inside radius of r centimeters and an outside radius of R centimeters (see figure). Find the area of one of the flat surfaces of the washer and write the area in factored form. $\pi(R - r)(R + r)$

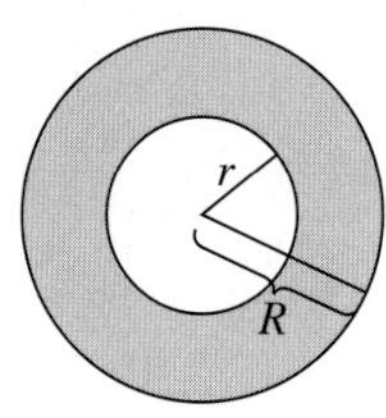

Revenue **The revenue from selling x units of a product at a price of p dollars per unit is given by $R = xp$. In Exercises 73 and 74, factor the expression for revenue to determine an expression that gives the price in terms of x.**

73. $R = 800x - 0.25x^2$
$p = 800 - 0.25x$

74. $R = 1000x - 0.4x^2$
$p = 1000 - 0.4x$

75. ***Simple Interest*** The total amount of money accrued from a principal of P dollars invested at a simple interest rate r (in decimal form) for t years is given by $P + Prt$. Factor this expression.
$P(1 + rt)$

76. ***Geometry*** The surface area of the rectangular solid shown below is given by $2x^2 + 4xh$. Factor this expression. $2x(x + 2h)$

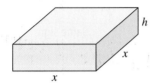

77. ***Geometry*** The cube shown has a volume of a^3 and is formed by the solids I, II, III, and IV.
See Additional Answers.

(a) Write an expression for the volume of the entire cube and for each of the solids I, II, III, and IV.

(b) Add the volumes of solids I, II, and III. Factor the result by factoring out the common binomial factor.

(c) Explain how the figure models the factoring pattern for the *difference of two cubes*.

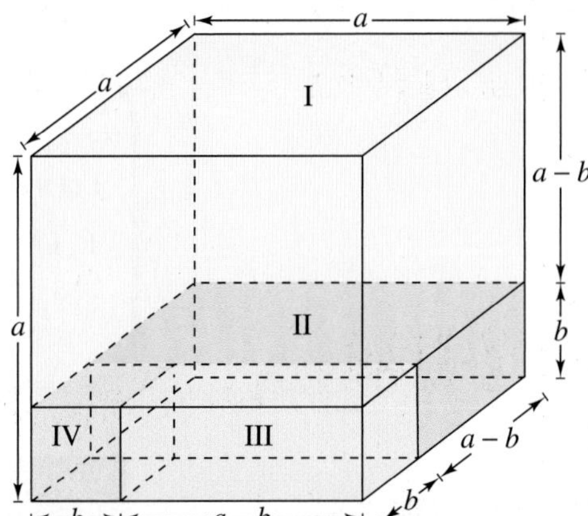

Explaining Concepts

78. ***Writing*** Explain how the word *factor* can be used as a noun or as a verb.
See Additional Answers.

79. ***Structure*** Explain what is meant by saying that a polynomial is in factored form.
The polynomial is written as a product of polynomials.

80. ***Writing*** Describe a method for finding the greatest common factor of two (or more) monomials.
See Additional Answers.

81. ***Writing*** How can you check your result after factoring a polynomial?
To check the result, multiply the factors to obtain the original polynomial.

82. ***Create an Example*** Give an example of a polynomial that is prime with respect to the integers.
$x^2 + 1$

83. ***Structure*** Which of the following polynomials *cannot* be factored by one of the formulas for special forms? Can this polynomial be factored at all? Explain why or why not. See Additional Answers.

a. $x^2 + y^2$ **b.** $x^2 - y^2$

c. $x^3 + y^3$ **d.** $x^3 - y^3$

84. ***Structure*** You can sometimes factor out a common binomial from a polynomial expression of four or more terms. What procedure can you use to do this?
Factor by grouping.

Cumulative Review

In Exercises 85–88, find the determinant of the matrix.

85. $\begin{bmatrix} 3 & 4 \\ 2 & 1 \end{bmatrix}$ -5

86. $\begin{bmatrix} -1 & 2 \\ -3 & 5 \end{bmatrix}$ 1

87. $\begin{bmatrix} -1 & 3 & 0 \\ -2 & 0 & 6 \\ 0 & 4 & 2 \end{bmatrix}$ 36

88. $\begin{bmatrix} 1 & 2 & 5 \\ -2 & 4 & 3 \\ -2 & 2 & 2 \end{bmatrix}$ 18

In Exercises 89–92, use a special product formula to find the product.

89. $(x + 7)(x - 7)$
$x^2 - 49$

90. $(2x + 3)^2$
$4x^2 + 12x + 9$

91. $(2x - 3)^2$
$4x^2 - 12x + 9$

92. $[(x - 2) - y][(x - 2) + y]$
$(x - 2)^2 - y^2 = x^2 - 4x - y^2 + 4$

5.5 Factoring Trinomials

- Recognize and factor perfect square trinomials.
- Factor trinomials of the forms $x^2 + bx + c$ and $ax^2 + bx + c$.
- Factor trinomials of the form $ax^2 + bx + c$ by grouping.
- Factor polynomials using the guidelines for factoring.

Perfect Square Trinomials

Perfect Square Trinomials

Let u and v represent real numbers, variables, or algebraic expressions.

1. $u^2 + 2uv + v^2 = (u + v)^2$ (Same sign)
2. $u^2 - 2uv + v^2 = (u - v)^2$ (Same sign)

EXAMPLE 1 Factoring Perfect Square Trinomials

a. $x^2 - 4x + 4 = x^2 - 2(x)(2) + 2^2$
$= (x - 2)^2$

b. $16y^2 + 24y + 9 = (4y)^2 + 2(4y)(3) + 3^2$
$= (4y + 3)^2$

c. $9x^2 - 30xy + 25y^2 = (3x)^2 - 2(3x)(5y) + (5y)^2$
$= (3x - 5y)^2$

EXAMPLE 2 Factoring Out a Common Monomial Factor First

a. $3x^2 - 30x + 75 = 3(x^2 - 10x + 25)$ Factor out common monomial factor.
$= 3(x - 5)^2$ Factor as perfect square trinomial.

b. $16y^3 + 80y^2 + 100y = 4y(4y^2 + 20y + 25)$ Factor out common monomial factor.
$= 4y(2y + 5)^2$ Factor as perfect square trinomial.

Exercises Within Reach®

Solutions in English & Spanish and tutorial videos at AlgebraWithinReach.com

Factoring a Perfect Square Trinomial **In Exercises 1–12, factor the perfect square trinomial.**

1. $x^2 + 4x + 4$ $(x + 2)^2$
2. $z^2 + 6z + 9$ $(z + 3)^2$
3. $25y^2 - 10y + 1$ $(5y - 1)^2$
4. $4z^2 + 28z + 49$ $(2z + 7)^2$
5. $9b^2 + 12b + 4$ $(3b + 2)^2$
6. $16a^2 - 24a + 9$ $(4a - 3)^2$
7. $36x^2 - 60xy + 25y^2$ $(6x - 5y)^2$
8. $4y^2 + 20yz + 25z^2$ $(2y + 5z)^2$
9. $3m^3 - 18m^2 + 27m$ $3m(m - 3)^2$
10. $4m^3 + 16m^2 + 16m$ $4m(m + 2)^2$
11. $20v^4 - 60v^3 + 45v^2$ $5v^2(2v - 3)^2$
12. $8y^3 + 24y^2 + 18y$ $2y(2y + 3)^2$

Factoring Trinomials

EXAMPLE 3 Factoring a Trinomial of the Form $x^2 + bx + c$

Factor the trinomial $x^2 + 3x - 4$.

SOLUTION

You need to find two numbers whose product is -4 and whose sum is 3. Using mental math, you can determine that the numbers are 4 and -1.

The product of 4 and –1 is –4.

$$x^2 + 3x - 4 = (x + 4)(x - 1)$$

The sum of 4 and –1 is 3.

Study Tip

Use a list to help you find the two numbers with the required product and sum. For Example 4(a):

Factors of -8	*Sum*
1, −8	−7
−1, 8	7
2, −4	−2
−2, 4	2

Because -2 is the required sum, the correct factorization is $x^2 - 2x - 8 = (x - 4)(x + 2)$.

EXAMPLE 4 Factoring Trinomials of the Form $x^2 + bx + c$

Factor each trinomial.

a. $x^2 - 2x - 8$

b. $x^2 - 5x + 6$

SOLUTION

a. You need to find two numbers whose product is -8 and whose sum is -2.

The product of -4 and 2 is –8.

$$x^2 - 2x - 8 = (x - 4)(x + 2)$$

The sum of -4 and 2 is -2.

b. You need to find two numbers whose product is 6 and whose sum is -5.

The product of -3 and -2 is 6.

$$x^2 - 5x + 6 = (x - 3)(x - 2)$$

The sum of -3 and -2 is -5.

Exercises Within Reach ®

Solutions in English & Spanish and tutorial videos at AlgebraWithinReach.com

Factoring a Trinomial of the Form $x^2 + bx + c$ **In Exercises 13–16, determine the missing factor.**

13. $a^2 + 6a + 8 = (a + 4)(\quad a + 2 \quad)$

14. $a^2 + 2a - 8 = (a + 4)(\quad a - 2 \quad)$

15. $y^2 - y - 20 = (y + 4)(\quad y - 5 \quad)$

16. $y^2 - 4y - 32 = (y + 4)(\quad y - 8 \quad)$

Factoring a Trinomial of the Form $x^2 + bx + c$ **In Exercises 17–24, factor the trinomial.**

17. $x^2 + 6x + 5$ $(x + 1)(x + 5)$

18. $x^2 + 7x + 10$ $(x + 2)(x + 5)$

19. $x^2 - 5x + 6$ $(x - 3)(x - 2)$

20. $x^2 - 10x + 24$ $(x - 4)(x - 6)$

21. $y^2 + 7y - 30$ $(y + 10)(y - 3)$

22. $m^2 - 3m - 10$ $(m - 5)(m + 2)$

23. $t^2 - 6t - 16$ $(t - 8)(t + 2)$

24. $x^2 + 4x - 12$ $(x + 6)(x - 2)$

Study Tip

With *any* factoring problem, remember that you can check your result by multiplying. For instance, in Example 5, you can check the result by multiplying $(x - 18)$ by $(x + 1)$ to see that you obtain $x^2 - 17x - 18$.

Remember that not all trinomials are factorable using integers. For instance, $x^2 - 2x - 4$ is not factorable using integers because there is no pair of factors of -4 whose sum is -2. Such non-factorable trinomials are called **prime polynomials**.

Additional Examples

Factor each trinomial.

a. $x^2 - 3x - 70$

b. $x^2 - 18x + 72$

Answers:

a. $(x + 7)(x - 10)$

b. $(x - 6)(x - 12)$

Guidelines for Factoring $x^2 + bx + c$

1. When c is *positive*, its factors have like signs that match the sign of b.
2. When c is *negative*, its factors have unlike signs.
3. When $|b|$ is small relative to $|c|$, first try those factors of c that are closest to each other in absolute value.
4. When $|b|$ is near $|c|$, first try those factors of c that are farthest from each other in absolute value.

EXAMPLE 5 Factoring a Trinomial of the Form $x^2 + bx + c$

Factor $x^2 - 17x - 18$.

SOLUTION

You need to find two numbers whose product is -18 and whose sum is -17. Because

$$|b| = |-17| = 17 \text{ and } c = |-18| = 18$$

are close in value, choose factors of -18 that are farthest from each other.

The product of -18 and 1 is -18.

$$x^2 - 17x - 18 = (x - 18)(x + 1)$$

The sum of -18 and 1 is -17.

Exercises Within Reach®

Solutions in English & Spanish and tutorial videos at AlgebraWithinReach.com

Factoring a Trinomial of the Form $x^2 + bx + c$ **In Exercises 25–28, determine the missing factor.**

25. $x^2 + 10x + 24 = (x + 4)(\quad x + 6 \quad)$

26. $x^2 + 7x + 12 = (x + 4)(\quad x + 3 \quad)$

27. $z^2 - 6z + 8 = (z - 4)(\quad z - 2 \quad)$

28. $z^2 + 2z - 24 = (z - 4)(\quad z + 6 \quad)$

Geometry **In Exercises 29 and 30, factor the polynomial expression for the area of the rectangle to determine an expression for the length of the rectangle.**

29. $x + 4$

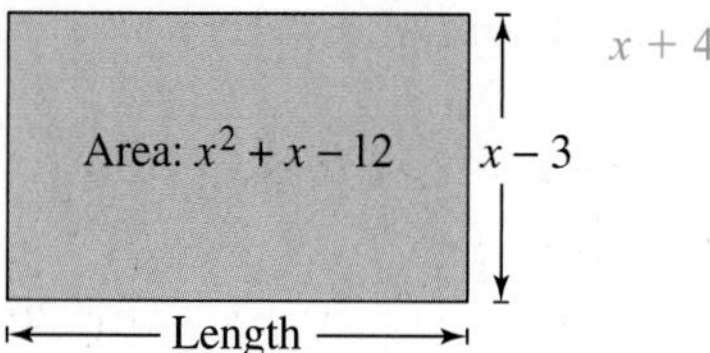

30. $x + 5$

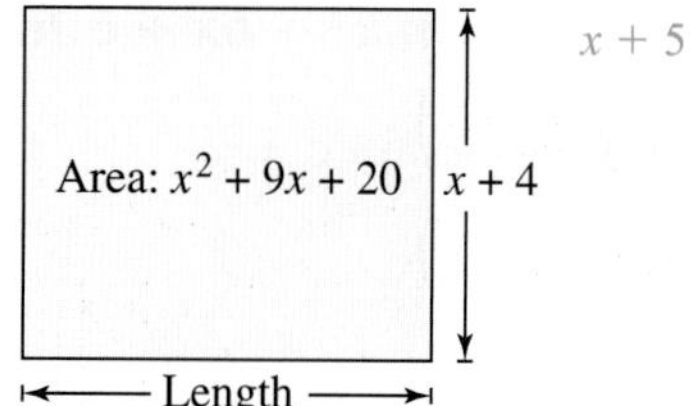

Factoring a Trinomial of the Form $x^2 + bx + c$ **In Exercises 31–36, factor the trinomial.**

31. $x^2 - 15x - 16$ $(x + 1)(x - 16)$

32. $x^2 + 17x - 18$ $(x - 1)(x + 18)$

33. $x^2 + x - 72$ $(x - 8)(x + 9)$

34. $x^2 - 2x - 120$ $(x + 10)(x - 12)$

35. $x^2 - 20x + 96$ $(x - 8)(x - 12)$

36. $y^2 - 35y + 300$ $(y - 15)(y - 20)$

EXAMPLE 6 Factoring a Trinomial of the Form $ax^2 + bx + c$

Factor the trinomial $4x^2 + 5x - 6$.

SOLUTION

First, observe that $4x^2 + 5x - 6$ has no common monomial factor. For this trinomial, $a = 4$, which factors as (1)(4) or (2)(2), and $c = -6$, which factors as $(-1)(6)$, $(1)(-6)$, $(-2)(3)$, or $(2)(-3)$. A test of the many possibilities is shown below.

Study Tip

When the original trinomial has no common monomial factor, its binomial factors cannot have common monomial factors. So, in Example 6, you do not have to test factors, such as $(4x - 6)$, that have a common monomial factor of 2. Which of the other factors in Example 6 did not need to be tested?

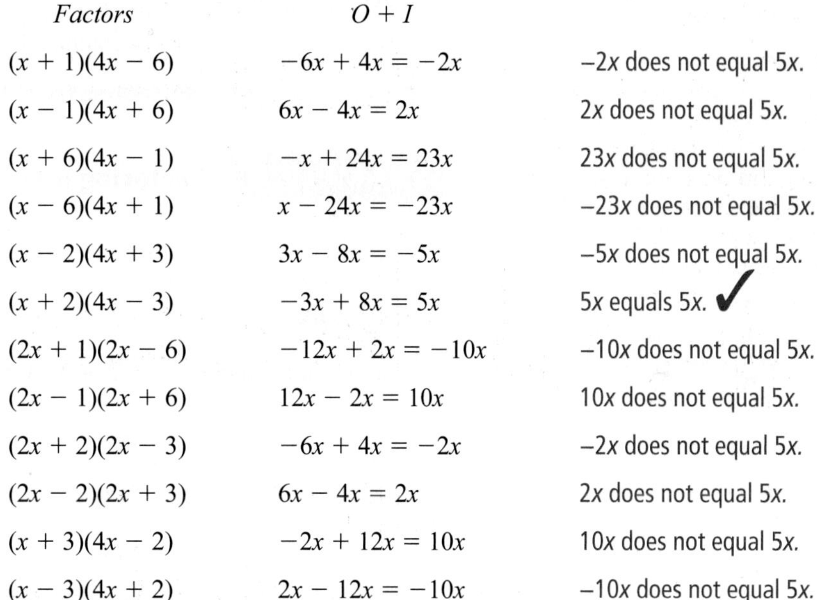
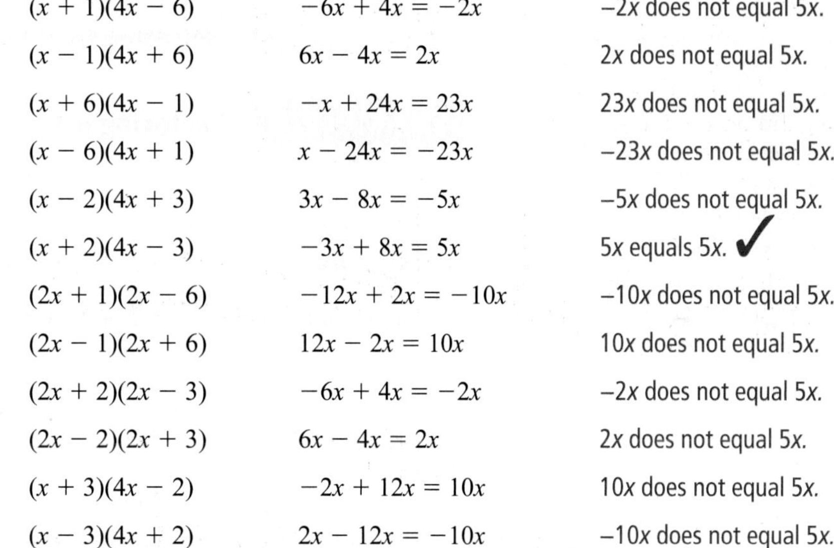

Factors	$O + I$	
$(x + 1)(4x - 6)$	$-6x + 4x = -2x$	$-2x$ does not equal $5x$.
$(x - 1)(4x + 6)$	$6x - 4x = 2x$	$2x$ does not equal $5x$.
$(x + 6)(4x - 1)$	$-x + 24x = 23x$	$23x$ does not equal $5x$.
$(x - 6)(4x + 1)$	$x - 24x = -23x$	$-23x$ does not equal $5x$.
$(x - 2)(4x + 3)$	$3x - 8x = -5x$	$-5x$ does not equal $5x$.
$(x + 2)(4x - 3)$	$-3x + 8x = 5x$	$5x$ equals $5x$. ✓
$(2x + 1)(2x - 6)$	$-12x + 2x = -10x$	$-10x$ does not equal $5x$.
$(2x - 1)(2x + 6)$	$12x - 2x = 10x$	$10x$ does not equal $5x$.
$(2x + 2)(2x - 3)$	$-6x + 4x = -2x$	$-2x$ does not equal $5x$.
$(2x - 2)(2x + 3)$	$6x - 4x = 2x$	$2x$ does not equal $5x$.
$(x + 3)(4x - 2)$	$-2x + 12x = 10x$	$10x$ does not equal $5x$.
$(x - 3)(4x + 2)$	$2x - 12x = -10x$	$-10x$ does not equal $5x$.

So, you can conclude that the correct factorization is

$$4x^2 + 5x - 6 = (x + 2)(4x - 3).$$

Check this result by multiplying $(x + 2)$ by $(4x - 3)$.

Exercises Within Reach®

Factoring a Trinomial of the Form $ax^2 + bx + c$ **In Exercises 37–40, determine the missing factor.**

37. $5x^2 + 18x + 9 = (x + 3)(\ 5x + 3\ )$

38. $5a^2 + 12a - 9 = (a + 3)(\ 5a - 3\ )$

39. $2y^2 - 3y - 27 = (y + 3)(\ 2y - 9\ )$

40. $3y^2 - y - 30 = (y + 3)(\ 3y - 10\ )$

Factoring a Trinomial of the Form $ax^2 + bx + c$ **In Exercises 41–46, factor the trinomial.**

41. $5z^2 + 2z - 3$ $(5z - 3)(z + 1)$

42. $3y^2 - 10y + 8$ $(3y - 4)(y - 2)$

43. $2t^2 - 7t - 4$ $(2t + 1)(t - 4)$

44. $3z^2 - z - 4$ $(3z - 4)(z + 1)$

45. $6x^2 - 5x - 25$ $(3x + 5)(2x - 5)$

46. $15x^2 + 4x - 3$ $(5x + 3)(3x - 1)$

Factoring Trinomials by Grouping

Study Tip

Factoring by grouping can be more efficient than the *guess, check*, and *revise* method, especially when the coefficients *a* and *c* have many factors.

Guidelines for Factoring $ax^2 + bx + c$ by Grouping

1. If necessary, write the trinomial in standard form.
2. Choose factors of the product ac that add up to b.
3. Use these factors to rewrite the middle term as a sum or difference.
4. Group and remove a common monomial factor from the first two terms and the last two terms.
5. If possible, factor out the common binomial factor.

EXAMPLE 7 Factoring a Trinomial by Grouping

Use factoring by grouping to factor the trinomial $3x^2 + 5x - 2$.

Additional Examples

Factor each trinomial.

a. $3x^2 + 11x - 4$

b. $24x^3 - 42x^2 + 9x$

Answers:

a. $(3x - 1)(x + 4)$

b. $3x(4x - 1)(2x - 3)$

SOLUTION

For the trinomial $3x^2 + 5x - 2$, $a = 3$ and $c = -2$, which implies that the product ac is -6. Now, because -6 factors as $(6)(-1)$, and $6 - 1 = 5 = b$, you can rewrite the middle term as $5x = 6x - x$. This produces the following result.

$$3x^2 + 5x - 2 = 3x^2 + (6x - x) - 2 \quad \text{Rewrite middle term.}$$

$$= (3x^2 + 6x) - (x + 2) \quad \text{Group terms.}$$

$$= 3x(x + 2) - (x + 2) \quad \text{Factor out common monomial factor in first group.}$$

$$= (x + 2)(3x - 1) \quad \text{Distributive Property}$$

So, the trinomial factors as $3x^2 + 5x - 2 = (x + 2)(3x - 1)$. Check this result as follows.

$$(x + 2)(3x - 1) = 3x^2 - x + 6x - 2 \quad \text{FOIL Method}$$

$$= 3x^2 + 5x - 2 \quad \text{Combine like terms.}$$

Exercises Within Reach®

Solutions in English & Spanish and tutorial videos at AlgebraWithinReach.com

Factoring a Trinomial by Grouping **In Exercises 47–56, use factoring by grouping to factor the trinomial.**

47. $3x^2 + 10x + 8$ $(3x + 4)(x + 2)$

48. $2x^2 + 9x + 9$ $(x + 3)(2x + 3)$

49. $5x^2 - 12x - 9$ $(5x + 3)(x - 3)$

50. $7x^2 - 13x - 2$ $(7x + 1)(x - 2)$

51. $15x^2 - 11x + 2$ $(3x - 1)(5x - 2)$

52. $12x^2 - 28x + 15$ $(2x - 3)(6x - 5)$

53. $10y^2 - 7y - 12$ $(5y + 4)(2y - 3)$

54. $6x^2 - x - 15$ $(3x - 5)(2x + 3)$

55. $2 + 5x - 12x^2$ $-(3x - 2)(4x + 1)$

56. $2 + x - 6x^2$ $-(3x - 2)(2x + 1)$

Summary of Factoring

Guidelines for Factoring Polynomials

1. Factor out any common factors.
2. Factor according to one of the special polynomial forms: difference of two squares, sum or difference of two cubes, or perfect square trinomials.
3. Factor trinomials, which have the form $ax^2 + bx + c$, using the methods for $a = 1$ and $a \neq 1$.
4. For polynomials with four terms, factor by grouping.
5. Check to see whether the factors themselves can be factored.
6. Check the results by multiplying the factors.

EXAMPLE 8 Factoring Polynomials

a. $3x^2 - 108 = 3(x^2 - 36)$ Factor out common factor.

$= 3(x + 6)(x - 6)$ Difference of two squares

b. $4x^3 - 32x^2 + 64x = 4x(x^2 - 8x + 16)$ Factor out common factor.

$= 4x(x - 4)^2$ Factor as perfect square trinomial.

c. $6x^3 + 27x^2 - 15x = 3x(2x^2 + 9x - 5)$ Factor out common factor.

$= 3x(2x - 1)(x + 5)$ Factor.

d. $x^3 - 3x^2 - 4x + 12 = (x^3 - 3x^2) + (-4x + 12)$ Group terms.

$= x^2(x - 3) - 4(x - 3)$ Factor out common factors.

$= (x - 3)(x^2 - 4)$ Distributive Property

$= (x - 3)(x + 2)(x - 2)$ Difference of two squares

Exercises Within Reach®

Solutions in English & Spanish and tutorial videos at AlgebraWithinReach.com

Factoring a Polynomial **In Exercises 57–64, factor the expression completely.**

57. $3x^3 - 3x$ $3x(x + 1)(x - 1)$

58. $20y^2 - 45$ $5(2y + 3)(2y - 3)$

59. $3x^3 - 18x^2 + 27x$ $3x(x - 3)^2$

60. $16z^3 - 56z^2 + 49z$ $z(4z - 7)^2$

61. $10t^3 + 2t^2 - 36t$ $2t(5t - 9)(t + 2)$

62. $8x^4 - 4x^3 - 24x^2$ $4x^2(2x + 3)(x - 2)$

63. $x^3 + 2x^2 - 16x - 32$ $(x + 2)(x + 4)(x - 4)$

64. $x^3 - 7x^2 - 4x + 28$ $(x + 2)(x - 2)(x - 7)$

Geometry **In Exercises 65 and 66, write an expression in factored form for the area of the shaded region of the figure.**

65.

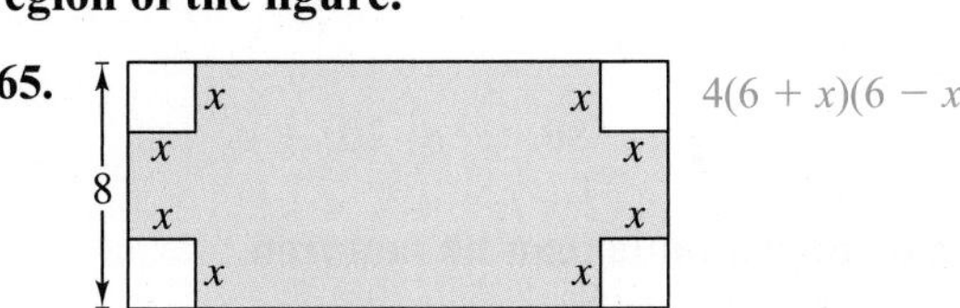

$4(6 + x)(6 - x)$

66.

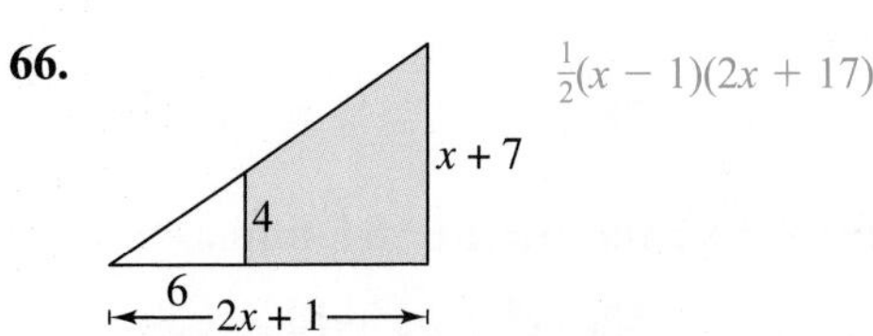

$\frac{1}{2}(x - 1)(2x + 17)$

Concept Summary: Factoring Trinomials

What

You can use some simple guidelines to help factor trinomials.

EXAMPLE

Factor the trinomial.

$x^2 + 5x + 6$

How

Guidelines for factoring $x^2 + bx + c$

1. When c is *positive*, its factors have like signs that match the sign of b.
2. When c is *negative*, its factors have unlike signs.
3. When $|b|$ is small relative to $|c|$, first try factors of c that are close to each other in absolute value.
4. When $|b|$ is near $|c|$, first try those factors of c that are farthest from each other in absolute value.

EXAMPLE

$x^2 + 5x + 6 = (x + 2)(x + 3)$

Why

You will need to know various factoring techniques when you learn how to solve polynomial equations. Here are some general guidelines for factoring polynomials.

1. Factor out any common factors.
2. Factor as a special form, if possible.
3. Factor using the methods for the forms $x^2 + bx + c$ or $ax^2 + bx + c$.
4. For polynomials with four terms, factor by grouping.
5. Factor the factors, if possible.
6. Check by multiplying factors.

Exercises Within Reach ®

Worked-out solutions to odd-numbered exercises at AlgebraWithinReach.com

Concept Summary Check

67. ***Guidelines for Factoring $x^2 + bx + c$*** When factoring $x^2 + 5x + 6$, how do you know that the signs of the factors of 6 are both positive?
See Additional Answers.

68. ***Special Polynomial Forms*** Name four special polynomial forms. See Additional Answers.

69. ***Guidelines for Factoring Polynomials*** By following the guidelines for factoring polynomials, what method should you use to factor $x^2 + 6x + 9$? The method for factoring a perfect square trinomial

70. ***Choosing a Factoring Method*** Why should you use factoring by grouping to factor $x^3 - 4x^2 - x - 4$?
When there are four terms you should factor by grouping.

Extra Practice

Geometric Modeling **In Exercises 71 and 72, write the factoring formula represented by the geometric model.**

71. 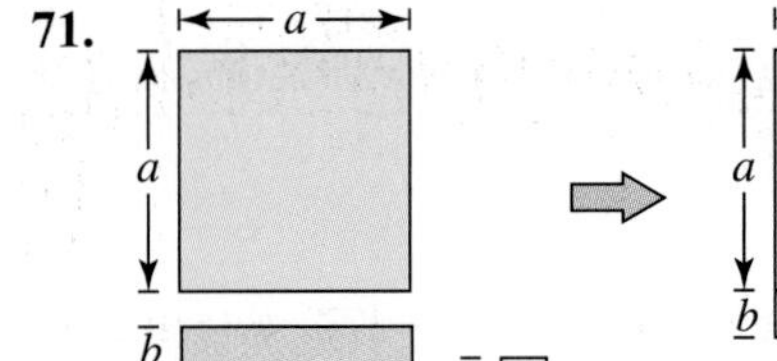

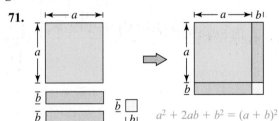

$a^2 + 2ab + b^2 = (a + b)^2$

72. 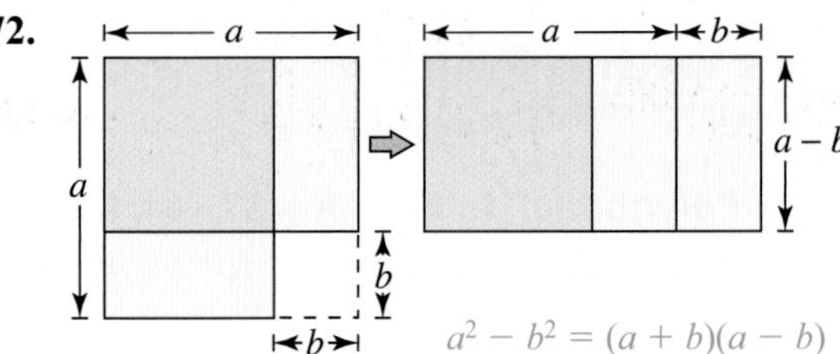

$a^2 - b^2 = (a + b)(a - b)$

Finding Coefficients **In Exercises 73–76, find two real numbers b such that the expression is a perfect square trinomial.**

73. $x^2 + bx + 81$ ± 18
74. $x^2 + bx + 49$ ± 14
75. $4x^2 + bx + 9$ ± 12
76. $16x^2 + bxy + 25y^2$ ± 40

Finding a Constant Term **In Exercises 77–80, find a real number c such that the expression is a perfect square trinomial.**

77. $x^2 + 8x + c$ 16
78. $x^2 + 12x + c$ 36
79. $y^2 - 6y + c$ 9
80. $z^2 - 20z + c$ 100

Finding Coefficients **In Exercises 81–84, find all integers b such that the trinomial can be factored.**

81. $x^2 + bx + 8$ $\pm 6, \pm 9$
82. $x^2 + bx + 10$ $\pm 7, \pm 11$
83. $x^2 + bx - 21$ $\pm 4, \pm 20$
84. $x^2 + bx - 7$ ± 6

Factoring a Polynomial **In Exercises 85–94, factor the expression, if possible.**

85. $4w^2 - 3w + 8$ Prime

86. $-6x^2 + 5x - 6$ Prime

87. $60y^3 + 35y^2 - 50y$ $5y(3y - 2)(4y + 5)$

88. $12x^2 + 42x^3 - 54x^4$ $-6x^2(x - 1)(9x + 2)$

89. $54x^3 - 2$ $2(3x - 1)(9x^2 + 3x + 1)$

90. $3t^3 - 24$ $3(t - 2)(t^2 + 2t + 4)$

91. $49 - (r - 2)^2$ $-(r + 5)(r - 9)$

92. $(x + 7y)^2 - 4a^2$ $(x + 7y + 2a)(x + 7y - 2a)$

93. $x^8 - 1$ $(x^4 + 1)(x^2 + 1)(x + 1)(x - 1)$

94. $x^4 - 16y^4$ $(x^2 + 4y^2)(x + 2y)(x - 2y)$

95. ***Number Problem*** For any integer n, the polynomial $8n^3 + 24n^2 + 16n$ represents the product of three consecutive even integers.

(a) Factor $8n^3 + 24n^2 + 16n$. $8n(n + 1)(n + 2)$

(b) Explain how to rewrite the factored form as $2n(2n + 2)(2n + 4)$. Factor 8 and distribute a factor of 2 into each binomial factor.

(c) When $n = 10$, what are the three integers? 20, 22, 24

96. ***Number Problem*** For any integer n, the polynomial $8n^3 + 12n^2 - 2n - 3$ represents the product of three consecutive odd integers.

(a) Factor $8n^3 + 12n^2 - 2n - 3$. $(2n - 1)(2n + 1)(2n + 3)$

(b) When $n = 15$, what are the three integers? 29, 31, 33

Explaining Concepts

97. ***Writing*** In your own words, explain how to factor $x^2 - 5x + 6$. Begin by finding the factors of 6 whose sum is -5. They are -2 and -3. The factorization is $(x - 2)(x - 3)$.

98. ***Writing*** Explain how you can check the factors of a trinomial. Give an example. To check the factors of a trinomial, multiply the factors to yield the original trinomial. The factors of $x^2 - 5x + 6$ are $x - 2$ and $x - 3$ because $(x - 2)(x - 3) = x^2 - 5x + 6$.

99. ***Structure*** Give an example of a prime trinomial. $x^2 + x + 1$

100. ***Error Analysis*** Describe and correct the error. See Additional Answers.

$$9x^2 - 9x - 54 = (3x + 6)(3x - 9)$$
$$= 3(x + 2)(x - 3)$$

101. ***Precision*** Is $x(x + 2) - 2(x + 2)$ completely factored? If not, show the complete factorization.

No. $x(x + 2) - 2(x + 2) = (x + 2)(x - 2)$

102. ***Precision*** Is $(2x - 4)(x + 1)$ completely factored? If not, show the complete factorization.

No. $(2x - 4)(x + 1) = 2(x - 2)(x + 1)$

103. ***Structure*** When factoring $x^2 - 7x + 12$, why is it unnecessary to test $(x - 3)(x + 4)$ or $(x + 3)(x - 4)$? For each pair of factors of 12, the signs must be the same to yield a positive product.

104. ***Creating a Test*** Create five factoring problems that you think represent a fair test of a person's factoring skills. Discuss how it is possible to *create* polynomials that are factorable.

104. Problems will vary. It is possible to create factorable polynomials by working backward: first list several factors, and then multiply them to form a single polynomial.

Cumulative Review

In Exercises 105–108, solve using a percent equation.

105. What is 125% of 340? 425

106. What is 68% of 250? 170

107. 725 is what percent of 2000? 36.25%

108. 34 is 5% of what number? 680

Mixture Problems **In Exercises 109–112, determine the numbers of units of solutions 1 and 2 needed to obtain a final solution of the specified amount and concentration.**

	Concentration of Solution 1	*Concentration of Solution 2*	*Concentration of Final Solution*	*Amount of Final Solution*
109.	20%	60%	30%	10 L
110.	30%	80%	50%	100 gal
111.	60%	90%	85%	120 gal
112.	25%	65%	45%	40 qt

109. 7.5 liters of 20% solution, 2.5 liters of 60% solution

110. 60 gallons of 30% solution, 40 gallons of 80% solution

111. 20 gallons of 60% solution, 100 gallons of 90% solution

112. 20 quarts of 25% solution, 20 quarts of 65% solution

5.6 Solving Polynomial Equations by Factoring

- Use the Zero-Factor Property to solve equations.
- Solve quadratic equations by factoring.
- Solve higher-degree polynomial equations by factoring.
- Solve application problems by factoring.

The Zero-Factor Property

Study Tip

The Zero-Factor Property is basically a formal way of saying that the only way the product of two or more factors can equal zero is if one (or more) of the factors equals zero.

Zero-Factor Property

Let a and b be real numbers, variables, or algebraic expressions. If a and b are factors such that

$$ab = 0$$

then $a = 0$ or $b = 0$. This property also applies to three or more factors.

EXAMPLE 1 Using the Zero-Factor Property

The Zero-Factor Property is the primary property for solving equations in algebra. For instance, to solve the equation

$$(x - 1)(x + 2) = 0 \qquad \text{Original equation}$$

you can use the Zero-Factor Property to conclude that either $(x - 1)$ or $(x + 2)$ equals 0. Setting the first factor equal to 0 implies that $x = 1$ is a solution.

$$x - 1 = 0 \;\Rightarrow\; x = 1 \qquad \text{First solution}$$

Similarly, setting the second factor equal to 0 implies that $x = -2$ is a solution.

$$x + 2 = 0 \;\Rightarrow\; x = -2 \qquad \text{Second solution}$$

So, the equation $(x - 1)(x + 2) = 0$ has exactly two solutions: $x = 1$ and $x = -2$. Check these solutions by substituting them into the original equation.

Exercises Within Reach®

Solutions in English & Spanish and tutorial videos at AlgebraWithinReach.com

Understanding the Zero-Factor Property **In Exercises 1–6, determine whether the equation is written in the correct form to apply the Zero-Factor Property.**

1. $x(x - 1) = 2$ No
2. $(x - 1)(x + 1) = 1$ No
3. $(x + 2) + (x - 1) = 0$ No
4. $x(x - 3) = 0$ Yes
5. $(x - 1)(x + 2) = 0$ Yes
6. $3(x^2 + x) = 0$ No

Using the Zero-Factor Property **In Exercises 7–16, use the Zero-Factor Property to solve the equation.**

7. $x(x - 4) = 0$ $0, 4$
8. $z(z + 6) = 0$ $-6, 0$
9. $(y - 3)(y + 10) = 0$ $-10, 3$
10. $(s - 7)(s + 4) = 0$ $-4, 7$
11. $25(a + 4)(a - 2) = 0$ $-4, 2$
12. $17(t - 3)(t + 8) = 0$ $-8, 3$
13. $(2t + 5)(3t + 1) = 0$ $-\frac{5}{2}, -\frac{1}{3}$
14. $(5x - 3)(2x - 8) = 0$ $\frac{3}{5}, 4$
15. $(x - 3)(2x + 1)(x + 4) = 0$ $-4, -\frac{1}{2}, 3$
16. $(y - 39)(2y + 7)(y + 12) = 0$ $-12, -\frac{7}{2}, 39$

Solving Quadratic Equations by Factoring

Definition of Quadratic Equation

A **quadratic equation** is an equation that can be written in the general form

$ax^2 + bx + c = 0$ Quadratic equation

where a, b, and c are real number with $a \neq 0$.

Study Tip

In Section 2.1, you learned that the basic idea in solving a linear equation is to *isolate the variable*. Notice in Example 2 that the basic idea in solving a quadratic equation is to factor the left side so that the equation can be converted into two linear equations.

EXAMPLE 2 Solving a Quadratic Equation by Factoring

Solve $x^2 - x - 6 = 0$.

SOLUTION

First, make sure that the right side of the equation is zero. Next, factor the left side of the equation. Finally, apply the Zero-Factor Property to find the solutions.

$x^2 - x - 6 = 0$ Write original equation.

$(x + 2)(x - 3) = 0$ Factor left side of equation.

$x + 2 = 0 \Rightarrow x = -2$ Set 1st factor equal to 0 and solve for x.

$x - 3 = 0 \Rightarrow x = 3$ Set 2nd factor equal to 0 and solve for x.

The solutions are $x = -2$ and $x = 3$.

Guidelines for Solving Quadratic Equations

1. Write the quadratic equation in general form.
2. Factor the left side of the equation.
3. Set each factor with a variable equal to zero.
4. Solve each linear equation.
5. Check each solution in the original equation.

Exercises Within Reach®

Solutions in English & Spanish and tutorial videos at AlgebraWithinReach.com

Solving a Quadratic Equation by Factoring **In Exercises 17–26, solve the equation by factoring.**

17. $x^2 + 2x = 0$ $-2, 0$

18. $x^2 - 5x = 0$ $0, 5$

19. $x^2 - 8x = 0$ $0, 8$

20. $x^2 + 16x = 0$ $-16, 0$

21. $x^2 - 25 = 0$ ± 5

22. $x^2 - 121 = 0$ ± 11

23. $x^2 - 3x - 10 = 0$ $-2, 5$

24. $x^2 - x - 12 = 0$ $-3, 4$

25. $x^2 - 10x + 24 = 0$ $4, 6$

26. $x^2 - 13x + 42 = 0$ $6, 7$

Study Tip

Be sure you see that one side of an equation must be zero to apply the Zero-Factor Property. For instance, in Example 3, you cannot simply factor the left side to obtain $x(2x + 5) = 12$ and assume that $x = 12$ and $2x + 5 = 12$ yield correct solutions. In fact, neither of the resulting solutions satisfies the original equation.

EXAMPLE 3 Solving a Quadratic Equation by Factoring

Solve $2x^2 + 5x = 12$.

SOLUTION

$2x^2 + 5x = 12$	Write original equation.
$2x^2 + 5x - 12 = 0$	Write in general form.
$(2x - 3)(x + 4) = 0$	Factor left side of equation.
$2x - 3 = 0 \Rightarrow x = \frac{3}{2}$	Set 1st factor equal to 0 and solve for x.
$x + 4 = 0 \Rightarrow x = -4$	Set 2nd factor equal to 0 and solve for x.

The solutions are $x = \frac{3}{2}$ and $x = -4$.

EXAMPLE 4 A Quadratic Equation with a Repeated Solution

Solve $x^2 - 2x + 16 = 6x$.

SOLUTION

$x^2 - 2x + 16 = 6x$	Write original equation.
$x^2 - 8x + 16 = 0$	Write in general form.
$(x - 4)^2 = 0$	Factor.
$x - 4 = 0$ or $x - 4 = 0$	Set factors equal to 0.
$x = 4$	Solve for x.

Note that even though the left side of this equation has two factors, the factors are the same. So, the only solution of the equation is

$$x = 4.$$

This solution is called a **repeated solution**.

You might tell students that Chapter 8 will introduce methods for solving quadratic equations that cannot be solved by factoring.

Exercises Within Reach®

Solutions in English & Spanish and tutorial videos at AlgebraWithinReach.com

Solving a Quadratic Equation by Factoring **In Exercises 27–40, solve the equation by factoring.**

27. $4x^2 + 15x = 25$ $-5, \frac{5}{4}$

28. $14x^2 + 9x = -1$ $-\frac{1}{2}, -\frac{1}{7}$

29. $7 + 13x - 2x^2 = 0$ $-\frac{1}{2}, 7$

30. $11 + 32y - 3y^2 = 0$ $-\frac{1}{3}, 11$

31. $3y^2 - 2 = -y$ $-1, \frac{2}{3}$

32. $-2x - 15 = -x^2$ $-3, 5$

33. $-13x + 36 = -x^2$ $4, 9$

34. $x^2 - 15 = -2x$ $-5, 3$

35. $m^2 - 8m + 18 = 2$ 4

36. $a^2 + 4a + 10 = 6$ -2

37. $x^2 + 16x + 57 = -7$ -8

38. $x^2 - 12x + 21 = -15$ 6

39. $4z^2 - 12z + 15 = 6$ $\frac{3}{2}$

40. $16t^2 + 48t + 40 = 4$ $-\frac{3}{2}$

Solving Higher-Degree Equations by Factoring

EXAMPLE 5 Solving a Polynomial Equation with Three Factors

$3x^3 = 15x^2 + 18x$	Original equation
$3x^3 - 15x^2 - 18x = 0$	Write in general form.
$3x(x^2 - 5x - 6) = 0$	Factor out common factor.
$3x(x - 6)(x + 1) = 0$	Factor.
$3x = 0 \Rightarrow x = 0$	Set 1st factor equal to 0.
$x - 6 = 0 \Rightarrow x = 6$	Set 2nd factor equal to 0.
$x + 1 = 0 \Rightarrow x = -1$	Set 3rd factor equal to 0.

The solutions are $x = 0, x = 6,$ and $x = -1$. Check these three solutions.

EXAMPLE 6 Solving a Polynomial Equation with Four Factors

$x^4 + x^3 - 4x^2 - 4x = 0$	Original equation
$x(x^3 + x^2 - 4x - 4) = 0$	Factor out common factor.
$x[(x^3 + x^2) + (-4x - 4)] = 0$	Group terms.
$x[x^2(x + 1) - 4(x + 1)] = 0$	Factor grouped terms.
$x[(x + 1)(x^2 - 4)] = 0$	Distributive Property
$x(x + 1)(x + 2)(x - 2) = 0$	Difference of two squares

$x = 0 \Rightarrow x = 0$

$x + 1 = 0 \Rightarrow x = -1$

$x + 2 = 0 \Rightarrow x = -2$

$x - 2 = 0 \Rightarrow x = 2$

The solutions are $x = 0, x = -1, x = -2,$ and $x = 2$. Check these four solutions.

Additional Examples

Solve each equation.

a. $2x^3 - 14x^2 = -20x$

b. $2x^4 - x^3 - 18x^2 + 9x = 0$

Answers:

a. $x = 0, x = 5, x = 2$

b. $x = 0, x = \frac{1}{2}, x = 3, x = -3$

Exercises Within Reach®

Solutions in English & Spanish and tutorial videos at AlgebraWithinReach.com

Solving a Polynomial Equation by Factoring **In Exercises 41–54, solve the equation by factoring.**

41. $x^3 - 19x^2 + 84x = 0$ 0, 7, 12

42. $x^3 + 18x^2 + 45x = 0$ −15, −3, 0

43. $6t^3 = t^2 + t$ $-\frac{1}{3}, 0, \frac{1}{2}$

44. $3u^3 = 5u^2 + 2u$ $-\frac{1}{3}, 0, 2$

45. $z^2(z + 2) - 4(z + 2) = 0$ ±2

46. $16(3 - u) - u^2(3 - u) = 0$ ±4, 3

47. $a^3 + 2a^2 - 9a - 18 = 0$ ±3, −2

48. $x^3 - 2x^2 - 4x + 8 = 0$ ±2

49. $c^3 - 3c^2 - 9c + 27 = 0$ ±3

50. $v^3 + 4v^2 - 4v - 16 = 0$ ±2, −4

51. $x^4 - 3x^3 - x^2 + 3x = 0$ ±1, 0, 3

52. $x^4 + 2x^3 - 9x^2 - 18x = 0$ ±3, −2, 0

53. $8x^4 + 12x^3 - 32x^2 - 48x = 0$ $\pm 2, -\frac{3}{2}, 0$

54. $9x^4 - 15x^3 - 9x^2 + 15x = 0$ $\pm 1, 0, \frac{5}{3}$

Applications

Application

EXAMPLE 7 Geometry: **Dimensions of a Room**

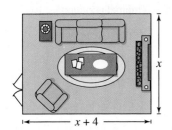

A rectangular room has an area of 192 square feet. The length of the room is 4 feet more than its width, as shown at the left. Find the dimensions of the room.

SOLUTION

Verbal Model: Length • Width = Area

Labels:
Length $= x + 4$ (feet)
Width $= x$ (feet)
Area $= 192$ (square feet)

Equation:

$$(x + 4)x = 192$$
$$x^2 + 4x - 192 = 0$$
$$(x + 16)(x - 12) = 0$$
$$x = -16 \quad \text{or} \quad x = 12$$

Because the negative solution does not make sense, choose the positive solution $x = 12$. When the width of the room is 12 feet, the length of the room is

$$\text{Length} = x + 4 = 12 + 4 = 16 \text{ feet.}$$

So, the dimensions of the room are 12 feet by 16 feet.

Exercises Within Reach®

Solutions in English & Spanish and tutorial videos at AlgebraWithinReach.com

55. ***Geometry*** The rectangular floor of a storage shed has an area of 540 square feet. The length of the floor is 7 feet more than its width (see figure). Find the dimensions of the floor. 20 feet × 27 feet

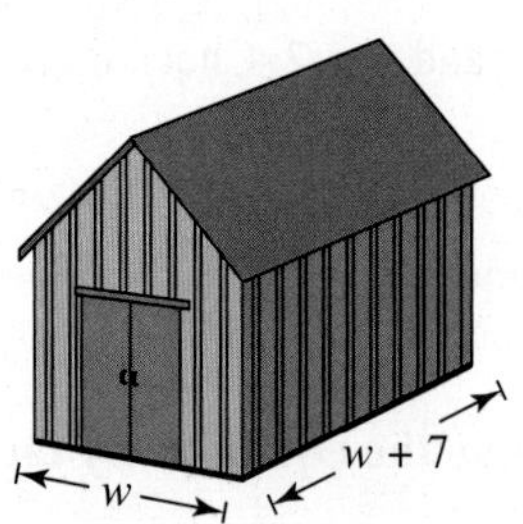

56. ***Geometry*** The outside dimensions of a picture frame are 28 centimeters and 20 centimeters (see figure). The picture alone has an area of 468 square centimeters. Find the width w of the frame. 1 centimeter

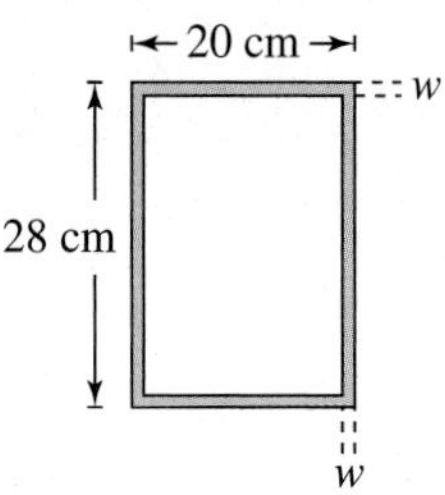

57. ***Geometry*** A triangle has an area of 27 square inches. The height of the triangle is $1\frac{1}{2}$ times its base (see figure). Find the base and height of the triangle.
Base: 6 inches; Height: 9 inches

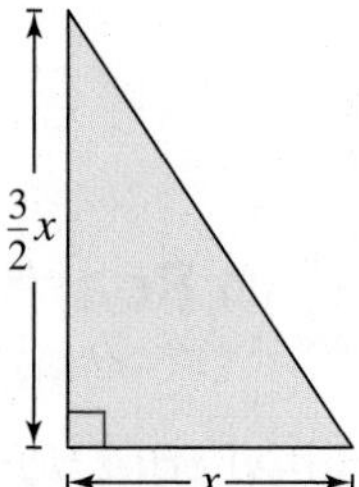

58. ***Geometry*** A triangle has an area of 60 square inches. The height of the triangle is 2 inches less than its base (see figure). Find the base and height of the triangle.
Base: 12 inches; Height: 10 inches

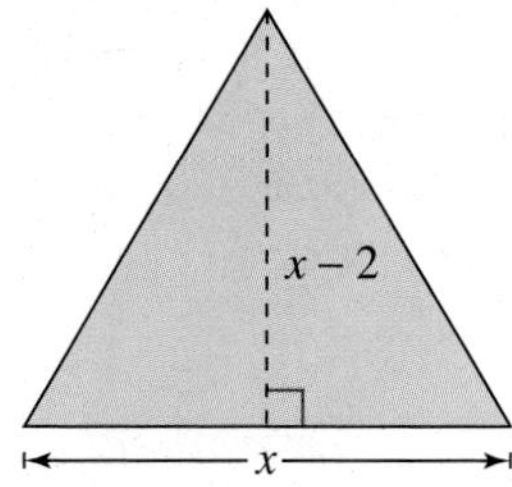

Application **EXAMPLE 8** **Free-Falling Object**

A rock is dropped into a well from a height of 64 feet above the water (see figure). The height h (in feet) of the rock relative to the surface of the water is modeled by the position function $h(t) = -16t^2 + 64$, where t is the time (in seconds) since the rock was dropped. How long does it take the rock to reach the water?

64 ft

SOLUTION

The surface of the water corresponds to a height of 0 feet. So, substitute 0 for $h(t)$ in the equation, and solve for t.

$0 = -16t^2 + 64$ Substitute 0 for $h(t)$.

$16t^2 - 64 = 0$ Write in general form.

$16(t^2 - 4) = 0$ Factor out common factor.

$16(t + 2)(t - 2) = 0$ Difference of two squares

$t = -2$ or $t = 2$ Solutions using Zero-Factor Property

Because a time of -2 seconds does not make sense, choose the positive solution $t = 2$, and conclude that the rock reaches the water 2 seconds after it is dropped.

Exercises Within Reach®

Solutions in English & Spanish and tutorial videos at AlgebraWithinReach.com

59. ***Free-Falling Object*** A tool is dropped from a construction project 400 feet above the ground. The height h (in feet) of the tool is modeled by the position function

$h(t) = -16t^2 + 400$

where t is the time in seconds. How long does it take for the tool to reach the ground? 5 seconds

60. ***Free-Falling Object*** A penny is dropped from the roof of a building 256 feet above the ground. The height h (in feet) of the penny is modeled by the position function

$h(t) = -16t^2 + 256$

where t is the time in seconds. How long does it take for the penny to reach the ground? 4 seconds

61. ***Free-Falling Object*** You throw a baseball upward with an initial velocity of 30 feet per second. The height h (in feet) of the baseball relative to your glove is modeled by the position function $h(t) = -16t^2 + 30t$, where t is the time in seconds. How long does it take for the ball to reach your glove? About 1.9 seconds

62. ***Free-Falling Object*** An object is thrown upward from the Royal Gorge Bridge in Colorado, 1053 feet above the Arkansas River, with an initial velocity of 48 feet per second. The height h (in feet) of the object is modeled by the position function $h(t) = -16t^2 + 48t + 1053$, where t is the time in seconds. How long does it take for the object to reach the river? 9.75 seconds

Dwight Smith/Shutterstock.com; Thomas W. Woodruff/Shutterstock.com

Concept Summary: *Solving Quadratic Equations by Factoring*

What

To solve a **quadratic equation** by factoring, use the guidelines for solving quadratic equations.

EXAMPLE

Solve $2x^2 + 18x = 0$.

How

- Write the quadratic equation in **general form**.
- Factor the left side of the equation.
- Set each factor with a variable equal to zero.
- Solve each linear equation.

EXAMPLE

$$2x^2 + 18x = 0$$
$$2x(x + 9) = 0$$
$$2x = 0 \Rightarrow x = 0$$
$$x + 9 = 0 \Rightarrow x = -9$$

Why

You can solve quadratic equations to answer many real-life problems, such as finding the length of time it will take a falling object to reach the ground. You can use the same guidelines to solve polynomial equations of higher degrees.

Exercises Within Reach®

Worked-out solutions to odd-numbered exercises at AlgebraWithinReach.com

Concept Summary Check

63. ***General Form*** Is the equation $2x^2 + 18x = 0$ in general form? Explain.
Yes. It has the form $ax^2 + bx + c = 0$ with $a \neq 0$.

64. ***Factoring the Left Side*** What are the factors of the left side of the equation $2x^2 + 18x = 0$?
$2x$ and $x + 9$

65. ***Using a Property*** How is the Zero-Factor Property used in the example above? The factors $2x$ and $x + 9$ are set equal to zero to form two linear equations.

66. ***Verifying Solutions*** How can you verify the solutions in the example above?
Check each solution in the original equation.

Extra Practice

Solving a Quadratic Equation by Factoring **In Exercises 67–78, solve the equation by factoring.**

67. $8x^2 = 5x$ $0, \frac{5}{8}$
68. $5x^2 = 7x$ $0, \frac{7}{5}$
69. $x(x - 5) = 36$ $-4, 9$
70. $s(s + 4) = 96$ $-12, 8$
71. $x(x + 2) - 10(x + 2) = 0$ $-2, 10$
72. $x(x - 15) + 3(x - 15) = 0$ $-3, 15$
73. $(x - 4)(x + 5) = 10$ $-6, 5$
74. $(u - 6)(u + 4) = -21$ $-1, 3$
75. $81 - (x + 4)^2 = 0$ $-13, 5$
76. $(s + 5)^2 - 49 = 0$ $-12, 2$
77. $(t - 2)^2 = 16$ $-2, 6$
78. $(s + 4)^2 = 49$ $-11, 3$

Graphical Reasoning **In Exercises 79–82, determine the x-intercepts of the graph and explain how the x-intercepts correspond to the solutions of the polynomial equation when $y = 0$.**

79. $y = x^2 - 9$

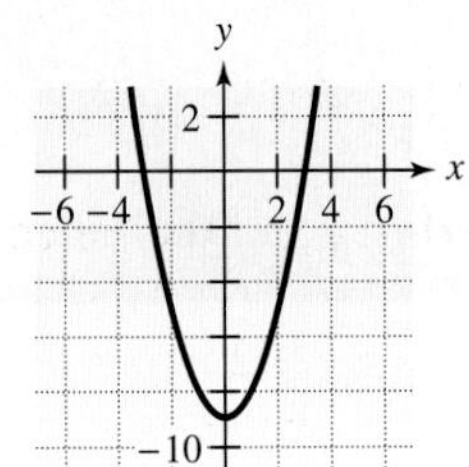

$(-3, 0), (3, 0)$; The x-intercepts are solutions of the polynomial equation.

80. 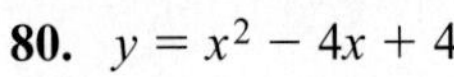
$y = x^2 - 4x + 4$

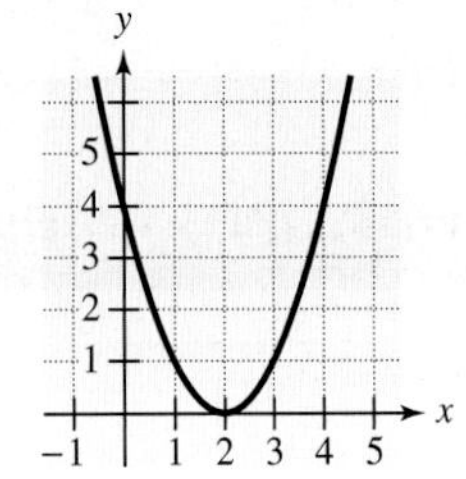

$(2, 0)$; The x-intercept is the solution of the polynomial equation.

81. 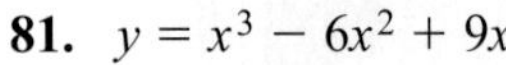
$y = x^3 - 6x^2 + 9x$

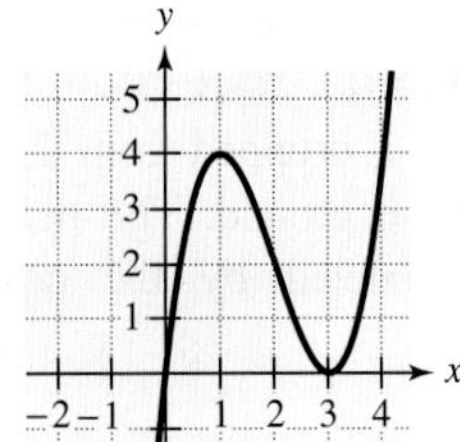

$(0, 0), (3, 0)$; The x-intercepts are solutions of the polynomial equation.

82. $y = x^3 - 3x^2 - x + 3$

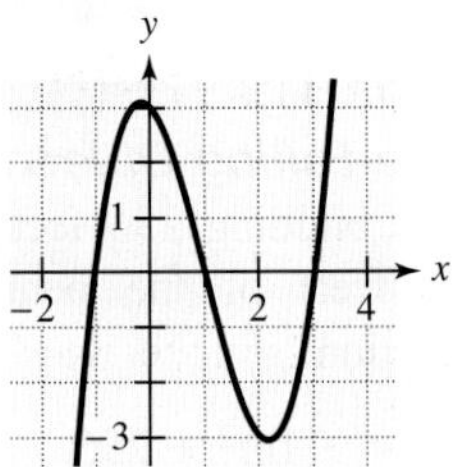

$(-1, 0), (1, 0), (3, 0)$; The x-intercepts are solutions of the polynomial equation.

Think About It **In Exercises 83 and 84, find a quadratic equation with the given solutions.**

83. $x = -2, \quad x = 6$ $x^2 - 4x - 12 = 0$

84. $x = -2, \quad x = 4$ $x^2 - 2x - 8 = 0$

85. ***Number Problem*** The sum of a positive number and its square is 240. Find the number. 15

86. ***Number Problem*** Find two consecutive positive integers whose product is 132. 11, 12

87. ***Free-Falling Object*** An object falls from the roof of a building 80 feet above the ground toward a balcony 16 feet above the ground. The object's height h (in feet) relative to the ground after t seconds is modeled by the equation $h = -16t^2 + 80$. How long does it take for the object to reach the balcony? 2 seconds

88. ***Free-Falling Object*** Your friend stands 96 feet above you on a cliff. You throw an object upward with an initial velocity of 80 feet per second. The object's height h (in feet) after t seconds is modeled by the equation $h = -16t^2 + 80t$. How long does it take for the object to reach your friend on the way up? On the way down?
2 seconds; 3 seconds

89. ***Break-Even Analysis*** The revenue R from the sale of x home theater systems is given by $R = 140x - x^2$. The cost of producing x systems is given by $C = 2000 + 50x$. How many home theater systems can be produced and sold in order to break even? 40 systems or 50 systems

90. ***Break-Even Analysis*** The revenue R from the sale of x digital cameras is given by $R = 120x - x^2$. The cost of producing x digital cameras is given by $C = 1200 + 40x$. How many cameras can be produced and sold in order to break even? 20 cameras or 60 cameras

Explaining Concepts

91. ***Structure*** What is the maximum number of solutions of an nth-degree polynomial equation? Give an example of a third-degree equation that has only one real number solution. Maximum number: n. The third-degree equation $(x + 1)^3 = 0$ has only one real solution: $x = -1$.

92. ***Structure*** What is the maximum number of first-degree factors that an nth-degree polynomial equation can have? Explain. See Additional Answers.

93. ***Think About It*** A quadratic equation has a repeated solution. Describe the x-intercept(s) of the graph of the equation formed by replacing 0 with y in the general form of the equation. See Additional Answers.

94. ***Reasoning*** A third-degree polynomial equation has two solutions. What must be special about one of the solutions? Explain. One of the solutions must be repeated because there are three factors that are set equal to zero.

95. ***Reasoning*** There are some polynomial equations that have real number solutions but cannot be solved by factoring. Explain how this can be. Many polynomial equations with irrational solutions cannot be factored.

96. ***Using a Graphing Calculator*** The polynomial equation $x^3 - x - 3 = 0$ *cannot* be solved algebraically using any of the techniques described in this book. It does, however, have one solution that is a real number. See Additional Answers.

(a) Use a graphing calculator to graph the related equation $y = x^3 - x - 3$. Use the graph to estimate the solution of the original equation.
$x \approx 1.67$

(b) Use the *table* feature of a graphing calculator to create a table and estimate the solution.
$x \approx 1.67$

Cumulative Review

In Exercises 97–100, find the unit price (in dollars per ounce) of the product.

97. A 12-ounce soda for \$0.75 \$0.0625 per ounce

98. A 12-ounce package of brown-and-serve rolls for \$1.89 \$0.1575 per ounce

99. A 30-ounce can of pumpkin pie filling for \$2.13
\$0.071 per ounce

100. Turkey meat priced at \$0.94 per pound
\$0.05875 per ounce

In Exercises 101–104, find the domain of the function.
See Additional Answers.

101. $f(x) = \dfrac{x + 3}{x + 1}$

102. $f(x) = \dfrac{12}{x - 2}$

103. $g(x) = \sqrt{3 - x}$

104. $h(x) = \sqrt{x^2 - 4}$

Kletr/Shutterstock.com

5 Chapter Summary

Section	What did you learn?	Explanation and Examples	Review Exercises
5.1	Use the rules of exponents to simplify expressions *(p. 230).*	$a^m \cdot a^n = a^{m+n}$ $(ab)^m = a^m \cdot b^m$ $(a^m)^n = a^{mn}$ $\frac{a^m}{a^n} = a^{m-n}$ $\left(\frac{a}{b}\right)^m = \frac{a^m}{b^m}$	1–14
	Rewrite exponential expressions involving negative and zero exponents *(p. 231).*	$a^0 = 1$ $a^{-m} = \frac{1}{a^m}$ $\left(\frac{a}{b}\right)^{-m} = \left(\frac{b}{a}\right)^m$	15–30
	Write very large and very small numbers in scientific notation *(p. 234).*	$1{,}230{,}000 = 1.23 \times 10^6$ $0.000123 = 1.23 \times 10^{-4}$	31–42
5.2	Identify the degrees and leading coefficients of polynomials *(p. 238).*	A polynomial in x is an expression of the form $a_nx^n + a_{n-1}x^{n-1} + \cdots + a_2x^2 + a_1x + a_0$ where $a_n \neq 0$. The polynomial is of degree n, and the number a_n is called the leading coefficient. The number a_0 is called the constant term.	43–46
	Add and subtract polynomials using a horizontal format and a vertical format *(p. 239).*	To *add* two polynomials, simply combine like terms. To *subtract* one polynomial from another, add the opposite.	47–62
	Use polynomials to model and solve real-life problems *(p. 242).*	Many real-life problems can be solved by finding a sum or difference of polynomial functions.	63–66
5.3	Use the Distributive Property and the FOIL Method to multiply polynomials *(p. 246).*	Distributive Property: $(2x - 7)(3x) = 2x(3x) - 7(3x)$ $= 6x^2 - 21x$ Foil Method: $(x + a)(x + b) = \underbrace{x \cdot x}_{\text{First}} + \underbrace{b \cdot x}_{\text{Outer}} + \underbrace{a \cdot x}_{\text{Inner}} + \underbrace{a \cdot b}_{\text{Last}}$	67–80
	Use special product formulas to multiply two binomials *(p. 249).*	Sum and difference of two terms: $(u + v)(u - v) = u^2 - v^2$ Square of a binomial: $(u + v)^2 = u^2 + 2uv + v^2$ $(u - v)^2 = u^2 - 2uv + v^2$	81–86
	Use multiplication of polynomials in application problems *(p. 250).*	Many real-life quantities can be modeled by products of polynomials.	87–90
5.4	Factor greatest common monomial factors from polynomials *(p. 256).*	The following are true for the greatest common monomial factor ax^n of a polynomial in x that has integer coefficients. **1.** The coefficient a must be the greatest integer that divides each of the coefficients in the polynomial. **2.** The variable factor x^n is the highest-powered variable factor that is common to all terms of the polynomial.	91–96

	What did you learn?	Explanation and Examples	Review Exercises
5.4	Factor polynomials by grouping terms *(p. 257).*	$x^3 - 5x^2 + x - 5 = (x^3 - 5x^2) + (x-5)$ $= x^2(x - 5) + 1(x - 5)$ $= (x - 5)(x^2 + 1)$	97–102
	Factor special products and factor completely *(pp. 258, 260).*	Difference of two squares: $u^2 - v^2 = (u + v)(u - v)$ Sum of two cubes: $u^3 + v^3 = (u + v)(u^2 - uv + v^2)$ Difference of two cubes: $u^3 - v^3 = (u - v)(u^2 + uv + v^2)$	103–114
5.5	Recognize and factor perfect square trinomials *(p. 264).*	Let u and v represent real numbers, variables, or algebraic expressions. **1.** $u^2 + 2uv + v^2 = (u + v)^2$ **2.** $u^2 - 2uv + v^2 = (u - v)^2$	115–118
	Factor trinomials of the forms $x^2 + bx + c$ and $ax^2 + bx + c$ *(pp. 265, 267).*	To factor $x^2 + bx + c$, find two factors of c whose sum is b. The product of 4 and −1 is −4. $x^2 + 3x - 4 = (x + 4)(x - 1)$ The sum of 4 and −1 is 3. To factor $ax^2 + bx + c$, find a combination of factors a and c such that the outer and inner products add up to the middle term bx.	119–124
	Factor trinomials of the form $ax^2 + bx + c$ by grouping *(p. 268).*	**1.** If necessary, write the trinomial in standard form. **2.** Choose factors of the product ac that add up to b. **3.** Use these factors to rewrite the middle term as a sum or difference. **4.** Group and remove a common monomial factor from the first two terms and the last two terms. **5.** If possible, factor out the common binomial factor.	125–130
	Factor polynomials using the guidelines for factoring *(p. 269).*	**1.** Factor out any common factors. **2.** Factor according to one of the special polynomial forms. **3.** Factor trinomials. **4.** For polynomials with four terms, factor by grouping. **5.** Check to see whether the factors can be factored. **6.** Check the results by multiplying the factors.	131–138
5.6	Use the Zero-Factor Property to solve equations *(p. 272).*	Let a and b be real numbers, variables, or algebraic expressions. If a and b are factors such that $ab = 0$, then $a = 0$ or $b = 0$. This property also applies to three or more factors.	139–144
	Solve quadratic equations by factoring *(p. 273).*	**1.** Write the quadratic equation in general form. **2.** Factor the left side of the equation. **3.** Set each factor with a variable equal to zero. **4.** Solve each linear equation. **5.** Check each solution in the original equation.	145–152
	Solve higher-degree polynomial equations by factoring *(p. 275).*	Use the same steps as for solving quadratic equations.	153–160
	Solve application problems by factoring *(p. 276).*	When you solve application problems by factoring, check your answers. Eliminate answers that are not appropriate in the context of the problem.	161–166

Review Exercises

Worked-out solutions to odd-numbered exercises at AlgebraWithinReach.com

5.1

Using Rules of Exponents **In Exercises 1–14, use the rules of exponents to simplify the expression.**

1. $x^4 \cdot x^5$ x^9
2. $-3y^2 \cdot y^4$ $-3y^6$
3. $(u^2)^3$ u^6
4. $(v^4)^2$ v^8
5. $(-2z)^3$ $-8z^3$
6. $(-3y)^2(2)$ $18y^2$
7. $-(u^2v)^2(-4u^3v)$ $4u^7v^3$
8. $(12x^2y)(3x^2y^4)^2$ $108x^6y^9$
9. $\dfrac{12z^5}{6z^2}$ $2z^3$
10. $\dfrac{15m^3}{25m}$ $\dfrac{3m^2}{5}$
11. $\dfrac{25g^4d^2}{80g^2d^2}$ $\dfrac{5g^2}{16}$
12. $\dfrac{-48u^8v^6}{(-2u^2v)^3}$ $-6u^2v^3$
13. $\left(\dfrac{72x^4}{6x^2}\right)^2$ $144x^4$
14. $\left(-\dfrac{y^2}{2}\right)^3$ $-\dfrac{y^6}{8}$

Using Rules of Exponents **In Exercises 15–18, use the rules of exponents to evaluate the expression.**

15. $(2^3 \cdot 3^2)^{-1}$ $\frac{1}{72}$
16. $(2^{-2} \cdot 5^2)^{-2}$ $\frac{16}{625}$
17. $\left(\dfrac{3}{4}\right)^{-3}$ $\frac{64}{27}$
18. $\left(\dfrac{1}{3^{-2}}\right)^2$ 81

Using Rules of Exponents **In Exercises 19–30, rewrite the expression using only positive exponents, and simplify. (Assume that any variables in the expression are nonzero.)**

19. $(6y^4)(2y^{-3})$ $12y$
20. $4(-3x)^{-3}$ $-\dfrac{4}{27x^3}$
21. $\dfrac{4x^{-2}}{2x}$ $\dfrac{2}{x^3}$
22. $\dfrac{15t^5}{24t^{-3}}$ $\dfrac{5t^8}{8}$
23. $(x^3y^{-4})^0$ 1
24. $(5x^{-2}y^4)^{-2}$ $\dfrac{x^4}{25y^8}$
25. $\dfrac{7a^6b^{-2}}{14a^{-1}b^4}$ $\dfrac{a^7}{2b^6}$
26. $\dfrac{2u^0v^{-2}}{10u^{-1}v^{-3}}$ $\dfrac{uv}{5}$
27. $\left(\dfrac{3x^{-1}y^2}{12x^5y^{-3}}\right)^{-1}$ $\dfrac{4x^6}{y^5}$
28. $\left(\dfrac{4x^{-3}z^{-1}}{8x^4z}\right)^{-2}$ $4x^{14}z^4$
29. $u^3(5u^0v^{-1})(9u)^2$ $\dfrac{405u^5}{v}$
30. $a^4(16a^{-2}b^4)(2b)^{-3}$ $2a^2b$

Writing in Scientific Notation **In Exercises 31–34, write the number in scientific notation.**

31. 0.0000319 3.19×10^{-5}
32. 0.0000008924 8.924×10^{-7}
33. 17,350,000 1.735×10^7
34. 849,600,000 8.496×10^8

Writing in Decimal Notation **In Exercises 35–38, write the number in decimal notation.**

35. 1.95×10^6 1,950,000
36. 7.025×10^4 70,250
37. 2.05×10^{-5} 0.0000205
38. 6.118×10^{-8} 0.00000006118

Using Scientific Notation **In Exercises 39–42, evaluate the expression without using a calculator.**

39. $(6 \times 10^3)^2$ 3.6×10^7
40. $(3 \times 10^{-3})(8 \times 10^7)$ 2.4×10^5
41. $\dfrac{3.5 \times 10^7}{7 \times 10^4}$ 500
42. $\dfrac{1}{(6 \times 10^{-3})^2}$ $\dfrac{250{,}000}{9}$

5.2

Determining the Degree and Leading Coefficient **In Exercises 43–46, write the polynomial in standard form, and determine its degree and leading coefficient.**

43. $6x^3 - 4x + 5x^2 - x^4$ $-x^4 + 6x^3 + 5x^2 - 4x$; 4; -1
44. $2x^6 - 5x^3 + x^5 - 7$ $2x^6 + x^5 - 5x^3 - 7$; 6; 2
45. $x^4 + 3x^5 - 4 - 6x$ $3x^5 + x^4 - 6x - 4$; 5; 3
46. $9x - 2x^3 + x^5 - 8x^7$ $-8x^7 + x^5 - 2x^3 + 9x$; 7; -8

Adding or Subtracting Polynomials **In Exercises 47–58, use a horizontal format to find the sum or difference.**

2 or 3

47. $(10x + 8) + (x^2 + 3x)$ $x^2 + 13x + 8$

48. $(6 - 4u) + (2u^2 - 3u)$ $2u^2 - 7u + 6$

49. $(5x^3 - 6x + 11) + (5 + 6x - x^2 - 8x^3)$
$-3x^3 - x^2 + 16$

50. $(7 - 12x^2 + 8x^3) + (x^4 - 6x^3 + 7x^2 - 5)$
$x^4 + 2x^3 - 5x^2 + 2$

51. $(3y - 4) - (2y^2 + 1)$ $-2y^2 + 3y - 5$

52. $(x^2 - 5) - (3 - 6x)$ $x^2 + 6x - 8$

53. $(-x^3 - 3x) - 4(2x^3 - 3x + 1)$ $-9x^3 + 9x - 4$

54. $(7z^2 + 6z) - 3(5z^2 + 2z)$ $-8z^2$

55. $3y^2 - [2y + 3(y^2 + 5)]$ $-2y - 15$

56. $(16a^3 + 5a) - 5[a + (2a^3 - 1)]$ $6a^3 + 5$

57. $(3x^5 + 4x^2 - 8x + 12) - (2x^5 + x) + (3x^2 - 4x^3 - 9)$
$x^5 - 4x^3 + 7x^2 - 9x + 3$

58. $(7x^4 - 10x^2 + 4x) + (x^3 - 3x) - (3x^4 - 5x^2 + 1)$
$4x^4 + x^3 - 5x^2 + x - 1$

Adding or Subtracting Polynomials **In Exercises 59–62, use a vertical format to find the sum or difference.**

59. $\begin{array}{r} 3x^2 + 5x \\ \underline{-4x^2 - x + 6} \end{array}$
$-x^2 + 4x + 6$

60. $\begin{array}{r} 6x + 1 \\ \underline{x^2 - 4x} \end{array}$
$x^2 + 2x + 1$

61. $\begin{array}{r} 3t - 5 \\ \underline{-(t^2 - t - 5)} \end{array}$
$-t^2 + 4t$

62. $\begin{array}{r} 10y^2 \quad + 3 \\ \underline{-(y^2 + 4y - 9)} \end{array}$
$9y^2 - 4y + 12$

Geometry **In Exercises 63 and 64, write and simplify an expression for the perimeter of the figure. Then find the perimeter of the figure for the given value of x.**

63. $x = \frac{2}{3}$

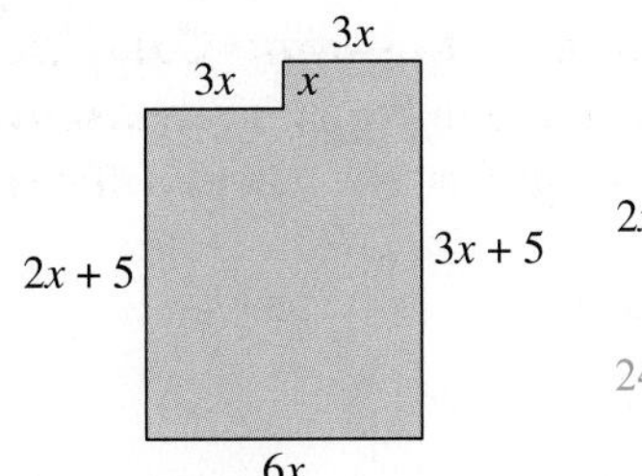

$18x + 10$; 22

64. $x = 3.5$

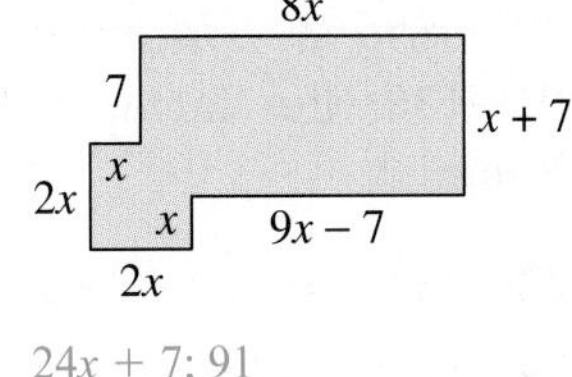

$24x + 7$; 91

65. ***Cost, Revenue, and Profit*** A manufacturer can produce and sell x backpacks per week. The total cost C (in dollars) of producing the backpacks is given by

$$C = 16x + 3000$$

and the total revenue R is given by $R = 35x$. Find the profit P obtained by selling 1200 backpacks per week.
\$19,800

66. ***Cost, Revenue, and Profit*** A manufacturer can produce and sell x notepads per week. The total cost C (in dollars) of producing the notepads is given by

$$C = 0.8x + 1000$$

and the total revenue R is given by $R = 1.6x$. Find the profit P obtained by selling 5000 notepads per week.
\$3000

5.3

Multiplying Polynomials **In Exercises 67–80, perform the multiplication and simplify.**

67. $-2x^3(x + 4)$ $-2x^4 - 8x^3$

68. $-4y^2(y - 2)$ $-4y^3 + 8y^2$

69. $3x(2x^2 - 5x + 3)$ $6x^3 - 15x^2 + 9x$

70. $-2y(5y^2 - y - 4)$ $-10y^3 + 2y^2 + 8y$

71. $(x - 2)(x + 7)$ $x^2 + 5x - 14$

72. $(u + 5)(u - 8)$ $u^2 - 3u - 40$

73. $(5x + 3)(3x - 4)$ $15x^2 - 11x - 12$

74. $(4x - 1)(2x - 5)$ $8x^2 - 22x + 5$

75. $(4x^2 + 3)(6x^2 + 1)$ $24x^4 + 22x^2 + 3$

76. $(3y^2 + 2)(4y^2 - 5)$ $12y^4 - 7y^2 - 10$

77. $(2x^2 - 3x + 2)(2x + 3)$ $4x^3 - 5x + 6$

78. $(5s^3 + 4s - 3)(4s - 5)$ $20s^4 - 25s^3 + 16s^2 - 32s + 15$

79. $2u(u - 7) - (u + 1)(u - 7)$ $u^2 - 8u + 7$

80. $(3v + 2)(-5v) + 5v(3v + 2)$ 0

Finding a Special Product **In Exercises 81–86, use a special product formula to find the product.**

81. $(4x - 7)^2$ $16x^2 - 56x + 49$

82. $(2x + 3y)^2$ $4x^2 + 12xy + 9y^2$

83. $(6v + 9)(6v - 9)$ $36v^2 - 81$

84. $(5x - 2y)(5x + 2y)$ $25x^2 - 4y^2$

85. $[(u - 3) + v][(u - 3) - v]$ $u^2 - v^2 - 6u + 9$

86. $[(m - 5) + n]^2$ $m^2 + n^2 - 10m + 2mn - 10n + 25$

Geometry **In Exercises 87 and 88, write an expression for the area of the shaded region of the figure. Then simplify the expression.**

87. 

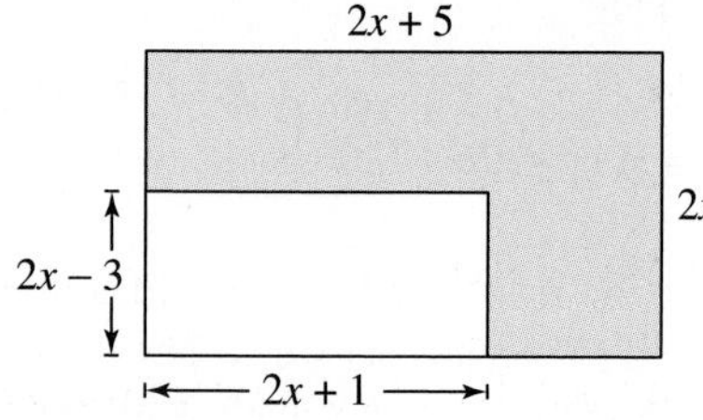

$2x(2x + 5) - (2x - 3)(2x + 1)$; $14x + 3$

88. $\frac{1}{2}(3x)(3x + 10) - \frac{1}{2}(2x)(3x)$; $\frac{3}{2}x^2 + 15x$

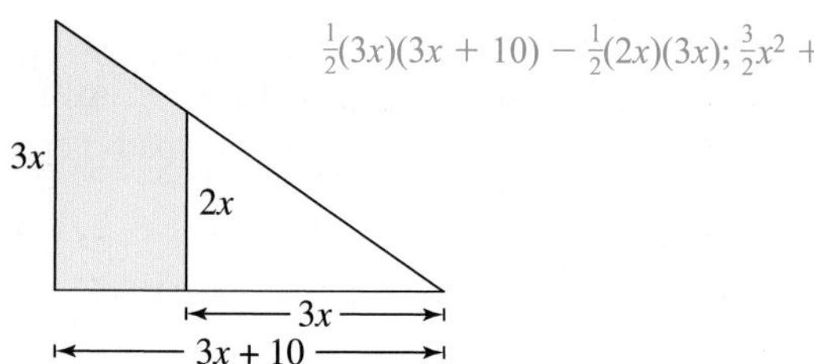

89. ***Compound Interest*** After 2 years, an investment of \$1000 compounded annually at an interest rate of 6% will yield the amount $1000(1 + 0.06)^2$. Find this product. \$1123.60

90. ***Compound Interest*** After 2 years, an investment of \$1500 compounded annually at an interest rate of r (in decimal form) will yield the amount $1500(1 + r)^2$. Find this product. $1500r^2 + 3000r + 1500$

5.4

Greatest Common Monomial Factor **In Exercises 91–96, factor out the greatest common monomial factor from the polynomial.**

91. $24x^2 - 18$ $6(4x^2 - 3)$ **92.** $14z^3 + 21$ $7(2z^3 + 3)$

93. $-3b^2 + b$ $-b(3b - 1)$ **94.** $-a^3 - 4a$ $-a(a^2 + 4)$

95. $6x^2 + 15x^3 - 3x$ $3x(2x + 5x^2 - 1)$ **96.** $8y - 12y^2 + 24y^3$ $4y(2 - 3y + 6y^2)$

A Common Binomial Factor **In Exercises 97 and 98, factor the expression by factoring out the common binomial factor.**

97. $28(x + 5) - 70(x + 5)$ $-42(x + 5)$

98. $(u - 9v)(u - v) + v(u - 9v)$ $u(u - 9v)$

Factoring by Grouping **In Exercises 99–102, factor the polynomial by grouping.**

99. $v^3 - 2v^2 - v + 2$ $(v + 1)(v - 1)(v - 2)$ **100.** $y^3 + 4y^2 - y - 4$ $(y + 1)(y - 1)(y + 4)$

101. $t^3 + 3t^2 + 3t + 9$ $(t + 3)(t^2 + 3)$ **102.** $x^3 + 7x^2 + 3x + 21$ $(x + 7)(x^2 + 3)$

Factoring the Difference of Two Squares **In Exercises 103–106, factor the difference of two squares.**

103. $x^2 - 36$ $(x + 6)(x - 6)$ **104.** $16y^2 - 49$ $(4y + 7)(4y - 7)$

105. $(u + 6)^2 - 81$ $(u - 3)(u + 15)$ **106.** $(y - 3)^2 - 16$ $(y + 1)(y - 7)$

Factoring the Sum or Difference of Two Cubes **In Exercises 107–110, factor the sum or difference of two cubes.**

107. $u^3 - 1$ $(u - 1)(u^2 + u + 1)$

108. $t^3 - 125$ $(t - 5)(t^2 + 5t + 25)$

109. $8x^3 + 27$ $(2x + 3)(4x^2 - 6x + 9)$

110. $64y^3 + 1$ $(4y + 1)(16y^2 - 4y + 1)$

Factoring Completely **In Exercises 111–114, factor the polynomial completely.**

111. $x^3 - x$ $x(x + 1)(x - 1)$

112. $y^4 - 4y^2$ $y^2(y + 2)(y - 2)$

113. $24 + 3u^3$ $3(u + 2)(u^2 - 2u + 4)$

114. $54 - 2x^3$ $-2(x - 3)(x^2 + 3x + 9)$

5.5

Factoring a Perfect Square Trinomial **In Exercises 115–118, factor the perfect square trinomial.**

115. $x^2 - 18x + 81$ $(x - 9)^2$ **116.** $y^2 + 16y + 64$ $(y + 8)^2$

117. $4s^2 + 40st + 100t^2$ $(2s + 10t)^2$

118. $9u^2 - 30uv + 25v^2$ $(3u - 5v)^2$

Factoring a Trinomial **In Exercises 119–124, factor the trinomial.**

119. $x^2 + 2x - 35$ $(x + 7)(x - 5)$ **120.** $x^2 - 12x + 32$ $(x - 8)(x - 4)$

121. $2x^2 - 7x + 6$ $(2x - 3)(x - 2)$ **122.** $5x^2 + 11x - 12$ $(x + 3)(5x - 4)$

123. $18x^2 + 27x + 10$ $(3x + 2)(6x + 5)$ **124.** $12x^2 - 13x - 14$ $(3x + 2)(4x - 7)$

Factoring a Trinomial by Grouping **In Exercises 125–130, use factoring by grouping to factor the trinomial.**

125. $4x^2 - 3x - 1$ $(4x + 1)(x - 1)$

126. $12x^2 - 7x + 1$ $(4x - 1)(3x - 1)$

127. $5x^2 - 12x + 7$ $(5x - 7)(x - 1)$

128. $3u^2 + 7u - 6$ $(3u - 2)(u + 3)$

129. $7s^2 + 10s - 8$ $(7s - 4)(s + 2)$

130. $3x^2 - 13x - 10$ $(3x + 2)(x - 5)$

Factoring a Polynomial **In Exercises 131–138, factor the expression completely.**

131. $4a - 64a^3$ $4a(1 + 4a)(1 - 4a)$

132. $3b + 27b^3$ $3b(1 + 9b^2)$

133. $z^3 + z^2 + 3z + 3$ $(z^2 + 3)(z + 1)$

134. $x^3 + 3x^2 - 4x - 12$ $(x + 3)(x + 2)(x - 2)$

135. $\frac{1}{4}x^2 + xy + y^2$ $\left(\frac{1}{2}x + y\right)^2$

136. $x^2 - \frac{2}{3}x + \frac{1}{9}$ $\left(x - \frac{1}{3}\right)^2$

137. $(x^2 - 10x + 25) - y^2$ $(x + y - 5)(x - y - 5)$

138. $u^6 - 8v^6$ $(u^2 - 2v^2)(u^4 + 2u^2v^2 + 4v^4)$

5.6

Using the Zero-Factor Property **In Exercises 139–144, use the Zero-Factor Property to solve the equation.**

139. $4x(x - 2) = 0$ $0, 2$

140. $-3x(2x + 6) = 0$ $-3, 0$

141. $(2x + 1)(x - 3) = 0$ $-\frac{1}{2}, 3$

142. $(x - 7)(3x - 8) = 0$ $\frac{8}{3}, 7$

143. $(x + 10)(4x - 1)(5x + 9) = 0$ $-10, -\frac{9}{5}, \frac{1}{4}$

144. $3x(x + 8)(2x - 7) = 0$ $-8, 0, \frac{7}{2}$

Solving a Quadratic Equation by Factoring **In Exercises 145–152, solve the quadratic equation by factoring.**

145. $3s^2 - 2s - 8 = 0$ $-\frac{4}{3}, 2$

146. $5v^2 - 12v - 9 = 0$ $-\frac{3}{5}, 3$

147. $m(2m - 1) + 3(2m - 1) = 0$ $-3, \frac{1}{2}$

148. $4w(2w + 8) - 7(2w + 8) = 0$ $-4, \frac{7}{4}$

149. $z(5 - z) + 36 = 0$ $-4, 9$

150. $(x + 3)^2 - 25 = 0$ $-8, 2$

151. $v^2 - 100 = 0$ ± 10

152. $x^2 - 121 = 0$ ± 11

Solving a Polynomial by Factoring **In Exercises 153–160, solve the equation by factoring.**

153. $2y^4 + 2y^3 - 24y^2 = 0$ $-4, 0, 3$

154. $9x^4 - 15x^3 - 6x^2 = 0$ $-\frac{1}{3}, 0, 2$

155. $x^3 - 11x^2 + 18x = 0$ $0, 2, 9$

156. $x^3 + 20x^2 + 36x = 0$ $-18, -2, 0$

157. $b^3 - 6b^2 - b + 6 = 0$ $\pm 1, 6$

158. $q^3 + 3q^2 - 4q - 12 = 0$ $\pm 2, -3$

159. $x^4 - 5x^3 - 9x^2 + 45x = 0$ $\pm 3, 0, 5$

160. $2x^4 + 6x^3 - 50x^2 - 150x = 0$ $\pm 5, -3, 0$

161. ***Number Problem*** Find two consecutive positive odd integers whose product is 99. 9, 11

162. ***Number Problem*** Find two consecutive positive even integers whose product is 168. 12, 14

163. ***Geometry*** A rectangle has an area of 900 square inches. The length of the rectangle is $2\frac{1}{4}$ times its width. Find the dimensions of the rectangle. 45 inches × 20 inches

164. ***Geometry*** A closed box with a square base stands 12 inches tall. The total surface area of the outside of the box is 512 square inches. What are the dimensions of the base? (*Hint:* The surface area is given by $S = 2x^2 + 4xh$.) 8 inches × 8 inches

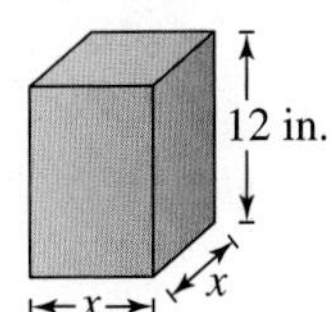

165. ***Free-Falling Object*** An object is dropped from a weather balloon 3600 feet above the ground. The height h (in feet) of the object is modeled by the position equation $h = -16t^2 + 3600$, where t is the time (in seconds). How long does it take the object to reach the ground? 15 seconds

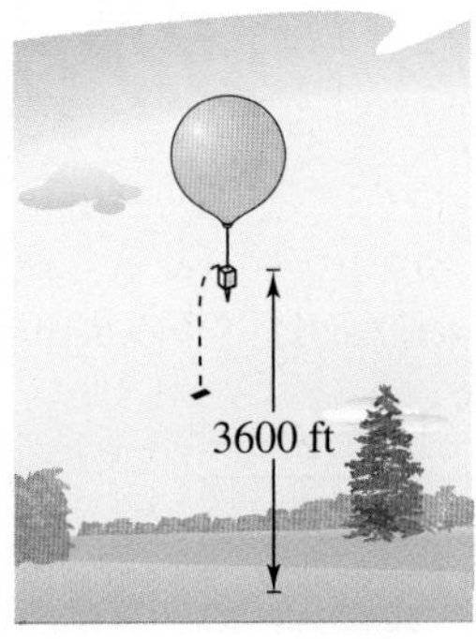

166. ***Free-Falling Object*** An object is thrown upward from the Trump Tower in New York City, which is 664 feet tall, with an initial velocity of 45 feet per second. The height h (in feet) of the object is modeled by the position equation $h = -16t^2 + 45t + 664$, where t is the time (in seconds). How long does it take the object to reach the ground? 8 seconds

Chapter Test

Solutions in English & Spanish and tutorial videos at AlgebraWithinReach.com

Take this test as you would take a test in class. After you are done, check your work against the answers in the back of the book.

1. Determine the degree and leading coefficient of $3 - 4.5x + 8.2x^3$.
   Degree: 3; Leading coefficient: 8.2
2. (a) Write 690,000,000 in scientific notation. 6.9×10^8
   (b) Write 4.72×10^{-5} in decimal notation. 0.0000472

In Exercises 3 and 4, rewrite each expression using only positive exponents, and simplify. (Assume that any variables in the expression are nonzero.)

3. (a) $\dfrac{2^{-1}x^5y^{-3}}{4x^{-2}y^2}$ $\dfrac{x^7}{8y^5}$ (b) $\left(\dfrac{-2x^2y}{z^{-3}}\right)^{-2}$ $\dfrac{1}{4x^4y^2z^6}$

4. (a) $\left(-\dfrac{2u^2}{v^{-1}}\right)^3\left(\dfrac{3v^2}{u^{-3}}\right)$ $-24u^9v^5$ (b) $\dfrac{(-3x^2y^{-1})^4}{6x^2y^0}$ $\dfrac{27x^6}{2y^4}$

In Exercises 5–9, perform the indicated operations and simplify.

5. (a) $(5a^2 - 3a + 4) + (a^2 - 4)$ $6a^2 - 3a$ (b) $(16 - y^2) - (16 + 2y + y^2)$ $-2y^2 - 2y$
6. (a) $-2(2x^4 - 5) + 4x(x^3 + 2x - 1)$ $8x^2 - 4x + 10$ (b) $4t - [3t - (10t + 7)]$ $11t + 7$
7. (a) $-3x(x - 4)$ $-3x^2 + 12x$ (b) $(2x - 3y)(x + 5y)$ $2x^2 + 7xy - 15y^2$
8. (a) $(x - 1)[2x + (x - 3)]$ $3x^2 - 6x + 3$ (b) $(2s - 3)(3s^2 - 4s + 7)$ $6s^3 - 17s^2 + 26s - 21$
9. (a) $(2w - 7)^2$ $4w^2 - 28w + 49$ (b) $[4 - (a + b)][(4 + (a + b)]$ $16 - a^2 - 2ab - b^2$

In Exercises 10–15, factor the expression completely.

10. $18y^2 - 12y$ $6y(3y - 2)$
11. $v^2 - \frac{16}{9}$ $\left(v - \frac{4}{3}\right)\left(v + \frac{4}{3}\right)$
12. $x^3 - 3x^2 - 4x + 12$ $(x + 2)(x - 2)(x - 3)$
13. $9u^2 - 6u + 1$ $(3u - 1)^2$
14. $6x^2 - 26x - 20$ $2(x - 5)(3x + 2)$
15. $x^3 + 27$ $(x + 3)(x^2 - 3x + 9)$

In Exercises 16–19, solve the equation.

16. $(x - 3)(x + 2) = 14$ $5, -4$
17. $(y + 2)^2 - 9 = 0$ $1, -5$
18. $12 + 5y - 3y^2 = 0$ $3, -\frac{4}{3}$
19. $2x^3 + 10x^2 + 8x = 0$ $-4, -1, 0$

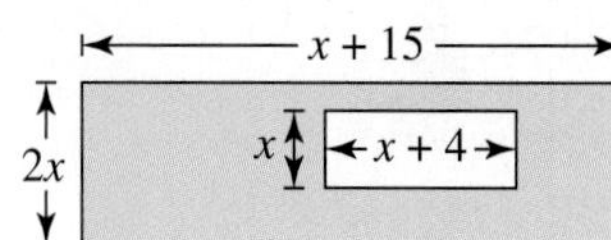

20. Write an expression for the area of the shaded region in the figure. Then simplify the expression. $x^2 + 26x$
21. The area of a rectangle is 54 square centimeters. The length of the rectangle is $1\frac{1}{2}$ times its width. Find the dimensions of the rectangle.
    6 centimeters $\times$ 9 centimeters
22. The area of a triangle is 20 square feet. The height of the triangle is 2 feet more than twice its base. Find the base and height of the triangle.
    Base: 4 feet; Height: 10 feet
23. The revenue R from the sale of x computer desks is given by $R = x^2 - 35x$. The cost C of producing x computer desks is given by $C = 150 + 12x$. How many computer desks must be produced and sold in order to break even? 50 computer desks

6 Rational Expressions, Equations, and Functions

MASTERY IS WITHIN REACH!

"No matter what I did, I kept getting really worked up about tests in my math class. Finally, I took a math study skills course the college offered. One of the most helpful strategies I learned was the ten steps to taking a math test. I went in to my math tests with a plan and felt more confident. I still get nervous, but it doesn't keep me from doing well on my tests anymore."

Kyle
Business

See page 321 for suggestions about using a test-taking strategy.

6.1 Rational Expressions and Functions

- Find the domain of a rational function.
- Simplify rational expressions.
- Use rational expressions to model and solve real-life problems.

The Domain of a Rational Function

Study Tip

Every polynomial is also a rational expression because you can consider the denominator to be 1. The domain of every polynomial is the set of all real numbers.

Definition of a Rational Expression

Let u and v be polynomials. The algebraic expression

$$\frac{u}{v}$$

is a **rational expression**. The **domain** of this rational expression is the set of all real numbers for which $v \neq 0$.

Definition of a Rational Function

Let $u(x)$ and $v(x)$ be polynomial functions. The function

$$f(x) = \frac{u(x)}{v(x)}$$

is a **rational function**. The **domain** of f is the set of all real numbers for which $v(x) \neq 0$.

A rational expression is undefined if the *denominator* is equal to 0. Students often incorrectly infer that this means that the expression is undefined if the *variable in the denominator* is equal to 0. Distinguish between these two ideas with the following examples.

1. $\frac{4-n}{n}$:
   domain = All real values of n such that $n \neq 0$.
2. $\frac{4-n}{n+6}$:
   domain = All real values of n such that $n \neq -6$.

EXAMPLE 1 Finding the Domain of a Rational Function

Find the domain of each rational function.

a. $f(x) = \frac{4}{x-2}$ **b.** $g(x) = \frac{5x}{x^2-16}$

SOLUTION

a. The denominator is 0 when $x - 2 = 0$ or $x = 2$. So, the domain is all real values of x such that $x \neq 2$.

b. The denominator is 0 when $x^2 - 16 = 0$. Solving this equation by factoring, you find that the denominator is 0 when $x = -4$ or $x = 4$. So, the domain is all real values of x such that $x \neq -4$ and $x \neq 4$.

Exercises Within Reach®

Solutions in English & Spanish and tutorial videos at AlgebraWithinReach.com

Finding the Domain of a Rational Function **In Exercises 1–6, find the domain of the rational function.** See Additional Answers.

1. $f(x) = \frac{4}{x-3}$ **2.** $g(x) = \frac{-2}{x-7}$ **3.** $f(z) = \frac{z+2}{z(z-4)}$

4. $f(x) = \frac{x^2}{x(x-1)}$ **5.** $f(t) = \frac{5t}{t^2-16}$ **6.** $f(x) = \frac{x}{x^2-4}$

Application

EXAMPLE 2 An Application Involving a Restricted Domain

You have started a small business that manufactures lamps. The initial investment for the business is \$120,000. The cost of manufacturing each lamp is \$15. So, your total cost of producing x lamps is

$C = 15x + 120{,}000.$ Cost function

Your average cost per lamp depends on the number of lamps produced. For instance, the average cost per lamp $\overline{C}$ of producing 100 lamps is

$\overline{C} = \dfrac{15(100) + 120{,}000}{100}$ Substitute 100 for x.

$= \dfrac{121{,}500}{100}$ Simplify.

$= \$1215.$ Average cost per lamp for 100 lamps

The average cost per lamp decreases as the number of lamps increases. For instance, the average cost per lamp $\overline{C}$ of producing 1000 lamps is

$\overline{C} = \dfrac{15(1000) + 120{,}000}{1000}$ Substitute 1000 for x.

$= \dfrac{135{,}000}{1000}$ Simplify.

$= \$135.$ Average cost per lamp for 1000 lamps

In general, the average cost of producing x lamps is

$\overline{C} = \dfrac{15x + 120{,}000}{x}.$ Average cost per lamp for x lamps

What is the domain of this rational function?

SOLUTION

If you were considering this function from only a mathematical point of view, you would say that the domain is all real values of x such that $x \neq 0$. However, because this function is a mathematical model representing a real-life situation, you must decide which values of x make sense in real life. For this model, the variable x represents the number of lamps that you produce. Assuming that you cannot produce a fractional number of lamps, you can conclude that the domain is the set of positive integers—that is,

$$\text{Domain} = \{1, 2, 3, 4, \ldots\}.$$

Study Tip

When a rational function is written, it is understood that the real numbers that make the denominator zero are excluded from the domain. These *implied* domain restrictions are generally not listed with the function. For instance, you know to exclude $x = 2$ and $x = -2$ from the function

$$f(x) = \frac{3x + 2}{x^2 - 4}$$

without having to list this information with the function.

Exercises Within Reach ®

Solutions in English & Spanish and tutorial videos at AlgebraWithinReach.com

Describing the Domain of a Rational Function **In Exercises 7–10, describe the domain.**

7. A rectangle of length x inches has an area of 500 square inches. The perimeter P of the rectangle is given by

$$P = 2\left(x + \frac{500}{x}\right).$$ $x > 0$

8. The cost C in millions of dollars for the government to seize p% of an illegal drug as it enters the country is given by

$$C = \frac{528p}{100 - p}.$$ $0 \le x < 100$

9. The inventory cost I when x units of a product are ordered from a supplier is given by

$$I = \frac{0.25x + 2000}{x}.$$ $\{1, 2, 3, 4, \ldots\}$

10. The average cost $\overline{C}$ for a manufacturer to produce x units of a product is given by

$$\overline{C} = \frac{1.35x + 4570}{x}.$$ $\{1, 2, 3, 4, \ldots\}$

Simplifying Rational Expressions

Simplifying Rational Expressions

Let u, v, and w represent real numbers, variables, or algebraic expressions such that $v \neq 0$ and $w \neq 0$. Then the following is valid.

$$\frac{uw}{vw} = \frac{u\cancel{w}}{v\cancel{w}} = \frac{u}{v}$$

EXAMPLE 3 Simplifying a Rational Expression

Simplify the rational expression $\dfrac{2x^3 - 6x}{6x^2}$.

SOLUTION

First note that the domain of the rational expression is all real values of x such that $x \neq 0$. Then, completely factor both the numerator and denominator.

$$\frac{2x^3 - 6x}{6x^2} = \frac{2x(x^2 - 3)}{2x(3x)} \qquad \text{Factor numerator and denominator.}$$

$$= \frac{\cancel{2x}(x^2 - 3)}{\cancel{2x}(3x)} \qquad \text{Divide out common factor } 2x.$$

$$= \frac{x^2 - 3}{3x} \qquad \text{Simplified form}$$

In simplified form, the domain of the rational expression is the same as that of the original expression—all real values of x such that $x \neq 0$.

EXAMPLE 4 Simplifying a Rational Expression

Simplify the rational expression $\dfrac{x^2 + 2x - 15}{3x - 9}$.

SOLUTION

The domain of the rational expression is all real values of x such that $x \neq 3$.

$$\frac{x^2 + 2x - 15}{3x - 9} = \frac{(x + 5)(x - 3)}{3(x - 3)} \qquad \text{Factor numerator and denominator.}$$

$$= \frac{(x + 5)\cancel{(x - 3)}}{3\cancel{(x - 3)}} \qquad \text{Divide out common factor } (x - 3).$$

$$= \frac{x + 5}{3}, \quad x \neq 3 \qquad \text{Simplified form}$$

Study Tip

Dividing out common factors can change the implied domain. In Example 4, the domain restriction $x \neq 3$ must be listed because it is no longer implied by the simplified expression.

Exercises Within Reach®

Solutions in English & Spanish and tutorial videos at AlgebraWithinReach.com

Simplifying a Rational Expression **In Exercises 11–16, simplify the rational expression.**

11. $\dfrac{3x^2 - 9x}{12x^2}$ $\dfrac{x - 3}{4x}$
12. $\dfrac{8x^3 + 4x^2}{20x}$ $\dfrac{x(2x + 1)}{5}, x \neq 0$
13. $\dfrac{x^2(x - 8)}{x(x - 8)}$ $x, x \neq 8, x \neq 0$
14. $\dfrac{a^2b(b - 3)}{b^3(b - 3)^2}$ $\dfrac{a^2}{b^2(b - 3)}$
15. $\dfrac{u^2 - 12u + 36}{u - 6}$ $u - 6, u \neq 6$
16. $\dfrac{z^2 + 22z + 121}{3z + 33}$ $\dfrac{z + 11}{3}, z \neq -11$

EXAMPLE 5 Simplifying a Rational Expression

Simplify the rational expression $\dfrac{x^3 - 16x}{x^2 - 2x - 8}$.

SOLUTION

The domain of the rational expression is all real values of x such that $x \neq -2$ and $x \neq 4$.

$$\frac{x^3 - 16x}{x^2 - 2x - 8} = \frac{x(x^2 - 16)}{(x + 2)(x - 4)} \quad \text{Partially factor.}$$

$$= \frac{x(x + 4)(x - 4)}{(x + 2)(x - 4)} \quad \text{Factor completely.}$$

$$= \frac{x(x + 4)\cancel{(x - 4)}}{(x + 2)\cancel{(x - 4)}} \quad \text{Divide out common factor } (x - 4).$$

$$= \frac{x(x + 4)}{x + 2},\ x \neq 4 \quad \text{Simplified form}$$

When you simplify a rational expression, keep in mind that you must list any domain restrictions that are no longer implied in the simplified expression. For instance, in Example 5 the restriction $x \neq 4$ is listed so that the domains agree for the original and simplified expressions. The example does not list $x \neq -2$ because this restriction is apparent by looking at either expression.

Study Tip

Be sure to *factor completely* the numerator and denominator of a rational expression in order to find any common factors. You may need to use a change in signs. Remember that the Distributive Property allows you to write $(b - a)$ as $-(a - b)$.

EXAMPLE 6 Simplification Involving a Change in Sign

Simplify the rational expression $\dfrac{2x^2 - 9x + 4}{12 + x - x^2}$.

SOLUTION

The domain of the rational expression is all real values of x such that $x \neq -3$ and $x \neq 4$.

$$\frac{2x^2 - 9x + 4}{12 + x - x^2} = \frac{(2x - 1)(x - 4)}{(4 - x)(3 + x)} \quad \text{Factor numerator and denominator.}$$

$$= \frac{(2x - 1)(x - 4)}{-(x - 4)(3 + x)} \quad (4 - x) = -(x - 4)$$

$$= \frac{(2x - 1)\cancel{(x - 4)}}{-\cancel{(x - 4)}(3 + x)} \quad \text{Divide out common factor } (x - 4).$$

$$= -\frac{2x - 1}{x + 3},\ x \neq 4 \quad \text{Simplified form}$$

Exercises Within Reach®

Solutions in English & Spanish and tutorial videos at AlgebraWithinReach.com

Simplifying a Rational Expression **In Exercises 17–22, simplify the rational expression.**

17. $\dfrac{y^3 - 4y}{y^2 + 4y - 12}$ $\dfrac{y(y + 2)}{y + 6}, y \neq 2$

18. $\dfrac{x^3 - 4x}{x^2 - 5x + 6}$ $\dfrac{x(x + 2)}{x - 3}, x \neq 2$

19. $\dfrac{3x^2 - 7x - 20}{12 + x - x^2}$ $-\dfrac{3x + 5}{x + 3}, x \neq 4$

20. $\dfrac{2x^2 + 3x - 5}{7 - 6x - x^2}$ $-\dfrac{2x + 5}{x + 7}, x \neq 1$

21. $\dfrac{2x^2 + 19x + 24}{2x^2 - 3x - 9}$ $\dfrac{x + 8}{x - 3}, x \neq -\dfrac{3}{2}$

22. $\dfrac{2y^2 + 13y + 20}{2y^2 + 17y + 30}$ $\dfrac{y + 4}{y + 6}, y \neq -\dfrac{5}{2}$

Study Tip

As you study the examples and work the exercises in this section and the next four sections, keep in mind that you are *rewriting expressions in simpler forms*. You are not solving equations. Equal signs are used in the steps of the simplification process only to indicate that the new form of the expression is *equivalent* to the original form.

Additional Examples

Simplify each rational expression.

a. $\dfrac{x^2 + 2x - 48}{3x + 24}$

b. $\dfrac{x^3 - 9x}{x^2 - 10x + 21}$

Answers:

a. $\dfrac{x - 6}{3}, x \neq -8$

b. $\dfrac{x(x + 3)}{x - 7}, x \neq 3$

EXAMPLE 7 Rational Expressions Involving Two Variables

a. $\dfrac{3xy + y^2}{2y} = \dfrac{y(3x + y)}{2y}$ Factor numerator.

$= \dfrac{\cancel{y}(3x + y)}{2\cancel{y}}$ Divide out common factor y.

$= \dfrac{3x + y}{2}, \ y \neq 0$ Simplified form

b. $\dfrac{4x^2y - y^3}{2x^2y - xy^2} = \dfrac{(4x^2 - y^2)y}{(2x - y)xy}$ Partially factor.

$= \dfrac{(2x - y)(2x + y)y}{(2x - y)xy}$ Factor completely.

$= \dfrac{\cancel{(2x - y)}(2x + y)\cancel{y}}{\cancel{(2x - y)}x\cancel{y}}$ Divide out common factors $(2x - y)$ and y.

$= \dfrac{2x + y}{x}, \ y \neq 0, \ y \neq 2x$ Simplified form

The domain of the original rational expression is all real values of x and y such that $x \neq 0$, $y \neq 0$, and $y \neq 2x$.

c. $\dfrac{2x^2 + 2xy - 4y^2}{5x^3 - 5xy^2} = \dfrac{2(x^2 + xy - 2y^2)}{5x(x^2 - y^2)}$ Partially factor.

$= \dfrac{2(x - y)(x + 2y)}{5x(x - y)(x + y)}$ Factor completely.

$= \dfrac{2\cancel{(x - y)}(x + 2y)}{5x\cancel{(x - y)}(x + y)}$ Divide out common factor $(x - y)$.

$= \dfrac{2(x + 2y)}{5x(x + y)}, \ x \neq y$ Simplified form

The domain of the original rational expression is all real values of x and y such that $x \neq 0$ and $x \neq \pm y$.

Exercises Within Reach®

Simplifying a Rational Expression **In Exercises 23–32, simplify the rational expression.**

23. $\dfrac{3xy^2}{xy^2 + x}$ $\dfrac{3y^2}{y^2 + 1}, x \neq 0$

24. $\dfrac{x + 3x^2y}{3xy + 1}$ $x, 3xy \neq -1$

25. $\dfrac{y^2 - 64x^2}{5(3y + 24x)}$ $\dfrac{y - 8x}{15}, y \neq -8x$

26. $\dfrac{x^2 - 25z^2}{2(3x + 15z)}$ $\dfrac{x - 5z}{6}, x \neq -5z$

27. $\dfrac{5xy + 3x^2y^2}{xy^3}$ $\dfrac{5 + 3xy}{y^2}, x \neq 0$

28. $\dfrac{4u^2v - 12uv^2}{18uv}$ $\dfrac{2(u - 3v)}{9}, u \neq 0, v \neq 0$

29. $\dfrac{u^2 - 4v^2}{u^2 + uv - 2v^2}$

$\dfrac{u - 2v}{u - v}, u \neq -2v$

30. $\dfrac{x^2 + 4xy}{x^2 - 16y^2}$

$\dfrac{x}{x - 4y}, x \neq -4y$

31. $\dfrac{3m^2 - 12n^2}{m^2 + 4mn + 4n^2}$

$\dfrac{3(m - 2n)}{m + 2n}$

32. $\dfrac{x^2 + xy - 2y^2}{x^2 + 3xy + 2y^2}$

$\dfrac{x - y}{x + y}, x \neq -2y$

Application

Application **EXAMPLE 8** Geometry: **Finding a Ratio**

Find the ratio of the area of the shaded portion of the triangle to the total area of the triangle.

SOLUTION

The area of the shaded portion of the triangle is given by

$$\begin{aligned}\text{Area} &= \frac{1}{2}(4x)(x+2)\\ &= \frac{1}{2}(4x^2+8x)\\ &= 2x^2+4x.\end{aligned}$$

The total area of the triangle is given by

$$\begin{aligned}\text{Area} &= \frac{1}{2}(4x+4x)(x+4)\\ &= \frac{1}{2}(8x)(x+4)\\ &= \frac{1}{2}(8x^2+32x)\\ &= 4x^2+16x.\end{aligned}$$

So, the ratio of the area of the shaded portion of the triangle to the total area of the triangle is

$$\begin{aligned}\frac{2x^2+4x}{4x^2+16x} &= \frac{2x(x+2)}{4x(x+4)}\\ &= \frac{x+2}{2(x+4)},\quad x>0.\end{aligned}$$

Exercises Within Reach®

Solutions in English & Spanish and tutorial videos at AlgebraWithinReach.com

Geometry **In Exercises 33–36, find the ratio of the area of the shaded portion to the total area of the figure.**

33. 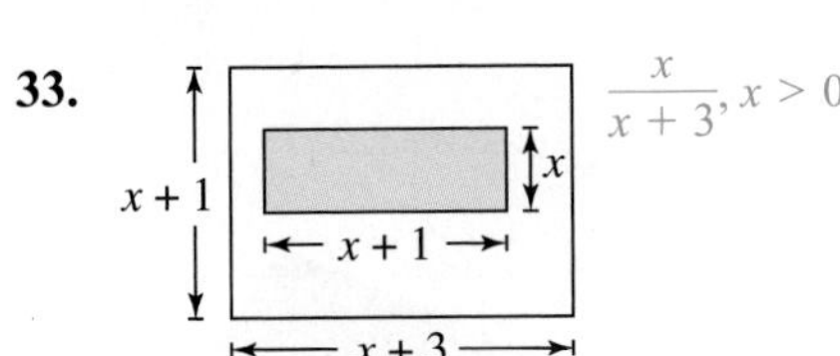

$\frac{x}{x+3}, x > 0$

34. 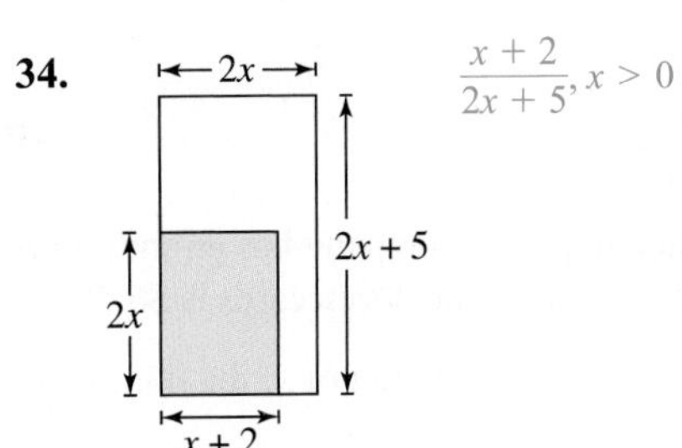

$\frac{x+2}{2x+5}, x > 0$

35. $\frac{1}{4}, x > 0$

36. 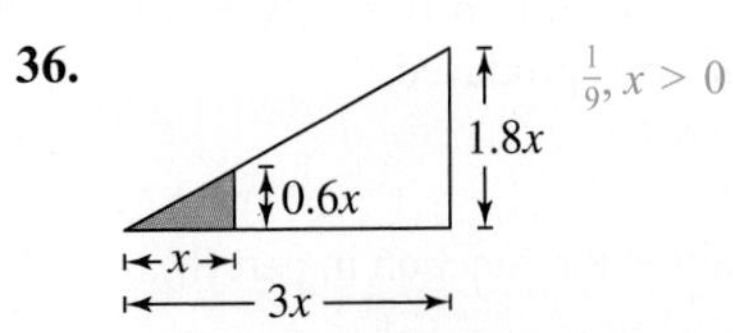

$\frac{1}{9}, x > 0$

Concept Summary: Simplifying Rational Expressions

What

A **rational expression** is in **simplified form** when its numerator and denominator have no common factors (other than ± 1).

EXAMPLE

Simplify $\frac{3x}{3x + 12}$.

How

Be sure to completely factor the numerator and denominator of a rational expression before concluding that there are no common factors.

EXAMPLE

$\frac{3x}{3x + 12}$ Original expression

$= \frac{3 \cdot x}{3(x + 4)}$ Factor completely.

$= \frac{\cancel{3} \cdot x}{\cancel{3}(x + 4)}$ Divide out factors.

$= \frac{x}{x + 4}$ Simplified form

Why

Simplifying a rational expression into a more usable form is a skill frequently used in algebra.

Exercises Within Reach ®

Worked-out solutions to odd-numbered exercises at AlgebraWithinReach.com

Concept Summary Check

37. ***Vocabulary*** How do you determine whether a rational expression is in simplified form? The rational expression is in simplified form when the numerator and denominator have no factors in common (other than ± 1).

38. ***Reasoning*** Can you divide out common terms from the numerator and denominator of a rational expression? Explain. No. Only common factors can be divided out of a rational expression.

39. ***Precision*** After factoring completely, what is one additional step that is sometimes needed to find common factors in the numerator and denominator of a rational expression? A change in signs is one additional step sometimes needed in order to find common factors.

40. ***Reasoning*** Is the following expression in simplified form? Explain your reasoning.

$$\frac{5 + x}{5 + (x + 2)}$$

Yes. You cannot divide out any common factors.

Extra Practice

Finding a Missing Factor **In Exercises 41–44, determine the missing factor.**

41. $\frac{(x + 5)(\;x(x - 2)\;)}{3x^2(x - 2)} = \frac{x + 5}{3x}, \; x \neq 2$

42. $\frac{(3y - 7)(\;y - 2\;)}{y^2 - 4} = \frac{3y - 7}{y + 2}, \; y \neq 2$

43. $\frac{(8x)(\;x + 3\;)}{x^2 - 2x - 15} = \frac{8x}{x - 5}, \; x \neq -3$

44. $\frac{(3 - z)(\;z + 2\;)}{z^3 + 2z^2} = \frac{3 - z}{z^2}, \; z \neq -2$

45. ***Average Cost*** A greeting card company has an initial investment of \$60,000. The cost of producing one dozen cards is \$6.50.

(a) Write the total cost C as a function of x, the number of dozens of cards produced. $C = 60{,}000 + 6.50x$

(b) Write the average cost per dozen $\overline{C} = C/x$ as a function of x, the number of dozens of cards produced.

$\overline{C} = \frac{60{,}000 + 6.50x}{x}$

(c) Determine the domain of the function in part (b). $\{1, 2, 3, 4, \ldots\}$

(d) Find the value of $\overline{C}(11{,}000)$. \$11.95

Tatiana Popova/Shutterstock.com

46. ***Distance Traveled*** A car starts on a trip and travels at an average speed of 55 miles per hour. Two hours later, a second car starts on the same trip and travels at an average speed of 65 miles per hour.

(a) Find the distance each vehicle has traveled when the second car has been on the road for t hours.
First car: $55(t + 2)$, Second car: $65t$

(b) Use the result of part (a) to write the distance between the first car and the second car as a function of t. $d = |10(11 - t)|$

(c) Write the ratio of the distance the second car has traveled to the distance the first car has traveled as a function of t.
$\frac{13t}{11(t + 2)}$

47. ***Geometry*** One swimming pool is circular and another is rectangular. The rectangular pool's width is three times its depth. Its length is 6 feet more than its width. The circular pool has a diameter that is twice the width of the rectangular pool, and it is 2 feet deeper. Find the ratio of the circular pool's volume to the rectangular pool's volume. $\pi, d > 0$

48. ***Geometry*** A circular pool has a radius five times its depth. A rectangular pool has the same depth as the circular pool. Its width is 4 feet more than three times its depth and its length is 2 feet less than six times its depth. Find the ratio of the rectangular pool's volume to the circular pool's volume.
$\frac{2(3d - 1)(3d + 4)}{25\pi d^2}, d > 0$

Explaining Concepts

49. ***Writing*** Describe the process for finding the implied domain restrictions of a rational function.
A rational expression is undefined when the denominator is equal to zero. To find the implied domain restriction of a rational function, find the values that make the denominator equal to zero (by setting the denominator equal to zero), and exclude those values from the domain.

50. ***Writing*** Describe a situation in which you would need to indicate a domain restriction to the right of a rational function.
A domain restriction would need to be indicated when a common factor (with at least one variable) is divided out of the rational expression.

51. ***Precision*** Give an example of a rational function whose domain is the set of all real numbers and whose denominator is a second-degree polynomial function.
$\frac{1}{x^2 + 1}$

52. ***Error Analysis*** Describe the error.

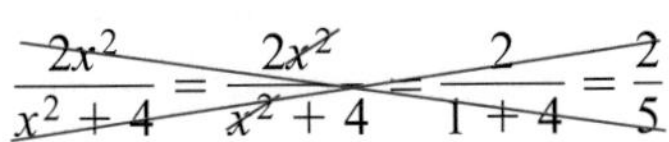

Only common factors (not terms) can be divided out.

53. ***Logic*** A student writes the following incorrect solution for simplifying a rational expression. Discuss the student's errors and misconceptions, and construct a correct solution.

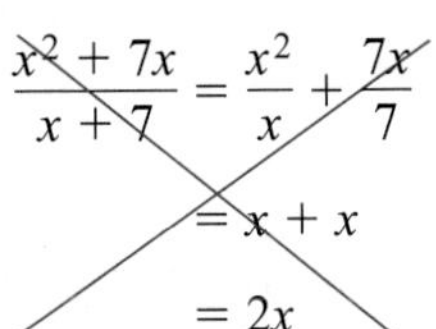

The student incorrectly divided (the denominator may not be split up) and the domain is not restricted.

Correct solution:
$\frac{x^2 + 7x}{x + 7} = \frac{x(x + 7)}{x + 7} = x, x \neq -7$

54. ***Reasoning*** Is the following statement true? Explain.

$$\frac{6x - 5}{5 - 6x} = -1$$

No. The domain must be restricted.
$\frac{6x - 5}{5 - 6x} = \frac{6x - 5}{-(6x - 5)} = -1, x \neq \frac{5}{6}$

55. ***Writing*** Explain how you can use a given polynomial function $f(x)$ to write a rational function $g(x)$ that is equivalent to $f(x), x \neq 2$. To write the polynomial $g(x)$, multiply $f(x)$ by $(x - 2)$ and divide by $(x - 2)$.

$g(x) = \frac{f(x)(x - 2)}{(x - 2)} = f(x), x \neq 2$

56. ***Think About It*** Is it possible for a rational function $f(x)$ (without added domain restrictions) to be undefined on an interval $[a, b]$, where a and b are real numbers such that $a < b$? Explain.
No. The rational function $f(x)$ can only be undefined at specific values of x, not on an interval.

Cumulative Review

In Exercises 57–60, find the product.

57. $\frac{1}{4}\left(\frac{3}{4}\right)$ $\frac{3}{16}$

58. $\frac{2}{3}\left(-\frac{5}{6}\right)$ $-\frac{5}{9}$

59. $\frac{1}{3}\left(\frac{3}{5}\right)(5)$ 1

60. $\left(-\frac{3}{7}\right)\left(\frac{2}{5}\right)\left(-\frac{1}{6}\right)$ $\frac{1}{35}$

In Exercises 61–64, perform the indicated multiplication.

61. $(-2a^3)(-2a)$ $4a^4$

62. $6x^2(-3x)$ $-18x^3$

63. $(-3b)(b^2 - 3b + 5)$ $-3b^3 + 9b^2 - 15b$

64. $ab^2(3a - 4ab + 6a^2b^2)$ $3a^2b^2 - 4a^2b^3 + 6a^3b^4$

6.2 Multiplying and Dividing Rational Expressions

- Multiply rational expressions and simplify.
- Divide rational expressions and simplify.

Multiplying Rational Expressions

Multiplying Rational Expressions

Let u, v, w, and z represent real numbers, variables, or algebraic expressions such that $v \neq 0$ and $z \neq 0$. Then the product of u/v and w/z is

$$\frac{u}{v} \cdot \frac{w}{z} = \frac{uw}{vz}.$$

EXAMPLE 1 Multiplying Rational Expressions

Remind students at the beginning that the implied domain restrictions on the original expression are $x \neq 0$ and $y \neq 0$.

Multiply the rational expressions.

$$\frac{4x^3y}{3xy^4} \cdot \frac{-6x^2y^2}{10x^4}$$

SOLUTION

$$\frac{4x^3y}{3xy^4} \cdot \frac{-6x^2y^2}{10x^4} = \frac{(4x^3y) \cdot (-6x^2y^2)}{(3xy^4) \cdot (10x^4)}$$ Multiply numerators and denominators.

$$= \frac{-24x^5y^3}{30x^5y^4}$$ Simplify.

$$= \frac{-4(\cancel{6})(\cancel{x^5})(\cancel{y^3})}{5(\cancel{6})(\cancel{x^5})(\cancel{y^3})(y)}$$ Factor and divide out common factors.

$$= -\frac{4}{5y}, \; x \neq 0$$ Simplified form

Exercises Within Reach®

Solutions in English & Spanish and tutorial videos at AlgebraWithinReach.com

Finding a Missing Factor **In Exercises 1–4, determine the missing factor.**

1. $\dfrac{7x^2}{3y(\quad)} = \dfrac{7}{3y}, \; x \neq 0$ x^2

2. $\dfrac{14x(x-3)^2}{(x-3)(\quad)} = \dfrac{2x}{x-3}$ $7(x-3)^2$

3. $\dfrac{3x(x+2)^2}{(x-4)(\quad)} = \dfrac{3x}{x-4}, \; x \neq -2$ $(x+2)^2$

4. $\dfrac{(x+1)^3}{x(\quad)} = \dfrac{x+1}{x}, \; x \neq -1$ $(x+1)^2$

Multiplying Rational Expressions **In Exercises 5–12, multiply and simplify.**

5. $4x \cdot \dfrac{7}{12x}$ $\frac{7}{3}, x \neq 0$

6. $\dfrac{8}{7y} \cdot (42y)$ $48, y \neq 0$

7. $\dfrac{8s^3}{9s} \cdot \dfrac{6s^2}{32s}$ $\frac{s^3}{6}, s \neq 0$

8. $\dfrac{3x^4}{7x} \cdot \dfrac{8x^2}{9}$ $\frac{8x^5}{21}, y \neq 0$

9. $16u^4 \cdot \dfrac{12}{8u^2}$ $24u^2, u \neq 0$

10. $18x^4 \cdot \dfrac{4}{15x}$ $\frac{24x^3}{5}, x \neq 0$

11. $\dfrac{8}{3+4x} \cdot (9+12x)$ $24, x \neq -\frac{3}{4}$

12. $(6-4x) \cdot \dfrac{10}{3-2x}$ $20, x \neq \frac{3}{2}$

EXAMPLE 2 Multiplying Rational Expressions

Multiply the rational expressions.

a. $\dfrac{x}{5x^2 - 20x} \cdot \dfrac{x-4}{2x^2 + x - 3}$

b. $\dfrac{4x^2 - 4x}{x^2 + 2x - 3} \cdot \dfrac{x^2 + x - 6}{4x}$

SOLUTION

a. $\dfrac{x}{5x^2 - 20x} \cdot \dfrac{x-4}{2x^2 + x - 3}$

$= \dfrac{x \cdot (x-4)}{(5x^2 - 20x) \cdot (2x^2 + x - 3)}$ Multiply numerators and denominators.

$= \dfrac{x(x-4)}{5x(x-4)(x-1)(2x+3)}$ Factor.

$= \dfrac{\cancel{x}\cancel{(x-4)}}{5\cancel{x}\cancel{(x-4)}(x-1)(2x+3)}$ Divide out common factors.

$= \dfrac{1}{5(x-1)(2x+3)},\ x \neq 0, x \neq 4$ Simplified form

b. $\dfrac{4x^2 - 4x}{x^2 + 2x - 3} \cdot \dfrac{x^2 + x - 6}{4x}$

$= \dfrac{4x(x-1)(x+3)(x-2)}{(x-1)(x+3)(4x)}$ Multiply and factor.

$= \dfrac{\cancel{4x}\cancel{(x-1)}\cancel{(x+3)}(x-2)}{\cancel{(x-1)}\cancel{(x+3)}\cancel{(4x)}}$ Divide out common factors.

$= x - 2,\ x \neq 0,\ x \neq 1,\ x \neq -3$ Simplified form

Exercises Within Reach®

Solutions in English & Spanish and tutorial videos at AlgebraWithinReach.com

Multiplying Rational Expressions **In Exercises 13–22, multiply and simplify.**

13. $\dfrac{8u^2v}{3u+v} \cdot \dfrac{u+v}{12u}$ $\dfrac{2uv(u+v)}{3(3u+v)}, u \neq 0$

14. $\dfrac{1-3xy}{4x^2y} \cdot \dfrac{46x^4y^2}{15-45xy}$ $\dfrac{23x^2y}{30}, x \neq 0, y \neq 0, 3xy \neq 1$

15. $\dfrac{12-r}{3} \cdot \dfrac{3}{r-12}$ $-1, r \neq 12$

16. $\dfrac{8-z}{8+z} \cdot \dfrac{z+8}{z-8}$ $-1, z \neq \pm 8$

17. $\dfrac{(2x-3)(x+8)}{x^3} \cdot \dfrac{x}{3-2x}$ $-\dfrac{x+8}{x^2}, x \neq \dfrac{3}{2}$

18. $\dfrac{x+14}{x^3(10-x)} \cdot \dfrac{x(x-10)}{5}$ $-\dfrac{x+14}{5x^2}, x \neq 10$

19. $\dfrac{4r-12}{r-2} \cdot \dfrac{r^2-4}{r-3}$ $4(r+2), r \neq 3, r \neq 2$

20. $\dfrac{5y-20}{5y+15} \cdot \dfrac{2y+6}{y-4}$ $2, y \neq -3, y \neq 4$

21. $\dfrac{2t^2-t-15}{t+2} \cdot \dfrac{t^2-t-6}{t^2-6t+9}$

$2t+5, t \neq 3, t \neq -2$

22. $\dfrac{y^2-16}{y^2+8y+16} \cdot \dfrac{3y^2-5y-2}{y^2-6y+8}$

$\dfrac{3y+1}{y+4}, y \neq 2, y \neq 4$

EXAMPLE 3 Multiplying Rational Expressions

Multiply the rational expressions.

a. $\frac{x-y}{y^2-x^2} \cdot \frac{x^2-xy-2y^2}{3x-6y}$

b. $\frac{x^2-3x+2}{x+2} \cdot \frac{3x}{x-2} \cdot \frac{2x+4}{x^2-5x}$

Additional Examples

Multiply the rational expressions.

a. $\frac{x^2+2x-3}{x^2-x} \cdot \frac{2x+3}{3x^2+5x-12}$

b. $\frac{x+y}{x^2+3xy+2y^2} \cdot \frac{x^2+xy-2y^2}{2x+4y}$

Answers:

a. $\frac{2x+3}{x(3x-4)},\ x \neq -3,\ x \neq 1$

b. $\frac{x-y}{2(x+2y)},\ x \neq -y$

SOLUTION

a. $\frac{x-y}{y^2-x^2} \cdot \frac{x^2-xy-2y^2}{3x-6y}$

$= \frac{(x-y)(x-2y)(x+y)}{(y+x)(y-x)(3)(x-2y)}$ Multiply and factor.

$= \frac{(x-y)(x-2y)(x+y)}{(y+x)(-1)(x-y)(3)(x-2y)}$ $(y-x) = -1(x-y)$

$= \frac{\cancel{(x-y)}\cancel{(x-2y)}\cancel{(x+y)}}{\cancel{(x+y)}(-1)\cancel{(x-y)}(3)\cancel{(x-2y)}}$ Divide out common factors.

$= -\frac{1}{3},\ x \neq y,\ x \neq -y,\ x \neq 2y$ Simplified form

b. $\frac{x^2-3x+2}{x+2} \cdot \frac{3x}{x-2} \cdot \frac{2x+4}{x^2-5x}$

$= \frac{(x-1)(x-2)(3)(x)(2)(x+2)}{(x+2)(x-2)(x)(x-5)}$ Multiply and factor.

$= \frac{(x-1)\cancel{(x-2)}(3)\cancel{(x)}(2)\cancel{(x+2)}}{\cancel{(x+2)}\cancel{(x-2)}\cancel{(x)}(x-5)}$ Divide out common factors.

$= \frac{6(x-1)}{x-5},\ x \neq 0,\ x \neq 2,\ x \neq -2$ Simplified form

Exercises Within Reach®

Solutions in English & Spanish and tutorial videos at AlgebraWithinReach.com

Multiplying Rational Expressions **In Exercises 23–30, multiply and simplify.**

23. $(4y^2-x^2) \cdot \frac{xy}{(x-2y)^2}$ $-\frac{xy(x+2y)}{x-2y}$

24. $(u-2v)^2 \cdot \frac{u+2v}{2v-u}$ $-(u-2v)(u+2v),\ u \neq 2v$

25. $\frac{x^2+2xy-3y^2}{(x+y)^2} \cdot \frac{x^2-y^2}{x+3y}$ $\frac{(x-y)^2}{x+y},\ x \neq -3y$

26. $\frac{(x-2y)^2}{x+2y} \cdot \frac{x^2+7xy+10y^2}{x^2-4y^2}$ $\frac{(x-2y)(x+5y)}{(x+2y)},\ x \neq 2y$

27. $\frac{x+5}{x-5} \cdot \frac{2x^2-9x-5}{3x^2+x-2} \cdot \frac{x^2-1}{x^2+7x+10}$

$\frac{(x-1)(2x+1)}{(3x-2)(x+2)},\ x \neq \pm 5,\ x \neq -1$

28. $\frac{t^2+4t+3}{2t^2-t-10} \cdot \frac{t}{t^2+3t+2} \cdot \frac{2t^2+4t^3}{t^2+3t}$

$\frac{2t^2(1+2t)}{(2t-5)(t+2)^2},\ t \neq -3,\ t \neq -1,\ t \neq 0$

29. $\frac{9-x^2}{2x+3} \cdot \frac{4x^2+8x-5}{4x^2-8x+3} \cdot \frac{6x^4-2x^3}{8x^2+4x}$

$-\frac{x^2(x+3)(x-3)(2x+5)(3x-1)}{2(2x+1)(2x+3)(2x-3)},\ x \neq 0,\ x \neq \frac{1}{2}$

30. $\frac{16x^2-1}{4x^2+9x+5} \cdot \frac{5x^2-9x-18}{x^2-12x+36} \cdot \frac{12+4x-x^2}{4x^2-13x+3}$

$-\frac{(4x+1)(5x+6)(x+2)}{(4x+5)(x+1)(x-6)},\ x \neq \frac{1}{4},\ x \neq 3$

Dividing Rational Expressions

Dividing Rational Expressions

Let u, v, w, and z represent real numbers, variables, or algebraic expressions such that $v \neq 0$, $w \neq 0$, and $z \neq 0$. Then the quotient of u/v and w/z is

$$\frac{u}{v} \div \frac{w}{z} = \frac{u}{v} \cdot \frac{z}{w} = \frac{uz}{vw}.$$

Study Tip

Don't forget to add domain restrictions as needed in division problems. In Example 4(a), an implied domain restriction in the original expression is $x \neq 1$. Because this restriction is not implied by the final expression, it must be added as a written restriction.

EXAMPLE 4 Dividing Rational Expressions

Divide the rational expressions.

a. $\dfrac{x}{x+3} \div \dfrac{4}{x-1}$ **b.** $\dfrac{2x}{3x-12} \div \dfrac{x^2-2x}{x^2-6x+8}$

SOLUTION

a. $\dfrac{x}{x+3} \div \dfrac{4}{x-1} = \dfrac{x}{x+3} \cdot \dfrac{x-1}{4}$ Invert divisor and multiply.

$= \dfrac{x(x-1)}{(x+3)(4)}$ Multiply numerators and denominators.

$= \dfrac{x(x-1)}{4(x+3)}, \ x \neq 1$ Simplified form

b. $\dfrac{2x}{3x-12} \div \dfrac{x^2-2x}{x^2-6x+8}$

$= \dfrac{2x}{3x-12} \cdot \dfrac{x^2-6x+8}{x^2-2x}$ Invert divisor and multiply.

$= \dfrac{(2)(x)(x-2)(x-4)}{(3)(x-4)(x)(x-2)}$ Factor.

$= \dfrac{(2)\cancel{(x)}\cancel{(x-2)}\cancel{(x-4)}}{(3)\cancel{(x-4)}\cancel{(x)}\cancel{(x-2)}}$ Divide out common factors.

$= \dfrac{2}{3}, \ x \neq 0, \ x \neq 2, \ x \neq 4$ Simplified form

Remember that the original expression is equivalent to $\frac{2}{3}$ except for $x = 0$, $x = 2$, and $x = 4$.

Exercises Within Reach®

Solutions in English & Spanish and tutorial videos at AlgebraWithinReach.com

Dividing Rational Expressions **In Exercises 31–38, divide and simplify.**

31. $\dfrac{x}{x+2} \div \dfrac{3}{x+1}$ $\dfrac{x(x+1)}{3(x+2)}, x \neq -1$

32. $\dfrac{x+3}{4} \div \dfrac{x-2}{x}$ $\dfrac{x(x+3)}{4(x-2)}, x \neq 0$

33. $x^2 \div \dfrac{3x}{4}$ $\dfrac{4x}{3}, x \neq 0$

34. $\dfrac{u}{10} \div u^2$ $\dfrac{6}{x}$

35. $\dfrac{2x}{5} \div \dfrac{x^2}{15}$ $\dfrac{1}{10u}$

36. $\dfrac{3y^2}{20} \div \dfrac{y}{15}$ $\dfrac{9y}{4}, y \neq 0$

37. $\dfrac{4x}{3x-3} \div \dfrac{x^2+2x}{x^2+x-2}$ $\frac{4}{3}, x \neq -2, x \neq 0, x \neq 1$

38. $\dfrac{5x+5}{2x} \div \dfrac{x^2-3x}{x^2-2x-3}$ $\dfrac{5(x+1)^2}{2x^2}, x \neq -1, x \neq 3$

EXAMPLE 5 Dividing Rational Expressions

Additional Example

Divide the rational expression.

$$\frac{4x}{10x+5} \div \frac{3x^2-8x}{2x+1}$$

Answer:

$$\frac{4}{5(3x-8)},\ x \neq 0,\ x \neq -\frac{1}{2}$$

a. $\dfrac{x^2-y^2}{2x+2y} \div \dfrac{2x^2-3xy+y^2}{6x+2y}$

$= \dfrac{x^2-y^2}{2x+2y} \cdot \dfrac{6x+2y}{2x^2-3xy+y^2}$ — Invert divisor and multiply.

$= \dfrac{(x+y)(x-y)(2)(3x+y)}{(2)(x+y)(2x-y)(x-y)}$ — Factor.

$= \dfrac{\cancel{(x+y)}\cancel{(x-y)}\cancel{(2)}(3x+y)}{\cancel{(2)}\cancel{(x+y)}(2x-y)\cancel{(x-y)}}$ — Divide out common factors.

$= \dfrac{3x+y}{2x-y},\ x \neq y,\ x \neq -y,\ y \neq -3x$ — Simplified form

b. $\dfrac{x^2-14x+49}{x^2-49} \div \dfrac{3x-21}{x^2+2x-35}$

$= \dfrac{x^2-14x+49}{x^2-49} \cdot \dfrac{x^2+2x-35}{3x-21}$ — Invert divisor and multiply.

$= \dfrac{(x-7)(x-7)(x+7)(x-5)}{(x-7)(x+7)(3)(x-7)}$ — Factor.

$= \dfrac{\cancel{(x-7)}\cancel{(x-7)}\cancel{(x+7)}(x-5)}{\cancel{(x-7)}\cancel{(x+7)}(3)\cancel{(x-7)}}$ — Divide out common factors.

$= \dfrac{x-5}{3},\ x \neq 5, x \neq 7, x \neq -7$ — Simplified form

Exercises Within Reach®

Solutions in English & Spanish and tutorial videos at AlgebraWithinReach.com

Dividing Rational Expressions **In Exercises 39–48, divide and simplify.**

39. $\dfrac{7xy^2}{10x^2y} \div \dfrac{21x^3}{45xy}$

$\dfrac{3y^2}{2x^3},\ y \neq 0$

40. $\dfrac{25x^2y}{60x^3y^2} \div \dfrac{5x^4y^3}{16x^2y}$

$\dfrac{4}{3x^3y^3}$

41. $\dfrac{3(a+b)}{4} \div \dfrac{(a+b)^2}{2}$

$\dfrac{3}{2(a+b)}$

42. $\dfrac{x^2+9}{5(x+2)} \div \dfrac{x+3}{5(x^2-4)}$

$\dfrac{(x^2+9)(x-2)}{x+3},\ x \neq \pm 2$

43. $\dfrac{2x+2y}{3} \div \dfrac{x^2-y^2}{x-y}$

$\dfrac{2}{3},\ x \neq y, x \neq -y$

44. $\dfrac{x^2+2x-xy-2y}{x^2-y^2} \div \dfrac{2x+4}{x+y}$

$\dfrac{1}{2},\ x \neq y, x \neq -y, x \neq -2$

45. $\dfrac{(x^3y)^2}{(x+2y)^2} \div \dfrac{x^2y}{(x+2y)^3}$

$x^4y(x+2y),\ x \neq 0, y \neq 0, x \neq -2y$

46. $\dfrac{x^2-y^2}{2x^2-8x} \div \dfrac{(x-y)^2}{2xy}$

$\dfrac{y(x+y)}{(x-4)(x-y)},\ x \neq 0, y \neq 0$

47. $\dfrac{x^2+2x-15}{x^2+11x+30} \div \dfrac{x^2-8x+15}{x^2+2x-24}$

$\dfrac{x-4}{x-5},\ x \neq -6, x \neq -5, x \neq 3$

48. $\dfrac{y^2+5y-14}{y^2+10y+21} \div \dfrac{y^2+5y+6}{y^2+7y+12}$

$\dfrac{(y-2)(y+4)}{(y+3)(y+2)},\ y \neq -7, y \neq -4$

Application **EXAMPLE 6** **Amount Spent on Meals and Beverages**

The annual amount A (in millions of dollars) Americans spent on meals and beverages purchased for the home, and the population P (in millions) of the United States, for the years 2003 through 2009 can be modeled by

$$A = \frac{32{,}684.374t + 396{,}404.73}{0.014t + 1}, \quad 3 \le t \le 9$$

and

$$P = 2.798t + 282.31, \quad 3 \le t \le 9$$

where t represents the year, with $t = 3$ corresponding to 2003. Find a model T for the amount Americans spent *per person* on meals and beverages for the home. (*Source:* U.S. Department of Agriculture and U.S. Census Bureau)

SOLUTION

To find a model T for the amount Americans spent per person on meals and beverages for the home, divide the total amount by the population.

$$T = \frac{32{,}684.374t + 396{,}404.73}{0.014t + 1} \div (2.798t + 282.31)$$ Divide amount spent by population.

$$= \frac{32{,}684.374t + 396{,}404.73}{0.014t + 1} \cdot \frac{1}{2.798t + 282.31}$$ Invert divisor and multiply.

$$= \frac{32{,}684.374t + 396{,}404.73}{(0.014t + 1)(2.798t + 282.31)}, \quad 3 \le t \le 9$$ Model

Exercises Within Reach®

Solutions in English & Spanish and tutorial videos at AlgebraWithinReach.com

49. ***Per Capita Income*** The total annual amount I (in millions of dollars) of personal income earned in Alabama, and its population P (in millions), for the years 2004 through 2009 can be modeled by

$I = 6591.43t + 102{,}139.0, \; 4 \le t \le 9$ and

$$P = \frac{0.184t + 4.3}{0.029t + 1}, \quad 4 \le t \le 9$$

where t represents the year, with $t = 4$ corresponding to 2004. Find a model Y for the annual per capita income for these years. (*Source:* U.S. Bureau of Economic Analysis and U.S. Census Bureau)

See Additional Answers.

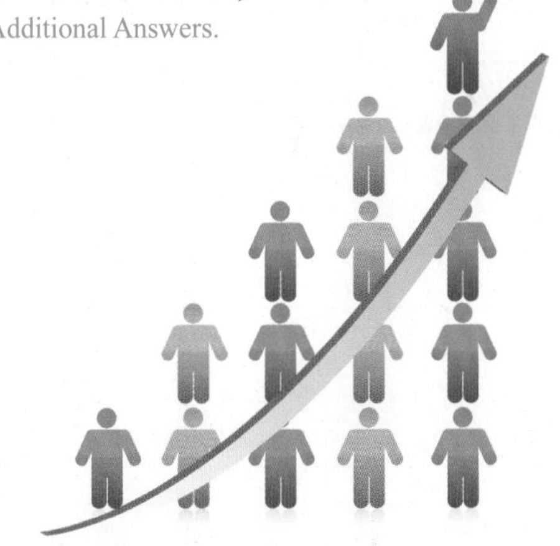

50. ***Per Capita Income*** The total annual amount I (in millions of dollars) of personal income earned in Montana, and its population P (in millions), for the years 2004 through 2009 can be modeled by

$$I = \frac{4651.460t + 14{,}528.41}{0.070t + 1}, \quad 4 \le t \le 9 \text{ and}$$

$$P = \frac{0.028t + 0.88}{0.018t + 1}, \quad 4 \le t \le 9$$

where t represents the year, with $t = 4$ corresponding to 2004. Find a model Y for the annual per capita income for these years. (*Source:* U.S. Bureau of Economic Analysis and U.S. Census Bureau) See Additional Answers.

alexmillos/Shutterstock.com; Nataliia Natykach/Shutterstock.com

Concept Summary: Multiplying and Dividing Rational Expressions

What

The rules for multiplying and dividing rational expressions are the same as the rules for multiplying and dividing numerical fractions.

EXAMPLE

Divide:

$$\frac{x^2 - 2x - 24}{x - 6} \div \frac{x^2 + 6x + 8}{x + 3}.$$

How

To divide two rational expressions, multiply the first expression by the reciprocal of the second.

EXAMPLE

$\dfrac{x^2 - 2x - 24}{x - 6} \div \dfrac{x^2 + 6x + 8}{x + 3}$ Original

$= \dfrac{x^2 - 2x - 24}{x - 6} \cdot \dfrac{x + 3}{x^2 + 6x + 8}$ Invert and multiply.

$= \dfrac{(x - 6)(x + 4)(x + 3)}{(x - 6)(x + 2)(x + 4)}$ Factor.

$= \dfrac{\cancel{(x - 6)}\cancel{(x + 4)}(x + 3)}{\cancel{(x - 6)}(x + 2)\cancel{(x + 4)}}$ Divide out common factors.

$= \dfrac{x + 3}{x + 2}, x \neq 6, x \neq -3, x \neq -4$ Simplified form

Why

Knowing how to multiply and divide rational expressions will help you solve rational equations.

Exercises Within Reach ®

Worked-out solutions to odd-numbered exercises at AlgebraWithinReach.com

Concept Summary Check

51. ***Writing*** Explain how to multiply two rational expressions. See Additional Answers.

52. ***Reasoning*** Why is factoring used in multiplying rational expressions? Factoring is used in multiplying rational expressions so that common factors in the numerator and denominator can be divided out.

53. ***Writing*** In your own words, explain how to divide rational expressions. To divide rational expressions, multiply the first expression by the reciprocal of the second expression.

54. ***Logic*** In dividing rational expressions, explain how you can lose implied domain restrictions when you invert the divisor. See Additional Answers.

Extra Practice

Using Order of Operations **In Exercises 55–62, perform the operations and simplify. (In Exercises 61 and 62, *n* is a positive integer.)**

55. $\left[\dfrac{x^2}{9} \cdot \dfrac{3(x + 4)}{x^2 + 2x}\right] \div \dfrac{x}{x + 2}$

$\dfrac{x + 4}{3}, x \neq -2, x \neq 0$

56. $\left(\dfrac{x^2 + 6x + 9}{x^2} \cdot \dfrac{2x + 1}{x^2 - 9}\right) \div \dfrac{4x^2 + 4x + 1}{x^2 - 3x}$

$\dfrac{x + 3}{x(2x + 1)}, x \neq \pm 3$

57. $\left[\dfrac{xy + y}{4x} \div (3x + 3)\right] \div \dfrac{y}{3x}$

$\dfrac{1}{4}, x \neq -1, x \neq 0, y \neq 0$

58. $\dfrac{3u^2 - u - 4}{u^2} \div \dfrac{3u^2 + 12u + 4}{u^4 - 3u^3}$

$\dfrac{u(3u - 4)(u + 1)(u - 3)}{3u^2 + 12u + 4}, u \neq 0, u \neq 3$

59. $\dfrac{2x^2 + 5x - 25}{3x^2 + 5x + 2} \cdot \dfrac{3x^2 + 2x}{x + 5} \div \left(\dfrac{x}{x + 1}\right)^2$

$\dfrac{(x + 1)(2x - 5)}{x}, x \neq -5, x \neq -1, x \neq -\dfrac{2}{3}$

60. $\dfrac{t^2 - 100}{4t^2} \cdot \dfrac{t^3 - 5t^2 - 50t}{t^4 + 10t^3} \div \dfrac{(t - 10)^2}{5t}$

$\dfrac{5(t + 5)}{4t^3}, t \neq \pm 10$

61. $x^3 \cdot \dfrac{x^{2n} - 9}{x^{2n} + 4x^n + 3} \div \dfrac{x^{2n} - 2x^n - 3}{x}$

$\dfrac{x^4}{(x^n + 1)^2}, x^n \neq -3, x^n \neq 3, x \neq 0$

62. $\dfrac{x^{n+1} - 8x}{x^{2n} + 2x^n + 1} \cdot \dfrac{x^{2n} - 4x^n - 5}{x} \div x^n$

$\dfrac{(x^n - 8)(x^n - 5)}{x^n(x^n + 1)}$

Probability **In Exercises 63–66, consider an experiment in which a marble is tossed into a rectangular box with dimensions $2x$ centimeters by $(4x + 2)$ centimeters. The probability that the marble will come to rest in the unshaded portion of the box is equal to the ratio of the unshaded area to the total area of the figure. Find the probability in simplified form.**

63. 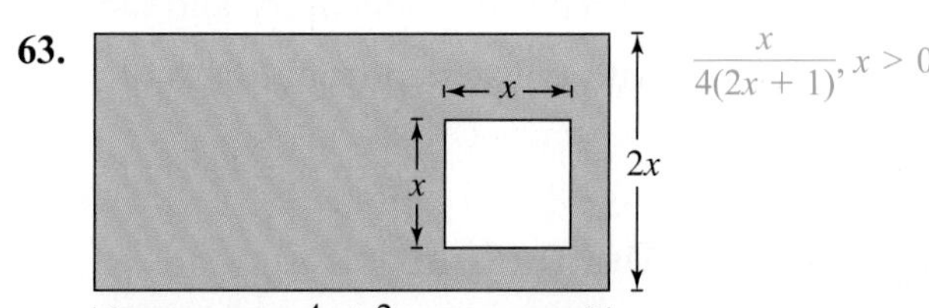

$\frac{x}{4(2x+1)}, x > 0$

64. 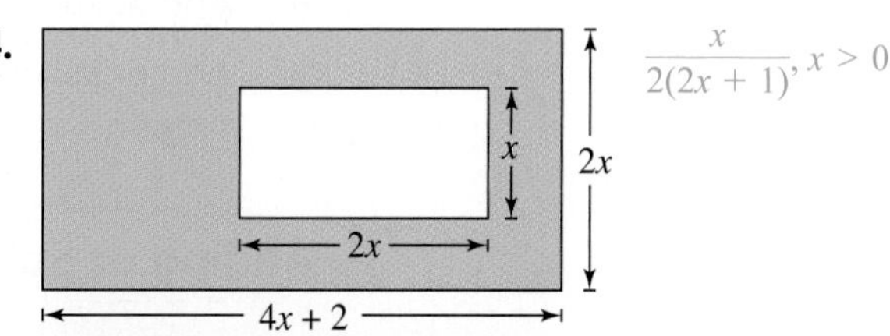

$\frac{x}{2(2x+1)}, x > 0$

65. 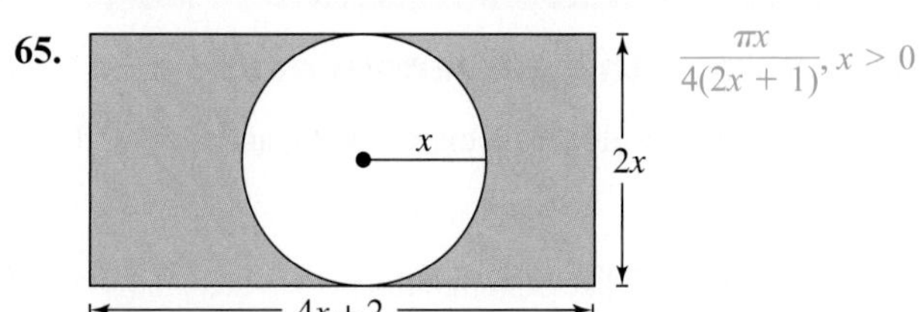

$\frac{\pi x}{4(2x+1)}, x > 0$

66. 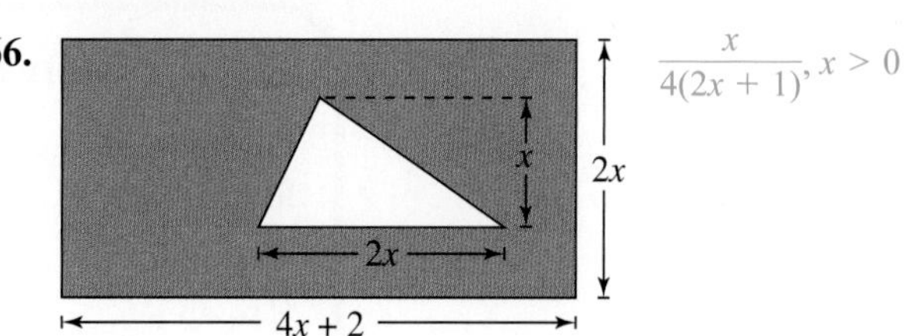

$\frac{x}{4(2x+1)}, x > 0$

Explaining Concepts

67. ***Writing*** Describe how the operation of division is used in the process of simplifying a product of rational expressions. See Additional Answers.

68. ***Writing*** In a quotient of two rational expressions, the denominator of the divisor is x. Describe a set of circumstances in which you will *not* need to list $x \neq 0$ as a domain restriction after dividing. See Additional Answers.

69. ***Logic*** Explain what is missing in the following statement.

$$\frac{x-a}{x-b} \div \frac{x-a}{x-b} = 1$$

The domain needs to be restricted, $x \neq a, x \neq b$.

70. ***Logic*** When two rational expressions are multiplied, the resulting expression is a polynomial. Explain how the total number of factors in the numerators of the expressions you multiplied compares to the total number of factors in the denominators. See Additional Answers.

71. ***Error Analysis*** Describe and correct the errors.

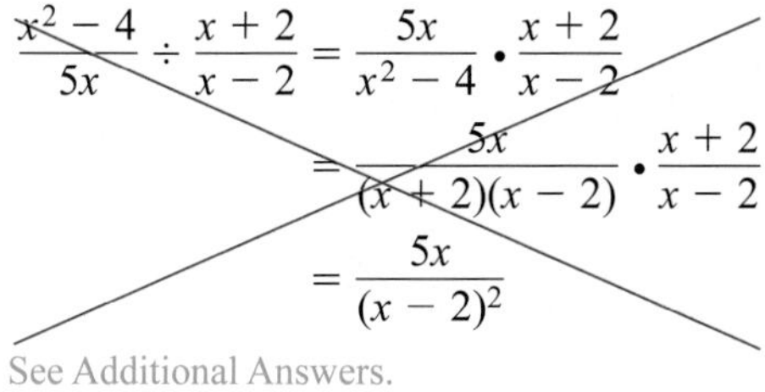

See Additional Answers.

72. ***Repeated Reasoning*** Complete the table for the given values of x. Round your answers to five decimal places.

x	60	100	1000
$\frac{x-10}{x+10}$	0.71429	0.81818	0.98020
$\frac{x+50}{x-50}$	11	3	1.10526
$\frac{x-10}{x+10} \cdot \frac{x+50}{x-50}$	7.85714	2.45455	1.08338

x	10,000	100,000	1,000,000
$\frac{x-10}{x+10}$	0.99800	0.99980	0.99998
$\frac{x+50}{x-50}$	1.01005	1.00100	1.00010
$\frac{x-10}{x+10} \cdot \frac{x+50}{x-50}$	1.00803	1.00080	1.00008

What kind of pattern do you see? Try to explain what is going on. Can you see why? See Additional Answers.

Cumulative Review

In Exercises 73–76, evaluate the expression.

73. $\frac{1}{8} + \frac{3}{8} + \frac{5}{8}$ $\frac{9}{8}$

74. $\frac{3}{7} - \frac{2}{7}$ $\frac{1}{7}$

75. $\frac{3}{5} + \frac{4}{15}$ $\frac{13}{15}$

76. $\frac{7}{6} - \frac{9}{7}$ $-\frac{5}{42}$

In Exercises 77 and 78, solve the equation by factoring.

77. $x^2 + 3x = 0$ $-3, 0$

78. $x^2 + 3x - 10 = 0$ $-5, 2$

6.3 Adding and Subtracting Rational Expressions

- Add or subtract rational expressions with like denominators, and simplify.
- Add or subtract rational expressions with unlike denominators, and simplify.

Adding or Subtracting with Like Denominators

Adding or Subtracting with Like Denominators

If u, v, and w are real numbers, variables, or algebraic expressions, and $w \neq 0$, the following rules are valid.

1. $\dfrac{u}{w} + \dfrac{v}{w} = \dfrac{u+v}{w}$ Add fractions with like denominators.

2. $\dfrac{u}{w} - \dfrac{v}{w} = \dfrac{u-v}{w}$ Subtract fractions with like denominators.

EXAMPLE 1 **Adding and Subtracting Rational Expressions**

a. $\dfrac{x}{4} + \dfrac{5-x}{4} = \dfrac{x+(5-x)}{4} = \dfrac{5}{4}$ Add numerators.

b. $\dfrac{7}{2x-3} - \dfrac{3x}{2x-3} = \dfrac{7-3x}{2x-3}$ Subtract numerators.

c. $\dfrac{x}{x^2-2x-3} - \dfrac{3}{x^2-2x-3} = \dfrac{x-3}{x^2-2x-3}$ Subtract numerators.

$= \dfrac{(1)(x-3)}{(x-3)(x+1)}$ Factor.

$= \dfrac{1}{x+1},\ x \neq 3$ Simplified form

Exercises Within Reach® Solutions in English & Spanish and tutorial videos at AlgebraWithinReach.com

Adding and Subtracting Rational Expressions **In Exercises 1–10, combine and simplify.**

1. $\dfrac{5x}{6} + \dfrac{4x}{6}$ $\dfrac{3x}{2}$

2. $\dfrac{7y}{12} + \dfrac{9y}{12}$ $\dfrac{4y}{3}$

3. $\dfrac{2}{3a} - \dfrac{11}{3a}$ $-\dfrac{3}{a}$

4. $\dfrac{6}{19x} - \dfrac{7}{19x}$ $-\dfrac{1}{19x}$

5. $\dfrac{x}{9} - \dfrac{x+2}{9}$ $-\dfrac{2}{9}$

6. $\dfrac{4-y}{4} + \dfrac{3y}{4}$ $\dfrac{y+2}{2}$

7. $\dfrac{2x-1}{x(x-3)} + \dfrac{1-x}{x(x-3)}$ $\dfrac{1}{x-3},\ x \neq 0$

8. $\dfrac{3-2n}{n(n+2)} - \dfrac{1-3n}{n(n+2)}$ $\dfrac{1}{n},\ n \neq -2$

9. $\dfrac{c}{c^2+3c-4} - \dfrac{1}{c^2+3c-4}$ $\dfrac{1}{c+4},\ c \neq 1$

10. $\dfrac{2v}{2v^2-5v-12} + \dfrac{3}{2v^2-5v-12}$ $\dfrac{1}{v-4},\ v \neq -\dfrac{3}{2}$

Adding or Subtracting with Unlike Denominators

The **least common multiple (LCM)** of two (or more) polynomials can be helpful when adding or subtracting rational expressions with *unlike* denominators. The least common multiple of two (or more) polynomials is the simplest polynomial that is a multiple of each of the original polynomials. This means that the LCM must contain all the *different* factors in each polynomial, with each factor raised to the greatest power of its occurrence in any one of the polynomials.

Additional Examples

Find the least common multiple of each pair of expressions.

a. $24x^3$, $36x$

b. $4x + 4$, $x^3 + x^2$

c. $5x - 25$, $2x^2 - 9x - 5$

Answers:

a. $72x^3$

b. $4x^2(x + 1)$

c. $5(x - 5)(2x + 1)$

EXAMPLE 2 Finding Least Common Multiples

Find the least common multiple of the expressions.

a. $6x, 2x^2, 9x^3$ **b.** $x^2 - x, 2x - 2$ **c.** $3x^2 + 6x, x^2 + 4x + 4$

SOLUTION

a. The least common multiple of

$$6x = 2 \cdot 3 \cdot x, \quad 2x^2 = 2 \cdot x^2, \quad \text{and} \quad 9x^3 = 3^2 \cdot x^3$$

is $2 \cdot 3^2 \cdot x^3 = 18x^3$.

b. The least common multiple of

$$x^2 - x = x(x - 1) \quad \text{and} \quad 2x - 2 = 2(x - 1)$$

is $2x(x - 1)$.

c. The least common multiple of

$$3x^2 + 6x = 3x(x + 2) \quad \text{and} \quad x^2 + 4x + 4 = (x + 2)^2$$

is $3x(x + 2)^2$.

To add or subtract rational expressions with *unlike* denominators, you must first rewrite the rational expressions so that they have a common denominator. You can always find a common denominator of two (or more) rational expressions by multiplying their denominators. However, if you use the **least common denominator (LCD)**, which is the least common multiple of the denominators, you may have less simplifying to do. After the rational expressions have been written with a common denominator, you can simply add or subtract using the rules given at the beginning of this section.

Exercises Within Reach®

Solutions in English & Spanish and tutorial videos at AlgebraWithinReach.com

Finding the Least Common Multiple **In Exercises 11–22, find the least common multiple of the expressions.**

11. $5x^2, 20x^3$ $20x^3$

12. $14t^2, 42t^5$ $42t^5$

13. $9y^3, 12y$ $36y^3$

14. $18m^2, 45m$ $90m^2$

15. $15x^2, 3(x + 5)$ $15x^2(x + 5)$

16. $6x^2, 15x(x - 1)$ $30x^2(x - 1)$

17. $63z^2(z + 1), 14(z + 1)^4$ $126z^2(z + 1)^4$

18. $18y^3, 27y(y - 3)^2$ $54y^3(y - 3)^2$

19. $8t(t + 2), 14(t^2 - 4)$ $56t(t + 2)(t - 2)$

20. $6(x^2 - 4), 2x(x + 2)$ $6x(x + 2)(x - 2)$

21. $2y^2 + y - 1, 4y^2 - 2y$ $2y(y + 1)(2y - 1)$

22. $t^3 + 3t^2 + 9t, 2t^2(t^2 - 9)$ $2t^2(t + 3)(t - 3)(t^2 + 3t + 9)$

EXAMPLE 3 Adding with Unlike Denominators

Add the rational expressions: $\frac{7}{6x} + \frac{5}{8x}$.

SOLUTION

By factoring the denominators, $6x = 2 \cdot 3 \cdot x$ and $8x = 2^3 \cdot x$, you can conclude that the least common denominator is $2^3 \cdot 3 \cdot x = 24x$.

$$\frac{7}{6x} + \frac{5}{8x} = \frac{7(4)}{6x(4)} + \frac{5(3)}{8x(3)}$$ Rewrite expressions using LCD of $24x$.

$$= \frac{28}{24x} + \frac{15}{24x}$$ Like denominators

$$= \frac{28 + 15}{24x} = \frac{43}{24x}$$ Add fractions and simplify.

EXAMPLE 4 Subtracting with Unlike Denominators

Subtract the rational expressions: $\frac{3}{x-3} - \frac{5}{x+2}$.

SOLUTION

The only factors of the denominators are $x - 3$ and $x + 2$. So, the least common denominator is $(x - 3)(x + 2)$.

$$\frac{3}{x-3} - \frac{5}{x+2}$$ Write original expressions.

$$= \frac{3(x+2)}{(x-3)(x+2)} - \frac{5(x-3)}{(x-3)(x+2)}$$ Rewrite expressions using LCD of $(x - 3)(x + 2)$.

$$= \frac{3x+6}{(x-3)(x+2)} - \frac{5x-15}{(x-3)(x+2)}$$ Distributive Property

$$= \frac{3x + 6 - 5x + 15}{(x-3)(x+2)}$$ Subtract fractions and use the Distributive Property.

$$= \frac{-2x+21}{(x-3)(x+2)}$$ Simplified form

Exercises Within Reach ®

Solutions in English & Spanish and tutorial videos at AlgebraWithinReach.com

Adding and Subtracting Rational Expressions **In Exercises 23–32, combine and simplify.**

23. $\frac{5}{4x} - \frac{3}{5}$ $\frac{-12x+25}{20x}$

24. $\frac{10}{b} + \frac{1}{10b}$ $\frac{101}{10b}$

25. $\frac{7}{a} + \frac{14}{a^2}$ $\frac{7(a+2)}{a^2}$

26. $\frac{1}{6u^2} - \frac{2}{9u}$ $\frac{3-4u}{18u^2}$

27. $25 + \frac{10}{x+4}$ $\frac{5(5x+22)}{x+4}$

28. $\frac{30}{x-6} - 4$ $-\frac{2(2x-27)}{x-6}$

29. $\frac{x}{x+3} - \frac{5}{x-2}$ $\frac{x^2-7x-15}{(x+3)(x-2)}$

30. $\frac{1}{x+4} - \frac{1}{x+2}$ $-\frac{2}{(x+4)(x+2)}$

31. $\frac{12}{x^2-9} - \frac{2}{x-3}$ $-\frac{2}{x+3}, x \neq 3$

32. $\frac{12}{x^2-4} - \frac{3}{x+2}$ $-\frac{3(x-6)}{(x+2)(x-2)}$

EXAMPLE 5 Adding with Unlike Denominators

Find the sum of the rational expressions.

$$\frac{6x}{x^2-4}+\frac{3}{2-x}$$

SOLUTION

$\frac{6x}{x^2-4}+\frac{3}{2-x}$ — Original expressions

$= \frac{6x}{(x+2)(x-2)}+\frac{3}{(-1)(x-2)}$ — Factor denominators.

$= \frac{6x}{(x+2)(x-2)}-\frac{3(x+2)}{(x+2)(x-2)}$ — Rewrite expressions using LCD of $(x+2)(x-2)$.

$= \frac{6x}{(x+2)(x-2)}-\frac{3x+6}{(x+2)(x-2)}$ — Distributive Property

$= \frac{6x-(3x+6)}{(x+2)(x-2)}$ — Subtract.

$= \frac{6x-3x-6}{(x+2)(x-2)}$ — Distributive Property

$= \frac{3x-6}{(x+2)(x-2)}$ — Simplify.

$= \frac{3(x-2)}{(x+2)(x-2)}$ — Factor numerator.

$= \frac{3\cancel{(x-2)}}{(x+2)\cancel{(x-2)}}$ — Divide out common factor.

$= \frac{3}{x+2},\ x \neq 2$ — Simplified form

Exercises Within Reach®

Solutions in English & Spanish and tutorial videos at AlgebraWithinReach.com

Adding and Subtracting Rational Expressions **In Exercises 33–42, combine and simplify.**

33. $\frac{20}{x-4}+\frac{20}{4-x}$ $\quad 0,\ x \neq 4$

34. $\frac{15}{2-t}-\frac{7}{t-2}$ $\quad -\frac{22}{t-2}$

35. $\frac{3x}{x-8}-\frac{6}{8-x}$ $\quad \frac{3(x+2)}{x-8}$

36. $\frac{1}{y-6}+\frac{y}{6-y}$ $\quad \frac{1-y}{y-6}$

37. $\frac{3x}{3x-2}+\frac{2}{2-3x}$ $\quad 1,\ x \neq \frac{2}{3}$

38. $\frac{y}{5y-3}-\frac{3}{3-5y}$ $\quad \frac{y+3}{5y-3}$

39. $\frac{3}{x-5}+\frac{2}{x+5}$ $\quad \frac{5(x+1)}{(x+5)(x-5)}$

40. $\frac{7}{2x-3}+\frac{3}{2x+3}$ $\quad \frac{4(5x+3)}{(2x-3)(2x+3)}$

41. $\frac{9}{5x}+\frac{3}{x-1}$ $\quad \frac{3(8x-3)}{5x(x-1)}$

42. $\frac{3}{x-1}+\frac{5}{4x}$ $\quad \frac{17x-5}{4x(x-1)}$

EXAMPLE 6 Subtracting with Unlike Denominators

$\frac{x}{x^2 - 5x + 6} - \frac{1}{x^2 - x - 2}$ Original expressions

$= \frac{x}{(x-3)(x-2)} - \frac{1}{(x-2)(x+1)}$ Factor denominators.

$= \frac{x(x+1)}{(x-3)(x-2)(x+1)} - \frac{1(x-3)}{(x-3)(x-2)(x+1)}$ Rewrite expressions using LCD of $(x-3)(x-2)(x+1)$.

$= \frac{x^2 + x}{(x-3)(x-2)(x+1)} - \frac{x-3}{(x-3)(x-2)(x+1)}$ Distributive Property

$= \frac{(x^2 + x) - (x - 3)}{(x-3)(x-2)(x+1)}$ Subtract fractions.

$= \frac{x^2 + x - x + 3}{(x-3)(x-2)(x+1)}$ Distributive Property

$= \frac{x^2 + 3}{(x-3)(x-2)(x+1)}$ Simplified form

Additional Examples

Perform the indicated operations and simplify.

a. $\frac{3}{x-1} + \frac{x+3}{x^2-1}$

b. $\frac{2x-5}{6x+9} - \frac{4}{2x^2+3x} + \frac{1}{x}$

Answers:

a. $\frac{2(2x+3)}{(x+1)(x-1)}$

b. $\frac{x-1}{3x}, x \neq -\frac{3}{2}$

EXAMPLE 7 Combining Rational Expressions

$$\frac{4x}{x^2 - 16} + \frac{x}{x+4} - \frac{2}{x} = \frac{4x}{(x+4)(x-4)} + \frac{x}{x+4} - \frac{2}{x}$$

$$= \frac{4x(x)}{x(x+4)(x-4)} + \frac{x(x)(x-4)}{x(x+4)(x-4)} - \frac{2(x+4)(x-4)}{x(x+4)(x-4)}$$

$$= \frac{4x^2 + x^2(x-4) - 2(x^2 - 16)}{x(x+4)(x-4)}$$

$$= \frac{4x^2 + x^3 - 4x^2 - 2x^2 + 32}{x(x+4)(x-4)}$$

$$= \frac{x^3 - 2x^2 + 32}{x(x+4)(x-4)}$$

Exercises Within Reach®

Solutions in English & Spanish and tutorial videos at AlgebraWithinReach.com

Combining Rational Expressions **In Exercises 43–52, combine and simplify.**

43. $\frac{x}{x^2 - x - 30} - \frac{1}{x+5}$ $\frac{6}{(x-6)(x+5)}$

44. $\frac{x}{x^2 - 9} + \frac{3}{x^2 - 5x + 6}$ $\frac{x^2 + x + 9}{(x+3)(x-3)(x-2)}$

45. $\frac{4}{x-4} + \frac{16}{(x-4)^2}$ $\frac{4x}{(x-4)^2}$

46. $\frac{3}{x-2} - \frac{1}{(x-2)^2}$ $\frac{3x-7}{(x-2)^2}$

47. $\frac{y}{x^2 + xy} - \frac{x}{xy + y^2}$ $\frac{y-x}{xy}, x \neq -y$

48. $\frac{5}{x+y} + \frac{5}{x^2 - y^2}$ $\frac{5(x-y+1)}{(x+y)(x-y)}$

49. $\frac{4}{x} - \frac{2}{x^2} + \frac{4}{x+3}$ $\frac{2(4x^2 + 5x - 3)}{x^2(x+3)}$

50. $\frac{5}{2} - \frac{1}{2x} - \frac{3}{x+1}$ $\frac{5x^2 - 2x - 1}{2x(x+1)}$

51. $\frac{3u}{u^2 - 2uv + v^2} + \frac{2}{u-v} - \frac{u}{u-v}$ $-\frac{u^2 - uv - 5u + 2v}{(u-v)^2}$

52. $\frac{1}{x-y} - \frac{3}{x+y} + \frac{3x-y}{x^2 - y^2}$ $\frac{x+3y}{(x+y)(x-y)}$

Application **EXAMPLE 8** **Marital Status**

The four types of martial status are never married, married, widowed, and divorced. For the years 2005 through 2010, the total number of people in the United States 18 years old or older P (in millions) and the total number of these people who were never married N (in millions) can be modeled by

$$P = \frac{10.443t + 201.35}{0.033t + 1} \quad \text{and} \quad N = \frac{-1.471t + 49.75}{-0.043t + 1}, \quad 5 \le t \le 10$$

where t represents the year, with $t = 5$ corresponding to 2005. Find a rational model T for the total number of people whose marital status was married, widowed, or divorced during this time period. (*Source:* U.S. Census Bureau)

SOLUTION

To find a model for T, find the difference of P and N.

$$T = \frac{10.443t + 201.35}{0.033t + 1} - \frac{-1.471t + 49.75}{-0.043t + 1}$$ Subtract N from P.

$$= \frac{(10.443t + 201.35)(-0.043t + 1) - (0.033t + 1)(-1.471t + 49.75)}{(0.033t + 1)(-0.043t + 1)}$$ Basic definition

$$= \frac{-0.400506t^2 + 1.6142t + 151.6}{(0.033t + 1)(-0.043t + 1)}$$ Simplify.

Exercises Within Reach®

Solutions in English & Spanish and tutorial videos at AlgebraWithinReach.com

Undergraduate Students **In Exercises 53 and 54, use the following models, which give the numbers (in millions) of males M and females F enrolled as undergraduate students from 2005 through 2010.**

$$M = \frac{-0.150t + 5.53}{-0.049t + 1}, \quad 5 \le t \le 10 \qquad F = \frac{-0.313t + 7.71}{-0.055t + 1}, \quad 5 \le t \le 10$$

In these models, t represents the year, with $t = 5$ corresponding to 2005. (*Source:* U.S. Department of Education)

53. Find a rational model T for the total number of undergraduate students (in millions) from 2005 through 2010.

$$T = \frac{0.023587t^2 - 1.14494t + 13.24}{(-0.049t + 1)(-0.055t + 1)}, \quad 5 \le t \le 10$$

54. Use the model you found in Exercise 53 to complete the table showing the total number of undergraduate students (rounded to the nearest million) each year from 2005 through 2010.

Year	2005	2006	2007
Undergraduates (in millions)	15	15	16

Year	2008	2009	2010
Undergraduates (in millions)	16	17	18

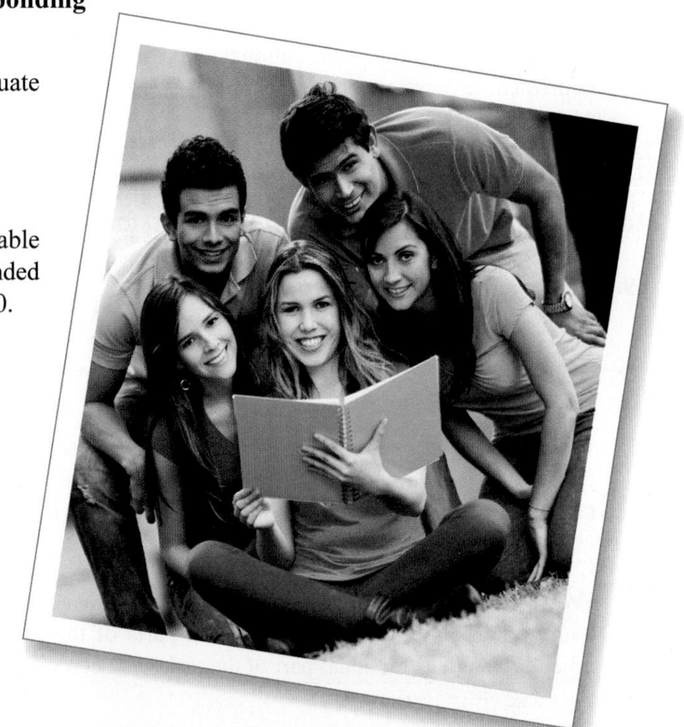

Andresr/Shutterstock.com

Concept Summary: Adding and Subtracting Rational Expressions

What

The rules for adding and subtracting rational expressions are the same as the rules for adding and subtracting numerical fractions.

EXAMPLE

Add: $\dfrac{7}{4x^2} + \dfrac{5}{2x}$.

How

To add (or subtract) rational expressions with unlike denominators do the following.

1. Use the **least common denominator** (LCD) to rewrite the fractions so they have like denominators.
2. Add (or subtract) the numerators.
3. Simplify the result.

EXAMPLE

$\dfrac{7}{4x^2} + \dfrac{5}{2x}$

$= \dfrac{7}{4x^2} + \dfrac{5(2x)}{2x(2x)}$ Rewrite.

$= \dfrac{7}{4x^2} + \dfrac{10x}{4x^2}$ Simplify.

$= \dfrac{10x + 7}{4x^2}$ Add.

Why

Knowing how to add and subtract rational expressions will help you solve rational equations.

Exercises Within Reach®

Worked-out solutions to odd-numbered exercises at AlgebraWithinReach.com

Concept Summary Check

55. ***True or False?*** Two rational expressions with *like* denominators have a common denominator. True

56. ***Logic*** When adding or subtracting rational expressions, how do you rewrite each rational expression as an equivalent expression whose denominator is the LCD? See Additional Answers.

57. ***Writing*** In your own words, describe how to add or subtract rational expressions with *like* denominators. See Additional Answers.

58. ***Writing*** In your own words, describe how to add or subtract rational expressions with *unlike* denominators. See Additional Answers.

Extra Practice

Adding and Subtracting Rational Expressions **In Exercises 59–62, combine and simplify.**

59. $\dfrac{4}{x^2} - \dfrac{4}{x^2 + 1}$ $\dfrac{4}{x^2(x^2 + 1)}$

60. $\dfrac{3}{y^2 - 3} + \dfrac{2}{3y^2}$ $\dfrac{11y^2 - 6}{3y^2(y^2 - 3)}$

61. $\dfrac{x + 2}{x - 1} - \dfrac{2}{x + 6} - \dfrac{14}{x^2 + 5x - 6}$ $\dfrac{x}{x - 1}, x \neq -6$

62. $\dfrac{-2x - 10}{x^2 + 8x + 15} + \dfrac{2}{x + 3} + \dfrac{x}{x + 5}$ $\dfrac{x}{x + 5}, x \neq -3$

63. ***Work Rate*** After working together for t hours on a common task, two workers have completed fractional parts of the job equal to $t/4$ and $t/6$. What fractional part of the task has been completed? $\dfrac{5t}{12}$

64. ***Work Rate*** After working together for t hours on a common task, two workers have completed fractional parts of the job equal to $t/3$ and $t/5$. What fractional part of the task has been completed? $\dfrac{8t}{15}$

65. ***Rewriting a Fraction*** The fraction $4/(x^3 - x)$ can be rewritten as a sum of three fractions, as follows.

$$\frac{4}{x^3 - x} = \frac{A}{x} + \frac{B}{x+1} + \frac{C}{x-1}$$

The numbers A, B, and C are the solutions of the system

$$\begin{cases} A + B + C = 0 \\ \quad -B + C = 0. \\ -A \qquad\quad = 4 \end{cases}$$

Solve the system and verify that the sum of the three resulting fractions is the original fraction.

$A = -4, B = 2, C = 2$

66. ***Rewriting a Fraction*** The fraction

$$\frac{x+1}{x^3 - x^2}$$

can be rewritten as a sum of three fractions, as follows.

$$\frac{x+1}{x^3 - x^2} = \frac{A}{x} + \frac{B}{x^2} + \frac{C}{x-1}$$

The numbers A, B, and C are the solutions of the system

$$\begin{cases} A \qquad\quad + C = 0 \\ -A + B \qquad = 1. \\ \quad -B \qquad = 1 \end{cases}$$

Solve the system and verify that the sum of the three resulting fractions is the original fraction.

$A = -2, B = -1, C = 2$

Explaining Concepts

67. ***Error Analysis*** Describe the error.

$$\frac{x-1}{x+4} - \frac{4x-11}{x+4} = \frac{x-1-4x-11}{x+4} = \frac{-3x-12}{x+4} = \frac{-3(x+4)}{x+4} = -3,\ x \neq -4$$

When the numerators are subtracted, the result should be $(x-1) - (4x-11) = x - 1 - 4x + 11$.

68. ***Error Analysis*** Describe the error.

$$\frac{2}{x} - \frac{3}{x+1} + \frac{x+1}{x^2} = \frac{2x(x+1) - 3x^2 + (x+1)^2}{x^2(x+1)} = \frac{2x^2 + x - 3x^2 + x^2 + 1}{x^2(x+1)} = \frac{x+1}{x^2(x+1)} = \frac{1}{x^2},\ x \neq -1$$

When the numerator is expanded in the second step, $(x+1)^2 \neq x^2 + 1$ and $2x(x+1) \neq 2x^2 + x$. Instead, the numerator should be $2x^2 + 2x - 3x^2 + x^2 + 2x + 1$.

69. ***Reasoning*** Is it possible for the least common denominator of two fractions to be the same as one of the fraction's denominators? If so, give an example.

Yes. The LCD of $4x + 1$ and $\frac{x}{x+2}$ is $x + 2$.

70. ***Precision*** Evaluate each expression at the given value of the variable in two different ways: (1) combine and simplify the rational expressions first and then evaluate the simplified expression at the given value of the variable, and (2) substitute the given value of the variable first and then simplify the resulting expression. Do you get the same result with each method? Discuss which method you prefer and why. List the advantages and/or disadvantages of each method.

(a) $\frac{1}{m-4} - \frac{1}{m+4} + \frac{3m}{m^2+16}$, $m = 2$ $\quad -\frac{7}{6}$

(b) $\frac{x-2}{x^2-9} + \frac{3x+2}{x^2-5x+6}$, $x = 4$ $\quad \frac{51}{7}$

(c) $\frac{3y^2+16y-8}{y^2+2y-8} - \frac{y-1}{y-2} + \frac{y}{y+4}$, $y = 3$ $\quad 8$

The results are the same. Answers will vary.

Cumulative Review

In Exercises 71–74, find the sum or difference.

71. $5v + (4 - 3v)$ $\quad 2v + 4$

72. $(2v + 7) + (9v + 8)$ $\quad 11v + 15$

73. $(x^2 - 4x + 3) - (6 - 2x)$ $\quad x^2 - 2x - 3$

74. $(5y + 2) - (2y^2 + 8y - 5)$ $\quad -2y^2 - 3y + 7$

In Exercises 75–78, factor the trinomial, if possible.

75. $x^2 - 7x + 12$ $\quad (x-3)(x-4)$

76. $c^2 + 6c + 10$ Not possible, prime

77. $2a^2 - 9a - 18$ $\quad (a-6)(2a+3)$

78. $6w^2 + 14w - 12$ $\quad 2(3w-2)(w+3)$

6.4 Complex Fractions

- Simplify complex fractions using rules for dividing rational expressions.
- Simplify complex fractions that have a sum or difference in the numerator and/or denominator.

Complex Fractions

Problems involving the division of two rational expressions are sometimes written as complex fractions. A **complex fraction** is a fraction that has a fraction in its numerator or denominator, or both.

$$\frac{\left\{\dfrac{x+2}{3}\right.}{\left\{\dfrac{x-2}{x}\right.}$$

Numerator fraction
→ Main fraction line
Denominator fraction

EXAMPLE 1 Simplifying a Complex Fraction

Simplify the complex fraction.

$$\frac{\left(\dfrac{5}{14}\right)}{\left(\dfrac{25}{8}\right)}$$

SOLUTION

$$\frac{\left(\dfrac{5}{14}\right)}{\left(\dfrac{25}{8}\right)} = \frac{5}{14} \cdot \frac{8}{25}$$ Invert divisor and multiply.

$$= \frac{5 \cdot 2 \cdot 2 \cdot 2}{2 \cdot 7 \cdot 5 \cdot 5}$$ Multiply and factor.

$$= \frac{\cancel{5} \cdot \cancel{2} \cdot 2 \cdot 2}{\cancel{2} \cdot 7 \cdot \cancel{5} \cdot 5}$$ Divide out common factors.

$$= \frac{4}{35}$$ Simplified form

Exercises Within Reach®

Solutions in English & Spanish and tutorial videos at AlgebraWithinReach.com

Simplifying a Complex Fraction **In Exercises 1–4, simplify the complex fraction.**

1. $\frac{\left(\dfrac{3}{16}\right)}{\left(\dfrac{9}{12}\right)}$ $\frac{1}{4}$

2. $\frac{\left(\dfrac{20}{21}\right)}{\left(\dfrac{8}{7}\right)}$ $\frac{5}{6}$

3. $\frac{\left(\dfrac{8x^2y}{3z^2}\right)}{\left(\dfrac{4xy}{9z^5}\right)}$ $6xz^3, x \neq 0, y \neq 0, z \neq 0$

4. $\frac{\left(\dfrac{36x^4}{5y^4z^5}\right)}{\left(\dfrac{9xy^2}{20z^5}\right)}$ $\frac{16x^3}{y^6}, x \neq 0, z \neq 0$

EXAMPLE 2 Simplifying Complex Fractions

Simplify each complex fraction.

a. $\dfrac{\left(\dfrac{4y^3}{(5x)^2}\right)}{\left(\dfrac{(2y)^2}{10x^3}\right)}$ **b.** $\dfrac{\left(\dfrac{x+1}{x+2}\right)}{\left(\dfrac{x+1}{x+5}\right)}$

Study Tip

Domain restrictions result from the values that make any denominator zero in a complex fraction. In Example 2(a), note that the original expression has three denominators: $(5x)^2$, $10x^3$, and $(2y)^2/10x^3$. The domain restrictions that result from these denominators are $x \neq 0$ and $y \neq 0$.

SOLUTION

a. $\dfrac{\left(\dfrac{4y^3}{(5x)^2}\right)}{\left(\dfrac{(2y)^2}{10x^3}\right)} = \dfrac{4y^3}{25x^2} \cdot \dfrac{10x^3}{4y^2}$ Invert divisor and multiply.

$= \dfrac{4y^2 \cdot y \cdot 2 \cdot 5x^2 \cdot x}{5 \cdot 5x^2 \cdot 4y^2}$ Multiply and factor.

$= \dfrac{\cancel{4y^2} \cdot y \cdot 2 \cdot \cancel{5x^2} \cdot x}{5 \cdot \cancel{5x^2} \cdot \cancel{4y^2}}$ Divide out common factors.

$= \dfrac{2xy}{5},\ x \neq 0,\ y \neq 0$ Simplified form

b. $\dfrac{\left(\dfrac{x+1}{x+2}\right)}{\left(\dfrac{x+1}{x+5}\right)} = \dfrac{x+1}{x+2} \cdot \dfrac{x+5}{x+1}$ Invert divisor and multiply.

$= \dfrac{(x+1)(x+5)}{(x+2)(x+1)}$ Multiply numerators and denominators.

$= \dfrac{\cancel{(x+1)}(x+5)}{(x+2)\cancel{(x+1)}}$ Divide out common factor.

$= \dfrac{x+5}{x+2},\ x \neq -1,\ x \neq -5$ Simplified form

Exercises Within Reach®

Solutions in English & Spanish and tutorial videos at AlgebraWithinReach.com

Simplifying a Complex Fraction **In Exercises 5–10, simplify the complex fraction.**

5. $\dfrac{\left(\dfrac{6x^3}{(5y)^2}\right)}{\left(\dfrac{(3x)^2}{15y^4}\right)}$ $\dfrac{2xy^2}{5}, x \neq 0, y \neq 0$

6. $\dfrac{\left(\dfrac{(3r)^3}{10t^4}\right)}{\left(\dfrac{9r}{(2t)^2}\right)}$ $\dfrac{6r^2}{5t^2}, r \neq 0$

7. $\dfrac{\left(\dfrac{y}{3-y}\right)}{\left(\dfrac{y^2}{y-3}\right)}$ $-\dfrac{1}{y}, y \neq 3$

8. $\dfrac{\left(\dfrac{x}{x-4}\right)}{\left(\dfrac{x}{4-x}\right)}$ $-1, x \neq 0, x \neq 4$

9. $\dfrac{\left(\dfrac{25x^2}{x-5}\right)}{\left(\dfrac{10x}{5+4x-x^2}\right)}$ $-\dfrac{5x(x+1)}{2}, x \neq -1, x \neq 0, x \neq 5$

10. $\dfrac{\left(\dfrac{5x}{x+7}\right)}{\left(\dfrac{10}{x^2+8x+7}\right)}$ $\dfrac{x(x+1)}{2}, x \neq -7, x \neq -1$

EXAMPLE 3 Simplifying a Complex Fraction

Simplify the complex fraction.

$$\frac{\left(\dfrac{x^2 + 4x + 3}{x - 2}\right)}{2x + 6}$$

SOLUTION

$$\frac{\left(\dfrac{x^2 + 4x + 3}{x - 2}\right)}{2x + 6} = \frac{\left(\dfrac{x^2 + 4x + 3}{x - 2}\right)}{\left(\dfrac{2x + 6}{1}\right)}$$ Rewrite denominator.

$$= \frac{x^2 + 4x + 3}{x - 2} \cdot \frac{1}{2x + 6}$$ Invert divisor and multiply.

$$= \frac{(x + 1)(x + 3)}{(x - 2)(2)(x + 3)}$$ Multiply and factor.

$$= \frac{(x + 1)\cancel{(x + 3)}}{(x - 2)(2)\cancel{(x + 3)}}$$ Divide out common factor.

$$= \frac{x + 1}{2(x - 2)}, \ x \neq -3$$ Simplified form

Exercises Within Reach ®

Solutions in English & Spanish and tutorial videos at AlgebraWithinReach.com

Simplifying a Complex Fraction **In Exercises 11–22, simplify the complex fraction.**

11. $\dfrac{\left(\dfrac{x^2 + 3x - 10}{x + 4}\right)}{3x - 6}$ $\quad \dfrac{x + 5}{3(x + 4)}, x \neq 2$

12. $\dfrac{\left(\dfrac{x^2 - 2x - 8}{x - 1}\right)}{5x - 20}$ $\quad \dfrac{x + 2}{5(x - 1)}, x \neq 4$

13. $\dfrac{2x - 14}{\left(\dfrac{x^2 - 9x + 14}{x + 3}\right)}$ $\quad \dfrac{2(x + 3)}{x - 2}, x \neq -3, x \neq 7$

14. $\dfrac{4x + 16}{\left(\dfrac{x^2 + 9x + 20}{x - 1}\right)}$ $\quad \dfrac{4(x - 1)}{x + 5}, x \neq -4, x \neq 1$

15. $\dfrac{\left(\dfrac{6x^2 - 17x + 5}{3x^2 + 3x}\right)}{\left(\dfrac{3x - 1}{3x + 1}\right)}$ $\quad \dfrac{(2x - 5)(3x + 1)}{3x(x + 1)}, x \neq \pm\dfrac{1}{3}$

16. $\dfrac{\left(\dfrac{6x^2 - 13x - 5}{5x^2 + 5x}\right)}{\left(\dfrac{2x - 5}{5x + 1}\right)}$ $\quad \dfrac{(3x + 1)(5x + 1)}{5x(x + 1)}, x \neq -\dfrac{1}{5}, x \neq \dfrac{5}{2}$

17. $\dfrac{\left(\dfrac{16x^2 + 8x + 1}{3x^2 + 8x - 3}\right)}{\left(\dfrac{4x^2 - 3x - 1}{x^2 + 6x + 9}\right)}$ $\quad \dfrac{(x + 3)(4x + 1)}{(3x - 1)(x - 1)}, x \neq -3, x \neq -\dfrac{1}{4}$

18. $\dfrac{\left(\dfrac{9x^2 - 24x + 16}{x^2 + 10x + 25}\right)}{\left(\dfrac{6x^2 - 5x - 4}{2x^2 + 3x - 35}\right)}$ $\quad \dfrac{(3x - 4)(2x - 7)}{(x + 5)(2x + 1)}, x \neq \dfrac{4}{3}, x \neq \dfrac{7}{2}$

19. $\dfrac{x^2 + x - 6}{x^2 - 4} \div \dfrac{x + 3}{x^2 + 4x + 4}$ $\quad x + 2, x \neq \pm 2, x \neq -3$

20. $\dfrac{t^3 + t^2 - 9t - 9}{t^2 - 5t + 6} \div \dfrac{t^2 + 6t + 9}{t - 2}$ $\quad \dfrac{t + 1}{t + 3}, t \neq 2, t \neq 3$

21. $\dfrac{\left(\dfrac{x^2 - 3x - 10}{x^2 - 4x + 4}\right)}{\left(\dfrac{21 + 4x - x^2}{x^2 - 5x - 14}\right)}$ $\quad -\dfrac{(x - 5)(x + 2)^2}{(x - 2)^2(x + 3)}, x \neq -2, x \neq 7$

22. $\dfrac{\left(\dfrac{x^2 + 5x + 6}{4x^2 - 20x + 25}\right)}{\left(\dfrac{x^2 - 5x - 24}{4x^2 - 25}\right)}$ $\quad \dfrac{(x + 2)(2x + 5)}{(2x - 5)(x - 8)}, x \neq -3, x \neq -\dfrac{5}{2}$

Complex Fractions with Sums or Differences

Complex fractions can have numerators and/or denominators that are sums or differences of fractions. One way to simplify such a complex fraction is to combine the terms so that the numerator and denominator each consist of a single fraction. Then divide by inverting the denominator and multiplying.

Study Tip

Another way of simplifying the complex fraction in Example 4 is to multiply the numerator and denominator by $3x$, the least common denominator of all the fractions in the numerator and denominator. This produces the same result, as shown below.

$$\frac{\left(\frac{x}{3}+\frac{2}{3}\right)}{\left(1-\frac{2}{x}\right)} = \frac{\left(\frac{x}{3}+\frac{2}{3}\right)}{\left(1-\frac{2}{x}\right)} \cdot \frac{3x}{3x}$$

$$= \frac{\frac{x}{3}(3x)+\frac{2}{3}(3x)}{(1)(3x)-\frac{2}{x}(3x)}$$

$$= \frac{x^2+2x}{3x-6}$$

$$= \frac{x(x+2)}{3(x-2)},\ x \neq 0$$

EXAMPLE 4 **Simplifying a Complex Fraction**

Simplify the complex fraction.

$$\frac{\left(\frac{x}{3}+\frac{2}{3}\right)}{\left(1-\frac{2}{x}\right)}$$

SOLUTION

$$\frac{\left(\frac{x}{3}+\frac{2}{3}\right)}{\left(1-\frac{2}{x}\right)} = \frac{\left(\frac{x}{3}+\frac{2}{3}\right)}{\left(\frac{x}{x}-\frac{2}{x}\right)}$$ Rewrite with least common denominator.

$$= \frac{\left(\frac{x+2}{3}\right)}{\left(\frac{x-2}{x}\right)}$$ Add fractions.

$$= \frac{x+2}{3} \cdot \frac{x}{x-2}$$ Invert divisor and multiply.

$$= \frac{x(x+2)}{3(x-2)},\ x \neq 0$$ Simplified form

Exercises Within Reach®

Solutions in English & Spanish and tutorial videos at AlgebraWithinReach.com

Simplifying a Complex Fraction **In Exercises 23–30, simplify the complex fraction.**

23. $\dfrac{\left(1+\frac{4}{y}\right)}{y}$ $\quad \frac{y+4}{y^2}$

24. $\dfrac{x}{\left(\frac{3}{x}+2\right)}$ $\quad \frac{x^2}{2x+3},\ x \neq 0$

25. $\dfrac{\left(\frac{4}{x}+3\right)}{\left(\frac{4}{x}-3\right)}$ $\quad -\frac{3x+4}{3x-4},\ x \neq 0$

26. $\dfrac{\left(\frac{1}{t}-1\right)}{\left(\frac{1}{t}+1\right)}$ $\quad \frac{1-t}{1+t},\ t \neq 0$

27. $\dfrac{\left(\frac{x}{2}\right)}{\left(2+\frac{3}{x}\right)}$ $\quad \frac{x^2}{2(2x+3)},\ x \neq 0$

28. $\dfrac{\left(1-\frac{2}{x}\right)}{\left(\frac{x}{2}\right)}$ $\quad \frac{2(x-2)}{x^2}$

29. $\dfrac{\left(3+\frac{9}{x-3}\right)}{\left(4+\frac{12}{x-3}\right)}$ $\quad \frac{3}{4},\ x \neq 0,\ x \neq 3$

30. $\dfrac{\left(4+\frac{16}{x-4}\right)}{\left(5+\frac{20}{x-4}\right)}$ $\quad \frac{4}{5},\ x \neq 0,\ x \neq 4$

Study Tip

In Example 5, you might wonder about the domain restrictions that result from the main denominator

$$\left(\frac{3}{x+2}+\frac{2}{x}\right)$$

of the original expression. By setting this expression equal to zero and solving for x, you can see that it leads to the domain restriction $x \neq -\frac{4}{5}$. Notice that this restriction is implied by the denominator of the simplified expression.

Encourage students to practice both of the methods shown for simplifying complex fractions.

Additional Examples

Simplify.

a. $\dfrac{\left(\dfrac{x^2+2x-3}{x-3}\right)}{4x+12}$

b. $\dfrac{\left(\dfrac{2}{x}-3\right)}{\left(1-\dfrac{1}{x-1}\right)}$

Answers:

a. $\dfrac{x-1}{4(x-3)}$, $x \neq -3$

b. $\dfrac{(2-3x)(x-1)}{x(x-2)}$, $x \neq 1$

EXAMPLE 5 Simplifying Complex Fractions

a. $\dfrac{\left(\dfrac{2}{x+2}\right)}{\left(\dfrac{3}{x+2}+\dfrac{2}{x}\right)} = \dfrac{\left(\dfrac{2}{x+2}\right)(x)(x+2)}{\left(\dfrac{3}{x+2}\right)(x)(x+2)+\left(\dfrac{2}{x}\right)(x)(x+2)}$ — $x(x + 2)$ is the least common denominator.

$= \dfrac{2x}{3x+2(x+2)}$ — Multiply and simplify.

$= \dfrac{2x}{3x+2x+4}$ — Distributive Property

$= \dfrac{2x}{5x+4},\ x \neq -2, x \neq 0$ — Simplify.

b. $\dfrac{5+x^{-2}}{8x^{-1}+x} = \dfrac{\left(5+\dfrac{1}{x^2}\right)}{\left(\dfrac{8}{x}+x\right)}$ — Rewrite with positive exponents.

$= \dfrac{\left(\dfrac{5x^2}{x^2}+\dfrac{1}{x^2}\right)}{\left(\dfrac{8}{x}+\dfrac{x^2}{x}\right)}$ — Rewrite with least common denominators.

$= \dfrac{\left(\dfrac{5x^2+1}{x^2}\right)}{\left(\dfrac{x^2+8}{x}\right)}$ — Add fractions.

$= \dfrac{5x^2+1}{x^2} \cdot \dfrac{x}{x^2+8}$ — Invert divisor and multiply.

$= \dfrac{x(5x^2+1)}{x^2(x^2+8)}$ — Multiply.

$= \dfrac{\cancel{x}(5x^2+1)}{\cancel{x}(x)(x^2+8)}$ — Divide out common factor.

$= \dfrac{5x^2+1}{x(x^2+8)}$ — Simplified form

Exercises Within Reach ®

Solutions in English & Spanish and tutorial videos at AlgebraWithinReach.com

Simplifying a Complex Fraction **In Exercises 31 and 32, simplify the complex fraction.**

31. $\dfrac{\left(\dfrac{1}{x}-\dfrac{1}{x+1}\right)}{\left(\dfrac{1}{x+1}\right)}$ $\quad \dfrac{1}{x}, x \neq -1$

32. $\dfrac{\left(\dfrac{5}{y}-\dfrac{6}{2y+1}\right)}{\left(\dfrac{5}{2y+1}\right)}$ $\quad \dfrac{4y+5}{5y}, y \neq -\dfrac{1}{2}$

Simplifying an Expression **In Exercises 33–36, simplify the expression.**

33. $\dfrac{2y-y^{-1}}{10-y^{-2}}$ $\quad \dfrac{y(2y^2-1)}{10y^2-1}, y \neq 0$

34. $\dfrac{9x-x^{-1}}{3+x^{-1}}$ $\quad 3x-1, x \neq -\dfrac{1}{3}, x \neq 0$

35. $\dfrac{7x^2+2x^{-1}}{5x^{-3}+x}$ $\quad \dfrac{x^2(7x^3+2)}{x^4+5}, x \neq 0$

36. $\dfrac{3x^{-2}-x}{4x^{-1}+6x}$ $\quad -\dfrac{x^3-3}{2x(3x^2+2)}$

Application EXAMPLE 6 Monthly Payment

The approximate annual percent interest rate r (in decimal form) of a monthly installment loan is

$$r = \frac{\left[\frac{24(MN - P)}{N}\right]}{\left(P + \frac{MN}{12}\right)}$$

where N is the total number of payments, M is the monthly payment, and P is the amount financed.

a. Simplify the expression.

b. Approximate the annual percent interest rate for a four-year home-improvement loan of $15,000 with monthly payments of $350.

SOLUTION

a. $$r = \frac{\left[\frac{24(MN - P)}{N}\right]}{\left(P + \frac{MN}{12}\right)} = \frac{24(MN - P)}{N} \div \frac{12P + MN}{12}$$

$$= \frac{24(MN - P)}{N} \cdot \frac{12}{12P + MN}$$

$$= \frac{288(MN - P)}{N(12P + MN)}$$

b. $$r = \frac{288(MN - P)}{N(12P + MN)} = \frac{288[350(48) - 15{,}000]}{48[12(15{,}000) + 350(48)]}$$

$$= \frac{288(1800)}{48(196{,}800)}$$

$$\approx 0.0549$$

The rate is about 5.5%.

Exercises Within Reach ®

Solutions in English & Spanish and tutorial videos at AlgebraWithinReach.com

37. ***Electronics*** When two resistors of resistance R_1 and R_2 (all in ohms) are connected in parallel, the total resistance (in ohms) is modeled by

$$\frac{1}{\left(\frac{1}{R_1} + \frac{1}{R_2}\right)}.$$

Simplify this complex fraction.

$\frac{R_1R_2}{R_1 + R_2}$

38. ***Using Results*** Use the simplified fraction in Exercise 37 to find the total resistance when $R_1 = 10$ ohms and $R_2 = 20$ ohms.

6.7 ohms

Maitree Laipitaksin/Shutterstock.com

Concept Summary: Simplifying Complex Fractions

What

When simplifying a **complex fraction**, the rules for dividing rational expressions still apply.

EXAMPLE

Simplify $\dfrac{\left(\frac{x+2}{3}\right)}{\left(\frac{x+2}{x}\right)}$.

How

To simplify a complex fraction, invert the denominator fraction and multiply.

EXAMPLE

$$\frac{\left(\frac{x+2}{3}\right)}{\left(\frac{x+2}{x}\right)} = \frac{x+2}{3} \cdot \frac{x}{x+2}$$

$$= \frac{x\cancel{(x+2)}}{3\cancel{(x+2)}}$$

$$= \frac{x}{3}, \ x \neq 0, \ x \neq -2$$

Why

Knowing how to simplify "messy" expressions, such as complex fractions, will help you as you continue your study of algebra.

A useful strategy in algebra is to *rewrite complicated problems into simpler forms.*

Exercises Within Reach®

Worked-out solutions to odd-numbered exercises at AlgebraWithinReach.com

Concept Summary Check

39. ***Vocabulary*** Define the term *complex fraction*. Give an example. See Additional Answers.

40. ***Precision*** Describe how to rewrite a complex fraction as a product. Invert the denominator fraction and multiply.

41. ***Writing*** Describe the method for simplifying complex fractions that involves the use of a least common denominator. See Additional Answers.

42. ***Reasoning*** Explain how you can find the implied domain restrictions for a complex fraction. See Additional Answers.

Extra Practice

Simplifying a Complex Fraction **In Exercises 43–48, simplify the complex fraction.**

43. $\dfrac{\left(\frac{y}{x}-\frac{x}{y}\right)}{\left(\frac{x+y}{xy}\right)}$ $y-x, x \neq 0, y \neq 0, x \neq -y$

44. $\dfrac{\left(\frac{x}{y}-\frac{y}{x}\right)}{\left(\frac{x-y}{xy}\right)}$ $x+y, x \neq 0, y \neq 0, x \neq y$

45. $\dfrac{\left(\frac{x}{x-3}-\frac{2}{3}\right)}{\left(\frac{10}{3x}+\frac{x^2}{x-3}\right)}$ $\dfrac{x(x+6)}{3x^3+10x-30}, x \neq 0, x \neq 3$

46. $\dfrac{\left(\frac{1}{2x}-\frac{6}{x+5}\right)}{\left(\frac{x}{x-5}+\frac{1}{x}\right)}$ $-\dfrac{(11x-5)(x-5)}{2(x+5)(x^2+x-5)}, x \neq 0, x \neq 5$

47. $\dfrac{\left(\frac{10}{x+1}\right)}{\left(\frac{1}{2x+2}+\frac{3}{x+1}\right)}$ $\dfrac{20}{7}, x \neq -1$

48. $\dfrac{\left(\frac{2}{x+5}\right)}{\left(\frac{2}{x+5}+\frac{1}{4x+20}\right)}$ $\dfrac{8}{9}, x \neq -5$

Simplifying an Expression **In Exercises 49–52, simplify the expression.**

49. $\dfrac{x^{-1}+y^{-1}}{x^{-1}-y^{-1}}$ $\dfrac{y+x}{y-x}, x \neq 0, y \neq 0$

50. $\dfrac{x^{-1}-y^{-1}}{x^{-2}-y^{-2}}$ $\dfrac{xy}{y+x}, x \neq 0, y \neq 0, x \neq y$

51. $\dfrac{x^{-2}-y^{-2}}{(x+y)^2}$ $\dfrac{y-x}{x^2y^2(y+x)}$

52. $\dfrac{x-y}{x^{-2}-y^{-2}}$ $-\dfrac{x^2y^2}{y+x}, x \neq 0, y \neq 0, x \neq y$

Simplifying an Expression **In Exercises 53 and 54, use the function to find and simplify the expression for $\dfrac{f(2+h)-f(2)}{h}$.**

53. $f(x) = \dfrac{1}{x}$ $-\dfrac{1}{2(h+2)}, h \neq 0$

54. $f(x) = \dfrac{x}{x-1}$ $-\dfrac{1}{h+1}, h \neq 0$

55. ***Average of Two Numbers*** Determine the average of two real numbers $x/5$ and $x/6$. $\frac{11x}{60}$

56. ***Average of Two Numbers*** Determine the average of two real numbers $2x/3$ and $3x/5$. $\frac{19x}{30}$

57. ***Average of Two Numbers*** Determine the average of two real numbers $2x/3$ and $x/4$. $\frac{11x}{24}$

58. ***Average of Two Numbers*** Determine the average of two real numbers $4/a^2$ and $2/a$. $\frac{a+2}{a^2}$

59. ***Average of Two Numbers*** Determine the average of two real numbers $(b + 5)/4$ and $2/b$. $\frac{b^2 + 5b + 8}{8b}$

60. ***Average of Two Numbers*** Determine the average of two real numbers $5/(2s)$ and $(s + 1)/5$. $\frac{2s^2 + 2s + 25}{20s}$

61. ***Number Problem*** Find three real numbers that divide the real number line between $x/9$ and $x/6$ into four equal parts (see figure).

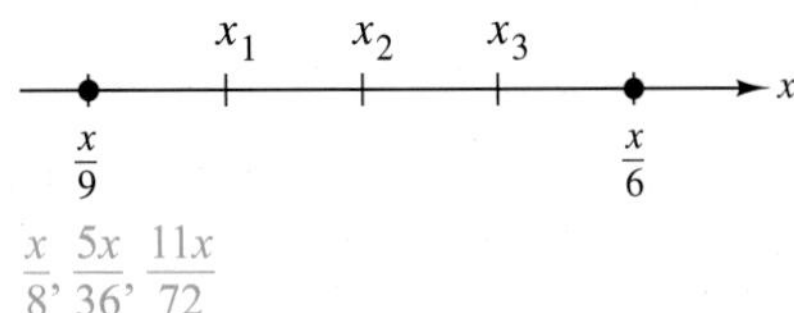

$\frac{x}{8}, \frac{5x}{36}, \frac{11x}{72}$

62. ***Number Problem*** Find two real numbers that divide the real number line between $x/3$ and $5x/4$ into three equal parts (see figure).

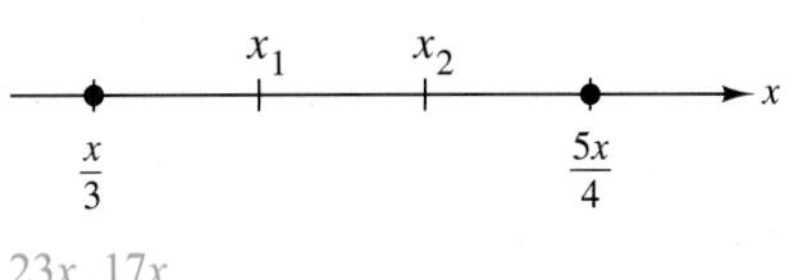

$\frac{23x}{36}, \frac{17x}{18}$

63. ***Electronics*** When three resistors of resistance R_1, R_2, and R_3 (all in ohms) are connected in parallel, the total resistance (in ohms) is modeled by

$$\frac{1}{\left(\frac{1}{R_1} + \frac{1}{R_2} + \frac{1}{R_3}\right)}.$$

Simplify this complex fraction.

$\frac{R_1R_2R_3}{R_1R_2 + R_1R_3 + R_2R_3}$

64. ***Using Results*** Use the simplified fraction found in Exercise 63 to determine the total resistance when $R_1 = 10$ ohms, $R_2 = 15$ ohms, and $R_3 = 20$ ohms.

About 4.615 ohms

Explaining Concepts

65. ***Reasoning*** Is the simplified form of a complex fraction a complex fraction? Explain.

No. A complex fraction can be written as the division of two rational expressions, so the simplified form will be a rational expression.

66. ***Writing*** Describe the effect of multiplying two rational expressions by their least common denominator. When two rational expressions are multiplied by their least common denominator, the denominator is divided out during the simplification.

Error Analysis **In Exercises 67 and 68, describe and correct the error.** See Additional Answers.

67. 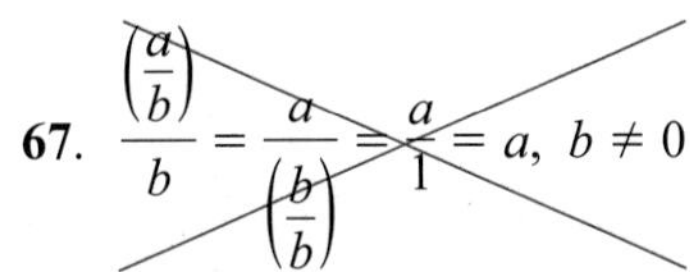

68. 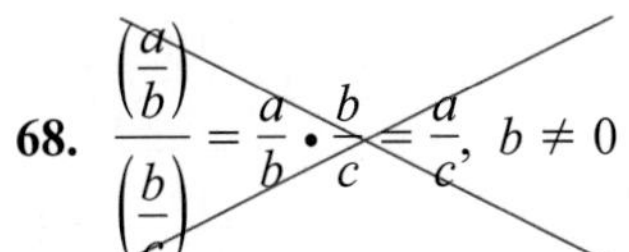

Cumulative Review

In Exercises 69 and 70, use the rules of exponents to simplify the expression.

69. $(2y)^3(3y)^2$ $72y^5$

70. $\frac{27x^4y^2}{9x^3y}$ $3xy$

In Exercises 71 and 72, factor the trinomial.

71. $3x^2 + 5x - 2$ $(3x - 1)(x + 2)$

72. $x^2 + xy - 2y^2$ $(x + 2y)(x - y)$

In Exercises 73–76, divide and simplify.

73. $\frac{x^2}{2} \div 4x$ $\frac{x}{8}, x \neq 0$

74. $\frac{4x^3}{3} \div \frac{2x^2}{9}$ $6x, x \neq 0$

75. $\frac{(x + 1)^2}{x + 2} \div \frac{x + 1}{(x + 2)^3}$ $(x + 1)(x + 2)^2, x \neq -2, x \neq -1$

76. $\frac{x^2 - 4x + 4}{x - 3} \div \frac{x^2 - 3x + 2}{x^2 - 6x + 9}$ $\frac{(x - 2)(x - 3)}{x - 1}, x \neq 2, x \neq 3$

Mid-Chapter Quiz: Sections 6.1–6.4

Solutions in English & Spanish and tutorial videos at AlgebraWithinReach.com

Take this quiz as you would take a quiz in class. After you are done, check your work against the answers in the back of the book.

In Exercises 1 and 2, find the domain of the rational function.

1. $f(x) = \dfrac{x}{x^2 + x}$ All real values of x such that $x \neq -1$ and $x \neq 0$.

2. $f(x) = \dfrac{x^2 + 4x}{x^2 - 4}$ All real values of x such that $x \neq -2$ and $x \neq 2$.

In Exercises 3–8, simplify the rational expression.

3. $\dfrac{9y^2}{6y}$ $\quad \dfrac{3}{2}y, y \neq 0$

4. $\dfrac{6u^4v^3}{15uv^3}$

5. $\dfrac{4x^2 - 1}{x - 2x^2}$

6. $\dfrac{(z + 3)^2}{2z^2 + 5z - 3}$

7. $\dfrac{5a^2b + 3ab^3}{a^2b^2}$

8. $\dfrac{2mn^2 - n^3}{2m^2 + mn - n^2}$

4. $\dfrac{2u^3}{5}, u \neq 0, v \neq 0$

5. $-\dfrac{2x + 1}{x}, x \neq \dfrac{1}{2}$

6. $\dfrac{z + 3}{2z - 1}, z \neq -3$

7. $\dfrac{5a + 3b^2}{ab}$

8. $\dfrac{n^2}{m + n}, 2m \neq n$

In Exercises 9–20, perform the indicated operations and simplify.

9. $\dfrac{11t^2}{6} \cdot \dfrac{9}{33t}$ $\quad \dfrac{t}{2}, t \neq 0$

10. $(x^2 + 2x) \cdot \dfrac{5}{x^2 - 4}$ $\quad \dfrac{5x}{x - 2}, x \neq -2$

11. $\dfrac{4}{3(x - 1)} \cdot \dfrac{12x}{6(x^2 + 2x - 3)}$

12. $\dfrac{32z^4}{5x^5y^5} \div \dfrac{80z^5}{25x^8y^6}$ $\quad \dfrac{2x^3y}{z}, x \neq 0, y \neq 0$

13. $\dfrac{a - b}{9a + 9b} \div \dfrac{a^2 - b^2}{a^2 + 2a + 1}$

14. $\dfrac{5u}{3(u + v)} \cdot \dfrac{2(u^2 - v^2)}{3v} \div \dfrac{25u^2}{18(u - v)}$

15. $\dfrac{5x - 6}{x - 2} + \dfrac{2x - 5}{x - 2}$ $\quad \dfrac{7x - 11}{x - 2}$

16. $\dfrac{x}{x^2 - 9} - \dfrac{4(x - 3)}{x + 3}$ $\quad -\dfrac{4x^2 - 25x + 36}{(x - 3)(x + 3)}$

17. $\dfrac{x^2 + 2}{x^2 - x - 2} + \dfrac{1}{x + 1} - \dfrac{x}{x - 2}$ $\quad 0, x \neq 2, x \neq -1$

18. $\dfrac{\left(\dfrac{9t^2}{3 - t}\right)}{\left(\dfrac{6t}{t - 3}\right)}$ $\quad -\dfrac{3t}{2}, t \neq 0, t \neq 3$

19. $\dfrac{\left(\dfrac{10}{x^2 + 2x}\right)}{\left(\dfrac{15}{x^2 + 3x + 2}\right)}$

20. $\dfrac{3x^{-1} - y^{-1}}{(x - y)^{-1}}$ $\quad \dfrac{(3y - x)(x - y)}{xy}, x \neq y$

11. $\dfrac{8x}{3(x + 3)(x - 1)^2}$

13. $\dfrac{(a + 1)^2}{9(a + b)^2}, a \neq b, a \neq -1$

14. $\dfrac{4(u - v)^2}{5uv}, u \neq \pm v$

19. $\dfrac{2(x + 1)}{3x}, x \neq -2, x \neq -1$

Applications

21. You open a floral shop with a setup cost of \$25,000. The cost of creating one dozen floral arrangements is \$144.

(a) Write the total cost C as a function of x, the number of dozens of floral arrangements created. $C = 25{,}000 + 144x$

(b) Write the average cost per dozen $\overline{C} = C/x$ as a function of x, the number of dozens of floral arrangements created.

(c) Find the value of $\overline{C}(500)$. \$194

21. (b) $\overline{C} = \dfrac{25{,}000 + 144x}{x}$

22. Determine the average of three real numbers x, $x/2$, and $2x/3$. $\quad \dfrac{13x}{18}$

Study Skills in Action

Using a Test-Taking Strategy

What do runners do before a race? They design a strategy for running their best. They make sure they get enough rest, eat sensibly, and get to the track early to warm up. In the same way, it is important for students to get a good night's sleep, eat a healthy meal, and get to class early to allow time to focus before a test.

The biggest difference between a runner's race and a math test is that a math student does not have to reach the finish line first! In fact, many students would increase their scores if they used all the test time instead of worrying about being the last student left in the class. This is why it is important to have a strategy for taking the test.

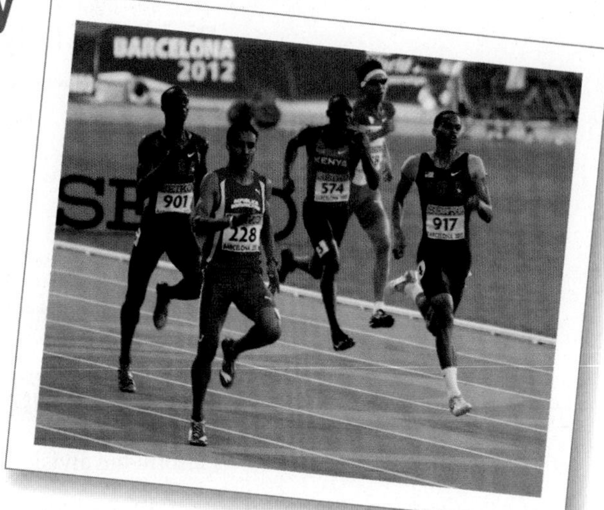

These runners are focusing on their techniques, not on whether other runners are ahead of or behind them.

Smart Study Strategy

Use Ten Steps for Test Taking

1. **Do a memory data dump.** As soon as you get the test, turn it over and write down anything that you still have trouble remembering sometimes (formulas, calculations, rules).
2. **Preview the test.** Look over the test and mark the questions you know how to do easily. These are the problems you should do first.
3. **Do a second memory data dump.** As you previewed the test, you may have remembered other information. Write this information on the back of the test.
4. **Develop a test progress schedule.** Based on how many points each question is worth, decide on a progress schedule. You should always have more than half the test done before half the time has elapsed.
5. **Answer the easiest problems first.** Solve the problems you marked while previewing the test.
6. **Skip difficult problems.** Skip the problems that you suspect will give you trouble.
7. **Review the skipped problems.** After solving all the problems that you know how to do easily, go back and reread the problems you skipped.
8. **Try your best at the remaining problems that confuse you.** Even if you cannot completely solve a problem, you may be able to get partial credit for a few correct steps.
9. **Review the test.** Look for any careless errors you may have made.
10. **Use all the allowed test time.** The test is not a race against the other students.

6.5 Dividing Polynomials and Synthetic Division

- Divide polynomials by monomials and write in simplest form.
- Use long division to divide polynomials by polynomials.
- Use synthetic division to divide and factor polynomials.

Dividing a Polynomial by a Monomial

Dividing a Polynomial by a Monomial

Let u, v, and w represent real numbers, variables, or algebraic expressions such that $w \neq 0$.

1. $\frac{u+v}{w} = \frac{u}{w} + \frac{v}{w}$ **2.** $\frac{u-v}{w} = \frac{u}{w} - \frac{v}{w}$

EXAMPLE 1 Dividing a Polynomial by a Monomial

Perform the division and simplify.

$$\frac{12x^2 - 20x + 8}{4x}$$

SOLUTION

$$\frac{12x^2 - 20x + 8}{4x} = \frac{12x^2}{4x} - \frac{20x}{4x} + \frac{8}{4x}$$ Divide each term in the numerator by $4x$.

$$= \frac{3(4x)(x)}{4x} - \frac{5(4x)}{4x} + \frac{2(4)}{4x}$$ Factor numerators.

$$= \frac{3(\cancel{4x})(x)}{\cancel{4x}} - \frac{5(\cancel{4x})}{\cancel{4x}} + \frac{2(\cancel{4})}{\cancel{4}x}$$ Divide out common factors.

$$= 3x - 5 + \frac{2}{x}$$ Simplified form

Exercises Within Reach®

Solutions in English & Spanish and tutorial videos at AlgebraWithinReach.com

Dividing a Polynomial by a Monomial **In Exercises 1–12, perform the division.**

1. $(7x^3 - 2x^2) \div x$ $7x^2 - 2x, x \neq 0$
2. $(3w^2 - 6w) \div w$ $3w - 6, w \neq 0$
3. $(m^4 + 2m^2 - 7) \div m$ $m^3 + 2m - \frac{7}{m}$
4. $(x^3 + x - 2) \div x$ $x^2 + 1 - \frac{2}{x}$
5. $(4x^2 - 2x) \div (-x)$ $-4x + 2, x \neq 0$
6. $(5y^3 + 6y^2 - 3y) \div (-y)$ $-5y^2 - 6y + 3, y \neq 0$
7. $\frac{50z^3 + 30z}{-5z}$ $-10z^2 - 6, z \neq 0$
8. $\frac{18c^4 - 24c^2}{-6c}$ $-3c^3 + 4c, c \neq 0$
9. $\frac{4v^4 + 10v^3 - 8v^2}{4v^2}$ $v^2 + \frac{5}{2}v - 2, v \neq 0$
10. $\frac{6x^4 + 8x^3 - 18x^2}{3x^2}$ $2x^2 + \frac{8}{3}x - 6, x \neq 0$
11. $(5x^2y - 8xy + 7xy^2) \div 2xy$ $\frac{5}{2}x - 4 + \frac{7}{2}y, x \neq 0, y \neq 0$
12. $(-14s^4t^2 + 7s^2t^2 - 18t) \div 2s^2t$ $-7s^2t + \frac{7}{2}t - \frac{9}{s^2}, t \neq 0$

Long Division

EXAMPLE 2 Long Division Algorithm for Positive Integers

Use the long division algorithm to divide 6584 by 28.

SOLUTION

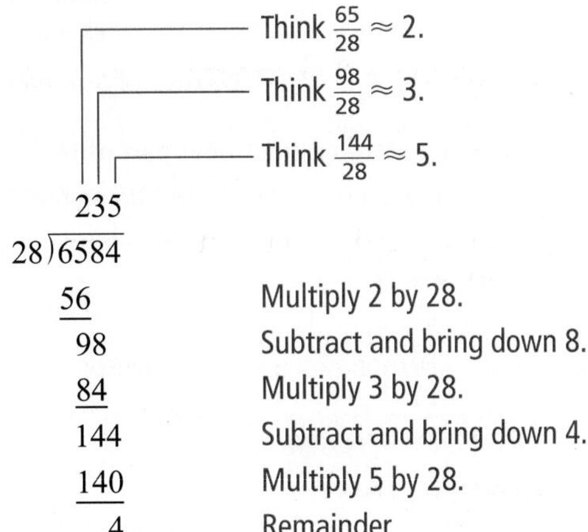

So, you have

$$6584 \div 28 = 235 + \frac{4}{28}$$

$$= 235 + \frac{1}{7}.$$

In Example 2, 6584 is the **dividend**, 28 is the **divisor**, 235 is the **quotient**, and 4 is the **remainder**.

Long Division of Polynomials

1. Write the dividend and divisor in descending powers of the variable.
2. Insert placeholders with zero coefficients for missing powers of the variable. (See Example 5.)
3. Perform the long division of the polynomials as you would with integers.
4. Continue the process until the degree of the remainder is less than that of the divisor.

Exercises Within Reach®

Solutions in English & Spanish and tutorial videos at AlgebraWithinReach.com

Using Long Division **In Exercises 13–18, use the long division algorithm to perform the division.**

13. Divide 1013 by 9. $112 + \frac{5}{9}$

14. Divide 3713 by 22. $168 + \frac{17}{22}$

15. $3235 \div 15$ $215 + \frac{2}{3}$

16. $6344 \div 28$ $226 + \frac{4}{7}$

17. $\frac{6055}{25}$ $242\frac{1}{5}$

18. $\frac{4160}{12}$ $346\frac{2}{3}$

Study Tip

Note that in Example 3, the division process requires $3x - 3$ to be subtracted from $3x + 4$. The difference

$$\begin{array}{r} 3x + 4 \\ \underline{-(3x - 3)} \end{array}$$

is implied and written simply as

$$\begin{array}{r} 3x + 4 \\ \underline{3x - 3} \\ 7. \end{array}$$

EXAMPLE 3 **Long Division Algorithm for Polynomials**

Think $x^2/x = x$.

Think $3x/x = 3$.

$$\begin{array}{r} x + 3 \\ x - 1 \overline{) x^2 + 2x + 4} \\ \underline{x^2 - x} \\ 3x + 4 \\ \underline{3x - 3} \\ 7 \end{array}$$

Multiply x by $(x - 1)$.
Subtract and bring down 4.
Multiply 3 by $(x - 1)$.
Subtract.

The remainder is a fractional part of the divisor, so you can write

$$\overbrace{\frac{x^2 + 2x + 4}{\underbrace{x - 1}_{\text{Divisor}}}}^{\text{Dividend}} = \overbrace{x + 3}^{\text{Quotient}} + \frac{\overset{\text{Remainder}}{7}}{\underbrace{x - 1}_{\text{Divisor}}}.$$

EXAMPLE 4 **Writing in Standard Form Before Dividing**

Divide $-13x^3 + 10x^4 + 8x - 7x^2 + 4$ by $3 - 2x$.

SOLUTION

First write the divisor and dividend in standard polynomial form.

$$\begin{array}{r} -5x^3 - x^2 + 2x - 1 \\ -2x + 3 \overline{) 10x^4 - 13x^3 - 7x^2 + 8x + 4} \\ \underline{10x^4 - 15x^3} \\ 2x^3 - 7x^2 \\ \underline{2x^3 - 3x^2} \\ -4x^2 + 8x \\ \underline{-4x^2 + 6x} \\ 2x + 4 \\ \underline{2x - 3} \\ 7 \end{array}$$

Multiply $-5x^3$ by $(-2x + 3)$.
Subtract and bring down $-7x^2$.
Multiply $-x^2$ by $(-2x + 3)$.
Subtract and bring down $8x$.
Multiply $2x$ by $(-2x + 3)$.
Subtract and bring down 4.
Multiply -1 by $(-2x + 3)$.
Subtract.

This shows that

$$\overbrace{\frac{10x^4 - 13x^3 - 7x^2 + 8x + 4}{\underbrace{-2x + 3}_{\text{Divisor}}}}^{\text{Dividend}} = \overbrace{-5x^3 - x^2 + 2x - 1}^{\text{Quotient}} + \frac{\overset{\text{Remainder}}{7}}{\underbrace{-2x + 3}_{\text{Divisor}}}.$$

Exercises Within Reach®

Solutions in English & Spanish and tutorial videos at AlgebraWithinReach.com

Using Long Division **In Exercises 19–26, use the long division algorithm to perform the division.**

19. $\dfrac{x^2 - 8x + 15}{x - 3}$ $x - 5, x \neq 3$

20. $\dfrac{t^2 - 18t + 72}{t - 6}$ $t - 12, t \neq 6$

21. Divide $21 - 4x - x^2$ by $3 - x$. $x + 7, x \neq 3$

22. Divide $5 + 4x - x^2$ by $1 + x$. $-x + 5, x \neq -1$

23. $(12 - 17t + 6t^2) \div (2t - 3)$ $3t - 4, t \neq \frac{3}{2}$

24. $(15 - 14u - 8u^2) \div (5 + 2u)$ $-4u + 3, u \neq -\frac{5}{2}$

25. $\dfrac{9x^3 - 3x^2 - 3x + 4}{3x + 2}$ $3x^2 - 3x + 1 + \dfrac{2}{3x + 2}$

26. $\dfrac{4y^3 + 12y^2 + 7y - 3}{2y + 3}$ $2y^2 + 3y - 1, y \neq -\frac{3}{2}$

EXAMPLE 5 Accounting for Missing Powers of x

Divide $x^3 - 2$ by $x - 1$.

SOLUTION

To account for the missing x^2- and x-terms, insert $0x^2$ and $0x$.

$$\begin{array}{r|l} x^2 + x + 1 & \\ x - 1 \overline{)x^3 + 0x^2 + 0x - 2} & \text{Insert } 0x^2 \text{ and } 0x. \\ \underline{x^3 - x^2} & \text{Multiply } x^2 \text{ by } (x-1). \\ x^2 + 0x & \text{Subtract and bring down } 0x. \\ \underline{x^2 - x} & \text{Multiply } x \text{ by } (x-1). \\ x - 2 & \text{Subtract and bring down } -2. \\ \underline{x - 1} & \text{Multiply 1 by } (x-1). \\ -1 & \text{Subtract.} \end{array}$$

So, you have

$$\frac{x^3 - 2}{x - 1} = x^2 + x + 1 - \frac{1}{x - 1}.$$

EXAMPLE 6 A Second-Degree Divisor

Study Tip

If the remainder of a division problem is zero, the divisor is said to *divide evenly* into the dividend.

Divide $x^4 + 6x^3 + 6x^2 - 10x - 3$ by $x^2 + 2x - 3$.

SOLUTION

$$\begin{array}{r|l} x^2 + 4x + 1 & \\ x^2 + 2x - 3 \overline{)x^4 + 6x^3 + 6x^2 - 10x - 3} & \\ \underline{x^4 + 2x^3 - 3x^2} & \text{Multiply } x^2 \text{ by } (x^2 + 2x - 3). \\ 4x^3 + 9x^2 - 10x & \text{Subtract and bring down } -10x. \\ \underline{4x^3 + 8x^2 - 12x} & \text{Multiply } 4x \text{ by } (x^2 + 2x - 3). \\ x^2 + 2x - 3 & \text{Subtract and bring down } -3. \\ \underline{x^2 + 2x - 3} & \text{Multiply 1 by } (x^2 + 2x - 3). \\ 0 & \text{Subtract.} \end{array}$$

So, $x^2 + 2x - 3$ divides evenly into $x^4 + 6x^3 + 6x^2 - 10x - 3$. That is,

$$\frac{x^4 + 6x^3 + 6x^2 - 10x - 3}{x^2 + 2x - 3} = x^2 + 4x + 1, \; x \neq -3, x \neq 1.$$

Exercises Within Reach®

Solutions in English & Spanish and tutorial videos at AlgebraWithinReach.com

Using Long Division **In Exercises 27–34, use the long division algorithm to perform the division.**

27. $\dfrac{x^2 + 16}{x + 4}$ $\quad x - 4 + \dfrac{32}{x + 4}$

28. $\dfrac{y^2 + 8}{y + 2}$ $\quad y - 2 + \dfrac{12}{y + 2}$

29. $\dfrac{x^3 + 125}{x + 5}$ $\quad x^2 - 5x + 25, x \neq -5$

30. $\dfrac{x^3 - 27}{x - 3}$ $\quad x^2 + 3x + 9, x \neq 3$

31. $(x^3 + 4x^2 + 7x + 7) \div (x^2 + 2x + 3)$

$x + 2 + \dfrac{1}{x^2 + 2x + 3}$

32. $(2x^3 + 2x^2 - 2x - 15) \div (2x^2 + 4x + 5)$

$x - 1 - \dfrac{3x + 10}{2x^2 + 4x + 5}$

33. $(4x^4 - 3x^2 + x - 5) \div (x^2 - 3x + 2)$

$4x^2 + 12x + 25 + \dfrac{52x - 55}{x^2 - 3x + 2}$

34. $(8x^5 + 6x^4 - x^3 + 1) \div (2x^3 - x^2 - 3)$

$4x^2 + 5x + 2 + \dfrac{14x^2 + 15x + 7}{2x^3 - x^2 - 3}$

Synthetic Division

Synthetic Division of a Third-Degree Polynomial

Use synthetic division to divide $ax^3 + bx^2 + cx + d$ by $x - k$, as follows.

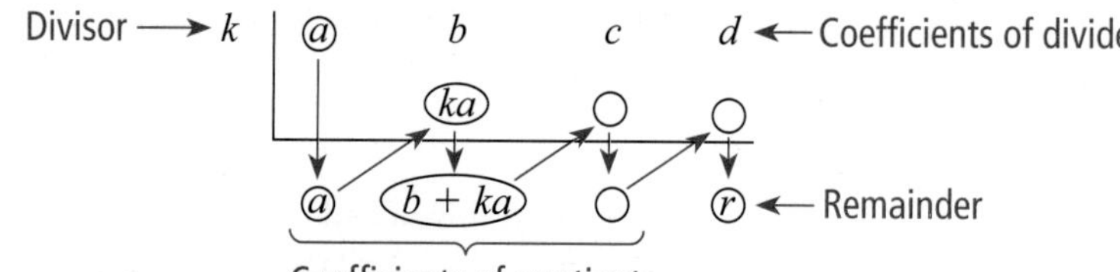

Vertical Pattern: Add terms.

Diagonal Pattern: Multiply by k.

EXAMPLE 7 Using Synthetic Division

Use synthetic division to divide $x^3 + 3x^2 - 4x - 10$ by $x - 2$.

SOLUTION

The coefficients of the dividend form the top of the synthetic division array. Because you are dividing by $x - 2$, write 2 at the top left of the array. To begin the algorithm, bring down the first coefficient. Then multiply this coefficient by 2, write the result in the second row, and add the two numbers in the second column. By continuing this pattern, you obtain the following.

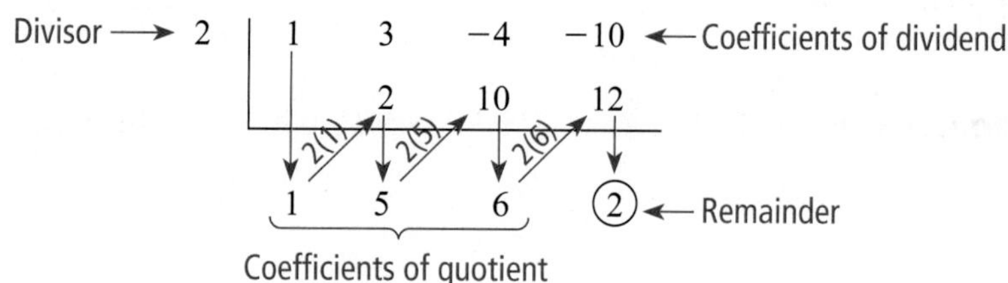

The bottom row shows the coefficients of the quotient. So, the quotient is

$$1x^2 + 5x + 6$$

and the remainder is 2. So, the result of the division problem is

$$\frac{x^3 + 3x^2 - 4x - 10}{x - 2} = x^2 + 5x + 6 + \frac{2}{x - 2}.$$

Additional Example

Use synthetic division to divide $x^4 - 10x^2 - 2x + 3$ by $x + 3$.

Answer:

−3	1	0	−10	−2	3
		−3	9	3	−3
	1	−3	−1	1	0

$x^3 - 3x^2 - x + 1,\ x \neq -3$

Exercises Within Reach®

Solutions in English & Spanish and tutorial videos at AlgebraWithinReach.com

Using Synthetic Division **In Exercises 35–42, use synthetic division to divide.**

35. $(x^2 + x - 6) \div (x - 2)$ $x + 3, x \neq 2$

36. $(x^2 + 5x - 6) \div (x + 6)$ $x - 1, x \neq -6$

37. $\dfrac{x^3 + 3x^2 - 1}{x + 4}$ $x^2 - x + 4 - \dfrac{17}{x + 4}$

38. $\dfrac{x^3 - 4x + 7}{x - 1}$ $x^2 + x - 3 + \dfrac{4}{x - 1}$

39. $\dfrac{5x^3 - 6x^2 + 8}{x - 4}$ $5x^2 + 14x + 56 + \dfrac{232}{x - 4}$

40. $\dfrac{5x^3 + 6x + 8}{x + 2}$ $5x^2 - 10x + 26 - \dfrac{44}{x + 2}$

41. $\dfrac{0.1x^2 + 0.8x + 1}{x - 0.2}$ $0.1x + 0.82 + \dfrac{1.164}{x - 0.2}$

42. $\dfrac{x^3 - 0.8x + 2.4}{x + 0.1}$ $x^2 - 0.1x - 0.79 + \dfrac{2.479}{x + 0.1}$

Synthetic division (or long division) can be used to factor polynomials. If the remainder in a synthetic division problem is zero, you know that the divisor divides *evenly* into the dividend.

EXAMPLE 8 Factoring a Polynomial

Completely factor the polynomial $x^3 - 7x + 6$ given that one of its factors is $x - 1$.

SOLUTION

The polynomial $x^3 - 7x + 6$ can be factored completely using synthetic division. Because $x - 1$ is a factor of the polynomial, you can divide as follows.

$$\begin{array}{r|rrrr} 1 & 1 & 0 & -7 & 6 \\ & & 1 & 1 & -6 \\ \hline & 1 & 1 & -6 & \boxed{0} \end{array} \leftarrow \text{Remainder}$$

Because the remainder is zero, the divisor divides evenly into the dividend:

$$\frac{x^3 - 7x + 6}{x - 1} = x^2 + x - 6.$$

From this result, you can factor the original polynomial as follows.

$$\begin{aligned} x^3 - 7x + 6 &= (x - 1)(x^2 + x - 6) \\ &= (x - 1)(x + 3)(x - 2) \end{aligned}$$

Study Tip

In Example 8, synthetic division is used to divide the polynomial by the factor $x - 1$. Long division could be used also.

Exercises Within Reach®

Solutions in English & Spanish and tutorial videos at AlgebraWithinReach.com

Factoring Completely **In Exercises 43–50, completely factor the polynomial given one of its factors.**

	Polynomial	*Factor*
43.	$x^3 - x^2 - 14x + 24$ $(x - 3)(x + 4)(x - 2)$	$x - 3$
44.	$x^3 + x^2 - 32x - 60$ $(x + 5)(x - 6)(x + 2)$	$x + 5$
45.	$4x^3 - 3x - 1$ $(x - 1)(2x + 1)^2$	$x - 1$
46.	$9x^3 + 51x^2 + 88x + 48$ $(x + 3)(3x + 4)^2$	$x + 3$
47.	$x^4 + 7x^3 + 3x^2 - 63x - 108$ $(x + 3)^2(x - 3)(x + 4)$	$x + 4$
48.	$x^4 - 6x^3 - 8x^2 + 96x - 128$ $(x - 4)^2(x + 4)(x - 2)$	$x - 4$
49.	$15x^2 - 2x - 8$ $5\left(x - \frac{4}{5}\right)(3x + 2)$	$x - \frac{4}{5}$
50.	$18x^2 - 9x - 20$ $6\left(x + \frac{5}{6}\right)(3x - 4)$	$x + \frac{5}{6}$

Finding a Constant **In Exercises 51 and 52, find the constant c such that the denominator divides evenly into the numerator.**

51. $\dfrac{x^3 + 2x^2 - 4x + c}{x - 2}$ -8

52. $\dfrac{x^4 - 3x^2 + c}{x + 6}$ -1188

Concept Summary: Dividing Polynomials

What

You can use long division to divide polynomials just as you do for integers.

EXAMPLE

Divide $(-4 + x^2)$ by $(2 + x)$.

How

1. Write the **dividend** and **divisor** in standard form.
2. Insert placeholders with zero coefficients for missing powers of the variable.
3. Use the long division algorithm.

EXAMPLE

$$\begin{array}{r} x - 2 \\ x + 2 \overline{)x^2 + 0x - 4} \\ \underline{x^2 + 2x} \\ -2x - 4 \\ \underline{-2x - 4} \\ 0 \end{array}$$

Why

You can use long division of polynomials to check the product of two polynomials.

Exercises Within Reach®

Worked-out solutions to odd-numbered exercises at AlgebraWithinReach.com

Concept Summary Check

53. ***Vocabulary*** Identify the dividend, divisor, quotient, and remainder of the equation $1253 \div 12 = 104 + \frac{5}{12}$.
Dividend: 1253, Divisor: 12, Quotient: 104, Remainder: 5

54. ***Inserting Placeholders*** What is the missing power of x in the dividend in the example above? Explain.
x; There is no term involving x in the dividend.

55. ***Reasoning*** Explain what it means for a divisor to divide *evenly* into a dividend. A divisor divides evenly into a dividend when the remainder is zero.

56. ***Reasoning*** Explain how you can check polynomial division. Polynomial division can be checked by multiplying the quotient and the remainder by the divisor.

Extra Practice

Using Long Division **In Exercises 57 and 58, use the long division algorithm to perform the division.**

57. $\dfrac{x^3 - 9x}{x - 2}$ $\quad x^2 + 2x - 5 - \dfrac{10}{x - 2}$

58. $\dfrac{2x^5 - 3x^3 + x}{x - 3}$ $\quad 2x^4 + 6x^3 + 15x^2 + 45x + 136 + \dfrac{408}{x - 3}$

Simplifying an Expression **In Exercises 59–62, simplify the expression.**

59. $\dfrac{8u^2v}{2u} + \dfrac{3(uv)^2}{uv}$ $\quad 7uv, u \neq 0, v \neq 0$

60. $\dfrac{15x^3y}{10x^2} + \dfrac{3xy^2}{2y}$ $\quad 3xy, x \neq 0, y \neq 0$

61. $\dfrac{x^2 + 3x + 2}{x + 2} + (2x + 3)$ $\quad 3x + 4, x \neq -2$

62. $\dfrac{x^2 + 2x - 3}{x - 1} - (3x - 4)$ $\quad -2x + 7, x \neq 1$

Dividing Polynomials **In Exercises 63 and 64, perform the division assuming that n is a positive integer.**

63. $\dfrac{x^{3n} + 3x^{2n} + 6x^n + 8}{x^n + 2}$ $\quad x^{2n} + x^n + 4, x^n \neq -2$

64. $\dfrac{x^{3n} - x^{2n} + 5x^n - 5}{x^n - 1}$ $\quad x^{2n} + 5, x^n \neq 1$

Think About It **In Exercises 65 and 66, the divisor, quotient, and remainder are given. Find the dividend.**

	Divisor	*Quotient*	*Remainder*
65.	$x - 6$	$x^2 + x + 1$	-4
	$x^3 - 5x^2 - 5x - 10$		
66.	$x + 3$	$x^2 - 2x - 5$	8
	$x^3 + x^2 - 11x - 7$		

67. ***Geometry*** The height of a cube is $x + 1$. The volume of the cube is $x^3 + 3x^2 + 3x + 1$. Use division to find the area of the base. $x^2 + 2x + 1$

68. ***Geometry*** A rectangular house has a volume of $(x^3 + 55x^2 + 650x + 2000)$ cubic feet (the space in the attic is not included). The height of the house is $(x + 5)$ feet (see figure). Find the number of square feet of floor space on the first floor of the house. $x^2 + 50x + 400$

Geometry **In Exercises 69 and 70, you are given the expression for the volume of the solid shown. Find the expression for the missing dimension.**

69. $V = x^3 + 18x^2 + 80x + 96$ $2x + 8$

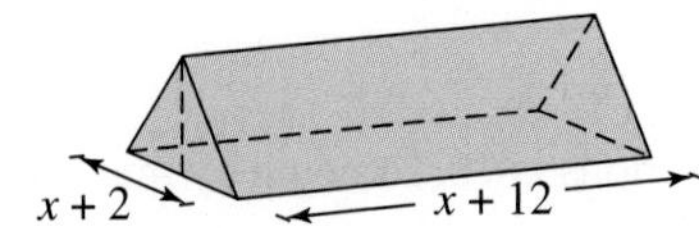

70. $V = 2h^3 + 3h^2 + h$ $2h + 1$

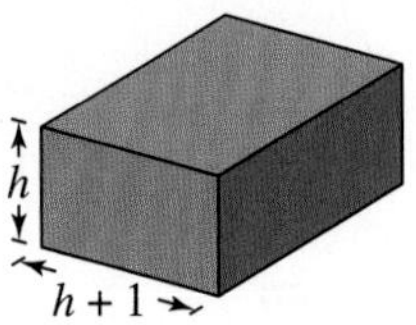

Explaining Concepts

71. ***Error Analysis*** Describe and correct the error.

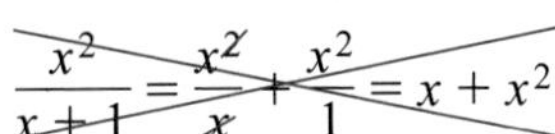

x is not a factor of the numerator.

$$\frac{6x + 5y}{x} = \frac{6x}{x} + \frac{5y}{x} = 6 + \frac{5y}{x}$$

72. ***Error Analysis*** Describe and correct the error.

$$\frac{x^2}{x+1} = \frac{x^2}{x} + \frac{x^2}{1} = x + x^2$$

Each term in the numerator can be divided by the denominator only if the denominator is a monomial.

$$\frac{x^2}{x+1} = x - 1 + \frac{1}{x+1}$$

73. ***Precision*** Create a polynomial division problem and identify the dividend, divisor, quotient, and remainder.

$$\frac{x^2 + 4}{x + 1} = x - 1 + \frac{5}{x + 1}$$

Dividend: $x^2 + 4$, Divisor: $x + 1$,
Quotient: $x - 1$, Remainder: 5

74. ***True or False?*** If the divisor divides evenly into the dividend, then the divisor and quotient are factors of the dividend. Justify your answer.

True. If $\frac{n(x)}{d(x)} = q(x)$, then $n(x) = d(x) \cdot q(x)$.

Cumulative Review

In Exercises 75–80, solve the inequality.

75. $7 - 3x > 4 - x$ $x < \frac{3}{2}$

76. $2(x + 6) - 20 < 2$ $x < 5$

77. $|x - 3| < 2$ $1 < x < 5$

78. $|x - 5| > 3$ $x < 2$ or $x > 8$

79. $\left|\frac{1}{4}x - 1\right| \geq 3$ $x \leq -8$ or $x \geq 16$

80. $\left|2 - \frac{1}{3}x\right| \leq 10$ $-24 \leq x \leq 36$

In Exercises 81 and 82, determine the quadrants in which the point must be located.

81. $(-3, y)$, y is a real number. Quadrant II or III

82. $(x, 7)$, x is a real number. Quadrant I or II

83. Describe the location of the set of points whose x-coordinates are 0. Located on the y-axis

84. Find the coordinates of the point five units to the right of the y-axis and seven units below the x-axis. $(5, -7)$

6.6 Solving Rational Equations

- Solve rational equations containing constant denominators.
- Solve rational equations containing variable denominators.
- Use rational equations to model and solve real-life problems.

Equations Containing Constant Denominators

EXAMPLE 1 Solving a Rational Equation

Solve $\frac{3}{5} = \frac{x}{2} + 1$.

Study Tip

A *rational equation* is an equation containing one or more rational expressions.

SOLUTION

The least common denominator of the fractions is 10, so begin by multiplying each side of the equation by 10.

$$10\left(\frac{3}{5}\right) = 10\left(\frac{x}{2} + 1\right)$$ Multiply each side by LCD of 10.

$$6 = 5x + 10$$ Distribute and simplify.

$$-4 = 5x \implies -\frac{4}{5} = x$$ Subtract 10 from each side, then divide each side by 5.

The solution is $x = -\frac{4}{5}$. You can check this in the original equation as follows.

Check

$$\frac{3}{5} \stackrel{?}{=} \frac{-4/5}{2} + 1$$ Substitute $-\frac{4}{5}$ for x in the original equation.

$$\frac{3}{5} \stackrel{?}{=} -\frac{4}{5} \cdot \frac{1}{2} + 1$$ Invert divisor and multiply.

$$\frac{3}{5} = -\frac{2}{5} + 1$$ Solution checks. ✓

Exercises Within Reach®

Solutions in English & Spanish and tutorial videos at AlgebraWithinReach.com

Checking Solutions **In Exercises 1 and 2, determine whether each value of x is a solution of the equation.**

	Equation	Values			Equation	Values	
1.	$\frac{x}{3} - \frac{x}{5} = \frac{4}{3}$	(a) $x = 0$	(b) $x = -2$	2.	$\frac{x}{4} + \frac{3}{4x} = 1$	(a) $x = -1$	(b) $x = 1$
		(c) $x = \frac{1}{8}$	(d) $x = 10$			(c) $x = 3$	(d) $x = \frac{1}{2}$

1. (a) Not a solution (b) Not a solution (c) Not a solution (d) Solution
2. (a) Not a solution (b) Solution (c) Solution (d) Not a solution

Solving a Rational Equation **In Exercises 3–8, solve the equation.**

3. $\frac{x}{6} - 1 = \frac{2}{3}$ 10
4. $\frac{y}{8} + 7 = -\frac{1}{2}$ -60
5. $\frac{1}{4} = \frac{z+1}{8}$ 1
6. $\frac{a}{2} = \frac{a+2}{3}$ 4
7. $\frac{x}{4} + \frac{x}{2} = \frac{2x}{3}$ 0
8. $\frac{x}{4} - \frac{x}{6} = \frac{1}{4}$ 3

EXAMPLE 2 Solving Rational Equations

Solve (a) $\frac{x-3}{6} = 7 - \frac{x}{12}$ and (b) $\frac{x^2}{3} + \frac{x}{2} = \frac{5}{6}$.

SOLUTION

a. The least common denominator of the fractions is 12, so begin by multiplying each side of the equation by 12.

$$\frac{x-3}{6} = 7 - \frac{x}{12}$$ Write original equation.

$$12\left(\frac{x-3}{6}\right) = 12\left(7 - \frac{x}{12}\right)$$ Multiply each side by LCD of 12.

$$2x - 6 = 84 - x$$ Distribute and simplify.

$$3x - 6 = 84$$ Add x to each side.

$$3x = 90 \quad \Rightarrow \quad x = 30$$ Add 6 to each side, then divide each side by 3.

The solution is $x = 30$. Check this in the original equation.

b. The least common denominator of the fractions is 6, so begin by multiplying each side of the equation by 6.

$$\frac{x^2}{3} + \frac{x}{2} = \frac{5}{6}$$ Write original equation.

$$6\left(\frac{x^2}{3} + \frac{x}{2}\right) = 6\left(\frac{5}{6}\right)$$ Multiply each side by LCD of 6.

$$\frac{6x^2}{3} + \frac{6x}{2} = \frac{30}{6}$$ Distributive Property

$$2x^2 + 3x = 5$$ Simplify.

$$2x^2 + 3x - 5 = 0$$ Subtract 5 from each side.

$$(2x + 5)(x - 1) = 0$$ Factor.

$$2x + 5 = 0 \quad \Rightarrow \quad x = -\frac{5}{2}$$ Set 1st factor equal to 0.

$$x - 1 = 0 \quad \Rightarrow \quad x = 1$$ Set 2nd factor equal to 0.

The solutions are $x = -\frac{5}{2}$ and $x = 1$. Check these in the original equation.

Additional Examples

Solve each equation.

a. $\frac{x+2}{6} - \frac{x-4}{8} = \frac{2}{3}$

b. $\frac{3x^2}{4} - \frac{x}{2} = \frac{1}{4}$

Answers:

a. $x = -4$

b. $x = -\frac{1}{3}, x = 1$

Exercises Within Reach®

Solutions in English & Spanish and tutorial videos at AlgebraWithinReach.com

Solving a Rational Equation **In Exercises 9–16, solve the equation.**

9. $\frac{z+2}{3} = 4 - \frac{z}{12}$ 8

10. $\frac{2y-9}{6} = 3y - \frac{3}{4}$ $-\frac{9}{32}$

11. $\frac{x-5}{5} + 3 = -\frac{x}{4}$ $-\frac{40}{9}$

12. $\frac{4x-2}{7} - \frac{5}{14} = 2x$ $-\frac{9}{20}$

13. $\frac{x^2}{2} - \frac{3x}{5} = -\frac{1}{10}$ $\frac{1}{5}, 1$

14. $\frac{x^2}{3} - \frac{x}{6} = \frac{1}{6}$ $-\frac{1}{2}, 1$

15. $\frac{t}{2} = 12 - \frac{3t^2}{2}$ $-3, \frac{8}{3}$

16. $\frac{x}{12} = \frac{1}{10} - \frac{x^2}{15}$ $-2, \frac{3}{4}$

Equations Containing Variable Denominators

EXAMPLE 3 An Equation Containing Variable Denominators

Solve $\frac{8}{3} = \frac{7}{x} - \frac{1}{3x}$.

SOLUTION

The least common denominator of the fractions is $3x$, so begin by multiplying each side of the equation by $3x$.

$\frac{8}{3} = \frac{7}{x} - \frac{1}{3x}$ Write original equation.

$3x\left(\frac{8}{3}\right) = 3x\left(\frac{7}{x} - \frac{1}{3x}\right)$ Multiply each side by LCD of $3x$.

$\frac{24x}{3} = \frac{21x}{x} - \frac{3x}{3x}$ Distributive Property

$8x = 21 - 1$ Simplify.

$x = \frac{20}{8} = \frac{5}{2}$ Subtract, divide each side by 8, and simplify.

The solution is $x = \frac{5}{2}$. Check this in the original equation.

Check

$\frac{8}{3} \stackrel{?}{=} \frac{7}{(5/2)} - \frac{1}{3(5/2)}$ Substitute $\frac{5}{2}$ for x.

$15\left(\frac{8}{3}\right) \stackrel{?}{=} 15\left(\frac{14}{5} - \frac{2}{15}\right)$ Invert divisors and multiply, and then multiply by LCD of 15.

$40 = 42 - 2$ Solution checks. ✓

Exercises Within Reach®

Solutions in English & Spanish and tutorial videos at AlgebraWithinReach.com

Solving a Rational Equation **In Exercises 17–30, solve the equation.**

17. $\frac{9}{25 - y} = -\frac{1}{4}$ 61

18. $-\frac{6}{u + 3} = \frac{2}{3}$ -12

19. $5 - \frac{12}{a} = \frac{5}{3}$ $\frac{18}{5}$

20. $\frac{5}{b} - 18 = 21$ $\frac{5}{39}$

21. $\frac{4}{x} - \frac{7}{5x} = -\frac{1}{2}$ $-\frac{26}{5}$

22. $\frac{5}{3} = \frac{6}{7x} + \frac{2}{x}$ $\frac{12}{7}$

23. $\frac{12}{y + 5} + \frac{1}{2} = 2$ 3

24. $\frac{7}{8} - \frac{16}{t - 2} = \frac{3}{4}$ 130

25. $\frac{5}{x} = \frac{25}{3(x + 2)}$ 3

26. $\frac{10}{x + 4} = \frac{15}{4(x + 1)}$ $\frac{4}{5}$

27. $\frac{8}{3x + 5} = \frac{1}{x + 2}$ $-\frac{11}{5}$

28. $\frac{500}{3x + 5} = \frac{50}{x - 3}$ 5

29. $\frac{3}{x + 2} - \frac{1}{x} = \frac{1}{5x}$ $\frac{4}{3}$

30. $\frac{12}{x + 5} + \frac{5}{x} = \frac{20}{x}$ -25

EXAMPLE 4 An Equation with No Solution

Solve $\dfrac{5x}{x-2} = 7 + \dfrac{10}{x-2}$.

SOLUTION

The least common denominator of the fractions is $x - 2$, so begin by multiplying each side of the equation by $x - 2$.

$\dfrac{5x}{x-2} = 7 + \dfrac{10}{x-2}$	Write original equation.
$(x-2)\left(\dfrac{5x}{x-2}\right) = (x-2)\left(7 + \dfrac{10}{x-2}\right)$	Multiply each side by $x - 2$.
$5x = 7(x-2) + 10$	Distribute and simplify.
$5x = 7x - 14 + 10$	Distributive Property
$5x = 7x - 4$	Combine like terms.
$-2x = -4$	Subtract $7x$ from each side.
$x = 2$	Divide each side by -2.

At this point, the solution appears to be $x = 2$. However, by performing a check, you can see that this "trial solution" is extraneous.

Check

$\dfrac{5x}{x-2} = 7 + \dfrac{10}{x-2}$	Write original equation.
$\dfrac{5(2)}{2-2} \stackrel{?}{=} 7 + \dfrac{10}{2-2}$	Substitute 2 for x.
$\dfrac{10}{0} \stackrel{?}{=} 7 + \dfrac{10}{0}$	Solution does not check. ✗

Because the check results in *division by zero*, you can conclude that 2 is extraneous. So, the original equation has no solution.

Exercises Within Reach®

Solutions in English & Spanish and tutorial videos at AlgebraWithinReach.com

Solving a Rational Equation **In Exercises 31–38, solve the equation.**

31. $\dfrac{1}{x-4} + 2 = \dfrac{2x}{x-4}$ No solution

32. $\dfrac{-2x}{x-1} + 2 = \dfrac{10}{x-1}$ No solution

33. $\dfrac{4}{x(x-1)} + \dfrac{3}{x} = \dfrac{4}{x-1}$ No solution

34. $\dfrac{10}{x(x-2)} + \dfrac{4}{x} = \dfrac{5}{x-2}$ No solution

35. $\dfrac{2}{x-10} - \dfrac{3}{x-2} = \dfrac{6}{x^2 - 12x + 20}$ 20

36. $\dfrac{5}{x+2} + \dfrac{2}{x^2 - 6x - 16} = -\dfrac{4}{x-8}$ $\frac{10}{3}$

37. $1 - \dfrac{6}{4-x} = \dfrac{x+2}{x^2 - 16}$ $-3, -2$

38. $\dfrac{4}{2x+3} + \dfrac{17}{5x-3} = 3$ $-\frac{11}{10}, 2$

Study Tip

Although cross-multiplication can be a little quicker than multiplying by the least common denominator, remember that *it can be used only with equations that have a single fraction on each side* of the equation.

EXAMPLE 5 **Cross-Multiplying**

$\frac{2x}{x+4} = \frac{3}{x-1}$ Original equation

$2x(x-1) = 3(x+4)$ Cross-multiply.

$2x^2 - 2x = 3x + 12$ Distributive Property

$2x^2 - 5x - 12 = 0$ Subtract $3x$ and 12 from each side.

$(2x+3)(x-4) = 0$ Factor.

$2x + 3 = 0 \Rightarrow x = -\frac{3}{2}$ Set 1st factor equal to 0.

$x - 4 = 0 \Rightarrow x = 4$ Set 2nd factor equal to 0.

The solutions are $x = -\frac{3}{2}$ and $x = 4$. Check these in the original equation.

EXAMPLE 6 **An Equation That Has Two Solutions**

$\frac{3x}{x+1} = \frac{12}{x^2-1} + 2$ Original equation

$(x^2-1)\left(\frac{3x}{x+1}\right) = (x^2-1)\left(\frac{12}{x^2-1} + 2\right)$ Multiply each side by LCD of $x^2 - 1$.

$(x-1)(3x) = 12 + 2(x^2-1)$ Distribute and simplify.

$3x^2 - 3x = 12 + 2x^2 - 2$ Distributive Property

$x^2 - 3x - 10 = 0$ Subtract $2x^2$ and 10 from each side.

$(x+2)(x-5) = 0$ Factor.

$x + 2 = 0 \Rightarrow x = -2$ Set 1st factor equal to 0.

$x - 5 = 0 \Rightarrow x = 5$ Set 2nd factor equal to 0.

The solutions are $x = -2$ and $x = 5$. Check these in the original equation.

Additional Examples

Solve each equation.

a. $\frac{3}{x+1} + \frac{5}{2} = 1$

b. $\frac{4}{x-2} + \frac{3x}{x+1} = 3$

Answers:

a. $x = -3$

b. $x = -10$

Exercises Within Reach®

Solutions in English & Spanish and tutorial videos at AlgebraWithinReach.com

Solving a Rational Equation **In Exercises 39–46, solve the equation.**

39. $\frac{2x}{5} = \frac{x^2 - 5x}{5x}$ -5

40. $\frac{3x}{4} = \frac{x^2 + 3x}{8x}$ $\frac{3}{5}$

41. $\frac{y+1}{y+10} = \frac{y-2}{y+4}$ 8

42. $\frac{x-3}{x+1} = \frac{x-6}{x+5}$ $\frac{9}{7}$

43. $\frac{x}{x-2} + \frac{3x}{x-4} = -\frac{2(x-6)}{x^2-6x+8}$ $3, -1$

44. $\frac{2(x+1)}{x^2-4x+3} + \frac{6x}{x-3} = \frac{3x}{x-1}$ $-1, -\frac{2}{3}$

45. $\frac{x}{3} = \frac{1 + \frac{4}{x}}{1 + \frac{2}{x}}$

$-3, 4$

46. $\frac{2x}{3} = \frac{1 + \frac{2}{x}}{1 + \frac{1}{x}}$

$-\frac{3}{2}, 2$

Application

Application

EXAMPLE 7 Finding a Save Percentage in Hockey

A hockey goalie faces 799 shots and saves 707 of them. How many additional consecutive saves does the goalie need to obtain a save percentage (in decimal form) of .900?

SOLUTION

Verbal Model: $\boxed{\text{Desired save percentage}} = \dfrac{\boxed{707 + (\text{Additional saves})}}{\boxed{799 + (\text{Additional saves})}}$

Labels: Desired save percentage $= 0.900$ (percentage in decimal form)

Additional saves $= x$ (saves)

Equation:

$$0.900 = \frac{707 + x}{799 + x}$$ Write equation.

$$0.900(799 + x) = 707 + x$$ Cross-multiply.

$$719.1 + 0.9x = 707 + x$$ Distributive Property

$$12.1 = 0.1x$$ Subtract 707 and $0.9x$ from each side.

$$121 = x$$ Divide each side by 0.1.

The goalie needs 121 additional consecutive saves to attain a save percentage of 0.900.

Exercises Within Reach®

Solutions in English & Spanish and tutorial videos at AlgebraWithinReach.com

47. ***Batting Average*** A softball player bats 47 times and hits the ball safely 8 times. How many additional consecutive times must the player hit the ball safely to obtain a batting average of .250?
5 hits

48. ***Batting Average*** A softball player bats 45 times and hits the ball safely 9 times. How many additional consecutive times must the player hit the ball safely to obtain a batting average of .280?
5 hits

49. ***Speed*** One person runs 1.5 miles per hour faster than a second person. The first person runs 4 miles in the same time the second person runs 3 miles. Find the speed of each person.
6 miles per hour, 4.5 miles per hour

50. ***Speed*** One person runs 2 miles per hour faster than a second person. The first person runs 5 miles in the same time the second person runs 4 miles. Find the speed of each person.
10 miles per hour, 8 miles per hour

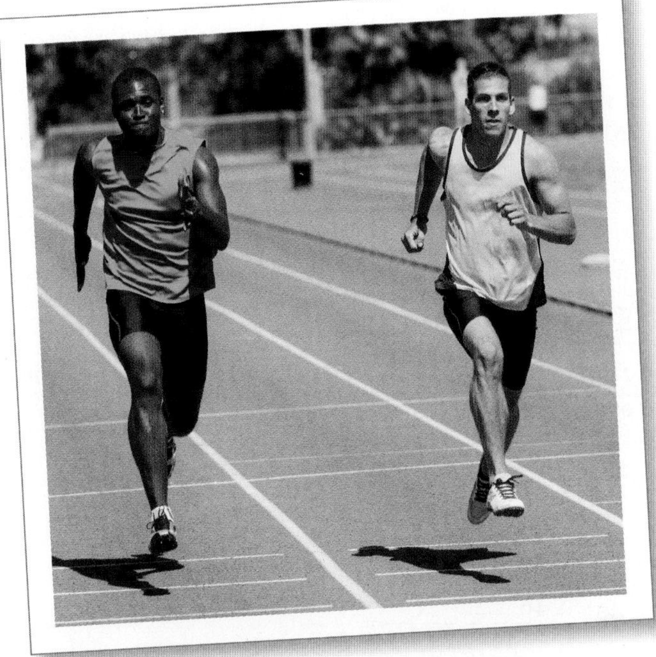

Concept Summary: Solving Rational Equations

What

The goal when solving rational equations is the same as when solving linear equations: You want to isolate the variable.

EXAMPLE

Solve $\frac{2}{x} - \frac{2}{5} = \frac{4}{x}$.

How

Use the following steps to solve rational equations.

1. Find the LCD of all fractions in the equation.
2. Multiply each side of the equation by the LCD.
3. Simplify each term.
4. Use the properties of equality to solve the resulting linear equation.

EXAMPLE

$\frac{2}{x} - \frac{2}{5} = \frac{4}{x}$ Original equation

$5x\left(\frac{2}{x} - \frac{2}{5}\right) = 5x\left(\frac{4}{x}\right)$ Multiply by LCD.

$\frac{10x}{x} - \frac{10x}{5} = \frac{20x}{x}$ Distributive Property

$10 - 2x = 20$ Simplify.

$-2x = 10$ Subtract 10.

$x = -5$ Divide by -2.

Why

You can use rational equations to model and solve many real-life problems. For example, knowing how to solve a rational equation can help you determine how many additional consecutive saves a goalie needs to obtain a specific save percentage.

Exercises Within Reach ®

Worked-out solutions to odd-numbered exercises at AlgebraWithinReach.com

Concept Summary Check

51. ***Writing*** What is a rational equation? A rational equation is an equation containing one or more rational expressions.

52. ***Writing*** Describe how to solve a rational equation. A rational equation can be solved by multiplying each side of the equation by the least common denominator.

53. ***Domain Restrictions*** Explain the domain restrictions that may exist for a rational equation. The domain of a rational equation must be restricted if any value of any of the variables makes any of the denominators zero.

54. ***Cross-Multiplication*** When can you use cross-multiplication to solve a rational equation? Explain. Cross-multiplication can be used only if the equation has a single fraction on each side of the equation.

Extra Practice

Think About It **In Exercises 55–62, if the exercise is an equation, solve it; if it is an expression, simplify it.**

55. $\frac{1}{2} = \frac{18}{x^2}$ ± 6

56. $\frac{1}{4} = \frac{16}{z^2}$ ± 8

57. $\frac{x+5}{4} - \frac{3x-8}{3} = \frac{4-x}{12}$ $\frac{43}{8}$

58. $\frac{2x-7}{10} - \frac{3x+1}{5} = \frac{6-x}{5}$ $-\frac{21}{2}$

59. $\frac{16}{x^2-16} + \frac{x}{2x-8} = \frac{1}{2}$ -12

60. $\frac{16}{x^2-16} + \frac{x}{2x-8} + \frac{1}{2}$ $\frac{x^2+2x+8}{(x+4)(x-4)}$

61. $\frac{5}{x+3} + \frac{5}{3} + 3$ $\frac{14x+57}{3(x+3)}$

62. $\frac{5}{x+3} + \frac{5}{3} = 3$ $\frac{3}{4}$

Using a Graph **In Exercises 63–66, (a) use the graph to determine any x-intercepts of the graph and (b) set $y = 0$ and solve the resulting rational equation to confirm the result of part (a).**

63. $y = \dfrac{x+2}{x-2}$

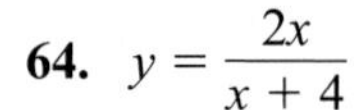

64. $y = \dfrac{2x}{x+4}$

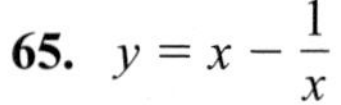

65. $y = x - \dfrac{1}{x}$

66. $y = x - \dfrac{2}{x} - 1$

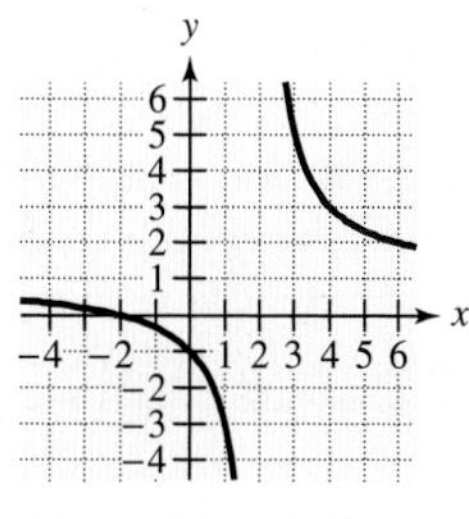

(a) and (b) $(-2, 0)$

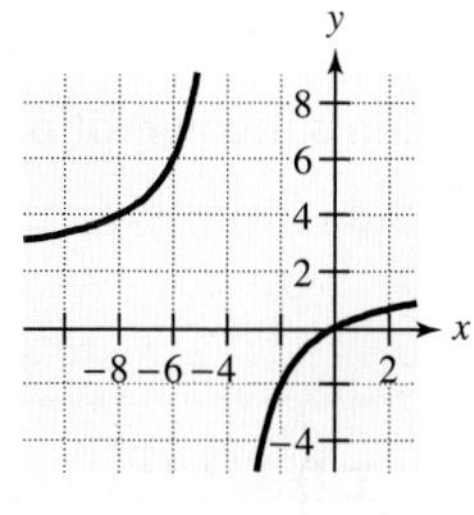

(a) and (b) $(0, 0)$

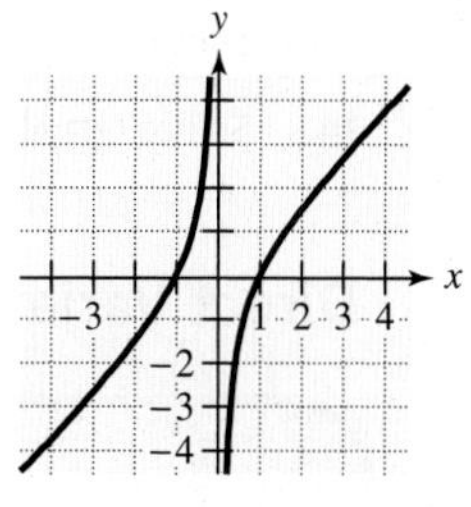

(a) and (b) $(-1, 0), (1, 0)$

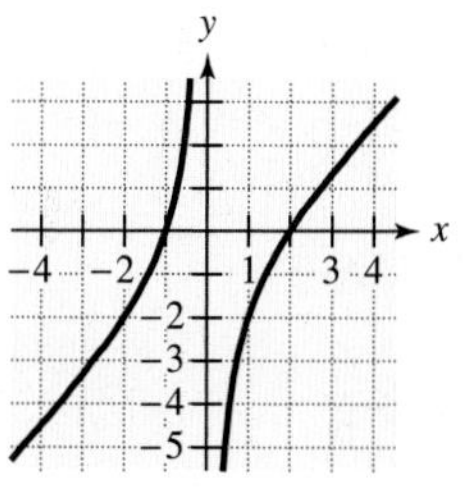

(a) and (b) $(-1, 0), (2, 0)$

67. ***Number Problem*** Find a number such that the sum of two times the number and three times its reciprocal is $\frac{203}{10}$. $\frac{3}{20}$, 10

68. ***Painting*** A painter can paint a room in 4 hours, while his partner can paint the room in 6 hours. How long would it take to paint the room if both worked together? 2.4 hours = 2 hours 24 minutes

69. ***Roofing*** A roofer requires 15 hours to shingle a roof, while an apprentice requires 21 hours. How long would it take to shingle the roof if both worked together? 8.75 hours = 8 hours, 45 minutes

70. ***Wind Speed*** A plane has a speed of 300 miles per hour in still air. The plane travels a distance of 680 miles with a tail wind in the same time it takes to travel 520 miles into a head wind. Find the speed of the wind. 40 miles per hour

Explaining Concepts

71. ***Writing*** Define the term *extraneous solution*. How do you identify an extraneous solution? An extraneous solution is a "trial solution" that does not satisfy the original equation. When you substitute this "trial solution" into the original equation, the result is false—the solution does not check, or is undefined.

72. ***Using a Graph*** Explain how you can use the graph of a rational equation to estimate the solution of the equation. Graph the rational equation and find any x-intercepts of the graph.

73. ***Precision*** Explain why the equation $\dfrac{n}{x} + n = \dfrac{n}{x}$ has no solution if n is any real nonzero number. When the equation is solved, the solution is $x = 0$. However, if $x = 0$, then there is division by zero, so the equation has no solution.

74. ***Reasoning*** Does multiplying a rational equation by its LCD produce an equivalent equation? Explain. Yes. By the multiplication property of equality, if $a = b$, then $ac = bc$.

Cumulative Review

In Exercises 75–78, factor the expression.

75. $x^2 - 81$
$(x + 9)(x - 9)$

76. $x^2 - 121$
$(x + 11)(x - 11)$

77. $4x^2 - \frac{1}{4}$
$\left(2x - \frac{1}{2}\right)\left(2x + \frac{1}{2}\right)$

78. $49 - (x - 2)^2$
$-(x + 5)(x - 9)$

In Exercises 79 and 80, find the domain of the rational function.

79. $f(x) = \dfrac{2x^2}{5}$
All real values of x

80. $f(x) = \dfrac{4}{x - 6}$
All real values of x such that $x \neq 6$

6.7 Variation

- Solve application problems involving direct variation.
- Solve application problems involving inverse variation.
- Solve application problems involving joint variation.

Direct Variation

Direct Variation

The following statements are equivalent.

1. y varies directly as x.

2. y is directly proportional to x.

3. $y = kx$ for some constant k.

The number k is called the **constant of proportionality**.

Application

EXAMPLE 1 Direct Variation

The total revenue R (in dollars) obtained from selling x ice show tickets is directly proportional to the number of tickets sold x. When 10,000 tickets are sold, the total revenue is \$142,500. Find a mathematical model that relates the total revenue R to the number of tickets sold x.

SOLUTION

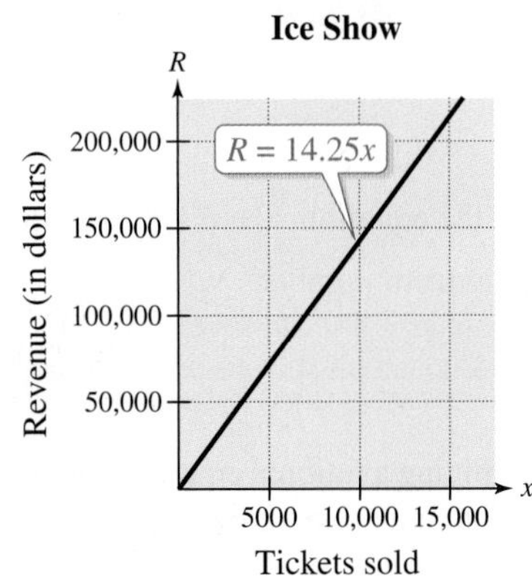

Because the total revenue is directly proportional to the number of tickets sold, the linear model is

$$R = kx.$$

To find the value of the constant k, use the fact that $R = 142{,}500$ when $x = 10{,}000$. Substituting these values into the model produces

$$142{,}500 = k(10{,}000) \qquad \text{Substitute 142,500 for } R \text{ and 10,000 for } x.$$

which implies that

$$k = \frac{142{,}500}{10{,}000} = 14.25.$$

So, the equation relating the total revenue to the total number of tickets sold is

$$R = 14.25x. \qquad \text{Direct variation model}$$

The graph of this equation is shown at the left.

Exercises Within Reach®

Solutions in English & Spanish and tutorial videos at AlgebraWithinReach.com

1. ***Revenue*** The total revenue R (in dollars) is directly proportional to the number of units sold x. When 500 units are sold, the total revenue is \$4825. Find a mathematical model that relates the total revenue R to the number of units sold x. $R = 9.65x$

2. ***Revenue*** The total revenue R (in dollars) is directly proportional to the number of units sold x. When 25 units are sold, the total revenue is \$300. Find a mathematical model that relates the total revenue R to the number of units sold x. $R = 12x$

Application **EXAMPLE 2** **Direct Variation**

Hooke's Law for springs states that the distance a spring is stretched (or compressed) is directly proportional to the force on the spring. A force of 20 pounds stretches a spring 5 inches.

a. Find a mathematical model that relates the distance the spring is stretched to the force applied to the spring.

b. How far will a force of 30 pounds stretch the spring?

SOLUTION

a. For this problem, let d represent the distance (in inches) that the spring is stretched and let F represent the force (in pounds) that is applied to the spring. Because the distance d is directly proportional to the force F, the model is

$$d = kF.$$

To find the value of the constant k, use the fact that $d = 5$ when $F = 20$. Substituting these values into the model produces

$$5 = k(20) \qquad \text{Substitute 5 for } d \text{ and 20 for } F.$$

$$\frac{5}{20} = k \qquad \text{Divide each side by 20.}$$

$$\frac{1}{4} = k. \qquad \text{Simplify.}$$

So, the equation relating distance to force is

$$d = \frac{1}{4}F. \qquad \text{Direct variation model}$$

b. When $F = 30$, the distance is

$$d = \frac{1}{4}(30) = 7.5 \text{ inches.}$$

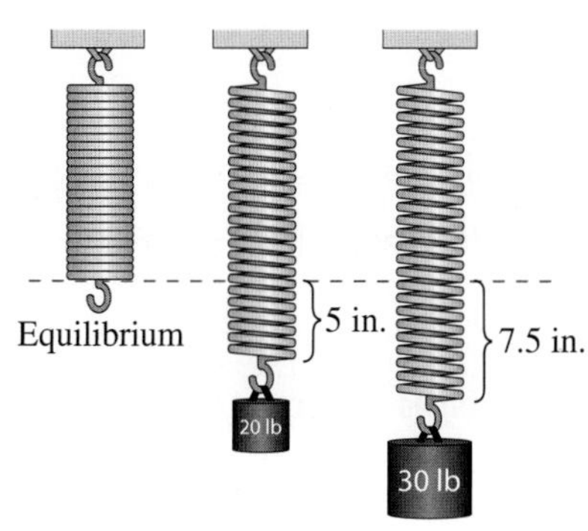

Exercises Within Reach®

Solutions in English & Spanish and tutorial videos at AlgebraWithinReach.com

3. ***Hooke's Law*** A force of 50 pounds stretches a spring 5 inches.
   (a) How far will a force of 20 pounds stretch the spring? 2 inches
   (b) What force is required to stretch the spring 1.5 inches? 15 pounds

4. ***Hooke's Law*** A force of 50 pounds stretches a spring 3 inches.
   (a) How far will a force of 20 pounds stretch the spring? 1.2 inches
   (b) What force is required to stretch the spring 1.5 inches? 25 pounds

5. ***Hooke's Law*** A baby weighing $10\frac{1}{2}$ pounds compresses the spring of a baby scale 7 millimeters. Determine the weight of a baby that compresses the spring 12 millimeters. 18 pounds

6. ***Hooke's Law*** An apple weighing 14 ounces compresses the spring of a produce scale 3 millimeters. Determine the weight of a grapefruit that compresses the spring 5 millimeters. $23\frac{1}{3}$ ounces

Vaclav Volrab/Shutterstock.com

Direct Variation as *n*th Power

The following statements are equivalent.

1. y varies directly as the nth power x.
2. y is directly proportional to the nth power of x.
3. $y = kx^n$ for some constant k.

Application

EXAMPLE 3 Direct Variation as a Power

The distance a ball rolls down an inclined plane is directly proportional to the square of the time it rolls. During the first second, a ball rolls down a plane a distance of 6 feet.

a. Find a mathematical model that relates the distance traveled to the time.

b. How far will the ball roll during the first 2 seconds?

SOLUTION

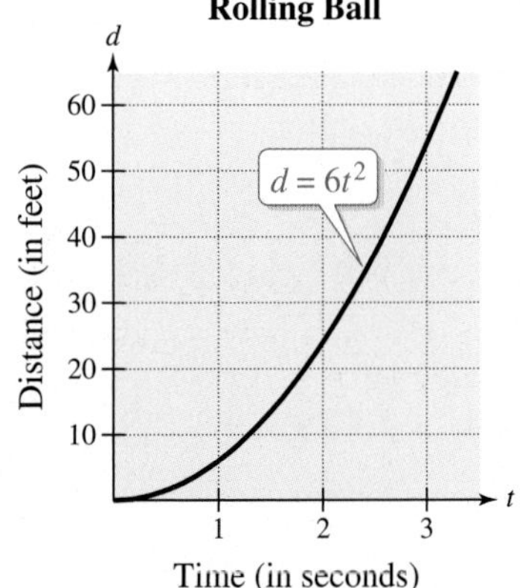

a. Letting d be the distance (in feet) that the ball rolls and letting t be the time (in seconds), you obtain the model

$$d = kt^2.$$

Because $d = 6$ when $t = 1$, you obtain

$d = kt^2$ Write original equation.

$6 = k(1)^2 \implies 6 = k$ Substitute 6 for d and 1 for t.

So, the equation relating distance to time is

$d = 6t^2$. Direct variation as 2nd power model

The graph of this equation is shown at the left.

b. When $t = 2$, the distance traveled is

$$d = 6(2)^2 = 6(4) = 24 \text{ feet}.$$

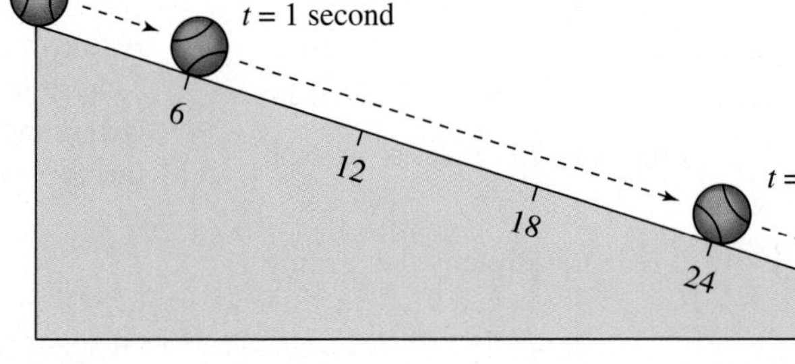

You can demonstrate direct variation and direct variation as nth power by having students consider the perimeter and area of a square. Point out that as the length of its sides increases, a square's area increases at a much greater rate than does its perimeter, as shown by the following pattern.

s	$P = 4s$	$A = s^2$
8	32	64
9	36	81
10	40	100
11	44	121
12	48	144

Exercises Within Reach®

Solutions in English & Spanish and tutorial videos at AlgebraWithinReach.com

7. ***Stopping Distance*** The stopping distance d of an automobile is directly proportional to the square of its speed s. On one road, a car requires 75 feet to stop from a speed of 30 miles per hour. How many feet does the car require to stop from a speed of 48 miles per hour on the same road? 192 feet

8. ***Frictional Force*** The frictional force F (between the tires of a car and the road) that is required to keep a car on a curved section of a highway is directly proportional to the square of the speed s of the car. By what factor does the force F change when the speed of the car is doubled on the same curve? 4

Inverse Variation

Inverse Variation

1. The following three statements are equivalent.

a. y varies inversely as x.

b. y is inversely proportional to x.

c. $y = \dfrac{k}{x}$ for some constant k.

2. If $y = \dfrac{k}{x^n}$, then y is inversely proportional to the nth power of x.

Application

EXAMPLE 4 Inverse Variation

The marketing department of a large company has found that the demand for one of its hand tools varies inversely as the price of the product. (When the price is low, more people are willing to buy the product than when the price is high.) When the price of the tool is \$7.50, the monthly demand is 50,000 tools. Approximate the monthly demand when the price is reduced to \$6.

SOLUTION

Let x represent the number of tools that are sold each month (the demand), and let p represent the price per tool (in dollars). Because the demand is inversely proportional to the price, the model is

$$x = \frac{k}{p}.$$

By substituting $x = 50{,}000$ when $p = 7.50$, you obtain

$$50{,}000 = \frac{k}{7.50} \qquad \text{Substitute 50,000 for } x \text{ and 7.50 for } p.$$

$$375{,}000 = k. \qquad \text{Multiply each side by 7.50.}$$

So, the inverse variation model is $x = \dfrac{375{,}000}{p}$.

Hand Tools

The graph of this equation is shown at the left. To find the demand that corresponds to a price of \$6, substitute 6 for p in the equation and obtain

$$x = \frac{375{,}000}{6} = 62{,}500 \text{ tools.}$$

So, when the price is lowered from \$7.50 per tool to \$6.00 per tool, you can expect the monthly demand to increase from 50,000 tools to 62,500 tools.

Exercises Within Reach®

Solutions in English & Spanish and tutorial videos at AlgebraWithinReach.com

9. ***Demand*** A company has found that the daily demand x for its boxes of chocolates is inversely proportional to the price p. When the price is \$5, the demand is 800 boxes. Approximate the demand when the price is increased to \$6.
667 boxes

10. ***Demand*** A company has found that the daily demand for its boxes of greeting cards is inversely proportional to the price. When the price is \$6, the demand is 600 boxes. Approximate the demand when the price is increased to \$7.
514 boxes

Application **EXAMPLE 5** **Direct Variation and Inverse Variation**

Remind students that direct variation is defined as $y = kx$. With inverse variation, $y = k/x$ can be written as $y = k(1/x)$, for some constant k.

A computer hardware manufacturer determines that the demand for its USB flash drive is directly proportional to the amount spent on advertising and inversely proportional to the price of the flash drive. When \$40,000 is spent on advertising and the price per unit is \$20, the monthly demand is 10,000 flash drives.

a. When the amount of advertising is increased to \$50,000, how much can the price increase and still maintain a monthly demand of 10,000 flash drives?

b. If you were in charge of the advertising department, would you recommend this increased expense in advertising?

SOLUTION

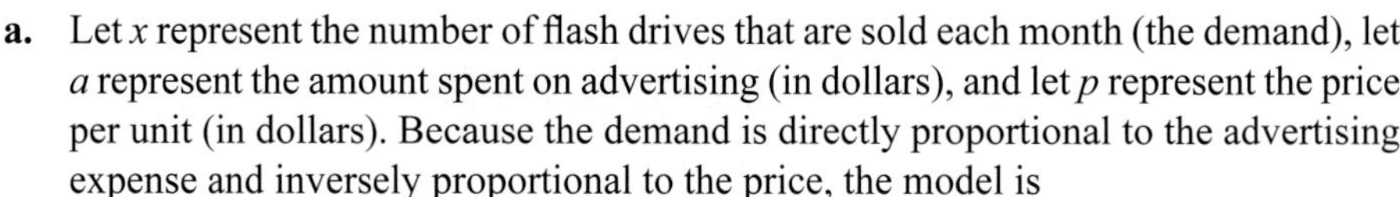

a. Let x represent the number of flash drives that are sold each month (the demand), let a represent the amount spent on advertising (in dollars), and let p represent the price per unit (in dollars). Because the demand is directly proportional to the advertising expense and inversely proportional to the price, the model is

$$x = \frac{ka}{p}.$$

By substituting 10,000 for x when $a = 40{,}000$ and $p = 20$, you obtain

$$10{,}000 = \frac{k(40{,}000)}{20}$$ Substitute 10,000 for x, 40,000 for a, and 20 for p.

$$200{,}000 = 40{,}000k$$ Multiply each side by 20.

$$5 = k.$$ Divide each side by 40,000.

So, the model is

$$x = \frac{5a}{p}.$$ Direct and inverse variation model

To find the price that corresponds to a demand of 10,000 and an advertising expense of \$50,000, substitute 10,000 for x and 50,000 for a in the model and solve for p.

$$10{,}000 = \frac{5(50{,}000)}{p} \quad \Longrightarrow \quad p = \frac{5(50{,}000)}{10{,}000} = \$25$$

So, the price can increase $\$25 - \$20 = \$5$.

b. The total revenue for selling 10,000 units at \$20 each is \$200,000, and the revenue for selling 10,000 units at \$25 each is \$250,000. Because increasing the advertising expense from \$40,000 to \$50,000 increases the revenue by \$50,000, you should recommend the increase in advertising expenses.

Additional Example

Have students write a mathematical model for the statement "N varies directly as the square of r and inversely as the cube of s."

Answer:

$N = (kr^2)/(s^3)$

Exercises Within Reach ®

Solutions in English & Spanish and tutorial videos at AlgebraWithinReach.com

11. ***Revenue*** The weekly demand for a company's frozen pizzas varies directly as the amount spent on advertising and inversely as the price per pizza. At \$5 per pizza, when \$500 is spent each week on ads, the demand is 2000 pizzas. When advertising is increased to \$600, what price yields a demand of 2000 pizzas? Is this increase worthwhile in terms of revenue?

\$6 per pizza; Answers will vary.

12. ***Revenue*** The monthly demand for a company's sports caps varies directly as the amount spent on advertising and inversely as the square of the price per cap. At \$15 per cap, when \$2500 is spent each week on ads, the demand is 300 caps. When advertising is increased to \$3000, what price yields a demand of 300 caps? Is this increase worthwhile in terms of revenue?

About \$16.43 per cap; Answers will vary.

Joint Variation

Joint Variation

1. The following three statements are equivalent.
   a. z varies jointly as x and y.
   b. z is jointly proportional to x and y.
   c. $z = kxy$ for some constant k.
2. If $z = kx^n y^m$, then z is jointly proportional to the nth power of x and the mth power of y.

Application **EXAMPLE 6** **Joint Variation**

The simple interest earned by a savings account is jointly proportional to the time and the principal. After one quarter (3 months), the interest for a principal of \$6000 is \$120. How much interest would a principal of \$7500 earn in 5 months?

SOLUTION

Let I represent the interest earned (in dollars), let P represent the principal (in dollars), and let t represent the time (in years). Because the interest is jointly proportional to the time and the principal, the model is

$$I = ktP.$$

Because $I = 120$ when $P = 6000$ and $t = \frac{1}{4}$, you have

$120 = k\left(\frac{1}{4}\right)(6000)$ Substitute 120 for I, $\frac{1}{4}$ for t, and 6000 for P.

$120 = 1500k$ Simplify.

$0.08 = k.$ Divide each side by 1500.

So, the model that relates interest to time and principal is

$I = 0.08tP.$ Joint variation model

To find the interest earned on a principal of \$7500 over a five-month period of time, substitute $P = 7500$ and $t = \frac{5}{12}$ into the model to obtain an interest of

$$I = 0.08\left(\frac{5}{12}\right)(7500)$$
$$= \$250.$$

Exercises Within Reach®

Solutions in English & Spanish and tutorial videos at AlgebraWithinReach.com

13. ***Simple Interest*** The simple interest earned by an account varies jointly as the time and the principal. A principal of \$600 earns \$10 interest in 4 months. How much would \$900 earn in 6 months? \$22.50

14. ***Simple Interest*** The simple interest earned by an account varies jointly as the time and the principal. In 2 years, a principal of \$5000 earns \$650 interest. How much would \$1000 earn in 1 year? \$65

Concept Summary: Variation

What

You can write mathematical models for three types of variation: **direct variation, inverse variation,** and **joint variation.**

EXAMPLE

Write a model for the statement.

1. y is directly proportional to x.
2. y is inversely proportional to x.
3. z is jointly proportional to x and y.

How

Use the definitions of direct, inverse, and joint variation to write the models. Let k be the **constant of proportionality**.

EXAMPLE

1. $y = kx$
2. $y = \frac{k}{x}$
3. $z = kxy$

Why

You can use variation models to solve many real-life problems. For instance, knowing how to use direct and inverse variation models can help you determine whether a company should increase the amount it spends on advertising.

Exercises Within Reach ®

Worked-out solutions to odd-numbered exercises at AlgebraWithinReach.com

Concept Summary Check

15. ***Direct Variation*** In a problem, y varies directly as x and the constant of proportionality is positive. If one of the variables increases, how does the other change? Explain. The other variable also increases because when one side of the equation increases, so must the other side.

16. ***Inverse Variation*** In a problem, y varies inversely as x and the constant of proportionality is positive. If one of the variables increases, how does the other change? Explain. The other variable decreases. The product of both variables is constant, so as one variable increases, the other one decreases.

17. ***Reasoning*** Are the following statements equivalent? Explain.

(a) y varies directly as x.

(b) y is directly proportional to the square of x.

No. The equation $y = kx$ is not equivalent to $y = kx^2$.

18. ***Complete the Sentence*** The direct variation model $y = kx$ can be described as "y varies directly as x," or "y is _______ _______ to x." directly proportional

Extra Practice

Writing a Model **In Exercises 19–24, write a model for the statement.**

19. I varies directly as V. $I = kV$

20. V is directly proportional to t. $V = kt$

21. p varies inversely as d. $p = k/d$

22. P is inversely proportional to the square root of $1 + r$. $P = k/\sqrt{1 + r}$

23. A varies jointly as l and w. $A = klw$

24. V varies jointly as h and the square of r. $V = khr^2$

Writing an Equation **In Exercises 25–30, find the constant of proportionality and write an equation that relates the variables.**

25. h is directly proportional to r, and $h = 28$ when $r = 12$. $h = \frac{7}{3}r$

26. n varies inversely as m, and $n = 32$ when $m = 1.5$. $n = 48/m$

27. F varies jointly as x and y, and $F = 500$ when $x = 15$ and $y = 8$. $F = \frac{25}{6}xy$

28. V varies jointly as h and the square of b, and $V = 288$ when $h = 6$ and $b = 12$. $V = \frac{1}{3}hb^2$

29. d varies directly as the square of x and inversely with r, and $d = 3000$ when $x = 10$ and $r = 4$. $d = 120x^2/r$

30. z is directly proportional to x and inversely proportional to the square root of y, and $z = 720$ when $x = 48$ and $y = 81$. $z = 135x/\sqrt{y}$

Determining the Type of Variation **In Exercises 31 and 32, determine whether the variation model is of the form $y = kx$ or $y = k/x$, and find k.**

31.

x	10	20	30	40	50
y	$\frac{2}{5}$	$\frac{1}{5}$	$\frac{2}{15}$	$\frac{1}{10}$	$\frac{2}{25}$

$y = k/x$ with $k = 4$

32.

x	10	20	30	40	50
y	-3	-6	-9	-12	-15

$y = kx$ with $k = -\frac{3}{10}$

33. ***Power Generation*** The power P generated by a wind turbine varies directly as the cube of the wind speed w. The turbine generates 400 watts of power in a 20-mile-per-hour wind. Find the power it generates in a 30-mile-per-hour wind. 1350 watts

34. ***Environment*** The graph shows the percent p of oil that remained in Chedabucto Bay, Nova Scotia, after an oil spill. The cleaning of the spill was left primarily to natural actions. After about a year, the percent that remained varied inversely as time. Find a model that relates p to t, where t is the number of years since the spill. Then use it to find the percent of oil that remained $6\frac{1}{2}$ years after the spill, and compare the result with the graph. $p = \frac{114}{t}$; 18%

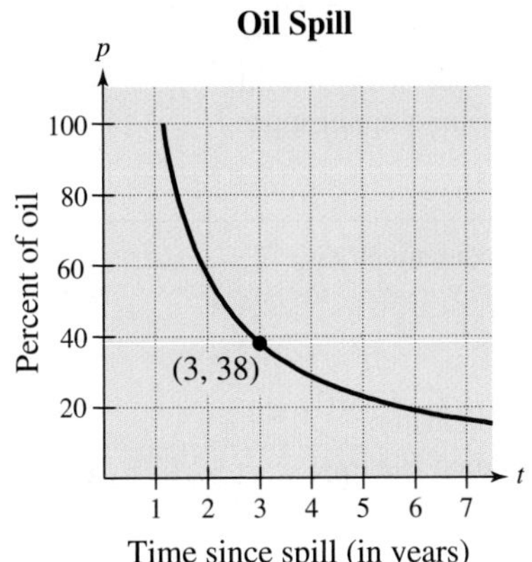

35. ***Engineering*** The load P that can be safely supported by a horizontal beam varies jointly as the product of the width W of the beam and the square of the depth D, and inversely as the length L (see figure).

(a) Write a model for the statement. $P = \frac{kWD^2}{L}$

(b) How does P change when the width and length of the beam are both doubled? Unchanged

(c) How does P change when the width and depth of the beam are doubled? Increases by a factor of 8

(d) How does P change when all three of the dimensions are doubled? Increases by a factor of 4

(e) How does P change when the depth of the beam is cut in half? Decreases by a factor of $\frac{1}{4}$

(f) A beam with width 3 inches, depth 8 inches, and length 120 inches can safely support 2000 pounds. Determine the safe load of a beam made from the same material if its depth is increased to 10 inches. 3125 pounds

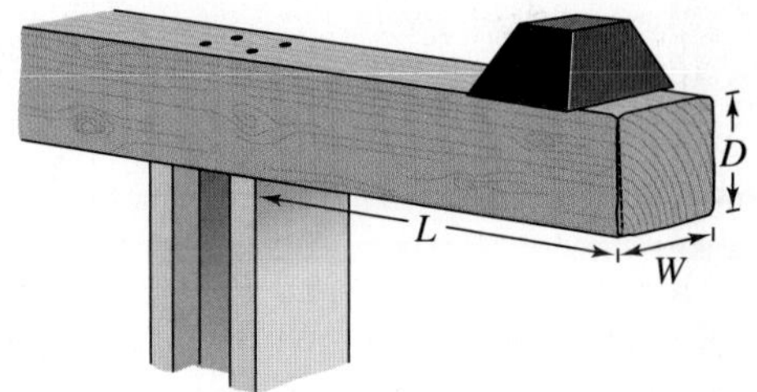

Explaining Concepts

True or False? **In Exercises 36 and 37, determine whether the statement is true or false. Justify your answer.**

36. In a situation involving both direct and inverse variation, y can vary directly as x and inversely as x at the same time. False. The equation would be $y = kx/x = k$, $x \neq 0$, and this is not a variation equation.

37. In a joint variation problem where z varies jointly as x and y, if x increases, then z and y must both increase. False. If x increases, then z and y do not both necessarily increase.

38. ***Precision*** If y varies directly as the square of x and x is doubled, how does y change? Use the rules of exponents to explain your answer. The variable y will quadruple. If $y = kx^2$ and x is replaced by $2x$, the result is $y = k(2x)^2 = 4kx^2$.

39. ***Precision*** If y varies inversely as the square of x and x is doubled, how does y change? Use the rules of exponents to explain your answer. The variable y will be one-fourth as great. If $y = k/x^2$ and x is replaced with $2x$, the result is $y = \frac{k}{(2x)^2} = \frac{k}{4x^2}$.

40. ***Think About It*** Describe a real-life problem for each type of variation (direct, inverse, and joint). Answers will vary.

Cumulative Review

In Exercises 41–44, write the expression using exponential notation.

41. $(6)(6)(6)(6)$ 6^4

42. $(-4)(-4)(-4)$ $(-4)^3$

43. $\left(\frac{1}{5}\right)\left(\frac{1}{5}\right)\left(\frac{1}{5}\right)\left(\frac{1}{5}\right)\left(\frac{1}{5}\right)$ $\left(\frac{1}{5}\right)^5$

44. $-\left(-\frac{3}{4}\right)\left(-\frac{3}{4}\right)\left(-\frac{3}{4}\right)$ $-\left(-\frac{3}{4}\right)^3$

In Exercises 45–48, use synthetic division to divide.

45. $(x^2 - 5x - 14) \div (x + 2)$ $x - 7, x \neq -2$

46. $(3x^2 - 5x + 2) \div (x + 1)$ $3x - 8 + \frac{10}{x+1}$

47. $\frac{4x^5 - 14x^4 + 6x^3}{x - 3}$ $4x^4 - 2x^3, x \neq 3$

48. $\frac{x^5 - 3x^2 - 5x + 1}{x - 2}$ $x^4 + 2x^3 + 4x^2 + 5x + 5 + \frac{11}{x - 2}$

6 Chapter Summary

	What did you learn?	Explanation and Examples	Review Exercises
6.1	Find the domain of a rational function *(p. 288)*.	The set of *usable* values of the variable (values that do not make the denominator zero) is called the domain of a rational function.	1–8
	Simplify rational expressions *(p. 290)*.	To simplify rational expressions, divide out common factors. $\frac{uw}{vw} = \frac{u\cancel{w}}{v\cancel{w}} = \frac{u}{v},\ w \neq 0$ Domain restrictions of the original expression that are not implied by the simplified form must be listed.	9–16
	Use rational expressions to model and solve real-life problems *(p. 293)*.	You can use rational expressions to model many real-life situations including geometry and business applications.	17, 18
6.2	Multiply rational expressions and simplify *(p. 296)*.	**1.** Multiply the numerators and the denominators. $\frac{u}{v} \cdot \frac{w}{z} = \frac{uw}{vz}$ **2.** Factor the numerator and the denominator. **3.** Simplify by dividing out the common factors.	19–26
	Divide rational expressions and simplify *(p. 299)*.	Invert the divisor and multiply using the steps for multiplying rational expressions. $\frac{u}{v} \div \frac{w}{z} = \frac{u}{v} \cdot \frac{z}{w} = \frac{uz}{vw}$	27–32
6.3	Add or subtract rational expressions with like denominators, and simplify *(p. 304)*.	**1.** Combine the numerators. $\frac{u}{w} \pm \frac{v}{w} = \frac{u \pm v}{w}$ **2.** Simplify the resulting rational expression.	33–42
	Add or subtract rational expressions with unlike denominators, and simplify *(p. 306)*.	**1.** Find the least common denominator (LCD) of the rational expressions. **2.** Rewrite each rational expression so that it has the LCD in its denominator. **3.** Combine these rational expressions with like denominators.	43–52
6.4	Simplify complex fractions using rules for dividing rational expressions *(p. 312)*.	When the numerator and denominator of the complex fraction each consist of a single fraction, use the rules for dividing rational expressions.	53–58
	Simplify complex fractions that have a sum or difference in the numerator and/or denominator *(p. 315)*.	When a sum or difference is present in the numerator and/or denominator of the complex fraction, first combine the terms so that the numerator and each denominator consist of a single fraction.	59–64

	What did you learn?	Explanation and Examples	Review Exercises
6.5	Divide polynomials by monomials and write in simplest form *(p. 322).*	Divide each term of the polynomial by the monomial, then simplify each fraction.	65–68
	Use long division to divide polynomials by polynomials *(p. 323).*	**1.** Write the dividend and divisor in descending powers of the variable. **2.** Insert placeholders with zero coefficients for missing powers of the variable. **3.** Perform the long division of the polynomials as you would with integers. **4.** Continue the process until the degree of the remainder is less than that of the divisor.	69–74
	Use synthetic division to divide and factor polynomials *(p. 326).*	Use synthetic division to divide $ax^3 + bx^2 + cx + d$ by $x - k$, as follows. Divisor ⟶ k | a b c d ⟵ Coefficients of dividend ka ○ ○ a $b + ka$ ○ r ⟵ Remainder Coefficients of quotient *Vertical Pattern:* Add terms. *Diagonal Pattern:* Multiply by k.	75–82
6.6	Solve rational equations containing constant denominators *(p. 330).*	**1.** Multiply each side of the equation by the LCD of all the fractions in the equation. **2.** Solve the resulting equation.	83–88
	Solve rational equations containing variable denominators *(p. 332).*	**1.** Determine the domain restrictions of the equation. **2.** Multiply each side of the equation by the LCD of all the fractions in the equation. **3.** Solve the resulting equation.	89–104
	Use rational equations to model and solve real-life problems *(p. 335).*	Many application problems require the use of rational equations. (See Example 7.)	105, 106
6.7	Solve application problems involving direct variation *(p. 338).*	Direct variation: $y = kx$ Direct variation as nth power: $y = kx^n$	107, 108
	Solve application problems involving inverse variation *(p. 341).*	Inverse variation: $y = \frac{k}{x}$ Inverse variation as nth power: $y = \frac{k}{x^n}$	109–111
	Solve application problems involving joint variation *(p. 343).*	Joint variation: $z = kxy$ Joint variation as nth and mth powers: $z = kx^ny^m$	112, 113

Review Exercises

Worked-out solutions to odd-numbered exercises at AlgebraWithinReach.com

6.1

Finding the Domain of a Rational Function **In Exercises 1–6, find the domain of the rational function.**

1. $f(y) = \dfrac{3y}{y-8}$ All real values of y such that $y \neq 8$

2. $g(t) = \dfrac{t+4}{t+12}$ All real values of t such that $t \neq -12$

3. $f(x) = \dfrac{2x}{x^2+1}$ All real values of x

4. $g(t) = \dfrac{t+2}{t^2+4}$ All real values of t

5. $g(u) = \dfrac{u}{u^2-7u+6}$ All real values of u such that $u \neq 1$ and $u \neq 6$

6. $f(x) = \dfrac{x-12}{x(x^2-16)}$ All real values of x such that $x \neq -4$ and $x \neq 4$

7. ***Geometry*** A rectangle of width w inches has an area of 36 square inches. The perimeter P of the rectangle is given by

$$P = 2\left(w + \frac{36}{w}\right).$$

Describe the domain of the function. $(0, \infty)$

8. ***Average Cost*** The average cost $\overline{C}$ for a manufacturer to produce x units of a product is given by

$$\overline{C} = \frac{15{,}000 + 0.75x}{x}.$$

Describe the domain of the function. $\{1, 2, 3, 4, \ldots\}$

Simplifying a Rational Expression **In Exercises 9–16, simplify the rational expression.**

9. $\dfrac{6x^4y^2}{15xy^2}$ $\quad \dfrac{2x^3}{5}, x \neq 0, y \neq 0$

10. $\dfrac{2(y^3z)^2}{28(yz^2)^2}$ $\quad \dfrac{y^4}{14z^2}, y \neq 0$

11. $\dfrac{5b-15}{30b-120}$ $\quad \dfrac{b-3}{6(b-4)}$

12. $\dfrac{4a}{10a^2+26a}$ $\quad \dfrac{2}{5a+13}, a \neq 0$

13. $\dfrac{9x-9y}{y-x}$ $\quad -9, x \neq y$

14. $\dfrac{x+3}{x^2-x-12}$ $\quad \dfrac{1}{x-4}, y \neq -3$

15. $\dfrac{x^2-5x}{2x^2-50}$ $\quad \dfrac{x}{2(x+5)}, x \neq 5$

16. $\dfrac{x^2+3x+9}{x^3-27}$ $\quad \dfrac{1}{x-3}$

Geometry **In Exercises 17 and 18, find the ratio of the area of the shaded portion to the total area of the figure.**

17. 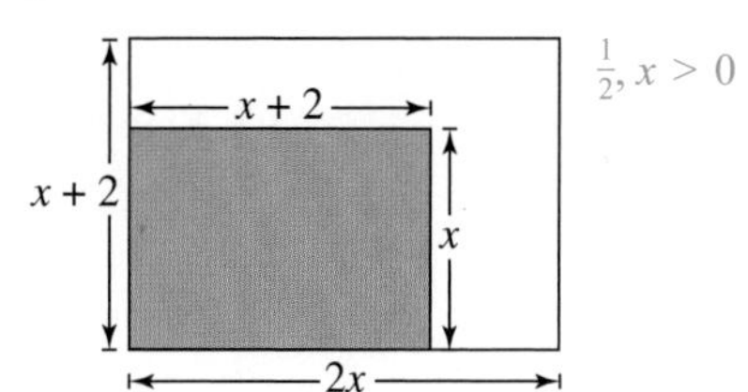

$\frac{1}{2}, x > 0$

18. 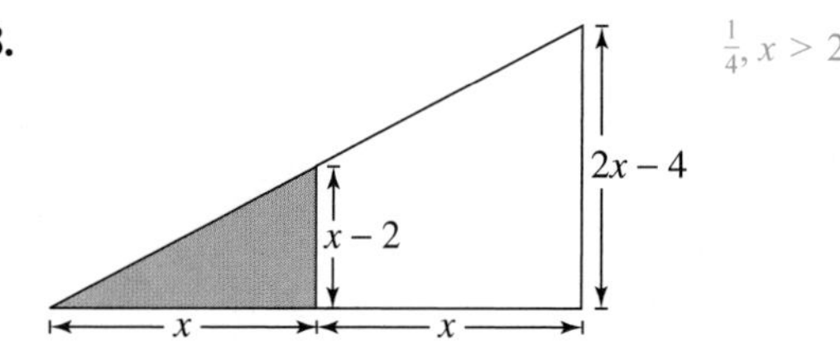

$\frac{1}{4}, x > 2$

6.2

Multiplying Rational Expressions **In Exercises 19–26, multiply and simplify.**

19. $\dfrac{4}{x} \cdot \dfrac{x^2}{12}$ $\quad \dfrac{x}{3}, x \neq 0$

20. $\dfrac{3}{y^3} \cdot 5y^3$ $\quad 15, y \neq 0$

21. $\dfrac{7}{8} \cdot \dfrac{2x}{y} \cdot \dfrac{y^2}{14x^2}$ $\quad \dfrac{y}{8x}, y \neq 0$

22. $\dfrac{15(x^2y)^3}{3y^3} \cdot \dfrac{12y}{x}$ $\quad 60x^5y, x \neq 0, y \neq 0$

23. $\dfrac{60z}{z+6} \cdot \dfrac{z^2-36}{5}$ $\quad 12z(z-6), z \neq -6$

24. $\dfrac{x^2-16}{6} \cdot \dfrac{3}{x^2-8x+16}$ $\quad \dfrac{x+4}{2(x-4)}$

25. $\dfrac{u}{u-3} \cdot \dfrac{3u-u^2}{4u^2}$ $\quad -\dfrac{1}{4}, u \neq 0, u \neq 3$

26. $x^2 \cdot \dfrac{x+1}{x^2-x} \cdot \dfrac{(5x-5)^2}{x^2+6x+5}$ $\quad \dfrac{25x(x-1)}{x+5}, x \neq \pm 1, x \neq 0$

Dividing Rational Expressions **In Exercises 27–32, divide and simplify.**

27. $24x^4 \div \dfrac{6x}{5}$ $\quad 20x^3, x \neq 0$

28. $\dfrac{8u^2}{3} \div \dfrac{u}{9}$ $\quad 24u, u \neq 0$

29. $25y^2 \div \dfrac{xy}{5}$ $\quad \dfrac{125y}{x}, y \neq 0$

30. $\dfrac{6}{z^2} \div 4z^2$ $\quad \dfrac{3}{2z^4}$

31. $\dfrac{x^2+3x+2}{3x^2+x-2} \div (x+2)$ $\quad \dfrac{1}{3x-2}, x \neq -2, x \neq -1$

32. $\dfrac{x^2-14x+48}{x^2-6x} \div (3x-24)$ $\quad \dfrac{1}{3x}, x \neq 6, x \neq 8$

6.3

Adding and Subtracting Rational Expressions **In Exercises 33–42, combine and simplify.**

33. $\dfrac{4x}{5}+\dfrac{11x}{5}$ $3x$

34. $\dfrac{7y}{12}-\dfrac{4y}{12}$ $\dfrac{y}{4}$

35. $\dfrac{15}{3x}-\dfrac{3}{3x}$ $\dfrac{4}{x}$

36. $\dfrac{4}{5x}+\dfrac{1}{5x}$ $\dfrac{1}{x}$

37. $\dfrac{8-x}{4x}+\dfrac{5}{4x}$ $-\dfrac{x-13}{4x}$

38. $\dfrac{3}{5x}-\dfrac{x-1}{5x}$ $-\dfrac{x-4}{5x}$

39. $\dfrac{2(3y+4)}{2y+1}+\dfrac{3-y}{2y+1}$ $\dfrac{5y+11}{2y+1}$

40. $\dfrac{4x-2}{3x+1}-\dfrac{x+1}{3x+1}$ $\dfrac{3(x-1)}{3x+1}$

41. $\dfrac{4x}{x+2}+\dfrac{3x-7}{x+2}-\dfrac{9}{x+2}$ $\dfrac{7x-16}{x+2}$

42. $\dfrac{3}{2y-3}-\dfrac{y-10}{2y-3}+\dfrac{5y}{2y-3}$ $\dfrac{4y+13}{2y-3}$

Adding and Subtracting Rational Expressions **In Exercises 43–52, combine and simplify.**

43. $\dfrac{3}{5x^2}+\dfrac{4}{10x}$ $\dfrac{2x+3}{5x^2}$

44. $\dfrac{3}{z}-\dfrac{5}{2z^2}$ $\dfrac{6z-5}{2z^2}$

45. $\dfrac{1}{x+5}+\dfrac{3}{x-12}$ $\dfrac{4x+3}{(x+5)(x-12)}$

46. $\dfrac{2}{x-10}+\dfrac{3}{4-x}$ $\dfrac{x-22}{(x-10)(4-x)}$

47. $5x+\dfrac{2}{x-3}-\dfrac{3}{x+2}$ $\dfrac{5x^3-5x^2-31x+13}{(x-3)(x+2)}$

48. $4-\dfrac{4x}{x+6}+\dfrac{7}{x-5}$ $\dfrac{31x-78}{(x+6)(x-5)}$

49. $\dfrac{6}{x-5}-\dfrac{4x+7}{x^2-x-20}$ $\dfrac{2x+17}{(x-5)(x+4)}$

50. $\dfrac{5}{x+2}+\dfrac{25-x}{x^2-3x-10}$ $\dfrac{4x}{(x+2)(x-5)}$

51. $\dfrac{5}{x+3}-\dfrac{4x}{(x+3)^2}-\dfrac{1}{x-3}$ $\dfrac{6(x-9)}{(x+3)^2(x-3)}$

52. $\dfrac{8}{y}-\dfrac{3}{y+5}+\dfrac{4}{y-2}$ $\dfrac{9y^2+50y-80}{y(y+5)(y-2)}$

6.4

Simplifying a Complex Fraction **In Exercises 53–58, simplify the complex fraction.**

53. $\dfrac{\left(\dfrac{6}{x}\right)}{\left(\dfrac{2}{x^3}\right)}$ $3x^2,\ x \neq 0$

54. $\dfrac{xy}{\left(\dfrac{5x^2}{2y}\right)}$ $\dfrac{2y^2}{5x},\ y \neq 0$

55. $\dfrac{\left(\dfrac{x}{x-2}\right)}{\left(\dfrac{2x}{2-x}\right)}$ $-\frac{1}{2},\ x \neq 0,\ x \neq 2$

56. $\dfrac{\left(\dfrac{y^2}{5-y}\right)}{\left(\dfrac{y}{y-5}\right)}$ $-y,\ y \neq 0,\ y \neq 5$

57. $\dfrac{\left(\dfrac{6x^2}{x^2+2x-35}\right)}{\left(\dfrac{x^3}{x^2-25}\right)}$ $\dfrac{6(x+5)}{x(x+7)},\ x \neq \pm 5$

58. $\dfrac{\left[\dfrac{24-18x}{(2-x)^2}\right]}{\left(\dfrac{60-45x}{x^2-4x+4}\right)}$ $\dfrac{2}{5},\ x \neq \dfrac{4}{3},\ x \neq 2$

Simplifying a Complex Fraction **In Exercises 59–64, simplify the complex fraction.**

59. $\dfrac{3t}{\left(5-\dfrac{2}{t}\right)}$ $\dfrac{3t^2}{5t-2},\ t \neq 0$

60. $\dfrac{\left(\dfrac{1}{x}-\dfrac{1}{2}\right)}{2x}$ $-\dfrac{x-2}{4x^2}$

61. $\dfrac{\left(x-3+\dfrac{2}{x}\right)}{\left(1-\dfrac{2}{x}\right)}$ $x-1,\ x \neq 0,\ x \neq 2$

62. $\dfrac{3x-1}{\left(\dfrac{2}{x^2}+\dfrac{5}{x}\right)}$ $\dfrac{x^2(3x-1)}{5x+2},\ x \neq 0$

63. $\dfrac{\left(\dfrac{1}{a^2-16}-\dfrac{1}{a}\right)}{\left(\dfrac{1}{a^2+4a}+4\right)}$ $\dfrac{-a^2+a+16}{(4a^2+16a+1)(a-4)},$ $a \neq -4,\ a \neq 0$

64. $\dfrac{\left(\dfrac{1}{x^2}-\dfrac{1}{y^2}\right)}{\left(\dfrac{1}{x}+\dfrac{1}{y}\right)}$ $\dfrac{y-x}{xy},\ x \neq -y$

6.5

Dividing a Polynomial by a Monomial **In Exercises 65–68, perform the division.**

65. $(4x^3 - x) \div (2x)$ $2x^2 - \frac{1}{2}, x \neq 0$

66. $(10x + 15) \div (5x)$ $2 + \frac{3}{x}$

67. $\frac{3x^3y^2 - x^2y^2 + x^2y}{x^2y}$ $3xy - y + 1, x \neq 0, y \neq 0$

68. $\frac{6a^3b^3 + 2a^2b - 4ab^2}{2ab}$ $3a^2b^2 + a - 2b, a \neq 0, b \neq 0$

Using Long Division **In Exercises 69–74, use the long division algorithm to perform the division.**

69. $\frac{5x^2 + 2x + 3}{x + 2}$ $5x - 8 + \frac{19}{x + 2}$

70. $\frac{4x^4 - x^3 - 7x^2 + 18x}{x - 2}$ $4x^3 + 7x^2 + 7x + 32 + \frac{64}{x - 2}$

71. $\frac{x^4 - 3x^2 + 2}{x^2 - 1}$ $x^2 - 2, x \neq \pm 1$

72. $\frac{x^4 - 4x^3 + 3x}{x^2 - 1}$ $x^2 - 4x + 1 - \frac{1}{x + 1}, x \neq 1$

73. $\frac{x^5 - 3x^4 + x^2 + 6}{x^3 - 2x^2 + x - 1}$ $x^2 - x - 3 - \frac{3x^2 - 2x - 3}{x^3 - 2x^2 + x - 1}$

74. $\frac{x^6 + 4x^5 - 3x^2 + 5x}{x^3 + x^2 - 4x + 3}$

$x^3 + 3x^2 + x + 8 - \frac{2(8x^2 - 17x + 12)}{x^3 + x^2 - 4x + 3}$

Using Synthetic Division **In Exercises 75–80, use synthetic division to divide.**

75. $\frac{x^2 + 3x + 5}{x + 1}$ $x + 2 + \frac{3}{x + 1}$

76. $\frac{2x^2 + x - 10}{x - 2}$ $2x + 5, x \neq 2$

77. $\frac{x^3 + 7x^2 + 3x - 14}{x + 2}$ $x^2 + 5x - 7, x \neq -2$

78. $\frac{x^4 - 2x^3 - 15x^2 - 2x + 10}{x - 5}$ $x^3 + 3x^2 - 2, x \neq 5$

79. $(x^4 - 3x^2 - 25) \div (x - 3)$

$x^3 + 3x^2 + 6x + 18 + \frac{29}{x - 3}$

80. $(2x^3 + 5x - 2) \div \left(x + \frac{1}{2}\right)$

$2x^2 - x + \frac{11}{2} - \frac{19}{4\left(x + \frac{1}{2}\right)}$

Factoring Completely **In Exercises 81 and 82, completely factor the polynomial given one of its factors.**

	Polynomial	*Factor*
81.	$x^3 + 2x^2 - 5x - 6$ $(x - 2)(x + 1)(x + 3)$	$x - 2$
82.	$2x^3 + x^2 - 2x - 1$ $(x + 1)(2x + 1)(x - 1)$	$x + 1$

6.6

Solving a Rational Equation **In Exercises 83–88, solve the equation.**

83. $\frac{x}{15} + \frac{3}{5} = 1$ 6

84. $\frac{x}{6} + \frac{5}{3} = 3$ 8

85. $\frac{3x}{8} = -15 + \frac{x}{4}$ -120

86. $\frac{t + 1}{6} = \frac{1}{2} - 2t$ $\frac{2}{13}$

87. $\frac{x^2}{6} - \frac{x}{12} = \frac{1}{2}$ $2, -\frac{3}{2}$

88. $\frac{x^2}{4} = -\frac{x}{12} + \frac{1}{6}$ $-1, \frac{2}{3}$

Solving a Rational Equation **In Exercises 89–104, solve the equation.**

89. $8 - \frac{12}{t} = \frac{1}{3}$ $\frac{36}{23}$

90. $5 + \frac{2}{x} = \frac{1}{4}$ $-\frac{8}{19}$

91. $\frac{2}{y} - \frac{1}{3y} = \frac{1}{3}$ 5

92. $\frac{7}{4x} - \frac{6}{8x} = 1$ 1

93. $r = 2 + \frac{24}{r}$ $-4, 6$

94. $\frac{2}{x} - \frac{x}{6} = \frac{2}{3}$ $-6, 2$

95. $\frac{t}{4} = \frac{4}{t}$ ± 4

96. $\frac{20}{u} = \frac{u}{5}$ ± 10

97. $\frac{3}{y+1} - \frac{8}{y} = 1$ $-4, -2$

98. $\frac{4x}{x-5} + \frac{2}{x} = -\frac{4}{x-5}$ $-\frac{5}{2}, 1$

99. $\frac{2x}{x-3} - \frac{3}{x} = 0$ No solution

100. $\frac{6x}{x-3} = 9 + \frac{18}{x-3}$ No solution

101. $\frac{12}{x^2+x-12} - \frac{1}{x-3} = -1$ $-2, 2$

102. $\frac{3}{x-1} + \frac{6}{x^2-3x+2} = 2$ $\frac{1}{2}, 4$

103. $\frac{5}{x^2-4} - \frac{6}{x-2} = -5$ $-\frac{9}{5}, 3$

104. $\frac{3}{x^2-9} + \frac{4}{x+3} = 1$ $0, 4$

105. ***Average Speeds*** You and a friend ride bikes for the same amount of time. You ride 24 miles and your friend rides 15 miles. Your friend's average speed is 6 miles per hour slower than yours. What are the average speeds of you and your friend?
16 miles per hour, 10 miles per hour

106. ***Average Speed*** You drive 220 miles to see a friend. The return trip takes 20 minutes less than the original trip, and your average speed is 5 miles per hour faster. What is your average speed on the return trip?
60 miles per hour

6.7

107. ***Hooke's Law*** A force of 100 pounds stretches a spring 4 inches. Find the force required to stretch the spring 6 inches. 150 pounds

108. ***Stopping Distance*** The stopping distance d of an automobile is directly proportional to the square of its speed s. How does the stopping distance change when the speed of the car is doubled?
The stopping distance d will increase by a factor of 4.

109. ***Travel Time*** The travel time between two cities is inversely proportional to the average speed. A train travels between the cities in 3 hours at an average speed of 65 miles per hour. How long would it take to travel between the cities at an average speed of 80 miles per hour? 2.44 hours

110. ***Demand*** A company has found that the daily demand x for its cordless telephones is inversely proportional to the price p. When the price is \$25, the demand is 1000 telephones. Approximate the demand when the price is increased to \$28. 893 telephones

111. ***Revenue*** The monthly demand for brand X athletic shoes varies directly as the amount spent on advertising and inversely as the square of the price per pair. When \$20,000 is spent on monthly advertising and the price per pair of shoes is \$55, the demand is 900 pairs. When advertising is increased to \$25,000, what price yields a demand of 900 pairs? Is this increase worthwhile in terms of revenue?
\$61.49; Answers will vary.

112. ***Simple Interest*** The simple interest earned by a savings account is jointly proportional to the time and the principal. After three quarters (9 months), the interest for a principal of \$12,000 is \$675. How much interest would a principal of \$8200 earn in 18 months?
\$922.50

113. ***Cost*** The cost of constructing a wooden box with a square base varies jointly as the height of the box and the square of the width of the box. A box of height 16 inches and of width 6 inches costs \$28.80. How much would a box of height 14 inches and of width 8 inches cost? \$44.80

Chapter Test

Solutions in English & Spanish and tutorial videos at AlgebraWithinReach.com

Take this test as you would take a test in class. After you are done, check your work against the answers in the back of the book.

1. Find the domain of $f(x) = \dfrac{x+1}{x^2 - 6x + 5}$. All real values of x such that $x \neq 1$ and $x \neq 5$

In Exercises 2 and 3, simplify the rational expression.

2. $\dfrac{4-2x}{x-2}$ $-2, x \neq 2$

3. $\dfrac{2a^2 - 5a - 12}{5a - 20}$ $\dfrac{2a+3}{5}, a \neq 4$

4. Find the least common multiple of x^2, $3x^3$, and $(x+4)^2$. $3x^3(x+4)^2$

In Exercises 5–18, perform the operation and simplify.

5. $\dfrac{4z^3}{5} \cdot \dfrac{25}{12z^2}$ $\dfrac{5z}{3}, z \neq 0$

6. $\dfrac{y^2 + 8y + 16}{2(y-2)} \cdot \dfrac{8y - 16}{(y+4)^3}$ $\dfrac{4}{y+4}, y \neq 2$

7. $\dfrac{(2xy^2)^3}{15} \div \dfrac{12x^3}{21}$ $\dfrac{14y^6}{15}, x \neq 0$

8. $(4x^2 - 9) \div \dfrac{2x+3}{2x^2 - x - 3}$

8. $(2x-3)^2(x+1)$, $x \neq -\frac{3}{2}, x \neq -1, x \neq \frac{3}{2}$

9. $\dfrac{3}{x-3} + \dfrac{x-2}{x-3}$ $\dfrac{x+1}{x-3}$

10. $2x + \dfrac{1 - 4x^2}{x+1}$ $\dfrac{-2x^2 + 2x + 1}{x+1}$

11. $\dfrac{5x}{x+2} - \dfrac{2}{x^2 - x - 6}$

11. $\dfrac{5x^2 - 15x - 2}{(x-3)(x+2)}$

12. $\dfrac{3}{x} - \dfrac{5}{x^2} + \dfrac{2x}{x^2 + 2x + 1}$

12. $\dfrac{5x^3 + x^2 - 7x - 5}{x^2(x+1)^2}$

13. $\dfrac{\left(\dfrac{3x}{x+2}\right)}{\left(\dfrac{12}{x^3 + 2x^2}\right)}$ $\dfrac{x^3}{4}, x \neq -2, x \neq 0$

14. $\dfrac{\left(9x - \dfrac{1}{x}\right)}{\left(\dfrac{1}{x} - 3\right)}$ $-(3x+1), x \neq 0, x \neq \frac{1}{3}$

15. $\dfrac{\left(\dfrac{3}{x^2} + \dfrac{1}{y}\right)}{\left(\dfrac{1}{x+y}\right)}$ $\dfrac{(3y + x^2)(x+y)}{x^2 y}, x \neq -y$

16. $\dfrac{6x^2 - 4x + 8}{2x}$ $3x - 2 + \dfrac{4}{x}$

17. $\dfrac{2x^4 - 15x^2 - 7}{x-3}$

17. $2x^3 + 6x^2 + 3x + 9 + \dfrac{20}{x-3}$

18. $\dfrac{t^4 + t^2 - 6t}{t^2 - 2}$ $t^2 + 3 - \dfrac{6t - 6}{t^2 - 2}$

In Exercises 19–21, solve the equation.

19. $\dfrac{3}{h+2} = \dfrac{1}{6}$ 16

20. $\dfrac{2}{x+5} - \dfrac{3}{x+3} = \dfrac{1}{x}$ $-1, -\frac{15}{2}$

21. $\dfrac{1}{x+1} + \dfrac{1}{x-1} = \dfrac{2}{x^2 - 1}$ No solution

22. Find a mathematical model that relates u to v when v varies directly as the square root of u, and $v = \frac{3}{2}$ when $u = 36$. $v = \frac{1}{4}\sqrt{u}$

23. When the temperature of a gas is not allowed to change, the absolute pressure P of the gas is inversely proportional to its volume V, according to Boyle's Law. A large balloon is filled with 180 cubic meters of helium at atmospheric pressure (1 atm) at sea level. What is the volume of the helium when the balloon rises to an altitude at which the atmospheric pressure is 0.75 atm? (Assume that the temperature does not change.) 240 cubic meters

7 Radicals and Complex Numbers

MASTERY IS WITHIN REACH!

"I learned about study groups by accident last semester. I started studying in the learning center. When I saw someone else from my class come into the center, I mentioned something to the tutor, and she suggested asking her over to study with us. We did, and it actually turned into a group of three or four students every session. I learned a lot more and enjoyed it more too."

Jennifer
Graphic Design

See page 379 for suggestions about studying in a group.

Rido/Shutterstock.com

7.1 Radicals and Rational Exponents

- Determine the nth roots of numbers and evaluate radical expressions.
- Use the rules of exponents to evaluate or simplify expressions with rational exponents.
- Evaluate radical functions and find the domains of radical functions.

Roots and Radicals

Definition of nth Root of a Number

Let a and b be real numbers and let n be an integer such that $n \geq 2$. If

$$a = b^n$$

then b is an **nth root of a**. When $n = 2$, the root is a **square root**. When $n = 3$, the root is a **cube root**.

Study Tip

In the definition at the right, "the nth root that has the same sign as a" means that the principal nth root of a is positive if a is positive and negative if a is negative. For instance, $\sqrt{4} = 2$ and $\sqrt[3]{-8} = -2$. Furthermore, to denote the negative square root of a number, you must use a negative sign in front of the radical. For instance, $-\sqrt{4} = -2$.

Principal nth Root of a Number

Let a be a real number that has at least one (real number) nth root. The **principal nth root of a** is the nth root that has the same sign as a, and it is denoted by the **radical**

$\sqrt[n]{a}$. Principal nth root

The positive integer n is the **index** of the radical, and the number a is the **radicand**. When $n = 2$, omit the index and write $\sqrt{a}$ rather than $\sqrt[2]{a}$.

EXAMPLE 1 Finding Roots of Numbers

a. $\sqrt{36} = 6$ because $6 \cdot 6 = 6^2 = 36$.

b. $-\sqrt{36} = -6$ because $6 \cdot 6 = 6^2 = 36$. So, $(-1)(\sqrt{36}) = (-1)(6) = -6$.

c. $\sqrt{-4}$ is not real because there is no real number that when multiplied by itself yields -4.

d. $\sqrt[3]{8} = 2$ because $2 \cdot 2 \cdot 2 = 2^3 = 8$.

e. $\sqrt[3]{-8} = -2$ because $(-2)(-2)(-2) = (-2)^3 = -8$.

Exercises Within Reach®

Solutions in English & Spanish and tutorial videos at AlgebraWithinReach.com

Finding a Root of a Number **In Exercises 1–16, find the root if it exists.**

1. $\sqrt{64}$ 8	2. $\sqrt{25}$ 5	3. $-\sqrt{100}$ -10	4. $-\sqrt{49}$ -7
5. $\sqrt{-25}$ Not a real number	6. $\sqrt{-1}$ Not a real number	7. $-\sqrt{-1}$ Not a real number	8. $-\sqrt{-4}$ Not a real number
9. $\sqrt[3]{27}$ 3	10. $\sqrt[3]{64}$ 4	11. $\sqrt[3]{-27}$ -3	12. $\sqrt[3]{-64}$ -4
13. $-\sqrt[3]{1}$ -1	14. $-\sqrt[3]{8}$ -2	15. $-\sqrt[3]{-27}$ 3	16. $-\sqrt[3]{-64}$ 4

Study Tip

The square roots of perfect squares are rational numbers, so $\sqrt{25}$, $\sqrt{49}$, and $\sqrt{\frac{4}{9}}$ are rational numbers. However, square roots such as $\sqrt{5}$, $\sqrt{6}$, and $\sqrt{\frac{2}{5}}$ are irrational numbers. Similarly, $\sqrt[3]{27}$ and $\sqrt[4]{16}$ are rational numbers, whereas $\sqrt[3]{6}$ and $\sqrt[4]{21}$ are irrational numbers.

Properties of *n*th Roots

Property	*Example*
1. If a is a positive real number and n is even, then a has exactly two (real) nth roots, which are denoted by $\sqrt[n]{a}$ and $-\sqrt[n]{a}$.	The two real square roots of 81 are $\sqrt{81} = 9$ and $-\sqrt{81} = -9$.
2. If a is any real number and n is odd, then a has only one (real) nth root, which is denoted by $\sqrt[n]{a}$.	$\sqrt[3]{27} = 3$ $\sqrt[3]{-64} = -4$
3. If a is a negative real number and n is even, then a has no (real) nth root.	$\sqrt{-64}$ is not a real number.

EXAMPLE 2 Classifying Perfect *n*th Powers

State whether each number is a perfect square, a perfect cube, both, or neither.

a. 81 **b.** -125 **c.** 64 **d.** 32

SOLUTION

a. 81 is a perfect square because $9^2 = 81$. It is not a perfect cube.

b. -125 is a perfect cube because $(-5)^3 = -125$. It is not a perfect square.

c. 64 is a perfect square because $8^2 = 64$, and it is also a perfect cube because $4^3 = 64$.

d. 32 is not a perfect square or a perfect cube. (It is, however, a perfect fifth power, because $2^5 = 32$.)

Exercises Within Reach®

Solutions in English & Spanish and tutorial videos at AlgebraWithinReach.com

Finding a Perfect Square **In Exercises 17–20, find the perfect square that represents the area of the square.**

17.
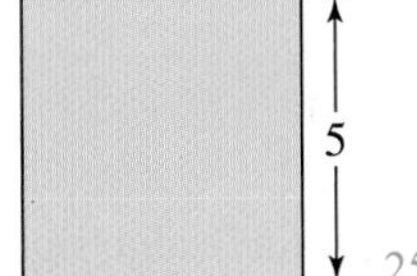

25

18.
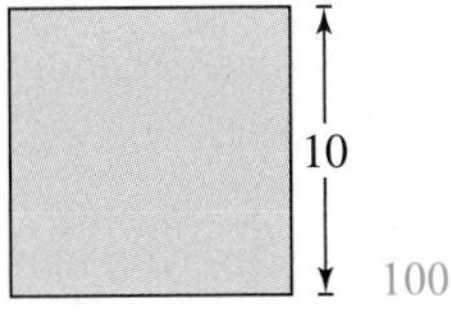

100

19.
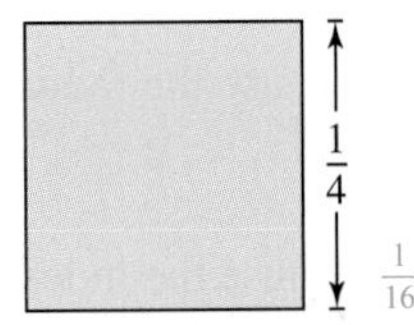

$\frac{1}{16}$

20.
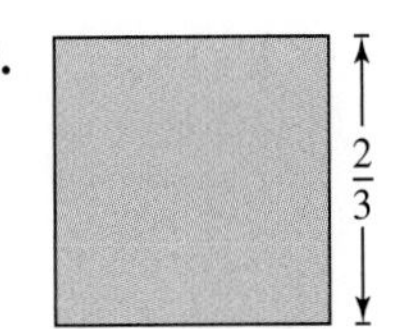

$\frac{4}{9}$

Classifying a Perfect nth Power **In Exercises 21–26, state whether the number is a perfect square, a perfect cube, or neither.**

21. 49 Perfect square
22. -27 Perfect cube
23. 1728 Perfect cube
24. 964 Neither
25. 96 Neither
26. 225 Perfect Square

Finding the Side Length **In Exercises 27–30, use the area A of the square or the volume V of the cube to find the side length l.**

27.
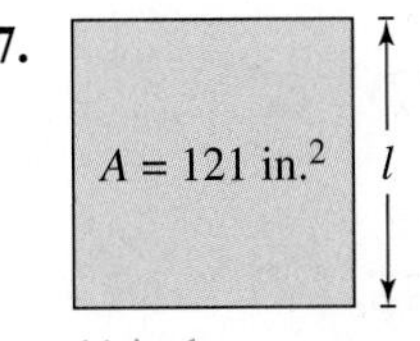

11 inches

28.
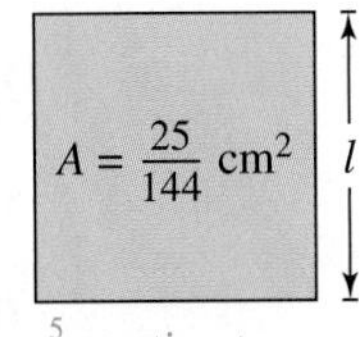

$\frac{5}{12}$ centimeter

29.
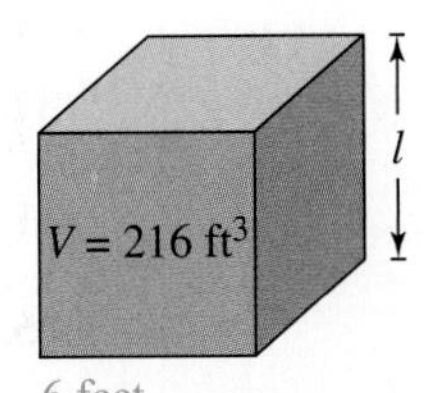

6 feet

30.
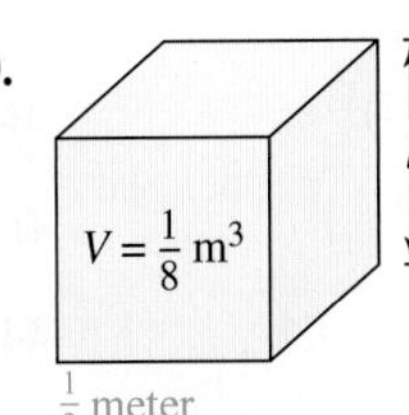

$\frac{1}{2}$ meter

Inverse Properties of *n*th Powers and *n*th Roots

Let a be a real number, and let n be an integer such that $n \geq 2$.

Property	*Example*
1. If a has a principal nth root, then $(\sqrt[n]{a})^n = a.$	$(\sqrt{5})^2 = 5$
2. If n is odd, then $\sqrt[n]{a^n} = a.$	$\sqrt[3]{5^3} = 5$
If n is even, then $\sqrt[n]{a^n} = \lvert a \rvert.$	$\sqrt{(-5)^2} = \lvert -5 \rvert = 5$

EXAMPLE 3 Inverse Operations

a. $(\sqrt{4})^2 = (2)^2 = 4$ and $\sqrt{4} = \sqrt{2^2} = 2$

b. $(\sqrt[3]{27})^3 = (3)^3 = 27$ and $\sqrt[3]{27} = \sqrt[3]{3^3} = 3$

c. $(\sqrt[4]{16})^4 = (2)^4 = 16$ and $\sqrt[4]{16} = \sqrt[4]{2^4} = 2$

d. $(\sqrt[5]{-243})^5 = (-3)^5 = -243$ and $\sqrt[5]{-243} = \sqrt[5]{(-3)^5} = -3$

EXAMPLE 4 Evaluating Radical Expressions

Evaluate each radical expression.

a. $\sqrt[3]{4^3}$ **b.** $\sqrt[3]{(-2)^3}$ **c.** $(\sqrt{7})^2$ **d.** $\sqrt{(-3)^2}$

SOLUTION

a. Because the index of the radical is odd, you can write

$$\sqrt[3]{4^3} = 4.$$

b. Because the index of the radical is odd, you can write

$$\sqrt[3]{(-2)^3} = -2.$$

c. Because the radicand is positive, $\sqrt{7}$ is real and you can write

$$(\sqrt{7})^2 = 7.$$

d. Because the index of the radical is even, you must include absolute value signs, and write

$$\sqrt{(-3)^2} = \lvert -3 \rvert = 3.$$

Exercises Within Reach®

Solutions in English & Spanish and tutorial videos at AlgebraWithinReach.com

Evaluating an Expression **In Exercises 31–46, evaluate the radical expression. If not possible, state the reason.**

31. $\sqrt{8^2}$ 8

32. $\sqrt{12^2}$ 12

33. $\sqrt[3]{5^3}$ 5

34. $\sqrt[3]{10^3}$ 10

35. $\sqrt[3]{(-7)^3}$ -7

36. $\sqrt[3]{\left(-\frac{1}{4}\right)^3}$ $-\frac{1}{4}$

37. $\sqrt{(-10)^2}$ 10

38. $\sqrt{(-12)^2}$ 12

39. $\sqrt{-\left(\frac{3}{10}\right)^2}$ Not a real number

40. $\sqrt{-9^2}$ Not a real number

41. $-\sqrt[3]{\left(\frac{1}{5}\right)^3}$ $-\frac{1}{5}$

42. $-\sqrt[3]{9^3}$ -9

43. $-\sqrt[4]{2^4}$ -2

44. $\sqrt[4]{3^4}$ 3

45. $-\sqrt[5]{7^5}$ -7

46. $\sqrt[5]{(-2)^5}$ -2

Rational Exponents

Study Tip

The numerator of a rational exponent denotes the *power* to which the base is raised, and the denominator denotes the *root* to be taken.

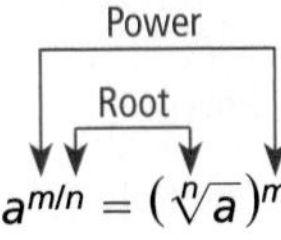

$a^{m/n} = (\sqrt[n]{a})^m$

Definition of Rational Exponents

Let a be a real number, and let n be an integer such that $n \geq 2$. If the principal nth root of a exists, then $a^{1/n}$ is defined as

$$a^{1/n} = \sqrt[n]{a}.$$

If m is a positive integer that has no common factor with n, then

$$a^{m/n} = (a^{1/n})^m = (\sqrt[n]{a})^m \quad \text{and} \quad a^{m/n} = (a^m)^{1/n} = \sqrt[n]{a^m}.$$

Summary of Rules of Exponents

Let r and s be rational numbers, and let a and b be real numbers, variables, or algebraic expressions. (All denominators and bases are nonzero.)

Product and Quotient Rules	*Example*
1. $a^r \cdot a^s = a^{r+s}$	$4^{1/2}(4^{1/3}) = 4^{5/6}$
2. $\dfrac{a^r}{a^s} = a^{r-s}$	$\dfrac{x^2}{x^{1/2}} = x^{2-(1/2)} = x^{3/2}$
Power Rules	
3. $(ab)^r = a^r \cdot b^r$	$(2x)^{1/2} = 2^{1/2}(x^{1/2})$
4. $(a^r)^s = a^{rs}$	$(x^3)^{1/2} = x^{3/2}$
5. $\left(\dfrac{a}{b}\right)^r = \dfrac{a^r}{b^r}$	$\left(\dfrac{x}{3}\right)^{2/3} = \dfrac{x^{2/3}}{3^{2/3}}$
Zero and Negative Exponent Rules	
6. $a^0 = 1$	$(3x)^0 = 1$
7. $a^{-r} = \dfrac{1}{a^r}$	$4^{-3/2} = \dfrac{1}{4^{3/2}} = \dfrac{1}{(2)^3} = \dfrac{1}{8}$
8. $\left(\dfrac{a}{b}\right)^{-r} = \left(\dfrac{b}{a}\right)^r$	$\left(\dfrac{x}{4}\right)^{-1/2} = \left(\dfrac{4}{x}\right)^{1/2} = \dfrac{2}{x^{1/2}}$

Exercises Within Reach ®

Solutions in English & Spanish and tutorial videos at AlgebraWithinReach.com

Using a Definition **In Exercises 47–50, determine the missing description.**

	Radical Form	*Rational Exponent Form*
47.	$\sqrt{36} = 6$	$36^{1/2} = 6$
48.	$\sqrt[3]{27^2} = 9$	$27^{2/3} = 9$
49.	$\sqrt[4]{256^3} = 64$	$256^{3/4} = 64$
50.	$\sqrt[3]{125} = 5$	$125^{1/3} = 5$

Writing in Radical Form **In Exercises 51–54, rewrite the expression in radical form.**

51. $x^{1/3}$ $\sqrt[3]{x}$ **52.** $n^{1/4}$ $\sqrt[4]{n}$ **53.** $y^{2/5}$ $(\sqrt[5]{y})^2$ or $\sqrt[5]{y^2}$ **54.** $u^{5/4}$ $(\sqrt[4]{u})^5$ or $\sqrt[4]{u^5}$

Remind students that perfect nth powers may not be obvious with some fractions until the fractions are reduced. For example,

$$\sqrt[3]{\frac{54}{128}} = \sqrt[3]{\frac{27}{64}} = \sqrt[3]{\frac{3^3}{4^3}} = \sqrt[3]{\left(\frac{3}{4}\right)^3} = \frac{3}{4}.$$

EXAMPLE 5 Evaluating Expressions with Rational Exponents

a. $8^{4/3} = (8^{1/3})^4 = (\sqrt[3]{8})^4 = 2^4 = 16$ Root is 3. Power is 4.

b. $(4^2)^{3/2} = 4^{2 \cdot (3/2)} = 4^{6/2} = 4^3 = 64$ Power-to-power rule

c. $25^{-3/2} = \frac{1}{25^{3/2}} = \frac{1}{(\sqrt{25})^3} = \frac{1}{5^3} = \frac{1}{125}$ Root is 2. Power is 3.

d. $\left(\frac{64}{125}\right)^{2/3} = \frac{64^{2/3}}{125^{2/3}} = \frac{(\sqrt[3]{64})^2}{(\sqrt[3]{125})^2} = \frac{4^2}{5^2} = \frac{16}{25}$ Root is 3. Power is 2.

e. $-16^{1/2} = -\sqrt{16} = -(4) = -4$ Root is 2. Power is 1.

f. $(-16)^{1/2} = \sqrt{-16}$ is not a real number. Root is 2. Power is 1.

Stress to students that the basic rules of exponents apply even if the exponents are negative numbers or fractions. Give comparison examples such as

$x^2 \cdot x^4 = x^6$

$x^{-2} \cdot x^4 = x^2$

$x^{1/2} \cdot x^{1/4} = x^{3/4}$.

EXAMPLE 6 Using Rules of Exponents

Use rational exponents to rewrite and simplify each expression.

a. $x\sqrt[4]{x^3} = x(x^{3/4}) = x^{1 + (3/4)} = x^{7/4}$

b. $\frac{\sqrt[3]{x^2}}{\sqrt{x^3}} = \frac{x^{2/3}}{x^{3/2}} = x^{(2/3) - (3/2)} = x^{-5/6} = \frac{1}{x^{5/6}}$

c. $\sqrt[3]{x^2y} = (x^2y)^{1/3} = (x^2)^{1/3}y^{1/3} = x^{2/3}y^{1/3}$

d. $\sqrt{\sqrt[3]{x}} = \sqrt{x^{1/3}} = (x^{1/3})^{1/2} = x^{(1/3)(1/2)} = x^{1/6}$

e. $\frac{(2x-1)^{4/3}}{\sqrt[3]{2x-1}} = \frac{(2x-1)^{4/3}}{(2x-1)^{1/3}} = (2x-1)^{(4/3)-(1/3)} = (2x-1)^{3/3} = 2x - 1$

Additional Examples

Simplify each expression.

a. $\frac{x\sqrt{x}}{\sqrt{x^3}}$ b. $(x^{3/4})^2 \cdot x^{-1/2}$

Answers:

a. 1 b. x

Exercises Within Reach®

Solutions in English & Spanish and tutorial videos at AlgebraWithinReach.com

Evaluating an Expression with a Rational Exponent **In Exercises 55–66, evaluate the expression.**

55. $27^{2/3}$ 9

56. $27^{4/3}$ 81

57. $(3^3)^{2/3}$ 9

58. $(8^2)^{3/2}$ 512

59. $32^{-2/5}$ $\frac{1}{4}$

60. $81^{-3/4}$ $\frac{1}{27}$

61. $\left(\frac{8}{27}\right)^{2/3}$ $\frac{4}{9}$

62. $\left(\frac{256}{625}\right)^{1/4}$ $\frac{4}{5}$

63. $-36^{1/2}$ -6

64. $-121^{1/2}$ -11

65. $(-27)^{-2/3}$ $\frac{1}{9}$

66. $(-243)^{-3/5}$ $-\frac{1}{27}$

Using Rules of Exponents **In Exercises 67–76, use rational exponents to rewrite and simplify the expression.**

67. $x\sqrt[3]{x^6}$ x^3

68. $t\sqrt[5]{t^2}$ $t^{7/5}$

69. $\frac{\sqrt[4]{t}}{\sqrt{t^5}}$ $t^{-9/4} = \frac{1}{t^{9/4}}$

70. $\frac{\sqrt[3]{x^4}}{\sqrt{x^3}}$ $x^{-1/6} = \frac{1}{x^{1/6}}$

71. $\sqrt[4]{x^3y}$ $x^{3/4}y^{1/4}$

72. $\sqrt[3]{u^4v^2}$ $u^{4/3}v^{2/3}$

73. $\sqrt{\sqrt[4]{y}}$ $y^{1/8}$

74. $\sqrt[3]{\sqrt{2x}}$ $2^{1/6}x^{1/6}$

75. $\frac{(x+y)^{3/4}}{\sqrt[4]{x+y}}$ $(x+y)^{1/2}$

76. $\frac{(a-b)^{1/3}}{\sqrt[3]{a-b}}$ 1

Radical Functions

EXAMPLE 7 Evaluating a Radical Function

Evaluate each radical function when $x = 4$.

a. $f(x) = \sqrt[3]{x - 31}$ **b.** $g(x) = \sqrt{16 - 3x}$

SOLUTION

a. $f(4) = \sqrt[3]{4 - 31} = \sqrt[3]{-27} = -3$

b. $g(4) = \sqrt{16 - 3(4)} = \sqrt{16 - 12} = \sqrt{4} = 2$

Domain of a Radical Function

Let n be an integer that is greater than or equal to 2.

1. If n is odd, the domain of $f(x) = \sqrt[n]{x}$ is the set of all real numbers.
2. If n is even, the domain of $f(x) = \sqrt[n]{x}$ is the set of all nonnegative real numbers.

EXAMPLE 8 Finding the Domain of a Radical Function

a. The domain of $f(x) = \sqrt[3]{x}$ is the set of all real numbers because for any real number x, the expression $\sqrt[3]{x}$ is a real number.

b. The domain of $f(x) = \sqrt{x^3}$ is the set of all nonnegative real numbers. For instance, 1 is in the domain but -1 is not because $\sqrt{(-1)^3} = \sqrt{-1}$ is not a real number.

EXAMPLE 9 Finding the Domain of a Radical Function

Find the domain of $f(x) = \sqrt{2x - 1}$.

SOLUTION

The domain of f consists of all x such that $2x - 1 \geq 0$. Using the methods described in Section 2.4, you can solve this inequality as follows.

$2x - 1 \geq 0$	Write original inequality.
$2x \geq 1$	Add 1 to each side.
$x \geq \frac{1}{2}$	Divide each side by 2.

So, the domain is the set of all real numbers x such that $x \geq \frac{1}{2}$.

Study Tip

In general, when the index n of a radical function is even, the domain of the function includes all real values for which the expression under the radical is greater than or equal to zero.

Exercises Within Reach®

Solutions in English & Spanish and tutorial videos at AlgebraWithinReach.com

Evaluating a Radical Function **In Exercises 77 and 78, evaluate the function for each indicated x-value, if possible, and simplify.**

77. $f(x) = \sqrt{2x + 9}$

(a) $f(0)$ (b) $f(8)$ (c) $f(-6)$ (d) $f(36)$

(a) 3 (b) 5 (c) Not a real number (d) 9

78. $f(x) = \sqrt[3]{2x - 1}$

(a) $f(0)$ (b) $f(-62)$ (c) $f(-13)$ (d) $f(63)$

(a) -1 (b) -5 (c) -3 (d) 5

Finding the Domain of a Radical Function **In Exercises 79–84, describe the domain of the function.**

79. $f(x) = \sqrt[5]{x}$ $-\infty < x < \infty$

80. $f(x) = \sqrt[3]{2x}$ $-\infty < x < \infty$

81. $f(x) = 3\sqrt{x}$ $0 \leq x < \infty$

82. $h(x) = \sqrt[4]{x}$ $0 \leq x < \infty$

83. $h(x) = \sqrt{2x + 9}$ $-\frac{9}{2} \leq x < \infty$

84. $f(x) = \sqrt{3x - 5}$ $\frac{5}{3} \leq x < \infty$

Concept Summary: Using Rational Exponents

What

You can use **rational exponents** and the rules of exponents to simplify some **radical** expressions.

EXAMPLE

Simplify the radical expression $\sqrt[3]{x} \cdot \sqrt[4]{x^3}$.

How

Use the definition of rational exponents to write the radicals in exponent form.

- $a^{1/n} = \sqrt[n]{a}$
- $a^{m/n} = (\sqrt[n]{a})^m = \sqrt[n]{a^m}$

Then use the rules of exponents to simplify the expression.

EXAMPLE

$$\sqrt[3]{x} \cdot \sqrt[4]{x^3} = x^{1/3} \cdot x^{3/4}$$
$$= x^{(1/3)+(3/4)}$$
$$= x^{(4/12)+(9/12)}$$
$$= x^{13/12}$$

Why

As you continue your study of mathematics, you will use and build upon the concepts that you learned in this section.

Exercises Within Reach ®

Worked-out solutions to odd-numbered exercises at AlgebraWithinReach.com

Concept Summary Check

85. *Structure* Write an expression that represents the principal cube root of x in radical form and in rational exponent form. $\sqrt[3]{x}, x^{1/3}$

86. *Structure* Write an expression that represents the principal fourth root of the cube of x in radical form and in rational exponent form. $\sqrt[4]{x^3}, x^{3/4}$

87. *Identifying a Rule* What rule of exponents is used to write $x^{1/3} \cdot x^{3/4}$ as $x^{1/3+3/4}$?
The product rule of exponents

88. *Writing the Radical Form* Write the simplified form of the expression $\sqrt[3]{x} \cdot \sqrt[4]{x^3}$ in radical form in two ways.
$\sqrt[12]{x^{13}}, (\sqrt[12]{x})^{13}$

Extra Practice

Finding Square Roots **In Exercises 89–92, find all of the square roots of the perfect square.**

89. $\frac{9}{16}$ $\pm\frac{3}{4}$ **90.** $\frac{25}{36}$ $\pm\frac{5}{6}$ **91.** 0.16 ± 0.4 **92.** 0.25 ± 0.5

Finding Cube Roots **In Exercises 93–96, find all of the cube roots of the perfect cube.**

93. $\frac{1}{1000}$ $\frac{1}{10}$ **94.** $-\frac{8}{125}$ $-\frac{2}{5}$ **95.** 0.001 0.1 **96.** -0.008 -0.2

Using Rules of Exponents **In Exercises 97–108, use rational exponents to rewrite and simplify the expression.**

97. $u^2\sqrt[3]{u}$ $u^{7/3}$

98. $y\sqrt[4]{y^2}$ $y^{3/2}$

99. $\dfrac{\sqrt{x}}{\sqrt{x^3}}$ $x^{-1} = \dfrac{1}{x}$

100. $\dfrac{\sqrt[3]{x^2}}{\sqrt[3]{x^4}}$ $x^{-2/3} = \dfrac{1}{x^{2/3}}$

101. $\sqrt[4]{y^3} \cdot \sqrt[3]{y}$ $y^{13/12}$

102. $\sqrt[6]{x^5} \cdot \sqrt[3]{x^4}$ $x^{13/6}$

103. $z^2\sqrt{y^5z^4}$ $y^{5/2}z^4$

104. $x^2\sqrt[3]{xy^4}$ $x^{7/3}y^{4/3}$

105. $\sqrt[4]{\sqrt{x^3}}$ $x^{3/8}$

106. $\sqrt[5]{\sqrt[3]{y^4}}$ $y^{4/15}$

107. $\dfrac{(3u-2v)^{2/3}}{\sqrt{(3u-2v)^3}}$ $\dfrac{1}{(3u-2v)^{5/6}}$

108. $\dfrac{\sqrt[4]{2x+y}}{(2x+y)^{3/2}}$ $\dfrac{1}{(2x+y)^{5/4}}$

Mathematical Modeling **In Exercises 109 and 110, use the formula for the *declining balances method***

$$r = 1 - \left(\frac{S}{C}\right)^{1/n}$$

to find the depreciation rate r (in decimal form). In the formula, n is the useful life of the item (in years), S is the salvage value (in dollars), and C is the original cost (in dollars).

109. A \$75,000 truck depreciates over an eight-year period, as shown in the graph. Find r. (Round your answer to three decimal places.) 0.128

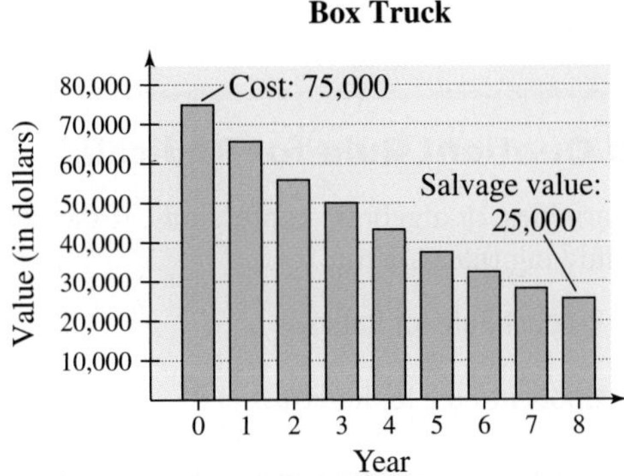

110. A \$125,000 stretch limousine depreciates over a 10-year period, as shown in the graph. Find r. (Round your answer to three decimal places.) 0.149

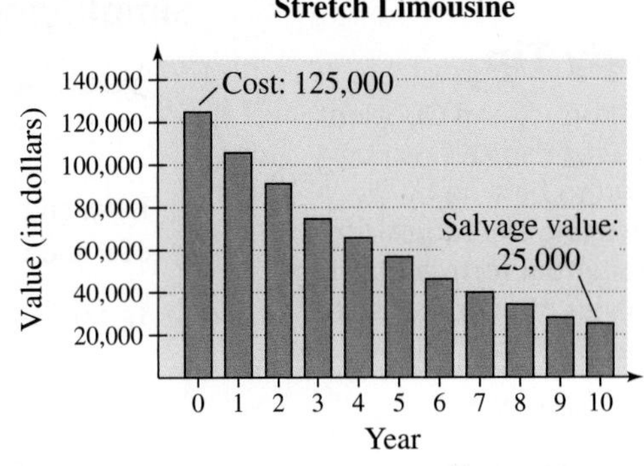

111. ***Geometry*** Find the dimensions of a piece of carpet for a square classroom with 529 square feet of floor space.
23 feet × 23 feet

112. ***Geometry*** Find the dimensions of a square mirror with an area of 1024 square inches.
32 inches × 32 inches

Explaining Concepts

113. ***Vocabulary*** What is the nth root of a number?
The nth root of a number a is the number b for which $a = b^n$.

114. ***Structure*** Explain how you can determine the domain of a function that has the form of a fraction with radical expressions in both the numerator and the denominator.
See Additional Answers.

115. ***Precision*** Is it true that $\sqrt{2} = 1.414$? Explain.
No. $\sqrt{2}$ is an irrational number. Its decimal representation is a nonterminating, nonrepeating decimal.

116. ***Number Sense*** Given that x represents a real number, state the conditions on n for each of the following.

(a) $\sqrt[n]{x^n} = x$ n is odd. (b) $\sqrt[n]{x^n} = |x|$ n is even.

117. ***Investigation*** Find all possible "last digits" of perfect squares. (For instance, the last digit of 81 is 1 and the last digit of 64 is 4.) Is it possible that 4,322,788,986 is a perfect square? 0, 1, 4, 5, 6, 9; Yes

118. ***Number Sense*** Use what you know about the domains of radical functions to write a set of rules for the domain of a rational exponent function of the form $f(x) = x^{1/n}$. Can the same rules be used for a function of the form $f(x) = x^{m/n}$? Explain.
For $f(x) = x^{1/n}$, when n is even, the domain is the set of all nonnegative real numbers. When n is odd, the domain is the set of all real numbers.
Yes. The denominator of the rational exponent determines any domain restrictions, not the numerator.

Cumulative Review

In Exercises 119–122, solve the equation.

119. $\dfrac{a}{5} = \dfrac{a-3}{2}$ 5

120. $\dfrac{x}{3} - \dfrac{3x}{4} = \dfrac{5x}{12}$ 0

121. $\dfrac{2}{u+4} = \dfrac{5}{8}$ $-\frac{4}{5}$

122. $\dfrac{6}{b} + 22 = 24$ 3

In Exercises 123–126, write a model for the statement.

123. s is directly proportional to the square of t. $s = kt^2$

124. r varies inversely as the fourth power of x. $r = \dfrac{k}{x^4}$

125. a varies jointly as b and c. $a = kbc$

126. x is directly proportional to y and inversely proportional to z. $x = \dfrac{ky}{z}$

7.2 Simplifying Radical Expressions

- Use the Product and Quotient Rules for Radicals to simplify radical expressions.
- Use rationalization techniques to simplify radical expressions.
- Use the Pythagorean Theorem in application problems.

Study Tip

The Product and Quotient Rules for Radicals can be shown to be true by converting the radicals to exponential form and using the rules of exponents on page 357.

Using Rule 3

$$\sqrt[n]{uv} = (uv)^{1/n} = u^{1/n}v^{1/n} = \sqrt[n]{u}\,\sqrt[n]{v}$$

Using Rule 5

$$\sqrt[n]{\frac{u}{v}} = \left(\frac{u}{v}\right)^{1/n} = \frac{u^{1/n}}{v^{1/n}} = \frac{\sqrt[n]{u}}{\sqrt[n]{v}}$$

Simplifying Radicals

Product and Quotient Rule for Radicals

Let u and v be real numbers, variables, or algebraic expressions. If the nth roots of u and v are real, then the following rules are true.

1. $\sqrt[n]{uv} = \sqrt[n]{u}\sqrt[n]{v}$ Product Rule for Radicals
2. $\sqrt[n]{\dfrac{u}{v}} = \dfrac{\sqrt[n]{u}}{\sqrt[n]{v}},\ v \neq 0$ Quotient Rule for Radicals

EXAMPLE 1 Removing Constant Factors from Radicals

Simplify each radical expression by removing as many factors as possible.

a. $\sqrt{75}$ **b.** $\sqrt{72}$ **c.** $\sqrt{162}$

SOLUTION

a. $\sqrt{75} = \sqrt{25 \cdot 3} = \sqrt{25}\sqrt{3} = 5\sqrt{3}$ 25 is a perfect square factor of 75.

b. $\sqrt{72} = \sqrt{36 \cdot 2} = \sqrt{36}\sqrt{2} = 6\sqrt{2}$ 36 is a perfect square factor of 72.

c. $\sqrt{162} = \sqrt{81 \cdot 2} = \sqrt{81}\sqrt{2} = 9\sqrt{2}$ 81 is a perfect square factor of 162.

Exercises Within Reach®

Solutions in English & Spanish and tutorial videos at AlgebraWithinReach.com

Removing a Constant Factor from a Radical **In Exercises 1 and 2, simplify the radical expression by removing as many factors as possible. Does the figure support your answer? Explain.**

1. $\sqrt{28}$

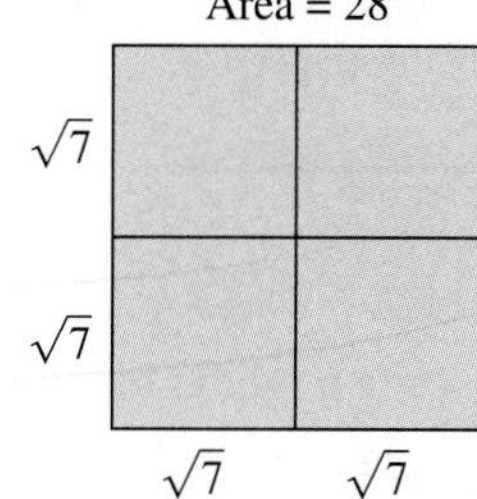

$2\sqrt{7}$; Yes. For a square, $A = s^2$, so $\sqrt{A} = s$. So, the figure shows that $\sqrt{A} = \sqrt{28} = s = 2\sqrt{7}$.

2. $\sqrt{54}$

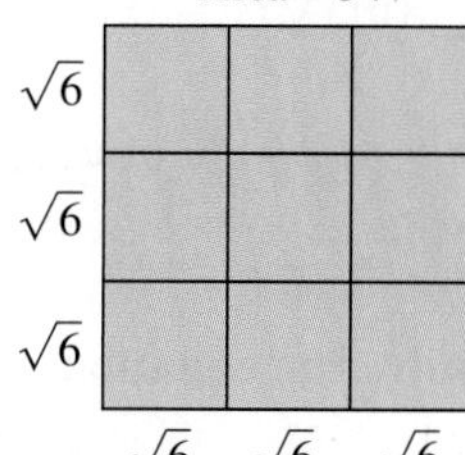

$3\sqrt{6}$; Yes. For a square, $A = s^2$, so $\sqrt{A} = s$. So, the figure shows that $\sqrt{A} = \sqrt{54} = s = 3\sqrt{6}$.

Removing a Constant Factor from a Radical **In Exercises 3–14, simplify the radical expression.**

3. $\sqrt{8}$ $2\sqrt{2}$ **4.** $\sqrt{12}$ $2\sqrt{3}$ **5.** $\sqrt{18}$ $3\sqrt{2}$ **6.** $\sqrt{27}$ $3\sqrt{3}$

7. $\sqrt{45}$ $3\sqrt{5}$ **8.** $\sqrt{125}$ $5\sqrt{5}$ **9.** $\sqrt{96}$ $4\sqrt{6}$ **10.** $\sqrt{84}$ $2\sqrt{21}$

11. $\sqrt{153}$ $3\sqrt{17}$ **12.** $\sqrt{147}$ $7\sqrt{3}$ **13.** $\sqrt{1183}$ $13\sqrt{7}$ **14.** $\sqrt{1176}$ $14\sqrt{6}$

EXAMPLE 2 Removing Variable Factors from Radicals

Simplify each radical expression.

a. $\sqrt{25x^2}$ b. $\sqrt{12x^3}$

c. $\sqrt{144x^4}$ d. $\sqrt{72x^3y^2}$

SOLUTION

a. $\sqrt{25x^2} = \sqrt{5^2x^2} = \sqrt{5^2}\sqrt{x^2}$ Product Rule for Radicals

$= 5|x|$ $\sqrt{x^2} = |x|$

b. $\sqrt{12x^3} = \sqrt{2^2x^2(3x)} = \sqrt{2^2}\sqrt{x^2}\sqrt{3x}$ Product Rule for Radicals

$= 2x\sqrt{3x}$ $\sqrt{2^2}\sqrt{x^2} = 2x,\ x \geq 0$

c. $\sqrt{144x^4} = \sqrt{12^2(x^2)^2} = \sqrt{12^2}\sqrt{(x^2)^2}$ Product Rule for Radicals

$= 12x^2$ $\sqrt{12^2}\sqrt{(x^2)^2} = 12|x^2| = 12x^2$

d. $\sqrt{72x^3y^2} = \sqrt{6^2x^2y^2} \cdot \sqrt{2x}$ Product Rule for Radicals

$= \sqrt{6^2}\sqrt{x^2}\sqrt{y^2} \cdot \sqrt{2x}$ Product Rule for Radicals

$= 6x|y|\sqrt{2x}$ $\sqrt{6^2}\sqrt{x^2}\sqrt{y^2} = 6x|y|,\ x \geq 0$

Additional Examples

Simplify each radical expression.

a. $\sqrt{48x^4y^3}$

b. $\sqrt[3]{54x^3y^5}$

Answers:

a. $4x^2y\sqrt{3y}$

b. $3xy\sqrt[3]{2y^2}$

EXAMPLE 3 Removing Factors from Radicals

Simplify each radical expression.

a. $\sqrt[3]{40}$

b. $\sqrt[4]{x^5}$

SOLUTION

a. $\sqrt[3]{40} = \sqrt[3]{8(5)} = \sqrt[3]{2^3} \cdot \sqrt[3]{5}$ Product Rule for Radicals

$= 2\sqrt[3]{5}$ $\sqrt[3]{2^3} = 2$

b. $\sqrt[4]{x^5} = \sqrt[4]{x^4(x)} = \sqrt[4]{x^4}\sqrt[4]{x}$ Product Rule for Radicals

$= x\sqrt[4]{x}$ $\sqrt[4]{x^4} = x,\ x \geq 0$

Exercises Within Reach®

Solutions in English & Spanish and tutorial videos at AlgebraWithinReach.com

Removing Factors from a Radical **In Exercises 15–34, simplify the radical expression.**

15. $\sqrt{4y^2}$ $2|y|$ **16.** $\sqrt{100x^2}$ $10|x|$ **17.** $\sqrt{9x^5}$ $3x^2\sqrt{x}$ **18.** $\sqrt{64x^3}$ $8x\sqrt{x}$

19. $\sqrt{48y^4}$ $4y^2\sqrt{3}$ **20.** $\sqrt{32x}$ $4\sqrt{2x}$ **21.** $\sqrt{117y^6}$ $3\sqrt{13}|y^3|$ **22.** $\sqrt{160x^8}$ $4\sqrt{10}x^4$

23. $\sqrt{120x^2y^3}$ $2|x|y\sqrt{30y}$ **24.** $\sqrt{125u^4v^6}$ $5\sqrt{5}u^2|v^3|$ **25.** $\sqrt{192a^5b^7}$ $8a^2b^3\sqrt{3ab}$ **26.** $\sqrt{363x^{10}y^9}$ $11|x^5|y^4\sqrt{3y}$

27. $\sqrt[3]{48}$ $2\sqrt[3]{6}$ **28.** $\sqrt[3]{54}$ $3\sqrt[3]{2}$ **29.** $\sqrt[3]{112}$ $2\sqrt[3]{14}$ **30.** $\sqrt[4]{112}$ $2\sqrt[4]{7}$

31. $\sqrt[4]{x^7}$ $x\sqrt[4]{x^3}$ **32.** $\sqrt[4]{x^9}$ $x^2\sqrt[4]{x}$ **33.** $\sqrt[4]{x^6}$ $|x|\sqrt[4]{x^2}$ **34.** $\sqrt[4]{x^{14}}$ $|x^3|\sqrt[4]{x^2}$

Study Tip

When you write a number as a product of prime numbers, you are writing its *prime factorization*. To find the perfect *n*th root factor of 486 in Example 4(a), you can write the prime factorization of 486.

$$486 = 2 \cdot 3 \cdot 3 \cdot 3 \cdot 3 \cdot 3$$
$$= 2 \cdot 3^5$$

From its prime factorization, you can see that 3^5 is a fifth root factor of 486.

$$\sqrt[5]{486} = \sqrt[5]{2 \cdot 3^5}$$
$$= \sqrt[5]{3^5}\sqrt[5]{2}$$
$$= 3\sqrt[5]{2}$$

Additional Examples

Simplify each radical expression.

a. $\sqrt{\dfrac{3x^2}{12y^4}}$ b. $\dfrac{\sqrt[3]{128x^2}}{4\sqrt[3]{2x}}$

Answers:

a. $\dfrac{|x|}{2y^2}$ b. $\sqrt[3]{x},\ x \neq 0$

EXAMPLE 4 Removing Factors from Radicals

a. $\sqrt[5]{486x^7} = \sqrt[5]{243x^5(2x^2)}$ Factor out perfect 5th powers.

$= \sqrt[5]{3^5x^5} \cdot \sqrt[5]{2x^2}$ Product Rule for Radicals

$= 3x\sqrt[5]{2x^2}$ $\sqrt[5]{3^5}\sqrt[5]{x^5} = 3x$

b. $\sqrt[3]{128x^3y^5} = \sqrt[3]{64x^3y^3(2y^2)}$ Factor out perfect 3rd powers.

$= \sqrt[3]{4^3x^3y^3} \cdot \sqrt[3]{2y^2}$ Product Rule for Radicals

$= 4xy\sqrt[3]{2y^2}$ $\sqrt[3]{4^3}\sqrt[3]{x^3}\sqrt[3]{y^3} = 4xy$

EXAMPLE 5 Removing Factors from Radicals

a. $\sqrt{\dfrac{81}{25}} = \dfrac{\sqrt{81}}{\sqrt{25}} = \dfrac{9}{5}$ Quotient Rule for Radicals

b. $\dfrac{\sqrt{56x^2}}{\sqrt{8}} = \sqrt{\dfrac{56x^2}{8}}$ Quotient Rule for Radicals

$= \sqrt{7x^2}$ Simplify.

$= \sqrt{7} \cdot \sqrt{x^2}$ Product Rule for Radicals

$= \sqrt{7}|x|$ $\sqrt{x^2} = |x|$

EXAMPLE 6 Removing Factors from Radicals

$-\sqrt[3]{\dfrac{y^5}{27x^3}} = -\dfrac{\sqrt[3]{y^3y^2}}{\sqrt[3]{27x^3}}$ Quotient Rule for Radicals

$= -\dfrac{\sqrt[3]{y^3} \cdot \sqrt[3]{y^2}}{\sqrt[3]{27} \cdot \sqrt[3]{x^3}}$ Product Rule for Radicals

$= -\dfrac{y\sqrt[3]{y^2}}{3x}$ Simplify.

Exercises Within Reach®

Solutions in English & Spanish and tutorial videos at AlgebraWithinReach.com

Removing Factors from a Radical **In Exercises 35–54, simplify the radical expression.**

35. $\sqrt[3]{40x^5}$ $2x\sqrt[3]{5x^2}$ **36.** $\sqrt[3]{81a^7}$ $3a^2\sqrt[3]{3a}$ **37.** $\sqrt[4]{324y^6}$ $3|y|\sqrt[4]{4y^2}$ **38.** $\sqrt[4]{160x^6}$ $2|x|\sqrt[4]{10x^2}$

39. $\sqrt[3]{x^4y^3}$ $xy\sqrt[3]{x}$ **40.** $\sqrt[3]{a^5b^6}$ $ab^2\sqrt[3]{a^2}$ **41.** $\sqrt[4]{4x^4y^6}$ $|xy|\sqrt[4]{4y^2}$ **42.** $\sqrt[4]{128u^4v^7}$ $2|u|v\sqrt[4]{8v^3}$

43. $\sqrt[5]{32x^5y^6}$ $2xy\sqrt[5]{y}$ **44.** $\sqrt[3]{16x^4y^5}$ $2xy\sqrt[3]{2xy^2}$ **45.** $\sqrt{\dfrac{16}{9}}$ $\dfrac{4}{3}$ **46.** $\sqrt{\dfrac{36}{49}}$ $\dfrac{6}{7}$

47. $\sqrt[3]{\dfrac{35}{64}}$ $\dfrac{\sqrt[3]{35}}{4}$ **48.** $\sqrt[4]{\dfrac{5}{16}}$ $\dfrac{\sqrt[4]{5}}{2}$ **49.** $\dfrac{\sqrt{39y^2}}{\sqrt{3}}$ $|y|\sqrt{13}$ **50.** $\dfrac{\sqrt{56w^3}}{\sqrt{2}}$ $2w\sqrt{7w}$

51. $\sqrt{\dfrac{32a^4}{b^2}}$ $\dfrac{4a^2\sqrt{2}}{|b|}$ **52.** $\sqrt{\dfrac{18x^2}{z^6}}$ $\dfrac{3|x|\sqrt{2}}{|z^3|}$ **53.** $\sqrt[5]{\dfrac{32x^2}{y^5}}$ $\dfrac{2\sqrt[5]{x^2}}{y}$ **54.** $\sqrt[3]{\dfrac{16z^3}{y^6}}$ $\dfrac{2z\sqrt[3]{2}}{y^2}$

Rationalization Techniques

Study Tip

When rationalizing a denominator, remember that for square roots, you want a perfect square in the denominator, for cube roots, you want a perfect cube, and so on. For instance, to find the rationalizing factor needed to create a perfect square in the denominator of Example 7(c), you can write the prime factorization of 18.

$$18 = 2 \cdot 3 \cdot 3 = 2 \cdot 3^2$$

From its prime factorization, you can see that 3^2 is a square root factor of 18. You need one more factor of 2 to create a perfect square in the denominator.

$$2 \cdot (2 \cdot 3^2) = 2 \cdot 2 \cdot 3^2 = 2^2 \cdot 3^2 = 4 \cdot 9 = 36.$$

Simplifying Radical Expressions

A radical expression is said to be in *simplest form* when all three of the statements below are true.

1. All possible nth-powered factors have been removed from each radical.
2. No radical contains a fraction.
3. No denominator of a fraction contains a radical.

EXAMPLE 7 Rationalizing Denominators

a. $\sqrt{\frac{3}{5}} = \frac{\sqrt{3}}{\sqrt{5}} = \frac{\sqrt{3}}{\sqrt{5}} \cdot \frac{\sqrt{5}}{\sqrt{5}} = \frac{\sqrt{15}}{\sqrt{5^2}} = \frac{\sqrt{15}}{5}$ Multiply by $\sqrt{5}/\sqrt{5}$ to create a perfect square in the denominator.

b. $\frac{4}{\sqrt[3]{9}} = \frac{4}{\sqrt[3]{9}} \cdot \frac{\sqrt[3]{3}}{\sqrt[3]{3}} = \frac{4\sqrt[3]{3}}{\sqrt[3]{27}} = \frac{4\sqrt[3]{3}}{3}$ Multiply by $\sqrt[3]{3}/\sqrt[3]{3}$ to create a perfect cube in the denominator.

c. $\frac{8}{3\sqrt{18}} = \frac{8}{3\sqrt{18}} \cdot \frac{\sqrt{2}}{\sqrt{2}} = \frac{8\sqrt{2}}{3\sqrt{36}} = \frac{8\sqrt{2}}{3\sqrt{6^2}}$ Multiply by $\sqrt{2}/\sqrt{2}$ to create a perfect square in the denominator.

$$= \frac{8\sqrt{2}}{3(6)} = \frac{4\sqrt{2}}{9}$$

EXAMPLE 8 Rationalizing Denominators

a. $\sqrt{\frac{8x}{12y^5}} = \sqrt{\frac{\cancel{(4)}(2)x}{\cancel{(4)}(3)y^5}} = \sqrt{\frac{2x}{3y^5}} = \frac{\sqrt{2x}}{\sqrt{3y^5}} \cdot \frac{\sqrt{3y}}{\sqrt{3y}} = \frac{\sqrt{6xy}}{\sqrt{3^2y^6}} = \frac{\sqrt{6xy}}{3|y^3|}$

b. $\sqrt[3]{\frac{54x^6y^3}{5z^2}} = \frac{\sqrt[3]{(3^3)(2)(x^6)(y^3)}}{\sqrt[3]{5z^2}} \cdot \frac{\sqrt[3]{25z}}{\sqrt[3]{25z}} = \frac{3x^2y\sqrt[3]{50z}}{\sqrt[3]{5^3z^3}} = \frac{3x^2y\sqrt[3]{50z}}{5z}$

Exercises Within Reach®

Solutions in English & Spanish and tutorial videos at AlgebraWithinReach.com

Rationalizing the Denominator **In Exercises 55–74, rationalize the denominator and simplify further, if possible.**

55. $\sqrt{\frac{1}{3}}$ $\frac{\sqrt{3}}{3}$
56. $\sqrt{\frac{1}{5}}$ $\frac{\sqrt{5}}{5}$
57. $\frac{1}{\sqrt{7}}$ $\frac{\sqrt{7}}{7}$
58. $\frac{12}{\sqrt{3}}$ $4\sqrt{3}$
59. $\sqrt[4]{\frac{5}{4}}$ $\frac{\sqrt[4]{20}}{2}$
60. $\sqrt[3]{\frac{9}{25}}$ $\frac{\sqrt[3]{45}}{5}$
61. $\frac{6}{\sqrt[3]{32}}$ $\frac{3\sqrt[3]{2}}{2}$
62. $\frac{10}{\sqrt[5]{16}}$ $5\sqrt[5]{2}$
63. $\frac{1}{\sqrt{y}}$ $\frac{\sqrt{y}}{y}$
64. $\frac{2}{\sqrt{3c}}$ $\frac{2\sqrt{3c}}{3c}$
65. $\sqrt{\frac{4}{x}}$ $\frac{2\sqrt{x}}{x}$
66. $\sqrt{\frac{4}{x^3}}$ $\frac{2\sqrt{x}}{x^2}$
67. $\frac{1}{x\sqrt{2}}$ $\frac{\sqrt{2}}{2x}$
68. $\frac{1}{3x\sqrt{x}}$ $\frac{\sqrt{x}}{3x^2}$
69. $\frac{6}{\sqrt{3b^3}}$ $\frac{2\sqrt{3b}}{b^2}$
70. $\frac{1}{\sqrt{xy}}$ $\frac{\sqrt{xy}}{xy}$
71. $\sqrt[3]{\frac{2x}{3y}}$ $\frac{\sqrt[3]{18xy^2}}{3y}$
72. $\sqrt[3]{\frac{20x^2}{9y^2}}$ $\frac{\sqrt[3]{60x^2y}}{3y}$
73. $\sqrt[3]{\frac{24x^3y^4}{25z}}$ $\frac{2xy\sqrt[3]{15yz^2}}{5z}$
74. $\sqrt[3]{\frac{3y^4z^6}{16x^5}}$ $\frac{yz^2\sqrt[3]{12xy}}{4x^2}$

Applications of Radicals

Radicals commonly occur in applications involving right triangles. Recall that a right triangle is one that contains a right (or 90°) angle, as shown below.

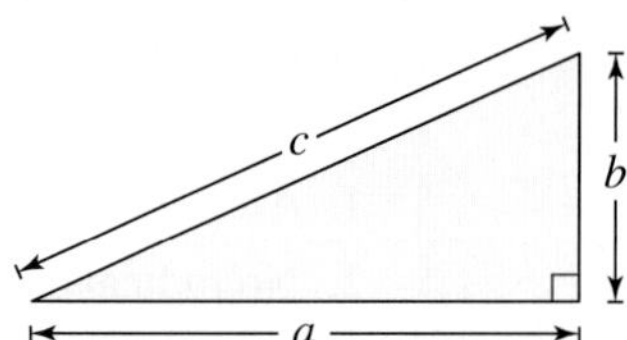

The relationship among the three sides of a right triangle is described by the **Pythagorean Theorem**, which states that if a and b are the lengths of the legs and c is the length of the hypotenuse, then

$$c = \sqrt{a^2 + b^2} \quad \text{and} \quad a = \sqrt{c^2 - b^2}. \qquad \text{Pythagorean Theorem: } a^2 + b^2 = c^2$$

Application **EXAMPLE 9** **Geometry: The Pythagorean Theorem**

Find the length of the hypotenuse of the right triangle shown below.

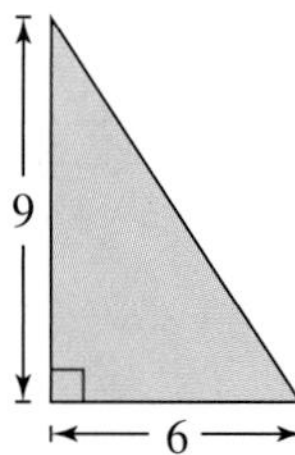

SOLUTION

Letting $a = 6$ and $b = 9$, use the Pythagorean Theorem to find c, as follows.

$c = \sqrt{a^2 + b^2}$ — Pythagorean Theorem

$= \sqrt{6^2 + 9^2}$ — Substitute 6 for a and 9 for b.

$= \sqrt{117}$ — Simplify.

$= \sqrt{9}\sqrt{13}$ — Product Rule for Radicals

$= 3\sqrt{13}$ — Simplify.

Exercises Within Reach®

Solutions in English & Spanish and tutorial videos at AlgebraWithinReach.com

Geometry **In Exercises 75–78, find the length of the hypotenuse of the right triangle.**

75.

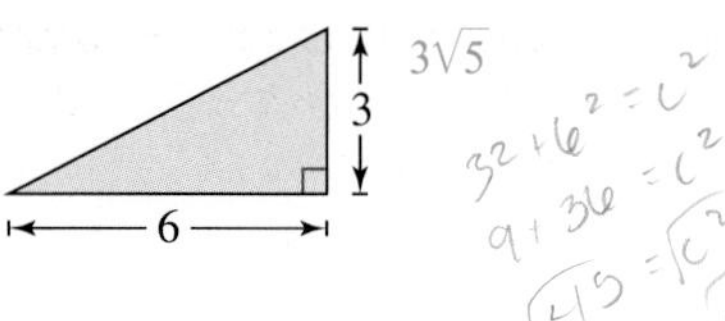

76.

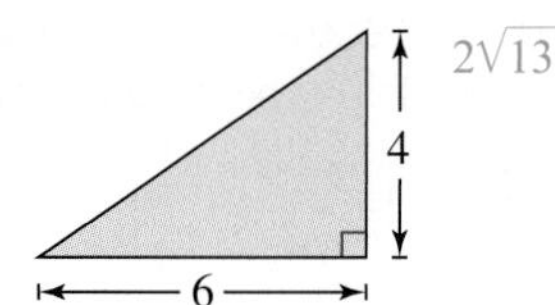

77.

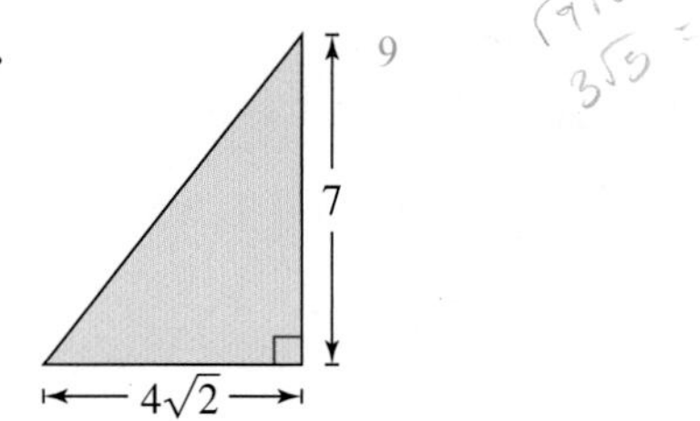

78.

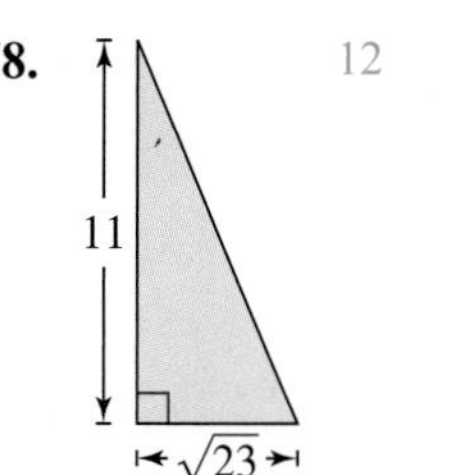

Application **EXAMPLE 10** **Dimensions of a Softball Diamond**

A softball diamond has the shape of a square with 60-foot sides, as shown below. The catcher is 5 feet behind home plate. How far does the catcher have to throw to reach second base?

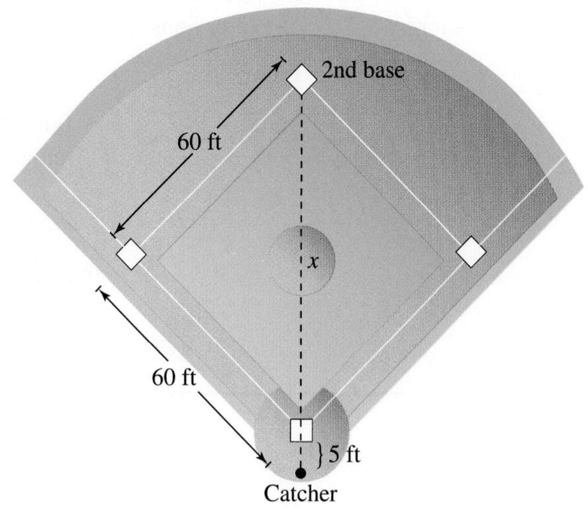

Let x be the hypotenuse of a right triangle with 60-foot sides. So, by the Pythagorean Theorem, you have the following.

$x = \sqrt{60^2 + 60^2}$	Pythagorean Theorem
$= \sqrt{7200}$	Simplify.
$= \sqrt{3600}\sqrt{2}$	Product Rule for Radicals
$= 60\sqrt{2}$	Simplify.
≈ 84.9 feet	Use a calculator.

So, the distance from home plate to second base is approximately 84.9 feet. Because the catcher is 5 feet behind home plate, the catcher must make a throw of

$$x + 5 \approx 84.9 + 5$$
$$= 89.9 \text{ feet.}$$

Exercises Within Reach®

Solutions in English & Spanish and tutorial videos at AlgebraWithinReach.com

79. *Geometry* A ladder is to reach a window that is 26 feet high. The ladder is placed 10 feet from the base of the wall (see figure). How long must the ladder be?

$2\sqrt{194} \approx 27.86$ feet

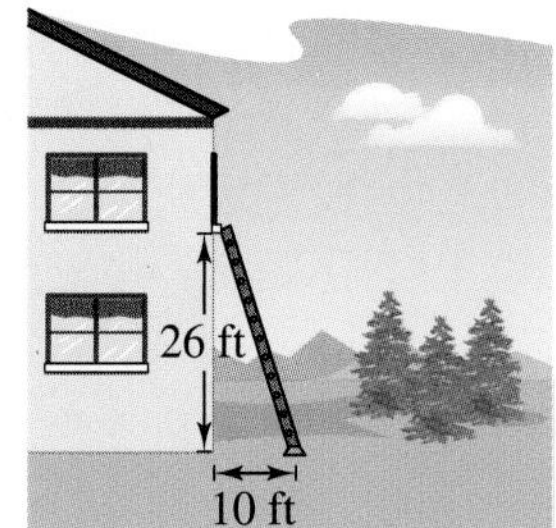

80. *Geometry* A string is attached to opposite corners of a piece of wood that is 6 inches wide and 14 inches long (see figure). How long must the string be?

$2\sqrt{58} \approx 15.23$ inches

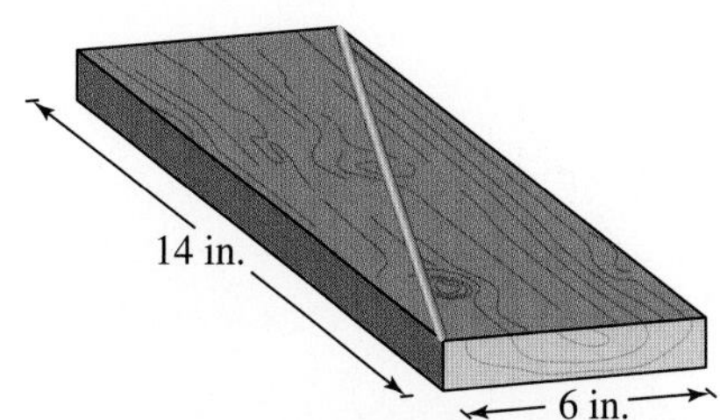

Wendy Nero/Shutterstock.com

Concept Summary: Simplifying Radical Expressions

What

You already knew how to simplify algebraic expressions. Now you can simplify radical expressions involving constants.

EXAMPLE

Simplify $\sqrt{\dfrac{54}{16}}$.

How

Use the Product and Quotient Rules for Radicals to simplify these types of expressions.

EXAMPLE

$\sqrt{\dfrac{54}{16}} = \dfrac{\sqrt{54}}{\sqrt{16}}$ Quotient Rule for Radicals

$= \dfrac{\sqrt{9} \cdot \sqrt{6}}{\sqrt{16}}$ Product Rule for Radicals

$= \dfrac{3\sqrt{6}}{4}$ Simplify.

Why

You can also use these rules to simplify radical expressions involving variables. It is trickier than simplifying radicals involving only constants. Just remember, $\sqrt{x^2} = |x|$.

Exercises Within Reach ®

Worked-out solutions to odd-numbered exercises at AlgebraWithinReach.com

Concept Summary Check

81. ***Stating a Rule*** State the Product Rule for Radicals in words and give an example.

If the nth roots of a and b are real, then

$\sqrt[n]{ab} = \sqrt[n]{a} \cdot \sqrt[n]{b}$.

$\sqrt[3]{108} = \sqrt[3]{27} \cdot \sqrt[3]{4} = 3\sqrt[3]{4}$

82. ***Stating a Rule*** State the Quotient Rule for Radicals in words and give an example.

If the nth roots of a and b are real, then

$\sqrt[n]{\dfrac{a}{b}} = \dfrac{\sqrt[n]{a}}{\sqrt[n]{b}}, b \neq 0.$

$\sqrt{\dfrac{16}{25}} = \dfrac{\sqrt{16}}{\sqrt{25}} = \dfrac{4}{5}$

83. ***Writing*** In your own words, describe the three conditions that must be true for a radical expression to be in simplest form.

- All possible nth-powered factors have been removed from each radical.
- No radical contains a fraction.
- No denominator of a fraction contains a radical.

84. ***Precision*** Explain why the Product Rule for Radicals cannot be applied to the expression $\sqrt{-8} \cdot \sqrt{-2}$.

In order to apply the Product Rule for Radicals, the nth roots of a and b must be real. $\sqrt{-8}$ and $\sqrt{-2}$ are not real.

Extra Practice

Simplifying a Radical Expression **In Exercises 85–98, simplify the radical expression.**

85. $\sqrt{0.04}$ 0.2

86. $\sqrt{0.25}$ 0.5

87. $\sqrt{0.0072}$ $0.06\sqrt{2}$

88. $\sqrt{0.0027}$ $0.03\sqrt{3}$

89. $\sqrt{\frac{60}{3}}$ $2\sqrt{5}$

90. $\sqrt{\frac{208}{4}}$ $2\sqrt{13}$

91. $\sqrt{\frac{13}{25}}$ $\dfrac{\sqrt{13}}{5}$

92. $\sqrt{\frac{15}{36}}$ $\dfrac{\sqrt{15}}{6}$

93. $\sqrt[3]{\dfrac{54a^4}{b^9}}$ $\dfrac{3a\sqrt[3]{2a}}{b^3}$

94. $\sqrt[4]{\dfrac{3u^2}{16v^8}}$ $\dfrac{\sqrt[4]{3u^2}}{2v^2}$

95. $\sqrt{4 \times 10^{-4}}$ $\frac{1}{50}$

96. $\sqrt{8.5 \times 10^3}$ $10\sqrt{85}$

97. $\sqrt[3]{2.4 \times 10^6}$ $20\sqrt[3]{300}$

98. $\sqrt[4]{3.2 \times 10^5}$ $20\sqrt[4]{2}$

99. ***Frequency*** The frequency f (in cycles per second) of a vibrating string is given by

$$f = \frac{1}{100}\sqrt{\frac{400 \times 10^6}{5}}.$$

Use a calculator to approximate this number. (Round the result to two decimal places.) 89.44 cycles per second

100. ***Period of a Pendulum*** The time t (in seconds) for a pendulum of length L (in feet) to go through one complete cycle (its period) is given by

$$t = 2\pi\sqrt{\frac{L}{32}}.$$

Find the period of a pendulum whose length is 4 feet. (Round your answer to two decimal places.) 2.22 seconds

101. ***Geometry*** The foundation of a house is 40 feet long and 30 feet wide. The roof rises 8 feet vertically and forms two congruent right triangles with the side of the house, as shown. (Assume there is no overhang.)

(a) Use the Pythagorean Theorem to find the length of the hypotenuse of each right triangle formed by the roof line. 17 feet

(b) Use the result of part (a) to determine the total area of the roof. 1360 square feet

Explaining Concepts

102. ***Number Sense*** When is $\sqrt{x^2} \neq x$? Explain. When x is negative, the expressions are not equal because the left side would be positive and the right side would be negative.

103. ***Writing*** Explain why $\sqrt{8}$ is not in simplest form.
The perfect square factor 4 needs to be removed from the radical. $\sqrt{8} = \sqrt{4 \cdot 2} = \sqrt{4} \cdot \sqrt{2} = 2\sqrt{2}$

104. ***Precision*** Describe how you would simplify $\frac{1}{\sqrt{3}}$.

Multiply both the denominator and numerator by $\sqrt{3}$ to rationalize the denominator.

105. ***Number Sense*** Enter any positive real number into your calculator and find its square root. Then repeatedly take the square root of the result.

$\sqrt{x}, \sqrt{\sqrt{x}}, \sqrt{\sqrt{\sqrt{x}}}, \ldots$

What real number do the results appear to be approaching?
1

106. ***Think About It*** Square the real number $\frac{5}{\sqrt{3}}$ and note that the radical is eliminated from the denominator. Is this equivalent to rationalizing the denominator? Why or why not? $\frac{25}{3}$; No. Rationalizing the denominator produces an equivalent expression, whereas squaring the number does not.

107. ***Writing*** Let u be a positive real number. Explain why $\sqrt[3]{u} \cdot \sqrt[4]{u} \neq \sqrt[12]{u}$.

Because $\sqrt[3]{u} = u^{1/3}$ and $\sqrt[4]{u} = u^{1/4}$, when $\sqrt[3]{u}$ and $\sqrt[4]{u}$ are multiplied, the rational exponents need to be added together. Therefore, $\sqrt[3]{u} \cdot \sqrt[4]{u} = u^{1/3} \cdot u^{1/4} = u^{7/12}$.

108. ***Writing*** Explain how to find a perfect nth root factor in the radicand of an nth root radical.
To find a perfect nth root factor, first factor the radicand completely. If the same factor appears at least n times, the perfect nth root factor is the common factor to the nth power.

Cumulative Review

In Exercises 109 and 110, solve the system of equations by graphing.

109. $\begin{cases} 2x + 3y = 12 \\ 4x - y = 10 \end{cases}$

(3, 2)

110. $\begin{cases} 3x + 2y = -4 \\ y = 3x + 7 \end{cases}$

(−2, 1)

In Exercises 111 and 112, solve the system of equations by the method of substitution.

111. $\begin{cases} y = x + 2 \\ y - x = 8 \end{cases}$ No solution

112. $\begin{cases} x - 3y = -2 \\ 7y - 4x = 6 \end{cases}$ $\left(-\frac{4}{5}, \frac{2}{5}\right)$

In Exercises 113 and 114, solve the system of equations by the method of elimination.

113. $\begin{cases} x + 4y + 3z = 2 \\ 2x + y + z = 10 \\ -x + y + 2z = 8 \end{cases}$

(4, −8, 10)

114. $\begin{cases} 1.5x - 3 = -2y \\ 3x + 4y = 6 \end{cases}$

Infinitely many solutions

7.3 Adding and Subtracting Radical Expressions

- Use the Distributive Property to add and subtract like radicals.
- Use radical expressions in application problems.

Adding and Subtracting Radical Expressions

Two or more radical expressions are called **like radicals** when they have the same index and the same radicand. For instance, the expressions $\sqrt{2}$ and $3\sqrt{2}$ are like radicals, whereas the expressions $\sqrt{3}$ and $\sqrt[3]{3}$ are not. Two radical expressions that are like radicals can be added or subtracted by adding or subtracting their coefficients.

EXAMPLE 1 Combining Radical Expressions

Simplify each expression by combining like radicals.

a. $\sqrt{7} + 5\sqrt{7} - 2\sqrt{7}$

b. $6\sqrt{x} - \sqrt[3]{4} - 5\sqrt{x} + 2\sqrt[3]{4}$

c. $3\sqrt[3]{x} + 2\sqrt[3]{x} + \sqrt{x} - 8\sqrt{x}$

SOLUTION

a. $\sqrt{7} + 5\sqrt{7} - 2\sqrt{7} = (1 + 5 - 2)\sqrt{7}$ Distributive Property

$= 4\sqrt{7}$ Simplify.

b. $6\sqrt{x} - \sqrt[3]{4} - 5\sqrt{x} + 2\sqrt[3]{4}$

$= (6\sqrt{x} - 5\sqrt{x}) + \left(-\sqrt[3]{4} + 2\sqrt[3]{4}\right)$ Group like radicals.

$= (6 - 5)\sqrt{x} + (-1 + 2)\sqrt[3]{4}$ Distributive Property

$= \sqrt{x} + \sqrt[3]{4}$ Simplify.

c. $3\sqrt[3]{x} + 2\sqrt[3]{x} + \sqrt{x} - 8\sqrt{x}$

$= (3 + 2)\sqrt[3]{x} + (1 - 8)\sqrt{x}$ Distributive Property

$= 5\sqrt[3]{x} - 7\sqrt{7}$ Simplify.

Consider an analogy. Simplifying the expression

$6\sqrt{x} - \sqrt[3]{4} - 5\sqrt{x} + 2\sqrt[3]{4}$
$= \sqrt{x} + \sqrt[3]{4}$

is similar to simplifying

$6a - b - 5a + 2b$
$= a + b$

where $a = \sqrt{x}$ and $b = \sqrt[3]{4}$. Stress that only radicals that are like radicals can be combined.

Exercises Within Reach®

Solutions in English & Spanish and tutorial videos at AlgebraWithinReach.com

Combining Radical Expressions **In Exercises 1–18, simplify the expression by combining like radicals.**

1. $3\sqrt{2} - \sqrt{2}$ $2\sqrt{2}$
2. $6\sqrt{5} - 2\sqrt{5}$ $4\sqrt{5}$
3. $4\sqrt[3]{y} + 9\sqrt[3]{y}$ $13\sqrt[3]{y}$
4. $3\sqrt[3]{3} + 6\sqrt[3]{3}$ $9\sqrt[3]{3}$
5. $\sqrt{7} + 3\sqrt{7} - 2\sqrt{7}$ $2\sqrt{7}$
6. $\sqrt{15} + 4\sqrt{15} - 2\sqrt{15}$ $3\sqrt{15}$
7. $8\sqrt{2} + 6\sqrt{2} - 5\sqrt{2}$ $9\sqrt{2}$
8. $2\sqrt{6} + 8\sqrt{6} - 3\sqrt{6}$ $7\sqrt{6}$
9. $2\sqrt{2} + 5\sqrt{2} - \sqrt{2} + 3\sqrt{2}$ $9\sqrt{2}$
10. $4\sqrt{5} - 3\sqrt{5} - 2\sqrt{5} + 12\sqrt{5}$ $11\sqrt{5}$
11. $\sqrt[4]{5} - 6\sqrt[4]{13} + 3\sqrt[4]{5} - \sqrt[4]{13}$ $4\sqrt[4]{5} - 7\sqrt[4]{13}$
12. $9\sqrt[3]{17} + 7\sqrt[3]{2} - 4\sqrt[3]{17} + \sqrt[3]{2}$ $5\sqrt[3]{17} + 8\sqrt[3]{2}$
13. $9\sqrt[3]{7} - \sqrt{3} + 4\sqrt[3]{7} + 2\sqrt{3}$ $13\sqrt[3]{7} + \sqrt{3}$
14. $5\sqrt{7} - 8\sqrt[4]{11} + \sqrt{7} + 9\sqrt[4]{11}$ $6\sqrt{7} + \sqrt[4]{11}$
15. $7\sqrt{x} + 5\sqrt[3]{9} - 3\sqrt{x} + 3\sqrt[3]{9}$ $8\sqrt[3]{9} + 4\sqrt{x}$
16. $4\sqrt{5} - 3\sqrt[4]{x} - 7\sqrt{5} - 16\sqrt[4]{x}$ $-19\sqrt[4]{x} - 3\sqrt{5}$
17. $3\sqrt[4]{x} + 5\sqrt[4]{x} - 3\sqrt{x} - 11\sqrt[4]{x}$ $-3\sqrt[4]{x} - 3\sqrt{x}$
18. $8\sqrt[3]{x} - 7\sqrt{x} - 6\sqrt[3]{x} + \sqrt[3]{x}$ $3\sqrt[3]{x} - 7\sqrt{x}$

EXAMPLE 2 Simplifying Before Combining Radical Expressions

Study Tip

It is important to realize that the expression $\sqrt{a} + \sqrt{b}$ is not equal to $\sqrt{a + b}$. For instance, you may be tempted to add $\sqrt{6} + \sqrt{3}$ and get $\sqrt{9} = 3$. But remember, you cannot add unlike radicals. So, $\sqrt{6} + \sqrt{3}$ cannot be simplified further.

Simplify each expression by combining like radicals.

a. $\sqrt{45x} + 3\sqrt{20x}$

b. $5\sqrt{x^3} - x\sqrt{4x}$

c. $6\sqrt{\dfrac{24}{x^4}} - 3\sqrt{\dfrac{54}{x^4}}$

d. $\sqrt{50y^5} - \sqrt{32y^5} + 3y^2\sqrt{2y}$

SOLUTION

a. $\sqrt{45x} + 3\sqrt{20x} = 3\sqrt{5x} + 6\sqrt{5x}$ — Simplify radicals.

$= 9\sqrt{5x}$ — Combine like radicals.

b. $5\sqrt{x^3} - x\sqrt{4x} = 5x\sqrt{x} - 2x\sqrt{x}$ — Simplify radicals.

$= 3x\sqrt{x}$ — Combine like radicals.

c. $6\sqrt{\dfrac{24}{x^4}} - 3\sqrt{\dfrac{54}{x^4}} = 6\dfrac{\sqrt{4 \cdot 6}}{x^2} - 3\dfrac{\sqrt{9 \cdot 6}}{x^2}$ — Inverse Property of Radicals

$= 6\dfrac{2\sqrt{6}}{x^2} - 3\dfrac{3\sqrt{6}}{x^2}$ — Inverse Property of Radicals

$= \dfrac{12\sqrt{6}}{x^2} - \dfrac{9\sqrt{6}}{x^2}$ — Multiply.

$= \dfrac{12\sqrt{6} - 9\sqrt{6}}{x^2}$ — Add fractions.

$= \dfrac{3\sqrt{6}}{x^2}$ — Combine like terms.

d. $\sqrt{50y^5} - \sqrt{32y^5} + 3y^2\sqrt{2y}$

$= \sqrt{5^2y^4(2y)} - \sqrt{4^2y^4(2y)} + 3y^2\sqrt{2y}$ — Factor radicands.

$= 5y^2\sqrt{2y} - 4y^2\sqrt{2y} + 3y^2\sqrt{2y}$ — Inverse Property of Radicals

$= 4y^2\sqrt{2y}$ — Combine like radicals.

Exercises Within Reach®

Solutions in English & Spanish and tutorial videos at AlgebraWithinReach.com

Combining Radical Expressions **In Exercises 19–34, simplify the expression by combining like radicals.**

19. $8\sqrt{27} - 3\sqrt{3}$ $21\sqrt{3}$

20. $9\sqrt{50} - 4\sqrt{2}$ $41\sqrt{2}$

21. $3\sqrt{45} + 7\sqrt{20}$ $23\sqrt{5}$

22. $5\sqrt{12} + 16\sqrt{27}$ $58\sqrt{3}$

23. $\sqrt{16x} - \sqrt{9x}$ $\sqrt{x}$

24. $\sqrt{25y} + \sqrt{64y}$ $13\sqrt{y}$

25. $5\sqrt{9x} - 3\sqrt{x}$ $12\sqrt{x}$

26. $4\sqrt{y} + 2\sqrt{16y}$ $12\sqrt{y}$

27. $\sqrt{18y} + 4\sqrt{72y}$ $27\sqrt{2y}$

28. $4\sqrt{12x} + \sqrt{75x}$ $13\sqrt{3x}$

29. $6\sqrt{x^3} - x\sqrt{9x}$ $3x\sqrt{x}$

30. $8\sqrt{x^3} + x\sqrt{81x}$ $17x\sqrt{x}$

31. $9\sqrt{\dfrac{40}{x^4}} - 2\sqrt{\dfrac{90}{x^4}}$ $\dfrac{12\sqrt{10}}{x^2}$

32. $11\sqrt{\dfrac{28}{x^8}} - 2\sqrt{\dfrac{112}{x^8}}$ $\dfrac{14\sqrt{7}}{x^4}$

33. $\sqrt{48x^5} + 2x\sqrt{27x^3} - 3x^2\sqrt{3x}$ $7x^2\sqrt{3x}$

34. $\sqrt{20x^5} - 4x\sqrt{80x^3} + 9x^2\sqrt{5x}$ $-5x^2\sqrt{5x}$

EXAMPLE 3 Simplifying Before Combining Radical Expressions

Simplify each expression by combining like radicals.

a. $\sqrt[3]{54y^5} + 4\sqrt[3]{2y^2}$ **b.** $\sqrt[3]{6x^4} + \sqrt[3]{48x} - \sqrt[3]{162x^4}$ **c.** $\sqrt[3]{\dfrac{6x}{16}} + 2\sqrt[3]{3x}$

SOLUTION

a. $\sqrt[3]{54y^5} + 4\sqrt[3]{2y^2} = 3y\sqrt[3]{2y^2} + 4\sqrt[3]{2y^2}$ Simplify radical.

$= (3y + 4)\sqrt[3]{2y^2}$ Distributive Property

b. $\sqrt[3]{6x^4} + \sqrt[3]{48x} - \sqrt[3]{162x^4}$ Write original expression.

$= x\sqrt[3]{6x} + 2\sqrt[3]{6x} - 3x\sqrt[3]{6x}$ Simplify radicals.

$= (x + 2 - 3x)\sqrt[3]{6x}$ Distributive Property

$= (2 - 2x)\sqrt[3]{6x}$ Combine like terms.

c. $\sqrt[3]{\dfrac{6x}{16}} + 2\sqrt[3]{3x} = \sqrt[3]{\dfrac{2 \cdot 3x}{2 \cdot 8}} + 2\sqrt[3]{3x}$ Factor numerator and denominator.

$= \sqrt[3]{\dfrac{\cancel{2} \cdot 3x}{\cancel{2} \cdot 8}} + 2\sqrt[3]{3x}$ Simplify fraction.

$= \sqrt[3]{\dfrac{3x}{2^3}} + 2\sqrt[3]{3x}$ Write 8 as 2^3.

$= \dfrac{1}{2}\sqrt[3]{3x} + 2\sqrt[3]{3x}$ Inverse Property of Radicals

$= \dfrac{5}{2}\sqrt[3]{3x}$ Combine like radicals.

Exercises Within Reach ®

Solutions in English & Spanish and tutorial videos at AlgebraWithinReach.com

Combining Radical Expressions **In Exercises 35–44, simplify the expression by combining like radicals.**

35. $2\sqrt[3]{54} + 12\sqrt[3]{16}$ $30\sqrt[3]{2}$

36. $4\sqrt[4]{48} - \sqrt[4]{243}$ $5\sqrt[4]{3}$

37. $\sqrt[3]{6x^4} + \sqrt[3]{48x}$ $(x + 2)\sqrt[3]{6x}$

38. $\sqrt[3]{54x} - \sqrt[3]{2x^4}$ $(3 - x)\sqrt[3]{2x}$

39. $5\sqrt[3]{24u^2} + 2\sqrt[3]{81u^5}$ $(10 + 6u)\sqrt[3]{3u^2}$

40. $3\sqrt[3]{16z^2} + 4\sqrt[3]{54z^5}$ $(6 + 12z)\sqrt[3]{2z^2}$

41. $\sqrt[3]{3x^4} + \sqrt[3]{81x} + \sqrt[3]{24x^4}$ $(3x + 3)\sqrt[3]{3x}$

42. $\sqrt[3]{2x^4} + \sqrt[3]{128x^4} - \sqrt[3]{54x}$ $(5x - 3)\sqrt[3]{2x}$

43. $\sqrt[3]{\dfrac{6x}{24}} + 4\sqrt[3]{2x}$ $\dfrac{9}{2}\sqrt[3]{2x}$

44. $\sqrt[3]{\dfrac{21x^2}{81}} + 3\sqrt[3]{7x^2}$ $\dfrac{10}{3}\sqrt[3]{7x^2}$

Geometry **In Exercises 45 and 46, the figure shows a stack of cube-shaped boxes. Use the volume of each box to find an expression for the height h of the stack.**

45.

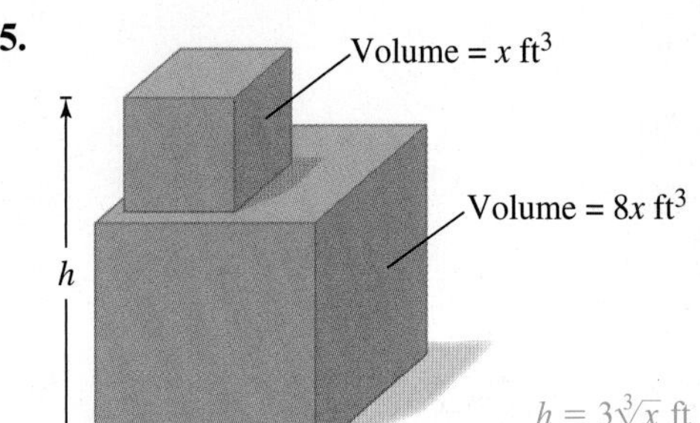

$h = 3\sqrt[3]{x}$ ft

46.

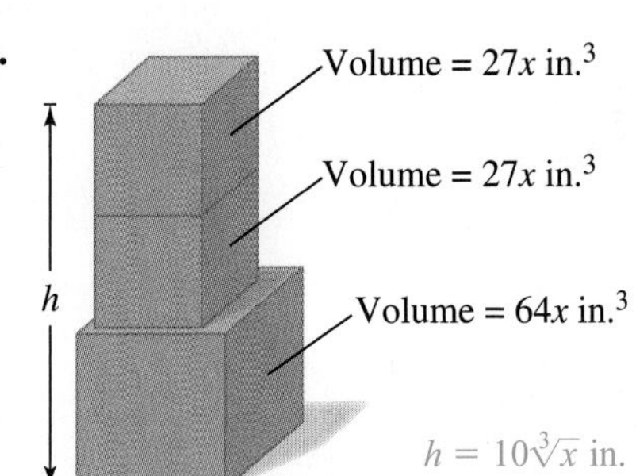

$h = 10\sqrt[3]{x}$ in.

EXAMPLE 4 Rationalizing Denominators Before Simplifying

Simplify each expression by combining like radicals.

a. $\sqrt{7} - \dfrac{5}{\sqrt{7}}$ **b.** $\sqrt{12y} - \dfrac{y}{\sqrt{3y}}$

SOLUTION

a. $\sqrt{7} - \dfrac{5}{\sqrt{7}} = \sqrt{7} - \left(\dfrac{5}{\sqrt{7}} \cdot \dfrac{\sqrt{7}}{\sqrt{7}}\right)$ Multiply by $\sqrt{7}/\sqrt{7}$ to remove the radical from the denominator.

$= \sqrt{7} - \dfrac{5\sqrt{7}}{7}$ Simplify.

$= \left(1 - \dfrac{5}{7}\right)\sqrt{7}$ Distributive Property

$= \dfrac{2}{7}\sqrt{7}$ Simplify.

b. $\sqrt{12y} - \dfrac{y}{\sqrt{3y}} = \sqrt{2^2(3y)} - \left(\dfrac{y}{\sqrt{3y}} \cdot \dfrac{\sqrt{3y}}{\sqrt{3y}}\right)$ Factor and multiply.

$= 2\sqrt{3y} - \dfrac{y\sqrt{3y}}{3y}$ Inverse Property of Radicals

$= 2\sqrt{3y} - \dfrac{\cancel{y}\sqrt{3y}}{3\cancel{y}}$ Divide out common factor.

$= 2\sqrt{3y} - \dfrac{\sqrt{3y}}{3}$ Simplify.

$= \dfrac{6\sqrt{3y}}{3} - \dfrac{\sqrt{3y}}{3}$ Rewrite with common denominator.

$= \dfrac{5\sqrt{3y}}{3},\ y > 0$ Combine like terms.

Exercises Within Reach ®

Solutions in English & Spanish and tutorial videos at AlgebraWithinReach.com

Rationalizing a Denominator Before Simplifying **In Exercises 47–56, perform the addition or subtraction and simplify your answer.**

47. $\sqrt{5} - \dfrac{3}{\sqrt{5}}$ $\dfrac{2\sqrt{5}}{5}$

48. $\sqrt{10} + \dfrac{5}{\sqrt{10}}$ $\dfrac{3\sqrt{10}}{2}$

49. $\sqrt{32} + \sqrt{\dfrac{1}{2}}$ $\dfrac{9\sqrt{2}}{2}$

50. $\sqrt{\dfrac{1}{5}} - \sqrt{45}$ $-\dfrac{14\sqrt{5}}{5}$

51. $\sqrt{18y} - \dfrac{y}{\sqrt{2y}}$ $\dfrac{5\sqrt{2y}}{2}$

52. $\dfrac{x}{\sqrt{3x}} + \sqrt{27x}$ $\dfrac{10\sqrt{3x}}{3}, x > 0$

53. $\dfrac{2}{\sqrt{3x}} + \sqrt{3x}$ $\dfrac{(3x + 2)\sqrt{3x}}{3x}$

54. $2\sqrt{7x} - \dfrac{4}{\sqrt{7x}}$ $\dfrac{(14x - 4)\sqrt{7x}}{7x}$

55. $\sqrt{7y^3} - \sqrt{\dfrac{9}{7y^3}}$ $\dfrac{(7y^3 - 3)\sqrt{7y}}{7y^2}$

56. $\sqrt{\dfrac{4}{3x^3}} + \sqrt{3x^3}$ $\dfrac{(3x^3 + 2)\sqrt{3x}}{3x^2}$

Applications

Application **EXAMPLE 5** **Geometry: Finding the Perimeter of a Triangle**

Write and simplify an expression for the perimeter of the triangle shown below.

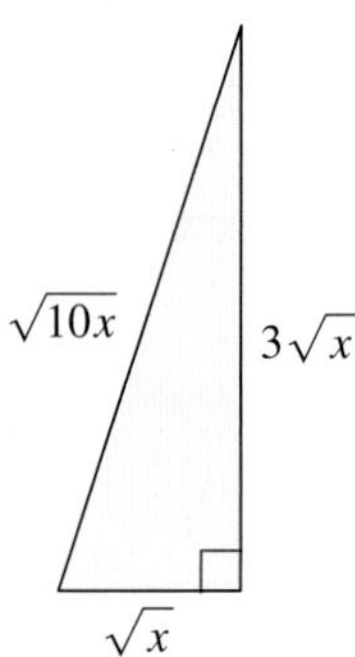

SOLUTION

$P = a + b + c$ Formula for perimeter of a triangle

$= \sqrt{x} + 3\sqrt{x} + \sqrt{10x}$ Substitute.

$= (1 + 3)\sqrt{x} + \sqrt{10x}$ Distributive Property

$= 4\sqrt{x} + \sqrt{10x}$ Simplify.

Exercises Within Reach®

Solutions in English & Spanish and tutorial videos at AlgebraWithinReach.com

Finding the Perimeter of a Figure **In Exercises 57–64, write and simplify an expression for the perimeter of the figure.**

57.

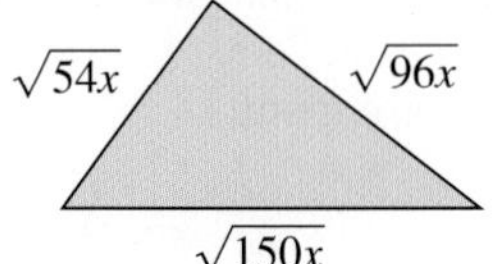

$12\sqrt{6x}$

58.

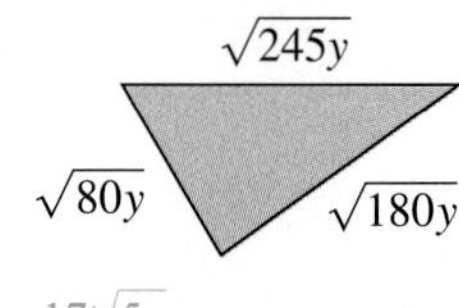

$17\sqrt{5y}$

59.

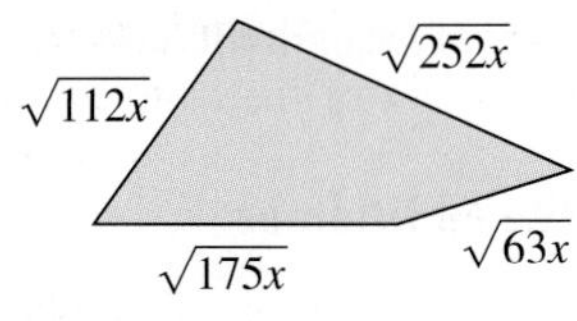

$18\sqrt{7x}$

60.

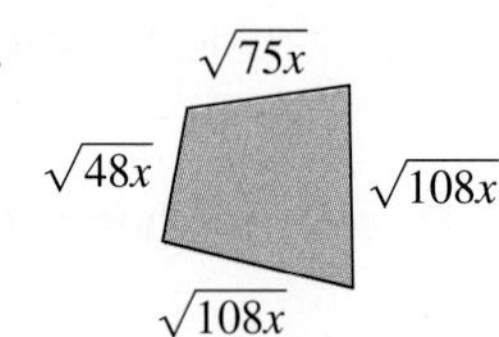

$21\sqrt{3x}$

61.

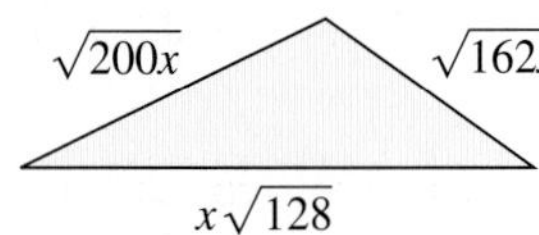

$19\sqrt{2x} + 8x\sqrt{2}$

62.

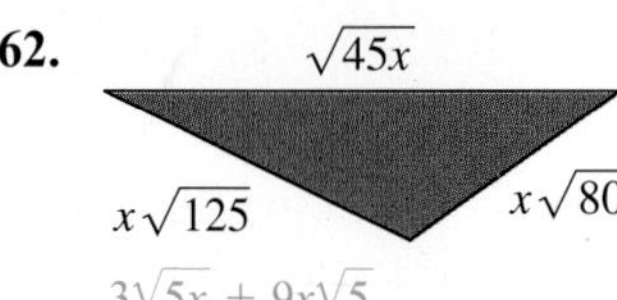

$3\sqrt{5x} + 9x\sqrt{5}$

63.

$x\sqrt{48}$ $\sqrt{12x}$ $\sqrt{27x}$ $x\sqrt{75}$

$9x\sqrt{3} + 5\sqrt{3x}$

64.

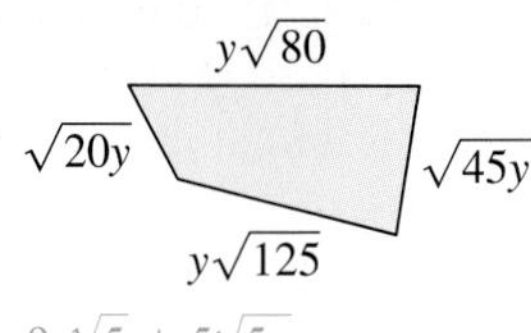

$9y\sqrt{5} + 5\sqrt{5y}$

Application **EXAMPLE 6** **Out-of-Pocket Expense**

Paying for College

The annual cost C of tuition and fees per student at a college and the average annual amount A of financial aid received per student at the college from 2005 through 2013 can be modeled by the functions

$$C = -186 - 180.2t + 3131.6\sqrt{t}, \quad 5 \le t \le 13 \qquad \text{Tuition and fees}$$

$$A = 4722 + 486.6t - 1695.9\sqrt{t}, \quad 5 \le t \le 13 \qquad \text{Aid amount}$$

where t represents the school year, with $t = 5$ corresponding to 2005–2006.

a. Find a function that models the average annual out-of-pocket expense E for tuition and fees at the college from 2005 through 2013.

b. Estimate a student's out-of-pocket expense for tuition and fees for the 2011–2012 school year.

SOLUTION

a. The difference of the cost and the aid gives the out-of-pocket expense.

$$C - A = (-186 - 180.2t + 3131.6\sqrt{t}) - (4722 + 486.6t - 1695.9\sqrt{t})$$

$$= -186 - 180.2t + 3131.6\sqrt{t} - 4722 - 486.6t + 1695.9\sqrt{t}$$

$$= -(186 + 4722) - (180.2t + 486.6t) + (3131.6\sqrt{t} + 1695.9\sqrt{t})$$

$$= -4908 - 666.8t + 4827.5\sqrt{t}$$

So, the average out-of-pocket expense E for tuition and fees is given by

$$E = -4908 - 666.8t + 4827.5\sqrt{t}.$$

b. Substitute 11 for t in the model for E.

$$E = -4908 - 666.8(11) + 4827.5\sqrt{11} \approx 3768.21$$

The average out-of-pocket expense for the 2011–2012 school year was \$3768.21.

Exercises Within Reach ®

Solutions in English & Spanish and tutorial videos at AlgebraWithinReach.com

65. ***Immigration*** The number of immigrants from Guatemala G and El Salvador S to become permanent residents of the United States from 2006 through 2011 can be modeled by the functions

$G = -3856 + 102{,}701\sqrt{t} - 58{,}994t + 8854\sqrt{t^3},$
$6 \le t \le 11$

$S = 2{,}249{,}527 - 2{,}230{,}479\sqrt{t} + 742{,}197t - 82{,}167\sqrt{t^3},$
$6 \le t \le 11$

where t represents the year, with $t = 6$ corresponding to 2006. (*Source:* U.S. Department of Homeland Security)

(a) Find a function that models the total number T of immigrants from Guatemala and El Salvador to become permanent residents of the United States from 2006 through 2011.

$T = 2{,}245{,}671 - 2{,}127{,}778\sqrt{t} + 683{,}203t - 73{,}313\sqrt{t^3}$

(b) Estimate T in 2009. 31,713 people

66. ***Immigration*** The number of immigrants from Oceania O and Australia A to become permanent residents of the United States from 2006 through 2011 can be modeled by the functions

$O = 339{,}061 - 334{,}718\sqrt{t} + 111{,}948t - 12{,}482\sqrt{t^3},$
$6 \le t \le 11$

$A = 224{,}864 - 228{,}070\sqrt{t} + 77{,}745t - 8807\sqrt{t^3},$
$6 \le t \le 11$

where t represents the year, with $t = 6$ corresponding to 2006. (*Source:* U.S. Department of Homeland Security)

(a) Find a function that models the number X of immigrants from Oceania excluding Australia to become permanent residents of the United States from 2006 through 2011.

$X = 114{,}197 - 106{,}648\sqrt{t} + 34{,}203t - 3675\sqrt{t^3}$

(b) Estimate X in 2011. 2644 people

Concept Summary: Adding and Subtracting Radical Expressions

What

When two or more radical expressions have the same index and the same radicand, they are called **like radicals**.

You can add and subtract radical expressions that contain like radicals.

EXAMPLE

Simplify the expression

$2\sqrt{54x} + \sqrt{6x} + \sqrt{24x^2} - 3\sqrt{6}$.

How

Use these steps to add or subtract radical expressions.

1. Write each radical in simplest form.
2. Combine like radicals.
   a. Use the Distributive Property to factor out any like radicals.
   b. Combine any like terms inside the parentheses.

EXAMPLE

$2\sqrt{54x} + \sqrt{6x} + \sqrt{24x^2} - 3\sqrt{6}$
$= 6\sqrt{6x} + \sqrt{6x} + 2x\sqrt{6} - 3\sqrt{6}$
$= (6 + 1)\sqrt{6x} + (2x - 3)\sqrt{6}$
$= 7\sqrt{6x} + (2x - 3)\sqrt{6}$

Why

You can use radical expressions to model real-life problems. Knowing how to add and subtract radical expressions will help you solve such problems. For example, you can find a radical function for the average annual out-of-pocket expenses for the students at a college.

Exercises Within Reach ®

Worked-out solutions to odd-numbered exercises at AlgebraWithinReach.com

Concept Summary Check

67. ***Writing Radicals in Simplest Form*** Which expression in the solution above is the result of writing the radicals in simplest form? $6\sqrt{6x} + \sqrt{6x} + 2x\sqrt{6} - 3\sqrt{6}$

68. ***Writing Radicals in Simplest Form*** What process is used to write the radicals in simplest form in the solution above? All possible perfect square factors are removed from each radical.

69. ***Identifying Like Radicals*** Identify the like radicals in the expression $6\sqrt{6x} + \sqrt{6x} + 2x\sqrt{6} - 3\sqrt{6}$. $6\sqrt{6x}$ and $\sqrt{6x}$, $2x\sqrt{6}$ and $-3\sqrt{6}$

70. ***Simplifying an Expression*** Explain why $(2x - 3)\sqrt{6}$ cannot be simplified further. The terms inside parentheses are not like terms, so they cannot be combined, and the radical is in simplest form.

Extra Practice

Combining Radical Expressions **In Exercises 71–86, simplify the expression by combining like radicals.**

71. $3\sqrt{x+1} + 10\sqrt{x+1}$ $13\sqrt{x+1}$
72. $7\sqrt{2a-3} - 4\sqrt{2a-3}$ $3\sqrt{2a-3}$
73. $\sqrt[3]{16t^4} - \sqrt[3]{54t^4}$ $-t\sqrt[3]{2t}$
74. $10\sqrt[3]{z} - \sqrt[3]{z^4}$ $(10 - z)\sqrt[3]{z}$
75. $\sqrt{5a} + 2\sqrt{45a^3}$ $(6a + 1)\sqrt{5a}$
76. $4\sqrt{3x^3} - \sqrt{12x}$ $(4x - 2)\sqrt{3x}$
77. $\sqrt{9x-9} + \sqrt{x-1}$ $4\sqrt{x-1}$
78. $\sqrt{4y+12} + \sqrt{y+3}$ $3\sqrt{y+3}$
79. $\sqrt{x^3 - x^2} + \sqrt{4x-4}$ $(x + 2)\sqrt{x-1}$
80. $\sqrt{9x-9} - \sqrt{x^3 - x^2}$ $(3 - x)\sqrt{x-1}$
81. $2\sqrt[3]{a^4b^2} + 3a\sqrt[3]{ab^2}$ $5a\sqrt[3]{ab^2}$
82. $3y\sqrt[4]{2x^5y^3} - x\sqrt[4]{162xy^7}$ 0
83. $\sqrt{4r^7s^5} + 3r^2\sqrt{r^3s^5} - 2rs\sqrt{r^5s^3}$ $3r^3s^2\sqrt{rs}$
84. $x\sqrt[3]{27x^5y^2} - x^2\sqrt[3]{x^2y^2} + z\sqrt[3]{x^8y^2}$ $x^2(2 + z)\sqrt[3]{x^2y^2}$
85. $\sqrt[3]{128x^9y^{10}} - 2x^2y\sqrt[3]{16x^3y^7}$ 0
86. $5\sqrt[3]{320x^5y^8} + 2x\sqrt[3]{135x^2y^8}$ $26xy^2\sqrt[3]{5x^2y^2}$

Comparing Radical Expressions **In Exercises 87–90, place the correct symbol (<, >, or =) between the expressions.**

87. $\sqrt{7} + \sqrt{18}$ $>$ $\sqrt{7 + 18}$
88. $\sqrt{10} - \sqrt{6}$ $<$ $\sqrt{10 - 6}$
89. 5 $<$ $\sqrt{9^2 - 4^2}$
90. 5 $=$ $\sqrt{3^2 + 4^2}$

91. ***Geometry*** The foundation of a house is 40 feet long and 30 feet wide. The roof rises 5 feet vertically and forms two congruent right triangles with the side of the house, as shown. (Assume there is no overhang.)

(a) Use the Pythagorean Theorem to find the length of the hypotenuse of each right triangle formed by the roof line. $5\sqrt{10} \approx 15.8$ feet

(b) You are replacing the drip edge around the entire perimeter of the roof. Find the total length of drip edge you need. $80 + 20\sqrt{10} \approx 143$ feet

92. ***Geometry*** The four corners are cut from a four-foot-by-eight-foot sheet of plywood, as shown in the figure. Find the perimeter of the remaining piece of plywood. $8 + 8\sqrt{2} \approx 19.3$ feet

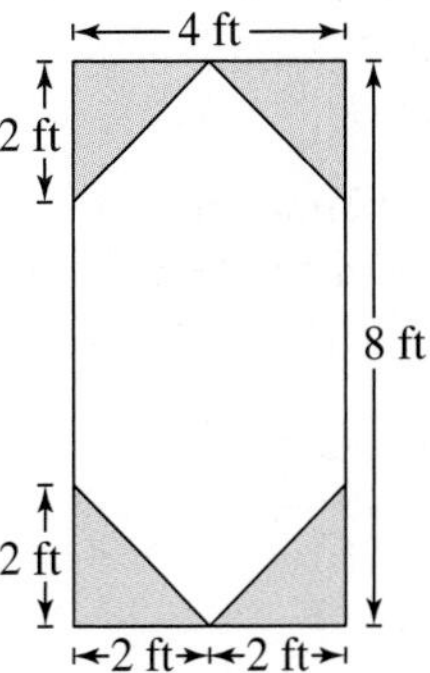

Explaining Concepts

93. ***Reasoning*** Will the sum of two radicals always be a radical? Give an example to support your answer.
No. $\sqrt{5} + (-\sqrt{5}) = 0$

94. ***Reasoning*** Will the difference of two radicals always be a radical? Give an example to support your answer.
No. $3x\sqrt{xy} - 3x\sqrt{xy} = 0$

95. ***Number Sense*** Is $\sqrt{2x} + \sqrt{2x}$ equal to $\sqrt{8x}$? Explain.
Yes. $\sqrt{2x} + \sqrt{2x} = 2\sqrt{2x} = \sqrt{4} \cdot \sqrt{2x} = \sqrt{8x}$

96. ***Structure*** Explain how adding two monomials compares to adding two radicals. You can add two monomials that are like terms or two radicals that are like radicals by using the Distributive Property to factor out the common monomial or radical, and then add the terms inside the parentheses.

97. ***Error Analysis*** Find and correct the error(s) in each solution.

(a) ~~$7\sqrt{3} + 4\sqrt{2} = 11\sqrt{5}$~~
The student combined terms with unlike radicands and added the radicands. The radical expression can be simplified no further.

(b) ~~$3\sqrt[3]{k} - 6\sqrt{k} = 3\sqrt{k}$~~
The student combined terms with unlike indices. The radical expressions can be simplified no further.

98. ***Structure*** Is $\sqrt{2} - \frac{1}{\sqrt{2}}$ in simplest form? Explain.
No. You can rationalize the denominator and combine like terms.
$\sqrt{2} - \frac{1}{\sqrt{2}} = \sqrt{2} - \frac{\sqrt{2}}{2} = \frac{\sqrt{2}}{2}$

Cumulative Review

In Exercises 99–104, combine the rational expressions and simplify.

99. $\frac{7z-2}{2z} - \frac{4z+1}{2z}$ $\quad \frac{3(z-1)}{2z}$

100. $\frac{2x+1}{3x} + \frac{3-4x}{3x}$ $\quad -\frac{2(x-2)}{3x}$

101. $\frac{2x+3}{x-3} + \frac{6-5x}{x-3}$ $\quad -3,\ x \neq 3$

102. $\frac{4m+6}{m+2} - \frac{3m+4}{m+2}$ $\quad 1,\ m \neq -2$

103. $\frac{2v}{v-5} - \frac{3}{5-v}$ $\quad \frac{2v+3}{v-5}$

104. $\frac{4}{x-4} + \frac{2x}{x+1}$ $\quad \frac{2(x^2-2x+2)}{(x-4)(x+1)}$

In Exercises 105–108, simplify the complex fraction.

105. $\frac{\left(\frac{2}{3}\right)}{\left(\frac{4}{15}\right)}$ $\quad \frac{5}{2}$

106. $\frac{\left(\frac{27a^3}{4b^2c}\right)}{\left(\frac{9ac^2}{10b^2}\right)}$ $\quad \frac{15a^2}{2c^3},\ a \neq 0,\ b \neq 0$

107. $\frac{3w-9}{\left(\frac{w^2-10w+21}{w+1}\right)}$ $\quad \frac{3(w+1)}{w-7},\ w \neq -1,\ w \neq 3$

108. $\frac{\left(\frac{x^2+2x-8}{x-8}\right)}{2x+8}$ $\quad \frac{x-2}{2(x-8)},\ x \neq -4$

Mid-Chapter Quiz: Sections 7.1–7.3

Solutions in English & Spanish and tutorial videos at AlgebraWithinReach.com

Take this quiz as you would take a quiz in class. After you are done, check your work against the answers in the back of the book.

In Exercises 1–4, evaluate the expression.

1. $\sqrt{255}$ 15
2. $\sqrt[4]{\frac{81}{16}}$ $\frac{3}{2}$
3. $49^{1/2}$ 7
4. $(-27)^{2/3}$ 9

In Exercises 5 and 6, evaluate the function for each indicated *x*-value, if possible, and simplify.

5. $f(x) = \sqrt{3x - 5}$
   (a) $f(0)$ (b) $f(2)$ (c) $f(10)$ (a) Not a real number (b) 1 (c) 5
6. $g(x) = \sqrt{9 - x}$
   (a) $g(-7)$ (b) $g(5)$ (c) $g(9)$ (a) 4 (b) 2 (c) 0

In Exercises 7 and 8, describe the domain of the function.

7. $g(x) = \dfrac{12}{\sqrt[3]{x}}$ $-\infty < x < 0$ and $0 < x < \infty$
8. $h(x) = \sqrt{3x + 10}$ $-\frac{10}{3} \le x < \infty$

In Exercises 9–14, simplify the radical expression.

9. $\sqrt{27x^2}$ $3|x|\sqrt{3}$
10. $\sqrt[4]{32x^8}$ $2x^2\sqrt[4]{2}$
11. $\sqrt{\dfrac{4u^3}{9}}$ $\dfrac{2u\sqrt{u}}{3}$
12. $\sqrt[3]{\dfrac{16}{u^6}}$ $\dfrac{2\sqrt[3]{2}}{u^2}$
13. $\sqrt{125x^3y^2z^4}$ $5x|y|z^2\sqrt{5x}$
14. $2a\sqrt[3]{16a^3b^5}$ $4a^2b\sqrt[3]{2b^2}$

In Exercises 15 and 16, rationalize the denominator and simplify further, if possible.

15. $\dfrac{24}{\sqrt{12}}$ $4\sqrt{3}$
16. $\dfrac{21x^2}{\sqrt{7x}}$ $3x\sqrt{7x}, x \ne 0$

In Exercises 17–22, simplify the expression by combining like radicals.

17. $2\sqrt{3} - 4\sqrt{7} + \sqrt{3}$ $3\sqrt{3} - 4\sqrt{7}$
18. $\sqrt{200y} - 3\sqrt{8y}$ $4\sqrt{2y}$
19. $5\sqrt{12} + 2\sqrt{3} - \sqrt{75}$ $7\sqrt{3}$
20. $\sqrt{25x + 50} - \sqrt{x + 2}$ $4\sqrt{x + 2}$
21. $6x\sqrt[3]{5x^2} + 2\sqrt[3]{40x^4}$
    $6x\sqrt[3]{5x^2} + 4x\sqrt[3]{5x}$
22. $3\sqrt{x^3y^4z^5} + 2xy^2\sqrt{xz^5} - xz^2\sqrt{xy^4z}$
    $4xy^2z^2\sqrt{xz}$

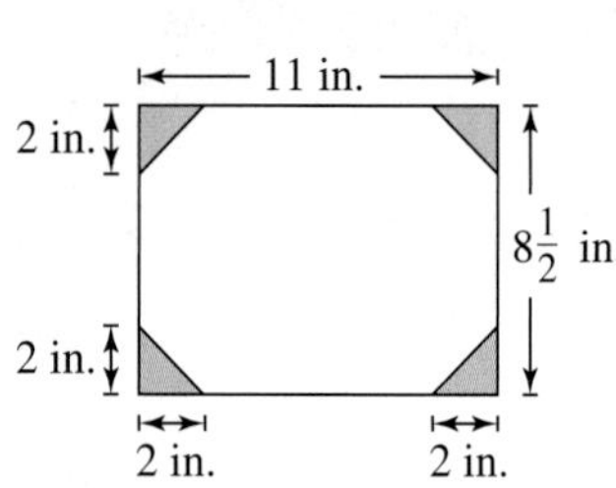

Application

23. The four corners are cut from an $8\frac{1}{2}$-inch-by-11-inch sheet of paper, as shown in the figure at the left. Find the perimeter of the remaining piece of paper.
    $23 + 8\sqrt{2} \approx 34.3$ inches

Study Skills in Action

Studying in a Group

Many students endure unnecessary frustration because they study by themselves. Studying in a group or with a partner has many benefits. First, the combined memory and comprehension of the members minimizes the likelihood of any member getting "stuck" on a particular problem. Second, discussing math often helps clarify unclear areas. Third, regular study groups keep many students from procrastinating. Finally, study groups often build a camaraderie that helps students stick with the course when it gets tough.

These students are keeping each other motivated.

Smart Study Strategy

Form a Weekly Study Group

1 ▶ Set up the group.

- Select students who are just as dedicated to doing well in the math class as you are.
- Find a regular meeting place on campus that has minimal distractions. Try to find a place that has a white board.
- Compare schedules, and select at least one time a week to meet, allowing at least 1.5 hours for study time.

2 ▶ Organize the study time. If you are unsure about how to structure your time during the first few study sessions, try using the guidelines at the right.

- *Review and compare notes - 20 minutes*
- *Identify and review the key rules, definitions, etc. - 20 minutes*
- *Demonstrate at least one homework problem for each key concept - 40 minutes*
- *Make small talk (saving this until the end improves your chances of getting through all the math) - 10 minutes*

3 ▶ Set up rules for the group. Consider using the following rules.

- Members must attend regularly, be on time, and participate.
- The sessions will focus on the key math concepts, not on the needs of one student.
- Students who skip classes will not be allowed to participate in the study group.
- Students who keep the group from being productive will be asked to leave the group.

4 ▶ Inform the instructor. Let the instructor know about your study group. Ask for advice about maintaining a productive group.

7.4 Multiplying and Dividing Radical Expressions

- Use the Distributive Property or the FOIL Method to multiply radical expressions.
- Determine the products of conjugates.
- Simplify quotients involving radicals by rationalizing the denominators.

Multiplying Rational Expressions

EXAMPLE 1 Multiplying Radical Expressions

Find each product and simplify.

a. $\sqrt{6} \cdot \sqrt{3}$ **b.** $\sqrt[3]{5} \cdot \sqrt[3]{16}$

SOLUTION

a. $\sqrt{6} \cdot \sqrt{3} = \sqrt{6 \cdot 3} = \sqrt{18} = \sqrt{9 \cdot 2} = 3\sqrt{2}$

b. $\sqrt[3]{5} \cdot \sqrt[3]{16} = \sqrt[3]{5 \cdot 16} = \sqrt[3]{80} = \sqrt[3]{8 \cdot 10} = 2\sqrt[3]{10}$

EXAMPLE 2 Multiplying Radical Expressions

Find each product and simplify.

a. $\sqrt{3}(2 + \sqrt{5})$ **b.** $\sqrt{2}(4 - \sqrt{8})$ **c.** $\sqrt{6}(\sqrt{12} - \sqrt{3})$

SOLUTION

a. $\sqrt{3}(2 + \sqrt{5}) = 2\sqrt{3} + \sqrt{3}\sqrt{5}$ Distributive Property

$= 2\sqrt{3} + \sqrt{15}$ Product Rule for Radicals

b. $\sqrt{2}(4 - \sqrt{8}) = 4\sqrt{2} - \sqrt{2}\sqrt{8}$ Distributive Property

$= 4\sqrt{2} - \sqrt{16} = 4\sqrt{2} - 4$ Product Rule for Radicals

c. $\sqrt{6}(\sqrt{12} - \sqrt{3}) = \sqrt{6}\sqrt{12} - \sqrt{6}\sqrt{3}$ Distributive Property

$= \sqrt{72} - \sqrt{18}$ Product Rule for Radicals

$= 6\sqrt{2} - 3\sqrt{2} = 3\sqrt{2}$ Find perfect square factors.

Exercises Within Reach®

Solutions in English & Spanish and tutorial videos at AlgebraWithinReach.com

Multiplying Radical Expression **In Exercises 1–20, multiply and simplify.**

1. $\sqrt{2} \cdot \sqrt{8}$ 4
2. $\sqrt{6} \cdot \sqrt{18}$ $6\sqrt{3}$
3. $\sqrt{3} \cdot \sqrt{15}$ $3\sqrt{5}$
4. $\sqrt{5} \cdot \sqrt{10}$ $5\sqrt{2}$
5. $\sqrt[3]{12} \cdot \sqrt[3]{6}$ $2\sqrt[3]{9}$
6. $\sqrt[3]{9} \cdot \sqrt[3]{3}$ 3
7. $\sqrt[4]{8} \cdot \sqrt[4]{2}$ 2
8. $\sqrt[4]{54} \cdot \sqrt[4]{3}$ $3\sqrt[4]{2}$
9. $\sqrt{7}(3 - \sqrt{7})$ $3\sqrt{7} - 7$
10. $\sqrt{3}(4 + \sqrt{3})$ $4\sqrt{3} + 3$
11. $\sqrt{2}(\sqrt{20} + 8)$ $2\sqrt{10} + 8\sqrt{2}$
12. $\sqrt{7}(\sqrt{14} + 3)$ $7\sqrt{2} + 3\sqrt{7}$
13. $\sqrt{6}(\sqrt{12} - \sqrt{3})$ $3\sqrt{2}$
14. $\sqrt{10}(\sqrt{5} + \sqrt{6})$ $5\sqrt{2} + 2\sqrt{15}$
15. $4\sqrt{3}(\sqrt{3} - \sqrt{5})$ $12 - 4\sqrt{15}$
16. $3\sqrt{5}(\sqrt{5} - \sqrt{2})$ $15 - 3\sqrt{10}$
17. $\sqrt{y}(\sqrt{y} + 4)$ $y + 4\sqrt{y}$
18. $\sqrt{x}(5 - \sqrt{x})$ $5\sqrt{x} - x$
19. $\sqrt{a}(4 - \sqrt{a})$ $4\sqrt{a} - a$
20. $\sqrt{z}(\sqrt{z} + 5)$ $z + 5\sqrt{z}$

Students sometimes have difficulty using the FOIL Method when radicals are involved. You may want to use an additional example such as the following.

$(4 + 2\sqrt{5})(3 - 4\sqrt{5})$

$= 12 - 16\sqrt{5} + 6\sqrt{5} - 8(5)$

$= -28 - 10\sqrt{5}$

Have students verify their results using their calculators.

In Section 5.3, you learned the FOIL Method for multiplying two binomials. This method can be used when the binomials involve radicals.

EXAMPLE 3 Using the FOIL Method to Multiply

Find each product and simplify.

a. $(2\sqrt{7} - 4)(\sqrt{7} + 1)$ **b.** $(3 - \sqrt{x})(1 + \sqrt{x})$

SOLUTION

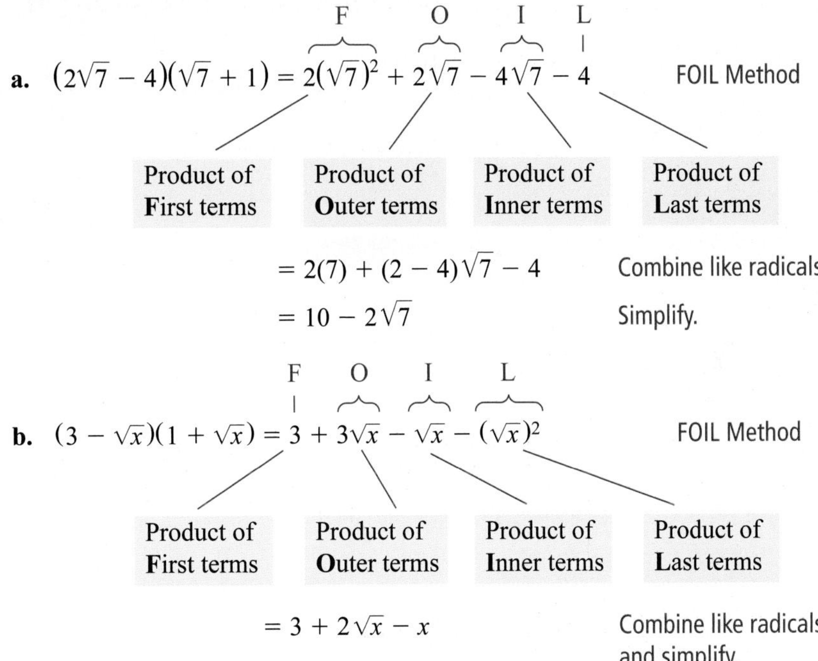

F O I L
a. $(2\sqrt{7} - 4)(\sqrt{7} + 1) = 2(\sqrt{7})^2 + 2\sqrt{7} - 4\sqrt{7} - 4$
FOIL Method
Product of First terms
Product of Outer terms
Product of Inner terms
Product of Last terms
$= 2(7) + (2 - 4)\sqrt{7} - 4$
Combine like radicals.
$= 10 - 2\sqrt{7}$
Simplify.
F O I L
b. $(3 - \sqrt{x})(1 + \sqrt{x}) = 3 + 3\sqrt{x} - \sqrt{x} - (\sqrt{x})^2$
FOIL Method
Product of First terms
Product of Outer terms
Product of Inner terms
Product of Last terms
$= 3 + 2\sqrt{x} - x$
Combine like radicals and simplify.

Exercises Within Reach®

Solutions in English & Spanish and tutorial videos at AlgebraWithinReach.com

Multiplying Radical Expressions **In Exercises 21–30, multiply and simplify.**

21. $(\sqrt{5} + 3)(\sqrt{3} - 5)$ $\sqrt{15} - 5\sqrt{5} + 3\sqrt{3} - 15$

22. $(\sqrt{7} + 6)(\sqrt{2} + 6)$ $\sqrt{14} + 6\sqrt{7} + 6\sqrt{2} + 36$

23. $(\sqrt{20} + 2)^2$ $8\sqrt{5} + 24$

24. $(4 - \sqrt{20})^2$ $36 - 16\sqrt{5}$

25. $(\sqrt{5} - \sqrt{3})(\sqrt{5} - \sqrt{3})$ $8 - 2\sqrt{15}$

26. $(\sqrt{2} + \sqrt{7})(\sqrt{2} + \sqrt{7})$ $9 + 2\sqrt{14}$

27. $(10 + \sqrt{2x})^2$ $100 + 20\sqrt{2x} + 2x$

28. $(5 - \sqrt{3y})^2$ $25 - 10\sqrt{3y} + 3y$

29. $(9\sqrt{x} + 2)(5\sqrt{x} - 3)$ $45x - 17\sqrt{x} - 6$

30. $(16\sqrt{u} - 3)(\sqrt{u} - 1)$ $16u - 19\sqrt{u} + 3$

Geometry **In Exercises 31–34, write and simplify an expression for the area of the rectangle.**

31.

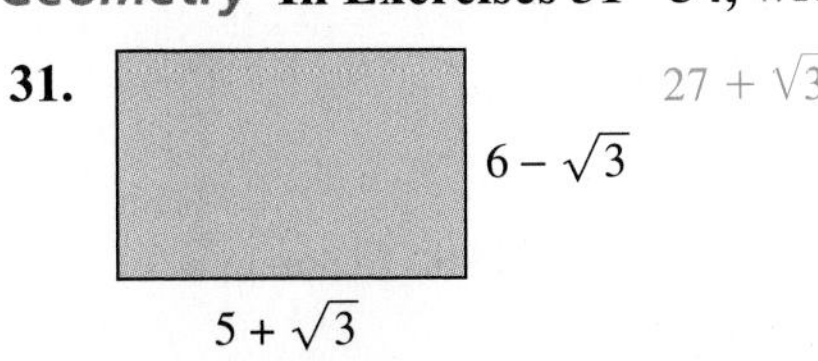

$27 + \sqrt{3}$

32.

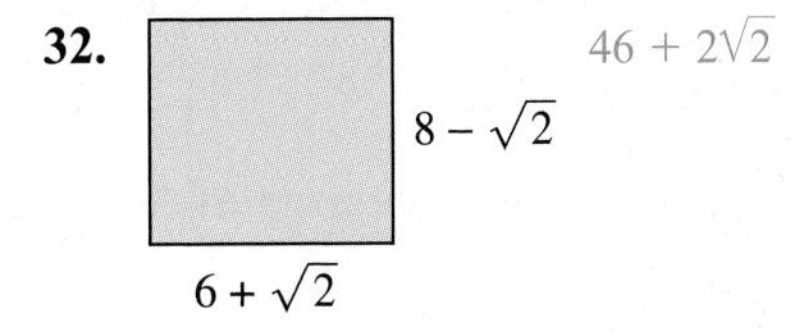

$46 + 2\sqrt{2}$

33.

$4 - \sqrt{x}$

$2 + \sqrt{x}$

$8 + 2\sqrt{x} - x$

34.

$8 - \sqrt{x}$

$5 + \sqrt{x}$

$40 + 3\sqrt{x} - x$

Conjugates

The expressions $3 + \sqrt{6}$ and $3 - \sqrt{6}$ are called **conjugates** of each other. Notice that they differ only in the sign between the terms. The product of two conjugates is the difference of two squares, which is given by the special product formula $(a + b)(a - b) = a^2 - b^2$. Here are some other examples.

EXAMPLE 4 Multiplying Conjugates

Expression	*Conjugate*	*Product*
$1 - \sqrt{3}$	$1 + \sqrt{3}$	$(1)^2 - (\sqrt{3})^2 = 1 - 3 = -2$
$\sqrt{5} + \sqrt{2}$	$\sqrt{5} - \sqrt{2}$	$(\sqrt{5})^2 - (\sqrt{2})^2 = 5 - 2 = 3$
$\sqrt{10} - 3$	$\sqrt{10} + 3$	$(\sqrt{10})^2 - (3)^2 = 10 - 9 = 1$
$\sqrt{x} + 2$	$\sqrt{x} - 2$	$(\sqrt{x})^2 - (2)^2 = x - 4,\ x \ge 0$

Looking ahead to solving equations involving radicals, have your students simplify $(\sqrt{3} + \sqrt{x})^2$.

EXAMPLE 5 Multiplying Conjugates

Find the conjugate of each expression and multiply each expression by its conjugate.

a. $2 - \sqrt{5}$ **b.** $\sqrt{3} + \sqrt{x}$

SOLUTION

a. The conjugate of $2 - \sqrt{5}$ is $2 + \sqrt{5}$.

$(2 - \sqrt{5})(2 + \sqrt{5}) = 2^2 - (\sqrt{5})^2$ Special product formula

$= 4 - 5 = -1$ Simplify.

b. The conjugate of $\sqrt{3} + \sqrt{x}$ is $\sqrt{3} - \sqrt{x}$.

$(\sqrt{3} + \sqrt{x})(\sqrt{3} - \sqrt{x}) = (\sqrt{3})^2 - (\sqrt{x})^2$ Special product formula

$= 3 - x,\ x \ge 0$ Simplify.

Exercises Within Reach®

Solutions in English & Spanish and tutorial videos at AlgebraWithinReach.com

Multiplying Conjugates **In Exercises 35–48, find the conjugate of the expression. Then multiply the expression by its conjugate and simplify.**

35. $2 + \sqrt{5}$ $2 - \sqrt{5}, -1$
36. $\sqrt{2} - 9$ $\sqrt{2} + 9, -79$
37. $\sqrt{11} - \sqrt{3}$ $\sqrt{11} + \sqrt{3}, 8$
38. $\sqrt{10} + \sqrt{7}$ $\sqrt{10} - \sqrt{7}, 3$
39. $\sqrt{15} + 3$ $\sqrt{15} - 3, 6$
40. $\sqrt{14} - 3$ $\sqrt{14} + 3, 5$
41. $\sqrt{x} - 3$ $\sqrt{x} + 3, x - 9$
42. $\sqrt{t} + 7$ $\sqrt{t} - 7, t - 49$
43. $\sqrt{2u} - \sqrt{3}$ $\sqrt{2u} + \sqrt{3}, 2u - 3$
44. $\sqrt{5a} + \sqrt{2}$ $\sqrt{5a} - \sqrt{2}, 5a - 2$
45. $2\sqrt{2} + \sqrt{4}$ $2\sqrt{2} - \sqrt{4}, 4$
46. $4\sqrt{3} + \sqrt{2}$ $4\sqrt{3} - \sqrt{2}, 46$
47. $\sqrt{x} + \sqrt{y}$ $\sqrt{x} - \sqrt{y}, x - y$
48. $3\sqrt{u} + \sqrt{3v}$ $3\sqrt{u} - \sqrt{3v}, 9u - 3v$

Geometry **In Exercises 49 and 50, find the area of the rectangle.**

49. 23

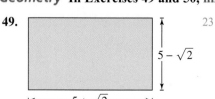

50. 58

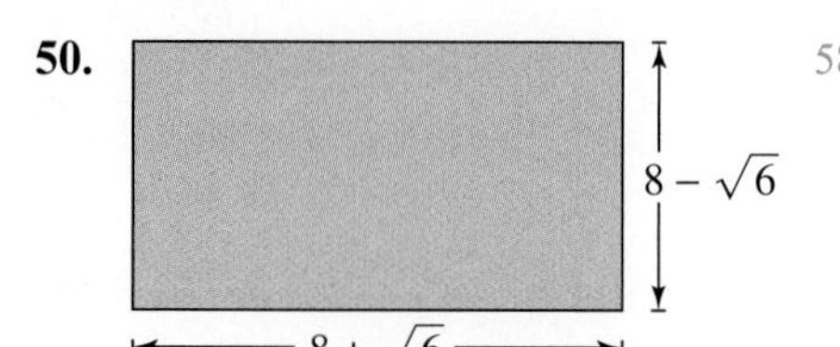

Dividing Radical Expressions

To simplify a quotient involving radicals, you rationalize the denominator. For single-term denominators, you can use the rationalization process described in Section 7.2. To rationalize a denominator involving two terms, multiply both the numerator and denominator by the *conjugate of the denominator*.

EXAMPLE 6 Simplifying Quotients Involving Radicals

Simplify (a) $\dfrac{\sqrt{3}}{1-\sqrt{5}}$ and (b) $\dfrac{4}{2-\sqrt{3}}$.

SOLUTION

a. $\dfrac{\sqrt{3}}{1-\sqrt{5}} = \dfrac{\sqrt{3}}{1-\sqrt{5}} \cdot \dfrac{1+\sqrt{5}}{1+\sqrt{5}}$ Multiply numerator and denominator by conjugate of denominator.

$= \dfrac{\sqrt{3}(1+\sqrt{5})}{1^2-(\sqrt{5})^2}$ Special product formula

$= \dfrac{\sqrt{3}+\sqrt{15}}{1-5}$ Simplify.

$= -\dfrac{\sqrt{3}+\sqrt{15}}{4}$ Simplify.

b. $\dfrac{4}{2-\sqrt{3}} = \dfrac{4}{2-\sqrt{3}} \cdot \dfrac{2+\sqrt{3}}{2+\sqrt{3}}$ Multiply numerator and denominator by conjugate of denominator.

$= \dfrac{4(2+\sqrt{3})}{2^2-(\sqrt{3})^2}$ Special product formula

$= \dfrac{8+4\sqrt{3}}{4-3}$ Simplify.

$= 8+4\sqrt{3}$ Simplify.

Exercises Within Reach®

Solutions in English & Spanish and tutorial videos at AlgebraWithinReach.com

Simplifying a Quotient Involving a Radical **In Exercises 51–60, simplify the expression.**

51. $\dfrac{1}{2+\sqrt{5}}$ $\sqrt{5}-2$

52. $\dfrac{1}{3+\sqrt{10}}$ $\sqrt{10}-3$

53. $\dfrac{\sqrt{2}}{2-\sqrt{6}}$ $-\sqrt{2}-\sqrt{3}$

54. $\dfrac{\sqrt{5}}{3-\sqrt{7}}$ $\dfrac{3\sqrt{5}+\sqrt{35}}{2}$

55. $\dfrac{5}{9-\sqrt{6}}$ $\dfrac{9+\sqrt{6}}{15}$

56. $\dfrac{8}{8-\sqrt{8}}$ $\dfrac{8+\sqrt{8}}{7}$

57. $\dfrac{6}{\sqrt{11}-2}$ $\dfrac{6(\sqrt{11}+2)}{7}$

58. $\dfrac{8}{\sqrt{7}+3}$ $4(3-\sqrt{7})$

59. $\dfrac{7}{\sqrt{3}+5}$ $\dfrac{7(5-\sqrt{3})}{22}$

60. $\dfrac{9}{\sqrt{7}-4}$ $-4-\sqrt{7}$

Additional Examples

Simplify each expression.

a. $\dfrac{6}{4+\sqrt{6}}$

b. $\dfrac{5}{\sqrt{3}+\sqrt{10}}$

Answers:

a. $\dfrac{12-3\sqrt{6}}{5}$

b. $-\dfrac{5\sqrt{3}-5\sqrt{10}}{7}$

EXAMPLE 7 Simplifying a Quotient Involving Radicals

$$\frac{5\sqrt{2}}{\sqrt{7}+\sqrt{2}} = \frac{5\sqrt{2}}{\sqrt{7}+\sqrt{2}} \cdot \frac{\sqrt{7}-\sqrt{2}}{\sqrt{7}-\sqrt{2}}$$ Multiply numerator and denominator by conjugate of denominator.

$$= \frac{5\sqrt{2}(\sqrt{7}-\sqrt{2})}{(\sqrt{7})^2-(\sqrt{2})^2}$$ Special product formula

$$= \frac{5(\sqrt{14}-\sqrt{4})}{7-2}$$ Simplify.

$$= \frac{5(\sqrt{14}-2)}{5}$$ Simplify.

$$= \frac{\cancel{5}(\sqrt{14}-2)}{\cancel{5}}$$ Divide out common factor.

$$= \sqrt{14}-2$$ Simplest form

EXAMPLE 8 Simplifying a Quotient Involving Radicals

Perform the division and simplify.

$$1 \div (\sqrt{x}-\sqrt{x+1})$$

SOLUTION

$$\frac{1}{\sqrt{x}-\sqrt{x+1}} = \frac{1}{\sqrt{x}-\sqrt{x+1}} \cdot \frac{\sqrt{x}+\sqrt{x+1}}{\sqrt{x}+\sqrt{x+1}}$$ Multiply numerator and denominator by conjugate of denominator.

$$= \frac{\sqrt{x}+\sqrt{x+1}}{(\sqrt{x})^2-(\sqrt{x+1})^2}$$ Special product formula

$$= \frac{\sqrt{x}+\sqrt{x+1}}{x-(x+1)}$$ Simplify.

$$= \frac{\sqrt{x}+\sqrt{x+1}}{-1}$$ Combine like terms.

$$= -\sqrt{x}-\sqrt{x+1}$$ Simplify.

Exercises Within Reach®

Solutions in English & Spanish and tutorial videos at AlgebraWithinReach.com

Simplifying a Quotient Involving Radicals **In Exercises 61–68, simplify the expression.**

61. $\dfrac{\sqrt{5}}{\sqrt{6}-\sqrt{5}}$ $\sqrt{30}+5$

62. $\dfrac{\sqrt{2}}{\sqrt{2}-\sqrt{3}}$ $-\sqrt{6}-2$

63. $\dfrac{4\sqrt{3}}{\sqrt{5}+\sqrt{3}}$ $2\sqrt{15}-6$

64. $\dfrac{5\sqrt{7}}{\sqrt{6}+\sqrt{7}}$ $35-5\sqrt{42}$

65. $1 \div (\sqrt{x}-\sqrt{x+3})$ $-\dfrac{\sqrt{x}+\sqrt{x+3}}{3}$

66. $1 \div (\sqrt{x}+\sqrt{x-2})$ $\dfrac{\sqrt{x}-\sqrt{x-2}}{2}$

67. $\dfrac{3x}{\sqrt{15}-\sqrt{3}}$ $\dfrac{(\sqrt{15}+\sqrt{3})x}{4}$

68. $\dfrac{5y}{\sqrt{12}+\sqrt{10}}$ $\dfrac{5y(2\sqrt{3}-\sqrt{10})}{2}$

At this point, students may still be tempted to rationalize $6/(\sqrt{x}-2)$ by multiplying by $\sqrt{x}/\sqrt{x}$. Ask them why this does not rationalize the denominator.

EXAMPLE 9 Simplifying Quotient Involving Radicals

Perform each division and simplify.

a. $6 \div (\sqrt{x}-2)$

b. $(2-\sqrt{3}) \div (\sqrt{6}+\sqrt{2})$

SOLUTION

a. $\dfrac{6}{\sqrt{x}-2} = \dfrac{6}{\sqrt{x}-2} \cdot \dfrac{\sqrt{x}+2}{\sqrt{x}+2}$ Multiply numerator and denominator by conjugate of denominator.

$= \dfrac{6(\sqrt{x}+2)}{(\sqrt{x})^2-2^2}$ Special product formula

$= \dfrac{6\sqrt{x}+12}{x-4}$ Simplify.

b. $\dfrac{2-\sqrt{3}}{\sqrt{6}+\sqrt{2}} = \dfrac{2-\sqrt{3}}{\sqrt{6}+\sqrt{2}} \cdot \dfrac{\sqrt{6}-\sqrt{2}}{\sqrt{6}-\sqrt{2}}$ Multiply numerator and denominator by conjugate of denominator.

$= \dfrac{2\sqrt{6}-2\sqrt{2}-\sqrt{18}+\sqrt{6}}{(\sqrt{6})^2-(\sqrt{2})^2}$ FOIL Method and special product formula

$= \dfrac{3\sqrt{6}-2\sqrt{2}-3\sqrt{2}}{6-2}$ Simplify.

$= \dfrac{3\sqrt{6}-5\sqrt{2}}{4}$ Simplify.

Exercises Within Reach®

Solutions in English & Spanish and tutorial videos at AlgebraWithinReach.com

Simplifying a Quotient Involving Radicals **In Exercises 69–78, simplify the expression.**

69. $8 \div (\sqrt{x}-3)$ $\dfrac{8\sqrt{x}+24}{x-9}$

70. $7 \div (\sqrt{x}-5)$ $\dfrac{7\sqrt{x}+35}{x-25}$

71. $\dfrac{\sqrt{5t}}{\sqrt{5}-\sqrt{t}}$ $\dfrac{5\sqrt{t}+t\sqrt{5}}{5-t}$

72. $\dfrac{\sqrt{2x}}{\sqrt{x}-\sqrt{2}}$ $\dfrac{x\sqrt{2}+2\sqrt{x}}{x-2}$

73. $\dfrac{8a}{\sqrt{3a}+\sqrt{a}}$ $4(\sqrt{3a}-\sqrt{a}),\ a \neq 0$

74. $\dfrac{7z}{\sqrt{5z}-\sqrt{z}}$ $\dfrac{7(\sqrt{5z}+\sqrt{z})}{4},\ z \neq 0$

75. $(\sqrt{7}+2) \div (\sqrt{7}-2)$ $\dfrac{4\sqrt{7}+11}{3}$

76. $(5-\sqrt{3}) \div (3+\sqrt{3})$ $\dfrac{9-4\sqrt{3}}{3}$

77. $(\sqrt{x}-5) \div (2\sqrt{x}-1)$ $\dfrac{2x-9\sqrt{x}-5}{4x-1}$

78. $(2\sqrt{t}+1) \div (2\sqrt{t}-1)$ $\dfrac{4t+4\sqrt{t}+1}{4t-1}$

Geometry **In Exercises 79 and 80, find the exact width w of the rectangle.**

79. Area = 10; width w; length $\sqrt{12}+\sqrt{5}$ $\dfrac{20\sqrt{3}-10\sqrt{5}}{7}$

80. Area = 32; width w; length $2\sqrt{5}+\sqrt{8}$ $\dfrac{16(\sqrt{5}-\sqrt{2})}{3}$

Concept Summary: Multiplying and Dividing Radical Expressions

What

You can use previously learned techniques to multiply and divide radical expressions.

EXAMPLE

Simplify each expression.

a. $(1 + \sqrt{2})(3 + \sqrt{2})$

b. $\dfrac{1}{\sqrt{2} - 1}$

How

- To multiply radical expressions, use the Distributive Property, FOIL Method, and special product formulas.
- To simplify quotients involving radicals, rationalize the denominators.

EXAMPLE

a. $(1 + \sqrt{2})(3 + \sqrt{2})$
$= 3 + \sqrt{2} + 3\sqrt{2} + 2$
$= 5 + 4\sqrt{2}$

b. $\dfrac{1}{\sqrt{2} - 1} \cdot \dfrac{\sqrt{2} + 1}{\sqrt{2} + 1} = \dfrac{\sqrt{2} + 1}{2 - 1}$
$= \sqrt{2} + 1$

Why

You can use radical expressions to model real-life problems. Knowing how to multiply and divide radical expressions will help you solve such problems. For example, you can multiply radical expressions to find the area of a cross section of the strongest beam that can be cut from a log.

Exercises Within Reach ®

Worked-out solutions to odd-numbered exercises at AlgebraWithinReach.com

Concept Summary Check

81. ***Multiplying Radical Expressions*** In the solution above, what technique is used to multiply $(1 + \sqrt{2})$ and $(3 + \sqrt{2})$? The FOIL Method

82. ***Product Rule for Radicals*** In the solution above, what term of the expression $3 + \sqrt{2} + 3\sqrt{2} + 2$ is obtained using the Product Rule for Radicals? 2

83. ***Describing a Relationship*** Describe the relationship between $\sqrt{2} + 1$ and $\sqrt{2} - 1$. They are conjugates.

84. ***Identifying a Process*** What process is used to rewrite $\dfrac{1}{\sqrt{2} - 1}$ as $\sqrt{2} + 1$ in the example above? Rationalizing the denominator

Extra Practice

Multiplying Radical Expressions **In Exercises 85–90, multiply and simplify.**

85. $(2\sqrt{2x} - \sqrt{5})(2\sqrt{2x} + \sqrt{5})$ $8x - 5$

86. $(\sqrt{7} - 3\sqrt{3t})(\sqrt{7} + 3\sqrt{3t})$ $7 - 27t$

87. $(\sqrt[3]{t} + 1)(\sqrt[3]{t^2} + 4\sqrt[3]{t} - 3)$ $t + 5\sqrt[3]{t^2} + \sqrt[3]{t} - 3$

88. $(\sqrt[3]{x} - 2)(\sqrt[3]{x^2} - 2\sqrt[3]{x} + 1)$ $x - 4\sqrt[3]{x^2} + 5\sqrt[3]{x} - 2$

89. $2\sqrt[3]{x^4y^5}(\sqrt[3]{8x^{12}y^4} + \sqrt[3]{16xy^9})$ $4xy^3(x^4\sqrt[3]{x} + y\sqrt[3]{2x^2y^2})$

90. $\sqrt[4]{8x^3y^5}(\sqrt[4]{4x^5y^7} - \sqrt[4]{3x^7y^6})$ $x^2y^2(2|y|\sqrt[4]{2} - \sqrt[4]{24x^2y^3})$

Evaluating a Function **In Exercises 91 and 92, evaluate the function as indicated.**

91. $f(x) = x^2 - 6x + 1$

(a) $f(2 - \sqrt{3})$ $2\sqrt{3} - 4$ (b) $f(3 - 2\sqrt{2})$ 0

92. $g(x) = x^2 + 8x + 11$

(a) $g(-4 + \sqrt{5})$ 0 (b) $g(-4\sqrt{2})$ $43 - 32\sqrt{2}$

Rationalizing the Numerator **In the study of calculus, students sometimes rewrite an expression by rationalizing the numerator. In Exercises 93–96, rationalize the numerator. (*Note:* The results will not be in simplest radical form.)**

93. $\dfrac{\sqrt{10}}{\sqrt{3x}}$ $\dfrac{10}{\sqrt{30x}}$

94. $\dfrac{\sqrt{5}}{\sqrt{7x}}$ $\dfrac{5}{\sqrt{35x}}$

95. $\dfrac{\sqrt{7} + \sqrt{3}}{5}$ $\dfrac{4}{5(\sqrt{7} - \sqrt{3})}$

96. $\dfrac{\sqrt{2} - \sqrt{5}}{4}$ $-\dfrac{3}{4(\sqrt{2} + \sqrt{5})}$

97. ***Geometry*** The width w and height h of the strongest rectangular beam that can be cut from a log with a diameter of 24 inches (see figure) are given by

$$w = 8\sqrt{3} \quad \text{and} \quad h = \sqrt{24^2 - (8\sqrt{3})^2}.$$

Find the area of the rectangular cross section of the beam, and write the area in simplest form.
$192\sqrt{2}$ square inches

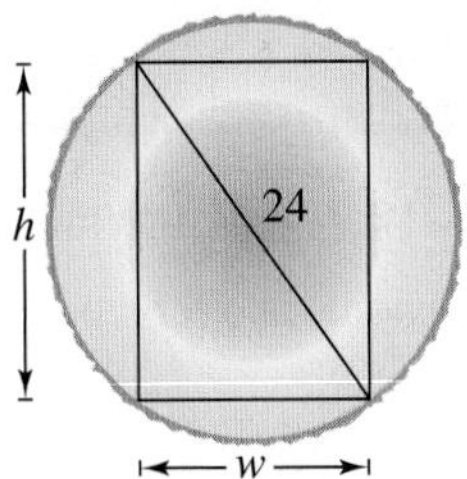

Figure for 97

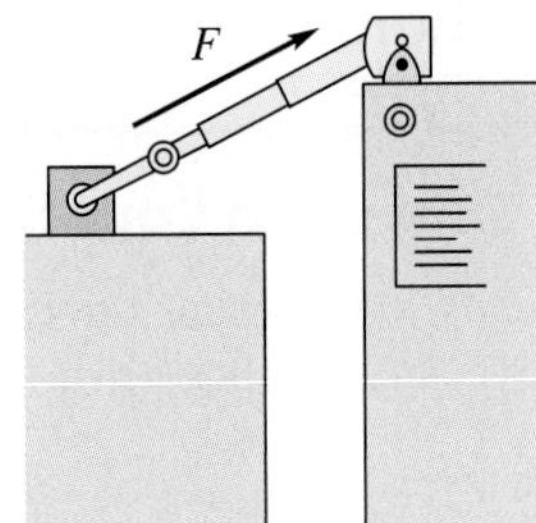

Figure for 98

98. ***Force*** The force required to slide a steel block weighing 500 pounds across a milling machine is

$$\frac{500k}{\dfrac{1}{\sqrt{k^2+1}} + \dfrac{k^2}{\sqrt{k^2+1}}}$$

where k is the friction constant (see figure). Simplify this expression.
$\dfrac{500k\sqrt{k^2+1}}{k^2+1}$

99. ***Basketball*** The area of the circular cross section of a basketball is 70 square inches. The area enclosed by a basketball hoop is about 254 square inches. Find the ratio of the diameter of the basketball to the diameter of the hoop. See Additional Answers.

100. ***Geometry*** The ratio of the width of the Temple of Hephaestus to its height (see figure) is approximately

$$\frac{w}{h} \approx \frac{2}{\sqrt{5}-1}.$$

This number is called the **golden section**. Early Greeks believed that the most aesthetically pleasing rectangles were those whose sides had this ratio.

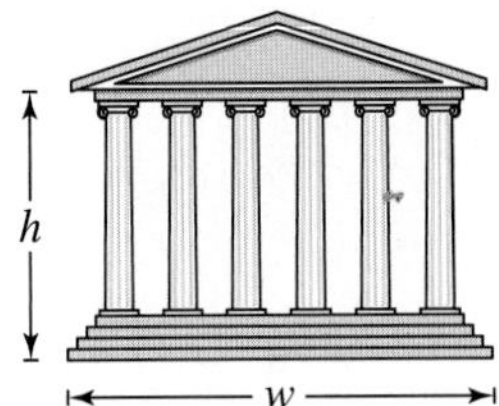

(a) Rationalize the denominator to simplify the expression. Then approximate the value of the expression. Round your answer to two decimal places.
$\dfrac{\sqrt{5}+1}{2} \approx 1.62$

(b) Use the Pythagorean Theorem, a straightedge, and a compass to construct a rectangle whose sides have the golden section as their ratio. Answers will vary.

Explaining Concepts

101. ***Number Sense*** Let a and b be integers, but not perfect squares. Describe the circumstances (if any) for which each expression represents a rational number. Explain.
See Additional Answers.

(a) $a\sqrt{b}$ (b) $\sqrt{a}\sqrt{b}$

102. ***Number Sense*** Given that a and b are positive integers, what type of number is the product of the expression $\sqrt{a} + \sqrt{b}$ and its conjugate? Explain.
Integer; Because a and b are integers, $(\sqrt{a}+\sqrt{b})(\sqrt{a}-\sqrt{b}) = a - b$ is an integer.

103. ***Exploration*** Find the conjugate of $\sqrt{a} + \sqrt{b}$. Multiply the conjugates. Next, find the conjugate of $\sqrt{b} + \sqrt{a}$. Multiply the conjugates. Explain how changing the order of the terms affects the conjugate and the product of the conjugates. See Additional Answers.

104. ***Exploration*** Rationalize the denominators of $\dfrac{1}{\sqrt{a}+\sqrt{b}}$ and $\dfrac{1}{\sqrt{b}+\sqrt{a}}$. Explain how changing the order of the terms in the denominator affects the rationalized form of the quotient. See Additional Answers.

Cumulative Review

In Exercises 105–108, solve the equation. If there is exactly one solution, check your solution. If not, justify your answer.

105. $3x - 18 = 0$ 6

106. $7t - 4 = 4t + 8$ 4

107. $3x - 4 = 3x$
No solution

108. $3(2x + 5) = 6x + 15$
Infinitely many solutions

In Exercises 109–112, solve the equation by factoring.

109. $x^2 - 144 = 0$ $-12, 12$

110. $4x^2 - 25 = 0$ $-\frac{5}{2}, \frac{5}{2}$

111. $x^2 + 2x - 15 = 0$ $-5, 3$

112. $6x^2 - x - 12 = 0$ $-\frac{4}{3}, \frac{3}{2}$

In Exercises 113–116, simplify the radical expression.

113. $\sqrt{32x^2y^5}$ $4|x|y^2\sqrt{2y}$

114. $\sqrt[3]{32x^2y^5}$ $2y\sqrt[3]{4x^2y^2}$

115. $\sqrt[4]{32x^2y^5}$ $2y\sqrt[4]{2x^2y}$

116. $\sqrt[5]{32x^2y^5}$ $2y\sqrt[5]{x^2}$

7.5 Radical Equations and Applications

- Solve a radical equation by raising each side to the nth power.
- Solve application problems involving radical equations.

Solving Radical Equations

Raising Each Side of an Equation to the *n*th Power

Let u and v be real numbers, variables, or algebraic expressions, and let n be a positive integer. If $u = v$, then it follows that

$$u^n = v^n.$$

This is called *raising each side of an equation to the nth power.*

EXAMPLE 1 Solving an Equation Having One Radical

$\sqrt{x} - 8 = 0$	Original equation
$\sqrt{x} = 8$	Isolate radical.
$(\sqrt{x})^2 = 8^2$	Square each side.
$x = 64$	Simplify.

Check

$\sqrt{64} - 8 \stackrel{?}{=} 0$	Substitute 64 for x in original equation.
$8 - 8 = 0$	Solution checks. ✓

So, the equation has one solution: $x = 64$.

Exercises Within Reach®

Solutions in English & Spanish and tutorial videos at AlgebraWithinReach.com

Checking Solutions of an Equation **In Exercises 1–4, determine whether each value of x is a solution of the equation.** See Additional Answers.

	Equation	*Values of x*			*Equation*	*Values of x*	
1.	$\sqrt{x} - 10 = 0$	(a) $x = -4$	(b) $x = -100$	**2.**	$\sqrt{3x} - 6 = 0$	(a) $x = \frac{2}{3}$	(b) $x = 2$
		(c) $x = \sqrt{10}$	(d) $x = 100$			(c) $x = 12$	(d) $x = -\frac{1}{3}\sqrt{6}$
3.	$\sqrt[3]{x - 4} = 4$	(a) $x = -60$	(b) $x = 68$	**4.**	$\sqrt[4]{2x} + 2 = 6$	(a) $x = 128$	(b) $x = 2$
		(c) $x = 20$	(d) $x = 0$			(c) $x = -2$	(d) $x = 0$

Solving an Equation **In Exercises 5–14, solve the equation and check your solution.**

5. $\sqrt{x} = 12$ 144
6. $\sqrt{x} = 5$ 25
7. $\sqrt{y} = 7$ 49
8. $\sqrt{t} = 4$ 16
9. $\sqrt[3]{z} = 3$ 27
10. $\sqrt[4]{x} = 3$ 81
11. $\sqrt{y} - 7 = 0$ 49
12. $\sqrt{t} - 13 = 0$ 169
13. $\sqrt{x} - 8 = 0$ 64
14. $\sqrt{x} - 10 = 0$ 100

Checking solutions of a radical equation is especially important because raising each side of an equation to the nth power to remove the radical(s) often introduces *extraneous* solutions.

EXAMPLE 2 Solving an Equation Having One Radical

$\sqrt{3x} + 6 = 0$	Original equation
$\sqrt{3x} = -6$	Isolate radical.
$(\sqrt{3x})^2 = (-6)^2$	Square each side.
$3x = 36$	Simplify.
$x = 12$	Divide each side by 3.

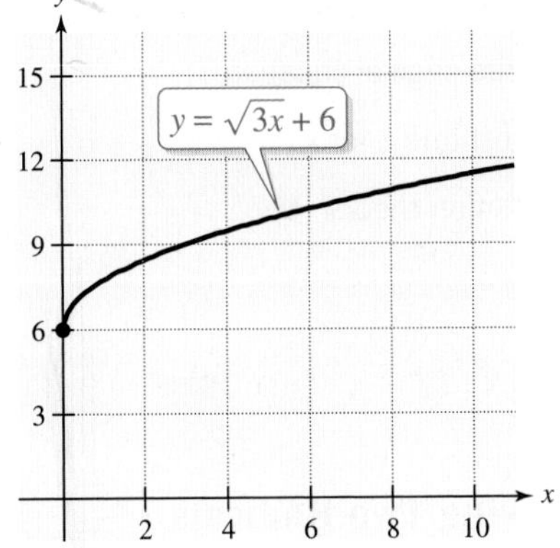

$\sqrt{3(12)} + 6 \stackrel{?}{=} 0$	Substitute 12 for x in original equation.
$6 + 6 \neq 0$	Solution does not check.

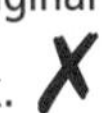

The solution $x = 12$ is an extraneous solution. So, the original equation has no solution. You can also check this graphically, as shown at the left. Notice that the graph does not cross the x-axis and so it has no x-intercept.

EXAMPLE 3 Solving an Equation Having One Radical

$\sqrt[3]{2x + 1} - 2 = 3$	Original equation
$\sqrt[3]{2x + 1} = 5$	Isolate radical.
$(\sqrt[3]{2x + 1})^3 = 5^3$	Cube each side.
$2x + 1 = 125$	Simplify.
$2x = 124$	Subtract 1 from each side.
$x = 62$	Divide each side by 2.

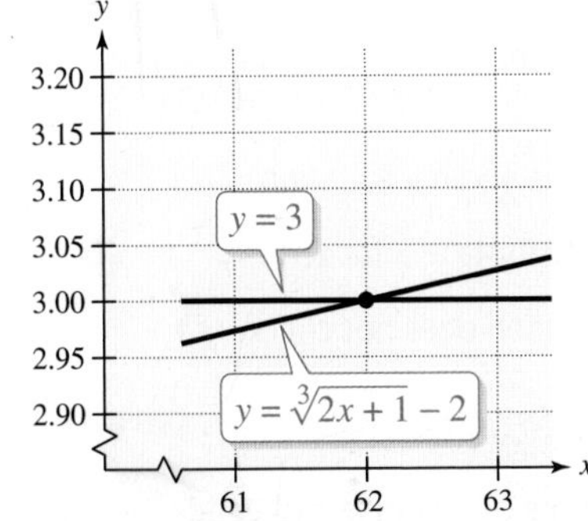

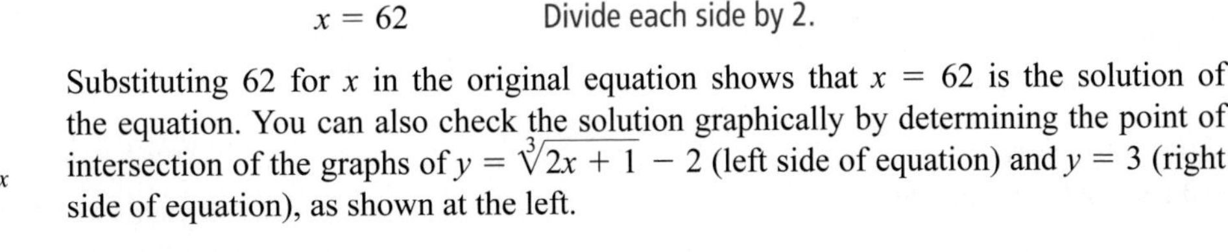

Substituting 62 for x in the original equation shows that $x = 62$ is the solution of the equation. You can also check the solution graphically by determining the point of intersection of the graphs of $y = \sqrt[3]{2x + 1} - 2$ (left side of equation) and $y = 3$ (right side of equation), as shown at the left.

Exercises Within Reach®

Solutions in English & Spanish and tutorial videos at AlgebraWithinReach.com

Solving an Equation **In Exercises 15–26, solve the equation and check your solution. (Some of the equations have no solution.)**

15. $\sqrt{u} + 13 = 0$ No solution

16. $\sqrt{y} + 15 = 0$ No solution

17. $\sqrt{10x} = 30$ 90

18. $\sqrt{8x} = 6$ $\frac{9}{2}$

19. $\sqrt{-3x} = 9$ -27

20. $\sqrt{-4y} = 4$ -4

21. $\sqrt{3y + 1} = 4$ 5

22. $\sqrt{3 - 2x} = 2$ $-\frac{1}{2}$

23. $\sqrt{9 - 2x} + 8 = -1$ No solution

24. $\sqrt{2t - 7} + 2 = -3$ No solution

25. $\sqrt[3]{y + 1} - 2 = 4$ 215

26. $4\sqrt[3]{x + 4} - 6 = 1$ $\frac{87}{64}$

EXAMPLE 4 Solving an Equation Having Two Radicals

$\sqrt{5x+3} = \sqrt{x+11}$	Original equation
$(\sqrt{5x+3})^2 = (\sqrt{x+11})^2$	Square each side.
$5x + 3 = x + 11$	Simplify.
$4x + 3 = 11$	Subtract x from each side.
$4x = 8$	Subtract 3 from each side.
$x = 2$	Divide each side by 4.

Check

$\sqrt{5x+3} = \sqrt{x+11}$	Write original equation.
$\sqrt{5(2)+3} \stackrel{?}{=} \sqrt{2+11}$	Substitute 2 for x.
$\sqrt{13} = \sqrt{13}$	Solution checks. ✓

So, the equation has one solution: $x = 2$.

EXAMPLE 5 Solving an Equation Having Two Radicals

$\sqrt[4]{3x} + \sqrt[4]{2x-5} = 0$	Original equation
$\sqrt[4]{3x} = -\sqrt[4]{2x-5}$	Isolate radicals.
$(\sqrt[4]{3x})^4 = (-\sqrt[4]{2x-5})^4$	Raise each side to fourth power.
$3x = 2x - 5$	Simplify.
$x = -5$	Subtract $2x$ from each side.

Ask students why it is necessary to isolate the radicals before raising each side to the fourth power in Example 5.

Try this simple illustration to show how extraneous solutions can be introduced. The equation $x = 2$ has only one solution. Trivially, x is 2. If each side of the equation is squared, you get the equation $x^2 = 4$, which has two solutions, $x = 2$ and $x = -2$. An "extra" solution has been introduced. Stress the checking of solutions in the *original* equation.

Check

$\sqrt[4]{3x} + \sqrt[4]{2x-5} = 0$	Write original equation.
$\sqrt[4]{3(-5)} + \sqrt[4]{2(-5)-5} \stackrel{?}{=} 0$	Substitute -5 for x.
$\sqrt[4]{-15} + \sqrt[4]{-15} \neq 0$	Solution does not check. 

The solution does not check because it yields fourth roots of negative radicands. So, this equation has no solution. Try checking this graphically. If you graph both sides of the equation, you will discover that the graphs do not intersect.

Exercises Within Reach®

Solutions in English & Spanish and tutorial videos at AlgebraWithinReach.com

Solving an Equation **In Exercises 27–36, solve the equation and check your solution. (Some of the equations have no solution.**

27. $\sqrt{x+3} = \sqrt{2x-1}$ 4

28. $\sqrt{3t+1} = \sqrt{t+15}$ 7

29. $\sqrt{3x+4} = \sqrt{4x+3}$ 1

30. $\sqrt{2x-7} = \sqrt{3x-12}$ 5

31. $\sqrt{3y-5} - 3\sqrt{y} = 0$ No solution

32. $\sqrt{2u+10} - 2\sqrt{u} = 0$ 5

33. $\sqrt[3]{3x-4} = \sqrt[3]{x+10}$ 7

34. $2\sqrt[3]{10-3x} = \sqrt[3]{2-x}$ $\frac{78}{23}$

35. $\sqrt[3]{2x+15} - \sqrt[3]{x} = 0$ −15

36. $\sqrt[4]{2x} + \sqrt[4]{x+3} = 0$ No solution

EXAMPLE 6 An Equation that Converts to a Quadratic Equation

$\sqrt{x} + 2 = x$	Original equation
$\sqrt{x} = x - 2$	Isolate radical.
$(\sqrt{x})^2 = (x-2)^2$	Square each side.
$x = x^2 - 4x + 4$	Simplify.
$-x^2 + 5x - 4 = 0$	Write in general form.
$(-1)(x-4)(x-1) = 0$	Factor.
$x - 4 = 0 \Rightarrow x = 4$	Set 1st factor equal to 0.
$x - 1 = 0 \Rightarrow x = 1$	Set 2nd factor equal to 0.

Check

First Solution	*Second Solution*
$\sqrt{4} + 2 \stackrel{?}{=} 4$	$\sqrt{1} + 2 \stackrel{?}{=} 1$
$2 + 2 = 4$ ✓	$1 + 2 \neq 1$ ✗

From the check you can see that $x = 1$ is an extraneous solution. So, the only solution is $x = 4$.

EXAMPLE 7 Repeatedly Squaring Each Side of an Equation

$\sqrt{3t+1} = 2 - \sqrt{3t}$	Original equation
$(\sqrt{3t+1})^2 = (2 - \sqrt{3t})^2$	Square each side (1st time).
$3t + 1 = 4 - 4\sqrt{3t} + 3t$	Simplify.
$-3 = -4\sqrt{3t}$	Isolate radical.
$(-3)^2 = (-4\sqrt{3t})^2$	Square each side (2nd time).
$9 = 16(3t)$	Simplify.
$\frac{3}{16} = t$	Divide each side by 48 and simplify.

The solution is $t = \frac{3}{16}$. Check this in the original equation.

Additional Examples

Solve each equation.

a. $4 - \sqrt{x-1} = \sqrt{3x+3}$

b. $\sqrt{x^2+11} - \sqrt{x^2-9} = 2$

Answers:

a. $x = 2$ ($x = 26$ is extraneous.)

b. $x = -5, x = 5$

Exercises Within Reach®

Solutions in English & Spanish and tutorial videos at AlgebraWithinReach.com

An Equation that Converts to a Quadratic Equation **In Exercises 37–42, solve the equation and check your solution(s).**

37. $\sqrt{x^2-2} = x + 4$ $-\frac{9}{4}$

38. $\sqrt{x^2-4} = x - 2$ 2

39. $\sqrt{2x} = x - 4$ 8

40. $\sqrt{x} = 6 - x$ 4

41. $\sqrt{8x+1} = x + 2$ 1, 3

42. $\sqrt{3x+7} = x + 3$ −2, −1

Repeatedly Squaring Each Side of an Equation **In Exercises 43–50, solve the equation and check your solution.**

43. $\sqrt{z+2} = 1 + \sqrt{z}$ $\frac{1}{4}$

44. $\sqrt{2x+5} = 7 - \sqrt{2x}$ $\frac{242}{49}$

45. $\sqrt{2t+3} = 3 - \sqrt{2t}$ $\frac{1}{2}$

46. $\sqrt{x} + \sqrt{x+2} = 2$ $\frac{1}{4}$

47. $\sqrt{x+5} - \sqrt{x} = 1$ 4

48. $\sqrt{x+1} = 2 - \sqrt{x}$ $\frac{9}{16}$

49. $\sqrt{x-6} + 3 = \sqrt{x+9}$ 7

50. $\sqrt{x+3} - \sqrt{x-1} = 1$ $\frac{13}{4}$

Applications

Application **EXAMPLE 8** **Electricity**

The amount of power consumed by an electrical appliance is given by $I = \sqrt{P/R}$, where I is the current measured in amps, R is the resistance measured in ohms, and P is the power measured in watts. Find the power used by an electric heater for which $I = 10$ amps and $R = 16$ ohms.

Study Tip

An alternative way to solve the problem in Example 8 would be first to solve the equation for P.

$$I = \sqrt{\frac{P}{R}}$$

$$I^2 = \left(\sqrt{\frac{P}{R}}\right)^2$$

$$I^2 = \frac{P}{R}$$

$$I^2R = P$$

At this stage, you can substitute the known values of I and R to obtain

$$P = (10)^2 16 = 1600.$$

SOLUTION

$$10 = \sqrt{\frac{P}{16}}$$ Substitute 10 for I and 16 for R in original equation.

$$10^2 = \left(\sqrt{\frac{P}{16}}\right)^2$$ Square each side.

$$100 = \frac{P}{16} \quad \Longrightarrow \quad 1600 = P$$ Simplify and multiply each side by 16.

So, the solution is $P = 1600$ watts. Check this in the original equation.

EXAMPLE 9 **Velocity of a Falling Object**

The velocity of a free-falling object can be determined from the equation $v = \sqrt{2gh}$, where v is the velocity measured in feet per second, $g = 32$ feet per second per second, and h is the distance (in feet) the object has fallen. Find the height from which a rock was dropped when it strikes the ground with a velocity of 50 feet per second.

SOLUTION

$$v = \sqrt{2gh}$$ Write original equation.

$$50 = \sqrt{2(32)h}$$ Substitute 50 for v and 32 for g.

$$50^2 = (\sqrt{64h})^2$$ Square each side.

$$2500 = 64h$$ Simplify.

$$39 \approx h$$ Divide each side by 64.

So, the rock was dropped from a height of about 39 feet. Check this in the original equation.

Exercises Within Reach®

Solutions in English & Spanish and tutorial videos at AlgebraWithinReach.com

Height **In Exercises 51 and 52, use the formula $t = \sqrt{d/16}$, which gives the time t (in seconds) for a free-falling object to fall d feet.**

51. A construction worker drops a nail from a building and observes it strike a water puddle after approximately 2 seconds. Estimate the height from which the nail was dropped. 64 feet

52. A farmer drops a stone down a well and hears it strike the water after approximately 4.5 seconds. Estimate the depth of the well. 324 feet

Free-Falling Object **In Exercises 53 and 54, use the equation for the velocity of a free-falling object, $v = \sqrt{2gh}$, as described in Example 9.**

53. A cliff diver strikes the water with a velocity of $32\sqrt{5}$ feet per second. Find the height from which the diver dove. 80 feet

54. A stone strikes the water with a velocity of 130 feet per second. Estimate to two decimal places the height from which the stone was dropped. 264.06 feet

Application EXAMPLE 10 Geometry: The Pythagorean Theorem

The distance between a house on shore and a playground on shore is 40 meters. The distance between the playground and a house on an island is 50 meters, as shown at the left. What is the distance between the two houses?

SOLUTION

You can see that the distances form a right triangle. So, you can use the Pythagorean Theorem to find the distance between the two houses.

$c = \sqrt{a^2 + b^2}$	Pythagorean Theorem
$50 = \sqrt{40^2 + b^2}$	Substitute 40 for a and 50 for c.
$50 = \sqrt{1600 + b^2}$	Simplify.
$50^2 = \left(\sqrt{1600 + b^2}\right)^2$	Square each side.
$2500 = 1600 + b^2$	Simplify.
$0 = b^2 - 900$	Write in general form.
$0 = (b + 30)(b - 30)$	Factor.
$b + 30 = 0 \Rightarrow b = -30$	Set 1st factor equal to 0.
$b + 30 = 0 \Rightarrow b = 30$	Set 2nd factor equal to 0.

Choose the positive solution to obtain a distance of 30 meters. Check this solution in the original equation.

Exercises Within Reach®

Solutions in English & Spanish and tutorial videos at AlgebraWithinReach.com

Geometry **In Exercises 55–58, find the length x of the unknown side of the right triangle. (Round your answer to two decimal places, if necessary.)**

55.

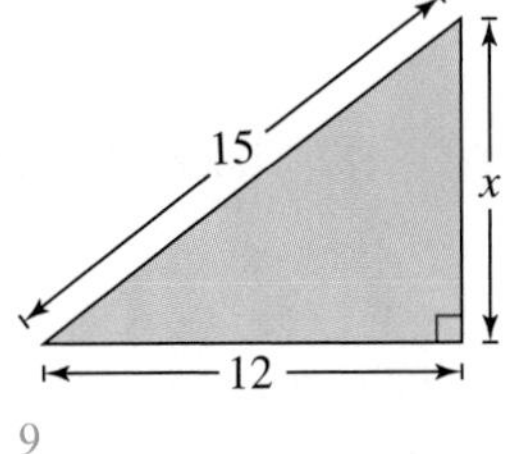

9

56.

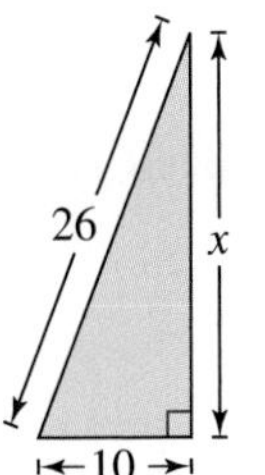

24

57.

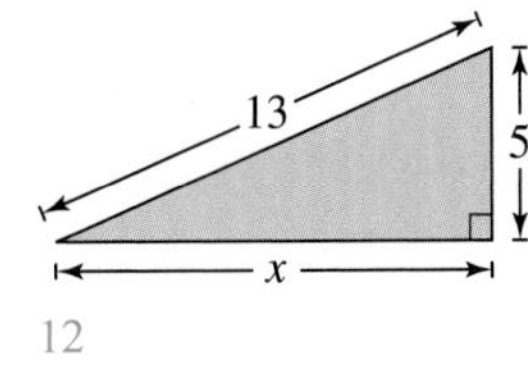

12

58.

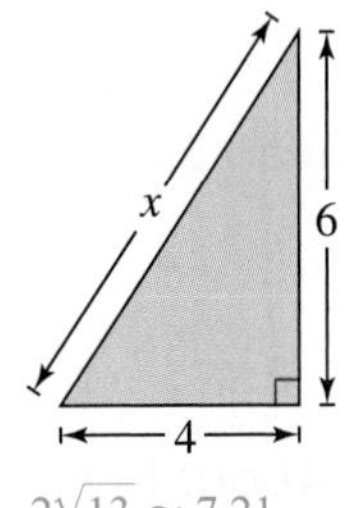

$2\sqrt{13} \approx 7.21$

59. ***Length of a Ramp*** A ramp is 20 feet long and rests on a porch that is 4 feet high (see figure). Find the distance x between the porch and the base of the ramp.

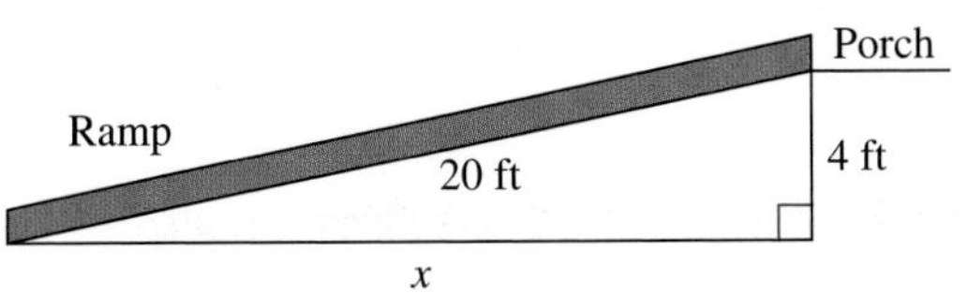

$8\sqrt{6} \approx 19.6$ feet

60. ***Ladder*** A ladder is 17 feet long, and the bottom of the ladder is 8 feet from the side of a house. How far does the ladder reach up the side of the house? 15 feet

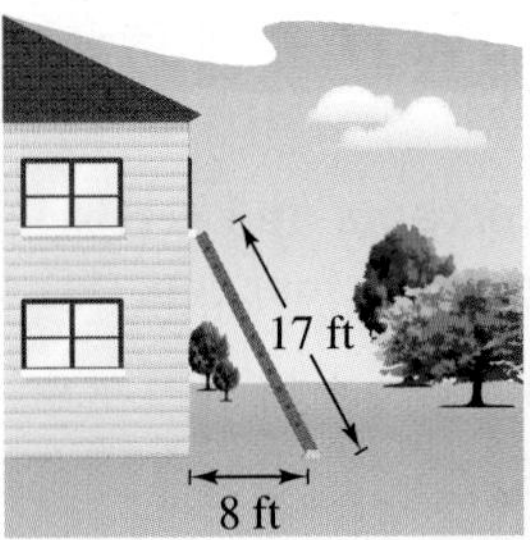

Concept Summary: Solving Radical Equations

What

Solving radical equations is somewhat like solving equations that contain fractions.

EXAMPLE

Solve $\sqrt{2x} + 1 = 5$.

How

First, try to get rid of the radicals and obtain a linear or polynomial equation. Then, solve the equation using the standard procedures.

EXAMPLE

$\sqrt{2x} + 1 = 5$ Original equation

$\sqrt{2x} = 4$ Isolate radical.

$(\sqrt{2x})^2 = (4)^2$ Square each side.

$2x = 16$ Simplify.

$x = 8$ Divide by 2.

Why

Knowing how to solve radical equations will help as you work through and solve problems involving the Pythagorean Theorem.

Exercises Within Reach®

Worked-out solutions to odd-numbered exercises at AlgebraWithinReach.com

Concept Summary Check

61. ***Solving an Equation*** In the example above, why is each side of the equation squared?
Each side is squared to eliminate the radical.

62. ***Solving an Equation*** Explain how to solve the equation $\sqrt[3]{2x} + 1 = 5$.
Isolate the radical, cube each side of the equation, and solve the resulting linear equation.

63. ***Checking a Solution*** One reason to check a solution in the original equation is the discover errors made in solving the equation. Describe another reason.
To determine whether the solution is extraneous.

64. ***Checking the Solution Graphically*** Explain how to check the solution in the example above graphically.
Graph $f(x) = \sqrt{2x} + 1$ and $g(x) = 5$, and determine whether they intersect at $x = 8$.

Extra Practice

Solving an Equation **In Exercises 65–68, solve the equation and check your solution(s).**

65. $3y^{1/3} = 18$ 216

66. $2x^{3/4} = 54$ 81

67. $(x + 4)^{2/3} = 4$ 4, −12

68. $(u - 2)^{4/3} = 81$ −25, 29

Think About It **In Exercises 69–72, use the given function to find the indicated value(s) of x.**

69. For $f(x) = \sqrt{x} - \sqrt{x - 9}$, find x such that $f(x) = 1$. 25

70. For $g(x) = \sqrt{x} + \sqrt{x - 5}$, find x such that $g(x) = 5$. 9

71. For $h(x) = \sqrt{x - 2} - \sqrt{4x + 1}$, find x such that $h(x) = -3$. 2, 6

72. For $f(x) = \sqrt{2x + 7} - \sqrt{x + 15}$, find x such that $f(x) = -1$. 1

Finding x-Intercepts **In Exercises 73–76, find the x-intercept(s) of the graph of the function without graphing the function.**

73. $f(x) = \sqrt{x + 5} - 3 + \sqrt{x}$ $\frac{4}{9}$

74. $f(x) = \sqrt{6x + 7} - 2 - \sqrt{2x + 3}$ 3

75. $f(x) = \sqrt{3x - 2} - 1 - \sqrt{2x - 3}$ 2, 6

76. $f(x) = \sqrt{5x + 6} - 1 - \sqrt{3x + 3}$ −1, 2

77. ***Airline Passengers*** An airline offers daily flights between Chicago and Denver. The total monthly cost C (in millions of dollars) of these flights is given by

$C = \sqrt{0.2x + 1}, x \geq 0$

where x is measured in thousands of passengers (see figure). The total cost of the flights for June is \$2.5 million. Approximately how many passengers flew in June? 26,250 passengers

Airline Costs

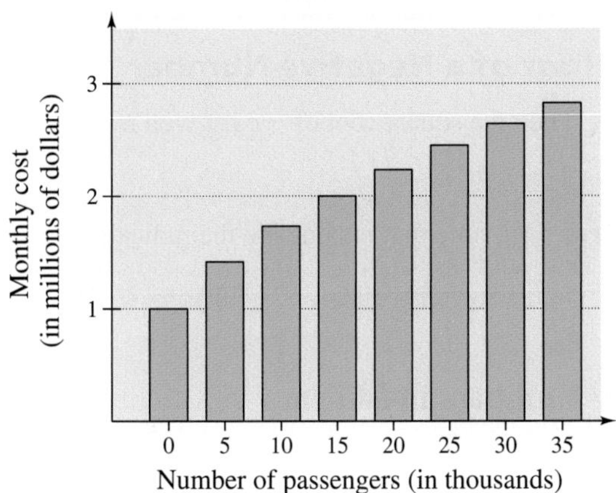

78. ***Killer Whales*** The weight w (in pounds) of a killer whale can be modeled by

$w = 280 + 325\sqrt{t}, \quad 0 \leq t \leq 144$

where t represents the age (in months) of the killer whale. At what age did the killer whale weigh about 3400 pounds? 92 months

79. ***Geometry*** The lateral surface area of a cone (see figure) is given by $S = \pi r\sqrt{r^2 + h^2}$. Solve the equation for h. Then find the height of a cone with a lateral surface area of $364\pi\sqrt{2}$ square centimeters and a radius of 14 centimeters. $h = \dfrac{\sqrt{S^2 - \pi^2 r^4}}{\pi r}$; 34 centimeters

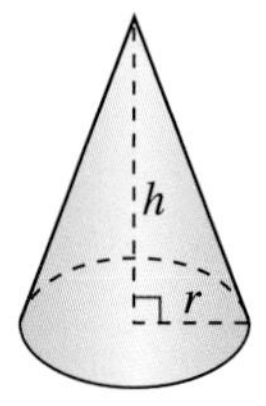

80. ***Geometry*** Determine the length and width of a rectangle with a perimeter of 92 inches and a diagonal of 34 inches. 30 inches × 16 inches

Explaining Concepts

81. ***Error Analysis*** Describe the error.

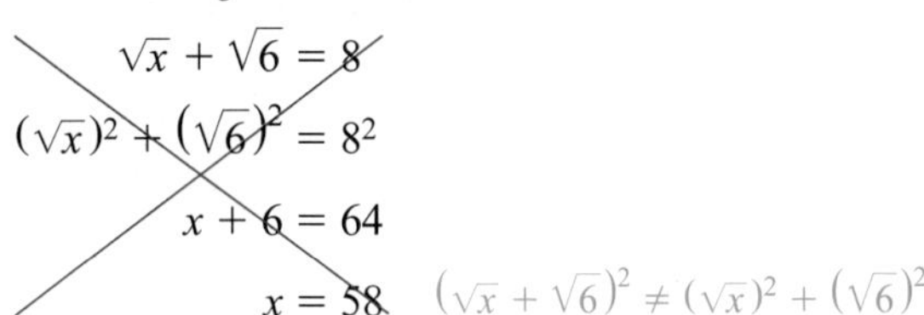

$(\sqrt{x} + \sqrt{6})^2 \neq (\sqrt{x})^2 + (\sqrt{6})^2$

82. ***Precision*** Does raising each side of an equation to the nth power always yield an equivalent equation? Explain.
No. Extraneous solutions may be introduced when each side of the equation is raised to the nth power.

83. ***Think About It*** Explain how to find the values of a and b for which the equation $x + \sqrt{x - a} = b$ has a solution of $x = 20$. (There are many correct values for a and b).
Substitute $x = 20$ into the equation, and then choose any value of a such that $a \leq 20$ and solve the resulting equation for b.

84. ***Number Sense*** Explain how you can tell that $\sqrt{x - 9} = -4$ has no solution without solving the equation.
Because the radicand of a rational expression is always nonnegative, the square root of the radicand must also be nonnegative.

Cumulative Review

In Exercises 85–88, determine whether the two lines are parallel, perpendicular, or neither.

85. L_1: $y = 4x + 2$
L_2: $y = 4x - 1$
Parallel

86. L_1: $y = 3x - 8$
L_2: $y = -3x - 8$
Neither

87. L_1: $y = -x + 5$
L_2: $y = x - 3$
Perpendicular

88. L_1: $y = 2x$
L_2: $y = \frac{1}{2}x + 4$
Neither

In Exercises 89 and 90, use matrices to solve the system of linear equations.

89. $\begin{cases} 4x - y = 10 \\ -7x - 2y = -25 \end{cases}$
(3, 2)

90. $\begin{cases} 3x - 2y = 5 \\ 6x - 5y = 14 \end{cases}$
(−1, −4)

In Exercises 91–94, simplify the expression.

91. $a^{3/5} \cdot a^{1/5}$ $a^{4/5}$

92. $\dfrac{m^2}{m^{2/3}}$ $m^{4/3}, m \neq 0$

93. $\left(\dfrac{x^{1/2}}{x^{1/8}}\right)^4$ $x^{3/2}, x \neq 0$

94. $\dfrac{(a + b)^{3/4}}{\sqrt[4]{a + b}}$ $(a + b)^{1/2}, a \neq -b$

7.6 Complex Numbers

- Perform operations on numbers in *i*-form.
- Add, subtract, and multiply complex numbers.
- Use complex conjugates to find the quotient of two complex numbers.

The Imaginary Unit *i*

The Square Root of a Negative Number

Let c be a positive real number. Then the square root of $-c$ is given by

$$\sqrt{-c} = \sqrt{c(-1)} = \sqrt{c}\sqrt{-1} = \sqrt{c}i$$

When writing $\sqrt{-c}$ in the ***i*-form**, $\sqrt{c}i$, note that i is outside the radical.

EXAMPLE 1 Writing Numbers in *i*-Form

Write each number in *i*-form.

a. $\sqrt{-36}$ **b.** $\sqrt{-\frac{16}{25}}$ **c.** $\sqrt{-54}$ **d.** $\frac{\sqrt{-48}}{\sqrt{-3}}$

SOLUTION

a. $\sqrt{-36} = \sqrt{36(-1)} = \sqrt{36}\sqrt{-1} = 6i$

b. $\sqrt{-\frac{16}{25}} = \sqrt{\frac{16}{25}(-1)} = \sqrt{\frac{16}{25}}\sqrt{-1} = \frac{4}{5}i$

c. $\sqrt{-54} = \sqrt{54(-1)} = \sqrt{54}\sqrt{-1} = 3\sqrt{6}i$

d. $\frac{\sqrt{-48}}{\sqrt{-3}} = \frac{\sqrt{48}\sqrt{-1}}{\sqrt{3}\sqrt{-1}} = \frac{\sqrt{48}i}{\sqrt{3}i} = \sqrt{\frac{48}{3}} = \sqrt{16} = 4$

Exercises Within Reach®

Solutions in English & Spanish and tutorial videos at AlgebraWithinReach.com

Writing a Number in i-Form **In Exercises 1–18, write the number in *i*-form.**

1. $\sqrt{-4}$ $2i$
2. $\sqrt{-9}$ $3i$
3. $-\sqrt{-144}$ $-12i$
4. $-\sqrt{-49}$ $-7i$
5. $\sqrt{-\frac{4}{25}}$ $\frac{2}{5}i$
6. $\sqrt{-\frac{9}{64}}$ $\frac{3}{8}i$
7. $-\sqrt{-\frac{36}{121}}$ $-\frac{6}{11}i$
8. $-\sqrt{-\frac{9}{25}}$ $-\frac{3}{5}i$
9. $\sqrt{-8}$ $2\sqrt{2}i$
10. $\sqrt{-75}$ $5\sqrt{3}i$
11. $\sqrt{-7}$ $\sqrt{7}i$
12. $\sqrt{-15}$ $\sqrt{15}i$
13. $\frac{\sqrt{-12}}{\sqrt{-3}}$ 2
14. $\frac{\sqrt{-45}}{\sqrt{-5}}$ 3
15. $\sqrt{-\frac{18}{25}}$ $\frac{3\sqrt{2}}{5}i$
16. $\sqrt{-\frac{20}{49}}$ $\frac{2\sqrt{5}}{7}i$
17. $\sqrt{-0.09}$ $0.3i$
18. $\sqrt{-0.0004}$ $0.02i$

To perform operations with square roots of negative numbers, you must *first* write the numbers in i-form. You can then add, subtract, and multiply as follows.

$ai + bi = (a + b)i$ Addition

$ai - bi = (a - b)i$ Subtraction

$(ai)(bi) = ab(i^2) = ab(-1) = -ab$ Multiplication

Study Tip

When performing operations with numbers in i-form, you sometimes need to be able to evaluate powers of the imaginary unit i. The first several powers of i are as follows.

$i^1 = i$
$i^2 = -1$
$i^3 = i(i^2) = i(-1) = -i$
$i^4 = (i^2)(i^2) = (-1)(-1) = 1$
$i^5 = i(i^4) = i(1) = i$
$i^6 = (i^2)(i^4) = (-1)(1) = -1$
$i^7 = (i^3)(i^4) = (-i)(1) = -i$
$i^8 = (i^4)(i^4) = (1)(1) = 1$

Note how the pattern of values i, -1, $-i$, and 1 repeats itself for powers greater than 4.

EXAMPLE 2 Operations with Square Roots of Negative Numbers

Perform each operation.

a. $\sqrt{-9} + \sqrt{-49}$ **b.** $\sqrt{-32} - 2\sqrt{-2}$

SOLUTION

a. $\sqrt{-9} + \sqrt{-49} = \sqrt{9}\sqrt{-1} + \sqrt{49}\sqrt{-1}$ Product Rule for Radicals

$= 3i + 7i$ Write in i-form.

$= 10i$ Simplify.

b. $\sqrt{-32} - 2\sqrt{-2} = \sqrt{32}\sqrt{-1} - 2\sqrt{2}\sqrt{-1}$ Product Rule for Radicals

$= 4\sqrt{2}i - 2\sqrt{2}i$ Write in i-form.

$= 2\sqrt{2}i$ Simplify.

EXAMPLE 3 Multiplying Square Roots of Negative Numbers

Find each product.

a. $\sqrt{-15}\sqrt{-15}$ **b.** $\sqrt{-5}(\sqrt{-45} - \sqrt{-4})$

SOLUTION

a. $\sqrt{-15}\sqrt{-15} = (\sqrt{15}i)(\sqrt{15}i)$ Write in i-form.

$= (\sqrt{15})^2 i^2$ Multiply.

$= 15(-1)$ $i^2 = -1$

$= -15$ Simplify.

b. $\sqrt{-5}(\sqrt{-45} - \sqrt{-4}) = \sqrt{5}i(3\sqrt{5}i - 2i)$ Write in i-form.

$= (\sqrt{5}i)(3\sqrt{5}i) - (\sqrt{5}i)(2i)$ Distributive Property

$= 3(5)(-1) - 2\sqrt{5}(-1)$ Multiply.

$= -15 + 2\sqrt{5}$ Simplify.

Exercises Within Reach®

Solutions in English & Spanish and tutorial videos at AlgebraWithinReach.com

Operations with Square Roots of Negative Numbers **In Exercises 19–34, perform the operation(s).** 27. -0.44 29. $-3 - 2\sqrt{3}$ 31. $4 + 3\sqrt{2}i$ 32. $-12 + 3i$

19. $\sqrt{-16} + \sqrt{-36}$ $10i$ **20.** $\sqrt{-25} - \sqrt{-9}$ $2i$ **21.** $\sqrt{-50} - \sqrt{-8}$ $3\sqrt{2}i$ **22.** $\sqrt{-500} + \sqrt{-45}$ $13\sqrt{5}i$

23. $\sqrt{-12}\sqrt{-2}$ $-2\sqrt{6}$ **24.** $\sqrt{-25}\sqrt{-6}$ $-5\sqrt{6}$ **25.** $\sqrt{-18}\sqrt{-3}$ $-3\sqrt{6}$ **26.** $\sqrt{-7}\sqrt{-7}$ -7

27. $\sqrt{-0.16}\sqrt{-1.21}$ **28.** $\sqrt{-0.49}\sqrt{-1.44}$ -0.84 **29.** $\sqrt{-3}(\sqrt{-3} + \sqrt{-4})$ **30.** $\sqrt{-12}(\sqrt{-3} - \sqrt{-12})$ 6

31. $\sqrt{-2}(3 - \sqrt{-8})$ **32.** $\sqrt{-9}(1 + \sqrt{-16})$ **33.** $(\sqrt{-16})^2$ -16 **34.** $(\sqrt{-2})^2$ -2

Complex Numbers

A number of the form $a + bi$, where a and b are real numbers, is called a **complex number**. The real number a is called the **real part** of the complex number $a + bi$, and the number bi is called the **imaginary part**.

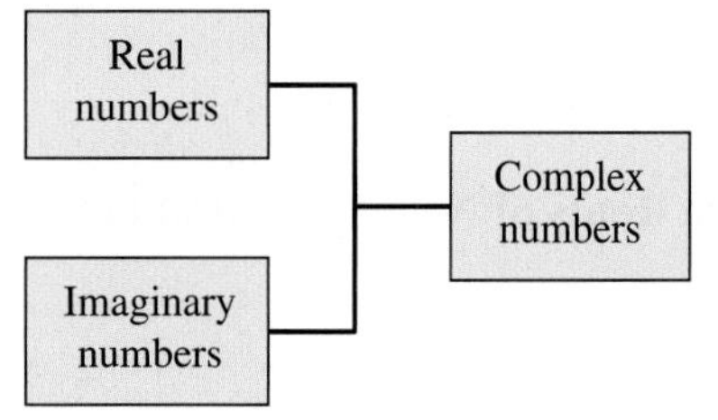

The real numbers and the imaginary numbers make up the complex numbers.

Definition of a Complex Number

If a and b are real numbers, the number $a + bi$ is a **complex number**, and it is said to be written in **standard form**. If $b = 0$, the number $a + bi = a$ is a real number. If $b \neq 0$, the number $a + bi$ is called an **imaginary number**. A number of the form bi, where $b \neq 0$, is called a **pure imaginary number**.

EXAMPLE 4 Equality of Two Complex Numbers

To determine whether the complex numbers $\sqrt{9} + \sqrt{-48}$ and $3 - 4\sqrt{3}i$ are equal, begin by writing the first number in standard form.

$$\sqrt{9} + \sqrt{-48} = \sqrt{3^2} + \sqrt{4^2(3)(-1)} = 3 + 4\sqrt{3}i$$

The two numbers are not equal because their imaginary parts differ in sign.

EXAMPLE 5 Equality of Two Complex Numbers

To find the values of x and y that satisfy the equation $3x - \sqrt{-25} = -6 + 3yi$, begin by writing the left side of the equation in standard form.

$$3x - 5i = -6 + 3yi$$ Each side is in standard form.

For these two numbers to be equal, their real parts must be equal to each other and their imaginary parts must be equal to each other.

Real Parts	*Imaginary Parts*
$3x = -6$	$3yi = -5i$
$x = -2$	$3y = -5$
	$y = -\frac{5}{3}$

So, $x = -2$ and $y = -\frac{5}{3}$.

Exercises Within Reach®

Solutions in English & Spanish and tutorial videos at AlgebraWithinReach.com

Equality of Two Complex Numbers **In Exercises 35–38, determine whether the complex numbers are equal.**

35. $\sqrt{1} + \sqrt{-25}$ and $1 + 5i$ Equal

36. $\sqrt{16} + \sqrt{-9}$ and $4 - 3i$ Not equal

37. $\sqrt{27} - \sqrt{-8}$ and $3\sqrt{3} + 2\sqrt{2}i$ Not equal

38. $\sqrt{18} - \sqrt{-12}$ and $3\sqrt{2} - 2\sqrt{3}i$ Equal

Equality of Two Complex Numbers **In Exercises 39–46, determine the values of a and b that satisfy the equation.**

39. $3 - 4i = a + bi$ $a = 3, b = -4$

40. $-8 + 6i = a + bi$ $a = -8, b = 6$

41. $5 - 4i = (a + 3) + (b - 1)i$ $a = 2, b = -3$

42. $-10 + 12i = 2a + (5b - 3)i$ $a = -5, b = 3$

43. $-4 - \sqrt{-8} = a + bi$ $a = -4, b = -2\sqrt{2}$

44. $\sqrt{-36} - 3 = a + bi$ $a = -3, b = 6$

45. $\sqrt{a} + \sqrt{-49} = 8 + bi$ $a = 64, b = 7$

46. $\sqrt{100} + \sqrt{b} = a + 2\sqrt{3}i$ $a = 10, b = -12$

To add or subtract two complex numbers, you add (or subtract) the real and imaginary parts separately. This is similar to combining like terms of a polynomial.

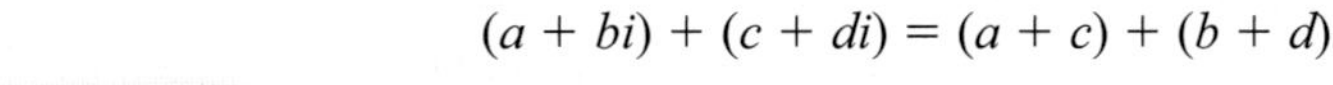

$(a + bi) + (c + di) = (a + c) + (b + d)i$ Addition of complex numbers

$(a + bi) - (c + di) = (a - c) + (b - d)i$ Subtraction of complex numbers

Study Tip

Note in part (b) of Example 6 that the sum of two complex numbers can be a real number.

EXAMPLE 6 Adding and Subtracting Complex Numbers

a. $(3 - i) + (-2 + 4i) = (3 - 2) + (-1 + 4)i = 1 + 3i$

b. $3i + (5 - 3i) = 5 + (3 - 3)i = 5$

c. $4 - (-1 + 5i) + (7 + 2i) = [4 - (-1) + 7] + (-5 + 2)i = 12 - 3i$

d. $(6 + 3i) + (2 - \sqrt{-8}) - \sqrt{-4} = (6 + 3i) + (2 - 2\sqrt{2}i) - 2i$

$= (6 + 2) + (3 - 2\sqrt{2} - 2)i$

$= 8 + (1 - 2\sqrt{2})i$

EXAMPLE 7 Multiplying Complex Numbers

a. $(7i)(-3i) = -21i^2$ Multiply.

$= -21(-1) = 21$ $i^2 = -1$

b. $(1 - i)(\sqrt{-9}) = (1 - i)(3i)$ Write in i-form.

$= 3i - 3i^2$ Distributive Property

$= 3i - 3(-1) = 3 + 3i$ $i^2 = -1$

c. $(2 - i)(4 + 3i) = 8 + 6i - 4i - 3i^2$ FOIL Method

$= 8 + 6i - 4i - 3(-1)$ $i^2 = -1$

$= 11 + 2i$ Combine like terms.

d. $(3 + 2i)(3 - 2i) = 3^2 - (2i)^2$ Special product formula

$= 9 - 4i^2$ Simplify.

$= 9 - 4(-1) = 13$ $i^2 = -1$

Additional Examples

Perform each operation and write the result in standard form.

a. $(7 + 4i) - (3 + \sqrt{-12})$

b. $(2 + 5i)^2$

Answers:

a. $4 + 2(2 - \sqrt{3})i$

b. $-21 + 20i$

Exercises Within Reach®

Solutions in English & Spanish and tutorial videos at AlgebraWithinReach.com

Adding and Subtracting Complex Numbers **In Exercises 47–54, perform the operation(s) and write the result in standard form.**

47. $(-4 - 7i) + (-10 - 33i)$ $-14 - 40i$

48. $(15 + 10i) - (2 + 10i)$ 13

49. $13i - (14 - 7i)$ $-14 + 20i$

50. $17i + (9 - 14i)$ $9 + 3i$

51. $6 - (3 - 4i) + 2i$ $3 + 6i$

52. $22 + (-5 + 8i) + 10i$ $17 + 18i$

53. $15i - (3 - 25i) + \sqrt{-81}$ $-3 + 49i$

54. $(-1 + i) - \sqrt{2} - \sqrt{-2}$ $(-1 - \sqrt{2}) + (1 - \sqrt{2})i$

Multiplying Complex Numbers **In Exercises 55–62, perform the operation and write the result in standard form.**

55. $(3i)(-8i)$ 24

56. $(-5i)(4i)$ 20

57. $(9 - 2i)(\sqrt{-4})$ $4 + 18i$

58. $(11 + 3i)(\sqrt{-25})$ 58. $-15 + 55i$

59. $(4 + 3i)(-7 + 4i)$ $-40 - 5i$

60. $(3 + 5i)(2 - 15i)$ $81 - 35i$

61. $(6 + 3i)(6 - 3i)$ 45

62. $(5 - 2i)(5 + 2i)$ 29

Complex Conjugates

Complex numbers of the form $a + bi$ and $a - bi$ are called **complex conjugates**. In general, the product of complex conjugates has the following form.

$$(a + bi)(a - bi) = a^2 - (bi)^2 = a^2 - b^2i^2 = a^2 - b^2(-1) = a^2 + b^2$$

EXAMPLE 8 Complex Conjugates

Complex Number	*Complex Conjugate*	*Product*
$4 - 5i$	$4 + 5i$	$4^2 + 5^2 = 41$
$3 + 2i$	$3 - 2i$	$3^2 + 2^2 = 13$
$-2 = -2 + 0i$	$-2 = -2 - 0i$	$(-2)^2 + 0^2 = 4$
$i = 0 + i$	$-i = 0 - i$	$0^2 + 1^2 = 1$

EXAMPLE 9 Writing Quotients in Standard Form

a. $\frac{2 - i}{4i} = \frac{2 - i}{4i} \cdot \frac{(-4i)}{(-4i)}$ Multiply numerator and denominator by complex conjugate of denominator.

$= \frac{-8i + 4i^2}{-16i^2}$ Multiply fractions.

$= \frac{-8i + 4(-1)}{-16(-1)}$ $i^2 = -1$

$= \frac{-8i - 4}{16}$ Simplify.

$= -\frac{1}{4} - \frac{1}{2}i$ Write in standard form.

b. $\frac{5}{3 - 2i} = \frac{5}{3 - 2i} \cdot \frac{3 + 2i}{3 + 2i}$ Multiply numerator and denominator by complex conjugate of denominator.

$= \frac{5(3 + 2i)}{(3 - 2i)(3 + 2i)}$ Multiply fractions.

$= \frac{5(3 + 2i)}{3^2 + 2^2}$ Product of complex conjugates

$= \frac{15 + 10i}{13}$ Simplify.

$= \frac{15}{13} + \frac{10}{13}i$ Write in standard form.

Exercises Within Reach®

Solutions in English & Spanish and tutorial videos at AlgebraWithinReach.com

Complex Conjugates **In Exercises 63–70, multiply the number by its complex conjugate and simplify.**

63. $-2 - 8i$ 68 **64.** $10 - 3i$ 109 **65.** $2 + i$ 5 **66.** $3 + 2i$ 13

67. $10i$ 100 **68.** 20 400 **69.** -12 144 **70.** $-12i$ 144

Writing a Quotient in Standard Form **In Exercises 71–74, write the quotient in standard form.**

71. $\frac{2 + i}{-5i}$ $-\frac{1}{5} + \frac{2}{5}i$ **72.** $\frac{1 + i}{3i}$ $\frac{1}{3} - \frac{1}{3}i$ **73.** $\frac{-12}{2 + 7i}$ $-\frac{24}{53} + \frac{84}{53}i$ **74.** $\frac{15}{2(1 - i)}$ $\frac{15}{4} + \frac{15}{4}i$

EXAMPLE 10 Writing a Quotient in Standard Form

Write $\frac{8-i}{8+i}$ in standard form.

SOLUTION

$$\frac{8-i}{8+i} = \frac{8-i}{8+i} \cdot \frac{8-i}{8-i}$$ Multiply numerator and denominator by complex conjugate of denominator.

$$= \frac{(8-i)(8-i)}{(8+i)(8-i)}$$ Multiply fractions.

$$= \frac{64-16i+i^2}{8^2+1^2}$$ FOIL Method and product of complex conjugates

$$= \frac{64-16i+(-1)}{8^2+1^2}$$ $i^2 = -1$

$$= \frac{63-16i}{65}$$ Simplify.

$$= \frac{63}{65} - \frac{16}{65}i$$ Write in standard form.

EXAMPLE 11 Writing a Quotient in Standard Form

Write $\frac{2+3i}{4-2i}$ in standard form.

SOLUTION

$$\frac{2+3i}{4-2i} = \frac{2+3i}{4-2i} \cdot \frac{4+2i}{4+2i}$$ Multiply numerator and denominator by complex conjugate of denominator.

$$= \frac{(2+3i)(4+2i)}{(4-2i)(4+2i)}$$ Multiply fractions.

$$= \frac{8+16i+6i^2}{4^2+2^2}$$ FOIL Method and product of complex conjugates

$$= \frac{8+16i+6(-1)}{4^2+2^2}$$ $i^2 = -1$

Some students incorrectly rewrite $(2+16i)/20$ as $(1+16i)/10$ or as $(2+4i)/5$.

$$= \frac{2+16i}{20}$$ Simplify.

$$= \frac{1}{10} + \frac{4}{5}i$$ Write in standard form.

Exercises Within Reach®

Solutions in English & Spanish and tutorial videos at AlgebraWithinReach.com

Writing a Quotient in Standard Form **In Exercises 75–80, write the quotient in standard form.**

75. $\frac{5-i}{5+i}$ $\frac{12}{13} - \frac{5}{13}i$

76. $\frac{9+i}{9-i}$ $\frac{40}{41} + \frac{9}{41}i$

77. $\frac{4-i}{3+i}$ $\frac{11}{10} - \frac{7}{10}i$

78. $\frac{2+i}{5-i}$ $\frac{9}{26} + \frac{7}{26}i$

79. $\frac{4+5i}{3-7i}$ $-\frac{23}{58} + \frac{43}{58}i$

80. $\frac{5+3i}{7-4i}$ $\frac{23}{65} + \frac{41}{65}i$

Concept Summary: Complex Numbers

What

The **imaginary unit** i is used to define the square root of a negative number.

- $i = \sqrt{-1}$
- $\sqrt{-c} = \sqrt{c}i$, where $c > 0$.

A number of the form $a + bi$, where a and b are real numbers, is called a **complex number**.

Performing operations with complex numbers is similar to performing operations with polynomials.

EXAMPLE

Perform each operation.

a. $(2 - i) + (4 + 2i)$

b. $(1 + i)(3 - 2i)$

c. $\frac{3}{i}$

How

To perform operations with complex numbers:

- Combine like terms to add and subtract.
- Use the Commutative, Associative, and Distributive Properties, along with the FOIL Method to multiply.
- Use **complex conjugates** to simplify quotients.

EXAMPLE

a. $(2 - i) + (4 + 2i)$

$= (2 + 4) + (-1 + 2)i = 6 + i$

b. $(1 + i)(3 - 2i)$

$= 3 - 2i + 3i - 2i^2$

$= 3 - 2i + 3i - 2(-1) = 5 + i$

c. $\frac{3}{i} = \frac{3}{i} \cdot \frac{-i}{-i} = \frac{-3i}{-(-1)} = -3i$

Why

The solutions of some quadratic equations are complex numbers. For instance, when a quadratic equation has no real solution, it has exactly two complex solutions. Knowing how to perform operations with complex numbers will help you identify such solutions.

Exercises Within Reach ®

Worked-out solutions to odd-numbered exercises at AlgebraWithinReach.com

Concept Summary Check

81. ***Writing a Number in i-Form*** Write $\sqrt{-2}$ in i-form. $\sqrt{2}i$

82. ***Identifying Like Terms*** Identify any like terms in the expression $(2 - i) + (4 + 2i)$. 2 and 4, $-i$ and $2i$

83. ***Multiplying Complex Numbers*** What method is used to multiply $(1 + i)$ and $(3 - 2i)$ in the solution above? The FOIL Method

84. ***Complex Conjugate*** What is the complex conjugate of i? What is the product of i and its complex conjugate? $-i$; 1

Extra Practice

Operations with Complex Numbers **In Exercises 85–96, perform the operation(s) and write the result in standard form.**

85. $\sqrt{-48} + \sqrt{-12} - \sqrt{-27}$ $3\sqrt{3}i$

86. $\sqrt{-32} - \sqrt{-18} + \sqrt{-50}$ $6\sqrt{2}i$

87. $(-5i)(-i)(\sqrt{-49})$ $-35i$

88. $(10i)(\sqrt{-36})(-5i)$ $300i$

89. $(-3i)^3$ $27i$

90. $(2i)^4$ 16

91. $(-2 + \sqrt{-5})(-2 - \sqrt{-5})$ 9

92. $(-3 - \sqrt{-12})(4 - \sqrt{-12})$ $-24 - 2\sqrt{3}i$

93. $(2 + 5i)^2$ $-21 + 20i$

94. $(8 - 3i)^2$ $55 - 48i$

95. $(3 + i)^3$ $18 + 26i$

96. $(2 - 2i)^3$ $-16 - 16i$

Evaluating Powers of i **In Exercises 97–104, simplify the expression.**

97. i^9 i

98. i^{11} $-i$

99. i^{42} -1

100. i^{24} 1

101. i^{35} $-i$

102. i^{64} 1

103. $(-i)^6$ -1

104. $(-i)^4$ 1

Operations with Complex Numbers **In Exercises 105–108, perform the operation by first writing each quotient in standard form.**

105. $\frac{5}{3+i} + \frac{1}{3-i}$ $\frac{9}{5} - \frac{2}{5}i$

106. $\frac{1}{1-2i} + \frac{4}{1+2i}$ $1 - \frac{6}{5}i$

107. $\frac{3i}{1+i} + \frac{2}{2+3i}$ $\frac{47}{26} + \frac{27}{26}i$

108. $\frac{i}{4-3i} - \frac{5}{2+i}$ $-\frac{53}{25} + \frac{29}{25}i$

Operations with Complex Conjugates **In Exercises 109–112, perform the operations.**

109. $(a + bi) + (a - bi)$ $2a$

110. $(a + bi)(a - bi)$ $a^2 + b^2$

111. $(a + bi) - (a - bi)$ $2bi$

112. $(a + bi)^2 + (a - bi)^2$ $2a^2 - 2b^2$

113. ***Cube Roots*** The principal cube root of 125, $\sqrt[3]{125}$, is 5. Evaluate the expression x^3 for each value of x.

(a) $x = \frac{-5 + 5\sqrt{3}i}{2}$ $\left(\frac{-5 + 5\sqrt{3}i}{2}\right)^3 = 125$

(b) $x = \frac{-5 - 5\sqrt{3}i}{2}$ $\left(\frac{-5 - 5\sqrt{3}i}{2}\right)^3 = 125$

114. ***Cube Roots*** The principal cube root of 27, $\sqrt[3]{27}$, is 3. Evaluate the expression x^3 for each value of x.

(a) $x = \frac{-3 + 3\sqrt{3}i}{2}$ $\left(\frac{-3 + 3\sqrt{3}i}{2}\right)^3 = 27$

(b) $\frac{-3 - 3\sqrt{3}i}{2}$ $\left(\frac{-3 - 3\sqrt{3}i}{2}\right)^3 = 27$

115. ***Pattern Recognition*** Compare the results of Exercises 113 and 114. Use the result to list the cube roots in simplest form of each number.

(a) 1 $1, \frac{-1 + \sqrt{3}i}{2}, \frac{-1 - \sqrt{3}i}{2}$

(b) 8 $2, -1 + \sqrt{3}i, -1 - \sqrt{3}i$

(c) 64 $4, -2 + 2\sqrt{3}i, -2 - 2\sqrt{3}i$

116. ***Algebraic Properties*** Consider the complex number $1 + 5i$.

(a) Find the additive inverse of the number. $-(1 + 5i) = -1 - 5i$

(b) Find the multiplicative inverse of the number. $\frac{1}{1 + 5i} = \frac{1}{26} - \frac{5}{26}i$

Explaining Concepts

117. ***Writing Rules*** Look back at Exercises 109–112. Based on your results, write a general rule for each exercise about operations on complex conjugates of the form $a + bi$ and $a - bi$. See Additional Answers.

118. ***True or False?*** Some numbers are both real and imaginary. Justify your answer. False. A number is either real or imaginary. For example, 2 is real, $2i$ is imaginary, and $2 + 2i$ is imaginary.

119. ***Error Analysis*** Describe and correct the error.

~~$\sqrt{-3}\sqrt{-3} = \sqrt{(-3)(-3)} = \sqrt{9} = 3$~~

The numbers must be written in i-form first.

$\sqrt{-3}\sqrt{-3} = (\sqrt{3}i)(\sqrt{3}i) = 3i^2 = -3$

120. ***Precision*** Explain why the Product Rule for Radicals cannot be used to produce the second expression in Exercise 119. The Product Rule for Radicals does not apply to $\sqrt{-3}\sqrt{-3}$ because $\sqrt{-3}$ is not real.

121. ***Number Sense*** The denominator of a quotient is a pure imaginary number of the form bi. How can you use the complex conjugate of bi to write the quotient in standard form? Can you use the number i instead of the conjugate of bi? Explain. See Additional Answers.

122. ***Number Sense*** The polynomial $x^2 + 1$ is prime *with respect to the integers*. It is not, however, prime *with respect to the complex numbers*. Show how $x^2 + 1$ can be factored using complex numbers. $x^2 + 1 = (x + i)(x - i)$

Cumulative Review

In Exercises 123–126, use the Zero-Factor Property to solve the equation.

123. $(x - 5)(x + 7) = 0$ $-7, 5$

124. $z(z - 2) = 0$ $0, 2$

125. $3y(y - 3)(y + 4) = 0$ $-4, 0, 3$

126. $(3x - 2)(4x + 1)(x + 9) = 0$ $-9, -\frac{1}{4}, \frac{2}{3}$

In Exercises 127–130, solve the equation and check your solution.

127. $\sqrt{x} = 9$ 81

128. $\sqrt[3]{t} = 8$ 512

129. $\sqrt{x} - 5 = 0$ 25

130. $\sqrt{2x + 3} - 7 = 0$ 23

7 Chapter Summary

	What did you learn?	Explanation and Examples	Review Exercises
7.1	Determine the nth roots of numbers and evaluate radical expressions *(p. 354).*	**1.** If a is a positive real number and n is even, then a has exactly two (real) nth roots, which are denoted by $\sqrt[n]{a}$ and $-\sqrt[n]{a}$. **2.** If a is any real number and n is odd, then a has only one (real) nth root, which is denoted by $\sqrt[n]{a}$. **3.** If a is a negative real number and n is even, then a has no (real) nth root.	1–12
	Use the rules of exponents to evaluate or simplify expressions with rational exponents *(p. 357).*	**1.** $a^{1/n} = \sqrt[n]{a}$ **2.** $a^{m/n} = (a^{1/n})^m = (\sqrt[n]{a})^m$ **3.** $a^{m/n} = (a^m)^{1/n} = \sqrt[n]{a^m}$ See page 357 for the rules of exponents as they apply to rational exponents.	13–34
	Evaluate radical functions and find the domains of radical functions *(p. 359).*	Let n be an integer that is greater than or equal to 2. **1.** If n is odd, the domain of $f(x) = \sqrt[n]{x}$ is the set of all real numbers. **2.** If n is even, the domain of $f(x) = \sqrt[n]{x}$ is the set of all nonnegative real numbers.	35–44
7.2	Use the Product and Quotient Rules for Radicals to simplify radical expressions *(p. 362).*	Let u and v be real numbers, variables, or algebraic expressions. If the nth roots of u and v are real, then the following rules are true. **1.** $\sqrt[n]{uv} = \sqrt[n]{u}\sqrt[n]{v}$ **2.** $\sqrt[n]{\frac{u}{v}} = \frac{\sqrt[n]{u}}{\sqrt[n]{v}},\ v \neq 0$	45–54
	Use rationalization techniques to simplify radical expressions *(p. 365).*	A radical expression is said to be in *simplest form* when all three of the statements below are true. **1.** All possible nth powered factors have been removed from each radical. **2.** No radical contains a fraction. **3.** No denominator of a fraction contains a radical.	55–58
	Use the Pythagorean Theorem in application problems *(p. 366).*	In a right triangle, if a and b are the lengths of the legs and c is the length of the hypotenuse, then $c = \sqrt{a^2 + b^2}$ and $a = \sqrt{c^2 - b^2}$. 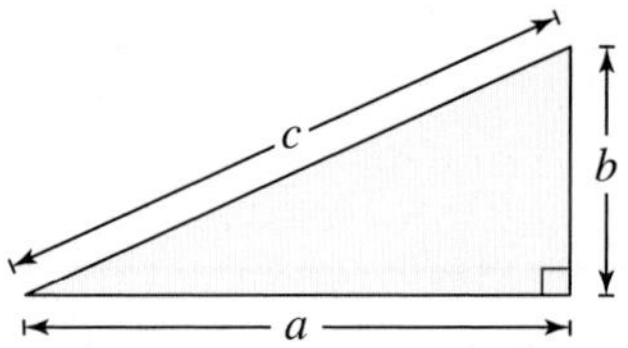 	59–62
7.3	Use the Distributive Property to add and subtract like radicals *(p. 370).*	**1.** Write each radical in simplest form. **2.** Combine like radicals: Use the Distributive Property to factor out the like radical, and then combine the terms inside the parentheses. $\sqrt{4x} + \sqrt{9x}$ $= 2\sqrt{x} + 3\sqrt{x}$ $= (2 + 3)\sqrt{x}$ $= 5\sqrt{x}$	63–70

	What did you learn?	Explanation and Examples	Review Exercises
7.3	Use radical expressions in application problems *(p. 374).*	Radical expressions can be used to model real-life quantities, such as the average cost of tuition (See Example 6).	71, 72
7.4	Use the Distributive Property or the FOIL Method to multiply radical expressions *(p. 380).*	**Distributive Property:** $\sqrt{2}(2+\sqrt{2}) = 2\sqrt{2} + (\sqrt{2})^2 = 2\sqrt{2} + 2$ **FOIL Method:** $(\sqrt{2}+1)(\sqrt{2}-3) = \overbrace{(\sqrt{2})^2}^{F} - \overbrace{3\sqrt{2}}^{O} + \overbrace{\sqrt{2}}^{I} - \overbrace{3}^{L} = 2 - 2\sqrt{2} - 3 = -1 - 2\sqrt{2}$	73–78
	Determine the products of conjugates *(p. 382).*	The product of two conjugates $(a+b)$ and $(a-b)$ is the difference of two squares. $(a+b)(a-b) = a^2 - b^2$	79–82
	Simplify quotients involving radicals by rationalizing the denominators *(p. 383).*	To simplify a quotient involving radicals, rationalize the denominator by multiplying both the numerator and denominator by the conjugate of the denominator.	83–86
7.5	Solve a radical equation by raising each side to the nth power *(p. 388).*	Let u and v be real numbers, variables, or algebraic expressions, and let n be a positive integer. If $u = v$, then it follows that $u^n = v^n$. $\sqrt{x} = 9$ Let $u = \sqrt{x}$, $v = 9$. $(\sqrt{x})^2 = 9^2$ $u^2 = v^2$ $x = 81$ Simplify.	87–96
	Solve application problems involving radical equations *(p. 392).*	Real-life problems involving electricity, the velocity of a falling object, and distances can result in radical equations (See Examples 8–10).	97–102
7.6	Perform operations on numbers in i-form *(p. 396).*	To perform operations with square roots of negative numbers, first write the numbers in i-form. Let c be a positive real number. Then the square root of $-c$ is given by $\sqrt{-c} = \sqrt{c(-1)} = \sqrt{c}\sqrt{-1} = \sqrt{c}\,i$. When writing $\sqrt{-c}$ in the i-form, $\sqrt{c}\,i$, note that i is outside the radical.	103–116
	Add, subtract, and multiply complex numbers *(p. 398).*	If a and b are real numbers, then $a + bi$ is a complex number in standard form. To add (or subtract) complex numbers, add (or subtract) the real and imaginary parts separately. To multiply complex numbers, use the Commutative, Associative, and Distributive Properties, along with the FOIL Method.	117–126
	Use complex conjugates to find the quotient of two complex numbers *(p. 400).*	Complex numbers of the form $(a+bi)$ and $(a-bi)$ are called complex conjugates. To write the quotient of two complex numbers in standard form, multiply the numerator and denominator by the complex conjugate of the denominator, and simplify.	127–132

Review Exercises

Worked-out solutions to odd-numbered exercises at AlgebraWithinReach.com

7.1

Evaluating an Expression **In Exercises 1–10, evaluate the radical expression. If not possible, state the reason.**

1. $-\sqrt{81}$ -9
2. $\sqrt{-16}$ Not a real number
3. $-\sqrt[3]{64}$ -4
4. $\sqrt[3]{-125}$ -5
5. $-\sqrt{\left(\frac{3}{4}\right)^2}$ $-\frac{3}{4}$
6. $\sqrt{\left(-\frac{9}{13}\right)^2}$ $\frac{9}{13}$
7. $\sqrt[3]{-\left(\frac{1}{5}\right)^3}$ $-\frac{1}{5}$
8. $-\sqrt[3]{\left(-\frac{27}{64}\right)^3}$ $\frac{27}{64}$
9. $\sqrt{-2^2}$ Not a real number
10. $-\sqrt{-3^2}$ Not a real number

Finding the Side Length **In Exercises 11 and 12, use the area A of the square or the volume V of the cube to find the side length l.**

11.

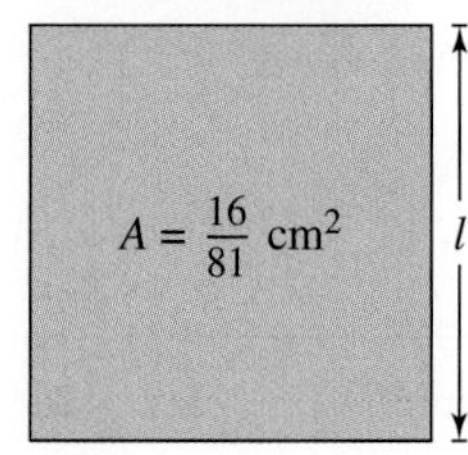

$\frac{4}{9}$ centimeter

12.

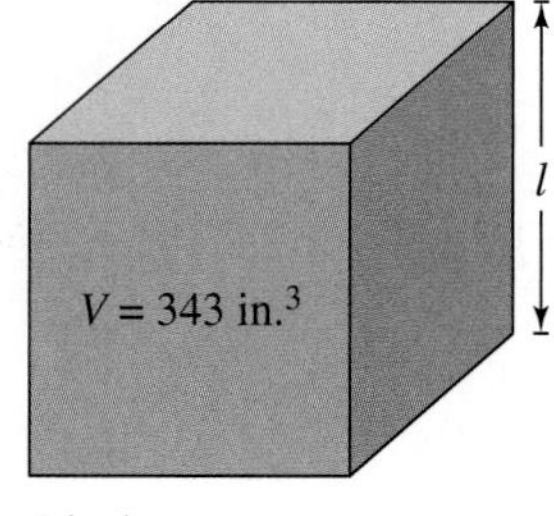

7 inches

Using a Definition **In Exercises 13–16, determine the missing description.**

	Radical Form	Rational Exponent Form
13.	$\sqrt[3]{27} = 3$	$27^{1/3} = 3$
14.	$\sqrt[3]{0.125} = 0.5$	$0.125^{1/3} = 0.5$
15.	$\sqrt[3]{216} = 6$	$216^{1/3} = 6$
16.	$\sqrt[4]{16} = 2$	$16^{1/4} = 2$

Evaluating an Expression with a Rational Exponent **In Exercises 17–22, evaluate the expression.**

17. $27^{4/3}$ 81
18. $16^{3/4}$ 8
19. $(-25)^{3/2}$ Not a real number
20. $-(4^3)^{2/3}$ -16
21. $8^{-4/3}$ $\frac{1}{16}$
22. $243^{-2/5}$ $\frac{1}{9}$

Using Rules of Exponents **In Exercises 23–34, use rational exponents to rewrite and simplify the expression.**

23. $x^{3/4} \cdot x^{-1/6}$ $x^{7/12}$
24. $a^{2/3} \cdot a^{3/5}$ $a^{19/15}$
25. $z\sqrt[3]{z^2}$ $z^{5/3}$
26. $x^2\sqrt[4]{x^3}$ $x^{11/4}$
27. $\dfrac{\sqrt[4]{x^3}}{\sqrt{x^4}}$ $\dfrac{1}{x^{5/4}}$
28. $\dfrac{\sqrt{x^3}}{\sqrt[3]{x^2}}$ $x^{5/6}$
29. $\sqrt[3]{a^3b^2}$ $ab^{2/3}$
30. $\sqrt[4]{m^3n^8}$ $m^{3/4}n^2$
31. $\sqrt[4]{\sqrt{x}}$ $x^{1/8}$
32. $\sqrt{\sqrt[3]{x^4}}$ $x^{2/3}$
33. $\dfrac{(3x+2)^{2/3}}{\sqrt[3]{3x+2}}$ $(3x+2)^{1/3}, x \neq -\frac{2}{3}$
34. $\dfrac{\sqrt[5]{3x+6}}{(3x+6)^{4/5}}$ $\dfrac{1}{(3x+6)^{3/5}}$

Evaluating a Radical Function **In Exercises 35–38, evaluate the function for each indicated x-value, if possible, and simplify.**

35. $f(x) = \sqrt{x - 2}$

(a) $f(-7)$ Not a real number

(b) $f(51)$ 7

36. $f(x) = \sqrt{6x - 5}$

(a) $f(5)$ 5

(b) $f(-1)$ Not a real number

37. $g(x) = \sqrt[3]{2x - 1}$

(a) $g(0)$ -1

(b) $g(14)$ 3

38. $g(x) = \sqrt[4]{x + 5}$

(a) $g(-4)$ 1

(b) $g(76)$ 3

Finding the Domain of a Radical Function **In Exercises 39–44, describe the domain of the function.**

39. $f(x) = \sqrt[7]{x}$ $-\infty < x < \infty$

40. $g(x) = \sqrt[3]{4x}$ $-\infty < x < \infty$

41. $g(x) = \sqrt{6x}$ $0 \le x < \infty$

42. $f(x) = \sqrt[4]{2x}$ $0 \le x < \infty$

43. $f(x) = \sqrt{9 - 2x}$ $-\infty < x \le \frac{9}{2}$

44. $g(x) = \sqrt[3]{x + 2}$ $-\infty < x < \infty$

7.2

Removing a Constant Factor from a Radical **In Exercises 45–48, simplify the radical expression.**

45. $\sqrt{63}$ $3\sqrt{7}$

46. $\sqrt{28}$ $2\sqrt{7}$

47. $\sqrt{242}$ $11\sqrt{2}$

48. $\sqrt{245}$ $7\sqrt{5}$

Removing Factors from a Radical **In Exercises 49–54, simplify the radical expression.**

49. $\sqrt{36u^5v^2}$ $6u^2|v|\sqrt{u}$

50. $\sqrt{24x^3y^4}$ $2xy^2\sqrt{6x}$

51. $\sqrt{0.25x^4y}$ $0.5x^2\sqrt{y}$

52. $\sqrt{0.16s^6t^3}$ $0.4|s^3|t\sqrt{t}$

53. $\sqrt[3]{48a^3b^4}$ $2ab\sqrt[3]{6b}$

54. $\sqrt[4]{48u^4v^6}$ $2|uv|\sqrt[4]{3v^2}$

Rationalizing the Denominator **In Exercises 55–58, rationalize the denominator and simplify further, if possible.**

55. $\sqrt{\dfrac{5}{6}}$ $\dfrac{\sqrt{30}}{6}$

56. $\dfrac{4y}{\sqrt{10z}}$ $\dfrac{2y\sqrt{10z}}{5z}$

57. $\dfrac{2}{\sqrt[3]{2x}}$ $\dfrac{\sqrt[3]{4x^2}}{x}$

58. $\sqrt[3]{\dfrac{16t}{s^2}}$ $\dfrac{2\sqrt[3]{2st}}{s}$

Geometry **In Exercises 59–62, find the length of the hypotenuse of the right triangle.**

59. $\sqrt{145}$

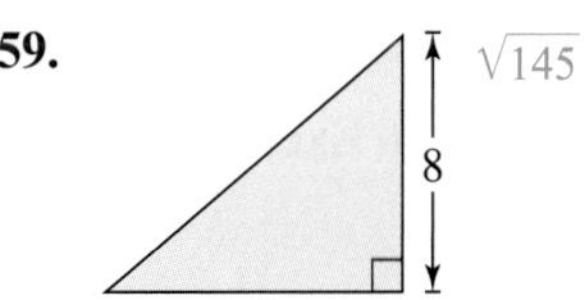

60. $\sqrt{769}$

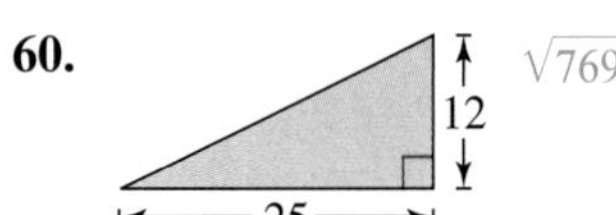

61. 17

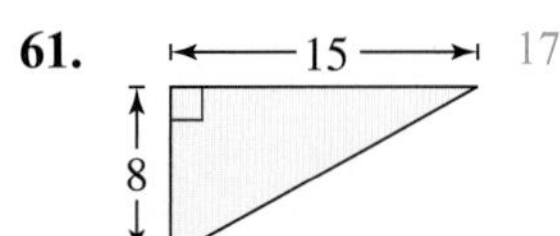

62. 8

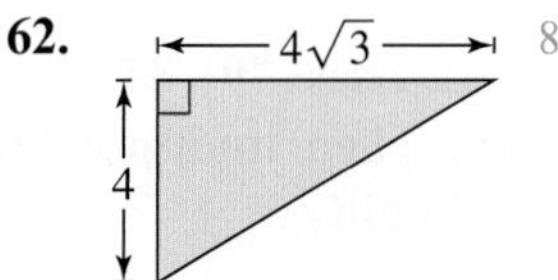

7.3

Combining Radical Expressions **In Exercises 63–70, simplify the expression by combining like radicals.**

63. $2\sqrt{24} + 7\sqrt{6} - \sqrt{54}$ $8\sqrt{6}$

64. $9\sqrt{50} - 5\sqrt{8} + \sqrt{48}$ $35\sqrt{2} + 4\sqrt{3}$

65. $5\sqrt{x} - \sqrt[3]{x} + 9\sqrt{x} - 8\sqrt[3]{x}$ $14\sqrt{x} - 9\sqrt[3]{x}$

66. $\sqrt{3x} - \sqrt[4]{6x^2} + 2\sqrt[4]{6x^2} - 4\sqrt{3x}$ $\sqrt[4]{6x^2} - 3\sqrt{3x}$

67. $10\sqrt[4]{y+3} - 3\sqrt[4]{y+3}$ $7\sqrt[4]{y+3}$

68. $5\sqrt[3]{x-3} + 4\sqrt[3]{x-3}$ $9\sqrt[3]{x-3}$

69. $2x\sqrt[3]{24x^2y} - \sqrt[3]{3x^5y}$ $3x\sqrt[3]{3x^2y}$

70. $4xy^2\sqrt[4]{243x} + 2y^2\sqrt[4]{48x^5}$ $16xy^2\sqrt[4]{3x}$

Dining Hall **In Exercises 71 and 72, a campus dining hall is undergoing renovations. The four corners of the hall are to be walled off and used as storage units (see figure.)**

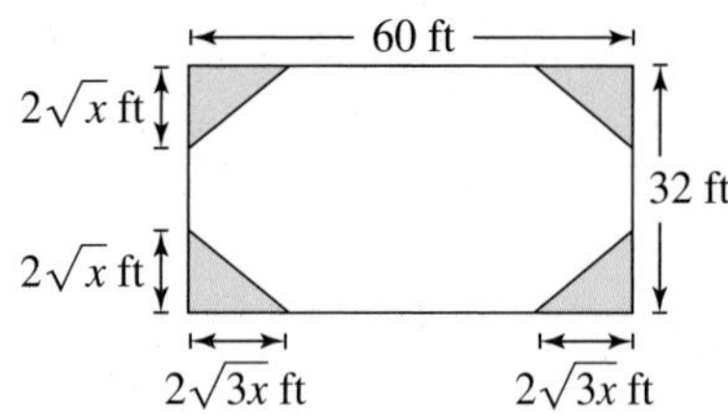

71. Find the perimeter of one of the storage units. $6\sqrt{x} + 2\sqrt{3x}$ feet

72. Find the perimeter of the newly designed dining hall. $184 + 8\sqrt{x} - 8\sqrt{3x}$ feet

7.4

Multiplying Radical Expressions **In Exercises 73–78, multiply and simplify.**

73. $\sqrt{15} \cdot \sqrt{20}$ $10\sqrt{3}$

74. $\sqrt{36} \cdot \sqrt{60}$ $12\sqrt{15}$

75. $\sqrt{10}(\sqrt{2} + \sqrt{5})$ $2\sqrt{5} + 5\sqrt{2}$

76. $\sqrt{12}(\sqrt{6} - \sqrt{8})$ $6\sqrt{2} - 4\sqrt{6}$

77. $(\sqrt{3} - \sqrt{x})(\sqrt{3} + \sqrt{x})$ $3 - x$

78. $(4 - 3\sqrt{2})^2$ $34 - 24\sqrt{2}$

Multiplying Conjugates **In Exercises 79–82, find the conjugate of the expression. Then multiply the expression by its conjugate and simplify.**

79. $3 - \sqrt{7}$ $3 + \sqrt{7}; 2$

80. $\sqrt{6} + 9$ $\sqrt{6} - 9; -75$

81. $\sqrt{x} + 20$ $\sqrt{x} - 20; x - 400$

82. $9 - \sqrt{2y}$ $9 + \sqrt{2y}; 81 - 2y$

Simplifying a Quotient Involving a Radical **In Exercises 83–86, simplify the expression.**

83. $\dfrac{\sqrt{2} - 1}{\sqrt{3} - 4}$ $-\dfrac{\sqrt{6} + 4\sqrt{2} - \sqrt{3} - 4}{13}$

84. $\dfrac{2 + \sqrt{20}}{3 + \sqrt{5}}$ $\sqrt{5} - 1$

85. $(\sqrt{x} + 10) \div (\sqrt{x} - 10)$ $\dfrac{x + 20\sqrt{x} + 100}{x - 100}$

86. $(3\sqrt{s} + 4) \div (\sqrt{s} + 2)$ $\dfrac{3s - 2\sqrt{s} - 8}{s - 4}$

7.5

Solving an Equation **In Exercises 87–96, solve the equation and check your solution(s). (Some of the equations have no solution.)**

87. $\sqrt{2x} - 8 = 0$ 32

88. $\sqrt{4x} + 6 = 9$ $\frac{9}{4}$

89. $\sqrt[4]{3x-1} + 6 = 3$ No real solution

90. $\sqrt[3]{5x-7} - 3 = -1$ 3

91. $\sqrt[3]{5x+2} - \sqrt[3]{7x-8} = 0$ 5

92. $\sqrt[4]{9x-2} - \sqrt[4]{8x} = 0$ 2

93. $\sqrt{2(x+5)} = x + 5$ $-5, -3$

94. $y - 2 = \sqrt{y+4}$ 5

95. $\sqrt{1+6x} = 2 - \sqrt{6x}$ $\frac{3}{32}$

96. $\sqrt{2+9b} + 1 = 3\sqrt{b}$ No real solution

97. ***Plasma TV*** The screen of a plasma television has a diagonal of 42 inches and a width of 36.8 inches. Find the height of the screen. $\sqrt{409.76} \approx 20.24$ inches

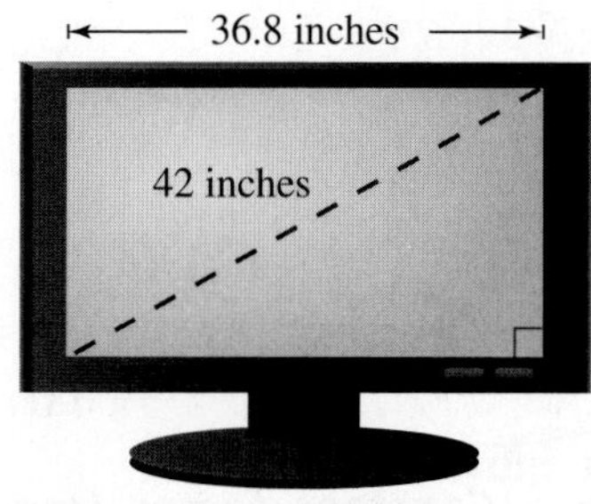

98. ***Geometry*** Determine the length and width of a rectangle with a perimeter of 82 inches and a diagonal of 29 inches. 20 inches × 21 inches

99. ***Period of a Pendulum*** The time t (in seconds) for a pendulum of length L (in feet) to go through one complete cycle (its period) is given by

$$t = 2\pi\sqrt{\frac{L}{32}}.$$

How long is the pendulum of a grandfather clock with a period of 1.9 seconds? 2.93 feet

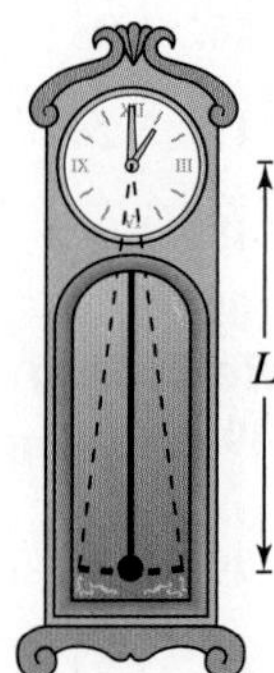

100. ***Height*** The time t (in seconds) for a free-falling object to fall d feet is given by

$$t = \sqrt{\frac{d}{16}}.$$

A child drops a pebble from a bridge and observes it strike the water after approximately 4 seconds. Estimate the height from which the pebble was dropped. 256 feet

Free-Falling Object **In Exercises 101 and 102, the velocity of a free-falling object can be determined from the equation**

$$v = \sqrt{2gh}$$

where v is the velocity (in feet per second), $g = 32$ feet per second per second, and h is the distance (in feet) the object has fallen.

101. Find the height from which a brick was dropped when it strikes the ground with a velocity of 64 feet per second. 64 feet

102. Find the height from which a wrench was dropped when it strikes the ground with a velocity of 112 feet per second. 196 feet

7.6

Writing a Number in i-Form **In Exercises 103–110, write the number in *i*-form.**

103. $\sqrt{-48}$ $4\sqrt{3}i$

104. $\sqrt{-0.16}$ $0.4i$

105. $10 - 3\sqrt{-27}$ $10 - 9\sqrt{3}i$

106. $3 + 2\sqrt{-500}$ $3 + 20\sqrt{5}i$

107. $\frac{3}{4} - 5\sqrt{-\frac{3}{25}}$ $\frac{3}{4} - \sqrt{3}i$

108. $\frac{2}{3} + 4\sqrt{-\frac{5}{16}}$ $\frac{2}{3} + \sqrt{5}i$

109. $8.4 + 20\sqrt{-0.81}$ $8.4 + 18i$

110. $-0.5 + 3\sqrt{-1.21}$ $-0.5 + 3.3i$

Operations with Square Roots of Negative Numbers **In Exercises 111–116, perform the operation(s) and write the result in standard form.**

111. $\sqrt{-9} - \sqrt{-1}$ $2i$

112. $\sqrt{-16} + \sqrt{-64}$ $12i$

113. $\sqrt{-81} + \sqrt{-36}$ $15i$

114. $\sqrt{-121} - \sqrt{-84}$ $(11 - 2\sqrt{21})i$

115. $\sqrt{-10}(\sqrt{-4} - \sqrt{-7})$ $\sqrt{70} - 2\sqrt{10}$

116. $\sqrt{-5}(\sqrt{-10} + \sqrt{-15})$ $-5\sqrt{2} - 5\sqrt{3}$

Equality of Two Complex Numbers **In Exercises 117–120, determine the values of a and b that satisfy the equation.**

117. $12 - 5i = (a + 2) + (b - 1)i$ $a = 10, b = -4$

118. $-48 + 9i = (a - 5) + (b + 10)i$ $a = -43, b = -1$

119. $\sqrt{-49} + 4 = a + bi$ $a = 4, b = 7$

120. $-3 - \sqrt{-4} = a + bi$ $a = -3, b = -2$

Operations with Complex Numbers **In Exercises 121–126, perform the operation and write the result in standard form.**

121. $(-4 + 5i) - (-12 + 8i)$ $8 - 3i$

122. $(-6 + 3i) + (-1 + i)$ $-7 + 4i$

123. $(4 - 3i)(4 + 3i)$ 25

124. $(12 - 5i)(2 + 7i)$ $59 + 74i$

125. $(6 - 5i)^2$ $11 - 60i$

126. $(2 - 9i)^2$ $-77 - 36i$

Writing a Quotient in Standard Form **In Exercises 127–132, write the quotient in standard form.**

127. $\frac{7}{3i}$ $-\frac{7}{3}i$

128. $\frac{4}{5i}$ $-\frac{4}{5}i$

129. $\frac{-3i}{4 - 6i}$ $\frac{9}{26} - \frac{3}{13}i$

130. $\frac{5i}{2 + 9i}$ $\frac{9}{17} + \frac{2}{17}i$

131. $\frac{3 - 5i}{6 + i}$ $\frac{13}{37} - \frac{33}{37}i$

132. $\frac{2 + i}{1 - 9i}$ $-\frac{7}{82} + \frac{19}{82}i$

Chapter Test

Solutions in English & Spanish and tutorial videos at AlgebraWithinReach.com

Take this test as you would take a test in class. After you are done, check your work against the answers in the back of the book.

In Exercises 1 and 2, evaluate each expression.

1. (a) $16^{3/2}$ 64
(b) $\sqrt{5}\sqrt{20}$ 10

2. (a) $125^{-2/3}$ $\frac{1}{25}$
(b) $\sqrt{3}\sqrt{12}$ 6

3. For $f(x) = \sqrt{9 - 5x}$, find $f(-8)$ and $f(0)$. $f(-8) = 7, f(0) = 3$

4. Describe the domain of $g(x) = \sqrt{7x - 3}$. $\left[\frac{3}{7}, \infty\right)$

In Exercises 5–7, simplify the expression.

5. (a) $\left(\dfrac{x^{1/2}}{x^{1/3}}\right)^2$ $x^{1/3}, x \neq 0$
(b) $5^{1/4} \cdot 5^{7/4}$ 25

6. (a) $\sqrt{\dfrac{32}{9}}$ $\dfrac{4\sqrt{2}}{3}$
(b) $\sqrt[3]{24}$ $2\sqrt[3]{3}$

7. (a) $\sqrt{24x^3}$ $2x\sqrt{6x}$
(b) $\sqrt[4]{16x^5y^8}$ $2xy^2\sqrt[4]{x}$

In Exercises 8 and 9, rationalize the denominator of the expression and simplify.

8. $\dfrac{2}{\sqrt[3]{9y}}$ $\dfrac{2\sqrt[3]{3y^2}}{3y}$

9. $\dfrac{10}{\sqrt{6} - \sqrt{2}}$ $\dfrac{5(\sqrt{6} + \sqrt{2})}{2}$

10. Subtract: $6\sqrt{18x} - 3\sqrt{32x}$ $6\sqrt{2x}$

11. Multiply and simplify: $\sqrt{5}(\sqrt{15x} + 3)$ $5\sqrt{3x} + 3\sqrt{5}$

12. Multiply and simplify: $(4 - \sqrt{2x})^2$ $16 - 8\sqrt{2x} + 2x$

13. Factor: $7\sqrt{27} + 14y\sqrt{12} = 7\sqrt{3}(\ 3 + 4y\ )$

In Exercises 14–16, solve the equation.

14. $\sqrt{6z} + 5 = 17$ 24

15. $\sqrt{x^2 - 1} = x - 2$ No solution

16. $\sqrt{x} - x + 6 = 0$ 9

In Exercises 17–20, perform the operation and write the result in standard form.

17. $(2 + 3i) - \sqrt{-25}$ $2 - 2i$

18. $(3 - 5i)^2$ $-16 - 30i$

19. $\sqrt{-16}(1 + \sqrt{-4})$ $-8 + 4i$

20. $(3 - 2i)(1 + 5i)$ $13 + 13i$

21. Write $\dfrac{5 - 2i}{3 + i}$ in standard form. $\frac{13}{10} - \frac{11}{10}i$

22. The velocity v (in feet per second) of an object is given by $v = \sqrt{2gh}$, where $g = 32$ feet per second per second and h is the distance (in feet) the object has fallen. Find the height from which a rock was dropped when it strikes the ground with a velocity of 96 feet per second. 144 feet

Cumulative Test: Chapters 5–7

Solutions in English & Spanish and tutorial videos at AlgebraWithinReach.com

Take this test as you would take a test in class. After you are done, check your work against the answers in the back of the book.

In Exercises 1–3, simplify the expression.

1. $(-2x^5y^{-2}z^0)^{-1}$ $-\frac{y^2}{2x^5}, y \neq 0, z \neq 0$

2. $\frac{12s^5t^{-2}}{20s^{-2}t^{-1}}$ $\frac{3s^7}{5t}, s \neq 0$

3. $\left(\frac{2x^{-4}y^3}{3x^5y^{-3}z^0}\right)^{-2}$ $\frac{9x^{18}}{4y^{12}}, x \neq 0, z \neq 0$

4. Evaluate $(5 \times 10^3)^2$ without using a calculator. 2.5×10^7

In Exercises 5–8, perform the operation(s) and simplify.

5. $(x^5 + 2x^3 + x^2 - 10x) - (2x^3 - x^2 + x - 4)$ $x^5 + 2x^2 - 11x + 4$

6. $-3(3x^3 - 4x^2 + x) + 3x(2x^2 + x - 1)$ $-3x^3 + 15x^2 - 6x$

7. $(x + 8)(3x - 2)$ $3x^2 + 22x - 16$

8. $(3x + 2)(3x^2 - x + 1)$ $9x^3 + 3x^2 + x + 2$

In Exercises 9–12, factor the expression completely.

9. $2x^2 - 11x + 15$ $(2x - 5)(x - 3)$

10. $9x^2 - 144$ $9(x + 4)(x - 4)$

11. $y^3 - 3y^2 - 9y + 27$ $(y - 3)^2(y + 3)$

12. $8t^3 - 40t^2 + 50t$ $2t(2t - 5)^2$

In Exercises 13 and 14, solve the equation.

13. $3x^2 + x - 24 = 0$ $-3, \frac{8}{3}$

14. $6x^3 - 486x = 0$ $0, \pm 9$

In Exercises 15–20, perform the operation(s) and simplify.

15. $\frac{x^2 + 8x + 16}{18x^2} \cdot \frac{2x^4 + 4x^3}{x^2 - 16}$ $\frac{x(x + 2)(x + 4)}{9(x - 4)}, x \neq -4, x \neq 0$

16. $\frac{x^2 + 4x}{2x^2 - 7x + 3} \div \frac{x^2 - 16}{x - 3}$ $\frac{x}{(2x - 1)(x - 4)}, x \neq -4, x \neq 3$

17. $\frac{5x}{x + 2} - \frac{2}{x^2 - x - 6}$ $\frac{5x^2 - 15x - 2}{(x + 2)(x - 3)}$

18. $\frac{2}{x} - \frac{x}{x^3 + 3x^2} + \frac{1}{x + 3}$ $\frac{3x + 5}{x(x + 3)}$

19. $\frac{\left(\frac{3x}{x + 2}\right)}{\left(\frac{12}{x^3 + 2x^2}\right)}$ $\frac{x^3}{4}, x \neq -2, x \neq 0$

20. $\frac{\left(\frac{x}{y} - \frac{y}{x}\right)}{\left(\frac{x - y}{xy}\right)}$ $x + y, x \neq 0, y \neq 0, x \neq y$

21. Use synthetic division to divide $2x^3 + 7x^2 - 5$ by $x + 4$. $2x^2 - x + 4 - \frac{21}{x + 4}$

22. Use the long division algorithm to divide $4x^4 - 6x^3 + x - 4$ by $2x - 1$.
$2x^3 - 2x^2 - x - \frac{4}{2x - 1}$

In Exercises 23 and 24, solve the equation.

23. $\dfrac{1}{x} + \dfrac{4}{10 - x} = 1$ 2, 5

24. $\dfrac{x - 3}{x} + 1 = \dfrac{x - 4}{x - 6}$ 2, 9

In Exercises 25–28, simplify the expression.

25. $\sqrt{24x^2y^3}$ $2|x|y\sqrt{6y}$

26. $\sqrt[3]{80a^{15}b^8}$ $2a^5b^2\sqrt[3]{10b^2}$

27. $(12a^{-4}b^6)^{1/2}$ $\dfrac{2|b^3|\sqrt{3}}{a^2}$

28. $\left(\dfrac{t^{1/2}}{t^{1/4}}\right)^2$ $\sqrt{t},\ t \neq 0$

29. Add: $10\sqrt{20x} + 3\sqrt{125x}$ $35\sqrt{5x}$

30. Multiply and simplify: $(\sqrt{2x} - 3)^2$ $2x - 6\sqrt{2x} + 9$

31. Simplify: $\dfrac{3}{\sqrt{10} - \sqrt{x}}$ $\dfrac{3(\sqrt{10} + \sqrt{x})}{10 - x}$

In Exercises 32–35, solve the equation.

32. $\sqrt{x - 5} - 6 = 0$ 41

33. $\sqrt{3 - x} + 10 = 11$ 2

34. $\sqrt{x + 5} - \sqrt{x - 7} = 2$ 11

35. $\sqrt{x - 4} = \sqrt{x + 7} - 1$ 29

In Exercises 36–38, perform the operation and write the result in standard form.

36. $\sqrt{-2}(\sqrt{-8} + 3)$ $-4 + 3\sqrt{2}i$

37. $(-4 + 11i) - (3 - 5i)$ $-7 + 16i$

38. $(5 + 2i)^2$ $21 + 20i$

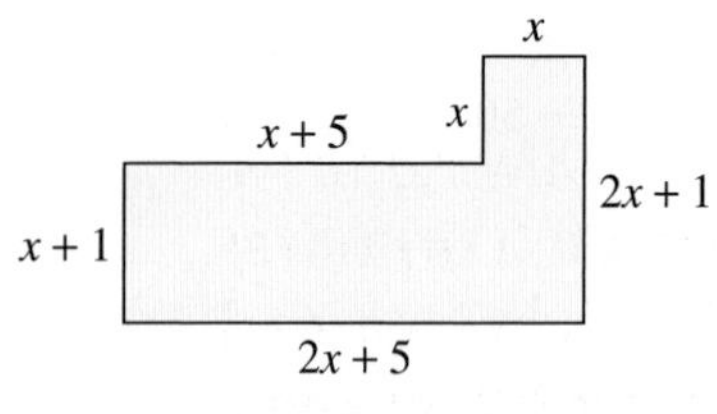

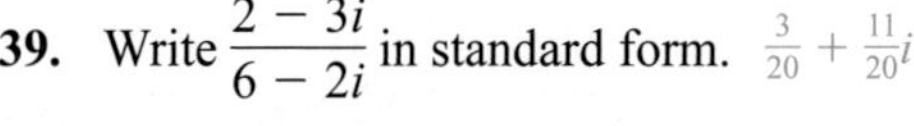

39. Write $\dfrac{2 - 3i}{6 - 2i}$ in standard form. $\frac{3}{20} + \frac{11}{20}i$

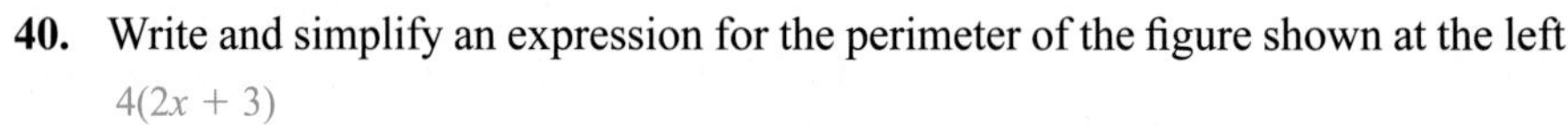

40. Write and simplify an expression for the perimeter of the figure shown at the left. $4(2x + 3)$

41. After working together for t hours on a common task, two workers have completed fractional parts of the job equal to $t/2$ and $2t/7$. What fractional part of the task has been completed? $\dfrac{11t}{14}$

42. A force of 50 pounds stretches a spring 7 centimeters.

(a) How far will a force of 20 pounds stretch the spring? 2.8 centimeters

(b) What force is required to stretch the spring 4.2 centimeters? 30 pounds

43. A guy wire on a 180-foot cell phone tower is attached to the top of the tower and to an anchor 90 feet from the base of the tower. Find the length of the guy wire. $90\sqrt{5} \approx 201.25$ feet

44. The time t (in seconds) for a free-falling object to fall d feet is given by

$$t = \sqrt{\frac{d}{16}}.$$

A construction worker drops a nail from a building and observes it strike a water puddle after approximately 5 seconds. Estimate the height from which the nail was dropped. 400 feet

Quadratic Equations, Functions, and Inequalities

MASTERY IS WITHIN REACH!

"I used to waste time reviewing for a test by just reading through my notes over and over again. Then I learned how to make note cards and keep working through them, saving the important ones for my mental cheat sheets. Now a couple of my friends and I go through our notes and make mental cheat sheets together. We are all doing well in our math class."

Steve
Art

See page 439 for suggestions about managing test anxiety.

8.1 Solving Quadratic Equations

- Solve quadratic equations by factoring.
- Solve quadratic equations by the Square Root Property.
- Use substitution to solve equations of quadratic form.

Solving Quadratic Equations by Factoring

From Section 5.6, remember that the first step in solving a quadratic equation by factoring is to write the equation in general form. Next, factor the left side. Finally, set each factor equal to zero and solve for x. Be sure to check each solution in the original equation.

EXAMPLE 1 Solving Quadratic Equations by Factoring

a. $x^2 + 5x = 24$ Original equation

$x^2 + 5x - 24 = 0$ Write in general form.

$(x + 8)(x - 3) = 0$ Factor.

$x + 8 = 0 \Rightarrow x = -8$ Set 1st factor equal to 0.

$x - 3 = 0 \Rightarrow x = 3$ Set 2nd factor equal to 0.

b. $3x^2 = 4 - 11x$ Original equation

$3x^2 + 11x - 4 = 0$ Write in general form.

$(3x - 1)(x + 4) = 0$ Factor.

$3x - 1 = 0 \Rightarrow x = \frac{1}{3}$ Set 1st factor equal to 0.

$x + 4 = 0 \Rightarrow x = -4$ Set 2nd factor equal to 0.

c. $9x^2 + 12 = 3 + 12x + 5x^2$ Original equation

$4x^2 - 12x + 9 = 0$ Write in general form.

$(2x - 3)(2x - 3) = 0$ Factor.

$2x - 3 = 0 \Rightarrow x = \frac{3}{2}$ Set factor equal to 0.

Check each solution in its original equation.

Study Tip

In Example 1(c), the quadratic equation produces two identical solutions. This is called a **double** or **repeated solution**.

Exercises Within Reach®

Solutions in English & Spanish and tutorial videos at AlgebraWithinReach.com

Solving a Quadratic Equation by Factoring **In Exercises 1–12, solve the equation by factoring.**

1. $x^2 - 15x + 54 = 0$ 6, 9
2. $x^2 + 15x + 44 = 0$ $-11, -4$
3. $x^2 - x - 30 = 0$ $-5, 6$
4. $x^2 - 2x - 48 = 0$ $-6, 8$
5. $x^2 + 4x = 45$ $-9, 5$
6. $x^2 - 7x = 18$ $-2, 9$
7. $x^2 - 16x + 64 = 0$ 8
8. $x^2 + 60x + 900 = 0$ -30
9. $9x^2 - 10x - 16 = 0$ $-\frac{8}{9}, 2$
10. $8x^2 - 10x + 3 = 0$ $\frac{1}{2}, \frac{3}{4}$
11. $u(u - 9) - 12(u - 9) = 0$ 9, 12
12. $16x(x - 8) - 12(x - 8) = 0$ $\frac{3}{4}, 8$

The Square Root Property

Square Root Property

The equation $u^2 = d$, where $d > 0$, has exactly two solutions:

$$u = \sqrt{d} \quad \text{and} \quad u = -\sqrt{d}.$$

These solutions can also be written as $u = \pm\sqrt{d}$. This solution process is also called **extracting square roots**.

EXAMPLE 2 Using the Square Root Property

a. $3x^2 = 15$ Original equation

$x^2 = 5$ Divide each side by 3.

$x = \pm\sqrt{5}$ Square Root Property

The solutions are $x = \sqrt{5}$ and $x = -\sqrt{5}$. Check these in the original equation.

b. $(x - 2)^2 = 10$ Original equation

$x - 2 = \pm\sqrt{10}$ Square Root Property

$x = 2 \pm \sqrt{10}$ Add 2 to each side.

The solutions are $x = 2 + \sqrt{10} \approx 5.16$ and $x = 2 - \sqrt{10} \approx -1.16$. Check these in the original equation.

c. $(3x - 6)^2 - 8 = 0$ Original equation

$(3x - 6)^2 = 8$ Add 8 to each side.

$3x - 6 = \pm 2\sqrt{2}$ Square Root Property and rewrite $\sqrt{8}$ as $2\sqrt{2}$.

$3x = 6 \pm 2\sqrt{2}$ Add 6 to each side.

$x = 2 \pm \dfrac{2\sqrt{2}}{3}$ Divide each side by 3.

The solutions are $x = 2 + \dfrac{2\sqrt{2}}{3} \approx 2.94$ and $x = 2 - \dfrac{2\sqrt{2}}{3} \approx 1.06$. Check these in the original equation.

Exercises Within Reach®

Solutions in English & Spanish and tutorial videos at AlgebraWithinReach.com

Using the Square Root Property **In Exercises 13–28, solve the equation by using the Square Root Property.**

13. $6x^2 = 54$ ± 3

14. $5t^2 = 5$ ± 1

15. $25x^2 = 16$ $\pm\frac{4}{5}$

16. $9z^2 = 121$ $\pm\frac{11}{3}$

17. $\dfrac{w^2}{4} = 49$ ± 14

18. $\dfrac{x^2}{6} = 24$ ± 12

19. $4x^2 - 25 = 0$ $\pm\frac{5}{2}$

20. $16y^2 - 121 = 0$ $\pm\frac{11}{4}$

21. $(x + 4)^2 = 64$ $-12, 4$

22. $(m - 12)^2 = 400$ $-8, 32$

23. $(x - 3)^2 = 0.25$ $2.5, 3.5$

24. $(x + 2)^2 = 0.81$ $-2.9, -1.1$

25. $(x - 2)^2 = 7$ $2 \pm \sqrt{7}$

26. $(y + 4)^2 = 27$ $-4 \pm 3\sqrt{3}$

27. $(2x + 1)^2 - 50 = 0$ $-\dfrac{1}{2} \pm \dfrac{5\sqrt{2}}{2}$

28. $(3x - 5)^2 - 48 = 0$ $\dfrac{5}{3} \pm \dfrac{4\sqrt{3}}{3}$

Square Root Property (Complex Square Root)

The equation $u^2 = d$, where $d < 0$, has exactly two solutions:

$$u = \sqrt{|d|}i \quad \text{and} \quad u = -\sqrt{|d|}\,i.$$

These solutions can also be written as $u = \pm\sqrt{|d|}i$.

EXAMPLE 3 Square Root Property (Complex Square Root)

a. $x^2 + 8 = 0$ — Original equation

$x^2 = -8$ — Subtract 8 from each side.

$x = \pm\sqrt{8}i = \pm 2\sqrt{2}i$ — Square Root Property

The solutions are $x = 2\sqrt{2}i$ and $x = -2\sqrt{2}i$. Check these in the original equation.

b. $(x - 4)^2 = -3$ — Original equation

$x - 4 = \pm\sqrt{3}i$ — Square Root Property

$x = 4 \pm \sqrt{3}i$ — Add 4 to each side.

The solutions are $x = 4 + \sqrt{3}i$ and $x = 4 - \sqrt{3}i$. Check these in the original equation.

c. $2(3x - 5)^2 + 32 = 0$ — Original equation

$2(3x - 5)^2 = -32$ — Subtract 32 from each side.

$(3x - 5)^2 = -16$ — Divide each side by 2.

$3x - 5 = \pm 4i$ — Square Root Property

$3x = 5 \pm 4i$ — Add 5 to each side.

$x = \frac{5}{3} \pm \frac{4}{3}i$ — Divide each side by 3.

The solutions are $x = \frac{5}{3} + \frac{4}{3}i$ and $x = \frac{5}{3} - \frac{4}{3}i$. Check these in the original equation.

Exercises Within Reach®

Solutions in English & Spanish and tutorial videos at AlgebraWithinReach.com

Using the Square Root Property **In Exercises 29–42, solve the equation by using the Square Root Property.**

29. $z^2 = -36$ $\pm 6i$
30. $x^2 = -16$ $\pm 4i$
31. $x^2 + 4 = 0$ $\pm 2i$
32. $p^2 + 9 = 0$ $\pm 3i$
33. $9u^2 + 17 = 0$ $\pm\frac{\sqrt{17}}{3}i$
34. $25x^2 + 4 = 0$ $\pm\frac{2}{5}i$
35. $(t - 3)^2 = -25$ $3 \pm 5i$
36. $(x + 5)^2 = -81$ $-5 \pm 9i$
37. $(3z + 4)^2 + 144 = 0$ $-\frac{4}{3} \pm 4i$
38. $(2y - 3)^2 + 25 = 0$ $\frac{3}{2} \pm \frac{5}{2}i$
39. $(4m + 1)^2 = -80$ $-\frac{1}{4} \pm \sqrt{5}i$
40. $(6y - 5)^2 = -8$ $\frac{5}{6} \pm \frac{\sqrt{2}}{3}i$
41. $36(t + 3)^2 = -100$ $-3 \pm \frac{5}{3}i$
42. $4(x - 4)^2 = -169$ $4 \pm \frac{13}{2}i$

Application EXAMPLE 4 Diameter of a Softball

The surface area of a sphere of radius r is given by

$$S = 4\pi r^2.$$

The surface area of a softball is

$$\frac{144}{\pi} \text{ square inches.}$$

Find the diameter d of the softball.

SOLUTION

$S = 4\pi r^2$	Write original equation.
$\frac{144}{\pi} = 4\pi r^2$	Substitute $\frac{144}{\pi}$ for S.
$\frac{144}{\pi(4\pi)} = \frac{4\pi r^2}{4\pi}$	Divide each side by 4π.
$\frac{36}{\pi^2} = r^2$	Simplify.
$\pm\sqrt{\frac{36}{\pi^2}} = r$	Square Root Property
$\pm\frac{6}{\pi} = r$	Simplify.

Choose the positive root to obtain $r = \frac{6}{\pi}$. Then,

$d = 2r$	Diameter is twice radius.
$= 2\left(\frac{6}{\pi}\right)$	Substitute $\frac{6}{\pi}$ for r.
$= \frac{12}{\pi}$	Simplify.
≈ 3.82 inches.	Use calculator to approximate.

So, the diameter of the softball is about 3.82 inches.

Exercises Within Reach®

Solutions in English & Spanish and tutorial videos at AlgebraWithinReach.com

43. ***Geometry*** The surface area S of a basketball is

$$\frac{900}{\pi} \text{ square inches.}$$

Find the radius r of the basketball. $\frac{15}{\pi} \approx 4.77$ inches

44. ***Unisphere*** The Unisphere is the world's largest man-made globe. It was built as the symbol of the 1964–1965 New York World's Fair. A sphere with the same diameter as the Unisphere globe would have a surface area of 45,239 square feet. What is the diameter of the Unisphere? (*Source:* The World's Fair and Exposition Information and Reference Guide)
120 feet

Rob Wilson/Shutterstock.com

Equations of Quadratic Form

Both the factoring method and the Square Root Property can be applied to nonquadratic equations that are of *quadratic form*. An equation is said to be of **quadratic form** if it has the form

$$au^2 + bu + c = 0$$

where u is an algebraic expression. Here are some examples.

Equation	*Written in Quadratic Form*
$x^4 + 5x^2 + 4 = 0$	$(x^2)^2 + 5(x^2) + 4 = 0$
$x - 5\sqrt{x} + 6 = 0$	$(\sqrt{x})^2 - 5(\sqrt{x}) + 6 = 0$
$2x^{2/3} + 5x^{1/3} - 3 = 0$	$2(x^{1/3})^2 + 5(x^{1/3}) - 3 = 0$
$18 + 2x^2 + (x^2 + 9)^2 = 8$	$(x^2 + 9)^2 + 2(x^2 + 9) - 8 = 0$

To solve an equation of quadratic form, it helps to make a substitution and rewrite the equation in terms of u, as demonstrated in Examples 5 and 6.

EXAMPLE 5 Solving an Equation of Quadratic Form

Solve $x^4 - 13x^2 + 36 = 0$.

SOLUTION

Begin by writing the original equation in quadratic form, as follows.

$x^4 - 13x^2 + 36 = 0$ Write original equation.

$(x^2)^2 - 13(x^2) + 36 = 0$ Write in quadratic form.

Next, let $u = x^2$ and substitute u into the equation written in quadratic form. Then, factor and solve the equation.

$u^2 - 13u + 36 = 0$ Substitute u for x^2.

$(u - 4)(u - 9) = 0$ Factor.

$u - 4 = 0 \Rightarrow u = 4$ Set 1st factor equal to 0.

$u - 9 = 0 \Rightarrow u = 9$ Set 2nd factor equal to 0.

At this point you have found the "u-solutions." To find the "x-solutions," replace u with x^2 and solve for x.

$u = 4 \Rightarrow x^2 = 4 \Rightarrow x = \pm 2$

$u = 9 \Rightarrow x^2 = 9 \Rightarrow x = \pm 3$

The solutions are $x = 2$, $x = -2$, $x = 3$, and $x = -3$. Check these in the original equation.

Exercises Within Reach®

Solutions in English & Spanish and tutorial videos at AlgebraWithinReach.com

Solving an Equation of Quadratic Form **In Exercises 45–50, solve the equation of quadratic form.**

45. $x^4 - 5x^2 + 4 = 0$ $\pm 1, \pm 2$

46. $x^4 - 10x^2 + 25 = 0$ $\pm\sqrt{5}$

47. $x^4 - 5x^2 + 6 = 0$ $\pm\sqrt{2}, \pm\sqrt{3}$

48. $x^4 - 10x^2 + 21 = 0$ $\pm\sqrt{3}, \pm\sqrt{7}$

49. $(x^2 - 4)^2 + 2(x^2 - 4) - 3 = 0$ $\pm 1, \pm\sqrt{5}$

50. $(x^2 - 1)^2 + (x^2 - 1) - 6 = 0$ $\pm\sqrt{3}, \pm\sqrt{2}i$

EXAMPLE 6 Solving Equations of Quadratic Form

Study Tip

When solving equations involving square roots, be sure to check for extraneous solutions.

a. $x - 5\sqrt{x} + 6 = 0$ Original equation

This equation is of quadratic form with $u = \sqrt{x}$.

$(\sqrt{x})^2 - 5(\sqrt{x}) + 6 = 0$ Write in quadratic form.

$u^2 - 5u + 6 = 0$ Substitute u for $\sqrt{x}$.

$(u - 2)(u - 3) = 0$ Factor.

$u - 2 = 0 \Rightarrow u = 2$ Set 1st factor equal to 0.

$u - 3 = 0 \Rightarrow u = 3$ Set 2nd factor equal to 0.

Now, using the u-solutions of 2 and 3, you obtain the x-solutions as follows.

$u = 2 \Rightarrow \sqrt{x} = 2 \Rightarrow x = 4$

$u = 3 \Rightarrow \sqrt{x} = 3 \Rightarrow x = 9$

b. $x^{2/3} - x^{1/3} - 6 = 0$ Original equation

This equation is of quadratic form with $u = x^{1/3}$.

$(x^{1/3})^2 - (x^{1/3}) - 6 = 0$ Write in quadratic form.

$u^2 - u - 6 = 0$ Substitute u for $x^{1/3}$.

$(u + 2)(u - 3) = 0$ Factor.

$u + 2 = 0 \Rightarrow u = -2$ Set 1st factor equal to 0.

$u - 3 = 0 \Rightarrow u = 3$ Set 2nd factor equal to 0.

Now, using the u-solutions of -2 and 3, you obtain the x-solutions as follows.

$u = -2 \Rightarrow x^{1/3} = -2 \Rightarrow x = -8$

$u = 3 \Rightarrow x^{1/3} = 3 \Rightarrow x = 27$

Additional Example

Solve $x - 13\sqrt{x} + 36 = 0$.

Answer:

The u-solutions are $u = 4$ and $u = 9$. Replacing u with $\sqrt{x}$, you obtain the x-solutions $x = 16$ and $x = 81$. (Note that some students may incorrectly conclude that $x = 2$ and $x = 3$.)

Exercises Within Reach®

Solutions in English & Spanish and tutorial videos at AlgebraWithinReach.com

Solving an Equation of Quadratic Form **In Exercises 51–64, solve the equation of quadratic form.**

51. $x - 3\sqrt{x} - 4 = 0$
16

52. $x - \sqrt{x} - 6 = 0$
9

53. $x - 7\sqrt{x} + 10 = 0$
4, 25

54. $x - 11\sqrt{x} + 24 = 0$
9, 64

55. $x^{2/3} - x^{1/3} - 6 = 0$
$-8, 27$

56. $x^{2/3} + 3x^{1/3} - 10 = 0$
$-125, 8$

57. $2x^{2/3} - 7x^{1/3} + 5 = 0$
$1, \frac{125}{8}$

58. $5x^{2/3} - 13x^{1/3} + 6 = 0$
$\frac{27}{125}, 8$

59. $x^{2/5} - 3x^{1/5} + 2 = 0$
1, 32

60. $x^{2/5} + 5x^{1/5} + 6 = 0$
$-243, -32$

61. $2x^{2/5} - 7x^{1/5} + 3 = 0$
$\frac{1}{32}, 243$

62. $2x^{2/5} + 3x^{1/5} + 1 = 0$
$-1, -\frac{1}{32}$

63. $x^{1/3} - x^{1/6} - 6 = 0$
729

64. $x^{1/3} + 2x^{1/6} - 3 = 0$
1

Concept Summary: Solving Quadratic Equations by Using the Square Root Property

What

Two ways to solve quadratic equations are (1) by factoring and (2) by using the **Square Root Property**.

EXAMPLE

Solve $x^2 = -144$ by using the Square Root Property.

How

To use the Square Root Property, take the square root of each side of the equation.

EXAMPLE

$x^2 = -144$ Original equation

$x = \pm 12i$ Square Root Property

The solutions are $x = -12i$ and $x = 12i$.

Notice that the solutions of the quadratic equation are complex numbers.

Why

Many real-life situations can be modeled by quadratic equations. Knowing how to solve these types of equations will help you answer many real-life problems.

Exercises Within Reach®

Worked-out solutions to odd-numbered exercises at AlgebraWithinReach.com

Concept Summary Check

65. ***Writing*** Describe how to solve a quadratic equation by using the Square Root Property. See Additional Answers.

66. ***True or False?*** Determine whether the following statement is true or false. Justify your answer.

The only solution of the equation $x^2 = 25$ is $x = 5$. False. Another solution is $x = -5$. $x^2 = (-5)^2 = 25$

67. ***Reasoning*** Does the equation $4x^2 + 9 = 0$ have two real solutions or two complex solutions? Explain your reasoning. Two complex solutions. When the squared expression is isolated on one side of the equation, the other side is negative. When the square root of each side is taken, the square root of the negative number is imaginary.

68. ***Writing*** Describe two ways to solve the equation $2x^2 - 18 = 0$. Factoring, Square Root Property

Extra Practice

Solving a Quadratic Equation **In Exercises 69–78, solve the equation.**

69. $x^2 + 900 = 0$ $\pm 30i$

70. $z^2 + 256 = 0$ $\pm 16i$

71. $\frac{2}{3}x^2 = 6$ ± 3

72. $\frac{1}{3}x^2 = 4$ $\pm 2\sqrt{3}$

73. $(p - 2)^2 - 108 = 0$ $2 \pm 6\sqrt{3}$

74. $(y + 12)^2 - 400 = 0$ $-32, 8$

75. $(p - 2)^2 + 108 = 0$ $2 \pm 6\sqrt{3}i$

76. $(y + 12)^2 + 400 = 0$ $-12 \pm 20i$

77. $(x + 2)^2 + 18 = 0$ $-2 \pm 3\sqrt{2}i$

78. $(x + 2)^2 - 18 = 0$ $-2 \pm 3\sqrt{2}$

Solving an Equation of Quadratic Form **In Exercises 79–88, solve the equation of quadratic form.**

79. $x^{1/2} - 3x^{1/4} + 2 = 0$ $1, 16$

80. $x^{1/2} - 5x^{1/4} + 6 = 0$ $16, 81$

81. $\frac{1}{x^2} - \frac{3}{x} + 2 = 0$ $\frac{1}{2}, 1$

82. $\frac{1}{x^2} - \frac{1}{x} - 6 = 0$ $-\frac{1}{2}, \frac{1}{3}$

83. $4x^{-2} - x^{-1} - 5 = 0$ $-1, \frac{4}{5}$

84. $2x^{-2} - x^{-1} - 1 = 0$ $-2, 1$

85. $(x^2 - 3x)^2 - 2(x^2 - 3x) - 8 = 0$ $\pm 1, 2, 4$

86. $(x^2 - 6x)^2 - 2(x^2 - 6x) - 35 = 0$ $\pm 1, 5, 7$

87. $16\left(\frac{x - 1}{x - 8}\right)^2 + 8\left(\frac{x - 1}{x - 8}\right) + 1 = 0$ $\frac{12}{5}$

88. $9\left(\frac{x + 2}{x + 3}\right)^2 - 6\left(\frac{x + 2}{x + 3}\right) + 1 = 0$ $-\frac{3}{2}$

Free-Falling Object **In Exercises 89–92, the height h (in feet) of a falling object at any time t (in seconds) is modeled by $h = -16t^2 + s_0$, where s_0 is the initial height (in feet). Use the model to find the time it takes for an object to fall to the ground given s_0.**

89. $s_0 = 256$
4 seconds

90. $s_0 = 48$
$\sqrt{3} \approx 1.73$ seconds

91. $s_0 = 128$
$2\sqrt{2} \approx 2.83$ seconds

92. $s_0 = 500$
$\frac{5\sqrt{5}}{2} \approx 5.59$ seconds

93. ***Free-Falling Object*** The height h (in feet) of an object thrown vertically upward from the top of a tower that is 144 feet tall is given by

$$h = 144 + 128t - 16t^2$$

where t measures the time (in seconds) from when the object is released. How long does it take for the object to reach the ground? 9 seconds

94. ***Profit*** The monthly profit P (in dollars) a company makes depends on the amount x (in dollars) the company spends on advertising according to the model

$$P = 800 + 120x - \frac{1}{2}x^2.$$

Find the amount the company must spend on advertising to make a monthly profit of \$8000. \$120

Compound Interest **The amount A in an account after 2 years, when a principal of P dollars is invested at annual interest rate r (in decimal form) compounded annually, is given by $A = P(1 + r)^2$. In Exercises 95 and 96, find r.**

95. $P = \$1500, A = \$1685.40$ 6%

96. $P = \$5000, A = \$5724.50$ 7%

Explaining Concepts

97. ***Reasoning*** For a quadratic equation $ax^2 + bx + c = 0$, where a, b, and c are real numbers with $a \neq 0$, explain why b and c can equal 0, but a cannot. A quadratic equation is of degree 2, so it must have an x^2-term, but it does not need an x-term or a constant.

98. ***Vocabulary*** Is the equation $x^6 - 6x^3 + 9 = 0$ of quadratic form? Explain your reasoning.
Yes. For an equation to be of quadratic form, the exponent of the algebraic expression in the first term must be twice the exponent of the same algebraic expression in the second term.

99. ***Logic*** Is it possible for a quadratic equation of the form $x^2 = m$ to have one real solution and one complex solution? Explain your reasoning. No. Complex solutions always occur in complex conjugate pairs.

100. ***Precision*** Describe a procedure for solving an equation of quadratic form. Give an example.
To solve an equation of quadratic form, determine an algebraic expression u such that substitution yields the quadratic equation $au^2 + bu + c = 0$. Solve this quadratic equation for u and then, through back-substitution, find the solution of the original equation.

Cumulative Review

In Exercises 101–104, solve the inequality and sketch the solution on the real number line.
See Additional Answers.

101. $3x - 8 > 4$ $x > 4$

102. $4 - 5x \geq 12$ $x \leq -\frac{8}{5}$

103. $2x - 6 \leq 9 - x$ $x \leq 5$

104. $x - 4 < 6$ or $x + 3 > 8$ $-\infty \leq x \leq \infty$

In Exercises 105 and 106, solve the system of linear equations.

105. $x + y - z = 4$
$2x + y + 2z = 10$
$x - 3y - 4z = -7$
(3, 2, 1)

106. $2x - y + z = -6$
$x + 5y - z = 7$
$-x - 2y - 3z = 8$
(−1, 1, −3)

In Exercises 107–112, combine the radical expressions, if possible, and simplify.

107. $5\sqrt{3} - 2\sqrt{3}$
$3\sqrt{3}$

108. $8\sqrt{27} + 4\sqrt{27}$
$36\sqrt{3}$

109. $16\sqrt[3]{y} - 9\sqrt[3]{x}$
$16\sqrt[3]{y} - 9\sqrt[3]{x}$

110. $12\sqrt{x-1} + 6\sqrt{x-1}$
$18\sqrt{x-1}$

111. $\sqrt{16m^4n^3} + m\sqrt{m^2n}$
$(4n + 1)m^2\sqrt{n}$

112. $x^2y\sqrt[4]{32x^2} + x\sqrt[4]{2x^6y^4} - y\sqrt[4]{162x^{10}}$
0

8.2 Completing the Square

- Rewrite quadratic expressions in completed square form.
- Solve quadratic equations by completing the square.

Constructing Perfect Square Trinomials

In this section, you will study a technique for rewriting an equation in a completed square form. This technique is called *completing the square*. Note that prior to completing the square, the coefficient of the second-degree term must be 1.

Emphasize that the coefficient of the second-degree term must be 1 before an equation can be rewritten by completing the square.

Completing the Square

To **complete the square** for the expression $x^2 + bx$, add $(b/2)^2$, which is the square of half the coefficient of x. Consequently,

$$x^2 + bx + \left(\frac{b}{2}\right)^2 = \left(x + \frac{b}{2}\right)^2.$$

(half)^2

Consider having students complete a pattern such as

$x^2 + 10x + \quad = (x + \quad)^2$
$x^2 + 11x + \quad = (x + \quad)^2$
$x^2 - 12x + \quad = (x - \quad)^2$

by asking what numbers should be added to create perfect square trinomials. Additional similar examples can be used as necessary to reinforce the pattern.

EXAMPLE 1 Constructing a Perfect Square Trinomial

What term should be added to $x^2 - 8x$ to make it a perfect square trinomial?

SOLUTION

For this expression, the coefficient of the x-term is -8. Add the square of half of this coefficient to the expression to make it a perfect square trinomial.

$$x^2 - 8x + \left(-\frac{8}{2}\right)^2 = x^2 - 8x + 16 \qquad \text{Add } \left(-\tfrac{8}{2}\right)^2 = 16 \text{ to the expression.}$$

You can then rewrite the expression as the square of a binomial, $(x - 4)^2$.

Exercises Within Reach®

Solutions in English & Spanish and tutorial videos at AlgebraWithinReach.com

Constructing a Perfect Square Trinomial **In Exercises 1–16, add a term to the expression to make it a perfect square trinomial.**

1. $x^2 + 8x +$ 16
2. $x^2 + 12x +$ 36
3. $y^2 - 20y +$ 100
4. $y^2 - 2y +$ 1
5. $x^2 + 14x +$ 49
6. $x^2 - 24x +$ 144
7. $t^2 + 5t +$ $\frac{25}{4}$
8. $u^2 + 7u +$ $\frac{49}{4}$
9. $x^2 - 9x +$ $\frac{81}{4}$
10. $y^2 - 11y +$ $\frac{121}{4}$
11. $a^2 - \frac{1}{3}a +$ $\frac{1}{36}$
12. $y^2 + \frac{4}{3}y +$ $\frac{4}{9}$
13. $y^2 + \frac{8}{5}y +$ $\frac{16}{25}$
14. $x^2 - \frac{9}{5}x +$ $\frac{81}{100}$
15. $r^2 - 0.4r +$ 0.04
16. $s^2 + 4.6s +$ 5.29

Solving Equations by Completing the Square

Completing the square can be used to solve quadratic equations. When using this procedure, remember to *preserve the equality* by adding the same constant to each side of the equation.

Study Tip

In Example 2, completing the square is used for the sake of illustration. This particular equation would be easier to solve by factoring. Try reworking the problem by factoring to see that you obtain the same two solutions.

EXAMPLE 2 Completing the Square: Leading Coefficient Is 1

Solve $x^2 + 12x = 0$ by completing the square.

SOLUTION

$x^2 + 12x = 0$ Write original equation.

$x^2 + 12x + 6^2 = 36$ Add $6^2 = 36$ to each side.

$\left(\frac{12}{2}\right)^2$

$(x + 6)^2 = 36$ Completed square form

$x + 6 = \pm\sqrt{36}$ Square Root Property

$x = -6 \pm 6$ Subtract 6 from each side.

$x = -6 + 6$ or $x = -6 - 6$ Separate solutions.

$x = 0$ $\quad x = -12$ Simplify.

The solutions are $x = 0$ and $x = -12$. Check these in the original equation.

Exercises Within Reach®

Solutions in English & Spanish and tutorial videos at AlgebraWithinReach.com

Solving an Equation **In Exercises 17–32, solve the equation first by completing the square and then by factoring.**

17. $x^2 - 20x = 0$
0, 20

18. $x^2 + 32x = 0$
−32, 0

19. $x^2 + 6x = 0$
−6, 0

20. $t^2 - 10t = 0$
0, 10

21. $y^2 - 5y = 0$
0, 5

22. $t^2 - 9t = 0$
0, 9

23. $t^2 - 8t + 7 = 0$
1, 7

24. $y^2 - 4y + 4 = 0$
2

25. $x^2 + 7x + 12 = 0$
−4, −3

26. $z^2 + 3z - 10 = 0$
−5, 2

27. $x^2 - 3x - 18 = 0$
−3, 6

28. $a^2 + 12a + 32 = 0$
−8, −4

29. $x^2 + 8x + 7 = 0$
−7, −1

30. $x^2 - 10x + 9 = 0$
−9, −1

31. $x^2 - 10x + 21 = 0$
3, 7

32. $x^2 - 10x + 24 = 0$
4, 6

EXAMPLE 3 Completing the Square: Leading Coefficient Is 1

$x^2 - 6x + 7 = 0$ Original equation

$x^2 - 6x = -7$ Subtract 7 from each side.

$x^2 - 6x + (-3)^2 = -7 + 9$ Add $(-3)^2 = 9$ to each side.

$\left(-\frac{6}{2}\right)^2$

$(x - 3)^2 = 2$ Completed square form

$x - 3 = \pm\sqrt{2}$ Square Root Property

$x = 3 \pm \sqrt{2}$ Add 3 to each side.

$x = 3 + \sqrt{2}$ or $x = 3 - \sqrt{2}$ Separate solutions.

The solutions are $x = 3 + \sqrt{2} \approx 4.41$ and $x = 3 - \sqrt{2} \approx 1.59$. Check these in the original equation.

EXAMPLE 4 Completing the Square: Leading Coefficient Is Not 1

$2x^2 - x - 2 = 0$ Original equation

$2x^2 - x = 2$ Add 2 to each side.

$x^2 - \frac{1}{2}x = 1$ Divide each side by 2.

$x^2 - \frac{1}{2}x + \left(-\frac{1}{4}\right)^2 = 1 + \frac{1}{16}$ Add $\left(-\frac{1}{4}\right)^2 = \frac{1}{16}$ to each side.

$\left(x - \frac{1}{4}\right)^2 = \frac{17}{16}$ Completed square form

$x - \frac{1}{4} = \pm\frac{\sqrt{17}}{4}$ Square Root Property

$x = \frac{1}{4} \pm \frac{\sqrt{17}}{4}$ Add $\frac{1}{4}$ to each side.

The solutions are $x = \frac{1}{4} + \frac{\sqrt{17}}{4} \approx 1.28$ and $x = \frac{1}{4} - \frac{\sqrt{17}}{4} \approx -0.78$. Check these in the original equation.

Exercises Within Reach ®

Solutions in English & Spanish and tutorial videos at AlgebraWithinReach.com

Completing the Square **In Exercises 33–38, solve the equation by completing the square. Give the solutions in exact form and in decimal form rounded to two decimal places.**

33. $x^2 - 4x - 3 = 0$
$2 + \sqrt{7} \approx 4.65$
$2 - \sqrt{7} \approx -0.65$

34. $x^2 - 6x + 7 = 0$
$3 + \sqrt{2} \approx 4.41$
$3 - \sqrt{2} \approx 1.59$

35. $x^2 + 4x - 3 = 0$
$-2 + \sqrt{7} \approx 0.65$
$-2 - \sqrt{7} \approx -4.65$

36. $x^2 + 6x + 7 = 0$
$-3 + \sqrt{2} \approx -1.59$
$-3 - \sqrt{2} \approx -4.41$

37. $3x^2 + 9x + 5 = 0$ $-\frac{3}{2} + \frac{\sqrt{21}}{6} \approx -0.74$
$-\frac{3}{2} - \frac{\sqrt{21}}{6} \approx -2.26$

38. $5x^2 - 15x + 7 = 0$ $\frac{3}{2} + \frac{\sqrt{85}}{10} \approx 2.42$
$\frac{3}{2} - \frac{\sqrt{85}}{10} \approx 0.58$

EXAMPLE 5 **Completing the Square: Leading Coefficient Is Not 1**

Solve $3x^2 - 6x + 1 = 0$ by completing the square.

SOLUTION

$3x^2 - 6x + 1 = 0$ Write original equation.

$3x^2 - 6x = -1$ Subtract 1 from each side.

$x^2 - 2x = -\frac{1}{3}$ Divide each side by 3.

$x^2 - 2x + (-1)^2 = -\frac{1}{3} + 1$ Add $(-1)^2 = 1$ to each side.

$(x - 1)^2 = \frac{2}{3}$ Completed square form

$x - 1 = \pm\sqrt{\frac{2}{3}}$ Square Root Property

$x - 1 = \pm\frac{\sqrt{6}}{3}$ Rationalize the denominator.

$x = 1 \pm \frac{\sqrt{6}}{3}$ Add 1 to each side.

The solutions are $x = 1 + \frac{\sqrt{6}}{3} \approx 1.82$ and $x = 1 - \frac{\sqrt{6}}{3} \approx 0.18$. Check these in the original equation.

Additional Examples

Solve each equation by completing the square.

a. $x^2 - 4x + 1 = 0$

b. $3x^2 - 2x - 4 = 0$

c. $x^2 - 5x + 10 = 0$

Answers:

a. $x = 2 \pm \sqrt{3}$

b. $x = \frac{1}{3} \pm \frac{\sqrt{13}}{3}$

c. $x = \frac{5}{2} \pm \frac{\sqrt{15}}{2}i$

Exercises Within Reach®

Solutions in English & Spanish and tutorial videos at AlgebraWithinReach.com

Completing the Square **In Exercises 39–46, solve the equation by completing the square. Give the solutions in exact form and in decimal form rounded to two decimal places.**

39. $2x^2 + 8x + 3 = 0$

$-2 + \frac{\sqrt{10}}{2} \approx -0.42$

$-2 - \frac{\sqrt{10}}{2} \approx -3.58$

40. $3x^2 - 24x - 5 = 0$

$4 + \frac{\sqrt{159}}{3} \approx 8.20$

$4 - \frac{\sqrt{159}}{3} \approx -0.20$

41. $4y^2 + 4y - 9 = 0$

$-\frac{1}{2} + \frac{\sqrt{10}}{2} \approx 1.08$

$-\frac{1}{2} - \frac{\sqrt{10}}{2} \approx -2.08$

42. $7u^2 - 8u - 3 = 0$

$\frac{4}{7} + \frac{\sqrt{37}}{7} \approx 1.44$

$\frac{4}{7} - \frac{\sqrt{37}}{7} \approx -0.30$

43. $x\left(x - \frac{2}{3}\right) = 14$

$\frac{1}{3} + \frac{\sqrt{127}}{3} \approx 4.09$

$\frac{1}{3} - \frac{\sqrt{127}}{3} \approx -3.42$

44. $2x\left(x + \frac{4}{3}\right) = 5$

$-\frac{2}{3} + \frac{\sqrt{106}}{6} \approx 1.05$

$-\frac{2}{3} - \frac{\sqrt{106}}{6} \approx -2.38$

45. $0.1x^2 + 0.5x = -0.2$

$-\frac{5}{2} + \frac{\sqrt{17}}{2} \approx -0.44$

$-\frac{5}{2} - \frac{\sqrt{17}}{2} \approx -4.56$

46. $0.2x^2 + 0.1x = 0.5$

$-\frac{1}{4} + \frac{\sqrt{41}}{4} \approx 1.35$

$-\frac{1}{4} - \frac{\sqrt{41}}{4} \approx -1.85$

EXAMPLE 6 A Quadratic Equation with Complex Solutions

Solve $x^2 - 4x + 8 = 0$ by completing the square.

SOLUTION

$x^2 - 4x + 8 = 0$	Write original equation.
$x^2 - 4x = -8$	Subtract 8 from each side.
$x^2 - 4x + (-2)^2 = -8 + 4$	Add $(-2)^2 = 4$ to each side.
$(x - 2)^2 = -4$	Completed square form
$x - 2 = \pm 2i$	Square Root Property
$x = 2 \pm 2i$	Add 2 to each side.

The solutions are $x = 2 + 2i$ and $x = 2 - 2i$. Check these in the original equation. Note that when a quadratic equation has no real solutions, the graph of the corresponding equation has no x-intercepts.

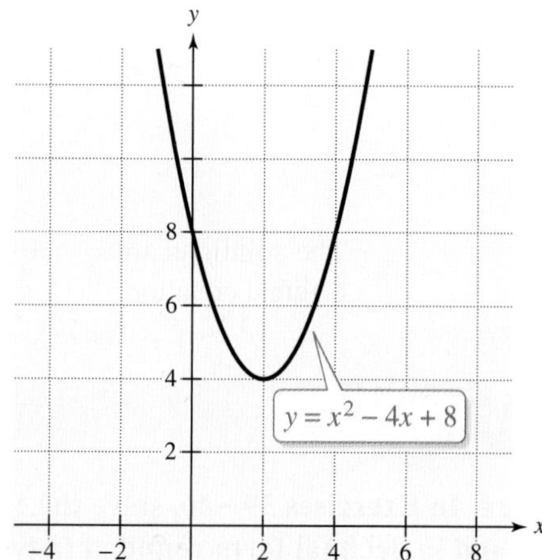

Exercises Within Reach®

Solutions in English & Spanish and tutorial videos at AlgebraWithinReach.com

Completing the Square **In Exercises 47–54, solve the equation by completing the square. Give the solutions in exact form and in decimal form rounded to two decimal places.**

47. $z^2 + 4z + 13 = 0$
$-2 \pm 3i$

48. $z^2 - 6z + 18 = 0$
$3 \pm 3i$

49. $x^2 + 9 = 4x$
$2 + \sqrt{5}i \approx 2 + 2.24i$
$2 - \sqrt{5}i \approx 2 - 2.24i$

50. $x^2 + 10 = 6x$
$3 \pm i$

51. $-x^2 + x - 1 = 0$
$\frac{1}{2} + \frac{\sqrt{3}}{2}i \approx 0.5 + 0.87i$
$\frac{1}{2} - \frac{\sqrt{3}}{2}i \approx 0.5 - 0.87i$

52. $-x^2 - x - 1 = 0$
$-\frac{1}{2} + \frac{\sqrt{3}}{2}i \approx -0.5 + 0.87i$
$-\frac{1}{2} - \frac{\sqrt{3}}{2}i \approx -0.5 - 0.87i$

53. $4z^2 - 3z + 2 = 0$
$\frac{3}{8} + \frac{\sqrt{23}}{8}i \approx 0.38 + 0.60i$
$\frac{3}{8} - \frac{\sqrt{23}}{8}i \approx 0.38 + 0.60i$

54. $5x^2 - 3x + 10 = 0$
$\frac{3}{10} + \frac{\sqrt{191}}{10}i \approx 0.30 + 1.38i$
$\frac{3}{10} - \frac{\sqrt{191}}{10}i \approx 0.30 - 1.38i$

Application EXAMPLE 7 Geometry: Dimensions of an iPhone

The first generation of the iPhone has an approximate volume of 4.968 cubic inches. Its width is 0.46 inch and its face has the dimensions x inches by $(x + 2.1)$ inches. Find the dimensions of the face in inches. (*Source:* Apple, Inc.)

SOLUTION

$lwh = V$	Formula for volume of a rectangular solid
$(x)(0.46)(x + 2.1) = 4.968$	Substitute 4.968 for V, x for l, 0.46 for w, and $x + 2.1$ for h.
$0.46x^2 + 0.966x = 4.968$	Multiply factors.
$x^2 + 2.1x = 10.8$	Divide each side by 0.46.
$x^2 + 2.1x + \left(\frac{2.1}{2}\right)^2 = 10.8 + 1.1025$	Add $\left(\frac{2.1}{2}\right)^2 = 1.1025$ to each side.
$(x + 1.05)^2 = 11.9025$	Completed square form
$x + 1.05 = \pm\sqrt{11.9025}$	Square Root Property
$x = -1.05 \pm \sqrt{11.9025}$	Subtract 1.05 from each side.

Choosing the positive root, you obtain

$x = -1.05 + 3.45 = 2.4$ inches — Length of face

and

$x + 2.1 = 2.4 + 2.1 = 4.5$ inches. — Height of face

Exercises Within Reach ®

Solutions in English & Spanish and tutorial videos at AlgebraWithinReach.com

55. ***Geometric Modeling*** Use the figure shown below.

(a) Find the area of the two adjoining rectangles and large square in the figure. $x^2 + 8x$

(b) Find the area of the small square in the lower right-hand corner of the figure and add it to the area found in part (a). $x^2 + 8x + 16$

(c) Find the dimensions and the area of the entire figure after adjoining the small square in the lower right-hand corner of the figure. Note that you have shown geometrically the technique of completing the square. $(x + 4)^2$

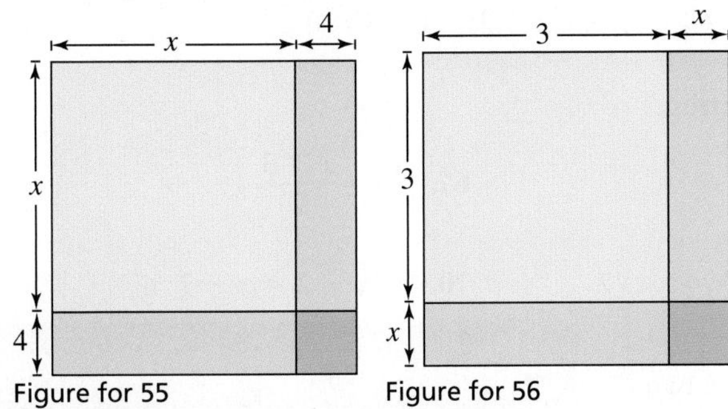

Figure for 55 Figure for 56

56. ***Geometric Modeling*** Repeat Exercise 55 for the figure shown above. (a) $6x + 9$ (b) $x^2 + 6x + 9$ (c) $(x + 3)^2$

57. ***Geometry*** An open box with a rectangular base of x inches by $(x + 4)$ inches has a height of 6 inches (see figure). The volume of the box is 840 cubic inches. Find the dimensions of the box.
6 inches × 10 inches × 14 inches

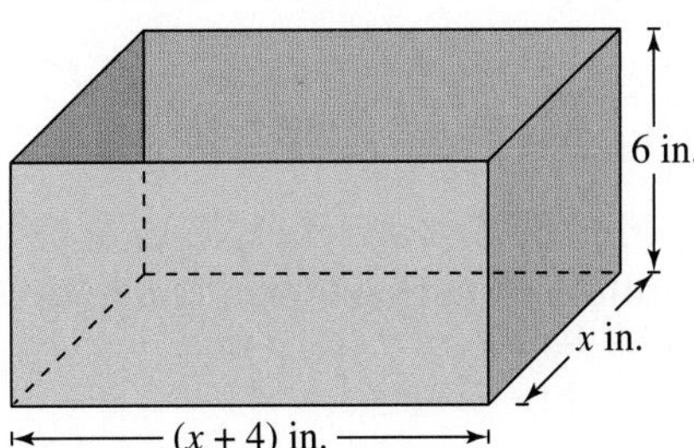

58. ***Geometry*** An open box with a rectangular base of $2x$ inches by $(6x - 2)$ inches has a height of 9 inches (see figure). The volume of the box is 1584 cubic inches. Find the dimensions of the box.
8 inches × 9 inches × 22 inches

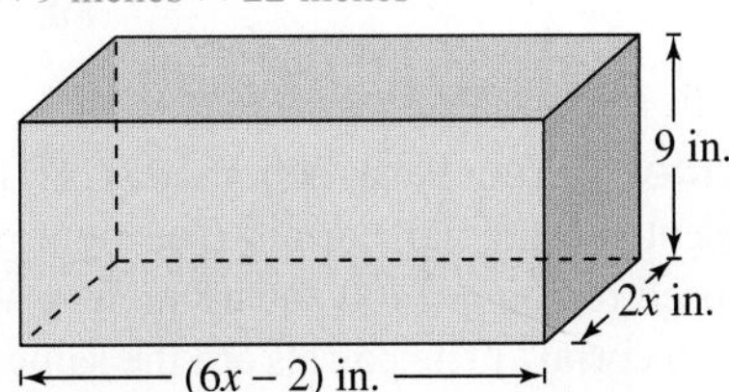

Concept Summary: Solving Quadratic Equations by Completing the Square

What

Another way to solve a quadratic equation is by **completing the square**.

EXAMPLE

Solve $2x^2 + 12x + 10 = 0$ by completing the square.

How

1. Write the equation in the form $x^2 + bx = c$.
2. Add $(b/2)^2$ to each side.
3. Write the equation in completed square form and use the Square Root Property to solve.

EXAMPLE

$2x^2 + 12x + 10 = 0$

$x^2 + 6x = -5$ Step 1

$x^2 + 6x + 3^2 = -5 + 9$ Step 2

$(x + 3)^2 = 4$

$x + 3 = \pm 2$

$x = -3 \pm 2$

$x = -1$ and $x = -5$ Step 3

Why

You can solve *any* quadratic equation by completing the square.

You can also use this method to identify quadratic equations that have no real solutions.

Exercises Within Reach ®

Worked-out solutions to odd-numbered exercises at AlgebraWithinReach.com

Concept Summary Check

59. ***Writing*** What is a perfect square trinomial?
A perfect square trinomial is one that can be written as $(x + k)^2$.

60. ***Completing the Square*** What term must be added to $x^2 + 5x$ to complete the square? Explain how you found the term.
$\frac{25}{4}$; The term is found by dividing the coefficient of the first-degree term by 2, and squaring the result to obtain $\left(\frac{5}{2}\right)^2 = \frac{25}{4}$.

61. ***Reasoning*** When using the method of completing the square to solve $2x^2 - 7x = 6$, what is the first step? Is the resulting equation equivalent to the original equation? Explain. Each side of the equation must be divided by 2 to obtain a leading coefficient of 1. The resulting equation is the same by the Multiplication Property of Equality.

62. ***Think About It*** Is it possible for a quadratic equation to have no real number solution? If so, give an example.
Yes. The solutions of $x^2 + 1 = 0$ are $\pm i$.

Extra Practice

Completing the Square **In Exercises 63 and 64, solve the equation by completing the square. Give the solutions in exact form and in decimal form rounded to two decimal places.**
See Additional Answers.

63. $0.2x^2 + 0.1x = -0.5$

64. $0.75x^2 + 1.25x + 1.5 = 0$

Solving an Equation **In Exercises 65–70, solve the equation.**

65. $\frac{x}{2} - \frac{1}{x} = 1$ $1 \pm \sqrt{3}$

66. $\frac{x}{2} + \frac{5}{x} = 4$ $4 \pm \sqrt{6}$

67. $\frac{x^2}{8} = \frac{x+3}{2}$ $-2, 6$

68. $\frac{x^2 + 2}{24} = \frac{x-1}{3}$ $4 \pm \sqrt{6}$

69. $\sqrt{2x + 1} = x - 3$ $4 \pm 2\sqrt{2}$

70. $\sqrt{3x - 2} = x - 2$ 6

71. ***Geometry*** You have 200 meters of fencing to enclose two adjacent rectangular corrals (see figure). The total area of the enclosed region is 1400 square meters. What are the dimensions of each corral? (The corrals are the same size.)
15 meters $\times$ $46\frac{2}{3}$ meters or 20 meters $\times$ 35 meters

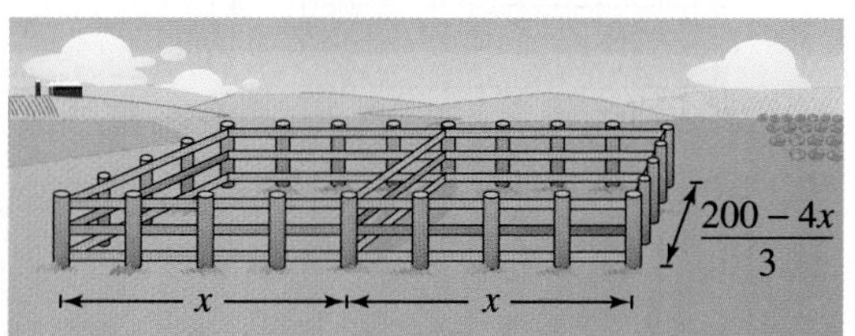

72. ***Geometry*** A kennel is adding a rectangular outdoor enclosure along one side of the kennel wall (see figure). The other three sides of the enclosure will be formed by a fence. The kennel has 111 feet of fencing and plans to use 1215 square feet of land for the enclosure. What are the dimensions of the enclosure?
30 feet × $40\frac{1}{2}$ feet or 15 feet × 81 feet

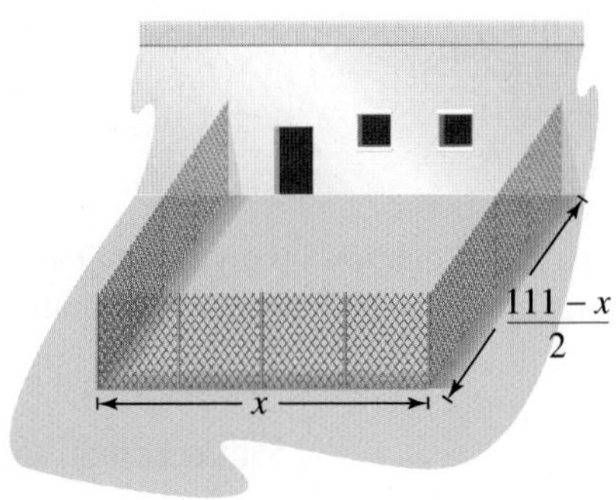

73. ***Revenue*** The revenue R (in dollars) from selling x pairs of running shoes is given by

$$R = x\left(80 - \frac{1}{2}x\right).$$

Find the number of pairs of running shoes that must be sold to produce a revenue of \$2750.
50 pairs, 110 pairs

74. ***Revenue*** The revenue R (in dollars) from selling x golf clubs is given by

$$R = x\left(150 - \frac{1}{10}x\right).$$

Find the number of golf clubs that must be sold to produce a revenue of \$15,033.60.
108 golf clubs, 1392 golf clubs

Explaining Concepts

75. ***Writing*** Explain the use of the Square Root Property when solving a quadratic equation by the method of completing the square. Use the method of completing the square to write the quadratic equation in the form $u^2 = d$. Then use the Square Root Property to simplify.

76. ***True or False?*** If you solve a quadratic equation by completing the square and obtain solutions that are rational numbers, then you could have solved the equation by factoring. Justify your answer.
True. Given the solutions $x = r_1$ and $x = r_2$, the quadratic equation can be written as $(x - r_1)(x - r_2) = 0$.

77. ***Think About It*** Consider the quadratic equation $(x - 1)^2 = d$.

(a) What value(s) of d will produce a quadratic equation that has exactly one (repeated) solution?
$d = 0$

(b) Describe the value(s) of d that will produce two different solutions, both of which are *rational* numbers. d is positive and is a perfect square.

(c) Describe the value(s) of d that will produce two different solutions, both of which are *irrational* numbers. d is positive and is not a perfect square.

(d) Describe the value(s) of d that will produce two different solutions, both of which are *complex* numbers. $d < 0$

78. ***Error Analysis*** You teach an algebra class and one of your students hands in the following solution. Describe and correct the error(s). See Additional Answers.

Solve $x^2 + 6x - 13 = 0$ by completing the square.

$$x^2 + 6x = 13$$
$$x^2 + 6x + \left(\frac{6}{2}\right)^2 = 13$$
$$(x + 3)^2 = 13$$
$$x + 3 = \pm\sqrt{13}$$
$$x = -3 \pm \sqrt{13}$$

Cumulative Review

In Exercises 79–86, perform the operation and simplify the expression.

79. $3\sqrt{5}\sqrt{500}$ 150

80. $2\sqrt{2x^2}\sqrt{27x}$ $6x\sqrt{6x}$

81. $(3 + \sqrt{2})(3 - \sqrt{2})$ 7

82. $(\sqrt[3]{6} - 2)(\sqrt[3]{4} + 1)$ $2\sqrt[3]{3} + \sqrt[3]{6} - 2\sqrt[3]{4} - 2$

83. $(3 + \sqrt{2})^2$ $11 + 6\sqrt{2}$

84. $(2\sqrt{x} - 5)^2$ $4x - 20\sqrt{x} + 25$

85. $\dfrac{8}{\sqrt{10}}$ $\dfrac{4\sqrt{10}}{5}$

86. $\dfrac{5}{\sqrt{12} - 2}$ $\dfrac{5(\sqrt{3} + 1)}{4}$

In Exercises 87 and 88, rewrite the expression using the specified rule, where *a* and *b* are nonnegative real numbers.

87. Product Rule: $\sqrt{ab} = \sqrt{a} \cdot \sqrt{b}$.

88. Quotient Rule: $\sqrt{\dfrac{a}{b}} = \dfrac{\sqrt{a}}{\sqrt{b}}$.

8.3 The Quadratic Formula

- Use the Quadratic Formula to solve quadratic equations.
- Determine the types of solutions of quadratic equations using the discriminant.
- Write quadratic equations from solutions of the equations.

The Quadratic Formula

Before completing the square for the *general* quadratic equation, consider solving a *specific* quadratic equation by completing the square and leaving it on the board for easy reference. Point out that the derivation of the Quadratic Formula is just a generalization of the technique presented in Section 8.2. Both the completing-the-square method and the Quadratic Formula can be used to solve *any* quadratic equation.

A fourth technique for solving a quadratic equation involves the *Quadratic Formula.* This formula is derived by completing the square for a general quadratic equation.

$$ax^2 + bx + c = 0 \qquad \text{General form, } a \neq 0$$

$$ax^2 + bx = -c \qquad \text{Subtract } c \text{ from each side.}$$

$$x^2 + \frac{b}{a}x = -\frac{c}{a} \qquad \text{Divide each side by } a.$$

$$x^2 + \frac{b}{a}x + \left(\frac{b}{2a}\right)^2 = -\frac{c}{a} + \left(\frac{b}{2a}\right)^2 \qquad \text{Add } \left(\frac{b}{2a}\right)^2 \text{ to each side.}$$

$$\left(x + \frac{b}{2a}\right)^2 = \frac{b^2 - 4ac}{4a^2} \qquad \text{Simplify.}$$

$$x + \frac{b}{2a} = \pm\sqrt{\frac{b^2 - 4ac}{4a^2}} \qquad \text{Square Root Property}$$

$$x = -\frac{b}{2a} \pm \frac{\sqrt{b^2 - 4ac}}{2|a|} \qquad \text{Subtract } \frac{b}{2a} \text{ from each side.}$$

$$x = \frac{-b \pm \sqrt{b^2 - 4ac}}{2a} \qquad \text{Simplify.}$$

Notice in the derivation of the Quadratic Formula that, because $\pm 2|a|$ represents the same numbers as $\pm 2a$, you can omit the absolute value bars.

Study Tip

The Quadratic Formula is one of the most important formulas in algebra, and you should memorize it. It helps to try to memorize a verbal statement of the rule. For instance, you might try to remember the following verbal statement of the Quadratic Formula: "The opposite of *b*, plus or minus the square root of *b* squared minus 4*ac*, all divided by 2*a*."

The Quadratic Formula

The solutions of $ax^2 + bx + c = 0$, $a \neq 0$, are given by the **Quadratic Formula**

$$x = \frac{-b \pm \sqrt{b^2 - 4ac}}{2a}.$$

Exercises Within Reach®

Solutions in English & Spanish and tutorial videos at AlgebraWithinReach.com

Writing a Quadratic Equation in General Form **In Exercises 1–4, write the quadratic equation in general form.**

1. $2x^2 = 7 - 2x$
$2x^2 + 2x - 7 = 0$

2. $7x^2 + 15x = 5$
$7x^2 + 15x - 5 = 0$

3. $x(10 - x) = 5$
$-x^2 + 10x - 5 = 0$

4. $x(2x + 9) = 12$
$2x^2 + 9x - 12 = 0$

EXAMPLE 1 The Quadratic Formula: Two Distinct Solutions

$x^2 + 6x = 16$	Original equation
$x^2 + 6x - 16 = 0$	Write in general form.
$x = \dfrac{-b \pm \sqrt{b^2 - 4ac}}{2a}$	Quadratic Formula
$x = \dfrac{-6 \pm \sqrt{6^2 - 4(1)(-16)}}{2(1)}$	Substitute 1 for a, 6 for b, and -16 for c.
$x = \dfrac{-6 \pm \sqrt{100}}{2}$	Simplify.
$x = \dfrac{-6 \pm 10}{2}$	Simplify.
$x = 2$ or $x = -8$	Solutions

The solutions are $x = 2$ and $x = -8$. Check these in the original equation.

Study Tip

In Example 1, the solutions are rational numbers, which means that the equation could have been solved by factoring. Try solving the equation by factoring.

Suggest that students write the Quadratic Formula for each homework problem before substituting specific values. Repeatedly writing the formula should help students learn it quickly.

EXAMPLE 2 The Quadratic Formula: Two Distinct Solutions

$-x^2 - 4x + 8 = 0$	Leading coefficient is negative.
$x^2 + 4x - 8 = 0$	Multiply each side by -1.
$x = \dfrac{-b \pm \sqrt{b^2 - 4ac}}{2a}$	Quadratic Formula
$x = \dfrac{-4 \pm \sqrt{4^2 - 4(1)(-8)}}{2(1)}$	Substitute 1 for a, 4 for b, and -8 for c.
$x = \dfrac{-4 \pm \sqrt{48}}{2}$	Simplify.
$x = \dfrac{-4 \pm 4\sqrt{3}}{2}$	Simplify.
$x = -2 \pm 2\sqrt{3}$	Solutions

The solutions are $x = -2 + 2\sqrt{3}$ and $x = -2 - 2\sqrt{3}$. Check these in the original equation.

Study Tip

If the leading coefficient of a quadratic equation is negative, you should begin by multiplying each side of the equation by -1, as shown in Example 2. This will produce a positive leading coefficient, which is easier to work with.

Some students may be tempted to reduce $(-4 \pm 4\sqrt{3})/2$ to $-2 \pm 4\sqrt{3}$. Ask them what is wrong with this reasoning.

Exercises Within Reach®

Solutions in English & Spanish and tutorial videos at AlgebraWithinReach.com

Solving a Quadratic Equation **In Exercises 5–8, solve the equation first by using the Quadratic Formula and then by factoring.**

5. $x^2 - 11x + 28 = 0$ 4, 7

6. $x^2 - 12x + 27 = 0$ 3, 9

7. $x^2 + 6x + 8 = 0$ $-2, -4$

8. $x^2 + 9x + 14 = 0$ $-7, -2$

Solving a Quadratic Equation **In Exercises 9–14, solve the equation by using the Quadratic Formula.**

9. $x^2 - 2x - 4 = 0$ $1 \pm \sqrt{5}$

10. $x^2 - 2x - 6 = 0$ $1 \pm \sqrt{7}$

11. $t^2 + 4t + 1 = 0$ $-2 \pm \sqrt{3}$

12. $y^2 + 6y - 8 = 0$ $-3 \pm \sqrt{17}$

13. $-x^2 + 10x - 23 = 0$ $5 \pm \sqrt{2}$

14. $-u^2 + 12u - 29 = 0$ $6 \pm \sqrt{7}$

Study Tip

Example 3 could have been solved as follows, without dividing each side by 2 in the first step.

$x =$

$$\frac{-(-24) \pm \sqrt{(-24)^2 - 4(18)(8)}}{2(18)}$$

$$x = \frac{24 \pm \sqrt{576 - 576}}{36}$$

$$x = \frac{24 \pm 0}{36}$$

$$x = \frac{2}{3}$$

While the result is the same, dividing each side by 2 simplifies the equation before the Quadratic Formula is applied. This allows you to work with smaller numbers.

Additional Examples

Solve each equation using the Quadratic Formula.

a. $3x^2 + 4x - 8 = 0$

b. $-x^2 - 2x - 4 = 0$

Answers:

a. $x = -\frac{2}{3} \pm \frac{2\sqrt{7}}{3}$

b. $x = -1 \pm \sqrt{3}i$

EXAMPLE 3 The Quadratic Formula: One Repeated Solution

$18x^2 - 24x + 8 = 0$	Original equation
$9x^2 - 12x + 4 = 0$	Divide each side by 2.
$x = \frac{-b \pm \sqrt{b^2 - 4ac}}{2a}$	Quadratic Formula
$x = \frac{-(-12) \pm \sqrt{(-12)^2 - 4(9)(4)}}{2(9)}$	Substitute 9 for a, -12 for b, and 4 for c.
$x = \frac{12 \pm \sqrt{144 - 144}}{18}$	Simplify.
$x = \frac{12 \pm \sqrt{0}}{18}$	Simplify.
$x = \frac{2}{3}$	Solution

The only solution is $x = \frac{2}{3}$. Check this in the original equation.

EXAMPLE 4 The Quadratic Formula: Complex Solutions

$2x^2 - 4x + 5 = 0$	Original equation
$x = \frac{-b \pm \sqrt{b^2 - 4ac}}{2a}$	Quadratic Formula
$x = \frac{-(-4) \pm \sqrt{(-4)^2 - 4(2)(5)}}{2(2)}$	Substitute 2 for a, -4 for b, and 5 for c.
$x = \frac{4 \pm \sqrt{-24}}{4}$	Simplify.
$x = \frac{4 \pm 2\sqrt{6}i}{4}$	Write in i-form.
$x = \frac{2(2 \pm \sqrt{6}i)}{2 \cdot 2}$	Factor numerator and denominator.
$x = \frac{\cancel{2}(2 \pm \sqrt{6}i)}{\cancel{2} \cdot 2}$	Divide out common factor.
$x = 1 \pm \frac{\sqrt{6}}{2}i$	Solutions

The solutions are $x = 1 + \frac{\sqrt{6}}{2}i$ and $x = 1 - \frac{\sqrt{6}}{2}i$. Check these in the original equation.

Exercises Within Reach®

Solutions in English & Spanish and tutorial videos at AlgebraWithinReach.com

Solving a Quadratic Equation **In Exercises 15 and 16, solve the equation first by using the Quadratic Formula and then by factoring.**

15. $16x^2 + 8x + 1 = 0$ $-\frac{1}{4}$

16. $9x^2 + 12x + 4 = 0$ $-\frac{2}{3}$

Solving a Quadratic Equation **In Exercises 17 and 18, solve the equation by using the Quadratic Formula.**

17. $2x^2 + 3x + 3 = 0$ $-\frac{3}{4} \pm \frac{\sqrt{15}}{4}i$

18. $2x^2 - 2x + 3 = 0$ $\frac{1}{2} \pm \frac{\sqrt{5}}{2}i$

The Discriminant

Using the Discriminant

Let a, b, and c be rational numbers such that $a \neq 0$. The **discriminant** of the quadratic equation $ax^2 + bx + c = 0$ is given by $b^2 - 4ac$, and can be used to classify the solutions of the equation as follows.

Discriminant	*Solution Type*
1. Perfect square	Two distinct rational solutions (Example 1)
2. Positive nonperfect square	Two distinct irrational solutions (Example 2)
3. Zero	One repeated rational solution (Example 3)
4. Negative number	Two distinct complex solutions (Example 4)

Study Tip

By reexamining Examples 1 through 4, you can see that the equations with rational or repeated solutions could have been solved by *factoring*. In general, quadratic equations (with integer coefficients) for which the discriminant is either zero or a perfect square are factorable using integer coefficients. Consequently, a quick test of the discriminant will help you decide which solution method to use to solve a quadratic equation.

EXAMPLE 5 Using the Discriminant

Determine the type of solution(s) for each quadratic equation.

a. $x^2 - x + 2 = 0$ **b.** $2x^2 - 3x - 2 = 0$

c. $x^2 - 2x + 1 = 0$ **d.** $x^2 - 2x - 1 = 9$

SOLUTION

Equation	*Discriminant*	*Solution Type*
a. $x^2 - x + 2 = 0$	$b^2 - 4ac = (-1)^2 - 4(1)(2)$ $= 1 - 8 = -7$	Two distinct complex solutions
b. $2x^2 - 3x - 2 = 0$	$b^2 - 4ac = (-3)^2 - 4(2)(-2)$ $= 9 + 16 = 25$	Two distinct rational solutions
c. $x^2 - 2x + 1 = 0$	$b^2 - 4ac = (-2)^2 - 4(1)(1)$ $= 4 - 4 = 0$	One repeated rational solution
d. $x^2 - 2x - 1 = 9$	$b^2 - 4ac = (-2)^2 - 4(1)(-10)$ $= 4 + 40 = 44$	Two distinct irrational solutions

Exercises Within Reach®

Solutions in English & Spanish and tutorial videos at AlgebraWithinReach.com

Using the Discriminant **In Exercises 19–26, use the discriminant to determine the type of solution(s) of the quadratic equation.**

19. $x^2 + x + 1 = 0$ Two distinct complex solutions

20. $x^2 + x - 1 = 0$ Two distinct irrational solutions

21. $3x^2 - 2x - 5 = 0$ Two distinct rational solutions

22. $5x^2 + 7x + 3 = 0$ Two distinct complex solutions

23. $9x^2 - 24x + 16 = 0$ One repeated rational solution

24. $2x^2 + 10x + 6 = 0$ Two distinct irrational solutions

25. $3x^2 - x = -2$ Two distinct complex solutions

26. $4x^2 - 16x = -16$ One repeated rational solution

Summary of Methods for Solving Quadratic Equations

Method	*Example*
1. Factoring	$3x^2 + x = 0$ $x(3x + 1) = 0 \Rightarrow x = 0$ and $x = -\frac{1}{3}$
2. Square Root Property	$(x + 2)^2 = 7$ $x + 2 = \pm\sqrt{7} \Rightarrow x = -2 + \sqrt{7}$ and $x = -2 - \sqrt{7}$
3. Completing the square	$x^2 + 6x = 2$ $x^2 + 6x + 3^2 = 2 + 9$ $(x + 3)^2 = 11 \Rightarrow x = -3 + \sqrt{11}$ and $x = -3 - \sqrt{11}$
4. Quadratic Formula	$3x^2 - 2x + 2 = 0 \Rightarrow x = \frac{-(-2) \pm \sqrt{(-2)^2 - 4(3)(2)}}{2(3)} = \frac{1}{3} \pm \frac{\sqrt{5}}{3}i$

Exercises Within Reach ®

Solutions in English & Spanish and tutorial videos at AlgebraWithinReach.com

Choosing a Method **In Exercises 27–42, solve the quadratic equation by using the most convenient method.**

27. $z^2 - 169 = 0$ ± 13

28. $t^2 = 144$ ± 12

29. $5y^2 + 15y = 0$ $-3, 0$

30. $12u^2 + 30u = 0$ $-\frac{5}{2}, 0$

31. $25(x - 3)^2 - 36 = 0$ $\frac{9}{5}, \frac{21}{5}$

32. $9(x + 4)^2 + 16 = 0$ $-4 \pm \frac{4}{3}i$

33. $2y(y - 18) + 3(y - 18) = 0$
$-\frac{3}{2}, 18$

34. $4y(y + 7) - 5(y + 7) = 0$
$-7, \frac{5}{4}$

35. $x^2 + 8x + 25 = 0$
$-4 \pm 3i$

36. $y^2 + 21y + 108 = 0$
$-12, -9$

37. $3x^2 - 13x + 169 = 0$
$\frac{13}{6} \pm \frac{13\sqrt{11}}{6}i$

38. $2x^2 - 15x + 225 = 0$
$\frac{15}{4} \pm \frac{15\sqrt{7}}{4}i$

39. $25x^2 + 80x + 61 = 0$
$-\frac{8}{5} \pm \frac{\sqrt{3}}{5}$

40. $14x^2 + 11x - 40 = 0$
$-\frac{11}{28} \pm \frac{\sqrt{2361}}{28}$

41. $7x(x + 2) + 5 = 3x(x + 1)$
$-\frac{11}{8} \pm \frac{\sqrt{41}}{8}$

42. $5x(x - 1) - 7 = 4x(x - 2)$
$-\frac{3}{2} \pm \frac{\sqrt{37}}{2}$

Using Technology and Algebra **In Exercises 43–46, use a graphing calculator to graph the function. Use the graph to approximate any x-intercepts of the graph. Set $y = 0$ and solve the resulting equation. Compare the result with the x-intercepts of the graph.** See Additional Answers.

43. $y = x^2 - 4x + 3$
$(1, 0), (3, 0)$
The result is the same.

44. $y = 5x^2 - 18x + 6$
$(3.23, 0), (0.37, 0)$
The result is the same.

45. $y = -0.03x^2 + 2x - 0.4$
$(0.20, 0), (66.47, 0)$
The result is the same.

46. $y = 3.7x^2 - 10.2x + 3.2$
$(2.40, 0), (0.36, 0)$
The result is the same.

Writing Quadratic Equations from Solutions

Using the Zero-Factor Property, you know that the equation $(x + 5)(x - 2) = 0$ has two solutions, $x = -5$ and $x = 2$. You can use the Zero-Factor Property in reverse to find a quadratic equation given its solutions. This process is demonstrated in Example 6.

Reverse of Zero-Factor Property

Let a and b be real numbers, variables, or algebraic expressions. If $a = 0$ or $b = 0$, then a and b are factors such that $ab = 0$.

EXAMPLE 6 Writing a Quadratic Equation from Its Solutions

Write a quadratic equation that has the solutions $x = 4$ and $x = -7$.

SOLUTION

Using the solutions $x = 4$ and $x = -7$, you can write the following.

$x = 4$	and	$x = -7$	Solutions
$x - 4 = 0$		$x + 7 = 0$	Obtain zero on one side of each equation.
		$(x - 4)(x + 7) = 0$	Reverse of Zero-Factor Property
		$x^2 + 3x - 28 = 0$	FOIL Method

So, a quadratic equation that has the solutions $x = 4$ and $x = -7$ is

$$x^2 + 3x - 28 = 0.$$

This is not the only quadratic equation with the solutions $x = 4$ and $x = -7$. You can obtain other quadratic equations with these solutions by multiplying $x^2 + 3x - 28 = 0$ by any nonzero real number.

Exercises Within Reach®

Solutions in English & Spanish and tutorial videos at AlgebraWithinReach.com

Writing a Quadratic Equation **In Exercises 47–58, write a quadratic equation that has the given solutions.**

47. $x = 5, x = -2$
$x^2 - 3x - 10 = 0$

48. $x = -2, x = 3$
$x^2 - x - 6 = 0$

49. $x = 1, x = 7$
$x^2 - 8x + 7 = 0$

50. $x = 2, x = 8$
$x^2 - 10x + 16 = 0$

51. $x = 1 + \sqrt{2}, x = 1 - \sqrt{2}$
$x^2 - 2x - 1 = 0$

52. $x = -3 + \sqrt{5}, x = -3 - \sqrt{5}$
$x^2 + 6x + 4 = 0$

53. $x = 5i, x = -5i$
$x^2 + 25 = 0$

54. $x = 2i, x = -2i$
$x^2 + 4 = 0$

55. $x = 12$
$x^2 - 24x + 144 = 0$

56. $x = -4$
$x^2 + 8x + 16 = 0$

57. $x = \frac{1}{2}$
$x^2 - x + \frac{1}{4} = 0$

58. $x = -\frac{3}{4}$
$x^2 + \frac{3}{2}x + \frac{9}{16} = 0$

Concept Summary: Solving Quadratic Equations by Using the Quadratic Formula

What

You can use the **Quadratic Formula** to solve quadratic equations.

EXAMPLE

Use the Quadratic Formula to solve $x^2 + 3x + 2 = 0$.

How

1. Identify the values of a, b, and c from the general form of the equation.
2. Substitute these values into the Quadratic Formula.
3. Simplify to obtain the solution.

EXAMPLE

$x = \dfrac{-b \pm \sqrt{b^2 - 4ac}}{2a}$ Quadratic Formula

$x = \dfrac{-3 \pm \sqrt{3^2 - 4(1)(2)}}{2(1)}$ Substitute.

$x = \dfrac{-3 \pm \sqrt{1}}{2} = \dfrac{-3 \pm 1}{2}$ Simplify.

$x = -1$ and $x = -2$

Why

You now know four methods for solving quadratic equations.

1. Factoring
2. Square Root Property
3. Completing the square
4. The Quadratic Formula

Remember that you can use the Quadratic Formula or completing the square to solve *any* quadratic equation.

Exercises Within Reach ®

Worked-out solutions to odd-numbered exercises at AlgebraWithinReach.com

Concept Summary Check

59. ***Vocabulary*** State the Quadratic Formula in words.
The opposite of b, plus or minus the square root of b squared minus $4ac$, all divided by $2a$.

60. ***Four Methods*** State the four methods used to solve quadratic equations.
The four methods are factoring, the Square Root Property, completing the square, and the Quadratic Formula.

61. ***Reasoning*** To solve the quadratic equation $3x^2 = 3 - x$ using the Quadratic Formula, what are the values of a, b, and c? The equation in general form is $3x^2 + x - 3 = 0$, so $a = 3$, $b = 1$, and $c = -3$.

62. ***Think About It*** The discriminant of a quadratic equation is -25. What type of solution(s) does the equation have? Two distinct complex solutions

Extra Practice

Solving an Equation **In Exercises 63–66, solve the equation.**

63. $\dfrac{x^2}{4} - \dfrac{2x}{3} = 1$
$\dfrac{4}{3} \pm \dfrac{2\sqrt{13}}{3}$

64. $\dfrac{x^2 - 9x}{6} = \dfrac{x - 1}{2}$
$6 \pm \sqrt{33}$

65. $\sqrt{x + 3} = x - 1$
$\dfrac{3}{2} + \dfrac{\sqrt{17}}{2}$

66. $\sqrt{2x - 3} = x - 2$
$3 + \sqrt{2}$

67. ***Geometry*** A rectangle has a width of x inches, a length of $(x + 6.3)$ inches, and an area of 58.14 square inches. Find its dimensions.
5.1 inches × 11.4 inches

68. ***Geometry*** A rectangle has a length of $(x + 1.5)$ inches, a width of x inches, and an area of 18.36 square inches. Find its dimensions.
3.6 inches × 5.1 inches

69. ***Depth of a River*** The depth d (in feet) of a river is given by

$$d = -0.25t^2 + 1.7t + 3.5, \quad 0 \le t \le 7$$

where t is the time (in hours) after a heavy rain begins. When is the river 6 feet deep?
2.15 or 4.65 hours

70. ***Free-Falling Object*** A stone is thrown vertically upward at a velocity of 20 feet per second from a bridge that is 40 feet above the level of the water. The height h (in feet) of the stone at time t (in seconds) after it is thrown is given by

$$h-16t^2 + 20t + 40.$$

(a) Find the time when the stone is again 40 feet above the water. 1.25 seconds

(b) Find the time when the stone strikes the water.
$\frac{5}{8} + \frac{\sqrt{185}}{8} \approx 2.33$ seconds

(c) Does the stone reach a height of 50 feet? Use the discriminant to justify your answer.
See Additional Answers.

71. ***Fuel Economy*** The fuel economy y (in miles per gallon) of a car is given by

$$y = -0.013x^2 + 1.25x + 5.6, \quad 5 \le x \le 75$$

where x is the speed (in miles per hour) of the car.

(a) Use a graphing calculator to graph the model.
See Additional Answers.

(b) Use the graph in part (a) to find the speeds at which you can travel and have a fuel economy of 32 miles per gallon. Verify your results algebraically.
31.3 or 64.8 miles per hour

72. (a) Determine the two solutions, x_1 and x_2, of each quadratic equation. Use the values of x_1 and x_2 to fill in the boxes.

Equation	x_1, x_2	$x_1 + x_2$	x_1x_2
(i) $x^2 - x - 6 = 0$	$-2, 3$	1	-6
(ii) $2x^2 + 5x - 3 = 0$	$-3, \frac{1}{2}$	$-\frac{5}{2}$	$-\frac{3}{2}$
(iii) $4x^2 - 9 = 0$	$-\frac{3}{2}, \frac{3}{2}$	0	$-\frac{9}{4}$
(iv) $x^2 - 10x + 34 = 0$	$5 + 3i$, $5 - 3i$	10	34

(b) Consider a general quadratic equation

$$ax^2 + bx + c = 0$$

whose solutions are x_1 and x_2. Use the results of part (a) to determine how the coefficients a, b, and c are related to both the sum $(x_1 + x_2)$ and the product (x_1x_2) of the solutions.
For the general quadratic equation $ax^2 + bx + c = 0$ with solutions x_1 and x_2,
$x_1 + x_2 = -\frac{b}{a}$ and $x_1x_2 = \frac{c}{a}$.

Explaining Concepts

Choosing a Method **In Exercises 73–76, determine the method of solving the quadratic equation that would be most convenient. Explain your reasoning.**

73. $(x - 3)^2 = 25$ The Square Root Property would be convenient because the equation is of the form $u^2 = d$.

74. $x^2 + 8x - 12 = 0$ Completing the square would be convenient because the expression cannot be factored and the leading coefficient is 1.

75. $2x^2 - 9x + 12 = 0$ The Quadratic Formula would be convenient because the equation is already in general form, the expression cannot be factored, and the leading coefficient is not 1.

76. $8x^2 - 40x = 0$ Factoring would be convenient because the expression can be easily factored.

77. ***Precision*** Explain how the discriminant of $ax^2 + bx + c = 0$ is related to the number of x-intercepts of the graph of $y = ax^2 + bx + c$.
See Additional Answers.

78. ***Error Analysis*** Describe and correct the student's error in writing a quadratic equation that has solutions $x = 2$ and $x = 4$. See Additional Answers.

$(x + 2)(x + 4) = 0$

$x^2 + 6x + 8 = 0$

Cumulative Review

In Exercises 79–82, find the distance between the points.

79. $(-1, 11), (2, 2)$ $3\sqrt{10}$

80. $(-2, 4), (3, -3)$ $\sqrt{74}$

81. $(-6, -2), (-3, -4)$ $\sqrt{13}$

82. $(-4, 7), (0, 4)$ 5

In Exercises 83–86, sketch the graph of the function.
See Additional Answers.

83. $f(x) = (x - 1)^2$

84. $f(x) = \frac{1}{2}x^2$

85. $f(x) = (x - 2)^2 + 4$

86. $f(x) = (x + 3)^2 - 1$

Mid-Chapter Quiz: Sections 8.1–8.3

Solutions in English & Spanish and tutorial videos at AlgebraWithinReach.com

Take this quiz as you would take a quiz in class. After you are done, check your work against the answers in the back of the book.

In Exercises 1–8, solve the quadratic equation by the specified method.

1. Factoring:
$2x^2 - 72 = 0$ ± 6

2. Factoring:
$2x^2 + 3x - 20 = 0$ $-4, \frac{5}{2}$

3. Square Root Property:
$3x^2 = 36$ $\pm 2\sqrt{3}$

4. Square Root Property
$(u - 3)^2 - 16 = 0$ $-1, 7$

5. Completing the square:
$m^2 + 7m + 2 = 0$ $-\frac{7}{2} \pm \frac{\sqrt{41}}{2}$

6. Completing the square:
$2y^2 + 6y - 5 = 0$ $-\frac{3}{2} \pm \frac{\sqrt{19}}{2}$

7. Quadratic Formula:
$x^2 + 4x - 6 = 0$ $-2 \pm \sqrt{10}$

8. Quadratic Formula:
$6v^2 - 3v - 4 = 0$ $\frac{1}{4} \pm \frac{\sqrt{105}}{12}$

In Exercises 9–16, solve the equation by using the most convenient method.

9. $x^2 + 5x + 7 = 0$ $-\frac{5}{2} \pm \frac{\sqrt{3}}{2}i$

10. $36 - (t - 4)^2 = 0$ $-2, 10$

11. $x(x - 10) + 3(x - 10) = 0$ $-3, 10$

12. $x(x - 3) = 10$ $-2, 5$

13. $4b^2 - 12b + 9 = 0$ $\frac{3}{2}$

14. $3m^2 + 10m + 5 = 0$ $-\frac{5}{3} \pm \frac{\sqrt{10}}{3}$

15. $x - 4\sqrt{x} - 21 = 0$ 49

16. $x^4 + 7x^2 + 12 = 0$ $\pm 2i, \pm\sqrt{3}i$

In Exercises 17 and 18, solve the equation of quadratic form.

17. $x - 4\sqrt{x} + 3 = 0$ $1, 9$

18. $x^4 - 14x^2 + 24 = 0$ $\pm\sqrt{2}, \pm 2\sqrt{3}$

Applications

19. The revenue R (in dollars) from selling x handheld video games is given by

$R = x(180 - 1.5x)$.

Find the number of handheld video games that must be sold to produce a revenue of $5400. 60 video games

20. A rectangle has a length of x meters, a width of $(100 - x)$ meters, and an area of 2275 square meters. Find its dimensions. 35 meters × 65 meters

21. The path of a baseball after it is hit is given by

$h = -0.003x^2 + 1.19x + 5.2$

where h is the height (in feet) of the baseball and x is the horizontal distance (in feet) of the ball from home plate. The ball hits the top of the outfield fence that is 10 feet high. How far is the outfield fence from home plate. 392.6 feet

Study Skills in Action

Managing Test Anxiety

Test anxiety is different from the typical nervousness that usually occurs during tests. It interferes with the thinking process. After leaving the classroom, have you suddenly been able to recall what you could not remember during the test? It is likely that this was a result of test anxiety. Test anxiety is a learned reaction or response—no one is born with it. The good news is that most students can learn to manage test anxiety.

It is important to get as much information as you can into your long-term memory and to practice retrieving the information before you take a test. The more you practice retrieving information, the easier it will be during the test.

Smart Study Strategy

Make Mental Cheat Sheets

No, we are not asking you to cheat! Just prepare as if you were going to and then memorize the information you've gathered.

1 ▶ Write down important information on note cards. This can include:

- formulas
- examples of problems you find difficult
- concepts that always trip you up

2 ▶ Memorize the information on the note cards. Flash through the cards, placing the ones containing information you know in one stack and the ones containing information you do not know in another stack. Keep working on the information you do not know.

3 ▶ As soon as you receive your test, turn it over and write down all the information you remember, starting with things you have the greatest difficulty remembering. Having this information available should boost your confidence and free up mental energy for focusing on the test.

Do not wait until the night before the test to make note cards. Make them after you study each section. Then review them two or three times a week.

Completing the square

To complete the square for the expression $x^2 + bx$, add $(b/2)^2$.

$x^2 + bx + (\frac{b}{2})^2 = (x + \frac{b}{2})^2$

Quadratic Formula:

$$x = \frac{-b \pm \sqrt{b^2 - 4ac}}{2a}$$

Use $x = -b/(2a)$ to find the vertex and axis of a parabola.

8.4 Graphs of Quadratic Functions

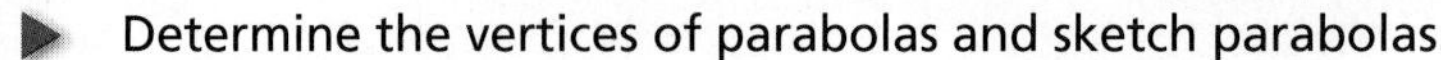

- Determine the vertices of parabolas and sketch parabolas.
- Write an equation of a parabola given the vertex and a point on the graph.
- Use parabolas to solve application problems.

Graphs of Quadratic Functions

Graphs of Quadratic Functions

The graph of $f(x) = ax^2 + bx + c$, $a \neq 0$, is a **parabola**. The completed square form

$$f(x) = a(x - h)^2 + k \qquad \text{Standard form}$$

is the **standard form** of the function. The **vertex** of the parabola occurs at the point (h, k), and the vertical line passing through the vertex is the **axis** of the parabola.

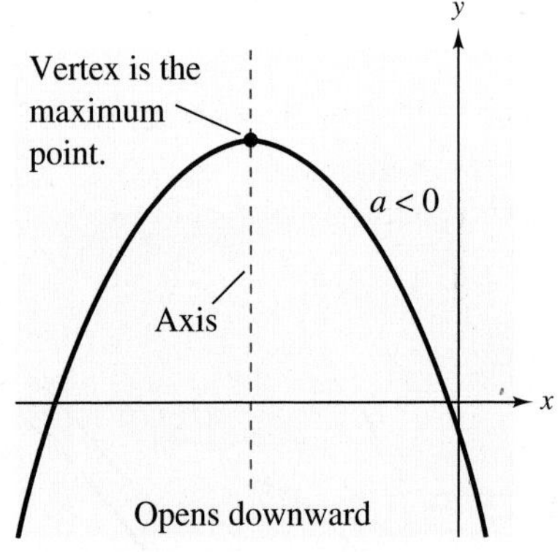

Every parabola is *symmetric* about its axis, which means that if it were folded along its axis, the two parts would match.

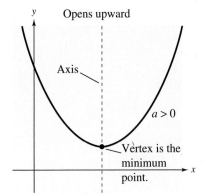

If a is positive, the graph of $f(x) = ax^2 + bx + c$ opens upward, and if a is negative, the graph opens downward, as shown at the left. Observe in the graphs that the y-coordinate of the vertex identifies the minimum function value when $a > 0$ and the maximum function value when $a < 0$.

EXAMPLE 1 Finding the Vertex by Completing the Square

Find the vertex of the parabola given by $f(x) = x^2 - 6x + 5$.

SOLUTION

Begin by writing the function in standard form.

$f(x) = x^2 - 6x + 5$	Write original function.
$f(x) = x^2 - 6x + (-3)^2 - (-3)^2 + 5$	Complete the square.
$f(x) = (x^2 - 6x + 9) - 9 + 5$	Group terms.
$f(x) = (x - 3)^2 - 4$	Standard form

From the standard form, you can see that the vertex of the parabola occurs at the point $(3, -4)$, as shown at the left. The minimum value of the function is $f(3) = -4$.

Exercises Within Reach®

Solutions in English & Spanish and tutorial videos at AlgebraWithinReach.com

Finding the Vertex of a Parabola **In Exercises 1–10, write the equation of the parabola in standard form, and find the vertex of its graph.**

1. $y = x^2 - 2x$ $y = (x - 1)^2 - 1, (1, -1)$
2. $y = x^2 + 2x$ $y = (x + 1)^2 - 1, (-1, -1)$
3. $y = x^2 - 4x + 7$ $y = (x - 2)^2 + 3, (2, 3)$
4. $y = x^2 + 6x - 5$ $y = (x + 3)^2 - 14, (-3, -14)$
5. $y = x^2 + 6x + 5$ $y = (x + 3)^2 - 4, (-3, -4)$
6. $y = x^2 - 4x + 5$ $y = (x - 2)^2 + 1, (2, 1)$
7. $y = -x^2 + 6x - 10$ $y = -(x - 3)^2 - 1, (3, -1)$
8. $y = -x^2 + 4x - 8$ $y = -(x - 2)^2 - 4, (2, -4)$
9. $y = 2x^2 + 6x + 2$ $y = 2\left(x + \frac{3}{2}\right)^2 - \frac{5}{2}, \left(-\frac{3}{2}, -\frac{5}{2}\right)$
10. $y = 3x^2 - 3x - 9$ $y = 3\left(x - \frac{1}{2}\right)^2 - \frac{39}{4}, \left(\frac{1}{2}, -\frac{39}{4}\right)$

In Example 1, the vertex of the graph was found by *completing the square*. Another approach to finding the vertex is to complete the square once for a general function and then use the resulting formula to find the vertex.

$$f(x) = ax^2 + bx + c \qquad \text{Quadratic function}$$

$$= a\left(x^2 + \frac{b}{a}x\right) + c \qquad \text{Factor } a \text{ out of first two terms.}$$

$$= a\left[x^2 + \frac{b}{a}x + \left(\frac{b}{2a}\right)^2\right] + c - \frac{b^2}{4a} \qquad \text{Complete the square.}$$

$$= a\left(x + \frac{b}{2a}\right)^2 + c - \frac{b^2}{4a} \qquad \text{Standard form}$$

From this form you can see that the vertex occurs when $x = -\dfrac{b}{2a}$.

When you ask students to find the vertex using $x = -\dfrac{b}{2a}$, remind them also to find the y-coordinate of the vertex.

EXAMPLE 2 Finding the Vertex Using a Formula

Find the vertex of the parabola given by $f(x) = 3x^2 - 9x$.

SOLUTION

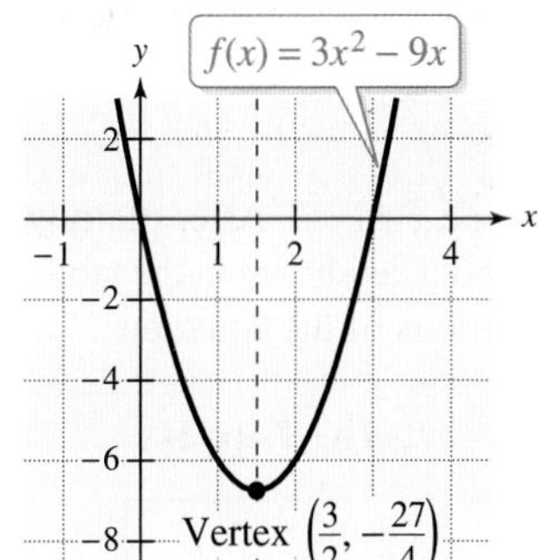

From the original function, it follows that $a = 3$ and $b = -9$. So, the x-coordinate of the vertex is

$$x = \frac{-b}{2a}$$

$$= \frac{-(-9)}{2(3)}$$

$$= \frac{3}{2}.$$

Substitute $\frac{3}{2}$ for x in the original equation to find the y-coordinate.

$$f\left(-\frac{b}{2a}\right) = f\left(\frac{3}{2}\right)$$

$$= 3\left(\frac{3}{2}\right)^2 - 9\left(\frac{3}{2}\right)$$

$$= -\frac{27}{4}$$

So, the vertex of the parabola is $\left(\frac{3}{2}, -\frac{27}{4}\right)$, the minimum value of the function is $f\left(\frac{3}{2}\right) = -\frac{27}{4}$, and the parabola opens upward, as shown in the figure.

Exercises Within Reach®

Solutions in English & Spanish and tutorial videos at AlgebraWithinReach.com

Finding the Vertex of a Parabola **In Exercises 11–16, find the vertex of the graph of the function by using the formula $x = -\dfrac{b}{2a}$.**

11. $f(x) = x^2 - 8x + 15$ $(4, -1)$

12. $f(x) = x^2 + 4x + 1$ $(-2, -3)$

13. $g(x) = -x^2 - 2x + 1$ $(-1, 2)$

14. $h(x) = -x^2 + 14x - 14$ $(7, 35)$

15. $y = 4x^2 + 4x + 4$ $\left(-\frac{1}{2}, 3\right)$

16. $y = 9x^2 - 12x$ $\left(\frac{2}{3}, -4\right)$

Study Tip

The x-and y-intercepts are useful points to plot. Another convenient fact is that the x-coordinate of the vertex lies halfway between the x-intercepts. Keep this in mind as you study the examples and do the exercises in this section.

Sketching a Parabola

1. Determine the vertex and axis of the parabola by completing the square or by using the formula $x = -\dfrac{b}{2a}$.
2. Plot the vertex, axis, x- and y-intercepts, and a few additional points on the parabola. (Using the symmetry about the axis can reduce the number of points you need to plot.)
3. Use the fact that the parabola opens *upward* when $a > 0$ and opens *downward* when $a < 0$ to complete the sketch.

EXAMPLE 3 Sketching a Parabola

To sketch the parabola given by $y = x^2 + 6x + 8$, begin by writing the equation in standard form.

$y = x^2 + 6x + 8$ — Write original equation.

$y = x^2 + 6x + 3^2 - 3^2 + 8$ — Complete the square.

$y = (x^2 + 6x + 9) - 9 + 8$ — Group terms.

$y = (x + 3)^2 - 1$ — Standard form

The vertex occurs at the point $(-3, -1)$ and the axis is the line $x = -3$. After plotting this information, calculate a few additional points on the parabola, as shown in the table. Note that the y-intercept is $(0, 8)$ and the x-intercepts are solutions of the equation

$$x^2 + 6x + 8 = (x + 4)(x + 2) = 0.$$

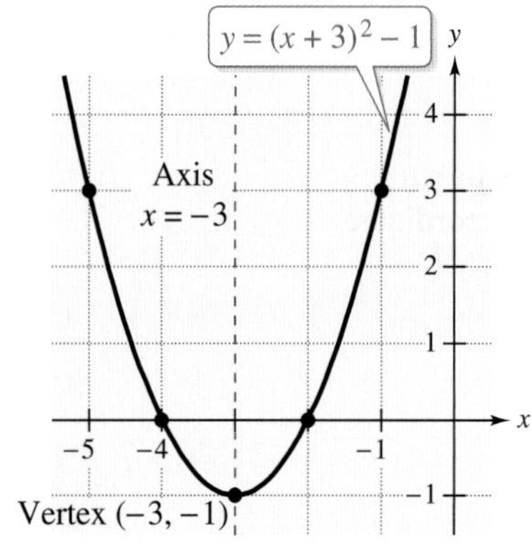

x	-5	-4	-3	-2	-1
$y = (x + 3)^2 - 1$	3	0	-1	0	3
Solution point	$(-5, 3)$	$(-4, 0)$	$(-3, -1)$	$(-2, 0)$	$(-1, 3)$

The graph of the parabola is shown at the left. Note that the parabola opens upward because the leading coefficient (in general form) is positive.

The graph of the parabola in Example 3 can also be obtained by shifting the graph of $y = x^2$ to the left three units and downward one unit, as discussed in Section 3.7.

Exercises Within Reach®

Solutions in English & Spanish and tutorial videos at AlgebraWithinReach.com

Sketching a Parabola **In Exercises 17–24, sketch the parabola. Identify the vertex and any x-intercepts.** See Additional Answers.

17. $g(x) = x^2 - 4$

18. $h(x) = x^2 - 9$

19. $f(x) = -x^2 + 4$

20. $f(x) = -x^2 + 9$

21. $y = (x - 4)^2$

22. $y = -(x + 4)^2$

23. $y = x^2 - 9x - 18$

24. $y = x^2 + 4x + 2$

Writing an Equation of a Parabola

To write an equation of a parabola with a vertical axis, use the fact that its standard equation has the form $y = a(x - h)^2 + k$, where (h, k) is the vertex.

EXAMPLE 4 Writing an Equation of a Parabola

Write an equation of the parabola with vertex $(-2, 1)$ and y-intercept $(0, -3)$, as shown at the left.

SOLUTION

Because the vertex occurs at $(h, k) = (-2, 1)$, the equation has the form

$y = a(x - h)^2 + k$	Standard form
$y = a[x - (-2)]^2 + 1$	Substitute -2 for h and 1 for k.
$y = a(x + 2)^2 + 1$	Simplify.

To find the value of a, use the fact that y-intercept is $(0, -3)$.

$y = a(x + 2)^2 + 1$	Write standard form.
$-3 = a(0 + 2)^2 + 1$	Substitute 0 for x and -3 for y.
$-1 = a$	Simplify.

So, the standard form of the equation of the parabola is $y = -(x + 2)^2 + 1$.

Exercises Within Reach®

Solutions in English & Spanish and tutorial videos at AlgebraWithinReach.com

Writing an Equation of a Parabola **In Exercises 25–30, write an equation of the parabola.**

25. $y = (x - 2)^2$

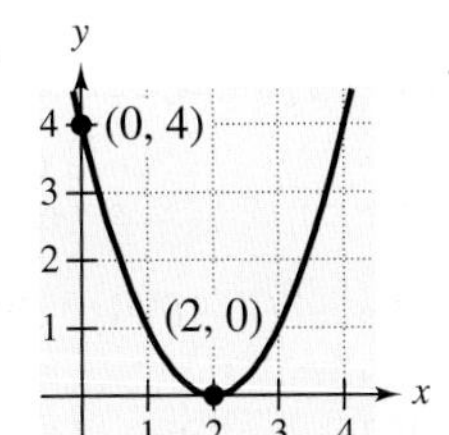

26. $y = -(x + 1)^2$

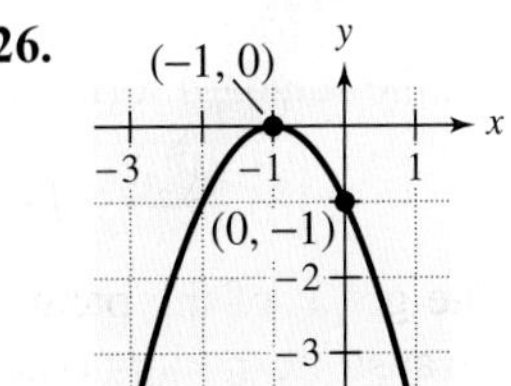

27. $y = -\frac{1}{2}(x - 2)^2 + 6$

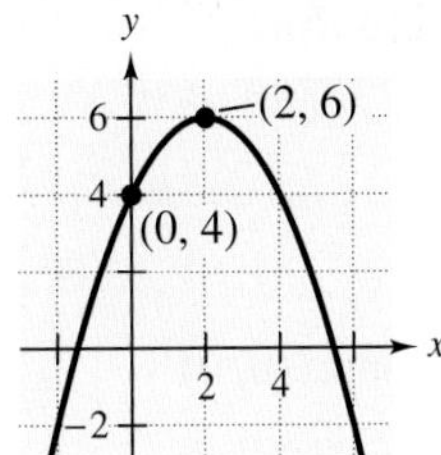

28. $y = 2(x + 1)^2 - 2$

29. $y = (x + 1)^2 - 3$

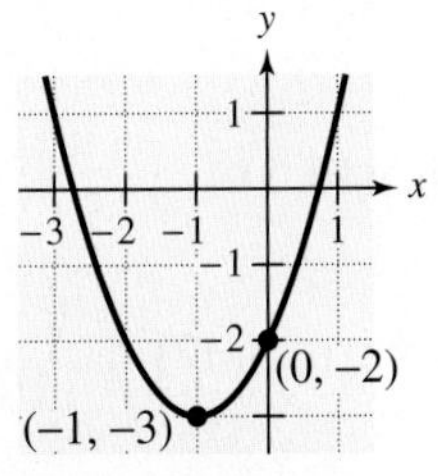

30. $y = -(x - 2)^2 + 2$

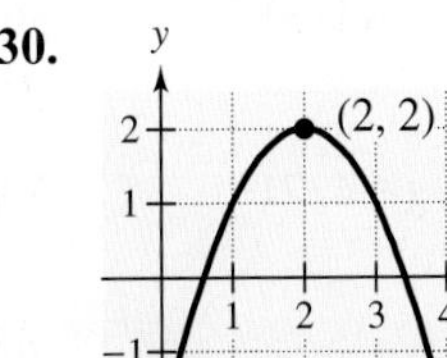

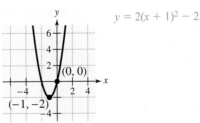

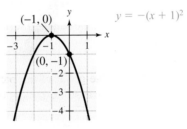

EXAMPLE 5 Writing an Equation of a Parabola

Write an equation of the parabola with vertex $(3, -4)$ that passes through the point $(5, -2)$, as shown below.

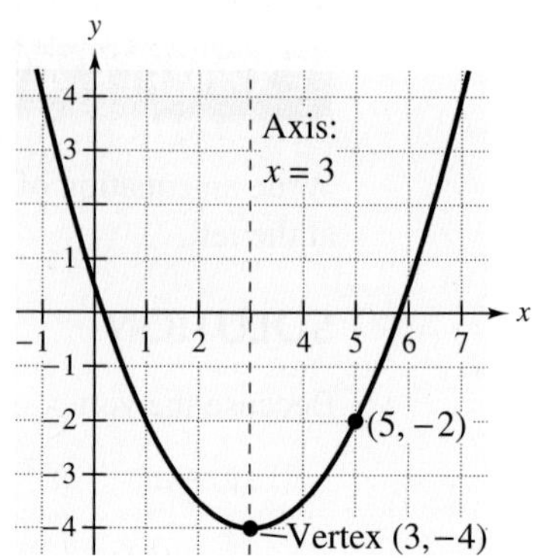

SOLUTION

Because the vertex occurs at $(h, k) = (3, -4)$, the equation has the form

$y = a(x - h)^2 + k$ Standard form

$y = a[x - 3)]^2 + (-4)$ Substitute 3 for h and -4 for k.

$y = a(x - 3)^2 - 4.$ Simplify.

To find the value of a, use the fact that the parabola passes through the point $(5, -2)$.

$y = a(x - 3)^2 - 4$ Write standard form.

$-2 = a(5 - 3)^2 - 4$ Substitute 5 for x and -2 for y.

$\frac{1}{2} = a$ Simplify.

So, the standard form of the equation of the parabola is $y = \frac{1}{2}(x - 3)^2 - 4$.

Exercises Within Reach®

Solutions in English & Spanish and tutorial videos at AlgebraWithinReach.com

Writing an Equation of a Parabola **In Exercises 31–40, write an equation of the parabola that satisfies the conditions.**

31. Vertex: $(2, 1)$; $a = 1$
$y = (x - 2)^2 + 1$

32. Vertex: $(-3, -3)$; $a = 1$
$y = (x + 3)^2 - 3$

33. Vertex: $(2, -4)$; Point on the graph: $(0, 0)$
$y = (x - 2)^2 - 4$

34. Vertex: $(-2, -4)$; Point on the graph: $(0, 0)$
$y = (x + 2)^2 - 4$

35. Vertex: $(-2, -1)$; Point on the graph: $(1, 8)$
$y = (x + 2)^2 - 1$

36. Vertex: $(4, 2)$; Point on the graph: $(2, -4)$
$y = -\frac{3}{2}(x - 4)^2 + 2$

37. Vertex: $(-1, 1)$; Point on the graph: $(-4, 7)$
$y = \frac{2}{3}(x + 1)^2 + 1$

38. Vertex: $(5, 2)$; Point on the graph: $(10, 3)$
$y = \frac{1}{25}(x - 5)^2 + 2$

39. Vertex: $(2, -2)$; Point on the graph: $(7, 8)$
$y = \frac{2}{5}(x - 2)^2 - 2$

40. Vertex: $(-4, 4)$; Point on the graph: $(-8, 12)$
$y = \frac{1}{2}(x + 4)^2 + 4$

Application

Application

EXAMPLE 6 **Bridge**

Each cable of a bridge is suspended (in the shape of a parabola) between two towers that are 1280 meters apart. The top of each tower is 152 meters above the roadway. The cables touch the roadway at the midpoint between the towers. Write an equation that models the cables of the bridge.

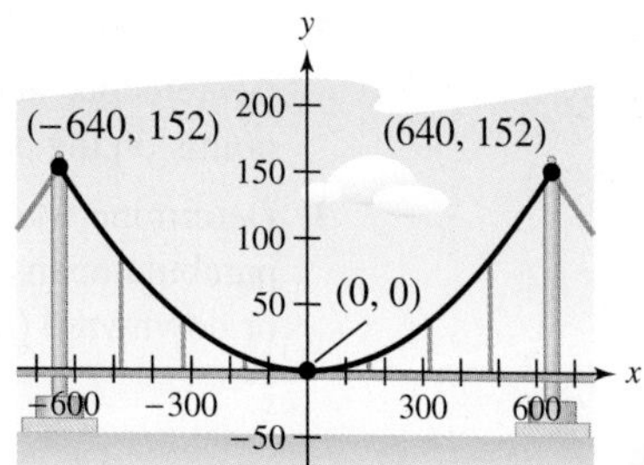

SOLUTION

From the figure, you can see that the vertex of the parabola occurs at (0, 0). So, the equation has the form

$y = a(x - h)^2 + k$ Standard form

$y = a(x - 0)^2 + 0$ Substitute 0 for h and 0 for k.

$y = ax^2.$ Simplify.

To find the value of a, use the fact that the parabola passes through the point (640, 152).

$y = ax^2$ Write standard form.

$152 = a(640)^2$ Substitute 640 for x and 152 for y.

$\frac{19}{51,200} = a$ Simplify.

So, an equation that models the cables of the bridge is $y = \frac{19}{51,200}x^2$.

Exercises Within Reach®

Solutions in English & Spanish and tutorial videos at AlgebraWithinReach.com

41. ***Roller Coaster Design*** A structural engineer must design a parabolic arc for the bottom of a roller coaster track. The vertex of the parabola is placed at the origin, and the parabola must pass through the points $(-30, 15)$ and $(30, 15)$ (see figure). Write an equation of the parabolic arc. $y = \frac{1}{60}x^2$

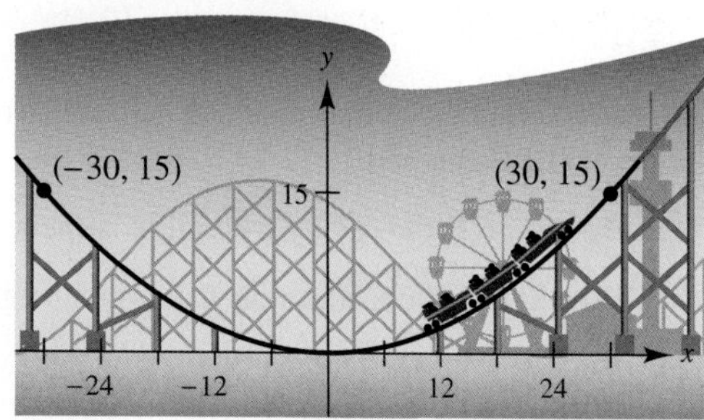

42. ***Highway Design*** A highway department engineer must design a parabolic arc to create a turn in a freeway around a park. The vertex of the parabola is placed at the origin, and the parabola must connect with roads represented by the equations $y = -0.4x - 100$ when $x < -500$ and $y = 0.4x - 100$ when $x > 500$ (see figure). Write an equation of the parabolic arc. $y = \frac{1}{2500}x^2$

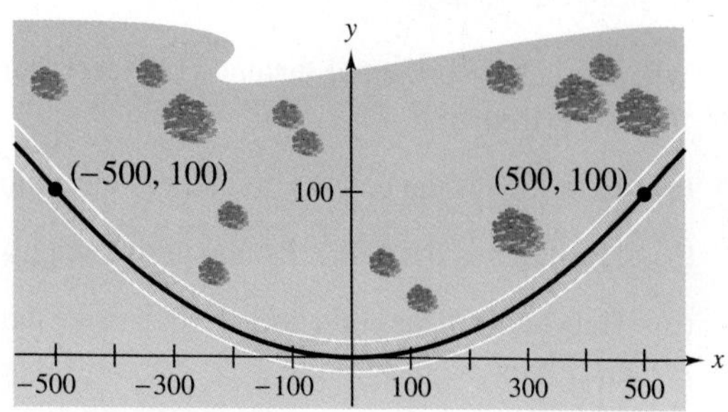

Concept Summary: Graphing Quadratic Functions

What

The graph of the quadratic function

$y = ax^2 + bx + c, a \neq 0$

is called a **parabola**.

EXAMPLE

Sketch the parabola given by $y = x^2 + 2x + 2$.

How

Use the following guidelines to sketch a parabola.

1. Determine the **vertex** and **axis** of the parabola.
2. Plot the vertex, axis, x- and y-intercepts, and a few additional points on the parabola.
3. Determine whether the parabola opens upward ($a > 0$) or downward ($a < 0$).

Why

You can use parabolas to model many real-life situations. For example, you can use a parabola to model the cables of a suspension bridge.

EXAMPLE

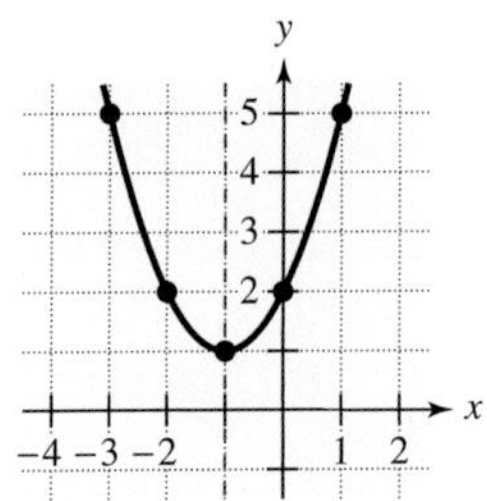

Exercises Within Reach ®

Worked-out solutions to odd-numbered exercises at AlgebraWithinReach.com

Concept Summary Check

43. ***Describing a Graph*** In your own words, describe the graph of the quadratic function $f(x) = ax^2 + bx + c$.
A parabola that opens upward when $a > 0$ and downward when $a < 0$.

44. ***Writing*** Explain how to find the vertex of the graph of a quadratic function. See Additional Answers.

45. ***Writing*** Explain how to find any x- or y-intercepts of the graph of a quadratic function.
See Additional Answers.

46. ***Reasoning*** Explain how to determine whether the graph of a quadratic function opens upward or downward.
If $a > 0$, the graph of $f(x) = ax^2 + bx + c$ opens upward, and if $a < 0$, the graph opens downward.

Extra Practice

Identifying a Transformation **In Exercises 47–54, identify the transformation of the graph of $f(x) = x^2$, and sketch the graph of h.** See Additional Answers.

47. $h(x) = x^2 - 1$ Vertical shift

48. $h(x) = x^2 + 3$ Vertical shift

49. $h(x) = (x + 2)^2$ Horizontal shift

50. $h(x) = (x - 4)^2$ Horizontal shift

51. $h(x) = -(x + 5)^2$ Horizontal shift and reflection in the x-axis

52. $h(x) = -x^2 - 6$ Vertical shift and reflection in the x-axis

53. $h(x) = -(x - 2)^2 - 3$
Horizontal and vertical shifts, reflection in the x-axis

54. $h(x) = -(x + 1)^2 + 5$
Horizontal and vertical shifts, reflection in the x-axis

55. ***Path of a Ball*** The height y (in feet) of a ball thrown by a child is given by

$y = \frac{1}{12}x^2 + 2x + 4$

where x is the horizontal distance (in feet) from where the ball is thrown.

(a) How high is the ball when it leaves the child's hand?

(b) How high is the ball at its maximum height?

(c) How far from the child does the ball strike the ground?

(a) 4 feet (b) 16 feet (c) $12 + 8\sqrt{3} \approx 25.9$ feet

56. ***Path of a Toy Rocket*** A child launches a toy rocket from a table. The height y (in feet) of the rocket is given by

$y = -\frac{1}{5}x^2 + 6x + 3$

where x is the horizontal distance (in feet) from where the rocket is launched.

(a) How high is the rocket when it is launched?

(b) How high is the rocket at its maximum height?

(c) How far from where it is launched does the rocket land?

(a) 3 feet (b) 48 feet (c) $15 + 4\sqrt{15} \approx 30.5$ feet

57. ***Path of a Golf Ball*** The height y (in yards) of a golf ball hit by a professional golfer is given by

$$y = \frac{1}{480}x^2 + \frac{1}{2}x$$

where x is the horizontal distance (in yards) from where you hit the ball.

(a) How high is the ball when it is hit? 0 yards

(b) How high is that ball at its maximum height? 30 yards

(c) How far from where the ball is hit does it strike the ground? 240 yards

58. ***Path of a Softball*** The height y (in feet) of a softball that you hit is given by

$$y = -\frac{1}{70}x^2 + 2x + 2$$

where x is the horizontal distance (in feet) from where you hit the ball.

(a) How high is the ball when you hit it? 2 feet

(b) How high is that ball at its maximum height? 72 feet

(c) How far from where you hit the ball does it strike the ground? $70 + 12\sqrt{35} \approx 141.0$ feet

59. ***Path of a Diver*** The path of a diver is given by

$$y = -\frac{4}{9}x^2 + \frac{24}{9}x + 10$$

where y is the height in feet and x is the horizontal distance from the end of the diving board in feet. What is the maximum height of the diver? 14 feet

60. ***Path of a Diver*** Repeat Exercise 59 when the path of the diver is modeled by

$$y = -\frac{4}{3}x^2 + \frac{10}{3}x + 10.$$ $\frac{145}{12} \approx 12.1$ feet

Explaining Concepts

61. ***Writing*** How is the discriminant related to the graph of a quadratic function? If the discriminant is positive, the parabola has two x-intercepts; if it is zero, the parabola has one x-intercept; and if it is negative, the parabola has no x-intercepts.

62. ***Think About It*** Is it possible for the graph of a quadratic function to have two y-intercepts? Explain.

No. The relation $f(x) = ax^2 + bx + c$ is a function, and so any vertical line will intersect the graph at most once.

63. ***Logic*** Explain how to determine the maximum (or minimum) value of a quadratic function. Find the y-coordinate of the vertex of the graph of the function.

64. ***Reasoning*** The domain of a quadratic function is the set of real numbers. Explain how to find the range.

If the graph of the quadratic function opens upward, the range is the set of all real numbers greater than or equal to the y-coordinate of the vertex. If the graph opens downward, the range is the set of all real numbers less than or equal to the y-coordinate of the vertex.

Cumulative Review

In Exercises 65–72, find the slope-intercept form of the equation of the line that passes through the two points.

65. $(0, 0), (4, -2)$ $y = -\frac{1}{2}x$

66. $(0, 0), (100, 75)$ $y = \frac{3}{4}x$

67. $(-1, -2), (3, 6)$ $y = 2x$

68. $(1, 5), (6, 0)$ $y = -x + 6$

69. $\left(\frac{3}{2}, 8\right), \left(\frac{11}{2}, \frac{5}{2}\right)$ $y = -\frac{11}{8}x + \frac{161}{16}$

70. $(0, 2), (7.3, 15.4)$ $y = \frac{134}{73}x + 2$

71. $(0, 8), (5, 8)$ $y = 8$

72. $(-3, 2), (-3, 5)$ $x = -3$

In Exercises 73–76, write the number in *i*-form.

73. $\sqrt{-64}$ $8i$

74. $\sqrt{-32}$ $4\sqrt{2}i$

75. $\sqrt{-0.0081}$ $0.09i$

76. $\sqrt{-\frac{20}{16}}$ $\frac{\sqrt{5}}{2}i$

8.5 Applications of Quadratic Equations

▶ Use quadratic equations to solve application problems.

Applications

Application

EXAMPLE 1 An Investment Problem

A car dealer buys a fleet of cars from a car rental agency for a total of $120,000. The dealer regains this $120,000 investment by selling all but 4 of the cars at an average profit of $2500 each. How many cars has the dealer sold, and what is the average price per car?

SOLUTION

Although this problem is stated in terms of average price and average profit per car, you can use a model that assumes that each car has sold for the same price.

Verbal Model: Selling price per car = Cost per car + Profit per car

Labels:
Number of cars sold $= x$ (cars)
Number of cars bought $= x + 4$ (cars)
Selling price per car $= 120{,}000/x$ (dollars per car)
Cost per car $= 120{,}000/(x + 4)$ (dollars per car)
Profit per car $= 2500$ (dollars per car)

Equation:

$$\frac{120{,}000}{x} = \frac{120{,}000}{x+4} + 2500$$

$$120{,}000(x + 4) = 120{,}000x + 2500x(x + 4), \quad x \neq 0, \; x \neq -4$$

$$120{,}000x + 480{,}000 = 120{,}000x + 2500x^2 + 10{,}000x$$

$$0 = 2500x^2 + 10{,}000x - 480{,}000$$

$$0 = x^2 + 4x - 192$$

$$0 = (x - 12)(x + 16)$$

$$x - 12 = 0 \quad \Rightarrow \quad x = 12$$

$$x + 16 = 0 \quad \Rightarrow \quad x = -16$$

By choosing the positive value, it follows that the dealer sold 12 cars at an average price of $120{,}000/12 = \$10{,}000$ per car.

Exercises Within Reach®

Solutions in English & Spanish and tutorial videos at AlgebraWithinReach.com

1. ***Selling Price*** A store owner buys a case of eggs for $21.60. The owner regains this investment by selling all but 6 dozen of the eggs at a profit of $0.30 per dozen. How many dozen eggs has the owner sold and what is the selling price per dozen? 18 dozen, $1.20 per dozen

2. ***Selling Price*** A computer store manager buys several computers of the same model for $12,600. The store can regain this investment by selling all but 4 of the computers at a profit of $360 per computer. To do this, how many computers must be sold, and at what price? 10 computers at $1260 per computer

Alexander Chaikin/Shutterstock.com

Application

EXAMPLE 2 Geometry: The Dimensions of a Picture

A picture is 6 inches taller than it is wide and has an area of 216 square inches, as shown at the left. What are the dimensions of the picture?

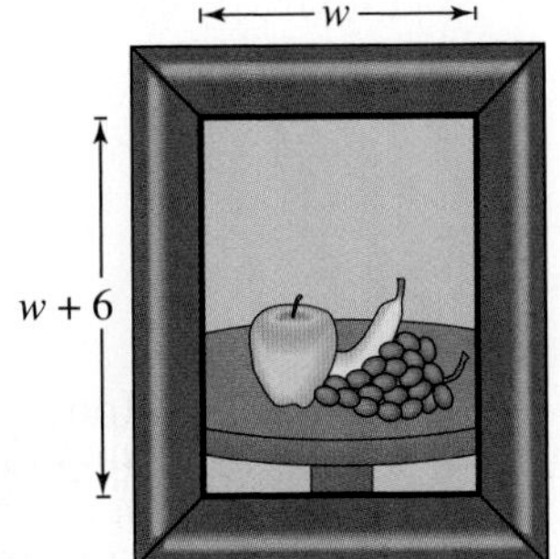

SOLUTION

Verbal Model: Area of picture = Width • Height

Labels:
Picture width = w (inches)
Picture height = $w + 6$ (inches)
Area = 216 (square inches)

Equation:

$$216 = w(w + 6)$$

$$0 = w^2 + 6w - 216$$

$$0 = (w + 18)(w - 12)$$

$$w + 18 = 0 \Rightarrow w = -18$$

$$w - 12 = 0 \Rightarrow w = 12$$

By choosing the positive value of w, you can conclude that the width of the picture is 12 inches and the height of the picture is $12 + 6 = 18$ inches.

Application

EXAMPLE 3 An Interest Problem

The amount A after 2 years in an account earning r percent (in decimal form) compounded annually is given by $A = P(1 + r)^2$, where P is the original investment. Find the interest rate when an investment of \$6000 increases to \$6933.75 over a two-year period.

SOLUTION

$$A = P(1 + r)^2$$ Write given formula.

$$6933.75 = 6000(1 + r)^2$$ Substitute 6933.75 for A and 6000 for P.

$$1.155625 = (1 + r)^2$$ Divide each side by 6000.

$$\pm 1.075 = 1 + r$$ Square Root Property

$$0.075 = r$$ Choose positive solution.

The annual interest rate is $r = 0.075 = 7.5\%$.

Exercises Within Reach®

Solutions in English & Spanish and tutorial videos at AlgebraWithinReach.com

3. ***Geometry*** A picture frame is 4 inches taller than it is wide and has an area of 192 square inches. What are the dimensions of the picture frame? 12 inches × 16 inches

4. ***Geometry*** The height of triangle is 8 inches less than its base. The area of the triangle is 192 square inches. Find the dimensions of the triangle. Base: 24 inches, Height: 16 inches

Compound Interest **The amount A after 2 years in an account earning r percent (in decimal form) compounded annually is given by $A = P(1 + r)^2$, where P is the original investment. In Exercises 5–8, find the interest rate r.**

5. $P = \$10{,}000$, $A = \$11{,}990.25$ 9.5%

6. $P = \$3000$, $A = \$3499.20$ 8%

7. $P = \$500$, $A = \$572.45$ 7%

8. $P = \$250$, $A = \$280.90$ 6%

Application **EXAMPLE 4** **Reduced Rates**

A ski club charters a bus for a ski trip at a cost of \$720. When four nonmembers accept invitations from the club to go on the trip, the bus fare per skier decreases by \$6. How many club members are going on the trip?

SOLUTION

Verbal Model: Fare per skier • Number of skiers = 720

Labels:
Number of ski club members $= x$ (people)
Number of skiers $= x + 4$ (people)
Original fare per skier $= \dfrac{720}{x}$ (dollars per person)
New fare per skier $= \dfrac{720}{x} - 6$ (dollars per person)

Equation:

$\left(\dfrac{720}{x} - 6\right)(x + 4) = 720$ Original equation.

$\left(\dfrac{720 - 6x}{x}\right)(x + 4) = 720$ Rewrite 1st factor.

$(720 - 6x)(x + 4) = 720x, \; x \neq 0$ Multiply each side by x.

$720x + 2880 - 6x^2 - 24x = 720x$ Multiply factors.

$-6x^2 - 24x + 2880 = 0$ Subtract $720x$ from each side.

$x^2 + 4x - 480 = 0$ Divide each side by -6.

$(x + 24)(x - 20) = 0$ Factor left side of equation.

$x + 24 = 0 \Rightarrow x = -24$ Set 1st factor equal to 0.

$x - 20 = 0 \Rightarrow x = 20$ Set 2nd factor equal to 0.

By choosing the positive value of x, you can conclude that 20 ski club members are going on the trip.

Exercises Within Reach®

Solutions in English & Spanish and tutorial videos at AlgebraWithinReach.com

9. ***Reduced Rates*** A service organization pays \$210 for a block of tickets to a baseball game. The block contains three more tickets than the organization needs for its member. By inviting 3 more people to attend (and share in the cost), the organization lowers the price per person by \$3.50. How many people are going to the game? 15 people

10. ***Reduced Fares*** A science club charters a bus to attend a science fair at a cost of \$480. To lower the bus fare per person, the club invites nonmembers to go along. When 3 nonmembers join the trip, the fare per person is decreased by \$1. How many people are going to the science fair? 32 people

Application **EXAMPLE 5** **Work-Rate Problem**

An office has two copy machines. Machine B is known to take 12 minutes longer than machine A to copy the company's monthly report. Using both machines together, it takes 8 minutes to copy the report. How long would it take each machine alone to copy the report?

SOLUTION

Verbal Model: [Work done by machine A] + [Work done by machine B] = [1 complete job]

[Rate for A] • [Time for both] + [Rate for B] • [Time for both] = 1

Labels:

Time for machine A $= t$	(minutes)
Rate for machine A $= 1/t$	(job per minute)
Time for machine B $= t + 12$	(minutes)
Rate for machine B $= 1/(t + 12)$	(job per minute)
Time for both machines $= 8$	(minutes)
Rate for both machines $= 1/8$	(job per minute)

Equation:

$$\frac{1}{t}(8) + \frac{1}{t+12}(8) = 1$$ Original equation

$$8\left(\frac{1}{t} + \frac{1}{t+12}\right) = 1$$ Distributive Property

$$8\left[\frac{t + 12 + t}{t(t+12)}\right] = 1$$ Rewrite with common denominator.

$$8t(t+12)\left[\frac{2t+12}{t(t+12)}\right] = t(t+12)$$ Multiply each side by $t(t + 12)$.

$$8(2t + 12) = t^2 + 12t$$ Simplify.

$$16t + 96 = t^2 + 12t$$ Distributive Property

$$0 = t^2 - 4t - 96$$ Subtract $16t + 96$ from each side.

$$0 = (t - 12)(t + 8)$$ Factor right side of equation.

$t - 12 = 0 \Rightarrow t = 12$ Set 1st factor equal to 0.

$t + 8 = 0 \Rightarrow t = -8$ Set 2nd factor equal to 0.

By choosing the positive value of t, you can conclude that machine A would take 12 minutes and machine B would take $12 + 12 = 24$ minutes.

Exercises Within Reach®

Solutions in English & Spanish and tutorial videos at AlgebraWithinReach.com

11. ***Work Rate*** An office has two printers. Machine B is known to take 3 minutes longer than machine A to produce the company's monthly financial report. Using both machines together, it takes 6 minutes to produce the report. How long would it take each machine alone to produce the report? 10.7 minutes, 13.7 minutes

12. ***Work Rate*** A builder works with two plumbing companies. Company A is known to take 3 days longer than Company B to install the plumbing in a particular style of house. Using both companies together, it takes 4 days to install the plumbing. How long would it take each company alone to install the plumbing? 6.8 days, 9.8 days

Application EXAMPLE 6 Geometry: The Pythagorean Theorem

An L-shaped sidewalk from the athletic center to the library on a college campus is 200 meters long, as shown at the left. By cutting diagonally across the grass, students shorten the walking distance to 150 meters. What are the lengths of the two legs of the sidewalk?

SOLUTION

Common Formula: $a^2 + b^2 = c^2$ Pythagorean Theorem

Labels: Length of one leg $= x$ (meters)
Length of other leg $= 200 - x$ (meters)
Length of diagonal $= 150$ (meters)

Equation:

$$x^2 + (200 - x)^2 = 150^2$$

$$x^2 + 40{,}000 - 400x + x^2 = 22{,}500$$

$$2x^2 - 400x + 40{,}000 = 22{,}500$$

$$2x^2 - 400x + 17{,}500 = 0$$

$$x^2 - 200x + 8750 = 0$$

Using the Quadratic Formula, you can find the solutions as follows.

$$x = \frac{-(-200) \pm \sqrt{(-200)^2 - 4(1)(8750)}}{2(1)}$$ Substitute 1 for *a*, -200 for *b*, and 8750 for *c*.

$$= \frac{200 \pm \sqrt{5000}}{2}$$

$$= \frac{200 \pm 50\sqrt{2}}{2}$$

$$= 100 \pm 25\sqrt{2}$$

Both solutions are positive, so it does not matter which you choose. When you let $x = 100 + 25\sqrt{2} \approx 135.4$ meters, the length of the other leg is $200 - x \approx 200 - 135.4 = 64.6$ meters.

Exercises Within Reach®

Solutions in English & Spanish and tutorial videos at AlgebraWithinReach.com

13. ***Geometry*** An L-shaped sidewalk from the library to the gym on a high school campus is 100 yards long, as shown in the figure. By cutting diagonally across the grass, students shorten the walking distance to 80 yards. What are the lengths of the two legs of the sidewalk?
23.5 yards, 76.5 yards

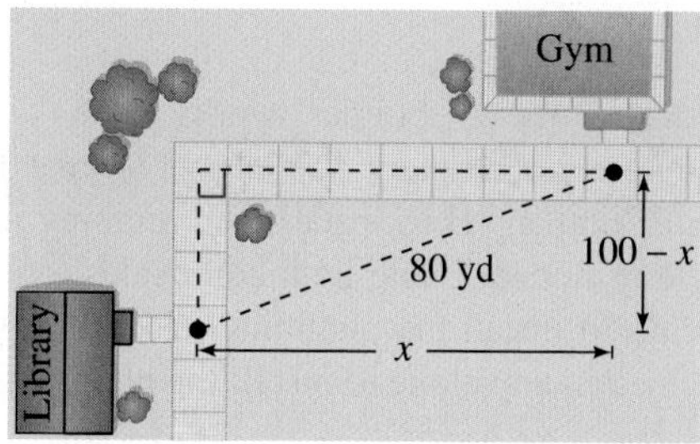

14. ***Delivery Route*** You deliver pizzas to an insurance office and an apartment complex (see figure). Your total mileage in driving to the insurance office and then the apartment complex is 12 miles. By using a direct route, you are able to drive just 9 miles to return to the pizza shop. Find the possible distances from the pizza shop to the insurance office. 3.9 miles or 8.1 miles

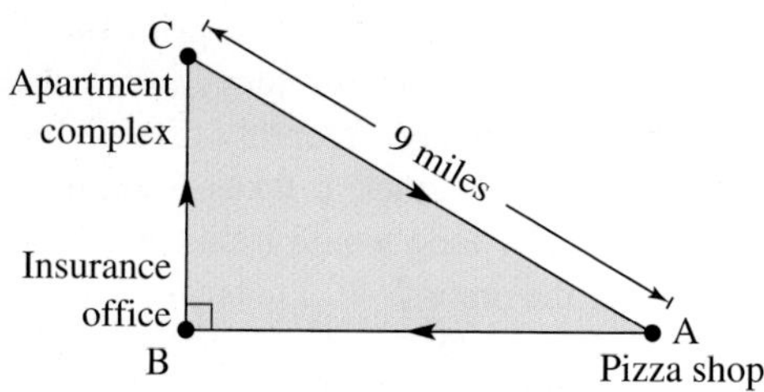

Application **EXAMPLE 7** **The Height of a Model Rocket**

A model rocket is projected straight upward from ground level according to the height equation

$$h = -16t^2 + 192t, \quad t \geq 0$$

where h is the height in feet and t is the time in seconds.

a. After how many seconds is the height 432 feet?

b. After how many seconds does the rocket hit the ground?

SOLUTION

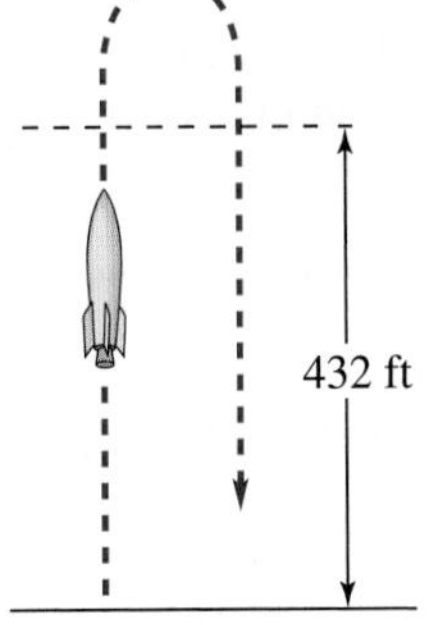

a.

$h = -16t^2 + 192t$	Write original equation.
$432 = -16t^2 + 192t$	Substitute 432 for h.
$16t^2 - 192t + 432 = 0$	Write in general form.
$t^2 - 12t + 27 = 0$	Divide each side by 16.
$(t - 3)(t - 9) = 0$	Factor left side of equation.
$t - 3 = 0 \Rightarrow t = 3$	Set 1st factor equal to 0.
$t - 9 = 0 \Rightarrow t = 9$	Set 2nd factor equal to 0.

The rocket attains a height of 432 feet at two different times—once (going up) after 3 seconds, and again (coming down) after 9 seconds.

b. To find the time it takes for the rocket to hit the ground, let the height be 0.

$0 = -16t^2 + 192t$	Substitute 0 for h in original equation.
$0 = t^2 - 12t$	Divide each side by -16.
$0 = t(t - 12)$	Factor right side of equation.
$t = 0$ or $t = 12$	Solutions

The rocket hits the ground after 12 seconds. (Note that the time of $t = 0$ seconds corresponds to the time of lift-off.)

Exercises Within Reach ®

Solutions in English & Spanish and tutorial videos at AlgebraWithinReach.com

15. ***Height*** You are hitting baseballs. When you toss the ball into the air, your hand is 5 feet above the ground (see figure). You hit the ball when it falls back to a height of 3.5 feet. You toss the ball with an initial velocity of 18 feet per second. The height h of the ball t seconds after leaving your hand is given by $h = 5 + 18t - 16t^2$. About how much time passes before you hit the ball?
1.2 seconds

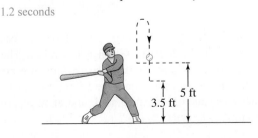

16. ***Height*** A model rocket is projected straight upward from ground level according to the height equation

$$h = -16t^2 + 160t$$

where h is the height of the rocket in feet and t is the time in seconds.

(a) After how many seconds is the height 336 feet?
3 seconds, 7 seconds

(b) After how many seconds does the rocket hit the ground?
10 seconds

(c) What is the maximum height of the rocket?
400 feet

Concept Summary: Using Quadratic Equations to Solve Problems

What

You can use quadratic equations to model many real-life problems.

EXAMPLE

A postcard is 2 inches wider than it is tall and has an area of 35 square inches. What are the dimensions of the postcard?

How

EXAMPLE

Here is one way to model this problem.

Create a Verbal Model:

Area of postcard = Height • Width

Assign labels:

Postcard height $= t$

Postcard width $= t + 2$

Area $= 35$

Write an equation:

$35 = t(t + 2)$

Why

Notice that the equation that models the problem is a quadratic equation. You can now solve the problem by solving the quadratic equation.

EXAMPLE

$35 = t(t + 2)$

$0 = t^2 + 2t - 35$

$0 = (t + 7)(t - 5)$

So, $t = -7$ and $t = 5$.

Because the height of the postcard cannot be negative, the height is 5 inches and the width is 7 inches.

Exercises Within Reach ®

Worked-out solutions to odd-numbered exercises at AlgebraWithinReach.com

Concept Summary Check

Problem Solving **In Exercises 17–20, a problem situation is given. Describe two quantities that can be set equal to each other to write an equation that can be used to solve the problem.**

17. You know the length of the hypotenuse and the sum of the lengths of the legs of a right triangle. You want to find the lengths of the legs. See Additional Answers.

18. You know the area of a rectangle and you know how many units longer the length is than the width. You want to find the length and width. The area is equal to the product of the width and the width plus the extra length.

19. You know the amount invested in an unknown number of product units. You know the number of units remaining when the investment is regained, and the profit per unit sold. You want to find the number of units sold and the price per unit. The quotient of the investment amount and the number of units is equal to the quotient of the investment amount and the number of units sold, plus the profit per unit sold.

20. You know the time in minutes for two machines to complete a task together and you know how many more minutes it takes one machine than the other to complete the task alone. You want to find the time to complete the task alone for each machine. See Additional Answers.

Extra Practice

Geometry **In Exercises 21–28, find the perimeter or area of the rectangle, as indicated.**

	Width	Length	Perimeter	Area
21.	$1.4l$	l	54 in.	177.19 in.^{2}
22.	w	$3.5w$	60 m	155.56 m^2
23.	w	$2.5w$	70 ft	250 ft^2
24.	w	$1.5w$	60 cm	216 cm^2
25.	w	$w + 3$	54 km	180 km^2
26.	$l - 6$	l	108 ft	720 ft^2
27.	$l - 20$	l	440 m	12,000 m^2
28.	w	$w + 5$	90 ft	500 ft^2

29. ***Selling Price*** A flea market vendor buys a box of DVD movies for \$50. The vendor regains this investment by selling all but 15 of the DVDs at a profit of \$3 each. How many DVDs has the vendor sold, and at what price? 10 DVDS at \$5 per DVD

30. ***Selling Price*** A running club buys a case of sweatshirts for \$750 to sell at a fundraiser. The club needs to sell all but 20 of the sweatshirts at a profit of \$10 per sweatshirt to regain the \$750 investment. How many sweatshirts must be sold, and at what price, to do this? 30 sweatshirts at \$25 per sweatshirt

Number Problem **In Exercises 31 and 32, find two positive integers that satisfy the requirement.**

31. The product of two consecutive integers is 182. 13, 14

32. The product of two consecutive odd integers is 323. 17, 19

33. ***Open Conduit*** An open-topped rectangular conduit for carrying water in a manufacturing process is made by folding up the edges of a sheet of aluminum that is 48 inches wide (see figure). A cross section of the conduit must have an area of 288 square inches. Find the width and height of the conduit.
Height: 12 inches; Width: 24 inches

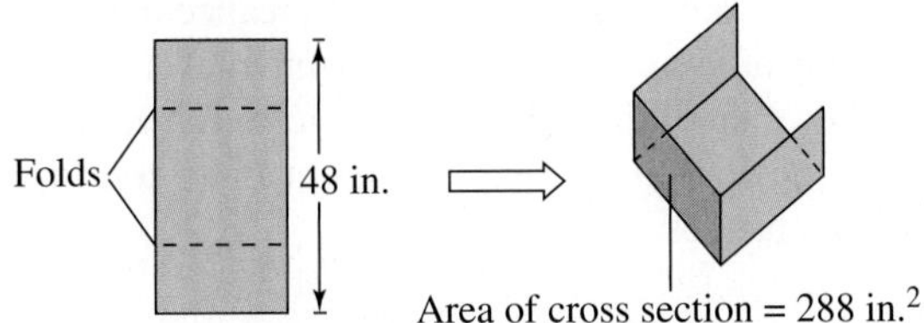

34. ***Speed*** A company uses a pickup truck for deliveries. The cost per hour for fuel is $C = v^2/300$, where v is the speed in miles per hour. The driver is paid \$15 per hour. The cost of wages and fuel for an 80-mile trip at constant speed is \$36. Find the possible speeds.
60 miles per hour or 75 miles per hour

35. ***Average Speed*** A truck traveled the first 100 miles of a trip at one speed and the last 135 miles at an average speed of 5 miles per hour less. The entire trip took 5 hours. What was the average speed for the first part of the trip? 50 miles per hour

36. ***Fenced Area*** A family builds a fence around three sides of their property (see figure). In total, they use 550 feet of fencing. By their calculations, the lot is 1 acre (43,560 square feet). Is this correct? Explain your reasoning. See Additional Answers.

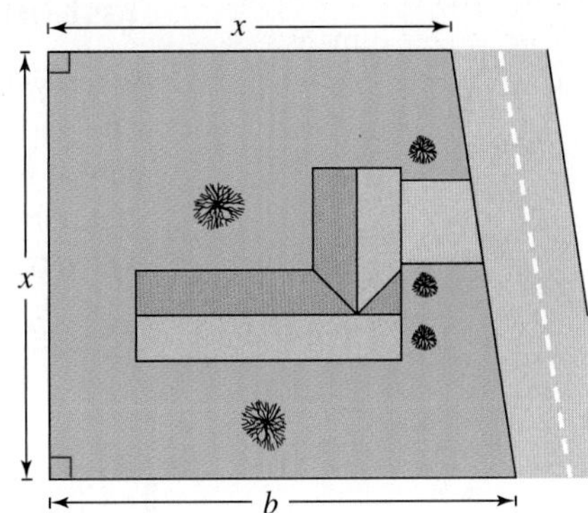

37. ***Distance*** Find any points on the line

$$y = 9$$

that are 10 units from the point (2, 3). (−6, 9), (10, 9)

38. ***Distance*** Find any points on the line

$$y = 14$$

that are 13 units from the point (1, 2). (−4, 14), (6, 14)

Explaining Concepts

39. ***Reasoning*** To solve some of the problems in this section, you wrote rational equations. Explain why these types of problems are included as applications of quadratic equations. To solve a rational equation, each side of the equation is multiplied by the LCD. The resulting equations in this section are quadratic equations.

40. ***Writing*** In Exercises 21–28 you are asked to find the perimeter or area of a rectangle. To do this, you must write an equation that can be solved for the length or width of the rectangle. Explain how you can tell when the equation will be a *quadratic equation* or a *linear equation*. See Additional Answers.

41. ***Think About It*** In a *reduced rates* problem such as Example 4, does the cost per person decrease by the same amount for each additional person? Explain.
No. For each additional person, the cost-per-person decrease gets smaller because the discount is distributed to more people

42. ***Think About It*** In a *height of an object* problem such as Example 7, suppose you try solving the height equation using a height greater than the maximum height reached by the object. What type of result will you get for t? Explain. The result will be a complex number because there is no real number value for the time that will yield a height greater than the maximum height.

Cumulative Review

In Exercises 43 and 44, solve the inequality and sketch the solution on the real number line.
See Additional Answers.

43. $5 - 3x > 17$ $x < -4$

44. $-3 < 2x + 3 < 5$ $-3 < x < 1$

In Exercises 45 and 46, solve the equation by completing the square.

45. $x^2 - 8x = 0$ 0, 8

46. $x^2 - 2x - 2 = 0$ $1 \pm \sqrt{3}$

8.6 Quadratic and Rational Inequalities

- Use test intervals to solve quadratic inequalities.
- Use test intervals to solve rational inequalities.
- Use inequalities to solve application problems.

Test Intervals and Quadratic Inequalities

When working with polynomial inequalities, it is important to realize that the value of a polynomial can change signs only at its **zeros**. That is, a polynomial can change signs only at the x-values for which the value of the polynomial is zero. When the real zeros of a polynomial are put in order, they divide the real number line into **test intervals** in which the polynomial has no sign changes.

Finding Test Intervals for a Polynomial

1. Find all real zeros of the polynomial, and arrange the zeros in increasing order. The zeros of a polynomial are called its **critical numbers**.
2. Use the critical numbers of the polynomial to determine its test intervals.
3. Choose a representative x-value in each test interval and evaluate the polynomial at that value. When the value of the polynomial is negative, the polynomial has negative values for *all* x-values in the interval. When the value of the polynomial is positive, the polynomial has positive values for *all* x-values in the interval.

EXAMPLE 1 Test Intervals for a Linear Polynomial

The first-degree polynomial $x + 2$ has a zero at $x = -2$, and it changes signs at that zero. You can picture this result on the real number line, as shown below.

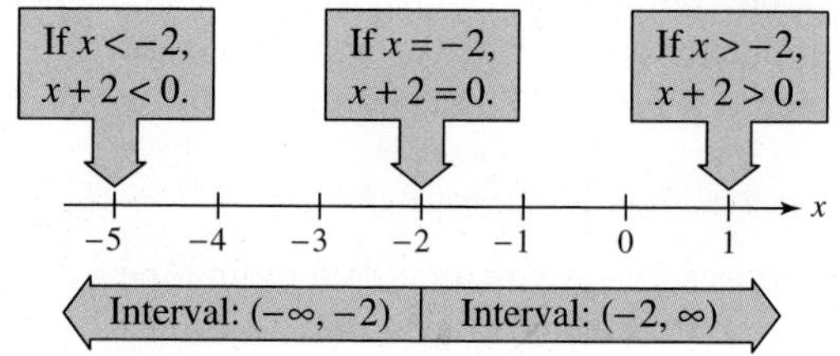

Note in the figure that the zero of the polynomial divides the real number line into two test intervals. The value of the polynomial is negative for every x-value in the first test interval $(-\infty, -2)$, and positive for every x-value in the second test interval $(-2, \infty)$.

Exercises Within Reach®

Solutions in English & Spanish and tutorial videos at AlgebraWithinReach.com

Finding Test Intervals **In Exercises 1–8, determine the intervals for which the polynomial is entirely negative and entirely positive.**

1. $x - 4$ Negative: $(-\infty, 4)$; Positive: $(4, \infty)$
2. $3 - x$ Negative: $(3, \infty)$; Positive: $(-\infty, 3)$
3. $3 - \frac{1}{2}x$ Negative: $(6, \infty)$; Positive: $(-\infty, 6)$
4. $\frac{2}{3}x - 8$ Negative: $(-\infty, 12)$; Positive: $(12, \infty)$
5. $4x(x - 5)$ Negative: $(0, 5)$; Positive: $(-\infty, 0)$ and $(5, \infty)$
6. $7x(3 - x)$ Negative: $(-\infty, 0)$ and $(3, \infty)$; Positive: $(0, 3)$
7. $4 - x^2$ Negative: $(-\infty, -2)$ and $(2, \infty)$; Positive: $(-2, 2)$
8. $x^2 - 36$ Negative: $(-6, 6)$; Positive: $(-\infty, -6)$ and $(6, \infty)$

EXAMPLE 2 Solving a Quadratic Inequality

Solve the inequality $x^2 - 5x < 0$.

SOLUTION

First find the critical numbers of $x^2 - 5x < 0$ by finding the solutions of the equation $x^2 - 5x = 0$.

$x^2 - 5x = 0$ Write corresponding equation.

$x(x - 5) = 0$ Factor.

$x = 0,\ x = 5$ Critical numbers

This implies that the test intervals are $(-\infty, 0)$, $(0, 5)$, and $(5, \infty)$. To test an interval, choose a convenient value in the interval and determine whether the value satisfies the inequality.

Study Tip

In Example 2, note that the original inequality contains a "less than" symbol. This means that the solution set does not contain the endpoints of the test interval (0, 5).

Test interval	Representative x-value	Is inequality satisfied?
$(-\infty, 0)$	$x = -1$	$(-1)^2 - 5(-1) \overset{?}{<} 0$ $6 \not< 0$
$(0, 5)$	$x = 1$	$1^2 - 5(1) \overset{?}{<} 0$ $-4 < 0$
$(5, \infty)$	$x = 6$	$6^2 - 5(6) \overset{?}{<} 0$ $6 \not< 0$

Of the three x-values tested above, only the value $x = 1$ satisfies the inequality $x^2 - 5x < 0$. So, you can conclude that the solution set of the inequality is $0 < x < 5$, as shown below.

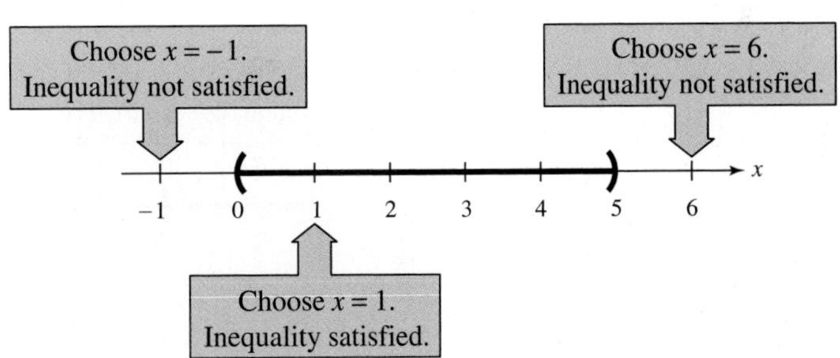

Exercises Within Reach®

Solutions in English & Spanish and tutorial videos at AlgebraWithinReach.com

Solving a Quadratic Inequality **In Exercises 9–16, solve the inequality and graph the solution on the real number line.** See Additional Answers.

9. $3x(x - 2) < 0$ $0 < x < 2$

10. $5x(x - 8) > 0$ $x < 0$ or $x > 8$

11. $3x(2 - x) \geq 0$ $0 \leq x \leq 2$

12. $5x(8 - x) > 0$ $0 < x < 8$

13. $x^2 + 4x > 0$ $x < -4$ or $x > 0$

14. $x^2 - 5x \geq 0$ $x \leq 0$ or $x \geq 5$

15. $x^2 - 3x - 10 \geq 0$ $x \leq -2$ or $x \geq 5$

16. $x^2 + 8x + 7 < 0$ $-7 < x < -1$

Some students may try to solve the quadratic inequality $(x + 4)(2x - 3) \geq 0$ by setting each factor greater than or equal to 0 ($x + 4 \geq 0$ and $2x - 3 \geq 0$), similar to the way they solve quadratic equations by factoring. Point out the error in this reasoning.

EXAMPLE 3 Solving a Quadratic Inequality

Solve the inequality $2x^2 + 5x \geq 12$.

SOLUTION

Begin by writing the inequality in the general form $2x^2 + 5x - 12 \geq 0$. Next, find the critical numbers by finding the solutions of the equation $2x^2 + 5x - 12 = 0$.

$2x^2 + 5x - 12 = 0$ Write corresponding equation.

$(x + 4)(2x - 3) = 0$ Factor.

$x = -4, \ x = \dfrac{3}{2}$ Critical numbers

This implies that the test intervals are $(-\infty, -4)$, $\left(-4, \frac{3}{2}\right)$, and $\left(\frac{3}{2}, \infty\right)$. To test an interval, choose a convenient value in the interval and determine whether the value satisfies the inequality.

Test interval	Representative *x*-value	Is inequality satisfied?
$(-\infty, -4)$	$x = -5$	$2(-5)^2 + 5(-5) \overset{?}{\geq} 12$ $25 \geq 12$
$\left(-4, \frac{3}{2}\right)$	$x = 0$	$2(0)^2 + 5(0) \overset{?}{\geq} 12$ $0 \not\geq 12$
$\left(\frac{3}{2}, \infty\right)$	$x = 2$	$2(2)^2 + 5(2) \overset{?}{\geq} 12$ $18 \geq 12$

From this table, you can see that the solution set of the inequality $2x^2 + 5x \geq 12$ is $x \leq -4$ or $x \geq \frac{3}{2}$, as shown below.

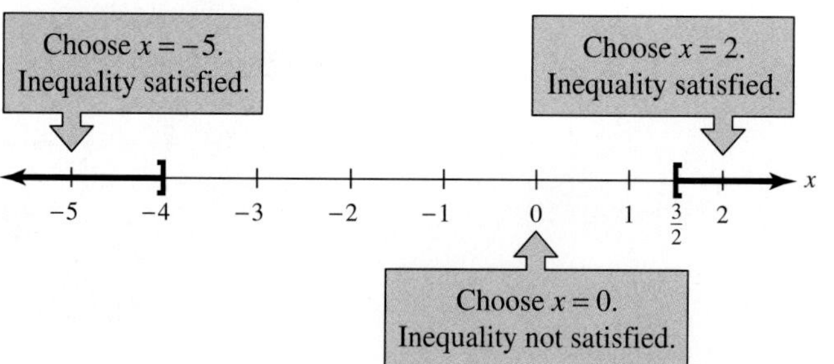

Study Tip

In Examples 2 and 3, the critical numbers are found by factoring. With quadratic polynomials that do not factor, you can use the Quadratic Formula to find the critical numbers. For instance, to solve the inequality

$x^2 - 2x - 1 \leq 0$

you can use the Quadratic Formula to determine that the critical numbers are

$1 - \sqrt{2} \approx -0.414$

and

$1 + \sqrt{2} \approx 2.414$.

Exercises Within Reach®

Solutions in English & Spanish and tutorial videos at AlgebraWithinReach.com

Solving a Quadratic Inequality **In Exercises 17–22, solve the inequality and graph the solution on the real number line.** See Additional Answers.

17. $x^2 > 4$
$x < -2$ or $x > 2$

18. $z^2 \leq 9$
$-3 \leq x \leq 3$

19. $x^2 + 5x \leq 36$
$-9 \leq x \leq 4$

20. $t^2 - 4t > 12$
$x < -2$ or $x > 6$

21. $u^2 + 2u - 2 > 1$
$x < -3$ or $x > 1$

22. $t^2 - 15t < -50$
$5 < x < 10$

EXAMPLE 4 Unusual Solution Sets

a. The solution set of the quadratic inequality

$$x^2 + 2x + 4 > 0$$

consists of the entire set of real numbers, $-\infty < x < \infty$. This is true because the value of the quadratic $x^2 + 2x + 4$ is positive for every real value of x. You can see in the graph at the left that the entire parabola lies above the x-axis.

b. The solution set of the quadratic inequality

$$x^2 + 2x + 1 \le 0$$

consists of the single number -1. This is true because $x^2 + 2x + 1 = (x + 1)^2$ has just one critical number, $x = -1$, and it is the only value that satisfies the inequality. You can see in the graph at the left that the parabola meets the x-axis only when $x = -1$.

c. The solution set of the quadratic inequality

$$x^2 + 3x + 5 < 0$$

is empty. This is true because the value of the quadratic $x^2 + 3x + 5$ is not less than zero for any value of x. No point on the parabola lies below the x-axis, as shown in the graph below on the left.

d. The solution set of the quadratic inequality

$$x^2 - 4x + 4 > 0$$

consists of all real numbers *except* the number 2. So, the solution set of the quadratic inequality is $x < 2$ or $x > 2$. You can see in the graph below on the right that the parabola lies above the x-axis *except* at $x = 2$, where it meets the x-axis.

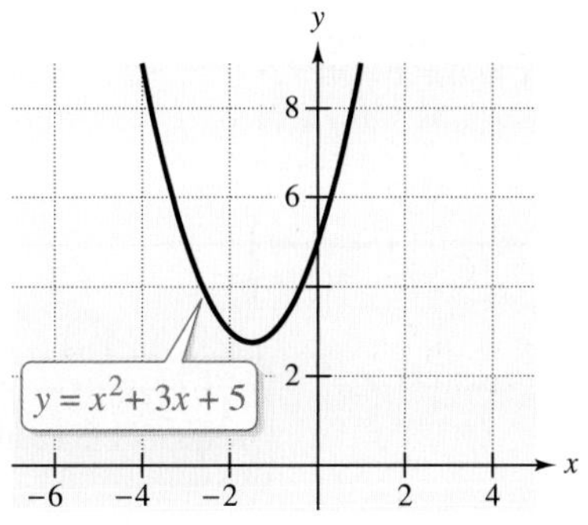

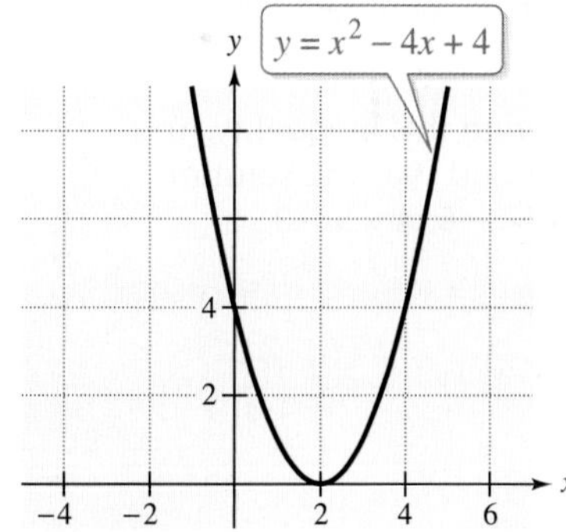

Exercises Within Reach®

Solutions in English & Spanish and tutorial videos at AlgebraWithinReach.com

Unusual Solution Sets **In Exercises 23–28, solve the inequality and graph the solution on the real number line. (Some of the inequalities have no solutions.)**

23. $x^2 + 4x + 5 < 0$
No solution

24. $x^2 + 14x + 49 < 0$
No solution

25. $x^2 + 6x + 10 > 0$
$-\infty < x < \infty$; See Additional Answers.

26. $x^2 + 2x + 1 \ge 0$
$-\infty < x < \infty$; See Additional Answers.

27. $y^2 + 16y + 64 \le 0$
$x = -8$; See Additional Answers.

28. $4x^2 + 28x + 49 \le 0$
$x = -\frac{7}{2}$; See Additional Answers.

Rational Inequalities

The concepts of critical numbers and test intervals can be extended to inequalities involving rational expressions. To do this, use the fact that the value of a rational expression can change sign only at its *zeros* (the x-values for which its numerator is zero) and its *undefined values* (the x-values for which its denominator is zero). These two types of numbers make up the **critical numbers** of a rational inequality.

Study Tip

When solving a rational inequality, you should begin by writing the inequality in general form, with the rational expression (as a single fraction) on the left and zero on the right. For instance, the first step in solving

$$\frac{2x}{x+3} < 4$$

is to write it as

$$\frac{2x}{x+3} - 4 < 0$$

$$\frac{2x - 4(x+3)}{x+3} < 0$$

$$\frac{-2x - 12}{x+3} < 0.$$

Try solving this inequality. You should find that the solution set is $x < -6$ or $x > -3$.

EXAMPLE 5 Solving a Rational Inequality

To solve the inequality $\frac{x}{x-2} > 0$, first find the critical numbers. The numerator is zero when $x = 0$, and the denominator is zero when $x = 2$. So, the two critical numbers are 0 and 2, which implies that the test intervals are $(-\infty, 0)$, $(0, 2)$, and $(2, \infty)$. To test an interval, choose a convenient value in the interval and determine whether the value satisfies the inequality.

Test interval	Representative x-value	Is inequality satisfied?
$(-\infty, 0)$	$x = -1$	$\frac{-1}{-1-2} \overset{?}{>} 0 \Rightarrow \frac{1}{3} > 0$
$(0, 2)$	$x = 1$	$\frac{1}{1-2} \overset{?}{>} 0 \Rightarrow -1 \not> 0$
$(2, \infty)$	$x = 3$	$\frac{3}{3-2} \overset{?}{>} 0 \Rightarrow 3 > 0$

From this table, you can see that the solution set of the inequality is $x < 0$ or $x > 2$, as shown below.

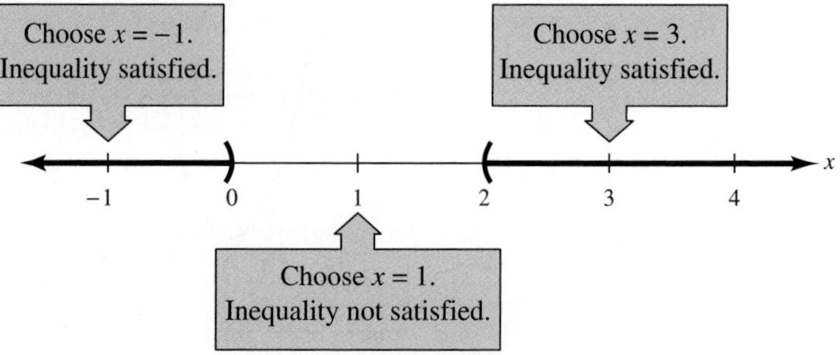

Exercises Within Reach ®

Solutions in English & Spanish and tutorial videos at AlgebraWithinReach.com

Solving a Rational Inequality **In Exercises 29–36, solve the inequality and graph the solution on the real number line.** See Additional Answers.

29. $\frac{5}{x-3} > 0$ $x > 3$

30. $\frac{3}{4-x} > 0$ $x < 4$

31. $\frac{-5}{x-3} > 0$ $x < 3$

32. $\frac{-3}{4-x} > 0$ $x > 4$

33. $\frac{x+4}{x-2} > 0$ $x < -4$ or $x > 2$

34. $\frac{x-5}{x+2} < 0$ $-2 < x < 5$

35. $\frac{3}{y-1} \le -1$ $-2 \le x < 1$

36. $\frac{2}{x-3} \ge -1$ $x \le 1$ or $x > 3$

Application

Application **EXAMPLE 6** **The Height of a Projectile**

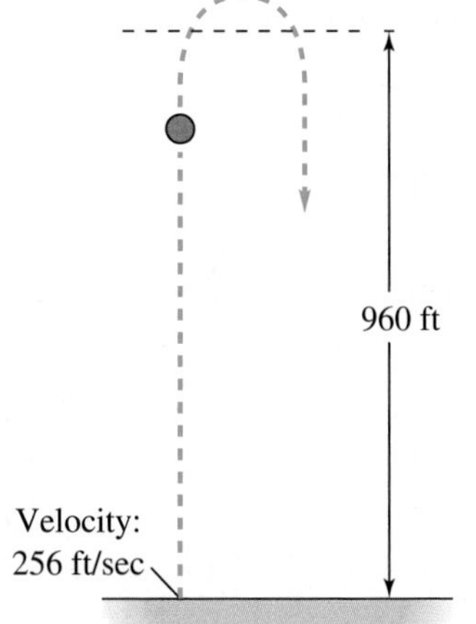

A projectile is fired straight upward from ground level with an initial velocity of 256 feet per second, as shown at the left. Its height h at any time t is given by

$$h = -16t^2 + 256t$$

where h is measured in feet and t is measured in seconds. During what interval of time will the height of the projectile exceed 960 feet?

SOLUTION

To solve this problem, begin by writing the inequality in general form.

$-16t^2 + 256t > 960$ Write original inequality.

$-16t^2 + 256t - 960 > 0$ Write in general form.

Next, find the critical numbers for $-16t^2 + 256t - 960 > 0$ by finding the solutions of the equation $-16t^2 + 256t - 960 = 0$.

$-16t^2 + 256t - 960 = 0$ Write corresponding equation.

$t^2 - 16t + 60 = 0$ Divide each side by -16.

$(t - 6)(t - 10) = 0$ Factor.

$t = 6, \ t = 10$ Critical numbers

This implies that the test intervals are

$(-\infty, 6), (6, 10)$, and $(10, \infty)$.   Test intervals

To test an interval, choose a convenient value in the interval and determine whether the value satisfies the inequality.

Test interval	Representative t-value	Is inequality satisfied?
$(-\infty, 6)$	$t = 0$	$-16(0)^2 + 256(0) \not> 960$
$(6, 10)$	$t = 7$	$-16(7)^2 + 256(7) > 960$
$(10, \infty)$	$t = 11$	$-16(11)^2 + 256(11) \not> 960$

So, the height will exceed 960 feet for values of t such that $6 < t < 10$.

Exercises Within Reach®

Solutions in English & Spanish and tutorial videos at AlgebraWithinReach.com

37. ***Height*** A projectile is fired straight upward from ground level with an initial velocity of 128 feet per second. Its height h at any time t is given by

$$h = -16t^2 + 128t$$

where h is measured in feet and t is measured in seconds. During what interval of time will the height of the projectile exceed 240 feet?

(3, 5)

38. ***Height*** A projectile is fired straight upward from ground level with an initial velocity of 88 feet per second. Its height h at any time t is given by

$$h = -16t^2 + 88t$$

where h is measured in feet and t is measured in seconds. During what interval of time will the height of the projectile exceed 50 feet?

$\left(\frac{11}{4} - \frac{\sqrt{71}}{4}, \frac{11}{4} + \frac{\sqrt{71}}{4}\right) \approx (0.64, 4.86)$

Concept Summary: Solving Quadratic and Rational Inequalities

What

You can use critical numbers and test intervals to solve quadratic and rational inequalities.

EXAMPLE

Solve the inequality

$x(x - 4) < 0$.

How

1. Find the critical numbers of the polynomial or rational inequality.
2. Use the critical numbers to determine the test intervals.
3. Choose a value in each interval and determine if the value satisfies the inequality.

EXAMPLE

$x(x - 4) = 0$

$x = 0,\ x = 4$

This implies that the test intervals are $(-\infty, 0)$, $(0, 4)$, and $(4, \infty)$.

By testing $x = -1$, $x = 1$, and $x = 5$, you can conclude that the solution set is $0 < x < 4$.

Why

By using the critical numbers to determine the test intervals, you need to test only a few values to find the solution set of the inequality.

Exercises Within Reach ®

Worked-out solutions to odd-numbered exercises at AlgebraWithinReach.com

Concept Summary Check

39. ***Vocabulary*** The test intervals of a polynomial are $(-\infty, -1)$, $(-1, 3)$, $(3, \infty)$. What are the critical numbers of the polynomial? The critical numbers are -1 and 3.
40. ***Writing*** In your own words, describe a procedure for solving quadratic inequalities. See Additional Answers.
41. ***Writing*** Is $x = 4$ a solution of the inequality $x(x - 4) < 0$? Explain. See Additional Answers.
42. ***Logic*** How is the procedure for finding the critical numbers of a quadratic inequality different from the procedure for finding the critical numbers of a rational inequality. See Additional Answers.

Extra Practice

Finding Critical Numbers **In Exercises 43–46, find the critical numbers.**

43. $x(2x - 5)$ $0, \frac{5}{2}$
44. $5x(x - 3)$ $0, 3$
45. $4x^2 - 81$ $\pm\frac{9}{2}$
46. $9y^2 - 16$ $\pm\frac{4}{3}$

Solving an Inequality **In Exercises 47–58, solve the inequality and graph the solution on the real number line.** See Additional Answers.

47. $6 - (x - 2)^2 < 0$ $x < 2 - \sqrt{6}$ or $x > 2 + \sqrt{6}$
48. $(y + 3)^2 - 6 \geq 0$ $x \leq -3 - \sqrt{6}$ or $x \geq -3 + \sqrt{6}$
49. $16 \leq (u + 5)^2$ $x \leq -9$ or $x \geq -1$
50. $25 \geq (x - 3)^2$ $-2 \leq x \leq 8$
51. $\dfrac{u - 6}{3u - 5} \leq 0$ $\frac{5}{3} < x \leq 6$
52. $\dfrac{3(u - 3)}{u + 1} < 0$ $-1 < x < 3$
53. $\dfrac{2(4 - t)}{4 + t} > 0$ $-4 < x < 4$
54. $\dfrac{2}{x - 5} \geq 3$ $5 < x \leq \frac{17}{3}$
55. $\dfrac{1}{x + 2} > -3$ $x < -\frac{7}{3}$ or $x > -2$
56. $\dfrac{4x}{x + 2} < -1$ $-2 < x < -\frac{2}{5}$
57. $\dfrac{6x}{x - 4} < 5$ $-20 < x < 4$
58. $\dfrac{x - 3}{x - 6} \leq 4$ $x < 6$ or $x \geq 7$

59. ***Compound Interest*** You are investing \$1000 in a certificate of deposit for 2 years, and you want the interest for that time period to exceed \$150. The interest is compounded annually. What interest rate should you have? [*Hint:* Solve the inequality $1000(1 + r)^2 > 1150$.]
$r > 7.24\%$

60. ***Compound Interest*** You are investing \$500 in a certificate of deposit for 2 years, and you want the interest for that time period to exceed \$50. The interest is compounded annually. What interest rate should you have? [*Hint:* Solve the inequality $500(1 + r)^2 > 550$.]
$r > 4.88\%$

61. ***Geometry*** You have 64 feet of fencing to enclose a rectangular region. Determine the interval for the length such that the area will exceed 240 square feet. $(12, 20)$

62. ***Cost, Revenue, and Profit*** The revenue and cost equations for a computer desk are given by

$$R = x(50 - 0.0002x) \text{ and } C = 12x + 150{,}000$$

where R and C are measured in dollars and x represents the number of desks sold. How many desks must be sold to obtain a profit of at least \$1,650,000?
$90{,}000 \le x \le 100{,}000$

63. ***Cost, Revenue, and Profit*** The revenue and cost equations for a digital camera are given by

$$R = x(125 - 0.0005x)$$

and

$$C = 3.5x + 185{,}000$$

where R and C are measured in dollars and x represents the number of cameras sold. How many cameras must be sold to obtain a profit of at least \$6,000,000?
$72{,}589 < x < 170{,}411$

64. ***Antibiotics*** The concentration C (in milligrams per liter) of an antibiotic t minutes after it is administered is given by

$$C(t) = \frac{21.9 - 0.043t}{1 + 0.005t}, \quad 30 \le t \le 500.$$

(a) Use a graphing calculator to graph the concentration function. See Additional Answers.

(b) How long does it take for the concentration of the antibiotic to fall below 5 milligrams per liter?
$\frac{4225}{17} \approx 248.5$ minutes

Explaining Concepts

65. ***Reasoning*** Explain why the critical numbers of a polynomial are not included in its test intervals.
The critical numbers of a polynomial are its zeros, so the value of the polynomial is zero at its critical numbers.

66. ***Precision*** Explain the difference in the solution sets of $x^2 - 4 < 0$ and $x^2 - 4 \le 0$. The solution set of $x^2 - 4 \le 0$ includes the values $x = -2$ and $x = 2$, and $x^2 - 4 < 0$ does not.

67. ***Reasoning*** The graph of a quadratic function g lies completely above the x-axis. What is the solution set of the inequality $g(x) < 0$? Explain your reasoning.
No solution. The value of the polynomial is positive for every real value of x, so there are no values that would make the polynomial negative.

68. ***Using a Graph*** Explain how you can use the graph of $f(x) = x^2 - x - 6$ to check the solution of $x^2 - x - 6 > 0$.
The solutions of the inequality are the values of x for which the parabola lies above the x-axis.

Cumulative Review

In Exercises 69–74, perform the operation and simplify.

69. $\dfrac{4xy^3}{x^2y} \cdot \dfrac{y}{8x}$ $\quad \dfrac{y^3}{2x^2}, y \ne 0$

70. $\dfrac{2x^2 - 2}{x^2 - 6x - 7} \cdot (x^2 - 10x + 21)$ $\quad 2(x - 1)(x - 3), x \ne -1, x \ne 7$

71. $\dfrac{x^2 - x - 6}{4x^3} \cdot \dfrac{x + 1}{x^2 + 5x + 6}$ $\quad \dfrac{(x - 3)(x + 1)}{4x^3(x + 3)}, x \ne -2$

72. $\dfrac{32x^3y}{y^9} \div \dfrac{8x^4}{y^6}$ $\quad \dfrac{4}{xy^2}$

73. $\dfrac{x^2 + 8x + 16}{x^2 - 6x} \div (3x - 24)$ $\quad \dfrac{(x + 4)^2}{3x(x - 6)(x - 8)}$

74. $\dfrac{x^2 + 6x - 16}{3x^2} \div \dfrac{x + 8}{6x}$ $\quad \dfrac{2(x - 2)}{x}, x \ne -8$

In Exercises 75–78, evaluate the expression for the specified value. Round your result to the nearest hundredth, if necessary.

75. $x^2; x = -\frac{1}{3}$ $\quad \frac{1}{9}$

76. $1000 - 20x^3; x = 4.02$ $\quad -299.30$

77. $\dfrac{100}{x^4}; x = 1.06$ $\quad 79.21$

78. $\dfrac{50}{1 - \sqrt{x}}; x = 0.1024$ $\quad 73.53$

8 Chapter Summary

	What did you learn?	Explanation and Examples	Review Exercises
8.1	Solve quadratic equations by factoring *(p. 414)*.	**1.** Write the equation in general form. **2.** Factor the left side. **3.** Set each factor equal to zero and solve for x.	1–10
	Solve quadratic equations by the Square Root Property *(p. 415)*.	**1.** The equation $u^2 = d$, where $d > 0$, has exactly two solutions: $u = \pm\sqrt{d}$. **2.** The equation $u^2 = d$, where $d < 0$, has exactly two solutions: $u = \pm\sqrt{\lvert d \rvert}i$.	11–22
	Use substitution to solve equations of quadratic form *(p. 418)*.	An equation is said to be of quadratic form if it has the form $au^2 + bu + c = 0$, where u is an algebraic expression. To solve an equation of quadratic form, it helps to make a substitution and rewrite the equation in terms of u.	23–30
8.2	Rewrite quadratic expressions in completed square form *(p. 422)*.	To complete the square for the expression $x^2 + bx$, add $(b/2)^2$, which is the square of half the coefficient of x. Consequently, $x^2 + bx + \left(\frac{b}{2}\right)^2 = \left(x + \frac{b}{2}\right)^2$.	31–36
	Solve quadratic equations by completing the square *(p. 423)*.	**1.** Prior to completing the square, the coefficient of the second-degree term must be 1. **2.** Preserve the equality by adding the same constant to each side of the equation. **3.** Use the Square Root Property to solve the quadratic equation.	37–42
8.3	Use the Quadratic Formula to solve quadratic equations *(p. 430)*.	The solutions of $ax^2 + bx + c = 0$, $a \neq 0$, are given by the Quadratic Formula $x = \dfrac{-b \pm \sqrt{b^2 - 4ac}}{2a}$.	43–48
	Determine the types of solutions of quadratic equations using the discriminant *(p. 433)*.	Let a, b, and c be rational numbers such that $a \neq 0$. The discriminant of the quadratic equation $ax^2 + bx + c = 0$ is given by $b^2 - 4ac$, and can be used to classify the solutions of the equation as follows. *Discriminant* — *Solution Type* **1.** Perfect square — Two distinct rational solutions **2.** Positive nonperfect square — Two distinct irrational solutions **3.** Zero — One repeated rational solution **4.** Negative number — Two distinct complex solutions	49–56

	What did you learn?	Explanation and Examples	Review Exercises
8.3	Write quadratic equations from solutions of the equations *(p. 435)*.	You can use the Zero-Factor Property in reverse to find a quadratic equation given its solutions.	57–62
8.4	Determine the vertices of parabolas and sketch parabolas *(p. 440)*.	**1.** Determine the vertex and axis of the parabola by completing the square or by using the formula $x = -b/(2a)$. **2.** Plot the vertex, axis, x- and y-intercepts, and a few additional points on the parabola. (Using the symmetry about the axis can reduce the number of points you need to plot.) **3.** Use the fact that the parabola opens upward when $a > 0$ and opens downward when $a < 0$ to complete the sketch.	63–70
	Write an equation of a parabola given the vertex and a point on the graph *(p. 443)*.	To write an equation of a parabola with a vertical axis, use the fact that its standard equation has the form $y = a(x - h)^2 + k$, where (h, k) is the vertex.	71–74
	Use parabolas to solve application problems *(p. 445)*.	You can solve applications involving the path of an object using parabolas.	75, 76
8.5	Use quadratic equations to solve application problems *(p. 448)*.	The following are samples of applications of quadratic equations. **1.** Investment **2.** Interest **3.** Height of a projectile **4.** Geometric dimensions **5.** Work rate **6.** Structural design **7.** Falling object **8.** Reduced rate	77–86
8.6	Use test intervals to solve quadratic inequalities *(p. 457)*.	**1.** For a polynomial expression, find all the real zeros. **2.** Arrange the zeros in increasing order. These numbers are called *critical numbers*. **3.** Use the critical numbers to determine the test intervals. **4.** Choose a representative x-value in each test interval and evaluate the expression at that value. When the value of the expression is negative, the expression has negative values for all x-values in the interval. When the value of the expression is positive, the expression has positive values for all x-values in the interval.	87–96
	Use test intervals to solve rational inequalities *(p. 460)*.	For a rational expression, find all the real zeros and those x-values for which the function is undefined. Then use Steps 2–4 as described above.	97–100
	Use inequalities to solve application problems *(p. 461)*.	Not only can you solve real-life applications using quadratic and rational equations, but you can also solve these types of problems using quadratic and rational inequalities.	101, 102

Review Exercises

Worked-out solutions to odd-numbered exercises at AlgebraWithinReach.com

8.1

Solving a Quadratic Equation by Factoring **In Exercises 1–10, solve the equation by factoring.**

1. $x^2 + 12x = 0$ $-12, 0$ **2.** $u^2 - 18u = 0$ $0, 18$

3. $3y^2 - 27 = 0$ ± 3 **4.** $2z^2 - 72 = 0$ ± 6

5. $4y^2 + 20y + 25 = 0$ $-\frac{5}{2}$

6. $x^2 + \frac{8}{3}x + \frac{16}{9} = 0$ $-\frac{4}{3}$

7. $2x^2 - 2x - 180 = 0$ $-9, 10$

8. $9x^2 + 18x - 135 = 0$ $-5, 3$

9. $2x^2 - 9x - 18 = 0$ $-\frac{3}{2}, 6$

10. $3x^2 - 19x + 20 = 0$ $\frac{4}{3}, 5$

Using the Square Root Property **In Exercises 11–22, solve the equation by using the Square Root Property.**

11. $z^2 = 144$ ± 12 **12.** $2x^2 = 98$ ± 7

13. $y^2 - 12 = 0$ $\pm 2\sqrt{3}$

14. $y^2 - 45 = 0$ $\pm 3\sqrt{5}$

15. $(x - 16)^2 = 400$ $-4, 36$

16. $(x + 3)^2 = 900$ $-33, 27$

17. $z^2 = -121$ $\pm 11i$

18. $u^2 = -225$ $\pm 15i$

19. $y^2 + 50 = 0$ $\pm 5\sqrt{2}i$

20. $x^2 + 48 = 0$ $\pm 4\sqrt{3}i$

21. $(y + 4)^2 + 18 = 0$ $-4 \pm 3\sqrt{2}i$

22. $(x - 2)^2 + 24 = 0$ $2 \pm 2\sqrt{6}i$

Solving an Equation of Quadratic Form **In Exercises 23–30, solve the equation of quadratic form.**

23. $x^4 - 4x^2 - 5 = 0$ $\pm\sqrt{5}, \pm i$

24. $x^4 - 10x^2 + 9 = 0$ $\pm 1, \pm 3$

25. $x - 4\sqrt{x} + 3 = 0$ $1, 9$

26. $x - 4\sqrt{x} + 13 = 0$ $-5 \pm 12i$

27. $(x^2 - 2x)^2 - 4(x^2 - 2x) - 5 = 0$ $1, 1 \pm \sqrt{6}$

28. $(\sqrt{x} - 2)^2 + 2(\sqrt{x} - 2) - 3 = 0$ 9

29. $x^{2/3} + 3x^{1/3} - 28 = 0$ $-343, 64$

30. $x^{2/5} + 4x^{1/5} + 3 - 0$ $243, -1$

8.2

Constructing a Perfect Square Trinomial **In Exercises 31–36, add a term to the expression to make it a perfect square trinomial.**

31. $z^2 + 18z +$ 81

32. $y^2 - 80y +$ 1600

33. $x^2 - 15x +$ $\frac{225}{4}$

34. $x^2 + 21x +$ $\frac{441}{4}$

35. $y^2 + \frac{2}{5}y +$ $\frac{1}{25}$

36. $x^2 - \frac{3}{4}x +$ $\frac{9}{64}$

Completing the Square **In Exercises 37–42, solve the equation by completing the square. Give the solutions in exact form and in decimal form rounded to two decimal places.**

37. $x^2 - 6x - 3 = 0$ $3 + 2\sqrt{3} \approx 6.46;\ 3 - 2\sqrt{3} \approx -0.46$

38. $x^2 + 12x + 6 = 0$
$-6 + \sqrt{30} \approx -0.52;\ -6 - \sqrt{30} \approx -11.48$

39. $v^2 + 5v + 4 = 0$ $-4, -1$ **40.** $u^2 - 5u + 6 = 0$ $2, 3$

41. $y^2 - \frac{2}{3}y + 2 = 0$
$\frac{1}{3} + \frac{\sqrt{17}}{3}i \approx 0.33 + 1.37i;\ \frac{1}{3} - \frac{\sqrt{17}}{3}i \approx 0.33 - 1.37i$

42. $t^2 + \frac{1}{2}t - 1 = 0$
$-\frac{1}{4} + \frac{\sqrt{17}}{4} \approx 0.78;\ -\frac{1}{4} - \frac{\sqrt{17}}{4} \approx -1.28$

8.3

Solving a Quadratic Equation **In Exercises 43–48, solve the equation by using the Quadratic Formula.**

43. $v^2 + v - 42 = 0$ $-7, 6$

44. $4x^2 + 12x + 9 = 0$ $-\frac{3}{2}$

45. $5x^2 - 16x + 2 = 0$ $\frac{8}{5} \pm \frac{3\sqrt{6}}{5}$

46. $3x^2 + 12x + 4 = 0$ $-2 \pm \frac{2\sqrt{6}}{3}$

47. $8x^2 - 6x + 2 = 0$ $\frac{3}{8} \pm \frac{\sqrt{7}}{8}i$

48. $x^2 - 4x + 8 = 0$ $2 \pm 2i$

Using the Discriminant **In Exercises 49–56, use the discriminant to determine the type of solution(s) of the quadratic equation.**

49. $x^2 + 4x + 4 = 0$
One repeated rational solution

50. $y^2 - 26y + 169 = 0$
One repeated rational solution

51. $s^2 - s - 20 = 0$
Two distinct rational solutions

52. $r^2 - 5r - 45 = 0$
Two distinct irrational solutions

53. $4t^2 + 16t + 10 = 0$
Two distinct irrational solutions

54. $8x^2 + 85x - 33 = 0$
Two distinct rational solutions

55. $v^2 - 6v + 21 = 0$
Two distinct complex solutions

56. $9y^2 + 1 = 0$
Two distinct complex solutions

Writing a Quadratic Equation **In Exercises 57–62, write a quadratic equation that has the given solutions.**

57. $x = 3, x = -7$
$x^2 + 4x - 21 = 0$

58. $x = -2, x = 8$
$x^2 - 6x - 16 = 0$

59. $x = 5 + \sqrt{7}, x = 5 - \sqrt{7}$
$x^2 - 10x + 18 = 0$

60. $x = 2 + \sqrt{2}, x = 2 - \sqrt{2}$
$x^2 - 4x + 2 = 0$

61. $x = 6 + 2i, x = 6 - 2i$
$x^2 - 12x + 40 = 0$

62. $x = 3 + 4i, x = 3 - 4i$
$x^2 - 6x + 25 = 0$

8.4

Finding the Vertex of a Parabola **In Exercises 63–66, write the equation of the parabola in standard form, and find the vertex of its graph.**

63. $y = x^2 - 8x + 3$
$y = (x - 4)^2 - 13$; Vertex: $(4, -13)$

64. $y = 8 - 8x - x^2$
$y = -(x + 4)^2 + 24$; Vertex: $(-4, 24)$

65. $y = 2x^2 - x + 3$
$y = 2\left(x - \frac{1}{4}\right)^2 + \frac{23}{8}$; Vertex: $\left(\frac{1}{4}, \frac{23}{8}\right)$

66. $y = 3x^2 + 2x - 6$
$y = 3\left(x + \frac{1}{3}\right)^2 - \frac{19}{3}$; Vertex: $\left(-\frac{1}{3}, -\frac{19}{3}\right)$

Sketching a Parabola **In Exercises 67–70, sketch the parabola. Identify the vertex and any x-intercepts.**
See Additional Answers.

67. $y = x^2 + 8x$

68. $y = -x^2 + 3x$

69. $f(x) = -x^2 - 2x + 4$

70. $f(x) = x^2 + 3x - 10$

Writing an Equation of a Parabola **In Exercises 71–74, write an equation of the parabola that satisfies the conditions.**

71. Vertex: $(2, -5)$
Point on the graph: $(0, 3)$
$y = 2(x - 2)^2 - 5$

72. Vertex: $(-4, 0)$
Point on the graph: $(0, -6)$
$y = -\frac{3}{8}(x + 4)^2$

73. Vertex: $(5, 0)$
Point on the graph: $(1, 1)$
$y = \frac{1}{16}(x - 5)^2$

74. Vertex: $(-2, 5)$
Point on the graph: $(-4, 11)$
$y = \frac{3}{2}(x + 2)^2 + 5$

75. ***Path of a Ball*** The height y (in feet) of a ball thrown by a child is given by $y = -\frac{1}{10}x^2 + 3x + 6$, where x is the horizontal distance (in feet) from where the ball is thrown.

(a) How high is the ball when it leaves the child's hand? 6 feet

(b) How high is the ball at its maximum height? 28.5 feet

(c) How far from the child does the ball strike the ground? 31.9 feet

MANDY GODBEHEAR/Shutterstock.com

76. ***Path of an Object*** You use a fishing rod to cast a lure into the water. The height y (in feet) of the lure is given by

$$y = -\frac{1}{90}x^2 + \frac{1}{5}x + 9,$$

where x is the horizontal distance (in feet) from where the lure is released.

(a) How high is the lure when it is released?
9 feet

(b) How high is the lure at its maximum height?
9.9 feet

(c) How far from its release point does the lure land?
$9 + 9\sqrt{11} \approx 38.8$ feet

8.5

77. ***Selling Price*** A car dealer buys a fleet of used cars for a total of \$80,000. The dealer regains this investment by selling all but 4 of these cars at a profit of \$1000 each. How many cars has the dealer sold, and at what price?
16 cars; \$5000

78. ***Selling Price*** A manager of a computer store buys several computers of the same model for \$27,000. The store can regain this investment by selling all but 5 of the computers at a profit of \$900 per computer. To do this, how many computers must be sold, and at what price?
10 computers; \$2700

79. ***Geometry*** The length of a rectangle is 12 inches greater than its width. The area of the rectangle is 85 square inches. Find the dimensions of the rectangle.
5 inches × 17 inches

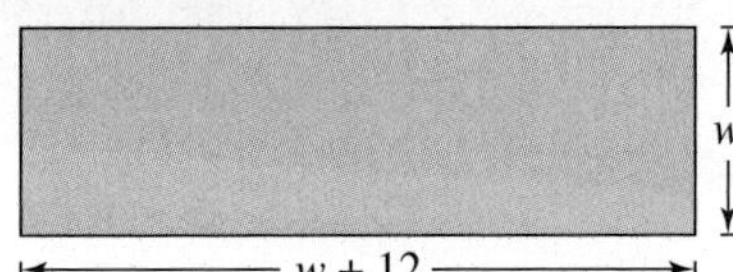

300dpi/Shutterstock.com

80. ***Geometry*** The height of a triangle is 3 inches greater than its base. The area of the triangle is 44 square inches. Find the base and height of the triangle.
Base: 8 inches, Height: 11 inches

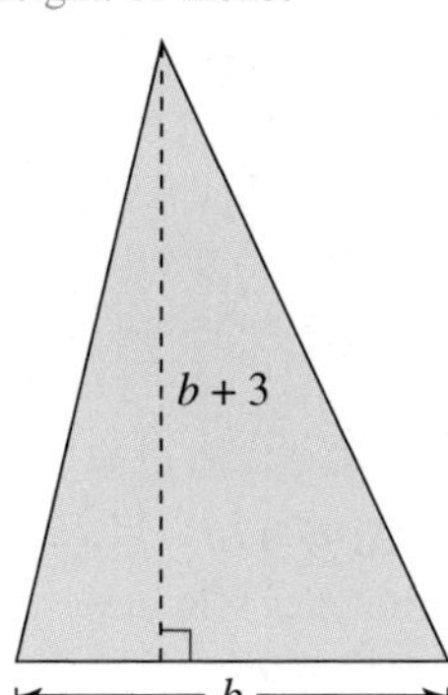

81. ***Reduced Rates*** A Little League baseball team pays \$96 for a block of tickets to a ball game. The block contains three more tickets than the team needs. By inviting 3 more people to attend (and share in the cost), the team lowers the price per ticket by \$1.60. How many people are going to the game?
15 people

82. ***Compound Interest*** You want to invest \$35,000 for 2 years at an annual interest rate of r (in decimal form). Interest on the account is compounded annually. Find the interest rate when a deposit of \$35,000 increases to \$40,221.44 over a two-year period.
7.2%

83. ***Geometry*** You leave campus to pick up two student council officers for a meeting. You drive a total of 20 miles to pick up the secretary and then the treasurer (see figure). By using a direct route, you are able to drive just 16 miles to return to campus. Find the possible distances from campus to the secretary's location. 4.7 miles or 15.3 miles

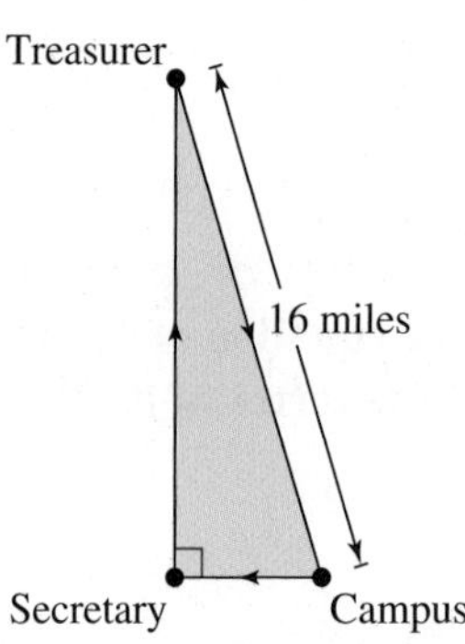

84. ***Geometry*** A corner lot has an L-shaped sidewalk along its sides. The total length of the sidewalk is 69 feet. By cutting diagonally across the lot, the walking distance is shortened to 51 feet. What are the lengths of the two legs of the sidewalk?
24 feet, 45 feet

85. ***Work-Rate Problem*** Working together, two people can complete a task in 10 hours. Working alone, one person takes 2 hours longer than the other. How long would it take each person to do the task alone?
$9 + \sqrt{101} \approx 19$ hours, $11 + \sqrt{101} \approx 21$ hours

86. ***Height*** A model rocket is projected straight upward at an initial velocity of 64 feet per second from a height of 192 feet. The height h at any time t is given by

$h = -16t^2 + 64t + 192$

where t is the time in seconds.

(a) After how many seconds is the height 256 feet?
2 seconds

(b) After how many seconds does the object hit the ground? 6 seconds

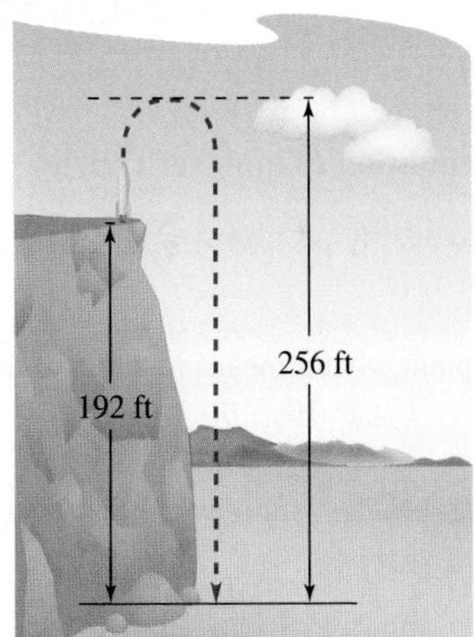

8.6

Finding Critical Numbers **In Exercises 87–90, find the critical numbers.**

87. $2x(x + 7)$ $-7, 0$

88. $x(x - 2) + 4(x - 2)$ $-4, 2$

89. $x^2 - 6x - 27$ $-3, 9$ **90.** $2x^2 + 11x + 5$ $-5, -\frac{1}{2}$

Solving a Quadratic Inequality **In Exercises 91–96, solve the inequality and graph the solution on the real number line.** See Additional Answers.

91. $5x(7 - x) > 0$ $0 < x < 7$

92. $-2x(x - 10) \le 0$ $x \le 0$ or $x \ge 10$

93. $16 - (x - 2)^2 \le 0$ $x \le -2$ or $x \ge 6$

94. $(x - 5)^2 - 36 > 0$ $x < -1$ or $x > 11$

95. $2x^2 + 3x - 20 < 0$ $-4 < x < \frac{5}{2}$

96. $-3x^2 + 10x + 8 \ge 0$ $-\frac{2}{3} \le x \le 4$

landysh/Shutterstock.com

Solving a Rational Inequality **In Exercises 97–100, solve the inequality and graph the solution on the real number line.** See Additional Answers.

97. $\dfrac{x + 3}{2x - 7} \ge 0$
$x \le -3$ or $x > \frac{7}{2}$

98. $\dfrac{3x + 2}{x - 3} > 0$
$x < -\frac{2}{3}$ or $x > 3$

99. $\dfrac{x + 4}{x - 1} < 0$
$-4 < x < 1$

100. $\dfrac{2x - 9}{x - 1} \le 0$
$1 < x \le \frac{9}{2}$

101. ***Height*** A projectile is fired straight upward from ground level with an initial velocity of 312 feet per second. Its height h at any time t is given by

$h = -16t^2 + 312t$

where h is measured in feet and t is measured in seconds. During what interval of time will the height of the projectile exceed 1200 feet?
(5.3, 14.2)

102. ***Average Cost*** The cost C of producing x notebooks is

$C = 100{,}000 + 0.9x, x > 0.$

Write the average cost

$$\overline{C} = \frac{C}{x}$$

as a function of x. Then determine how many notebooks must be produced for the average cost per unit to be less than \$2.

$\overline{C} = \dfrac{100{,}000}{x} + 0.9, x > 0; x > 90{,}910$ notebooks

Chapter Test

Solutions in English & Spanish and tutorial videos at AlgebraWithinReach.com

Take this test as you would take a test in class. After you are done, check your work against the answers in the back of the book.

In Exercises 1–6, solve the equation by the specified method.

1. Factoring:
$x(x - 3) - 10(x - 3) = 0$ 3, 10

2. Factoring:
$6x^2 - 34x - 12 = 0$ $-\frac{1}{3}, 6$

3. Square Root Property:
$(x - 2)^2 = 0.09$ 1.7, 2.3

4. Square Root Property
$(x + 4)^2 + 100 = 0$ $-4 \pm 10i$

5. Completing the square:
$2x^2 - 6x + 3 = 0$ $\frac{3}{2} \pm \frac{\sqrt{3}}{2}$

6. Completing the square:
$2y(y - 2) = 7$ $1 \pm \frac{3\sqrt{2}}{2}$

In Exercises 7 and 8, solve the equation of quadratic form.

7. $\frac{1}{x^2} - \frac{6}{x} + 4 = 0$ $\frac{3}{4} \pm \frac{\sqrt{5}}{4}$

8. $x^{2/3} - 9x^{1/3} + 8 = 0$ 1, 512

9. Find the discriminant and explain what it means in terms of the type of solutions of the quadratic equation $5x^2 - 12x + 10 = 0$.
-56; A negative discriminant tells us the equation has two imaginary solutions.

10. Write a quadratic equation that has the solutions -7 and -3.
$x^2 + 10x + 21 = 0$

In Exercises 11 and 12, sketch the parabola. Identify the vertex and any *x*-intercepts. See Additional Answers.

11. $y = -x^2 + 2x - 4$

12. $y = x^2 - 2x - 15$

In Exercises 13–15, solve the inequality and graph the solution on the real number line. See Additional Answers.

13. $16 \le (x - 2)^2$
$x \le -2$ or $x \ge 6$

14. $2x(x - 3) < 0$
$0 < x < 3$

15. $\frac{x + 1}{x - 5} \le 0$
$-1 \le x < 5$

16. The width of a rectangle is 22 feet less than its length. The area of the rectangle is 240 square feet. Find the dimensions of the rectangle. 8 feet × 30 feet

17. An English club charters a bus trip to a Shakespearean festival. The cost of the bus is \$1250. To lower the bus fare per person, the club invites nonmembers to go along. When 10 nonmembers join the trip, the fare per person is decreased by \$6.25. How many club members are going on the trip? 40 members

18. An object is dropped from a height of 75 feet. Its height h (in feet) at any time t is given by $h = -16t^2 + 75$, where t is measured in seconds. Find the time required for the object to fall to a height of 35 feet. $\frac{\sqrt{10}}{2} \approx 1.58$ seconds

19. Two buildings are connected by an L-shaped walkway. The total length of the walkway is 155 feet. By cutting diagonally across the grass, the walking distance is shortened to 125 feet. What are the lengths of the two legs of the walkway?
35 feet, 120 feet

Exponential and Logarithmic Functions

MASTERY IS WITHIN REACH!

"I failed my first college math class because, for some reason, I thought just showing up and listening would be enough. I was wrong. Now, I get to class a little early to review my notes, sit where I can see the instructor, and ask questions. I try to learn and remember as much as possible in class because I am so busy juggling work and college."

Avonne
Business

See page 499 for suggestions about making the most of class time.

9.1 Exponential Functions

- Evaluate and graph exponential functions.
- Evaluate the natural base e and graph natural exponential functions.
- Use exponential functions to solve application problems.

Exponential Functions

Whereas polynomial and rational functions have terms with variable bases and constant exponents, **exponential functions** have terms with constant bases and variable exponents.

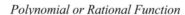

Polynomial or Rational Function	*Exponential Function*
Constant Exponents	Variable Exponents
$f(x) = x^2$, $f(x) = x^{-3}$	$f(x) = 2^x$, $f(x) = 3^{-x}$
Variable Bases	Constant Bases

Study Tip

Rule 4 of the rules of exponential functions indicates that 2^{-x} can be written as

$$2^{-x} = \frac{1}{2^x}$$

Similarly, $\frac{1}{3^{-x}}$ can be written as

$$\frac{1}{3^{-x}} = 3^x.$$

In other words, you can move a *factor* from the numerator to the denominator (or from the denominator to the numerator) by changing the sign of its exponent.

Definition of Exponential Function

The **exponential function f with base a** is denoted by $f(x) = a^x$, where $a > 0$, $a \neq 1$, and x is any real number.

Rules of Exponential Functions

Let a be a positive real number, and let x and y be real numbers, variables, or algebraic expressions.

1. $a^x \cdot a^y = a^{x+y}$ Product rule

2. $\frac{a^x}{a^y} = a^{x-y}$ Quotient rule

3. $(a^x)^y = a^{xy}$ Power-to-power rule

4. $a^{-x} = \frac{1}{a^x} = \left(\frac{1}{a}\right)^x$ Negative exponent rule

EXAMPLE 1 Using a Calculator

To evaluate exponential functions with a calculator, you can use the exponential key y^x or ∧. For example, to evaluate $3^{-1.3}$, you can use the following keystrokes.

Keystrokes	*Display*	
3 y^x 1.3 +/− =	0.239741	Scientific
3 ∧ ((−) 1.3) ENTER	0.239741	Graphing

Exercises Within Reach®

Solutions in English & Spanish and tutorial videos at AlgebraWithinReach.com

Using a Calculator **In Exercises 1–6, use a calculator to evaluate the expression. (Round your answer to three decimal places.)**

1. $2^{-2.3}$ 0.203

2. $5^{1.4}$ 9.518

3. $5^{\sqrt{2}}$ 9.739

4. $4^{-\pi}$ 0.013

5. $6^{1/3}$ 1.817

6. $6^{-1/3}$ 0.550

EXAMPLE 2 Evaluating Exponential Functions

Evaluate each function. Use a calculator only if it is necessary or more efficient.

	Function	*Values*
a.	$f(x) = 2^x$	$x = 3, x = -4, x = \pi$
b.	$g(x) = 12^x$	$x = 3, x = -0.1, x = \frac{5}{7}$
c.	$h(x) = (1.04)^{2x}$	$x = 0, x = -2, x = \sqrt{2}$

SOLUTION

	Evaluation	*Comment*
a.	$f(3) = 2^3 = 8$	Calculator is not necessary.
	$f(-4) = 2^{-4} = \dfrac{1}{2^4} = \dfrac{1}{16}$	Calculator is not necessary.
	$f(\pi) = 2^{\pi} \approx 8.825$	Calculator is necessary.
b.	$g(3) = 12^3 \approx 1728$	Calculator is more efficient
	$g(-0.1) = 12^{-0.1} \approx 0.780$	Calculator is necessary.
	$g\left(\frac{5}{7}\right) = 12^{5/7} \approx 5.900$	Calculator is necessary.
c.	$h(0) = (1.04)^{2 \cdot 0} = (1.04)^0 = 1$	Calculator is not necessary.
	$h(-2) = (1.04)^{2(-2)} \approx 0.855$	Calculator is more efficient.
	$h(\sqrt{2}) = (1.04)^{2\sqrt{2}} \approx 1.117$	Calculator is necessary.

Additional Examples

Evaluate each expression.

a. 3^3

b. $7^{-1/3}$

c. $1.085^{\sqrt{3}}$

Answers:

a. 27

b. About 0.523

c. About 1.152

Exercises Within Reach®

Solutions in English & Spanish and tutorial videos at AlgebraWithinReach.com

Evaluating an Exponential Function **In Exercises 7–16, evaluate the function as indicated. Use a calculator only if it is necessary or more efficient. (Round your answers to three decimal places.)**

7. $f(x) = 3^x$
(a) $x = -2$ $\frac{1}{9}$
(b) $x = 0$ 1
(c) $x = 1$ 3

8. $F(x) = 3^{-x}$
(a) $x = -2$ 9
(b) $x = 0$ 1
(c) $x = 1$ $\frac{1}{3}$

9. $g(x) = 2.2^{-x}$
(a) $x = 1$ 0.455
(b) $x = 3$ 0.094
(c) $x = \sqrt{6}$ 0.145

10. $G(x) = 4.2^x$
(a) $x = -1$ 0.238
(b) $x = -2$ 0.057
(c) $x = \sqrt{2}$ 7.610

11. $f(t) = 500\left(\frac{1}{2}\right)^t$
(a) $t = 0$ 500
(b) $t = 1$ 250
(c) $t = \pi$ 56.657

12. $g(s) = 1200\left(\frac{2}{3}\right)^s$
(a) $s = 0$ 1200
(b) $s = 2$ 533.333
(c) $s = \sqrt{2}$ 676.317

13. $f(x) = 1000(1.05)^{2x}$
(a) $x = 0$ 1000
(b) $x = 5$ 1628.895
(c) $x = 10$ 2653.298

14. $g(t) = 10{,}000(1.03)^{4t}$
(a) $t = 1$ 11,255.088
(b) $t = 3$ 14,257.609
(c) $t = 5.5$ 19,161.034

15. $h(x) = \dfrac{5000}{(1.06)^{8x}}$
(a) $x = 5$ 486.111
(b) $x = 10$ 47.261
(c) $x = 20$ 0.447

16. $P(t) = \dfrac{10{,}000}{(1.01)^{12t}}$
(a) $t = 2$ 7875.661
(b) $t = 10$ 3029.948
(c) $t = 20$ 918.058

EXAMPLE 3 Graphing Exponential Functions

In the same coordinate plane, sketch the graph of each function. Determine the domain and range of each function.

a. $f(x) = 2^x$ **b.** $g(x) = 4^x$

SOLUTION

The table lists some values of each function, and the figure at the left shows the graph of each function. From the graphs, you can see that the domain of each function is the set of all real numbers and that the range of each function is the set of all positive real numbers.

x	-2	-1	0	1	2	3
2^x	$\frac{1}{4}$	$\frac{1}{2}$	1	2	4	8
4^x	$\frac{1}{16}$	$\frac{1}{4}$	1	4	16	64

EXAMPLE 4 Graphing Exponential Functions

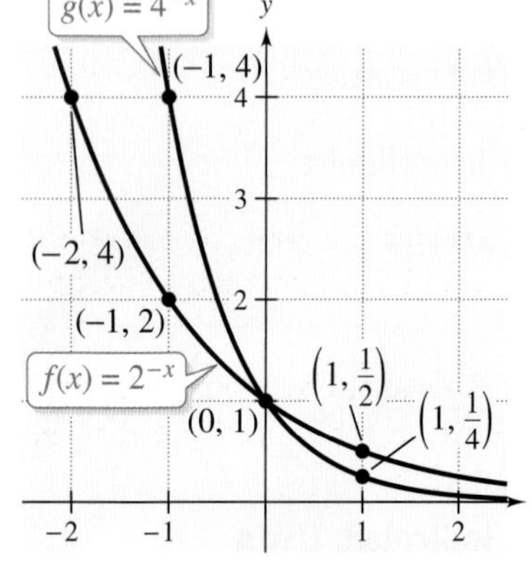

In the same coordinate plane, sketch the graph of each function.

a. $f(x) = 2^{-x}$ **b.** $g(x) = 4^{-x}$

SOLUTION

The table lists some values of each function, and the figure at the left shows the graph of each function.

x	-3	-2	-1	0	1	2
2^{-x}	8	4	2	1	$\frac{1}{2}$	$\frac{1}{4}$
4^{-x}	64	16	4	1	$\frac{1}{4}$	$\frac{1}{16}$

Exercises Within Reach®

Solutions in English & Spanish and tutorial videos at AlgebraWithinReach.com

Matching **In Exercises 17–20, match the function with its graph.**

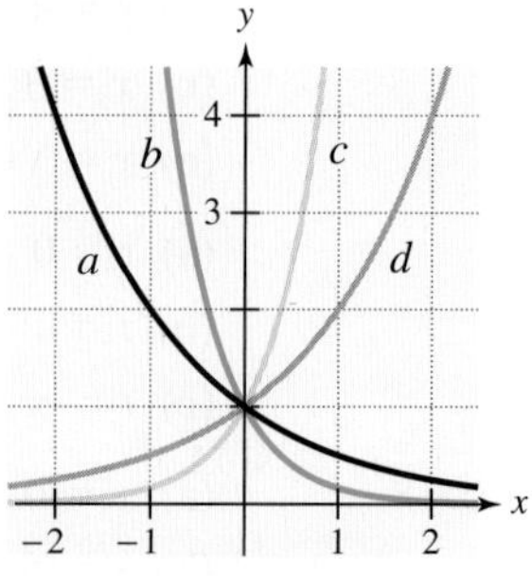

17. $f(x) = 2^x$ d

18. $g(x) = 6^x$ c

19. $h(x) = 2^{-x}$ a

20. $k(x) = 6^{-x}$ b

Graphing an Exponential Function **In Exercises 21–24, sketch the graph of the function. Determine the domain and range.** See Additional Answers.

21. $f(x) = 3^x$ Domain: $-\infty < x < \infty$, Range: $x > 0$

22. $h(x) = \frac{1}{2}(3^x)$ Domain: $-\infty < x < \infty$, Range: $x > 0$

23. $f(x) = 3^{-x} = \left(\frac{1}{3}\right)^x$ Domain: $-\infty < x < \infty$, Range: $x > 0$

24. $h(x) = \frac{1}{2}(3^{-x})$ Domain: $-\infty < x < \infty$, Range: $x > 0$

Study Tip

An **asymptote** of a graph is a line to which the graph becomes arbitrarily close as $|x|$ or $|y|$ increases without bound. In other words, when a graph has an asymptote, it is possible to move far enough out on the graph so that there is almost no difference between the graph and the asymptote.

Graph of $y = a^x$

- Domain: $(-\infty, \infty)$
- Range: $(0, \infty)$
- Intercept: $(0, 1)$
- Increasing (moves up to the right)
- Asymptote: x-axis

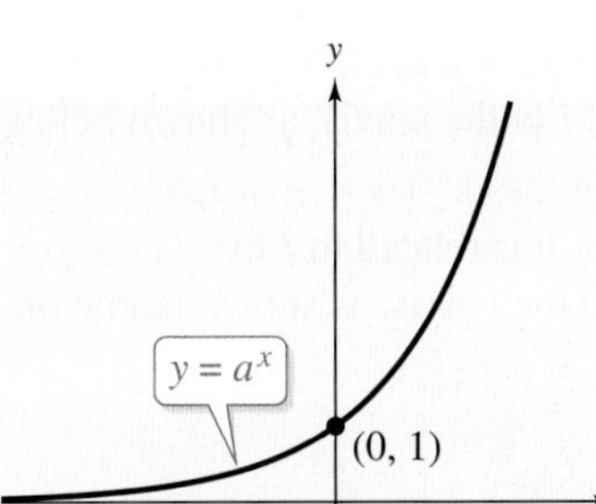

Graph of $y = a^{-x} = \left(\frac{1}{a}\right)^x$

- Domain: $(-\infty, \infty)$
- Range: $(0, \infty)$
- Intercept: $(0, 1)$
- Decreasing (moves down to the right)
- Asymptote: x-axis

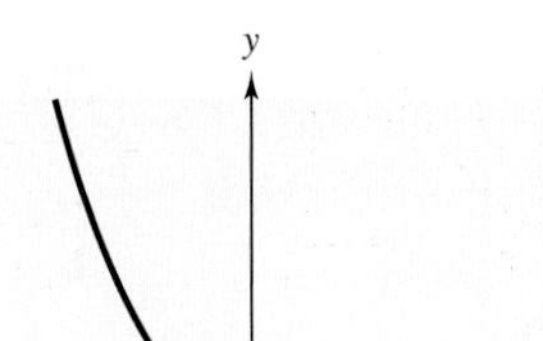

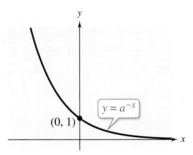

EXAMPLE 5 Transformations of Graphs

Use transformations to analyze and sketch the graph of each function.

a. $g(x) = 3^{x+1}$ **b.** $h(x) = 3^x - 2$

SOLUTION

Consider the function $f(x) = 3^x$.

a. The function g is related to f by $g(x) = f(x + 1)$. To sketch the graph of g, shift the graph of f one unit to the left, as shown below on the left. Note that the y-intercept of g is $(0, 3)$.

b. The function h is related to f by $h(x) = f(x) - 2$. To sketch the graph of g, shift the graph of f two units downward, as shown below on the right. Note that the y-intercept of h is $(0, -1)$ and the horizontal asymptote is $y = -2$.

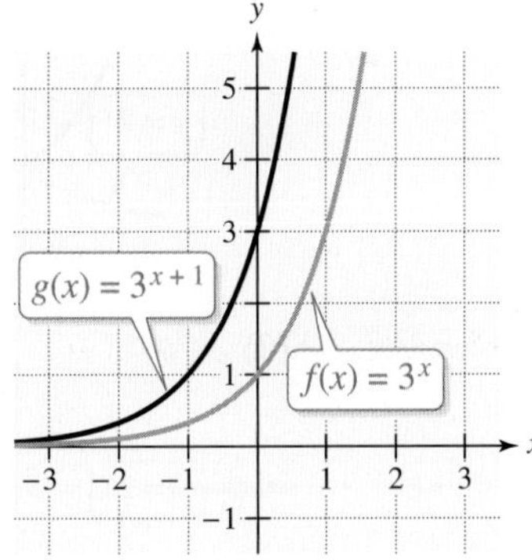

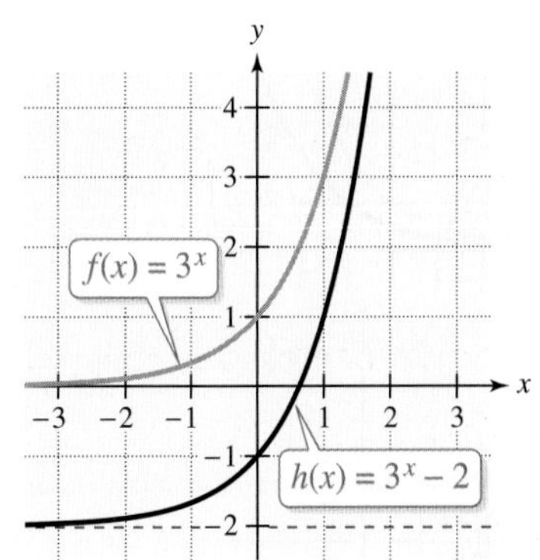

Exercises Within Reach ®

Solutions in English & Spanish and tutorial videos at AlgebraWithinReach.com

Transformations of Graphs **In Exercises 25–32, use transformations to analyze and sketch the graph of the function.** See Additional Answers.

25. $g(x) = 3^x - 1$ **26.** $g(x) = 3^x + 1$ **27.** $g(x) = 5^{x-1}$ **28.** $g(x) = 5^{x+3}$

29. $g(x) = 2^x + 3$ **30.** $g(x) = 2^{x+3}$ **31.** $g(x) = 2^{x-4}$ **32.** $g(x) = 2^x - 4$

EXAMPLE 6 Reflections of Graphs

Additional Examples

Sketch the graph of each function.

a. $f(x) = -2^x$

b. $g(x) = 2^x - 3$

Answers:

a.

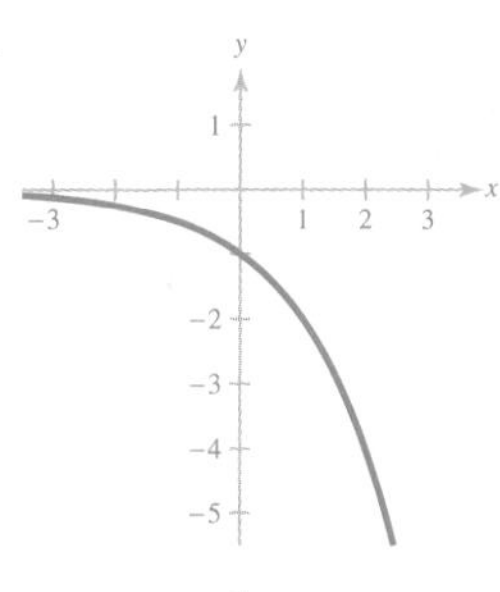

b.

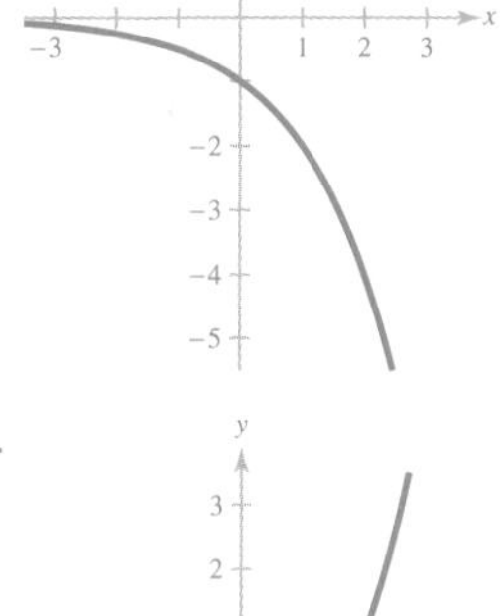

Use transformations to analyze and sketch the graph of each function.

a. $g(x) = -3^x$ **b.** $h(x) = 3^{-x}$

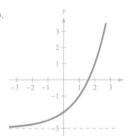

SOLUTION

Consider the function $f(x) = 3^x$.

a. The function g is related to f by $g(x) = -f(x)$. To sketch the graph of g, reflect the graph of f in the x-axis, as shown below on the left. Note that the y-intercept of g is $(0, -1)$.

b. The function h is related to f by $h(x) = f(-x)$. To sketch the graph of h, reflect the graph of f in the y-axis, as shown below on the right.

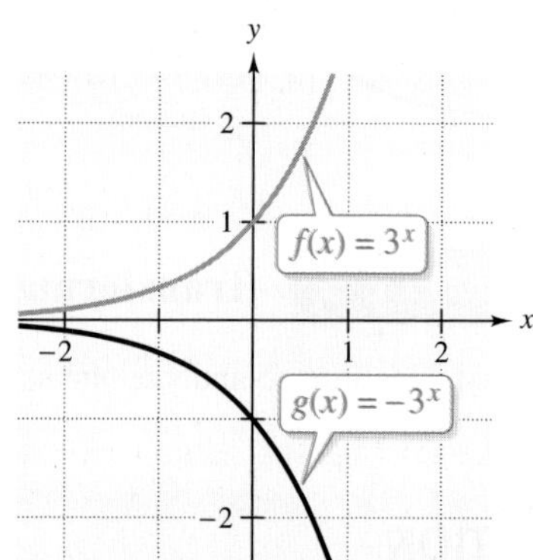

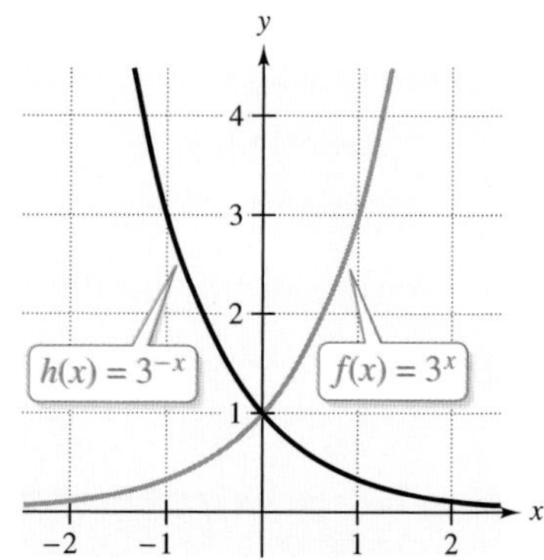

Exercises Within Reach ®

Solutions in English & Spanish and tutorial videos at AlgebraWithinReach.com

Matching **In Exercises 33–36, match the function with its graph.**

(a)

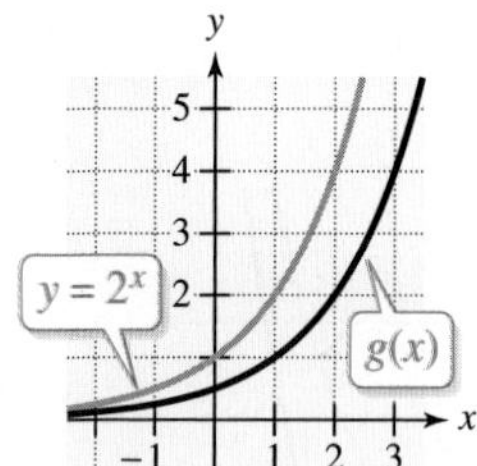

(b)

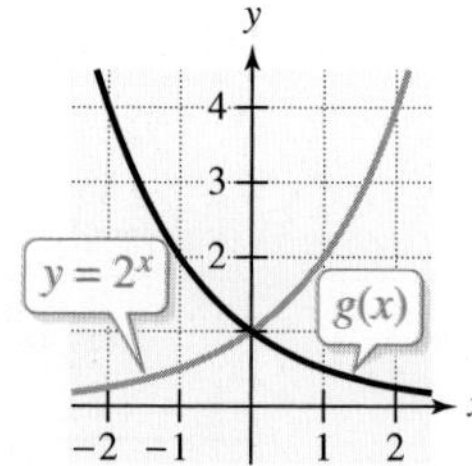

(c)

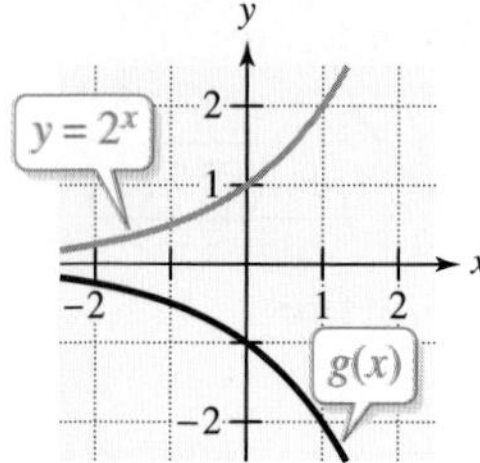

(d)

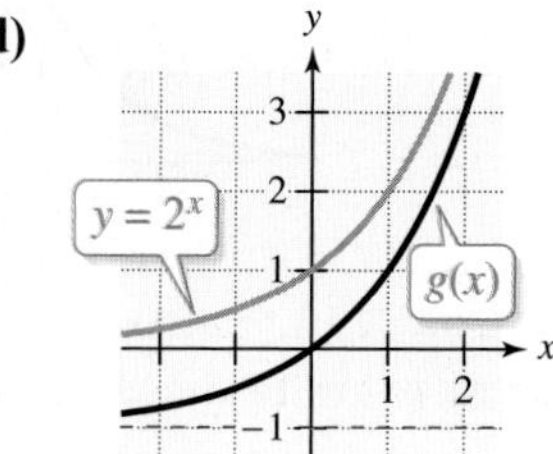

33. $g(x) = 2^{-x}$ b **34.** $g(x) = 2^x - 1$ d **35.** $g(x) = 2^{x-1}$ a **36.** $g(x) = -2^x$ c

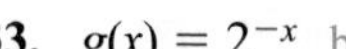

Reflections of Graphs **In Exercises 37–40, use transformations to analyze and sketch the graph of the function.** See Additional Answers.

37. $g(x) = -4^x$ **38.** $g(x) = 4^{-x}$ **39.** $g(x) = 5^{-x}$ **40.** $g(x) = -5^x$

The Natural Exponential Function

So far, integers or rational numbers have been used as bases of exponential functions. In many applications of exponential functions, the convenient choice for a base is the following irrational number, denoted by the letter "e."

$e \approx 2.71828\ldots$ Natural base

This number is called the **natural base**. The function

$f(x) = e^x$ Natural exponential function

is called the **natural exponential function**. To evaluate the natural exponential function, you need a calculator, preferably one with a natural exponential key (e^x). Here are some examples.

Value	*Keystrokes*	*Display*	
e^2	2 (e^x)	7.3890561	Scientific
e^2	(e^x) 2 ( )) (ENTER)	7.3890561	Graphing
e^{-3}	3 (+/−) (e^x)	0.0497871	Scientific
e^{-3}	(e^x) ((−)) 3 ( )) (ENTER)	0.0497871	Graphing

EXAMPLE 7 Graphing the Natural Exponential Function

When evaluating the natural exponential function, remember that e is the constant number $2.71828\ldots$ and x is a variable. After evaluating this function at several values, you can sketch its graph, as shown at the left.

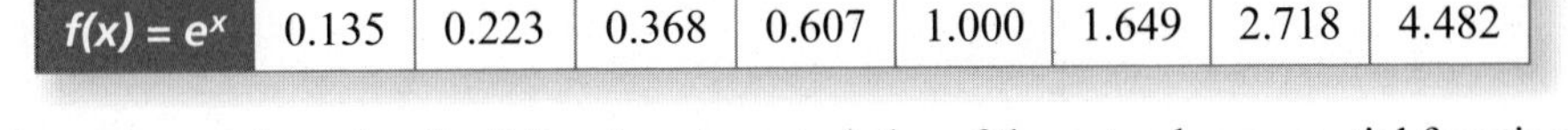

x	−2	−1.5	−1	−0.5	0	0.5	1	1.5
$f(x) = e^x$	0.135	0.223	0.368	0.607	1.000	1.649	2.718	4.482

From the graph, notice the following characteristics of the natural exponential function.

- Domain: $(-\infty, \infty)$
- Range: $(0, \infty)$
- Intercept: $(0, 1)$
- Increasing (moves up to the right)
- Asymptote: x-axis

Exercises Within Reach®

Solutions in English & Spanish and tutorial videos at AlgebraWithinReach.com

Using a Calculator **In Exercises 41–44, use a calculator to evaluate the expression. (Round your answer to three decimal places.)**

41. $e^{1/3}$ 1.396 **42.** $e^{-1/3}$ 0.717 **43.** $3(2e^{1/2})^3$ 107.561 **44.** $(9e^2)^{3/2}$ 542.309

Evaluating an Exponential Function **In Exercises 45 and 46, use a calculator to evaluate the function as indicated. (Round your answers to three decimal places.)**

45. $g(x) = 10e^{-0.5x}$
(a) $x = -4$ 73.891
(b) $x = 4$ 1.353
(c) $x = 8$ 0.183

46. $A(t) = 200e^{0.1t}$
(a) $t = 10$ 543.656
(b) $t = 20$ 1477.811
(c) $t = 40$ 10,919.630

Transformations of Graphs **In Exercises 47–50, use transformations to analyze and sketch the graph of the function.** See Additional Answers.

47. $g(x) = -e^x$ **48.** $g(x) = e^{-x}$ **49.** $g(x) = e^x + 1$ **50.** $g(x) = e^{x+1}$

Applications

Application **EXAMPLE 8** **Radioactive Decay**

A particular radioactive element has a half-life of 25 years. For an initial mass of 10 grams, the mass y (in grams) that remains after t years is given by

$$y = 10\left(\frac{1}{2}\right)^{t/25}, \quad t \geq 0$$

How much of the initial mass remains after 120 years?

SOLUTION

When t = 120, the mass is given by

$$y = 10\left(\frac{1}{2}\right)^{120/25}$$ Substitute 120 for t.

$$= 10\left(\frac{1}{2}\right)^{4.8}$$ Simplify.

$$\approx 0.359.$$ Use a calculator.

So, after 120 years, the mass has decayed from an initial amount of 10 grams to only 0.359 gram. Note that the graph of the function shows the 25-year half-life. That is, after 25 years the mass is 5 grams (half of the original), after another 25 years the mass is 2.5 grams, and so on.

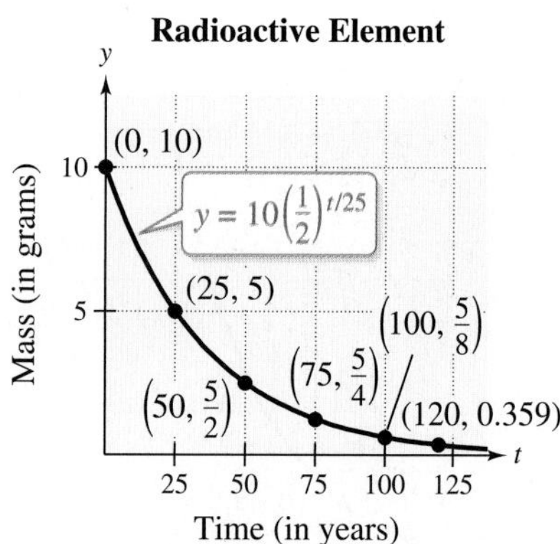

Exercises Within Reach®

Solutions in English & Spanish and tutorial videos at AlgebraWithinReach.com

51. ***Radioactive Decay*** Cesium-137 is a man-made radioactive element with a half-life of 30 years. A large amount of cesium-137 was introduced into the atmosphere by nuclear weapons testing in the 1950s and 1960s. By now, much of that substance has decayed. Most of the cesium-137 currently in the atmosphere is from nuclear accidents like those at Chernobyl in the Ukraine and Fukushima in Japan. Cesium-137 is one of the most dangerous substances released during those events.

    After t years, 16 grams of cesium-137 decays to a mass y (in grams) given by

    $$y = 16\left(\frac{1}{2}\right)^{t/30}, \quad t \geq 0.$$

    How much of the initial mass remains after 80 years?
    2.520 grams

52. ***Radioactive Substance*** In July of 1999, an individual bought several leaded containers from a metals recycler and found two of them labeled "radioactive." An investigation showed that the containers, originally obtained from Ohio State University, apparently had been used to store iodine-131 starting in January of 1999. Because iodine-131 has a half-life of only 8 days, no elevated radiation levels were detected. (*Source:* United States Nuclear Regulatory Commission)

    Suppose 6 grams of iodine-131 is stored in January. The mass y (in grams) that remains after t days is given by

    $$y = 6\left(\tfrac{1}{2}\right)^{t/8}, t \geq 0.$$

    How much of the substance is left in July, after 180 days have passed?
    About $1.01 \times 10^{-6} = 0.00000101$ gram

Formulas for Compound Interest

After t years, the balance A in an account with principal P and annual interest rate r (in decimal form) is given by one of the following formulas.

1. For n compoundings per year: $A = P\left(1 + \frac{r}{n}\right)^{nt}$

2. For continuous compounding: $A = Pe^{rt}$

Application

EXAMPLE 9 Comparing Three Types of Compounding

A total of \$15,000 is invested at an annual interest rate of 8%. Find the balance after 6 years for each type of compounding.

a. Quarterly **b.** Monthly **c.** Continuous

SOLUTION

a. Letting $P = 15{,}000$, $r = 0.08$, $n = 4$, and $t = 6$, the balance after 6 years when compounded quarterly is

$$A = 15{,}000\left(1 + \frac{0.08}{4}\right)^{4(6)}$$

$$\approx \$24{,}126.56.$$

b. Letting $P = 15{,}000$, $r = 0.08$, $n = 12$, and $t = 6$, the balance after 6 years when compounded monthly is

$$A = 15{,}000\left(1 + \frac{0.08}{12}\right)^{12(6)}$$

$$\approx \$24{,}202.53.$$

c. Letting $P = 15{,}000$, $r = 0.08$, and $t = 6$, the balance after 6 years when compounded continuously is

$$A = 15{,}000e^{0.08(6)}$$

$$\approx \$24{,}241.12.$$

Note that the balance is greater with continuous compounding than with quarterly or monthly compounding.

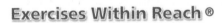

Exercises Within Reach®

Solutions in English & Spanish and tutorial videos at AlgebraWithinReach.com

Compound Interest **In Exercises 53–56, complete the table to determine the balance A for P dollars invested at rate r for t years, compounded n times per year.** See Additional Answers.

n	1	4	12	365	Continuous compounding
A					

	Principal	*Rate*	*Time*
53.	$P = \$100$	$r = 7\%$	$t = 15$ years
54.	$P = \$600$	$r = 4\%$	$t = 5$ years
55.	$P = \$2000$	$r = 9.5\%$	$t = 10$ years
56.	$P = \$1500$	$r = 6.5\%$	$t = 20$ years

michaeljung/Shutterstock.com

Concept Summary: Graphing Exponential Functions

What

The **exponential function f with base a** is

$f(x) = a^x$

where $a > 0$, $a \neq 1$, and x is any real number.

You can use the graph of $y = a^x$ to sketch the graphs of functions of the form $f(x) = b \pm a^{x+c}$.

EXAMPLE

Sketch the graph of $h(x) = -2^x + 1$.

How

One way to sketch the graph of h is to relate it to the graph of $f(x) = 2^x$.

EXAMPLE

The function h is related to f by $h(x) = -f(x) + 1$. To sketch the graph of h, reflect the graph of f in the x-axis and shift it one unit upward.

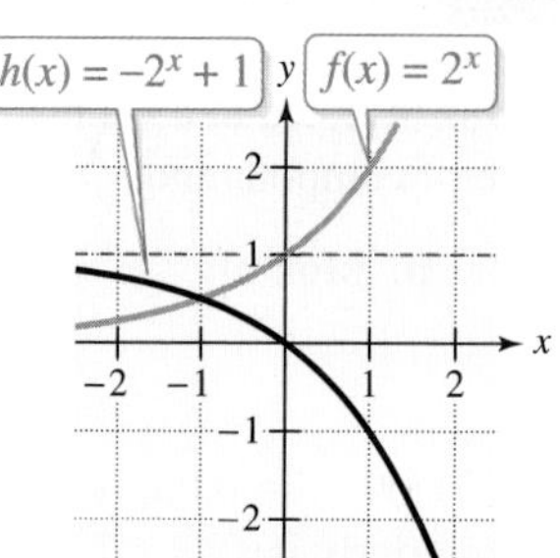

Why

Many real-life situations are modeled by exponential functions. Two such situations are radioactive decay and compound interest.

Exercises Within Reach®

Worked-out solutions to odd-numbered exercises at AlgebraWithinReach.com

Concept Summary Check

57. ***Describing a Transformation*** Describe how to transform the graph of $f(x) = 2^x$ to obtain the graph of $g(x) = -2^x$. Reflect the graph of f in the x-axis.

58. ***Describing a Transformation*** Describe how to transform the graph of $g(x) = -2^x$ to obtain the graph of $h(x) = -2^x + 1$. Shift the graph of g 1 unit upward.

59. ***Analyzing Behavior*** How does a reflection in the x-axis affect the behavior of an exponential function? The behavior is reversed. When the graph of the original function is increasing, the reflected graph is decreasing. When the graph of the original function is decreasing, the reflected graph is increasing.

60. ***Analyzing the Asymptote*** How does a shift upward affect the asymptote of an exponential function? The asymptote shifts 1 unit upward.

Extra Practice

Using Rules of Exponential Functions **In Exercises 61–68, simplify the expression.**

61. $3^x \cdot 3^{x+2}$ 3^{2x+2}

62. $e^{3x} \cdot e^{-x}$ e^{2x}

63. $3(e^x)^{-2}$ $\frac{3}{e^{2x}}$

64. $4(e^{2x})^{-1}$ $\frac{4}{e^{2x}}$

65. $\frac{e^{x+2}}{e^x}$ e^2

66. $\frac{3^{2x+3}}{3^{x+1}}$ 3^{x+2}

67. $\sqrt[3]{-8e^{3x}}$ $-2e^x$

68. $\sqrt{4e^{6x}}$ $2e^{3x}$

Compound Interest **In Exercises 69 and 70, complete the table to determine the principal P that will yield a balance of A dollars when invested at rate r for t years, compounded n times per year.**

See Additional Answers.

	Balance	*Rate*	*Time*
69.	$A = \$5000$	$r = 7\%$	$t = 10$ years
70.	$A = \$100{,}000$	$r = 9\%$	$t = 20$ years

n	1	4	12	365	Continuous compounding
P					

71. ***Demand*** The daily demand x and the price p for a collectible are related by $p = 25 - 0.4e^{0.02x}$. Find the prices for demands of (a) $x = 100$ units and (b) $x = 125$ units. (a) \$22.04 (b) \$20.13

72. ***Compound Interest*** A sum of \$5000 is invested at an annual interest rate of 6%, compounded monthly. Find the balance in the account after 5 years. \$6744.25

73. ***Property Value*** The value of a piece of property doubles every 15 years. You buy the property for \$64,000. Its value t years after the date of purchase should be $V(t) = 64{,}000(2)^{t/15}$. Use the model to approximate the values of the property (a) 5 years and (b) 20 years after its purchase. (a) \$80,634.95 (b) \$161,269.89

74. ***Depreciation*** The value of a car that originally cost \$16,000 depreciates so that each year it is worth $\frac{3}{4}$ of its value from the previous year. Find a model for $V(t)$, the value of the car after t years. Sketch a graph of the model, and determine the value of the car 2 years and 4 years after its purchase.
$V(t) = 16{,}000\left(\frac{3}{4}\right)^t$; See Additional Answers.
\$9000 after 2 years, \$5062.50 after 4 years

75. ***Savings Plan*** You decide to start saving pennies according to the following pattern. You save 1 penny the first day, 2 pennies the second day, 4 the third day, 8 the fourth day, and so on. Each day you save twice the number of pennies you saved on the previous day. Write an exponential function that models this problem. How many pennies do you save on the thirtieth day?
$f(t) = 2^{t-1}$; $f(30) = 536{,}870{,}912$ pennies

76. ***Exploration*** Consider the function

$$f(x) = \left(1 + \frac{1}{x}\right)^x.$$

(a) Use a calculator to complete the table.

x	1	10	100	1000	10,000
$f(x)$	2	2.5937	2.7048	2.7169	2.7181

(b) Use the table to sketch the graph of f. Does the graph appear to approach a horizontal asymptote?
See Additional Answers.
Yes.

(c) From parts (a) and (b), what conclusions can you make about the value of f as x gets larger and larger?
The value approaches e.

Explaining Concepts

77. ***Reasoning*** Explain why $y = 1^x$ is not an exponential function. By definition, the base of an exponential function must be positive and not equal to 1. The function $y = 1^x$ simplifies to the constant function $y = 1$.

78. ***Structure*** Compare the graphs of $f(x) = 3^x$ and $g(x) = \left(\frac{1}{3}\right)^x$. f is an increasing function and g is a decreasing function. The graph of g is a reflection of f in the y-axis.

79. ***Number Sense*** Does e equal $\dfrac{271{,}801}{99{,}990}$? Explain.
No. e is an irrational number.

80. ***Analyzing Graphs*** Consider the graphs of the functions $f(x) = 3^x$, $g(x) = 4^x$, and $h(x) = 4^{-x}$.

(a) What point do the graphs of f, g, and h have in common? (0, 1)

(b) Describe the asymptote for each graph.
Horizontal asymptote: $y = 0$

(c) State whether each graph increases or decreases as x increases.
f and g increase and h decreases as x increases.

(d) Compare the graphs of f and g.
g increases at a faster rate than f.

81. ***Reasoning*** Use the characteristics of the exponential function with base 2 to explain why $2^{\sqrt{2}}$ is greater than 2 but less than 4. Because $1 < \sqrt{2} < 2$ and $2 > 0$, $2^1 < 2^{\sqrt{2}} < 2^2$. So, $2 < 2^{\sqrt{2}} < 4$.

82. ***Reasoning*** Identify the graphs of $y_1 = e^{0.2x}$, $y_2 = e^{0.5x}$, and $y_3 = e^x$ in the figure. Describe the effect on the graph of $y = e^{kx}$ when $k > 0$ is changed. y_1: a, y_2: b, y_3: c;
The graph increases at a greater rate as k increases.

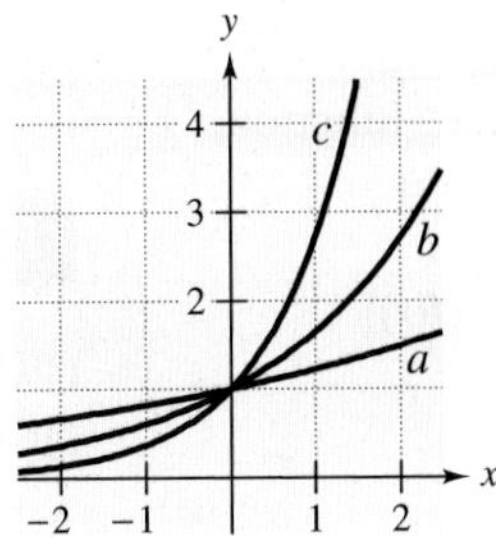

83. ***Reasoning*** Consider functions of the form $f(x) = k^x$, where k is positive. Describe the real values of k for which the values of f will *increase*, *decrease*, and *remain constant* as x increases. When $k > 1$, the values of f will increase. When $0 < k < 1$, the values of f will decrease. When $k = 1$, the values of f will remain constant.

84. ***Repeated Reasoning*** Look back at your answers to the compound interest problems in Exercises 53–56. In terms of the interest earned, would you say that the difference between quarterly compounding and annual compounding is greater than the difference between *hourly* compounding and daily compounding? Explain.
See Additional Answers.

Cumulative Review

In Exercises 85 and 86, find the domain of the function.

85. $g(s) = \sqrt{s - 4}$ $x \geq 4$

86. $h(t) = \dfrac{\sqrt{t^2 - 1}}{t - 2}$ $x \leq -1$ or $1 \leq x < 2$ or $x > 2$

In Exercises 87 and 88, sketch the graph of the equation. Use the Vertical Line Test to determine whether y is a function of x. See Additional Answers.

87. $y^2 = x - 1$ y is not a function of x.

88. $x = y^4 + 1$ y is not a function of x.

9.2 Composite and Inverse Functions

- Form composite functions and find the domains of composite functions.
- Find inverse functions algebraically.
- Compare the graph of a function with the graph of its inverse.

Composite Functions

Definition of Composition of Two Functions

The **composition** of the functions f and g is given by $(f \circ g)(x) = f(g(x))$. The domain of the **composite function** $(f \circ g)$ is the set of all x in the domain of g such that $g(x)$ is in the domain of f.

Study Tip

A composite function can be viewed as a function within a function, where the composition

$$(f \circ g)(x) = f(g(x))$$

has f as the "outer" function and g as the "inner" function. This is reversed in the composition

$$(g \circ f)(x) = g(f(x)).$$

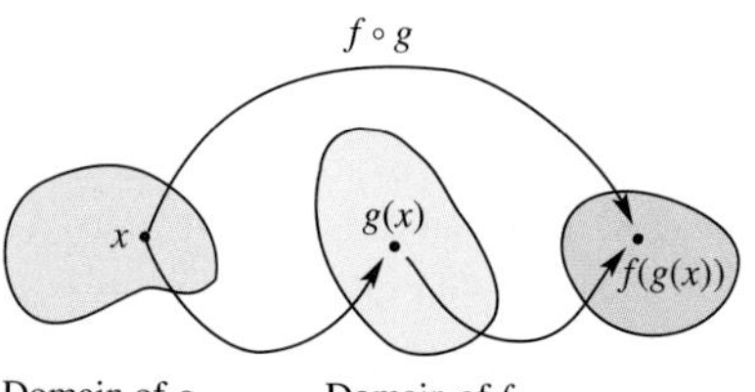

EXAMPLE 1 **Finding the Composition of Two Functions**

Given $f(x) = 2x + 4$ and $g(x) = 3x - 1$, find the composition of f with g. Then evaluate the composite function when $x = 1$ and when $x = -3$.

SOLUTION

$(f \circ g)(x) = f(g(x))$	Definition of $f \circ g$
$= f(3x - 1)$	$g(x) = 3x - 1$ is the inner function.
$= 2(3x - 1) + 4$	Input $3x - 1$ into the outer function f.
$= 6x - 2 + 4$	Distributive Property
$= 6x + 2$	Simplify.

When $x = 1$, the value of this composite function is

$$(f \circ g)(1) = 6(1) + 2 = 8.$$

When $x = -3$, the value of this composite function is

$$(f \circ g)(-3) = 6(-3) + 2 = -16.$$

Exercises Within Reach®

Solutions in English & Spanish and tutorial videos at AlgebraWithinReach.com

Finding the Composition of Two Functions **In Exercises 1 and 2, find the compositions.**

1. $f(x) = 2x + 3$, $g(x) = x - 6$

(a) $(f \circ g)(x)$ (b) $(g \circ f)(x)$

(c) $(f \circ g)(4)$ (d) $(g \circ f)(7)$

(a) $2x - 9$ (b) $2x - 3$ (c) -1 (d) 11

2. $f(x) = x - 5$, $g(x) = 3x + 2$

(a) $(f \circ g)(x)$ (b) $(g \circ f)(x)$

(c) $(f \circ g)(3)$ (d) $(g \circ f)(3)$

(a) $3x - 3$ (b) $3x - 13$ (c) 6 (d) -4

EXAMPLE 2 Comparing the Compositions of Functions

Given $f(x) = 2x - 3$ and $g(x) = x^2 + 1$, find each composition.

a. $(f \circ g)(x)$ **b.** $(g \circ f)(x)$

Additional Example:

Given $f(x) = x^2 - 9$ and $g(x) = \sqrt{9 - x^2}$, find the composition of f with g. Then find the domain of the composition.

Answer:

$(f \circ g)(x) = -x^2$

Domain: $-3 \le x \le 3$

SOLUTION

a. $(f \circ g)(x) = f(g(x))$ Definition of $f \circ g$

$= f(x^2 + 1)$ $g(x) = x^2 + 1$ is the inner function.

$= 2(x^2 + 1) - 3$ Input $x^2 + 1$ into the outer function f.

$= 2x^2 + 2 - 3$ Distributive Property

$= 2x^2 - 1$ Simplify.

b. $(g \circ f)(x) = g(f(x))$ Definition of $g \circ f$

$= g(2x - 3)$ $f(x) = 2x - 3$ is the inner function.

$= (2x - 3)^2 + 1$ Input $2x - 3$ into the outer function g.

$= 4x^2 - 12x + 9 + 1$ Expand.

$= 4x^2 - 12x + 10$ Simplify.

Note that $(f \circ g)(x) \neq (g \circ f)(x)$

EXAMPLE 3 Finding the Domain of a Composite Function

Find the domain of the composition of f with g when $f(x) = x^2$ and $g(x) = \sqrt{x}$.

Study Tip

To determine the domain of a composite function, first write the composite function in simplest form. Then use the fact that its domain either is equal to or is a restriction of the domain of the "inner" function.

SOLUTION

$(f \circ g)(x) = f(g(x))$ Definition of $f \circ g$

$= f(\sqrt{x})$ $g(x) = \sqrt{x}$ is the inner function.

$= (\sqrt{x})^2$ Input $\sqrt{x}$ into the outer function f.

$= x,\ x \ge 0$ Domain of $f \circ g$ is all $x \ge 0$.

The domain of the inner function $g(x) = \sqrt{x}$ is the set of all nonnegative real numbers. The simplified form of $f \circ g$ has no restriction on this set of numbers. So, the restriction $x \ge 0$ must be added to the composition of this function. The domain of $f \circ g$ is the set of all nonnegative real numbers.

Exercises Within Reach®

Solutions in English & Spanish and tutorial videos at AlgebraWithinReach.com

Finding the Composition of Two Functions **In Exercises 3 and 4, find the compositions.**

3. $f(x) = x^2 + 3,\ g(x) = x + 2$

(a) $(f \circ g)(x)$ $x^2 + 4x + 7$ (b) $(g \circ f)(x)$ $x^2 + 5$

(c) $(f \circ g)(2)$ 19 (d) $(g \circ f)(-3)$ 14

4. $f(x) = 2x + 1,\ g(x) = x^2 - 5$

(a) $(f \circ g)(x)$ $2x^2 - 9$ (b) $(g \circ f)(x)$ $4x^2 + 4x - 4$

(c) $(f \circ g)(-1)$ -7 (d) $(g \circ f)(3)$ 44

Finding the Domains of Composite Functions **In Exercises 5 and 6, find the compositions (a) $f \circ g$ and (b) $g \circ f$. Then find the domain of each composition.** See Additional Answers.

5. $f(x) = \sqrt{x + 2}$

$g(x) = x - 4$

6. $f(x) = \sqrt{x - 5}$

$g(x) = x + 3$

Inverse Functions

Let f be a function from the set $A = \{1, 2, 3, 4\}$ to the set $B = \{3, 4, 5, 6\}$ such that

$$f(x) = x + 2: \quad \{(1, 3), (2, 4), (3, 5), (4, 6)\}.$$

By interchanging the first and second coordinates of each of these ordered pairs, you can form another function that is called the **inverse function** of f, denoted by f^{-1}. It is a function from the set B to the set A, and can be written as follows.

$$f^{-1}(x) = x - 2: \quad \{(3, 1), (4, 2), (5, 3), (6, 4)\}.$$

Interchanging the ordered pairs of a function f produces a function only when f is one-to-one. A function f is **one-to-one** when each value of the dependent variable corresponds to exactly one value of the independent variable.

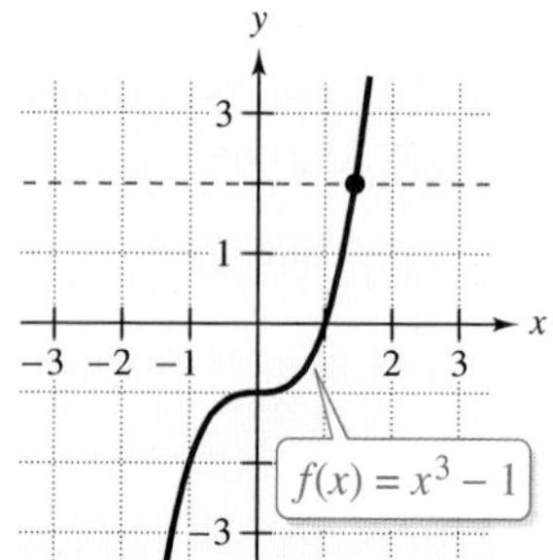

Horizontal Line Test for Inverse Functions

A function f has an inverse function f^{-1} if and only if f is one-to-one. Graphically, a function f has an inverse function f^{-1} if and only if no *horizontal* line intersects the graph of f at more than one point.

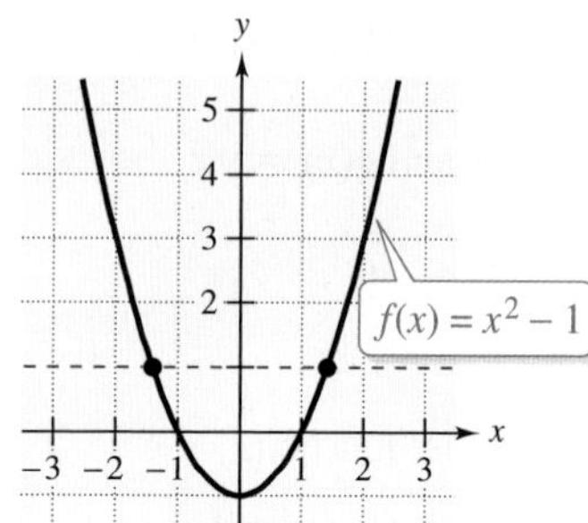

EXAMPLE 4 Applying the Horizontal Line Test

Use the Horizontal Line Test to determine whether the function is one-to-one and so has an inverse function.

a. The graph of the function $f(x) = x^3 - 1$ is shown at the top left. Because no horizontal line intersects the graph of f at more than one point, you can conclude that f *is* a one-to-one function and *does* have an inverse function.

b. The graph of the function $f(x) = x^2 - 1$ is shown at the bottom left. Because it is possible to find a horizontal line that intersects the graph of f at more than one point, you can conclude that f *is not* a one-to-one function and *does not* have an inverse function.

Point out that when students look at this definition of inverse function, they are seeing an application of composition. It can be used to verify that two functions are inverse functions of each other.

Definition of Inverse Function

Let f and g be two functions such that

$$f(g(x)) = x \quad \text{for every } x \text{ in the domain of } g$$

and

$$g(f(x)) = x \quad \text{for every } x \text{ in the domain of } f.$$

The function g is called the **inverse function** of the function f, and is denoted by f^{-1} (read "f-inverse"). So, $f(f^{-1}(x)) = x$ and $f^{-1}(f(x)) = x$. The domain of f must be equal to the range of f^{-1}, and vice versa.

Exercises Within Reach®

Solutions in English & Spanish and tutorial videos at AlgebraWithinReach.com

Applying the Horizontal Line Test **In Exercises 7–12, sketch the graph of the function. Then use the Horizontal Line Test to determine whether the function is one-to-one and so has an inverse function.**

See Additional Answers.

7. $f(x) = x^2 - 2$ No
8. $f(x) = \frac{1}{5}x$ Yes
9. $f(x) = x^2, \; x \geq 0$ Yes
10. $f(x) = \sqrt{-x}$ Yes
11. $g(x) = \sqrt{25 - x^2}$ No
12. $g(x) = |x - 4|$ No

Study Tip

You can graph a function and use the Horizontal Line Test to see whether the function is one-to-one before trying to find its inverse function.

Finding an Inverse Function Algebraically

1. In the equation for $f(x)$, replace $f(x)$ with y.
2. Interchange x and y.
3. Solve the new equation for y. (If the new equation does not represent y as a function of x, the function f does not have an inverse function.)
4. Replace y with $f^{-1}(x)$.
5. Verify that f and f^{-1} are inverse functions of each other by showing that $f(f^{-1}(x)) = x = f^{-1}(f(x))$.

EXAMPLE 5 Finding Inverse Functions

a.

$f(x) = 2x + 3$	Original function
$y = 2x + 3$	Replace $f(x)$ with y.
$x = 2y + 3$	Interchange x and y.
$y = \dfrac{x-3}{2}$	Solve for y.
$f^{-1}(x) = \dfrac{x-3}{2}$	Replace y with $f^{-1}(x)$.

You can verify that $f(f^{-1}(x)) = x = f^{-1}(f(x))$, as follows.

$$f(f^{-1}(x)) = f\left(\frac{x-3}{2}\right) = 2\left(\frac{x-3}{2}\right) + 3 = (x-3) + 3 = x$$

$$f^{-1}(f(x)) = f^{-1}(2x+3) = \frac{(2x+3)-3}{2} = \frac{2x}{2} = x$$

Emphasize that *both* conditions, $f(f^{-1}(x)) = x$ and $f^{-1}(f(x)) = x$, must be satisfied in order for two functions to be inverse functions of each other.

b.

$f(x) = x^3 + 3$	Original function
$y = x^3 + 3$	Replace $f(x)$ with y.
$x = y^3 + 3$	Interchange x and y.
$y = \sqrt[3]{x-3}$	Solve for y.
$f^{-1}(x) = \sqrt[3]{x-3}$	Replace y with $f^{-1}(x)$.

You can verify that $f(f^{-1}(x)) = x = f^{-1}(f(x))$, as follows.

$$f(f^{-1}(x)) = f\left(\sqrt[3]{x-3}\right) = \left(\sqrt[3]{x-3}\right)^3 + 3 = (x-3) + 3 = x$$

$$f^{-1}(f(x)) = f^{-1}(x^3+3) = \sqrt[3]{(x^3+3)-3} = \sqrt[3]{x^3} = x$$

Additional Examples

Determine whether each function has an inverse function. If it does, find its inverse function.

a. $f(x) = 5x - 7$

b. $f(x) = 3x^3$

Answers:

a. $f^{-1}(x) = \dfrac{x+7}{5}$

b. $f^{-1}(x) = \sqrt[3]{\dfrac{x}{3}}$

Exercises Within Reach®

Solutions in English & Spanish and tutorial videos at AlgebraWithinReach.com

Finding an Inverse Function **In Exercises 13–16, find the inverse function of f.**

13. $f(x) = 3 - 4x$ $\quad f^{-1}(x) = \dfrac{3-x}{4}$

14. $f(t) = 6t + 1$ $\quad f^{-1}(t) = \dfrac{t-1}{6}$

15. $f(t) = t^3 - 1$ $\quad f^{-1}(t) = \sqrt[3]{t+1}$

16. $f(t) = t^5 + 8$ $\quad f^{-1}(t) = \sqrt[5]{t-8}$

Graphs of Inverse Functions

The graphs of f and f^{-1} are related to each other in the following way. If the point (a, b) lies on the graph of f, then the point (b, a) must lie on the graph of f^{-1}, and vice versa. This means that the graph of f^{-1} is a reflection of the graph of f in the line $y = x$, as shown in the figure below.

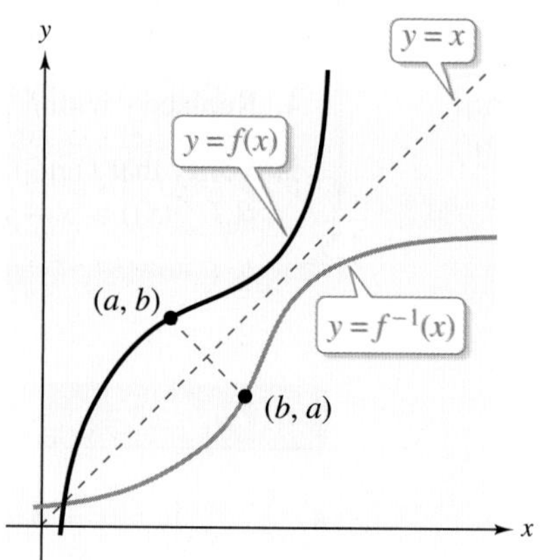

EXAMPLE 6 The Graphs of f and f^{-1}

Sketch the graphs of the inverse functions $f(x) = 2x - 3$ and $f^{-1}(x) = \frac{1}{2}(x + 3)$ on the same rectangular coordinate system, and show that the graphs are reflections of each other in the line $y = x$.

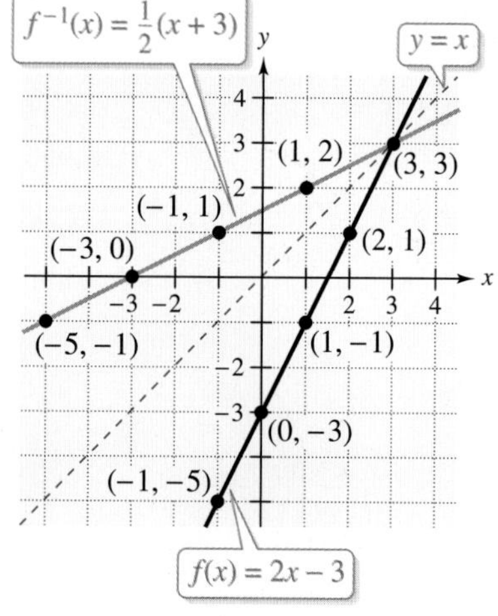

SOLUTION

The graphs of f and f^{-1} are shown at the left. Visually, it appears that the graphs are reflections of each other. You can verify this reflective property by testing a few points on each graph. Note in the following list that if the point (a, b) is on the graph of f, then the point (b, a) is on the graph of f^{-1}.

$f(x) = 2x - 3$	$f^{-1}(x) = \frac{1}{2}(x + 3)$
$(-1, -5)$	$(-5, -1)$
$(0, -3)$	$(-3, 0)$
$(1, -1)$	$(-1, 1)$
$(2, 1)$	$(1, 2)$
$(3, 3)$	$(3, 3)$

Exercises Within Reach®

Solutions in English & Spanish and tutorial videos at AlgebraWithinReach.com

The Graphs of f and f^{-1} **In Exercises 17–22, sketch the graphs of f and f^{-1} on the same rectangular coordinate system. Show that the graphs are reflections of each other in the line $y = x$.** See Additional Answers.

17. $f(x) = x + 4$
$f^{-1}(x) = x - 4$

18. $f(x) = x - 7$
$f^{-1}(x) = x + 7$

19. $f(x) = 3x - 1$
$f^{-1}(x) = \frac{1}{3}(x + 1)$

20. $f(x) = 5 - 4x$
$f^{-1}(x) = -\frac{1}{4}(x - 5)$

21. $f(x) = x^2 - 1,\ x \ge 0$
$f^{-1}(x) = \sqrt{x + 1}$

22. $f(x) = (x + 2)^2,\ x \ge -2$
$f^{-1}(x) = \sqrt{x} - 2$

EXAMPLE 7 Verifying Inverse Functions Graphically

Graphically verify that f and g are inverse functions of each other.

$$f(x) = x^2, \; x \geq 0 \quad \text{and} \quad g(x) = \sqrt{x}$$

SOLUTION

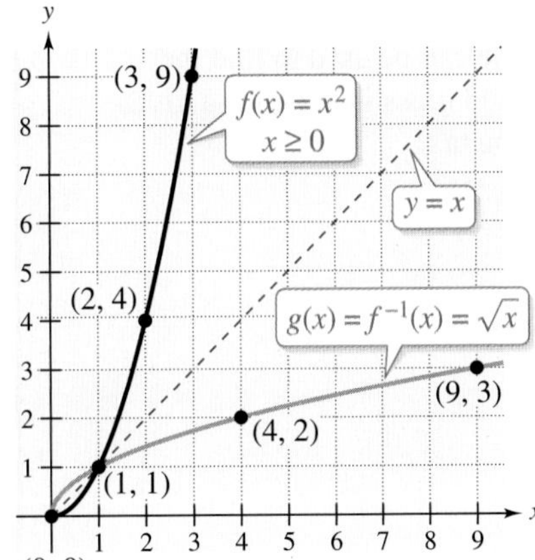

You can graphically verify that f and g are inverse functions of each other by graphing the functions on the same rectangular coordinate system, as shown at the left. Visually, it appears that the graphs are reflections of each other in the line $y = x$. You can verify this reflective property by testing a few points on each graph. Note in the following list that if the point (a, b) is on the graph of f, then the point (b, a) is on the graph of g.

$f(x) = x^2, \; x \geq 0$	$g(x) = f^{-1}(x) = \sqrt{x}$
$(0, 0)$	$(0, 0)$
$(1, 1)$	$(1, 1)$
$(2, 4)$	$(4, 2)$
$(3, 9)$	$(9, 3)$

So, f and g are inverse functions of each other.

Exercises Within Reach®

Matching **In Exercises 23–26, match the graph with the graph of its inverse function.**

(a)

(b)

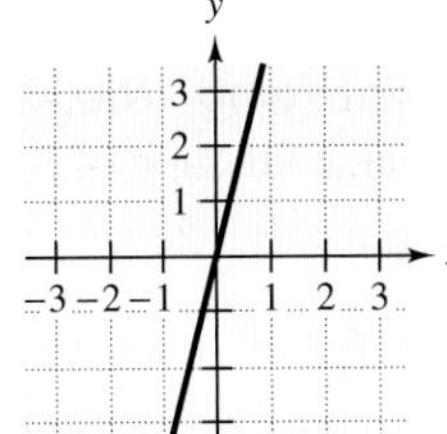

(c)

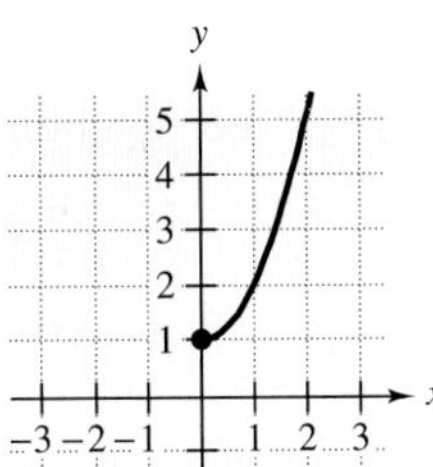

(d)

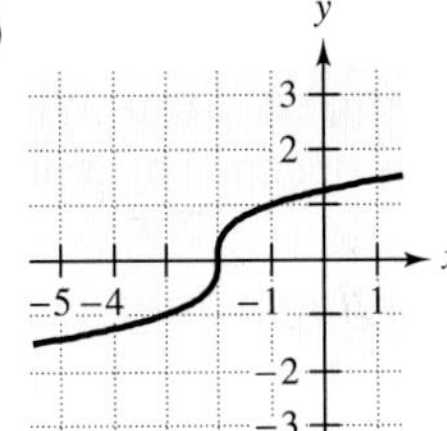

23.

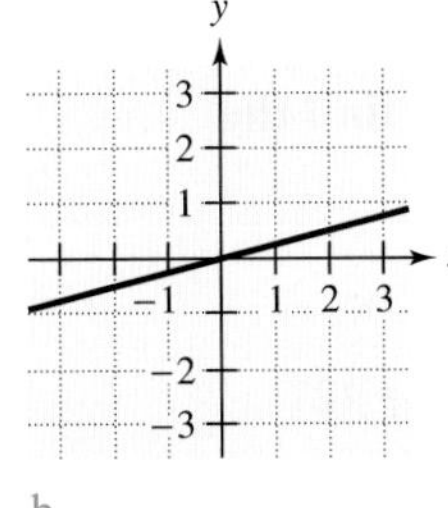

b

24.

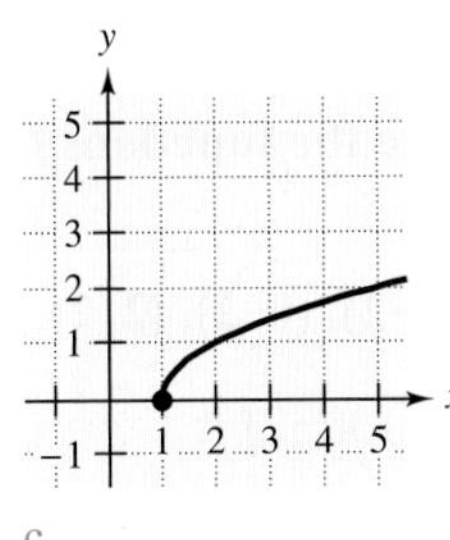

c

25.

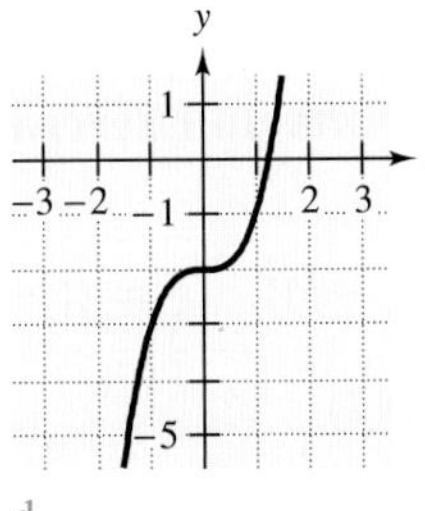

d

26.

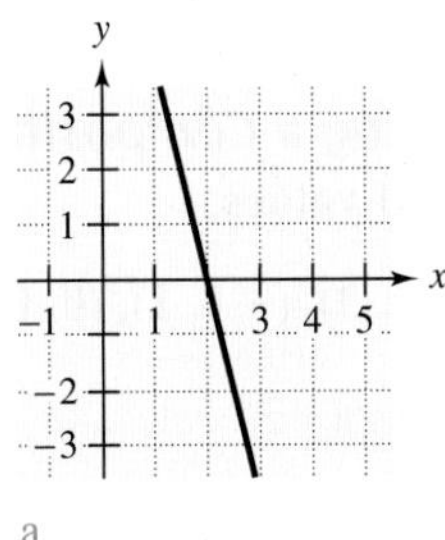

a

Verifying Inverse Functions Graphically **In Exercises 27–30, sketch the graphs of f and g on the same rectangular coordinate system. Plot the indicated points on the graph of f. Then plot the reflections of these points in the line $y = x$ to verify that f and g are inverse functions.** See Additional Answers.

27. $f(x) = \frac{1}{3}x$

$g(x) = 3x$

Points: $(0, f(0))$, $(3, f(3))$, and $(6, f(6))$

28. $f(x) = \frac{1}{5}x - 1$

$g(x) = 5x + 5$

Points: $(0, f(0))$, $(5, f(5))$, and $(10, f(10))$

29. $f(x) = \sqrt{x - 4}$

$g(x) = x^2 + 4, \; x \geq 0$

Points: $(4, f(4))$, $(5, f(5))$, and $(8, f(8))$

30. $f(x) = \sqrt{4 - x}$

$g(x) = 4 - x^2, \; x \geq 0$

Points: $(0, f(0))$, $(3, f(3))$, and $(4, f(4))$

Concept Summary: Finding Inverse Functions

What

You can form functions using one or more other functions by creating **composite functions** and **inverse functions**.

You can think of a composite function as a "function within a function." You can think of inverse functions as functions that "undo" each other.

EXAMPLE

Find the inverse function of

$f(x) = 2x + 6$.

How

Given an equation for $f(x)$, here is how to find the inverse function f^{-1}.

1. Replace $f(x)$ with y.
2. Interchange x and y.
3. Solve for y.
4. Replace y with $f^{-1}(x)$.
5. Verify that f and f^{-1} are inverse functions.

EXAMPLE

$$f(x) = 2x + 6$$
$$y = 2x + 6$$
$$x = 2y + 6$$
$$y = \frac{x - 6}{2}$$
$$f^{-1}(x) = \frac{x - 6}{2}$$

Why

Inverse functions are used in many real-life problems. For instance, the inverse of a function used to find income tax based on adjusted gross income can be used to find the adjusted gross income for a given income tax.

Exercises Within Reach®

Worked-out solutions to odd-numbered exercises at AlgebraWithinReach.com

Concept Summary Check

31. *Applying a Step* Which equation in the solution above is obtained by replacing $f(x)$ with y?
$y = 2x + 6$

32. *Identifying a Step* In the solution above, what step is performed after obtaining the equation $x = 2y + 6$?
Solve for y.

33. *Identifying the Inverse Function* What is the inverse function of $f(x) = 2x + 6$?
$f^{-1}(x) = \frac{x - 6}{2}$

34. *Verifying Inverse Functions* To show that two functions f and g are inverse functions, you must show that both $f(g(x))$ and $g(f(x))$ are equal to what? x

Extra Practice

Evaluating a Composition of Functions **In Exercises 35–38, use the functions f and g to find the indicated values.**

$f = \{(-2, 3), (-1, 1), (0, 0), (1, -1), (2, -3)\}$, $g = \{(-3, 1), (-1, -2), (0, 2), (2, 2), (3, 1)\}$

35. $(f \circ g)(-3)$ -1

36. $(f \circ g)(2)$ -3

37. $(g \circ f)(-2)$ 1

38. $(g \circ f)(2)$ 1

Verifying Inverse Functions **In Exercises 39 and 40, verify algebraically that the functions f and g are inverse functions of each other.** See Additional Answers.

39. $f(x) = 1 - 2x$, $g(x) = \frac{1}{2}(1 - x)$

40. $f(x) = 2x - 1$, $g(x) = \frac{1}{2}(x + 1)$

Finding an Inverse Function **In Exercises 41–44, find the inverse function (if it exists).**

41. $g(x) = x^2 + 4$
Inverse does not exist.

42. $h(x) = (4 - x)^2$
Inverse does not exist.

43. $h(x) = \sqrt{x}$
$h^{-1}(x) = x^2, x \ge 0$

44. $h(x) = \sqrt{x + 5}$
$h^{-1}(x) = x^2 - 5, x \ge 0$

Restricting the Domain **In Exercises 45 and 46, restrict the domain of f so that f is a one-to-one function. Then find the inverse function f^{-1} and state its domain. (Note: There is more than one correct answer.)** See Additional Answers.

45. $f(x) = (x - 2)^2$

46. $f(x) = 9 - x^2$

47. ***Ripples*** You are standing on a bridge over a calm pond and drop a pebble, causing ripples of concentric circles in the water. The radius (in feet) of the outermost ripple is given by $r(t) = 0.6t$, where t is time in seconds after the pebble hits the water. The area of the circle is given by the function $A(r) = \pi r^2$. Find an equation for the composition $A(r(t))$. Describe the input and output of this composite function. What is the area of the circle after 3 seconds? $A(r(t)) = 0.36\pi t^2$;
Input: time, Output: area; $A(r(3)) = 10.2$ square feet

48. ***Daily Production Cost*** The daily cost of producing x units in a manufacturing process is

$C(x) = 8.5x + 300.$

The number of units produced in t hours during a day is given by

$x(t) = 12t, \ 0 \le t \le 8.$

Find, simplify, and interpret $(C \circ x)(t)$.
$(C \circ x)(t) = 102t + 300, 0 \le t \le 8$
This function represents the production cost after t hours of operation.

49. ***Hourly Wage*** Your wage is \$9.00 per hour plus \$0.65 for each unit produced per hour. So, your hourly wage y in terms of the number of units produced x is $y = 9 + 0.65x$.

(a) Find the inverse function. $y = \frac{20}{13}(x - 9)$

(b) Determine the number of units produced when your hourly wage averages \$14.20. 8 units

50. ***Federal Income Tax*** In 2012, the function $T = 0.15(x - 8700) + 870$ represented the federal income tax owed by a single person whose adjusted gross income x was between \$8700 and \$35,350. (*Source:* Internal Revenue Service)

(a) Find the inverse function. $T = \dfrac{x - 870}{0.15} + 8700$

(b) What does each variable represent in the inverse function? x: income tax owed, T: adjusted gross income

(c) Use the context of the problem to determine the domain of the inverse function. $870 \le x \le 4867.5$

(d) Determine the adjusted gross income for a single person who owed \$3315 in federal income taxes in 2012. \$25,000

Explaining Concepts

True or False? **In Exercises 51–54, decide whether the statement is true or false. If true, explain your reasoning. If false, give an example.**

51. If the inverse function of f exists, then the y-intercept of f is an x-intercept of f^{-1}. Explain. True. If the point (a, b) lies on the graph of f, then the point (b, a) must lie on the graph of f^{-1}, and vice versa.

52. There exists no function f such that $f = f^{-1}$.
False. $f(x) = \dfrac{1}{x} = f^{-1}(x)$

53. If the inverse function of f exists, then the domains of f and f^{-1} are the same.
False. $f(x) = \sqrt{x - 1}$, Domain: $x \ge 1$
$f^{-1}(x) = x^2 + 1$, Domain: $x \ge 0$

54. If the inverse function of f exists and the graph of f passes through the point (2, 2), then the graph of f^{-1} also passes through the point (2, 2).
True. If the point (a, b) lies on the graph of f, then the point (b, a) must lie on the graph of f^{-1}, and vice versa.

55. ***Writing*** Describe how to find the inverse of a function given by a set of ordered pairs. Give an example.
Interchange the coordinates of each ordered pair. The inverse of the function defined by $\{(3, 6), (5, -2)\}$ is $\{(6, 3), (-2, 5)\}$.

56. ***Reasoning*** Why must a function be one-to-one in order for its inverse to be a function? A function that is not one-to-one has two inputs with the same output, so its inverse has two outputs for the same input and cannot be a function.

Cumulative Review

In Exercises 57–60, identify the transformation of the graph of $f(x) = x^2$.

57. $h(x) = -x^2$ Reflection in the x-axis

58. $g(x) = (x - 4)^2$ Horizontal shift

59. $k(x) = (x + 3)^2 - 5$ Horizontal and vertical shifts

60. $v(x) = x^2 + 1$ Vertical shift

In Exercises 61–64, factor the expression completely.

61. $16 - (y + 2)^2$ $(6 + y)(2 - y)$

62. $2x^3 - 6x$ $2x(x^2 - 3)$

63. $5 - u + 5u^2 - u^3$ $-(u^2 + 1)(u - 5)$

64. $t^2 + 10t + 25$ $(t + 5)^2$

In Exercises 65–68, graph the equation.
See Additional Answers.

65. $3x - 4y = 6$

66. $y = 3 - \frac{1}{2}x$

67. $y = -(x - 2)^2 + 1$

68. $y = x^2 - 6x + 5$

9.3 Logarithmic Functions

- Evaluate and graph logarithmic functions.
- Evaluate and graph natural logarithmic functions.
- Use the change-of-base formula to evaluate logarithms.

Logarithmic Functions

Additional Examples
Evaluate each logarithm.
a. $\log_3 27$
b. $\log_2 2$
c. $\log_{25} 5$
Answers:
a. 3
b. 1
c. $\frac{1}{2}$

Definition of Logarithmic Function

Let a and x be positive real numbers such that $a \neq 1$. The **logarithm of x with base a** is denoted by $\log_a x$ and is defined as follows.

$$y = \log_a x \quad \text{if and only if} \quad x = a^y$$

The function $f(x) = \log_a x$ is the **logarithmic function with base a.** Note that the inverse function of $f(x) = a^x$ is $f^{-1}(x) = \log_a x$.

EXAMPLE 1 Evaluating Logarithms

Evaluate each logarithm.

a. $\log_3 9$ **b.** $\log_4 2$ **c.** $\log_5 1$ **d.** $\log_{10} \frac{1}{10}$ **e.** $\log_3(-1)$

SOLUTION

In each case you should answer the question, "To what power must the base be raised to obtain the given number?"

a. The power to which 3 must be raised to obtain 9 is 2. That is,

$$3^2 = 9 \quad \Rightarrow \quad \log_3 9 = 2.$$

b. The power to which 4 must be raised to obtain 2 is $\frac{1}{2}$. That is,

$$4^{1/2} = 2 \quad \Rightarrow \quad \log_4 2 = \frac{1}{2}.$$

c. The power to which 5 must be raised to obtain 1 is 0. That is,

$$5^0 = 1 \quad \Rightarrow \quad \log_5 1 = 0.$$

d. The power to which 10 must be raised to obtain $\frac{1}{10}$ is -1. That is,

$$10^{-1} = \frac{1}{10} \quad \Rightarrow \quad \log_{10} \frac{1}{10} = -1.$$

e. There is no power to which 3 can be raised to obtain -1. The reason for this is that for any value of x, 3^x is a positive number. So, $\log_3(-1)$ is undefined.

Study Tip

Study the results of parts (c), (d), and (e) carefully. Each of the logarithms illustrates an important special property of logarithms that you should know.

Exercises Within Reach®

Solutions in English & Spanish and tutorial videos at AlgebraWithinReach.com

Evaluating a Logarithm **In Exercises 1–12, evaluate the logarithm. (If not possible, state the reason.)**

1. $\log_2 8$ 3 **2.** $\log_3 27$ 3 **3.** $\log_9 3$ $\frac{1}{2}$ **4.** $\log_{125} 5$ $\frac{1}{3}$

5. $\log_{16} 8$ $\frac{3}{4}$ **6.** $\log_{81} 9$ $\frac{1}{2}$ **7.** $\log_4 1$ 0 **8.** $\log_3 1$ 0

9. $\log_2 \frac{1}{16}$ -4 **10.** $\log_3 \frac{1}{9}$ -2 **11.** $\log_2(-3)$ Undefined **12.** $\log_4(-4)$ Undefined

Properties of Logarithms

Let a and x be positive real numbers such that $a \neq 1$. Then the following properties are true.

1. $\log_a 1 = 0$ because $a^0 = 1$.
2. $\log_a a = 1$ because $a^1 = a$.
3. $\log_a a^x = x$ because $a^x = a^x$.

The logarithmic function with base 10 is called the **common logarithmic function**. On most calculators, this function can be evaluated with the common logarithmic key [LOG], as illustrated in the next example.

EXAMPLE 2 Evaluating Common Logarithms

Evaluate each logarithm. Use a calculator only if necessary.

a. $\log_{10} 100$ **b.** $\log_{10} 0.01$ **c.** $\log_{10} 5$ **d.** $\log_{10} 2.5$

SOLUTION

a. The power to which 10 must be raised to obtain 100 is 2. That is,

$$10^2 = 100 \implies \log_{10} 100 = 2.$$

b. The power to which 10 must be raised to obtain 0.01 or $\frac{1}{100}$ is -2. That is,

$$10^{-2} = \frac{1}{100} \implies \log_{10} 0.01 = -2.$$

c. There is no simple power to which 10 can be raised to obtain 5, so you should use a calculator to evaluate $\log_{10} 5$.

Keystrokes	*Display*	
5 [LOG]	0.69897	Scientific
[LOG] 5 [)] [ENTER]	0.69897	Graphing

So, rounded to three decimal places, $\log_{10} 5 \approx 0.699$.

d. There is no simple power to which 10 can be raised to obtain 2.5, so you should use a calculator to evaluate $\log_{10} 2.5$.

Keystrokes	*Display*	
2.5 [LOG]	0.39794	Scientific
[LOG] 2.5 [)] [ENTER]	0.39794	Graphing

So, rounded to three decimal places, $\log_{10} 2.5 \approx 0.398$.

Study Tip

Be sure you see that the value of a logarithm can be zero or negative, as in Example 2(b), but you *cannot* take the logarithm of zero or a negative number. This means that the logarithms $\log_{10}(-10)$ and $\log_5 0$ are undefined.

Exercises Within Reach®

Solutions in English & Spanish and tutorial videos at AlgebraWithinReach.com

Evaluating a Common Logarithm **In Exercises 13–24, evaluate the logarithm. Use a calculator only if necessary.**

13. $\log_{10} 1000$ 3
14. $\log_{10} 10{,}000$ 4
15. $\log_{10} 0.1$ -1
16. $\log_{10} 0.00001$ -5
17. $\log_{10} \frac{1}{10{,}000}$ -4
18. $\log_{10} \frac{1}{100}$ -2
19. $\log_{10} 42$ 1.6232
20. $\log_{10} 7561$ 3.8786
21. $\log_{10} 0.023$ -1.6383
22. $\log_{10} 0.149$ -0.8268
23. $\log_{10}(\sqrt{5} + 3)$ 0.7190
24. $\log_{10} \frac{\sqrt{3}}{2}$ -0.0625

EXAMPLE 3 Graphing a Logarithmic Function

On the same rectangular coordinate system, sketch the graph of each function.

a. $f(x) = 2^x$

b. $g(x) = \log_2 x$

SOLUTION

a. Begin by making a table of values for $f(x) = 2^x$.

x	-2	-1	0	1	2	3
$f(x) = 2^x$	$\frac{1}{4}$	$\frac{1}{2}$	1	2	4	8

By plotting these points and connecting them with a smooth curve, you obtain the graph shown at the left.

b. Because $g(x) = \log_2 x$ is the inverse function of $f(x) = 2^x$, the graph of g is obtained by reflecting the graph of f in the line $y = x$, as shown at the left.

Study Tip

In Example 3, the inverse property of logarithmic functions is used to sketch the graph of $g(x) = \log_2 x$. You could also use a standard point-plotting approach or a graphing calculator.

Notice from the graph of $g(x) = \log_2 x$, shown above, that the domain of the function is the set of positive numbers and the range is the set of all real numbers. The basic characteristics of the graph of a logarithmic function are summarized in the figure below. Note that the graph has one x-intercept at $(1, 0)$. Also note that $x = 0$ (y-axis) is a vertical asymptote of the graph.

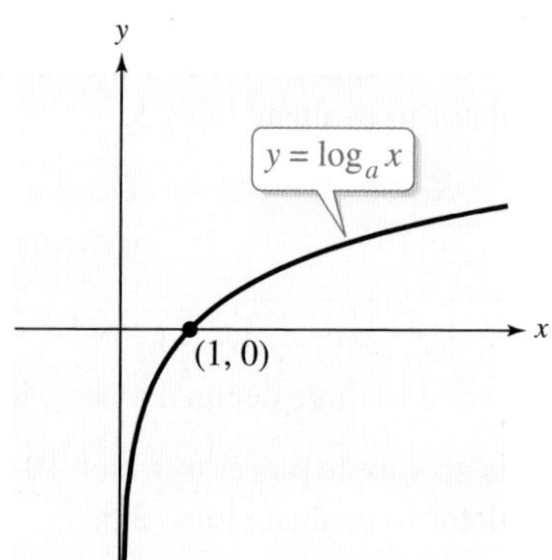

Graph of $y = \log_a x$, $a > 1$

- Domain: $(0, \infty)$
- Range: $(-\infty, \infty)$
- Intercept: $(1, 0)$
- Increasing (moves up to the right)
- Asymptote: y-axis

Exercises Within Reach®

Solutions in English & Spanish and tutorial videos at AlgebraWithinReach.com

Graphing a Logarithmic Function **In Exercises 25–30, sketch the graph of f. Then use the graph of f to sketch the graph of g.** See Additional Answers.

25. $f(x) = 3^x$
$g(x) = \log_3 x$

26. $f(x) = 4^x$
$g(x) = \log_4 x$

27. $f(x) = 6^x$
$g(x) = \log_6 x$

28. $f(x) = 5^x$
$g(x) = \log_5 x$

29. $f(x) = \left(\frac{1}{2}\right)^x$
$g(x) = \log_{1/2} x$

30. $f(x) = \left(\frac{1}{3}\right)^x$
$g(x) = \log_{1/3} x$

EXAMPLE 4 Sketching the Graphs of Logarithmic Functions

The graph of each function is similar to the graph of $f(x) = \log_{10} x$, as shown in the figures. From the graph, you can determine the domain of the function.

a. Because $g(x) = \log_{10}(x - 1) = f(x - 1)$, the graph of g can be obtained by shifting the graph of f one unit to the right. The vertical asymptote of the graph of g is $x = 1$. The domain of g is $(1, \infty)$.

b. Because $h(x) = 2 + \log_{10} x = 2 + f(x)$, the graph of h can be obtained by shifting the graph of f two units upward. The vertical asymptote of the graph of h is $x = 0$. The domain of h is $(0, \infty)$.

c. Because $k(x) = -\log_{10} x = -f(x)$, the graph of k can be obtained by reflecting the graph of f in the x-axis. The vertical asymptote of the graph of k is $x = 0$. The domain of k is $(0, \infty)$.

d. Because $j(x) = \log_{10}(-x) = f(-x)$, the graph of j can be obtained by reflecting the graph of f in the y-axis. The vertical asymptote of the graph of j is $x = 0$. The domain of j is $(-\infty, 0)$.

f(x) = log10 x
(1, 0)
(2, 0)
g(x) = log10(x − 1)

(a)

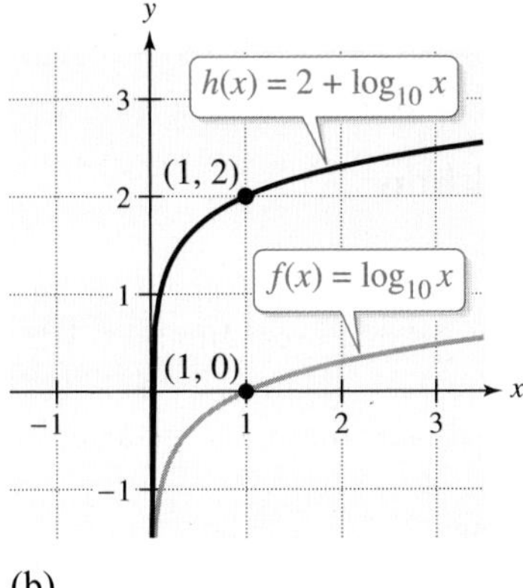

(b)

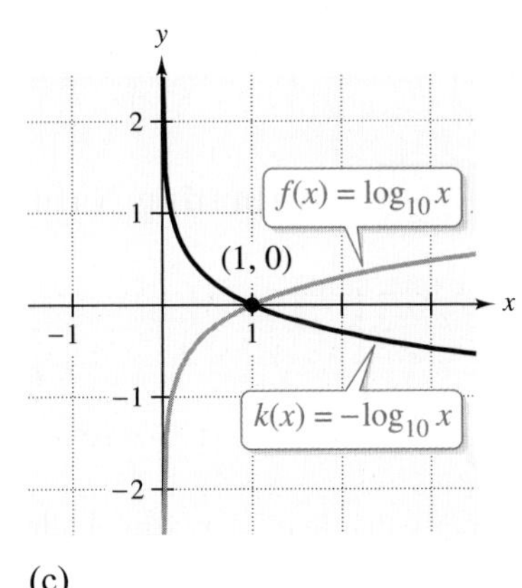

(c)

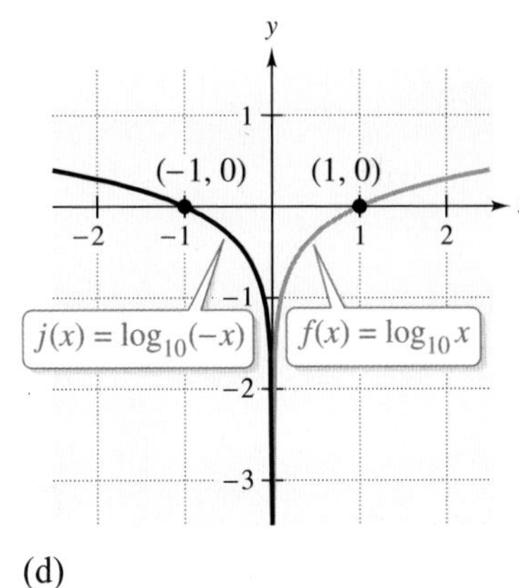

(d)

Exercises Within Reach®

Solutions in English & Spanish and tutorial videos at AlgebraWithinReach.com

Matching **In Exercises 31–34, match the function with its graph.**

(a)

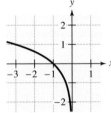

(b)

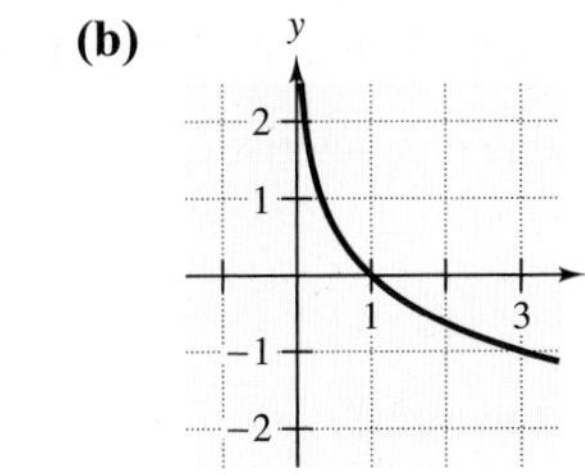

(c)

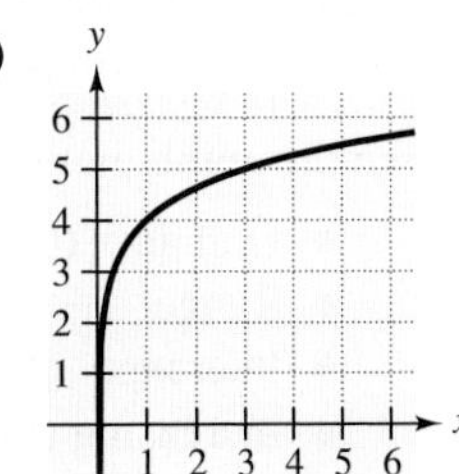

(d)

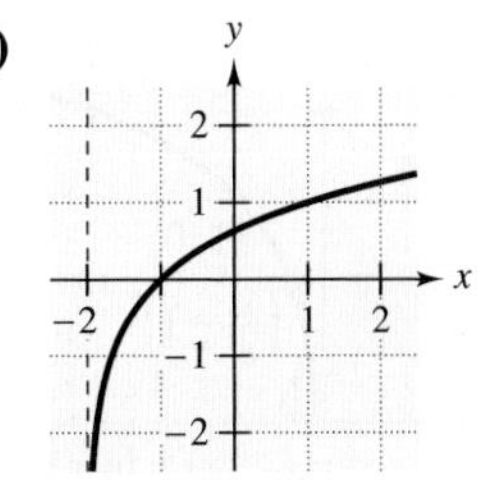

31. $f(x) = 4 + \log_3 x$ c

32. $f(x) = -\log_3 x$ b

33. $f(x) = \log_3(-x)$ a

34. $f(x) = \log_3(x + 2)$ d

Identifying a Transformation **In Exercises 35–40, identify the transformation of the graph of $f(x) = \log_2 x$. Then sketch the graph of h. Determine the vertical asymptote and the domain of the graph of h.** See Additional Answers.

35. $h(x) = 3 + \log_2 x$
Vertical shift two units upward; $x = 0$; $(0, \infty)$

36. $h(x) = -5 + \log_2 x$
Vertical shift five units downward; $x = 0$; $(0, \infty)$

37. $h(x) = \log_2(x - 2)$
Horizontal shift two units to the right; $x = 2$; $(2, \infty)$

38. $h(x) = \log_2(x + 5)$
Horizontal shift five units to the left; $x = -5$; $(-5, \infty)$

39. $h(x) = \log_2(-x)$
Reflection in the y-axis; $x = 0$; $(-\infty, 0)$

40. $h(x) = -\log_2 x$
Reflection in the x-axis; $x = 0$; $(0, \infty)$

The Natural Logarithmic Function

The Natural Logarithmic Function

The function defined by

$$f(x) = \log_e x = \ln x$$

where $x > 0$, is called the **natural logarithmic function**.

Properties of Natural Logarithms

Let x be a positive real number. Then the following properties are true.

1. $\ln 1 = 0$ because $e^0 = 1$.

2. $\ln e = 1$ because $e^1 = e$.

3. $\ln e^x = x$ because $e^x = e^x$.

EXAMPLE 5 Evaluating Natural Logarithms

Evaluate each expression.

a. $\ln e^2$ **b.** $\ln \frac{1}{e}$

SOLUTION

Using the property that $\ln e^x = x$, you obtain the following.

a. $\ln e^2 = 2$ **b.** $\ln \frac{1}{e} = \ln e^{-1} = -1$

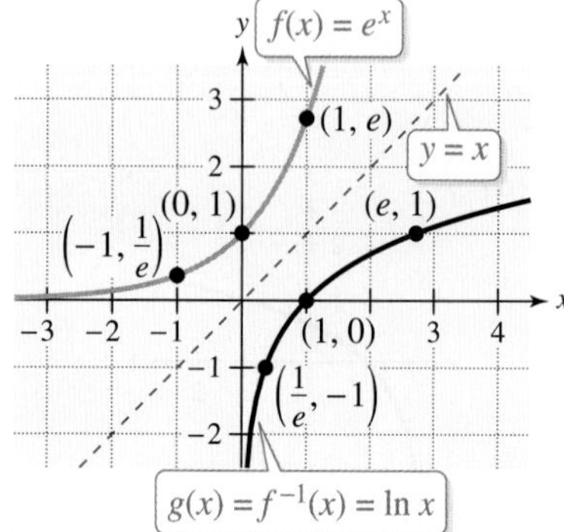

EXAMPLE 6 Graphing the Natural Logarithmic Function

Because the functions $f(x) = e^x$ and $g(x) = \ln x$ are inverse functions of each other, their graphs are reflections of each other in the line $y = x$, as shown in the figure.

From the graph, notice the following characteristics of the natural logarithmic function.

- Domain: $(0, \infty)$
- Range: $(-\infty, \infty)$
- Intercept: $(1, 0)$
- Increasing (moves up to the right)
- Asymptote: y-axis

Exercises Within Reach®

Solutions in English & Spanish and tutorial videos at AlgebraWithinReach.com

Evaluating a Natural Logarithm **In Exercises 41–44, evaluate the expression.**

41. $\ln e^3$ 3 **42.** $\ln e^6$ 6 **43.** $\ln \frac{1}{e^2}$ −2 **44.** $\ln \frac{1}{e^4}$ −4

Graphing a Natural Logarithmic Function **In Exercises 45–48, sketch the graph of the function.**
See Additional Answers.

45. $f(x) = 3 + \ln x$ **46.** $h(x) = 2 + \ln x$ **47.** $f(x) = -\ln x$ **48.** $f(x) = \ln(-x)$

Change of Base

Change-of-Base Formula

Let a, b, and x be positive real numbers such that $a \neq 1$ and $b \neq 1$. Then $\log_a x$ is given as follows.

$$\log_a x = \frac{\log_b x}{\log_b a} \quad \text{or} \quad \log_a x = \frac{\ln x}{\ln a}$$

EXAMPLE 7 Changing Bases to Evaluate Logarithms

a. Use *common* logarithms to evaluate $\log_3 5$.

b. Use *natural* logarithms to evaluate $\log_6 2$.

SOLUTION

Using the change-of-base formula, you can convert to common and natural logarithms by writing

$$\log_3 5 = \frac{\log_{10} 5}{\log_{10} 3} \quad \text{and} \quad \log_6 2 = \frac{\ln 2}{\ln 6}.$$

Now, use the following keystrokes.

a.

Keystrokes	*Display*	
5 [LOG] [÷] 3 [LOG] [=]	1.4649735	Scientific
[LOG] 5 [)] [÷] [LOG] 3 [)] [ENTER]	1.4649735	Graphing

So, $\log_3 5 \approx 1.465$.

b.

Keystrokes	*Display*	
2 [LN] [÷] 6 [LN] [=]	0.3868528	Scientific
[LN] 2 [)] [÷] [LN] 6 [)] [ENTER]	0.3868528	Graphing

So, $\log_6 2 \approx 0.387$.

Study Tip

In Example 7(a), $\log_3 5$ could have been evaluated using natural logarithms in the change-of-base formula.

$$\log_3 5 = \frac{\ln 5}{\ln 3} \approx 1.465$$

Notice that you get the same answer whether you use natural logarithms or common logarithms in the change-of-base formula.

Exercises Within Reach®

Using the Change-of-Base Formula **In Exercises 49–62, use (a) common logarithms and (b) natural logarithms to evaluate the expression. (Round your answer to four decimal places.)**

49. $\log_9 36$ 1.6309

50. $\log_7 411$ 3.0929

51. $\log_5 14$ 1.6397

52. $\log_6 9$ 1.2263

53. $\log_2 0.72$ −0.4739

54. $\log_{12} 0.6$ −0.2056

55. $\log_{15} 1250$ 2.6332

56. $\log_{20} 125$ 1.6117

57. $\log_{1/4} 16$ −2

58. $\log_{1/3} 18$ −2.6309

59. $\log_4 \sqrt{42}$ 1.3481

60. $\log_5 \sqrt{21}$ 0.9458

61. $\log_2(1 + e)$ 1.8946

62. $\log_4(2 + e^3)$ 2.2325

Concept Summary: Evaluating Logarithms

What

The function $f(x) = \log_a x$, where a and x are positive real numbers such that $a \neq 1$, is the **logarithmic function with base a.**

The exponential function with base a and the logarithmic function with base a are inverse functions.

$f(x) = a^x \qquad f^{-1}(x) = \log_a x$

EXAMPLE

Evaluate $\log_3 81$.

How

To evaluate a logarithm, you should answer the question, " To what power must the base be raised to obtain the given number?"

EXAMPLE

The power to which 3 must be raised to obtain 81 is 4. That is.

$3^4 = 81 \Rightarrow \log_3 81 = 4.$

Why

Logarithmic functions are used in many scientific and business applications. For instance, you can use a logarithmic function to determine the length of a home mortgage given any monthly payment.

Exercises Within Reach®

Worked-out solutions to odd-numbered exercises at AlgebraWithinReach.com

Concept Summary Check

63. ***Inverse Functions*** What is the inverse of the function $y = a^x$? $y = \log_a x$ or $x = a^y$

64. ***The Base of a Logarithm*** What is the base of the logarithm $\log_3 81$? 3

65. ***The Base of a Natural Logarithm*** What is the base of the logarithm ln 5? e

66. ***The Change-of-Base Formula*** Write an expression involving natural logarithms that is equivalent to $\log_3 81$. $\frac{\ln 81}{\ln 3}$

Extra Practice

Writing Exponential Form **In Exercises 67–74, write the logarithmic equation in exponential form.**

67. $\log_7 49 = 2$ $7^2 = 49$
68. $\log_{11} 121 = 2$ $11^2 = 121$
69. $\log_2 \frac{1}{32} = -5$ $2^{-5} = \frac{1}{32}$
70. $\log_3 \frac{1}{27} = -3$ $3^{-3} = \frac{1}{27}$
71. $\log_{36} 6 = \frac{1}{2}$ $36^{1/2} = 6$
72. $\log_{64} 4 = \frac{1}{3}$ $64^{1/3} = 4$
73. $\log_8 4 = \frac{2}{3}$ $8^{2/3} = 4$
74. $\log_{16} 8 = \frac{3}{4}$ $16^{3/4} = 8$

Writing Logarithmic Form **In Exercises 75–78, write the exponential equation in logarithmic form.**

75. $6^2 = 36$ $\log_6 36 = 2$
76. $3^5 = 243$ $\log_3 243 = 5$
77. $8^{2/3} = 4$ $\log_8 4 = \frac{2}{3}$
78. $81^{3/4} = 27$ $\log_{81} 27 = \frac{3}{4}$

Graphing a Logarithmic Function **In Exercises 79–86, sketch the graph of the function. Identify the vertical asymptote.** See Additional Answers.

79. $h(s) = -2 \log_3 s$ $s = 0$
80. $g(x) = -\frac{1}{2} \log_3 x$ $x = 0$
81. $f(x) = \log_{10}(10x)$ $x = 0$
82. $g(x) = \log_4(4x)$ $x = 0$
83. $y = \log_4(x - 1) - 2$ $x = 1$
84. $y = \log_4(x - 2) + 3$ $x = 2$
85. $y = -\log_3 x + 2$ $x = 0$
86. $f(x) = -\log_6(x + 2)$ $x = -2$

87. ***American Elk*** The antler spread a (in inches) and shoulder height h (in inches) of an adult male American elk are related by the model

$h = 116 \log_{10}(a + 40) - 176.$

Approximate to one decimal place the shoulder height of a male American elk with an antler spread of 55 inches. 53.4 inches

Delmas Lehman/Shutterstock.com

88. ***Meteorology*** Most tornadoes last less than 1 hour and travel about 20 miles. The speed of the wind S (in miles per hour) near the center of a tornado and the distance d (in miles) the tornado travels are related by the model $S = 93 \log_{10} d + 65$. On March 18, 1925, a large tornado struck portions of Missouri, Illinois, and Indiana, covering a distance of 220 miles. Approximate to one decimal place the speed of the wind near the center of this tornado.
282.8 miles per hour

89. ***Tractrix*** A person walking along a dock (the y-axis) drags a boat by a 10-foot rope (see figure). The boat travels along a path known as a tractrix. The equation of the path is

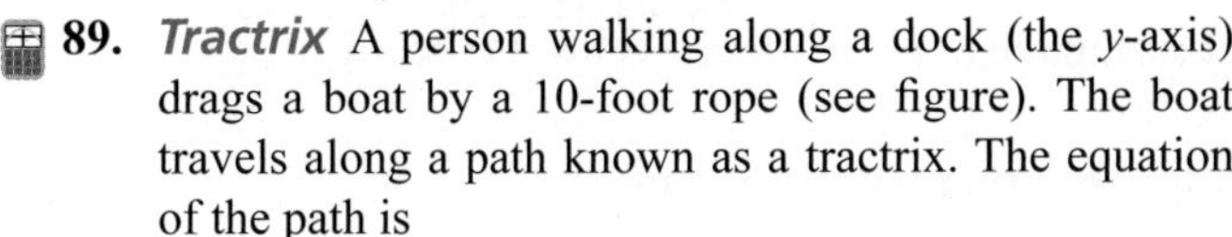

$$y = 10 \ln\left(\frac{10 + \sqrt{100 - x^2}}{x}\right) - \sqrt{100 - x^2}.$$

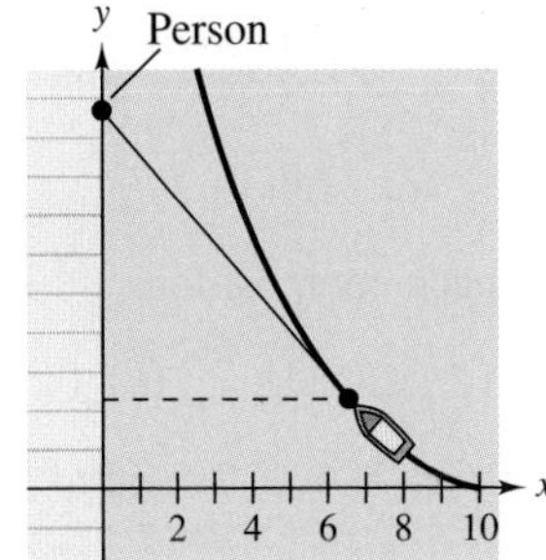

(a) Use a graphing calculator to graph the function. What is the domain of the function?
See Additional Answers; Domain: $0 < x \le 10$

(b) Identify any asymptotes. $x = 0$

(c) Determine the position of the person when the x-coordinate of the position of the boat is $x = 2$.
(0, 22.9)

90. ***Home Mortgage*** The model

$$t = 10.042 \ln\left(\frac{x}{x - 1250}\right), \quad x > 1250$$

approximates the length t (in years) of a home mortgage of \$150,000 at 10% interest in terms of the monthly payment x.

(a) Use a graphing calculator to graph the model. Describe the change in the length of the mortgage as the monthly payment increases.
See Additional Answers; As the monthly payment increases, the length of the loan decreases.

(b) Use the graph in part (a) to approximate the length of the mortgage when the monthly payment is \$1316.35. 30 years

(c) Use the result of part (b) to find the total amount paid over the term of the mortgage. What amount of the total is interest costs? \$473,886; \$323,886

Think About It **In Exercises 91–96, answer the question for the function $f(x) = \log_{10} x$. (Do not use a calculator.)**

91. What is the domain of f? $(0, \infty)$

92. Find the inverse function of f. $f^{-1}(x) = 10^x$

93. Describe the values of $f(x)$ for $1000 \le x \le 10{,}000$.
$3 \le f(x) \le 4$

94. Describe the values of x, given that $f(x)$ is negative.
$0 < x < 1$

95. By what amount will x increase, given that $f(x)$ is increased by 1 unit? A factor of 10

96. Find the ratio of a to b when $f(a) = 3 + f(b)$.
$10^3 = 1000$

Explaining Concepts

True or False? **In Exercises 97 and 98, determine whether the statement is true or false. Justify your answer.**

97. The statement $8 = 2^3$ is equivalent to $2 = \log_8 3$.
False. $8 = 2^3$ is equivalent to $3 = \log_2 8$.

98. The graph of $f(x) = \ln x$ is the reflection of the graph of $f(x) = e^x$ in the x-axis.
False. The graphs are reflections in the line $y = x$.

99. ***Vocabulary*** Explain the difference between common logarithms and natural logarithms. See Additional Answers.

100. ***Structure*** Explain the relationship between the domain of the graph of $f(x) = \log_5 x$ and the range of the graph of $g(x) = 5^x$. See Additional Answers.

101. ***Think About It*** Discuss how shifting or reflecting the graph of a logarithmic function affects the domain and the range. See Additional Answers.

102. ***Number Sense*** Explain why $\log_a x$ is defined only when $0 < a < 1$ and $a > 1$. See Additional Answers.

Cumulative Review

In Exercises 103–106, use the rules of exponents to simplify the expression.

103. $(-m^6n)(m^4n^3)$ $-m^{10}n^4$

104. $(m^2n^4)^3(mn^2)$ m^7n^{14}

105. $\dfrac{36x^4y}{8xy^3}$ $\dfrac{9x^3}{2y^2}, x \ne 0$

106. $-\left(\dfrac{3x}{5y}\right)^5$ $-\dfrac{243x^5}{3125y^5}$

In Exercises 107–110, perform the indicated operation(s) and simplify. (Assume all variables are positive.)

107. $25\sqrt{3x} - 3\sqrt{12x}$ $19\sqrt{3x}$

108. $(\sqrt{x} + 3)(\sqrt{x} - 3)$ $x - 9$

109. $\sqrt{u}(\sqrt{20} - \sqrt{5})$ $\sqrt{5u}$

110. $(2\sqrt{t} + 3)^2$ $4t + 12\sqrt{t} + 9$

Mid-Chapter Quiz: Sections 9.1–9.3

Solutions in English & Spanish and tutorial videos at AlgebraWithinReach.com

Take this quiz as you would take a quiz in class. After you are done, check your work against the answers in the back of the book.

1. (a) $\frac{16}{9}$ (b) 1 (c) $\frac{3}{4}$ (d) 1.540

1. Given $f(x) = \left(\frac{4}{3}\right)^x$, find (a) $f(2)$, (b) $f(0)$, (c) $f(-1)$, and (d) $f(1.5)$.

2. Identify the horizontal asymptote of the graph of $g(x) = 3^{x-1}$.
Horizontal asymptote: $y = 0$

In Exercises 3–6, match the function with its graph.

3. $f(x) = 2^x$ b

4. $g(x) = -2^x$ d

5. $h(x) = 2^{-x}$ a

6. $s(x) = 2^x - 2$ c

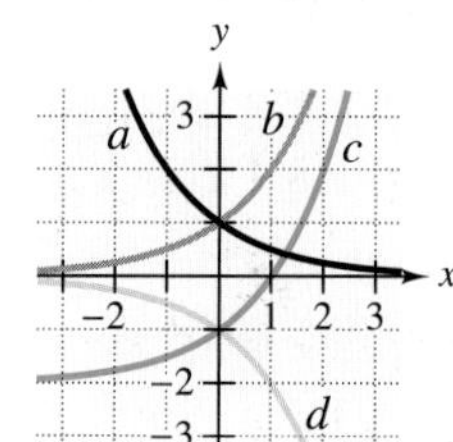

7. (a) $2x^3 - 3$
(b) $(2x-3)^3 = 8x^3 - 36x^2 + 54x - 27$
(c) -19
(d) 125

7. Given $f(x) = 2x - 3$ and $g(x) = x^3$, find each composition.

(a) $(f \circ g)(x)$ (b) $(g \circ f)(x)$ (c) $(f \circ g)(-2)$ (d) $(g \circ f)(4)$

8. $f(g(x)) = 5 - 2\left[\frac{1}{2}(5 - x)\right]$
$= 5 - 5 + x = x$
$g(f(x)) = \frac{1}{2}[5 - (5 - 2x)]$
$= \frac{1}{2}(2x) = x$
See Additional Answers.

8. Verify algebraically and graphically that $f(x) = 5 - 2x$ and $g(x) = \frac{1}{2}(5 - x)$ are inverse functions of each other.

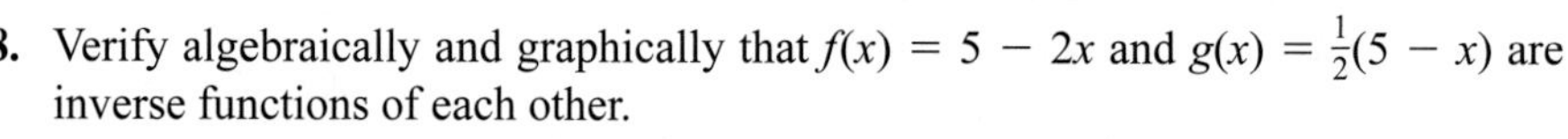

In Exercises 9 and 10, find the inverse function of *f*.

9. $f(x) = 10x + 3$ $f^{-1}(x) = \dfrac{x-3}{10}$ **10.** $f(t) = \frac{1}{2}t^3 + 2$ $f^{-1}(t) = \sqrt[3]{2t - 4}$

11. Write the logarithmic equation $\log_9 \frac{1}{81} = -2$ in exponential form. $9^{-2} = \frac{1}{81}$

12. Write the exponential equation $2^6 = 64$ in logarithmic form. $\log_2 64 = 6$

13. Evaluate $\log_5 125$ without a calculator. 3

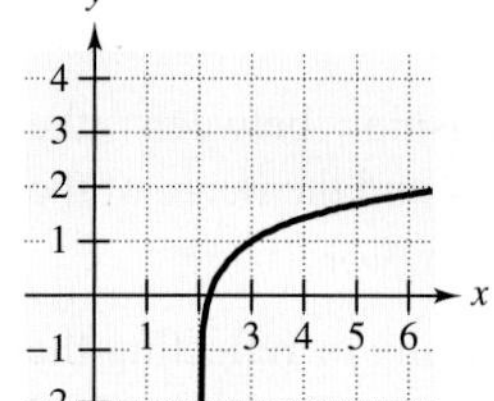

In Exercises 14 and 15, sketch the graph of the function. Identify the vertical asymptote. See Additional Answers.

14. $f(t) = \ln(t + 3)$ $t = -3$ **15.** $h(x) = 1 + \ln x$ $x = 0$

16. Use the graph of f shown at the left to determine h and k in $f(x) = \log_5(x - h) + k$. $h = 2, k = 1$

17. Use a calculator and the change-of-base formula to evaluate $\log_3 782$. 6.0639

Applications

18. You deposit \$1200 in an account at an annual interest rate of $6\frac{1}{4}\%$. Complete the table to determine the balance A in the account after 15 years when the account is compounded n times per year.

n	1	4	12	365	Continuous compounding
A	\$2979.31	\$3042.18	\$3056.86	\$3064.06	\$3064.31

19. After t years, 14 grams of a radioactive element with a half-life of 40 years decays to a mass y (in grams) given by $y = 14\left(\frac{1}{2}\right)^{t/40}$, $t \geq 0$. How much of the initial mass remains after 125 years? 1.60 grams

Study Skills in Action

Making the Most of Class Time

Have you ever slumped at your desk while in class and thought, "I'll just get the notes down and study later—I'm too tired"? Learning math in college is a team effort, between instructor and student. The more you understand in class, the more you will be able to learn while studying outside of class.

Approach math class with the intensity of a navy pilot during a mission briefing. The pilot has strategic plans to learn during the briefing. He or she listens intensely, takes notes, and memorizes important information. The goal is for the pilot to leave the briefing with a clear picture of the mission. It is the same with a student in a math class.

These students are sitting in the front row, where they are more likely to pay attention.

Smart Study Strategy

Take Control of Your Class Time

1 ▸ **Sit where you can easily see and hear the instructor, and the instructor can see you.** The instructor may be able to tell when you are confused just by the look on your face, and may adjust the lesson accordingly. In addition, sitting in this strategic place will keep your mind from wandering.

2 ▸ **Pay attention to what the instructor says about the math, not just what is written on the board.** Write problems on the left side of your notes and what the instructor says about the problems on the right side.

3 ▸ **If the instructor is moving through the material too fast, ask a question.** Questions help to slow the pace for a few minutes and also to clarify what is confusing to you.

4 ▸ **Try to memorize new information while learning it.** Repeat in your head what you are writing in your notes. That way you are reviewing the information twice.

5 ▸ **Ask for clarification.** If you don't understand something at all and don't even know how to phrase a question, just ask for clarification. You might say something like, "Could you please explain the steps in this problem one more time?"

6 ▸ **Think as intensely as if you were going to take a quiz on the material at the end of class.** This kind of mindset will help you to process new information.

7 ▸ **If the instructor asks for someone to go up to the board, volunteer.** The student at the board often receives additional attention and instruction to complete the problem.

8 ▸ **At the end of class, identify concepts or problems on which you still need clarification.** Make sure you see the instructor or a tutor as soon as possible.

9.4 Properties of Logarithms

- Use the properties of logarithms to evaluate logarithms.
- Rewrite, expand, or condense logarithmic expressions.
- Use the properties of logarithms to solve application problems.

Properties of Logarithms

Properties of Logarithms

Let a be a positive real number such that $a \neq 1$, and let n be a real number. If u and v are real numbers, variables, or algebraic expressions such that $u > 0$ and $v > 0$, then the following properties are true.

	Logarithm with Base a	*Natural Logarithm*
1. Product Property:	$\log_a(uv) = \log_a u + \log_a v$	$\ln(uv) = \ln u + \ln v$
2. Quotient Property:	$\log_a \frac{u}{v} = \log_a u - \log_a v$	$\ln \frac{u}{v} = \ln u - \ln v$
3. Power Property:	$\log_a u^n = n \log_a u$	$\ln u^n = n \ln u$

EXAMPLE 1 Using Properties of Logarithms

Use $\ln 2 \approx 0.693$, $\ln 3 \approx 1.099$, and $\ln 5 \approx 1.609$ to approximate each expression.

a. $\ln \frac{2}{3}$ **b.** $\ln 10$ **c.** $\ln 30$

SOLUTION

a. $\ln \frac{2}{3} = \ln 2 - \ln 3$ Quotient Property

$\approx 0.693 - 1.099 = -0.406$ Substitute for ln 2 and ln 3.

b. $\ln 10 = \ln(2 \cdot 5)$ Factor.

$= \ln 2 + \ln 5$ Product Property

$\approx 0.693 + 1.609$ Substitute for ln 2 and ln 5.

$= 2.302$ Simplify.

c. $\ln 30 = \ln(2 \cdot 3 \cdot 5)$ Factor.

$= \ln 2 + \ln 3 + \ln 5$ Product Property

$\approx 0.693 + 1.099 + 1.609$ Substitute for ln 2, ln 3, and ln 5.

$= 3.401$ Simplify.

Exercises Within Reach®

Using Properties of Logarithms **In Exercises 1–8, use $\ln 3 \approx 1.0986$ and $\ln 5 \approx 1.6094$ to approximate the expression. Use a calculator to verify your result.**

1. $\ln \frac{5}{3}$ 0.5108 **2.** $\ln \frac{3}{5}$ −0.5108 **3.** $\ln 9$ 2.1972 **4.** $\ln 15$ 2.7080

5. $\ln 75$ 4.3174 **6.** $\ln 45$ 3.8066 **7.** $\ln \sqrt{45}$ 1.9033 **8.** $\ln \sqrt[3]{25}$ 1.0729

When using the properties of logarithms, it helps to state the properties *verbally*. For instance, the verbal form of the Product Property

$$\ln(uv) = \ln u + \ln v$$

is: *The log of a product is the sum of the logs of the factors*. Similarly, the verbal form of the Quotient Property

$$\ln \frac{u}{v} = \ln u - \ln v$$

is: *The log of a quotient is the difference of the logs of the numerator and denominator.*

Study Tip

Remember that you can verify results, such as those given in Examples 1 and 2, with a calculator.

EXAMPLE 2 Using Properties of Logarithms

Use the properties of logarithms to verify that $-\ln 2 = \ln \frac{1}{2}$.

SOLUTION

Using the Power Property, you can write the following.

$-\ln 2 = (-1) \ln 2$ Rewrite coefficient as -1.

$= \ln 2^{-1}$ Power Property

$= \ln \frac{1}{2}$ Rewrite 2^{-1} as $\frac{1}{2}$.

Exercises Within Reach®

Solutions in English & Spanish and tutorial videos at AlgebraWithinReach.com

Using Properties of Logarithms **In Exercises 9–14, use the properties of logarithms to verify the statement.**

9. $-3 \log_4 2 = \log_4 \frac{1}{8}$ $\quad -3 \log_4 2 = \log_4 2^{-3} = \log_4 \frac{1}{8}$

10. $-2 \ln \frac{1}{3} = \ln 9$ $\quad -2 \ln \frac{1}{3} = \ln\left(\frac{1}{3}\right)^{-2} = \ln 3^2 = \ln 9$

11. $-3 \log_{10} 3 + \log_{10} \frac{3}{2} = \log_{10} \frac{1}{18}$ $\quad -3 \log_{10} 3 + \log_{10} \frac{3}{2} = \log_{10} 3^{-3} + \log_{10} \frac{3}{2} = \log_{10} \frac{1}{27} + \log_{10} \frac{3}{2} = \log_{10}\left(\frac{1}{27} \cdot \frac{3}{2}\right) = \log_{10} \frac{1}{18}$

12. $-2 \ln 2 + \ln 24 = \ln 6$ $\quad -2 \ln 2 + \ln 24 = \ln 2^{-2} + \ln 24 = \ln \frac{1}{4} + \ln 24 = \ln\left(\frac{1}{4} \cdot 24\right) = \ln 6$

13. $-\ln \frac{1}{7} = \ln 56 - \ln 8$ $\quad -\ln \frac{1}{7} = \ln\left(\frac{1}{7}\right)^{-1} = \ln 7 = \ln \frac{56}{8} = \ln 56 - \ln 8$

14. $-\log_5 10 = \log_5 10 - \log_5 100$ $\quad -\log_5 10 = \log_5 10^{-1} = \log_5 \frac{1}{10} = \log_5 \frac{10}{100} = \log_5 10 - \log_5 100$

Using Properties of Logarithms **In Exercises 15–34, use the properties of logarithms to evaluate the expression.**

15. $\log_{12} 12^3$ $\;3$
16. $\log_3 81$ $\;4$
17. $\log_4\left(\frac{1}{16}\right)^2$ $\;-4$
18. $\log_7\left(\frac{1}{49}\right)^3$ $\;-6$
19. $\log_5 \sqrt[3]{5}$ $\;\frac{1}{3}$
20. $\ln \sqrt{e}$ $\;\frac{1}{2}$
21. $\ln 14^0$ $\;0$
22. $\ln\left(\frac{7.14}{7.14}\right)$ $\;0$
23. $\ln e^{-9}$ $\;-9$
24. $\ln e^7$ $\;7$
25. $\log_8 4 + \log_8 16$ $\;2$
26. $\log_{10} 5 + \log_{10} 20$ $\;2$
27. $\log_3 54 - \log_3 2$ $\;3$
28. $\log_3 324 - \log_3 4$ $\;4$
29. $\log_2 5 - \log_2 40$ $\;-3$
30. $\log_4\left(\frac{3}{16}\right) + \log_4\left(\frac{1}{3}\right)$ $\;-2$
31. $\ln e^8 + \ln e^4$ $\;12$
32. $\ln e^5 - \ln e^2$ $\;3$
33. $\ln \frac{e^3}{e^2}$ $\;1$
34. $\ln(e^2 \cdot e^4)$ $\;6$

Rewriting Logarithmic Expressions

In Examples 1 and 2, the properties of logarithms were used to rewrite logarithmic expressions involving the log of a *constant*. A more common use of these properties is to rewrite the log of a *variable expression*.

EXAMPLE 3 Expanding Logarithmic Expressions

Use the properties of logarithms to expand each expression.

a. $\log_{10} 7x^3$ **b.** $\log_6 \frac{8x^3}{y}$ **c.** $\ln \frac{\sqrt{3x-5}}{7}$

SOLUTION

a. $\log_{10} 7x^3 = \log_{10} 7 + \log_{10} x^3$ Product Property

$= \log_{10} 7 + 3 \log_{10} x$ Power Property

b. $\log_6 \frac{8x^3}{y} = \log_6 8x^3 - \log_6 y$ Quotient Property

$= \log_6 8 + \log_6 x^3 - \log_6 y$ Product Property

$= \log_6 8 + 3 \log_6 x - \log_6 y$ Power Property

c. $\ln \frac{\sqrt{3x-5}}{7} = \ln\left[\frac{(3x-5)^{1/2}}{7}\right]$ Rewrite using rational exponent.

$= \ln(3x-5)^{1/2} - \ln 7$ Quotient Property

$= \frac{1}{2}\ln(3x-5) - \ln 7$ Power Property

Exercises Within Reach®

Solutions in English & Spanish and tutorial videos at AlgebraWithinReach.com

Expanding a Logarithmic Expression **In Exercises 35–58, use the properties of logarithms to expand the expression. (Assume all variables are positive.)**

35. $\log_3 11x$ $\log_3 11 + \log_3 x$
36. $\log_2 3x$ $\log_2 3 + \log_2 x$
37. $\ln 3y$ $\ln 3 + \ln y$
38. $\ln 5x$ $\ln 5 + \ln x$
39. $\log_7 x^2$ $2 \log_7 x$
40. $\log_3 x^3$ $3 \log_3 x$
41. $\log_4 x^{-3}$ $-3 \log_4 x$
42. $\log_2 s^{-4}$ $-4 \log_2 s$
43. $\log_4 \sqrt{3x}$ $\frac{1}{2}(\log_4 3 + \log_4 x)$
44. $\log_3 \sqrt[3]{5y}$ $\frac{1}{3}(\log_3 5 + \log_3 y)$
45. $\log_2 \frac{z}{17}$ $\log_2 z - \log_2 17$
46. $\log_{10} \frac{7}{y}$ $\log_{10} 7 - \log_{10} y$
47. $\log_9 \frac{\sqrt{x}}{12}$ $\frac{1}{2}\log_9 x - \log_9 12$
48. $\ln \frac{\sqrt{x}}{x+9}$ $\frac{1}{2}\ln x - \ln(x+9)$
49. $\ln x^2(y+2)$ $2 \ln x + \ln(y+2)$
50. $\ln y(y+1)^2$ $\ln y + 2\ln(y+1)$
51. $\log_4[x^6(x+7)^2]$ $6 \log_4 x + 2\log_4(x+7)$
52. $\log_8[(x-y)^3z^6]$ $3\log_8(x-y) + 6\log_8 z$
53. $\log_3 \sqrt[3]{x+1}$ $\frac{1}{3}\log_3(x+1)$
54. $\log_5 \sqrt{xy}$ $\frac{1}{2}(\log_5 x + \log_5 y)$
55. $\ln \sqrt{x(x+2)}$ $\frac{1}{2}[\ln x + \ln(x+2)]$
56. $\ln \sqrt[3]{x(x+5)}$ $\frac{1}{3}[\ln x + \ln(x+5)]$
57. $\ln \frac{xy^2}{z^3}$ $\ln x + 2\ln y - 3\ln z$
58. $\log_5 \frac{x^2y^5}{z^7}$ $2\log_5 x + 5\log_5 y - 7\log_5 z$

When you rewrite a logarithmic expression as in Example 3, you are *expanding* the expression. The reverse procedure is demonstrated in Example 4, and is called *condensing* a logarithmic expression.

Be aware that some students confuse the property of logarithms that states $\log_b u - \log_b v = \log_b(u/v)$ with $\log_v u = \log_b u/\log_b v$. So, they may be tempted to rewrite $\log_a x = \log_b x/\log_b a$ incorrectly as $\log_b x - \log_b a$.

Additional Examples

a. Expand $\ln \sqrt{x}y^3$.

b. Condense $3 \log_4 x + 5 \log_4 y$.

Answers:

a. $\frac{1}{2} \ln x + 3 \ln y$

b. $\log_4 x^3y^5,\ x > 0,\ y > 0$

EXAMPLE 4 Condensing Logarithmic Expressions

Use the properties of logarithms to condense each expression.

a. $\ln x - \ln 3$ **b.** $\frac{1}{2}\log_3 x + \log_3 5$ **c.** $3(\ln 4 + \ln x)$

SOLUTION

a. $\ln x - \ln 3 = \ln \frac{x}{3}$ Quotient Property

b. $\frac{1}{2}\log_3 x + \log_3 5 = \log_3 x^{1/2} + \log_3 5$ Power Property

$= \log_3 5\sqrt{x}$ Product Property

c. $3(\ln 4 + \ln x) = 3(\ln 4x)$ Product Property

$= \ln(4x)^3$ Power Property

$= \ln 64x^3$ Simplify.

When you expand or condense a logarithmic expression, it is possible to change the domain of the expression. For instance, the domain of the function

$f(x) = 2 \ln x$ Domain is the set of positive real numbers.

is the set of positive real numbers, whereas the domain of

$g(x) = \ln x^2$ Domain is the set of nonzero real numbers.

is the set of nonzero real numbers. So, when you expand or condense a logarithmic expression, you should check to see whether the rewriting has changed the domain of the expression. In such cases, you should restrict the domain appropriately. For instance, you can write

$$f(x) = 2 \ln x$$
$$= \ln x^2,\ x > 0.$$

Exercises Within Reach®

Solutions in English & Spanish and tutorial videos at AlgebraWithinReach.com

Condensing a Logarithmic Expression **In Exercises 59–68, use the properties of logarithms to condense the expression.**

59. $\log_{12} x - \log_{12} 3$ $\log_{12} \frac{x}{3}$

60. $\log_6 12 - \log_6 y$ $\log_6 \frac{12}{y}$

61. $\log_3 5 + \log_3 x$ $\log_3 5x$

62. $\log_5 2x + \log_5 3y$ $\log_5 6xy$

63. $7 \log_2 x + 3 \log_2 z$ $\log_2 x^7z^3$

64. $2 \log_{10} x + \frac{1}{2}\log_{10} y$ $\log_{10} x^2\sqrt{y}$

65. $4(\ln x + \ln y)$ $\ln x^4y^4$

66. $\frac{1}{3}(\ln 10 + \ln 4x)$ $\ln 2\sqrt[3]{5x}$

67. $\log_4(x + 8) - 3 \log_4 x$ $\log_4 \frac{x+8}{x^3}$

68. $5 \log_3 x + \log_3(x - 6)$ $\log_3 x^5(x - 6)$

Applications

Application

EXAMPLE 5 Human Memory Model

In an experiment, students attended several lectures on a subject. Every month for a year after that, the students were tested to see how much of the material they remembered. The average scores for the group are given by the human memory model

Be aware that some students may incorrectly simplify $80 - 9 \ln 9$ as $71 \ln 9$. Review the order of operations.

$$f(t) = 80 - \ln(t + 1)^9, \quad 0 \le t \le 12$$

where t is the time in months. Find the average score for the group after 8 months.

SOLUTION

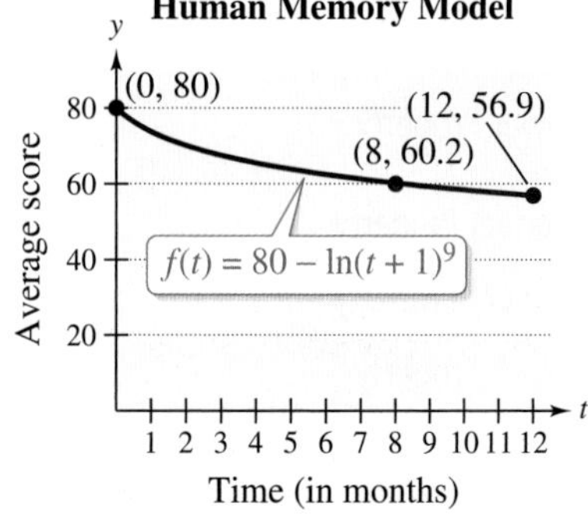

To make the calculations easier, rewrite the model using the Power Property, as follows.

$$f(t) = 80 - 9 \ln(t + 1), \quad 0 \le t \le 12$$

After 8 months, the average score was

$$f(8) = 80 - 9 \ln(8 + 1) \qquad \text{Substitute 8 for } t.$$

$$= 80 - 9 \ln 9 \qquad \text{Add.}$$

$$\approx 80 - 19.8 \qquad \text{Simplify.}$$

$$= 60.2. \qquad \text{Average score after 8 months}$$

The graph of the function is shown at the left.

Exercises Within Reach®

Solutions in English & Spanish and tutorial videos at AlgebraWithinReach.com

69. ***Human Memory Model*** Students participating in an experiment attended several lectures on a subject. Every month for a year after that, the students were tested to see how much of the material they remembered. The average scores for the group are given by the human memory model

$$f(t) = 80 - \log_{10}(t + 1)^{12}, \quad 0 \le t \le 12$$

where t is the time in months. (See figure.) Use the Power Property to rewrite the model. Then find the average scores for the group after 2 months and 8 months.

$f(t) = 80 - 12 \log_{10}(t + 1)$; $f(2) \approx 74.27$; $f(8) \approx 68.55$

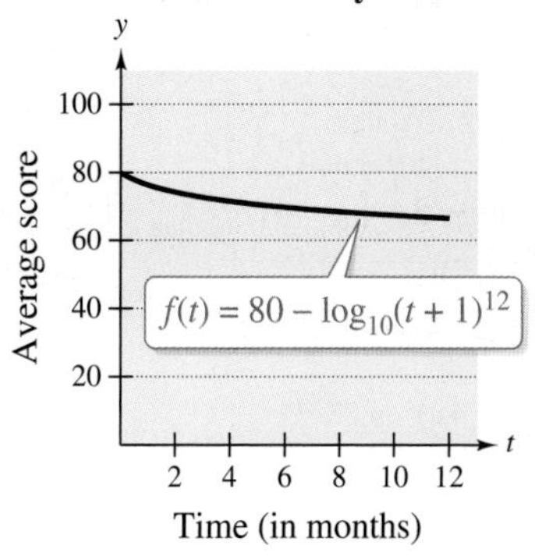

70. ***Human Memory Model*** Students participating in an experiment attended several lectures on a subject. Every month for a year after that, the students were tested to see how much of the material they remembered. The average scores for the group are given by the human memory model

$$f(t) = 75 - \ln(t + 1)^6, \quad 0 \le t \le 12$$

where t is the time in months. (See figure.) Use the Power Property to rewrite the model. Then find the average scores for the group after 1 month and 11 months.

$f(t) = 75 - 6 \ln(t + 1)$; $f(1) \approx 70.84$, $f(11) \approx 60.09$

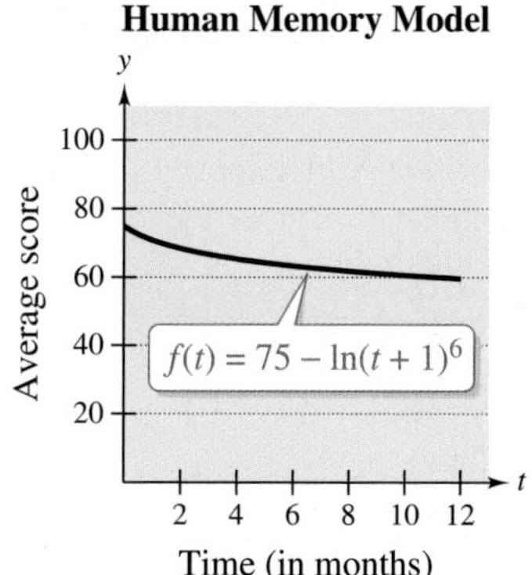

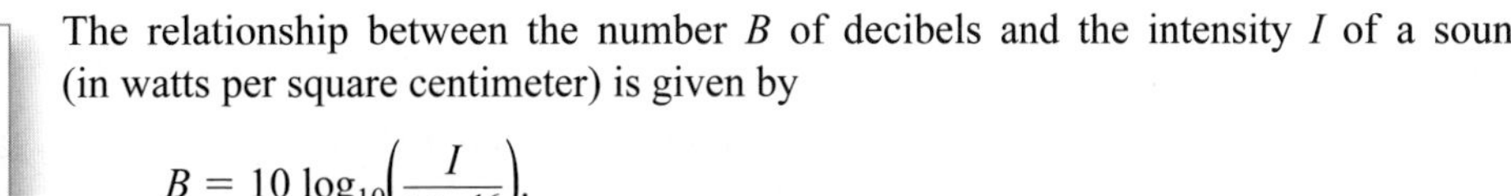

Application EXAMPLE 6 Sound Intensity

The relationship between the number B of decibels and the intensity I of a sound (in watts per square centimeter) is given by

$$B = 10 \log_{10}\left(\frac{I}{10^{-16}}\right).$$

Use properties of logarithms to write the formula in simpler form, and determine the number of decibels of a thunderclap with an intensity of 10^{-1} watt per square centimeter.

SOLUTION

$B = 10 \log_{10}\left(\frac{I}{10^{-16}}\right)$	Write formula for decibels.
$= 10(\log_{10} I - \log_{10} 10^{-16})$	Quotient Property
$= 10 \log_{10} I - 10(-16 \log_{10} 10)$	Power Property
$= 10 \log_{10} I - 10(-16)$	$\log_{10} 10 = 1$
$= 10 \log_{10} I + 160$	Simplify.

When the intensity is $I = 10^{-1}$ watt per square centimeter, the number of decibels is

$B = 10 \log_{10} I + 160$	Write formula for decibels.
$= 10 \log_{10}(10^{-1}) + 160$	Substitute 10^{-1} for I.
$= 10(-1) + 160$	$\log_{10}(10^{-1}) = -1$
$= -10 + 160$	Multiply.
$= 150$ decibels.	Add.

Exercises Within Reach ®

Solutions in English & Spanish and tutorial videos at AlgebraWithinReach.com

71. ***Sound Intensity*** The relationship between the number B of decibels and the intensity I of a sound (in watts per square meter) is given by

$$B = 10 \log_{10}\left(\frac{I}{10^{-12}}\right).$$

Use properties of logarithms to write the formula in simpler form, and determine the number of decibels in the front row of the concert when the intensity is 10^{-1} watt per square meter.

$B = 10 \log_{10} I + 120$; 110 decibels

72. ***Sound Intensity*** The relationship between the number B of decibels and the intensity I of a sound (in watts per square centimeter) is given by

$$B = 10 \log_{10}\left(\frac{I}{10^{-16}}\right).$$

Find the number of decibels in the middle of an arena during a concert when the intensity is 10^{-6} watt per square centimeter.

100 decibels

Concept Summary: Using Properties of Logarithms

What

Here are three properties of logarithms.

1. Product Property
2. Quotient Property
3. Power Property

You can use these properties to evaluate, rewrite, expand, or condense logarithmic expressions.

EXAMPLE

Show that $2(\log_2 3 + \log_2 x - \log_2 y)$ and $\log_2\left(\frac{3x}{y}\right)^2$ are equivalent.

How

Use the Product Property, the Quotient Property, and the Power Property.

EXAMPLE

$2(\log_2 3 + \log_2 x - \log_2 y)$

$= 2(\log_2 3x - \log_2 y)$ Product Property

$= 2\left(\log_2 \frac{3x}{y}\right)$ Quotient Property

$= \log_2\left(\frac{3x}{y}\right)^2$ Power Property

Why

Knowing how to use the properties of logarithms correctly can help you rewrite logarithmic expressions into forms that are easier to evaluate. Just be careful that when doing so, you identify whether you change the domain of the expression.

Exercises Within Reach®

Worked-out solutions to odd-numbered exercises at AlgebraWithinReach.com

Concept Summary Check

73. ***Rewriting an Expression*** In the solution above, is the final expression obtained by expanding or by condensing the original expression? Condensing

74. ***Using the Product Property*** Use the Product Property of Logarithms to condense the expression $\log_2 3 + \log_2 x$. $\log_2 3x$

75. ***Condensing an Expression*** What property can you use to condense the expression $2\left(\log_2 \frac{3x}{y}\right)$?
The Power Property of Logarithms

76. ***Using the Quotient Property*** Use the Quotient Property of Logarithms to expand the expression $\log_2\left(\frac{3x}{y}\right)$. $\log_2 3x - \log_2 y$

Extra Practice

Condensing a Logarithmic Expression **In Exercises 77–86, use the properties of logarithms to condense the expression.**

77. $3 \ln x + \ln y - 2 \ln z$ $\ln \frac{x^3y}{z^2}$

78. $4 \ln 2 + 2 \ln x - \frac{1}{2} \ln y$ $\ln \frac{16x^2}{\sqrt{y}}$

79. $2[\ln x - \ln(x + 1)]$ $\ln\left(\frac{x}{x+1}\right)^2$

80. $5\left[\ln x - \frac{1}{2}\ln(x + 4)\right]$ $\ln\left(\frac{x}{\sqrt{x+4}}\right)^5$

81. $\frac{1}{3}\log_5(x + 3) - \log_5(x - 6)$ $\log_5 \frac{\sqrt[3]{x+3}}{x-6}$

82. $\frac{1}{4}\log_6(x + 1) - 5\log_6(x - 4)$ $\log_6 \frac{\sqrt[4]{x+1}}{(x-4)^5}$

83. $5\log_6(c + d) - \frac{1}{2}\log_6(m - n)$ $\log_6 \frac{(c+d)^5}{\sqrt{m-n}}$

84. $2\log_5(x + y) + 3\log_5 w$ $\log_5(x + y)^2w^3$

85. $\frac{1}{5}(3\log_2 x - 4\log_2 y)$ $\log_2 \sqrt[5]{\frac{x^3}{y^4}}$

86. $\frac{1}{3}[\ln(x - 6) - 4\ln y - 2\ln z]$ $\ln \sqrt[3]{\frac{x-6}{y^4z^2}}$

Simplifying a Logarithmic Expression **In Exercises 87–92, simplify the expression.**

87. $\ln 3e^2$ $2 + \ln 3$

88. $\log_3(3^2 \cdot 4)$ $2 + \log_3 4$

89. $\log_5 \sqrt{50}$ $1 + \frac{1}{2}\log_5 2$

90. $\log_2 \sqrt{22}$ $\frac{1}{2} + \frac{1}{2}\log_2 11$

91. $\log_8 \frac{8}{x^3}$ $1 - 3\log_8 x$

92. $\ln \frac{6}{e^5}$ $\ln 6 - 5$

Molecular Transport **In Exercises 93 and 94, use the following information. The energy E (in kilocalories per gram molecule) required to transport a substance from the outside to the inside of a living cell is given by**

$$E = 1.4(\log_{10} C_2 - \log_{10} C_1)$$

where C_1 and C_2 are the concentrations of the substance outside and inside the cell, respectively.

93. Condense the expression. $E = \log_{10}\left(\frac{C_2}{C_1}\right)^{1.4}$

94. The concentration of a substance inside a cell is twice the concentration outside the cell. How much energy is required to transport the substance from outside to inside the cell? 0.4214 kilocalorie

Explaining Concepts

True or False? **In Exercises 95–100, use properties of logarithms to determine whether the equation is true or false. Justify your answer.**

95. $\log_2 8x = 3 + \log_2 x$
True. $\log_2 8x = \log_2 8 + \log_2 x = 3 + \log_2 x$

96. $\log_3(u + v) = \log_3 u + \log_3 v$
False. $\log_3(u + v)$ does not simplify.

97. $\log_3(u + v) = \log_3 u \bullet \log_3 v$
False. $\log_3(u + v)$ does not simplify.

98. $\dfrac{\log_6 10}{\log_6 3} = \log_6 10 - \log_6 3$
False. $\log_6 \frac{10}{3} = \log_6 10 - \log_6 3$

99. If $f(x) = \log_a x$, then $f(ax) = 1 + f(x)$.
See Additional Answers.

100. If $f(x) = \log_a x$, then $f(a^n) = n$.
True. $f(a^n) = \log_a a^n = n$

True or False? **In Exercises 101–104, determine whether the statement is true or false given that $f(x) = \ln x$. Justify your answer.**

101. $f(0) = 0$ False. 0 is not in the domain of f.

102. $f(2x) = \ln 2 + \ln x$ True. $f(2x) = \ln 2x = \ln 2 + \ln x$

103. $f(x - 3) = \ln x - \ln 3,\ x > 3$ False. $f(x - 3) = \ln(x - 3)$

104. $\sqrt{f(x)} = \frac{1}{2} \ln x$ False. $\sqrt{f(x)} = \sqrt{\ln x} \neq \ln \sqrt{x}$

105. ***Think About It*** Explain how you can show that $\dfrac{\ln x}{\ln y} \neq \ln \dfrac{x}{y}$. Evaluate when $x = e$ and $y = e$.

106. ***Think About It*** Without a calculator, approximate the natural logarithms of as many integers as possible between 1 and 20 using $\ln 2 \approx 0.6931$, $\ln 3 \approx 1.0986$, $\ln 5 \approx 1.6094$, and $\ln 7 \approx 1.9459$. Explain the method you used. Then verify your results with a calculator and explain any differences in the results. See Additional Answers.

Cumulative Review

In Exercises 107–112, solve the equation.

107. $\frac{2}{3}x + \frac{2}{3} = 4x - 6$ 2

108. $x^2 - 10x + 17 = 0$ $5 \pm 2\sqrt{2}$

109. $\dfrac{5}{2x} - \dfrac{4}{x} = 3$ $-\frac{1}{2}$

110. $\dfrac{1}{x} + \dfrac{2}{x - 5} = 0$ $\frac{5}{3}$

111. $|x - 4| = 3$ 1, 7

112. $\sqrt{x + 2} = 7$ 47

In Exercises 113–116, sketch the parabola. Identify the vertex and any x-intercepts. See Additional Answers.

113. $g(x) = -(x + 2)^2$

114. $f(x) = x^2 - 16$

115. $g(x) = -2x^2 + 4x - 7$

116. $h(x) = x^2 + 6x + 14$

In Exercises 117–120, find the compositions (a) $f \circ g$ and (b) $g \circ f$. Then find the domain of each composition.
See Additional Answers.

117. $f(x) = 4x + 9$
$g(x) = x - 5$

118. $f(x) = \sqrt{x}$
$g(x) = x - 3$

119. $f(x) = \dfrac{1}{x}$
$g(x) = x + 2$

120. $f(x) = \dfrac{5}{x^2 - 4}$
$g(x) = x + 1$

9.5 Solving Exponential and Logarithmic Equations

- Use one-to-one properties to solve exponential and logarithmic equations.
- Use inverse properties to solve exponential and logarithmic equations.
- Use exponential or logarithmic equations to solve application problems.

Using One-to-One Properties

One-to-One Properties of Exponential and Logarithmic Equations

Let a be a positive real number such that $a \neq 1$, and let x and y be real numbers. Then the following properties are true.

1. $a^x = a^y$ if and only if $x = y$.

2. $\log_a x = \log_a y$ if and only if $x = y$ $(x > 0,\ y > 0)$.

EXAMPLE 1 Solving Exponential and Logarithmic Equations

Solve each equation.

a. $4^{x+2} = 64$

b. $\ln(2x - 3) = \ln 11$

SOLUTION

a. $4^{x+2} = 64$ Write original equation.

$4^{x+2} = 4^3$ Rewrite with like bases.

$x + 2 = 3$ One-to-one property

$x = 1$ Subtract 2 from each side.

The solution is $x = 1$. Check this in the original equation.

b. $\ln(2x - 3) = \ln 11$ Write original equation.

$2x - 3 = 11$ One-to-one property

$2x = 14$ Add 3 to each side.

$x = 7$ Divide each side by 2.

The solution is $x = 7$. Check this in the original equation.

Exercises Within Reach® Solutions in English & Spanish and tutorial videos at AlgebraWithinReach.com

Using a One-to-One Property **In Exercises 1–14, solve the equation.**

1. $7^x = 7^3$ 3

2. $4^x = 4^6$ 6

3. $e^{1-x} = e^4$ -3

4. $e^{x+3} = e^8$ 5

5. $3^{2-x} = 81$ -2

6. $4^{2x-1} = 64$ 2

7. $6^{2x} = 36$ 1

8. $5^{3x} = 25$ $\frac{2}{3}$

9. $\ln 5x = \ln 22$ $\frac{22}{5}$

10. $\ln 4x = \ln 30$ $\frac{15}{2}$

11. $\ln(3 - x) = \ln 10$ -7

12. $\ln(2x - 3) = \ln 17$ 10

13. $\log_3(4 - 3x) = \log_3(2x + 9)$ -1

14. $\log_4(2x - 6) = \log_4(5x + 6)$ -4

Using Inverse Properties

Solving Exponential Equations

To solve an exponential equation, first isolate the exponential expression. Then *take the logarithm* of each side of the equation and solve for the variable.

Inverse Properties of Exponents and Logarithms

	Base a	*Natural Base e*
1.	$\log_a(a^x) = x$	$\ln(e^x) = x$
2.	$a^{(\log_a x)} = x$	$e^{(\ln x)} = x$

Study Tip

Remember that to evaluate a logarithm such as $\log_2 7$, you need to use the change-of-base formula.

$$\log_2 7 = \frac{\ln 7}{\ln 2} \approx 2.807$$

Similarly,

$$\log_4 9 + 3 = \frac{\ln 9}{\ln 4} + 3$$
$$\approx 1.585 + 3$$
$$= 4.585.$$

EXAMPLE 2 Solving Exponential Equations

Solve each exponential equation.

a. $2^x = 7$ **b.** $4^{x-3} = 9$ **c.** $2e^x = 10$

SOLUTION

a.

$2^x = 7$ Write original equation.

$\log_2(2^x) = \log_2 7$ Take logarithm of each side.

$x = \log_2 7$ Inverse property

The solution is $x = \log_2 7 \approx 2.807$. Check this in the original equation.

b.

$4^{x-3} = 9$ Write original equation.

$\log_4(4^{x-3}) = \log_4 9$ Take logarithm of each side.

$x - 3 = \log_4 9$ Inverse property

$x = \log_4 9 + 3$ Add 3 to each side.

The solution is $x = \log_4 9 + 3 \approx 4.585$. Check this in the original equation.

c.

$2e^x = 10$ Write original equation.

$e^x = 5$ Divide each side by 2.

$\ln(e^x) = \ln 5$ Take logarithm of each side.

$x = \ln 5$ Inverse property

The solution is $x = \ln 5 \approx 1.609$. Check this in the original equation.

Exercises Within Reach®

Solutions in English & Spanish and tutorial videos at AlgebraWithinReach.com

Solving an Exponential Equation **In Exercises 15–26, solve the exponential equation. (Round your answer to two decimal places, if necessary.)**

15. $3^x = 91$ 4.11 **16.** $4^x = 40$ 2.66 **17.** $5^x = 8.2$ 1.31 **18.** $2^x = 3.6$ 1.85

19. $3^{2-x} = 8$ 0.11 **20.** $5^{3-x} = 15$ 1.32 **21.** $10^{x+6} = 250$ −3.60 **22.** $12^{x-1} = 324$ 3.33

23. $\frac{1}{4}e^x = 5$ 3 **24.** $\frac{2}{3}e^x = 1$ 0.41 **25.** $4e^{-x} = 24$ −1.79 **26.** $6e^{-x} = 3$ 0.69

Additional Examples

Solve each equation.

a. $2^x = 32$

b. $3(2^x) = 42$

c. $4e^{2x} = 5$

Answers:

a. $x = 5$

b. $x = \frac{\ln 14}{\ln 2} \approx 3.807$

c. $x = \frac{1}{2} \ln \frac{5}{4} \approx 0.112$

EXAMPLE 3 Solving Exponential Equations

Solve each exponential equation.

a. $5 + e^{x+1} = 20$ **b.** $23 - 5e^{x+1} = 3$

SOLUTION

a.

$5 + e^{x+1} = 20$	Write original equation.
$e^{x+1} = 15$	Subtract 5 from each side.
$\ln e^{x+1} = \ln 15$	Take the logarithm of each side.
$x + 1 = \ln 15$	Inverse Property
$x = -1 + \ln 15$	Subtract 1 from each side.

The solution is $x = -1 + \ln 15 \approx 1.708$. You can check this as follows.

Check

$5 + e^{x+1} = 20$	Write original equation.
$5 + e^{-1+\ln 15+1} \stackrel{?}{=} 20$	Substitute $-1 + \ln 15$ for x.
$5 + e^{\ln 15} \stackrel{?}{=} 20$	Simplify.
$5 + 15 = 20$	Solution checks. ✓

b.

$23 - 5e^{x+1} = 3$	Write original equation.
$-5e^{x+1} = -20$	Subtract 23 from each side.
$e^{x+1} = 4$	Divide each side by -5.
$\ln e^{x+1} = \ln 4$	Take the logarithm of each side.
$x + 1 = \ln 4$	Inverse property
$x = \ln 4 - 1$	Subtract 1 from each side.

The solution is $x = \ln 4 - 1 \approx 0.386$. You can check this as follows.

Check

$23 - 5e^{x+1} = 3$	Write original equation.
$23 - 5e^{(\ln 4 - 1)+1} \stackrel{?}{=} 3$	Substitute $\ln 4 - 1$ for x.
$23 - 5e^{\ln 4} \stackrel{?}{=} 3$	Simplify.
$23 - 5(4) = 3$	Solution checks. 

Exercises Within Reach®

Solutions in English & Spanish and tutorial videos at AlgebraWithinReach.com

Solving an Exponential Equation **In Exercises 27–36, solve the exponential equation. (Round your answer to two decimal places.)**

27. $7 + e^{2-x} = 28$ -1.04

28. $5^{x+6} - 4 = 12$ -4.28

29. $4 + e^{2x} = 10$ 0.90

30. $10 + e^{4x} = 18$ 0.52

31. $17 - e^{x/4} = 14$ 4.39

32. $50 - e^{x/2} = 35$ 5.42

33. $8 - 12e^{-x} = 7$ 2.48

34. $6 - 3e^{-x} = -15$ -1.95

35. $4(1 + e^{x/3}) = 84$ 8.99

36. $50(3 - e^{2x}) = 125$ -0.35

Solving Logarithmic Equations

To solve a logarithmic equation, first isolate the logarithmic expression. Then *exponentiate* each side of the equation and solve for the variable.

EXAMPLE 4 Solving Logarithmic Equations

Solve each logarithmic equation.

a. $2 \log_4 x = 5$ **b.** $\frac{1}{4} \log_2 x = \frac{1}{2}$

SOLUTION

a. $2 \log_4 x = 5$ Write original equation.

$\log_4 x = \frac{5}{2}$ Divide each side by 2.

$4^{\log_4 x} = 4^{5/2}$ Exponentiate each side.

$x = 4^{5/2}$ Inverse Property

$x = 32$ Simplify.

The solution is $x = 32$. Check this in the original equation, as follows.

Check

$2 \log_4 x = 5$ Write original equation.

$2 \log_4(32) \stackrel{?}{=} 5$ Substitute 32 for x.

$2(2.5) \stackrel{?}{=} 5$ Use a calculator.

$5 = 5$ Solution checks. ✓

b. $\frac{1}{4} \log_2 x = \frac{1}{2}$ Write original equation.

$\log_2 x = 2$ Multiply each side by 4.

$2^{\log_2 x} = 2^2$ Exponentiate each side.

$x = 4$ Inverse property

The solution is $x = 4$. Check this in the original equation.

Exercises Within Reach®

Solutions in English & Spanish and tutorial videos at AlgebraWithinReach.com

Solving a Logarithmic Equation **In Exercises 37–46, solve the logarithmic equation. (Round your answer to two decimal places, if necessary.)**

37. $\log_{10} x = -1$ 0.1

38. $\log_{10} x = 3$ 1000

39. $4 \log_3 x = 28$ 2187

40. $6 \log_2 x = 18$ 8

41. $\frac{1}{6} \log_3 x = \frac{1}{3}$ 9

42. $\frac{1}{8} \log_5 x = \frac{1}{2}$ 625

43. $\log_{10} 4x = 2$ 27

44. $\log_3 6x = 4$ 13.5

45. $2 \log_4(x + 5) = 3$ 3

46. $5 \log_{10}(x + 2) = 15$ 998

EXAMPLE 5 **Solving Logarithmic Equations**

Study Tip

When checking approximate solutions to exponential and logarithmic equations, be aware that the check will not be exact because the solutions are approximate.

a. $3 \log_{10} x = 6$ Original equation

$\log_{10} x = 2$ Divide each side by 3.

$x = 10^2$ Exponential form

$x = 100$ Simplify.

The solution is $x = 100$. Check this in the original equation.

b. $20 \ln 0.2x = 30$ Original equation

$\ln 0.2x = 1.5$ Divide each side by 20.

$e^{\ln 0.2x} = e^{1.5}$ Exponentiate each side.

$0.2x = e^{1.5}$ Inverse property

$x = 5e^{1.5}$ Divide each side by 0.2.

The solution is $x = 5e^{1.5} \approx 22.408$. Check this in the original equation.

Additional Examples

Solve each equation.

a. $2 \log_5 3x = 4$

b. $\ln(x - 2) + \ln(2x - 3) = 2 \ln x$

Answers:

a. $x = \frac{25}{3}$

b. $x = 6$

EXAMPLE 6 **Checking for Extraneous Solutions**

$\log_6 x + \log_6(x - 5) = 2$ Original equation

$\log_6[x(x - 5)] = 2$ Condense the left side.

$x(x - 5) = 6^2$ Exponential form

$x^2 - 5x - 36 = 0$ Write in general form.

$(x - 9)(x + 4) = 0$ Factor.

$x - 9 = 0 \Rightarrow x = 9$ Set 1st factor equal to 0.

$x + 4 = 0 \Rightarrow x = -4$ Set 2nd factor equal to 0.

Check the possible solutions $x = 9$ and $x = -4$ in the original equation.

First Solution

$\log_6(9) + \log_6(9 - 5) \stackrel{?}{=} 2$

$\log_6(9 \cdot 4) \stackrel{?}{=} 2$

$\log_6 36 = 2$ ✓

Second Solution

$\log_6(-4) + \log_6(-4 - 5) \stackrel{?}{=} 2$

$\log_6(-4) + \log_6(-9) \neq 2$ ✗

Of the two possible solutions, only $x = 9$ checks. So, $x = -4$ is extraneous.

Exercises Within Reach®

Solutions in English & Spanish and tutorial videos at AlgebraWithinReach.com

Solving a Logarithmic Equation **In Exercises 47–52, solve the logarithmic equation. (Round your answer to two decimal places, if necessary.)**

47. $2 \log_{10} x = 10$ 100,000

48. $7 \log_2 x = 35$ 32

49. $3 \ln 0.1x = 4$ 37.94

50. $16 \ln 0.25x = 48$ 80.34

51. $\log_{10} x + \log_{10}(x - 3) = 1$ 5

52. $\log_{10} x + \log_{10}(x + 1) = 0$ 0.62

Application

Application **EXAMPLE 7** **Compound Interest**

A deposit of \$5000 is placed in a savings account for 2 years. The interest is compounded continuously. At the end of 2 years, the balance in the account is \$5416.44. What is the annual interest rate for this account?

SOLUTION

Formula: $A = Pe^{rt}$

Labels:
Principal $= P = 5000$ (dollars)
Amount $= A = 5416.44$ (dollars)
Time $= t = 2$ (years)
Annual interest rate $= r$ (percent in decimal form)

Equation:
$5416.44 = 5000e^{2r}$ Substitute for A, P, and t.
$1.083288 = e^{2r}$ Divide each side by 5000.
$\ln 1.083288 = \ln(e^{2r})$ Take logarithm of each side.
$0.08 \approx 2r \Rightarrow 0.04 \approx r$ Inverse property

The interest rate is about 4%. Check this solution.

Exercises Within Reach ®

Solutions in English & Spanish and tutorial videos at AlgebraWithinReach.com

Compound Interest **In Exercises 53 and 54, use the formula for continuous compounding $A = Pe^{rt}$, where A is the account balance after t years for the principal P and annual interest rate r (in decimal form).**

53. A deposit of \$10,000 is placed in a savings account for 2 years. The interest is compounded continuously. At the end of 2 years, the balance in the account is \$11,051.71. What is the annual interest rate for this account? 5%

54. A deposit of \$2500 is placed in a savings account for 2 years. The interest is compounded continuously. At the end of 2 years, the balance in the account is \$2847.07. What is the annual interest rate for this account? 6.5%

55. ***Friction*** In order to restrain an untrained horse, a trainer partially wraps a rope around a cylindrical post in a corral (see figure). The horse is pulling on the rope with a force of 200 pounds. The force F (in pounds) needed to hold back the horse is $F = 200e^{-0.5\pi\theta/180}$, where θ is the angle of wrap (in degrees). Find the smallest value of θ for which a force of 80 pounds will hold the horse by solving for θ when $F = 80$. 105°

56. ***Online Retail*** The projected online retail sales S (in billions of dollars) in the United States for the years 2009 through 2014 are modeled by the equation $S = 67.2e^{0.0944t}$, for $9 \le t \le 14$, where t is the time in years, with $t = 9$ corresponding to 2009. Find the year when S is about \$230 billion by solving for t when $S = 230$. (*Source:* Forrester Research, Inc.) 2013

Concept Summary: Solving Exponential and Logarithmic Equations

What

You can use the one-to-one properties and the inverse properties to solve exponential and logarithmic equations.

EXAMPLE

Solve each equation.

a. $3^{x-2} = 3^3$

b. $\log_6 3x = 2$

How

One-to-One Properties Let a, x, and y be real numbers ($a > 0, a \neq 1$).

1. $a^x = a^y$ if and only if $x = y$.

2. $\log_a x = \log_a y$ if and only if $x = y$ ($x > 0, y > 0$).

Inverse Properties

1. $\log_a(a^x) = x$ $\quad \ln(e^x) = x$

2. $a^{(\log_a x)} = x$ $\quad e^{(\ln x)} = x$

EXAMPLE

a. $3^{x-2} = 3^3$

$x - 2 = 3$ One-to-one property

$x = 5$ Add 2 to each side.

b. $\log_6 3x = 2$

$6^{\log_6 3x} = 6^2$ Exponentiate each side.

$3x = 6^2$ Inverse property

$x = 12$ Divide each side by 3.

Why

You will use exponential and logarithmic equations often in your study of mathematics and its applications. For instance, you will solve a logarithmic equation to determine the intensity of a sound.

Exercises Within Reach®

Worked-out solutions to odd-numbered exercises at AlgebraWithinReach.com

Concept Summary Check

57. ***Applying a One-to-One Property*** Explain why you can apply a one-to-one property to solve the equation $3^{x-2} = 3^3$. Each side of the equation is in exponential form with the same base.

58. ***Applying a One-to-One Property*** Explain how to apply a one-to-one property to the equation $3^{x-2} = 3^3$. Set the exponents equal to each other.

59. ***Applying an Inverse Property*** Explain how to apply an inverse property to the equation $\log_6 3x = 2$. Exponentiate each side of the equation using the base 6, and then rewrite the left side of the equation as $3x$.

60. ***Applying an Inverse Property*** Explain how to apply an inverse property to solve an exponential equation. When you have an exponential expression on each side of the equation, take the logarithm of each side.

Extra Practice

Solving an Exponential or a Logarithmic Equation **In Exercises 61–80, solve the equation, if possible. (Round your answer to two decimal places, if necessary.)**

61. $5^x = \frac{1}{125}$ -3

62. $3^x = \frac{1}{243}$ -5

63. $2^{x+2} = \frac{1}{16}$ -6

64. $3^{x+2} = \frac{1}{27}$ -5

65. $\log_6 3x = \log_6 18$ 6

66. $\log_5 2x = \log_5 36$ 18

67. $\log_4(x - 8) = \log_4(-4)$ No solution

68. $\log_5(2x - 3) = \log_5(4x - 5)$ No solution

69. $\frac{1}{5}(4^{x+2}) = 300$ 3.28

70. $3(2^{t+4}) = 350$ 2.87

71. $6 + 2^{x-1} = 1$ No solution

72. $24 + e^{4-x} = 22$ No solution

73. $\log_5(x + 3) - \log_5 x = 1$ 0.75

74. $\log_3(x - 2) + \log_3 5 = 3$ 7.40

75. $\log_2(x - 1) + \log_2(x + 3) = 3$ 2.46

76. $\log_6(x - 5) + \log_6 x = 2$ 9.00

77. $\log_{10} 4x - \log_{10}(x - 2) = 1$ 3.33

78. $\log_2 3x - \log_2(x + 4) = 3$ No solution

79. $\log_2 x + \log_2(x + 2) - \log_2 3 = 4$ 6.00

80. $\log_3 2x + \log_3(x - 1) - \log_3 4 = 1$ 3.00

81. ***Doubling Time*** Solve the exponential equation

$$5000 = 2500e^{0.09t}$$

for t to determine the number of years for an investment of \$2500 to double in value when compounded continuously at the rate of 9%. 7.7 years

82. ***Doubling Rate*** Solve the exponential equation

$$10{,}000 = 5000e^{10r}$$

for r to determine the interest rate required for an investment of \$5000 to double in value when compounded continuously for 10 years. 6.9%

83. ***Sound Intensity*** The relationship between the number B of decibels and the intensity I of a sound in watts per square centimeter is given by

$$B = 10 \log_{10}\left(\frac{I}{10^{-16}}\right).$$

Determine the intensity of a sound that registers 80 decibels on a decibel meter. 10^{-8} watt per square centimeter

84. ***Sound Intensity*** The relationship between the number B of decibels and the intensity I of a sound in watts per square centimeter is given by

$$B = 10 \log_{10}\left(\frac{I}{10^{-16}}\right).$$

Determine the intensity of a sound that registers 110 decibels on a decibel meter.
10^{-5} watt per square centimeter

Newton's Law of Cooling **In Exercises 85 and 86, use Newton's Law of Cooling**

$$kt = \ln \frac{T - S}{T_0 - S}$$

where T is the temperature of a body (in °F), t is the number of hours elapsed, S is the temperature of the environment, and T_0 is the initial temperature of the body.

85. A corpse is discovered in a motel room at 10 P.M., and its temperature is 85°F. Three hours later, the temperature of the corpse is 78°F. The temperature of the motel room is a constant 65°F.

(a) What is the constant k? -0.144

(b) Find the time of death assuming the body temperature is 98.6°F at the time of death. 6:24 P.M.

(c) What is the temperature of the corpse two hours after death? 90.2°F

86. A corpse is discovered in the bedroom of a home at 7 A.M., and its temperature is 92°F. Two hours later, the temperature of the corpse is 88°F. The temperature of the bedroom is a constant 68°F.

(a) What is the constant k? -0.091

(b) Find the time of death assuming the body temperature is 98.6°F at the time of death. 4:20 A.M.

(c) What is the temperature of the corpse three hours after death? 91.3°F

Explaining Concepts

87. ***Think About It*** Which equation can be solved without logarithms, $2^{x-1} = 32$ or $2^{x-1} = 30$? Explain.

$2^{x-1} = 32$, because you can write 32 as 2^5 and then apply the one-to-one property of exponential equations.

88. ***Writing*** Explain how to solve $10^{2x-1} = 5316$.

Take the common logarithm of each side of the equation and solve the resulting linear equation for x.

89. ***Writing*** In your own words, state the guidelines for solving exponential and logarithmic equations.

See Additional Answers.

90. ***Think About It*** Why is it possible for a logarithmic equation to have an extraneous solution?

Exponentiating each side of the equation may introduce extraneous solutions.

Cumulative Review

In Exercises 91–94, solve the equation by using the Square Root Property.

91. $x^2 = -25$ $\pm 5i$

92. $x^2 - 49 = 0$ ± 7

93. $9n^2 - 16 = 0$ $\pm\frac{4}{3}$

94. $(2a + 3)^2 = 18$ $-\frac{3}{2} \pm \frac{3\sqrt{2}}{2}$

In Exercises 95 and 96, solve the equation of quadratic form.

95. $t^4 - 13t^2 + 36 = 0$ $\pm 2, \pm 3$

96. $u + 2\sqrt{u} - 15 = 0$ 9

In Exercises 97–100, find the perimeter or area of the rectangle, as indicated.

	Width	*Length*	*Perimeter*	*Area*
97.	$2.5x$	x	42 in.	90 in.^{2}
98.	w	$1.6w$	78 ft	360 ft^2
99.	w	$w + 4$	56 km	192 km^2
100.	$x - 3$	x	66 cm	270 cm^2

9.6 Applications

- Use exponential equations to solve compound interest problems.
- Use exponential equations to solve growth and decay problems.
- Use logarithmic equations to solve intensity problems.

Notice that the formulas for periodic compounding and continuous compounding have five variables and four variables, respectively. Using basic algebra skills and the properties of exponents and logarithms, you are now able to solve for A, P, r, or t in either formula, given the values of all the other variables in the formula.

Compound Interest

In Section 9.1, you were introduced to two formulas for compound interest. Recall that in these formulas, A is the balance, P is the principal, r is the annual interest rate (in decimal form), and t is the time in years.

n Compoundings per Year	*Continuous Compounding*
$A = P\left(1 + \frac{r}{n}\right)^{nt}$	$A = Pe^{rt}$

Application

EXAMPLE 1 Finding the Annual Interest Rate

An investment of \$50,000 is made in an account that compounds interest quarterly. After 4 years, the balance in the account is \$71,381.07. What is the annual interest rate for this account?

Study Tip

To remove an exponent from one side of an equation, you can often raise each side of the equation to the *reciprocal* power. For instance, in Example 1, the exponent 16 is eliminated from the right side by raising each side to the reciprocal power $\frac{1}{16}$.

SOLUTION

Formula: $A = P\left(1 + \frac{r}{n}\right)^{nt}$

Labels:

Principal $= P = 50{,}000$	(dollars)
Amount $= A = 71{,}381.07$	(dollars)
Time $= t = 4$	(years)
Number of compoundings per year $= n = 4$	
Annual interest rate $= r$	(percent in decimal form)

Equation:

$71{,}381.07 = 50{,}000\left(1 + \frac{r}{4}\right)^{(4)(4)}$	Substitute for A, P, n, and t.
$1.42762 \approx \left(1 + \frac{r}{4}\right)^{16}$	Divide each side by 50,000.
$(1.42762)^{1/16} \approx 1 + \frac{r}{4}$	Raise each side to $\frac{1}{16}$ power.
$1.0225 \approx 1 + \frac{r}{4}$	Simplify.
$0.09 \approx r$	Subtract 1 from each side and then multiply each side by 4.

The annual interest rate is about 9%. Check this in the original problem.

Exercises Within Reach®

Solutions in English & Spanish and tutorial videos at AlgebraWithinReach.com

Finding an Annual Interest Rate **In Exercises 1 and 2, find the annual interest rate.**

	Principal	*Balance*	*Time*	*Compounding*	
1.	\$500	\$1004.83	10 years	Monthly	7%
2.	\$3000	\$21,628.70	20 years	Quarterly	10%

Application

EXAMPLE 2 Doubling Time for Continuous Compounding

An investment is made in a trust fund at an annual interest rate of 8.75%, compounded continuously. How long will it take for the investment to double?

Study Tip

In "doubling time" problems, you do not need to know the value of the principal P to find the doubling time. As shown in Example 2, the factor P divides out the equation and so does not affect the doubling time.

SOLUTION

$A = Pe^{rt}$ Formula for continuous compounding

$2P = Pe^{0.0875t}$ Substitute known values.

$2 = e^{0.0875t}$ Divide each side by P.

$\ln 2 = 0.0875t$ Inverse property

$\frac{\ln 2}{0.0875} = t$ Divide each side by 0.0875.

$7.92 \approx t$ Use a calculator.

It will take about 7.92 years for the investment to double.

Application

EXAMPLE 3 Finding the Type of Compounding

You deposit \$1000 in an account. At the end of 1 year, your balance is \$1077.63. The bank tells you that the annual interest rate for the account is 7.5%. How was the interest compounded?

SOLUTION

If the interest had been compounded continuously at 7.5%, then the balance would have been $A = 1000e^{(0.075)(1)} = \$1077.88$. Because the actual balance is slightly less than this, you should use the formula for interest that is compounded n times per year.

$$1077.63 = 1000\left(1 + \frac{0.075}{n}\right)^n$$

At this point, it is not clear what you should do to solve the equation for n. However, by completing a table like the one shown below, you can see that $n = 12$. So, the interest was compounded monthly.

n	1	4	12	365
$1000\left(1 + \frac{0.075}{n}\right)^n$	1075	1077.14	1077.63	1077.88

Exercises Within Reach®

Solutions in English & Spanish and tutorial videos at AlgebraWithinReach.com

Doubling Time **In Exercises 3–6, find the time for an investment to double at the given annual interest rate, compounded continuously.**

3. 8% 8.66 years **4.** 5% 13.86 years **5.** 6.75% 10.27 years **6.** 9.75% 7.11 years

Compound Interest **In Exercises 7 and 8, determine the type of compounding. Solve the problem by trying the more common types of compounding.**

	Principal	*Balance*	*Time*	*Rate*
7.	\$5000 (Yearly)	\$8954.24	10 years	6%
8.	\$5000 (Monthly)	\$9096.98	10 years	6%

In Example 3, notice that an investment of $1000 compounded monthly produced a balance of $1077.63 at the end of 1 year. Because $77.63 of this amount is interest, the **effective yield** for the investment is

$$\text{Effective yield} = \frac{\text{Year's interest}}{\text{Amount invested}} = \frac{77.63}{1000}$$
$$= 0.07763$$
$$= 7.763\%.$$

In other words, the effective yield for an investment collecting compound interest is the *simple interest rate* that would yield the same balance at the end of 1 year.

Application

EXAMPLE 4 Finding the Effective Yield

An investment is made in an account that pays 6.75% interest, compounded continuously. What is the effective yield for this investment?

SOLUTION

Notice that you do not have to know the principal or the time that the money will be left in the account. Instead, you can choose an arbitrary principal, such as $1000. Then, because effective yield is based on the balance at the end of 1 year, you can use the following formula.

$$A = Pe^{rt}$$
$$= 1000e^{0.0675(1)}$$
$$\approx 1069.83$$

Now, because the account would earn $69.83 in interest after 1 year for a principal of $1000, you can conclude that the effective yield is

$$\text{Effective yield} = \frac{69.83}{1000} \qquad \text{Divide interest by principal.}$$
$$= 0.06983 \qquad \text{Simplify.}$$
$$= 6.983\%. \qquad \text{Write in percent form.}$$

Exercises Within Reach®

Solutions in English & Spanish and tutorial videos at AlgebraWithinReach.com

Finding an Effective Yield **In Exercises 9–16, find the effective yield.**

	Rate	*Compounding*		*Rate*	*Compounding*
9.	8%	Continuously 8.33%	**10.**	9.5%	Continuously 9.97%
11.	7%	Monthly 7.23%	**12.**	8%	Yearly 8%
13.	6%	Quarterly 6.14%	**14.**	9%	Quarterly 9.31%
15.	$5\frac{1}{4}\%$	Daily 5.39%	**16.**	8%	Monthly 8.30%

17. ***Finding Effective Yield*** Is it necessary to know the principal P to find the effective yield in Exercises 9–16? Explain.

No. The effective yield is the ratio of the year's interest to the amount invested. The ratio will remain the same regardless of the amount invested.

18. ***Compounding and Effective Yield*** Consider the results of Exercises 9, 12, and 16. When the interest is compounded more frequently, what inference can you make about the difference between the effective yield and the stated annual percentage rate?

The difference increases.

Growth and Decay

Exponential Growth and Decay

The mathematical model for exponential growth or decay is given by

$$y = Ce^{kt}.$$

For this model, C is the original amount, and y is the amount after time t. The number k is a constant that is determined by the rate of growth (or decay). When $k > 0$, the model represents **exponential growth**, and when $k < 0$, it represents **exponential decay**.

Application

EXAMPLE 5 Website Growth

A college created an algebra tutoring website in 2010. The number of hits per year at the website has grown exponentially. The website had 4080 hits in 2010 and 8568 hits in 2013. Predict the number of hits in 2018.

SOLUTION

t (year)	Ce^{kt} (hits)
0	$Ce^{k(0)} = 4080$
3	$Ce^{k(3)} = 8568$
8	$Ce^{k(8)} = ?$

In the exponential growth model $y = Ce^{kt}$, let $t = 0$ represent 2010. Next, use the information given in the problem to set up the table shown. Because $Ce^{k(0)} = Ce^0 = 4080$, you can conclude that $C = 4080$. Then, using this value of C, you can solve for k as follows.

$Ce^{k(3)} = 8568$ From table

$4080e^{3k} = 8568$ Substitute 4080 for C.

$e^{3k} = 2.1$ Divide each side by 4080.

$3k = \ln 2.1$ Inverse property

$k = \frac{1}{3}\ln 2.1 \approx 0.2473$ Divide each side by 3 and simplify.

Finally, you can use this value of k in the model from the table to predict the number of hits in 2018 to be $4080e^{0.2473(8)} \approx 29{,}503$.

Exercises Within Reach®

Solutions in English & Spanish and tutorial videos at AlgebraWithinReach.com

19. ***Computer Virus*** In 2005, a computer worm called "Samy" interrupted the operations of a social networking website by inserting the payload message "but most of all, Samy is my hero" in the personal profile pages of the website's users. It is said that the "Samy" worm's message spread from 73 users to 1 million users within 20 hours.

(a) Find the constants C and k to obtain an exponential growth model $y = Ce^{kt}$ for the "Samy" worm.
$y = 73e^{0.4763t}$

(b) Use your model from part (a) to estimate how long it took the "Samy" worm to drop its payload message in 5300 personal profile pages.
9 hours

20. ***Album Downloads*** In 2006, about 27.6 million albums were purchased through downloading in the United States. In 2011, the number had increased to about 104.8 million. Use an exponential growth model to predict the number of albums that will be purchased through downloading in 2015. (*Source:* Recording Industry Association of America)
304.6 million albums

Application

EXAMPLE 6 Radioactive Decay

Radioactive iodine-125 is a by-product of some types of nuclear reactors. Its half-life is 60 days. That is, after 60 days, a given amount of radioactive iodine-125 will have decayed to half the original amount. A nuclear accident occurs and releases 20 grams of radioactive iodine-125. How long will it take for the radioactive iodine to decay to 1 gram?

SOLUTION

t (days)	Ce^{kt} (grams)
0	$Ce^{k(0)} = 20$
60	$Ce^{k(60)} = 10$
?	$Ce^{k(t)} = 1$

Use the model for exponential decay, $y = Ce^{kt}$, and the information given in the problem to set up the table shown. Because

$$Ce^{k(0)} = Ce^0 = 20$$

you can conclude that

$$C = 20.$$

Then, using this value of C, you can solve for k as follows.

$$Ce^{k(60)} = 10 \quad \text{From table}$$

$$20e^{60k} = 10 \quad \text{Substitute 20 for } C.$$

$$e^{60k} = \frac{1}{2} \quad \text{Divide each side by 20.}$$

$$60k = \ln\frac{1}{2} \quad \text{Inverse property}$$

$$k = \frac{1}{60}\ln\frac{1}{2} \quad \text{Divide each side by 60.}$$

$$k \approx -0.01155 \quad \text{Simplify.}$$

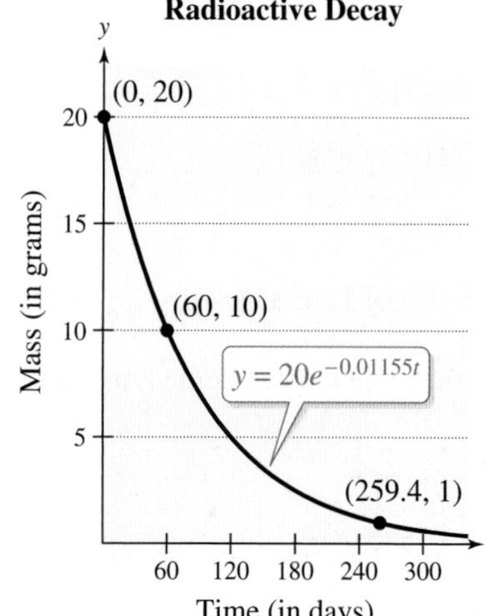

Finally, you can use this value of k in the model from the table to find the time when the amount is 1 gram, as follows.

$$Ce^{kt} = 1 \quad \text{From table}$$

$$20e^{-0.01155t} = 1 \quad \text{Substitute for } C \text{ and } k.$$

$$e^{-0.01155t} = \frac{1}{20} \quad \text{Divide each side by 20.}$$

$$-0.01155t = \ln\frac{1}{20} \quad \text{Inverse property}$$

$$t = \frac{1}{-0.01155}\ln\frac{1}{20} \quad \text{Divide each side by } -0.01155.$$

$$t \approx 259.4 \text{ days} \quad \text{Simplify.}$$

So, 20 grams of radioactive iodine-125 will decay to 1 gram after about 259.4 days. This solution is shown graphically at the left.

Exercises Within Reach®

Solutions in English & Spanish and tutorial videos at AlgebraWithinReach.com

21. ***Radioactive Decay*** Radioactive radium (^{226}Ra) has a half-life of 1620 years. Starting with 5 grams of this substance, how much will remain after 1000 years?
3.3 grams

22. ***Radioactive Decay*** Carbon 14 (^{14}C) has a half-life of 5730 years. Starting with 5 grams of this substance, how much will remain after 1000 years?
4.4 grams

Intensity Models

On the Richter scale, the magnitude R of an earthquake is given by the intensity model

$$R = \log_{10} I$$

where I is the intensity of the shock wave.

Application

EXAMPLE 7 Earthquake Intensity

On March 11, 2011, an earthquake of magnitude 9.0 in Honshu, Japan, had devastating results. It left more than 330,000 buildings destroyed or damaged, 130,900 people displaced, and 15,700 people dead.

On March 9, 2011, an earthquake near the east coast of Honshu, Japan, measured 7.3 on the Richter scale. Two days later, an earthquake in the same region measured 9.0 on the Richter scale. Compare the intensities of the two earthquakes.

SOLUTION

The intensity of the earthquake on March 9 is given as follows.

$R = \log_{10} I$ Intensity model

$7.3 = \log_{10} I$ Substitute 7.3 for R.

$10^{7.3} = I$ Inverse property

The intensity of the earthquake on March 11 can be found in a similar way.

$R = \log_{10} I$ Intensity model

$9.0 = \log_{10} I$ Substitute 9.0 for R.

$10^{9.0} = I$ Inverse property

The ratio of these two intensities is

$$\frac{I \text{ for March 11}}{I \text{ for March 9}} = \frac{10^{9.0}}{10^{7.3}}$$

$$= 10^{9.0 - 7.3}$$

$$= 10^{1.7}$$

$$\approx 50.$$

So, the earthquake on March 11 had an intensity that was about 50 times greater than the intensity of the earthquake on March 9.

Exercises Within Reach ®

Solutions in English & Spanish and tutorial videos at AlgebraWithinReach.com

Earthquake Intensity **In Exercises 23–26, compare the intensities of the two earthquakes.**

	Location	*Date*	*Magnitude*
23.	Chile	5/22/1960	9.5
	Chile	2/11/2011	6.8

The earthquake of 1960 was about 501 times as intense.

	Location	*Date*	*Magnitude*
24.	Alaska	3/28/1964	9.2
	Alaska	9/2/2011	6.8

The earthquake of 1964 was about 251 times as intense.

25.	New Zealand	6/13/2011	6.0
	New Zealand	7/6/2011	7.6

The earthquake on July 6 was about 40 times as intense.

26.	India	9/18/2011	6.9
	India	10/29/2011	3.5

The earthquake on September 18 was about 2512 times as intense.

Concept Summary: Using Exponential and Logarithmic Equations to Solve Problems

What

Many real-life situations involve exponential and logarithmic equations.

- investing (compound interest)
- population growth (**exponential growth**)
- radioactive decay (**exponential decay**)
- earthquake intensity (logarithmic equation)

EXAMPLE

How long will it take an investment to double at an annual interest rate of 6%, compounded continuously?

How

You can use a formula containing an exponential equation to solve problems involving continuous compounding.

EXAMPLE

$$A = Pe^{rt}$$
$$2P = Pe^{0.06t}$$
$$2 = e^{0.06t}$$
$$\ln 2 = 0.06t$$
$$11.55 \approx t$$

Why

Knowing the correct formula to use is the first step in solving many real-life problems involving exponential and logarithmic equations. Here are a few other formulas you should know.

Compound interest (n compoundings per year):

$$A = P\left(1 + \frac{r}{n}\right)^{nt}$$

Exponential growth or decay:

$y = Ce^{kt}$

Earthquake magnitude: $R = \log_{10} I$

Exercises Within Reach®

Worked-out solutions to odd-numbered exercises at AlgebraWithinReach.com

Concept Summary Check

27. ***Choosing a Formula*** Why is the formula $A = Pe^{rt}$ used in the solution above? It is the formula for continuously compounded interest.

28. ***Using a Formula*** Why is $2P$ substituted for A in the solution above? To determine when the balance of the investment is twice the original principal P.

29. ***Using a Formula*** Why is 0.06 substituted for r in the solution above? The variable r represents the interest rate in decimal form.

30. ***Interpreting an Equation*** Does the equation $120 = 16e^{0.2t}$ represent a situation involving exponential growth or exponential decay? Exponential growth

Extra Practice

Doubling Time **In Exercises 31–34, find the time for the investment to double.**

	Principal	*Rate*	*Compounding*		*Principal*	*Rate*	*Compounding*
31.	\$2500	7.5%	Monthly 9.27 years	**32.**	\$250	6.5%	Yearly 11.01 years
33.	\$900	$5\frac{3}{4}\%$	Quarterly 12.14 years	**34.**	\$1500	$7\frac{1}{4}\%$	Monthly 9.59 years

Exponential Growth and Decay **In Exercises 35–38, find the constant k such that the graph of $y = Ce^{kt}$ passes through the points.**

35. 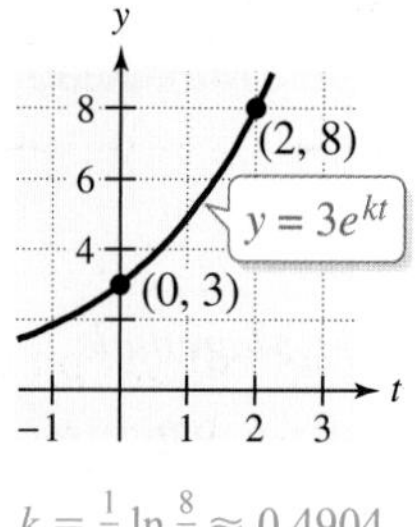

$k = \frac{1}{2}\ln\frac{8}{3} \approx 0.4904$

36. 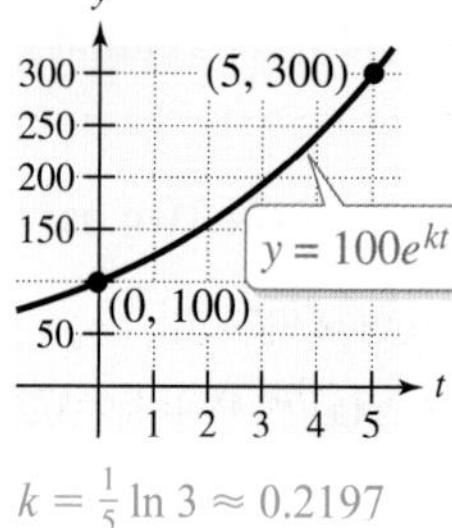

$k = \frac{1}{5}\ln 3 \approx 0.2197$

37. 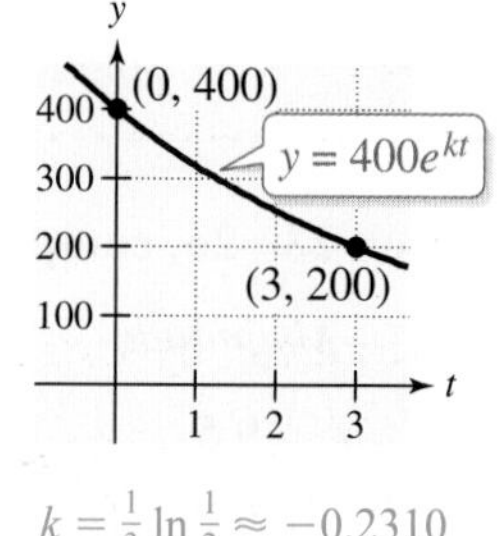

$k = \frac{1}{3}\ln\frac{1}{2} \approx -0.2310$

38. 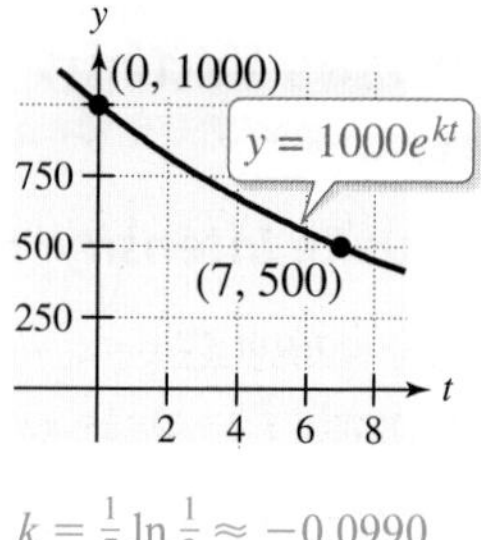

$k = \frac{1}{7}\ln\frac{1}{2} \approx -0.0990$

Acidity **In Exercises 39–42, use the acidity model $\text{pH} = -\log_{10}[H^+]$, where acidity (pH) is a measure of the hydrogen ion concentration $[H^+]$ (measured in moles of hydrogen per liter) of a solution.**

39. Find the pH of a solution that has a hydrogen ion concentration of 9.2×10^{-8}. 7.04

40. Compute the hydrogen ion concentration of a solution that has a pH of 4.7. 2.0×10^{-5}

41. A blueberry has a pH of 2.5 and a liquid antacid has a pH of 9.5. The hydrogen ion concentration of the blueberry is how many times the concentration of the antacid? 10^7 times

42. When pH of a solution decreases by 1 unit, the hydrogen ion concentration increases by what factor? 10

43. ***World Population*** Projections of the world population P (in billions) every 5 years from 2010 through 2050 can be modeled by

$$P = \frac{10.9}{1 + 0.80e^{-0.031t}}, \quad 10 \le t \le 50$$

where t represents the year, with $t = 10$ corresponding to 2010 (see figure). Use the model to predict the world population in 2022. (*Source:* U.S. Census Bureau) 7.761 billion people

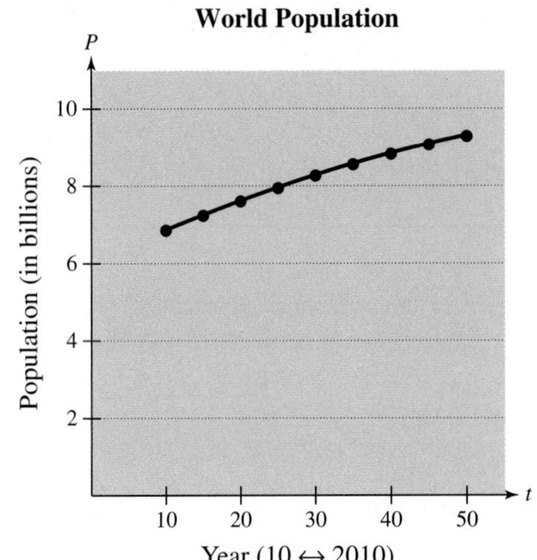

44. ***World Population*** Use the model P given in Exercise 43 to predict the world population in 2038. 8.746 billion people

45. ***Population Growth*** The population p of a species of wild rabbit t years after it is introduced into a new habitat is given by

$$p(t) = \frac{5000}{1 + 4e^{-t/6}}.$$

(a) Determine the size of the population of rabbits that was introduced into the habitat. 1000 rabbits

(b) Determine the size of the population of rabbits after 9 years. 2642 rabbits

(c) After how many years will the size of the population of rabbits be 2000? 5.88 years

46. ***Advertising Effect*** The sales S (in thousands of units) of a brand of jeans after the company spends x hundred dollars in advertising are given by

$$S = 10(1 - e^{kx}).$$

(a) Write S as a function of x given that 2500 pairs of jeans are sold when \$500 is spent on advertising. $S = 10(1 - e^{-0.0575x})$

(b) How many pairs of jeans are sold when \$700 is spent on advertising? 3314 pairs

Explaining Concepts

47. ***Structure*** Explain how to determine whether an exponential model of the form $y = Ce^{kt}$ models growth or decay. When $k > 0$, the model represents exponential growth, and when $k < 0$, the model represents exponential decay.

48. ***Structure*** The formulas for periodic and continuous compounding have the four variables A, P, r, and t in common. Explain what each variable measures. A is the balance, P is the principal, r is the annual interest rate (in decimal form), and t is the time in years.

49. ***Think About It*** For what types of compounding is the effective yield on an investment greater than the annual interest rate? Explain. When the investment is compounded more than once in a year (quarterly, monthly, daily, continuously), the effective yield is greater than the interest rate.

50. ***Number Sense*** If the reading on the Richter scale is increased by 1, the intensity of the earthquake is increased by what factor? Explain. The intensity is increased by 10 because $10^{x+1} = 10^x \cdot 10^1$.

Cumulative Review

In Exercises 51–54, solve the equation by using the Quadratic Formula.

51. $x^2 - 7x - 5 = 0$ $\frac{7}{2} \pm \frac{\sqrt{69}}{2}$

52. $x^2 + 5x - 3 = 0$ $-\frac{5}{2} \pm \frac{\sqrt{37}}{2}$

53. $3x^2 + 9x + 4 = 0$ $-\frac{3}{2} \pm \frac{\sqrt{33}}{6}$

54. $3x^2 + 4x = -2x + 5$ $-1 \pm \frac{2\sqrt{6}}{3}$

In Exercises 55–58, solve the inequality and graph the solution on the real number line. See Additional Answers.

55. $\frac{4}{x - 4} > 0$ $x > 4$

56. $\frac{x - 1}{x + 2} < 0$ $-2 < x < 1$

57. $\frac{2x}{x - 3} > 1$ $x < -3$ or $x > 3$

58. $\frac{x - 5}{x + 2} \le -1$ $-2 < x \le \frac{3}{2}$

9 Chapter Summary

	What did you learn?	Explanation and Examples	Review Exercises
9.1	Evaluate and graph exponential functions *(p. 472)*.	*Graph of* $y = a^x$ Domain: $(-\infty, \infty)$ Range: $(0, \infty)$ Intercept: $(0, 1)$ Increases from left to right. The x-axis is a horizontal asymptote.	1–10
	Evaluate the natural base e and graph natural exponential functions *(p. 477)*.	The natural exponential function is simply an exponential function with a special base, the natural base $e \approx 2.71828$.	11–16
	Use exponential functions to solve application problems *(p. 478)*.	*Formulas for compound interest:* **1.** For n compoundings per year: $A = P\left(1 + \frac{r}{n}\right)^{nt}$ **2.** For continuous compounding: $A = Pe^{rt}$	17–20
9.2	Form composite functions and find the domains of composite functions *(p. 482)*.	$(f \circ g)(x) = f(g(x))$ The domain of $(f \circ g)$ is the set of all x in the domain of g such that $g(x)$ is in the domain of f.	21–26
	Find inverse functions algebraically *(p. 485)*.	**1.** In the equation for $f(x)$, replace $f(x)$ with y. **2.** Interchange x and y. **3.** Solve for y. (If y is not a function of x, the original equation does not have an inverse.) **4.** Replace y with $f^{-1}(x)$. **5.** Verify that $f(f^{-1}(x)) = x = f^{-1}(f(x))$.	27–34
	Compare the graph of a function with the graph of its inverse *(p. 486)*.		35, 36
9.3	Evaluate and graph logarithmic functions *(p. 490)*.	Let a and x be positive real numbers such that $a \neq 1$. $y = \log_a x$ if and only if $x = a^y$ *Graph of* $y = \log_a x$ Domain: $(0, \infty)$ Range: $(-\infty, \infty)$ Intercept: $(1, 0)$ Increases from left to right. The y-axis is a vertical asymptote.	37–50

	What did you learn?	Explanation and Examples	Review Exercises
9.3	Evaluate and graph natural logarithmic functions *(p. 494)*.	The natural logarithmic function $\ln x$ is defined by $f(x) = \log_e x = \ln x$.	51–56
	Use the change-of-base formula to evaluate logarithms *(p. 495)*.	$\log_a x = \dfrac{\log_b x}{\log_b a}$ or $\log_a x = \dfrac{\ln x}{\ln a}$	57–60
9.4	Use the properties of logarithms to evaluate logarithms *(p. 500)*.	Let a be a positive real number such that $a \neq 1$, and let n be a real number. If u and v are real numbers, variables, or algebraic expressions such that $u > 0$ and $v > 0$, then the following properties are true. *Logarithm with base a* / *Natural logarithm* **1.** $\log_a(uv) = \log_a u + \log_a v$ / $\ln(uv) = \ln u + \ln v$ **2.** $\log_a \dfrac{u}{v} = \log_a u - \log_a v$ / $\ln \dfrac{u}{v} = \ln u - \ln v$ **3.** $\log_a u^n = n \log_a u$ / $\ln u^n = n \ln u$	61–66
	Rewrite, expand, or condense logarithmic expressions *(p. 502)*.	In the properties of logarithms stated above, the left side of each equation gives the condensed form and the right side gives the expanded form.	67–86
	Use the properties of logarithms to solve application problems *(p. 504)*.	The relationship between the number B of decibels and the intensity I of a sound (in watts per square centimeter) is given by $B = 10 \log_{10}\left(\dfrac{I}{10^{-16}}\right)$.	87, 88
9.5	Use one-to-one properties to solve exponential and logarithmic equations *(p. 508)*.	Let a be a positive real number such that $a \neq 1$, and let x and y be real numbers. Then the following properties are true. **1.** $a^x = a^y$ if and only if $x = y$. **2.** $\log_a x = \log_a y$ if and only if $x = y$ $(x > 0, y > 0)$.	89–94
	Use inverse properties to solve exponential and logarithmic equations *(p. 509)*.	*Base a* / *Natural base e* **1.** $\log_a(a^x) = x$ / $\ln(e^x) = x$ **2.** $a^{(\log_a x)} = x$ / $e^{(\ln x)} = x$	95–110
	Use exponential or logarithmic equations to solve application problems *(p. 513)*.	Exponential or logarithmic equations can be used to solve many real-life problems involving compound interest and the physical sciences.	111, 112
9.6	Use exponential equations to solve compound interest problems *(p. 516)*.	In the following compound interest formulas, A is the balance, P is the principal, r is the annual interest rate (in decimal form), and t is the time in years. *n Compoundings per Year:* $A = P\left(1 + \dfrac{r}{n}\right)^{nt}$ *Continuous Compounding:* $A = Pe^{rt}$	113–124
	Use exponential equations to solve growth and decay problems *(p. 519)*.	The mathematical model for exponential growth or decay is $y = Ce^{kt}$ where C is the original amount, y is the amount after time t, and k is a constant determined by the rate of growth (or decay). The model represents growth when $k > 0$ and decay when $k < 0$.	125, 126
	Use logarithmic problems to solve intensity problems *(p. 521)*.	On the Richter scale, the magnitude R of an earthquake is given by the intensity model $R = \log_{10} I$, where I is the intensity of the shock wave.	127, 128

Review Exercises

Worked-out solutions to odd-numbered exercises at AlgebraWithinReach.com

9.1

Evaluating an Exponential Function **In Exercises 1–4, evaluate the function as indicated. Use a calculator only if it is necessary or more efficient. (Round your answers to three decimal places.)**

1. $f(x) = 4^x$

(a) $x = -3$ $\frac{1}{64}$

(b) $x = 1$ 4

(c) $x = 2$ 16

2. $g(x) = 4^{-x}$

(a) $x = -2$ 16

(b) $x = 0$ 1

(c) $x = 2$ $\frac{1}{16}$

3. $g(t) = 5^{-t/3}$

(a) $t = -3$ 5

(b) $t = \pi$ 0.185

(c) $t = 6$ $\frac{1}{25}$

4. $h(s) = 1 - 3^{0.2s}$

(a) $s = 0$ 0

(b) $s = 2$ -0.552

(c) $s = \sqrt{10}$ -1.003

Graphing an Exponential Function **In Exercises 5–10, sketch the graph of the function.**
See Additional Answers.

5. $f(x) = 3^x$

6. $f(x) = 3^{-x}$

7. $f(x) = 3^x - 3$

8. $f(x) = 3^x + 5$

9. $f(x) = 3^{x+1}$

10. $f(x) = 3^{x-1}$

Evaluating an Exponential Function **In Exercises 11 and 12, use a calculator to evaluate the exponential function as indicated. (Round your answers to three decimal places, if necessary.)**

11. $f(x) = 3e^{-2x}$

(a) $x = 3$ 0.007

(b) $x = 0$ 3

(c) $x = -19$ 9.557×10^{16}

12. $g(x) = e^{x/5} + 11$

(a) $x = 12$ 22.023

(b) $x = -8$ 11.202

(c) $x = 18.4$ 50.646

Transformations of Graphs **In Exercises 13–16, use transformations to analyze and sketch the graph of the function.** See Additional Answers.

13. $y = e^{-x} + 1$

14. $y = -e^x + 1$

15. $g(x) = e^{x+2}$

16. $h(t) = e^{x-2}$

Compound Interest **In Exercises 17 and 18, complete the table to determine the balance A for P dollars invested at rate r for t years, compounded n times per year.** See Additional Answers.

n	1	4	12	365	Continuous compounding
A					

	Principal	*Rate*	*Time*
17.	$P = \$5000$	$r = 10\%$	$t = 40$ years
18.	$P = \$10{,}000$	$r = 9.5\%$	$t = 30$ years

19. ***Radioactive Decay*** After t years, 21 grams of a radioactive element with a half-life of 25 years decays to a mass y (in grams) given by $y = 21\left(\frac{1}{2}\right)^{t/25}$, $t \geq 0$. How much of the initial mass remains after 58 years?
4.21 grams

20. ***Depreciation*** The value of a truck that originally cost \$38,000 depreciates so that each year it is worth $\frac{2}{3}$ of its value from the previous year. Find a model for $V(t)$, the value of the truck after t years. Sketch a graph of the model, and determine the value of the truck 6 years after its purchase.
$V(t) = 38{,}000\left(\frac{2}{3}\right)^t$; See Additional Answers; \$3336.08

9.2

Finding the Composition of Two Functions **In Exercises 21–24, find the compositions.**

21. $f(x) = x + 2$, $g(x) = x^2$

(a) $(f \circ g)(2)$ 6

(b) $(g \circ f)(-1)$ 1

22. $f(x) = \sqrt[3]{x}$, $g(x) = x + 2$

(a) $(f \circ g)(6)$ 2 (b) $(g \circ f)(64)$ 6

23. $f(x) = \sqrt{x+1}$, $g(x) = x^2 - 1$

(a) $(f \circ g)(5)$ 5 (b) $(g \circ f)(-1)$ −1

24. $f(x) = \dfrac{1}{x-4}$, $g(x) = \dfrac{x+1}{2x}$

(a) $(f \circ g)(1)$ $-\frac{1}{3}$ (b) $(g \circ f)\left(\frac{1}{5}\right)$ $-\frac{7}{5}$

Finding the Domains of Composite Functions **In Exercises 25 and 26, find the compositions (a) $f \circ g$ and (b) $g \circ f$. Then find the domain of each composition.**

25. $f(x) = \sqrt{x+6}$, $g(x) = 2x$

(a) $(f \circ g)(x) = \sqrt{2x+6}$
Domain: $[-3, \infty)$

(b) $(g \circ f)(x) = 2\sqrt{x+6}$
Domain: $[-6, \infty)$

26. $f(x) = \dfrac{2}{x-4}$, $g(x) = x^2$

(a) $(f \circ g)(x) = \dfrac{2}{x^2-4}$
Domain: All real numbers such that $x \neq -2$ and $x \neq 2$

(b) $(g \circ f)(x) = \dfrac{4}{(x-4)^2}$
Domain: All real numbers such that $x \neq 4$

Applying the Horizontal Line Test **In Exercises 27 and 28, use the Horizontal Line Test to determine whether the function is one-to-one and so has an inverse function.**

27. $f(x) = x^2 - 25$ No

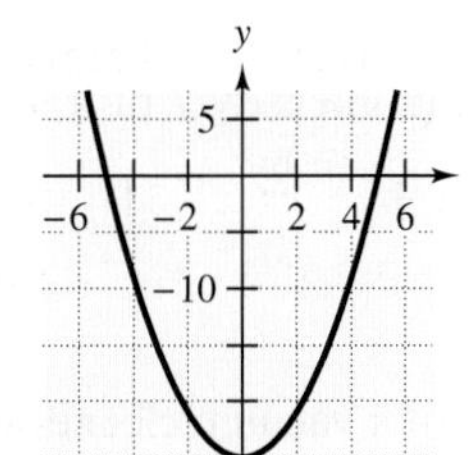

28. $f(x) = \frac{1}{4}x^3$ Yes

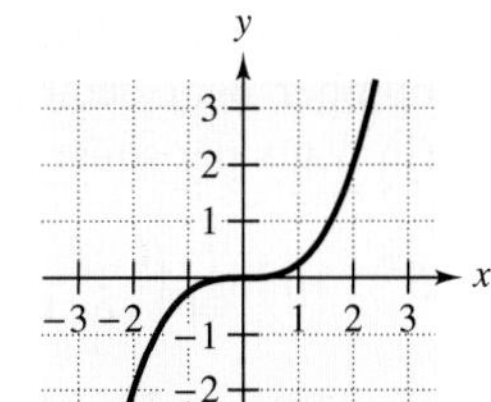

Finding an Inverse Function **In Exercises 29–34, find the inverse function.**

29. $f(x) = 3x + 4$ $f^{-1}(x) = \frac{1}{3}(x - 4)$

30. $f(x) = 2x - 3$ $f^{-1}(x) = \frac{1}{2}(x + 3)$

31. $h(x) = \sqrt{5x}$ $h^{-1}(x) = \frac{1}{5}x^2, x \geq 0$

32. $g(x) = x^2 + 2, x \geq 0$ $g^{-1}(x) = \sqrt{x-2}$

33. $f(t) = t^3 + 4$ $f^{-1}(t) = \sqrt[3]{t-4}$

34. $h(t) = \sqrt[3]{t-1}$ $h^{-1}(t) = t^3 + 1$

Verifying Inverse Functions Graphically **In Exercises 35 and 36, sketch the graphs of f and g on the same rectangular coordinate system. Plot the indicated points on the graph of f. Then plot the reflections of these points in the line $y = x$ to verify that f and g are inverse functions.**

See Additional Answers.

35. $f(x) = 3x + 4$

$g(x) = \frac{1}{3}(x - 4)$

Points: $(-1, f(-1)), (0, f(0)), (1, f(1))$

36. $f(x) = \frac{1}{3}\sqrt[3]{x}$

$g(x) = 27x^3$

Points: $(-1, f(-1)), (0, f(0)), (1, f(1))$

9.3

Evaluating a Logarithm **In Exercises 37–44, evaluate the logarithm.**

37. $\log_{10} 1000$ 3

38. $\log_{27} 3$ $\frac{1}{3}$

39. $\log_3 \frac{1}{9}$ −2

40. $\log_4 \frac{1}{16}$ −2

41. $\log_2 64$ 6

42. $\log_{10} 0.01$ −2

43. $\log_3 1$ 0

44. $\log_2 \sqrt{4}$ 1

Graphing a Logarithmic Function **In Exercises 45–50, sketch the graph of the function. Identify the vertical asymptote.** See Additional Answers.

45. $f(x) = \log_3 x$ $x = 0$

46. $f(x) = -\log_3 x$ $x = 0$

47. $f(x) = -1 + \log_3 x$ $x = 0$

48. $f(x) = 1 + \log_3 x$ $x = 0$

49. $f(x) = \log_2(x - 4)$ $x = 4$

50. $f(x) = \log_4(x + 1)$ $x = -1$

Evaluating a Natural Logarithm **In Exercises 51 and 52, evaluate the expression.**

51. $\ln e^7$ 7

52. $\ln \dfrac{1}{e^3}$ −3

Graphing a Natural Logarithmic Function **In Exercises 53–56, sketch the graph of the function.**

See Additional Answers.

53. $y = \ln(x - 3)$

54. $y = -\ln(x + 2)$

55. $y = 5 - \ln x$

56. $y = 3 + \ln x$

Using the Change-of-Base Formula **In Exercises 57–60, use (a) common logarithms and (b) natural logarithms to evaluate the expression. (Round your answer to four decimal places.)**

57. $\log_4 9$ 1.5850

58. $\log_{1/2} 5$ −2.3219

59. $\log_8 160$ 2.4406

60. $\log_3 0.28$ −1.1587

9.4

Using Properties of Logarithms **In Exercises 61–66, use $\log_5 2 \approx 0.4307$ and $\log_5 3 \approx 0.6826$ to approximate the expression. (Round your answer to four decimal places.)**

61. $\log_5 18$ 1.7959

62. $\log_5 \sqrt{6}$ 0.5567

63. $\log_5 \frac{1}{2}$ −0.4307

64. $\log_5 \frac{2}{3}$ −0.2519

65. $\log_5(12)^{2/3}$ 1.0293

66. $\log_5(5^2 \cdot 6)$ 3.1133

Expanding a Logarithmic Expression **In Exercises 67–72, use the properties of logarithms to expand the expression. (Assume all variables are positive.)**

67. $\log_4 6x^4$

$\log_4 6 + 4 \log_4 x$

68. $\log_{12} 2x^{-5}$

$\log_{12} 2 - 5 \log_{12} x$

69. $\log_5 \sqrt{x + 2}$

$\frac{1}{2} \log_5(x + 2)$

70. $\ln \sqrt[3]{\frac{x}{5}}$

$\frac{1}{3}(\ln x - \ln 5)$

71. $\ln \frac{x + 2}{x + 3}$

$\ln(x + 2) - \ln(x + 3)$

72. $\ln x(x + 4)^2$ $\ln x + 2 \ln(x + 4)$

Condensing a Logarithmic Expression **In Exercises 73–80, use the properties of logarithms to condense the expression.**

73. $5 \log_2 y$ $\log_2 y^5$

74. $-\frac{2}{3} \ln 3y$ $\ln\left(\frac{1}{3y}\right)^{2/3}, y > 0$

75. $\log_8 16x + \log_8 2x^2$ $\log_8 32x^3$

76. $\log_4 6x - \log_4 10$ $\log_4 \frac{3x}{5}$

77. $-2(\ln 2x - \ln 3)$ $\ln \frac{9}{4x^2}, x > 0$

78. $5(1 + \ln x + \ln 2)$ $5 + \ln 32x^5$

79. $4[\log_2 k - \log_2(k - t)]$ $\log_2\left(\frac{k}{k - t}\right)^4, k > t, k > 0$

80. $\frac{1}{3}(\log_8 a + 2 \log_8 b)$ $\log_8 \sqrt[3]{ab^2}$

True or False? **In Exercises 81–86, use the properties of logarithms to determine whether the equation is true or false. If it is false, state why or give an example to show that it is false.**

81. $\log_2 4x = 2 \log_2 x$

False. $\log_2 4x = \log_2 4 + \log_2 x = 2 + \log_2 x$

82. $\frac{\ln 5x}{\ln 10x} = \ln \frac{1}{2}$

False. $\ln \frac{5x}{10x} = \ln \frac{1}{2}$

83. $\log_{10} 10^{2x} = 2x$ True

84. $e^{\ln t} = t, t > 0$ True

85. $\log_4 \frac{16}{x} = 2 - \log_4 x$ True

86. $6 \ln x + 6 \ln y = \ln(xy)^6, x > 0, y > 0$ True

87. ***Sound Intensity*** The relationship between the number B of decibels and the intensity I of a sound (in watts per square meter) is given by

$$B = 10 \log_{10}\left(\frac{I}{10^{-12}}\right).$$

Find the number of decibels of a vacuum cleaner with an intensity of 10^{-4} watt per square meter.

80 decibels

88. ***Human Memory Model*** A psychologist finds that the percent p of retention in a group of subjects can be modeled by

$$p = \frac{\log_{10}(10^{68})}{\log_{10}(t + 1)^{20}}$$

where t is the time in months after the subjects' initial testing. Use properties of logarithms to write the formula in simpler form, and determine the percent of retention after 5 months.

$P = \frac{17}{5 \log_{10}(t + 1)}$; 4.37% retention

9.5

Using a One-to-one Property **In Exercises 89–94, solve the equation.**

89. $2^x = 64$ 6

90. $6^x = 216$ 3

91. $4^{x-3} = \frac{1}{16}$ 1

92. $3^{x-2} = 81$ 6

93. $\log_7(x + 6) = \log_7 12$ 6

94. $\ln(8 - x) = \ln 3$ 5

Solving an Exponential Equation **In Exercises 95–100, solve the exponential equation, if possible. (Round your answer to two decimal places.)**

95. $3^x = 500$ 5.66

96. $8^x = 1000$ 3.32

97. $2e^{0.5x} = 45$ 6.23

98. $125e^{-0.4x} = 40$ 2.85

99. $12(1 - 4^x) = 18$ No solution

100. $25(1 - e^t) = 12$ −0.65

Solving a Logarithmic Equation **In Exercises 101–110, solve the logarithmic equation. (Round your answer to two decimal places.)**

101. $\ln x = 7.25$ 1408.10

102. $\ln x = -0.5$ 0.61

103. $\log_{10} 4x = 2.1$ 31.47

104. $\log_2 2x = -0.65$ 0.32

105. $\log_3(2x + 1) = 2$ 4

106. $\log_5(x - 10) = 2$ 35

107. $\frac{1}{3}\log_2 x + 5 = 7$ 64

108. $4 \log_5(x + 1) = 4.8$ 5.90

109. $\log_3 x + \log_3 7 = 4$ 11.57

110. $2 \log_4 x - \log_4(x - 1) = 1$ 2

111. ***Compound Interest*** A deposit of \$5000 is placed in a savings account for 2 years. The interest is compounded continuously. At the end of 2 years, the balance in the account is \$5751.37. What is the annual interest rate for this account?
7%

112. ***Sound Intensity*** The relationship between the number B of decibels and the intensity I of a sound (in watts per square centimeter) is given by

$$B = 10 \log_{10}\left(\frac{I}{10^{-16}}\right).$$

Determine the intensity of a firework display I that registers 130 decibels on a decibel meter.
10^{-3} watt per square centimeter

9.6

Finding an Annual Interest Rate **In Exercises 113–118, find the annual interest rate.**

	Principal	Balance	Time	Compounding
113.	\$250 5%	\$410.90	10 years	Quarterly
114.	\$1000 6%	\$1348.85	5 years	Monthly
115.	\$5000 7.5%	\$15,399.30	15 years	Daily
116.	\$10,000 6.5%	\$35,236.45	20 years	Yearly
117.	\$1800 6.5%	\$46,422.61	50 years	Continuous
118.	\$7500 5%	\$15,877.50	15 years	Continuous

Finding an Effective Yield **In Exercises 119–124, find the effective yield.**

	Rate	Compounding
119.	5.5% 5.65%	Daily
120.	6% 6.17%	Monthly
121.	7.5% 7.71%	Quarterly
122.	8% 8%	Yearly
123.	7.5% 7.79%	Continuously
124.	3.75% 3.82%	Continuously

Radioactive Decay **In Exercises 125 and 126, determine the amount of the radioactive isotope that will remain after 1000 years.**

	Isotope	Half-Life (Years)	Initial Quantity	Amount After 1000 Years
125.	^{226}Ra	1620	3.5 g	2.282 g
126.	^{14}C	5730	10 g	8.861 g

Earthquake Intensity **In Exercises 127 and 128, compare the intensities of the two earthquakes.**
See Additional Answers.

	Location	Date	Magnitude
127.	San Francisco, California	4/18/1906	8.3
	San Francisco, California	3/5/2012	4.0
128.	Virginia	8/23/2011	5.8
	Colorado	8/23/2011	5.3

Chapter Test

Solutions in English & Spanish and tutorial videos at AlgebraWithinReach.com

Take this test as you would take a test in class. After you are done, check your work against the answers in the back of the book.

1. $f(-1) = 81$
   $f(0) = 54$
   $f\left(\frac{1}{2}\right) = 18\sqrt{6} \approx 44.09$
   $f(2) = 24$

3. (a) $(f \circ g)(x) = 18x^2 - 63x + 55$
   Domain: $(-\infty, \infty)$
   (b) $(g \circ f)(x) = -6x^2 - 3x + 5$
   Domain: $(-\infty, \infty)$

5. $(f \circ g)(x) = -\frac{1}{2}(-2x + 6) + 3$
   $= (x - 3) + 3$
   $= x$
   $(g \circ f)(x) = -2\left(-\frac{1}{2}x + 3\right) + 6$
   $= (x - 6) + 6$
   $= x$

1. Evaluate $f(t) = 54\left(\frac{2}{3}\right)^t$ when $t = -1, 0, \frac{1}{2}$, and 2.
2. Sketch a graph of the function $f(x) = 2^{x-5}$ and identify the horizontal asymptote. See Additional Answers; $y = 0$
3. Find the compositions (a) $(f \circ g)$ and (b) $(g \circ f)$. Then find the domain of each composition.
   $f(x) = 2x^2 + x \qquad g(x) = 5 - 3x$
4. Find the inverse function of $f(x) = 9x - 4$. $f^{-1}(x) = \frac{1}{9}(x + 4)$
5. Verify algebraically that the functions f and g are inverse functions of each other.
   $f(x) = -\frac{1}{2}x + 3 \qquad g(x) = -2x + 6$
6. Evaluate $\log_4 \frac{1}{256}$ without a calculator. -4
7. Describe the relationship between the graphs of $f(x) = \log_5 x$ and $g(x) = 5^x$. The graph of g is a reflection in the line $y = x$ of the graph of f.
8. Use the properties of logarithms to expand $\log_8(4\sqrt{x}/y^4)$. $\frac{2}{3} + \frac{1}{2}\log_8 x - 4\log_8 y$
9. Use the properties of logarithms to condense $\ln x - 4\ln y$. $\ln\frac{x}{y^4}, y > 0$

In Exercises 10–17, solve the equation. Round your answer to two decimal places, if necessary.

10. $\log_2 x = 5$ 32
11. $9^{2x} = 182$ 1.18
12. $400e^{0.08t} = 1200$ 13.73
13. $3\ln(2x - 3) = 10$ 15.52
14. $12(7 - 2^x) = -300$ 5
15. $\log_2 x + \log_2 4 = 5$ 8
16. $\ln x - \ln 2 = 4$ 109.20
17. $30(e^x + 9) = 300$ 0
18. Determine the balance after 20 years when \$2000 is invested at 7% compounded (a) quarterly and (b) continuously. (a) \$8012.78 (b) \$8110.40
19. Determine the principal that will yield \$100,000 when invested at 9% compounded quarterly for 25 years. \$10,806.08
20. A principal of \$500 yields a balance of \$1006.88 in 10 years when the interest is compounded continuously. What is the annual interest rate? 7%
21. A car that originally cost \$20,000 has a depreciated value of \$15,000 after 1 year. Find the value of the car when it is 5 years old by using the exponential model $y = Ce^{kt}$. \$4746.09

In Exercises 22–24, the population p of a species of fox t years after it is introduced into a new habitat is given by

$$p(t) = \frac{2400}{1 + 3e^{-t/4}}.$$

22. Determine the size of the population that was introduced into the habitat. 600 foxes
23. Determine the size of the population after 4 years. 1141 foxes
24. After how many years will the size of the population be 1200? 4.4 years

10

Conics

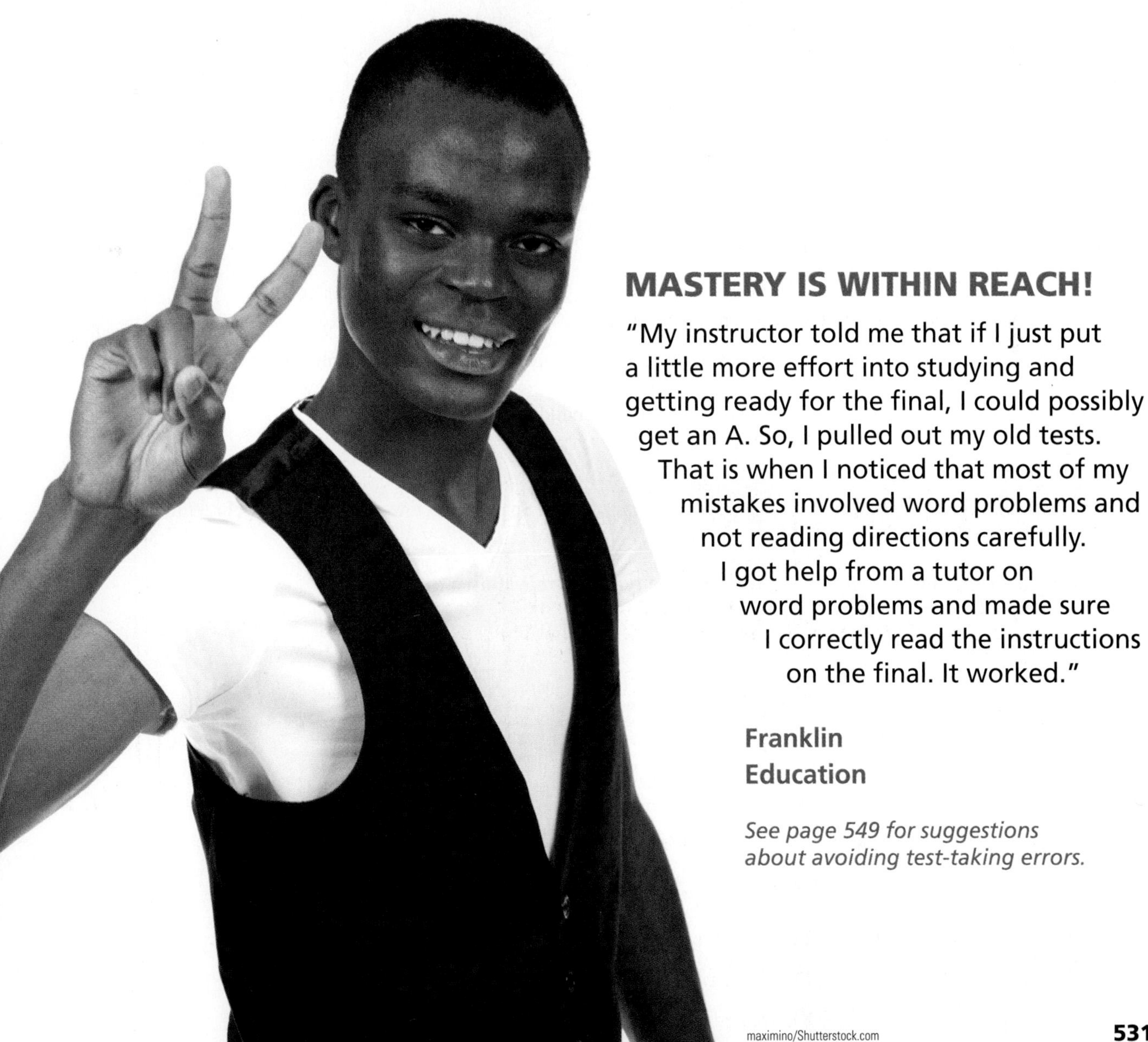

MASTERY IS WITHIN REACH!

"My instructor told me that if I just put a little more effort into studying and getting ready for the final, I could possibly get an A. So, I pulled out my old tests. That is when I noticed that most of my mistakes involved word problems and not reading directions carefully. I got help from a tutor on word problems and made sure I correctly read the instructions on the final. It worked."

Franklin
Education

See page 549 for suggestions about avoiding test-taking errors.

maximino/Shutterstock.com

10.1 Circles and Parabolas

- Recognize the four basic conics: circles, parabolas, ellipses, and hyperbolas.
- Graph and write equations of circles.
- Graph and write equations of parabolas.

The Conics

In Section 8.4, you saw that the graph of a second-degree equation of the form $y = ax^2 + bx + c$ is a parabola. A parabola is one of four types of **conics** or **conic sections**. The other three types are circles, ellipses, and hyperbolas. All four types have equations of second degree. Each figure can be obtained by intersecting a plane with a double-napped cone, as shown below.

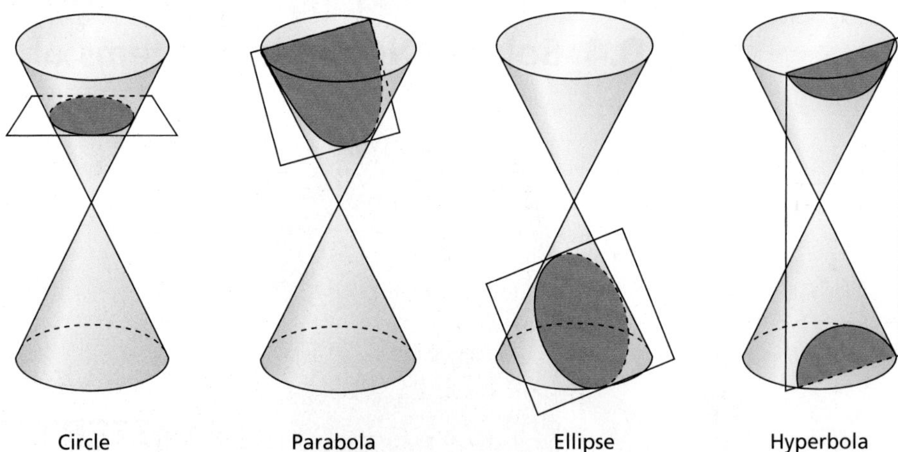

Circle Parabola Ellipse Hyperbola

Conics occur in many practical applications. Reflective surfaces in satellite dishes, flashlights, and telescopes often have a parabolic shape. The orbits of planets are elliptical, and the orbits of comets are usually elliptical or hyperbolic. Ellipses and parabolas are also used in building archways and bridges.

Exercises Within Reach ®

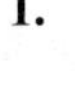

Identifying a Conic **In Exercises 1–4, Identify the type of conic shown in the graph.**

1.

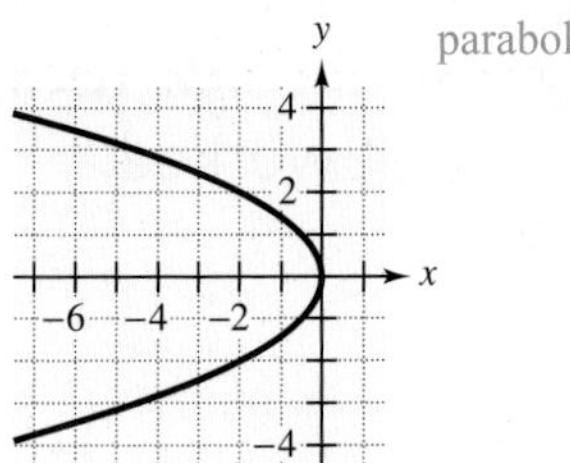

parabola

2.

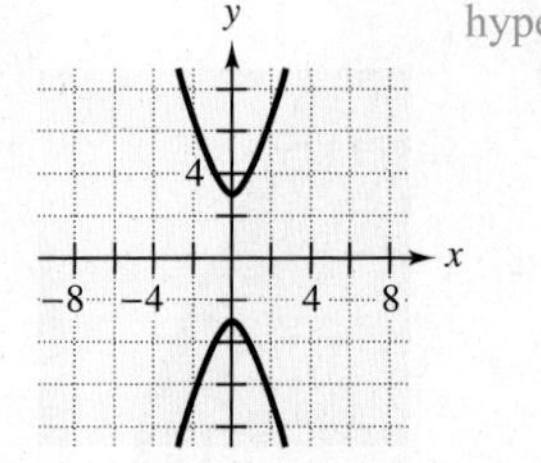

hyperbola

3.

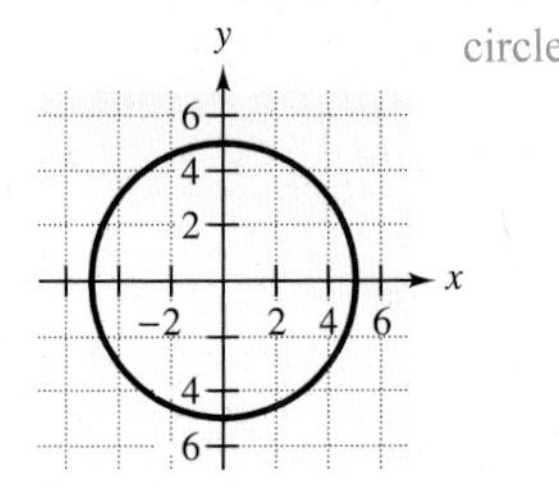

circle

4.

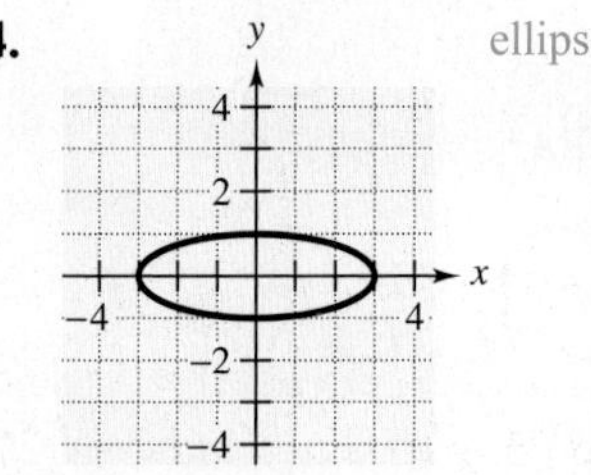

ellipse

Circles

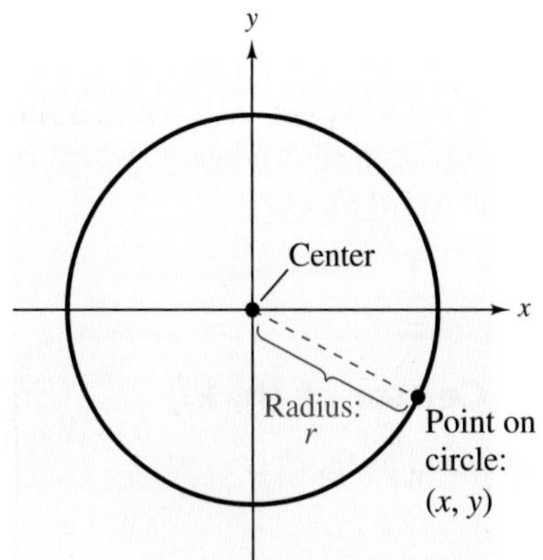

A **circle** in the rectangular coordinate system consists of all points (x, y) that are a given positive distance r from a fixed point, called the **center** of the circle. The distance r is called the **radius** of the circle.

> **Standard Equation of a Circle (Center at Origin)**
>
> The **standard form of the equation of a circle centered at the origin** is
>
> $x^2 + y^2 = r^2$. Circle with center at (0, 0)
>
> The positive number r is called the **radius** of the circle.

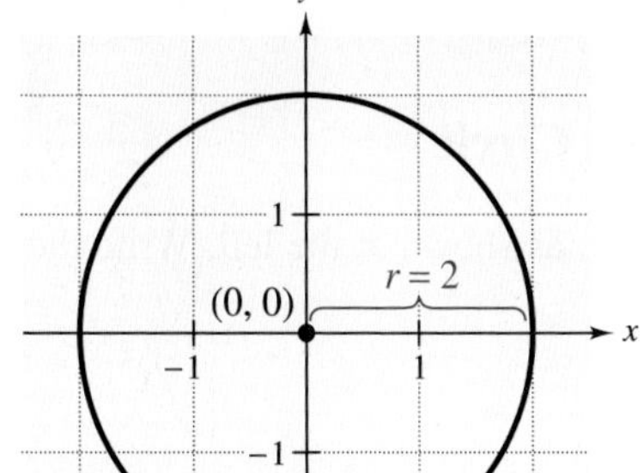

EXAMPLE 1 Writing an Equation of a Circle

Write an equation of the circle that is centered at the origin and has a radius of 2, as shown at the left.

SOLUTION

$x^2 + y^2 = r^2$	Standard form with center at (0, 0)
$x^2 + y^2 = 2^2$	Substitute 2 for r.
$x^2 + y^2 = 4$	Equation of circle

EXAMPLE 2 Sketching a Circle

Identify the radius of the circle given by $4x^2 + 4y^2 - 25 = 0$. Then sketch the circle.

SOLUTION

Begin by writing the equation in standard form.

$4x^2 + 4y^2 - 25 = 0$	Write original equation.
$4x^2 + 4y^2 = 25$	Add 25 to each side.
$x^2 + y^2 = \frac{25}{4}$	Divide each side by 4.
$x^2 + y^2 = \left(\frac{5}{2}\right)^2$	Standard form

From the standard form of the equation of this circle centered at the origin, you can see that the radius is $\frac{5}{2}$. The graph of the circle is shown at the left.

Exercises Within Reach ®

Solutions in English & Spanish and tutorial videos at AlgebraWithinReach.com

Writing an Equation of a Circle **In Exercises 5–8, write the standard form of the equation of the circle centered at the origin.**

5. Radius: 5 $x^2 + y^2 = 25$ **6.** Radius: 9 $x^2 + y^2 = 81$ **7.** Radius: $\frac{2}{3}$ $x^2 + y^2 = \frac{4}{9}$ **8.** Radius: $\frac{5}{2}$ $x^2 + y^2 = \frac{25}{4}$

Sketching a Circle **In Exercises 9–12, identify the center and radius of the circle, and sketch its graph.** See Additional Answers.

9. $x^2 + y^2 = 16$ **10.** $x^2 + y^2 = 1$ **11.** $x^2 + y^2 = 36$ **12.** $x^2 + y^2 = 15$

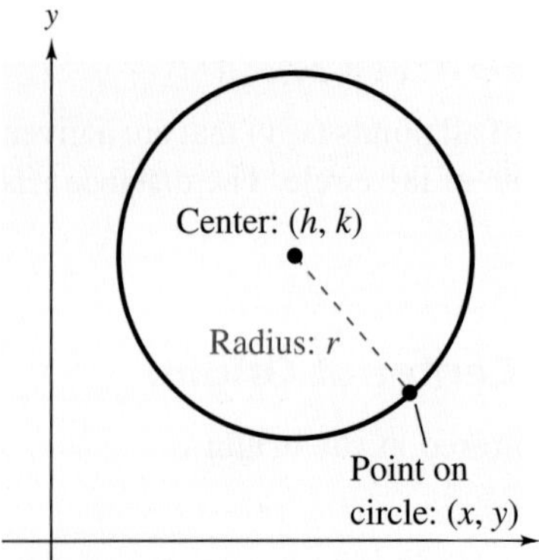

Consider a circle whose radius is r and whose center is the point (h, k), as shown at the left. Let (x, y) be any point on the circle. To find an equation for this circle, you can use a variation of the Distance Formula and write

$$\text{Radius} = r = \sqrt{(x-h)^2 + (y-k)^2}.$$ Distance Formula (See Section 3.1.)

By squaring each side of this equation, you obtain the equation shown below, which is called the *standard form of the equation of a circle centered at (h, k).*

Standard Equation of a Circle [Center at (h, k)]

The **standard form of the equation of a circle centered at (h, k)** is

$$(x-h)^2 + (y-k)^2 = r^2.$$

EXAMPLE 3 Writing an Equation of a Circle

The point $(2, 5)$ lies on a circle whose center is $(5, 1)$, as shown at the left. Write the standard form of the equation of this circle.

SOLUTION

The radius r of the circle is the distance between $(2, 5)$ and $(5, 1)$.

$r = \sqrt{(2-5)^2 + (5-1)^2}$ Distance Formula

$= \sqrt{(-3)^2 + 4^2}$ Simplify.

$= \sqrt{9 + 16}$ Simplify.

$= \sqrt{25}$ Simplify.

$= 5$ Radius

Using $(h, k) = (5, 1)$ and $r = 5$, the equation of the circle is

$(x-h)^2 + (y-k)^2 = r^2$ Standard form

$(x-5)^2 + (y-1)^2 = 5^2$ Substitute for h, k, and r.

$(x-5)^2 + (y-1)^2 = 25.$ Equation of circle

From the graph, you can see that the center of the circle is shifted five units to the right and one unit upward from the origin.

Exercises Within Reach®

Solutions in English & Spanish and tutorial videos at AlgebraWithinReach.com

Writing an Equation of a Circle **In Exercises 13–18, write the standard form of the equation of the circle centered at (h, k).**

13. Center: $(4, 3)$ $(x-4)^2 + (y-3)^2 = 100$
Radius: 10

14. Center: $(-4, 8)$ $(x+4)^2 + (y-8)^2 = 49$
Radius: 7

15. Center: $(6, -5)$ $(x-6)^2 + (y+5)^2 = 9$
Radius: 3

16. Center: $(-5, -2)$ $(x+5)^2 + (y+2)^2 = \frac{25}{4}$
Radius: $\frac{5}{2}$

17. Center: $(-2, 1)$ $(x+2)^2 + (y-1)^2 = 4$
Passes through the point $(0, 1)$

18. Center: $(8, 2)$ $(x-8)^2 + (y-2)^2 = 4$
Passes through the point $(8, 0)$

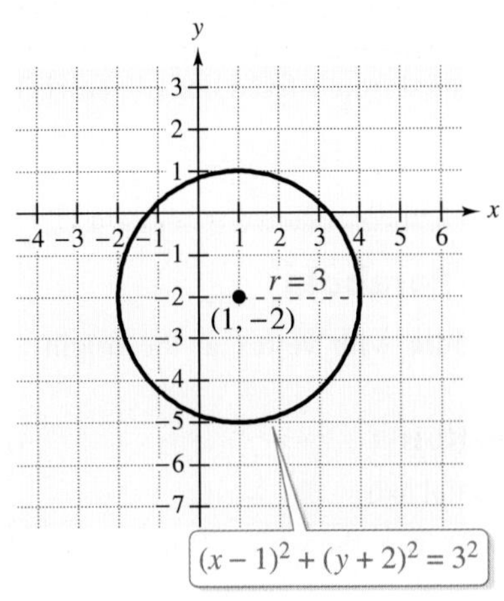

EXAMPLE 4 Writing an Equation in Standard Form

Write the equation $x^2 + y^2 - 2x + 4y - 4 = 0$ in standard form. Then sketch the circle represented by the equation.

SOLUTION

$x^2 + y^2 - 2x + 4y - 4 = 0$	Write original equation.
$(x^2 - 2x + \square) + (y^2 + 4y + \square) = 4$	Group terms.
$[x^2 - 2x + (-1)^2] + (y^2 + 4y + 2^2) = 4 + 1 + 4$	Complete the squares.
$(x - 1)^2 + (y + 2)^2 = 3^2$	Standard form

The circle is centered at $(1, -2)$ with a radius of 3, as shown at the left.

Application

EXAMPLE 5 Mechanical Drawing

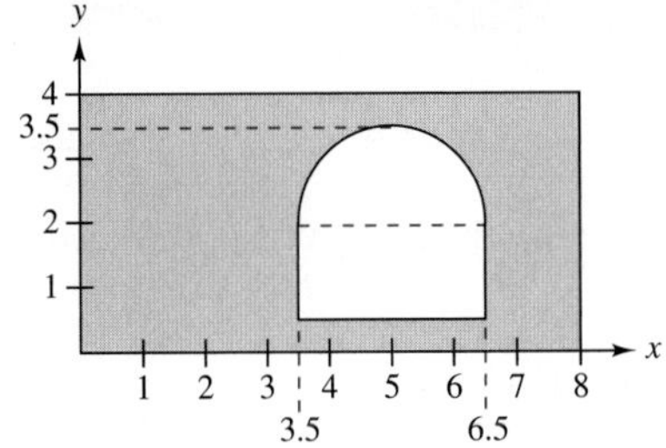

In a mechanical drawing class, you have to program a computer to model the metal piece shown at the left. Find an equation that represents the semicircular portion of the hole in the metal piece.

SOLUTION

The center of the circle is $(h, k) = (5, 2)$ and the radius of the circle is $r = 1.5$. This implies that the equation of the entire circle is

$(x - h)^2 + (y - k)^2 = r^2$	Standard form
$(x - 5)^2 + (y - 2)^2 = 1.5^2$	Substitute for h, k, and r.
$(x - 5)^2 + (y - 2)^2 = 2.25.$	Equation of circle

To find the equation of the upper portion of the circle, solve this standard equation for y.

$$(x - 5)^2 + (y - 2)^2 = 2.25$$

$$(y - 2)^2 = 2.25 - (x - 5)^2$$

$$y - 2 = \pm\sqrt{2.25 - (x - 5)^2}$$

$$y = 2 \pm \sqrt{2.25 - (x - 5)^2}$$

Finally, take the positive square root to obtain the equation of the upper portion of the circle.

$$y = 2 + \sqrt{2.25 - (x - 5)^2}$$

Exercises Within Reach ®

Solutions in English & Spanish and tutorial videos at AlgebraWithinReach.com

Sketching a Circle **In Exercises 19 and 20, identify the center and radius of the circle, and sketch its graph.** See Additional Answers.

19. $x^2 + y^2 + 2x + 6y + 6 = 0$ Center: $(-1, -3)$, $r = 2$

20. $x^2 + y^2 + 6x - 4y - 3 = 0$ Center: $(-3, 2)$, $r = 4$

Dog Leash **A dog is leashed to a side of a house. The boundary has a diameter of 80 feet.**

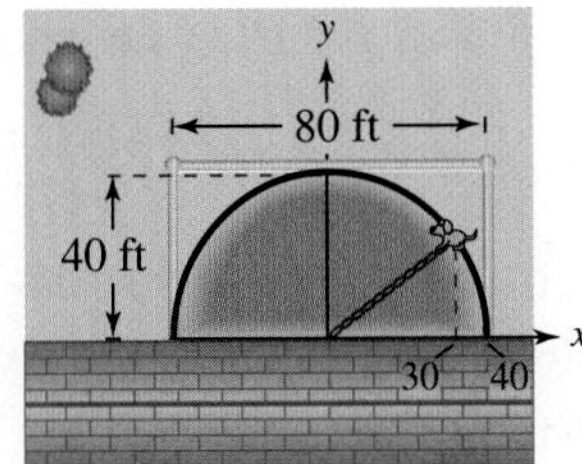

21. Write an equation that represents the semicircle. $y = \sqrt{1600 - x^2}$

22. The dog stands on the semicircle, 10 feet from the fence, as shown at the right. Use the equation you found in Exercise 21 to find how far the dog is from the house. $10\sqrt{7} \approx 26.5$ feet

Parabolas

A **parabola** is the set of all points (x, y) that are equidistant from a fixed line (**directrix**) and a fixed point (**focus**) not on the line.

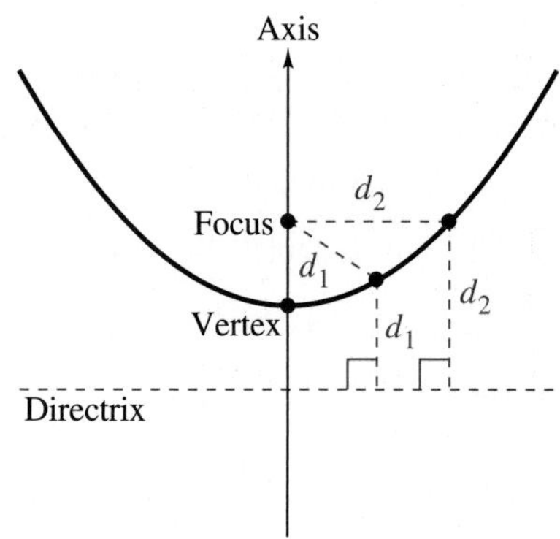

Standard Equation of a Parabola

The **standard form of the equation of a parabola** with vertex at the origin $(0, 0)$ is

$x^2 = 4py, \ p \neq 0$ Vertical axis

$y^2 = 4px, \ p \neq 0.$ Horizontal axis

The focus lies on the axis p units (*directed distance*) from the vertex. If the vertex is at (h, k), then the standard form of the equation is

$(x - h)^2 = 4p(y - k), \ p \neq 0$ Vertical axis; directrix: $y = k - p$

$(y - k)^2 = 4p(x - h), \ p \neq 0.$ Horizontal axis; directrix: $x = h - p$

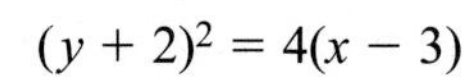

EXAMPLE 6 Writing an Equation of a Parabola

Write the standard form of the equation of the parabola with the given vertex and focus.

a. Vertex: $(0, 0)$
Focus: $(0, -2)$

b. Vertex: $(3, -2)$
Focus: $(4, -2)$

SOLUTION

a.
$$x^2 = 4py$$
$$x^2 = 4(-2)y$$
$$x^2 = -8y$$

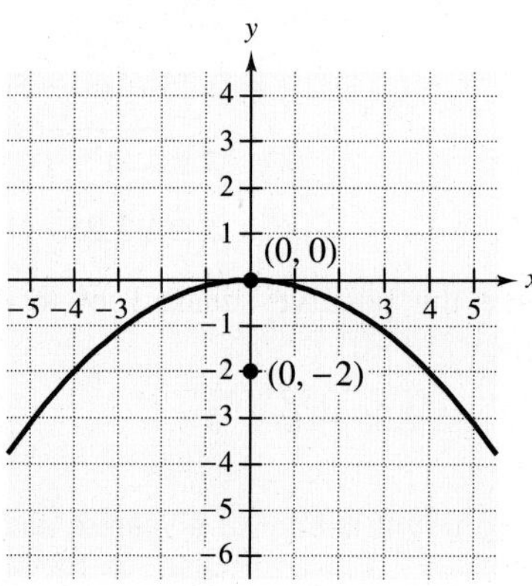

b.
$$(y - k)^2 = 4p(x - h)$$
$$[y - (-2)]^2 = 4(1)(x - 3)$$
$$(y + 2)^2 = 4(x - 3)$$

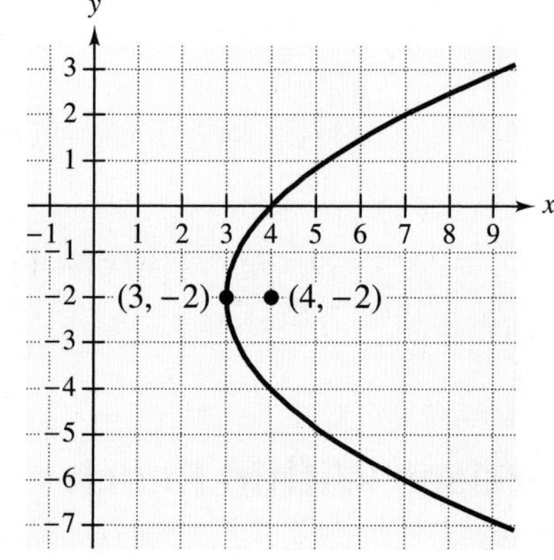

Study Tip

If the focus of a parabola is above or to the right of the vertex, p is positive. If the focus is below or to the left of the vertex, p is negative.

Exercises Within Reach®

Solutions in English & Spanish and tutorial videos at AlgebraWithinReach.com

Writing an Equation of a Parabola **In Exercises 23–26, write the standard form of the equation of the parabola with its vertex at the origin.**

23. Focus: $\left(0, -\frac{3}{2}\right)$ $x^2 = -6y$

24. Focus: $\left(\frac{5}{4}, 0\right)$ $y^2 = 5x$

25. Focus: $(-2, 0)$ $y^2 = -8x$

26. Focus: $(0, -2)$ $x^2 = -8y$

Writing an Equation of a Parabola **In Exercises 27 and 28, write the standard form of the equation of the parabola with its vertex at (h, k).**

27. Vertex: $(3, 2)$, Focus: $(1, 2)$ $(y - 2)^2 = -8(x - 3)$

28. Vertex: $(-1, 2)$, Focus: $(-1, 0)$ $(x + 1)^2 = -8(y - 2)$

EXAMPLE 7 Sketching a Parabola

Sketch the parabola given by $y = \frac{1}{8}x^2$, and identify its vertex and focus.

SOLUTION

Because the equation can be written in the standard form $x^2 = 4py$, it is a parabola whose vertex is at the origin. You can identify the focus of the parabola by writing its equation in standard form.

$y = \frac{1}{8}x^2$	Write original equation.
$\frac{1}{8}x^2 = y$	Interchange sides of the equation.
$x^2 = 8y$	Multiply each side by 8.
$x^2 = 4(2)y$	Rewrite 8 in the form $4p$.

From this standard form, you can see that $p = 2$. Because the parabola opens upward, as shown at the left, you can conclude that the focus lies $p = 2$ units above the vertex. So, the focus is (0, 2).

Parabolas occur in a wide variety of applications. For instance, a parabolic reflector can be formed by revolving a parabola around its axis. The light rays emanating from the focus of a parabolic reflector used in a flashlight are all parallel to one another, as shown at the right.

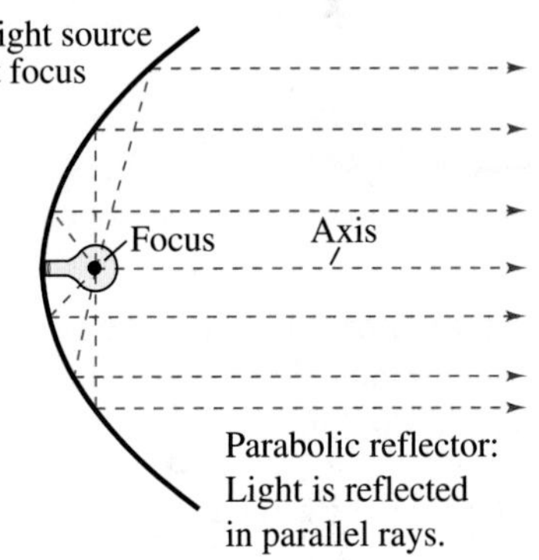

Exercises Within Reach®

Solutions in English & Spanish and tutorial videos at AlgebraWithinReach.com

Sketching a Parabola **In Exercises 29–42, sketch the parabola, and identify its vertex and focus.**

See Additional Answers.

29. $y = \frac{1}{2}x^2$
Vertex: (0, 0), Focus: $\left(0, \frac{1}{2}\right)$

30. $y = 2x^2$
Vertex: (0, 0), Focus: $\left(0, \frac{1}{8}\right)$

31. $y^2 = -10x$
Vertex: (0, 0), Focus: $\left(-\frac{5}{2}, 0\right)$

32. $y^2 = 3x$
Vertex: (0, 0), Focus: $\left(\frac{3}{4}, 0\right)$

33. $x^2 + 8y = 0$
Vertex: (0, 0), Focus: (0, −2)

34. $x + y^2 = 0$
Vertex: (0, 0), Focus: $\left(-\frac{1}{4}, 0\right)$

35. $(x - 1)^2 + 8(y + 2) = 0$
Vertex: (1, −2), Focus: (1, −4)

36. $(x + 3) + (y - 2)^2 = 0$
Vertex: (−3, 2), Focus: $\left(-\frac{13}{4}, 2\right)$

37. $\left(y + \frac{1}{2}\right)^2 = 2(x - 5)$
Vertex: $\left(5, -\frac{1}{2}\right)$, Focus: $\left(\frac{11}{2}, -\frac{1}{2}\right)$

38. $\left(x + \frac{1}{2}\right)^2 = 4(y - 3)$
Vertex: $\left(-\frac{1}{2}, 3\right)$, Focus: $\left(-\frac{1}{2}, 4\right)$

39. $y = \frac{1}{3}(x^2 - 2x + 10)$
Vertex: (1, 3), Focus: $\left(1, \frac{15}{4}\right)$

40. $4x - y^2 - 2y - 33 = 0$
Vertex: (8, −1), Focus: (9, −1)

41. $y^2 + 6y + 8x + 25 = 0$
Vertex: (−2, −3), Focus: (−4, −3)

42. $y^2 - 4y - 4x = 0$
Vertex: (−1, 2), Focus: (0, 2)

Concept Summary: Writing and Graphing Equations of Circles

What

The standard form of the equation of a **circle** with **center** (h, k) and **radius** r is

$(x - h)^2 + (y - k)^2 = r^2.$

EXAMPLE

Identify the center and radius of the circle given by

$x^2 + y^2 - 4x - 2y + 1 = 0.$

Then sketch the graph.

How

To identify the center and radius of the circle, write the equation in standard form.

EXAMPLE

$x^2 + y^2 - 4x - 2y + 1 = 0$ Original

$(x - 2)^2 + (y - 1)^2 = 4$ Standard form

From the equation, you can identify the center and radius.

Center: $(2, 1)$

Radius: $\sqrt{4} = 2$

Why

Knowing how to write equations of circles in standard form allows you to easily sketch the graphs of circles. For example, here is the graph of the circle $(x - 2)^2 + (y - 1)^2 = 4$.

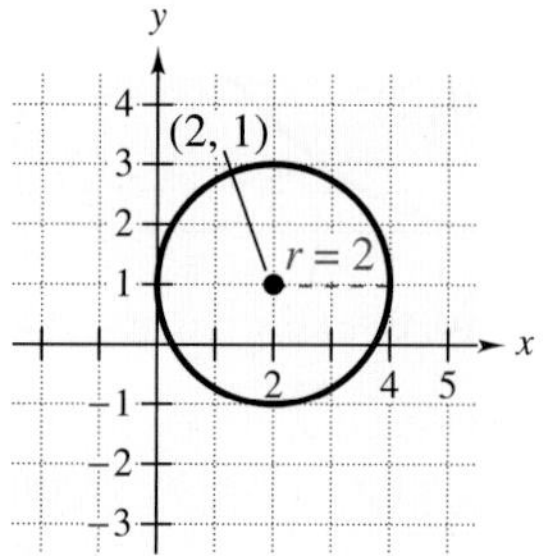

You can use a similar process to sketch parabolas.

Exercises Within Reach®

Worked-out solutions to odd-numbered exercises at AlgebraWithinReach.com

Concept Summary Check

43. ***Reasoning*** Explain how to identify the center and radius of the circle given by the equation $x^2 + y^2 - 36 = 0$. See Additional Answers.

44. ***Translation*** Is the center of the circle given by the equation $(x + 2)^2 + (y + 4)^2 = 20$ shifted two units to the right and four units upward from the origin? Explain your reasoning. See Additional Answers.

45. ***Standard Form*** Which standard form of the equation of a parabola should you use to write an equation for a parabola with vertex $(2, -3)$ and focus $(2, 1)$? Explain your reasoning. See Additional Answers.

46. ***Think About It*** Given the equation of a parabola, explain how to determine whether the parabola opens upward, downward, to the right, or to the left.
See Additional Answers.

Extra Practice

Sketching a Circle **In Exercises 47–50, identify the center and radius of the circle, and sketch its graph.**
See Additional Answers.

47. $\left(x + \frac{9}{4}\right)^2 + (y - 4)^2 = 16$ Center: $\left(-\frac{9}{4}, 4\right)$, $r = 4$

48. $(x - 5)^2 + \left(y + \frac{3}{4}\right)^2 = 1$ Center: $\left(5, -\frac{3}{4}\right)$, $r = 1$

49. $x^2 + y^2 + 10x - 4y - 7 = 0$ Center: $(-5, 2)$, $r = 6$

50. $x^2 + y^2 - 14x + 8y + 56 = 0$ Center: $(7, -4)$, $r = 3$

51. ***Observation Wheel*** Write an equation that represents the circular wheel of the Singapore Flyer in Singapore, which has a diameter of 150 meters. Place the origin of the rectangular coordinate system at the center of the wheel.
$x^2 + y^2 = 75^2$

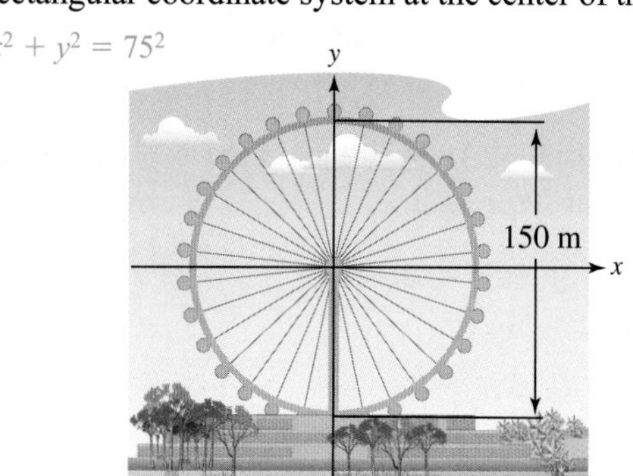

52. ***Mirror*** Write an equation that represents the circular mirror, with a diameter of 3 feet, shown in the figure. The wall hangers of the mirror are shown as two points on the circle. Use the equation to determine the height of the left wall hanger.
$(x - 10)^2 + (y - 5.5)^2 = 2.25$; 6.8 feet

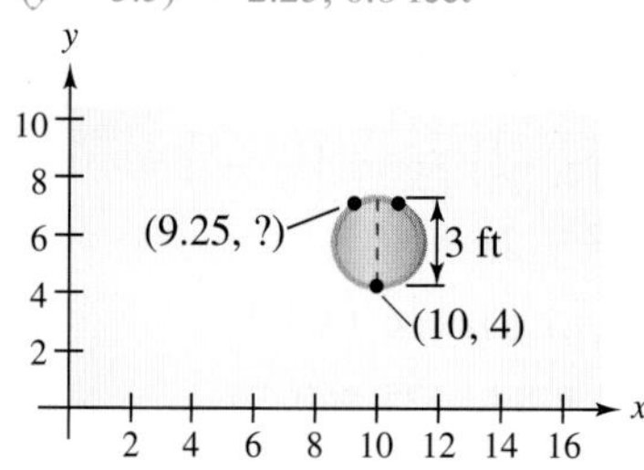

53. ***Suspension Bridge*** Each cable of a suspension bridge is suspended (in the shape of a parabola) between two towers that are 120 meters apart, and the top of each tower is 20 meters above the roadway. The cables touch the roadway at the midpoint between the two towers (see figure).

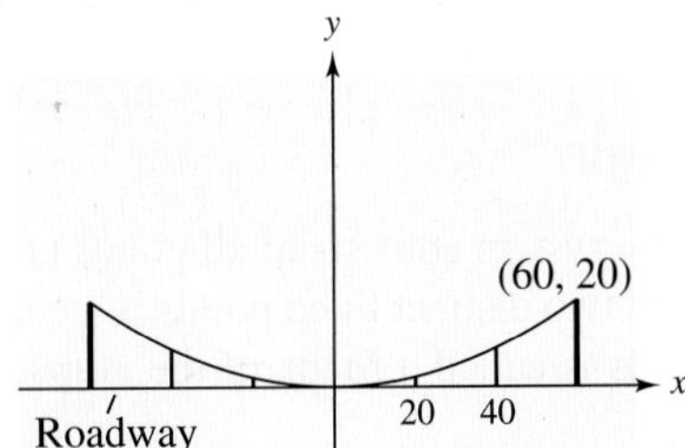

(a) Write an equation that represents the parabolic shape of each cable. $x^2 = 180y$

(b) Complete the table by finding the height of the suspension cables y over the roadway at a distance of x meters from the center of the bridge.

x	0	20	40	60
y	0	$2\frac{2}{9}$	$8\frac{8}{9}$	20

54. ***Beam Deflection*** A simply supported beam is 16 meters long and has a load at the center (see figure). The deflection of the beam at its center is 3 centimeters. Assume that the shape of the deflected beam is parabolic.

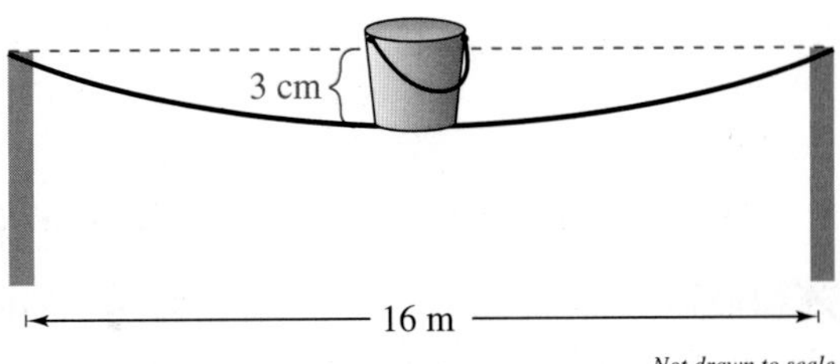

(a) Write an equation of the parabola. (Assume that the origin is at the center of the deflected beam.) $x^2 = \frac{640,000}{3}y$

(b) How far from the center of the beam is the deflection equal to 1 centimeter? 462 centimeters

Explaining Concepts

55. ***Precision*** The point $(-4, 3)$ lies on a circle with center $(-1, 1)$. does the point $(3, 2)$ lie on the same circle? Explain your reasoning. See Additional Answers.

56. ***Logic*** A student claims that

$$x^2 + y^2 - 6y = -5$$

does not represent a circle. Is the student correct? Explain your reasoning. See Additional Answers.

57. ***Reasoning*** Is y a function of x in the equation $y^2 = 6x$? Explain.
No. For each $x > 0$, there correspond two values of y.

58. ***Reasoning*** Is it possible for a parabola to intersect its directrix? Explain. See Additional Answers.

59. ***Think About It*** If the vertex and focus of a parabola are on a horizontal line, is the directrix of the parabola vertical? Explain. Yes. The directrix of a parabola is perpendicular to the line through the vertex and focus.

Cumulative Review

In Exercises 60–65, solve the equation by completing the square.

60. $x^2 + 4x = 6$ $-2 \pm \sqrt{10}$

61. $x^2 + 6x = -4$ $-3 \pm \sqrt{5}$

62. $x^2 - 2x - 3 = 0$ $-1, 3$

63. $4x^2 - 12x - 10 = 0$ $\frac{3}{2} \pm \frac{\sqrt{19}}{2}$

64. $2x^2 + 5x - 8 = 0$ $-\frac{5}{4} \pm \frac{\sqrt{89}}{4}$

65. $9x^2 - 12x = 14$ $\frac{2}{3} \pm \sqrt{2}$

In Exercises 66–69, use the properties of logarithms to expand the expression. (Assume all variables are positive.)

66. $\log_8 x^{10}$ $10 \log_8 x$

67. $\log_{10} \sqrt{xy^3}$ $\frac{1}{2}(\log_{10} x + 3 \log_{10} y)$

68. $\ln 5x^2y$ $\ln 5 + 2 \ln x + \ln y$

69. $\ln \frac{x}{y^4}$ $\ln x - 4 \ln y$

In Exercises 70–73, use the properties of logarithms to condense the expression.

70. $\log_{10} x + \log_{10} 6$ $\log_{10} 6x$

71. $2 \log_3 x - \log_3 y$ $\log_3 \frac{x^2}{y}$

72. $3 \ln x + \ln y - \ln 9$ $\ln \frac{x^3y}{9}$

73. $4(\ln x + \ln y) - \ln(x^4 + y^4)$ $\ln \frac{x^4y^4}{x^4 + y^4}$

10.2 Ellipses

- Graph and write equations of ellipses centered at the origin.
- Graph and write equations of ellipses centered at (h, k).

Ellipses Centered at the Origin

An **ellipse** in the rectangular coordinate system consists of all points (x, y) such that the sum of the distances between (x, y) and two distinct fixed points is a constant, as shown at the left. Each of the two fixed points is called a **focus** of the ellipse. (The plural of focus is *foci*.)

$d_1 + d_2$ is constant.

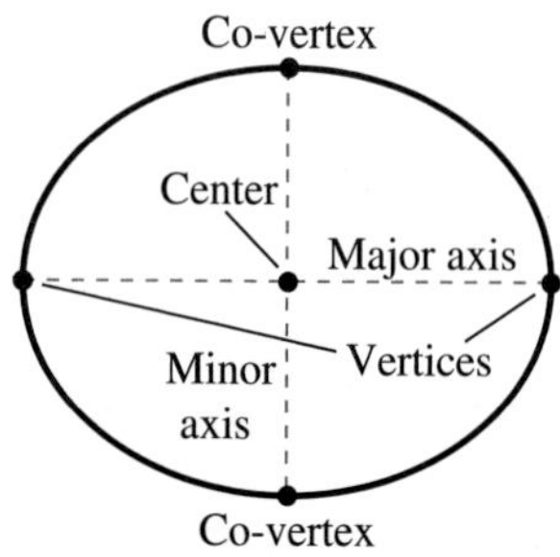

Mention that the major axis is the longer axis.

Standard Equation of an Ellipse (Center at Origin)

The **standard form of the equation of an ellipse centered at the origin** with major and minor axes of lengths $2a$ and $2b$ is

$$\frac{x^2}{a^2} + \frac{y^2}{b^2} = 1 \quad \text{or} \quad \frac{x^2}{b^2} + \frac{y^2}{a^2} = 1, \qquad 0 < b < a.$$

The vertices lie on the major axis, a units from the center, and the co-vertices lie on the minor axis, b units from the center, as shown below.

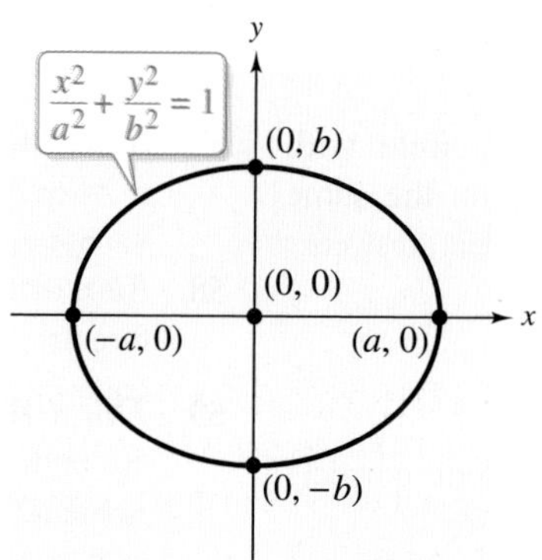

Major axis is horizontal.
Minor axis is vertical.

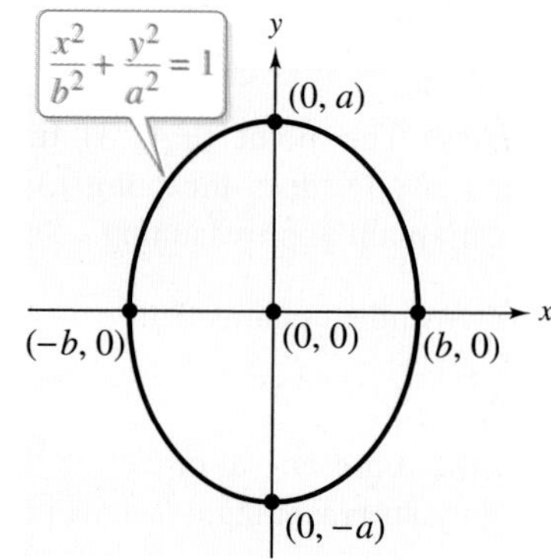

Major axis is vertical.
Minor axis is horizontal.

Exercises Within Reach®

Solutions in English & Spanish and tutorial videos at AlgebraWithinReach.com

Identifying the Standard Form **In Exercises 1–4, determine whether the standard form of the equation of the ellipse is $\frac{x^2}{a^2} + \frac{y^2}{b^2} = 1$ or $\frac{x^2}{b^2} + \frac{y^2}{a^2} = 1$.**

1. 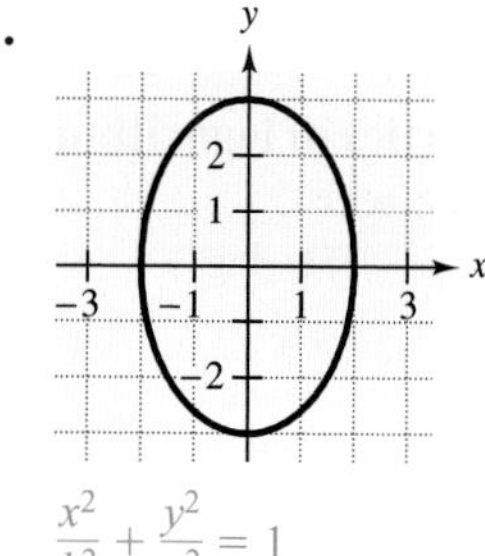

$\frac{x^2}{b^2} + \frac{y^2}{a^2} = 1$

2. 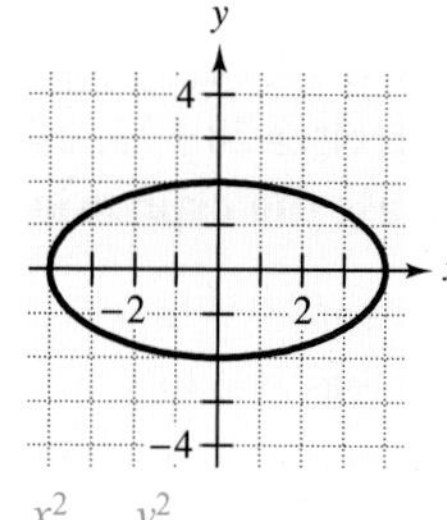

$\frac{x^2}{a^2} + \frac{y^2}{b^2} = 1$

3. 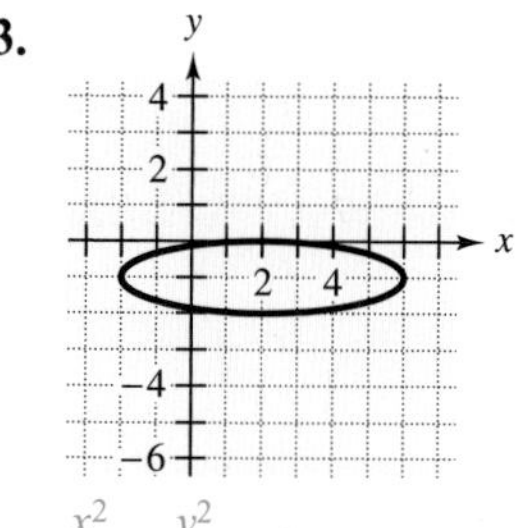

$\frac{x^2}{a^2} + \frac{y^2}{b^2} = 1$

4. $\frac{x^2}{b^2} + \frac{y^2}{a^2} = 1$

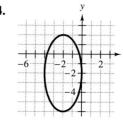

EXAMPLE 1 Writing the Standard Equation of an Ellipse

Write an equation of the ellipse that is centered at the origin, with vertices $(-3, 0)$ and $(3, 0)$ and co-vertices $(0, -2)$ and $(0, 2)$.

SOLUTION

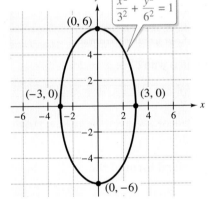

Begin by plotting the vertices and co-vertices, as shown at the left. The center of the ellipse is $(0, 0)$ and the major axis is horizontal. So, the standard form of the equation is

$$\frac{x^2}{a^2} + \frac{y^2}{b^2} = 1 \qquad \text{Major axis is horizontal.}$$

The distance from the center to either vertex is $a = 3$, and the distance from the center to either co-vertex is $b = 2$. So, the standard form of the equation of the ellipse is

$$\frac{x^2}{3^2} + \frac{y^2}{2^2} = 1. \qquad \text{Standard form}$$

EXAMPLE 2 Sketching an Equation

Sketch the ellipse given by $4x^2 + y^2 = 36$. Identify the vertices and co-vertices.

SOLUTION

To sketch an ellipse, it helps first to write its equation in standard form.

$$4x^2 + y^2 = 36 \qquad \text{Write original equation.}$$

$$\frac{x^2}{9} + \frac{y^2}{36} = 1 \qquad \text{Divide each side by 36 and simplify.}$$

$$\frac{x^2}{3^2} + \frac{y^2}{6^2} = 1 \qquad \text{Standard form}$$

Because the denominator of the y^2-term is larger than the denominator of the x^2-term, you can conclude that the major axis is vertical. Moreover, because $a = 6$, the vertices are $(0, -6)$ and $(0, 6)$. Finally, because $b = 3$, the co-vertices are $(-3, 0)$ and $(3, 0)$, as shown at the left.

Exercises Within Reach®

Solutions in English & Spanish and tutorial videos at AlgebraWithinReach.com

Writing the Standard Equation of an Ellipse **In Exercises 5–8, write the standard form of the equation of the ellipse centered at the origin.**

	Vertices	*Co-vertices*			*Vertices*	*Co-vertices*	
5.	$(-4, 0), (4, 0)$	$(0, -3), (0, 3)$	$\frac{x^2}{16} + \frac{y^2}{9} = 1$	**6.**	$(-2, 0), (2, 0)$	$(0, -1), (0, 1)$	$\frac{x^2}{4} + \frac{y^2}{1} = 1$
7.	$(0, -6), (0, 6)$	$(-3, 0), (3, 0)$	$\frac{x^2}{9} + \frac{y^2}{36} = 1$	**8.**	$(0, -5), (0, 5)$	$(-1, 0), (1, 0)$	$\frac{x^2}{1} + \frac{y^2}{25} = 1$

Sketching an Ellipse **In Exercises 9–12, sketch the ellipse. Identify the vertices and co-vertices.**

See Additional Answers.

9. $\frac{x^2}{16} + \frac{y^2}{4} = 1$ Vertices: $(\pm 4, 0)$, Co-vertices: $(0, \pm 2)$

10. $\frac{x^2}{9} + \frac{y^2}{25} = 1$ Vertices: $(0, \pm 5)$, Co-vertices: $(\pm 3, 0)$

11. $4x^2 + y^2 = 4$ Vertices: $(0, \pm 2)$, Co-vertices: $(\pm 1, 0)$

12. $4x^2 + 9y^2 = 36$ Vertices: $(\pm 3, 0)$, Co-vertices: $(0, \pm 2)$

Ellipses Centered at (h, k)

Standard Equation of an Ellipse [Center at (h, k)]

The **standard form of the equation of an ellipse centered at (h, k)** with major and minor axes of lengths $2a$ and $2b$, where $0 < b < a$, is

$$\frac{(x-h)^2}{a^2} + \frac{(y-k)^2}{b^2} = 1 \qquad \text{Major axis is horizontal.}$$

or

$$\frac{(x-h)^2}{b^2} + \frac{(y-k)^2}{a^2} = 1. \qquad \text{Major axis is vertical.}$$

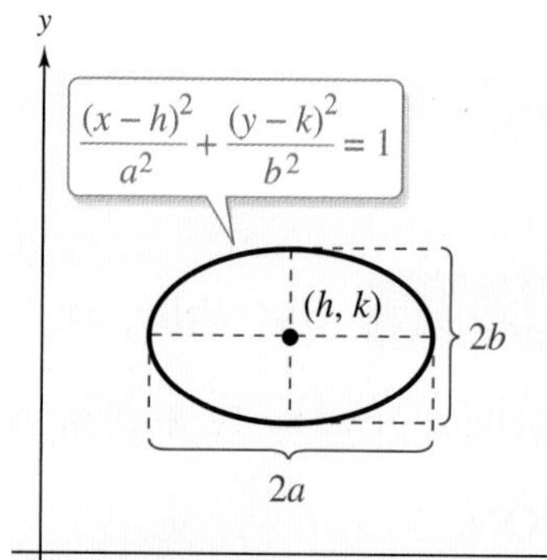

Major axis is horizontal.
Minor axis is vertical.

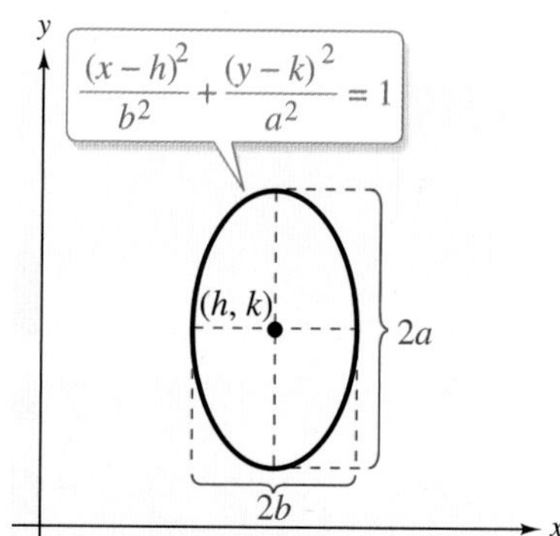

Major axis is vertical.
Minor axis is horizontal.

The foci lie on the major axis, c units form the center, with $c^2 = a^2 - b^2$.

Exercises Within Reach ®

Solutions in English & Spanish and tutorial videos at AlgebraWithinReach.com

Identifying the Standard Form **In Exercises 13–16, determine whether the standard form of the equation of the ellipse is $\frac{(x-h)^2}{a^2} + \frac{(y-k)^2}{b^2} = 1$ or $\frac{(x-h)^2}{b^2} + \frac{(y-k)^2}{a^2} = 1$.**

13.

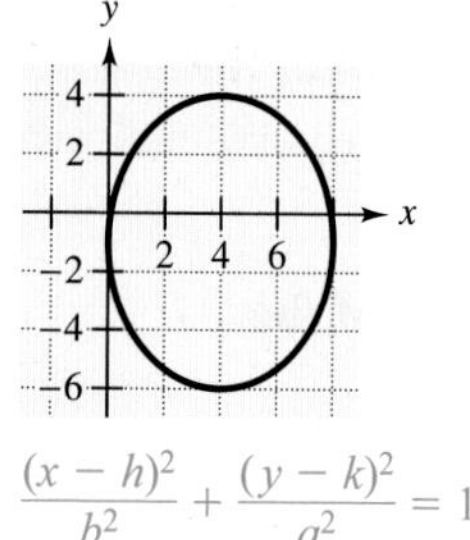

$\frac{(x-h)^2}{b^2} + \frac{(y-k)^2}{a^2} = 1$

14.

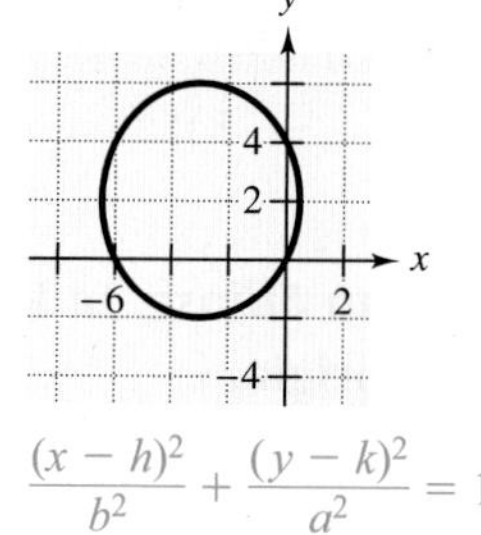

$\frac{(x-h)^2}{b^2} + \frac{(y-k)^2}{a^2} = 1$

15.

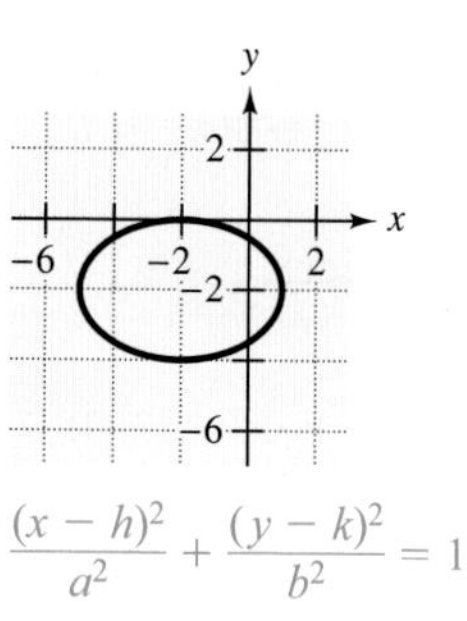

$\frac{(x-h)^2}{a^2} + \frac{(y-k)^2}{b^2} = 1$

16.

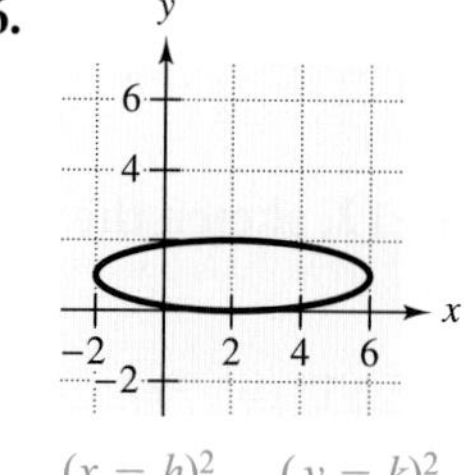

$\frac{(x-h)^2}{a^2} + \frac{(y-k)^2}{b^2} = 1$

EXAMPLE 3 Writing the Standard Equation of an Ellipse

Write the standard form of the equation of the ellipse with vertices $(-2, 2)$ and $(4, 2)$ and co-vertices $(1, 3)$ and $(1, 1)$, as shown below.

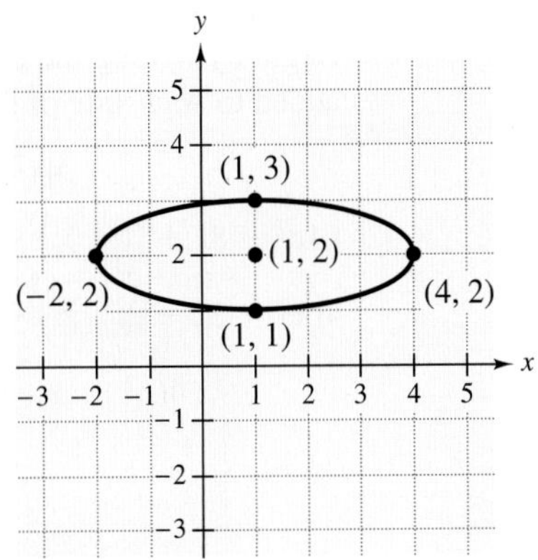

SOLUTION

Because the vertices are $(-2, 2)$ and $(4, 2)$, the center of the ellipse is $(h, k) = (1, 2)$. The distance from the center to either vertex is $a = 3$, and the distance from the center to either co-vertex is $b = 1$. Because the major axis is horizontal, the standard form of the equation is

$$\frac{(x-h)^2}{a^2} + \frac{(y-k)^2}{b^2} = 1. \qquad \text{Major axis is horizontal.}$$

Substitute the values of h, k, a, and b to obtain

$$\frac{(x-1)^2}{3^2} + \frac{(y-2)^2}{1^2} = 1. \qquad \text{Standard form}$$

From the graph, you can see that the center of the ellipse is shifted one unit to the right and two units upward from the origin.

Exercises Within Reach®

Solutions in English & Spanish and tutorial videos at AlgebraWithinReach.com

Writing the Standard Equation of an Ellipse **In Exercises 17–20, write the standard form of the equation of the ellipse.**

17.

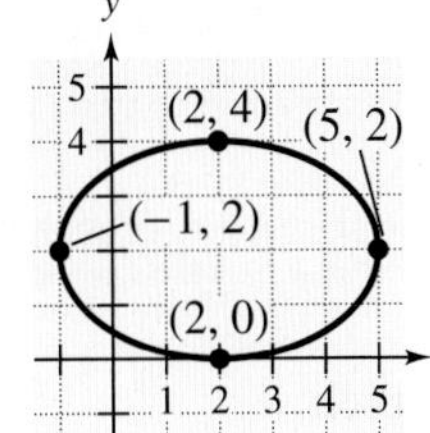

$\frac{(x-2)^2}{9} + \frac{(y-2)^2}{4} = 1$

18.

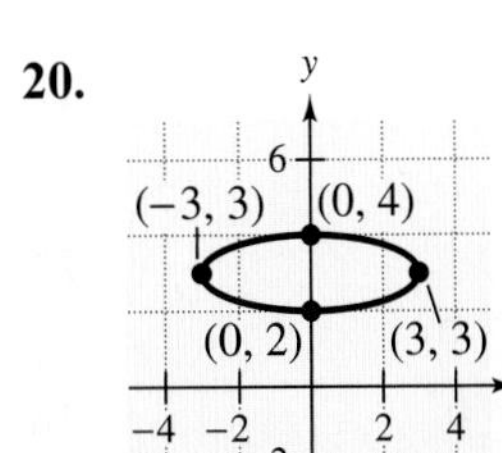

$\frac{(x+5)^2}{4} + \frac{(y+2)^2}{16} = 1$

19.

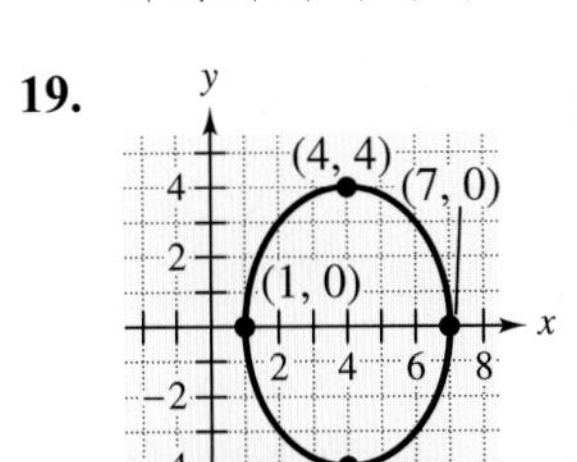

$\frac{(x-4)^2}{9} + \frac{y^2}{16} = 1$

20.

$\frac{x^2}{9} + (y-3)^2 = 1$

EXAMPLE 4 Sketching an Ellipse

Sketch the ellipse given by $4x^2 + y^2 - 8x + 6y + 9 = 0$.

SOLUTION

Begin by writing the equation in standard form. In the fourth step, note that 9 and 4 are added to *each* side of the equation.

$$4x^2 + y^2 - 8x + 6y + 9 = 0 \quad \text{Write original equation.}$$

$$(4x^2 - 8x + \square) + (y^2 + 6y + \square) = -9 \quad \text{Group terms.}$$

$$4(x^2 - 2x + \square) + (y^2 + 6y + \square) = -9 \quad \text{Factor 4 out of } x\text{-terms.}$$

$$4(x^2 - 2x + 1) + (y^2 + 6y + 9) = -9 + 4(1) + 9 \quad \text{Complete the squares.}$$

$$4(x - 1)^2 + (y + 3)^2 = 4 \quad \text{Simplify.}$$

$$\frac{(x - 1)^2}{1} + \frac{(y + 3)^2}{4} = 1 \quad \text{Divide each side by 4.}$$

$$\frac{(x - 1)^2}{1^2} + \frac{(y + 3)^2}{2^2} = 1 \quad \text{Standard form}$$

You can see that the center of the ellipse is $(h, k) = (1, -3)$. Because the denominator of the y^2-term is larger than the denominator of the x^2-term, you can conclude that the major axis is vertical. Because the denominator of the x^2-term is $b^2 = 1^2$, you can locate the endpoints of the minor axis one unit to the right of the center and one unit to the left of the center. Because the denominator of the y^2-term is $a^2 = 2^2$, you can locate the endpoints of the major axis two units up from the center and two units down from the center, as shown below on the left. To complete the graph, sketch an oval shape that is determined by the vertices and co-vertices, as shown below on the right.

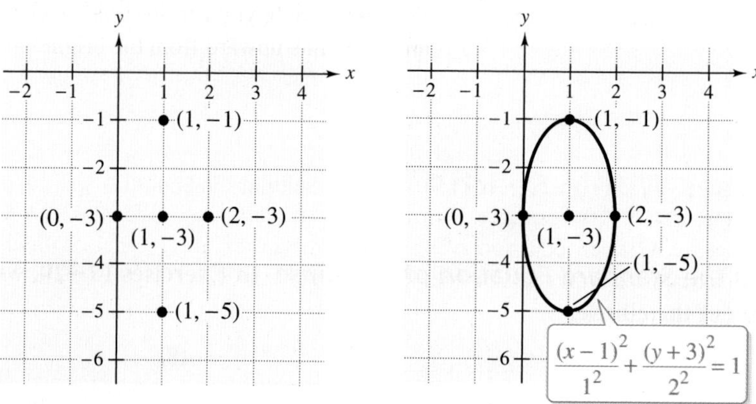

Exercises Within Reach®

Solutions in English & Spanish and tutorial videos at AlgebraWithinReach.com

Sketching an Ellipse **In Exercises 21–26, find the center and vertices of the ellipse, and sketch its graph.** See Additional Answers.

21. $4(x - 2)^2 + 9(y + 2)^2 = 36$
Center: $(2, -2)$, Vertices: $(-1, -2), (5, -2)$

22. $2(x + 5)^2 + 8(y - 2)^2 = 72$
Center: $(-5, 2)$, Vertices: $(-11, 2), (1, 2)$

23. $9x^2 + 4y^2 + 36x - 24y + 36 = 0$
Center: $(-2, 3)$, Vertices: $(-2, 6), (-2, 0)$

24. $9x^2 + 4y^2 - 36x + 8y + 31 = 0$
Center: $(2, -1)$, Vertices: $\left(2, -\frac{5}{2}\right), \left(2, \frac{1}{2}\right)$

25. $25x^2 + 9y^2 - 200x + 54y + 256 = 0$
Center: $(4, -3)$, Vertices: $(4, -8), (4, 2)$

26. $25x^2 + 16y^2 - 150x - 128y + 81 = 0$
Center: $(3, 4)$, Vertices: $(3, -1), (3, 9)$

Application EXAMPLE 5 Semielliptical Archway

You are responsible for designing a semielliptical archway, as shown at the left. The height of the archway is 10 feet, and its width is 30 feet. Write an equation that represents the archway and use the equation to sketch an accurate diagram of the archway.

SOLUTION

To make the equation simple, place the origin at the center of the ellipse. This means that the standard form of the equation is

$$\frac{x^2}{a^2} + \frac{y^2}{b^2} = 1. \qquad \text{Major axis is horizontal.}$$

Because the major axis is horizontal, it follows that $a = 15$ and $b = 10$, which implies that the equation of the ellipse is

$$\frac{x^2}{15^2} + \frac{y^2}{10^2} = 1. \qquad \text{Standard form}$$

To write an equation that represents the archway, solve this equation for y.

$$\frac{x^2}{225} + \frac{y^2}{100} = 1 \qquad \text{Simplify denominators.}$$

$$\frac{y^2}{100} = 1 - \frac{x^2}{225} \qquad \text{Subtract } \frac{x^2}{225} \text{ from each side.}$$

$$y^2 = 100\left(1 - \frac{x^2}{225}\right) \qquad \text{Multiply each side by 100.}$$

$$y = 10\sqrt{1 - \frac{x^2}{225}} \qquad \text{Take the positive square root of each side.}$$

To make an accurate sketch of the archway, calculate several y-values, as shown in the table. Then use the values in the table to sketch the archway, as shown at the left.

x	y
± 15	0
± 12.5	5.53
± 10	7.45
± 7.5	8.66
± 5	9.43
± 2.5	9.86
0	10

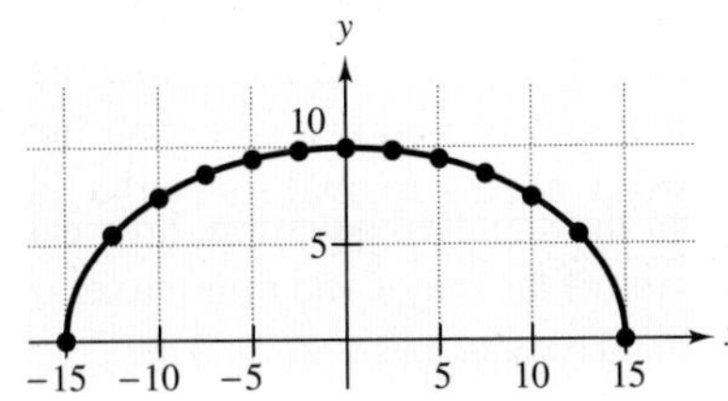

Exercises Within Reach®

Solutions in English & Spanish and tutorial videos at AlgebraWithinReach.com

27. ***Motorsports*** Most sprint car dirt tracks are elliptical in shape. Write an equation of an elliptical race track with a major axis that is 1230 feet long and a minor axis that is 580 feet long.
$\frac{x^2}{615^2} + \frac{y^2}{290^2} = 1$ or $\frac{x^2}{290^2} + \frac{y^2}{615^2} = 1$

Jeff Schultes/Shutterstock.com

28. ***Bicycle Chainwheel*** The pedals of a bicycle drive a chainwheel, which drives a smaller sprocket wheel on the rear axle (see figure). Many chainwheels are circular. Some, however, are slightly elliptical, which tends to make pedaling easier. Write an equation of an elliptical chainwheel with a major axis that is 8 inches long and a minor axis that is $7\frac{1}{2}$ inches long.
$\frac{x^2}{16} + \frac{y^2}{225/16} = 1$ or $\frac{x^2}{225/16} + \frac{y^2}{16} = 1$

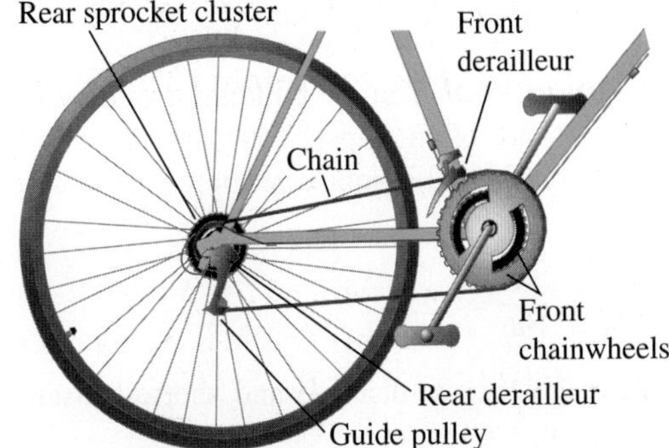

Concept Summary: Writing Equations of Ellipses

What

The standard form of the equation of an **ellipse** centered at the origin is

$$\frac{x^2}{a^2} + \frac{y^2}{b^2} = 1 \text{ or } \frac{x^2}{b^2} + \frac{y^2}{a^2} = 1,$$

where $0 < b < a$.

The **vertices** lie on the **major axis**, a units from the **center**. The **co-vertices** lie on the **minor axis**, b units from the center.

EXAMPLE

Write an equation of the ellipse centered at the origin, with vertices $(0, -6)$ and $(0, 6)$ and co-vertices $(-3, 0)$ and $(3, 0)$.

How

To write the standard form of the equation of an ellipse centered at $(0, 0)$, follow these steps.

1. Plot the vertices and co-vertices.
2. Decide whether the major axis is horizontal or vertical.
3. Identify the values of a and b.

EXAMPLE

From the graph of the ellipse (see figure), you can see the vertices and co-vertices. You can also see that the major axis is vertical, $a = 6$, and $b = 3$. So an equation of the ellipse is $\frac{x^2}{3^2} + \frac{y^2}{6^2} = 1$.

Why

Understanding how to write an equation of an ellipse centered at the origin will help you when you learn how to write an equation of an ellipse *not* centered at the origin.

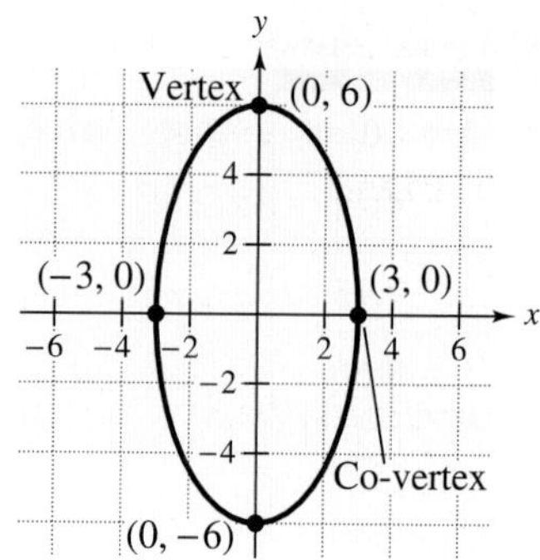

Exercises Within Reach®

Worked-out solutions to odd-numbered exercises at AlgebraWithinReach.com

Concept Summary Check

29. ***Vocabulary*** Define an ellipse and write the standard form of the equation of an ellipse centered at the origin. See Additional Answers.

30. ***Writing an Equation*** What points do you need to know in order to write the equation of an ellipse? The center, vertices, and co-vertices are needed to write the equation of an ellipse.

31. ***Think About It*** From the standard equation, how can you determine the lengths of the major and minor axes of an ellipse? The length of the major axis is $2a$, and the length of the minor axis is $2b$.

32. ***Writing*** From the standard equation, how can you determine the orientation of the major and minor axes of an ellipse? See Additional Answers.

Extra Practice

Writing the Standard Equation of an Ellipse **In Exercises 33 and 34, write the standard form of the equation of the ellipse centered at the origin.**

33. Major axis (vertical) 10 units, minor axis 6 units
$\frac{x^2}{9} + \frac{y^2}{25} = 1$

34. Major axis (horizontal) 24 units, minor axis 10 units
$\frac{x^2}{144} + \frac{y^2}{25} = 1$

Sketching an Ellipse **In Exercises 35 and 36, sketch the ellipse. Identify the vertices and co-vertices.**
See Additional Answers.

35. $16x^2 + 25y^2 - 9 = 0$

36. $64x^2 + 36y^2 - 49 = 0$

37. ***Wading Pool*** You are building a wading pool that is in the shape of an ellipse. Your plans give the following equation for the elliptical shape of the pool, measured in feet.

$$\frac{x^2}{324} + \frac{y^2}{196} = 1$$

Find the longest distance and shortest distance across the pool. 36 feet, 28 feet

38. ***Oval Office*** In the White House, the Oval Office is in the shape of an ellipse. The perimeter (in meters) of the floor can be modeled by the equation

$$\frac{x^2}{19.36} + \frac{y^2}{30.25} = 1.$$

Find the longest distance and shortest distance across the office. 11 meters, 8.8 meters

Airplane **In Exercises 39 and 40, an airplane with enough fuel to fly 800 miles safely will take off from airport A and land at airport B. Answer the following questions given the situation in each exercise.**

(a) Explain why the region in which the airplane can fly is bounded by an ellipse (see figure).

(b) Let (0, 0) represent the center of the ellipse. Find the coordinates of each airport.

(c) Suppose the plane flies from airport A straight past airport B to a vertex of the ellipse, and then straight back to airport B. How far does the plane fly? Use your answer to find the coordinates of the vertices.

(d) Write an equation of the ellipse. (*Hint:* $c^2 = a^2 - b^2$)

(e) The area of an ellipse is given by $A = \pi ab$. Find the area of the ellipse.

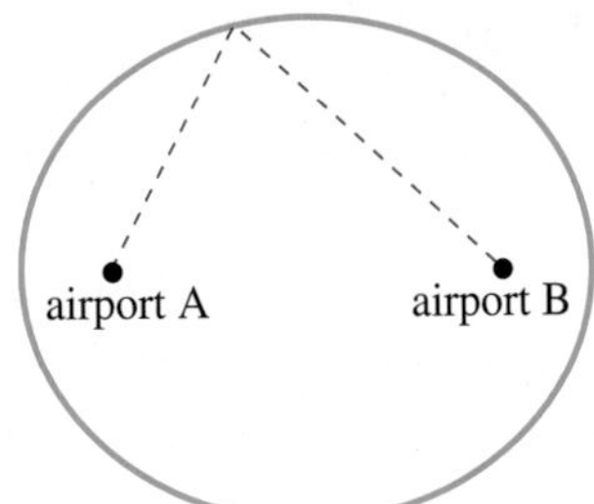

39. Airport A is 500 miles from airport B.
See Additional Answers.

40. Airport A is 650 miles form airport B.
See Additional Answers.

Explaining Concepts

41. ***Vocabulary*** Describe the relationship between circles and ellipses. How are they similar? How do they differ?
A circle is an ellipse in which the major axis and the minor axis have the same length. Both circles and ellipses have foci. However, a circle has a single focus located at the center and an ellipse has two foci that lie on the major axis.

42. ***Area*** The area of an ellipse is given by $A = \pi ab$. Explain how this area is related to the area of a circle.
An ellipse is a circle when $a = b$, and this distance represents the radius of the circle. Therefore, $A = \pi ab$ is equivalent to $A = \pi r^2$ when $a = b$.

43. ***Vocabulary*** Explain the significance of the foci in an ellipse. The sum of the distances between each point on the ellipse and the two foci is a constant.

44. ***Reasoning*** Explain how to write an equation of an ellipse when you know the coordinates of the vertices and co-vertices. Plot the vertices and the co-vertices. Find the center and determine the major and minor axes. a and b are the distances from the center to the vertices and co-vertices, respectively.

45. ***Logic*** From the standard form of the equation, explain how you can determine if the graph of an ellipse intersects the x- or y-axis. See Additional Answers.

Cumulative Review

In Exercises 46–53, evaluate the function as indicated and sketch the graph of the function.
See Additional Answers.

46. $f(x) = 4^x$
(a) $x = 3$
(b) $x = -1$
(a) $f(3) = 64$
(b) $f(-1) = \frac{1}{4}$

47. $f(x) = 3^{-x}$
(a) $x = -2$
(b) $x = 2$
(a) $f(-2) = 9$
(b) $f(2) = \frac{1}{9}$

48. $g(x) = 5^{x-1}$
(a) $x = 4$
(b) $x = 0$
(a) $g(4) = 125$
(b) $g(0) = \frac{1}{5}$

49. $g(x) = 6e^{0.5x}$
(a) $x = -1$
(b) $x = 2$
(a) $g(-1) \approx 3.639$
(b) $g(2) \approx 16.310$

50. $h(x) = \log_{10} 2x$
(a) $x = 5$
(b) $x = 500$
(a) $h(5) = 1$
(b) $h(500) = 3$

51. $h(x) = \log_{16} 4x$
(a) $x = 4$
(b) $x = 64$
(a) $h(4) = 1$
(b) $h(64) = 2$

52. $f(x) = \ln(-x)$
(a) $x = -6$
(b) $x = 3$
(a) $f(-6) \approx 1.79$
(b) $f(3)$ does not exist.

53. $f(x) = \log_4(x - 3)$
(a) $x = 3$
(b) $x = 35$
(a) $f(3)$ does not exist.
(b) $f(35) = \frac{5}{2}$

Mid-Chapter Quiz: Sections 10.1–10.2

Solutions in English & Spanish and tutorial videos at AlgebraWithinReach.com

Take this quiz as you would take a quiz in class. After you are done, check your work against the answers in the back of the book.

In Exercises 1 and 2, write the standard form of the equation of the circle centered at (h, k).

1. Center: $(0, 0)$
   Radius: 5
   $x^2 + y^2 = 25$

2. Center: $(3, -5)$
   Passes through the point $(0, -1)$
   $(x - 3)^2 + (y + 5)^2 = 25$

In Exercises 3 and 4, write the standard form of the equation of the parabola with its vertex at (h, k).

3. Vertex: $(-2, 1)$ $(y - 1)^2 = 8(x + 2)$
   Focus: $(0, 1)$

4. Vertex: $(2, 3)$ $(x - 2)^2 = -8(y - 3)$
   Focus: $(2, 1)$

5. Write the standard form of the equation of the ellipse shown in the figure.

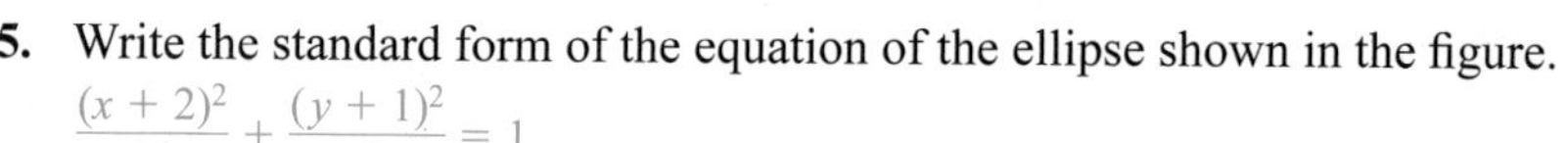

$\frac{(x + 2)^2}{16} + \frac{(y + 1)^2}{4} = 1$

6. Write the standard form of the equation of the ellipse with vertices $(0, -10)$ and $(0, 10)$ and co-vertices $(-6, 0)$ and $(6, 0)$.

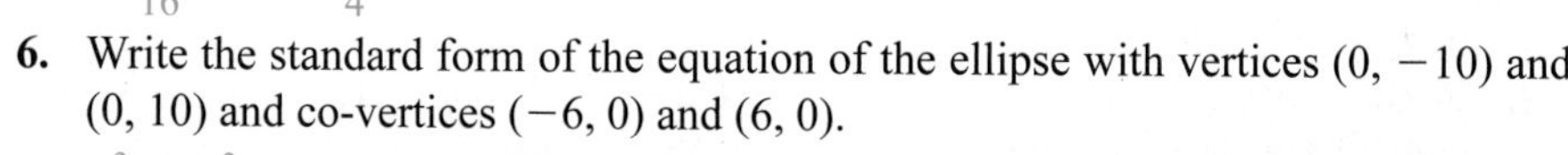

$\frac{x^2}{36} + \frac{y^2}{100} = 1$

In Exercises 7 and 8, write the equation of the circle in standard form. Then find the center and the radius of the circle.

7. $x^2 + y^2 + 6y - 7 = 0$ $x^2 + (y + 3)^2 = 16$; Center: $(0, -3)$, $r = 4$

8. $x^2 + y^2 + 2x - 4y + 4 = 0$ $(x + 1)^2 + (y - 2)^2 = 1$; Center: $(-1, 2)$, $r = 1$

In Exercises 9 and 10, write the equation of the parabola in standard form. Then find the vertex and the focus of the parabola.

9. $x = y^2 - 6y - 7$ $(y - 3)^2 = x + 16$; Vertex: $(-16, 3)$, Focus: $\left(-\frac{63}{4}, 3\right)$

10. $x^2 - 8x + y + 12 = 0$ $(x - 4)^2 = -(y - 4)$; Vertex: $(4, 4)$, Focus: $\left(4, \frac{15}{4}\right)$

In Exercises 11 and 12, write the equation of the ellipse in standard form. Then find the center and the vertices of the ellipse.

11. $4x^2 + y^2 - 16x - 20 = 0$
    $\frac{(x - 2)^2}{9} + \frac{y^2}{36} = 1$
    Center: $(2, 0)$
    Vertices: $(2, -6)$, $(2, 6)$

12. $4x^2 + 9y^2 - 48x + 36y + 144 = 0$
    $\frac{(x - 6)^2}{9} + \frac{(y + 2)^2}{4} = 1$
    Center: $(6, -2)$
    Vertices: $(3, -2)$, $(9, -2)$

In Exercises 13–18, sketch the graph of the equation.
See Additional Answers.

13. $(x + 5)^2 + (y - 1)^2 = 9$

14. $9x^2 + y^2 = 81$

15. $x = -y^2 - 4y$

16. $x^2 + (y + 4)^2 = 1$

17. $y = x^2 - 2x + 1$

18. $4(x + 3)^2 + (y - 2)^2 = 16$

Study Skills in Action

Avoiding Test-Taking Errors

For some students, the day they get their math tests back is just as nerve-racking as the day they take the test. Do you look at your grade, sigh hopelessly, and stuff the test in your book bag? This kind of response is not going to help you to do better on the next test. When professional football players lose a game, the coach does not let them just forget about it. They review all their mistakes and discuss how to correct them. That is what you need to do with every math test.

There are six types of test errors, as listed below. Look at your test and see what types of errors you make. Then decide what you can do to avoid making them again. Many students need to do this with a tutor or instructor the first time through.

Smart Study Strategy

Analyze Your Errors

Type of error	Corrective action
1 ▶ **Misreading Directions:** You do not correctly read or understand directions.	Read the instructions in the textbook exercises at least twice and make sure you understand what they mean. Make this a habit in time for the next test.
2 ▶ **Careless Errors:** You understand how to do a problem but make careless errors, such as not carrying a sign, miscopying numbers, and so on.	Pace yourself during a test to avoid hurrying. Also, make sure you write down every step of the solution neatly. Use a finger to move from one step to the next, looking for errors.
3 ▶ **Concept Errors:** You do not understand how to apply the properties and rules needed to solve a problem.	Find a tutor who will work with you on the next chapter. Visit the instructor to make sure you understand the math.
4 ▶ **Application Errors:** You can do numerical problems that are similar to your homework problems but struggle with problems that vary, such as application problems.	Do not just mimic the steps of solving an application problem. Explain out loud why you are doing each step. Ask the instructor or tutor for different types of problems.
5 ▶ **Test-Taking Errors:** You hurry too much, do not use all of the allowed time, spend too much time on one problem, and so on.	Refer to the *Ten Steps for Test-Taking* on page 321.
6 ▶ **Study Errors:** You do not study the right material or do not learn it well enough to remember it on a test without resources such as notes.	Take a practice test. Work with a study group. Confer with your instructor. Do not try to learn a whole chapter's worth of material in one night—cramming does not work in math!

10.3 Hyperbolas

- Analyze hyperbolas centered at the origin.
- Analyze hyperbolas centered at (h, k).

Hyperbolas Centered at the Origin

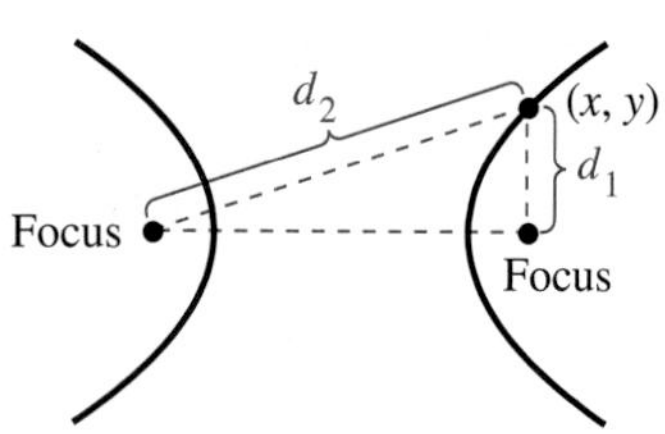

$d_2 - d_1$ is a positive constant.

A **hyperbola** in the rectangular coordinate system consists of all points (x, y) such that the difference of the distances between (x, y) and two fixed points is a positive constant, as shown at the left. The two fixed points are called the **foci** of the hyperbola. The line on which the foci lie is called the **transverse axis** of the hyperbola.

Standard Equation of a Hyperbola (Center at Origin)

The **standard form of the equation of a hyperbola centered at the origin** is

$$\frac{x^2}{a^2} - \frac{y^2}{b^2} = 1 \qquad \text{or} \qquad \frac{y^2}{a^2} - \frac{x^2}{b^2} = 1$$

Transverse axis is horizontal. Transverse axis is vertical.

where a and b are positive real numbers. The **vertices** of the hyperbola lie on the transverse axis, a units from the center, as shown below.

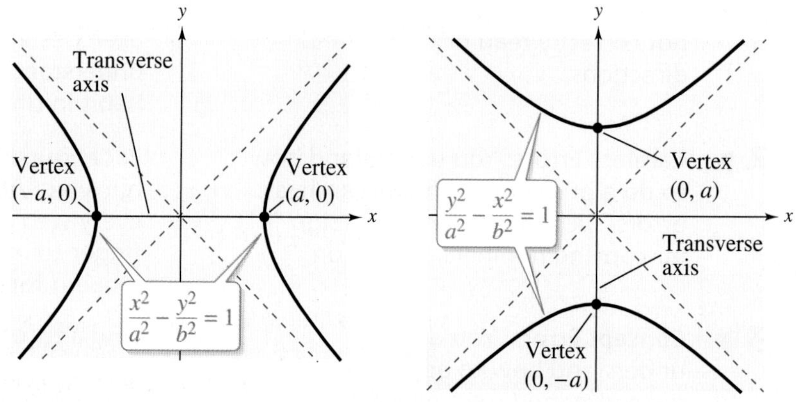

Exercises Within Reach®

Solutions in English & Spanish and tutorial videos at AlgebraWithinReach.com

Identifying the Standard Form **In Exercises 1–4, determine whether the standard form of the equation of the hyperbola is $\frac{x^2}{a^2} - \frac{y^2}{b^2} = 1$ or $\frac{y^2}{a^2} - \frac{x^2}{b^2} = 1$.** See Additional Answers.

1.

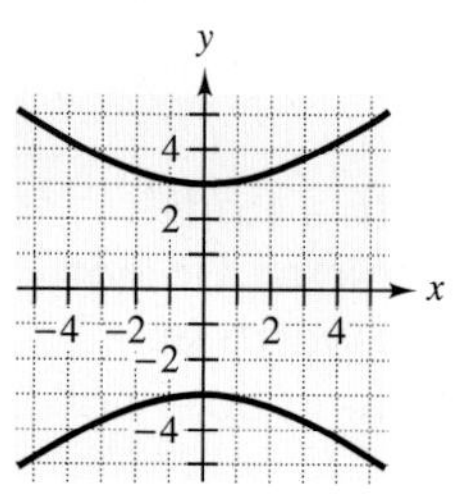

2.

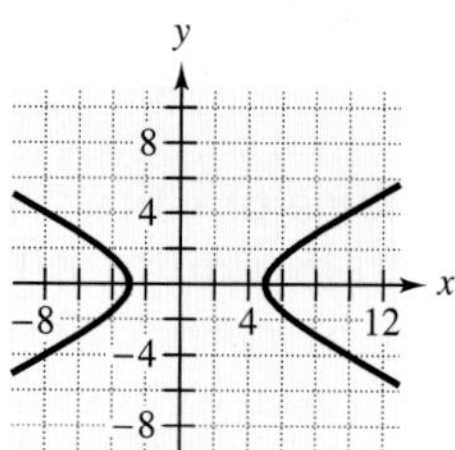

3.

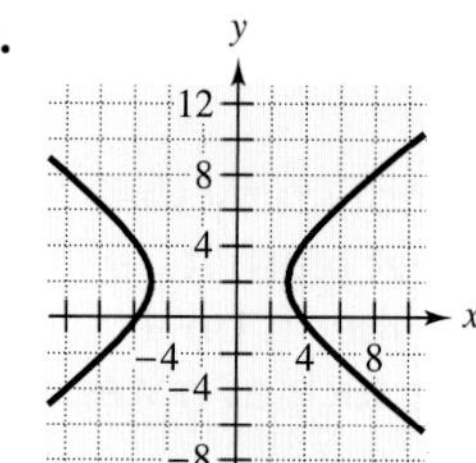

4.

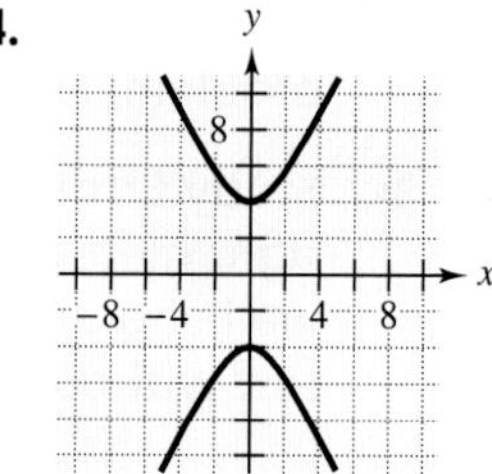

A hyperbola has two disconnected parts, each of which is called a **branch** of the hyperbola. The two branches approach a pair of intersecting lines called the **asymptotes** of the hyperbola. The two asymptotes intersect at the center of the hyperbola.

Study Tip

To sketch a hyperbola, form a **central rectangle** that is centered at the origin and has side lengths of $2a$ and $2b$. The asymptotes pass through the corners of the central rectangle, and the vertices of the hyperbola lie at the centers of opposite sides of the central rectangle.

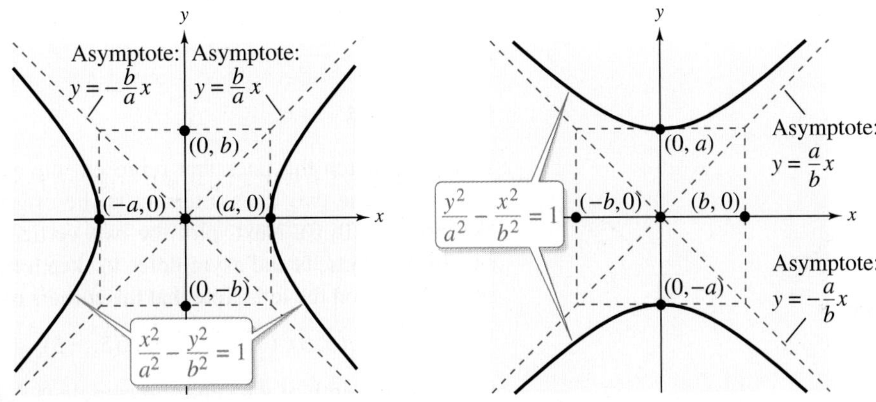

Transverse axis is horizontal.

Transverse axis is vertical.

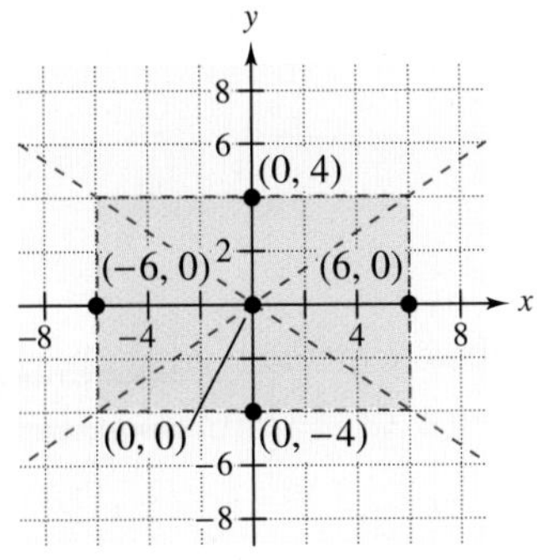

EXAMPLE 1 Sketching a Hyperbola

Identify the vertices of the hyperbola given by the equation, and sketch the hyperbola.

$$\frac{x^2}{36} - \frac{y^2}{16} = 1$$

SOLUTION

From the standard form of the equation

$$\frac{x^2}{6^2} - \frac{y^2}{4^2} = 1$$

you can see that the center of the hyperbola is the origin and the transverse axis is horizontal. So, the vertices lie six units to the left and right of the center at the points

$$(-6, 0) \text{ and } (6, 0).$$

Because $a = 6$ and $b = 4$, you can sketch the hyperbola by first drawing a central rectangle with a width of $2a = 12$ and a height of $2b = 8$, as shown in the top left figure. Next, draw the asymptotes of the hyperbola through the corners of the central rectangle, and plot the vertices. Finally, draw the hyperbola, as shown in the bottom left figure.

Exercises Within Reach®

Solutions in English & Spanish and tutorial videos at AlgebraWithinReach.com

Sketching a Hyperbola **In Exercises 5–10, identify the vertices of the hyperbola, and sketch its graph.** See Additional Answers.

5. $\frac{x^2}{9} - \frac{y^2}{25} = 1$
Vertices: $(\pm 3, 0)$

6. $\frac{x^2}{4} - \frac{y^2}{9} = 1$
Vertices: $(\pm 2, 0)$

7. $\frac{y^2}{9} - \frac{x^2}{25} = 1$
Vertices: $(0, \pm 3)$

8. $\frac{y^2}{4} - \frac{x^2}{9} = 1$
Vertices: $(0, \pm 2)$

9. $y^2 - x^2 = 9$
Vertices: $(0, \pm 3)$

10. $y^2 - x^2 = 1$
Vertices: $(0, \pm 1)$

EXAMPLE 2 Writing an Equation of a Hyperbola

Write the standard form of the equation of the hyperbola with a vertical transverse axis and vertices (0, 3) and (0, −3). The equations of the asymptotes of the hyperbola are $y = \frac{3}{5}x$ and $y = -\frac{3}{5}x$.

SOLUTION

To begin, sketch the lines that represent the asymptotes, as shown below on the left. Note that these two lines intersect at the origin, which implies that the center of the hyperbola is (0, 0). Next, plot the two vertices at the points (0, 3) and (0, −3). You can use the vertices and asymptotes to sketch the central rectangle of the hyperbola, as shown below on the left. Note that the corners of the central rectangle occur at the points

$$(-5, 3), (5, 3), (-5, -3), \text{ and } (5, -3).$$

Because the width of the central rectangle is $2b = 10$, it follows that $b = 5$. Similarly, because the height of the central rectangle is $2a = 6$, it follows that $a = 3$. Now that you know the values of a and b, you can use the standard form of the equation of a hyperbola to write an equation.

$\frac{y^2}{a^2} - \frac{x^2}{b^2} = 1$ Transverse axis is vertical.

$\frac{y^2}{3^2} - \frac{x^2}{5^2} = 1$ Substitute 3 for a and 5 for b.

$\frac{y^2}{9} - \frac{x^2}{25} = 1$ Simplify.

The graph is shown below on the right.

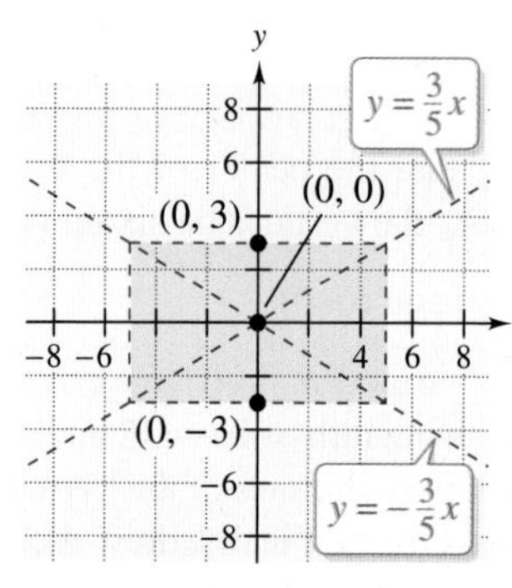

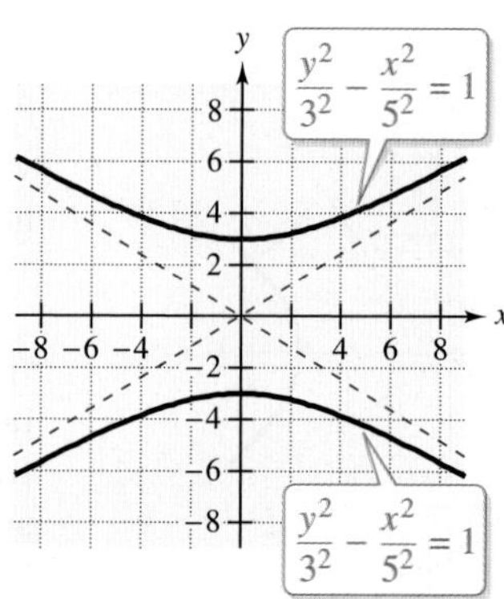

Study Tip

For a hyperbola, note that a and b are not determined in the same way as for an ellipse, where a is always greater than b. In the standard form of the equation of a hyperbola, a^2 is always the denominator of the positive term.

Exercises Within Reach®

Solutions in English & Spanish and tutorial videos at AlgebraWithinReach.com

Writing an Equation of a Hyperbola **In Exercises 11–16, write the standard form of the equation of the hyperbola centered at the origin.**

	Vertices	*Asymptotes*			*Vertices*	*Asymptotes*	
11.	(−4, 0), (4, 0)	$y = 2x$	$y = -2x$	**12.**	(−2, 0), (2, 0)	$y = \frac{1}{3}x$	$y = -\frac{1}{3}x$
	$\frac{x^2}{16} - \frac{y^2}{64} = 1$				$\frac{x^2}{4} - \frac{y^2}{4/9} = 1$		
13.	(0, −4), (0, 4)	$y = \frac{1}{2}x$	$y = -\frac{1}{2}x$	**14.**	(0, −2), (0, 2)	$y = 3x$	$y = -3x$
	$\frac{y^2}{16} - \frac{x^2}{64} = 1$				$\frac{y^2}{4} - \frac{x^2}{4/9} = 1$		
15.	(−9, 0), (9, 0)	$y = \frac{2}{3}x$	$y = -\frac{2}{3}x$	**16.**	(−1, 0), (1, 0)	$y = \frac{1}{2}x$	$y = -\frac{1}{2}x$
	$\frac{x^2}{81} - \frac{y^2}{36} = 1$				$\frac{x^2}{1} - \frac{y^2}{1/4} = 1$		

Hyperbolas Centered at (h, k)

Standard Equation of a Hyperbola [Center at (h, k)]

The **standard form of the equation of a hyperbola centered at (h, k)** is

$$\frac{(x-h)^2}{a^2} - \frac{(y-k)^2}{b^2} = 1 \quad \text{Transverse axis is horizontal.}$$

or

$$\frac{(y-k)^2}{a^2} - \frac{(x-h)^2}{b^2} = 1 \quad \text{Transverse axis is vertical.}$$

where a and b are positive real numbers. The vertices lie on the transverse axis, a units from the center, as shown below.

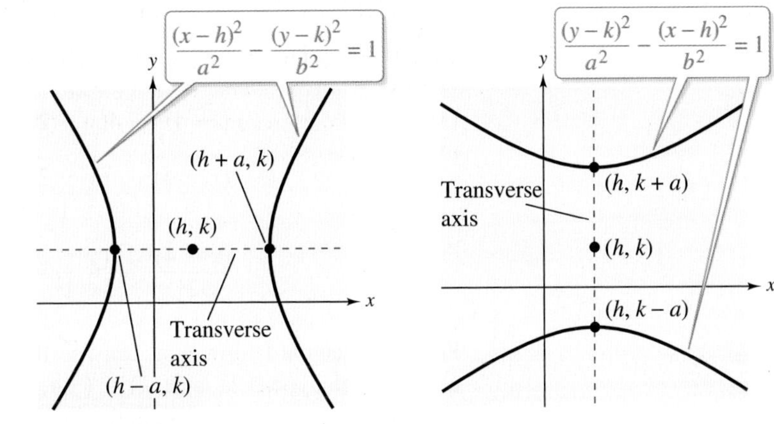

EXAMPLE 3 **Sketching a Hyperbola**

Sketch the hyperbola given by $\frac{(y-1)^2}{9} - \frac{(x+2)^2}{4} = 1$.

SOLUTION

From the form of the equation, you can see that the transverse axis is vertical, and the center of the hyperbola is $(h, k) = (-2, 1)$. Because $a = 3$ and $b = 2$, you can begin by sketching a central rectangle that is six units high and four units wide, centered at $(-2, 1)$. Then, sketch the asymptotes by drawing lines through the corners of the central rectangle. Sketch the hyperbola, as shown at the left. From the graph, you can see that the center of the hyperbola is shifted two units to the left and one unit upward from the origin.

Exercises Within Reach®

Solutions in English & Spanish and tutorial videos at AlgebraWithinReach.com

Sketching a Hyperbola **In Exercises 17–20, find the center and vertices of the hyperbola, and sketch its graph.** See Additional Answers.

17. $\frac{(x-1)^2}{4} - \frac{(y+2)^2}{1} = 1$

Center: $(1, -2)$, Vertices: $(-1, -2), (3, -2)$

18. $\frac{(x-2)^2}{4} - \frac{(y-3)^2}{9} = 1$

Center: $(2, 3)$, Vertices: $(0, 3), (4, 3)$

19. $(y+4)^2 - (x-3)^2 = 25$

Center: $(3, -4)$, Vertices: $(3, 1), (3, -9)$

20. $(y+6)^2 - (x-2)^2 = 1$

Center: $(2, -6)$, Vertices: $(2, -5), (2, -7)$

EXAMPLE 4 Sketching a Hyperbola

Sketch the hyperbola given by

$$x^2 - 4y^2 + 8x + 16y - 4 = 0.$$

SOLUTION

Complete the square to write the equation in standard form.

$x^2 - 4y^2 + 8x + 16y - 4 = 0$	Write original equation.
$(x^2 + 8x + \square) - (4y^2 - 16y + \square) = 4$	Group terms.
$(x^2 + 8x + \square) - 4(y^2 - 4y + \square) = 4$	Factor 4 out of y-terms.
$(x^2 + 8x + 16) - 4(y^2 - 4y + 4) = 4 + 16 - 4(4)$	Complete the squares.
$(x + 4)^2 - 4(y - 2)^2 = 4$	Simplify.
$\dfrac{(x + 4)^2}{4} - \dfrac{(y - 2)^2}{1} = 1$	Divide each side by 4.
$\dfrac{(x + 4)^2}{2^2} - \dfrac{(y - 2)^2}{1^2} = 1$	Standard form

From the standard form, you can see that the transverse axis is horizontal, and the center of the hyperbola is $(h, k) = (-4, 2)$. Because $a = 2$ and $b = 1$, you can begin by sketching a central rectangle that is four units wide and two units high, centered at $(-4, 2)$. Then, sketch the asymptotes by drawing lines through the corners of the central rectangle. Sketch the hyperbola, as shown below. From the graph, you can see that the center of the hyperbola is shifted four units to the left and two units upward from the origin.

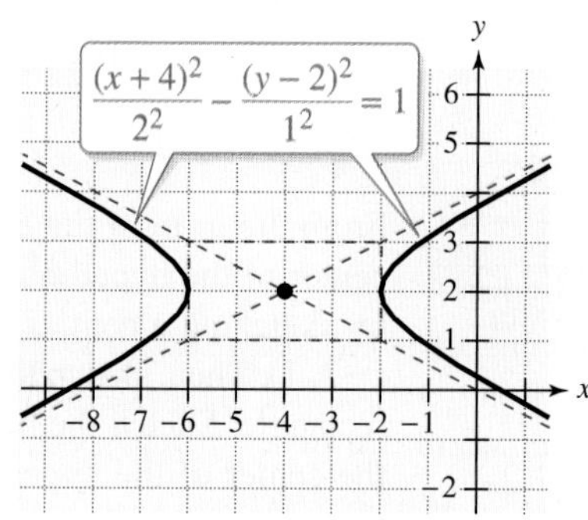

Exercises Within Reach®

Solutions in English & Spanish and tutorial videos at AlgebraWithinReach.com

Sketching a Hyperbola **In Exercises 21–24, find the center and vertices of the hyperbola, and sketch its graph.** See Additional Answers.

21. $9x^2 - y^2 - 36x - 6y + 18 = 0$
Center: $(2, -3)$, Vertices: $(1, -3)$, $(3, -3)$

22. $x^2 - 9y^2 + 36y - 72 = 0$
Center: $(0, 2)$, Vertices: $(-6, 2)$, $(6, 2)$

23. $4x^2 - y^2 + 24x + 4y + 28 = 0$
Center: $(-3, 2)$, Vertices: $(-4, 2)$, $(-2, 2)$

24. $25x^2 - 4y^2 + 100x + 8y + 196 = 0$
Center: $(-2, 1)$, Vertices: $(-2, -4)$, $(-2, 6)$

Application EXAMPLE 5 Navigation

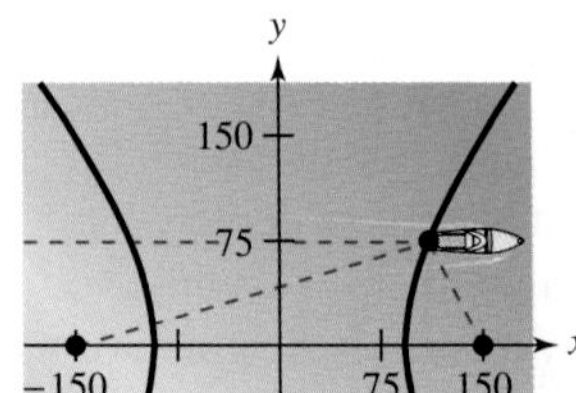

Long-distance radio navigation for aircraft and ships uses synchronized pulses transmitted by widely separated transmitting stations. The locations of two transmitting stations that are 300 miles apart are represented by the points $(-150, 0)$ and $(150, 0)$ (see figure). A ship's location is given by $(x, 75)$. The difference in the arrival times of pulses transmitted simultaneously to the ship from the two stations is constant at any point on the hyperbola given by

$$\frac{x^2}{8649} - \frac{y^2}{13{,}851} = 1$$

which passes through the ship's location and has the two stations as foci. Use the equation to find the x-coordinate of the ship's location.

SOLUTION

To find the x-coordinate of the ship's location, substitute $y = 75$ into the equation for the hyperbola and solve for x.

$$\frac{x^2}{8649} - \frac{y^2}{13{,}851} = 1$$ Write equation for hyperbola.

$$\frac{x^2}{8649} - \frac{75^2}{13{,}851} = 1$$ Substitute 75 for y.

$$x^2 = 8649\left[1 + \frac{75^2}{13{,}851}\right]$$ Isolate the variable.

$$x^2 \approx 12{,}161.4$$ Use a calculator.

$$x \approx 110.3$$ Take positive square root of each side.

The ship is about 110.3 miles to the right of the y-axis.

Exercises Within Reach®

Solutions in English & Spanish and tutorial videos at AlgebraWithinReach.com

25. ***Optics*** Hyperbolic mirrors are used in some telescopes. The figure shows a cross section of a hyperbolic mirror as the right branch of a hyperbola. A property of the mirror is that a light ray directed at the focus $(48, 0)$ is reflected to the other focus $(0, 0)$. Use the equation of the hyperbola

$$89x^2 - 55y^2 - 4272x + 31{,}684 = 0$$

to find the coordinates of the mirror's vertex.

$\left(24 + \sqrt{220}, 0\right) \approx (38.8, 0)$

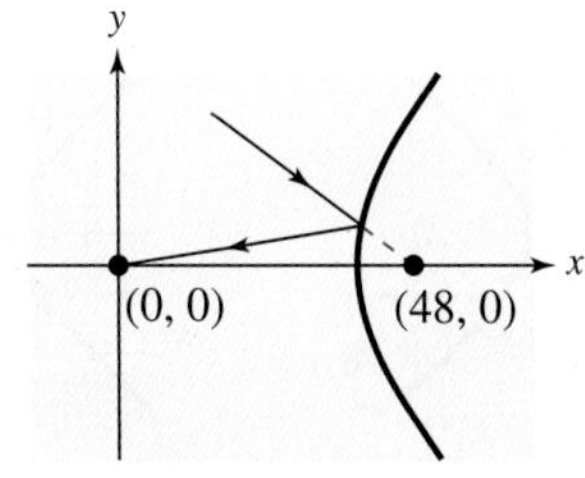

26. ***Art*** A sculpture has a hyperbolic cross section (see figure). Write an equation that models the curved sides of the sculpture.

$\frac{x^2}{1} - \frac{3y^2}{169} = 1$

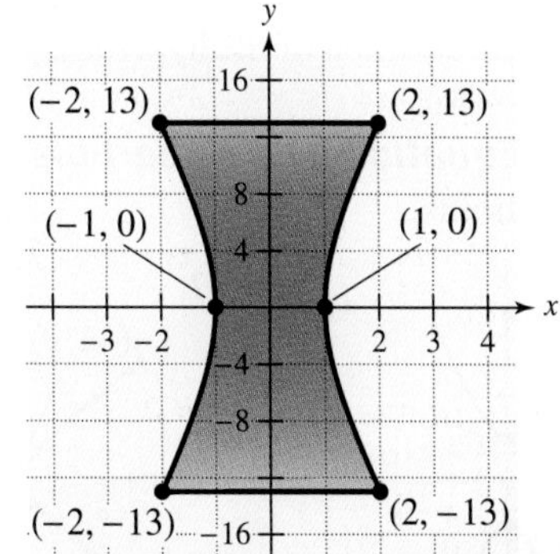

Concept Summary: Sketching Hyperbolas

What

You can use the standard form of the equation of a **hyperbola** to sketch its graph.

EXAMPLE

Sketch the hyperbola.

$$\frac{x^2}{9} - \frac{y^2}{4} = 1$$

How

1. Write the equation in standard form.
2. Determine the center and the values of a and b.
3. Use the values of $2a$ and $2b$ to sketch the **central rectangle**.
4. Draw the **asymptotes** through the corners of the central rectangle.
5. Plot the **vertices** and finally sketch the hyperbola.

EXAMPLE

$$\frac{x^2}{3^2} - \frac{y^2}{2^2} = 1 \qquad \text{Standard form}$$

The center of the hyperbola is the origin. Because $a = 3$ and $b = 2$, the central rectangle has a width of $2a = 2(3) = 6$ units and a height of $2b = 2(2) = 4$ units. The graph is shown at the right.

Why

Understanding how to sketch the graph of a hyperbola centered at the origin will help you when you learn how to sketch the graph of a hyperbola *not* centered at the origin.

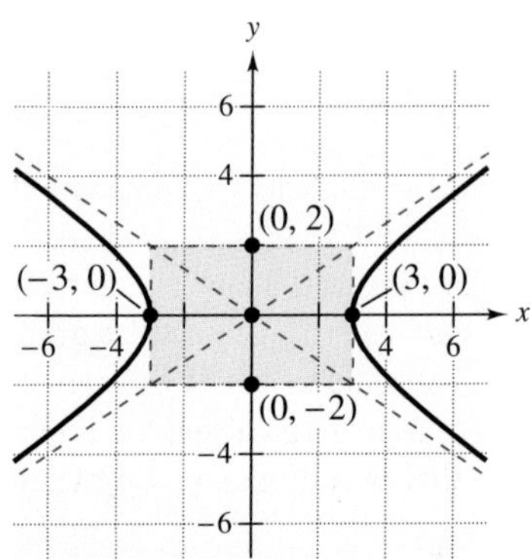

Exercises Within Reach®

Worked-out solutions to odd-numbered exercises at AlgebraWithinReach.com

Concept Summary Check

27. ***Vocabulary*** You are given the equation of a hyperbola in the standard form

$$\frac{x^2}{a^2} - \frac{y^2}{b^2} = 1.$$

Explain how you can sketch the central rectangle for the hyperbola. Explain how you can use the central rectangle to sketch the asymptotes of the hyperbola.
See Additional Answers.

28. ***Reasoning*** You are given the vertices and the equations of the asymptotes of a hyperbola. Explain how you can determine the values of a and b in the standard form of the equation of the hyperbola.
See Additional Answers.

29. ***Logic*** What are the dimensions of the central rectangle and the coordinates of the center of the hyperbola whose equation in standard form is

$$\frac{(y-k)^2}{a^2} - \frac{(x-h)^2}{b^2} = 1?$$

Central rectangle dimensions: $2a \times 2b$, Center: (h, k)

30. ***Reasoning*** Given the equation of a hyperbola in the general polynomial form

$$ax^2 - by^2 + cx + dy + e = 0$$

what process can you use to find the center of the hyperbola? The method of completing the square can be used to determine the center of the hyperbola.

Writing an Equation of a Hyperbola **In Exercises 31–34, write the standard form of the equation of the hyperbolas.**

31.

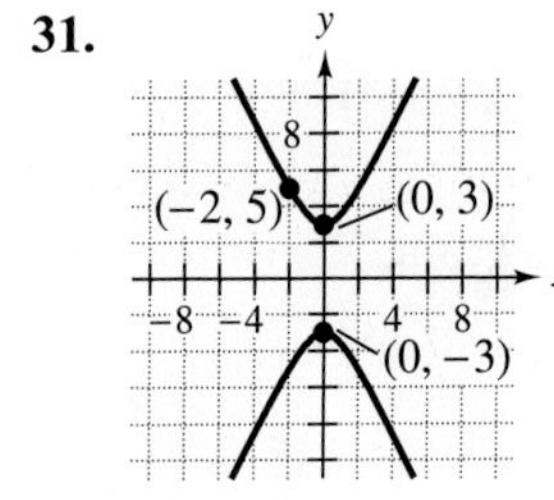

$\frac{y^2}{9} - \frac{x^2}{9/4} = 1$

32.

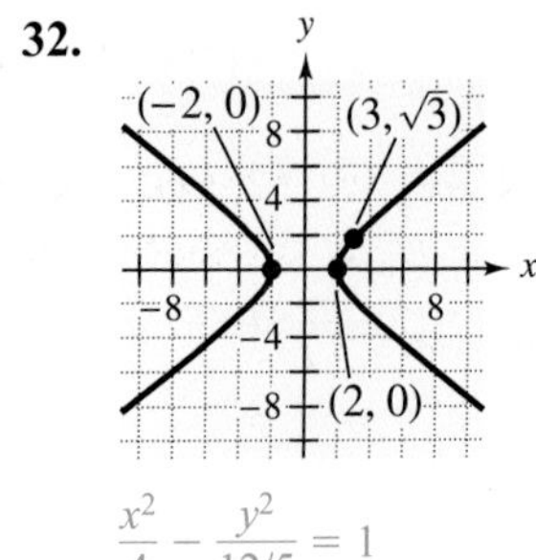

$\frac{x^2}{4} - \frac{y^2}{12/5} = 1$

33.

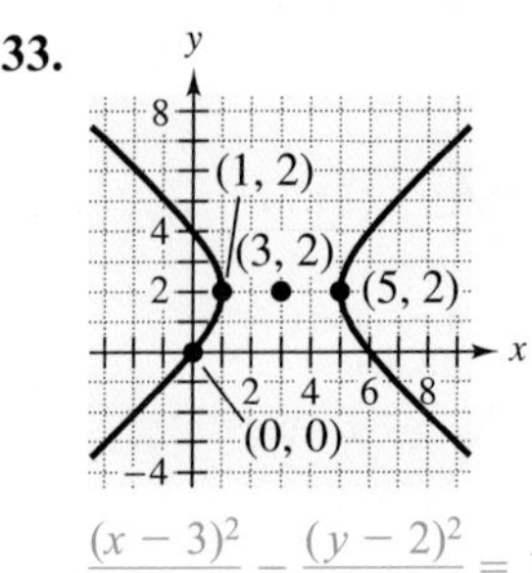

$\frac{(x-3)^2}{4} - \frac{(y-2)^2}{16/5} = 1$

34.

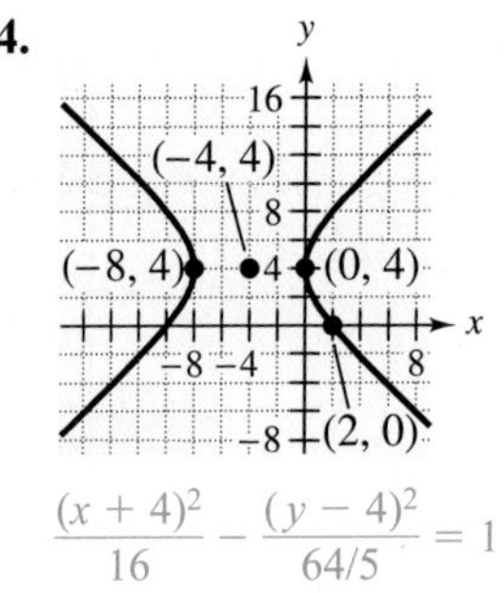

$\frac{(x+4)^2}{16} - \frac{(y-4)^2}{64/5} = 1$

Sketching a Hyperbola **In Exercises 35–38, identify the vertices and asymptotes of the hyperbola, and sketch its graph.** See Additional Answers.

35. $\frac{x^2}{1} - \frac{y^2}{9/4} = 1$ Vertices: $(\pm 1, 0)$, Asymptotes: $y = \pm\frac{3}{2}x$

36. $\frac{y^2}{1/4} - \frac{x^2}{25/4} = 1$ Vertices: $\left(0, \pm\frac{1}{2}\right)$, Asymptotes: $y = \pm\frac{1}{5}x$

37. $4y^2 - x^2 + 16 = 0$

Vertices: $(\pm 4, 0)$, Asymptotes: $y = \pm\frac{1}{2}x$

38. $4y^2 - 9x^2 - 36 = 0$

Vertices: $(0, \pm 3)$, Asymptotes: $y = \pm\frac{3}{2}x$

Identifying a Conic **In Exercises 39–44, determine whether the graph represented by the equation is a circle, a parabola, an ellipse, or a hyperbola.**

39. $\frac{(x-3)^2}{4^2} + \frac{(y-4)^2}{6^2} = 1$ Ellipse

40. $\frac{(x+2)^2}{25} + \frac{(y-2)^2}{25} = 1$ Circle

41. $x^2 - y^2 = 1$ Hyperbola

42. $2x + y^2 = 0$ Parabola

43. $y^2 - x^2 - 2y + 8x - 19 = 0$ Hyperbola

44. $9x^2 + y^2 - 18x - 8y + 16 = 0$ Ellipse

45. ***Aeronautics*** When an airplane travels faster than the speed of sound, the sound waves form a cone behind the airplane. When the airplane is flying parallel to the ground, the sound waves intersect the ground in a hyperbola with the airplane directly above its center (see figure). A sonic boom can be heard along the hyperbola. You hear a sonic boom that is audible along a hyperbola with the equation

$$\frac{x^2}{100} - \frac{y^2}{4} = 1$$

where x and y are measured in miles. What is the shortest horizontal distance you could be from the airplane? 10 miles

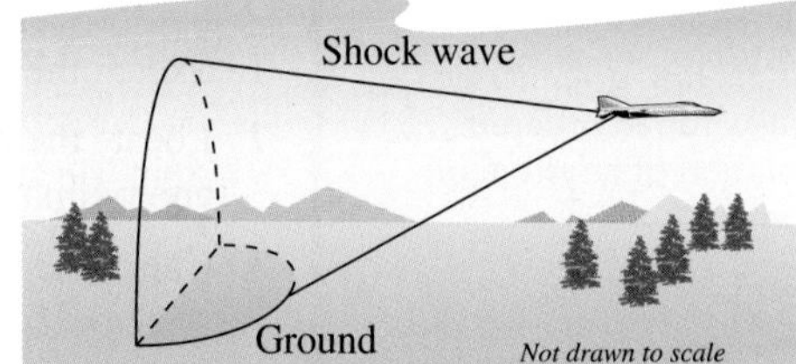

Explaining Concepts

46. ***Think About It*** Describe the part of the hyperbola

$$\frac{(x-3)^2}{4} - \frac{(y-1)^2}{9} = 1$$

given by each equation.

(a) $x = 3 - \frac{2}{3}\sqrt{9 + (y-1)^2}$ Left half

(b) $y = 1 + \frac{3}{2}\sqrt{(x-3)^2 - 4}$ Top half

47. ***Reasoning*** Consider the definition of a hyperbola. How many hyperbolas have a given pair of points as foci? Explain your reasoning.

Infinitely many. The constant difference of the distances can be different for an infinite number of hyperbolas that have the same set of foci.

48. ***Precision*** How many hyperbolas pass through a given point and have a given pair of points as foci? Explain your reasoning.

One. The difference in the distances between the given point and the given foci is constant, so only one branch of one hyperbola can pass through the point.

49. ***Project*** Cut cone-shaped pieces of styrofoam to demonstrate how to obtain each type of conic section: circle, parabola, ellipse, and hyperbola. Discuss how you could write directions for someone else to form each conic section. Compile a list of real-life situations and/or everyday objects in which conic sections may be seen.

Answers will vary.

Cumulative Review

In Exercises 50 and 51, solve the system of equations by graphing.

50. $\begin{cases} -x + 3y = 8 \\ 4x - 12y = -32 \end{cases}$

Infinitely many solutions

51. $\begin{cases} x - 3y = 5 \\ 2x - 6y = -5 \end{cases}$

No solution

In Exercises 52 and 53, solve the system of linear equations by the method of elimination.

52. $\begin{cases} x + y = 3 \\ x - y = 2 \end{cases}$

$\left(\frac{5}{2}, \frac{1}{2}\right)$

53. $\begin{cases} 4x + 3y = 3 \\ x - 2y = 9 \end{cases}$

$(3, -3)$

10.4 Solving Nonlinear Systems of Equations

- Solve nonlinear systems of equations graphically.
- Solve nonlinear systems of equations by substitution.
- Solve nonlinear systems of equations by elimination.

Solving Nonlinear Systems of Equations by Graphing

A **nonlinear system of equations** is a system that contains at least one nonlinear equation. Nonlinear systems of equations can have no solution, one solution, or two or more solutions.

Review what it means, graphically and algebraically, for an ordered pair to be a solution of a system of linear equations.

Solving a Nonlinear System Graphically

1. Sketch the graph of each equation in the system.
2. Locate the point(s) of intersection of the graphs (if any) and graphically approximate the coordinates of the point(s).
3. Check the coordinates by substituting them into each equation in the original system. If the coordinates do not check, you may have to use an algebraic approach, as discussed later in this section.

EXAMPLE 1 Solving a Nonlinear System Graphically

Find all solutions of the nonlinear system of equations.

$$\begin{cases} x = (y-3)^2 & \text{Equation 1} \\ x + y = 5 & \text{Equation 2} \end{cases}$$

SOLUTION

Begin by sketching the graph of each equation. Solve the first equation for y.

$x = (y-3)^2$ Write original equation.

$\pm\sqrt{x} = y - 3$ Take the square root of each side.

$3 \pm \sqrt{x} = y$ Add 3 to each side.

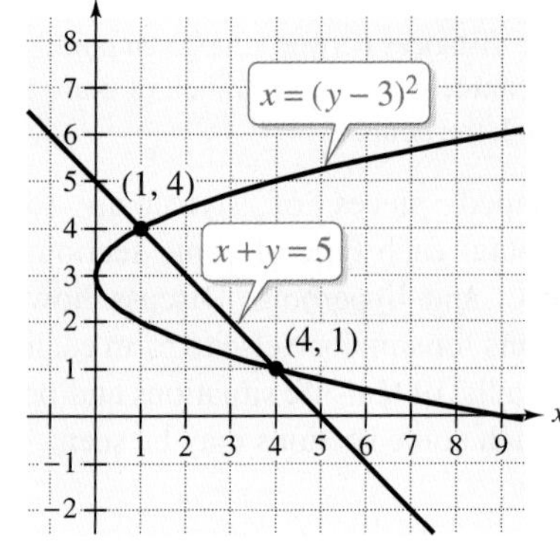

The graph of $y = 3 \pm \sqrt{x}$ is a parabola with its vertex at $(0, 3)$. The graph of the second equation is a line with a slope of -1 and a y-intercept of $(0, 5)$. The system appears to have two solutions: $(4, 1)$ and $(1, 4)$, as shown at the left. Check these solutions in the original system.

Exercises Within Reach®

Solutions in English & Spanish and tutorial videos at AlgebraWithinReach.com

Solving a System Graphically **In Exercises 1–4, graph the equations to determine whether the system has any solutions. Find any solutions that exist.** See Additional Answers.

1. $\begin{cases} x + y = 2 \\ x^2 - y = 0 \end{cases}$ $(-2, 4), (1, 1)$

2. $\begin{cases} y = 4 \\ x^2 - y = 0 \end{cases}$ $(-2, 4), (2, 4)$

3. $\begin{cases} x^2 + y = 9 \\ x - y = -3 \end{cases}$ $(2, 5), (-3, 0)$

4. $\begin{cases} x - y^2 = 0 \\ x - y = 2 \end{cases}$ $(4, 2), (1, -1)$

EXAMPLE 2 Solving a Nonlinear System Graphically

Find all solutions of the nonlinear system of equations.

$$\begin{cases} x^2 + y^2 = 25 & \text{Equation 1} \\ x - y = 1 & \text{Equation 2} \end{cases}$$

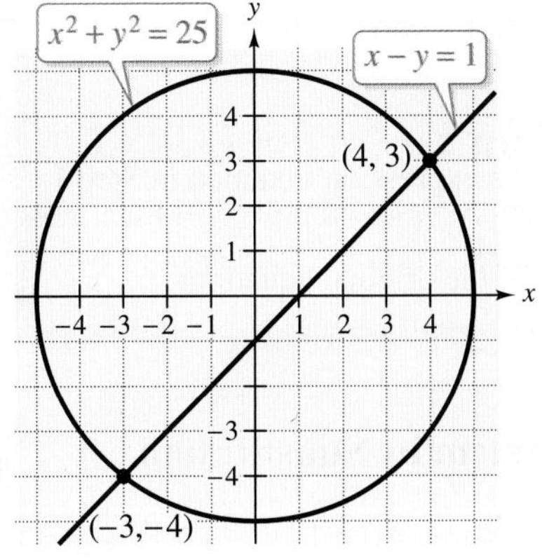

SOLUTION

Begin by sketching the graph of each equation. The graph of the first equation is a circle centered at the origin with a radius of 5. The graph of the second equation is a line with a slope of 1 and a y-intercept of $(0, -1)$. The system appears to have two solutions: $(-3, -4)$ and $(4, 3)$, as shown at the left.

Check

To check $(-3, -4)$, substitute -3 for x and -4 for y in each equation.

$(-3)^2 + (-4)^2 \stackrel{?}{=} 25$ Substitute -3 for x and -4 for y in Equation 1.

$9 + 16 = 25$ Solution checks in Equation 1. ✓

$(-3) - (-4) \stackrel{?}{=} 1$ Substitute -3 for x and -4 for y in Equation 2.

$-3 + 4 = 1$ Solution checks in Equation 2. ✓

To check $(4, 3)$, substitute 4 for x and 3 for y in each equation.

$4^2 + 3^2 \stackrel{?}{=} 25$ Substitute 4 for x and 3 for y in Equation 1

$16 + 9 = 25$ Solution checks in Equation 1. ✓

$4 - 3 \stackrel{?}{=} 1$ Substitute 4 for x and 3 for y in Equation 2

$1 = 1$ Solution checks in Equation 2. ✓

Exercises Within Reach®

Solutions in English & Spanish and tutorial videos at AlgebraWithinReach.com

Solving a System Graphically **In Exercises 5–12, graph the equations to determine whether the system has any solutions. Find any solutions that exist.** See Additional Answers.

5. $\begin{cases} x^2 + y^2 = 100 \\ x + y = 2 \end{cases}$

$(-6, 8), (8, -6)$

6. $\begin{cases} x^2 + y^2 = 169 \\ x + y = 7 \end{cases}$

$(-5, 12), (12, -5)$

7. $\begin{cases} x^2 + y^2 = 25 \\ 2x - y = -5 \end{cases}$

$(0, 5), (-4, -3)$

8. $\begin{cases} x = 0 \\ x^2 + y^2 = 9 \end{cases}$

$(0, 3), (0, -3)$

9. $\begin{cases} x - 2y = 4 \\ x^2 - y = 0 \end{cases}$

No real solution

10. $\begin{cases} y = \sqrt{x - 2} \\ x - 2y = 1 \end{cases}$

$(3, 1)$

11. $\begin{cases} x^2 - y^2 = 16 \\ 3x - y = 12 \end{cases}$

$(5, 3), (4, 0)$

12. $\begin{cases} 9x^2 - 4y^2 = 36 \\ 5x - 2y = 0 \end{cases}$

No real solution

Solving Nonlinear Systems of Equations by Substitution

Method of Substitution

To solve a system of two equations in two variables, use the steps below.

1. Solve one of the equations for one variable in terms of the other.
2. Substitute the expression obtained in Step 1 into the other equation to obtain an equation in one variable.
3. Solve the equation obtained in Step 2.
4. Back-substitute the solution from Step 3 into the expression obtained in Step 1 to find the value of the other variable.
5. Check the solution to see that it satisfies *both* of the original equations.

EXAMPLE 3 Solving a Nonlinear System by Substitution

Solve the nonlinear system of equations.

$$\begin{cases} 4x^2 + y^2 = 4 & \text{Equation 1} \\ -2x + y = 2 & \text{Equation 2} \end{cases}$$

SOLUTION

Begin by solving for y in Equation 2 to obtain $y = 2x + 2$. Next, substitute this expression for y into Equation 1.

$4x^2 + y^2 = 4$	Write Equation 1.
$4x^2 + (2x + 2)^2 = 4$	Substitute $2x + 2$ for y.
$4x^2 + 4x^2 + 8x + 4 = 4$	Expand.
$8x^2 + 8x = 0$	Simplify.
$8x(x + 1) = 0$	Factor.
$8x = 0 \Rightarrow x = 0$	Set 1st factor equal to 0.
$x + 1 = 0 \Rightarrow x = -1$	Set 2nd factor equal to 0.

Finally, back-substitute these values of x into the revised Equation 2 to solve for y.

For $x = 0$: $y = 2(0) + 2 = 2$

For $x = -1$: $y = 2(-1) + 2 = 0$

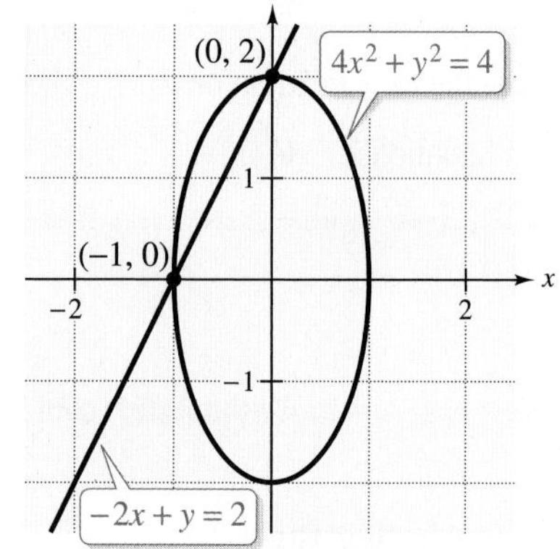

Remind students that *both* solutions must be checked in *both* equations of the system.

So, the system of equations has two solutions: $(0, 2)$ and $(-1, 0)$. The graph of the system is shown in the figure. Check these solutions in the original system.

Exercises Within Reach®

Solutions in English & Spanish and tutorial videos at AlgebraWithinReach.com

Solving a System by Substitution **In Exercises 13–16, solve the system by the method of substitution.**

13. $\begin{cases} y = 2x^2 \\ y = 6x - 4 \end{cases}$ (1, 2), (2, 8)

14. $\begin{cases} y = 5x^2 \\ y = -5x + 10 \end{cases}$ (−2, 20), (1, 5)

15. $\begin{cases} x^2 + y^2 = 4 \\ x + y = 2 \end{cases}$ (0, 2), (2, 0)

16. $\begin{cases} 2x^2 - y^2 = -8 \\ x - y = 6 \end{cases}$ (−14, −20), (2, −4)

EXAMPLE 4 Solving a Nonlinear System: The No Solution Case

Solve the nonlinear system of equations.

$$\begin{cases} x^2 - y = 0 & \text{Equation 1} \\ x \;\; - y = 1 & \text{Equation 2} \end{cases}$$

SOLUTION

Begin by solving for y in Equation 2 to obtain $y = x - 1$. Next, substitute this expression for y into Equation 1.

$x^2 - y = 0$ Write Equation 1.

$x^2 - (x - 1) = 0$ Substitute $x - 1$ for y.

$x^2 - x + 1 = 0$ Distributive Property

Use the Quadratic Formula, because this equation cannot be factored.

$$x = \frac{-(-1) \pm \sqrt{(-1)^2 - 4(1)(1)}}{2(1)}$$ Use Quadratic Formula.

$$= \frac{1 \pm \sqrt{1-4}}{2} = \frac{1 \pm \sqrt{-3}}{2}$$ Simplify.

Now, because the Quadratic Formula yields a negative number inside the radical, you can conclude that the equation $x^2 - x + 1 = 0$ has no real solution. So, the system has no real solution. The graph of the system is shown in the figure. From the graph, you can see that the parabola and the line have no point of intersection.

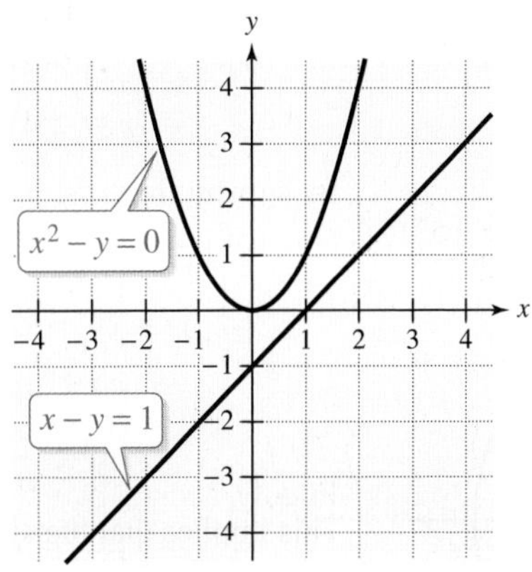

Exercises Within Reach®

Solutions in English & Spanish and tutorial videos at AlgebraWithinReach.com

Solving a System by Substitution **In Exercises 17–24, solve the system by the method of substitution.**

17. $\begin{cases} x^2 - 4y^2 = 16 \\ x^2 + y^2 = 1 \end{cases}$ No real solution

18. $\begin{cases} 2x^2 - y^2 = 12 \\ 3x^2 - y^2 = -4 \end{cases}$ No real solution

19. $\begin{cases} y = x^2 - 3 \\ x^2 + y^2 = 9 \end{cases}$ $(\pm\sqrt{5}, 2), (0, -3)$

20. $\begin{cases} x^2 + y^2 = 25 \\ x - 3y = -5 \end{cases}$ $(-5, 0), (4, 3)$

21. $\begin{cases} 16x^2 + 9y^2 = 144 \\ 4x + 3y = 12 \end{cases}$ $(0, 4), (3, 0)$

22. $\begin{cases} 4x^2 + 16y^2 = 64 \\ x + 2y = 4 \end{cases}$ $(4, 0), (0, 2)$

23. $\begin{cases} x^2 - y^2 = 9 \\ x^2 + y^2 = 1 \end{cases}$ No real solution

24. $\begin{cases} x^2 + y^2 = 1 \\ x + y = 7 \end{cases}$ No real solution

Solving Nonlinear Systems of Equations by Elimination

EXAMPLE 5 Solving a Nonlinear System by Elimination

Solve the nonlinear system of equations.

$$\begin{cases} 4x^2 + y^2 = 64 & \text{Equation 1} \\ x^2 + y^2 = 52 & \text{Equation 2} \end{cases}$$

SOLUTION

Both equations have y^2 as a term (and no other terms containing y). To eliminate y, multiply Equation 2 by -1 and then add.

$$\begin{array}{rcl} 4x^2 + y^2 &=& 64 \\ -x^2 - y^2 &=& -52 \\ \hline 3x^2 \quad &=& 12 \end{array} \qquad \text{Add equations.}$$

After eliminating y, solve the remaining equation for x.

$3x^2 = 12$ Write resulting equation.

$x^2 = 4$ Divide each side by 3.

$x = \pm 2$ Take square root of each side.

By substituting $x = 2$ into Equation 2, you obtain

$x^2 + y^2 = 52$ Write Equation 2.

$(2)^2 + y^2 = 52$ Substitute 2 for x.

$y^2 = 48$ Subtract 4 from each side.

$y = \pm 4\sqrt{3}.$ Take square root of each side and simplify.

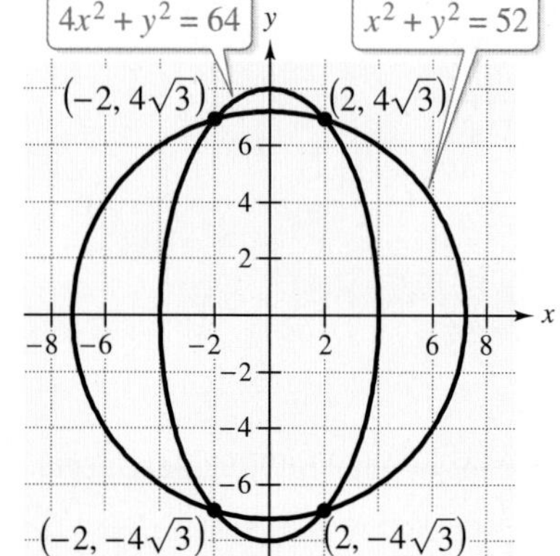

By substituting $x = -2$, you obtain the same values of y, as follows.

$x^2 + y^2 = 52$ Write Equation 2.

$(-2)^2 + y^2 = 52$ Substitute -2 for x.

$y^2 = 48$ Subtract 4 from each side.

$y = \pm 4\sqrt{3}.$ Take square root of each side and simplify.

This implies that the system has four solutions:

$$(2, 4\sqrt{3}),\quad (2, -4\sqrt{3}),\quad (-2, 4\sqrt{3}),\quad (-2, -4\sqrt{3}).$$

Check these solutions in the original system. The graph of the system is shown in the figure. Notice that the graph of Equation 1 is an ellipse and the graph of Equation 2 is a circle.

Exercises Within Reach®

Solving a System by Elimination **In Exercises 25–28, solve the system by the method of elimination.**

25. $\begin{cases} x^2 + 2y = 1 \\ x^2 + y^2 = 4 \end{cases}$ $(\pm\sqrt{3}, -1)$

26. $\begin{cases} x + y^2 = 5 \\ 2x^2 + y^2 = 6 \end{cases}$ $(1, \pm 2), \left(-\frac{1}{2}, \pm\frac{\sqrt{22}}{2}\right)$

27. $\begin{cases} -x + y^2 = 10 \\ x^2 - y^2 = -8 \end{cases}$ $(2, \pm 2\sqrt{3}), (-1, \pm 3)$

28. $\begin{cases} x^2 + y = 9 \\ x^2 - y^2 = 7 \end{cases}$ $(\pm\sqrt{11}, -2), (\pm 2\sqrt{2}, 1)$

Application EXAMPLE 6 Avalanche Rescue System

RECCO technology is often used in conjunction with other rescue methods such as avalanche dogs, transceivers, and probe lines.

RECCO® is an avalanche rescue system utilized by rescue organizations worldwide. RECCO technology enables quick directional pinpointing of a victim's exact location using harmonic radar. The two-part system consists of a detector used by rescuers and reflectors that are integrated into apparel, helmets, protection gear, or boots. The range of the detector through snow is 30 meters. Two rescuers are 30 meters apart on the surface. What is the maximum depth of a reflection that is in range of both rescuers?

SOLUTION

Let the first rescuer be located at the origin and let the second rescuer be located 30 meters (units) to the right. The range of each detector is circular and can be modeled by the following equations.

$$x^2 + y^2 = 30^2 \qquad \text{Range of first rescuer}$$

$$(x - 30)^2 + y^2 = 30^2 \qquad \text{Range of second rescuer}$$

Using methods demonstrated earlier in this section, you will find that these two equations intersect when

$x = 15$ and $y \approx \pm 25.98$.

You are concerned only about the lower portions of the circles. So, the maximum depth in range of both rescuers is point R, as shown below, which is about 26 meters beneath the surface.

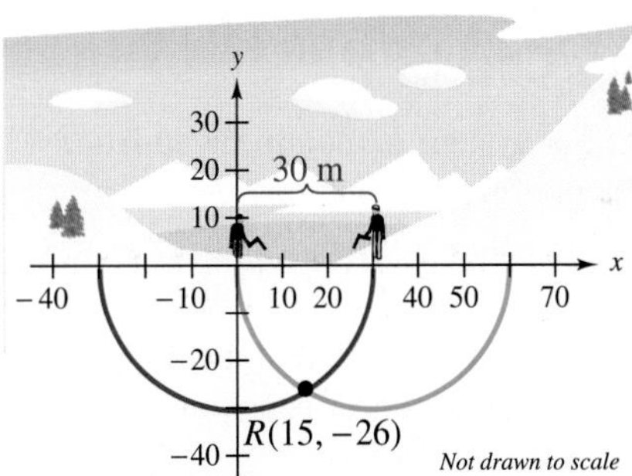

Exercises Within Reach®

Solutions in English & Spanish and tutorial videos at AlgebraWithinReach.com

29. ***Busing Boundary*** To be eligible to ride the school bus to East High School, a student must live at least 1 mile from the school (see figure). Describe the portion of Clarke Street for which the residents are *not* eligible to ride the school bus. Use a coordinate system in which the school is at (0, 0) and each unit represents 1 mile.

Between points $\left(-\frac{3}{5}, -\frac{4}{5}\right)$ and $\left(\frac{4}{5}, -\frac{3}{5}\right)$

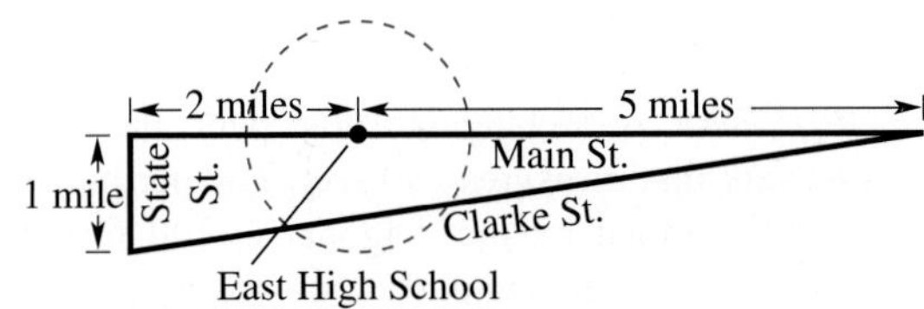

30. ***Search Team*** A search team of three members splits to search an area of the woods. Each member carries a family service radio with a circular range of 3 miles. The team members agree to communicate from their bases every hour. The second member sets up base 3 miles north of the first member. Where should the third member set up base to be as far east as possible but within direct communication range of each of the other two searchers? Use a coordinate system in which the first member is at (0, 0) and each unit represents 1 mile.

$\left(\frac{3\sqrt{3}}{2}, \frac{3}{2}\right)$

Concept Summary: Solving Nonlinear Systems of Equations

What

One way to solve a **nonlinear system of equations** is by graphing.

EXAMPLE

Find all solutions of the nonlinear system of equations.

$$\begin{cases} y = 2x - 1 & \text{Equation 1} \\ y = 2x^2 - 1 & \text{Equation 2} \end{cases}$$

How

Sketch the graph of each equation. Then approximate and check the point(s) of intersection (if any).

EXAMPLE

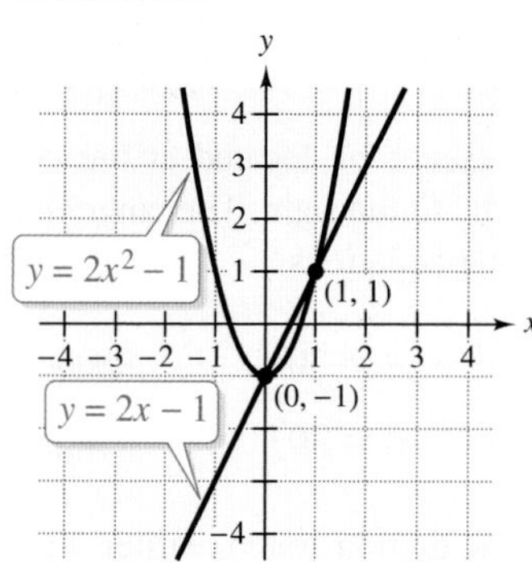

The system has two solutions: $(0, -1), (1, 1)$.

Why

Not all systems of equations are linear. There are many examples of nonlinear systems of equations in real-life applications. You can use graphing, substitution, and elimination to solve these systems just as you did with linear systems of equations.

Exercises Within Reach®

Worked-out solutions to odd-numbered exercises at AlgebraWithinReach.com

Concept Summary Check

31. ***Vocabulary*** How is a system of nonlinear equations different from a system of linear equations?
In addition to zero, one, or infinitely many solutions, a system of nonlinear equations can have two or more solutions.

32. ***Solving a System*** Identify the different methods you can use to solve a system of nonlinear equations.
Graphically, by substitution, by elimination

33. ***Choosing a Method*** If one of the equations in a system is linear, which algebraic method usually works best for solving the system? Substitution

34. ***Choosing a Method*** If both of the equations in a system are conics, which algebraic method usually works best for solving the system? Elimination

Extra Practice

Solving a System by Substitution **In Exercises 35–38, solve the system by the method of substitution.**

35. $\begin{cases} y = x^2 - 5 \\ 3x + 2y = 10 \end{cases}$
$(-4, 11), \left(\frac{5}{2}, \frac{5}{4}\right)$

36. $\begin{cases} x + y = 4 \\ x^2 - y^2 = 4 \end{cases}$
$\left(\frac{5}{2}, \frac{3}{2}\right)$

37. $\begin{cases} y = \sqrt{4 - x} \\ x + 3y = 6 \end{cases}$
$(0, 2), (3, 1)$

38. $\begin{cases} y = \sqrt{25 - x^2} \\ x + y = 7 \end{cases}$
$(3, 4), (4, 3)$

Solving a System by Elimination **In Exercises 39–42, solve the system by the method of elimination.**

39. $\begin{cases} \dfrac{x^2}{4} + y^2 = 1 \\ x^2 + \dfrac{y^2}{4} = 1 \end{cases}$
$\left(\pm\frac{2\sqrt{5}}{5}, \pm\frac{2\sqrt{5}}{5}\right)$

40. $\begin{cases} x^2 - y^2 = 1 \\ \dfrac{x^2}{2} + y^2 = 1 \end{cases}$
$\left(\pm\frac{2\sqrt{3}}{3}, \pm\frac{\sqrt{3}}{3}\right)$

41. $\begin{cases} y^2 - x^2 = 10 \\ x^2 + y^2 = 16 \end{cases}$
$(\pm\sqrt{3}, \pm\sqrt{13})$

42. $\begin{cases} x^2 + y^2 = 25 \\ x^2 + 2y^2 = 36 \end{cases}$
$(\pm\sqrt{14}, \pm\sqrt{11})$

43. ***Sailboat*** A sail for a sailboat is shaped like a right triangle that has a perimeter of 36 meters and a hypotenuse of 15 meters. Find the dimensions of the sail.
9 meters × 12 meters

44. ***Dog Park*** A rectangular dog park has a diagonal sidewalk that measures 290 feet. The perimeter of each triangle formed by the diagonal is 700 feet. Find the dimensions of the dog park. 200 feet × 210 feet

45. ***Hyperbolic Mirror*** In a hyperbolic mirror, light rays directed to one focus are reflected to the other focus. The mirror in the figure has the equation

$$\frac{x^2}{9} - \frac{y^2}{16} = 1.$$

At which point on the mirror will light from the point (0, 10) reflect to the focus? (3.633, 2.733)

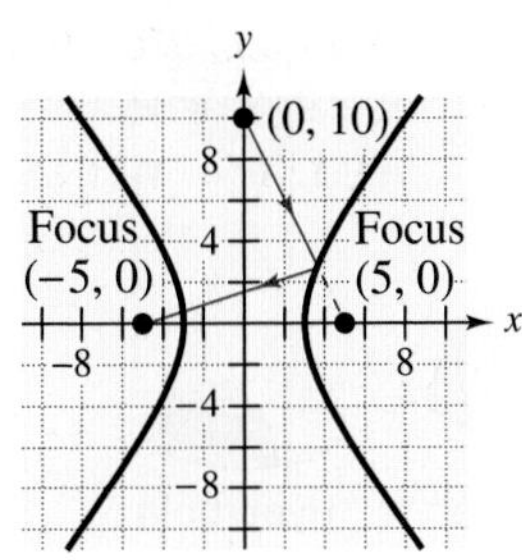

46. ***Miniature Golf*** You are playing miniature golf and your golf ball is at (−15, 25) (see figure). A wall at the end of the enclosed area is part of a hyperbola whose equation is

$$\frac{x^2}{19} - \frac{y^2}{81} = 1.$$

Using the reflective property of hyperbolas given in Exercise 45, at which point on the wall must your ball hit for it to go into the hole? (The ball bounces off the wall only once.) (4.989, 5.011)

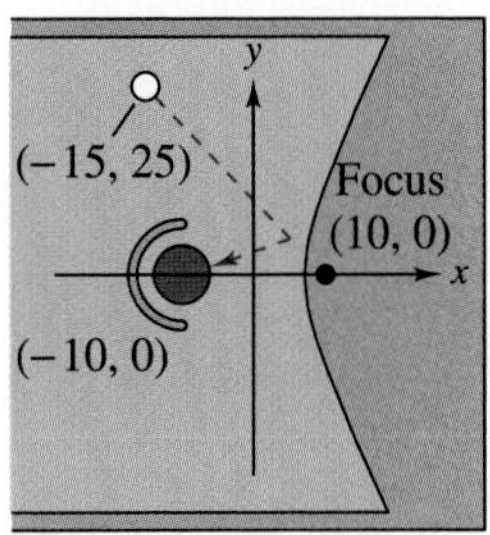

Explaining Concepts

47. ***Writing*** Explain how to solve a nonlinear system of equations using the method of substitution.

Solve one of the equations for one variable in terms of the other. Substitute that expression into the other equation and solve. Back-substitute the solution into the first equation to find the value of the other variable. Check the solution to see that it satisfies both of the original equations.

48. ***Writing*** Explain how to solve a nonlinear system of equations using the method of elimination.

Multiply Equation 2 by a factor that makes the coefficients of one variable equal. Subtract Equation 2 from Equation 1. Write the resulting equation and solve. Back-substitute the solution into either equation to find the value of the other variable. Check the solution to see that it satisfies both of the original equations.

49. ***Precision*** What is the maximum number of points of intersection of a line and a hyperbola? Explain.

Two. The line can intersect a branch of the hyperbola at most twice, and it can intersect only one point on each branch at the same time.

50. ***Precision*** A circle and a parabola can have 0, 1, 2, 3, or 4 points of intersection. Sketch the circle given by $x^2 + y^2 = 4$. Discuss how this circle could intersect a parabola with an equation of the form $y = x^2 + C$. Then find the values of C for each of the five cases described below.

(a) No points of intersection
$C < -\frac{17}{4}$ or $C > 2$

(b) One point of intersection
$C = 2$

(c) Two points of intersection
$C = -\frac{17}{4}$ or $-2 < C < 2$

(d) Three points of intersection
$C = -2$

(e) Four points of intersection
$-\frac{17}{4} < C < -2$

Cumulative Review

In Exercises 51–62, solve the equation and check your solution(s).

51. $\sqrt{6 - 2x} = 4$ -5

52. $\sqrt{x + 3} = -9$ No real solution

53. $\sqrt{x} = x - 6$ 9

54. $\sqrt{x + 14} = \sqrt{x} + 3$ $\frac{25}{36}$

55. $3^x = 243$ 5

56. $4^x = 256$ 4

57. $5^{x-1} = 310$ 4.564

58. $e^{0.5x} = 8$ 4.1589

59. $\log_{10} x = 0.01$ 1.023

60. $\log_4 8x = 3$ 8

61. $2 \ln(x + 1) = -2$ -0.632

62. $\ln(x + 3) - \ln x = \ln 1$ No real solution

10 Chapter Summary

	What did you learn?	Explanation and Examples	Review Exercises
10.1	Recognize the four basic conics: circles, parabolas, ellipses, and hyperbolas *(p. 532)*.	Circle Parabola Ellipse Hyperbola	1–10
	Graph and write equations of circles *(p. 533)*.	*Standard equation of a circle with radius r and center (0, 0):* $x^2 + y^2 = r^2$ *Standard equation of a circle with radius r and center (h, k):* $(x - h)^2 + (y - k)^2 = r^2$	11–18
	Graph and write equations of parabolas *(p. 536)*.	*Standard equation of a parabola with vertex (0, 0):* $x^2 = 4py$ Vertical axis $y^2 = 4px$ Horizontal axis *Standard equation of a parabola with vertex at (h, k):* $(x - h)^2 = 4p(y - k)$ Vertical axis $(y - k)^2 = 4p(x - h)$ Horizontal axis The focus of a parabola lies on the axis, a directed distance of p units from the vertex.	19–24
10.2	Graph and write equations of ellipses centered at the origin *(p. 540)*.	*Standard equation of an ellipse centered at the origin:* $\frac{x^2}{a^2} + \frac{y^2}{b^2} = 1$ Major axis is horizontal. $\frac{x^2}{b^2} + \frac{y^2}{a^2} = 1$ Major axis is vertical.	25–32
	Graph and write equations of ellipses centered at (h, k) *(p. 542)*.	*Standard equation of an ellipse centered at (h, k):* $\frac{(x - h)^2}{a^2} + \frac{(y - k)^2}{b^2} = 1$ Major axis is horizontal. $\frac{(x - h)^2}{b^2} + \frac{(y - k)^2}{a^2} = 1$ Major axis is vertical. In all of the standard equations for ellipses, $0 < b < a$. The vertices of the ellipse lie on the major axis, a units from the center (the major axis has length $2a$). The co-vertices lie on the minor axis, b units from the center (the minor axis has length $2b$).	33–40

	What did you learn?	Explanation and Examples	Review Exercises
10.3	Analyze hyperbolas centered at the origin *(p. 550)*.	*Standard form of a hyperbola centered at the origin:* $\dfrac{x^2}{a^2} - \dfrac{y^2}{b^2} = 1$ Transverse axis is horizontal. $\dfrac{y^2}{a^2} - \dfrac{x^2}{b^2} = 1$ Transverse axis is vertical.	41–48
	Analyze hyperbolas centered at (h, k) *(p. 553)*.	*Standard equation of a hyperbola centered at (h, k):* $\dfrac{(x-h)^2}{a^2} - \dfrac{(y-k)^2}{b^2} = 1$ Transverse axis is horizontal. $\dfrac{(y-k)^2}{a^2} - \dfrac{(x-h)^2}{b^2} = 1$ Transverse axis is vertical. A hyperbola's vertices lie on the transverse axis, a units from the center. A hyperbola's central rectangle has side lengths of $2a$ and $2b$. A hyperbola's asymptotes pass through opposite corners of its central rectangle.	49–54
10.4	Solve nonlinear systems of equations graphically *(p. 558)*.	**1.** Sketch the graph of each equation in the system. **2.** Graphically approximate the coordinates of any points of intersection of the graphs. **3.** Check the coordinates by substituting them into each equation in the original system. If the coordinates do not check, you may have to use an algebraic approach such as substitution or elimination.	55–58
	Solve nonlinear systems of equations by substitution *(p. 560)*.	**1.** Solve one equation for one variable in terms of the other. **2.** Substitute the expression obtained in Step 1 into the other equation to obtain an equation in one variable. **3.** Solve the equation obtained in Step 2. **4.** Back-substitute the solution from Step 3 into the expression obtained in Step 1 to find the value of the other variable. **5.** Check that the solution satisfies *both* of the original equations.	59–66, 77–82
	Solve nonlinear systems of equations by elimination *(p. 562)*.	**1.** Obtain coefficients for x (or y) that are opposites by multiplying one or both equations by suitable constants. **2.** Add the equations to eliminate one variable. Solve the resulting equation. **3.** Back-substitute the value from Step 2 into either of the original equations and solve for the other variable. **4.** Check the solution in both of the original equations.	67–76

Review Exercises

Worked-out solutions to odd-numbered exercises at AlgebraWithinReach.com

10.1

Identifying a Conic **In Exercises 1–10, identify the type of conic shown in the graph.**

1.
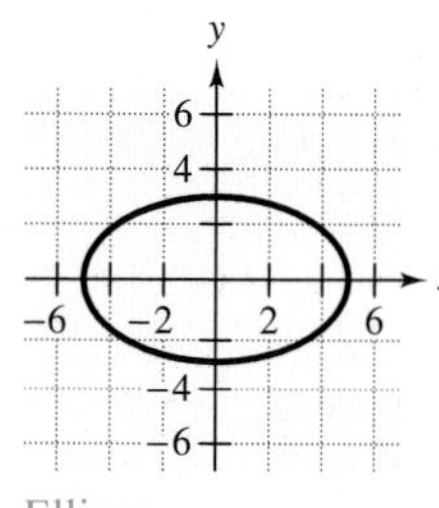
Ellipse

2.
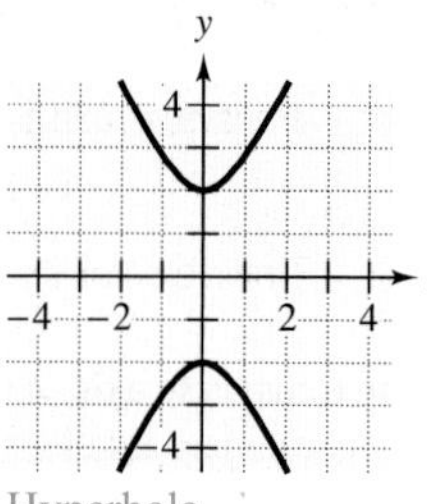
Hyperbola

3.
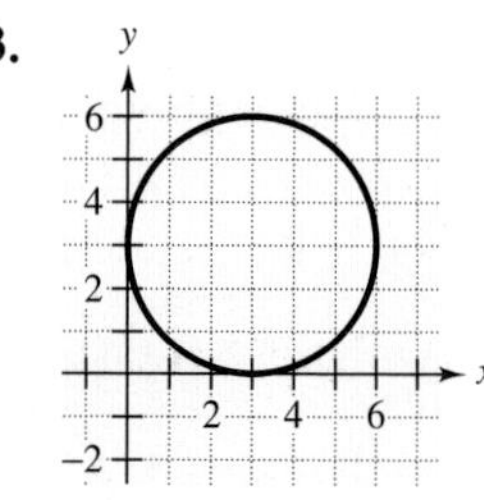
Circle

4.
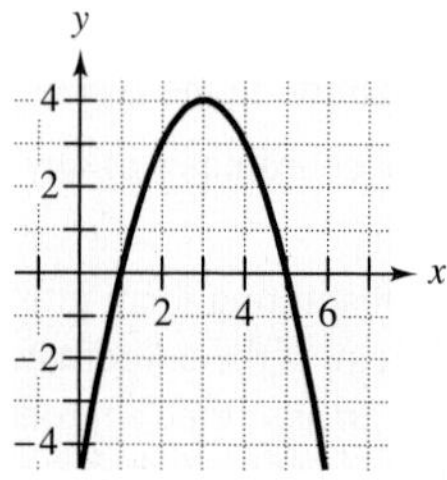
Parabola

5.
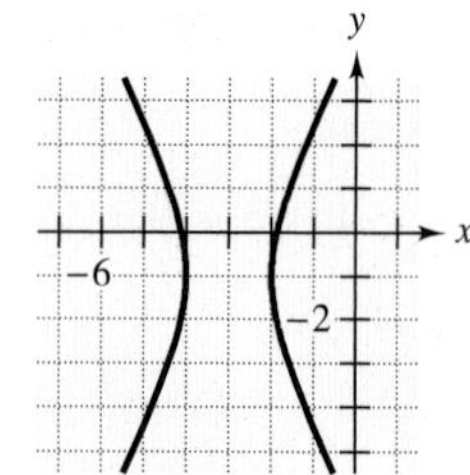
Hyperbola

6.
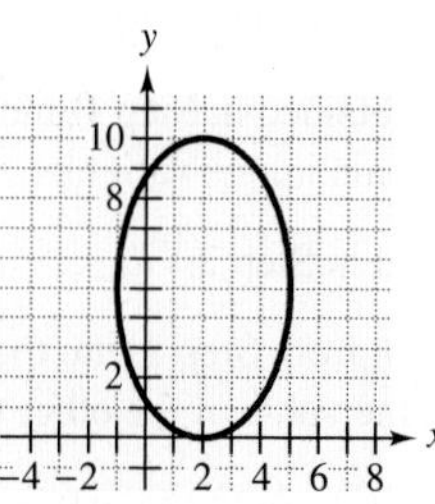
Ellipse

7.
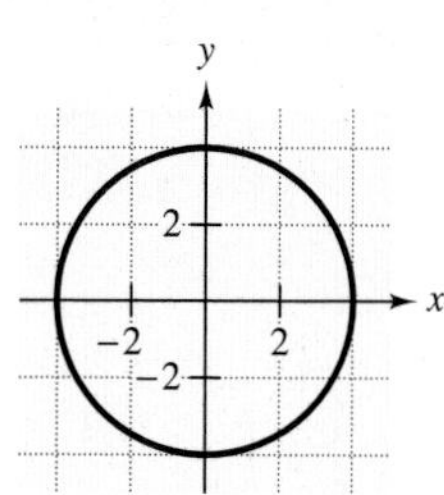
Circle

8.
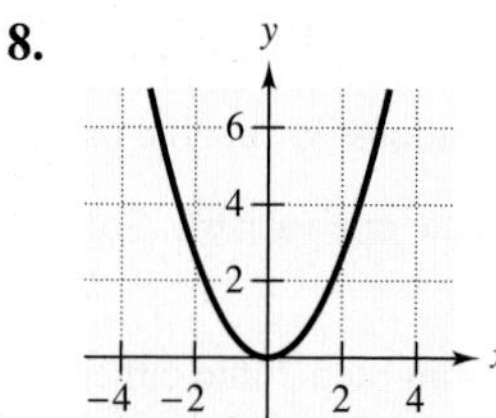
Parabola

9.
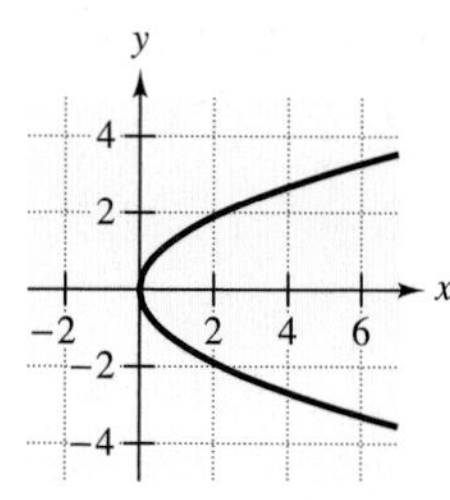
Parabola

10.
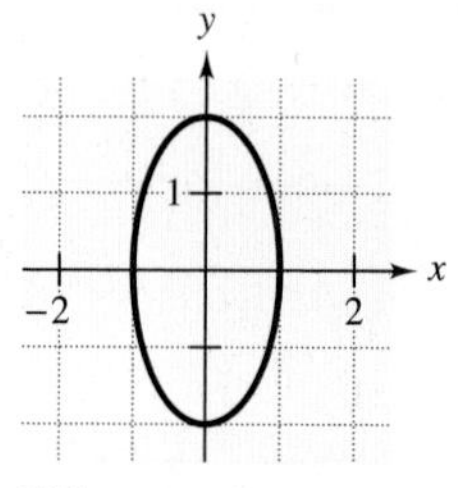
Ellipse

Writing an Equation of a Circle **In Exercises 11 and 12, write the standard form of the equation of the circle centered at the origin.**

11. Radius: 6
$x^2 + y^2 = 36$

12. Passes through the point $(-1, 3)$
$x^2 + y^2 = 10$

Sketching a Circle **In Exercises 13 and 14, identify the center and radius of the circle, and sketch its graph.** See Additional Answers.

13. $x^2 + y^2 = 64$
Center: $(0, 0)$, $r = 8$

14. $9x^2 + 9y^2 - 49 = 0$
Center: $(0, 0)$, $r = \frac{7}{3}$

Writing an Equation of a Circle **In Exercises 15 and 16, write the standard form of the equation of the circle centered at (h, k).**

15. Center: $(2, 6)$, Radius: 3
$(x - 2)^2 + (y - 6)^2 = 9$

16. Center: $(-2, 3)$, Passes through the point $(1, 1)$
$(x + 2)^2 + (y - 3)^2 = 13$

Sketching a Circle **In Exercises 17 and 18, identify the center and radius of the circle, and sketch its graph.**
See Additional Answers.

17. $x^2 + y^2 + 6x + 8y + 21 = 0$
Center: $(-3, -4)$, $r = 2$

18. $x^2 + y^2 - 8x + 16y + 75 = 0$
Center: $(4, -8)$, $r = \sqrt{5}$

Writing an Equation of a Parabola **In Exercises 19–22, write the standard form of the equation of the parabola. Then sketch the parabola.**
See Additional Answers.

19. Vertex: $(0, 0)$, Focus: $(6, 0)$
$y^2 = 24x$

20. Vertex: $(0, 0)$, Focus: $(0, 1)$
$x^2 = 4y$

21. Vertex: $(0, 5)$, Focus: $(2, 5)$
$(y - 5)^2 = 8x$

22. Vertex: $(-3, 2)$, Focus: $(-3, 0)$
$(x + 3)^2 = -8(y - 2)$

Sketching a Parabola **In Exercises 23 and 24, sketch the parabola, and identify its vertex and focus.**
See Additional Answers.

23. $y = \frac{1}{2}x^2 - 8x + 7$
Vertex: $(8, -25)$, Focus: $\left(8, -\frac{49}{2}\right)$

24. $x = y^2 + 10y - 4$
Vertex: $(-29, -5)$, Focus: $\left(-\frac{115}{4}, -5\right)$

10.2

Writing the Standard Equation of an Ellipse **In Exercises 25–28, write the standard form of the equation of the ellipse centered at the origin.**

25. Vertices: $(0, -5)$, $(0, 5)$

Co-vertices: $(-2, 0)$, $(2, 0)$

$\frac{x^2}{4} + \frac{y^2}{25} = 1$

26. Vertices: $(-8, 0)$, $(8, 0)$

Co-vertices: $(0, -3)$, $(0, 3)$

$\frac{x^2}{64} + \frac{y^2}{9} = 1$

27. Major axis (vertical) 6 units, minor axis 4 units

$\frac{x^2}{4} + \frac{y^2}{9} = 1$

28. Major axis (horizontal) 12 units, minor axis 2 units

$\frac{x^2}{36} + y^2 = 1$

Sketching an Ellipse **In Exercises 29–32, sketch the ellipse. Identify the vertices and co-vertices.**
See Additional Answers.

29. $\frac{x^2}{64} + \frac{y^2}{16} = 1$
Vertices: $(\pm 8, 0)$, Co-vertices: $(0, \pm 4)$

30. $\frac{x^2}{9} + y^2 = 1$
Vertices: $(\pm 3, 0)$, Co-vertices: $(0, \pm 1)$

31. $36x^2 + 9y^2 - 36 = 0$
Vertices: $(0, \pm 2)$, Co-vertices: $(\pm 1, 0)$

32. $100x^2 + 4y^2 - 4 = 0$
Vertices: $(0, \pm 1)$, Co-vertices: $\left(\pm\frac{1}{5}, 0\right)$

Writing the Standard Equation of an Ellipse **In Exercises 33–36, write the standard form of the equation of the ellipse.**

33. Vertices: $(-2, 4)$, $(8, 4)$

Co-vertices: $(3, 0)$, $(3, 8)$

$\frac{(x-3)^2}{25} + \frac{(y-4)^2}{16} = 1$

34. Vertices: $(0, 5)$, $(12, 5)$

Co-vertices: $(6, 2)$, $(6, 8)$

$\frac{(x-6)^2}{36} + \frac{(y-5)^2}{9} = 1$

35. Vertices: $(0, 0)$, $(0, 8)$

Co-vertices: $(-3, 4)$, $(3, 4)$

$\frac{x^2}{9} + \frac{(y-4)^2}{16} = 1$

36. Vertices: $(5, -3)$, $(5, 13)$

Co-vertices: $(3, 5)$, $(7, 5)$

$\frac{(x-5)^2}{4} + \frac{(y-5)^2}{64} = 1$

Sketching an Ellipse **In Exercises 37–40, find the center and vertices of the ellipse, and sketch its graph.**
See Additional Answers.

37. $9(x + 1)^2 + 4(y - 2)^2 = 144$
Center: $(-1, 2)$
Vertices: $(-1, -4)$, $(-1, 8)$

38. $x^2 + 25y^2 - 4x - 21 = 0$
Center: $(2, 0)$
Vertices: $(-3, 0)$, $(7, 0)$

39. $16x^2 + y^2 + 6y - 7 = 0$
Center: $(0, -3)$
Vertices: $(0, -7)$, $(0, 1)$

40. $x^2 + 4y^2 + 10x - 24y + 57 = 0$
Center: $(-5, 3)$
Vertices: $(-7, 3)$, $(-3, 3)$

10.3

Sketching a Hyperbola **In Exercises 41–44, identify the vertices and asymptotes of the hyperbola, and sketch its graph.**
See Additional Answers.

41. $x^2 - y^2 = 25$
Vertices: $(\pm 5, 0)$
Asymptotes: $y = \pm x$

42. $y^2 - x^2 = 16$
Vertices: $(0, \pm 4)$
Asymptotes: $y = \pm x$

43. $\frac{y^2}{25} - \frac{x^2}{4} = 1$
Vertices: $(0, \pm 5)$
Asymptotes: $y = \pm\frac{5}{2}x$

44. $\frac{x^2}{16} - \frac{y^2}{25} = 1$
Vertices: $(\pm 4, 0)$
Asymptotes: $y = \pm\frac{5}{4}x$

Writing an Equation of a Hyperbola **In Exercises 45–48, write the standard form of the equation of the hyperbola centered at the origin.**

	Vertices	*Asymptotes*	
45.	$(-2, 0), (2, 0)$	$y = \frac{3}{2}x$	$y = -\frac{3}{2}x$
	$\frac{x^2}{4} - \frac{y^2}{9} = 1$		
46.	$(0, -6), (0, 6)$	$y = 3x$	$y = -3x$
	$\frac{y^2}{36} - \frac{x^2}{4} = 1$		
47.	$(0, -8), (0, 8)$	$y = \frac{4}{5}x$	$y = -\frac{4}{5}x$
	$\frac{y^2}{64} - \frac{x^2}{100} = 1$		
48.	$(-3, 0), (3, 0)$	$y = \frac{4}{3}x$	$y = -\frac{4}{3}x$
	$\frac{x^2}{9} - \frac{y^2}{16} = 1$		

Sketching a Hyperbola **In Exercises 49–52, find the center and vertices of the hyperbola, and sketch its graph.** See Additional Answers.

49. $\frac{(x-3)^2}{9} - \frac{(y+1)^2}{4} = 1$

Center: $(3, -1)$, Vertices: $(0, -1), (6, -1)$

50. $\frac{(x+4)^2}{25} - \frac{(y-7)^2}{64} = 1$

Center: $(-4, 7)$, Vertices: $(-9, 7), (1, 7)$

51. $8y^2 - 2x^2 + 48y + 16x + 8 = 0$

Center: $(4, -3)$, Vertices: $(4, -1), (4, -5)$

52. $25x^2 - 4y^2 - 200x - 40y = 0$

Center: $(4, -5)$, Vertices: $\left(4 \pm 2\sqrt{3}, -5\right)$

Writing an Equation of a Hyperbola **In Exercises 53 and 54, write the standard form of the equation of the hyperbola.**

53.

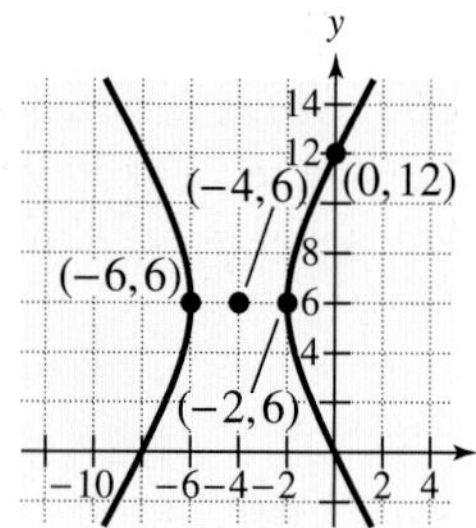

$\frac{(x+4)^2}{4} - \frac{(y-6)^2}{12} = 1$

54.

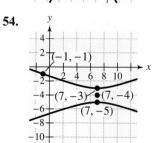

$(y+4)^2 - \frac{(x-7)^2}{8} = 1$

10.4

Solving a System Graphically **In Exercises 55–58, graph the equations to determine whether the system has any solutions. Find any solutions that exist.** See Additional Answers.

55. $\begin{cases} y = x^2 \\ y = 3x \end{cases}$

$(0, 0), (3, 9)$

56. $\begin{cases} y = 2 + x^2 \\ y = 8 - x \end{cases}$

$(-3, 11), (2, 6)$

57. $\begin{cases} x^2 + y^2 = 16 \\ -x + y = 4 \end{cases}$

$(-4, 0), (0, 4)$

58. $\begin{cases} 2x^2 - y^2 = -8 \\ y = x + 6 \end{cases}$

$(-2, 4), (14, 20)$

Solving a System by Substitution **In Exercises 59–66, solve the system by the method of substitution.**

59. $\begin{cases} y = 5x^2 \\ y = -15x - 10 \end{cases}$

$(-1, 5), (-2, 20)$

60. $\begin{cases} y^2 = 16x \\ 4x - y = -24 \end{cases}$

No real solution

61. $\begin{cases} x^2 + y^2 = 1 \\ x + y = -1 \end{cases}$

$(-1, 0), (0, -1)$

62. $\begin{cases} y^2 - x^2 = 9 \\ x + y = 1 \end{cases}$

$(-4, 5)$

63. $\begin{cases} 4x + y^2 = 2 \\ 2x - y = -11 \end{cases}$

$\left(-\frac{17}{2}, -6\right), \left(-\frac{7}{2}, 4\right)$

64. $\begin{cases} x^2 + y^2 = 10 \\ 2x - y = 5 \end{cases}$

$(1, -3), (3, 1)$

65. $\begin{cases} x^2 + y^2 = 9 \\ x + 2y = 3 \end{cases}$

$\left(-\frac{9}{5}, \frac{12}{5}\right), (3, 0)$

66. $\begin{cases} x^2 + y^2 = 4 \\ x - 2y = 4 \end{cases}$

$(0, -2), \left(\frac{8}{5}, -\frac{6}{5}\right)$

Solving a System by Elimination **In Exercises 67–76, solve the system by the method of elimination.**

67. $\begin{cases} 6x^2 - y^2 = 15 \\ x^2 + y^2 = 13 \end{cases}$

$(\pm 2, \pm 3)$

68. $\begin{cases} x^2 + y^2 = 16 \\ -x^2 + \dfrac{y^2}{16} = 1 \end{cases}$

$(0, \pm 4)$

69. $\begin{cases} x^2 + y^2 = 7 \\ x^2 - y^2 = 1 \end{cases}$

$(\pm 2, \pm\sqrt{3})$

70. $\begin{cases} x^2 + y^2 = 25 \\ y^2 - x^2 = 7 \end{cases}$

$(\pm 3, \pm 4)$

71. $\begin{cases} x^2 - y^2 = 4 \\ x^2 + y^2 = 4 \end{cases}$

$(\pm 2, 0)$

72. $\begin{cases} x^2 + y^2 = 25 \\ x^2 - y^2 = -36 \end{cases}$

No real solution

73. $\begin{cases} x^2 + y^3 = 13 \\ 2x^2 + 3y^2 = 30 \end{cases}$

$(\pm 3, \pm 2)$

74. $\begin{cases} 3x^2 - y^2 = 4 \\ x^2 + 4y^2 = 10 \end{cases}$

$(\pm\sqrt{2}, \pm\sqrt{2})$

75. $\begin{cases} 4x^2 + 9y^2 = 36 \\ 2x^2 - 9y^2 = 18 \end{cases}$

$(\pm 3, 0)$

76. $\begin{cases} 5x^2 - 2y^2 = -13 \\ 3x^2 + 4y^2 = 39 \end{cases}$

$(\pm 1, \pm 3)$

77. ***Geometry*** A circuit board has a perimeter of 28 centimeters and a diagonal of 10 centimeters. Find the dimensions of the circuit board.

6 centimeters × 8 centimeters

78. ***Geometry*** A ceramic tile has a perimeter of 6 inches and a diagonal of $\sqrt{5}$ inches. Find the dimensions of the tile.

1 inch × 2 inches

79. ***Ice Rink*** A rectangular ice rink has an area of 3000 square feet. The diagonal across the rink is 85 feet. Find the dimensions of the rink.

40 feet × 75 feet

80. ***Cell Phone*** A cell phone has a rectangular external display that contains 19,200 pixels with a diagonal of 200 pixels. Find the resolution (dimensions in pixels) of the external display.

120 pixels × 160 pixels

81. ***Geometry*** A piece of wire 100 inches long is cut into two pieces. Each of the two pieces is then bent into a square. The area of one square is 144 square inches greater than the area of the other square. Find the length of each piece of wire.

Piece 1: 38.48 inches
Piece 2: 61.52 inches

82. ***Geometry*** You have 250 feet of fencing to enclose two corrals of equal size (see figure). The combined area of the corrals is 2400 square feet. Find the dimensions of each corral.

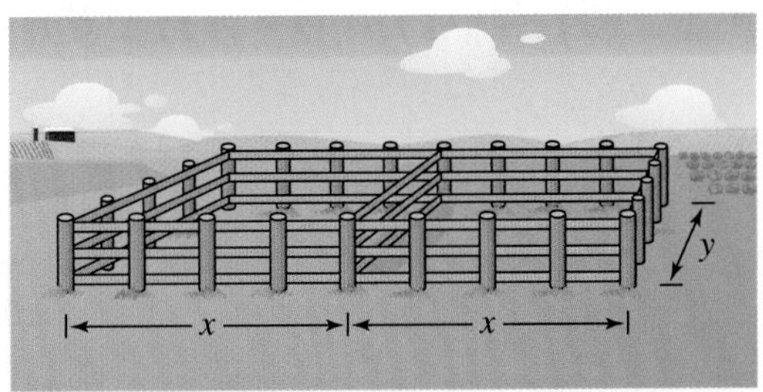

40 feet × 30 feet or $22\frac{1}{2}$ feet × $53\frac{1}{3}$ feet

Chapter Test

Solutions in English & Spanish and tutorial videos at AlgebraWithinReach.com

Take this test as you would take a test in class. After you are done, check your work against the answers in the back of the book.

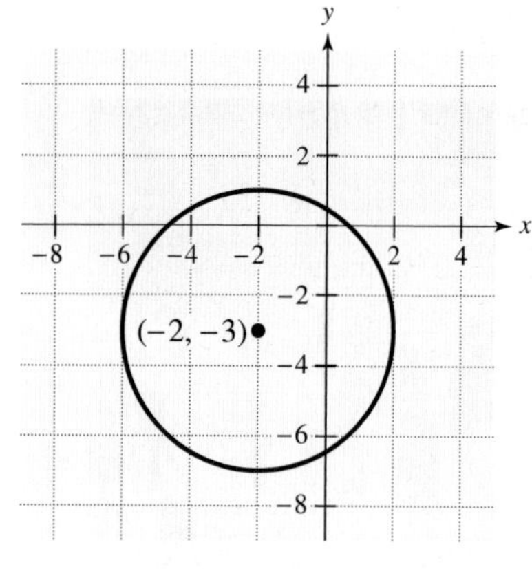

1. Write the standard form of the equation of the circle shown at the left.
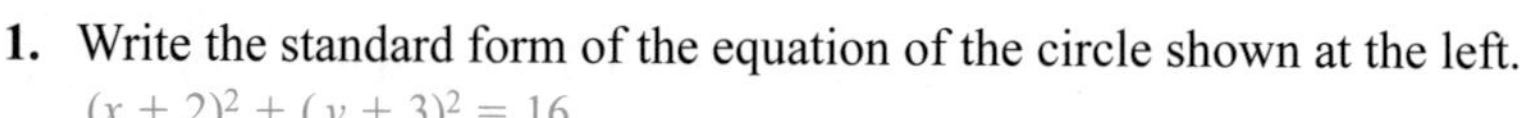
$(x + 2)^2 + (y + 3)^2 = 16$

In Exercises 2 and 3, write the standard form of the equation of the circle, and sketch its graph. See Additional Answers.

2. $x^2 + y^2 - 2x - 6y + 1 = 0$ $(x - 1)^2 + (y - 3)^2 = 9$

3. $x^2 + y^2 + 4x - 6y + 4 = 0$ $(x + 2)^2 + (y - 3)^2 = 9$

4. Identify the vertex and the focus of the parabola $x = -3y^2 + 12y - 8$, and sketch its graph. Vertex: $(4, 2)$, Focus: $\left(\frac{47}{12}, 2\right)$; See Additional Answers.

5. Write the standard form of the equation of the parabola with vertex $(7, -2)$ and focus $(7, 0)$. $(x - 7)^2 = 8(y + 2)$

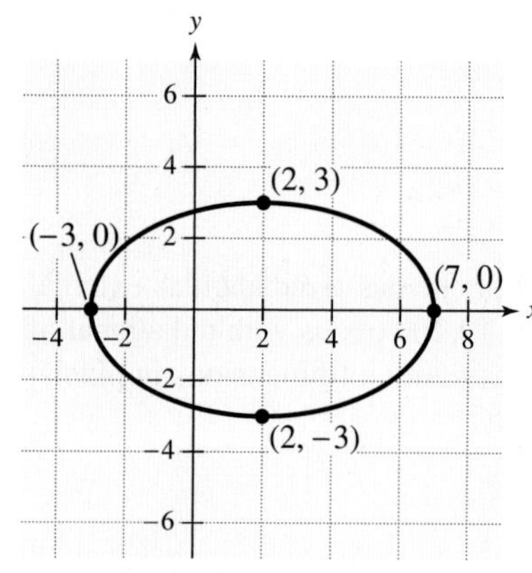

6. Write the standard form of the equation of the ellipse shown at the left.
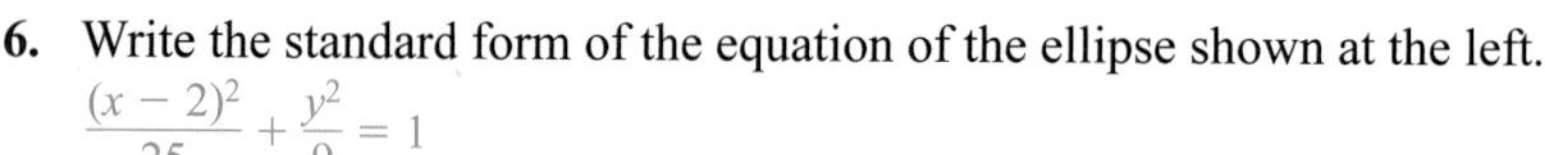
$\frac{(x - 2)^2}{25} + \frac{y^2}{9} = 1$

In Exercises 7 and 8, find the center and vertices of the ellipse, and sketch its graph. See Additional Answers.

7. $16x^2 + 4y^2 = 64$ Center: $(0, 0)$, Vertices: $(0, \pm 4)$

8. $25x^2 + 4y^2 - 50x - 24y - 39 = 0$ Center: $(1, 3)$, Vertices: $(1, -2)$, $(1, 8)$

In Exercises 9 and 10, write the standard form of the equation of the hyperbola.

9. Vertices: $(-3, 0)$, $(3, 0)$; Asymptotes: $y = \pm\frac{2}{3}x$ $\frac{x^2}{9} - \frac{y^2}{4} = 1$

10. Vertices: $(0, -5)$, $(0, 5)$; Asymptotes: $y = \pm\frac{5}{2}x$ $\frac{y^2}{25} - \frac{x^2}{4} = 1$

In Exercises 11 and 12, find the center and vertices of the hyperbola, and sketch its graph. See Additional Answers.

11. $4x^2 - 2y^2 - 24x + 20 = 0$ Center: $(3, 0)$, Vertices: $(1, 0)$, $(5, 0)$

12. $16y^2 - 25x^2 + 64y + 200x - 736 = 0$ Center: $(4, -2)$, Vertices: $(4, -7)$; $(4, 3)$

In Exercises 13–15, solve the nonlinear system of equations.

13. $\begin{cases} \frac{x^2}{16} + \frac{y^2}{9} = 1 \\ 3x + 4y = 12 \end{cases}$ $(0, 3)$, $(4, 0)$

14. $\begin{cases} x^2 + y^2 = 16 \\ \frac{x^2}{16} - \frac{y^2}{9} = 1 \end{cases}$ $(\pm 4, 0)$

15. $\begin{cases} x^2 + y^2 = 10 \\ x^2 = y^2 + 2 \end{cases}$ $(\sqrt{6}, 2)$, $(\sqrt{6}, -2)$, $(-\sqrt{6}, 2)$, $(-\sqrt{6}, -2)$

16. Write the equation of the circular orbit of a satellite 1000 miles above the surface of Earth. Place the origin of the rectangular coordinate system at the center of Earth and assume the radius of Earth is 4000 miles. $x^2 + y^2 = 5000^2$

17. A rectangle has a perimeter of 56 inches and a diagonal of 20 inches. Find the dimensions of the rectangle. 16 inches $\times$ 12 inches

Cumulative Test: Chapters 8–10

Solutions in English & Spanish and tutorial videos at AlgebraWithinReach.com

Take this test as you would take a test in class. After you are done, check your work against the answers in the back of the book.

In Exercises 1–4, solve the equation by the specified method.

1. Factoring:

$4x^2 - 9x - 9 = 0$ $-\frac{3}{4}, 3$

2. Square Root Property:

$(x - 5)^2 - 64 = 0$ $-3, 13$

3. Completing the square:

$x^2 - 10x - 25 = 0$ $5 \pm 5\sqrt{2}$

4. Quadratic Formula:

$3x^2 + 6x + 2 = 0$ $-1 \pm \frac{\sqrt{3}}{3}$

5. Solve the equation of quadratic form: $x^4 - 8x^2 + 15 = 0$

$\pm\sqrt{3}, \pm\sqrt{5}$

In Exercises 6 and 7, solve the inequality and graph the solution on the real number line. See Additional Answers.

6. $3x^2 + 8x \leq 3$ $\left[-3, \frac{1}{3}\right]$

7. $\frac{3x + 4}{2x - 1} < 0$ $\left(-\frac{4}{3}, \frac{1}{2}\right)$

8. Find a quadratic equation having the solutions -2 and 6. $x^2 - 4x - 12 = 0$

9. Find the compositions (a) $f \circ g$ and (b) $g \circ f$ for $f(x) = 2x^2 - 3$ and $g(x) = 5x - 1$. Then find the domain of each composition.

(a) $(f \circ g)(x) = 50x^2 - 20x - 1$; Domain: $(-\infty, \infty)$
(b) $(g \circ f)(x) = 10x^2 - 16$; Domain: $(-\infty, \infty)$

10. Find the inverse function of $f(x) = \frac{5 - 3x}{4}$. $f^{-1}(x) = -\frac{4}{3}x + \frac{5}{3}$

11. Evaluate $f(x) = 7 + 2^{-x}$ when $x = 1, 0.5$, and 3.

$f(1) = \frac{15}{2}, f(0.5) \approx 7.707, f(3) = \frac{57}{8}$

12. Sketch the graph of $f(x) = 4^{x-1}$ and identify the horizontal asymptote.

See Additional Answers; $y = 0$

13. Describe the relationship between the graphs of $f(x) = e^x$ and $g(x) = \ln x$.

The graphs are reflections of each other in the line $y = x$.

14. Sketch the graph of $\log_3(x - 1)$ and identify the vertical asymptote.

See Additional Answers; $x = 1$

15. Evaluate $\log_4 \frac{1}{16}$. -2

16. Use the properties of logarithms to condense $3(\log_2 x + \log_2 y) - \log_2 z$.

$\log_2 \frac{x^3y^3}{z}$

17. Use the properties of logarithms to expand $\log_{10} \frac{\sqrt{x + 1}}{x^4}$.

$\frac{1}{2}\log_{10}(x + 1) - 4\log_{10} x$

In Exercises 18–21, solve the equation.

18. $\log_x\left(\frac{1}{9}\right) = -2$ 3

19. $4 \ln x = 10$ 12.182

20. $500(1.08)^t = 2000$ 18.013

21. $3(1 + e^{2x}) = 20$ 0.867

22. If the inflation rate averages 2.8% over the next 5 years, the approximate cost C of goods and services t years from now is given by

$C(t) = P(1.028)^t,\ 0 \le t \le 5$

where P is the percent cost. The price of an oil change is presently \$29.95. Estimate the price 5 years from now. \$34.38

23. Determine the effective yield of an 8% interest rate compounded continuously.
8.33%

24. Determine the length of time for an investment of \$1500 to quadruple in value when the investment earns 7% compounded continuously. 19.8 years

25. Write the standard form of the equation of the circle, and sketch its graph.

$x^2 + y^2 - 6x + 14y - 6 = 0$ $(x - 3)^2 + (y + 7)^2 = 64$; See Additional Answers.

26. Identify the vertex and focus of the parabola, and sketch its graph.

$y = 2x^2 - 20x + 5$ Vertex: $(5, -45)$, Focus: $\left(5, -\frac{359}{8}\right)$ See Additional Answers.

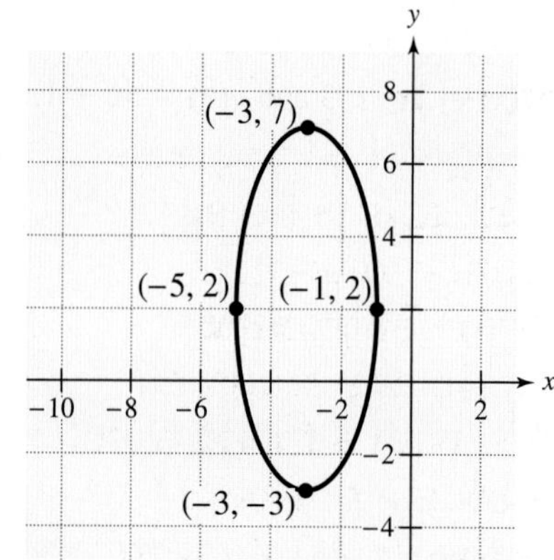

27. Write the standard form of the equation of the ellipse shown at the left.
$\frac{(x + 3)^2}{4} + \frac{(y - 2)^2}{25} = 1$

28. Find the center and vertices of the ellipse, and sketch its graph.

$4x^2 + y^2 = 4$ Center: $(0, 0)$, Vertices: $(0, \pm 2)$ See Additional Answers.

29. Write the standard form of the equation of the hyperbola with vertices $(0, -3)$ and $(0, 3)$ and asymptotes $y = \pm 3x$.
$\frac{y^2}{9} - x^2 = 1$

30. Find the center and vertices of the hyperbola, and sketch its graph.

$x^2 - 9y^2 + 18y = 153$ Center: $(0, 1)$, Vertices: $(\pm 12, 1)$ See Additional Answers.

In Exercises 31 and 32, solve the nonlinear system of equations.

31. $\begin{cases} y = x^2 - x - 1 \\ 3x - y = 4 \end{cases}$ $(1, -1), (3, 5)$

32. $\begin{cases} x^2 + 5y^2 = 21 \\ -x + y^2 = 5 \end{cases}$ $(-4, \pm 1), (-1, \pm 2)$

33. A rectangle has an area of 32 square feet and a perimeter of 24 feet. Find the dimensions of the rectangle.
8 feet × 4 feet

34. A rectangle has an area of 21 square feet and a perimeter of 20 feet. Find the dimensions of the rectangle.
7 feet × 3 feet

11

Sequences, Series, and the Binomial Theorem

MASTERY IS WITHIN REACH!

"When I was in my math courses, I had to get an early start on getting ready for my finals. I had to get a good grade in the last math course because I was competing with other students to get into the dental hygiene program. I studied with a friend from class a couple of weeks before the final. I stuck to studying for the final more since we did it together. We both did just fine on the final."

Mindy
Dental Hygiene

See page 593 for suggestions about preparing for the final exam.

11.1 Sequences and Series

- Write the terms of sequences.
- Find the apparent nth term of a sequence.
- Sum the terms of a sequence to obtain a series.

Sequences

A mathematical **sequence** is simply an ordered list of numbers. Each number in the list is a **term** of the sequence. A sequence can have a finite number of terms or an infinite number of terms.

> **Sequences**
>
> An **infinite sequence** $a_1, a_2, a_3, \ldots, a_n, \ldots$ is a function whose domain is the set of positive integers.
>
> A **finite sequence** $a_1, a_2, a_3, \ldots, a_n$ is a function whose domain is the finite set $\{1, 2, 3, \ldots, n\}$.

On occasion it is convenient to begin subscripting a sequence with 0 instead of 1 so that the terms of the sequence become $a_0, a_1, a_2, a_3, \ldots$. When this is the case, the domain includes 0.

EXAMPLE 1 Writing the Terms of a Sequence

Write the first six terms of the sequence with the given nth term.

a. $a_n = n^2 - 1$ Begin sequence with $n = 1$.

b. $a_n = 3(2^n)$ Begin sequence with $n = 0$.

SOLUTION

a. $a_1 = (1)^2 - 1 = 0$ $a_2 = (2)^2 - 1 = 3$ $a_3 = (3)^2 - 1 = 8$

$a_4 = (4)^2 - 1 = 15$ $a_5 = (5)^2 - 1 = 24$ $a_6 = (6)^2 - 1 = 35$

The sequence can be written as $0, 3, 8, 15, 24, 35, \ldots, n^2 - 1, \ldots$.

b. $a_0 = 3(2^0) = 3 \cdot 1 = 3$ $a_1 = 3(2^1) = 3 \cdot 2 = 6$

$a_2 = 3(2^2) = 3 \cdot 4 = 12$ $a_3 = 3(2^3) = 3 \cdot 8 = 24$

$a_4 = 3(2^4) = 3 \cdot 16 = 48$ $a_5 = 3(2^5) = 3 \cdot 32 = 96$

The sequence can be written as $3, 6, 12, 24, 48, 96, \ldots, 3(2^n), \ldots$.

Exercises Within Reach®

Solutions in English & Spanish and tutorial videos at AlgebraWithinReach.com

Writing the Terms of a Sequence **In Exercises 1–10, write the first five terms of the sequence with the given nth term. (Assume that n begins with 1.)**

1. $a_n = 2n$ 2, 4, 6, 8, 10
2. $a_n = 3n$ 3, 6, 9, 12, 15
3. $a_n = \left(\frac{1}{4}\right)^n$ $\frac{1}{4}, \frac{1}{16}, \frac{1}{64}, \frac{1}{256}, \frac{1}{1024}$
4. $a_n = \left(\frac{1}{3}\right)^n$ $\frac{1}{3}, \frac{1}{9}, \frac{1}{27}, \frac{1}{81}, \frac{1}{243}$
5. $a_n = 5n - 2$ 3, 8, 13, 18, 23
6. $a_n = 2n + 3$ 5, 7, 9, 11, 13
7. $a_n = \dfrac{4}{n + 3}$ $1, \frac{4}{5}, \frac{2}{3}, \frac{4}{7}, \frac{1}{2}$
8. $a_n = \dfrac{9}{5 + n}$ $\frac{3}{2}, \frac{9}{7}, \frac{9}{8}, 1, \frac{9}{10}$
9. $a_n = \dfrac{3n}{5n - 1}$ $\frac{3}{4}, \frac{2}{3}, \frac{9}{14}, \frac{12}{19}, \frac{5}{8}$
10. $a_n = \dfrac{2n}{6n - 3}$ $\frac{2}{3}, \frac{4}{9}, \frac{2}{5}, \frac{8}{21}, \frac{10}{27}$

EXAMPLE 2 A Sequence Whose Terms Alternate in Sign

Write the first six terms of the sequence whose nth term is

$$a_n = \frac{(-1)^n}{2n-1}. \qquad \text{Begin sequence with } n = 1.$$

SOLUTION

$$a_1 = \frac{(-1)^1}{2(1)-1} = -\frac{1}{1} \qquad a_2 = \frac{(-1)^2}{2(2)-1} = \frac{1}{3} \qquad a_3 = \frac{(-1)^3}{2(3)-1} = -\frac{1}{5}$$

$$a_4 = \frac{(-1)^4}{2(4)-1} = \frac{1}{7} \qquad a_5 = \frac{(-1)^5}{2(5)-1} = -\frac{1}{9} \qquad a_6 = \frac{(-1)^6}{2(6)-1} = \frac{1}{11}$$

The sequence can be written as $-1, \frac{1}{3}, -\frac{1}{5}, \frac{1}{7}, -\frac{1}{9}, \frac{1}{11}, \ldots, \frac{(-1)^n}{2n-1}, \ldots$.

Some very important sequences in mathematics involve terms that are defined with special types of products call *factorials*.

Definition of Factorial

When n is a positive integer, n **factorial** is defined as

$$n! = 1 \cdot 2 \cdot 3 \cdot 4 \cdot \cdots \cdot (n-1) \cdot n.$$

As a special case, zero factorial is defined as $0! = 1$.

EXAMPLE 3 Writing Factorials

The first several factorial values are as follows.

$0! = 1$ $\qquad$ $1! = 1$

$2! = 1 \cdot 2 = 2$ $\qquad$ $3! = 1 \cdot 2 \cdot 3 = 6$

$4! = 1 \cdot 2 \cdot 3 \cdot 4 = 24$ $\qquad$ $5! = 1 \cdot 2 \cdot 3 \cdot 4 \cdot 5 = 120$

Exercises Within Reach®

Solutions in English & Spanish and tutorial videos at AlgebraWithinReach.com

Writing the Terms of a Sequence **In Exercises 11–16, write the first five terms of the sequence with the given nth term. (Assume that n begins with 1.)**

11. $a_n = (-1)^n 2n$ $\quad -2, 4, -6, 8, -10$

12. $a_n = (-1)^{n+1} 3n$ $\quad 3, -6, 9, -12, 15$

13. $a_n = \left(-\frac{1}{2}\right)^{n+1}$ $\quad \frac{1}{4}, -\frac{1}{8}, \frac{1}{16}, -\frac{1}{32}, \frac{1}{64}$

14. $a_n = \left(-\frac{2}{3}\right)^{n-1}$ $\quad 1, -\frac{2}{3}, \frac{4}{9}, -\frac{8}{27}, \frac{16}{81}$

15. $a_n = \frac{(-1)^n}{n^2}$ $\quad -1, \frac{1}{4}, -\frac{1}{9}, \frac{1}{16}, -\frac{1}{25}$

16. $a_n = \frac{(-1)^{n+1}}{n}$ $\quad 1, -\frac{1}{2}, \frac{1}{3}, -\frac{1}{4}, \frac{1}{5}$

Writing a Factorial **In Exercises 17–20, write the product represented by the factorial. Then evaluate the product.**

17. $6!$ $\quad 1 \cdot 2 \cdot 3 \cdot 4 \cdot 5 \cdot 6; 720$

18. $7!$ $\quad 1 \cdot 2 \cdot 3 \cdot 4 \cdot 5 \cdot 6 \cdot 7; 5040$

19. $9!$ $\quad 1 \cdot 2 \cdot 3 \cdot 4 \cdot 5 \cdot 6 \cdot 7 \cdot 8 \cdot 9; 362,880$

20. $10!$ $\quad 1 \cdot 2 \cdot 3 \cdot 4 \cdot 5 \cdot 6 \cdot 7 \cdot 8 \cdot 9 \cdot 10; 3,628,800$

Many calculators have a factorial key, denoted by $\boxed{n!}$. If your calculator has such a key, try using it to evaluate $n!$ for several values of n. You will see that as n increases, the value of $n!$ becomes very large. For instance, $10! = 3{,}628{,}800$.

EXAMPLE 4 Writing Sequences Involving Factorials

Additional Example:

Write the first six terms of the sequence whose nth term is

$$a_n = \frac{2n!}{(2n)!}.$$

Answer:

$a_0 = 2,\ a_1 = 1,\ a_2 = \frac{1}{6},\ a_3 = \frac{1}{60},$

$a_4 = \frac{1}{840},\ a_5 = \frac{1}{15{,}120}$

Write the first six terms of the sequence with the given nth term.

a. $a_n = \frac{1}{n!}$ Begin sequence with $n = 0$.

b. $a_n = \frac{2^n}{n!}$ Begin sequence with $n = 0$.

SOLUTION

a. $a_0 = \frac{1}{0!} = \frac{1}{1} = 1$ $\qquad a_1 = \frac{1}{1!} = \frac{1}{1} = 1$

$a_2 = \frac{1}{2!} = \frac{1}{1 \cdot 2} = \frac{1}{2}$ $\qquad a_3 = \frac{1}{3!} = \frac{1}{1 \cdot 2 \cdot 3} = \frac{1}{6}$

$a_4 = \frac{1}{4!} = \frac{1}{1 \cdot 2 \cdot 3 \cdot 4} = \frac{1}{24}$ $\qquad a_5 = \frac{1}{5!} = \frac{1}{1 \cdot 2 \cdot 3 \cdot 4 \cdot 5} = \frac{1}{120}$

b. $a_0 = \frac{2^0}{0!} = \frac{1}{1} = 1$ $\qquad a_1 = \frac{2^1}{1!} = \frac{2}{1} = 2$

$a_2 = \frac{2^2}{2!} = \frac{2 \cdot 2}{1 \cdot 2} = \frac{4}{2} = 2$ $\qquad a_3 = \frac{2^3}{3!} = \frac{2 \cdot 2 \cdot 2}{1 \cdot 2 \cdot 3} = \frac{8}{6} = \frac{4}{3}$

$a_4 = \frac{2^4}{4!} = \frac{2 \cdot 2 \cdot 2 \cdot 2}{1 \cdot 2 \cdot 3 \cdot 4} = \frac{2}{3}$ $\qquad a_5 = \frac{2^5}{5!} = \frac{2 \cdot 2 \cdot 2 \cdot 2 \cdot 2}{1 \cdot 2 \cdot 3 \cdot 4 \cdot 5} = \frac{4}{15}$

EXAMPLE 5 Simplifying Expressions Involving Factorials

a. $\frac{7!}{5!} = \frac{\cancel{1 \cdot 2 \cdot 3 \cdot 4 \cdot 5} \cdot 6 \cdot 7}{\cancel{1 \cdot 2 \cdot 3 \cdot 4 \cdot 5}} = 6 \cdot 7 = 42$

b. $\frac{n!}{(n-1)!} = \frac{\cancel{1 \cdot 2 \cdot 3 \cdot \ldots \cdot (n-1)} \cdot n}{\cancel{1 \cdot 2 \cdot 3 \cdot \ldots \cdot (n-1)}} = n$

Exercises Within Reach®

Solutions in English & Spanish and tutorial videos at AlgebraWithinReach.com

Writing the Terms of a Sequence **In Exercises 21–24, write the first five terms of the sequence with the given *n*th term. (Assume that *n* begins with 1.)**

21. $a_n = \frac{n!}{n}$ 1, 1, 2, 6, 24

22. $a_n = \frac{n!}{n+1}$ $\frac{1}{2}, \frac{2}{3}, \frac{3}{2}, \frac{24}{5}, 20$

23. $a_n = \frac{(n+1)!}{n!}$ 2, 3, 4, 5, 6

24. $a_n = \frac{n!}{(n-1)!}$ 1, 2, 3, 4, 5

Simplifying an Expression Involving a Factorial **In Exercises 25–32, simplify the expression.**

25. $\frac{5!}{4!}$ 5

26. $\frac{6!}{8!}$ $\frac{1}{56}$

27. $\frac{25!}{20!5!}$ 53,130

28. $\frac{20!}{15!5!}$ 15,504

29. $\frac{n!}{(n+1)!}$ $\frac{1}{n+1}$

30. $\frac{(n+2)!}{n!}$ $(n+1)(n+2)$

31. $\frac{(n+1)!}{(n-1)!}$ $n(n+1)$

32. $\frac{(2n)!}{(2n-1)!}$ $2n$

Finding the *n*th Term of a Sequence

Sometimes you have the first several terms of a sequence and need to find a formula (the *n*th term) to generate those terms. Pattern recognition is crucial in finding a form for the *n*th term.

EXAMPLE 6 Finding the *n*th Term of a Sequence

Write an expression for the *n*th term of each sequence.

a. $\frac{1}{2}, \frac{1}{4}, \frac{1}{8}, \frac{1}{16}, \frac{1}{32}, \ldots$ **b.** $1, -4, 9, -16, 25, \ldots$

SOLUTION

a.

n:	1	2	3	4	5	...	n
Terms:	$\frac{1}{2}$	$\frac{1}{4}$	$\frac{1}{8}$	$\frac{1}{16}$	$\frac{1}{32}$	...	a_n

Pattern: The numerators are 1 and the denominators are increasing powers of 2.

So, an expression for the *n*th term is $\frac{1}{2^n}$.

b.

n:	1	2	3	4	5	...	n
Terms:	1	-4	9	-16	25	...	a_n

Pattern: The terms have alternating signs, with those in the even positions being negative. The absolute value of each term is the square of *n*.

So, an expression for the *n*th term is $(-1)^{n+1}n^2$.

Exercises Within Reach®

Solutions in English & Spanish and tutorial videos at AlgebraWithinReach.com

Matching **In Exercises 33–36, match the sequence with the graph of its first 10 terms.**

(a) 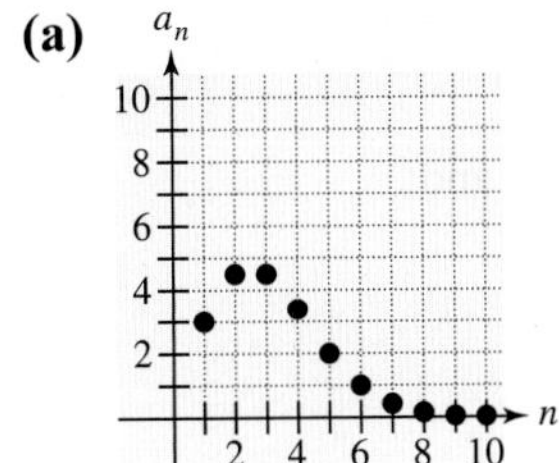

(b) 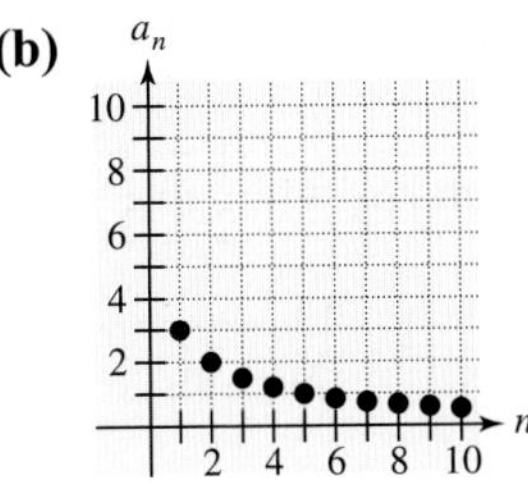

(c) 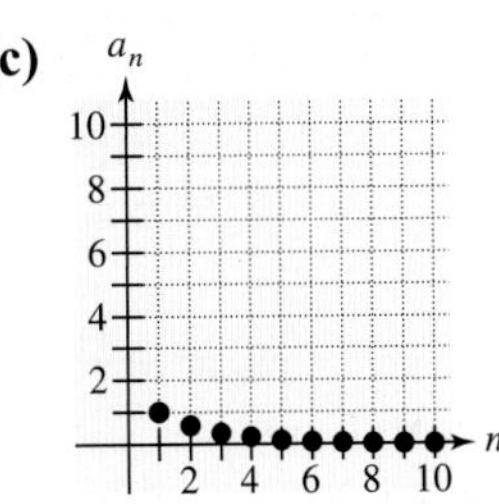

(d) 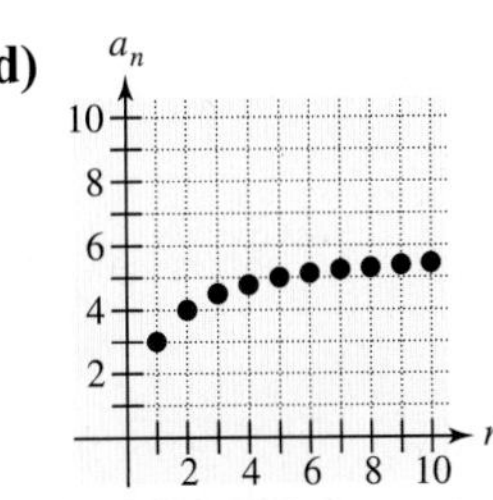

33. $\frac{6}{n+1}$ b

34. $a_n = \frac{6n}{n+1}$ d

35. $a_n = (0.6)^{n-1}$ c

36. $a_n = \frac{3^n}{n!}$ a

Finding the nth Term **In Exercises 37–42, write an expression for the *n*th term of the sequence. (Assume that *n* begins with 1.)**

37. $1, 3, 5, 7, 9, \ldots$ $a_n = 2n - 1$

38. $2, -4, 6, -8, 10, \ldots$ $a_n = (-1)^{n+1}2n$

39. $0, 3, 8, 15, 24, \ldots$ $a_n = n^2 - 1$

40. $1, 8, 27, 64, 125, \ldots$ $a_n = n^3$

41. $-\frac{1}{5}, \frac{1}{25}, -\frac{1}{125}, \frac{1}{625}, -\frac{1}{3125}, \ldots$ $a_n = \left(-\frac{1}{5}\right)^n$

42. $1, \frac{1}{4}, \frac{1}{9}, \frac{1}{16}, \frac{1}{25}, \ldots$ $a_n = \frac{1}{n^2}$

Series

Definition of Series

For an infinite sequence $a_1, a_2, a_3, \ldots, a_n, \ldots$

1. the sum of the first n terms

$$S_n = a_1 + a_2 + a_3 + \cdots + a_n$$

is called a **partial sum**, and

2. the sum of all the terms

$$a_1 + a_2 + a_3 + \cdots + a_n + \cdots$$

is called an **infinite series**, or simply a **series**.

EXAMPLE 7 Finding Partial Sums

Find the indicated partial sums for each sequence.

a. Find S_1, S_2, and S_5 for $a_n = 3n - 1$.

b. Find S_2, S_3, and S_4 for $a_n = \dfrac{(-1)^n}{n+1}$.

SOLUTION

a. The first five terms of the sequence $a_n = 3n - 1$ are

$$a_1 = 2,\ a_2 = 5,\ a_3 = 8,\ a_4 = 11, \text{ and } a_5 = 14$$

So, the partial sums are

$$S_1 = 2,\ S_2 = 2 + 5 = 7, \text{ and } S_5 = 2 + 5 + 8 + 11 + 14 = 40.$$

b. The first four terms of the sequence $a_n = \dfrac{(-1)^n}{n+1}$ are

$$a_1 = -\frac{1}{2},\ a_2 = \frac{1}{3},\ a_3 = -\frac{1}{4}, \text{ and } a_4 = \frac{1}{5}.$$

So, the partial sums are

$$S_2 = -\frac{1}{2} + \frac{1}{3} = -\frac{1}{6}, \text{ and } S_3 = -\frac{1}{2} + \frac{1}{3} - \frac{1}{4} = -\frac{5}{12}, \text{ and}$$

$$S_4 = -\frac{1}{2} + \frac{1}{3} - \frac{1}{4} + \frac{1}{5} = -\frac{13}{60}.$$

Exercises Within Reach®

Solutions in English & Spanish and tutorial videos at AlgebraWithinReach.com

Finding Partial Sums **In Exercises 43–46, find the indicated partial sums for the sequence.**

43. Find S_1, S_2, and S_6 for $a_n = 2n + 5$. $S_1 = 7$; $S_2 = 16$; $S_6 = 72$

44. Find S_3, S_4, and S_{10} for $a_n = n^3 - 1$. $S_3 = 33$; $S_4 = 96$; $S_{10} = 3015$

45. Find S_2, S_3, and S_9 for $a_n = \dfrac{1}{n}$. $S_2 = \frac{3}{2}$; $S_3 = \frac{11}{6}$; $S_9 = \frac{7129}{2520}$

46. Find S_1, S_2, and S_5 for $a_n = \dfrac{(-1)^{n+1}}{n+1}$. $S_1 = \frac{1}{2}$; $S_3 = \frac{5}{12}$; $S_5 = \frac{23}{60}$

Sigma notation expresses a compound sum more concisely. Use several different examples to help students become more comfortable with it. As a first example, consider placing $a_1, a_2, a_3, \ldots, a_n$ above each of the corresponding values of the sequence to help reinforce the meaning of the notation.

Definition of Sigma Notation

The sum of the first n terms of the sequence whose nth term is a_n is

$$\sum_{i=1}^{n} a_i = a_1 + a_2 + a_3 + a_4 + \ldots + a_n$$

where i is the **index of summation**, n is the **upper limit of summation**, and 1 is the **lower limit of summation**.

Study Tip

In Example 8(a), the index of summation is i and the summation begins with $i = 1$. Any letter can be used as the index of summation, and the summation can begin with any integer. For instance, in Example 8(b), the index of summation is k and the summation begins with $k = 0$.

EXAMPLE 8 Finding Sums in Sigma Notation

a. $\displaystyle\sum_{i=1}^{6} 2i = 2(1) + 2(2) + 2(3) + 2(4) + 2(5) + 2(6)$

$= 2 + 4 + 6 + 8 + 10 + 12$

$= 42$

b. $\displaystyle\sum_{k=0}^{8} \frac{1}{k!} = \frac{1}{0!} + \frac{1}{1!} + \frac{1}{2!} + \frac{1}{3!} + \frac{1}{4!} + \frac{1}{5!} + \frac{1}{6!} + \frac{1}{7!} + \frac{1}{8!}$

$\displaystyle = 1 + 1 + \frac{1}{2} + \frac{1}{6} + \frac{1}{24} + \frac{1}{120} + \frac{1}{720} + \frac{1}{5040} + \frac{1}{40,320}$

≈ 2.71828

Note that this sum is approximately $e = 2.71828\ldots$.

EXAMPLE 9 Writing a Sum in Sigma Notation

Write the sum in sigma notation: $\frac{2}{2} + \frac{2}{3} + \frac{2}{4} + \frac{2}{5} + \frac{2}{6}$.

SOLUTION

To write this sum in sigma notation, you must find a pattern for the terms. You can see that the terms have numerators of 2 and denominators that range over the integers from 2 to 6. So, one possible sigma notation is

$$\sum_{i=1}^{5} \frac{2}{i+1} = \frac{2}{2} + \frac{2}{3} + \frac{2}{4} + \frac{2}{5} + \frac{2}{6}.$$

Exercises Within Reach®

Solutions in English & Spanish and tutorial videos at AlgebraWithinReach.com

Finding a Partial Sum **In Exercises 47–54, find the partial sum.**

47. $\displaystyle\sum_{k=1}^{5} 6$ 30

48. $\displaystyle\sum_{k=1}^{4} 5k$ 50

49. $\displaystyle\sum_{i=0}^{6} (2i+5)$ 77

50. $\displaystyle\sum_{i=0}^{4} (2i+3)$ 35

51. $\displaystyle\sum_{j=0}^{3} \frac{1}{j^2+1}$ $\frac{9}{5}$

52. $\displaystyle\sum_{j=1}^{5} \frac{(-1)^{j+1}}{j^2}$ $\frac{3019}{3600}$

53. $\displaystyle\sum_{n=0}^{5} \left(-\frac{1}{3}\right)^n$ $\frac{182}{243}$

54. $\displaystyle\sum_{k=1}^{4} \left(\frac{5}{3}\right)^{k-1}$ $\frac{272}{27}$

Writing a Sum in Sigma Notation **In Exercises 55–58, write the sum in sigma notation. (Begin with $k = 1$.)**

55. $1 + 2 + 3 + 4 + 5$ $\displaystyle\sum_{k=1}^{5} k$

56. $5 + 10 + 15 + 20 + 25 + 30$ $\displaystyle\sum_{k=1}^{6} 5k$

57. $\frac{1}{2} + \frac{2}{3} + \frac{3}{4} + \frac{4}{5} + \frac{5}{6} + \cdots + \frac{11}{12}$ $\displaystyle\sum_{k=1}^{11} \frac{k}{k+1}$

58. $\frac{2}{4} + \frac{4}{5} + \frac{6}{6} + \frac{8}{7} + \cdots + \frac{40}{23}$ $\displaystyle\sum_{k=1}^{20} \frac{2k}{k+3}$

Concept Summary: Finding the nth Term of a Sequence

What

A **sequence** is an ordered list of numbers. You can use the first several **terms** of a sequence to write an expression for the nth term a_n of the sequence.

EXAMPLE

Write an expression for the nth term of the sequence $\frac{2}{3}, \frac{2}{6}, \frac{2}{9}, \frac{2}{12}, \ldots$.

How

Use pattern recognition to find a form for the nth term.

EXAMPLE

n:	1	2	3	4	. . .	n
Terms:	$\frac{2}{3}$	$\frac{2}{6}$	$\frac{2}{9}$	$\frac{2}{12}$	. . .	a_n

Pattern: Each numerator is 2 and each denominator is three times n.

So, an expression for the nth term is $a_n = \frac{2}{3n}$.

Why

You can use the expression for the nth term of a sequence to represent **partial sums** for the sequence in **sigma notation.**

For instance, for the sequence whose nth term is $a_n = \frac{2}{3n}$, the partial sum S_5 is

$$S_5 = \sum_{n=1}^{5} \frac{2}{3n}.$$

Exercises Within Reach®

Worked-out solutions to odd-numbered exercises at AlgebraWithinReach.com

Concept Summary Check

59. ***Identifying a Term*** What is the third term of the sequence in the example above?
$\frac{2}{9}$

60. ***Identifying a Domain Value*** In the example above, what domain value of the sequence corresponds to the term $\frac{2}{12}$? $n = 4$

61. ***Finding a Term of a Sequence*** Explain how to find the 6th term of the sequence in the example above.
Substitute 6 for n in a_n: $a_6 = \frac{2}{3(6)} = \frac{1}{9}$

62. ***Vocabulary*** Explain the difference between a *sequence* and a *series*. A sequence is an ordered list of numbers called terms, and a series is the sum of the terms of an infinite sequence.

Extra Practice

63. ***Compound Interest*** A deposit of \$500 is made in an account that earns 7% interest compounded yearly. The balance in the account after N years is given by

$$A_N = 500(1 + 0.07)^N, \quad N = 1, 2, 3, \ldots.$$

(a) Compute the first eight terms of the sequence.
\$535, \$572.45, \$612.52, \$655.40, \$701.28, \$750.37, \$802.89, \$859.09

(b) Find the balance in this account after 40 years by computing A_{40}.
\$7487.23

(c) Use a graphing calculator to graph the first 40 terms of the sequence.
See Additional Answers.

(d) The graph shows how the terms of the sequence increase. Do the terms keep increasing by the same amount? Explain.
No; For increasing values of N, the values of the terms A_N increase by greater amounts.

64. ***Soccer Ball Design*** The number of degrees a_n in each angle of a regular n-sided polygon is

$$a_n = \frac{180(n-2)}{n}, \quad n \geq 3.$$

The surface of a soccer ball is made of regular hexagons and pentagons. When a soccer ball is taken apart and flattened, as shown in the figure, the sides do not meet each other. Use the terms a_5 and a_6 to explain why there are gaps between adjacent hexagons.

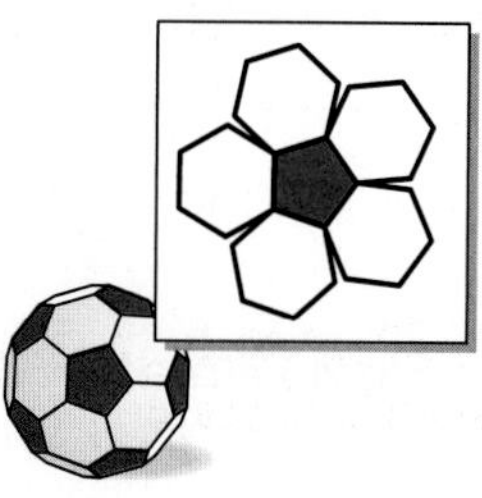

$a_5 = 108°$, $a_6 = 120°$; At the point where any two hexagons and a pentagon meet, the sum of the three angles is $a_5 + 2a_6 = 348° < 360°$. Therefore, there is a gap of 12°.

65. ***Stars*** Stars are formed by placing n equally spaced points on a circle and connecting each point with a second point on the circle (see figure). The measure of degrees d_n of the angle at each tip of the star is given by

$$d_n = \frac{180(n-4)}{n}, \quad n \geq 5.$$

Write the first six terms of this sequence.
36°, 60°, 77.1°, 90°, 100°, 108°

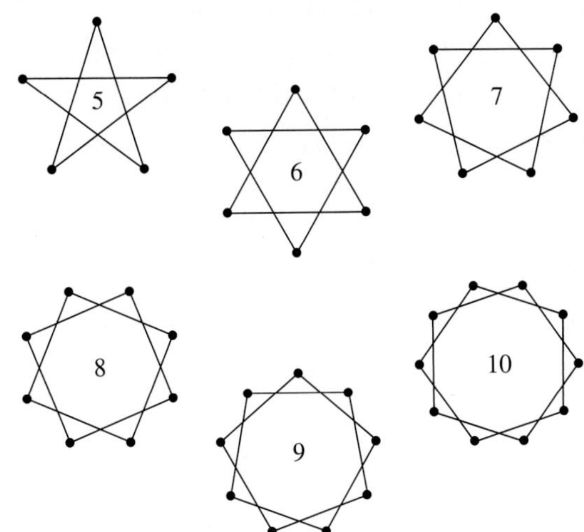

66. ***Stars*** The stars in Exercise 65 were formed by placing n equally spaced points on a circle and connecting each point with the second point from it on the circle. The stars in the figure for this exercise were formed in a similar way except that each point was connected with the third point from it. For these stars, the measure in degrees d_n of the angle at each point is given by

$$d_n = \frac{180(n-6)}{n}, \quad n \geq 7.$$

Write the first five terms of this sequence.
25.7°, 45°, 60°, 72°, 81.8°

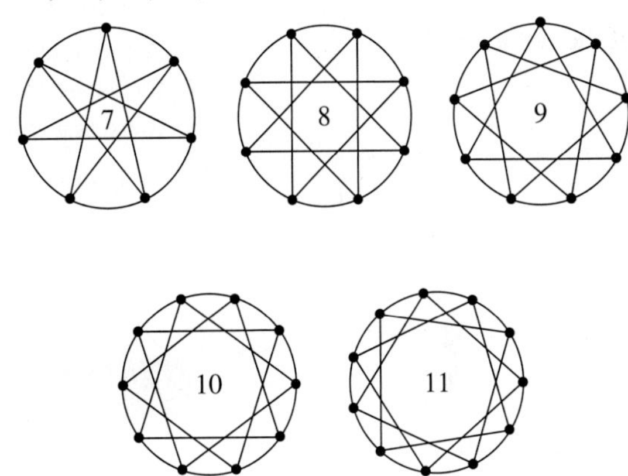

Explaining Concepts

67. ***Structure*** In your own words, explain why a sequence is a function. A sequence is a function because there is only one value for each term of the sequence.

68. ***Number Sense*** The nth term of a sequence is $a_n = (-1)^n n$. Which terms of the sequence are negative? Explain. The terms are negative when n is odd because $(-1)^n = -1$ when n is odd and $(-1)^n = 1$ when n is even.

69. ***Number Sense*** Explain the difference between $a_n = 4n!$ and $a_n = (4n)!$. See Additional Answers.

True or False? **In Exercises 70–72, determine whether the statement is true or false. Justify your answer.**
See Additional Answers.

70. $\sum_{i=1}^{4} (i^2 + 2i) = \sum_{i=1}^{4} i^2 + \sum_{i=1}^{4} 2i$ True

71. $\sum_{k=1}^{4} 3k = 3\sum_{k=1}^{4} k$ True

72. $\sum_{j=1}^{4} 2^j = \sum_{j=3}^{6} 2^{j-2}$ True

Cumulative Review

In Exercises 73–76, evaluate the expression for the specified value of the variable.

73. $-2n + 15; n = 3$ 9 **74.** $-20n + 100; n = 4$ 20

75. $25 - 3(n + 4); n = 8$ -11

76. $-\frac{3}{2}(n - 1) + 6; n = 10$ $-\frac{15}{2}$

In Exercises 77–80, identify the center and radius of the circle, and sketch its graph. See Additional Answers.

77. $x^2 + y^2 = 36$ Center: $(0, 0)$, $r = 6$

78. $4x^2 + 4y^2 = 9$ Center: $(0, 0)$, $r = \frac{3}{2}$

79. $x^2 + y^2 + 4x - 12 = 0$ Center: $(-2, 0)$, $r = 4$

80. $x^2 + y^2 - 10x - 2y - 199 = 0$ Center: $(5, 1)$, $r = 15$

In Exercises 81–84, identify the vertex and focus of the parabola, and sketch its graph. See Additional Answers.

81. $x^2 = 6y$ Vertex: $(0, 0)$, Focus: $\left(0, \frac{3}{2}\right)$

82. $y^2 = 9x$ Vertex: $(0, 0)$, Focus: $\left(\frac{9}{4}, 0\right)$

83. $x^2 + 8y + 32 = 0$ Vertex: $(0, -4)$, Focus: $(0, -6)$

84. $y^2 - 10x + 6y + 29 = 0$
Vertex: $(2, -3)$, Focus: $\left(\frac{9}{2}, -3\right)$

11.2 Arithmetic Sequences

- Recognize, write, and find the nth terms of arithmetic sequences.
- Find the nth partial sum of an arithmetic sequence.
- Use arithmetic sequences to solve application problems.

Arithmetic Sequences

Definition of Arithmetic Sequence

A sequence is called **arithmetic** when the differences between consecutive terms are the same. So, the sequence $a_1, a_2, a_3, a_4, \ldots, a_n, \ldots$ is arithmetic when there is a number d such that

$$a_2 - a_1 = d, a_3 - a_2 = d, a_4 - a_3 = d$$

and so on. The number d is the **common difference** of the sequence.

Additional Example

Have students decide whether the sequence 1, 7, 13, 19, 26, 33, 39, . . . is an arithmetic sequence.

Answer: No

EXAMPLE 1 Examples of Arithmetic Sequences

a. The sequence whose nth term is $3n + 2$ is arithmetic. For this sequence, the common difference between consecutive terms is 3.

$$5, 8, 11, 14, \ldots, 3n + 2, \ldots$$ Begin with $n = 1$.

$8 - 5 = 3$

b. The sequence whose nth term is $7 - 5n$ is arithmetic. For this sequence, the common difference between consecutive terms is -5.

$$2, -3, -8, -13, \ldots, 7 - 5n, \ldots$$ Begin with $n = 1$.

$-3 - 2 = -5$

c. The sequence whose nth term is $\frac{1}{4}(n + 3)$ is arithmetic. For this sequence, the common difference between consecutive terms is $\frac{1}{4}$.

$$1, \frac{5}{4}, \frac{3}{2}, \frac{7}{4}, \ldots, \frac{1}{4}(n + 3), \ldots$$ Begin with $n = 1$.

$\frac{5}{4} - 1 = \frac{1}{4}$

Exercises Within Reach®

Solutions in English & Spanish and tutorial videos at AlgebraWithinReach.com

Finding the Common Difference **In Exercises 1–6, find the common difference of the arithmetic sequence.**

1. 2, 5, 8, 11, . . . 3
2. −8, 0, 8, 16, . . . 8
3. 100, 94, 88, 82, . . . −6
4. 3200, 2800, 2400, 2000, . . . −400
5. $4, \frac{9}{2}, 5, \frac{11}{2}, 6, \ldots$ $\frac{1}{2}$
6. $1, \frac{5}{3}, \frac{7}{3}, 3, \ldots$ $\frac{2}{3}$

Finding the Common Difference **In Exercises 7–12, find the common difference of the arithmetic sequence with the given nth term.**

7. $a_n = 4n + 5$ 4
8. $a_n = 7n + 6$ 7
9. $a_n = 8 - 3n$ −3
10. $a_n = 12 - 4n$ −4
11. $a_n = \frac{1}{2}(n + 1)$ $\frac{1}{2}$
12. $a_n = \frac{1}{3}(n + 4)$ $\frac{1}{3}$

The *n*th Term of an Arithmetic Sequence

The *n*th term of an arithmetic sequence has the form

$$a_n = a_1 + (n - 1)d$$

where d is the common difference between the terms of the sequence, and a_1 is the first term.

Study Tip

The *n*th term of an arithmetic sequence can be derived from the following pattern.

$a_1 = a_1$	1st term
$a_2 = a_1 + d$	2nd term
$a_3 = a_1 + 2d$	3rd term
$a_4 = a_1 + 3d$	4th term
$a_5 = a_1 + 4d$ (1 less)	5th term
⋮	⋮
$a_n = a_1 + (n - 1)d$ (1 less)	*n*th term

EXAMPLE 2 Finding the *n*th Term of an Arithmetic Sequence

a. Find a formula for the *n*th term of the arithmetic sequence whose common difference is 2 and whose first term is 5.

b. Find a formula for the *n*th term of the arithmetic sequence whose common difference is 3 and whose first term is 2.

SOLUTION

a. You know that the formula for the *n*th term is of the form $a_n = a_1 + (n - 1)d$. Moreover, because the common difference is $d = 2$ and the first term is $a_1 = 5$, the formula must have the form

$$a_n = 5 + 2(n - 1).$$ Substitute 5 for a_1 and 2 for d.

So, the formula for the *n*th term is $a_n = 2n + 3$, and the sequence has the following form.

$$5, 7, 9, 11, 13, \ldots, 2n - 3, \ldots$$

b. You know that the formula for the *n*th term is of the form $a_n = a_1 + (n - 1)d$. Moreover, because the common difference is $d = 3$ and the first term is $a_1 = 2$, the formula must have the form

$$a_n = 2 + 3(n - 1).$$ Substitute 2 for a_1 and 3 for d.

So, the formula for the *n*th term is $a_n = 3n - 1$, and the sequence has the following form.

$$2, 5, 8, 11, 14, \ldots, 3n - 1, \ldots$$

Exercises Within Reach®

Solutions in English & Spanish and tutorial videos at AlgebraWithinReach.com

Finding the nth Term **In Exercises 13–22, find a formula for the *n*th term of the arithmetic sequence.**

13. $a_1 = 4,\ d = 3$
$a_n = 3n + 1$

14. $a_1 = 7,\ d = 2$
$a_n = 2n + 5$

15. $a_1 = \frac{1}{2},\ d = \frac{3}{2}$
$a_n = \frac{3}{2}n - 1$

16. $a_1 = \frac{5}{3},\ d = \frac{1}{3}$
$a_n = \frac{1}{3}n + \frac{4}{3}$

17. $a_1 = 100,\ d = -5$
$a_n = -5n + 105$

18. $a_1 = -6,\ d = -1$
$a_n = -n - 5$

19. $a_3 = 6,\ d = \frac{3}{2}$
$a_n = \frac{3}{2}n + \frac{3}{2}$

20. $a_6 = 5,\ d = \frac{3}{2}$
$a_n = \frac{3}{2}n - 4$

21. $a_1 = 5,\ a_5 = 15$
$a_n = \frac{5}{2}n + \frac{5}{2}$

22. $a_2 = 93,\ a_6 = 65$
$a_n = -7n + 107$

EXAMPLE 3 **Finding a Term**

Find the ninth term of the arithmetic sequence that begins with 2 and 9.

SOLUTION

For this sequence, the common difference is $d = 9 - 2 = 7$. Because the common difference is $d = 7$ and the first term is $a_1 = 2$, the formula must have the form

$$a_n = 2 + 7(n - 1).$$ Substitute 2 for a_1 and 7 for d.

So, a formula for the nth term is

$$a_n = 7n - 5$$

which implies that the ninth term is

$$a_n = 7(9) - 5 = 58.$$

When you know the nth term and the common difference of an arithmetic sequence, you can find the $(n + 1)$th term by using the **recursion formula**

$$a_{n+1} = a_n + d.$$

EXAMPLE 4 **Using a Recursion Formula**

The 12th term of an arithmetic sequence is 52 and the common difference is 3.

a. What is the 13th term of the sequence? **b.** What is the first term?

SOLUTION

a. You know that $a_{12} = 52$ and $d = 3$. So, using the recursion formula $a_{13} = a_{12} + d$, you can determine that the 13th term of sequence is

$$a_{13} = 52 + 3 = 55.$$

b. Using $n = 12$, $d = 3$, and $a_{12} = 52$ in the formula $a_n = a_1 + (n - 1)d$ yields

$$52 = a_1 + (12 - 1)(3)$$

$$19 = a_1.$$

Exercises Within Reach ®

Solutions in English & Spanish and tutorial videos at AlgebraWithinReach.com

Finding a Term **In Exercises 23 and 24, the first two terms of an arithmetic sequence are given. Find the missing term.**

23. $a_1 = 5, a_2 = 11, a_{10} =$ 59

24. $a_1 = 3, a_2 = 13, a_9 =$ 83

Using a Recursion Formula **In Exercises 25–28, write the first five terms of the arithmetic sequence defined recursively.**

25. $a_1 = 14$
$a_{k+1} = a_k + 6$
14, 20, 26, 32, 38

26. $a_1 = 3$
$a_{k+1} = a_k - 2$
3, 1, −1, −3, −5

27. $a_1 = 23$
$a_{k+1} = a_k - 5$
23, 18, 13, 8, 3

28. $a_1 = -16$
$a_{k+1} = a_k + 5$
−16, −11, −6, −1, 4

Using a Recursion Formula **In Exercises 29 and 30, find (a) a_1 and (b) a_5.**

29. $a_4 = 23,\ d = 6$ $a_1 = 5, a_5 = 29$

30. $a_4 = \frac{15}{4},\ d = -\frac{3}{4}$ $a_1 = 6, a_5 = 3$

The Partial Sum of an Arithmetic Sequence

Study Tip

You can use the formula for the nth partial sum of an arithmetic sequence to find the sum of consecutive numbers. For instance, the sum of the integers from 1 to 100 is

$$\sum_{i=1}^{100} i = \frac{100}{2}(1 + 100)$$
$$= 50(101)$$
$$= 5050.$$

The nth Partial Sum of an Arithmetic Sequence

The nth partial sum of the arithmetic sequence whose nth term is a_n is

$$\sum_{i=1}^{n} a_i = a_1 + a_2 + a_3 + a_4 + \cdots + a_n$$
$$= \frac{n}{2}(a_1 + a_n).$$

Or, equivalently, you can find the sum of the first n terms of an arithmetic sequence by multiplying the average of the first and nth terms by n.

EXAMPLE 5 Finding the nth Partial Sum

Find the sum of the first 20 terms of the arithmetic sequence whose nth term is $4n + 1$.

SOLUTION

The first term of this sequence is $a_1 = 4(1) + 1 = 5$ and the 20th term is $a_{20} = 4(20) + 1 = 81$. So, the sum of the first 20 terms is given by

$\sum_{i=1}^{n} a_i = \frac{n}{2}(a_1 + a_n)$	nth partial sum formula
$\sum_{i=1}^{20} (4i + 1) = \frac{20}{2}(a_1 + a_{20})$	Substitute 20 for n.
$= 10(5 + 81)$	Substitute 5 for a_1 and 81 for a_{20}.
$= 10(86)$	Simplify.
$= 860.$	nth partial sum

Exercises Within Reach®

Solutions in English & Spanish and tutorial videos at AlgebraWithinReach.com

Finding the nth Partial Sum **In Exercises 31–40, find the partial sum.**

31. $\sum_{k=1}^{20} k$ 210

32. $\sum_{k=1}^{30} 4k$ 1860

33. $\sum_{k=1}^{50} (k + 3)$ 1425

34. $\sum_{n=1}^{30} (n + 2)$ 525

35. $\sum_{k=1}^{10} (5k - 2)$ 255

36. $\sum_{k=1}^{100} (4k - 1)$ 20,100

37. $\sum_{n=1}^{500} \frac{n}{2}$ 62,625

38. $\sum_{n=1}^{300} \frac{n}{3}$ 15,050

39. $\sum_{n=1}^{30} \left(\frac{1}{3}n - 4\right)$ 35

40. $\sum_{n=1}^{75} (0.3n + 5)$ 1230

EXAMPLE 6 Finding the *n*th Partial Sum

Find the sum of the even integers from 2 to 100.

SOLUTION

Because the integers

2, 4, 6, 8, . . . , 100

form an arithmetic sequence, you can find the sum as follows.

$$\sum_{i=1}^{n} a_i = \frac{n}{2}(a_1 + a_n) \qquad \textit{n}\text{th partial sum formula}$$

$$\sum_{i=1}^{50} 2i = \frac{50}{2}(a_1 + a_{50}) \qquad \text{Substitute 50 for } n.$$

$$= 25(2 + 100) \qquad \text{Substitute 2 for } a_1 \text{ and 100 for } a_{50}.$$

$$= 25(102) \qquad \text{Simplify.}$$

$$= 2550 \qquad \textit{n}\text{th partial sum}$$

EXAMPLE 7 Finding the *n*th Partial Sum

Find the sum.

$1 + 3 + 5 + 7 + 9 + 11 + 13 + 15 + 17 + 19$

SOLUTION

To begin, notice that the sequence is arithmetic (with a common difference of 2). Moreover, the sequence has 10 terms. So, the sum of the sequence is

$$\sum_{i=1}^{n} a_i = \frac{n}{2}(a_1 + a_n) \qquad \textit{n}\text{th partial sum formula}$$

$$\sum_{i=1}^{10} (2i - 1) = \frac{10}{2}(a_1 + a_{10}) \qquad \text{Substitute 10 for } n.$$

$$= 5(1 + 19) \qquad \text{Substitute 1 for } a_1 \text{ and 19 for } a_{10}.$$

$$= 5(20) \qquad \text{Simplify.}$$

$$= 100. \qquad \textit{n}\text{th partial sum}$$

Exercises Within Reach®

Finding the nth Partial Sum **In Exercises 41–48, find the *n*th partial sum of the arithmetic sequence.**

41. 5, 12, 19, 26, 33, . . . , $n = 12$ 522

42. 2, 12, 22, 32, 42, . . . , $n = 20$ 1940

43. −50, −38, −26, −14, −2, . . . , $n = 50$ 12,200

44. −16, −8, 0, 8, 16, . . . , $n = 30$ 3000

45. 1, 4.5, 8, 11.5, 15, . . . , $n = 12$ 243

46. 2.2, 2.8, 3.4, 4.0, 4.6, . . . , $n = 12$ 66

47. $a_1 = 0.5, a_4 = 1.7, \ldots,$ $n = 10$ 23

48. $a_1 = 15, a_{100} = 307, \ldots,$ $n = 100$ 16,100

49. ***Number Problem*** Find the sum of the first 75 positive integers. 2850

50. ***Number Problem*** Find the sum of the first 50 positive odd integers. 2500

Application

Application

EXAMPLE 8 Total Sales

Your business sells \$100,000 worth of handmade furniture during its first year. You have a goal of increasing annual sales by \$25,000 each year for 9 years. If you meet this goal, how much will you sell during your first 10 years of business?

SOLUTION

The annual sales during the first 10 years form the following arithmetic sequence.

\$100,000, \$125,000, \$150,000, \$175,000, \$200,000, \$225,000, \$250,000, \$275,000, \$300,000, \$325,000

Using the formula for the nth partial sum of an arithmetic sequence, you can find the total sales during the first 10 years as follows.

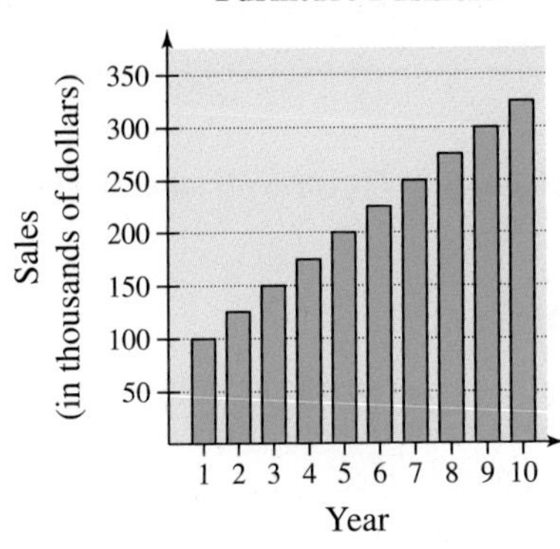

$$\text{Total sales} = \frac{n}{2}(a_1 + a_n)$$ nth partial sum formula

$$= \frac{10}{2}(100{,}000 + 325{,}000)$$ Substitute for n, a_1 and a_n.

$$= 5(425{,}000)$$ Simplify.

$$= \$2{,}125{,}000$$ Simplify.

From the bar graph shown at the left, notice that the annual sales for your company follow a *linear growth* pattern. In other words, saying that a quantity increases arithmetically is the same as saying that it increases linearly.

Exercises Within Reach ®

Solutions in English & Spanish and tutorial videos at AlgebraWithinReach.com

51. ***Salary*** In your new job as an actuary, your starting salary will be \$54,000 with an increase of \$3000 at the end of each of the first 5 years. How much will you be paid through the end of your first 6 years of employment with the company? \$369,000

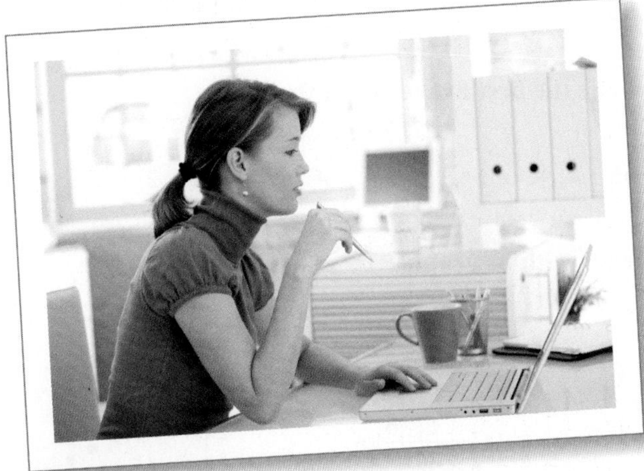

52. ***Wages*** You earn 5 dollars on the first day of the month, 10 dollars on the second day, 15 dollars on the third day, and so on. Determine the total amount that you will earn during a 30-day month. \$2325

StockLite/Shutterstock.com

53. ***Baling Hay*** In the first two trips baling hay around a large field (see figure), a farmer obtains 93 bales and 89 bales, respectively. The farmer estimates that the number will continue to decrease in a linear pattern. Estimate the total number of bales obtained after completing the six remaining trips around the field. 632 bales

First trip
Second trip
Third trip
Fourth trip
Fifth trip
Sixth trip
Seventh trip
Eighth trip

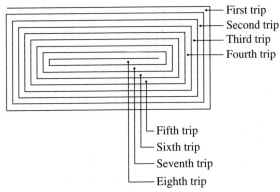

54. ***Ticket Prices*** There are 20 rows of seats on the main floor of an outdoor arena: 20 seats in the first row, 21 seats in the second row, 22 seats in the third row, and so on. How much should you charge per ticket to obtain \$15,000 for the sale of all the seats on the main floor? \$25.43

Concept Summary: Writing Formulas for Arithmetic Sequences

What

An **arithmetic sequence** is a sequence whose consecutive terms have a **common difference**. You can use a formula to represent an arithmetic sequence.

EXAMPLE

Find a formula for the nth term of the arithmetic sequence.

8, 12, 16, 20, 24, . . .

How

The formula for the nth term of an arithmetic sequence is of the form $a_n = a_1 + (n - 1)d$.

EXAMPLE

From the sequence, you can see that the common difference is $d = 4$ and the first term is $a_1 = 8$.

$a_n = a_1 + (n - 1)d$ Write formula.

$= 8 + 4(n - 1)$ Substitute.

$= 4n + 4$ Simplify.

Why

Once you find a formula for the nth term of an arithmetic sequence, you can use this formula to find any term in the sequence.

Exercises Within Reach®

Worked-out solutions to odd-numbered exercises at AlgebraWithinReach.com

Concept Summary Check

55. ***Vocabulary*** In an arithmetic sequence, the common difference between consecutive terms is 3. How can you use the value of one term to find the value of the next term in the sequence? Add 3 to the value of the term.

56. ***Writing*** Explain how you can use the first two terms of an arithmetic sequence to write a formula for the nth term of the sequence. Find the difference, $d = a_2 - a_1$, between the two terms. The nth term is $a_n = a_1 + (n - 1)d$.

57. ***Reasoning*** Explain how you can use the average of the first term and nth term of an arithmetic sequence to find the nth partial sum of the sequence. The nth partial sum can be found by multiplying the number of terms by the average of the first term and the nth term.

58. ***Logic*** In an arithmetic sequence, you know the common difference d between consecutive terms. What else do you need to know to find a_6? You need to know the value of another term a_n of the sequence for a known value of n.

Extra Practice

Writing the Terms of a Sequence **In Exercises 59–62, write the first five terms of the arithmetic sequence.**

59. $a_1 = 7,\ d = 5$ 7, 12, 17, 22, 27

60. $a_1 = 8,\ d = 3$ 8, 11, 14, 17, 20

61. $a_1 = 11,\ d = 4$ 11, 15, 19, 23, 27

62. $a_1 = 18,\ d = 10$ 18, 28, 38, 48, 58

Writing the Terms of a Sequence **In Exercises 63–66, write the first five terms of the arithmetic sequence. (Assume that n begins with 1.)**

63. $a_n = 3n + 4$ 7, 10, 13, 16, 19

64. $a_n = 5n - 4$ 1, 6, 11, 16, 21

65. $a_n = \frac{5}{2}n - 1$ $\frac{3}{2}, 4, \frac{13}{2}, 9, \frac{23}{2}$

66. $a_n = \frac{2}{3}n + 2$ $\frac{8}{3}, \frac{10}{3}, 4, \frac{14}{3}, \frac{16}{3}$

Matching **In Exercises 67–70, match the arithmetic sequence with its graph.**

(a)

a_n 14 12 10 8 6 4 2 −2 2 4 6 10 12 14 n

(b)

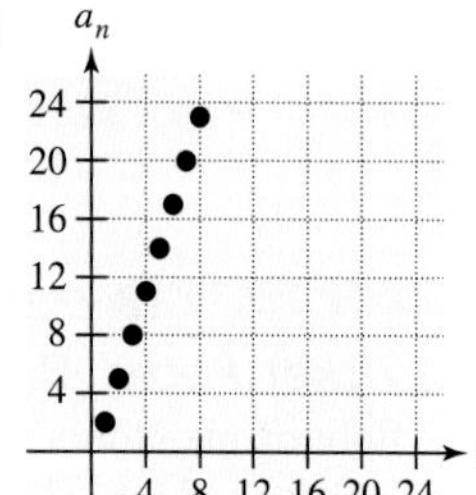

(c)

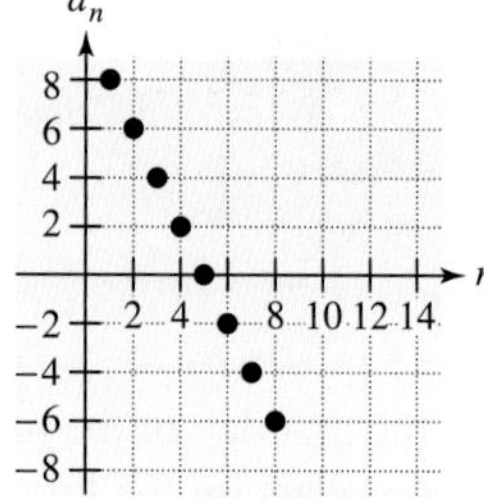

(d)

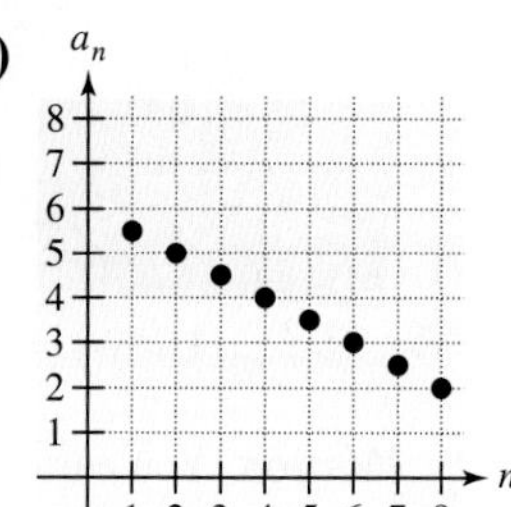

67. $a_n = -\frac{1}{2}n + 6$ d

68. $a_n = -2n + 10$ c

69. $a_1 = 12$
$a_{n+1} = a_n - 2$ a

70. $a_1 = 2$
$a_{n+1} = a_n + 3$ b

71. ***Free-Falling Object*** A free-falling object falls 16 feet during the first second, 48 feet during the second second, 80 feet during the third second, and so on. What total distance does the object fall in 8 seconds?
1024 feet

72. ***Free-Falling Object*** A free-falling object falls 4.9 meters during the first second, 14.7 meters during the second second, 24.5 meters during the third second, and so on. What total distance does the object fall in 5 seconds? 122.5 meters

73. ***Clock Chimes*** A clock chimes once at 1:00, twice at 2:00, three times at 3:00, and so on. The clock also chimes once at 15-minute intervals that are not on the hour. How many times does the clock chime in a 12-hour period?
114 times

74. ***Clock Chimes*** A clock chimes once at 1:00, twice at 2:00, three times at 3:00, and so on. The clock also chimes once on the half-hour. How many times does the clock chime in a 12-hour period? 90 times

Explaining Concepts

75. ***Vocabulary*** Explain what a recursion formula does.
A recursion formula gives the relationship between the terms a_{n+1} and a_n.

76. ***Reasoning*** Explain how to use the nth term a_n and the common difference d of an arithmetic sequence to write a recursion formula for the term a_{n+2} of the sequence.
Because a_{n+2} is two terms away from a_n, add twice the difference d to a_n.
$a_{n+2} = a_n + 2d$

77. ***Writing*** Is it possible to use the nth term a_n and the common difference d of an arithmetic sequence to write a recursion formula for the term a_{2n}? Explain.
Yes. Because a_{2n} is n terms away from a_n, add n times the difference d to a_n.
$a_{2n} = a_n + nd$

78. ***Reasoning*** Each term of an arithmetic sequence is multiplied by a constant C. Is the resulting sequence arithmetic? If so, how does the common difference compare with the common difference of the original sequence? Yes. The new common difference is C times the common difference of the original sequence.

79. ***Pattern Recognition***

(a) Compute the sums of positive odd integers.

$1 + 3 = 4$

$1 + 3 + 5 = 9$

$1 + 3 + 5 + 7 = 16$

$1 + 3 + 5 + 7 + 9 = 25$

$1 + 3 + 5 + 7 + 9 + 11 = 36$

(b) Do the partial sums of the positive odd integers form an arithmetic sequence? Explain.
No. There is no common difference between consecutive terms of the sequence.

(c) Use the sums in part (a) to make a conjecture about the sums of positive odd integers. Check your conjecture for the sum below.

$1 + 3 + 5 + 7 + 9 + 11 + 13 = 49$.

$\sum_{k=1}^{n} (2k - 1) = n^2$

Cumulative Review

In Exercises 80–83, find the center and vertices of the ellipse.

80. $\dfrac{(x-4)^2}{25} + \dfrac{(y+5)^2}{9} = 1$ Center: $(4, -5)$ Vertices: $(-1, -5), (9, -5)$

81. $\dfrac{(x+2)}{4} + (y-8)^2 = 1$ Center: $(-2, 8)$ Vertices: $(-4, 8), (0, 8)$

82. $9x^2 + 4y^2 - 18x + 24y + 9 = 0$
Center: $(1, -3)$, Vertices: $(1, -6), (1, 0)$

83. $x^2 + 4y^2 - 8x + 12 = 0$ Center: $(4, 0)$ Vertices: $(2, 0), (6, 0)$

In Exercises 84–87, write the sum in sigma notation. (Begin with $k = 1$.)

84. $3 + 4 + 5 + 6 + 7 + 8 + 9$ $\sum_{k=1}^{7} (k+2)$

85. $3 + 6 + 9 + 12 + 15$ $\sum_{k=1}^{5} 3k$

86. $12 + 15 + 18 + 21 + 24$ $\sum_{k=1}^{5} (3k+9)$

87. $2 + 2^2 + 2^3 + 2^4 + 2^5$ $\sum_{k=1}^{5} 2^k$

Mid-Chapter Quiz: Sections 11.1–11.2

Solutions in English & Spanish and tutorial videos at AlgebraWithinReach.com

Take this quiz as you would take a quiz in class. After you are done, check your work against the answers in the back of the book.

In Exercises 1–4, write the first five terms of the sequence. (Assume that *n* begins with 1.)

1. $a_n = 4n$ 4, 8, 12, 16, 20

2. $a_n = 2n + 5$ 7, 9, 11, 13, 15

3. $a_n = 32\left(\frac{1}{4}\right)^{n-1}$ 32, 8, 2, $\frac{1}{2}$, $\frac{1}{8}$

4. $a_n = \frac{(-3)^n n}{n+4}$ $-\frac{3}{5}$, 3, $-\frac{81}{7}$, $\frac{81}{2}$, -135

In Exercises 5–10, find the partial sum.

5. $\sum_{k=1}^{4} 10k$ 100

6. $\sum_{i=1}^{10} 4$ 40

7. $\sum_{j=1}^{5} \frac{60}{j+1}$ 87

8. $\sum_{n=1}^{4} \frac{12}{n}$ 25

9. $\sum_{n=1}^{5} (3n-1)$ 40

10. $\sum_{k=1}^{4} (k^2-1)$ 26

In Exercises 11–14, write the sum in sigma notation. (Begin with $k = 1$.)

11. $\frac{2}{3(1)} + \frac{2}{3(2)} + \frac{2}{3(3)} + \cdots + \frac{2}{3(20)}$ $\sum_{k=1}^{20} \frac{2}{3k}$

12. $\frac{1}{1^3} - \frac{1}{2^3} + \frac{1}{3^3} - \cdots + \frac{1}{25^3}$ $\sum_{k=1}^{25} \frac{(-1)^{k+1}}{k^3}$

13. $0 + \frac{1}{2} + \frac{2}{3} + \frac{3}{4} + \cdots + \frac{19}{20}$ $\sum_{k=1}^{20} \frac{k-1}{k}$

14. $\frac{1}{2} + \frac{4}{2} + \frac{9}{2} + \cdots + \frac{100}{2}$ $\sum_{k=1}^{10} \frac{k^2}{2}$

In Exercises 15 and 16, find the common difference of the arithmetic sequence.

15. $1, \frac{3}{2}, 2, \frac{5}{2}, 3, \ldots$ $\frac{1}{2}$

16. 100, 94, 88, 82, 76, . . . -6

In Exercises 17 and 18, find a formula for the *n*th term of the arithmetic sequence.

17. $a_1 = 20$, $a_4 = 11$ $a_n = -3n + 23$

18. $a_1 = 32$, $d = -4$ $a_n = -4n + 36$

19. Find the sum of the first 200 positive even numbers. 40,200

20. You save \$0.50 on one day, \$1.00 the next day, \$1.50 the next day, and so on. How much will you have accumulated at the end of one year (365 days)?
\$33,397.50

Study Skills in Action

Preparing for the Final Exam

At the end of the semester, most students are inundated with projects, papers, and tests. Instructors may speed up the pace in lectures to get through all the material. If something unexpected is going to happen to a student, it often happens during this time.

Getting through the last couple of weeks of a math course can be challenging. This is why it is important to plan your review time for the final exam at least three weeks before the exam.

These students are planning how they will study for the final exam.

Smart Study Strategy

Form a Final Exam Study Group

1 ▶ **Form a study group of three or four students several weeks before the final exam.** The intent of this group is to review what you have already learned while continuing to learn new material.

2 ▶ **Find out what material you must know for the final, even if the instructor has not yet covered it.** As a group, meet with the instructor outside of class. A group is likely to receive more attention and can ask more questions.

3 ▶ **Ask for or create a practice final and have the instructor look at it.** Make sure the problems are at an appropriate level of difficulty. Look for sample problems in old tests and in cumulative tests in the textbook. Review what the textbook and your notes say as you look for problems. This will refresh your memory.

4 ▶ **Have each group member take the practice final exam.** Then have each member identify what he or she needs to study. Make sure you can complete the problems with the speed and accuracy that are necessary to complete the real final exam.

5 ▶ **Decide when the group is going to meet during the next couple of weeks and what you will cover during each session.** The tutoring or learning center on campus is an ideal setting in which to meet. Many libraries have small study rooms that study groups can reserve. Set up several study times for each week. If you live at home, make sure your family knows that this is a busy time.

6 ▶ **During the study group sessions, make sure you stay on track.** Prepare for each study session by knowing what material in the textbook you are going to cover and having the class notes for that material. When you have questions, assign a group member to go to the instructor for answers. Then this member can relay the correct information to the other group members. Save socializing for after the final exam.

11.3 Geometric Sequences

- Recognize, write, and find the nth terms of geometric sequences.
- Find the sums of finite and infinite geometric sequences.
- Use geometric sequences to solve application problems.

Geometric Sequences

Definition of Geometric Sequence

A sequence is called **geometric** when the ratios of consecutive terms are the same. So, the sequence $a_1, a_2, a_3, a_4, \ldots, a_n, \ldots$ is geometric when there is a number r, with $r \neq 0$, such that

$$\frac{a_2}{a_1} = r, \frac{a_3}{a_2} = r, \frac{a_4}{a_3} = r$$

and so on. The number r is the **common ratio** of the sequence.

EXAMPLE 1 Examples of Geometric Sequences

a. The sequence whose nth term is 2^n is geometric. For this sequence, the common ratio between consecutive terms is 2.

$$2, 4, 8, 16, \ldots, 2^n, \ldots \qquad \text{Begin with } n = 1.$$

$$\frac{4}{2} = 2$$

b. The sequence whose nth term is $4(3^n)$ is geometric. For this sequence, the common ratio between consecutive terms is 3.

$$12, 36, 108, 324, \ldots, 4(3^n), \ldots \qquad \text{Begin with } n = 1.$$

$$\frac{36}{12} = 3$$

c. The sequence whose nth term is $\left(-\frac{1}{3}\right)^n$ is geometric. For this sequence, the common ratio between consecutive terms is $-\frac{1}{3}$.

$$-\frac{1}{3}, \frac{1}{9}, -\frac{1}{27}, \frac{1}{81}, \ldots, \left(-\frac{1}{3}\right)^n, \ldots \qquad \text{Begin with } n = 1.$$

$$\frac{1/9}{-1/3} = -\frac{1}{3}$$

Exercises Within Reach ®

Solutions in English & Spanish and tutorial videos at AlgebraWithinReach.com

Identifying a Geometric Sequence **In Exercises 1–10, determine whether the sequence is geometric. If so, find the common ratio.**

1. $3, 6, 12, 24, \ldots$ Geometric, 2
2. $2, 6, 18, 54, \ldots$ Geometric, 3
3. $1, \pi, \pi^2, \pi^3, \ldots$ Geometric, π
4. $e, e^2, e^3, e^4, \ldots$ Geometric, e
5. $10, 15, 20, 25, \ldots$ Not geometric
6. $1, 8, 27, 64, 125, \ldots$ Not geometric
7. $64, 32, 16, 8, \ldots$ Geometric, $\frac{1}{2}$
8. $10, 20, 40, 80, \ldots$ Geometric, 2
9. $a_n = 4\left(-\frac{1}{2}\right)^n$ Geometric, $-\frac{1}{2}$
10. $a_n = -2\left(\frac{1}{3}\right)^n$ Geometric, $\frac{1}{3}$

The nth Term of a Geometric Sequence

The nth term of a geometric sequence has the form

$$a_n = a_1 r^{n-1}$$

where r is the common ratio of consecutive terms of the sequence. So, every geometric sequence can be written in the following form.

$$a_1, a_1 r, a_1 r^2, a_1 r^3, a_1 r^4, \ldots, a_1 r^{n-1}, \ldots$$

Study Tip

When you know the nth term of a geometric sequence, you can multiply that term by r to find the $(n + 1)$th term. That is, $a_{n+1} = ra_n$.

EXAMPLE 2 Finding the nth Term of a Geometric Sequence

Find a formula for the nth term of the geometric sequence whose common ratio is 3 and whose first term is 1. Then find the eighth term of the sequence.

SOLUTION

Because the common ratio is 3 and the first term is 1, the formula for the nth term must be

$$a_n = a_1 r^{n-1} = (1)(3)^{n-1} = 3^{n-1}.$$ Substitute 1 for a_1 and 3 for r.

The sequence has the form $1, 3, 9, 27, 81, \ldots, 3^{n-1}, \ldots$.

The eighth term of the sequence is $a_8 = 3^{8-1} = 3^7 = 2187$.

Have students compare Examples 2 and 3. Note that in Example 2, $r > 1$ and the sequence increases, whereas in Example 3, $r < 1$ and the sequence decreases. Ask students to explain why.

EXAMPLE 3 Finding the nth Term of a Geometric Sequence

Find a formula for the nth term of the geometric sequence whose first two terms are 4 and 2.

SOLUTION

Because the common ratio is

$$r = \frac{a_2}{a_1} = \frac{2}{4} = \frac{1}{2}$$

the formula for the nth term must be

$$a_n = a_1 r^{n-1}$$ Formula for geometric sequence

$$= 4\left(\frac{1}{2}\right)^{n-1}.$$ Substitute 4 for a_1 and $\frac{1}{2}$ for r.

The sequence has the form $4, 2, 1, \frac{1}{2}, \frac{1}{4}, \ldots, 4\left(\frac{1}{2}\right)^{n-1}, \ldots$.

Exercises Within Reach®

Solutions in English & Spanish and tutorial videos at AlgebraWithinReach.com

Finding the nth Term **In Exercises 11–18, find a formula for the nth term of the geometric sequence. Then find a_7.**

11. $a_1 = 1,\ r = 2$ $a_n = 2^{n-1}$; $a_7 = 64$

12. $a_1 = 5,\ r = 4$ $a_n = 5(4)^{n-1}$; $a_7 = 20{,}480$

13. $a_1 = 9,\ r = \frac{2}{3}$ $a_n = 9\left(\frac{2}{3}\right)^{n-1}$; $a_7 = \frac{64}{81}$

14. $a_1 = 10,\ r = -\frac{1}{5}$ $a_n = 10\left(-\frac{1}{5}\right)^{n-1}$; $a_7 = \frac{2}{3125}$

15. $a_1 = 2,\ a_2 = 4$ $a_n = 2(2)^{n-1}$; $a_7 = 128$

16. $a_1 = 25,\ a_2 = 125$ $a_n = 25(5)^{n-1}$; $a_7 = 390{,}625$

17. $a_1 = 8,\ a_2 = 2$ $a_n = 8\left(\frac{1}{4}\right)^{n-1}$; $a_7 = \frac{1}{512}$

18. $a_1 = 18,\ a_2 = 8$ $a_n = 18\left(\frac{4}{9}\right)^{n-1}$; $a_7 = \frac{8192}{59{,}049}$

The Sum of a Geometric Sequence

The nth Partial Sum of a Geometric Sequence

The nth partial sum of the geometric sequence whose nth term is $a_n = a_1 r^{n-1}$ is given by

$$\sum_{i=1}^{n} a_1 r^{i-1} = a_1 + a_1 r + a_1 r^2 + a_1 r^3 + \cdots + a_1 r^{n-1} = a_1\left(\frac{r^n - 1}{r - 1}\right).$$

EXAMPLE 4 Finding the nth Partial Sum

Find the sum $1 + 2 + 4 + 8 + 16 + 32 + 64 + 128$.

SOLUTION

This is a geometric sequence whose common ratio is $r = 2$. Because the first term of the sequence is $a_1 = 1$, follows that the sum is

$$\sum_{i=1}^{8} 2^{i-1} = (1)\left(\frac{2^8 - 1}{2 - 1}\right) = \frac{256 - 1}{2 - 1} = 255.$$ Substitute 1 for a_1 and 2 for r.

Additional Example

Find the sum

$$\sum_{i=1}^{7} \frac{3^{i-1}}{2}.$$

Answer: 546.5

EXAMPLE 5 Finding the nth Partial Sum

Find the sum of the first five terms of the geometric sequence whose nth term is $a_n = \left(\frac{2}{3}\right)^n$.

SOLUTION

$$\sum_{i=1}^{5} \left(\frac{2}{3}\right)^i = \frac{2}{3}\left[\frac{(2/3)^5 - 1}{(2/3) - 1}\right]$$ Substitute $\frac{2}{3}$ for a_1 and $\frac{2}{3}$ for r.

$$= \frac{2}{3}\left[\frac{(32/243) - 1}{-1/3}\right]$$ Simplify.

$$= \frac{422}{243} \approx 1.737$$ Use a calculator to simplify.

Exercises Within Reach®

Solutions in English & Spanish and tutorial videos at AlgebraWithinReach.com

Finding the nth Partial Sum **In Exercises 19–26, find the *n*th partial sum of the geometric sequence. Round to the nearest hundredth, if necessary.**

19. $4, 12, 36, 108, \ldots, \quad n = 8$ 13,120

20. $5, 10, 20, 40, 80, \ldots, \quad n = 10$ 5115

21. $1, -3, 9, -27, 81, \ldots, \quad n = 10$ −14,762

22. $3, -6, 12, -24, 48, \ldots, \quad n = 12$ −4095

23. $8, 4, 2, 1, \frac{1}{2}, \ldots, \quad n = 15$ 16.00

24. $9, 6, 4, \frac{8}{3}, \frac{16}{9}, \ldots, \quad n = 10$ 26.53

25. $a_n = \left(\frac{3}{4}\right)^n, \quad n = 6$ 2.47

26. $a_n = \left(\frac{5}{6}\right)^n, \quad n = 4$ 2.59

Using Sigma Notation **In Exercises 27–30, find the partial sum. Round to the nearest hundredth, if necessary.**

27. $\sum_{i=1}^{10} 2^{i-1}$ 1023

28. $\sum_{i=1}^{6} 3^{i-1}$ 364

29. $\sum_{i=1}^{12} 3\left(\frac{3}{2}\right)^{i-1}$ 772.48

30. $\sum_{i=1}^{20} 12\left(\frac{2}{3}\right)^{i-1}$ 35.99

Sum of an Infinite Geometric Series

If $a_1, a_1r, a_1r^2, \ldots, a_1r^n, \ldots$ is an infinite geometric sequence and $|r| < 1$, then the sum of the terms of the corresponding infinite geometric series is

$$S = \sum_{i=0}^{\infty} a_1 r^i = \frac{a_1}{1-r}.$$

EXAMPLE 6 Finding the Sum of an Infinite Geometric Series

Find each sum.

a. $\sum_{i=1}^{\infty} 5\left(\frac{3}{4}\right)^{i-1}$ **b.** $\sum_{n=0}^{\infty} 4\left(\frac{3}{10}\right)^{n}$ **c.** $\sum_{i=0}^{\infty} \left(-\frac{3}{5}\right)^{i}$

SOLUTION

a. The series is geometric, with $a_1 = 5\left(\frac{3}{4}\right)^{1-1} = 5$ and $r = \frac{3}{4}$. So,

$$\sum_{i=1}^{\infty} 5\left(\frac{3}{4}\right)^{i-1} = \frac{5}{1-(3/4)} = \frac{5}{1/4} = 20.$$

b. The series is geometric, with $a_1 = 4\left(\frac{3}{10}\right)^{0} = 4$ and $r = \frac{3}{10}$. So,

$$\sum_{n=0}^{\infty} 4\left(\frac{3}{10}\right)^{n} = \frac{4}{1-(3/10)} = \frac{4}{7/10} = \frac{40}{7}.$$

c. The series is geometric, with $a_1 = \left(-\frac{3}{5}\right)^{0} = 1$ and $r = -\frac{3}{5}$. So,

$$\sum_{i=0}^{\infty} \left(-\frac{3}{5}\right)^{i} = \frac{1}{1-(-3/5)} = \frac{1}{1+(3/5)} = \frac{5}{8}.$$

Exercises Within Reach®

Finding the Sum of an Infinite Geometric Series **In Exercises 31–42, find the sum.**

31. $\sum_{n=1}^{\infty} 8\left(\frac{3}{4}\right)^{n-1}$ 32

32. $\sum_{n=1}^{\infty} 10\left(\frac{5}{6}\right)^{n-1}$ 60

33. $\sum_{n=0}^{\infty} 2\left(\frac{2}{3}\right)^{n}$ 6

34. $\sum_{n=0}^{\infty} 4\left(\frac{1}{4}\right)^{n}$ $\frac{16}{3}$

35. $\sum_{n=0}^{\infty} \left(-\frac{3}{7}\right)^{n}$ $\frac{7}{10}$

36. $\sum_{n=0}^{\infty} \left(-\frac{5}{9}\right)^{n}$ $\frac{9}{14}$

37. $\sum_{n=1}^{\infty} \left(\frac{1}{2}\right)^{n-1}$ 2

38. $\sum_{n=1}^{\infty} \left(\frac{1}{3}\right)^{n-1}$ $\frac{3}{2}$

39. $\sum_{n=1}^{\infty} \left(-\frac{1}{2}\right)^{n-1}$ $\frac{2}{3}$

40. $\sum_{n=1}^{\infty} \left(-\frac{1}{3}\right)^{n-1}$ $\frac{3}{4}$

41. $\sum_{n=0}^{\infty} \left(\frac{1}{10}\right)^{n}$ $\frac{10}{9}$

42. $\sum_{n=0}^{\infty} \left(\frac{1}{8}\right)^{n}$ $\frac{8}{7}$

Applications

Application **EXAMPLE 7** **A Lifetime Salary**

You accept a job as a meteorologist that pays a salary of \$45,000 the first year. During the next 39 years, you receive a 6% raise each year. What is your total salary over the 40-year period?

SOLUTION

Using a geometric sequence, your salary during the first year is $a_1 = 45{,}000$. Then, with a 6% raise each year, your salary for the next 2 years will be as follows.

$$a_2 = 45{,}000 + 45{,}000(0.06) = 45{,}000(1.06)^1$$

$$a_3 = 45{,}000(1.06) + 45{,}000(1.06)(0.06) = 45{,}000(1.06)^2$$

From this pattern, you can see that the common ratio of the geometric sequence is $r = 1.06$. Using the formula for the nth partial sum of a geometric sequence, you can find the total salary over the 40-year period.

$$\text{Total salary} = a_1\left(\frac{r^n - 1}{r - 1}\right)$$

$$= 45{,}000\left[\frac{(1.06)^{40} - 1}{1.06 - 1}\right]$$

$$= 45{,}000\left[\frac{(1.06)^{40} - 1}{0.06}\right]$$

$$\approx \$6{,}964{,}288$$

Exercises Within Reach®

Solutions in English & Spanish and tutorial videos at AlgebraWithinReach.com

43. ***Salary*** You accept a job as an archaeologist that pays a salary of \$30,000 the first year. During the next 39 years, you receive a 5% raise each year. What is your total salary over the 40-year period? \$3,623,993.23

44. ***Salary*** You accept a job as marine biologist that pays a salary of \$45,000 the first year. During the next 39 years, you receive a 5.5% raise each year. What is your total salary over the 40-year period? \$6,147,252.63

45. ***Wages*** You work at a company that pays \$0.01 for the first day, \$0.02 for the second day, \$0.04 for the third day, and so on. The daily wage keeps doubling. What is your total income for working (a) 29 days and (b) 30 days?
(a) \$5,368,709.11 (b) \$10,737,418.23

46. ***Wages*** You work at a company that pays \$0.01 for the first day, \$0.03 for the second day, \$0.09 for the third day, and so on. The daily wage keeps tripling. What is your total income for working (a) 25 days and (b) 26 days?
(a) \$4,236,443,047.21 (b) \$12,709,329,141.60

Application EXAMPLE 8 **Increasing Annuity**

You deposit \$100 in an account each month for 2 years. The account pays an annual interest rate of 9%, compounded monthly. What is your balance at the end of 2 years? (This type of savings plan is called an **increasing annuity**.)

SOLUTION

The first deposit would earn interest for the full 24 months, the second deposit would earn interest for 23 months, the third deposit would earn interest for 22 months, and so on. Using the formula for compound interest, you can find the balance of the account at the end of 2 years.

$$\begin{aligned}\text{Total} &= a_1 + a_2 + \cdots + a_{24}\\ &= 100\left(1+\frac{0.09}{12}\right)^1 + 100\left(1+\frac{0.09}{12}\right)^2 + \cdots + 100\left(1+\frac{0.09}{12}\right)^{24}\\ &= 100(1.0075)^1 + 100(1.0075)^2 + \cdots + 100(1.0075)^{24}\\ &= 100(100.75)\left(\frac{1.0075^{24}-1}{1.0075-1}\right)\\ &= \$2638.49\end{aligned}$$

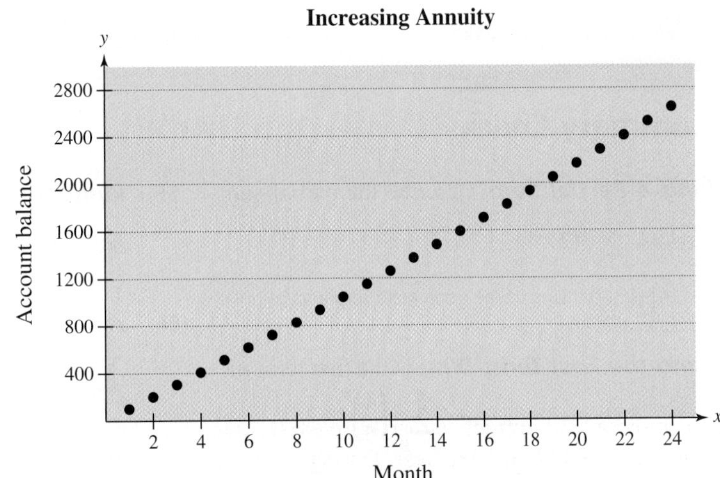

Exercises Within Reach ®

Solutions in English & Spanish and tutorial videos at AlgebraWithinReach.com

Increasing Annuity **In Exercises 47–52, find the balance A in an increasing annuity in which a principal of P dollars is invested each month for t years, compounded monthly at rate r.**

47. $P = \$50$ $t = 10$ years $r = 9\%$ \$9748.28

48. $P = \$50$ $t = 5$ years $r = 7\%$ \$3600.53

49. $P = \$30$ $t = 40$ years $r = 8\%$ \$105,428.44

50. $P = \$200$ $t = 30$ years $r = 10\%$ \$455,865.06

51. $P = \$100$ $t = 30$ years $r = 6\%$ \$100,953.76

52. $P = \$100$ $t = 25$ years $r = 8\%$ \$95,736.66

Concept Summary: Finding Sums of Geometric Sequences

What

A **geometric sequence** is a sequence whose consecutive terms have a **common ratio.**

You can use formulas to find partial sums of geometric sequences and sums of infinite geometric series.

EXAMPLE

Find the sum.

a. $\sum_{i=1}^{5} 3(2)^{i-1}$

b. $\sum_{n=0}^{\infty} 5\left(\frac{4}{5}\right)^n$

How

The nth partial sum of a geometric sequence:

$$\sum_{i=1}^{n} a_1 r^{i-1} = a_1\left(\frac{r^n - 1}{r - 1}\right)$$

Sum of an infinite geometric series:

$$\sum_{i=0}^{\infty} a_1 r^i = \frac{a_1}{1 - r}$$

EXAMPLE

a. $\sum_{i=1}^{5} 3(2)^{i-1} = 3\left(\frac{2^5 - 1}{2 - 1}\right)$

$= 3(31) = 93$

b. $\sum_{n=0}^{\infty} 5\left(\frac{4}{5}\right)^n = \frac{5}{1 - (4/5)} = \frac{5}{1/5} = 25$

Why

Many physics problems and compound interest problems involve sums of geometric sequences. You can use the formulas to find these sums efficiently.

Exercises Within Reach®

Worked-out solutions to odd-numbered exercises at AlgebraWithinReach.com

Concept Summary Check

53. ***Describing a Partial Sum*** Describe the partial sum represented by $\sum_{i=1}^{6} 3(2)^{i-1}$.

The sum of the first 6 terms of the geometric sequence whose nth term is $a_n = 3(2)^{i-1}$

54. ***Identifying the First Term*** What is the first term of the geometric series with the sum $\sum_{n=0}^{\infty} 5\left(\frac{4}{5}\right)^n$? 5

55. ***Identifying the Common Ratio*** What is the common ratio of the geometric series with the sum $\sum_{n=0}^{\infty} 5\left(\frac{4}{5}\right)^n$?

$\frac{4}{5}$

56. ***An Infinite Geometric Series*** Explain why the sum $\sum_{n=0}^{\infty} 5\left(\frac{4}{5}\right)^n$ is finite. The common ratio of $\frac{4}{5}$ is less than 1.

Extra Practice

Writing the Terms of a Sequence **In Exercises 57–60, write the first five terms of the geometric sequence.**

57. $a_1 = 5,\ r = -2$ $5, -10, 20, -40, 80$

58. $a_1 = -12,\ r = -1$ $-12, 12, -12, 12, -12$

59. $a_1 = -4,\ r = -\frac{1}{2}$ $-4, 2, -1, \frac{1}{2}, -\frac{1}{4}$

60. $a_1 = 3,\ r = -\frac{3}{2}$ $3, -\frac{9}{2}, \frac{27}{4}, -\frac{81}{8}, \frac{243}{16}$

Finding the Sum of a Geometric Series **In Exercises 61 and 62, find the sum.**

61. $8 + 6 + \frac{9}{2} + \frac{27}{8} + \cdots$ 32

62. $3 - 1 + \frac{1}{3} - \frac{1}{9} + \cdots$ $\frac{9}{4}$

63. ***Geometry*** An equilateral triangle has an area of 1 square unit. The triangle is divided into four smaller triangles and the center triangle is shaded (see figure). Each of the three unshaded triangles is then divided into four smaller triangles and each center triangle is shaded. This process is repeated one more time. What is the total area of the shaded regions?

$\frac{37}{64} \approx 0.578$ square unit

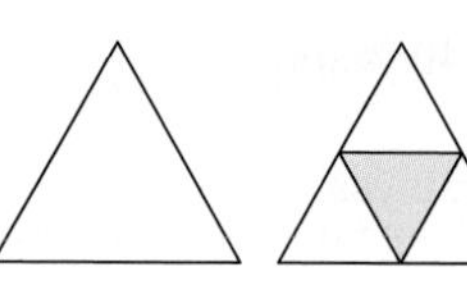

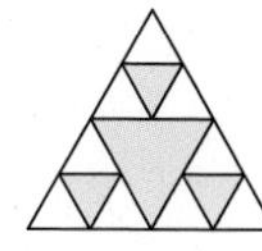

64. ***Geometry*** A square has an area of 1 square unit. The square is divided into nine smaller squares and the center square is shaded (see figure). Each of the eight unshaded squares is then divided into nine smaller squares and each center square is shaded. This process is repeated one more time. What is the total area of the shaded regions? $\frac{217}{729} \approx 0.298$ square unit

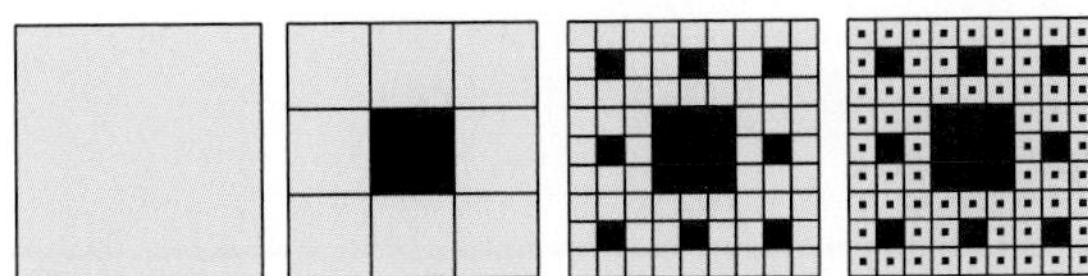

65. ***Cooling*** The temperature of water in an ice cube tray is 70°F when it is placed in a freezer. Its temperature n hours after being placed in the freezer is 20% less than 1 hour earlier. Find a formula for the nth term of the geometric sequence that gives the temperature of the water after n hours in the freezer. Then find the temperature after 6 hours in the freezer. $a_n = 70(0.8)^n$; 18.4°F

66. ***Bungee Jumping*** A bungee jumper drops 100 feet from a bridge and rebounds 75% of that distance for a total distance of 175 feet (see figure). Each successive drop and rebound covers 75% of the distance of the previous drop and rebound. Evaluate the expression below for the total distance traveled during 10 drops and rebounds.

$175 + 175(0.75) + \cdots + 175(0.75)^9$ 660.58 feet

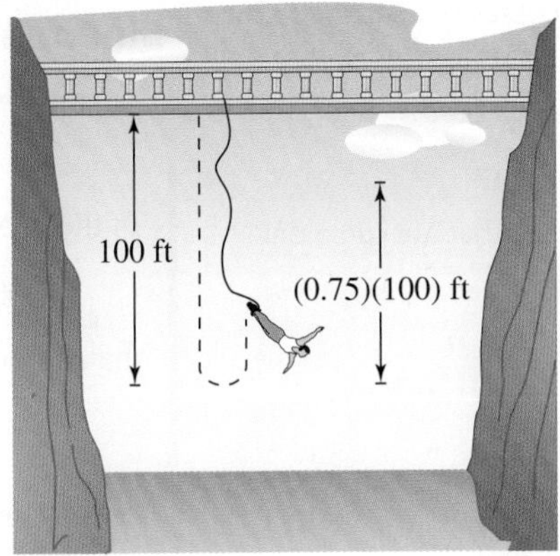

Explaining Concepts

67. ***Structure*** What is the general formula for the nth term of a geometric sequence? $a_n = a_1 r^{n-1}$

68. ***Writing*** How can you determine whether a sequence is geometric? A sequence is geometric when the ratios between consecutive terms are the same.

69. ***Writing*** Explain the difference between an arithmetic sequence and a geometric sequence. An arithmetic sequence has a common difference between consecutive terms whereas a geometric sequence has a common ratio between consecutive terms.

70. ***Structure*** Give an example of a geometric sequence whose terms alternate is sign. $a_n = \left(-\frac{2}{3}\right)^{n-1}$

71. ***Think About It*** Explain why the terms of a geometric sequence decrease when $a_1 > 0$ and $0 < r < 1$. When a positive number is multiplied by a number between 0 and 1, the result is a smaller positive number, so the terms of the sequence decrease.

72. ***Geometry*** A unit square is divided into two equal rectangles. One of the resulting rectangles is then divided into two equal rectangles, as shown in the figure. This process is repeated indefinitely. See Additional Answers.

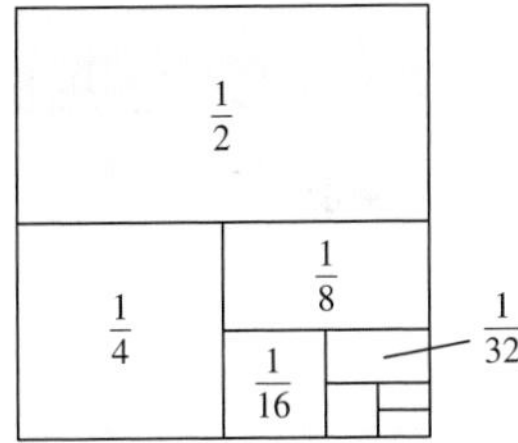

(a) Explain why the areas of the rectangles (from largest to smallest) form a geometric sequence.

(b) Find a formula for the nth term of the geometric sequence.

(c) Use the formula for the sum of an infinite geometric series to show that the combined area of the rectangles is 1.

Cumulative Review

In Exercises 73 and 74, solve the system of equations by the method of substitution.

73. $\begin{cases} y = 2x^2 \\ y = 2x + 4 \end{cases}$ $(-1, 2), (2, 8)$

74. $\begin{cases} x^2 + y^2 = 1 \\ x^2 + y = 1 \end{cases}$ $(\pm 1, 0), (0, 1)$

In Exercises 75–78, find the annual interest rate.

	Principal	*Balance*	*Time*	*Compounding*
75.	\$1000	\$2219.64	10 years	Monthly
76.	\$2000	\$3220.65	8 years	Quarterly
77.	\$2500	\$10,619.63	20 years	Yearly
78.	\$3500	\$25,861.70	40 years	Continuously

75. 8% 76. 6% 77. 7.5% 78. 5%

In Exercises 79 and 80, sketch the hyperbola. Identify the vertices and asymptotes. See Additional Answers.

79. $\frac{x^2}{16} - \frac{y^2}{9} = 1$

80. $\frac{y^2}{1} - \frac{x^2}{4} = 1$

11.4 The Binomial Theorem

- Use the Binomial Theorem to calculate binomial coefficients.
- Use Pascal's Triangle to calculate binomial coefficients.
- Expand binomial expressions.

Binomial Coefficients

Other notations that are commonly used for ${}_nC_r$ are $\binom{n}{r}$ and $C(n, r)$.

The Binomial Theorem

In the expansion of $(x + y)^n$

$$(x + y)^n = x^n + nx^{n-1}y + \cdots + {}_nC_r x^{n-r}y^r + \cdots + nxy^{n-1} + y^n$$

the coefficient of $x^{n-r}y^r$ is given by

$${}_nC_r = \frac{n!}{(n-r)!r!}.$$

EXAMPLE 1 Finding Binomial Coefficients

a. ${}_8C_2 = \dfrac{8!}{6! \cdot 2!} = \dfrac{(8 \cdot 7) \cdot \cancel{6!}}{\cancel{6!} \cdot 2!} = \dfrac{8 \cdot 7}{2 \cdot 1} = 28$

b. ${}_{10}C_3 = \dfrac{10!}{7! \cdot 3!} = \dfrac{(10 \cdot 9 \cdot 8) \cdot \cancel{7!}}{\cancel{7!} \cdot 3!} = \dfrac{10 \cdot 9 \cdot 8}{3 \cdot 2 \cdot 1} = 120$

c. ${}_7C_0 = \dfrac{7!}{7! \cdot 0!} = \dfrac{\cancel{7!}}{\cancel{7!} \cdot 1} = 1$

d. ${}_8C_8 = \dfrac{8!}{0! \cdot 8!} = \dfrac{\cancel{8!}}{1 \cdot \cancel{8!}} = 1$

e. ${}_7C_3 = \dfrac{7!}{4! \cdot 3!} = \dfrac{(7 \cdot 6 \cdot 5) \cdot \cancel{4!}}{\cancel{4!} \cdot 3!} = \dfrac{7 \cdot 6 \cdot 5}{3 \cdot 2 \cdot 1} = 35$

f. ${}_7C_4 = \dfrac{7!}{3! \cdot 4!} = \dfrac{(7 \cdot 6 \cdot 5) \cdot \cancel{4!}}{3! \cdot \cancel{4!}} = \dfrac{7 \cdot 6 \cdot 5}{3 \cdot 2 \cdot 1} = 35$

g. ${}_{12}C_1 = \dfrac{12!}{11! \cdot 1!} = \dfrac{(12) \cdot \cancel{11!}}{\cancel{11!} \cdot 1!} = \dfrac{12}{1} = 12$

h. ${}_{12}C_{11} = \dfrac{12!}{1! \cdot 11!} = \dfrac{(12) \cdot \cancel{11!}}{1! \cdot \cancel{11!}} = \dfrac{12}{1} = 12$

Study Tip

In Example 1, note that the answers to parts (e) and (f) are the same and that the answers to parts (g) and (h) are the same. In general ${}_nC_r = {}_nC_{n-r}$.

Exercises Within Reach ®

Solutions in English & Spanish and tutorial videos at AlgebraWithinReach.com

Finding a Binomial Coefficient **In Exercises 1–10, evaluate the binomial coefficient ${}_nC_r$.**

1. ${}_6C_4$ 15
2. ${}_9C_3$ 84
3. ${}_{10}C_5$ 252
4. ${}_{12}C_9$ 220
5. ${}_{12}C_{12}$ 1
6. ${}_8C_1$ 8
7. ${}_{20}C_6$ 38,760
8. ${}_{15}C_{10}$ 3003
9. ${}_{20}C_{14}$ 38,760
10. ${}_{15}C_5$ 3003

Pascal's Triangle

There is a convenient way to remember a pattern for binomial coefficients. By arranging the coefficients in a triangular pattern, you obtain the following array, which is called **Pascal's Triangle**. This triangle is named after the famous French mathematician Blaise Pascal (1623–1662).

Pascal's Triangle

```
                    1
                  1   1
                1   2   1
              1   3   3   1            1 + 2 = 3
            1   4   6   4   1
          1   5  10  10   5   1
        1   6  15  20  15   6   1      10 + 5 = 15
      1   7  21  35  35  21   7   1
```

Study Tip

The top row in Pascal's Triangle is called the *zeroth row* because it corresponds to the binomial expansion

$(x + y)^0 = 1.$

Similarly, the next row is called the *first row* because it corresponds to the binomial expansion

$(x + y)^1 = 1(x) + 1(y).$

In general, the *nth row* in Pascal's Triangle gives the coefficients of $(x + y)^n$.

The first and last numbers in each row of Pascal's Triangle are 1. Every other number in each row is formed by adding the two numbers immediately above the number. Pascal noticed that numbers in this triangle are precisely the same numbers that are the coefficients of binomial expansions, as follows.

$$(x + y)^0 = 1 \qquad \text{0th row}$$
$$(x + y)^1 = 1x + 1y \qquad \text{1st row}$$
$$(x + y)^2 = 1x^2 + 2xy + 1y^2 \qquad \text{2nd row}$$
$$(x + y)^3 = 1x^3 + 3x^2y + 3xy^2 + 1y^3 \qquad \text{3rd row}$$
$$(x + y)^4 = 1x^4 + 4x^3y + 6x^2y^2 + 4xy^3 + 1y^4 \qquad \vdots$$
$$(x + y)^5 = 1x^5 + 5x^4y + 10x^3y^2 + 10x^2y^3 + 5xy^4 + 1y^5$$
$$(x + y)^6 = 1x^6 + 6x^5y + 15x^4y^2 + 20x^3y^3 + 15x^2y^4 + 6xy^5 + 1y^6$$
$$(x + y)^7 = 1x^7 + 7x^6y + 21x^5y^2 + 35x^4y^3 + 35x^3y^4 + 21x^2y^5 + 7xy^6 + 1y^7$$

Exercises Within Reach®

Solutions in English & Spanish and tutorial videos at AlgebraWithinReach.com

11. ***Pascal's Triangle*** Find the eighth row of Pascal's Triangle using the diagram below.

```
                    1
                  1   1
                1   2   1
              1   3   3   1
            1   4   6   4   1
          1   5  10  10   5   1
        1   6  15  20  15   6   1
      1   7  21  35  35  21   7   1
    1   8  28  56  70  56  28   8   1
```

12. ***Pascal's Triangle*** Use your answer to Exercise 11 to expand $(x + y)^8$.

$(x + y)^8 = 1x^8 + 8x^7y + 28x^6y^2 + 56x^5y^3 + 70x^4y^4 + 56x^3y^5 + 28x^2y^6 + 8xy^7 + 1y^8$

EXAMPLE 2 **Using Pascal's Triangle**

Use the fifth row of Pascal's Triangle to evaluate ${}_5C_2$.

SOLUTION

1	5	10	10	5	1
${}_5C_0$	${}_5C_1$	${}_5C_2$	${}_5C_3$	${}_5C_4$	${}_5C_5$

So, ${}_5C_2 = 10$.

Check

$${}_5C_2 = \frac{5!}{3! \cdot 2!} = \frac{(5 \cdot 4) \cdot \cancel{3!}}{\cancel{3!} \cdot 2!} = \frac{5 \cdot 4}{2 \cdot 1} = 10 \checkmark$$

EXAMPLE 3 **Using Pascal's Triangle**

Use the sixth row of Pascal's Triangle to evaluate ${}_6C_4$.

SOLUTION

1	6	15	20	15	6	1
${}_6C_0$	${}_6C_1$	${}_6C_2$	${}_6C_3$	${}_6C_4$	${}_6C_5$	${}_6C_6$

So, ${}_6C_4 = 15$.

Check

$${}_6C_4 = \frac{6!}{2! \cdot 4!} = \frac{(6 \cdot 5) \cdot \cancel{4!}}{2! \cdot \cancel{4!}} = \frac{6 \cdot 5}{2 \cdot 1} = 15 \checkmark$$

Exercises Within Reach®

Solutions in English & Spanish and tutorial videos at AlgebraWithinReach.com

Using Pascal's Triangle **In Exercises 13–22, use Pascal's Triangle to evaluate ${}_nC_r$.**

13. ${}_6C_2$ 15

14. ${}_9C_3$ 84

15. ${}_7C_3$ 35

16. ${}_9C_5$ 126

17. ${}_8C_4$ 70

18. ${}_{10}C_6$ 210

19. ${}_5C_3$ 10

20. ${}_8C_6$ 28

21. ${}_7C_4$ 35

22. ${}_{10}C_2$ 45

Binomial Expansions

EXAMPLE 4 Expanding a Binomial

Write the expansion of the expression $(x + 1)^5$.

SOLUTION

The binomial coefficients from the fifth row of Pascal's Triangle are

1, 5, 10, 10, 5, 1.

So, the expansion is as follows.

$$(x + 1)^5 = (1)x^5 + (5)x^4(1) + (10)x^3(1^2) + (10)x^2(1^3) + (5)x(1^4) + (1)(1^5)$$
$$= x^5 + 5x^4 + 10x^3 + 10x^2 + 5x + 1$$

To expand binomials representing *differences*, rather than sums, you alternate signs. For example, $(x - 1)^3 = x^3 - 3x^2 + 3x - 1$.

EXAMPLE 5 Expanding a Binomial

Write the expansion of each expression.

a. $(x - 3)^4$ **b.** $(2x - 1)^3$

SOLUTION

a. The binomial coefficients from the fourth row of Pascal's Triangle are

1, 4, 6, 4, 1.

So, the expansion is as follows.

$$(x - 3)^4 = (1)x^4 - (4)x^3(3) + (6)x^2(3^2) - (4)x(3^3) + (1)(3^4)$$
$$= x^4 - 12x^3 + 54x^2 - 108x + 81$$

b. The binomial coefficients from the third row of Pascal's Triangle are

1, 3, 3, 1.

So, the expansion is as follows.

$$(2x - 1)^3 = (1)(2x)^3 - (3)(2x)^2(1) + (3)(2x)(1^2) - (1)(1^3)$$
$$= 8x^3 - 12x^2 + 6x - 1$$

Exercises Within Reach®

Solutions in English & Spanish and tutorial videos at AlgebraWithinReach.com

Using Pascal's Triangle **In Exercises 23–26, use Pascal's Triangle to expand the expression.**

23. $(t + 5)^3$ $t^3 + 15t^2 + 75t + 125$

24. $(y + 2)^4$ $y^4 + 8y^3 + 24y^2 + 32y + 16$

25. $(m - n)^5$ $m^5 - 5m^4n + 10m^3n^2 - 10m^2n^3 + 5mn^4 - n^5$

26. $(r - s)^7$ $r^7 - 7r^6s + 21r^5s^2 - 35r^4s^3 + 35r^3s^4 - 21r^2s^5 + 7rs^6 - s^7$

Using the Binomial Theorem **In Exercises 27–30, use the Binomial Theorem to expand the expression.**

27. $(x + 3)^6$ $x^6 + 18x^5 + 135x^4 + 540x^3 + 1215x^2 + 1458x + 729$

28. $(m - 4)^4$ $m^4 - 16m^3 + 96m^2 - 256m + 256$

29. $(u - v)^3$ $u^3 - 3u^2v + 3uv^2 - v^3$

30. $(x - y)^4$ $x^4 - 4x^3y + 6x^2y^2 - 4xy^3 + y^4$

EXAMPLE 6 Expanding a Binomial

Write the expansion of the expression $(x - 2y)^4$.

SOLUTION

Use the fourth row of Pascal's Triangle, as follows.

$$(x - 2y)^4 = (1)x^4 - (4)x^3(2y) + (6)x^2(2y)^2 - (4)x(2y)^3 + (1)(2y)^4$$
$$= x^4 - 8x^3y + 24x^2y^2 - 32xy^3 + 16y^4$$

EXAMPLE 7 Expanding a Binomial

Write the expansion of the expression $(x^2 + 4)^3$.

SOLUTION

Use the third row of Pascal's Triangle, as follows.

$$(x^2 + 4)^3 = (1)(x^2)^3 + (3)(x^2)^2(4) + (3)x^2(4^2) + (1)(4^3)$$
$$= x^6 + 12x^4 + 48x^2 + 64$$

EXAMPLE 8 Expanding a Binomial

Write the expansion of the expression $(x^2 - 4)^3$.

SOLUTION

Use the third row of Pascal's Triangle, as follows.

$$(x^2 - 4)^3 = (1)(x^2)^3 - (3)(x^2)^2(4) + (3)x^2(4^2) - (1)(4^3)$$
$$= x^6 - 12x^4 + 48x^2 - 64$$

Exercises Within Reach®

Solutions in English & Spanish and tutorial videos at AlgebraWithinReach.com

Using the Binomial Theorem **In Exercises 31–42, use the Binomial Theorem to expand the expression.**

31. $(3a - 1)^5$ $243a^5 - 405a^4 + 270a^3 - 90a^2 + 15a - 1$

32. $(1 - 4b)^3$ $1 - 12b + 48b^2 - 64b^3$

33. $(2y + z)^6$ $64y^6 + 192y^5z + 240y^4z^2 + 160y^3z^3 + 60y^2z^4 + 12yz^5 + z^6$

34. $(3c + d)^6$ $729c^6 + 1458c^5d + 1215c^4d^2 + 540c^3d^3 + 135c^2d^4 + 18cd^5 + d^6$

35. $(x^2 + 2)^4$
$x^8 + 8x^6 + 24x^4 + 32x^2 + 16$

36. $(5 + y^2)^5$
$3125 + 3125y^2 + 1250y^4 + 250y^6 + 25y^8 + y^{10}$

37. $(3a + 2b)^4$
$81a^4 + 216a^3b + 216a^2b^2 + 96ab^3 + 16b^4$

38. $(4u - 3v)^3$
$64u^3 - 144u^2v + 108uv^2 - 27v^3$

39. $\left(x + \frac{2}{y}\right)^4$ $x^4 + \frac{8x^3}{y} + \frac{24x^2}{y^2} + \frac{32x}{y^3} + \frac{16}{y^4}$

40. $\left(s + \frac{1}{t}\right)^5$ $s^5 + \frac{5s^4}{t} + \frac{10s^3}{t^2} + \frac{10s^2}{t^3} + \frac{5s}{t^4} + \frac{1}{t^5}$

41. $(2x^2 - y)^5$
$32x^{10} - 80x^8y + 80x^6y^2 - 40x^4y^3 + 10x^2y^4 - y^5$

42. $(x - 4y^3)^4$
$x^4 - 16x^3y^3 + 96x^2y^6 - 256xy^9 + 256y^{12}$

Sometimes you will need to find a specific term in a binomial expansion. Instead of writing out the entire expansion, you can use the fact that from the Binomial Theorem, the $(r + 1)$th term is

$${}_nC_r x^{n-r} y^r.$$

EXAMPLE 9 **Finding a Term in a Binomial Expansion**

a. Find the sixth term in the expansion of $(a + 2b)^8$.

b. Find the coefficient of the term a^6b^5 in the expansion of $(3a - 2b)^{11}$.

SOLUTION

a. In this case, $6 = r + 1$ means that $r = 5$. Because $n = 8$, $x = a$, and $y = 2b$, the sixth term in the binomial expansion is

$$\begin{aligned} {}_nC_r x^{n-r} y^r &= {}_8C_5 a^{8-5}(2b)^5 \\ &= 56 \cdot a^3 \cdot (2b)^5 \\ &= 56(2^5)\, a^3b^5 \\ &= 1792a^3b^5. \end{aligned}$$

b. In this case, $n = 11$, $r = 5$, $x = 3a$, and $y = -2b$. Substitute these values to obtain

$$\begin{aligned} {}_nC_r x^{n-r} y^r &= {}_{11}C_5(3a)^6(-2b)^5 \\ &= 462(729a^6)(-32b^5) \\ &= -10{,}777{,}536a^6b^5. \end{aligned}$$

So, the coefficient is $-10{,}777{,}536$.

Additional Examples

a. Expand $(x + 5)^5$.

b. Find the fifth term in the expansion of $(2x + 1)^9$.

Answers:

a. $x^5 + 25x^4 + 250x^3 + 1250x^2 + 3125x + 3125$

b. $4032x^5$

Exercises Within Reach®

Solutions in English & Spanish and tutorial videos at AlgebraWithinReach.com

Finding a Specific Term **In Exercises 43–48, find the specified term in the expansion of the binomial.**

43. $(x + y)^{10}$, 4th term $120x^7y^3$

44. $(x - y)^6$, 7th term y^6

45. $(a + 6b)^9$, 5th term $163{,}296a^5b^4$

46. $(3a - b)^{12}$, 10th term $-5940a^3b^9$

47. $(4x + 3y)^9$, 8th term $1{,}259{,}712x^2y^7$

48. $(5a + 6b)^5$, 5th term $32{,}400ab^4$

Finding the Coefficient of a Term **In Exercises 49–52, find the coefficient of the given term in the expansion of the binomial.**

	Expression	*Term*
49.	$(x + 1)^{10}$ 120	x^7
50.	$(x^2 - 3)^4$ 54	x^4
51.	$(x + 3)^{12}$ 5940	x^9
52.	$(3 - y^3)^5$ -90	x^9

Concept Summary: Using Pascal's Triangle to Expand Binomials

What

You can use **Pascal's Triangle** to **expand a binomial**.

EXAMPLE

Use Pascal's Triangle to expand $(x - 2)^3$.

```
            1
          1   1
        1   2   1
      1   3   3   1
    1   4   6   4   1
  1   5  10  10   5   1
1   6  15  20  15   6   1
```

How

Because the exponent is 3, use the third row of Pascal's Triangle.

EXAMPLE

The **binomial coefficients** from the third row of Pascal's Triangle are

1, 3, 3, 1.

So, the expansion is as follows.

$$(x - 2)^3 = (1)x^3 - (3)x^2(2) + (3)x(2)^2 - (1)(2)^3$$
$$= x^3 - 6x^2 + 12x - 8$$

Why

Knowing how to form Pascal's Triangle will allow you to expand complex binomials easily.

Using Pascal's Triangle to find

$(x - 2)^3$

is easier than finding the product

$(x - 2)(x - 2)(x - 2)$.

Exercises Within Reach®

Worked-out solutions to odd-numbered exercises at AlgebraWithinReach.com

Concept Summary Check

53. ***Writing*** In your own words, explain how to form the rows in Pascal's Triangle. The first and last numbers in each row are 1. Every other number in the row is formed by adding the two numbers immediately above the number.

54. ***Reasoning*** How many terms are in the expansion of $(x + y)^{10}$? 11

55. ***Writing*** How do the expansions of $(x + y)^n$ and $(x - y)^n$ differ? The signs of the terms alternate in the expansion of $(x - y)^n$.

56. ***Reasoning*** Which row of Pascal's Triangle would you use to evaluate ${}_{10}C_3$? Because $n = 10$, the tenth row of Pascal's Triangle should be used.

Extra Practice

Using the Binomial Theorem **In Exercises 57–62, use the Binomial Theorem to expand the expression.**

57. $(\sqrt{x} + 5)^3$ $x^{3/2} + 15x + 75x^{1/2} + 125$

58. $(2\sqrt{t} - 1)^3$ $8t^{3/2} - 12t + 6t^{1/2} - 1$

59. $(x^{2/3} - y^{1/3})^3$ $x^2 - 3x^{4/3}y^{1/3} + 3x^{2/3}y^{2/3} - y$

60. $(u^{3/5} + 2)^5$ $u^3 + 10u^{12/5} + 40u^{9/5} + 80u^{6/5} + 80u^{3/5} + 32$

61. $(3\sqrt{t} + \sqrt[4]{t})^4$ $81t^2 + 108t^{7/4} + 54t^{3/2} + 12t^{5/4} + t$

62. $(x^{3/4} - 2x^{5/4})^4$ $16x^5 - 32x^{9/2} + 24x^4 - 8x^{7/2} + x^3$

Evaluating ${}_nC_r$ **In Exercises 63–68, use a graphing calculator to evaluate ${}_nC_r$.**

63. ${}_{30}C_6$ 593,775

64. ${}_{40}C_8$ 76,904,685

65. ${}_{52}C_5$ 2,598,960

66. ${}_{100}C_4$ 3,921,225

67. ${}_{800}C_{797}$ 85,013,600

68. ${}_{1000}C_2$ 499,500

Using the Binomial Theorem **In Exercises 69 and 70, use the Binomial Theorem to expand the complex number.**

69. $(1 + i)^4$ -4

70. $(2 - i)^5$ $-38 - 41i$

Using the Binomial Theorem **In Exercises 71–74, use the Binomial Theorem to approximate the quantity rounded to three decimal places. For example:**

$$(1.02)^{10} = (1 + 0.02)^{10} \approx 1 + 10(0.02) + 45(0.02)^2.$$

71. $(1.02)^8$ 1.172

72. $(2.005)^{10}$ 1049.890

73. $(2.99)^{12}$ 510,568.785

74. $(1.98)^9$ 467.721

Probability **In the study of probability, it is sometimes necessary to use the expansion $(p + q)^n$, where $p + q = 1$. In Exercises 75–78, use the Binomial Theorem to expand the expression.**

75. $\left(\frac{1}{2} + \frac{1}{2}\right)^5$ $\frac{1}{32} + \frac{5}{32} + \frac{10}{32} + \frac{10}{32} + \frac{5}{32} + \frac{1}{32}$

76. $\left(\frac{2}{3} + \frac{1}{3}\right)^4$ $\frac{16}{81} + \frac{32}{81} + \frac{24}{81} + \frac{8}{81} + \frac{1}{81}$

77. $\left(\frac{1}{4} + \frac{3}{4}\right)^4$ $\frac{1}{256} + \frac{12}{256} + \frac{54}{256} + \frac{108}{256} + \frac{81}{256}$

78. $\left(\frac{2}{5} + \frac{3}{5}\right)^3$ $\frac{8}{125} + \frac{36}{125} + \frac{54}{125} + \frac{27}{125}$

79. ***Pascal's Triangle*** Rows 0 through 6 of Pascal's Triangle are shown. Find the sum of the numbers in each row. Describe the pattern.

1
1 1
1 2 1
1 3 3 1
1 4 6 4 1
1 5 10 10 5 1
1 6 15 20 15 6 1

The sum of the numbers in each row is a power of 2. Because the sum of the numbers in Row 2 is $1 + 2 + 1 = 4 = 2^2$, the sum of the numbers in Row n is 2^n.

80. ***Pascal's Triangle*** Use each encircled group of numbers to form a 2×2 matrix. Find the determinant of each matrix. Describe the pattern.

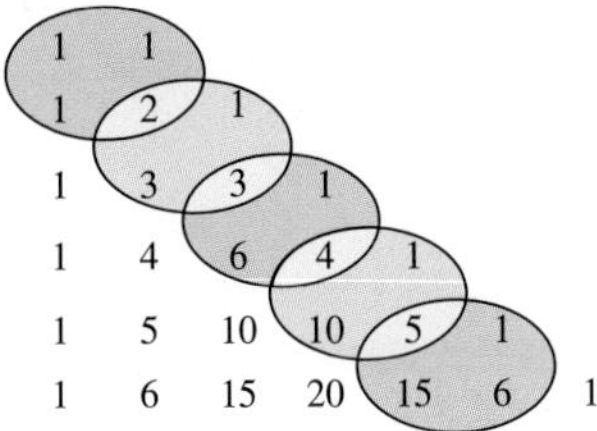

1, 3, 6, 10, 15; The difference between consecutive determinants increases by 1.

Explaining Concepts

81. ***Precision*** In the expansion of $(x + y)^{10}$, is ${}_6C_4$ the coefficient of the x^4y^6 term? Explain. No. The coefficient of x^4y^6 is ${}_{10}C_6$ because $n = 10$ and $r = 6$.

82. ***Number Sense*** Which of the following is equal to ${}_{11}C_5$? Explain.

(a) $\dfrac{11 \cdot 10 \cdot 9 \cdot 8 \cdot 7}{5 \cdot 4 \cdot 3 \cdot 2 \cdot 1}$

(b) $\dfrac{11 \cdot 10 \cdot 9 \cdot 8 \cdot 7}{6 \cdot 5 \cdot 4 \cdot 3 \cdot 2 \cdot 1}$

a; ${}_{11}C_5 = \dfrac{11!}{6!5!} = \dfrac{11 \cdot 10 \cdot 9 \cdot 8 \cdot 7}{5 \cdot 4 \cdot 3 \cdot 2 \cdot 1}$

83. ***Reasoning*** When finding the seventh term of a binomial expansion by evaluating ${}_nC_r x^{n-r}y^r$, what value should you substitute for r? Explain. The value for r is 6 because $6 + 1 = 7$.

84. ***Reasoning*** In the expansion of $(x + 2)^9$, are the coefficients of the x^3-term and the x^6-term identical? Explain. No. The binomial coefficients are the same, but the coefficients have a different number of factors of 2.

x^3-term: $84(x)^3(2)^6 = 5376x^3$

x^6-term: $84(x)^6(2)^3 = 672x^6$

Cumulative Review

In Exercises 85 and 86, find the partial sum of the arithmetic sequence.

85. $\displaystyle\sum_{i=1}^{15} (2 + 3i)$ 390

86. $\displaystyle\sum_{k=1}^{25} (9k - 5)$ 2800

In Exercises 87 and 88, find the partial sum of the geometric sequence. Round to the nearest hundredth, if necessary.

87. $\displaystyle\sum_{k=1}^{8} 5^{k-1}$ 97,656

88. $\displaystyle\sum_{i=1}^{20} 10\left(\tfrac{3}{4}\right)^{i-1}$ 39.87

11 Chapter Summary

	What did you learn?	Explanation and Examples	Review Exercises
11.1	Write the terms of sequences *(p. 576)*.	An infinite sequence $a_1, a_2, a_3, \ldots, a_n, \ldots$ is a function whose domain is the set of positive integers. A finite sequence $a_1, a_2, a_3, \ldots, a_n$ is a function whose domain is the finite set $\{1, 2, 3, \ldots, n\}$. When n is a positive integer, n factorial is defined as $n! = 1 \cdot 2 \cdot 3 \cdot 4 \cdot \cdots \cdot (n-1) \cdot n.$ As a special case, zero factorial is defined as $0! = 1$.	1–8
	Find the apparent nth term of a sequence *(p. 579)*.	Sometimes you have the first several terms of a sequence and need to find a formula (the nth term) to generate those terms. Pattern recognition is crucial in finding a form for the nth term.	9–16
	Sum the terms of a sequence to obtain a series *(p. 580)*.	For an infinite sequence $a_1, a_2, a_3, \ldots, a_n, \ldots$ **1.** the sum of the first n terms is called a partial sum, and **2.** the sum of all the terms is called an infinite series, or simply a series. The sum of the first n terms of the sequence whose nth term is a_n is $\sum_{i=1}^{n} a_i = a_1 + a_2 + a_3 + a_4 + \ldots + a_n$ where i is the index of summation, n is the upper limit of summation, and 1 is the lower limit of summation.	17–24
11.2	Recognize, write, and find the nth terms of arithmetic sequences *(p. 584)*.	A sequence is called arithmetic when the differences between consecutive terms are the same. The difference d between each pair of consecutive terms is called the common difference of the sequence. The nth term of an arithmetic sequence has the form $a_n = a_1 + (n-1)d$ where d is the common difference between the terms of the sequence, and a_1 is the first term.	25–36
	Find the nth partial sum of an arithmetic sequence *(p. 587)*.	The nth partial sum of the arithmetic sequence whose nth term is a_n is $\sum_{i=1}^{n} a_i = a_1 + a_2 + a_3 + a_4 + \ldots + a_n$ $= \frac{n}{2}(a_1 + a_n).$	37–42
	Use arithmetic sequences to solve application problems *(p. 589)*.	Saying that a quantity increases arithmetically is the same as saying that it increases linearly.	43–46

	What did you learn?	Explanation and Examples	Review Exercises
11.3	Recognize, write, and find the nth terms of geometric sequences *(p. 594).*	The nth term of a geometric sequence has the form $a_n = a_1 r^{n-1}$ where r is the common ratio of consecutive terms of the sequence. So, every geometric sequence can be written in the following form. $a_1, a_1 r, a_1 r^2, a_1 r^3, a_1 r^4, \ldots, a_1 r^{n-1}, \ldots$	47–60
	Find the sums of finite and infinite geometric sequences *(p. 596).*	The nth partial sum of the geometric sequence whose nth term is $a_n = a_1 r^{n-1}$ is given by $\sum_{i=1}^{n} a_1 r^{i-1} = a_1 + a_1 r + a_1 r^2 + a_1 r^3 + \cdots + a_1 r^{n-1} = a_1\left(\frac{r^n - 1}{r - 1}\right).$ If $a_1, a_1 r, a_1 r^2, \ldots, a_1 r^n, \ldots$ is an infinite geometric sequence and $\lvert r \rvert < 1$, then the sum of the terms of the corresponding infinite geometric series is $S = \sum_{i=0}^{\infty} a_1 r^i = \frac{a_1}{1 - r}.$	61–72
	Use geometric sequences to solve application problems *(p. 598).*	You can use a partial sum of a geometric sequence to find the value of an increasing annuity.	73–76
11.4	Use the Binomial Theorem to calculate binomial coefficients *(p. 602).*	In the expansion of $(x + y)^n$ $(x + y)^n = x^n + nx^{n-1}y + \cdots + {}_nC_r x^{n-r}y^r + \cdots + nxy^{n-1} + y^n$ the coefficient of $x^{n-r}y^r$ is given by ${}_nC_r = \frac{n!}{(n-r)!r!}.$	77–84
	Use Pascal's Triangle to calculate binomial coefficients *(p. 603).*	The first and last numbers in each row of Pascal's Triangle are 1. Every other number in each row is formed by adding the two numbers immediately above the number. 1 1 1 1 2 1 1 3 3 1 1 4 6 4 1 1 5 10 10 5 1 1 6 15 20 15 6 1	85–88
	Expand binomial expressions *(p. 605).*	When you write out the coefficients of a binomial raised to a power, you are expanding a binomial. The formulas for binomial coefficients give you an easy way to expand binomials.	89–102

Review Exercises

Worked-out solutions to odd-numbered exercises at AlgebraWithinReach.com

11.1

Writing the Terms of a Sequence **In Exercises 1–8, write the first five terms of the sequence with the given *n*th term. (Assume that *n* begins with 1.)**

1. $a_n = 3n + 5$

$8, 11, 14, 17, 20$

2. $a_n = \frac{1}{2}n - 4$

$-\frac{7}{2}, -3, -\frac{5}{2}, -2, -\frac{3}{2}$

3. $a_n = \dfrac{n}{3n - 1}$

$\frac{1}{2}, \frac{2}{5}, \frac{3}{8}, \frac{4}{11}, \frac{5}{14}$

4. $a_n = 3^n + n$

$4, 11, 30, 85, 248$

5. $a_n = (n + 1)!$

$2, 6, 24, 120, 720$

6. $a_n = (-1)^n n!$

$-1, 2, -6, 24, -120$

7. $a_n = \dfrac{n!}{2n}$

$\frac{1}{2}, \frac{1}{2}, 1, 3, 12$

8. $a_n = \dfrac{(n + 1)!}{(2n)!}$

$1, \frac{1}{4}, \frac{1}{30}, \frac{1}{336}, \frac{1}{5040}$

Finding the nth Term **In Exercises 9–16, write an expression for the *n*th term of the sequence. (Assume that *n* begins with 1.)**

9. $4, 7, 10, 13, 16, \ldots$

$a_n = 3n + 1$

10. $3, -6, 9, -12, 15, \ldots$

$a_n = (-1)^{n-1}3n$

11. $\frac{1}{2}, \frac{1}{5}, \frac{1}{10}, \frac{1}{17}, \frac{1}{26}, \ldots$

$a_n = \dfrac{1}{n^2 + 1}$

12. $\frac{0}{2}, \frac{1}{3}, \frac{2}{4}, \frac{3}{5}, \frac{4}{6}, \ldots$

$a_n = \dfrac{n - 1}{n + 1}$

13. $3, 1, -1, -3, -5, \ldots$

$a_n = -2n + 5$

14. $3, 7, 11, 15, 19, \ldots$

$a_n = 4n - 1$

15. $\frac{3}{2}, \frac{12}{5}, \frac{27}{10}, \frac{48}{17}, \frac{75}{26}, \ldots$

$a_n = \dfrac{3n^2}{n^2 + 1}$

16. $-1, \frac{1}{2}, -\frac{1}{4}, \frac{1}{8}, -\frac{1}{16}, \ldots$

$a_n = \dfrac{(-1)^n}{2^{n-1}}$

Finding a Partial Sum **In Exercises 17–20, find the partial sum.**

17. $\displaystyle\sum_{k=1}^{4} 7$

28

18. $\displaystyle\sum_{k=1}^{4} \frac{(-1)^k}{k}$

$-\frac{7}{12}$

19. $\displaystyle\sum_{i=1}^{5} \frac{i - 2}{i + 1}$

$\frac{13}{20}$

20. $\displaystyle\sum_{n=1}^{4} \left(\frac{1}{n} - \frac{1}{n + 2}\right)$

$\frac{17}{15}$

Writing a Sum in Sigma Notation **In Exercises 21–24, write the sum in sigma notation. (Begin with $k = 1$.)**

21. $[5(1) - 3] + [5(2) - 3] + [5(3) - 3] + [5(4) - 3]$

$\displaystyle\sum_{k=1}^{4} (5k - 3)$

22. $1(1 - 5) + 2(2 - 5) + 3(3 - 5) + 4(4 - 5) + 5(5 - 5)$

$\displaystyle\sum_{k=1}^{5} k(k - 5)$

23. $\dfrac{1}{3(1)} + \dfrac{1}{3(2)} + \dfrac{1}{3(3)} + \dfrac{1}{3(4)} + \dfrac{1}{3(5)} + \dfrac{1}{3(6)}$

$\displaystyle\sum_{k=1}^{6} \frac{1}{3k}$

24. $\left(-\frac{1}{3}\right)^0 + \left(-\frac{1}{3}\right)^1 + \left(-\frac{1}{3}\right)^2 + \left(-\frac{1}{3}\right)^3 + \left(-\frac{1}{3}\right)^4$

$\displaystyle\sum_{k=1}^{5} \left(-\frac{1}{3}\right)^{k-1}$

11.2

Finding the Common Difference **In Exercises 25 and 26, find the common difference of the arithmetic sequence.**

25. $50, 44.5, 39, 33.5, 28, \ldots$ -5.5

26. $9, 12, 15, 18, 21, \ldots$ 3

Writing the Terms of a Sequence **In Exercises 27–30, write the first five terms of the arithmetic sequence with the given *n*th term. (Assume that *n* begins with 1.)**

27. $a_n = 132 - 5n$

$127, 122, 117, 112, 107$

28. $a_n = 2n + 3$

$5, 7, 9, 11, 13$

29. $a_n = \frac{1}{3}n + \frac{5}{3}$

$2, \frac{7}{3}, \frac{8}{3}, 3, \frac{10}{3}$

30. $a_n = -\frac{3}{5}n + 1$

$\frac{2}{5}, -\frac{1}{5}, -\frac{4}{5}, -\frac{7}{5}, -2$

Using a Recursion Formula **In Exercises 31 and 32, write the first five terms of the arithmetic sequence defined recursively.**

31. $a_1 = 80$

$a_{k+1} = a_k - \frac{5}{2}$

$80, \frac{155}{2}, 75, \frac{145}{2}, 70$

32. $a_1 = 30$

$a_{k+1} = a_k - 12$

$30, 18, 6, -6, -18$

Finding the nth Term **In Exercises 33–36, find a formula for the *n*th term of the arithmetic sequence.**

33. $a_1 = 10,\ d = 4$ $a_n = 4n + 6$

34. $a_1 = 32,\ d = -2$ $a_n = -2n + 34$

35. $a_1 = 1000,\ a_2 = 950$ $a_n = -50n + 1050$

36. $a_2 = 150,\ a_5 = 201$ $a_n = 17n + 116$

Finding the nth Partial Sum **In Exercises 37–40, find the partial sum.**

37. $\sum_{k=1}^{12} (7k - 5)$ 486

38. $\sum_{k=1}^{10} (100 - 10k)$ 450

39. $\sum_{j=1}^{120} \left(\frac{1}{4}j + 1\right)$ 1935

40. $\sum_{j=1}^{50} \frac{3j}{2}$ $\frac{3825}{2}$

Finding the nth Partial Sum **In Exercises 41 and 42, find the *n*th partial sum of the arithmetic sequence.**

41. 5.25, 6.5, 7.75, 9, . . . , $n = 60$ 2527.5

42. $\frac{5}{2}, 3, \frac{7}{2}, 4, \frac{9}{2}, \ldots, n = 150$ 5962.5

43. ***Number Problem*** Find the sum of the first 50 positive integers that are multiples of 4. 5100

44. ***Number Problem*** Find the sum of the integers from 225 to 300. 19,950

45. ***Auditorium Seating*** Each row in a small auditorium has three more seats than the preceding row. The front row seats 22 people and there are 12 rows of seats. Find the seating capacity of the auditorium. 462 seats

46. ***Wages*** You earn \$25 on the first day of the month and \$100 on the last day of the month. Each day you are paid \$2.50 more than the previous day. How much do you earn in a 31-day month? 1937.50

11.3

Identifying a Geometric Sequence **In Exercises 47 and 48, determine whether the sequence is geometric. If so, find the common ratio.**

47. $8, 20, 50, 125, \frac{625}{2}, \ldots$ Geometric, $\frac{5}{2}$

48. $27, -18, 12, -8, \frac{16}{3}, \ldots$ Geometric, $-\frac{2}{3}$

Writing the Terms of a Sequence **In Exercises 49–54, write the first five terms of the geometric sequence.**

49. $a_1 = 10,\ r = 3$
10, 30, 90, 270, 810

50. $a_1 = 2,\ r = -5$
2, −10, 50, −250, 1250

51. $a_1 = 100,\ r = -\frac{1}{2}$
$100, -50, 25, -\frac{25}{2}, \frac{25}{4}$

52. $a_1 = 20,\ r = \frac{1}{5}$
$20, 4, \frac{4}{5}, \frac{4}{25}, \frac{4}{125}$

53. $a_1 = 4,\ r = \frac{3}{2}$
$4, 6, 9, \frac{27}{2}, \frac{81}{4}$

54. $a_1 = 32,\ r = -\frac{3}{4}$
$32, -24, 18, -\frac{27}{2}, \frac{81}{8}$

Finding the nth Term **In Exercises 55–60, find a formula for the *n*th term of the geometric sequence.**

55. $a_1 = 1,\ r = -\frac{2}{3}$ $a_n = \left(-\frac{2}{3}\right)^{n-1}$

56. $a_1 = 100,\ r = 1.07$ $a_n = 100(1.07)^{n-1}$

57. $a_1 = 24,\ a_2 = 72$ $a_n = 24(3)^{n-1}$

58. $a_1 = 16,\ a_2 = -4$ $a_n = 16\left(-\frac{1}{4}\right)^{n-1}$

59. $a_1 = 12,\ a_4 = -\frac{3}{2}$ $a_n = 12\left(-\frac{1}{2}\right)^{n-1}$

60. $a_2 = 1,\ a_3 = \frac{1}{3}$ $a_n = 3\left(\frac{1}{3}\right)^{n-1}$

Finding the nth Partial Sum **In Exercises 61–64, find the *n*th partial sum of the geometric sequence. Round to the nearest hundredth, if necessary.**

61. 200, 280, 392, 548.8, . . . , $n = 12$ 27,846.96

62. 25, 22.5, 20.25, 18.225, . . . , $n = 20$ 219.61

63. 27, −36, 48, −64, . . . , $n = 14$ −637.85

64. −1024, 512, −256, 128, . . . , $n = 9$ −684

Using Sigma Notation **In Exercises 65–68, find the partial sum. Round to the nearest thousandth, if necessary.**

65. $\sum_{n=1}^{12} 2^n$ 8190

66. $\sum_{n=1}^{12} (-2)^n$ 2730

67. $\sum_{k=1}^{8} 5\left(-\frac{3}{4}\right)^k$ −1.928

68. $\sum_{k=1}^{12} (-0.6)^{k-1}$ 0.624

Finding the Sum of an Infinite Geometric Series **In Exercises 69–72, find the sum**

69. $\sum_{i=1}^{\infty} \left(\frac{7}{8}\right)^{i-1}$ 8

70. $\sum_{i=1}^{\infty} \left(\frac{3}{5}\right)^{i-1}$ $\frac{5}{2}$

71. $\sum_{k=0}^{\infty} 4\left(\frac{2}{3}\right)^k$ 12

72. $\sum_{k=0}^{\infty} 1.3\left(\frac{1}{10}\right)^k$ $\frac{13}{9}$

73. ***Depreciation*** A company pays \$120,000 for a machine. During the next 5 years, the machine depreciates at the rate of 30% per year. (That is, at the end of each year, the depreciated value is 70% of what it was at the beginning of the year.)

(a) Find a formula for the nth term of the geometric sequence that gives the value of the machine n full years after it was purchased.

$a_n = 120{,}000(0.70)^n$

(b) Find the depreciated value of the machine at the end of 5 full years.

\$20,168.40

74. ***Population Increase*** A city of 85,000 people is growing at the rate of 1.2% per year. (That is, at the end of each year, the population is 1.012 times what it was at the beginning of the year.)

(a) Find a formula for the nth term of the geometric sequence that gives the population after n years.

$a_n = 85{,}000(1.012)^n$

(b) Estimate the population after 50 years.

154,328 people

75. ***Internet*** On its first day, a website has 1000 visits. During the next 89 days, the number of visits increases by 12.5% each day. What is the total number of visits during the 90-day period?

321,222,672 visits

76. ***Increasing Annuity*** You deposit \$200 in an account each month for 10 years. The account pays an annual interest rate of 8%, compounded monthly. What is your balance at the end of 10 years?

\$36,833.14

11.4

Finding a Binomial Coefficient **In Exercises 77–84, evaluate the binomial coefficient $_nC_r$.**

77. $_8C_3$ 56

78. $_{12}C_2$ 66

79. $_{15}C_4$ 1365

80. $_{100}C_1$ 100

81. $_{40}C_4$ 91,390

82. $_{32}C_8$ 10,518,300

83. $_{25}C_6$ 177,100

84. $_{48}C_5$ 1,712,304

Using Pascal's Triangle **In Exercises 85–88, use Pascal's Triangle to evaluate $_nC_r$.**

85. $_4C_2$ 6

86. $_9C_9$ 1

87. $_{10}C_3$ 120

88. $_6C_3$ 20

Using Pascal's Triangle **In Exercises 89–92, use Pascal's Triangle to expand the expression.**

89. $(x - 5)^4$

$x^4 - 20x^3 + 150x^2 - 500x + 625$

90. $(x + y)^7$

$x^7 + 7x^6y + 21x^5y^2 + 35x^4y^3 + 35x^3y^4 + 21x^2y^5 + 7xy^6 + y^7$

91. $(5x + 2)^3$

$125x^3 + 150x^2 + 60x + 8$

92. $(x - 3y)^4$

$x^4 - 12x^3y + 54x^2y^2 - 108xy^3 + 81y^4$

Using the Binomial Theorem **In Exercises 93–98, use the Binomial Theorem to expand the expression.**

93. $(x + 1)^{10}$

$x^{10} + 10x^9 + 45x^8 + 120x^7 + 210x^6 + 252x^5 + 210x^4 + 120x^3 + 45x^2 + 10x + 1$

94. $(y - 2)^6$

$y^6 - 12y^5 + 60y^4 - 160y^3 + 240y^2 - 192y + 64$

95. $(3x - 2y)^4$

$81x^4 - 216x^3y + 216x^2y^2 - 96xy^3 + 16y^4$

96. $(4u + v)^5$

$1024u^5 + 1280u^4v + 640u^3v^2 + 160u^2v^3 + 20uv^4 + v^5$

97. $(u^2 + v^3)^5$

$u^{10} + 5u^8v^3 + 10u^6v^6 + 10u^4v^9 + 5u^2v^{12} + v^{15}$

98. $(x^4 + y^5)^4$

$x^{16} - 4x^{12}y^5 + 6x^8y^{10} - 4x^4y^{15} + y^{20}$

Finding a Specific Term **In Exercises 99 and 100, find the specified term in the expansion of the binomial.**

99. $(x + 2)^{10}$, 7th term $13{,}440x^4$

100. $(2x - 3y)^5$, 4th term $-1080x^2y^3$

Finding the Coefficient of a Term **In Exercises 101 and 102, find the coefficient of the given term in the expansion of the binomial.**

	Expression	*Term*
101.	$(x - 3)^{10}$	x^5 −61,236
102.	$(3x + 4y)^6$	x^2y^4 34,560

Chapter Test

Solutions in English & Spanish and tutorial videos at AlgebraWithinReach.com

Take this test as you would take a test in class. After you are done, check your work against the answers in the back of the book.

1. Write the first five terms of the sequence whose nth term is $a_n = \left(-\frac{3}{5}\right)^{n-1}$. (Assume that n begins with 1.) $1, -\frac{3}{5}, \frac{9}{25}, -\frac{27}{125}, \frac{81}{625}$

2. Write the first five terms of the sequence whose nth term is $a_n = 3n^2 - n$. (Assume that n begins with 1.) 2, 10, 24, 44, 70

In Exercises 3–5, find the partial sum.

3. $\sum_{n=1}^{12} 5$ 60

4. $\sum_{k=0}^{8} (2k - 3)$ 45

5. $\sum_{n=1}^{5} (3 - 4n)$ -45

6. Write the sum in sigma notation: $\frac{2}{3(1)+1} + \frac{2}{3(2)+1} + \cdots + \frac{2}{3(12)+1}$. $\sum_{k=1}^{12} \frac{2}{3k+1}$

7. Write the sum in sigma notation:
$\left(\frac{1}{2}\right)^0 + \left(\frac{1}{2}\right)^2 + \left(\frac{1}{2}\right)^4 + \left(\frac{1}{2}\right)^6 + \left(\frac{1}{2}\right)^8 + \left(\frac{1}{2}\right)^{10}$. $\sum_{k=1}^{6} \left(\frac{1}{2}\right)^{2k-2}$

8. Write the first five terms of the arithmetic sequence whose first term is $a_1 = 12$ and whose common difference is $d = 4$. 12, 16, 20, 24, 28

9. Find a formula for the nth term of the arithmetic sequence whose first term is $a_1 = 5000$ and whose common difference is $d = -100$. $a_n = -100n + 5100$

10. Find the sum of the first 50 positive integers that are multiples of 3. 3825

11. Find the common ratio of the geometric sequence: $-4, 3, -\frac{9}{4}, \frac{27}{16}, \ldots$ $-\frac{3}{4}$

12. Find a formula for the nth term of the geometric sequence whose first term is $a_1 = 4$ and whose common ratio is $r = \frac{1}{2}$. $a_n = 4\left(\frac{1}{2}\right)^{n-1}$

In Exercises 13 and 14, find the partial sum.

13. $\sum_{n=1}^{8} 2(2^n)$ 1020

14. $\sum_{n=1}^{10} 3\left(\frac{1}{2}\right)^n$ $\frac{3069}{1024}$

In Exercises 15 and 16, find the sum.

15. $\sum_{i=1}^{\infty} \left(\frac{1}{2}\right)^i$ 1

16. $\sum_{i=1}^{\infty} 10(0.4)^{i-1}$ $\frac{50}{3}$

17. Evaluate the binomial coefficient ${}_{20}C_3$. 1140

18. Use Pascal's Triangle to expand the expression $(x - 2)^5$.
$x^5 - 10x^4 + 40x^3 - 80x^2 + 80x - 32$

19. Find the coefficient of the term x^3y^5 in the expansion of the expression $(x + y)^8$. 56

20. A free-falling object falls 4.9 meters during the first second, 14.7 more meters during the second second, 24.5 more meters during the third second, and so on. Assume the pattern continues. What is the total distance the object falls in 10 seconds? 490 meters

21. You deposit \$80 each month in an increasing annuity that pays 4.8% compounded monthly. What is the balance after 45 years? \$153,287.87

Additional Answers

CHAPTER 1

Section 1.1 *(pp. 2–9)*

1. (a) $\{1, 2, 6\}$ (b) $\{-6, 0, 1, 2, 6\}$
(c) $\left\{-6, -\frac{4}{3}, 0, \frac{5}{8}, 1, 2, 6\right\}$ (d) $\left\{-\sqrt{6}, \sqrt{2}, \pi\right\}$

2. (a) $\{5, 101\}$ (b) $\{-1, 0, 5, 101\}$
(c) $\left\{-\frac{10}{3}, -1, 0, \frac{2}{5}, \frac{5}{2}, 5, 101\right\}$ (d) $\left\{-\pi, -\sqrt{3}, \sqrt{3}\right\}$

3. (a) $\{\sqrt{4}\}$ (b) $\{\sqrt{4}, 0\}$
(c) $\left\{-4.2, \sqrt{4}, -\frac{1}{9}, 0, \frac{3}{11}, 5.\overline{5}, 5.543\right\}$ (d) $\{\sqrt{11}\}$

4. (a) $\{3, 110\}$ (b) $\{-\sqrt{25}, 0, 3, 110\}$
(c) $\left\{-\sqrt{25}, -0.\overline{1}, -\frac{5}{3}, 0, 0.85, 3, 110\right\}$ (d) $\{-\sqrt{6}\}$

49. The fractions are converted to decimals and plotted on a number line to determine the order.

50. Rewrite the fractions with the same denominator 6, then plot on a number line or compare the numerators. Because the fractions are negative, the greater fraction has the lesser numerator.

65.
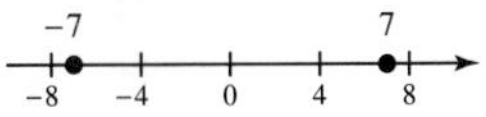

66.
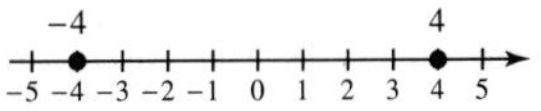

67.
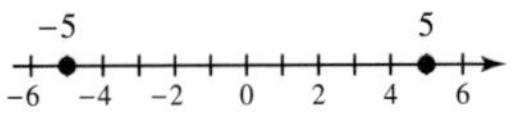

68.
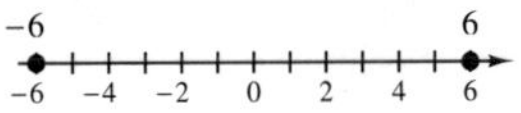

69.
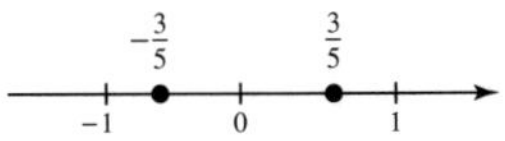

70.
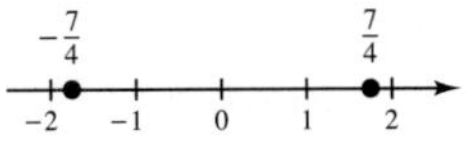

71.
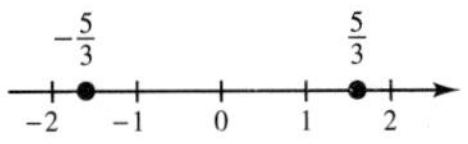

72.
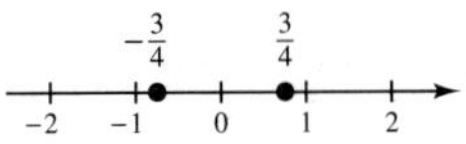

73.
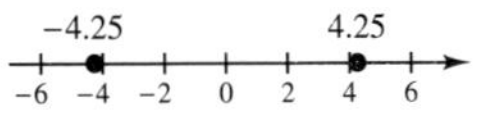

74.
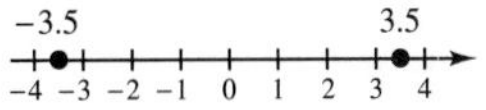

Section 1.3 *(pp. 18–25)*

23.

$ac = bc, c \neq 0$	Write original equation.
$\frac{1}{c}(ac) = \frac{1}{c}(bc)$	Multiplication Property of Equality
$\frac{1}{c}(ca) = \frac{1}{c}(cb)$	Commutative Property of Multiplication
$\left(\frac{1}{c} \cdot c\right)a = \left(\frac{1}{c} \cdot c\right)b$	Associative Property of Multiplication
$1 \cdot a = 1 \cdot b$	Multiplicative Inverse Property
$a = b$	Multiplicative Identity Property

24.

$a \cdot 1 = 1 \cdot a$	Write original equation.
$a = a$	Multiplicative Identity Property

25.

$a = (a + b) + (-b)$	Write original equation.
$a = a + [b + (-b)]$	Associative Property of Addition
$a = a + 0$	Additive Inverse Property
$a = a$	Additive Identity Property

26.

$a + (-a) = 0$	Write original equation.
$0 = 0$	Additive Inverse Property

41.

$x + 5 = 3$	Original equation
$(x + 5) + (-5) = 3 + (-5)$	Addition Property of Equality
$x + [5 + (-5)] = -2$	Associative Property of Addition
$x + 0 = -2$	Additive Inverse Property
$x = -2$	Additive Identity Property

42.

$x - 8 = 20$	Original equation
$(x - 8) + 8 = 20 + 8$	Addition Property of Equality
$x + (-8 + 8) = 28$	Associative Property of Addition
$x + 0 = 28$	Additive Inverse Property
$x = 28$	Additive Identity Property

43.

$2x - 5 = 6$	Original equation
$(2x - 5) + 5 = 6 + 5$	Addition Property of Equality
$2x + (-5 + 5) = 11$	Associative Property of Addition
$2x + 0 = 11$	Additive Inverse Property
$2x = 11$	Additive Identity Property
$\frac{1}{2}(2x) = \frac{1}{2}(11)$	Multiplication Property of Equality
$\left(\frac{1}{2} \cdot 2\right)x = \frac{11}{2}$	Associative Property of Multiplication
$1 \cdot x = \frac{11}{2}$	Multiplicative Inverse Property
$x = \frac{11}{2}$	Multiplicative Identity Property

44.

$3x + 4 = 10$	Original equation
$(3x + 4) + (-4) = 10 + (-4)$	Addition Property of Equality
$3x + [4 + (-4)] = 6$	Associative Property of Addition
$3x + 0 = 6$	Additive Inverse Property
$3x = 6$	Additive Identity Property
$\frac{1}{3}(3x) = \frac{1}{3}(6)$	Multiplication Property of Equality
$\left(\frac{1}{3} \cdot 3\right)x = 2$	Associative Property of Multiplication
$1 \cdot x = 2$	Multiplicative Inverse Property
$x = 2$	Multiplicative Identity Property

45.

$-4x - 4 = 0$	Original equation
$(-4x - 4) + 4 = 0 + 4$	Addition Property of Equality
$-4x + (-4 + 4) = 4$	Associative Property of Addition
$-4x + 0 = 4$	Additive Inverse Property
$-4x = 4$	Additive Identity Property
$-\frac{1}{4}(-4x) = -\frac{1}{4}(4)$	Multiplication Property of Equality
$\left[-\frac{1}{4}\cdot(-4)\right]x = -1$	Associative Property of Multiplication
$1 \cdot x = -1$	Multiplicative Inverse Property
$x = -1$	Multiplicative Identity Property

46.

$-5x + 25 = 5$	Original equation
$(-5x + 25) + (-25) = 5 + (-25)$	Addition Property of Equality
$-5x + [25 + (-25)] = -20$	Associative Property of Addition
$-5x + 0 = -20$	Additive Inverse Property
$-5x = -20$	Additive Identity Property
$-\frac{1}{5}(-5x) = -\frac{1}{5}(-20)$	Multiplication Property of Equality
$\left[-\frac{1}{5}\cdot(-5)\right]x = 4$	Associative Property of Multiplication
$1 \cdot x = 4$	Multiplicative Inverse Property
$x = 4$	Multiplicative Identity Property

47. Every real number except zero has an additive inverse. The additive inverse (or opposite) of a number is the same distance from zero as that number. Because there is no distance from zero to zero, zero does not have an additive inverse.

48. Every real number except zero has a multiplicative inverse. The multiplicative inverse of every number is 1 over that number. Because zero cannot be in the denominator, zero does not have a multiplicative inverse.

49. No.

Subtraction: $7 - 4 = 3 \neq -3 = 4 - 7$

Division: $18 \div 6 = 3 \neq \frac{1}{3} = 6 \div 18$

50. No.

Subtraction: $(10 - 2) - 9 = -1 \neq 17 = 10 - (2 - 9)$

Division: $(40 \div 4) \div 2 = 5 \neq 20 = 40 \div (4 \div 2)$

Mid-Chapter Quiz *(p. 26)*

1.

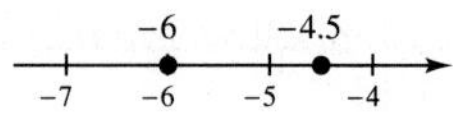

2.

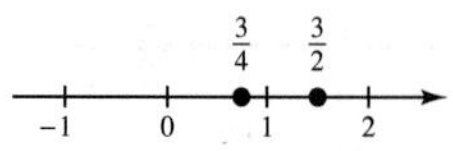

Section 1.4 *(pp. 28–35)*

72. To simplify $5x + 3x$ using the Distributive Property, factor the x out of each term, then add the numbers inside the parentheses. $5x + 3x = (5 + 3)x = 8x$

73. To combine like terms in an algebraic expression, first determine which terms are like terms. Then add the coefficients of the like terms and attach the common variable factor. For example, $5x^4 + (-2x^4) = [5 + (-2)]x^4 = 3x^4$.

74. To simplify an algebraic expression, algebraically remove the parentheses and combine like terms. To evaluate an algebraic expression, substitute values of the variables into the expression, and then simplify.

103. (a) Square: $\dfrac{4(4 - 3)}{2} = 2$ diagonals

Pentagon: $\dfrac{5(5 - 3)}{2} = 5$ diagonals

Hexagon: $\dfrac{6(6 - 3)}{2} = 9$ diagonals

(b) The numerator of the formula will always be even because it is the product of an even natural number and an odd natural number. Because the numerator is an even natural number and the denominator is 2, the formula will always yield a natural number.

Section 1.5 *(pp. 36–43)*

53. Perimeter: $2(2w) + 2w = 6w$; Area: $2w(w) = 2w^2$

54. Perimeter: $2(l) + 2(0.5l) = 3l$

Area: $(l)(0.5l) = 0.5l^2$

55. Perimeter: $6 + 2x + 3 + x + 3 + x = 4x + 12$

Area: $3x + 6x = 9x$ or $6(2x) - 3(x) = 9x$

56. Perimeter: $5 + (x + 2) + 1 + x + 3 + x + 1 + (x + 2) = 4x + 14$

Area: $5(x + 2) - 3x = 2x + 10$ or $5(2) + x + x = 2x + 10$

Review Exercises *(pp. 46–47)*

5.

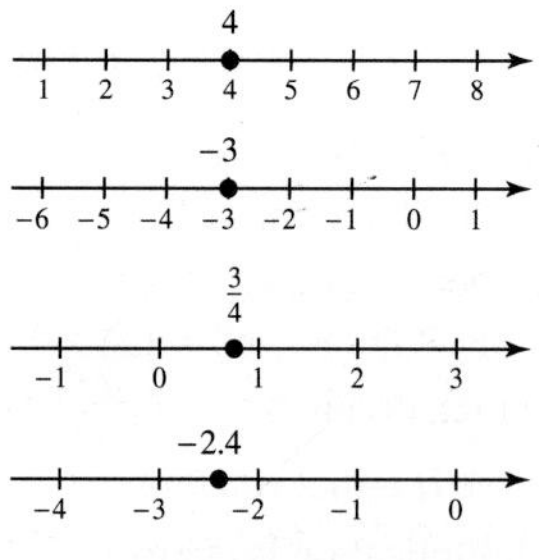

6.

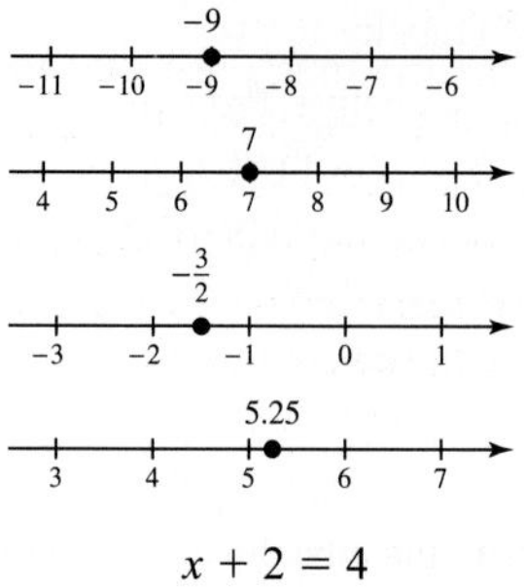

57.

$x + 2 = 4$	Original equation
$(x + 2) + (-2) = 4 + (-2)$	Addition Property of Equality
$x + [2 + (-2)] = 2$	Associative Property of Addition
$x + 0 = 2$	Additive Inverse Property
$x = 2$	Additive Identity Property

58.

$3x - 8 = 1$	Original equation
$(3x - 8) + 8 = 1 + 8$	Addition Property of Equality
$3x + (-8 + 8) = 9$	Associative Property of Addition
$3x + 0 = 9$	Additive Inverse Property
$3x = 9$	Additive Identity Property
$\frac{1}{3}(3x) = \frac{1}{3}(9)$	Multiplication Property of Equality
$\left(\frac{1}{3} \cdot 3\right)x = 3$	Associative Property of Multiplication
$1 \cdot x = 3$	Multiplicative Inverse Property
$x = 3$	Multiplicative Identity Property

CHAPTER 2

Section 2.1 *(pp. 50–57)*

68. (a)

t	0	1	2
Width	15	14.5	14
Length	15	15.5	16
Area	225	224.75	224

t	3	5	10
Width	13.5	12.5	10
Length	16.5	17.5	20
Area	222.75	218.75	200

(b) Because the length is t times the width and the perimeter is fixed, as t increases, the length increases and the width and area decrease. The maximum area occurs when the length and width are equal.

Section 2.2 *(pp. 58–65)*

1. 26(biweekly pay) + (bonus) = 37,120
Equation: $26x + 2800 = 37{,}120$; Solution: \$1320

2. 8(wage per hour) + 0.75(number of units) = 146
Equation: $8(10) + 0.75x = 146$; Solution: 88 units

85. To convert a percent to a decimal, divide by 100. For example: $24\% = \frac{24}{100} = 0.24$.
To convert a decimal to a percent, multiply by 100. For example: $0.18 = 0.18(100)\% = 18\%$.

87. Mathematical modeling is the use of mathematics to solve problems that occur in real-life situations. For examples, review the real-life problems in the exercise set.

88. The percent increase required to raise your salary after a 7% reduction is 7.53%. The percents are not the same because the base number is smaller for the percent increase.

Section 2.3 *(pp. 66–75)*

49. Perimeter: linear units—inches and meters
Area: square units—square feet and square centimeters
Volume: cubic units—cubic inches and cubic feet

Section 2.4 *(pp. 78–85)*

7. 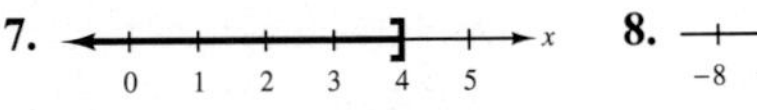

8. 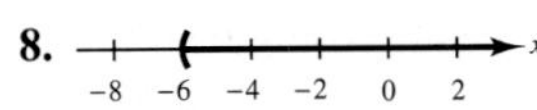

9. 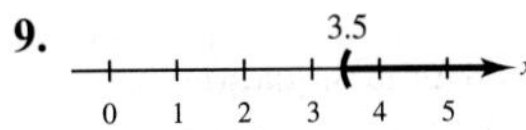

10. 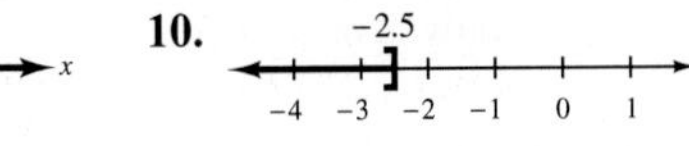

11. 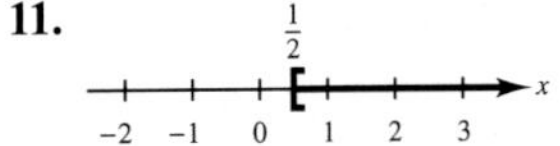

12. 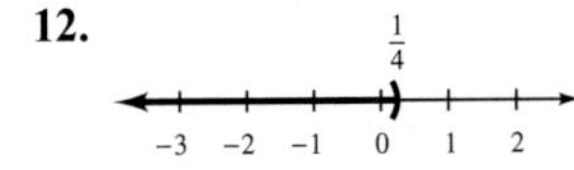

13.

14. 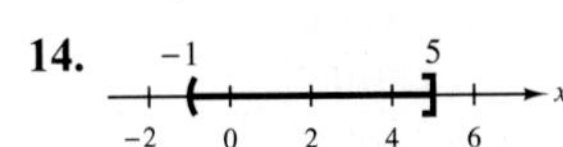

15.

16. 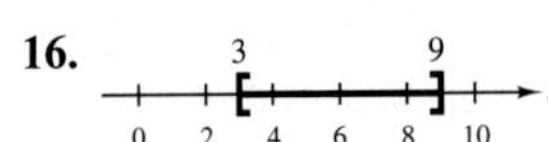

17. 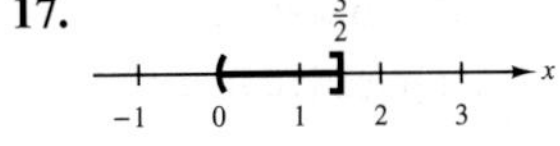

18. 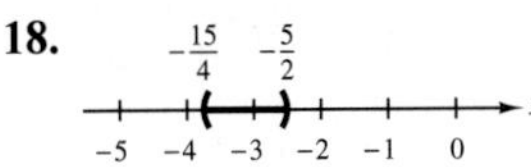

19. 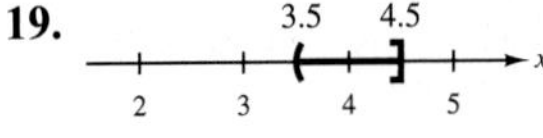

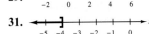

20. 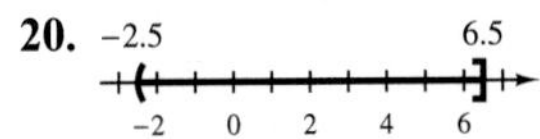

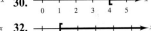

25.

26. 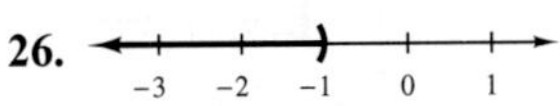

27.

28. 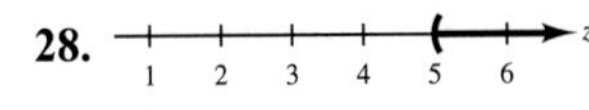

29.

30.

31.

32.

33. 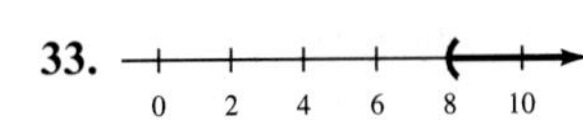

34. 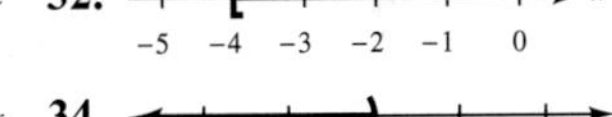

35. 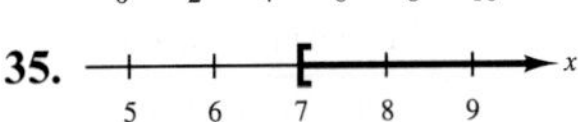

36. 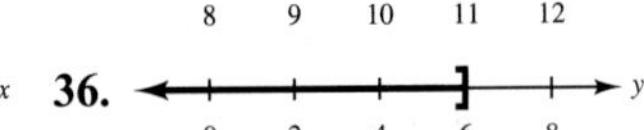

37.

38. 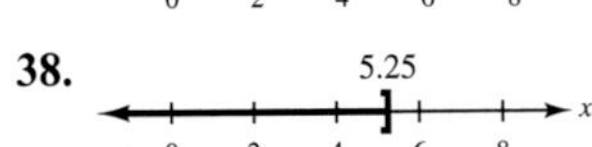

39. 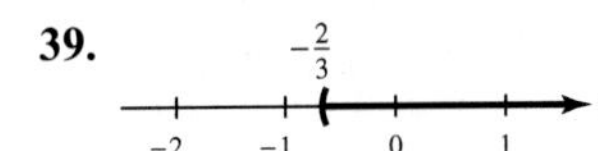

40. 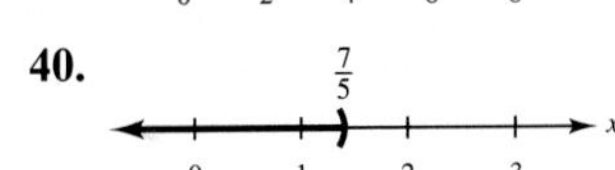

41. 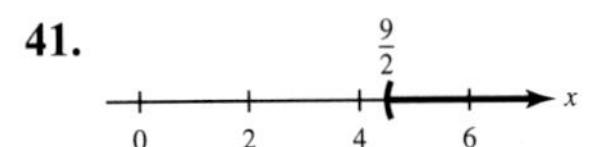

42. 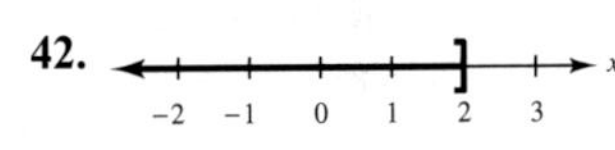

43. 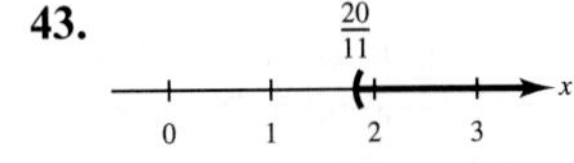

44. 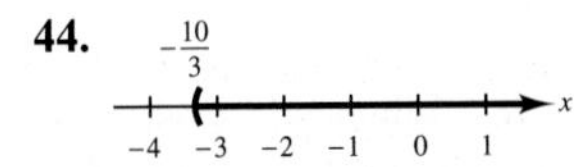

45. 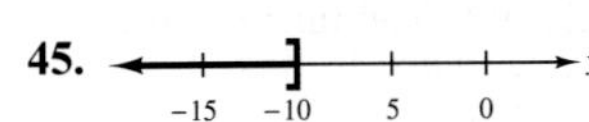

46. 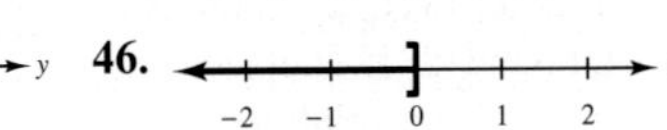

47. 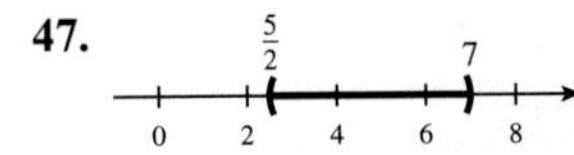

48. 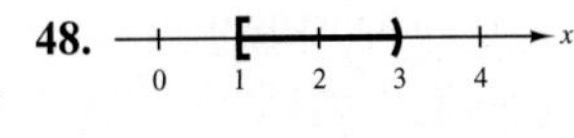

49. 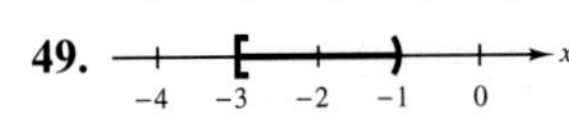

50. 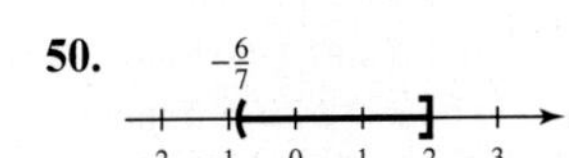

51. 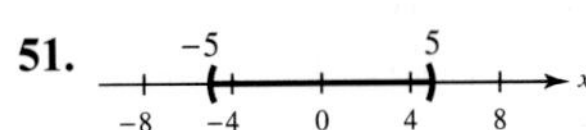

52. 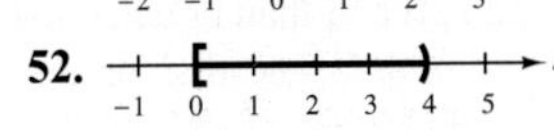

53. 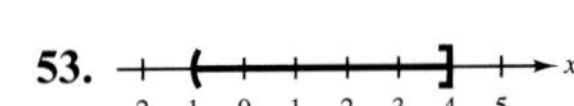

54. 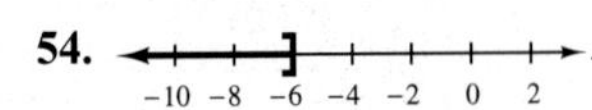

57. 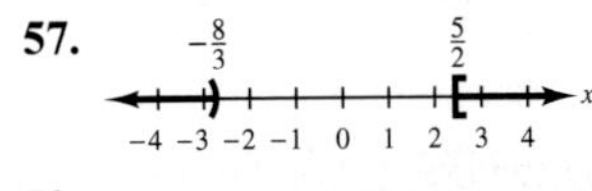

58. 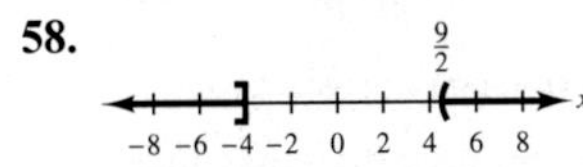

59. 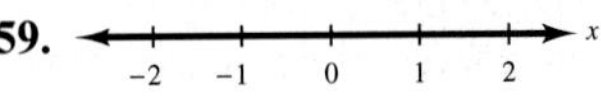

60. 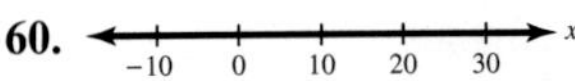

66. (a) Not equivalent. $x < 3$ represents all values of x less than 3, whereas $x > 3$ represents all values of x greater than 3.
(b) Equivalent. $3 < x$ represents all values of x greater than 3, and so does $x > 3$.
(c) Equivalent. When each side of an inequality is multiplied or divided by a negative number, the inequality symbol is reversed. So, $-x < -3$ is equivalent to $x > 3$.
(d) Not equivalent. $-3 < x$ represents all values of x greater than -3, which is not the same as $x > 3$, which represents all values of x greater than 3.

75. 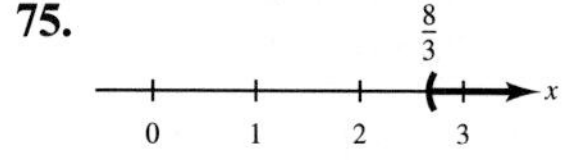

76. 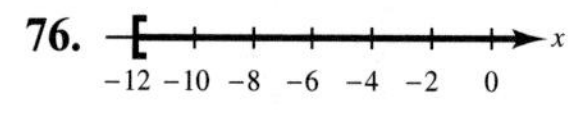

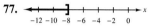

77. −12 −10 −8 −6 −4 −2 0

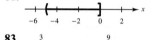

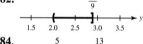

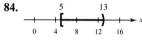

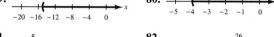

78. $\frac{12}{7}$ 0 1 2 3

79. −15 −20 −16 −12 −8 −4 0

80. −5 −4 −3 −2 −1 0

81. −5 −6 −4 −2 0 2

82. $\frac{26}{9}$ 1.5 2.0 2.5 3.0 3.5

83. $-\frac{3}{2}$ $\frac{9}{2}$ −2 0 2 4 6

84. 5 13 0 4 8 12 16

85. 1 10 0 3 6 9 12

86. 0 2 4 6 8 10

93. The multiplication and division properties differ. The inequality symbol is reversed if both sides of an inequality are multiplied or divided by a negative real number.

94. $-10 \le -x \le 3$; If x is multiplied by -1, then -3 and 10 must also be multiplied by -1 and both inequality symbols must be reversed.

95. The solution set of a linear inequality is bounded if the solution is written as a double inequality. Otherwise, the solution set is unbounded.

96. The solution set of two linear inequalities joined by the word *or* will never be a bounded interval. One linear inequality is unbounded. If another inequality is joined to it by the word *or*, either no solutions will be added or many solutions will be added, but no solutions will be taken away.

98. $x < a$ or $x > b$; Both of the intervals are unbounded ($x < a$ goes to $-\infty$ and $x > b$ goes to ∞) and there is no overlap between a and b, so there are two distinct intervals.

99. $x > a$ or $x < b$; $x > a$ includes all real numbers between a and b and greater than or equal to b, all the way to ∞. $x < b$ adds all real numbers less than or equal to a, all the way to $-\infty$.

Section 2.5 *(pp. 86–93)*

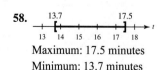

57. 41.826 42.65 (40 41 42 43 44 45)
Fastest: 41.826 seconds
Slowest: 42.65 seconds

58. 13.7 17.5 (13 14 15 16 17 18)
Maximum: 17.5 minutes
Minimum: 13.7 minutes

59. To solve an absolute value equation, first write the equation in standard form. Because the solutions of $|x| = a$ are $x = a$ and $x = -a$, rewrite the equation as two equivalent equations. Then, solve for each value of x. For example: $|x - 3| + 2 = 7$ in standard form is $|x - 3| = 5$, which means that $x - 3 = 5$ or $x - 3 = -5$. So, $x = 8$ or $x = -2$.

84. $|x - b| > |a|$; For example, if $a = -3$ and $b = -10$, you have the set of all real numbers that are more than $|-3|$ units from -10, or

$$|x - (-10)| > |-3|$$
$$x + 10 > 3 \quad \text{or} \quad x + 10 < -3$$
$$x > -7 \qquad\qquad x < -13.$$

86. The error is that when the inequality is rewritten as two inequalities, the inequality symbols are reversed. The second line should be $3x - 4 \ge -5$ or $3x - 4 \le 5$. Because $|3x - 4|$ is always nonnegative, you can recognize without solving that the inequality is true for all real values of x, $-\infty < x < \infty$.

Review Exercises *(pp. 96–99)*

31. (Total pay) = (Weekly wage)(Number of weeks) + (Training session pay)
Equation: $2635 = 320x + 75$; Solution: 8 weeks

32. (Earnings) = (Hours worked)(Hourly wage) + 1.25(Number of sales)
Equation: $88 = 8 \cdot 6 + 1.25x$; Solution: 32 sales

87. 88. 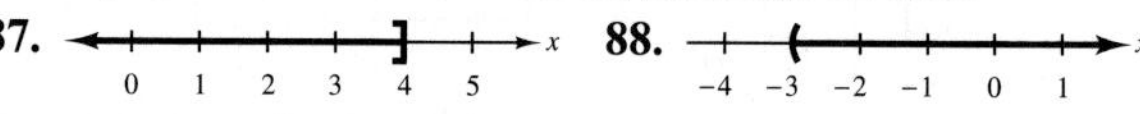

89. 1 2 3 4 5 6 7 8

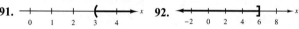

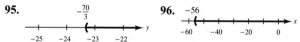

90. 0 1 2 3 4 5 6

91. 0 1 2 3 4

92. −2 0 2 4 6 8

93. −2 0 2 4 6 8

94. 0 1 2 3 4

95. $-\frac{70}{3}$ −25 −24 −23 −22

96. −56 −60 −40 −20 0

97. −5 −4 −3 −2 −1 0 1 2

98. $\frac{4}{5}$ −2 −1 0 1 2

99. −7 −8 −6 −4 −2 0

100. $-\frac{5}{2}$ −4 −2 0 2 4

101. −1 −20 −16 −12 −8 −4 0

102. 5 27 0 10 20 30

103. −3 −4 −2 0 2

104. 1 2 3 4 5

125. 116.6 0 20 40 60 80 100 120

126. 104 40 50 60 70 80 90 100 110

CHAPTER 3

Section 3.1 *(pp. 102–109)*

1.

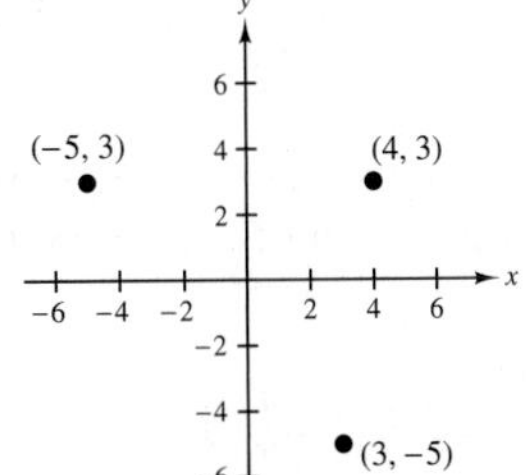

2.

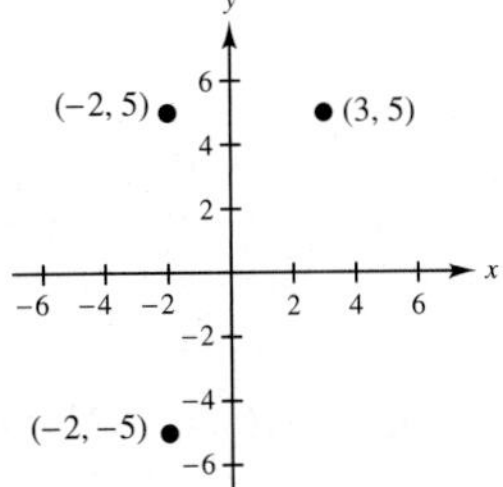

3.

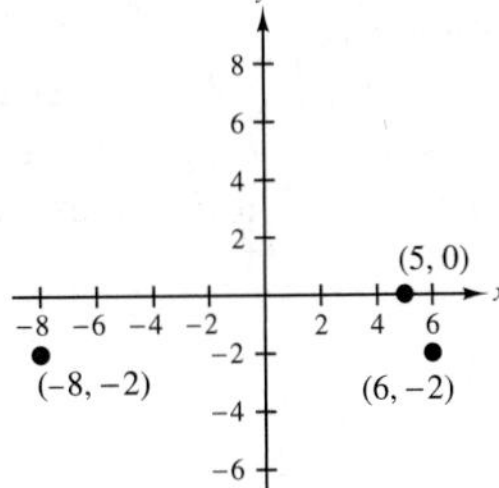

4.

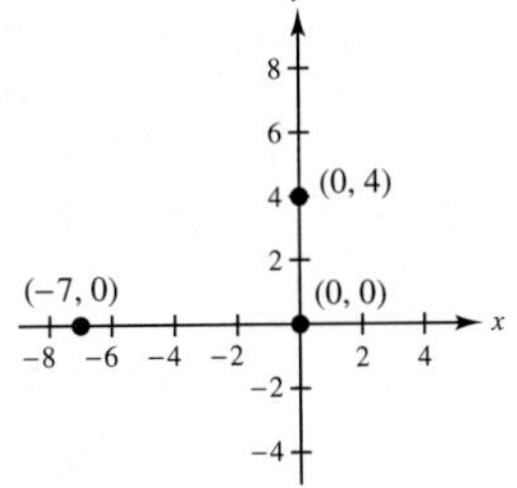

5.

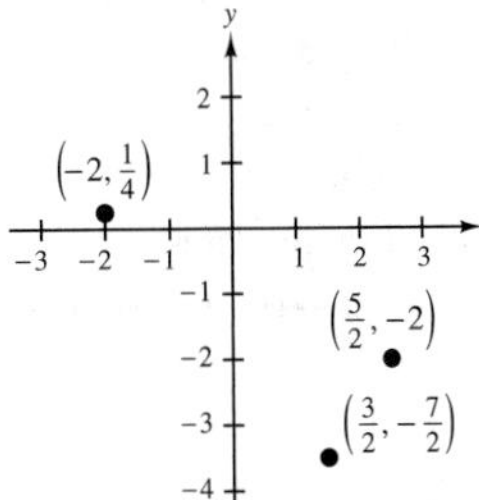

6.

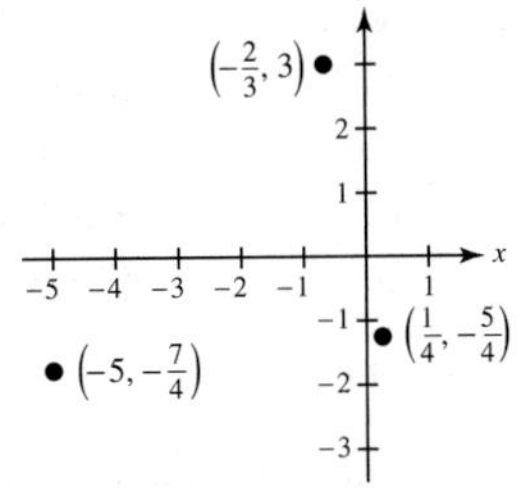

7. A: $(-2, 4)$, B: $(0, -2)$, C: $(4, -2)$

8. A: $(4, 2)$, B: $(-1, -2)$, C: $(0, 0)$

9. A: $(4, -2)$, B: $\left(-3, -\frac{5}{2}\right)$, C: $(3, 0)$

10. A: $\left(-\frac{7}{2}, -2\right)$, B: $(5, 1)$, C: $(0, 4)$

11.

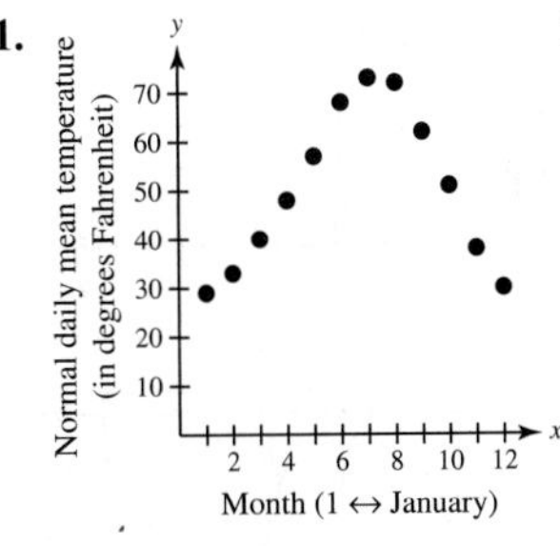

12.

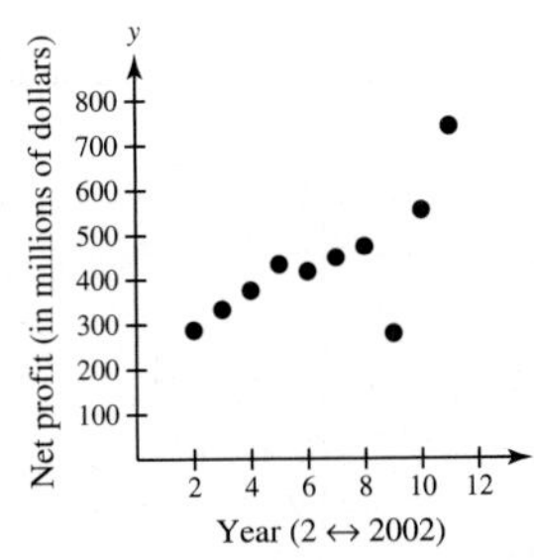

13.

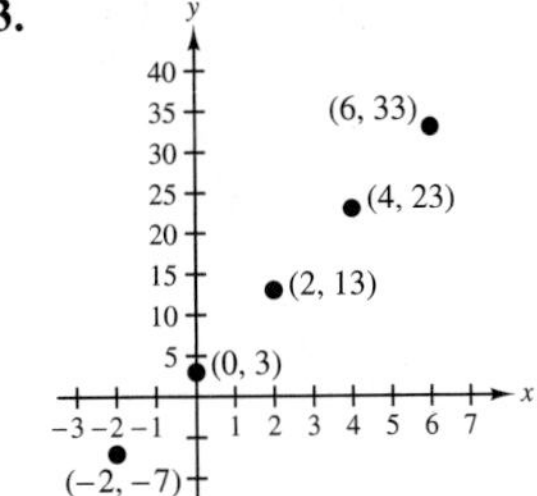
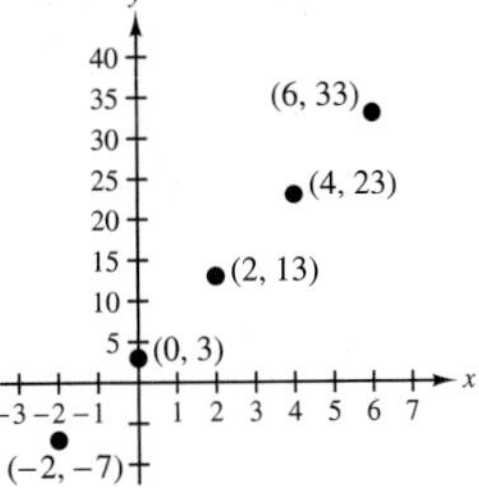

14.

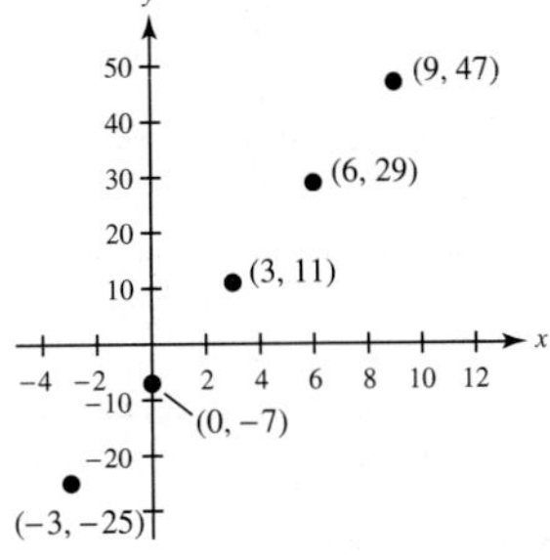
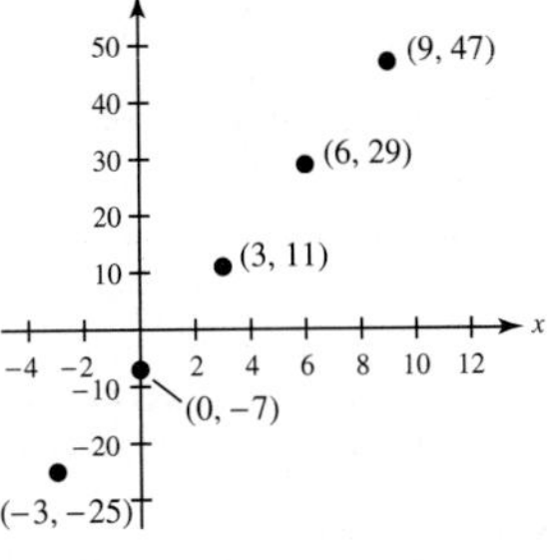

15.

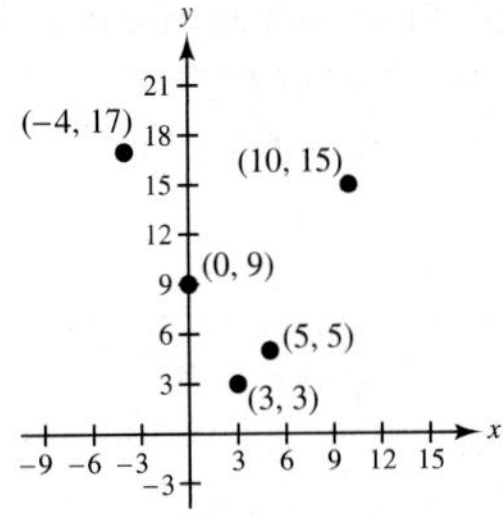

16.

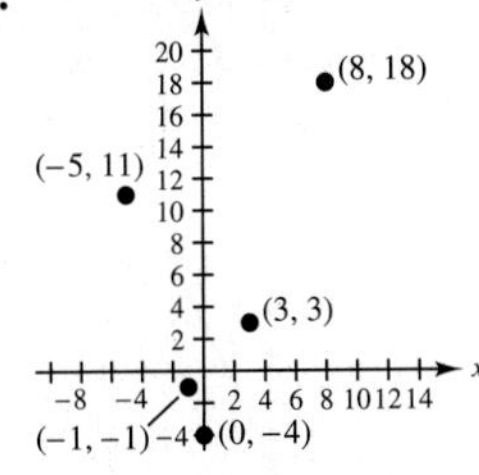

45.

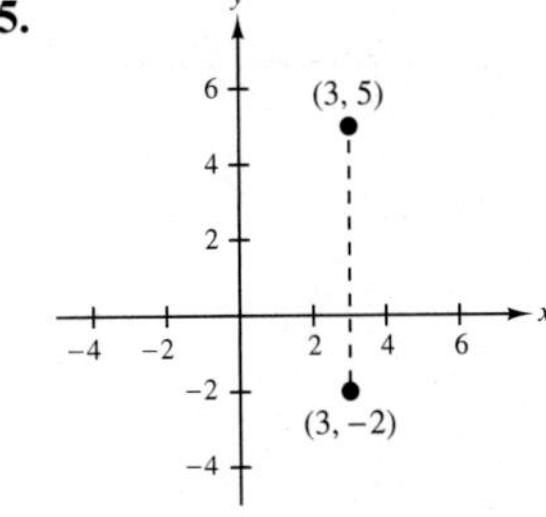

46.

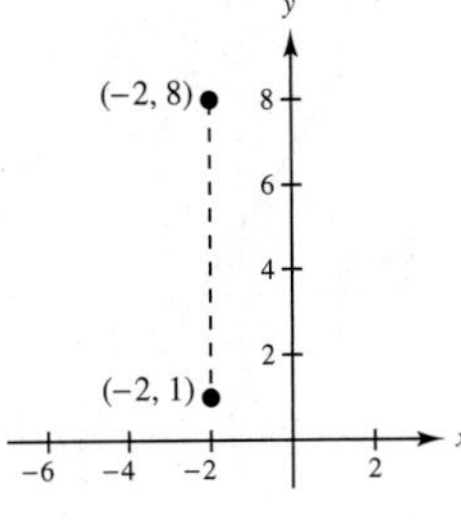

47.

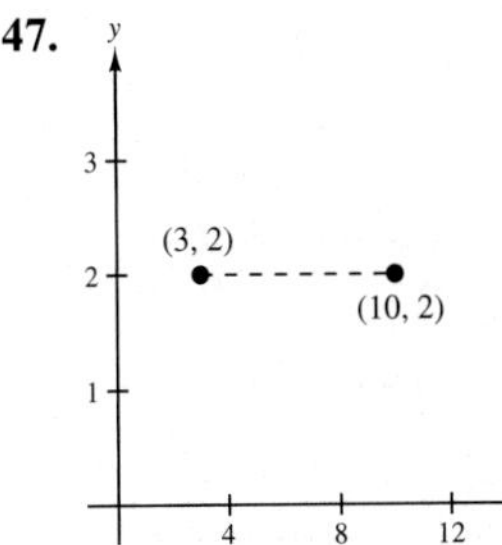

48.

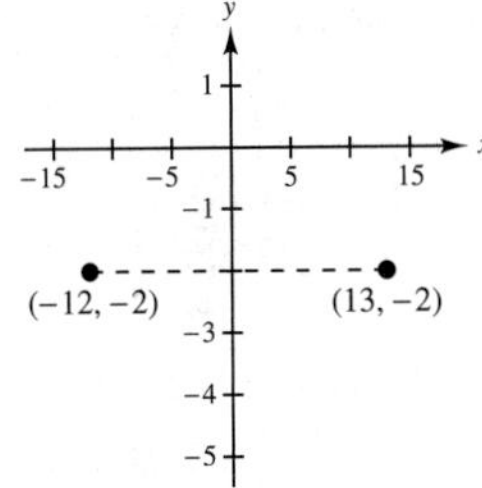

57.

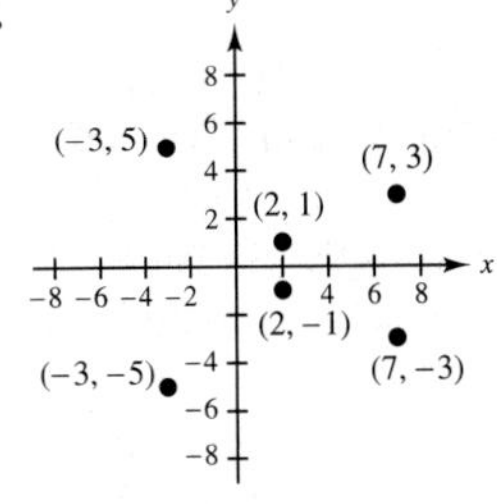

58.

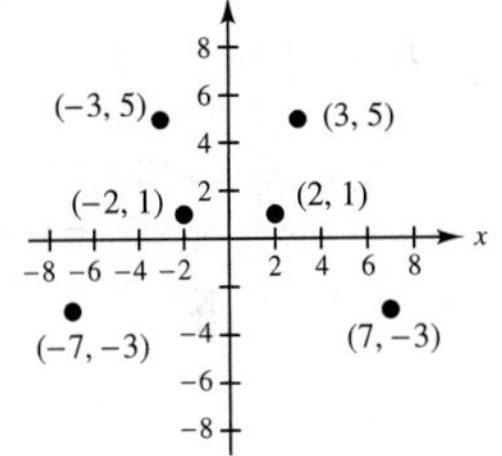

Section 3.2 *(pp. 110–117)*

1.

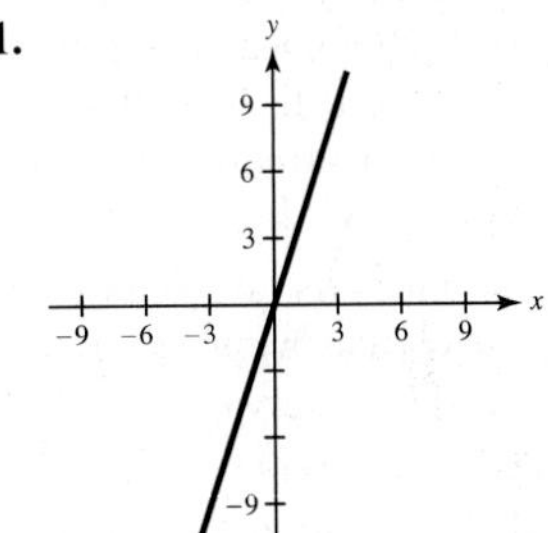

2.

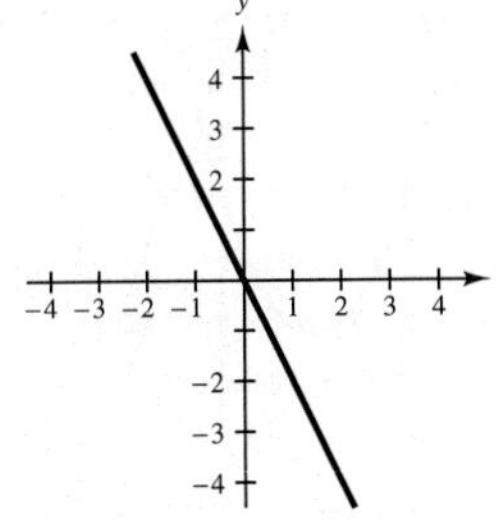

3.

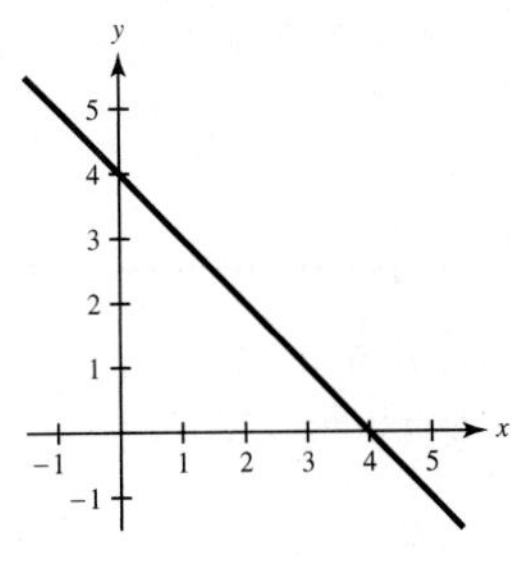

4.

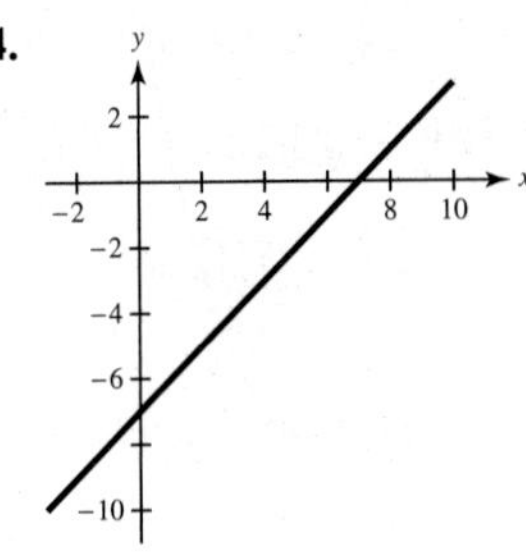

ADDITIONAL ANSWERS

5.

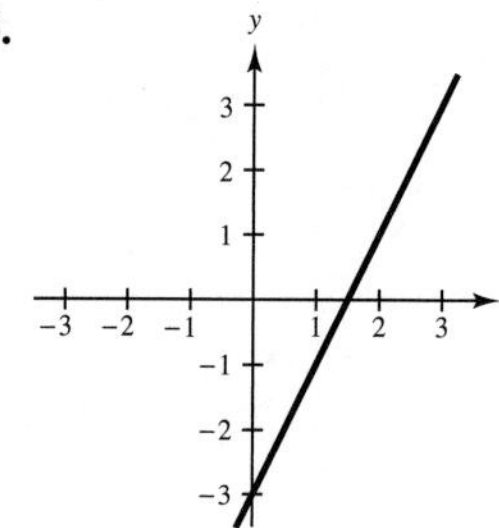

6.

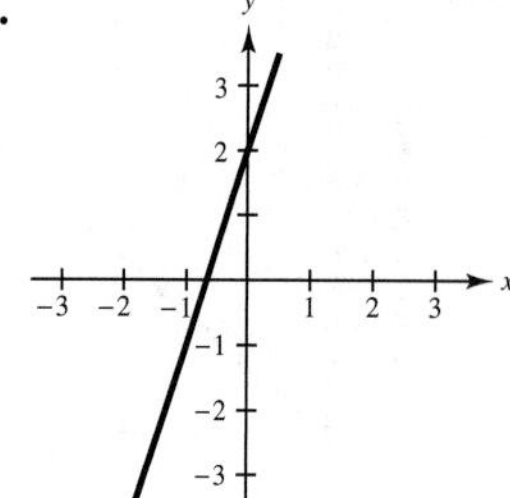

7.

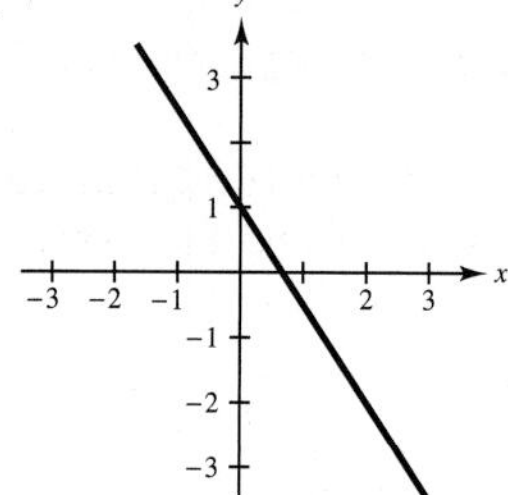

8.

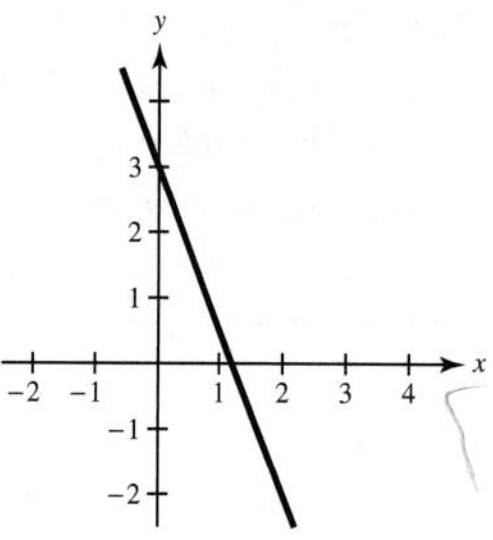

9.

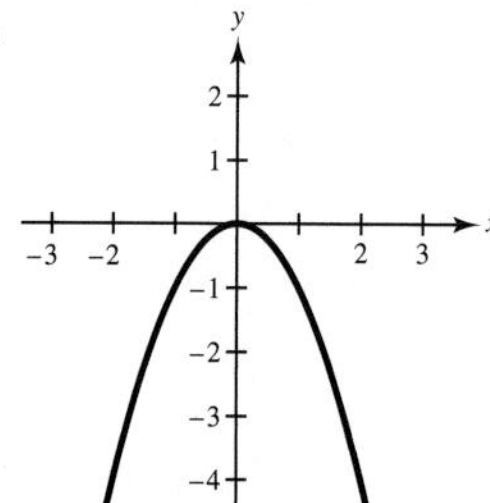

10.

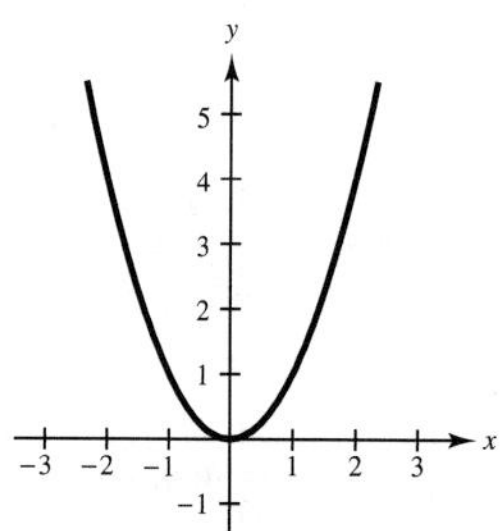

11.

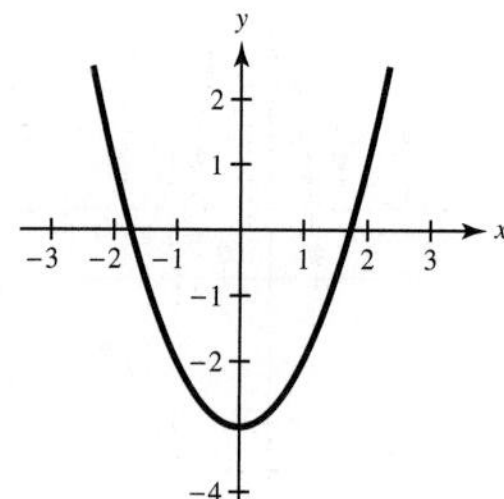

12.

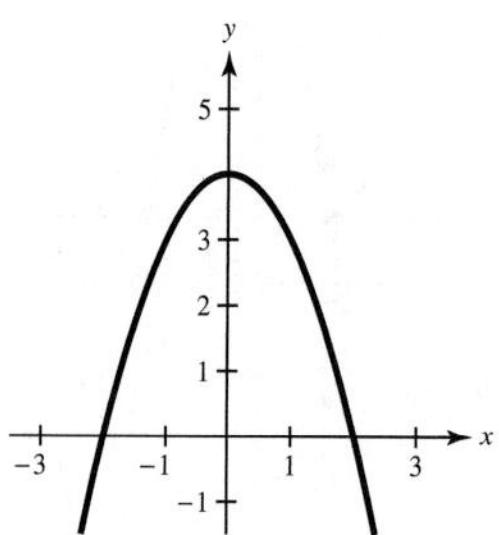

13.

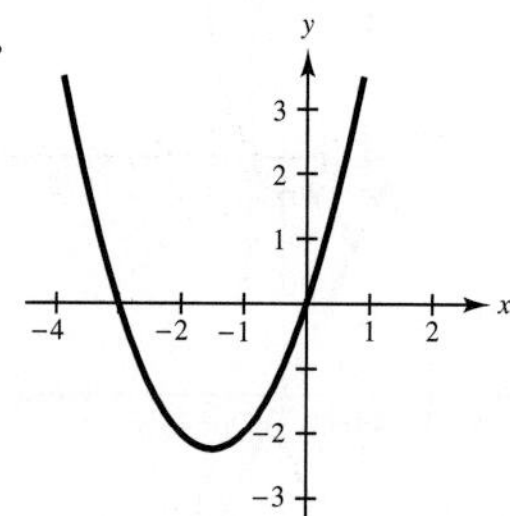

14.

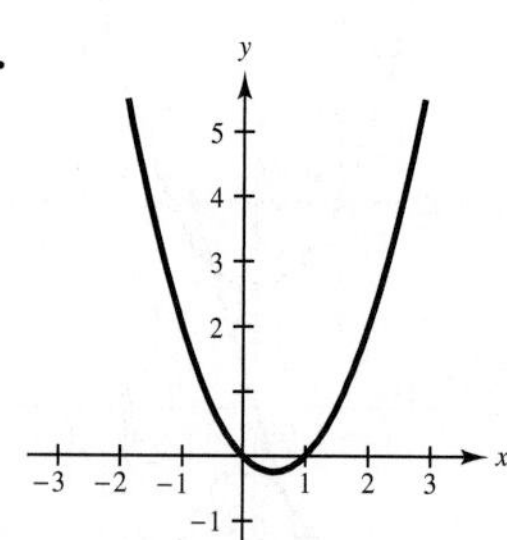

15.

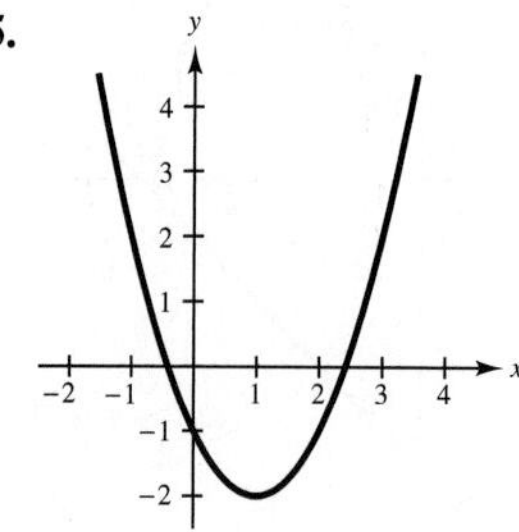

16.

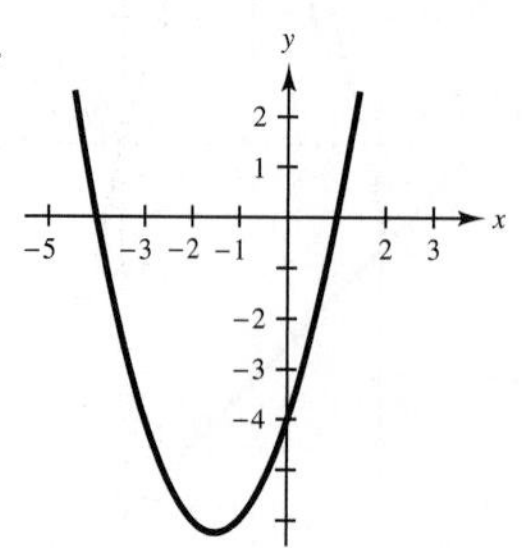

17.

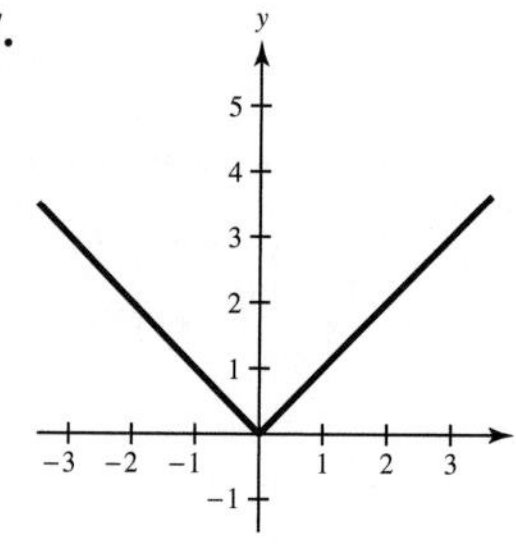

18.

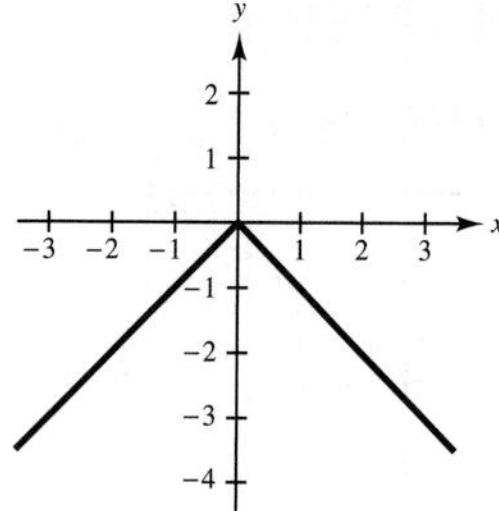

19.

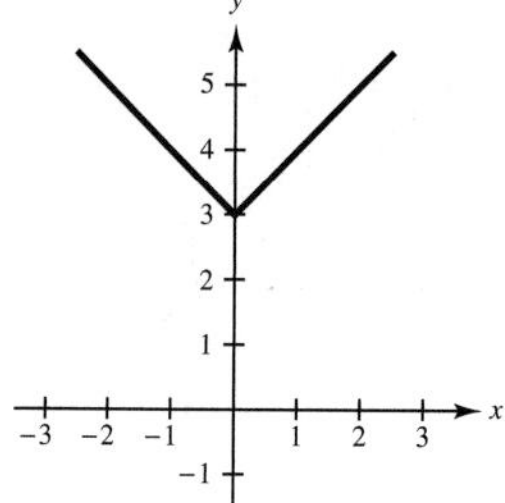

20.

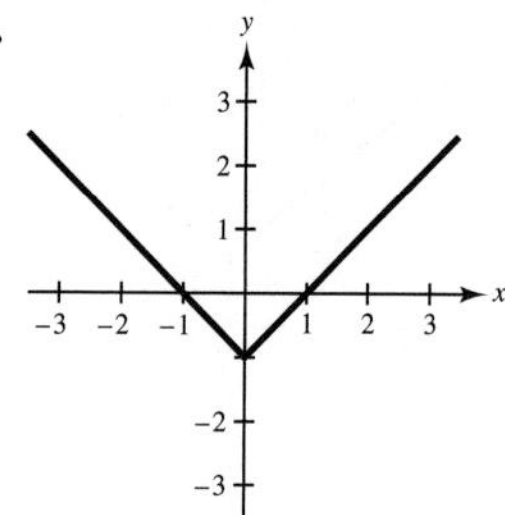

21.

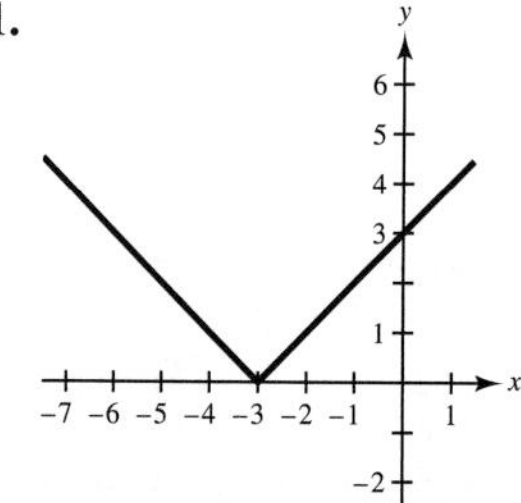

22.

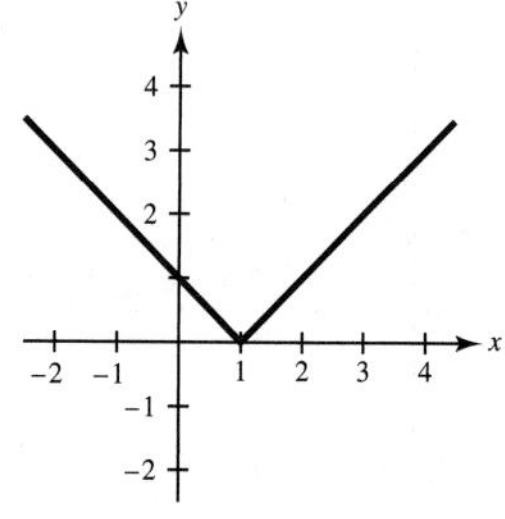

23.

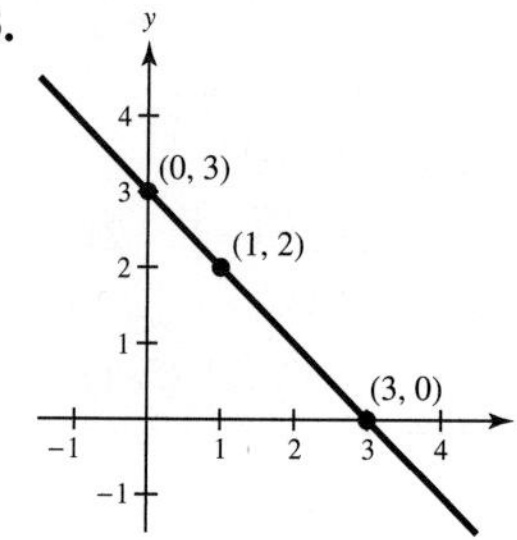

24.

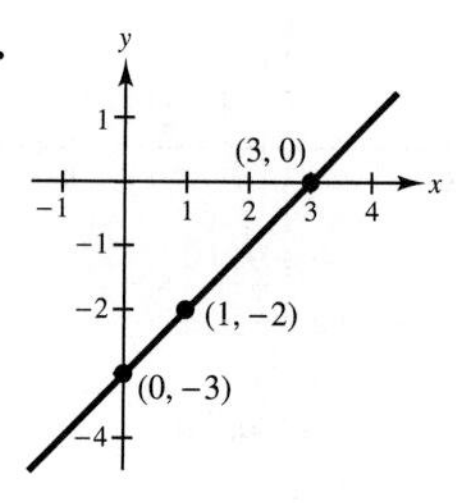

25.

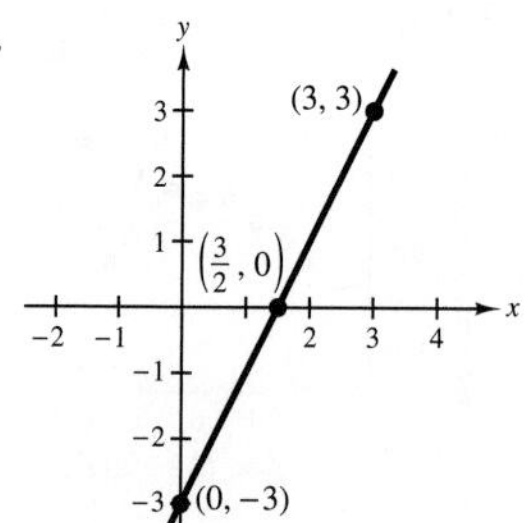

26.

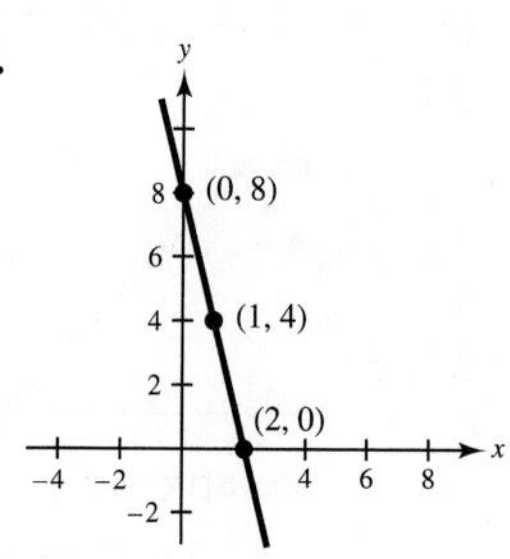

27.

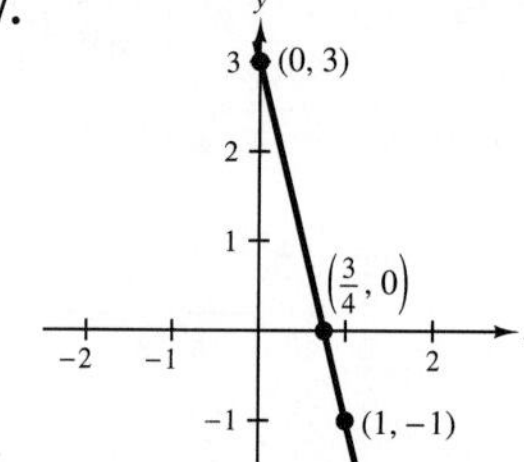

28.

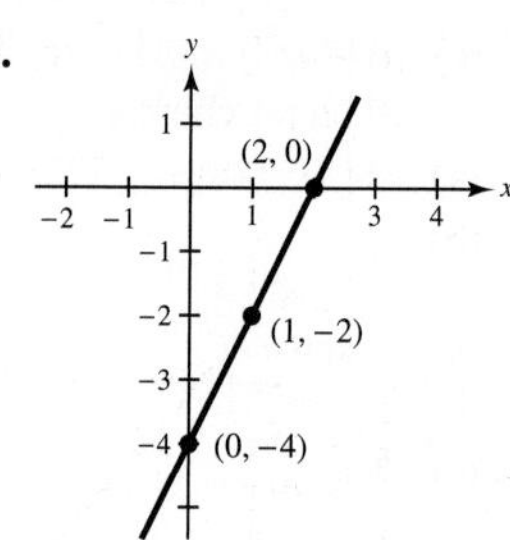

29.

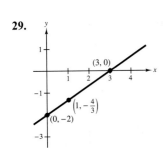

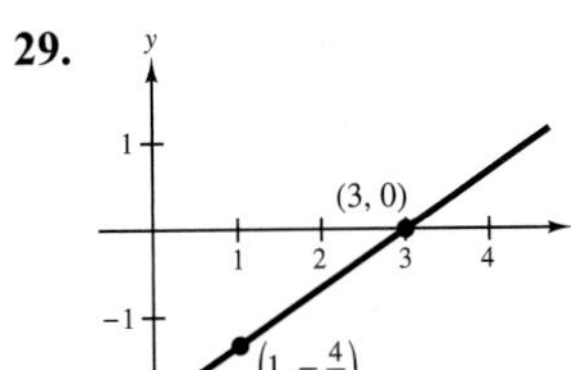

30.

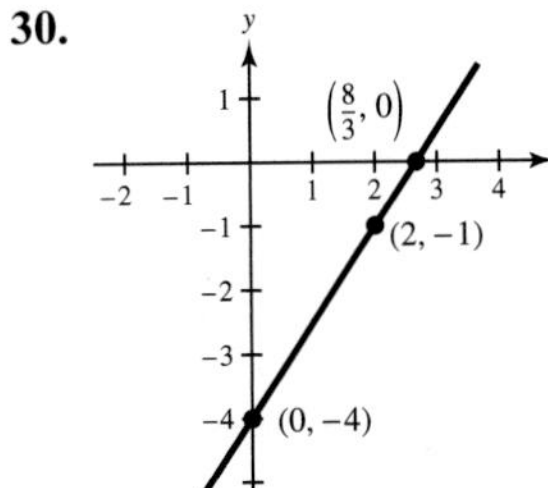

31.

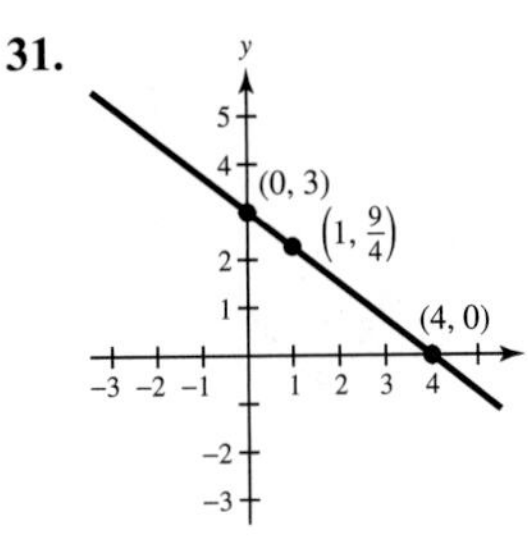

32.

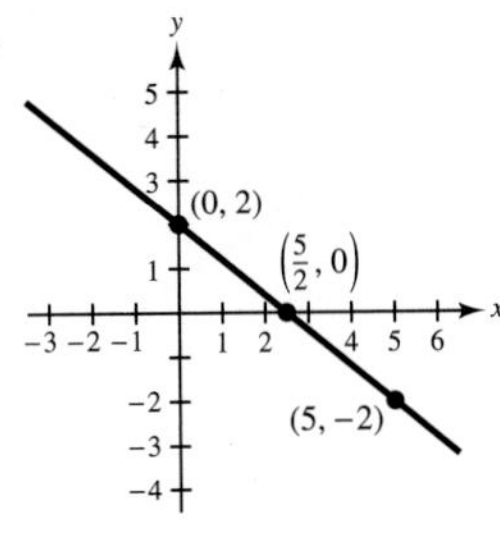

33.

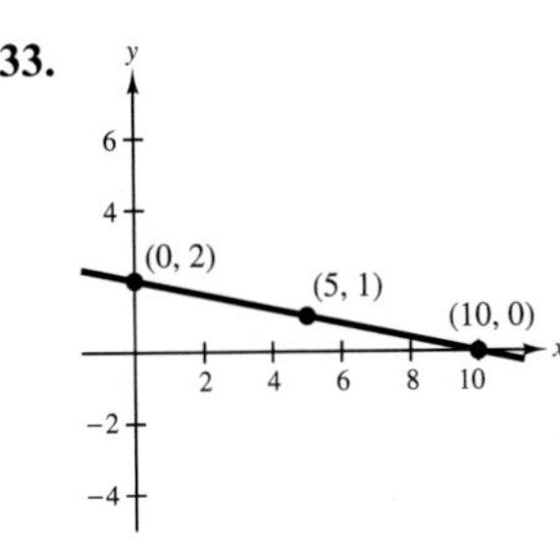

34.

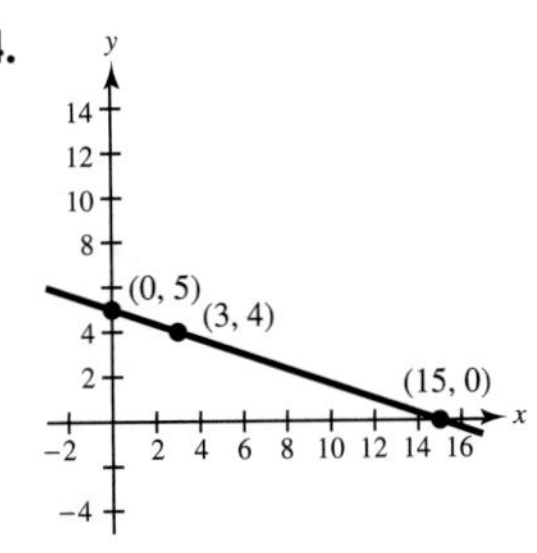

35.

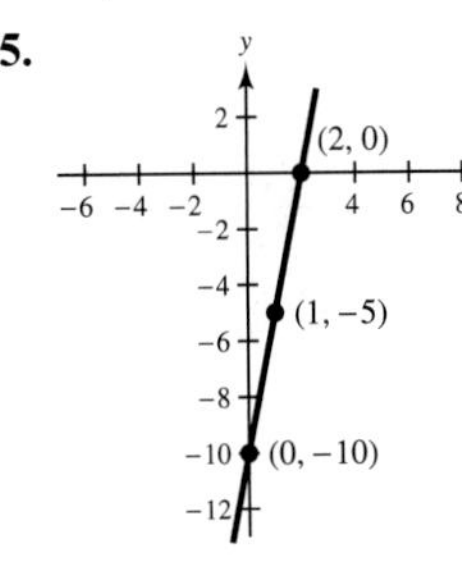

36.

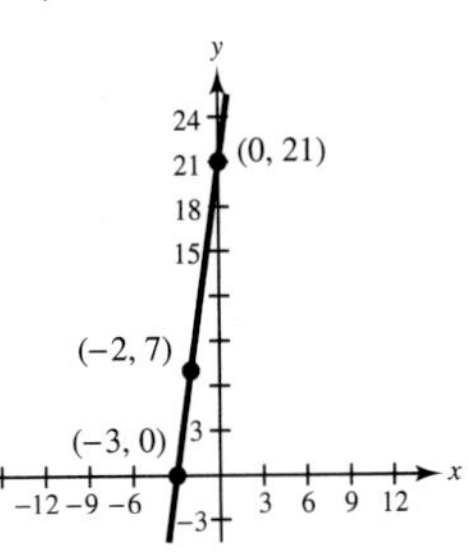

37. (a) $y = 40{,}000 - 5000t, 0 \le t \le 7$

(b)

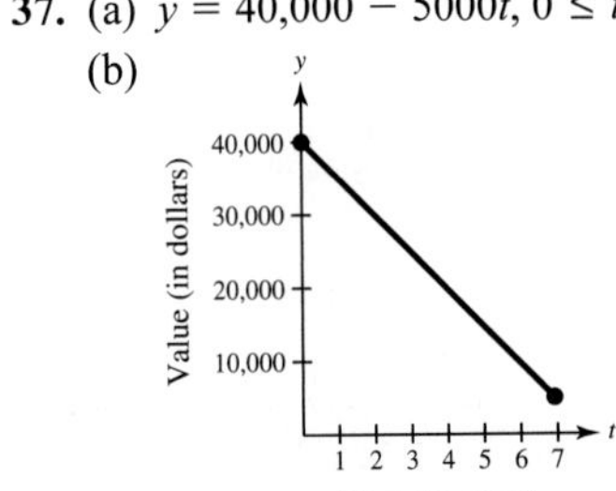

(c) (0, 40,000); This represents the value of the delivery van when purchased.

38. (a) $y = 65{,}000 - 5500t, 0 \le t \le 10$

(b)

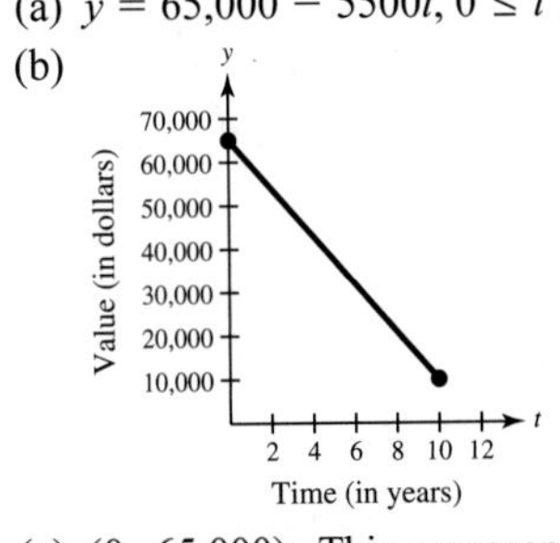

(c) (0, 65,000); This represents the value of the limousine when purchased.

39. (b)

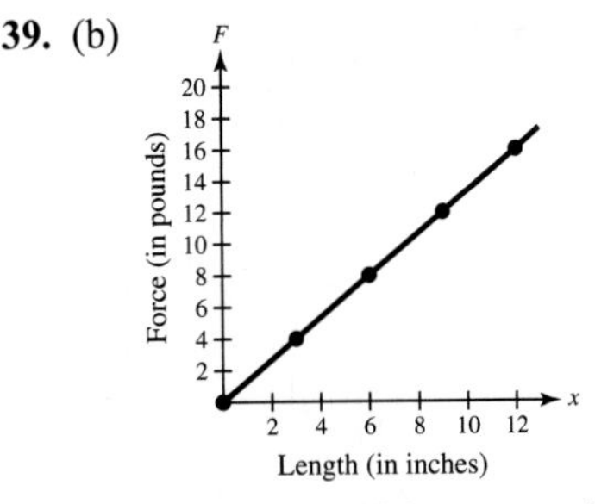

42. Yes. The graph will have two x-intercepts and one y-intercept. From the figure, it is clear that the x-intercepts are $(-2, 0)$ and $(0, 0)$. The y-intercept is $(0, 0)$.

43. False. The x-intercept is the point at which the graph crosses the x-axis, so the value of y is 0. To find the x-intercept, substitute 0 for y in the equation and solve for x.

45.

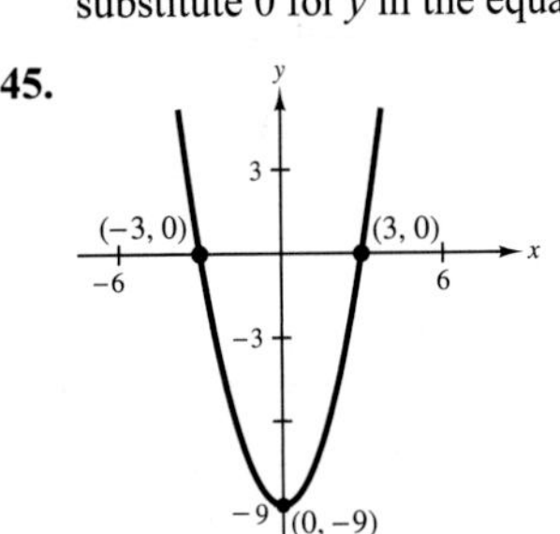

46.

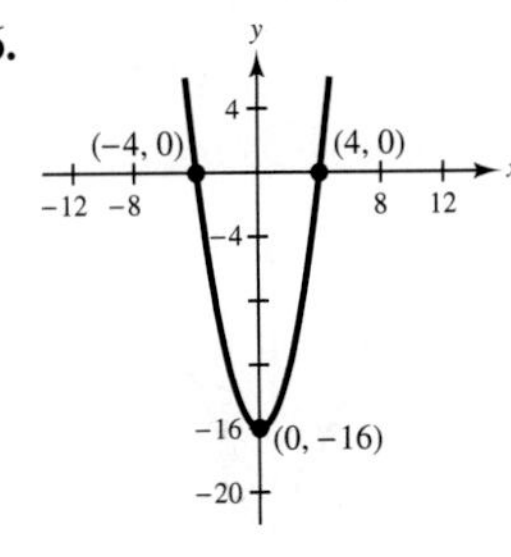

47.

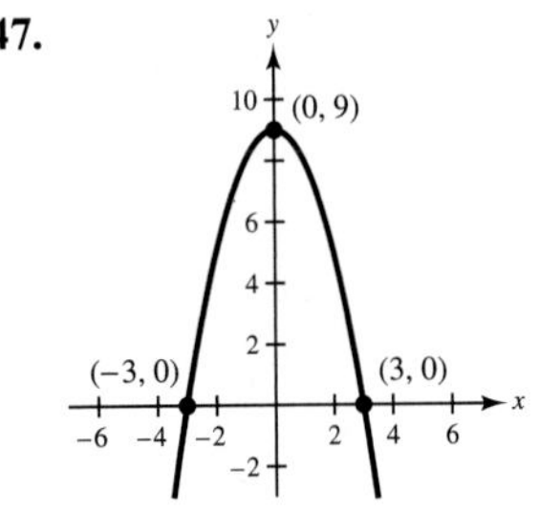

48.

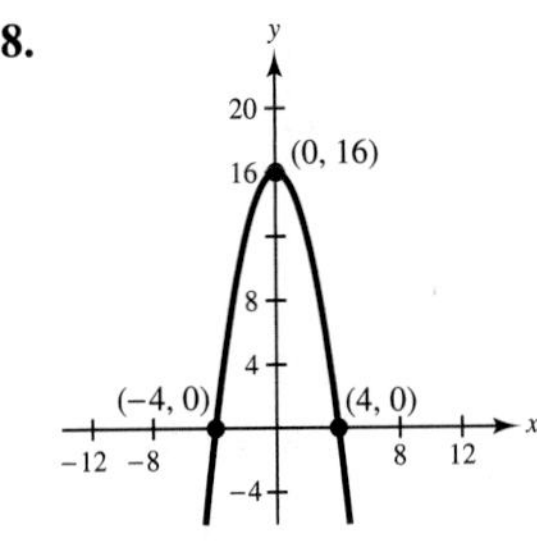

49.

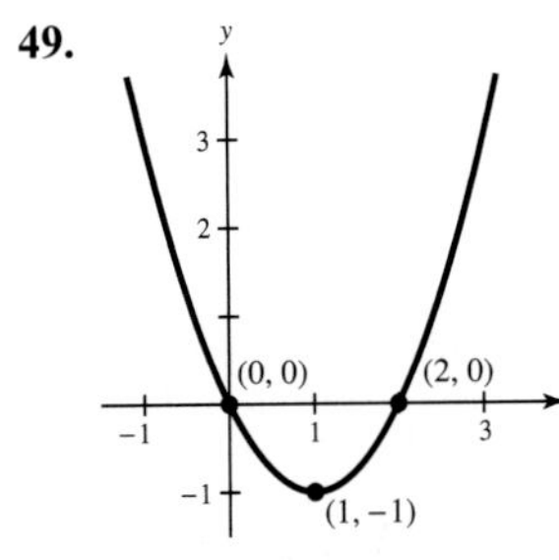

50.

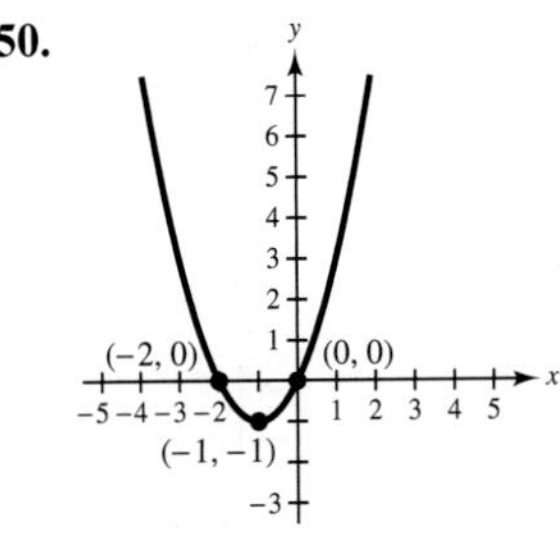

51.

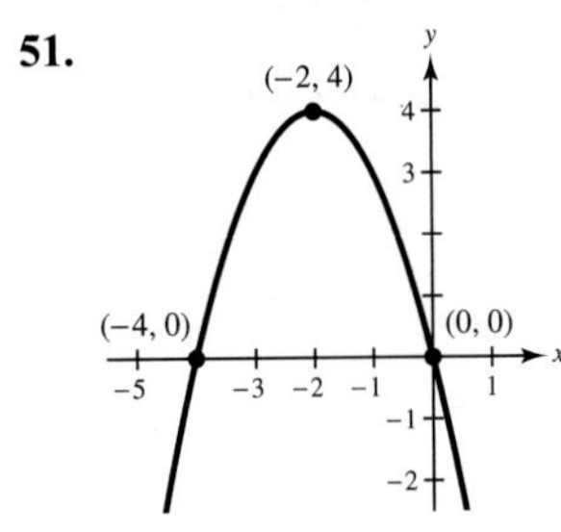

52.

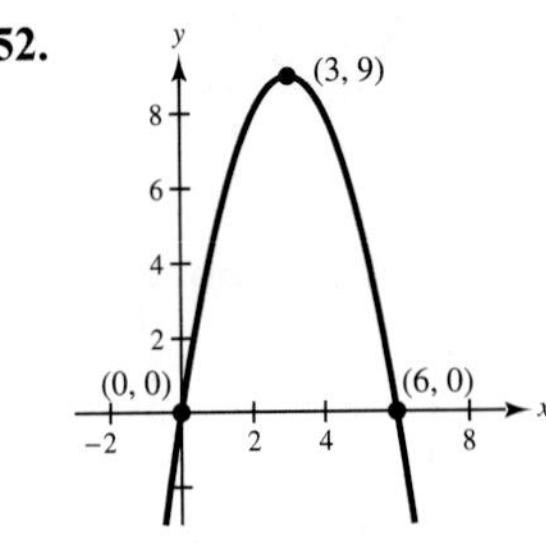

53.

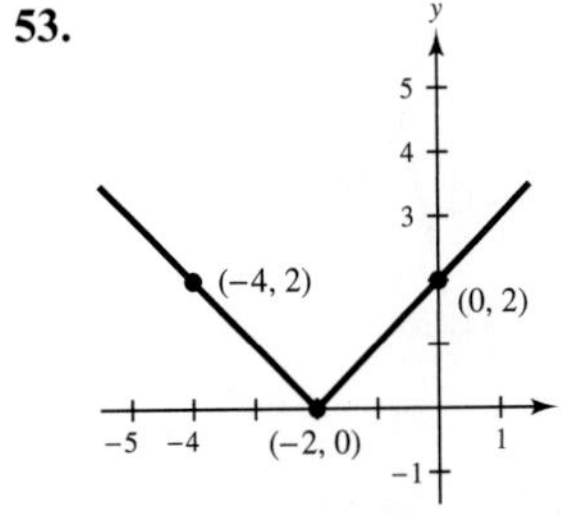

54.

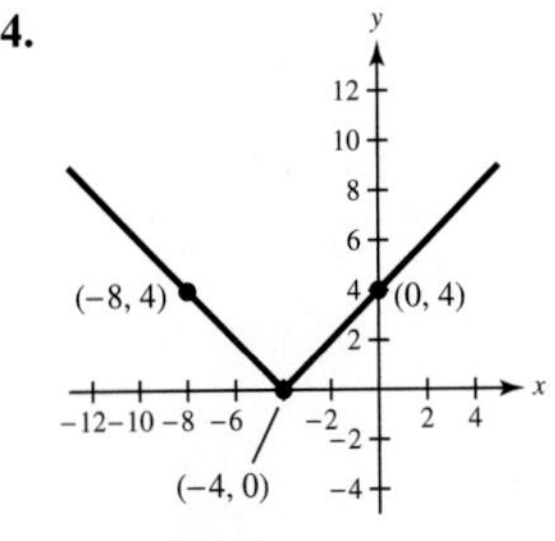

55.

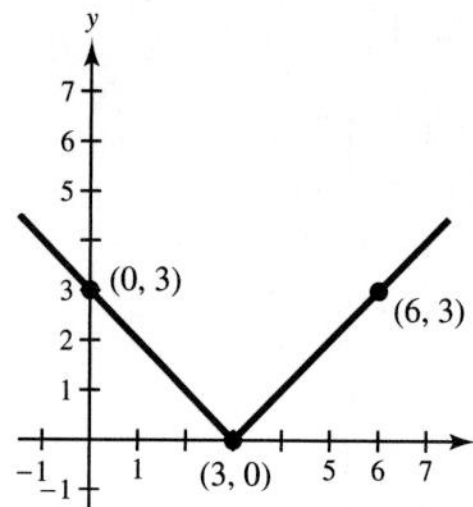

56.

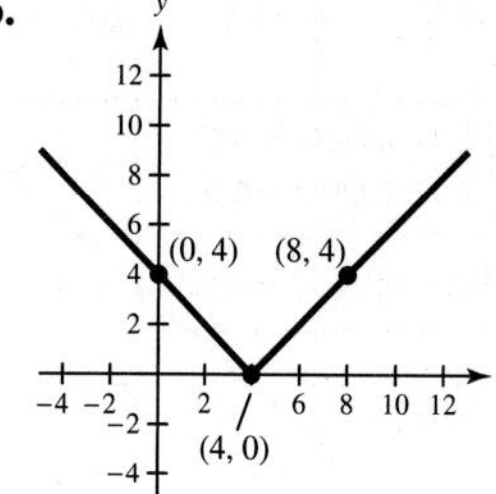

58.

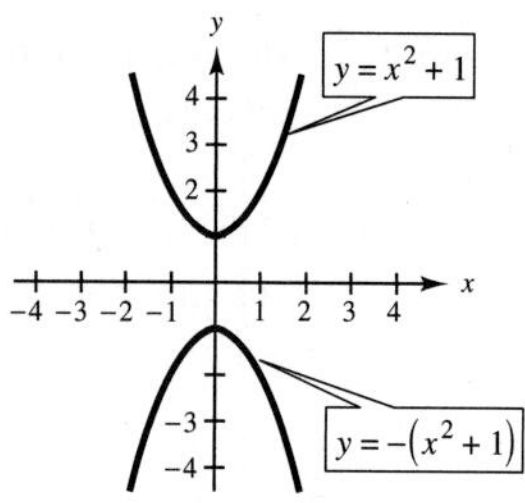

Additional Example:

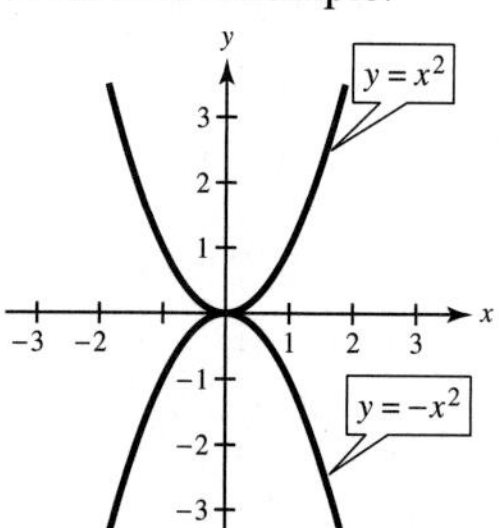

59.

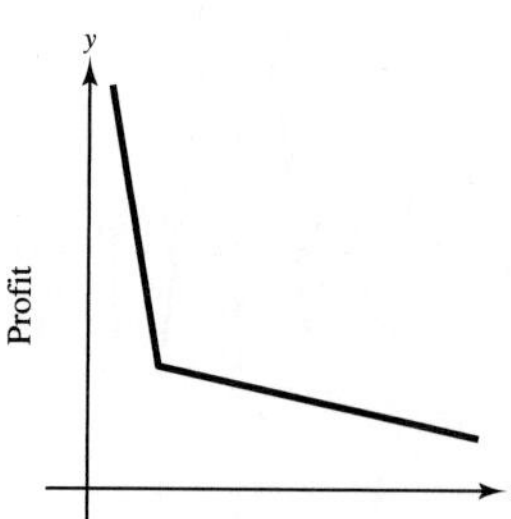

71.

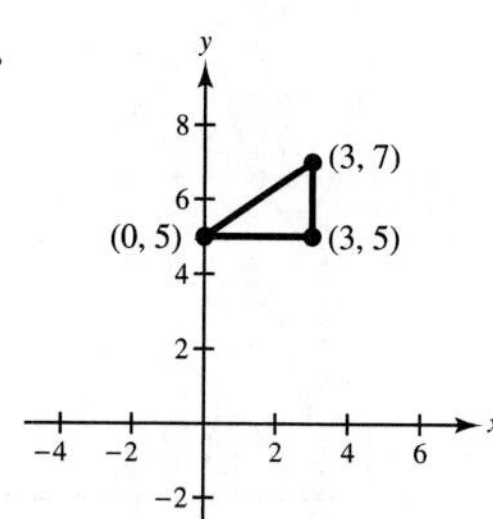

72.

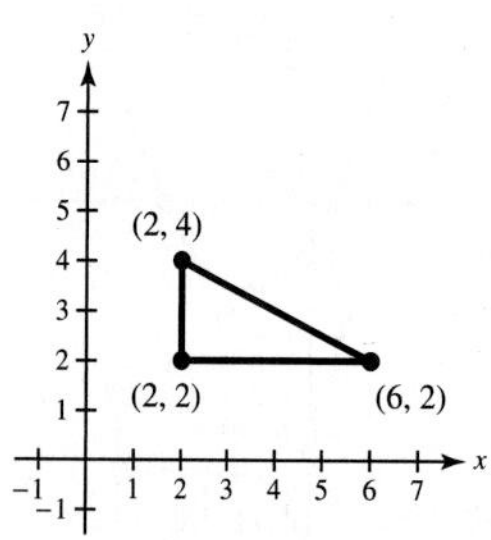

Section 3.3 *(pp. 118–125)*

1.

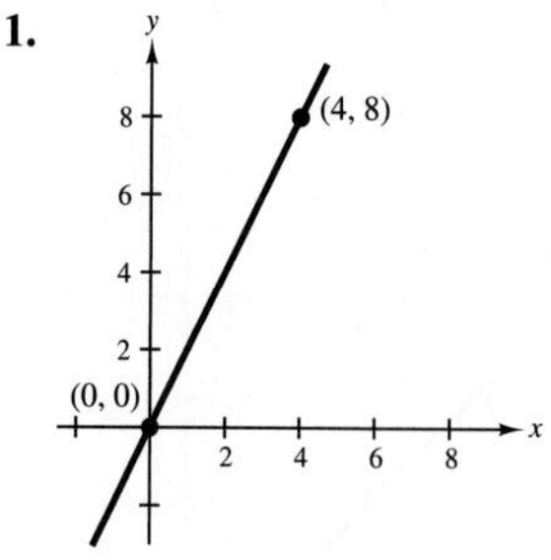

2.

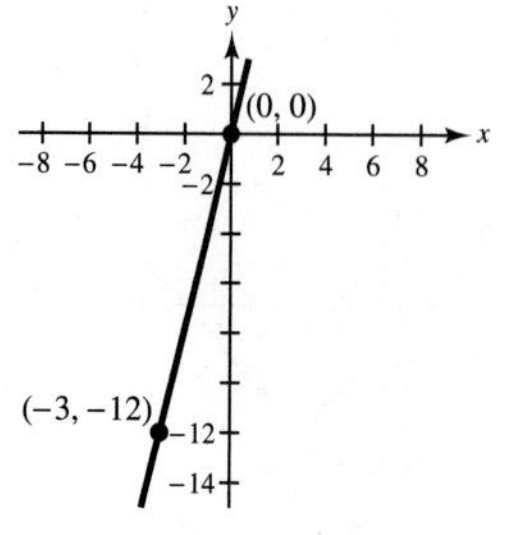

3.

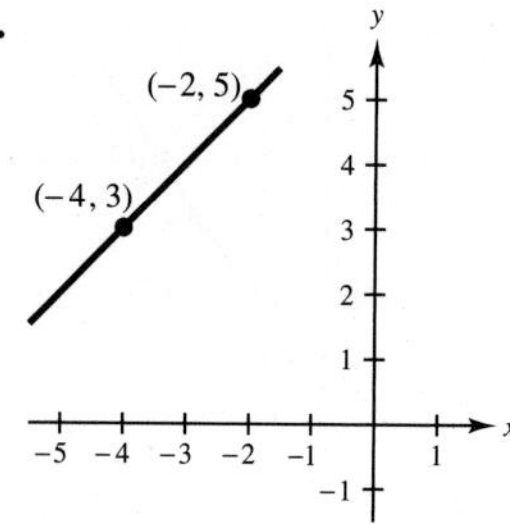

4.

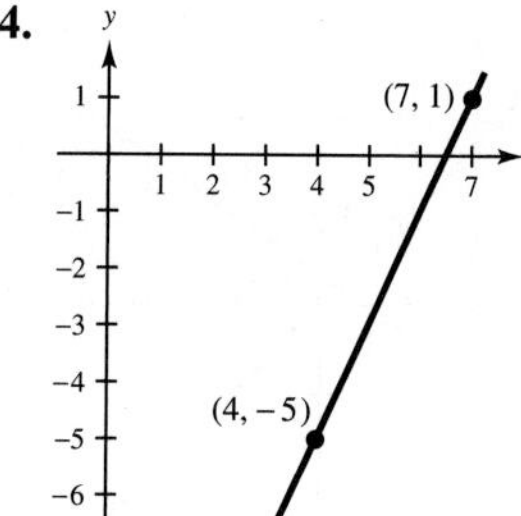

5.

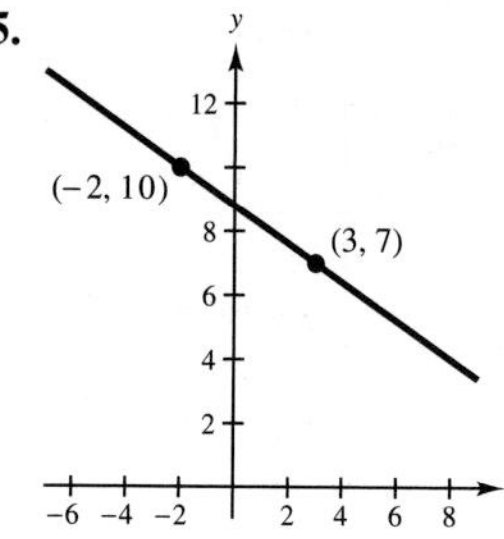

6.

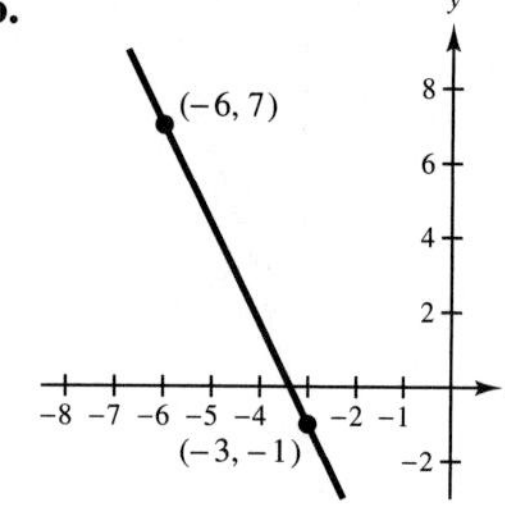

7.

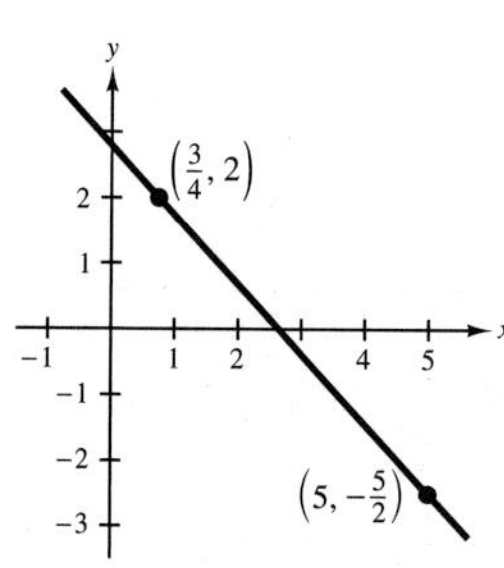

8.

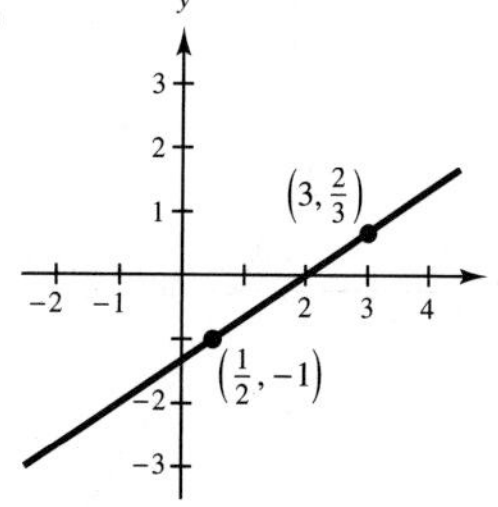

17.

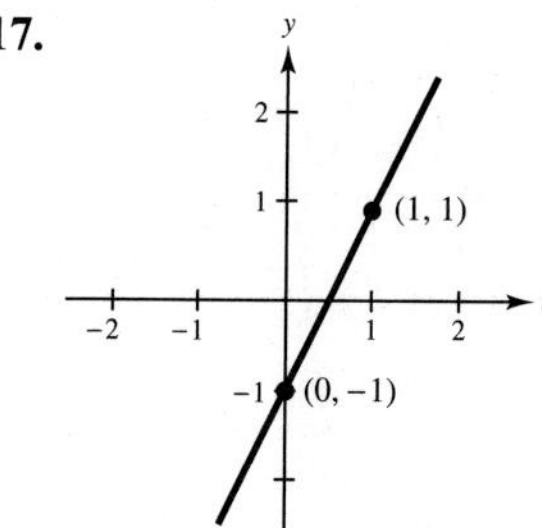

18.

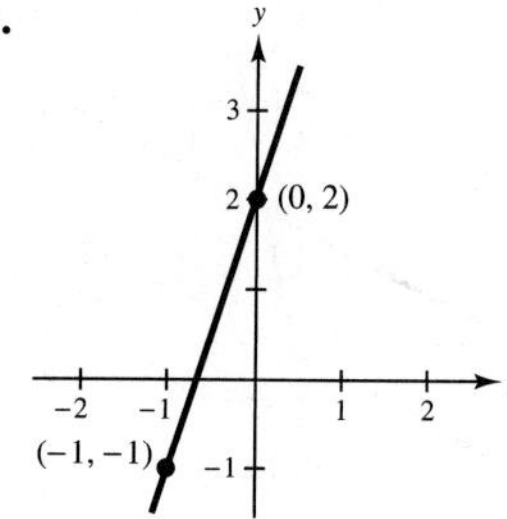

19.

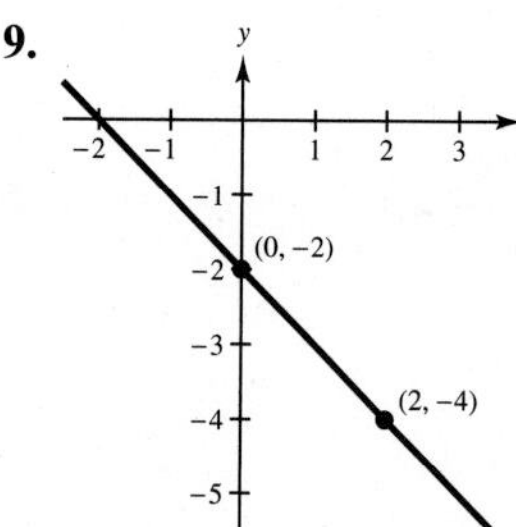

20.

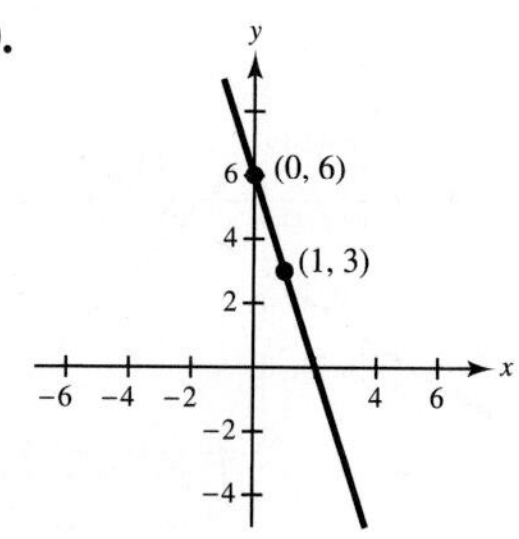

21.

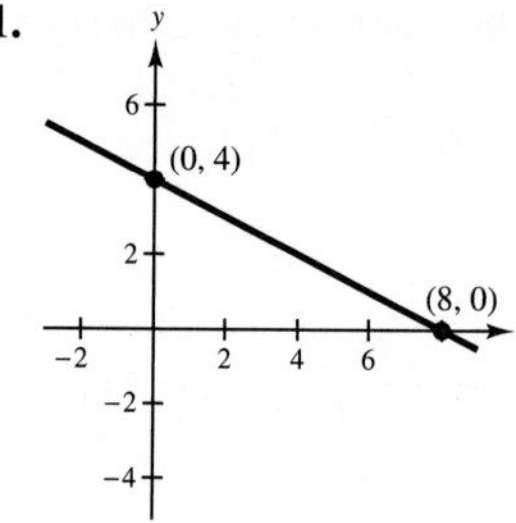

22.

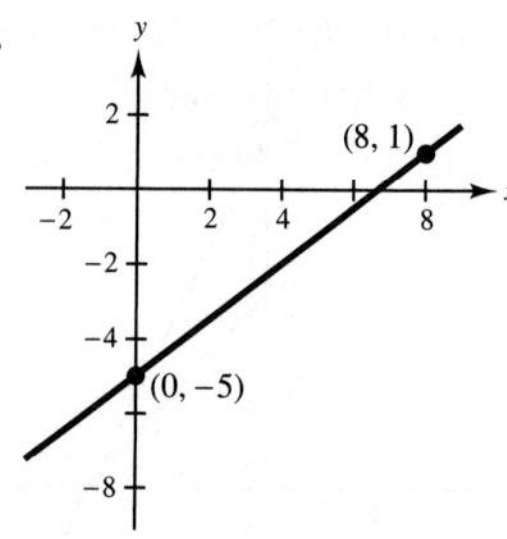

27.

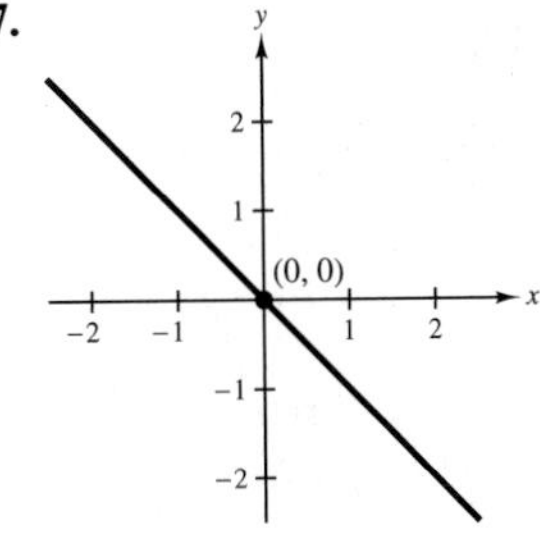

28.

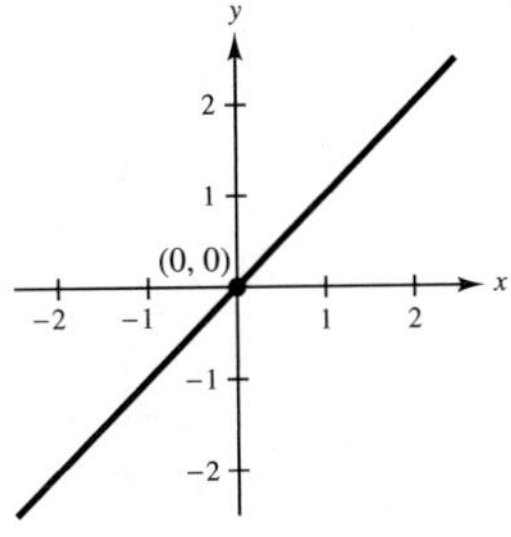

29.

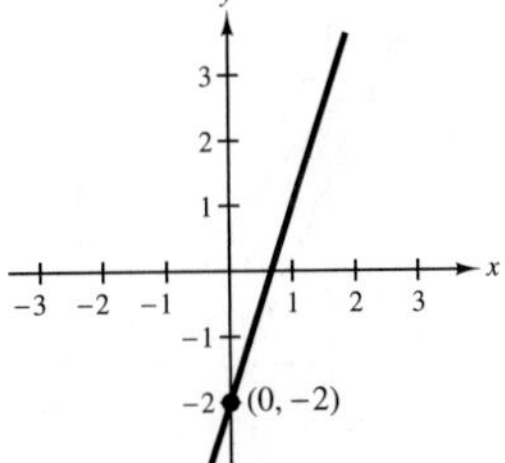

30.

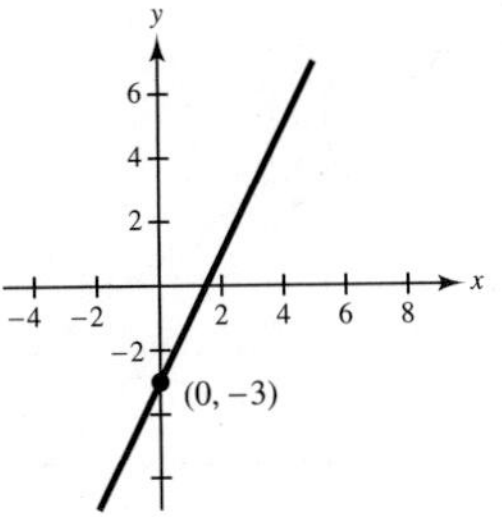

31.

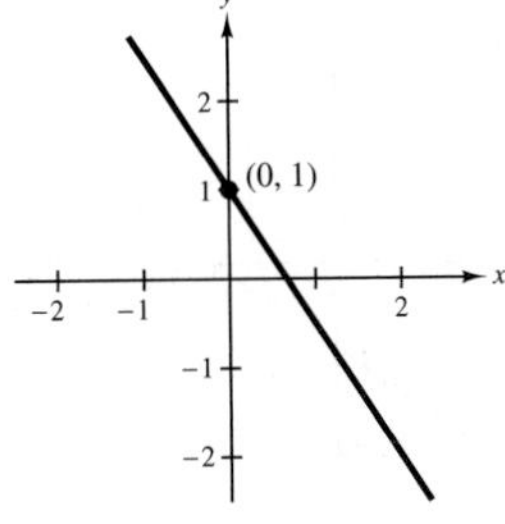

32.

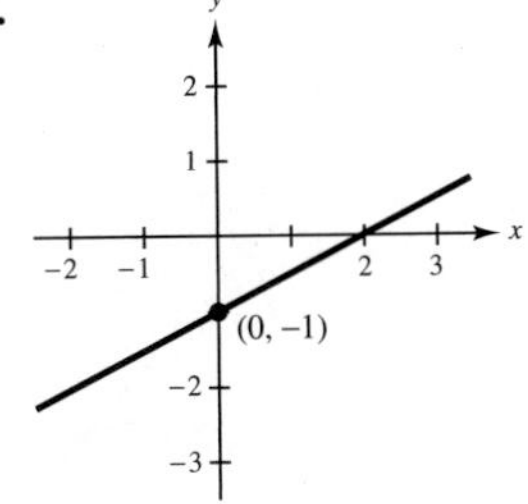

33.

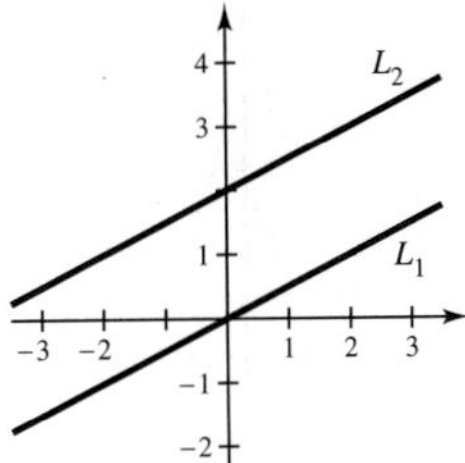

34.

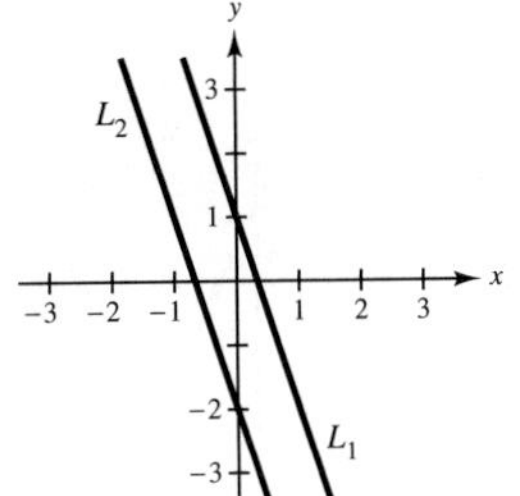

59.

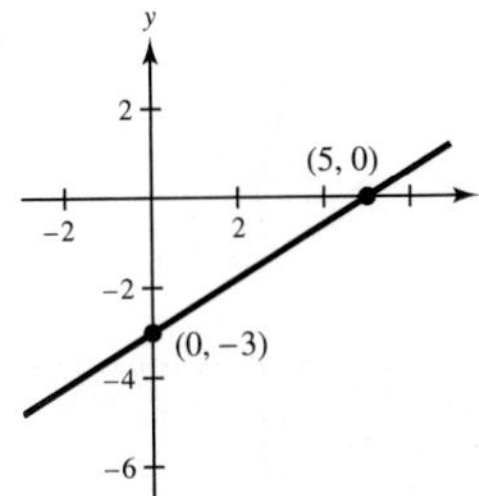

60.

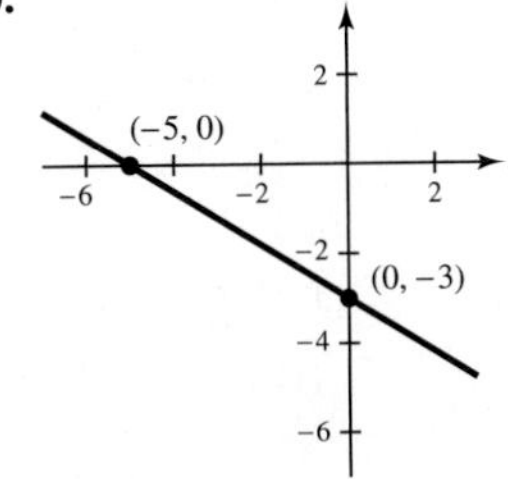

61.

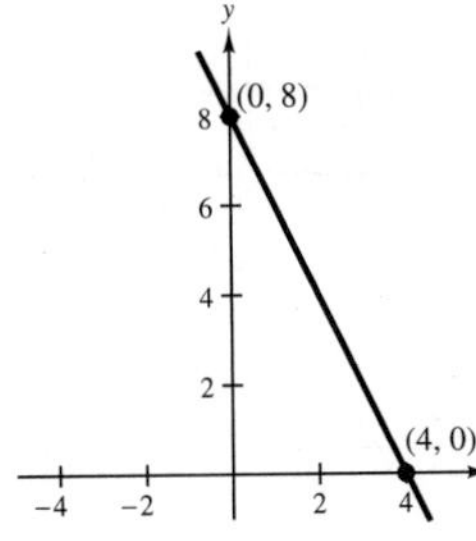

62.

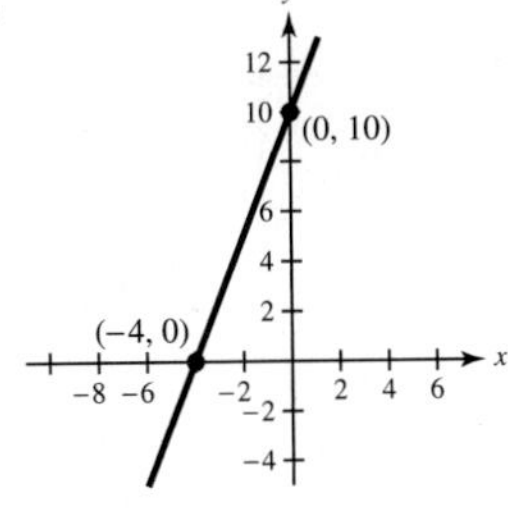

Section 3.4 *(pp. 126–133)*

39. $\frac{x}{3} + \frac{y}{2} = 1$ **40.** $-\frac{x}{6} + \frac{y}{2} = 1$

41. $-\frac{6x}{5} - \frac{3y}{7} = 1$ **42.** $-\frac{3x}{8} - \frac{y}{4} = 1$

47.

Distance from the deep end	0	8	16	24	32	40
Depth of water	9	8	7	6	5	4

48.

Distance from the tall end	0	3	6	9	12
Height of block	6	5	4	3	2

53.

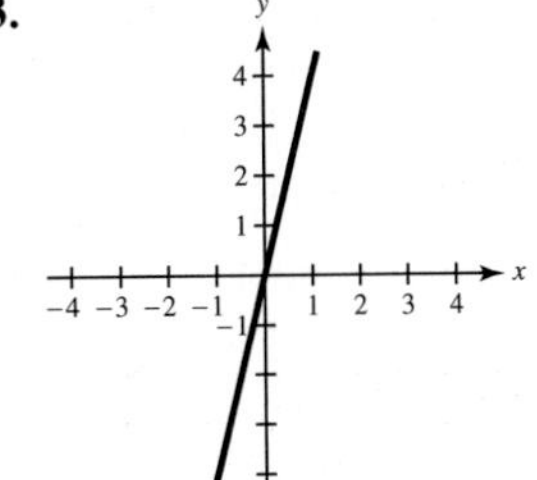

54.

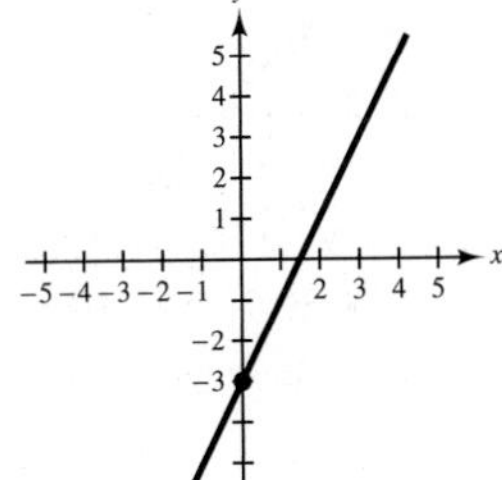

55.

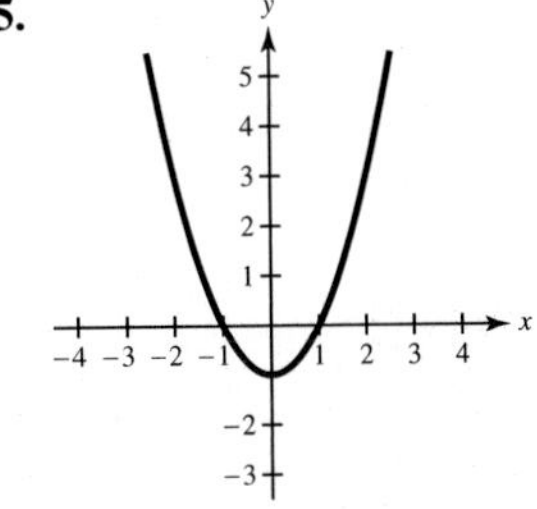

56.

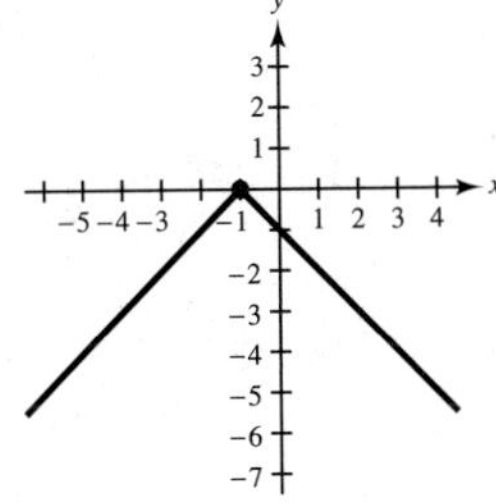

Mid-Chapter Quiz *(p. 134)*

3.

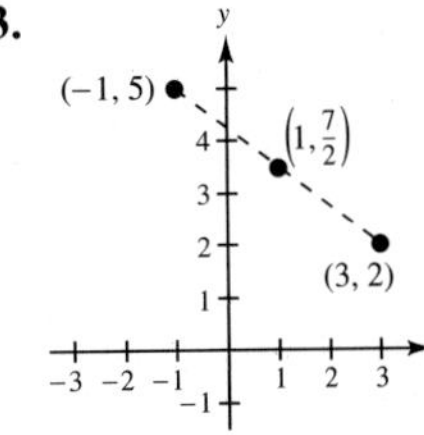

4.

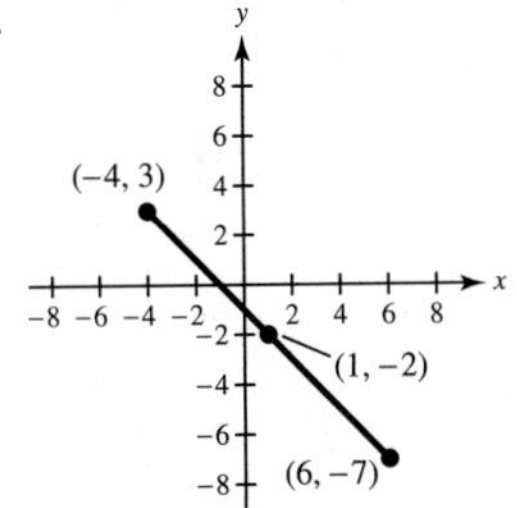

5.

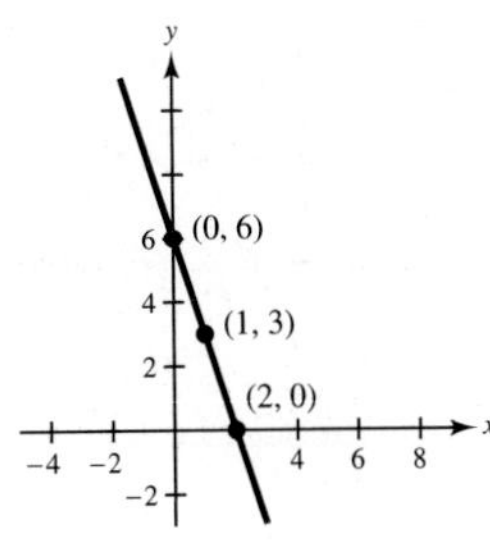

6.

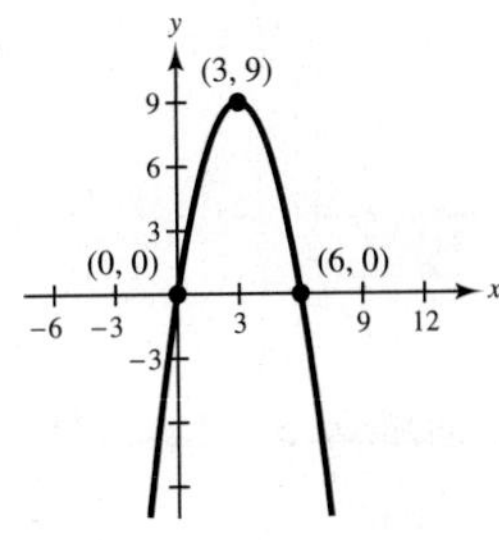

7.

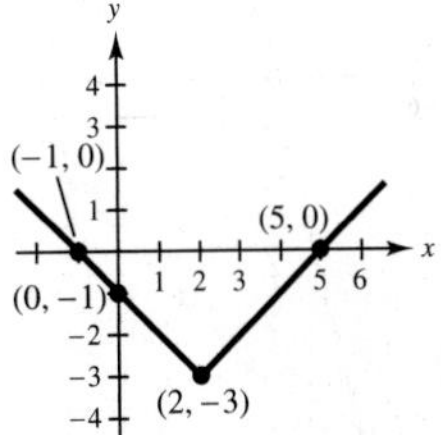

ADDITIONAL ANSWERS

11.

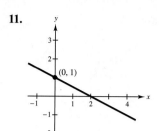

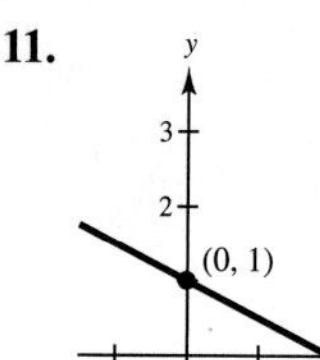

12.

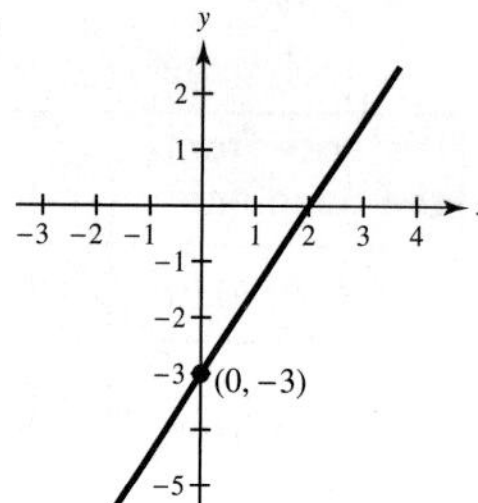

16.

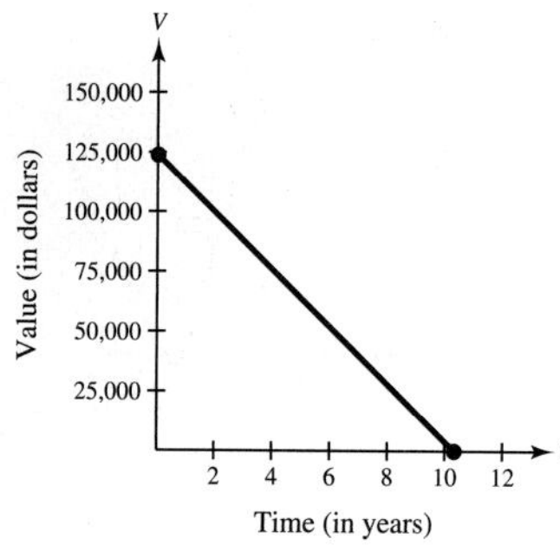

Section 3.5 *(pp. 136–143)*

7.

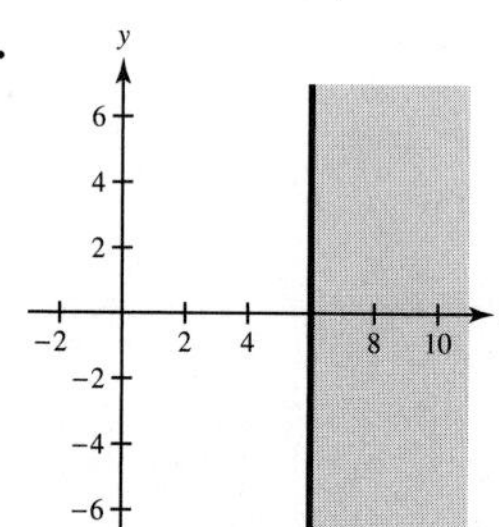

8.

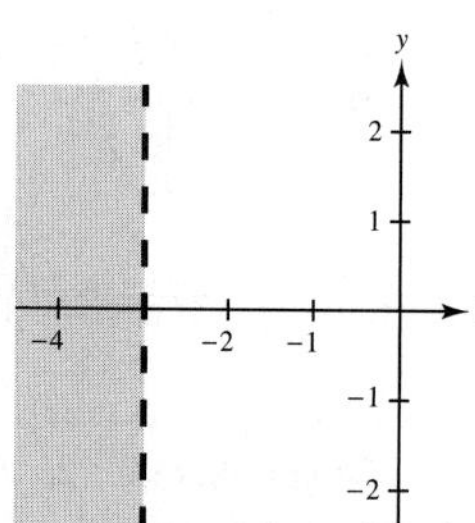

9.

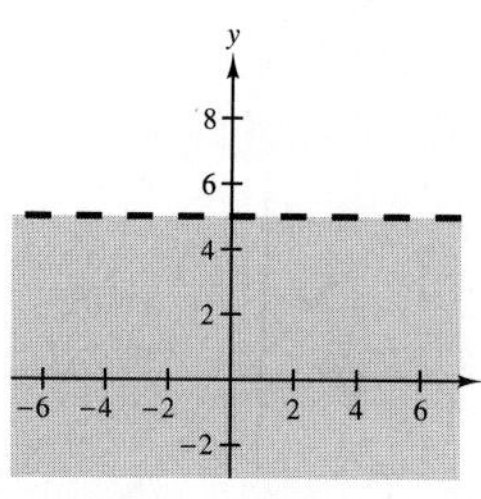

10.

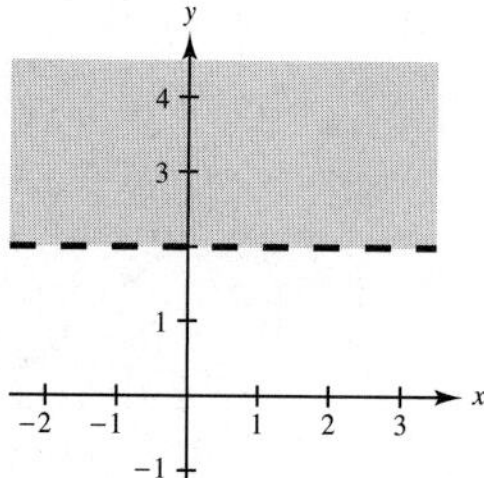

11.

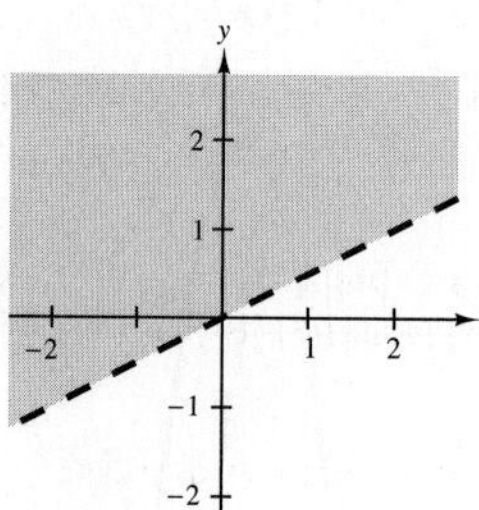

12.

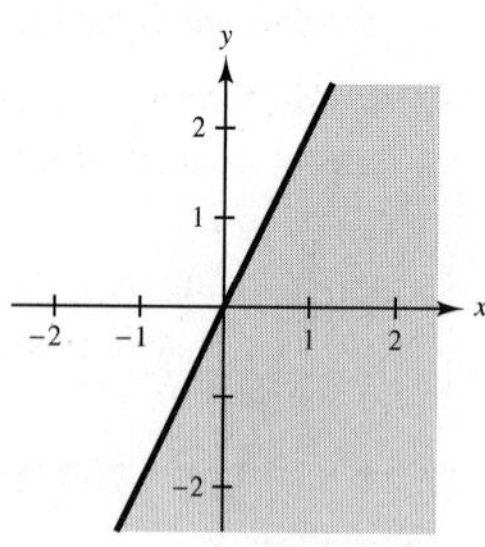

19.

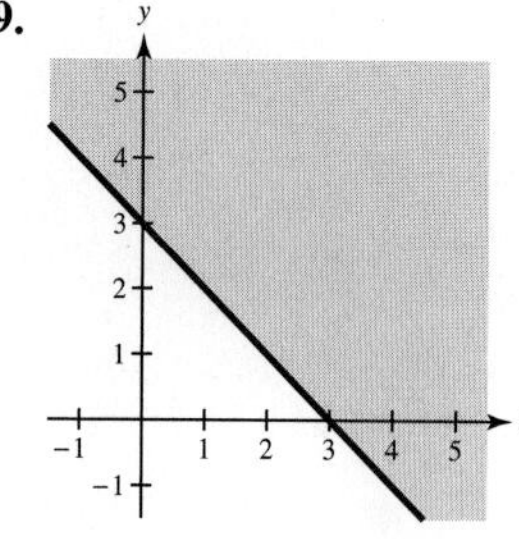

20.

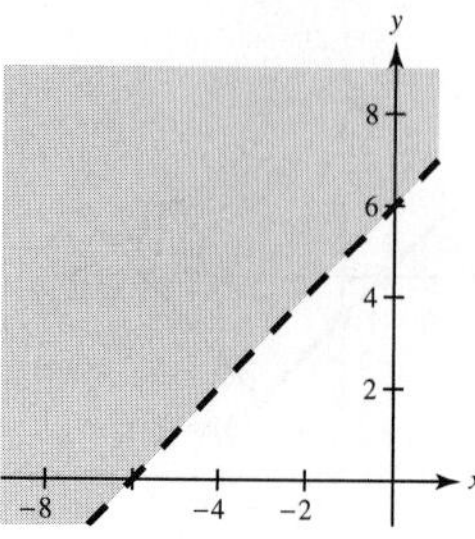

21.

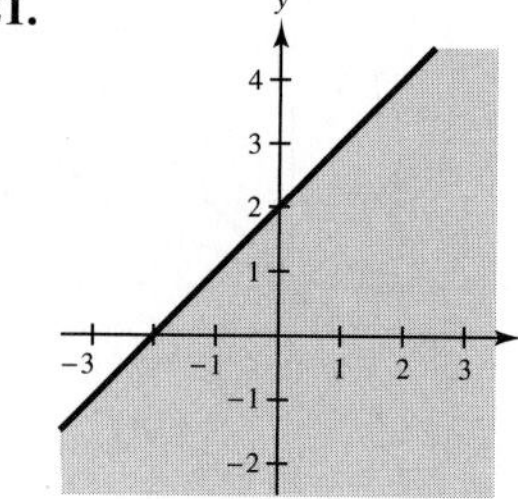

22.

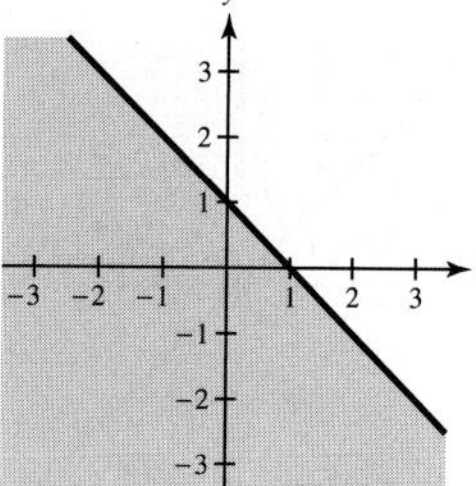

23.

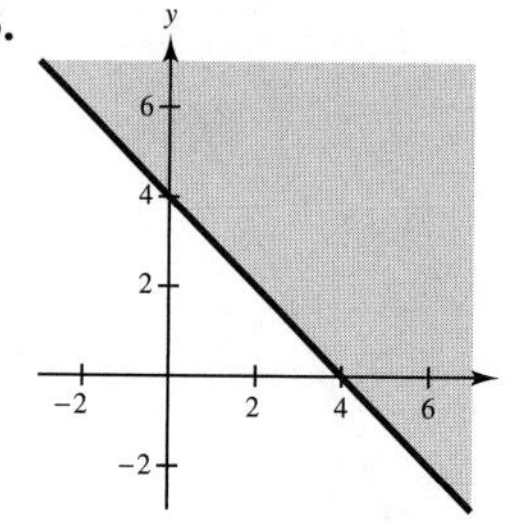

24.

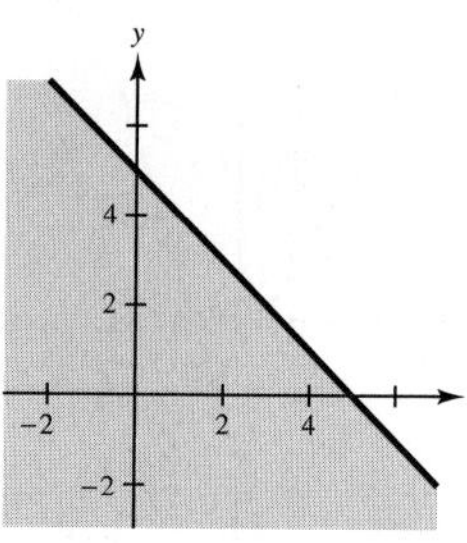

25.

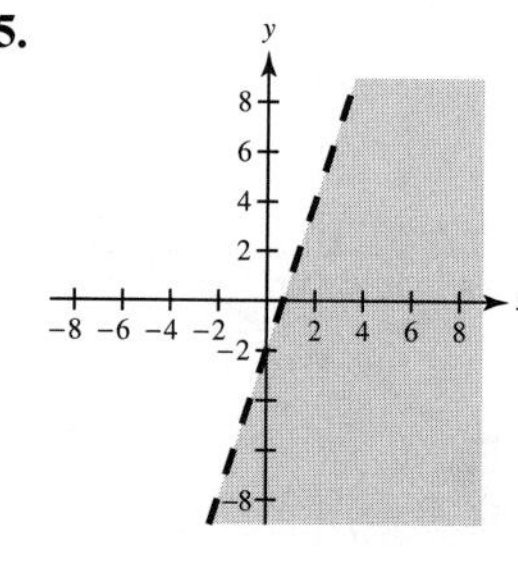

26.

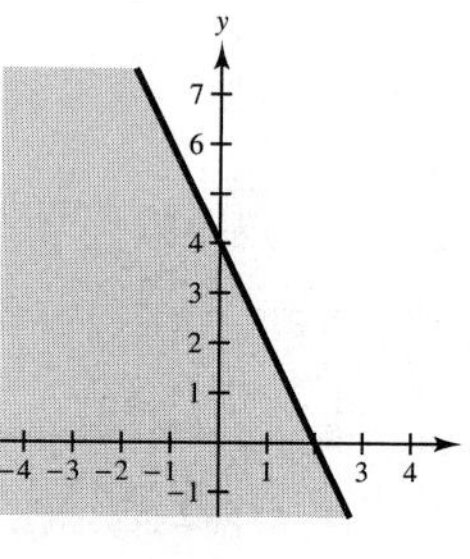

27.

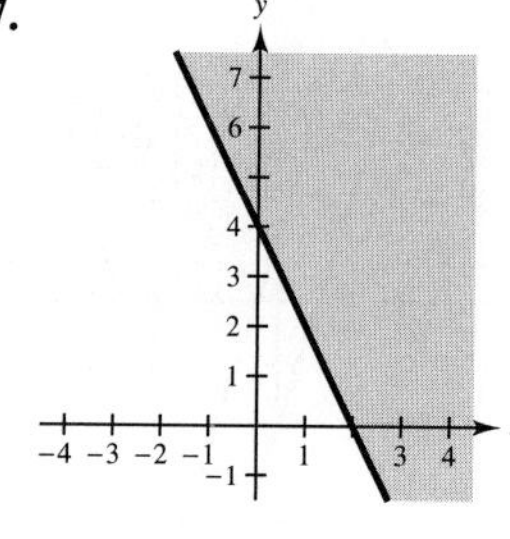

28.

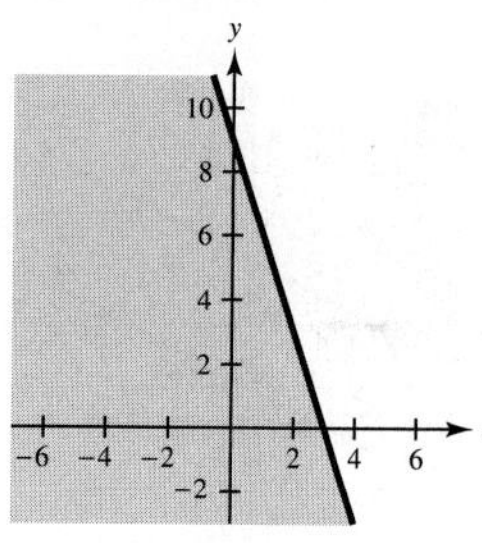

29.

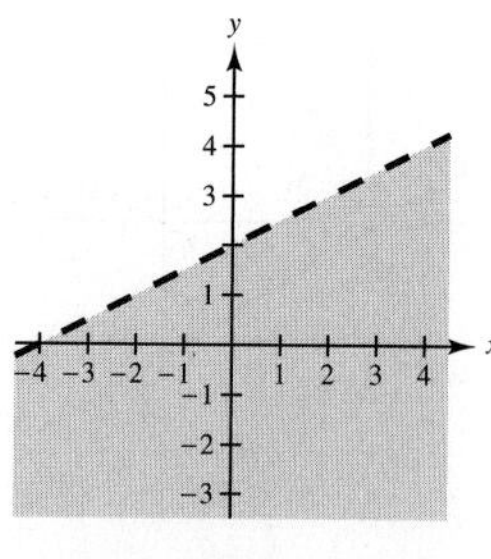

30.

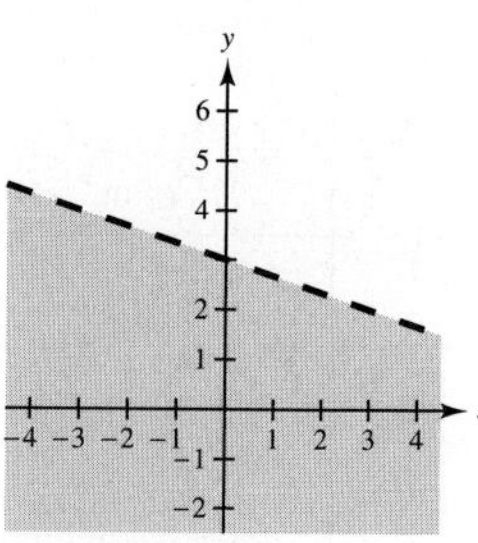

31.

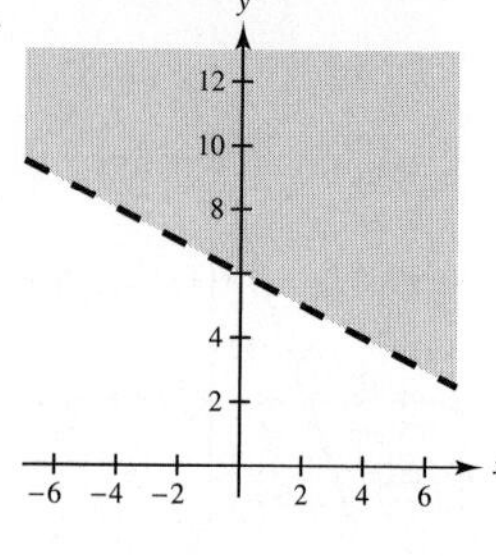

32.

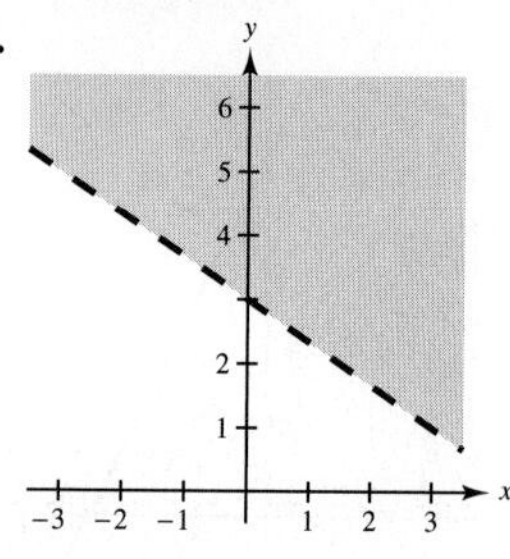

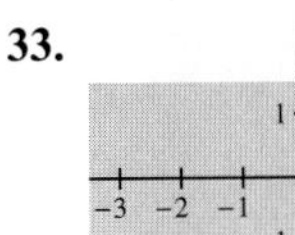

33.
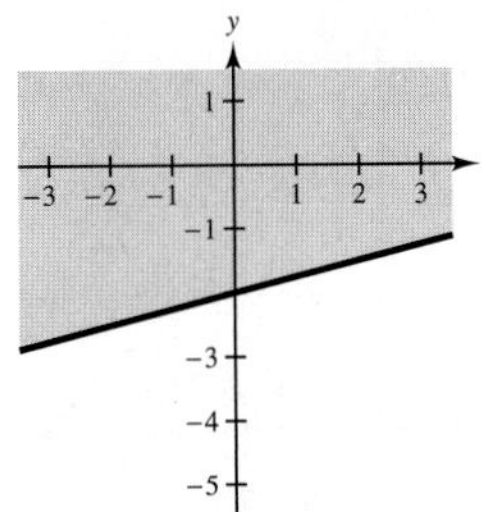

34.
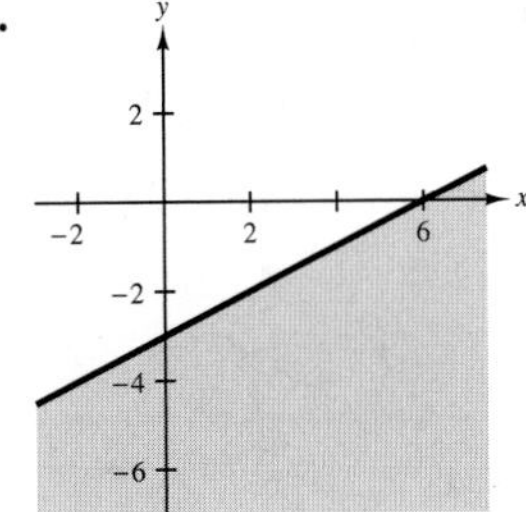

35.
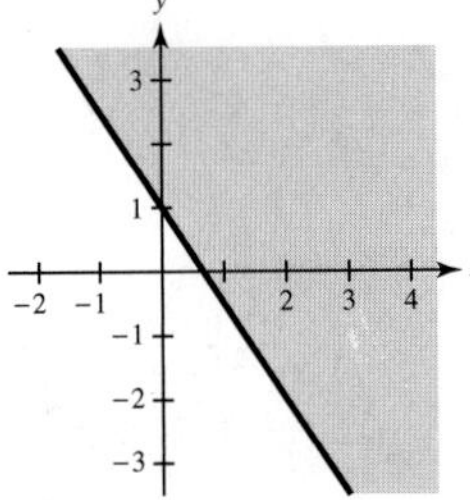

36.
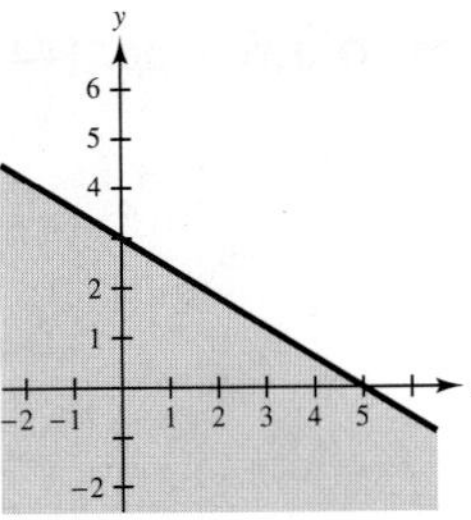

37.
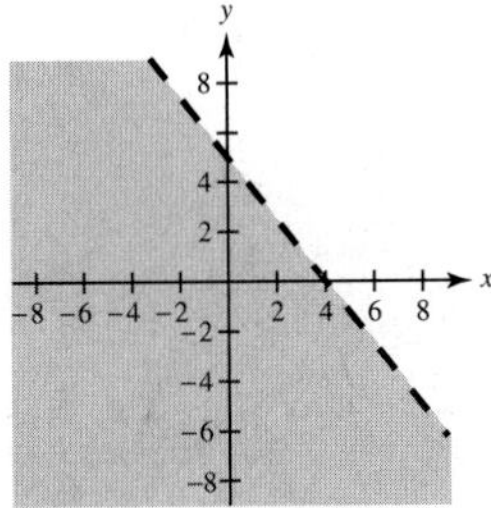

38.
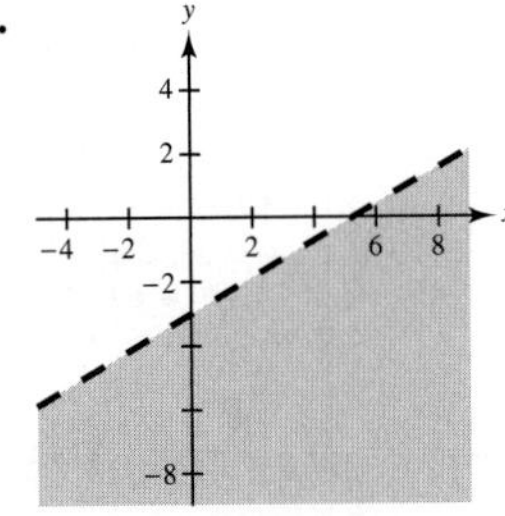

39.
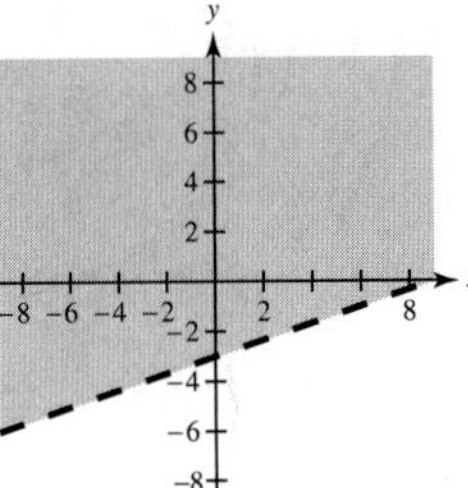

40.
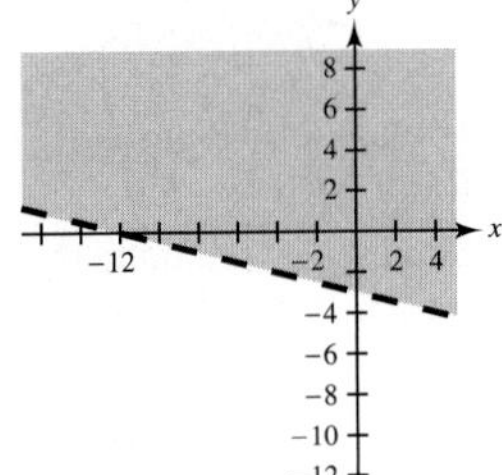

41.
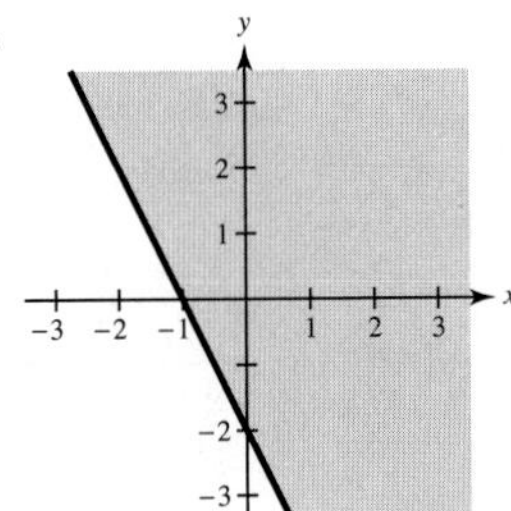

42.
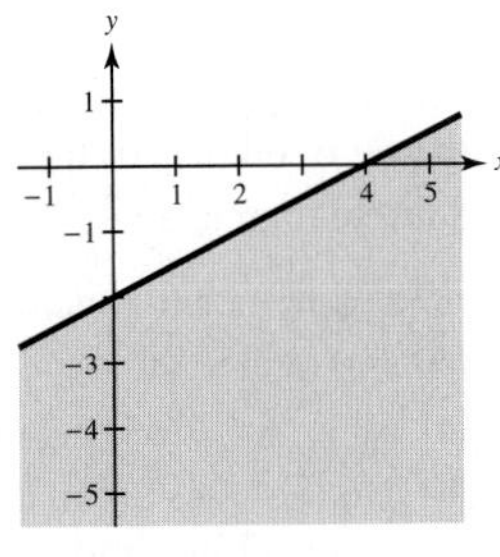

43.
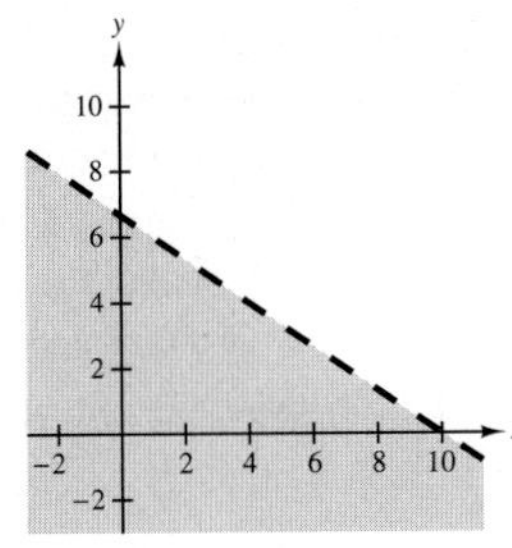

44.
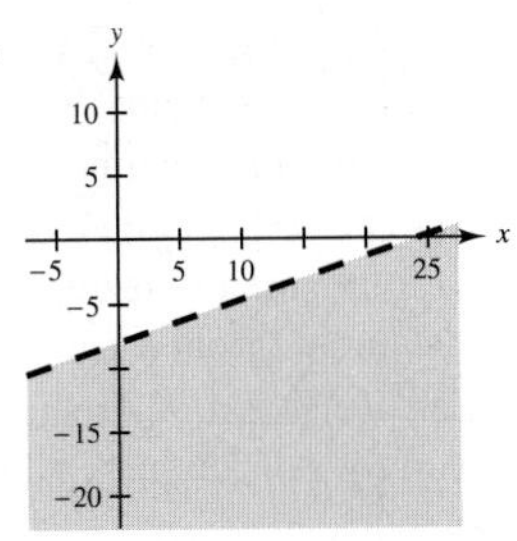

45.
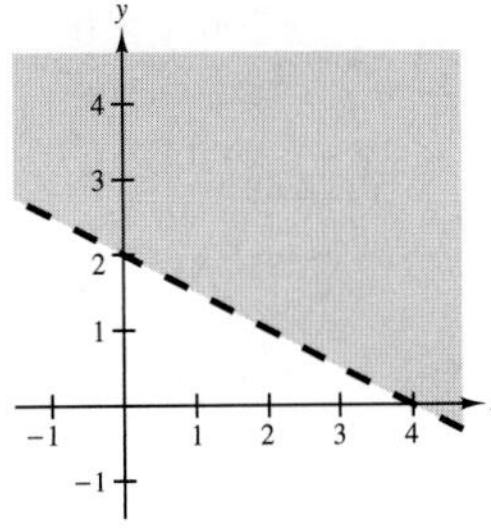

46.
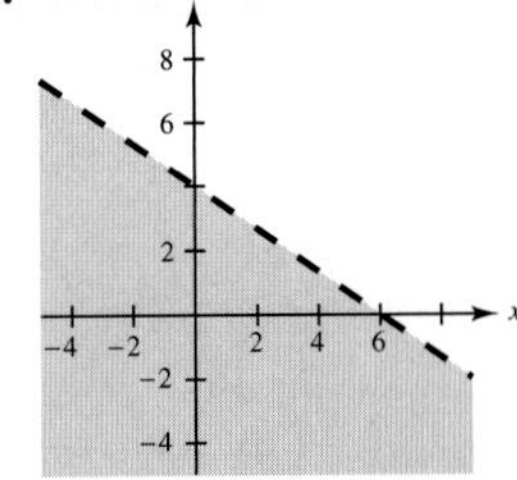

47.
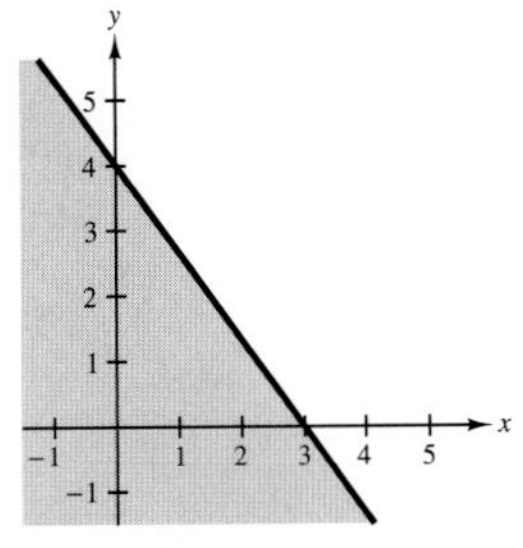

48.
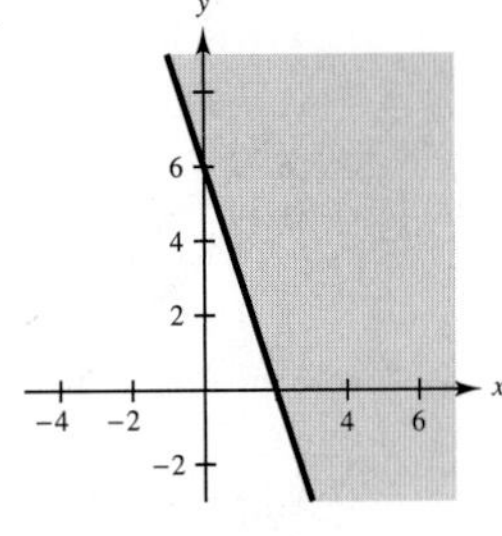

49. (a) $11x + 9y \geq 240; x \geq 0, y \geq 0$

(b)
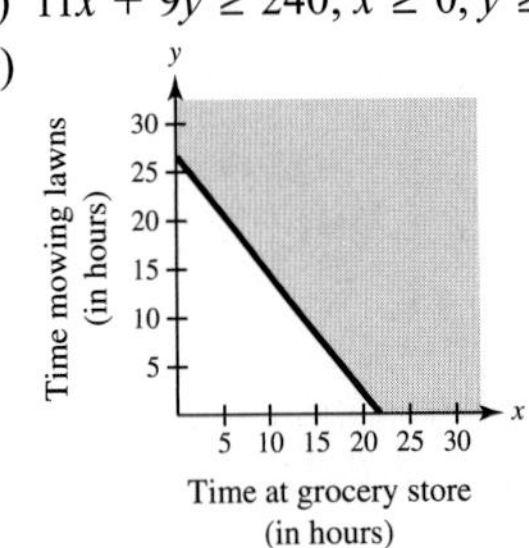

Sample answer: (x, y): (2, 25), (4, 22), (10, 15)

50. (a) $8x + 10y \geq 160; x \geq 0, y \geq 0$

(b)
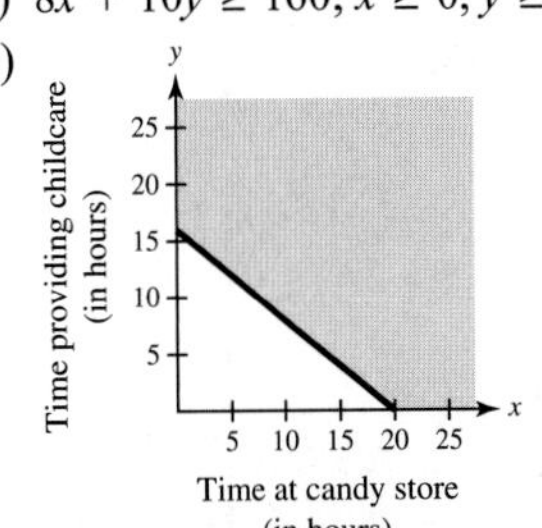

Sample answers: (x, y): (6, 20), (11, 12), (18, 8)

51. False. Another way to determine if (2, 4) is a solution of the inequality is to substitute $x = 2$ and $y = 4$ into the inequality $2x + 3y > 12$.

52. False. A point on the boundary line cannot be used to determine the half-plane that represents the solution of the linear inequality.

59. (a) $10x + 15y \leq 1000$ or $y \leq -\frac{2}{3}x + \frac{200}{3}; x \geq 0, y \geq 0$

(b)
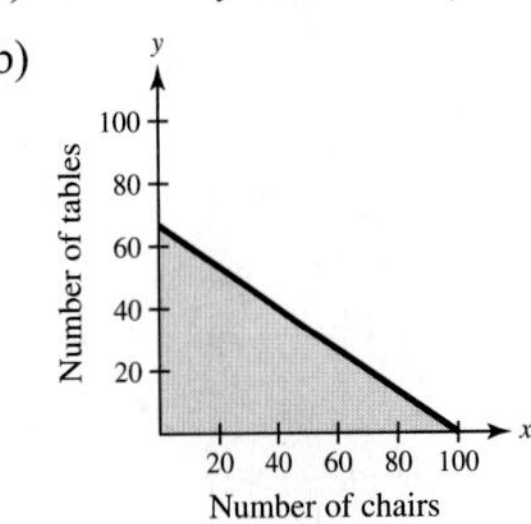

60. (a) $15x + 6y \le 2000$ or $y \le -\frac{5}{2}x + \frac{1000}{3}$; $x \ge 0, y \ge 0$

(b)

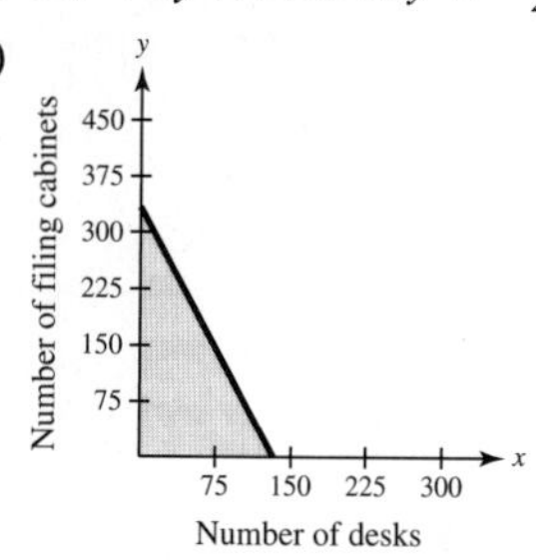

61. (b)

62. (b)

63. (b)

67. To write a double inequality such that the solution is the graph of a line, use the same real number as the bounds of the inequality, and use $\le$ for one inequality symbol and $\le$ for the other inequality symbol. For example, the solution of $2 \le x + y \le 2$ is $x + y = 2$. Graphically, $x + y = 2$ is a line in the plane.

69. The boundary lines of the inequalities are parallel lines. Therefore, $a = c$. Because every point in the solution set of $y \le ax + b$ is a solution of $y < cx + d$, the inequality $y \le ax + b$ lies in the half-plane of $y < cx + d$. Therefore, $y = ax + b$ lies under $y = cx + d$ and $b < d$ (assuming $b > 0$ and $d > 0$).

74. **75.**

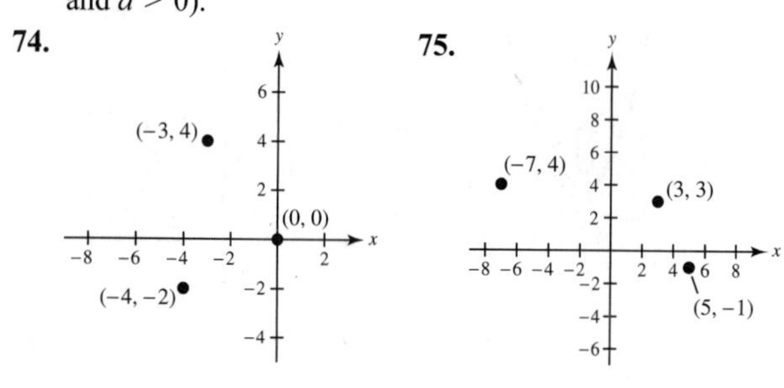

76. **77.**

Section 3.6 *(pp. 144–151)*

1. **2.**

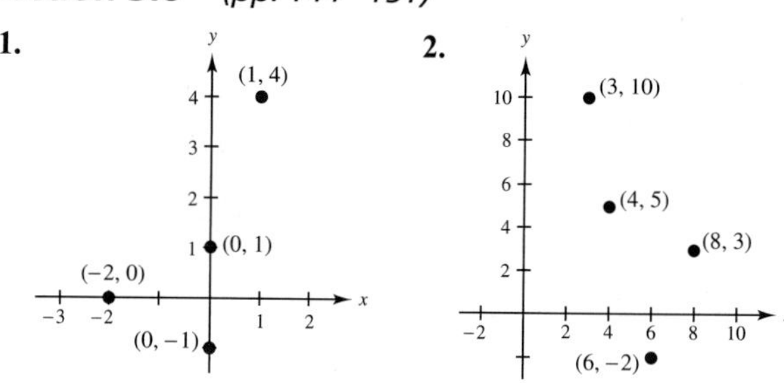

3. **4.**

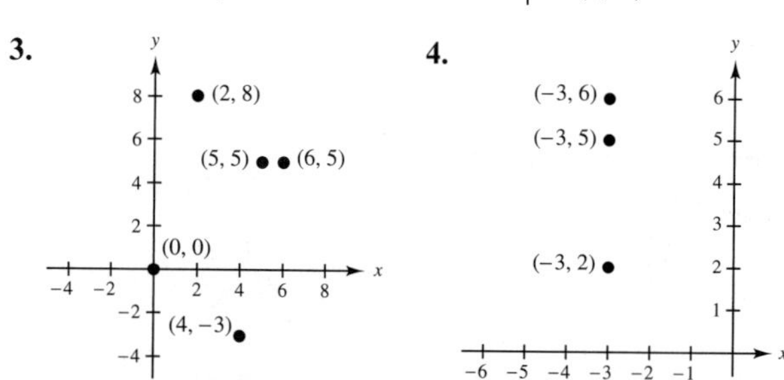

31. Domain: All real numbers r such that $r > 0$
Range: All real numbers C such that $C > 0$

32. Domain: All real numbers s such that $s > 0$
Range: All real numbers A such that $A > 0$

33. Domain: All real numbers r such that $r > 0$
Range: All real numbers A such that $A > 0$

34. Domain: All real numbers r such that $r > 0$
Range: All real numbers V such that $V > 0$

35. The domain is the set of inputs of the function, and the range is the set of outputs of the function.

45. All real numbers x such that $x \ge \frac{1}{2}$

46. All real numbers x such that $x \le \frac{8}{3}$

47. All real numbers t **48.** All real numbers x

54. (a) \$360, \$480, \$570, \$660
(b) No. $h < 0$ is not in the domain of W.

55. (a) \$240, \$300, \$420, \$750
(b) When the worker works between 0 and 40 hours, the wage is \$12 per hour. When more than 40 hours are worked, the worker earns time-and-a-half for the additional hours.

57. Not correct. The price of the shoes varies. The store could sell a lot of pairs of inexpensive shoes one day but make less than the amount from a few pairs of more expensive shoes.

58. Yes. If a subset of A includes only input values that are mapped to only one output in the range, then the subset of A is a function.

59. No. If set B is a function, no element of the domain is matched with two different elements in the range. A subset of B would have fewer elements in the domain, and therefore, a subset of B would be a function.

60. *Sample answer*: Statement: The number of ceramic tiles required to floor a kitchen is a function of the area of the floor.
Domain: The area of the floor
Range: The number of ceramic tiles
This is a function.

61. *Sample answer*: Statement: The distance driven on a trip is a function of the number of hours spent in the car.
Domain: The number of hours in the car
Range: The distance driven
The function has an infinite number of ordered pairs.

66.
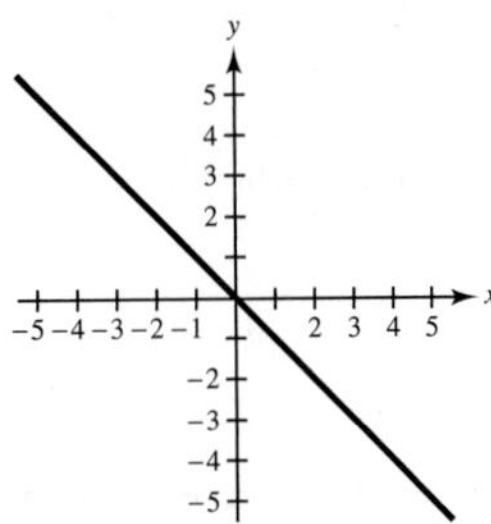

67.
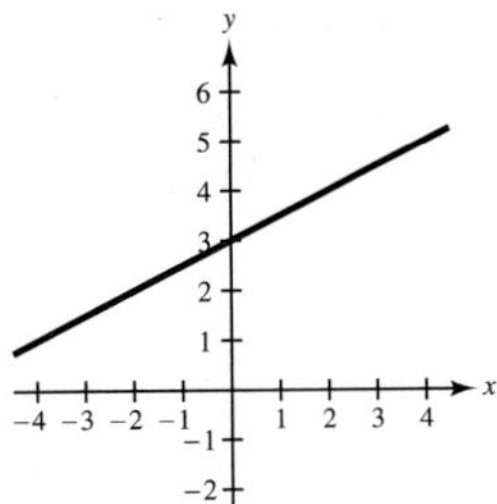

68.
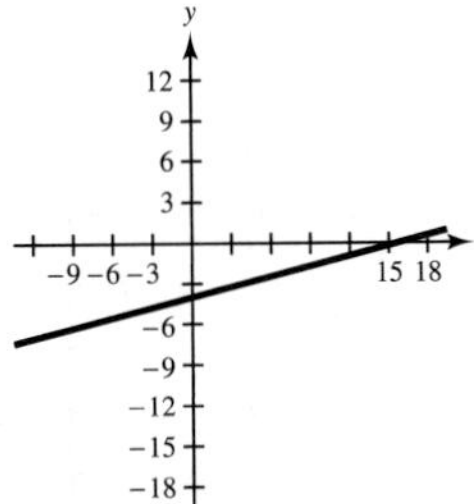

69.
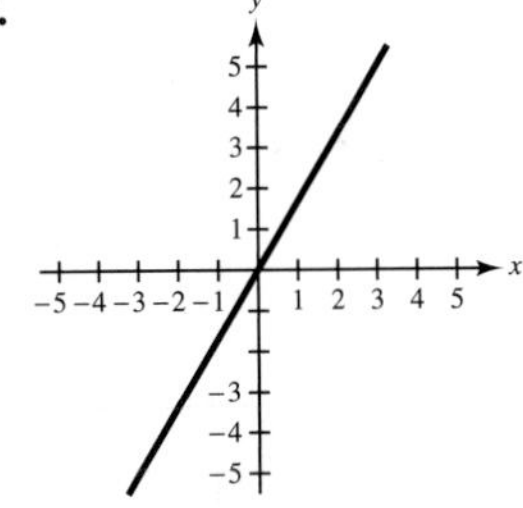

70.
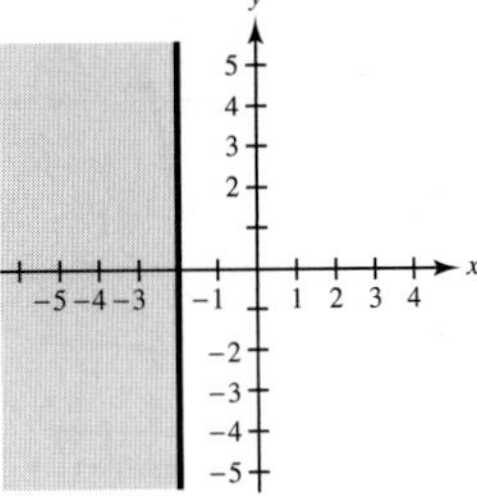

71.
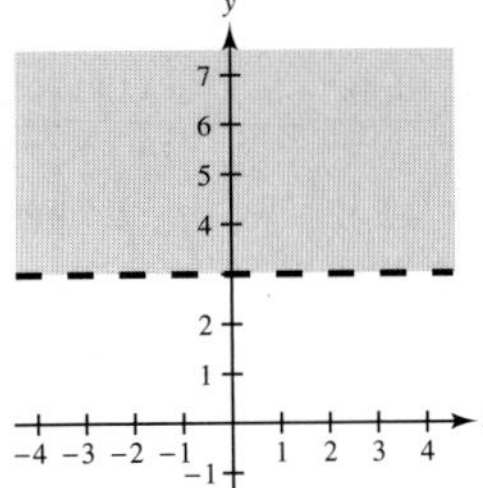

72.
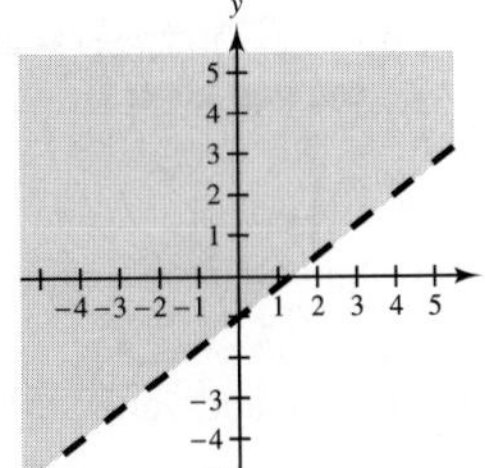

73.
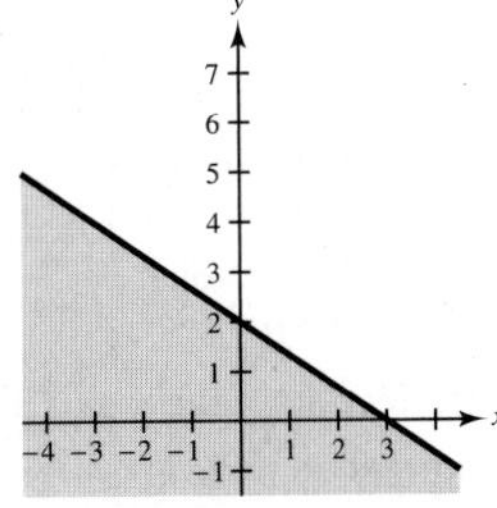

Section 3.7 *(pp. 152–159)*

1.
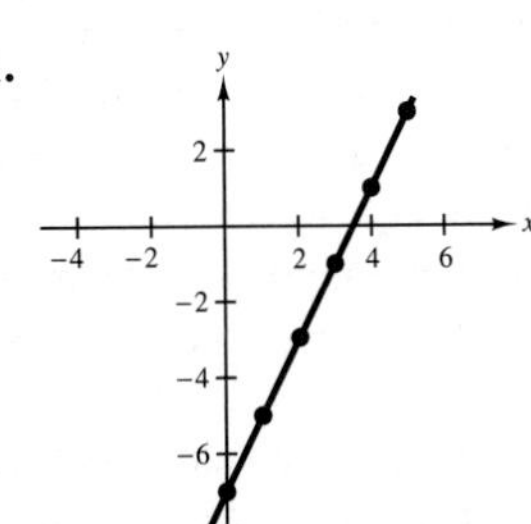

2.
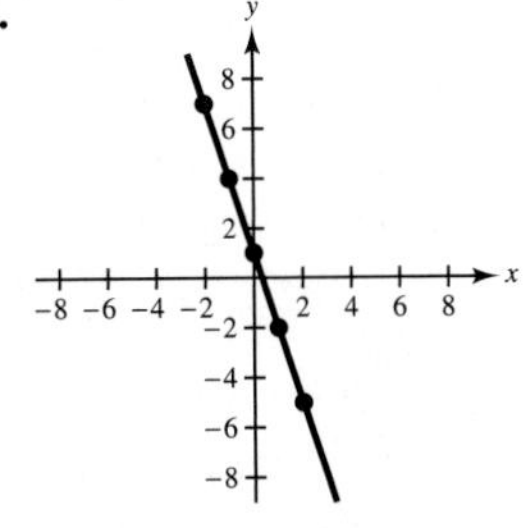

3.
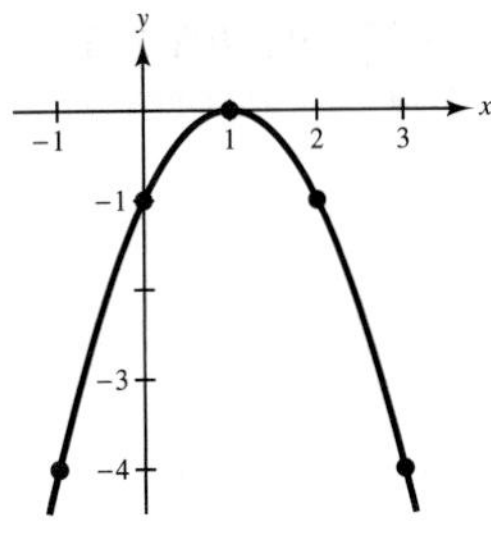

4.
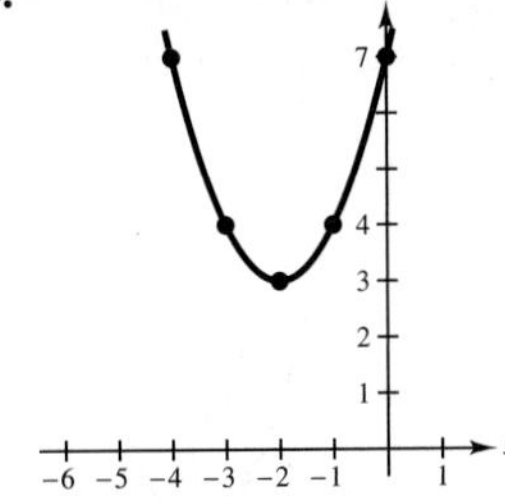

5.
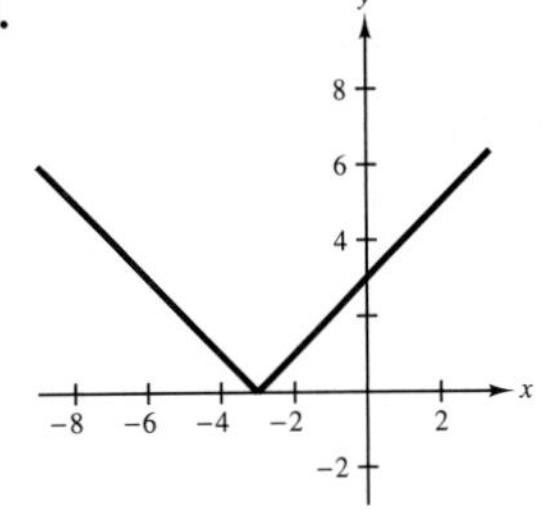

6.
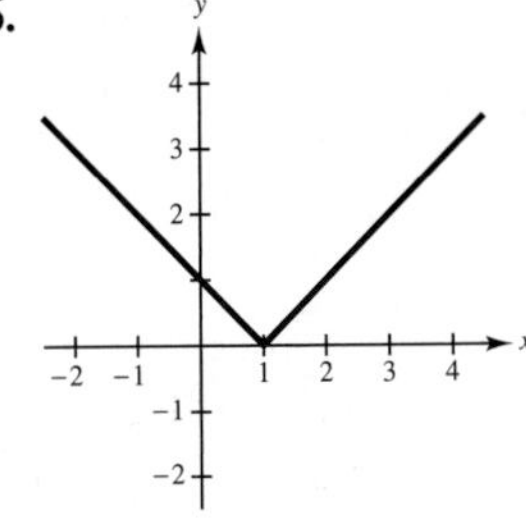

7.
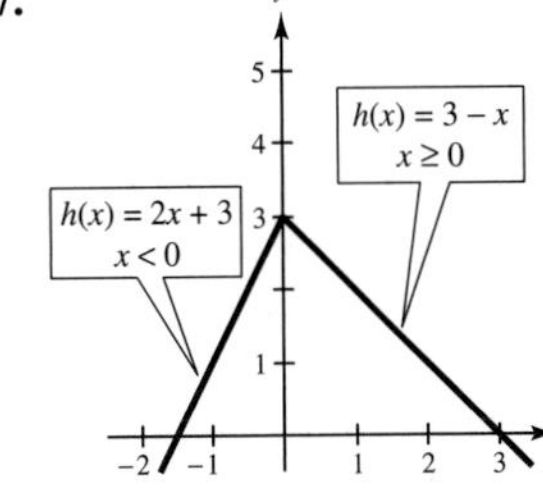

8.
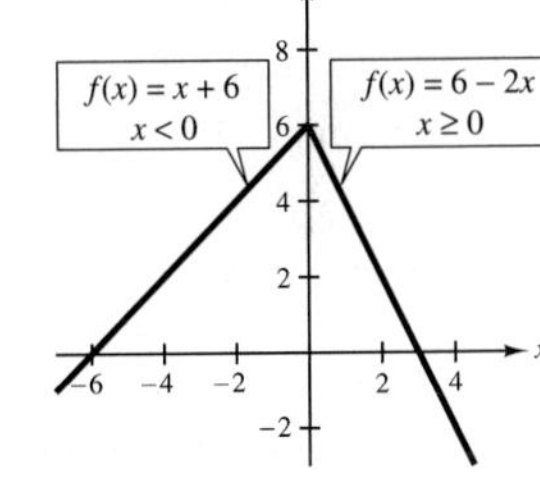

9.
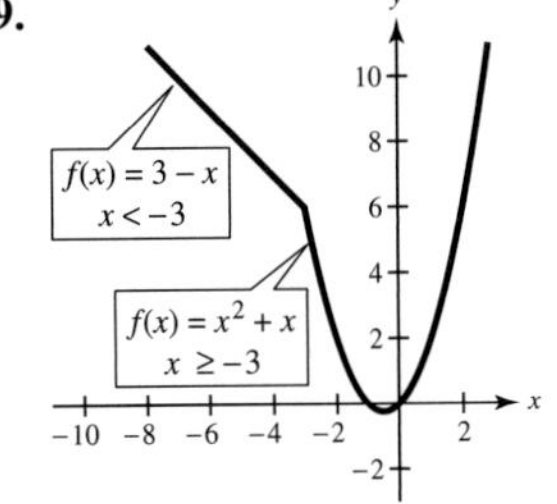

10.
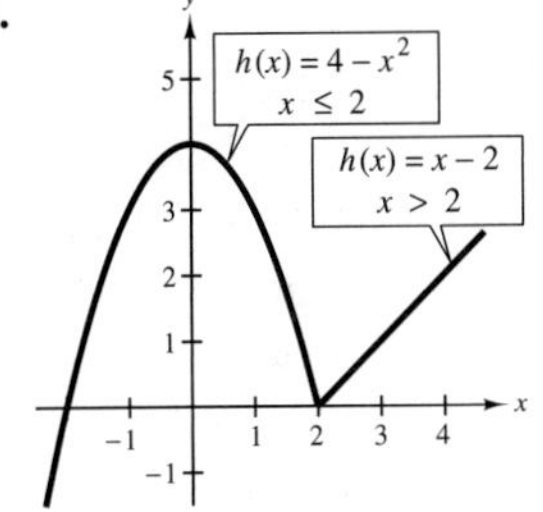

11.
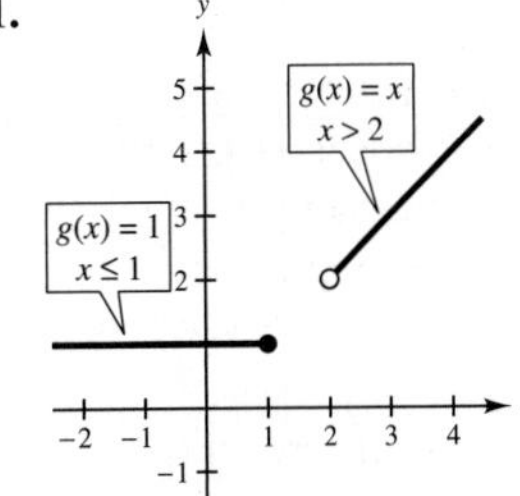

12.
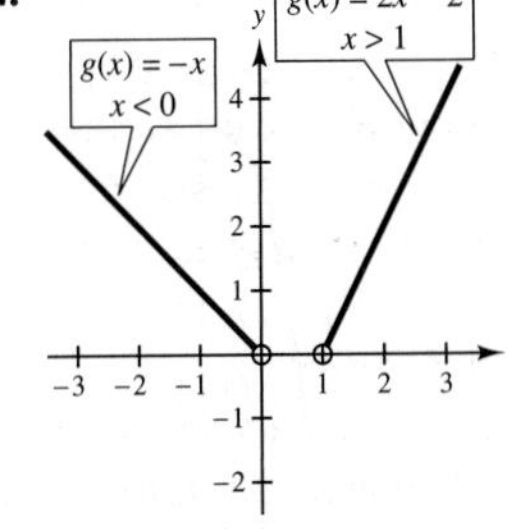

23. (a) Vertical shift 2 units upward

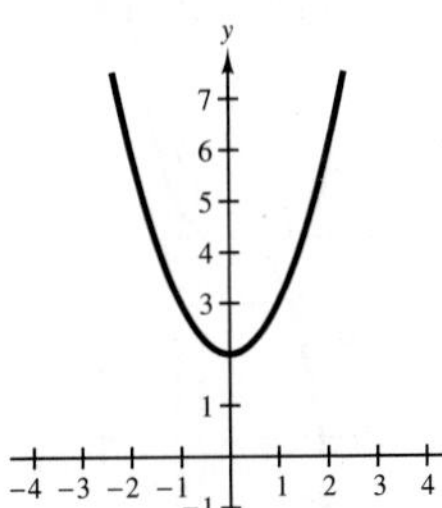

(b) Vertical shift 4 units downward

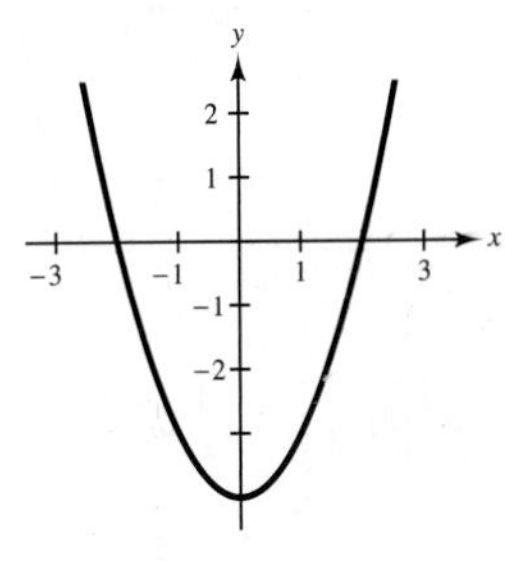

(c) Horizontal shift 2 units to the left

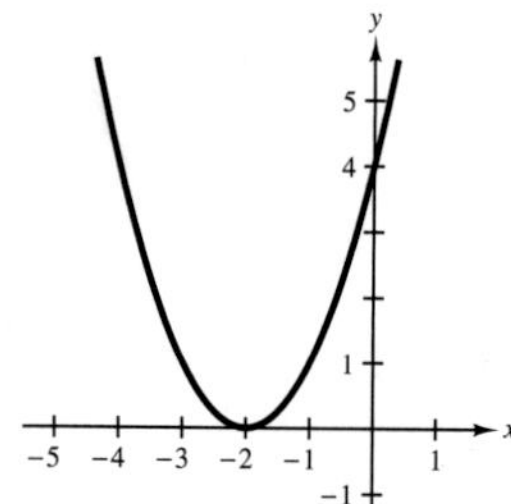

(d) Horizontal shift 4 units to the right

24. (a) Horizontal shift 3 units to the right

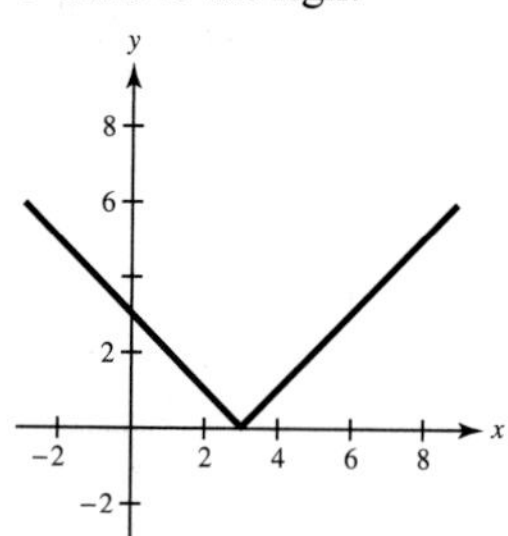

(b) Horizontal shift 3 units to the left

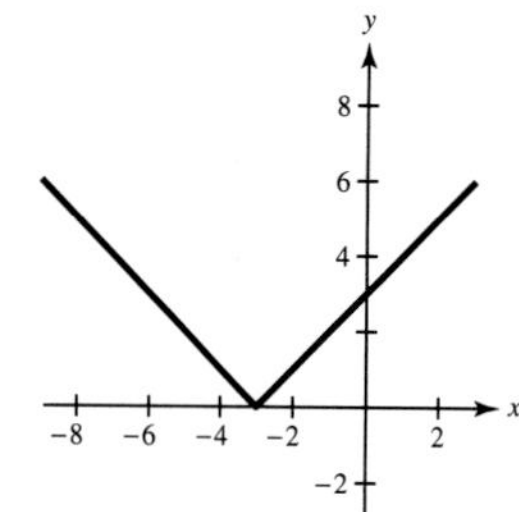

(c) Vertical shift 4 units downward

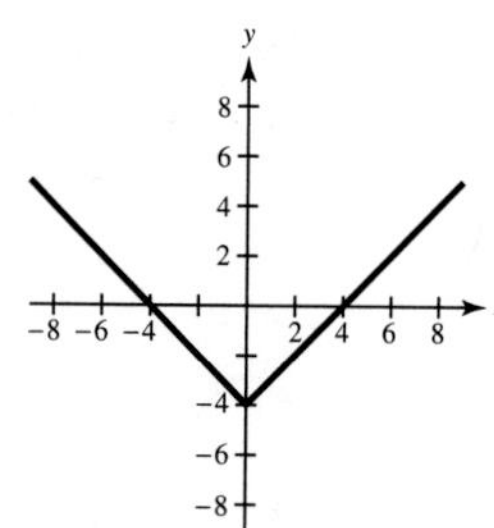

(d) Vertical shift 6 units upward

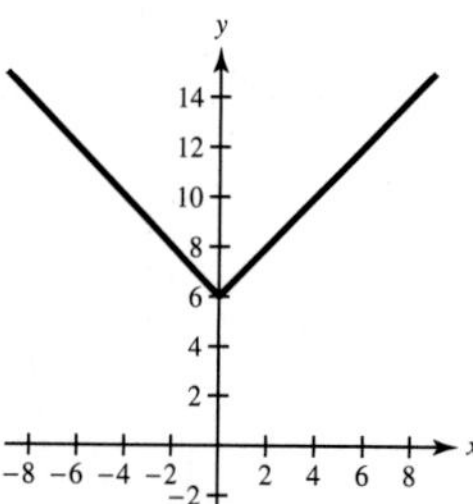

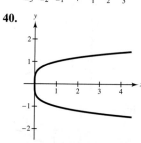

25. (a) Vertical shift 3 units upward

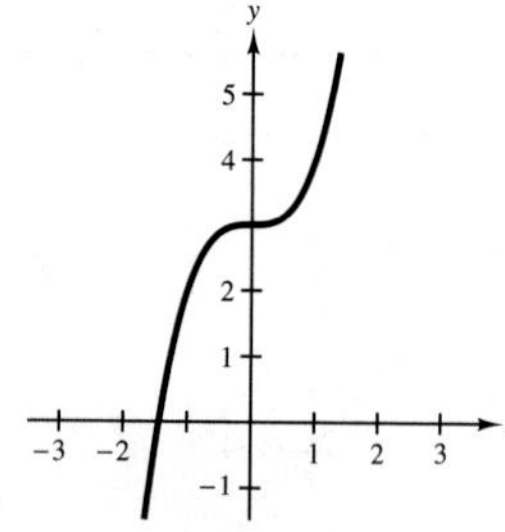

(b) Vertical shift 5 units downward

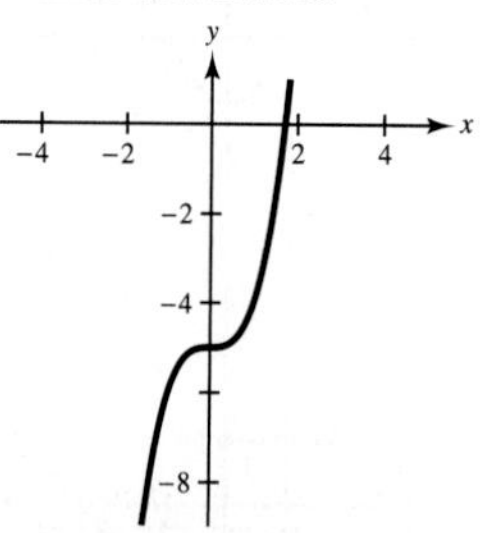

(c) Horizontal shift 3 units to the right

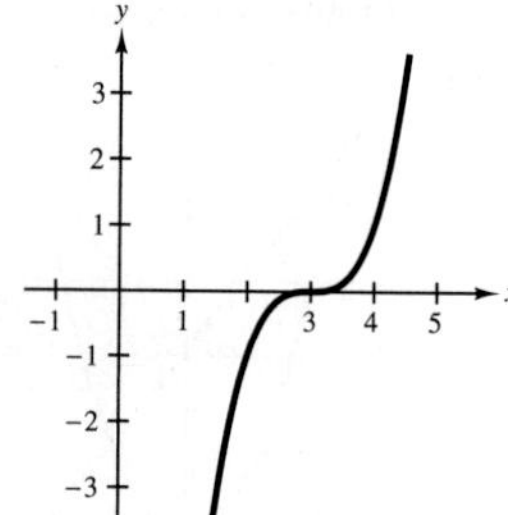

(d) Horizontal shift 2 units to the left

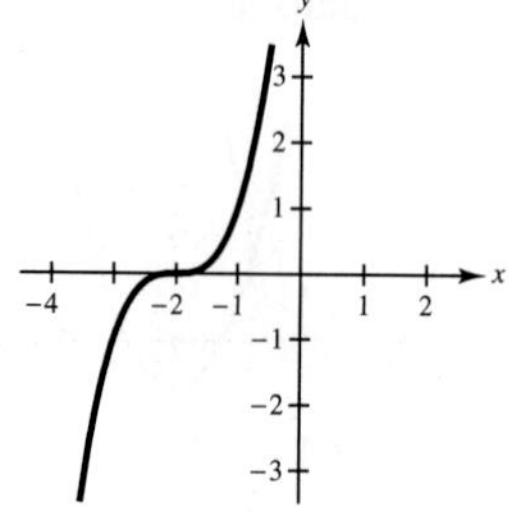

26. (a) Horizontal shift 2 units to the right

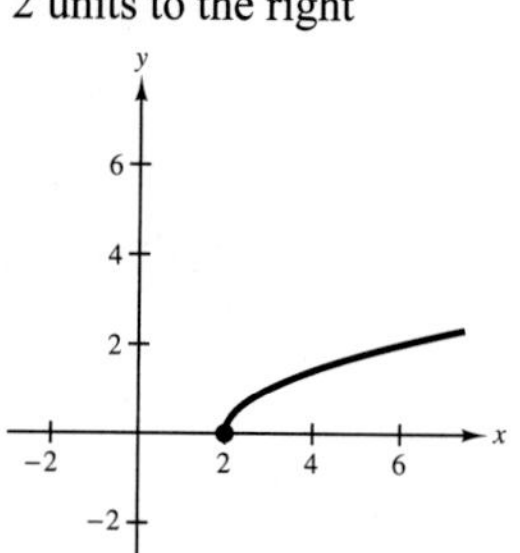

(b) Vertical shift 2 units upward

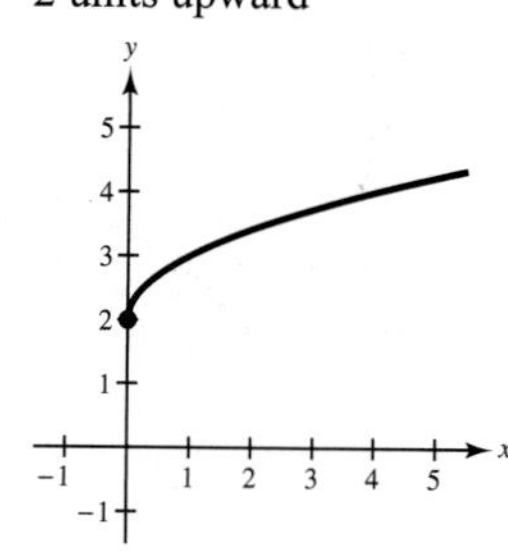

(c) Vertical shift 4 units upward

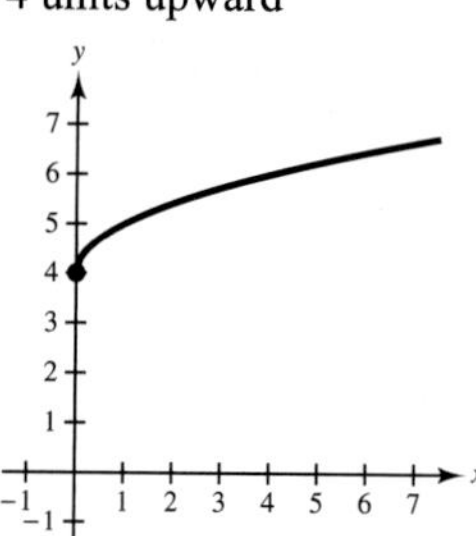

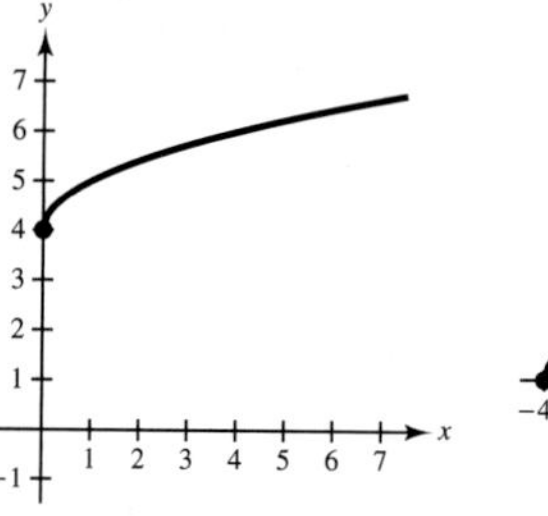

(d) Horizontal shift 4 units to the left

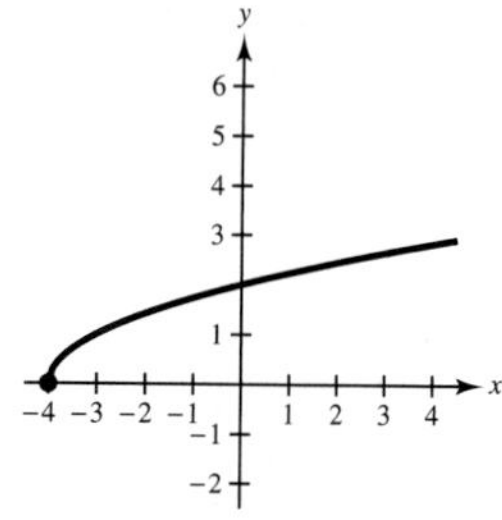

37.

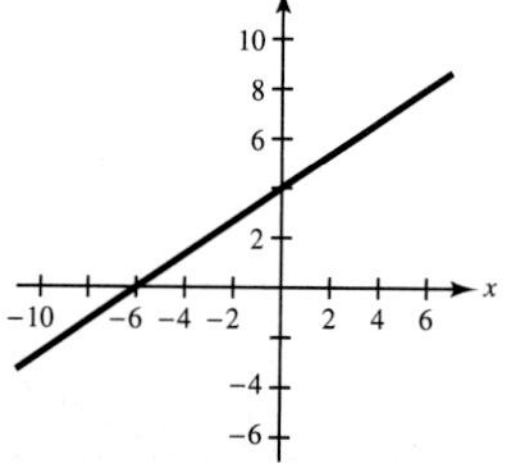

38.

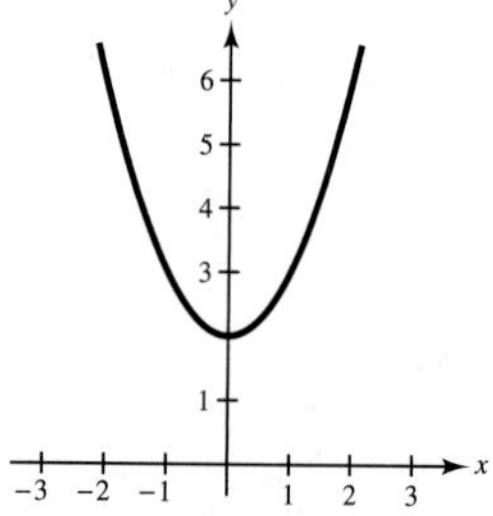

39.

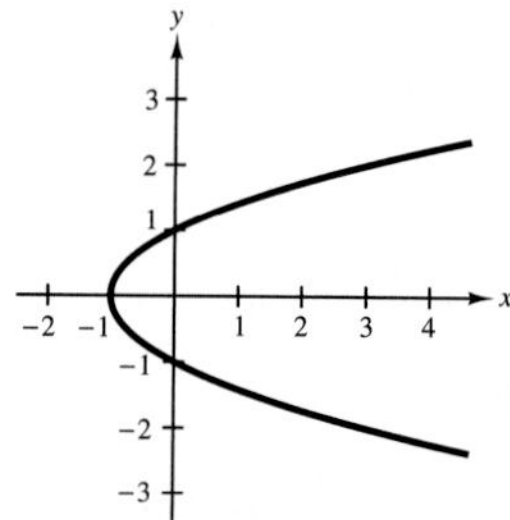

40.

41. Horizontal shift of $y = x^3$ (cubing function)
$y = (x - 2)^3$

42. Vertical shift of $y = |x|$ (absolute value function)
$y = |x| + 3$

43. Reflection in the x-axis and horizontal and vertical shifts of $y = x^2$ (squaring function)
$y = -(x + 1)^2 + 1$

44. Horizontal and vertical shift of $y = \sqrt{x}$ (square root function)
$y = \sqrt{x - 1} + 1$

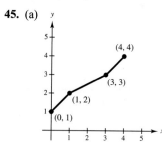

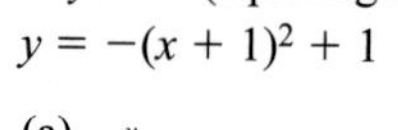

45. (a)

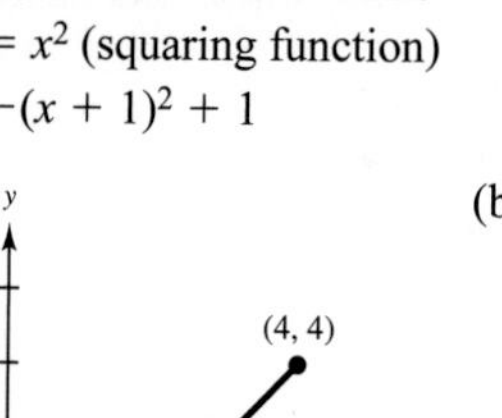

(b)

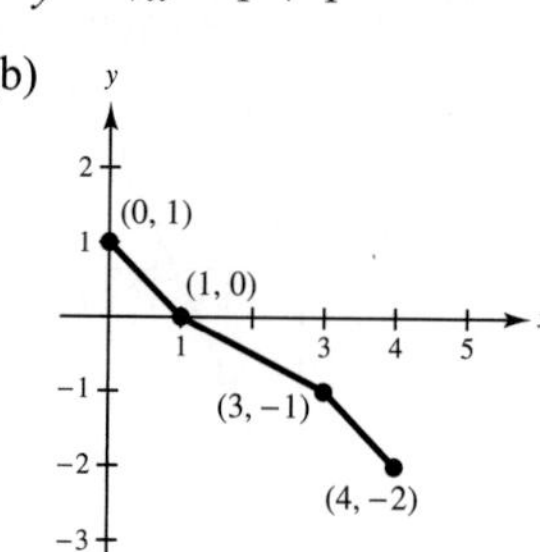

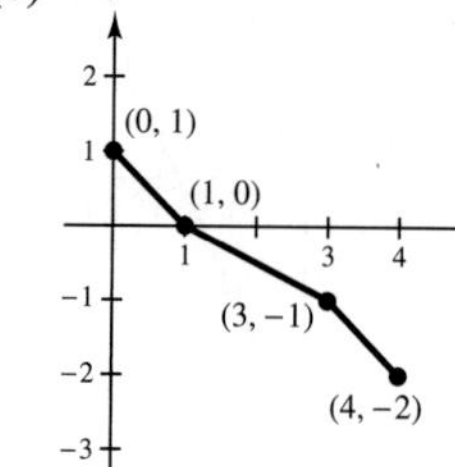

(c)

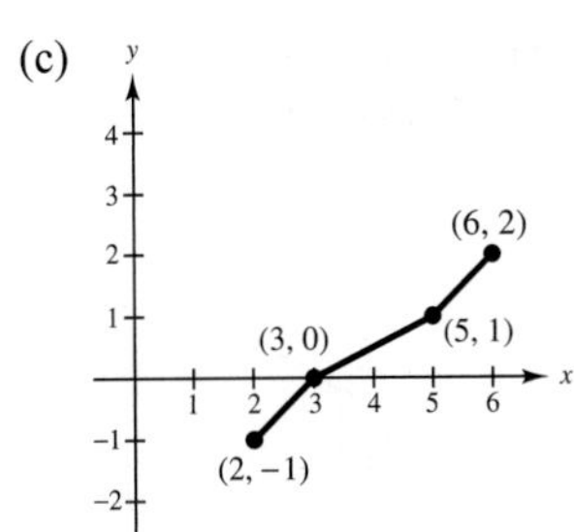

(d)

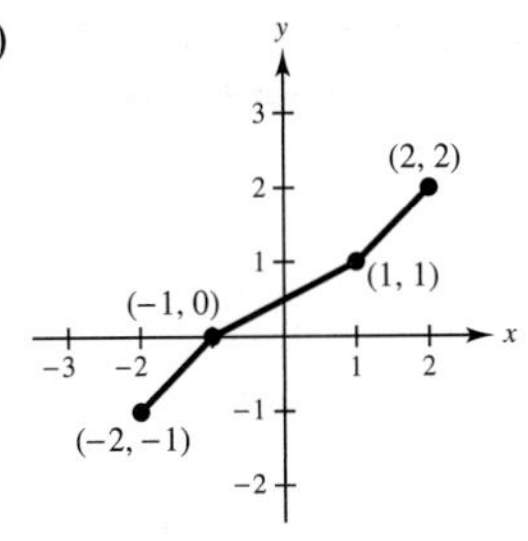

(e)

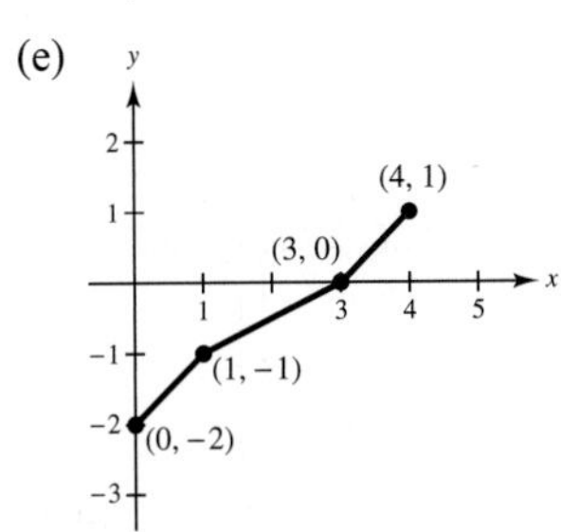

(f)

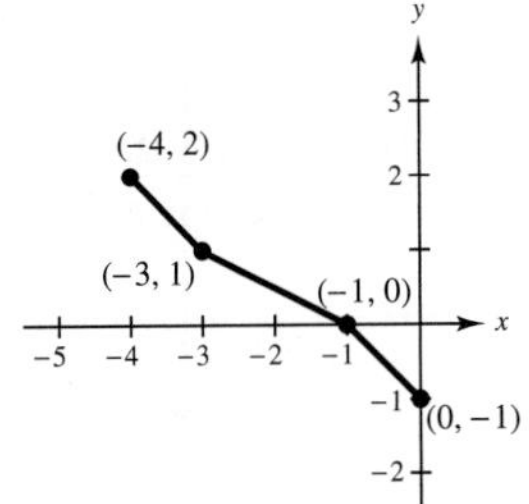

46. (a)

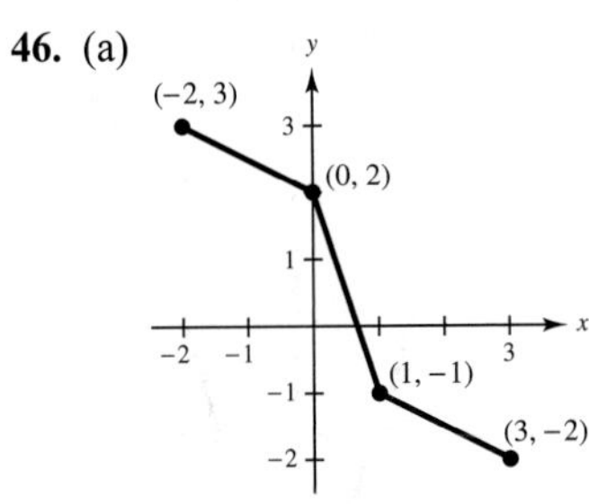

(b)

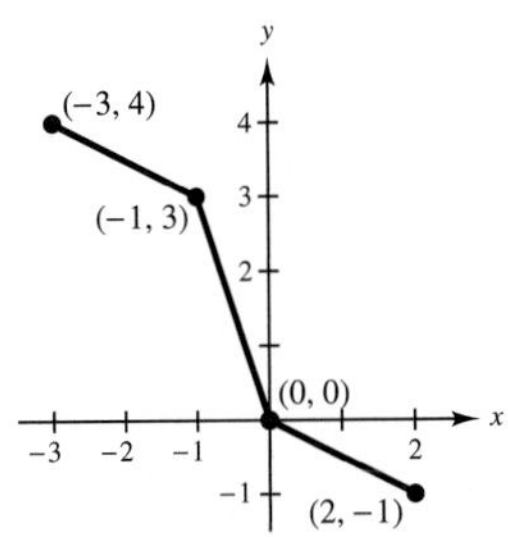

(c)

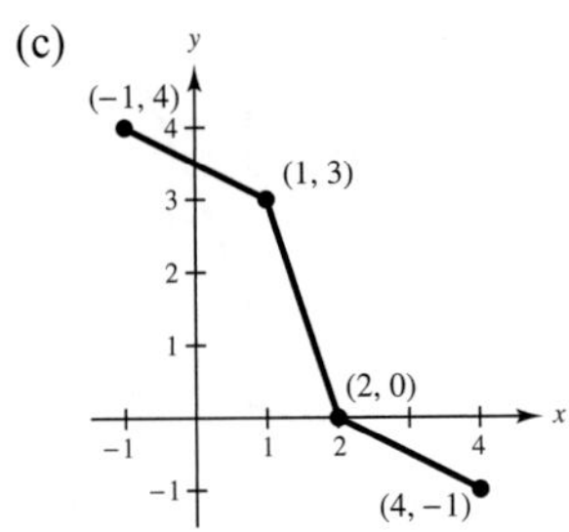

(d)

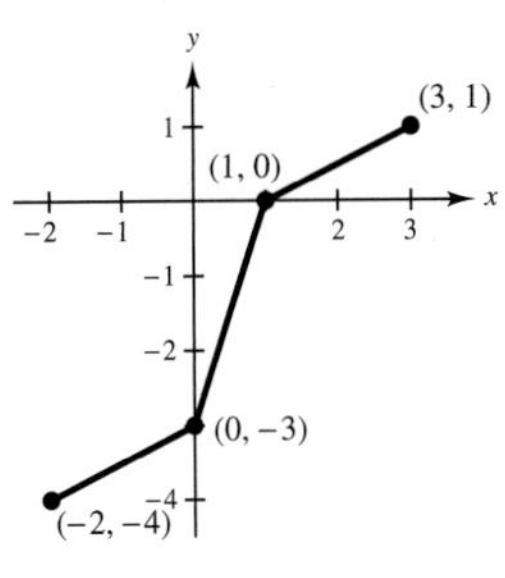

(e)

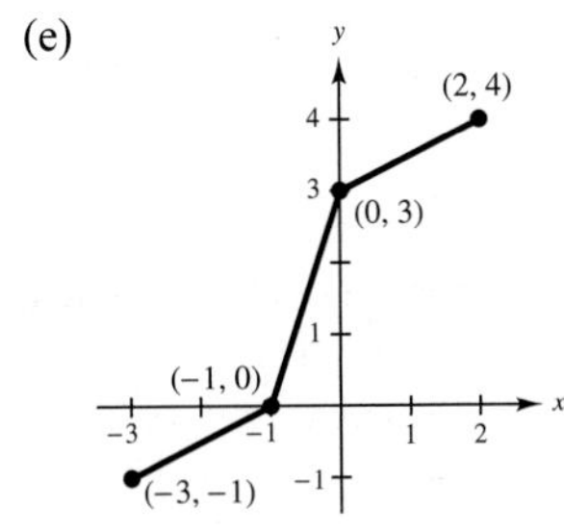

(f)

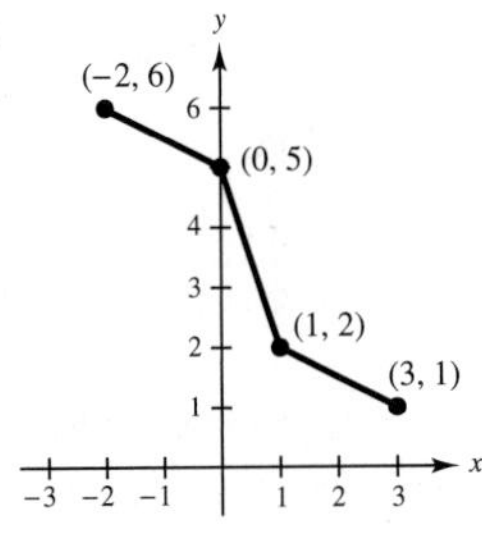

47. (a) Perimeter: $P = 2l + 2w$

$$200 = 2l + 2w$$
$$200 - 2l = 2w$$
$$100 - l = w$$

Area: $A = lw$

$$A = l(100 - l)$$

(b)

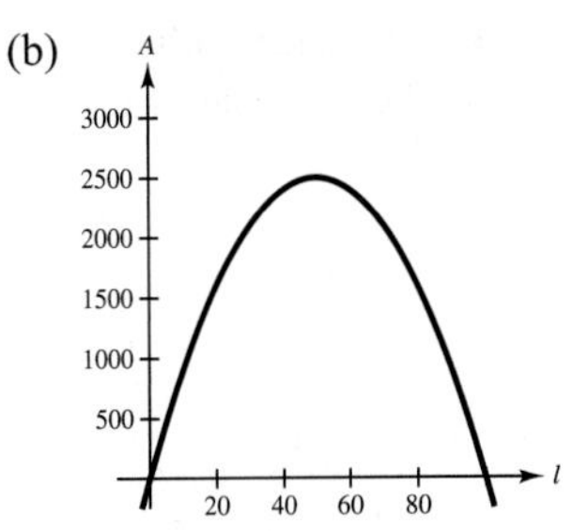

(c) $l = 50$. The figure is a square.

Review Exercises *(pp. 162–165)*

1.

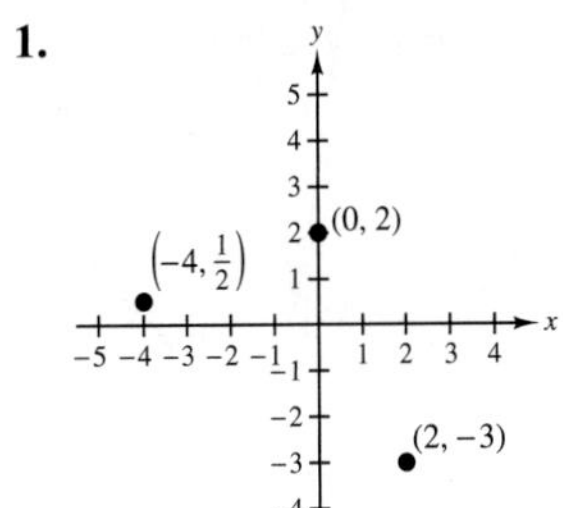

2.

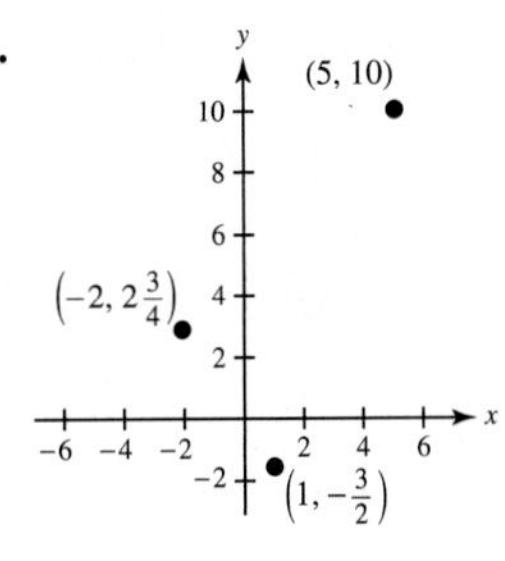

17.

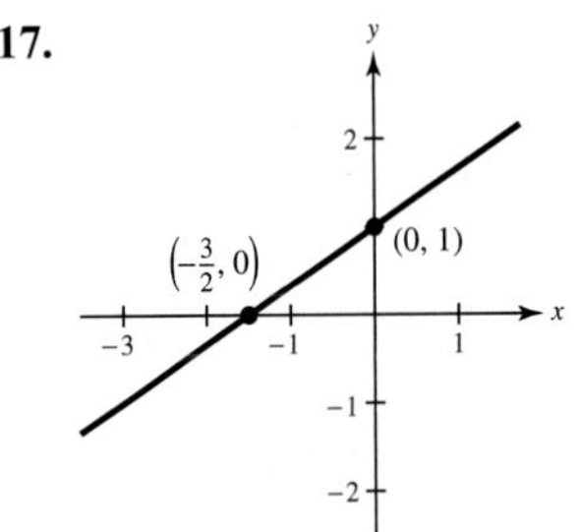

18.

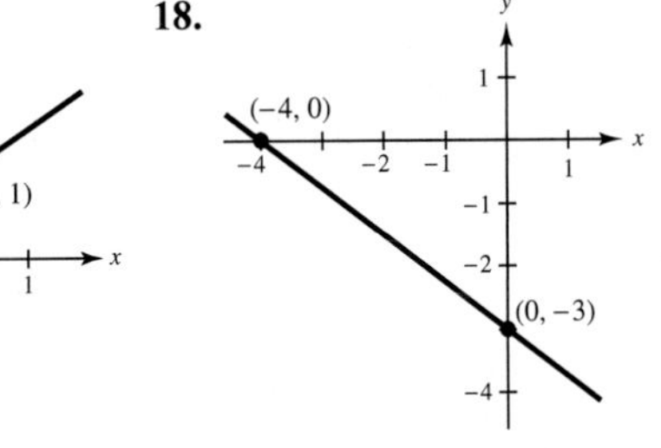

19.

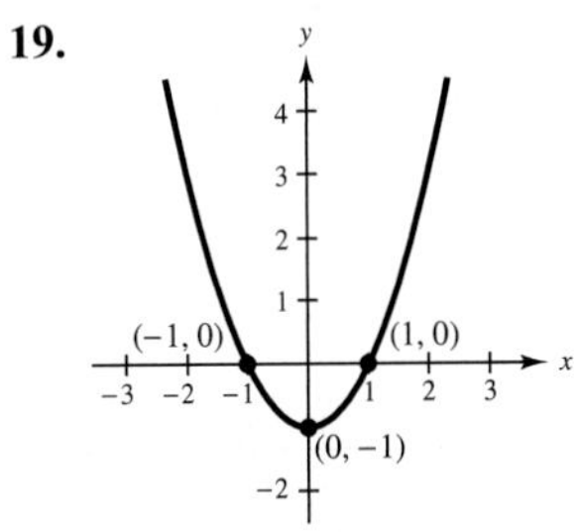

20.

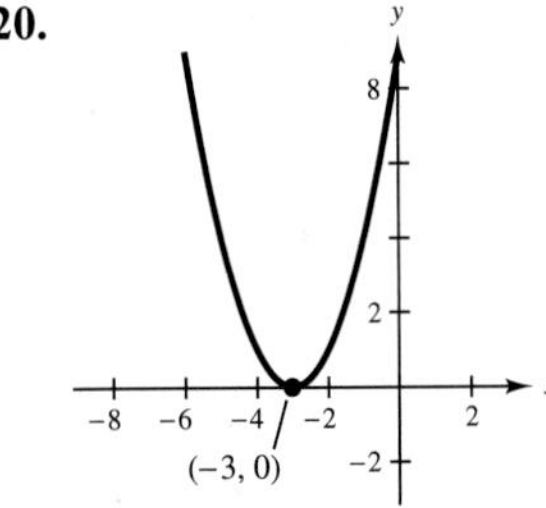

21.

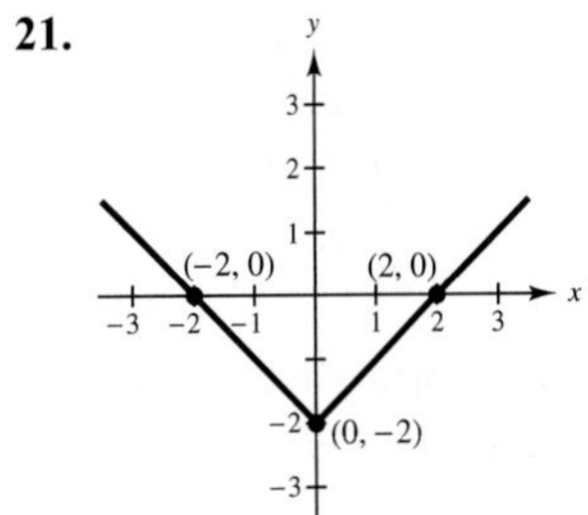

22.

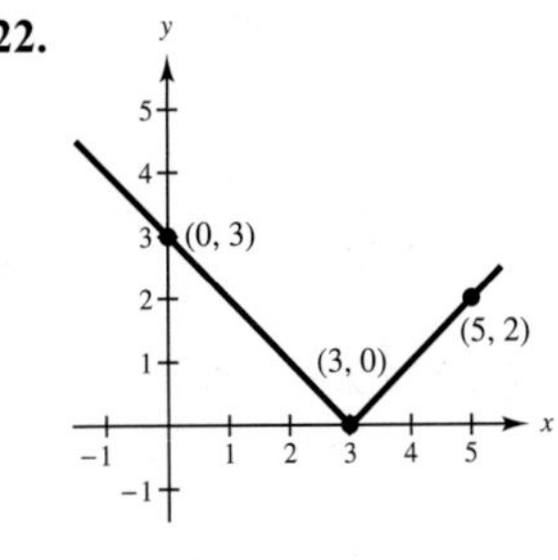

23.

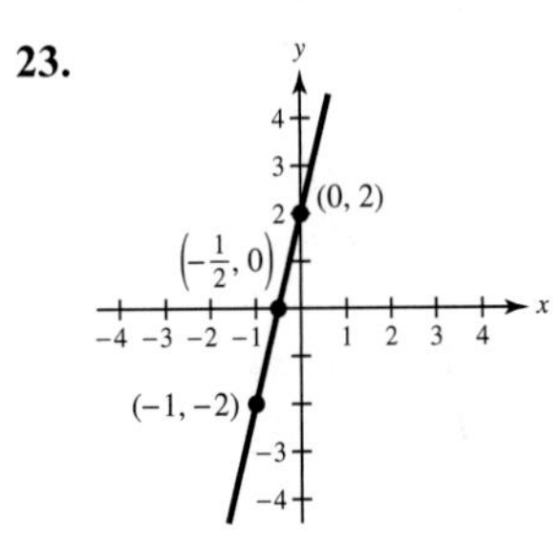

24.

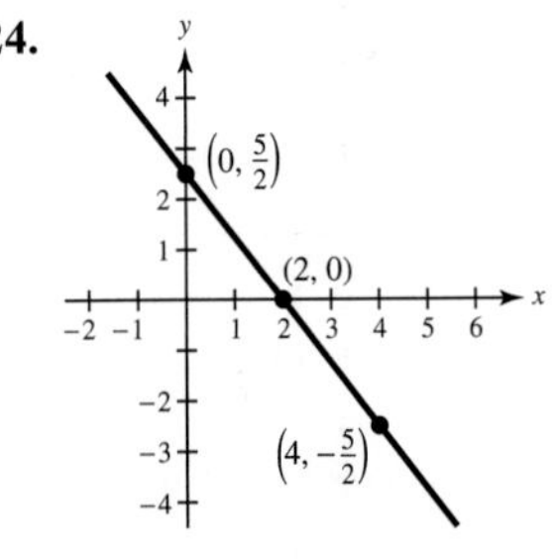

25.

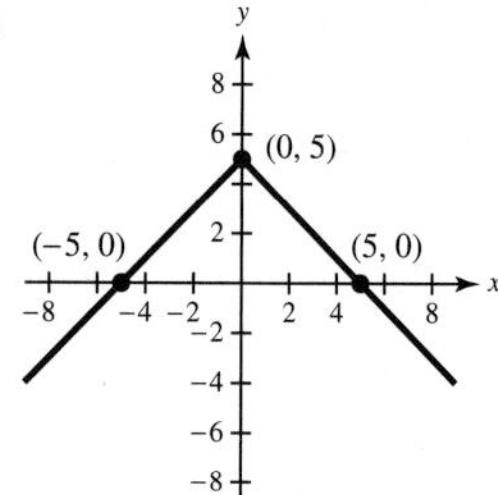

26.

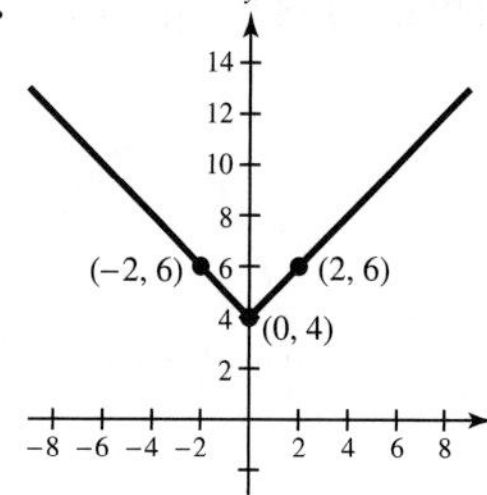
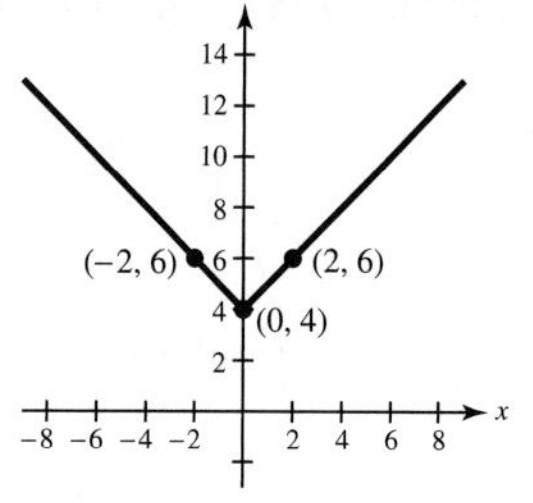

27.

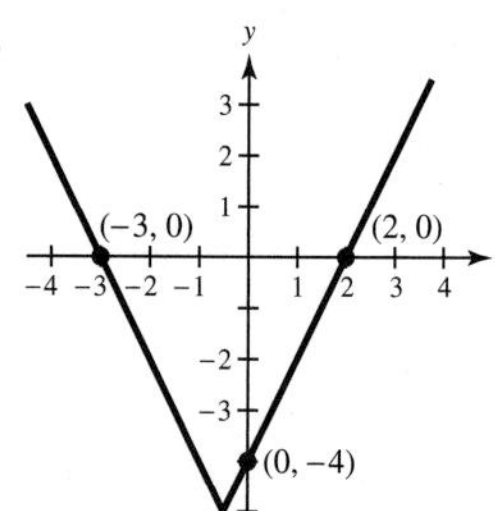

28.

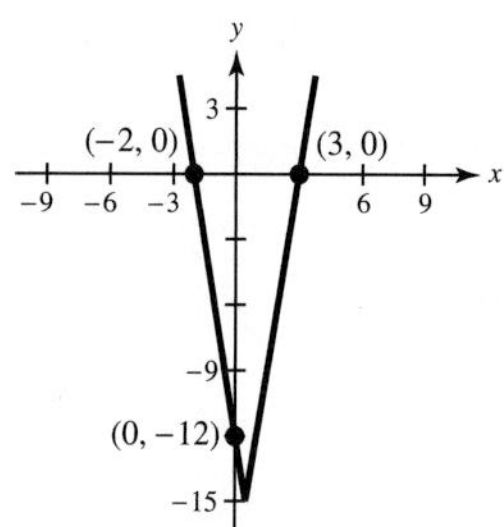

29. (b)

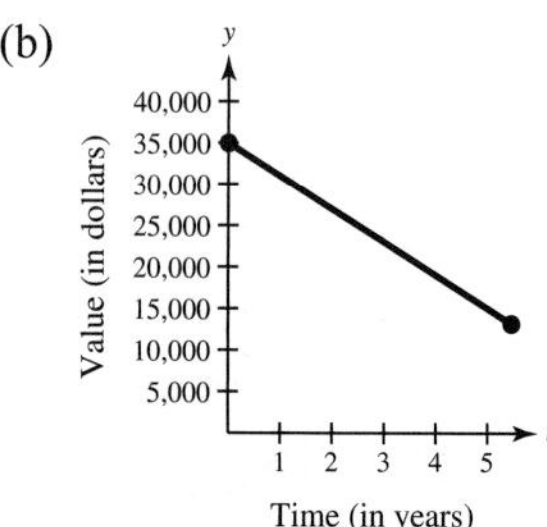

30. (b)

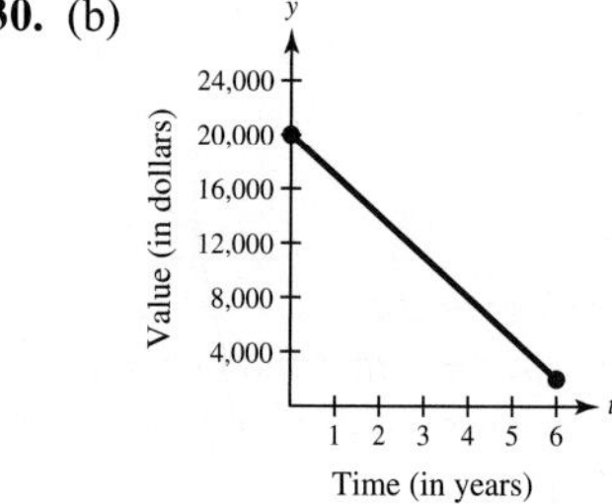

37.

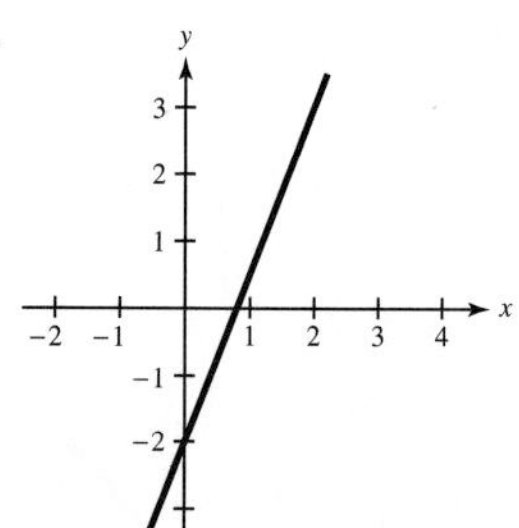

38.

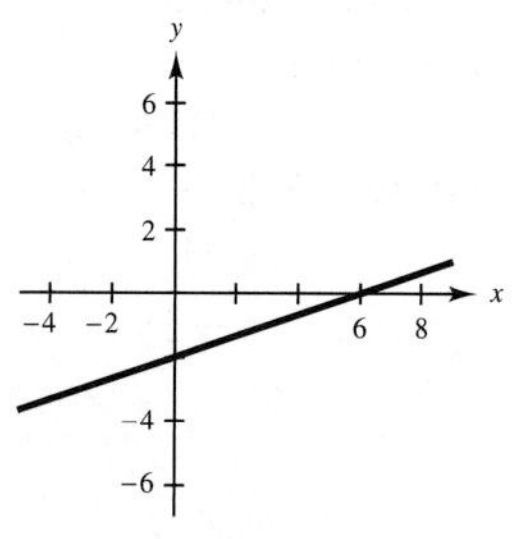

39.

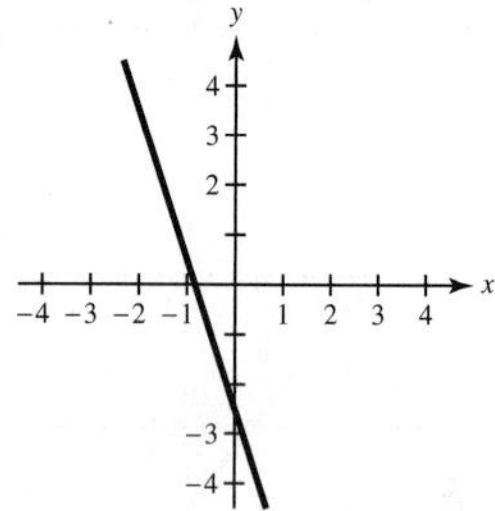

40.

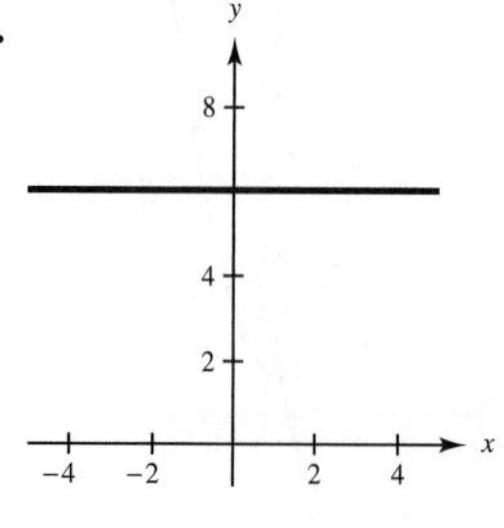

59. (a) $y = -3x + 1$ (b) $y = \frac{1}{3}x - 1$

60. (a) $y = -\frac{1}{2}x + \frac{9}{2}$ (b) $y = 2x + 7$

65.

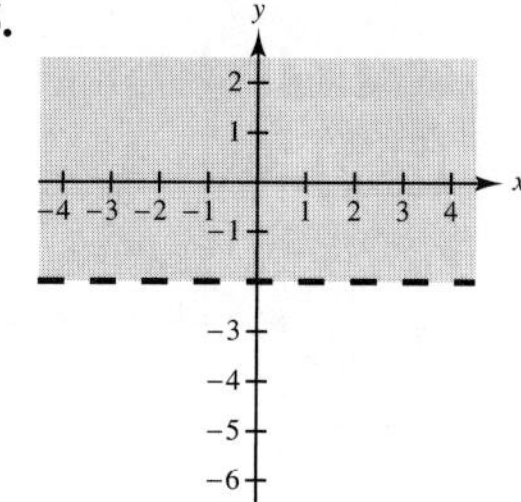

66.

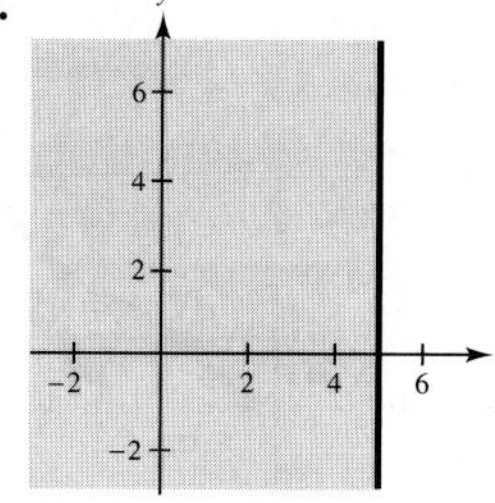

67.

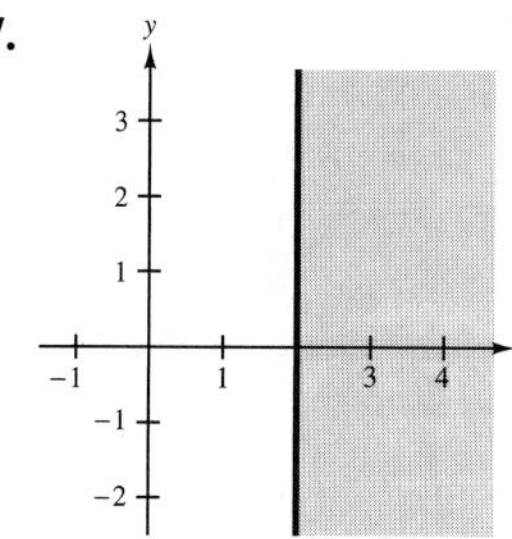

68.

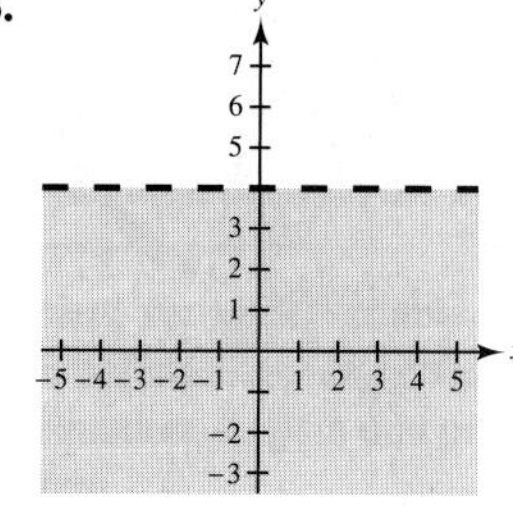

69.

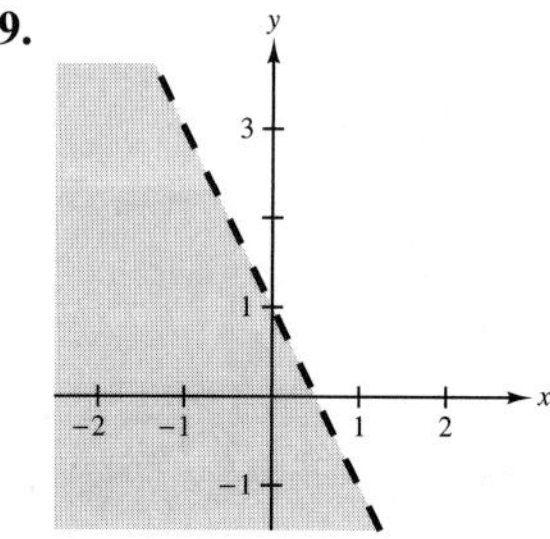

70.

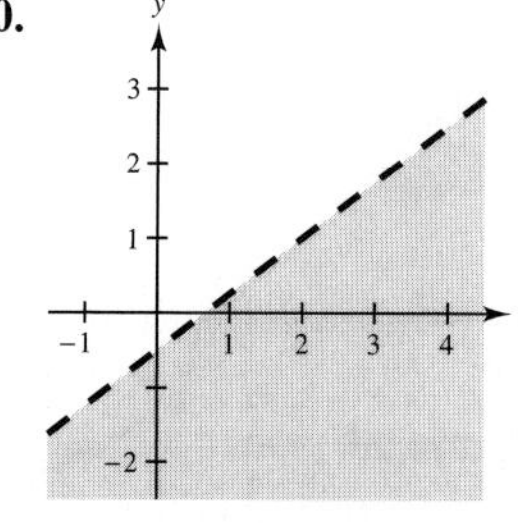

71.

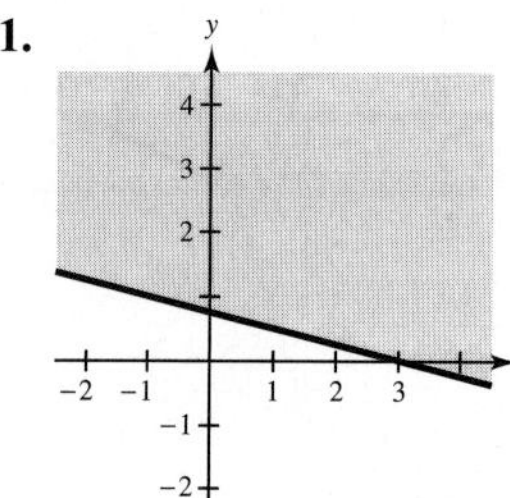

72.

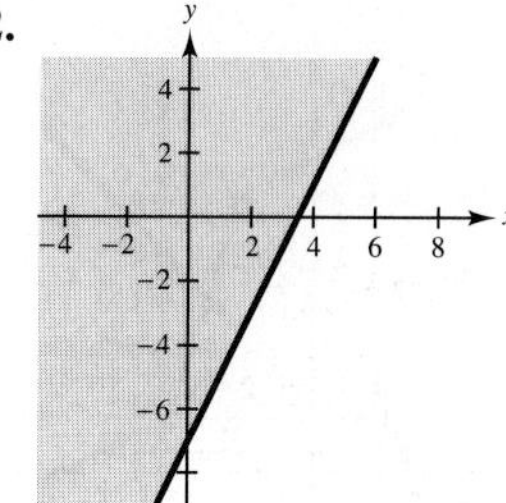

73. (b)

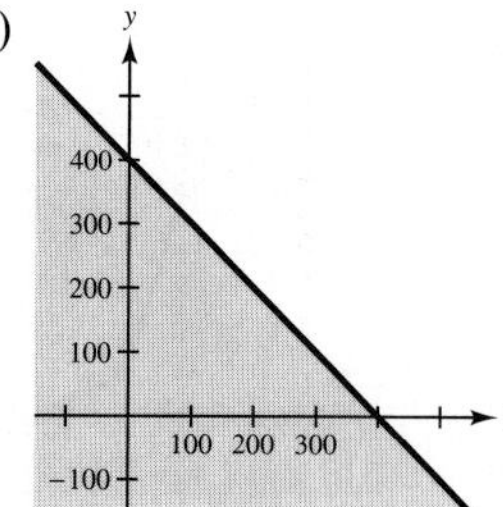

74. (b)

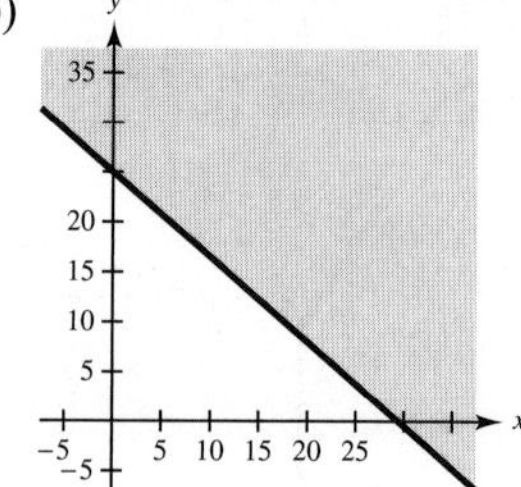

Sample answer: (x, y): (10, 17), (12, 15), (15, 13)

75.

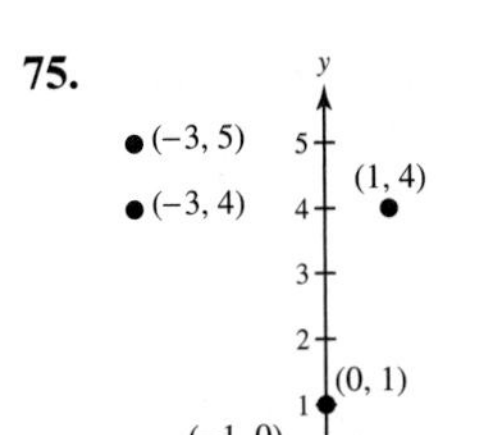

76.

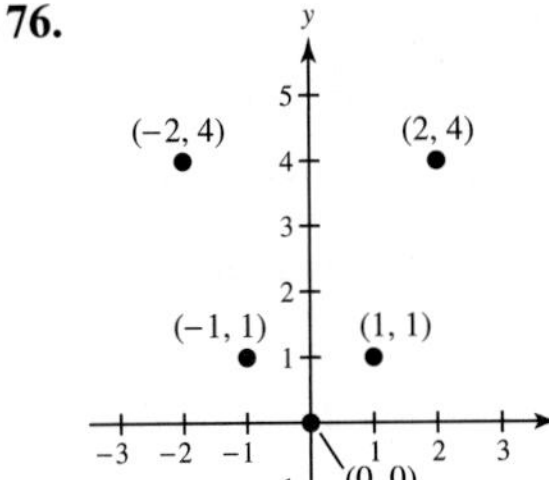

91.

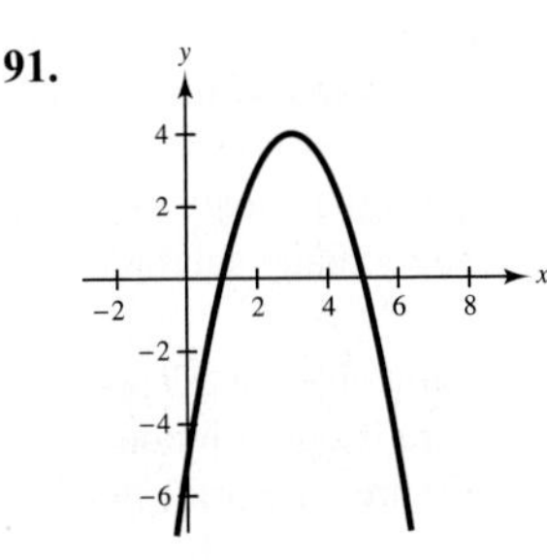

92.

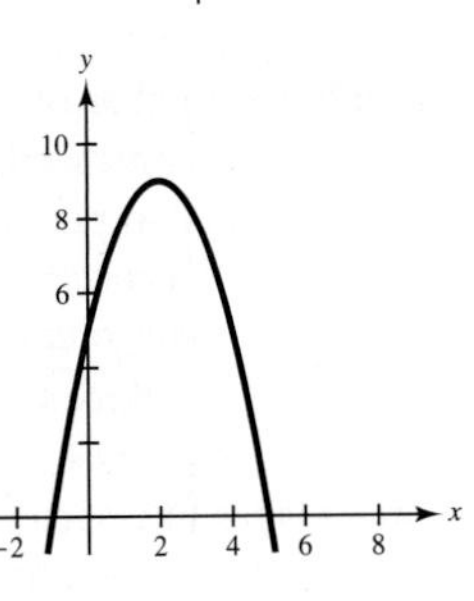

93.

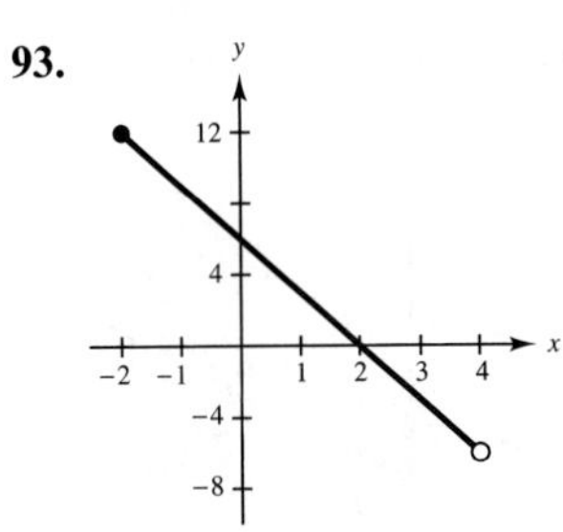

94.

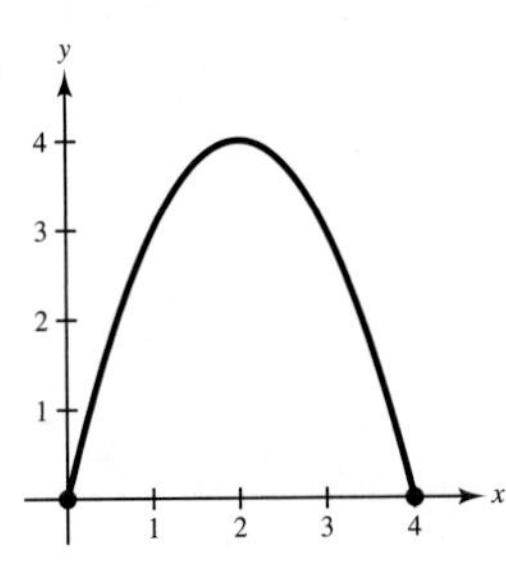

95.

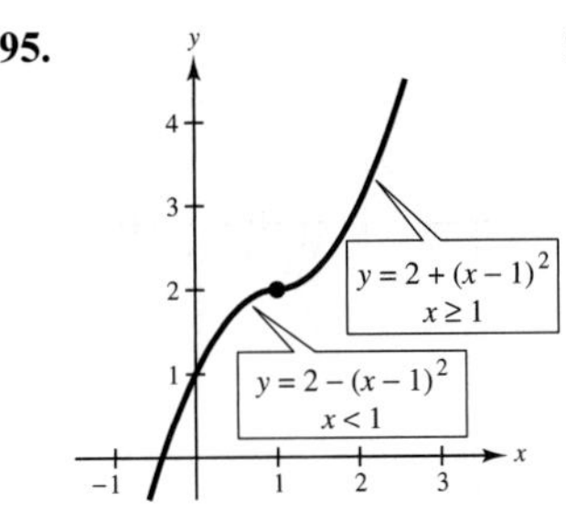

96.

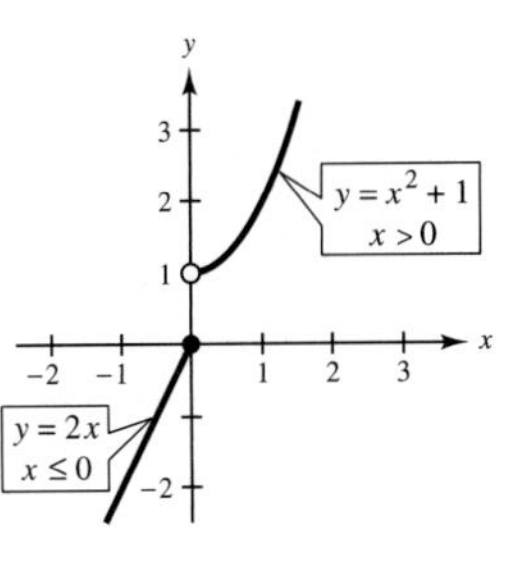

103.

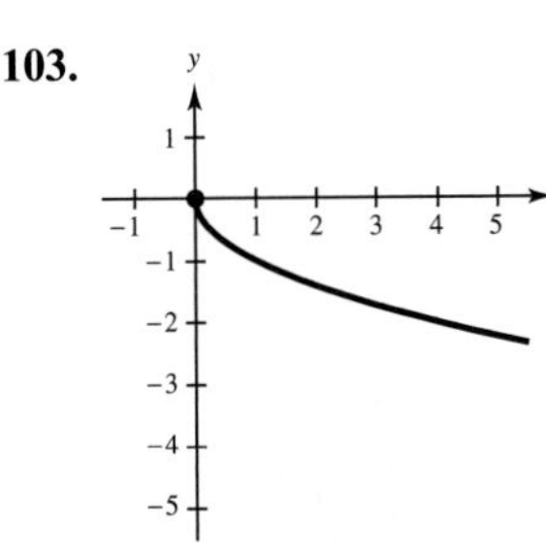

104.

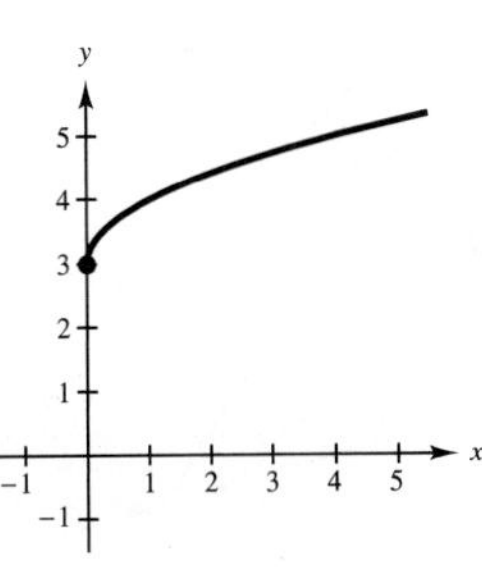

105. **106.**

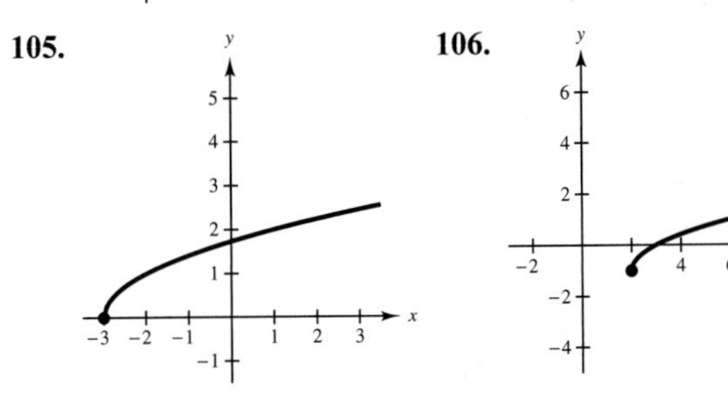

Chapter Test *(p. 166)*

2.

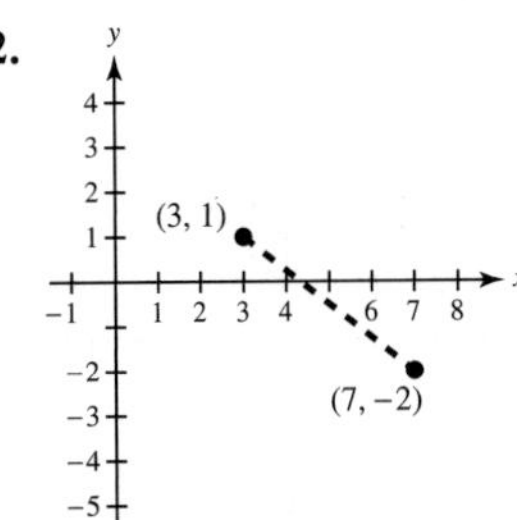

4.

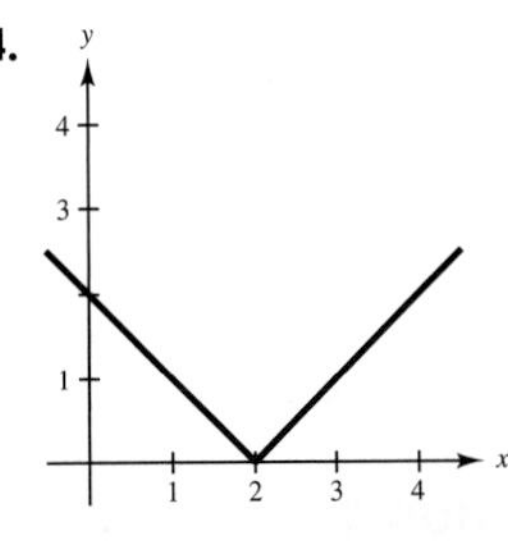

6.

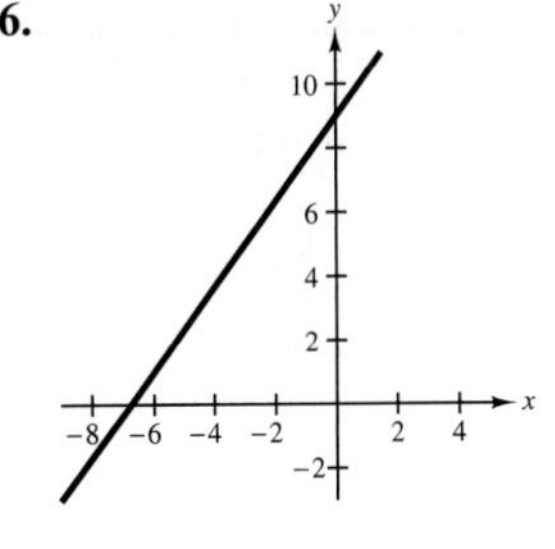

7.

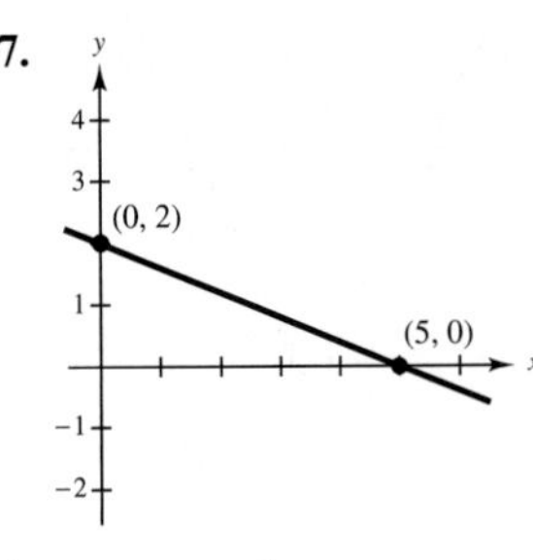

11.

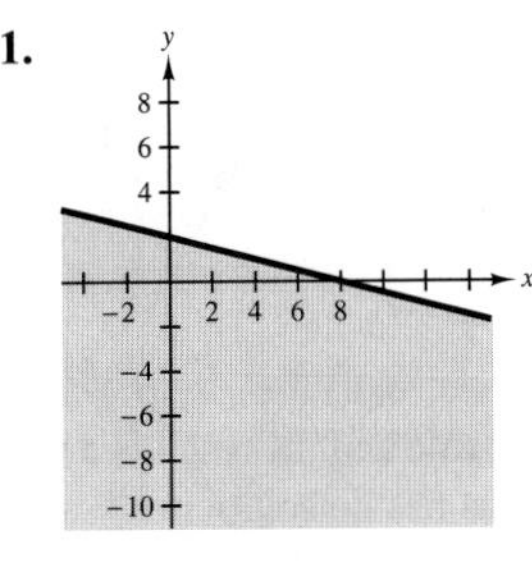

16.

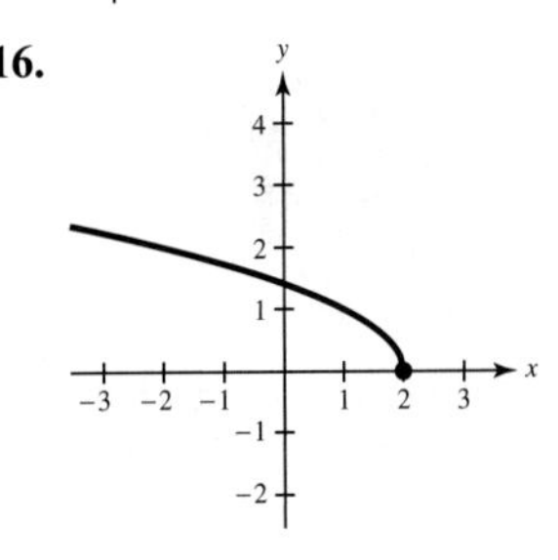

CHAPTER 4

Section 4.1 *(pp. 168–175)*

65. *Sample answer:* Pick any coefficient for each variable, then substitute the desired solution for the corresponding variable to find the right-hand side of the equation. Repeat this process for the second equation in the system.

Example: Solution (1, 4)

Equation 1: $-x + 3y$

$$-(1) + 3(4) = 11$$

Equation 2: $2x - y$

$$2(1) - (4) = -2$$

System: $\begin{cases} -x + 3y = 11 \\ 2x - y = -2 \end{cases}$

70.

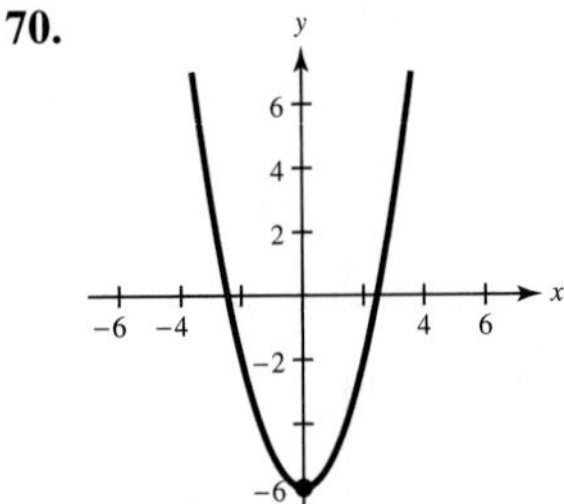

71.

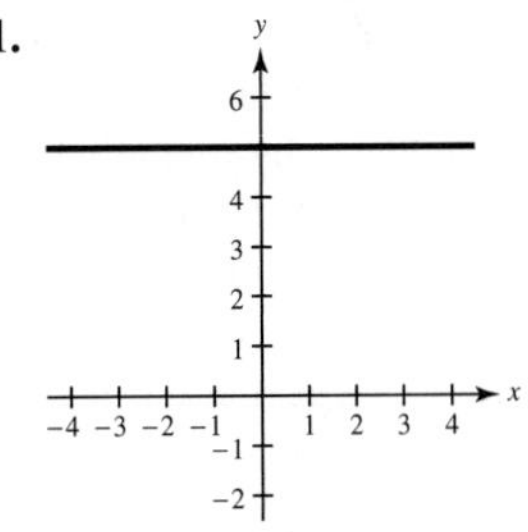

72.

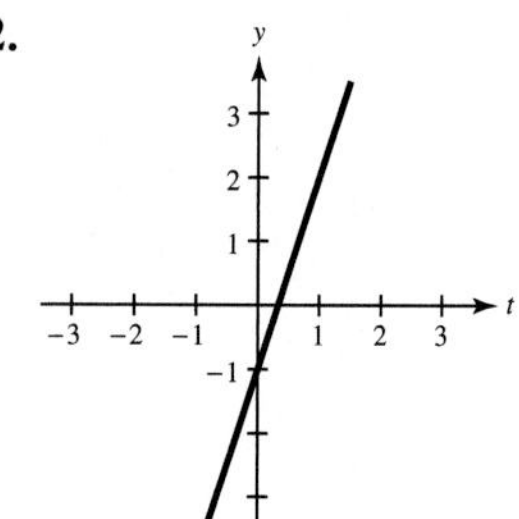

73.

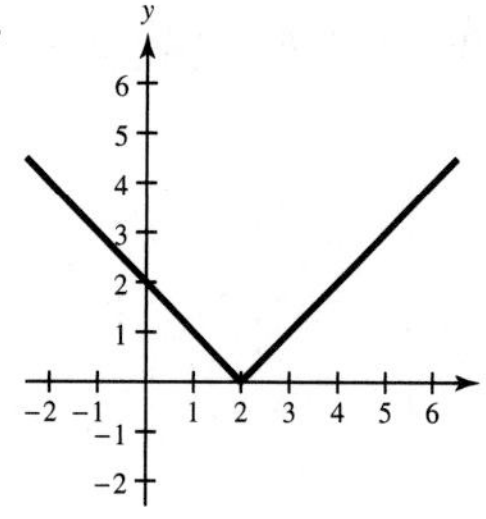

Section 4.2 (pp. 176–183)

25. 550 miles per hour is the speed of the plane; 50 miles per hour is the wind speed.
26. 540 miles per hour is the speed of the plane; 60 miles per hour is the wind speed.

29. (b)

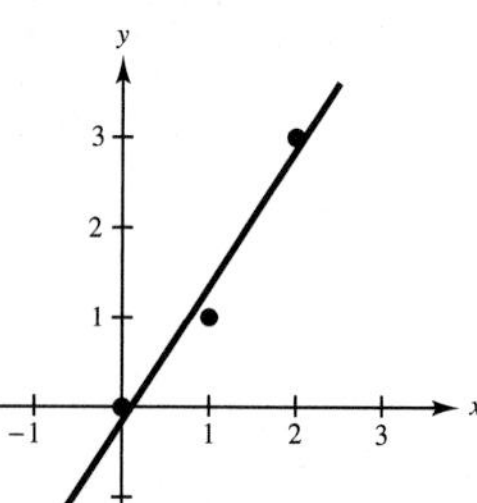

30. (b)

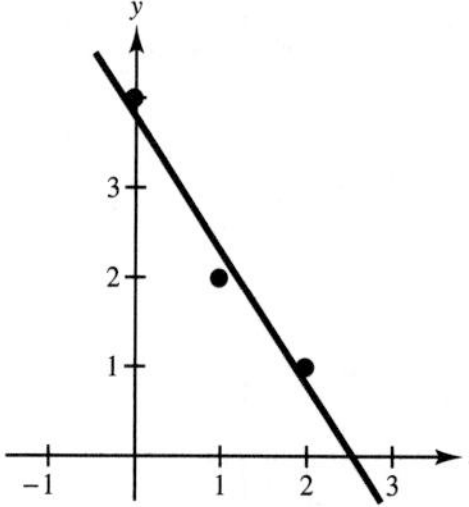

52. When a nonzero multiple of one equation is added to another equation to eliminate a variable and the result is a false statement, such as $0 = 5$, the system has no solution.

$$\begin{cases} x - y = 3 \\ x - y = 8 \end{cases} \Rightarrow \begin{array}{r} -x + y = -3 \\ x - y = 8 \\ \hline 0 = 5 \end{array}$$

53. (1) Obtain coefficients for x (or y) that are opposites by multiplying all terms of one or both equations by suitable constants.
    (2) Add the equations to eliminate one variable, and solve the resulting equation.
    (3) Back-substitute the value obtained in Step (2) into either of the original equations and solve for the other variable.
    (4) Check your solution in both of the original equations.

Section 4.3 (pp. 184–191)

29. Gaussian elimination enables you to obtain an equivalent system in row-echelon form. The next steps involve the use of back-substitution to find the solution.
30. An equivalent system can be produced by interchanging two equations, by multiplying one of the equations by a nonzero constant, or by adding a multiple of one of the equations to another equation to replace the latter equation.
33. Yes. The first equation was multiplied by -2 and added to the second equation. Then the first equation was multiplied by -3 and added to the third equation.
34. Yes. The first equation was added to the second equation. Then the first equation was multiplied by -2 and added to the third equation.
35. $\begin{cases} x + 2y - z = -4 \\ y + 2z = 1 \\ 3x + y + 3z = 15 \end{cases}$
36. $\begin{cases} x - y + z = -12 \\ x + y + z = 2 \\ -2x - 4y + 3z = -68 \end{cases}$
43. The solution is apparent because the row-echelon form is $\begin{cases} x = 1 \\ y = -3. \\ z = 4 \end{cases}$
46. If two equations in the system are dependent, either there are an infinite number of solutions and the graph of the solution is two planes intersecting at a line, or there is no solution and the graph is two parallel planes. If all three equations in the system are dependent, there are infinitely many solutions and the graph of the solution is one plane.
47. The graphs are three planes with three possible situations. If all three planes intersect in one point, there is one solution. If all three planes intersect in one line, there are an infinite number of solutions. If each pair of planes intersects in a line, but the three lines of intersection are all parallel, there is no solution.

Mid-Chapter Quiz (p. 192)

2.

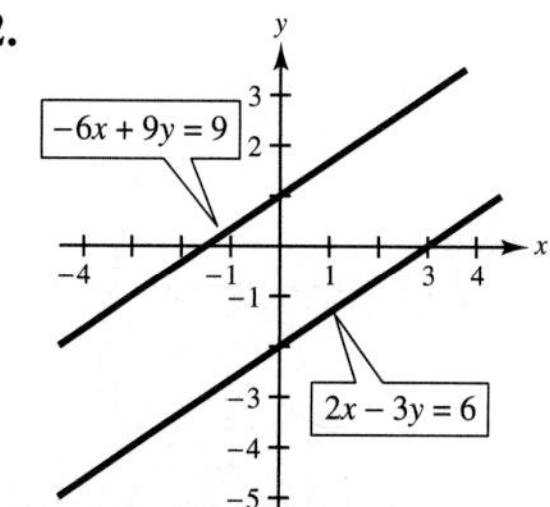

3.

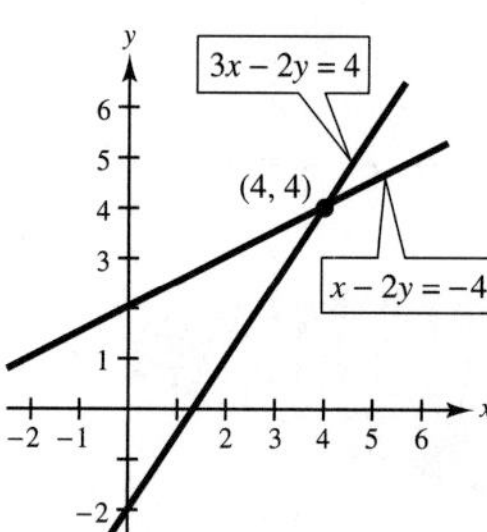

4.

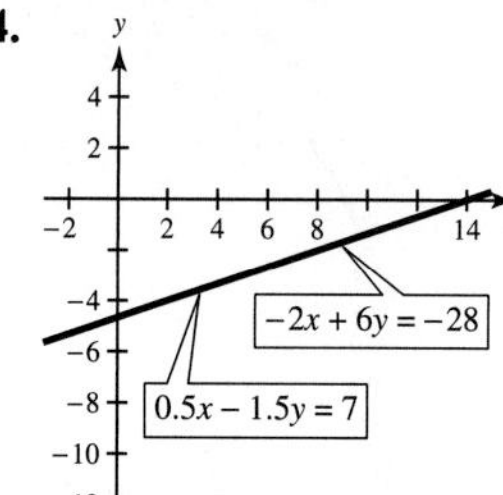

5.

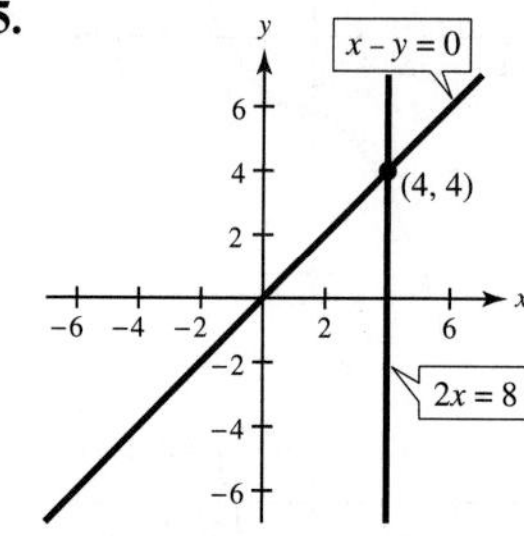

6.

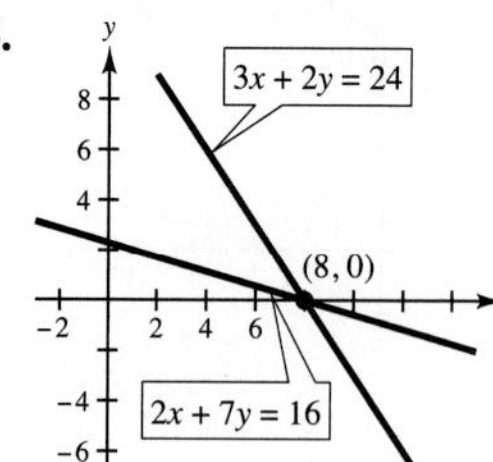

7.

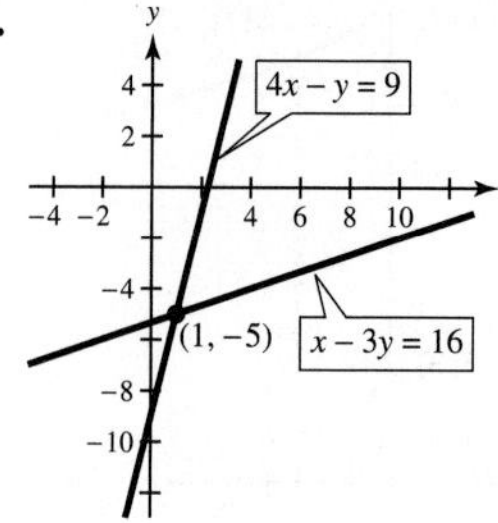

15. $\begin{cases} x + y = -2 \\ 2x - y = 32 \end{cases}$
16. $\begin{cases} x + y + z = 7 \\ 2x - y = 9 \\ -2x + y + 3z = 21 \end{cases}$

Section 4.4 (pp. 194–203)

13. (a) $\begin{bmatrix} 1 & 1 & 0 \\ 5 & -2 & -2 \\ 2 & 4 & 1 \end{bmatrix}$ (b) $\left[\begin{array}{rrr|r} 1 & 1 & 0 & 0 \\ 5 & -2 & -2 & 12 \\ 2 & 4 & 1 & 5 \end{array}\right]$

14. (a) $\begin{bmatrix} 9 & -3 & 1 \\ 12 & 0 & -8 \\ 3 & 4 & -1 \end{bmatrix}$ (b) $\left[\begin{array}{ccc:c} 9 & -3 & 1 & 13 \\ 12 & 0 & -8 & 5 \\ 3 & 4 & -1 & 6 \end{array}\right]$

41. 1×4 (one row and four columns), 2×2 (two rows and two columns), 4×1 (four rows and one column)

43. Row operations are performed on systems of equations and elementary row operations are performed on matrices.

59. The first entry in the first column is 1, and the other two are zero. In the second column, the first entry is a nonzero real number, the second number is 1, and the third number is zero. In the third column, the first two entries are nonzero real numbers and the third entry is 1.

Section 4.5 *(pp. 204–213)*

81.

82.

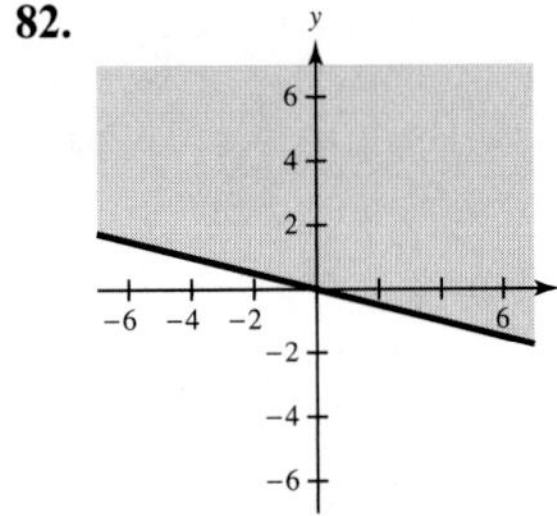

83.

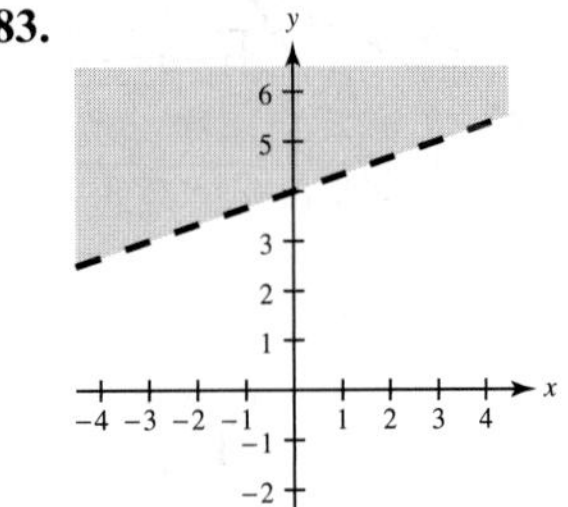

84.

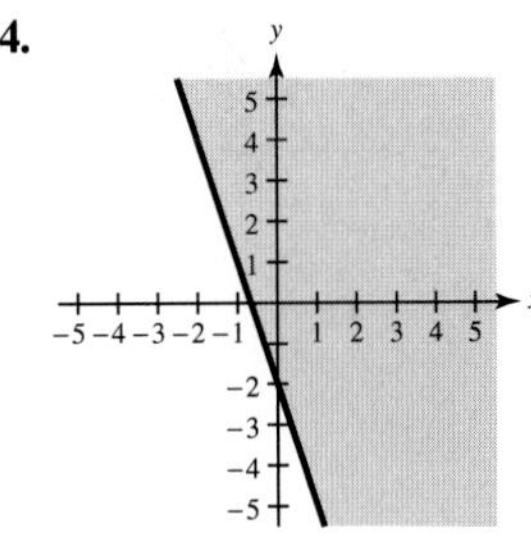

87.

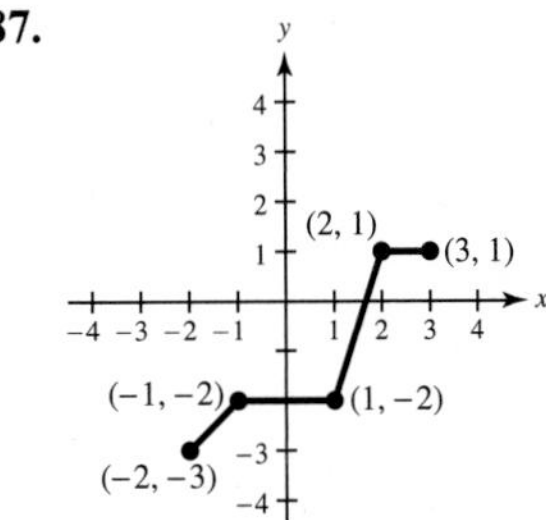

88.

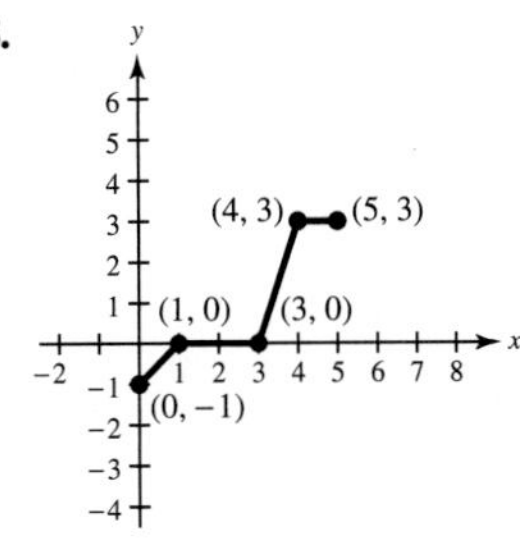

89.

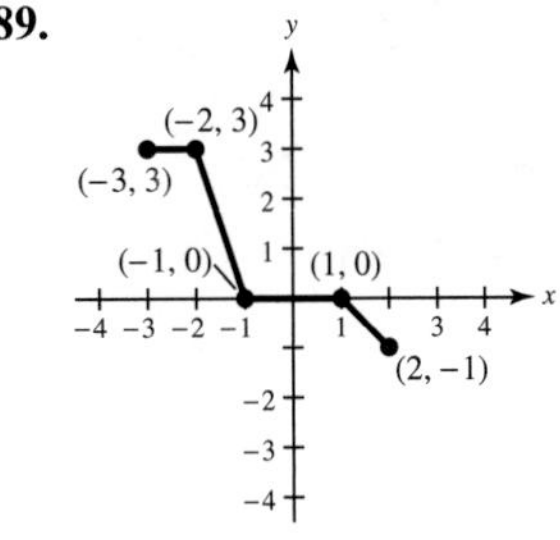

90.

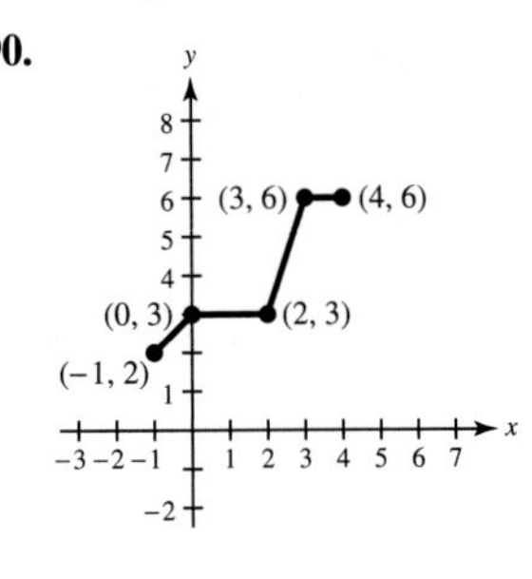

Section 4.6 *(pp. 214–221)*

1.

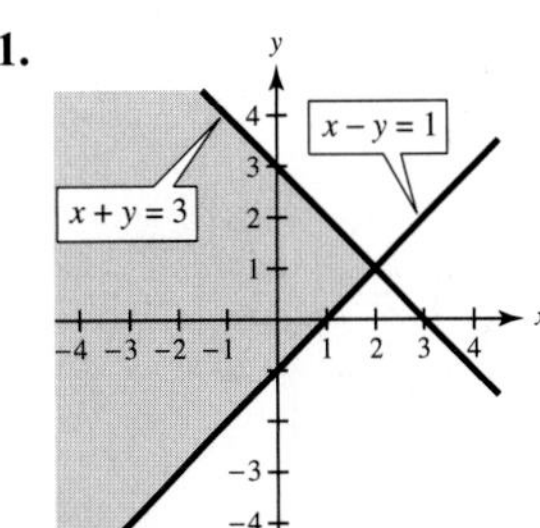

2.

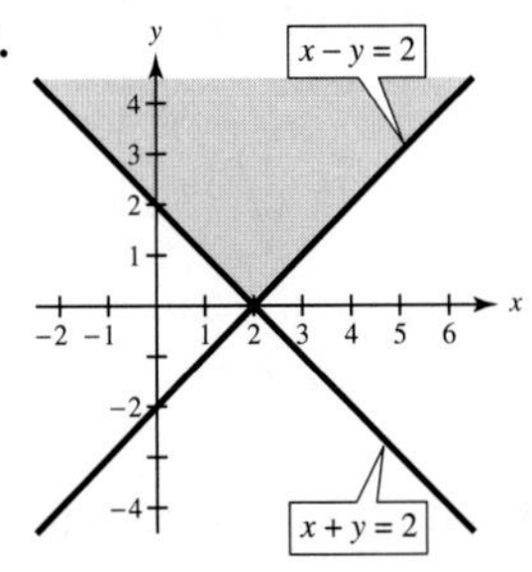

3.

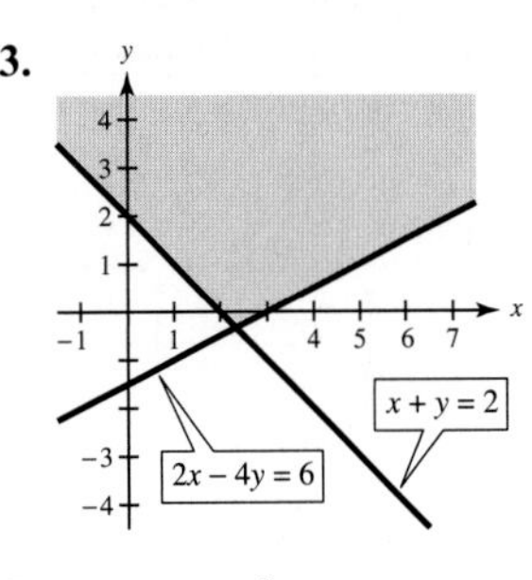

4.

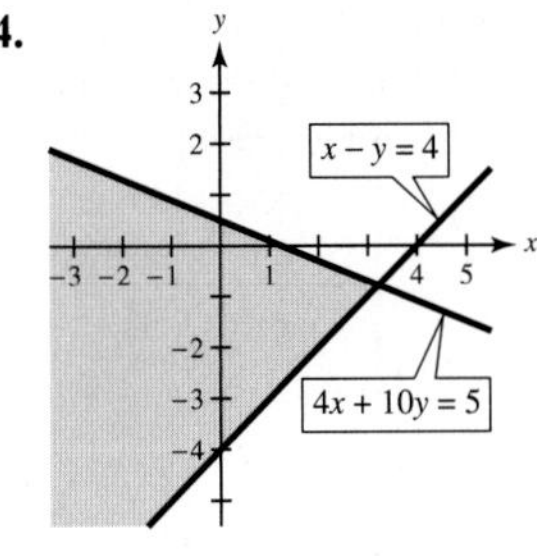

5.

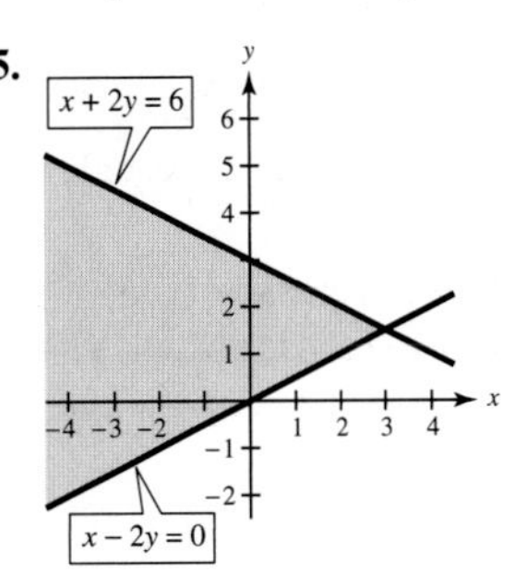

6.

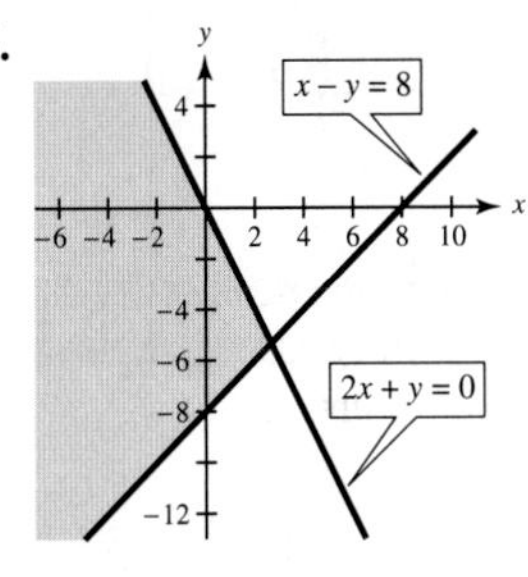

7.

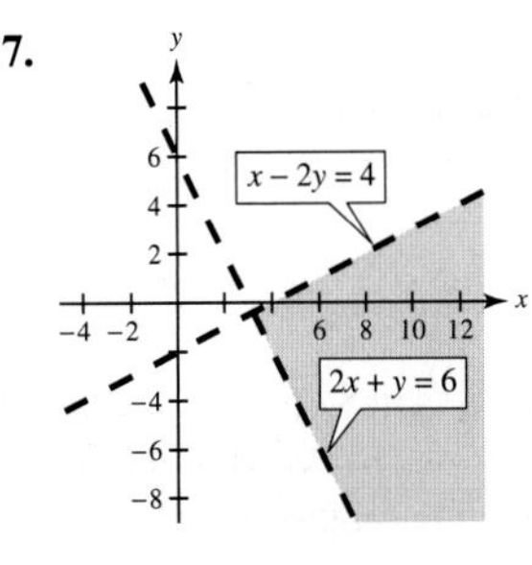

8.

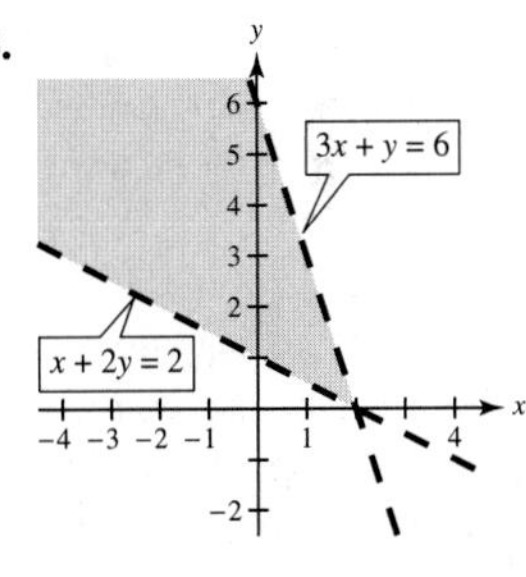

9.

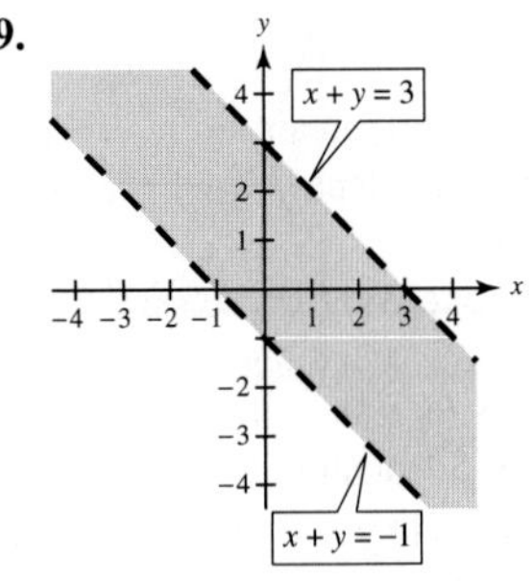

10. No solution

11.

12.

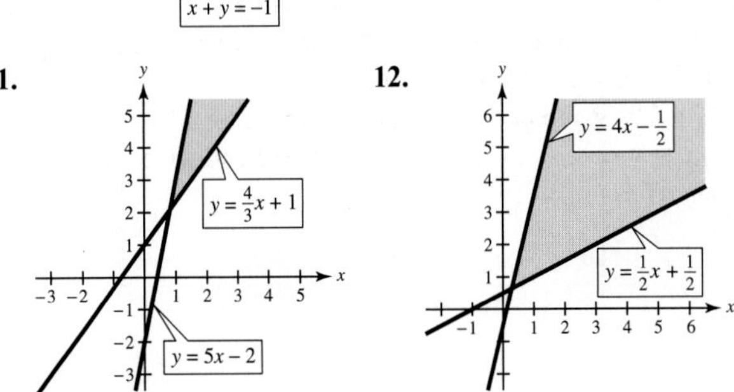

13.

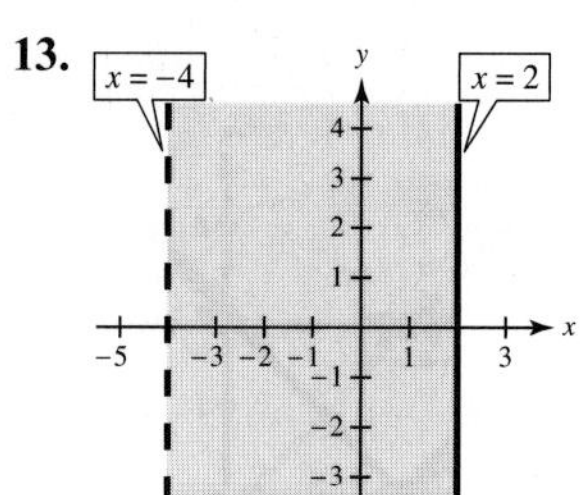

14.

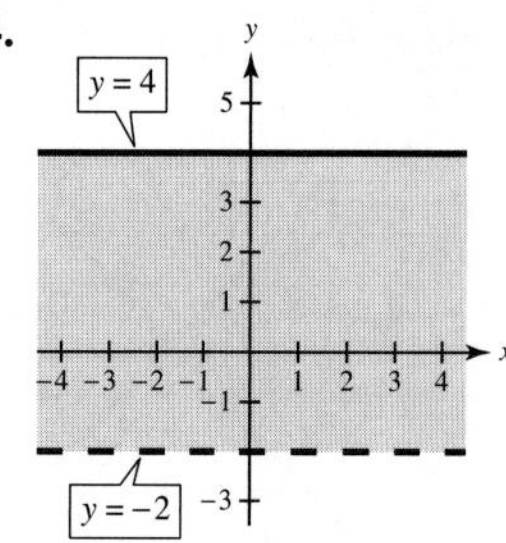

15.

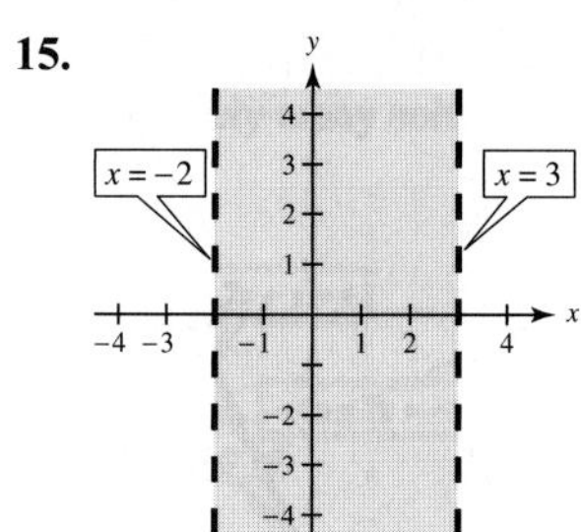

16.

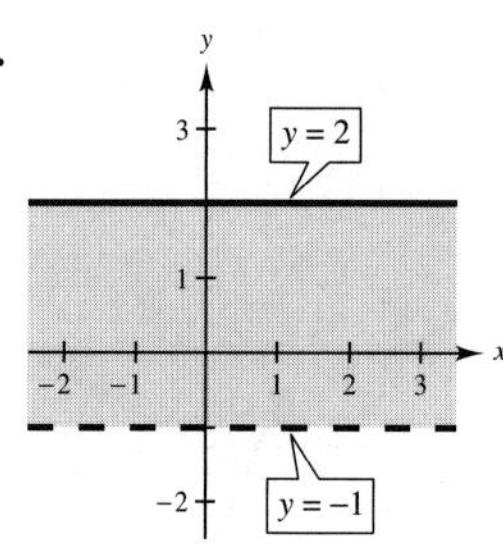

17.

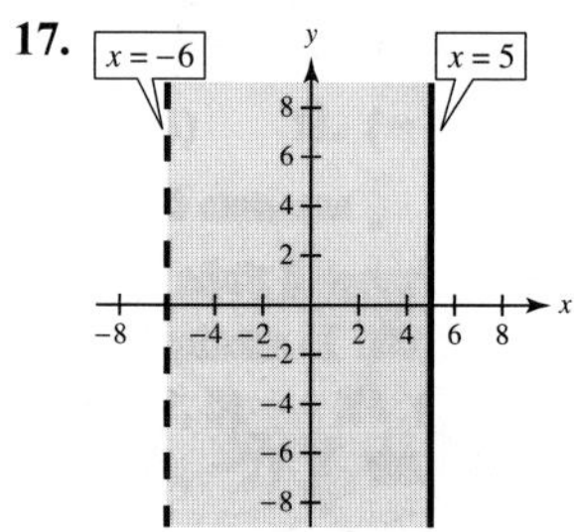

18.

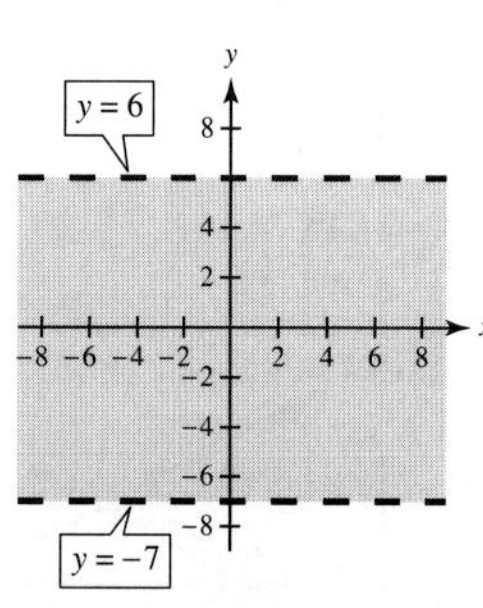

19.

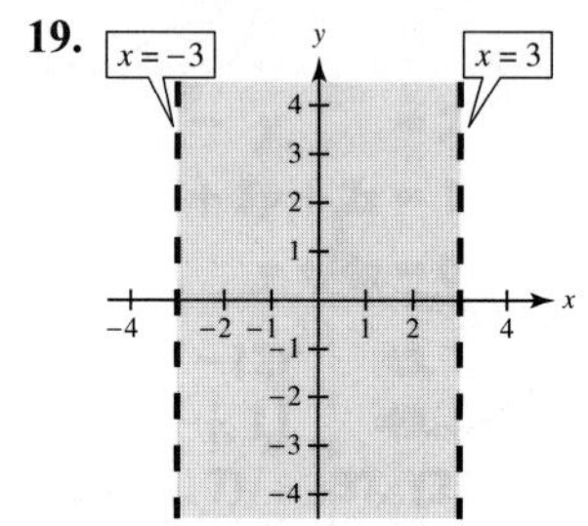

20.

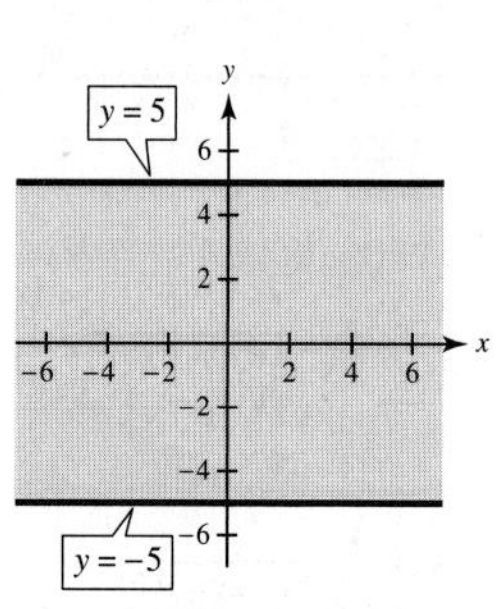

21.

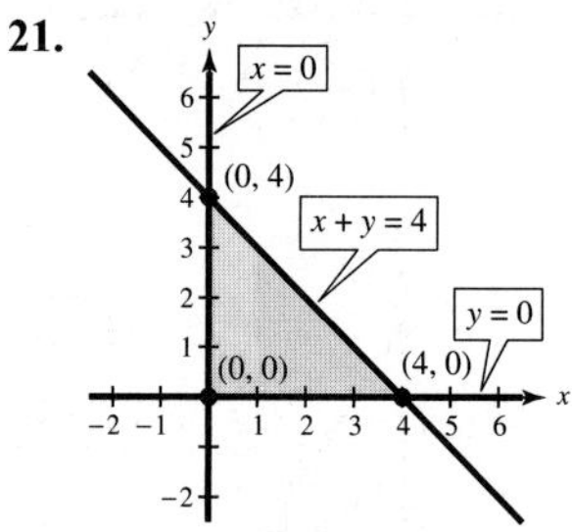

22.

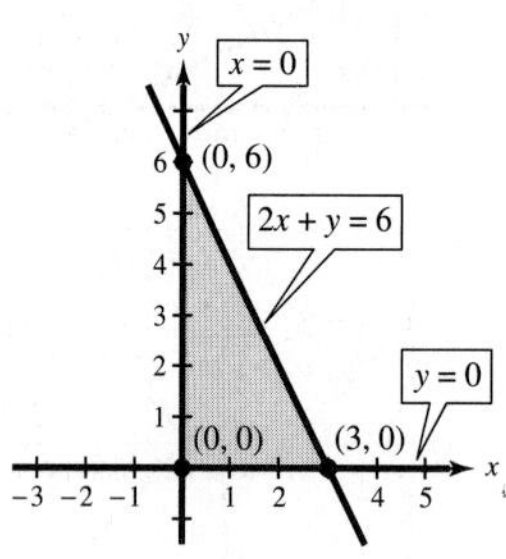

23.

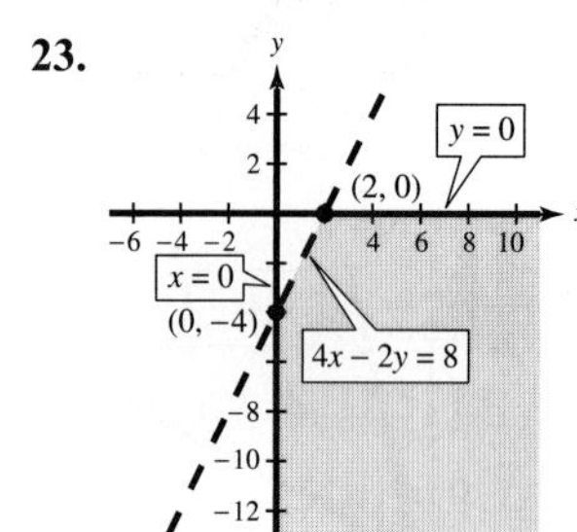

24.

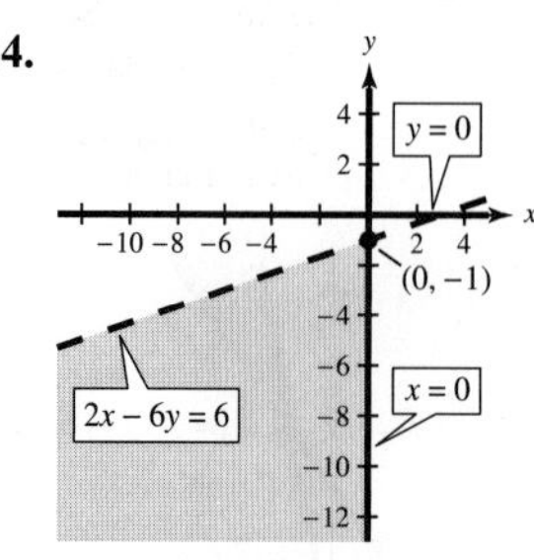

25. 26.

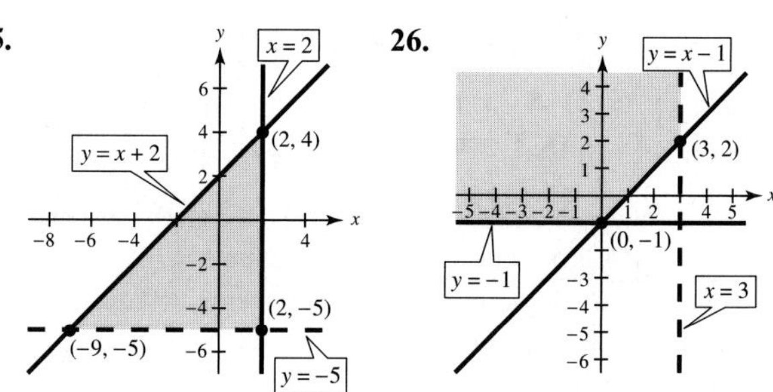

27. 28.

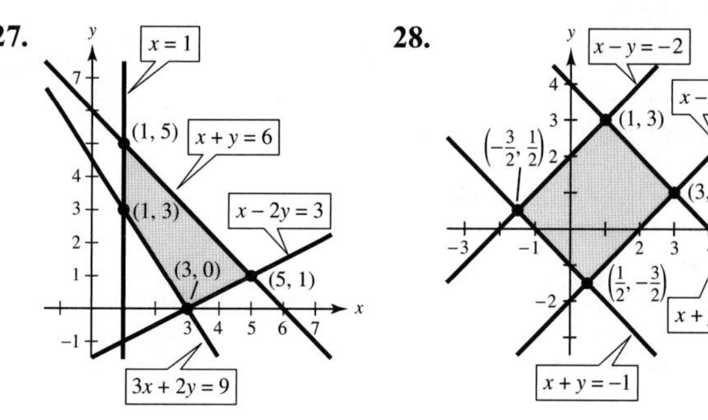

29. 30.

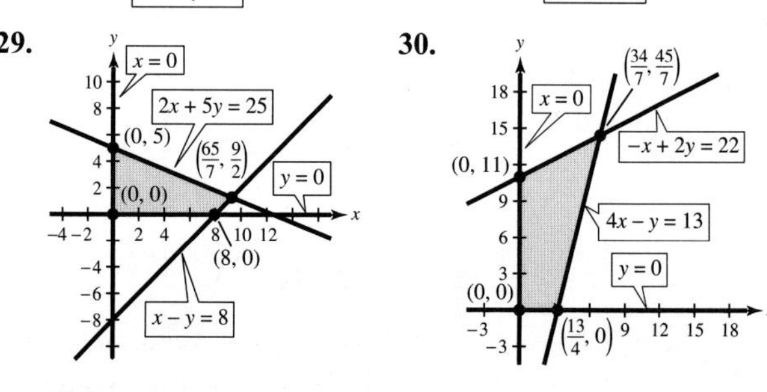

35. $\begin{cases} x + \frac{3}{2}y \le 12 \\ \frac{4}{3}x + \frac{3}{4}y \le 16 \\ x \ge 0 \\ y \ge 0 \end{cases}$

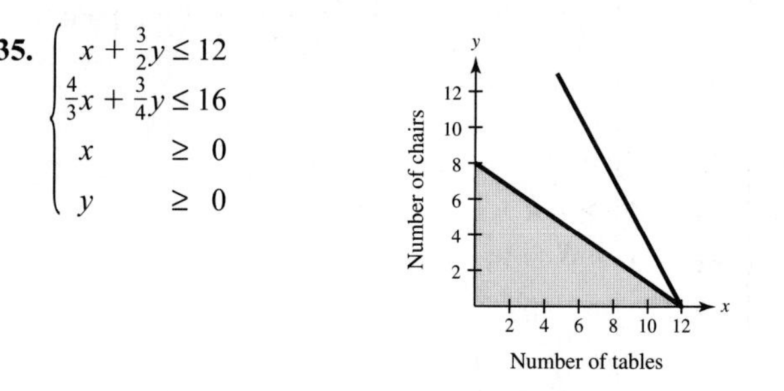

36. $\begin{cases} 3x + \frac{5}{2}y \le 20 \\ \frac{5}{4}x + y \le 16 \\ x \ge 0 \\ y \ge 0 \end{cases}$

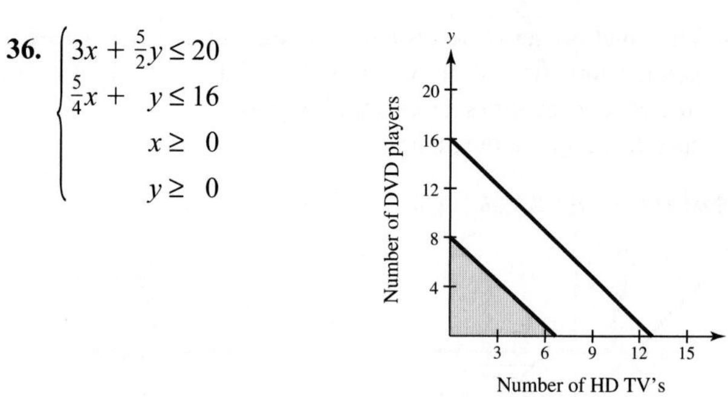

37. A system of linear inequalities in two variables consists of two or more linear inequalities in two variables.

38. A dashed line should be used when the inequality contains $<$ or $>$ and a solid line should be used when the inequality contains $\le$ or $\ge$.

39. Not necessarily. Two boundary lines can intersect outside the solution region at a point that is not a vertex of the region.

40. Yes. If the half-planes of any two graphs in the system do not intersect, the system has no solution.

41.

$x + y = 1$ $-x + y = 1$ $y = 0$

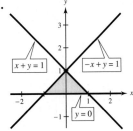

42.

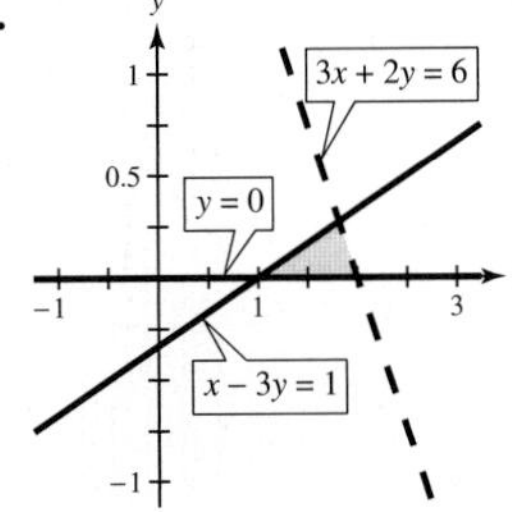

43. No solution

44.

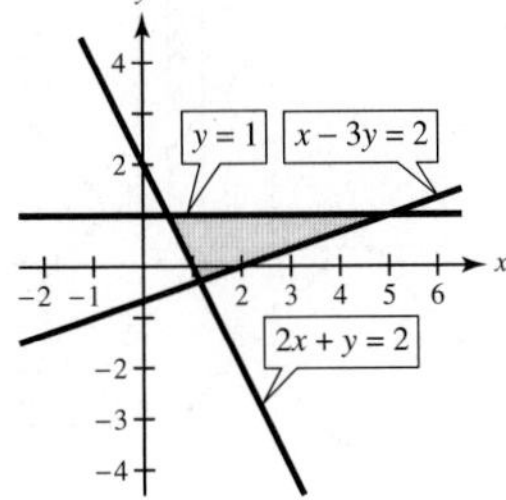

45.

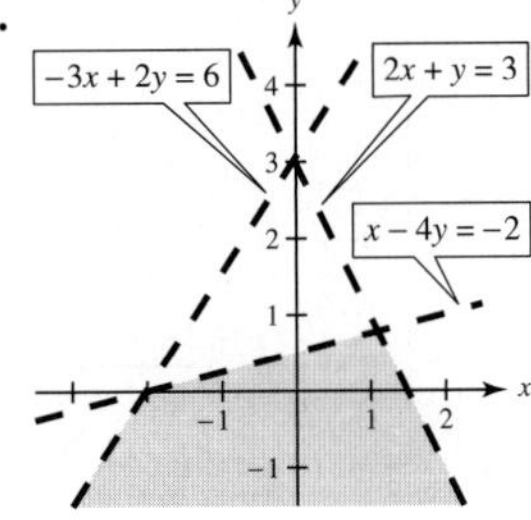

46. No solution

47.

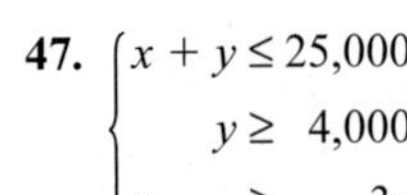

$$\begin{cases} x + y \le 25{,}000 \\ y \ge 4{,}000 \\ x \ge 3y \end{cases}$$

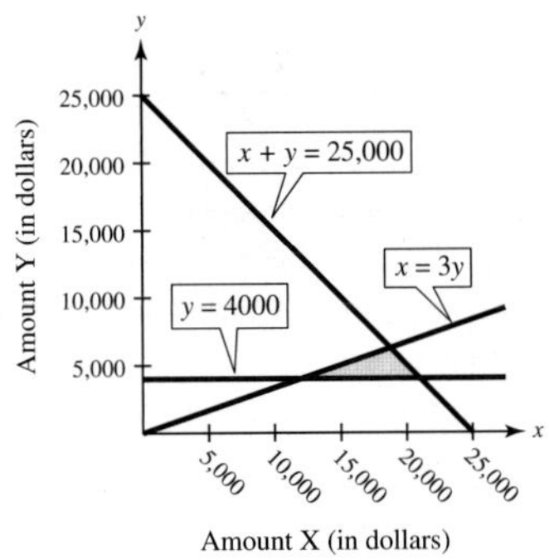

48. $$\begin{cases} 12x + 10y \ge 200 \\ 8x + 10y \ge 100 \\ 6x + 8y \ge 120 \\ x \ge 0 \\ y \ge 0 \end{cases}$$

55. The solution set of a system of linear equations is either a point, a line, or a plane, whereas the solution set of a system of linear inequalities is an infinite number of solutions in a specific region of the plane.

Review Exercises *(pp. 224–226)*

3.

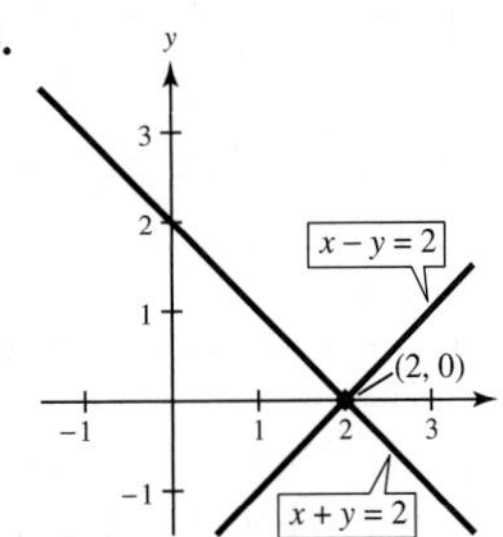

4.

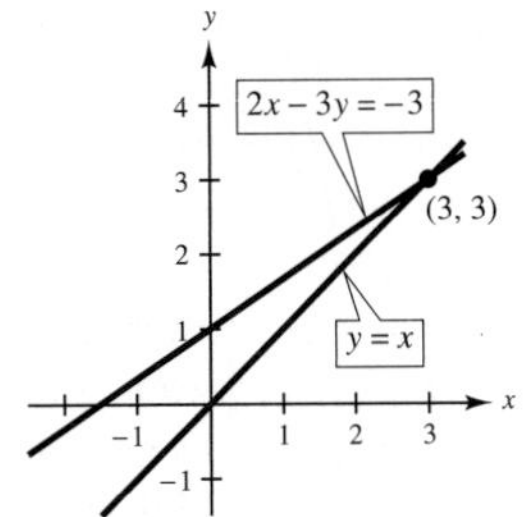

5.

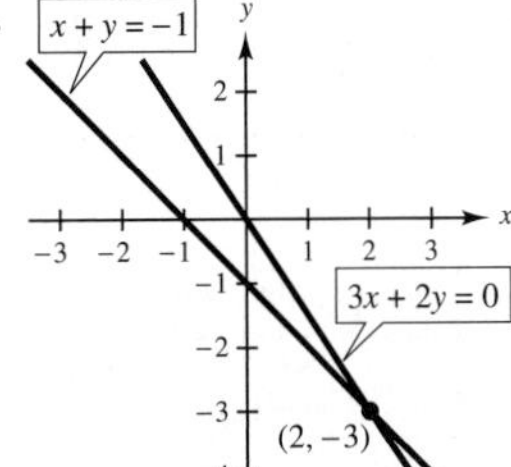

6.

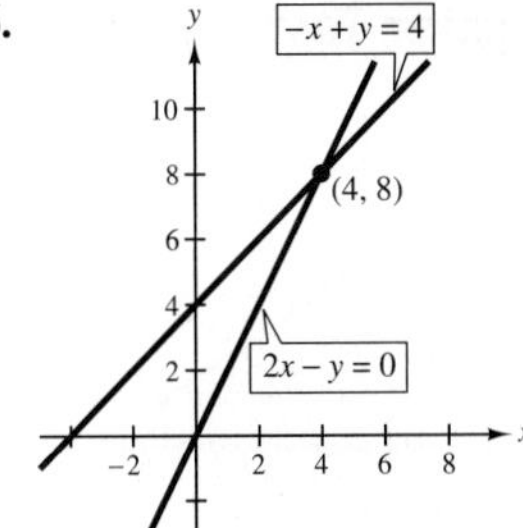

7.

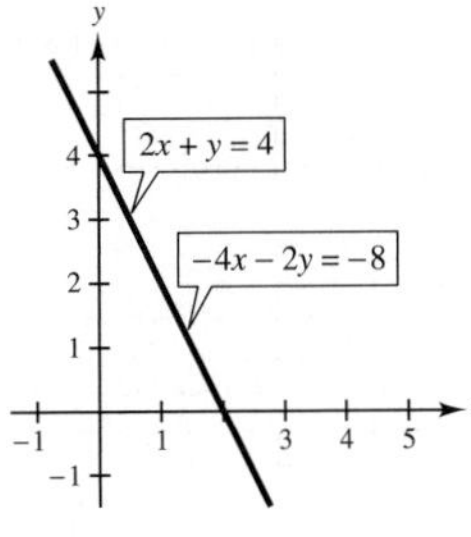

8.

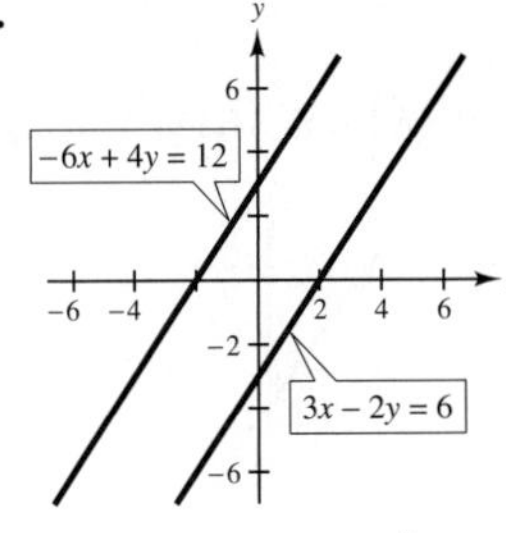

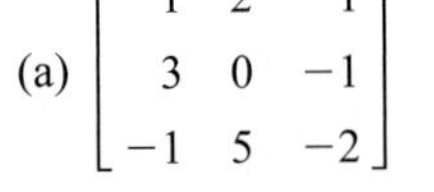

38. (a) $\begin{bmatrix} 1 & 2 & 1 \\ 3 & 0 & -1 \\ -1 & 5 & -2 \end{bmatrix}$ (b) $\left[\begin{array}{ccc|c} 1 & 2 & 1 & 4 \\ 3 & 0 & -1 & 2 \\ -1 & 5 & -2 & -6 \end{array}\right]$

67.

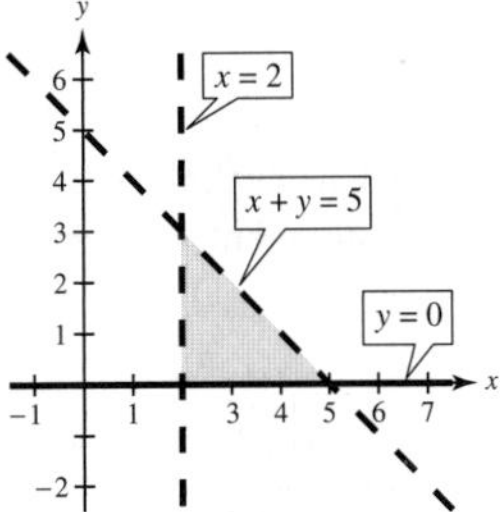

68.

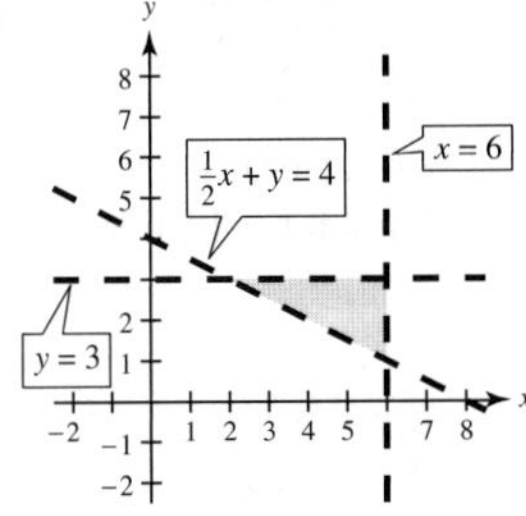

69.

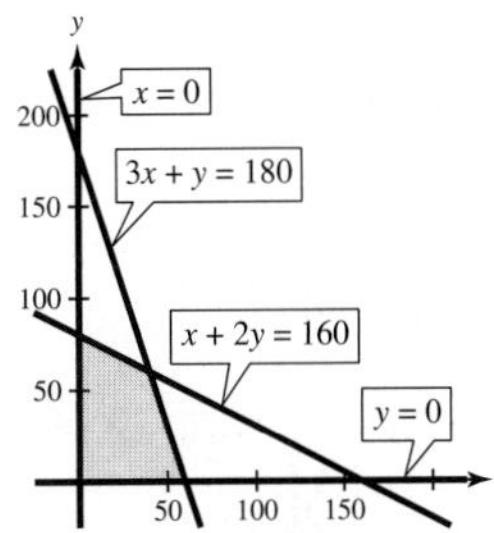

70.

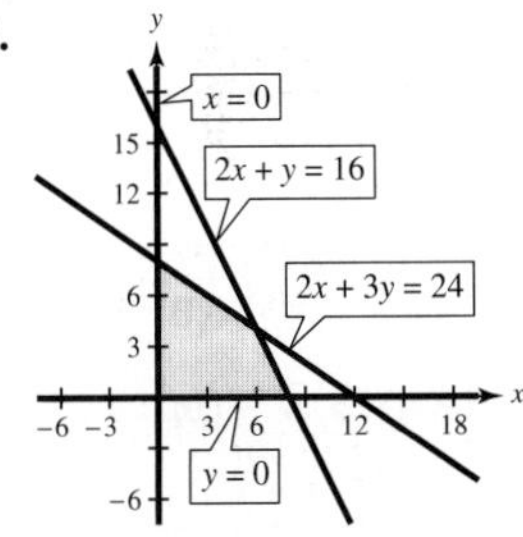

71.

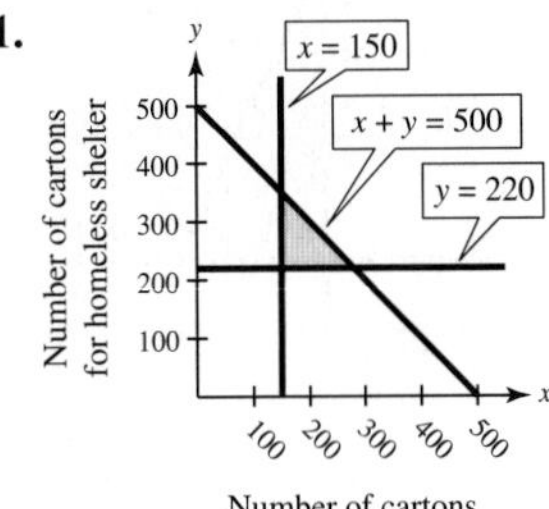

72.

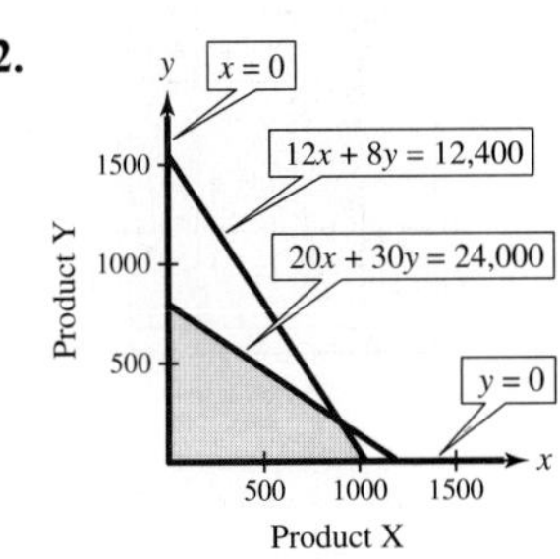

ADDITIONAL ANSWERS

Chapter Test (p. 227)

14.

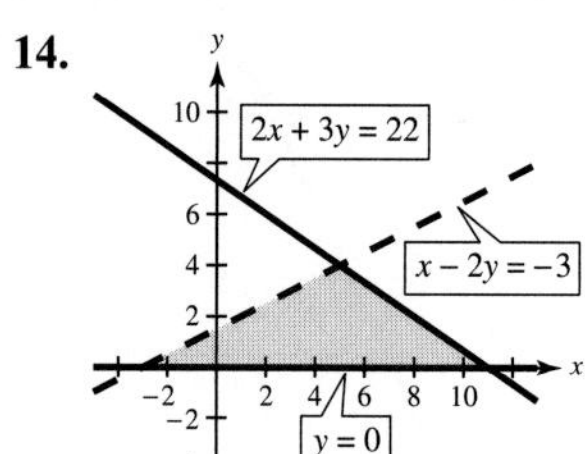

17.

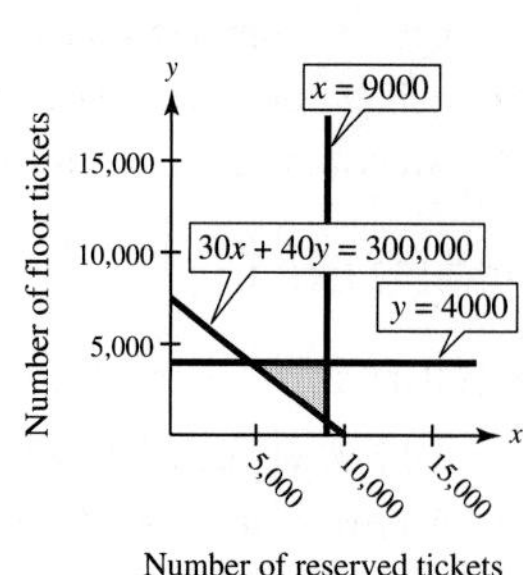

Cumulative Test: Chapters 1–4 (p. 228)

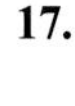

17.

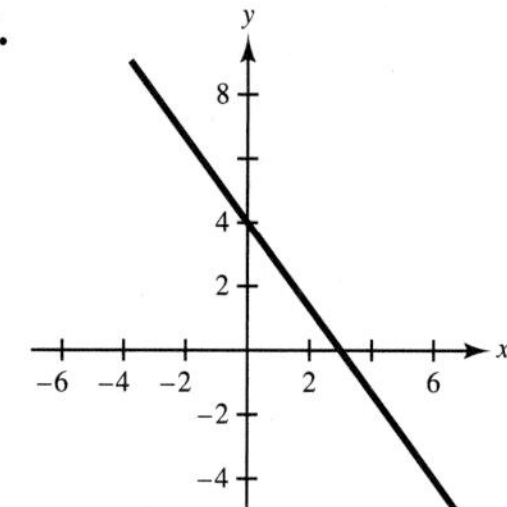

18.

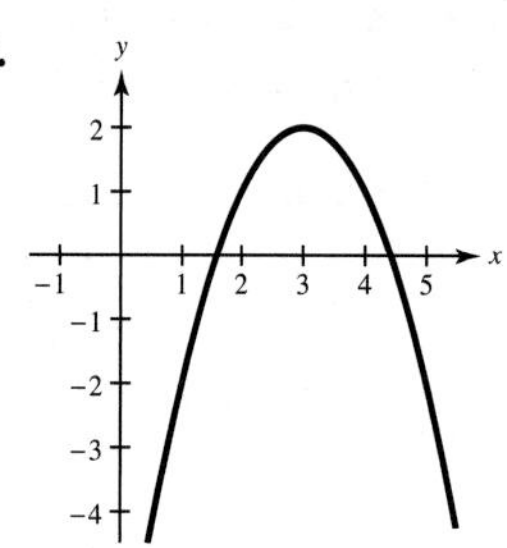

CHAPTER 5

Section 5.1 (pp. 230–237)

104. False. Because 0.142 is not between 1 and 10, the expression is not in scientific notation. The equivalent expression in scientific notation is 1.42×10^9.

114.

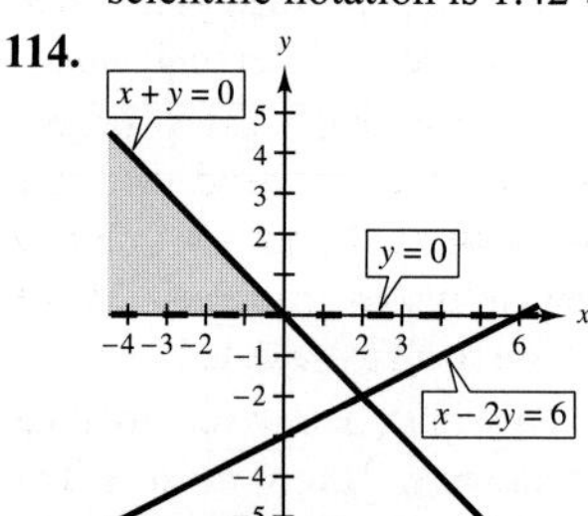

Section 5.3 (pp. 246–253)

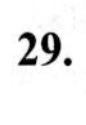

29.

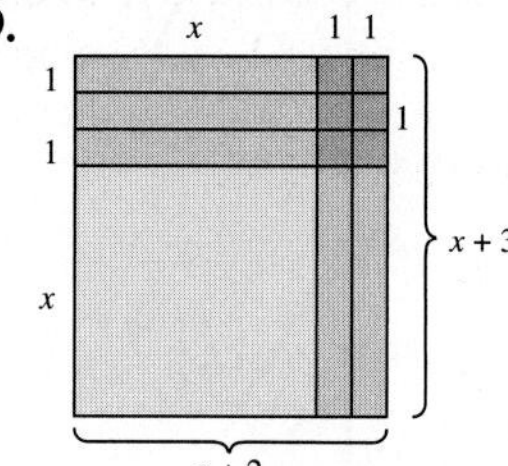

30.

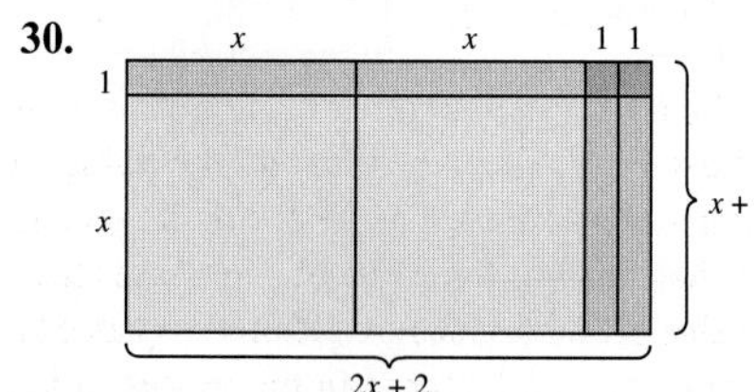

31.

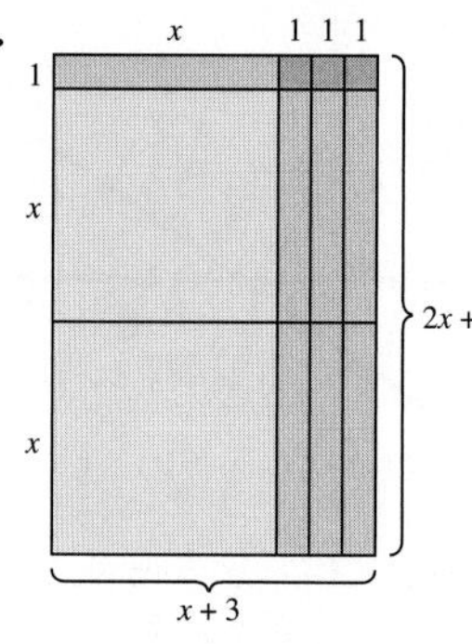

32.

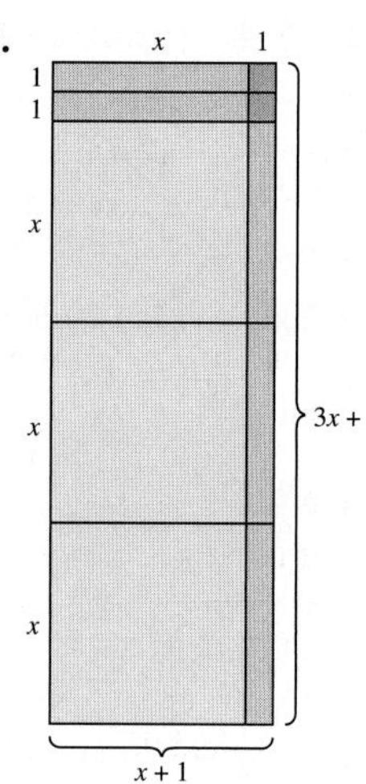

76.

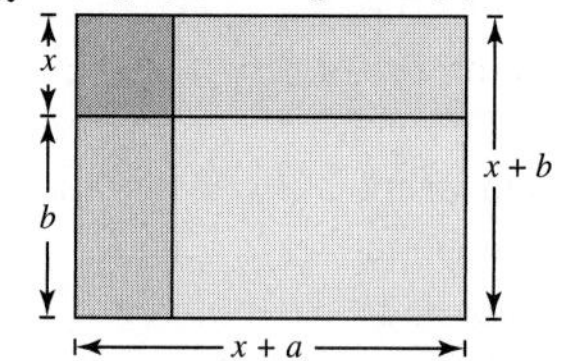

83. Multiplying a polynomial by a monomial is an application of the Distributive Property. Polynomial multiplication requires repeated use of this property.

84. When the FOIL Method is used to square a binomial there are four terms, but the outer and inner terms are like terms. Combining these like terms leaves a total of three terms.

87.

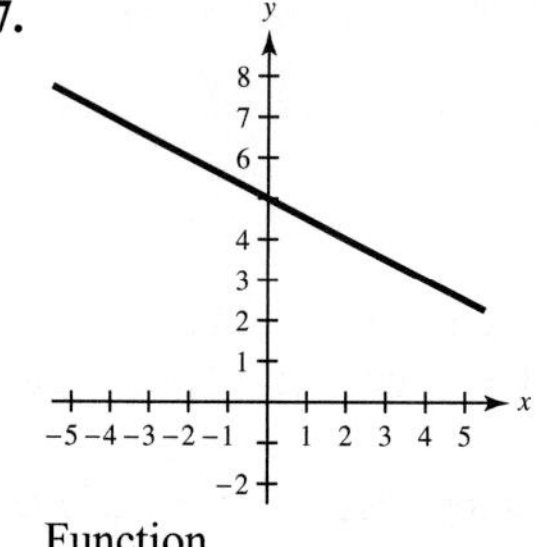

Function

88.

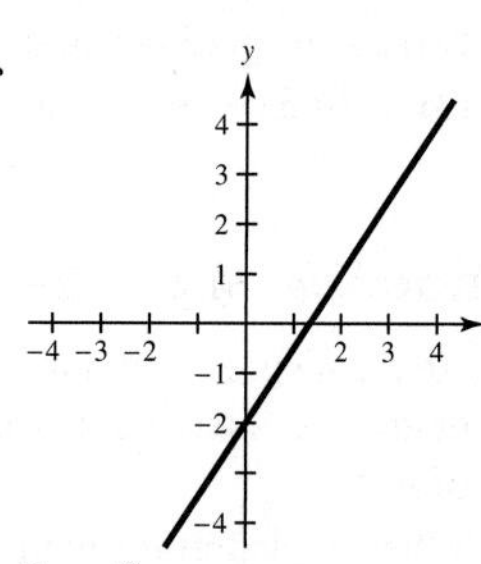

Function

89.

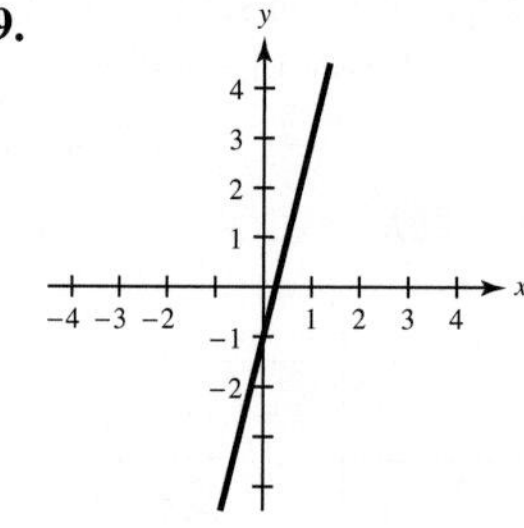

Function

90.

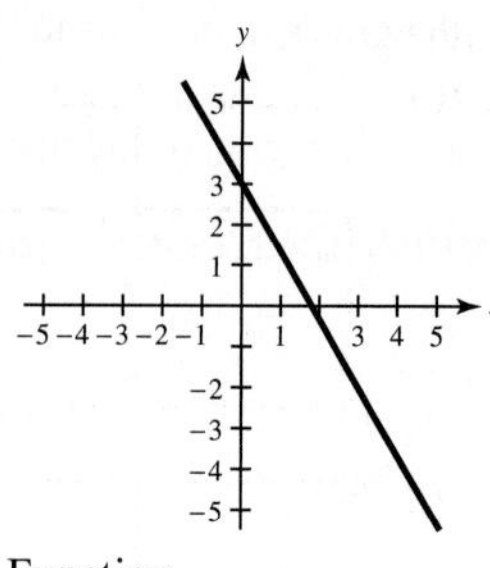

Function

91.

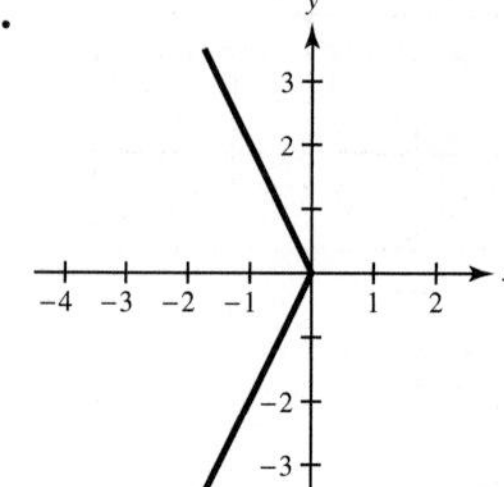

Not a function

92.

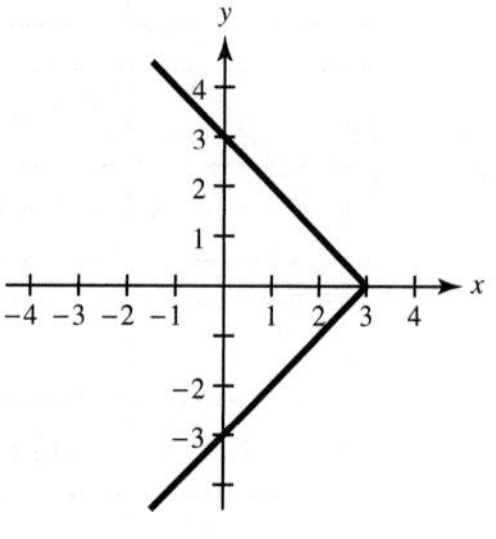

Not a function

Section 5.4 *(pp. 256–263)*

69. $x^2(3x + 4) - (3x + 4) = (3x + 4)(x + 1)(x - 1)$
$3x(x^2 - 1) + 4(x^2 - 1) = (x + 1)(x - 1)(3x + 4)$

70. $2x^2(3x - 4) + 3(3x - 4) = (3x - 4)(2x^2 + 3)$
$3x(2x^2 + 3) - 4(2x^2 + 3) = (2x^2 + 3)(3x -)$

77. (a) Entire cube: a^3, Solid I: $a^2(a - b)$, Solid II: $ab(a - b)$, Solid III: $b^2(a - b)$, Solid IV: b^3
(b) $a^2(a - b) + ab(a - b) + b^2(a - b)$
$= (a - b)(a^2 + ab + b^2)$
(c) When the smaller cube (solid IV) is removed, the remaining solid has a volume of $a^3 - b^3$ which is equal to $(a - b)(a^2 + ab + b^2)$, the sum of the volumes of solids I, II, and III.

78. Noun: any one of the expressions that, when multiplied together, yield the product
Verb: the process of finding the expressions that, when multiplied together, yield the given product

80. To find the greatest common factor, first determine the prime factorization of each monomial. The greatest common factor is the product of each common prime factor raised to its lowest power in any one of the monomials.

83. The polynomial (a) $x^2 + y^2$ cannot be factored at all, because a sum of two squares with no common monomial factors is prime with respect to the integers.

Section 5.5 *(pp. 264–271)*

67. Because 6 is positive, its factors have like signs that match the sign of 5 which is positive.

68. Difference of two squares, sum of two cubes, difference of two cubes, perfect square trinomial

100. $(3x + 6)(3x - 9) = 3(x + 2)3(x - 3)$
$= 9(x + 2)(x - 3)$

Section 5.6 *(pp. 272–279)*

92. An nth-degree polynomial can have at most n factors of first degree because the degree of the product of n linear factors of n.

93. When a quadratic equation has a repeated solution, the graph of the equation has one x-intercept, which is the vertex of the graph.

96. (a)

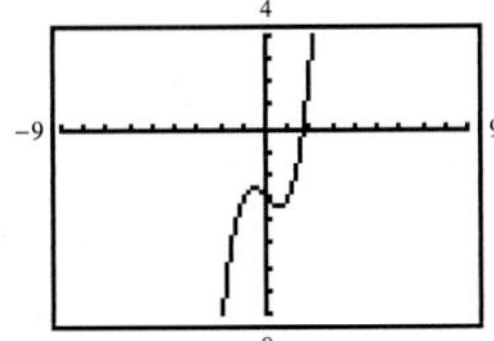

(b)

x	1.0	1.1	1.2	1.3
$x^3 - x - 3$	−3.0	−2.769	−2.472	−2.103

x	1.4	1.5	1.6	1.7
$x^3 - x - 3$	−1.656	−1.125	−0.504	0.213

x	1.8	1.9	2.0
$x^3 - x - 3$	1.032	1.959	3.0

101. All real numbers x such that $x \neq -1$

102. All real numbers x such that $x \neq 2$

103. All real numbers x such that $x \leq 3$

104. All real numbers x such that $x \leq -2$ or $x \geq 2$

CHAPTER 6

Section 6.1 *(pp. 288–295)*

1. All real values of x such that $x \neq 3$

2. All real values of x such that $x \neq 7$

3. All real values of z such that $z \neq 0$ and $z \neq 4$

4. All real values of x such that $x \neq 0$ and $x \neq 1$

5. All real values of t such that $t \neq -4$ and $t \neq 4$

6. All real values of x such that $x \neq -2$ and $x \neq 2$

Section 6.2 *(pp. 296–303)*

49. $Y = \dfrac{(6591.43t + 102{,}139.0)(0.029t + 1)}{0.184t + 4.3}, 4 \leq t \leq 9$

50. $Y = \dfrac{(4651.460t + 14{,}528.41)(0.018t + 1)}{(0.070t + 1)(0.028t + 0.88)}, 4 \leq t \leq 9$

51. To multiply two rational expressions, let a, b, c, and d represent real numbers, variables, or algebraic expressions such that $b \neq 0$ and $d \neq 0$. Then the product of $\frac{a}{b}$ and $\frac{c}{d}$ is $\frac{a}{b} \cdot \frac{c}{d} = \frac{ac}{bd}$.

54. If the second expression has any implied domain restrictions, when it is inverted during division, the denominator becomes the numerator, so any restrictions might be lost.

67. In simplifying a product of rational expressions, you divide the common factors out of the numerator and denominator.

68. *Sample answer:* If there is a power of x of second degree or higher in the denominator of the dividend and the numerators of the dividend and divisor do not include any powers of x, then $x \neq 0$ will be an implied restriction in the resulting quotient and so it will not need to be listed.

70. Multiplying two rational expressions involves dividing common factors out of the numerators and denominators. The result is the constant polynomial 1 when the factors in the numerators and denominators are identical. The result can be a polynomial other than 1 only when there are additional factors in the numerator after dividing out common factors.

71. The first expression needs to be multiplied by the reciprocal of the second expression (not the second by the reciprocal of the first), and the domain needs to be restricted.

$$\frac{x^2 - 4}{5x} \div \frac{x + 2}{x - 2} = \frac{x^2 - 4}{5x} \cdot \frac{x - 2}{x + 2} = \frac{(x - 2)^2(x + 2)}{5x(x + 2)} = \frac{(x - 2)^2}{5x}, x \neq \pm 2$$

72. The values in the first row get larger and closer to 1 as the value of x increases (as x becomes larger, the value of 10 becomes smaller in comparison). The values in the second row get smaller and closer to 1 as the value of x increases (as x becomes larger, the value of 50 becomes smaller in comparison). The values in the third row are larger than the values in the first row and smaller than the values in the second row and get closer to 1 as the value of x increases.

Section 6.3 *(pp. 304–311)*

56. To rewrite each expression, multiply the numerator and the denominator by a factor (or factors) of the LCM of the individual denominators such that the denominator of the expression is the LCD.

57. To add or subtract rational expressions with like denominators, simply add (or subtract) the terms in the numerators and keep the common denominator.

58. To add or subtract rational expressions with unlike denominators, rewrite each expression as an equivalent expression whose denominator is the LCD and then add (or subtract) the numerators and keep the common denominator.

Section 6.4 *(pp. 312–319)*

39. A complex fraction is a fraction that has a fraction it its numerator or denominator, or both.

$$\frac{\left(\frac{x-3}{x}\right)}{\left(\frac{3x-9}{4}\right)}$$

41. To simplify a complex fraction, multiply the numerator and denominator by the least common denominator of all of the fractions in the numerator and denominator.

42. To find the implied domain restrictions for a complex fraction, set each factor in each denominator equal to zero.

67. In the second step, the set of parentheses cannot be moved because division is not associative.

$$\frac{(a/b)}{b} = \frac{a}{b} \cdot \frac{1}{b} = \frac{a}{b^2}$$

68. In the second step, the denominator fraction should have been inverted before being multiplied by the numerator.

$$\frac{(a/b)}{(b/c)} = \frac{a}{b} \cdot \frac{c}{b} = \frac{ac}{b^2}, c \neq 0$$

CHAPTER 7

Section 7.1 *(pp. 354–361)*

114. For each radical expression in the numerator and the denominator, if the index of the radical expression is even, the domain of the expression includes all real values for which the radicand is greater than or equal to zero. If the index of the radical expression is odd, the domain of the expression includes all real values. The domain of the function excludes all real values for which the radical expression in the denominator is zero.

Section 7.4 *(pp. 380–387)*

99. $\frac{\sqrt{4445}}{127}$

101. (a) If either a or b (or both) equal zero, the expression is zero and therefore rational.

(b) If the product of a and b is a perfect square, then the expression is rational.

103. $\sqrt{a} - \sqrt{b}$; $(\sqrt{a} + \sqrt{b})(\sqrt{a} - \sqrt{b}) = a - b$;

$\sqrt{b} - \sqrt{a}$; $(\sqrt{b} + \sqrt{a})(\sqrt{b} - \sqrt{a}) = b - a$;

When the order of the terms is changed, the conjugate and the product both change by a factor of -1.

104. $\frac{1}{\sqrt{a} + \sqrt{b}} = \frac{\sqrt{a} - \sqrt{b}}{a - b}$ and

$\frac{1}{\sqrt{b} + \sqrt{a}} = \frac{\sqrt{b} - \sqrt{a}}{b - a} = \frac{\sqrt{a} - \sqrt{b}}{a - b}$

Changing the order of the terms in the denominator does not affect the rationalized form of the quotient.

Section 7.5 *(pp. 388–395)*

1. (a) Not a solution (b) Not a solution
(c) Not a solution (d) Solution

2. (a) Not a solution (b) Not a solution
(c) Solution (d) Not a solution

3. (a) Not a solution (b) Solution
(c) Not a solution (d) Not a solution

4. (a) Solution (b) Not a solution
(c) Not a solution (d) Not a solution

Section 7.6 *(pp. 396–403)*

117. Exercise 109: The sum of complex conjugates of the form $a + bi$ and $a - bi$ is twice the real number of a, or $2a$.

Exercise 110: The product of complex conjugates of the form $a + bi$ and $a - bi$ is the sum of the squares of a and b, or $a^2 + b^2$.

Exercise 111: The difference of complex conjugates of the form $a + bi$ and $a - bi$ is twice the imaginary number bi, or $2bi$.

Exercise 112: The sum of the squares of complex conjugates of the form $a + bi$ and $a - bi$ is the difference of twice the squares of a and b, or $2a^2 - 2b^2$.

121. To simplify the quotient, multiply the numerator and the denominator by $-bi$. This will yield a positive real number in the denominator. The number i can also be used to simplify the quotient. The denominator will be the opposite of b, but the resulting number will be the same.

CHAPTER 8

Section 8.1 *(pp. 414–421)*

65. Write the equation in the form $u^2 = d$, where u is an algebraic expression and d is a positive constant. Take the square root of each side of the equation to obtain the solutions $u = \pm\sqrt{d}$.

101.

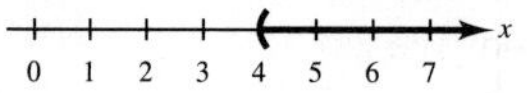

102.

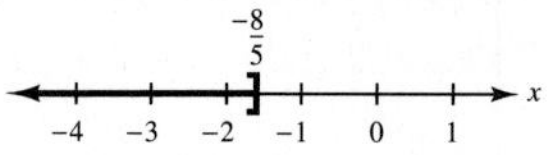

103.

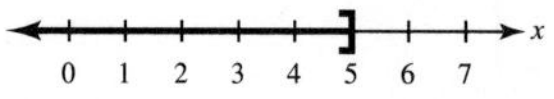

104.

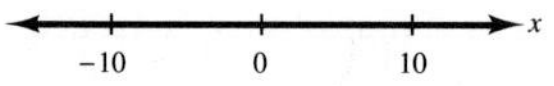

Section 8.2 *(pp. 422–429)*

63. $-\frac{1}{4} + \frac{\sqrt{39}}{4}i \approx -0.25 + 1.56i$

$-\frac{1}{4} - \frac{\sqrt{39}}{4}i \approx -0.25 - 1.56i$

64. $-\frac{5}{6} + \frac{\sqrt{47}}{6}i \approx -0.83 + 1.14i$

$-\frac{5}{6} - \frac{\sqrt{47}}{6}i \approx -0.83 - 1.14i$

78. The student forgot to add $\left(\frac{6}{2}\right)^2$ to the right side of the equation.

Correct solution: $x^2 + 6x = 13$

$$x^2 + 6x + \left(\tfrac{6}{2}\right)^2 = 13 + \left(\tfrac{6}{2}\right)^2$$
$$(x + 3)^2 = 22$$
$$x + 3 = \pm\sqrt{22}$$
$$x = -3 \pm \sqrt{22}$$

Section 8.3 *(pp. 430–437)*

43.

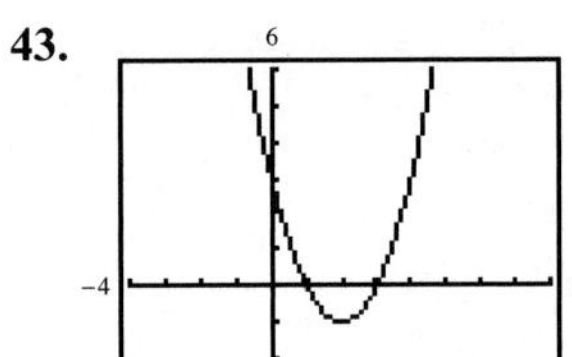

44.

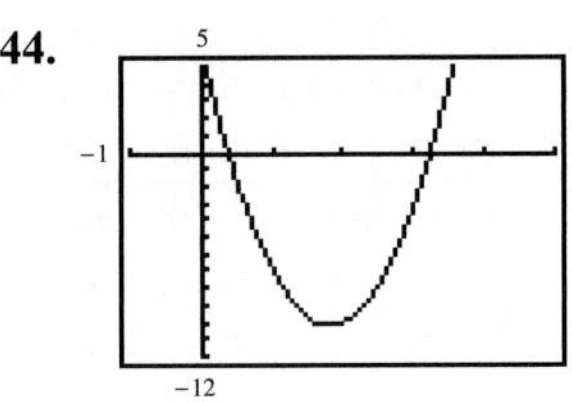

45.

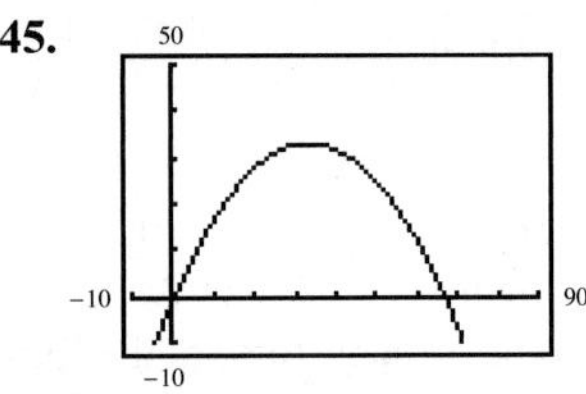

46.

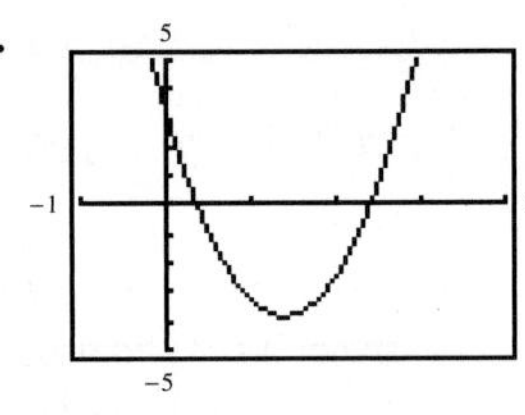

70. (c) No. In order for the discriminant to be greater than or equal to zero, the value of c must be greater than or equal to -6.25. Therefore, the height cannot exceed 46.25 feet or the value of c would be less than -6.25 when the equation is set equal to zero.

71. (a)

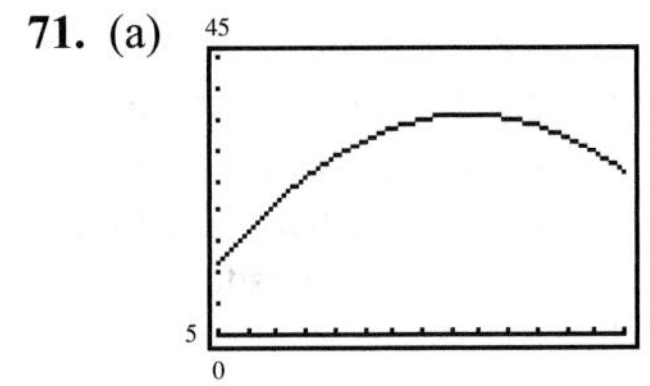

77. When the Quadratic Formula is applied to $ax^2 + bx + c = 0$, the square root of the discriminant is evaluated. When the discriminant is positive, the square root of the discriminant is positive and will yield two real solutions (or x-intercepts). When the discriminant is zero, the equation has one real solution (or x-intercept). When the discriminant is negative, the square root of the discriminant is negative and will yield two complex solutions (no x-intercepts).

78. The solutions $x = 2$ and $x = 4$ can be rewritten as $x - 2 = 0$ and $x - 4 = 0$, so the factors are $x - 2$ and $x - 4$, respectively.

$$(x - 2)(x - 4) = 0$$
$$x^2 - 6x + 8 = 0$$

83.

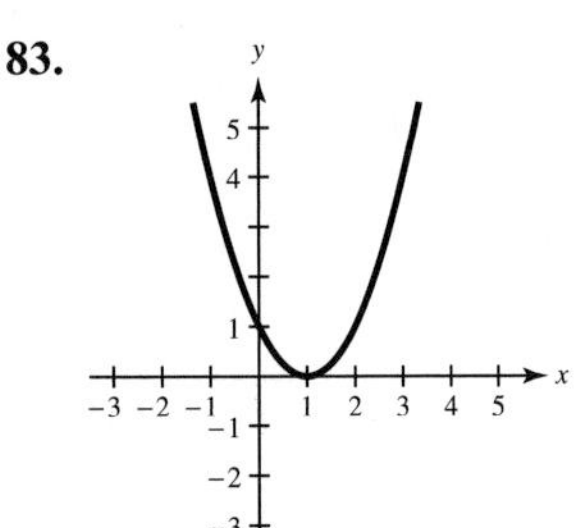

84.

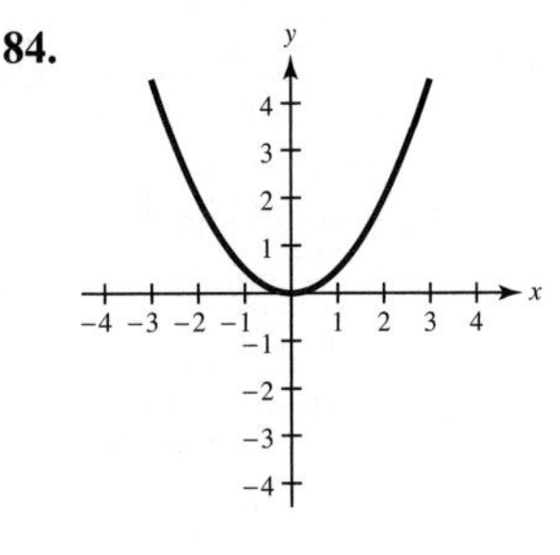

85.

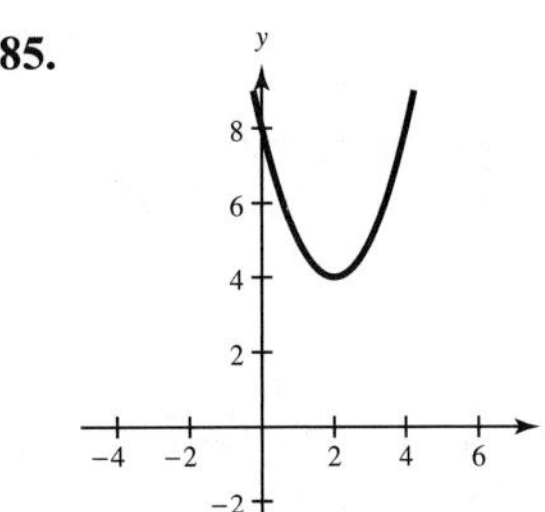

86.

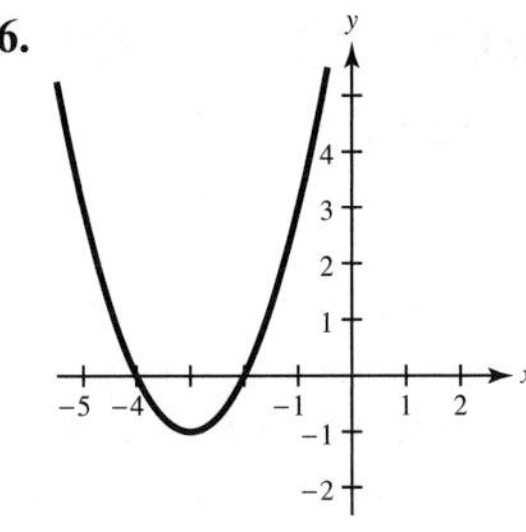

Section 8.4 *(pp. 440–447)*

17.

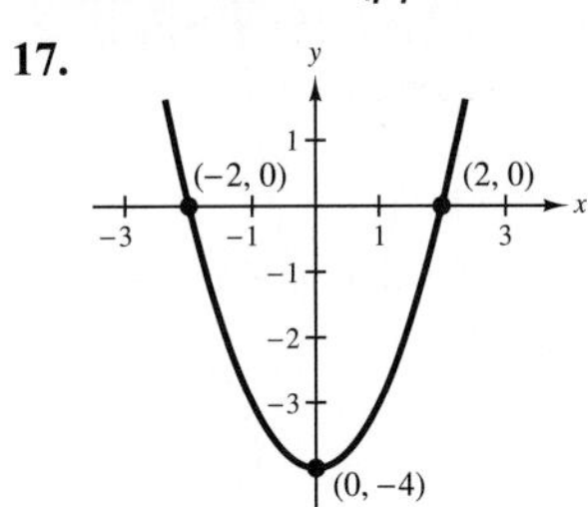

18.

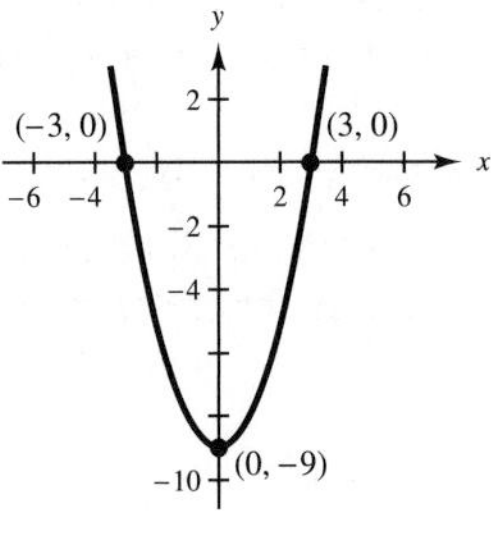

19.

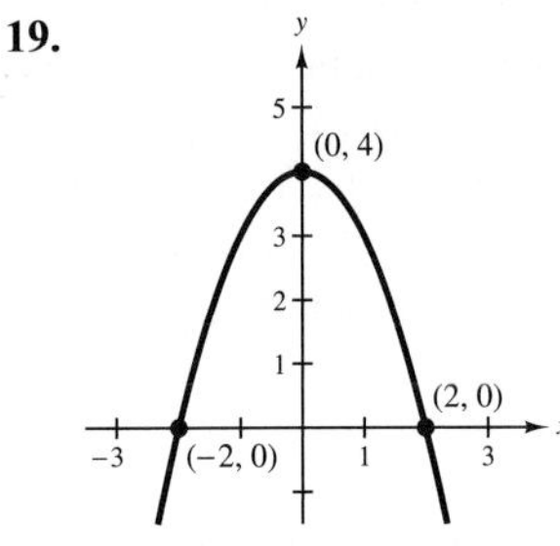

20.

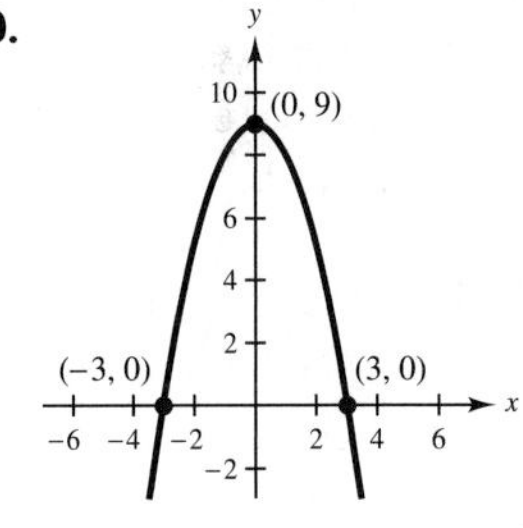

21.

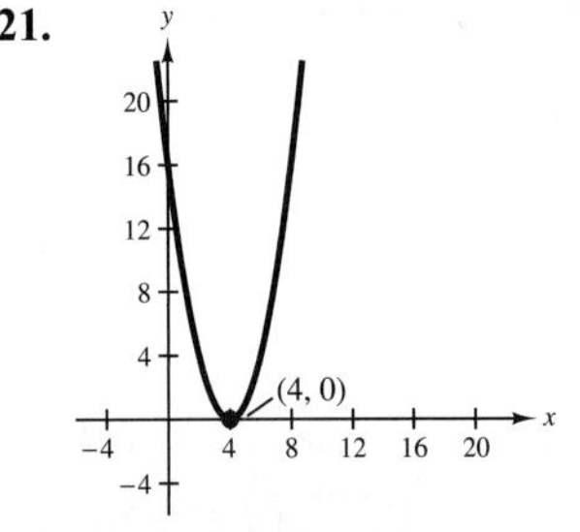

22.

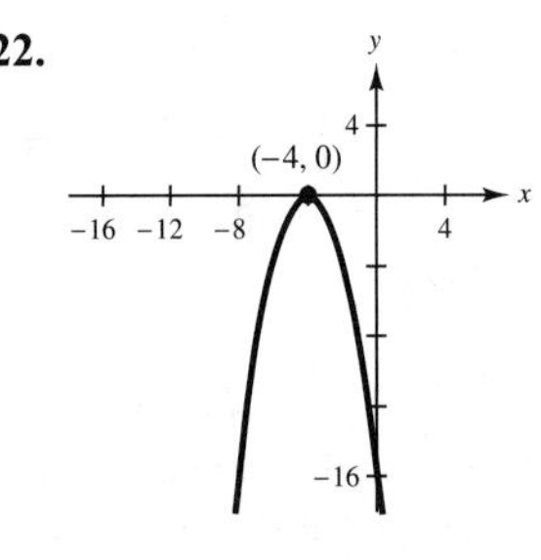

23.

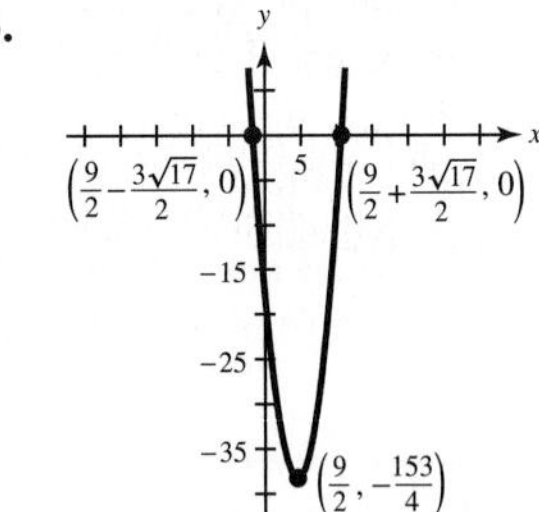

24.

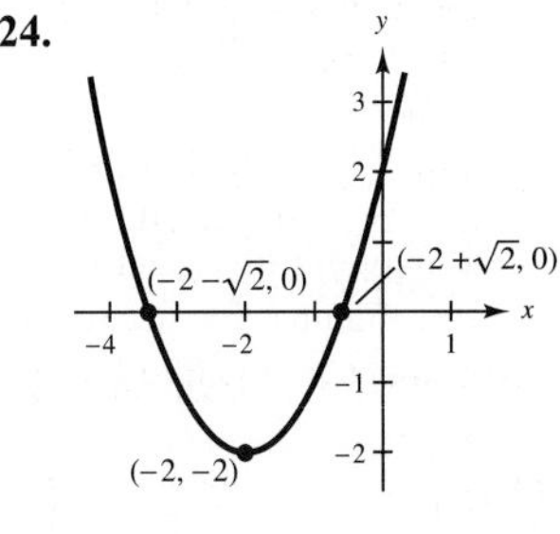

44. The x-coordinate of the vertex of a quadratic function written in the general form $f(x) = ax^2 + bx + c$ can be found by using $x = -b/(2a)$. Substitute the value found for x into the equation to find the y-coordinate of the vertex. The vertex can also be found by using the method of completing the square to write the quadratic function in the standard form $y = a(x - h)^2 + k$. The vertex is located at the point (h, k).

45. To find any x-intercepts, set $y = 0$ and solve the resulting equation for x. To find any y-intercepts, set $x = 0$ and solve the resulting equation for y.

47.
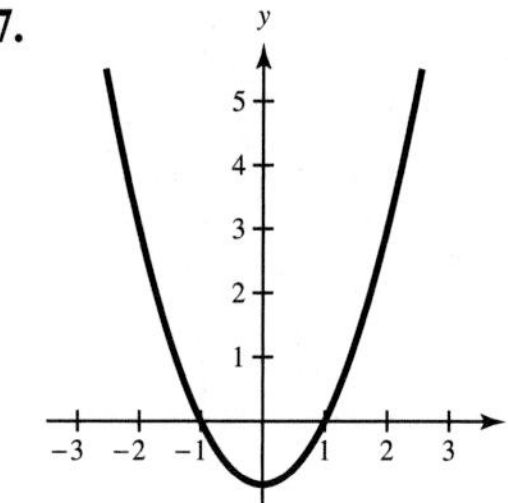

48.
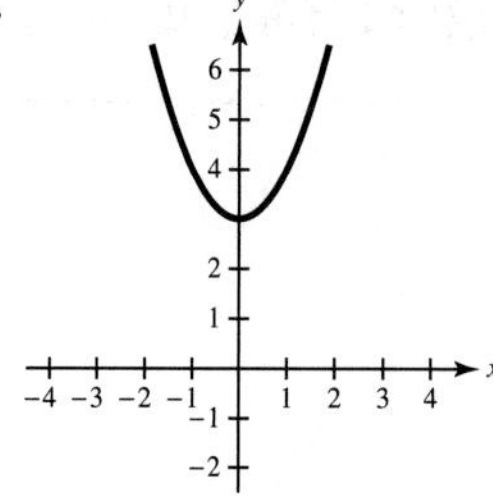

49.
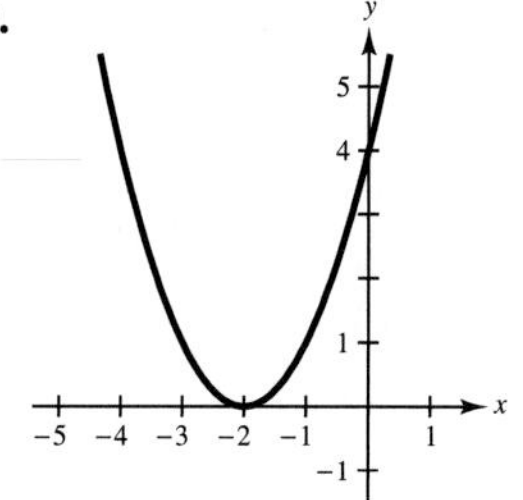

50.
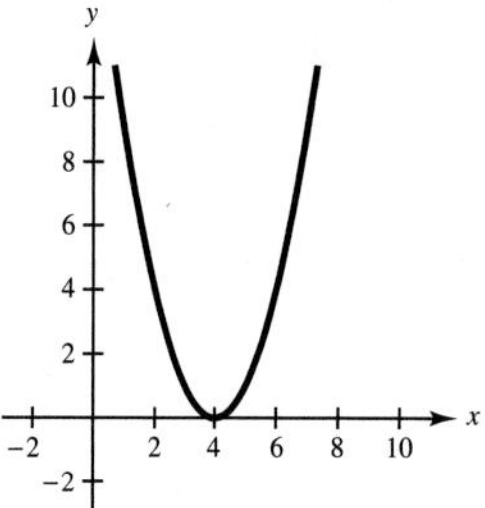

51.
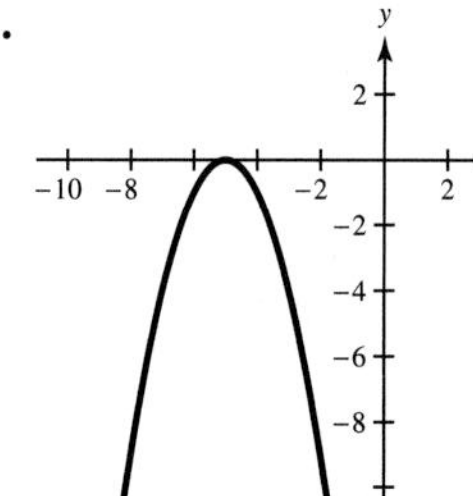

52.
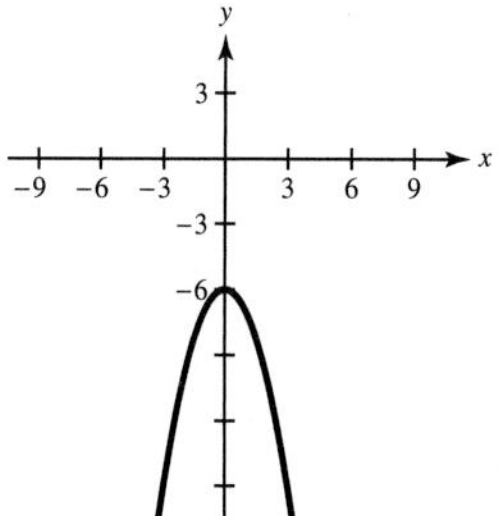

53.
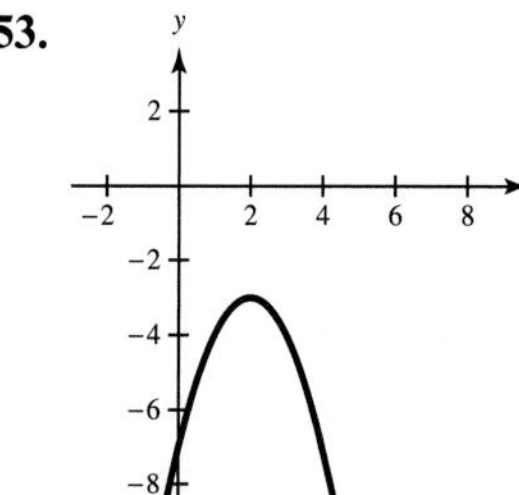

54.
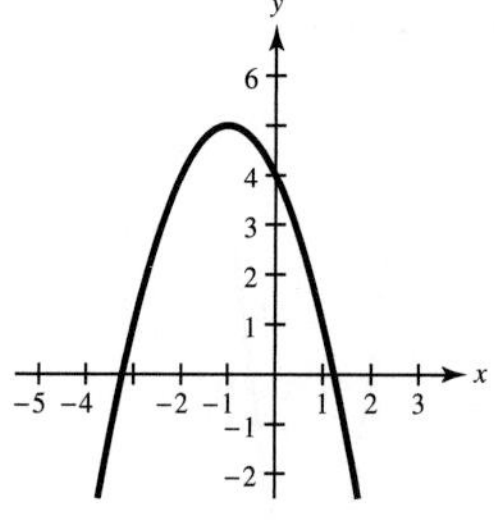

Section 8.5 *(pp. 448–455)*

17. The Pythagorean Theorem can be used to set the sum of the square of a missing length and the square of the difference of the sum of the lengths and the missing length equal to the square of the length of the hypotenuse.

20. The product of the quotient of 1 and the time for the first machine and the time for both machines, plus the product of the quotient of 1, and the time for the first machine plus the extra minutes for the second machine and the time for both machines, is equal to 1.

36. No.

Area $= \frac{1}{2}(b_1 + b_2)h = \frac{1}{2}x[x + (550 - 2x)] = 43{,}560$

This equation has no real solution.

40. If the problem involves finding the area, the length and width can be found by using the formula for the perimeter of a rectangle, which is a linear equation. If the problem involves finding the perimeter, the length and width can be found by using the formula for the area of a rectangle, which is a quadratic equation.

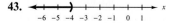

43.
−6 −5 −4 −3 −2 −1 0 1 x

44.
−6 −5 −4 −3 −2 −1 0 1 2 x

Section 8.6 *(pp. 456–463)*

9.
−1 0 1 2 3 x

10.
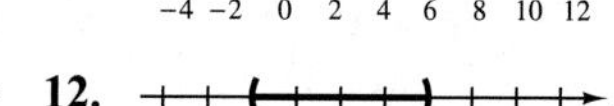

11.
−1 0 1 2 3 x

12.
−4 −2 0 2 4 6 8 10 12 14

13.
−8 −6 −4 −2 0 2 4 6 x

14.
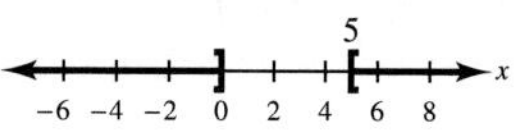

15.
5
−6 −4 −2 0 2 4 6 8 x

16.
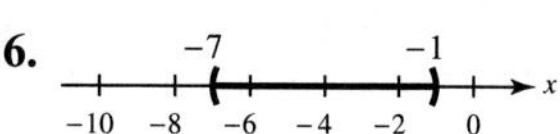

17.
−4 −2 0 2 4 x

18.
−3 −2 −1 0 1 2 3 z

19.
−9
−10 −8 −6 −4 −2 0 2 4 6 x

20.
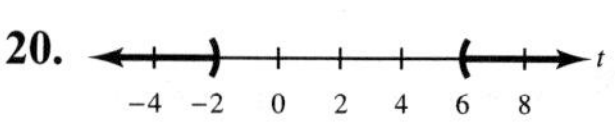

21.
−4 −3 −2 −1 0 1 2 u

22.
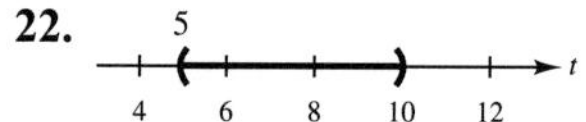

25.
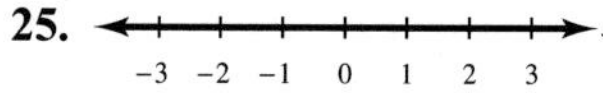

26.
−3 −2 −1 0 1 2 3 x

27.
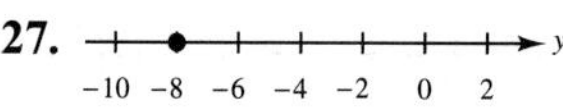

28.
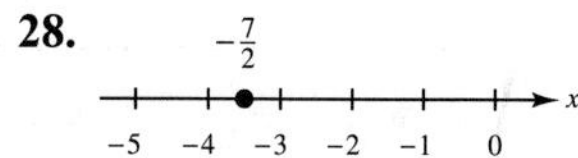

29.
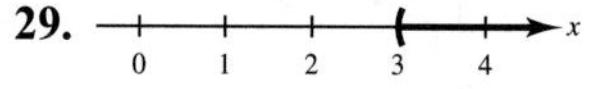

30.
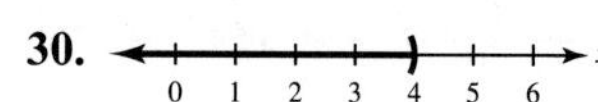

31.
0 1 2 3 4 5 6 x

32.
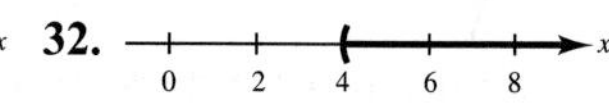

33.
−6 −4 −2 0 2 4 x

34.
5
−4 −2 0 2 4 6 x

35. **36.**
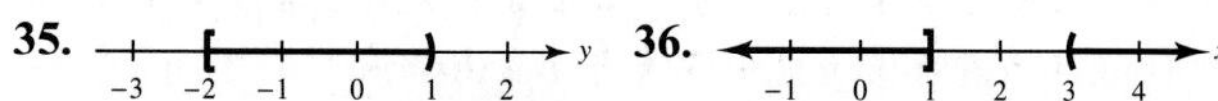

40. To solve a quadratic inequality, first write the inequality in general form. Then solve the corresponding equation. The solutions of the equation are the critical numbers, which determine the test intervals. Choose a representative x-value in each test interval and evaluate the polynomial at that value. If the value of the polynomial satisfies the inequality, then that interval is part of the solution set.

41. No. $x = 4$ is not in the solution set $0 < x < 4$.

42. The critical numbers of a quadratic inequality are the zeros of the polynomial, and the critical numbers of a rational inequality are the zeros and the undefined values.

47.
$2-\sqrt{6}$ $2+\sqrt{6}$
−2 0 2 4 6 8 x

48.
$-3-\sqrt{6}$ $-3+\sqrt{6}$
−8 −6 −4 −2 0 2 4 y

49.
−9 −1
−10 −8 −6 −4 −2 0 u

50.
−4 −2 0 2 4 6 8 x

51. 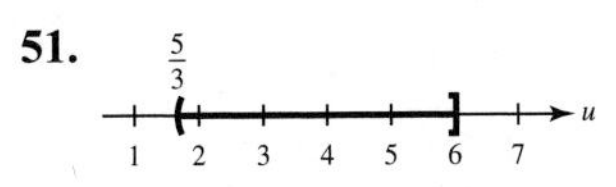

52.

53.

54.

55.

56.

57.

58.

64. (a) 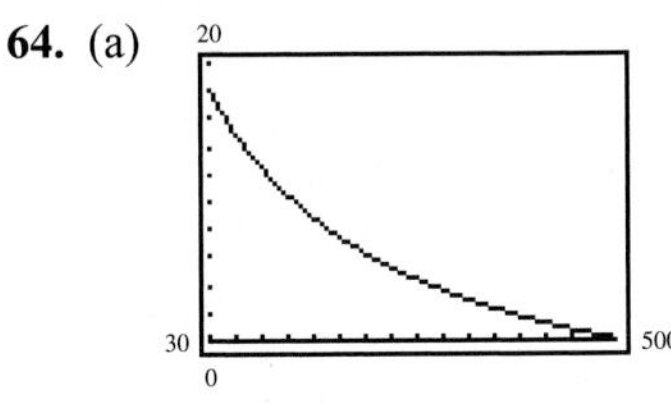

Review Exercises (pp. 466–469)

67.

68. 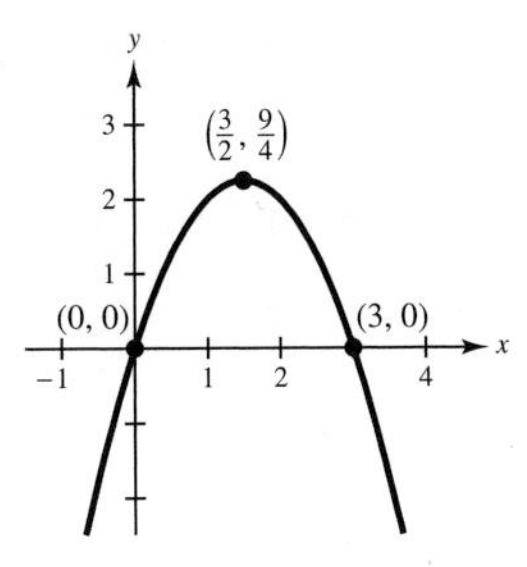

69. 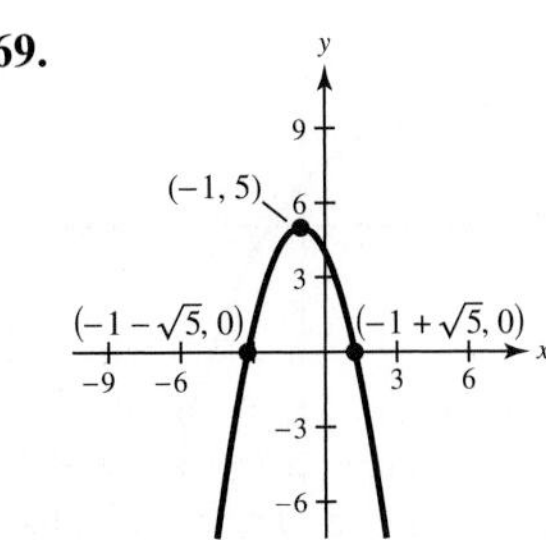

70. 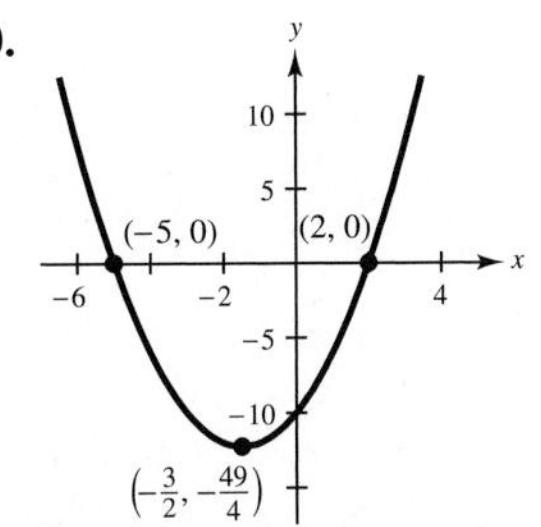

91. 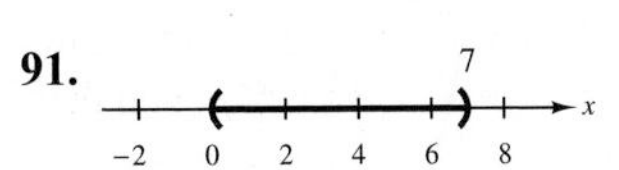

92. 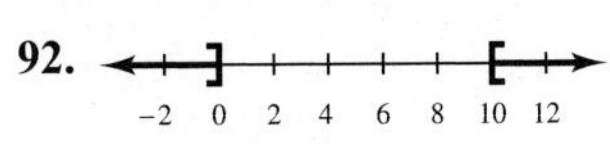

93. 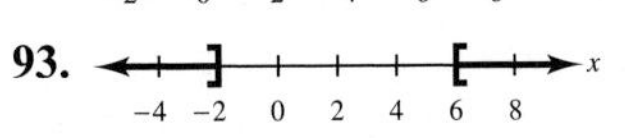

94. 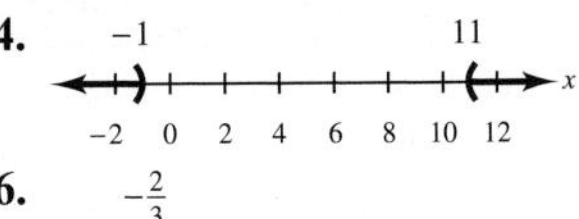

95. 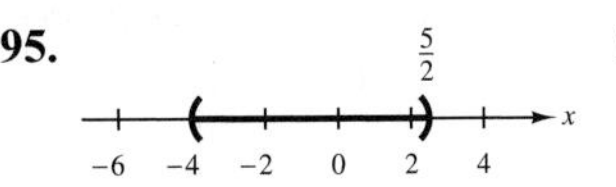

96. 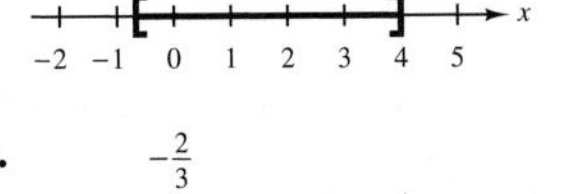

97.

98.

99.

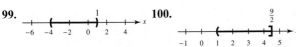

100. 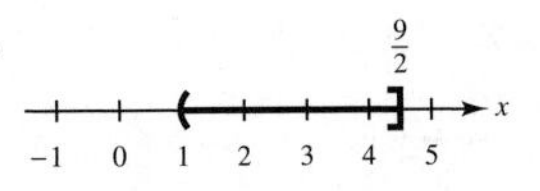

Chapter Test (p. 470)

11.

12. 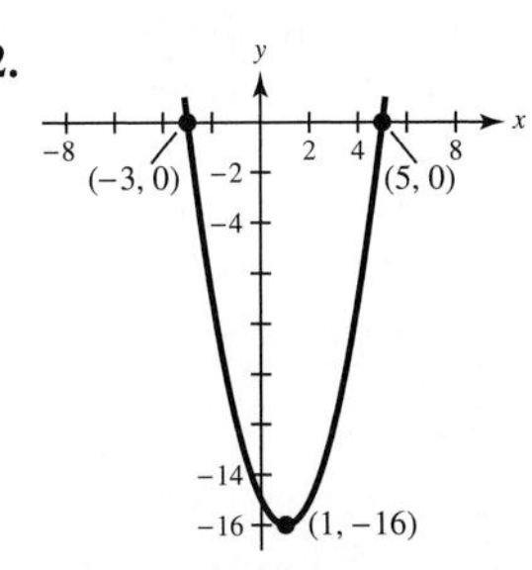

13.

14. 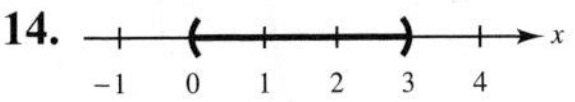

15. 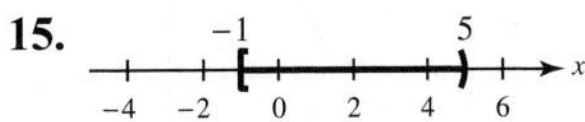

CHAPTER 9

Section 9.1 (pp. 472–481)

21. 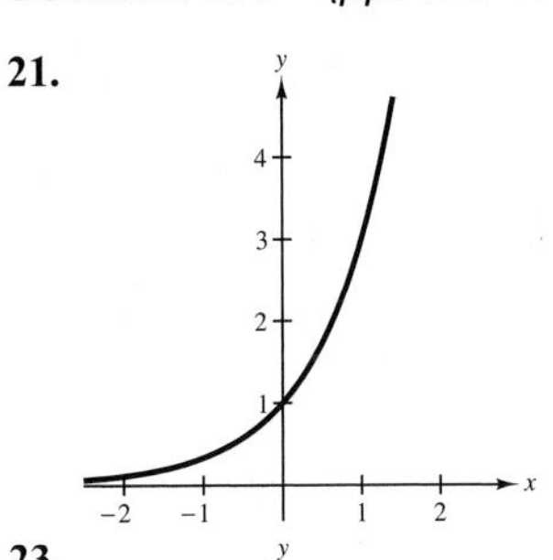

22. 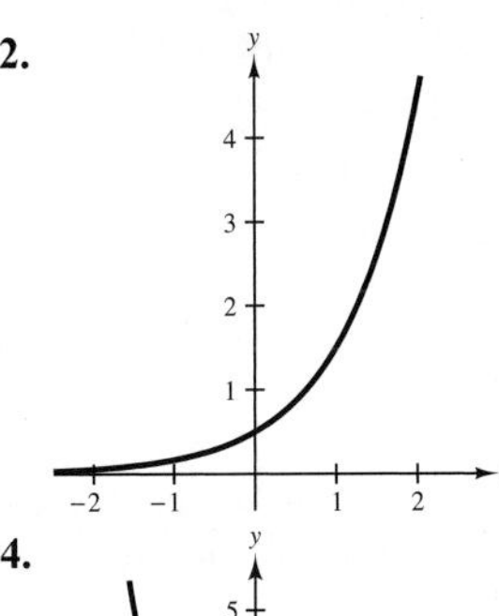

23. 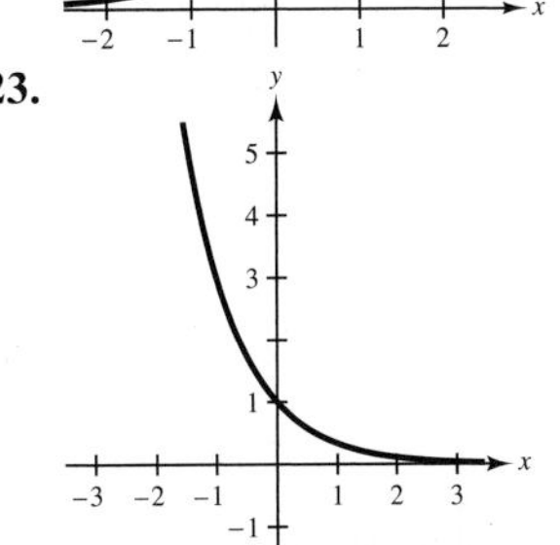

24. 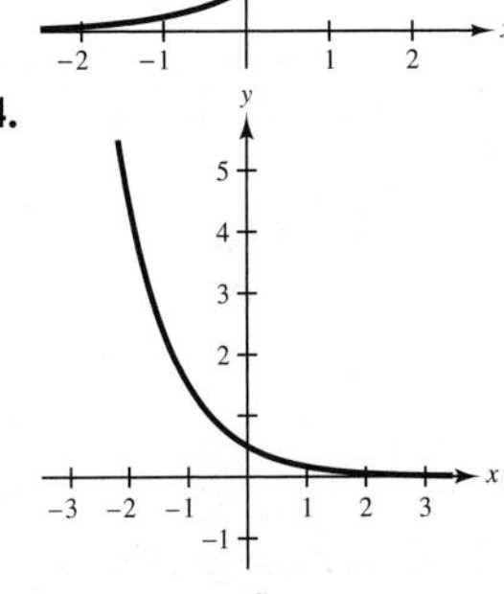

25. The function g is related to $f(x) = 3^x$ by $g(x) = f(x) - 1$. To sketch g, shift the graph of f one unit downward.
y-intercept: $(0, 0)$
Asymptote: $y = -1$

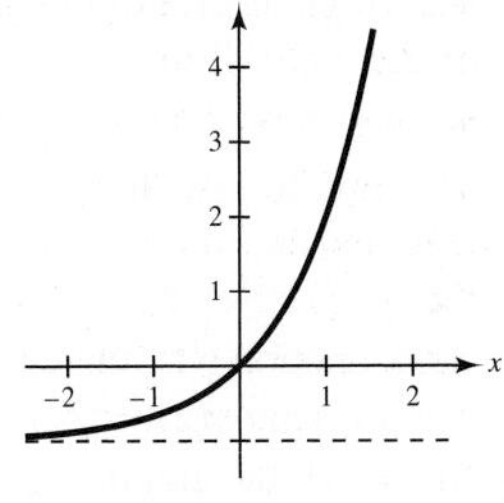

26. The function g is related to $f(x) = 3^x$ by $g(x) = f(x) + 1$. To sketch the graph of g, shift the graph of f one unit upward.
y-intercept: $(0, 2)$
Asymptote: $y = 1$

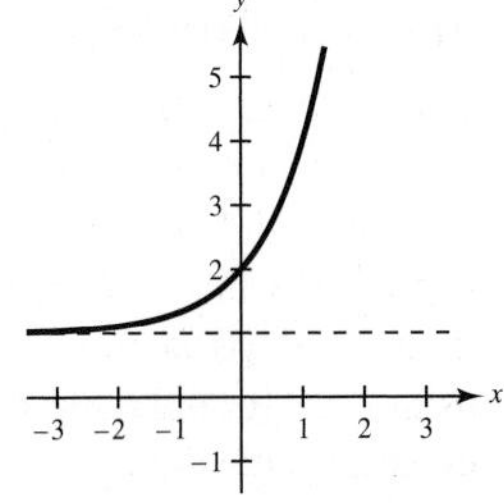

27. The function g is related to $f(x) = 5^x$ by $g(x) = f(x - 1)$. To sketch the graph of g, shift the graph of f one unit to the right.
y-intercept: $\left(0, \frac{1}{5}\right)$
Asymptote: x-axis

28. The function g is related to $f(x) = 5^x$ by $g(x) = f(x + 3)$. To sketch the graph of g, shift the graph of f three units to the left.
y-intercept: $(0, 125)$
Asymptote: x-axis

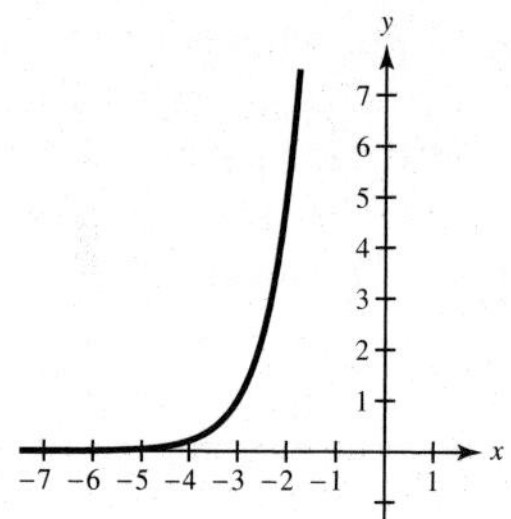

29. The function g is related to $f(x) = 2^x$ by $g(x) = f(x) + 3$. To sketch the graph of g, shift the graph of f three units upward.
y-intercept: $(0, 4)$
Asymptote: $y = 3$

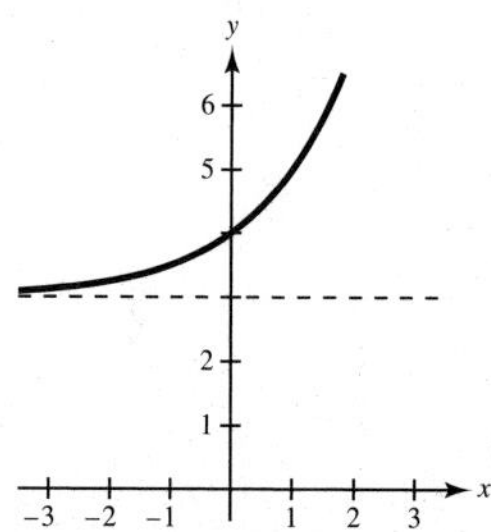

30. The function g is related to $f(x) = 2^x$ by $g(x) = f(x + 3)$. To sketch the graph of g, shift the graph of f three units to the left.
y-intercept: $(0, 8)$
Asymptote: x-axis

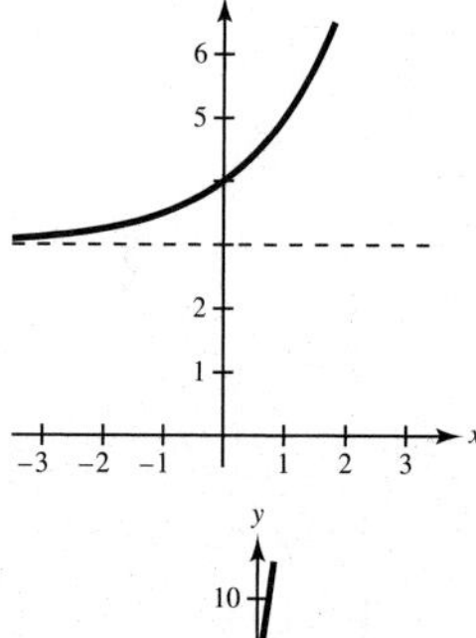

31. The function g is related to $f(x) = 2^x$ by $g(x) = f(x - 4)$. To sketch the graph of g, shift the graph of f four units to the right.
y-intercept: $\left(0, \frac{1}{16}\right)$
Asymptote: x-axis

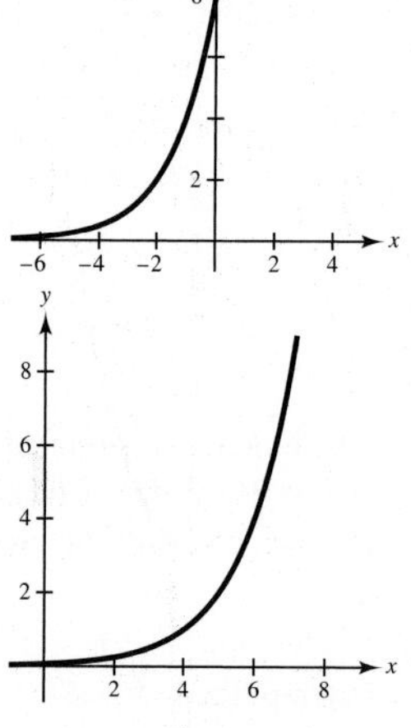

32. The function g is related to $f(x) = 2^x$ by $g(x) = f(x) - 4$. To sketch the graph of g, shift the graph of f four units downward.
y-intercept: $(0, -3)$
Asymptote: $y = -4$

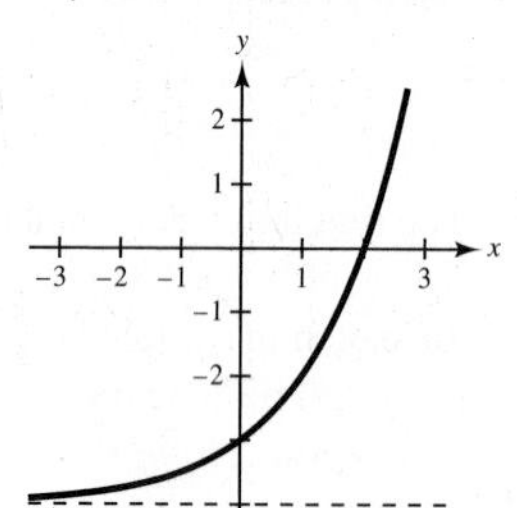

37. The function g is related to $f(x) = 4^x$ by $g(x) = -f(x)$. To sketch the graph of g, reflect the graph of f in the x-axis.

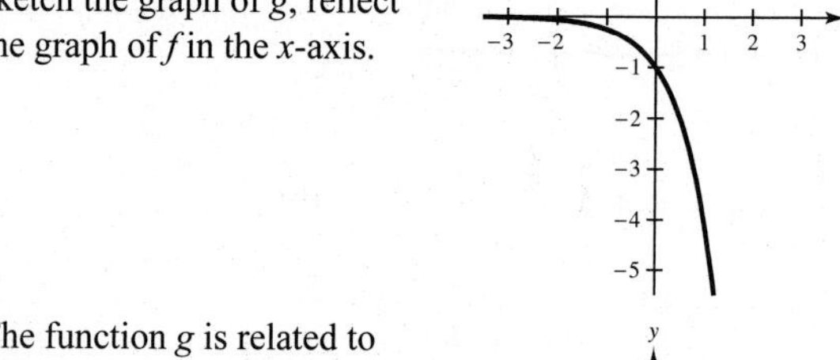

38. The function g is related to $f(x) = 4^x$ by $g(x) = f(-x)$. To sketch the graph of g, reflect the graph of f in the y-axis.

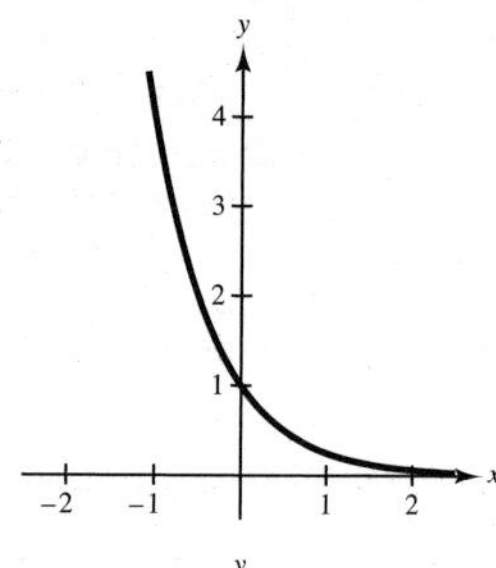

39. The function g is related to $f(x) = 5^x$ by $g(x) = f(-x)$. To sketch the graph of g, reflect the graph of f in the y-axis.

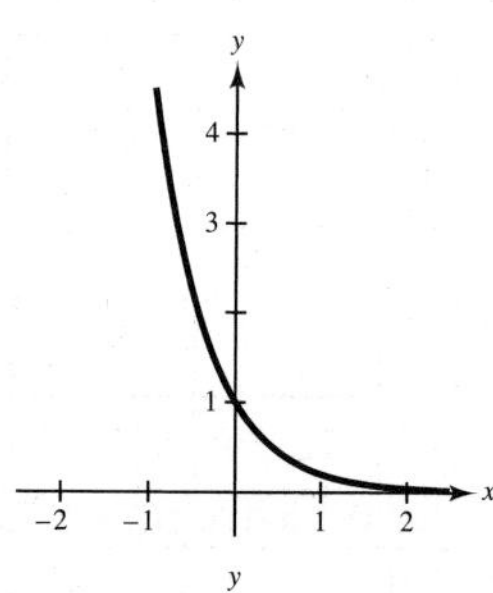

40. The function g is related to $f(x) = 5^x$ by $g(x) = -f(x)$. To sketch the graph of g, reflect the graph of f in the x-axis.

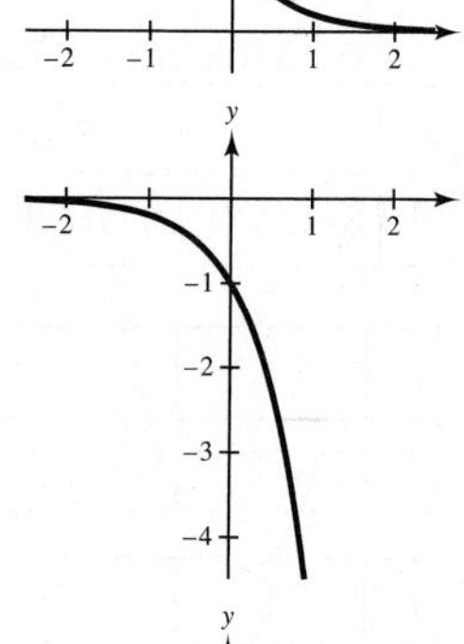

47. The function g is related to $f(x) = e^x$ by $g(x) = -f(x)$. To sketch the graph of g, reflect the graph of f in the x-axis.

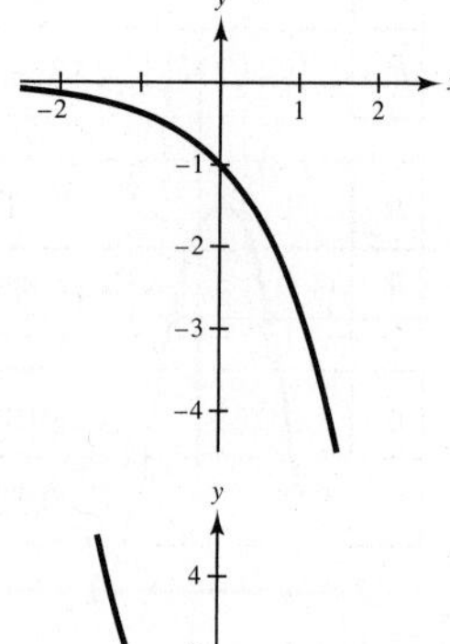

48. The function g is related to $f(x) = e^x$ by $g(x) = f(-x)$. To sketch the graph of g, reflect the graph of f in the x-axis.

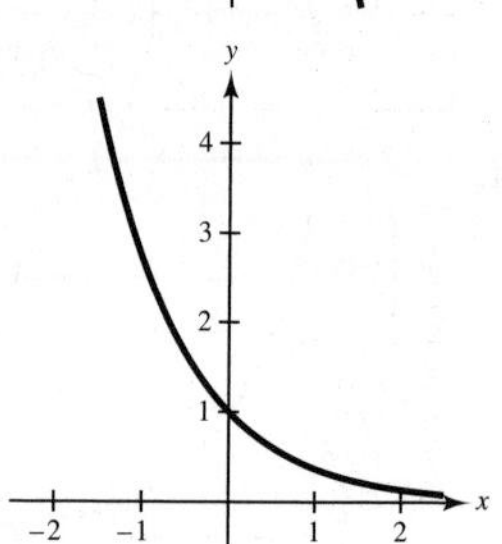

49. The function g is related to $f(x) = e^x$ by $g(x) = f(x) + 1$. To sketch the graph of g, shift the graph of f one unit upward.

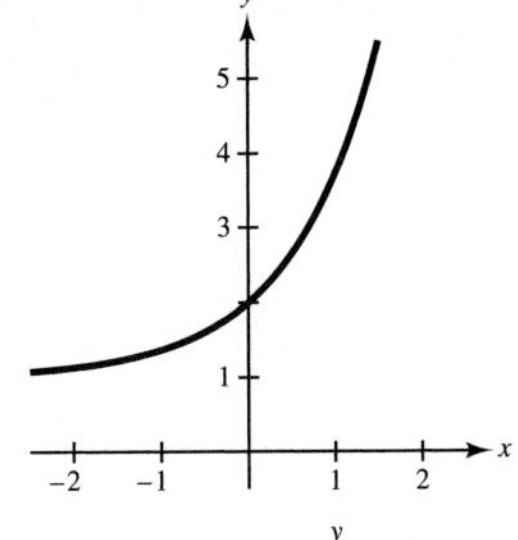

50. The function g is related to $f(x) = e^x$ by $g(x) = f(x + 1)$. To sketch the graph of g, shift the graph of f one unit to the left.

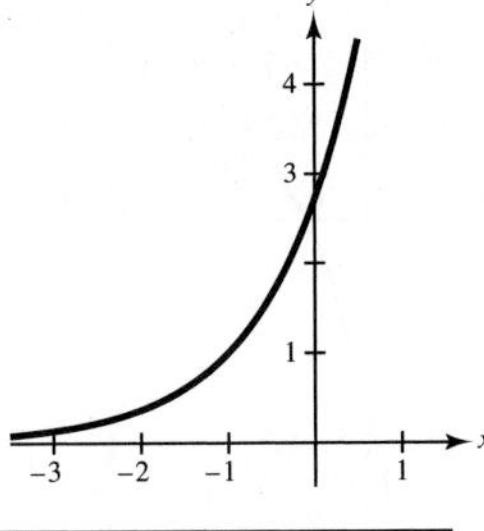

53.

n	1	4	12	365	Continuous
A	\$275.90	\$283.18	\$284.89	\$285.74	\$285.77

54.

n	1	4	12	365	Continuous
A	\$729.99	\$732.11	\$732.60	\$732.83	\$732.84

55.

n	1	4	12	365	Continuous
A	\$4956.46	\$5114.30	\$5152.11	\$5170.78	\$5171.42

56.

n	1	4	12	365	Continuous
A	\$5285.47	\$5446.73	\$5484.67	\$5503.31	\$5503.95

69.

n	1	4	12	365	Continuous
P	\$2541.75	\$2498.00	\$2487.98	\$2483.09	\$2482.93

70.

n	1	4	12
P	\$17,843.09	\$16,862.99	\$16,641.28

n	365	Continuous
P	\$16,533.56	\$16,529.89

74.

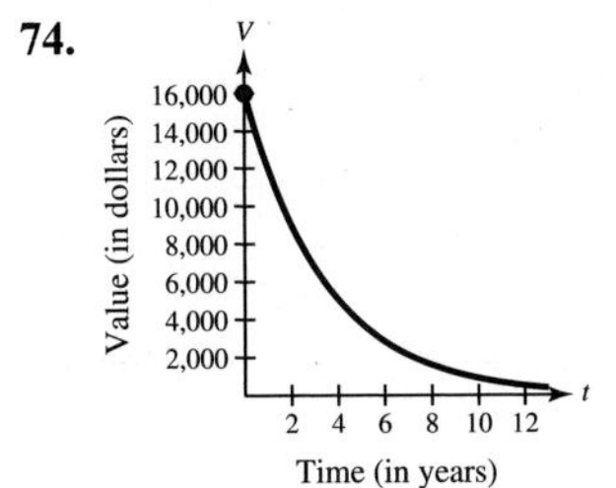

76. (b)

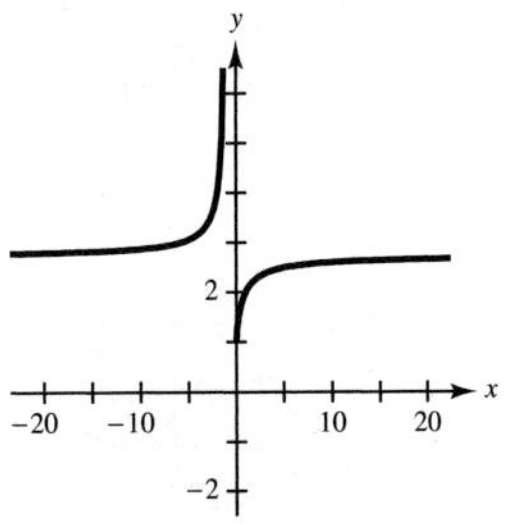

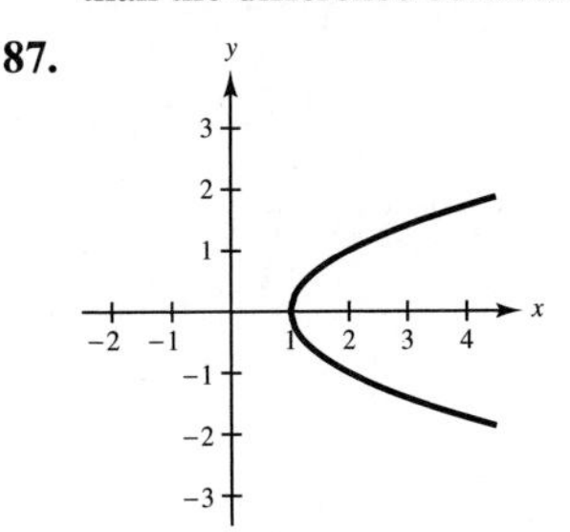

84. Yes. In terms of interest earned, the difference between quarterly and annual compounding tends to be greater than the difference between continuous and daily compounding, which is greater than the difference between hourly and daily compounding.

87.

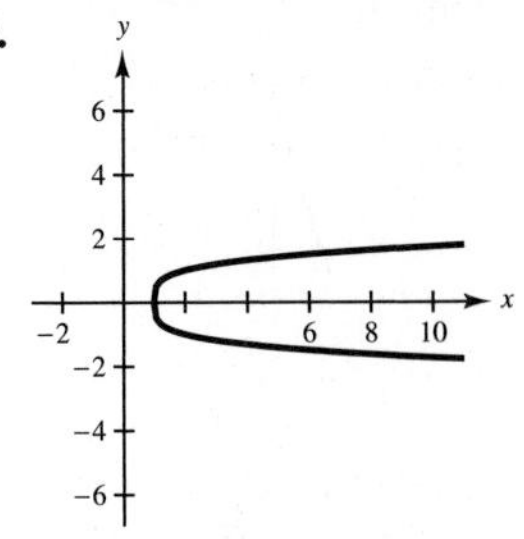

88.

Section 9.2 *(pp. 482–489)*

5. (a) $(f \circ g)(x) = \sqrt{x - 2}$
Domain: $x \geq 2$

(b) $(g \circ f)(x) = \sqrt{x + 2} - 4$
Domain: $x \geq -2$

6. (a) $(f \circ g)(x) = \sqrt{x - 2}$
Domain: $x \geq 2$

(b) $(g \circ f)(x) = \sqrt{x - 5} + 3$
Domain: $x \geq 5$

7.

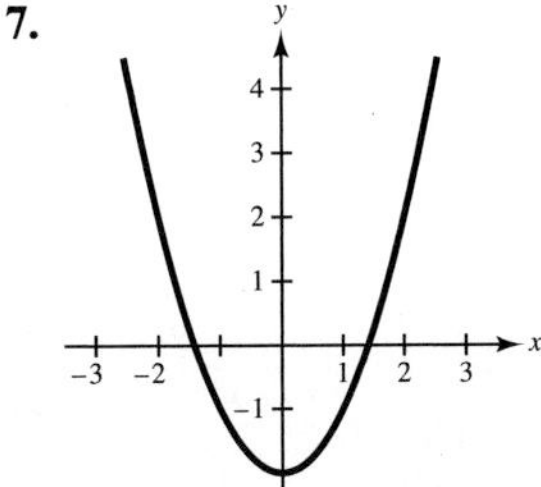

8.

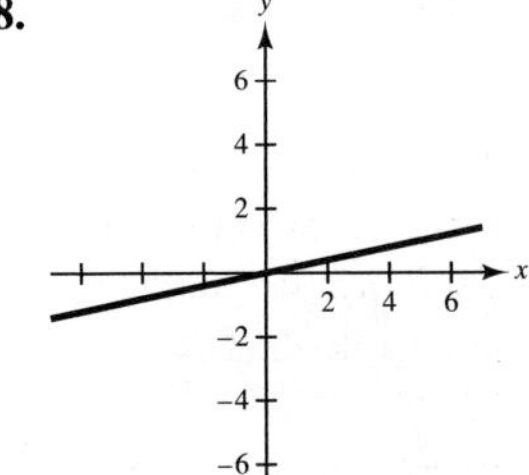

9.

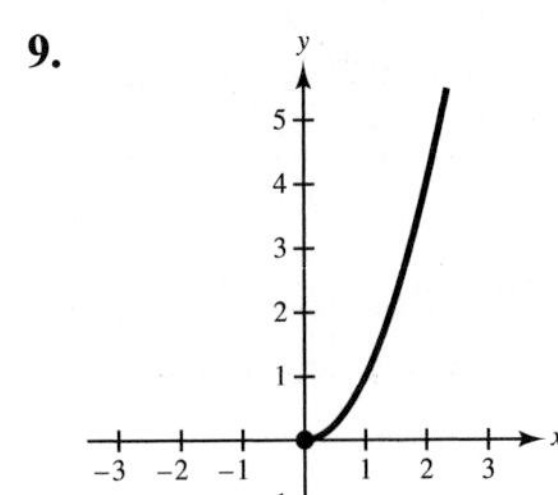

10.

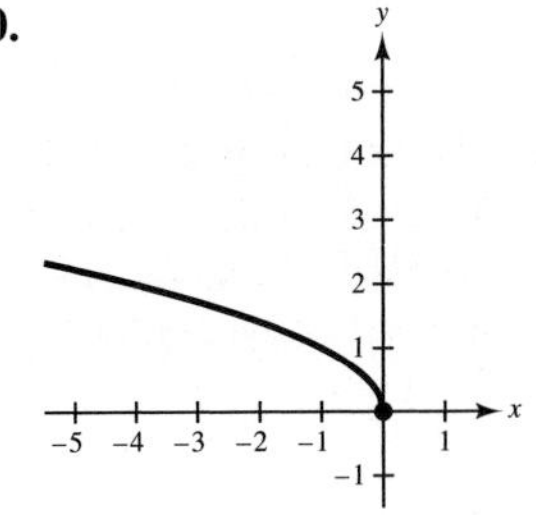

11.

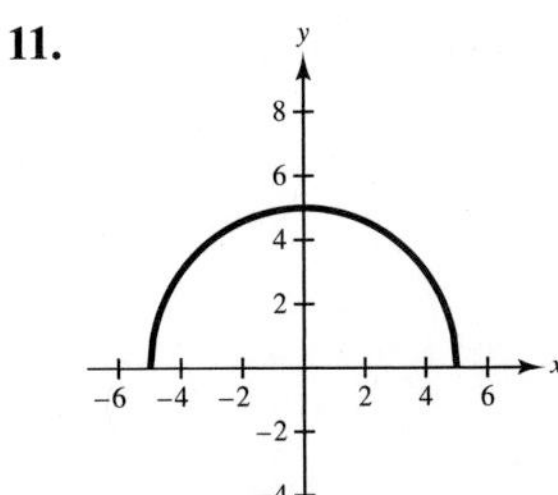

12.

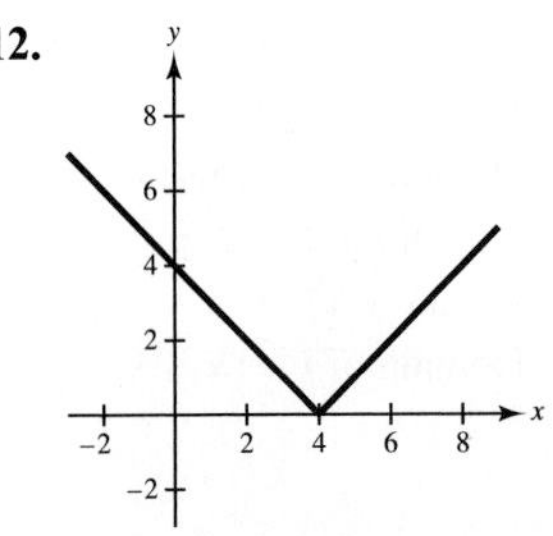

17.

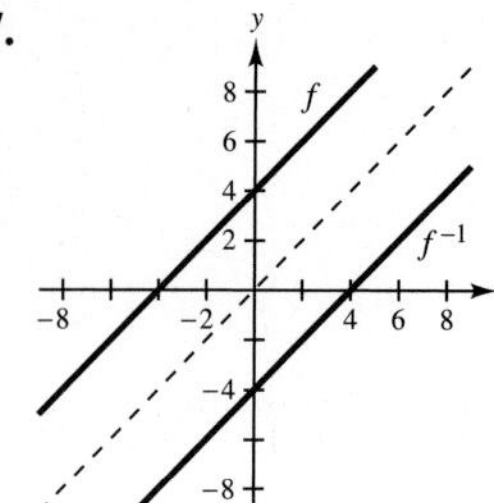

18.

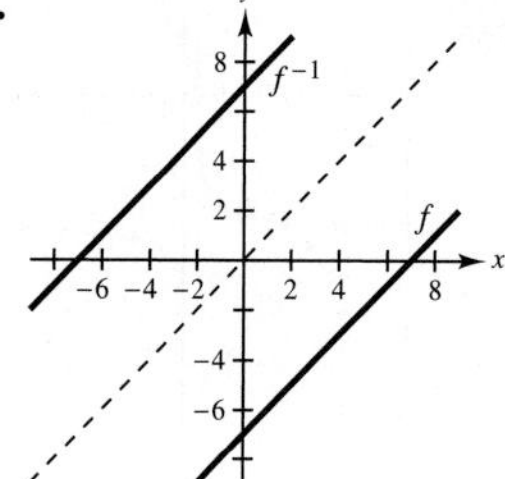

19.

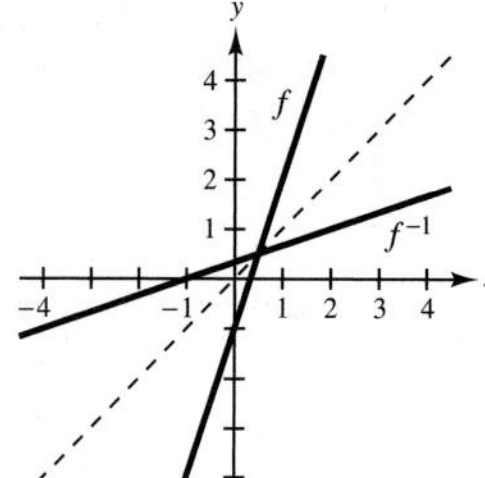

20.

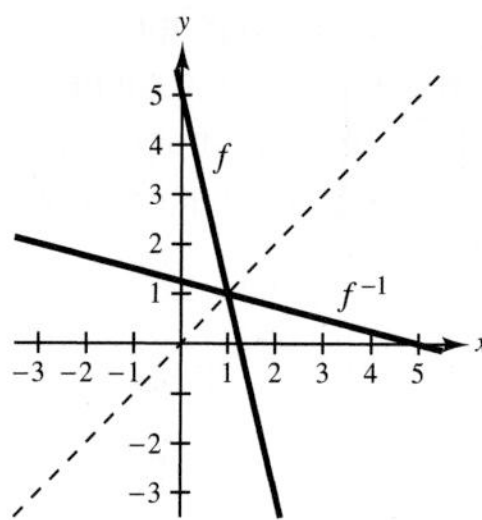

21.

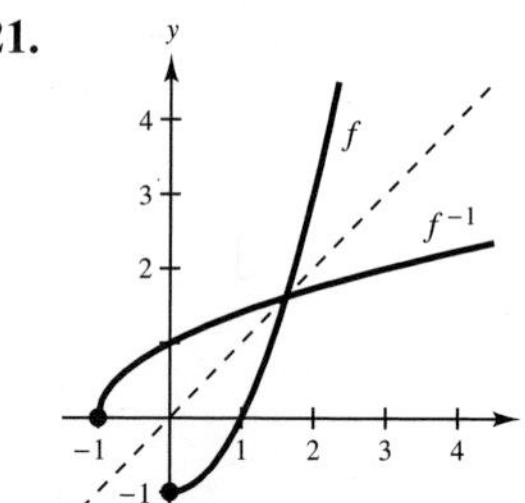

22.

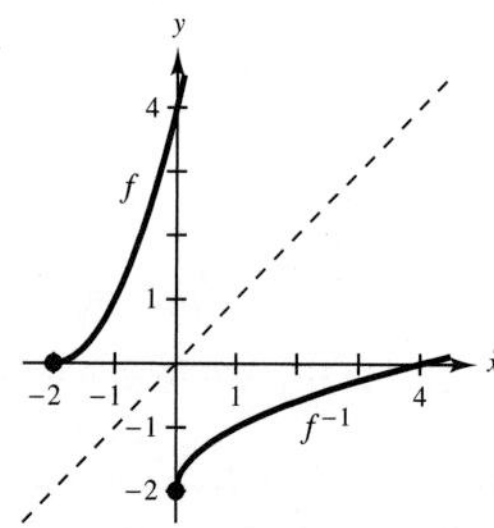

27.

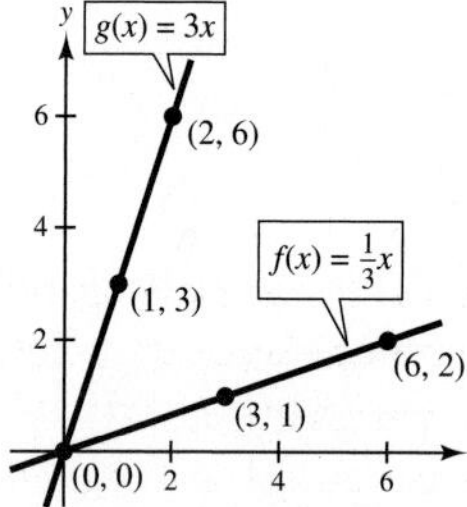

28.

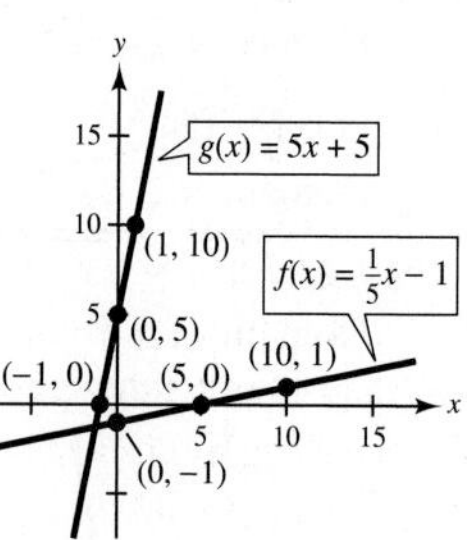

29.

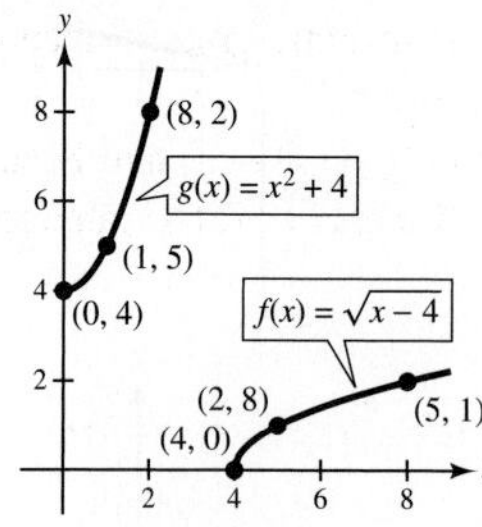

30.

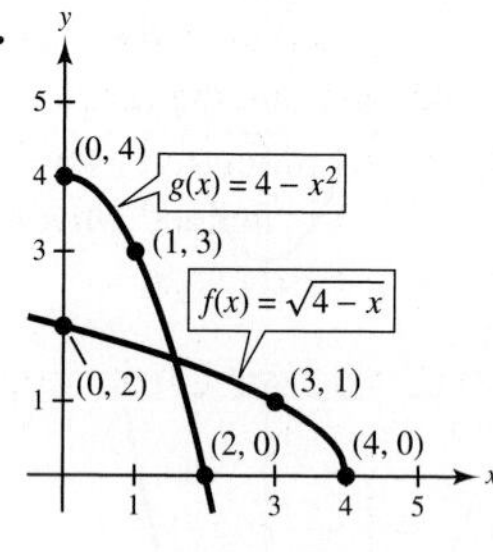

39. $f(g(x)) = f\left[\frac{1}{2}(1 - x)\right] = 1 - 2\left[\frac{1}{2}(1 - x)\right] = 1 - (1 - x) = x$

$g(f(x)) = g(1 - 2x) = \frac{1}{2}\left[1 - (1 - 2x)\right] = \frac{1}{2}(2x) = x$

40. $f(g(x)) = f\left[\frac{1}{2}(x + 1)\right] = 2\left[\frac{1}{2}(x + 1)\right] - 1 = (x + 1) - 1 = x$

$g(f(x)) = g(2x - 1) = \frac{1}{2}\left[(2x - 1) + 1\right] = \frac{1}{2}(2x) = x$

45. Domain of f: $x \geq 2$

$f^{-1}(x) = \sqrt{x} + 2$

Domain of f^{-1}: $x \geq 0$

46. Domain of f: $x \geq 0$

$f^{-1}(x) = \sqrt{9 - x}$

Domain of f^{-1}: $x \leq 9$

65.

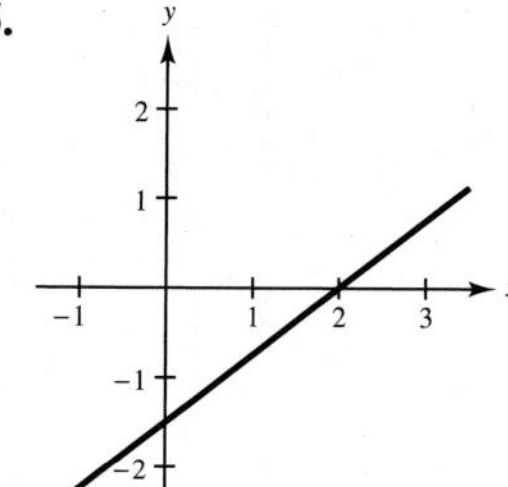

66.

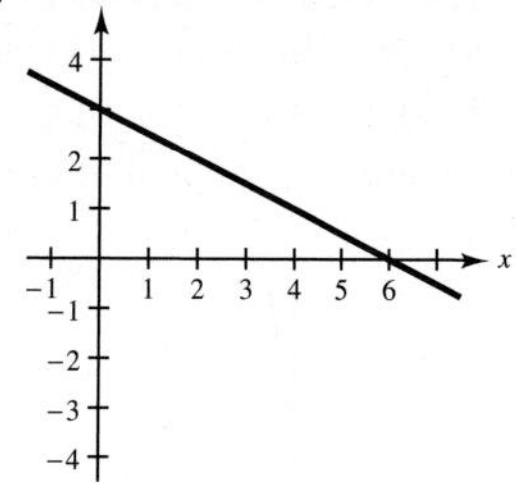

67.

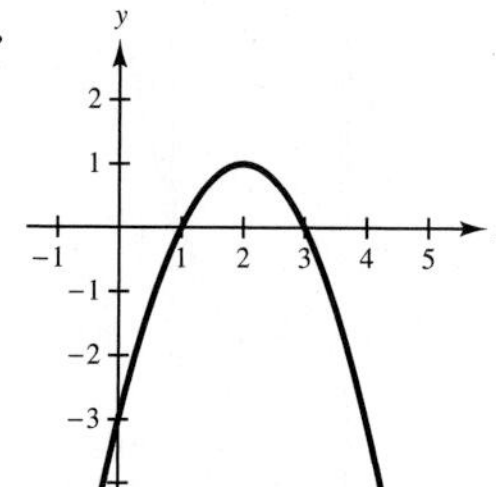

68.

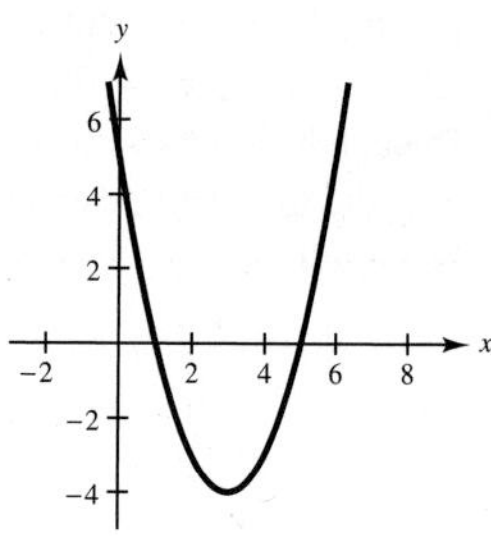

Section 9.3 *(pp. 490–497)*

25.

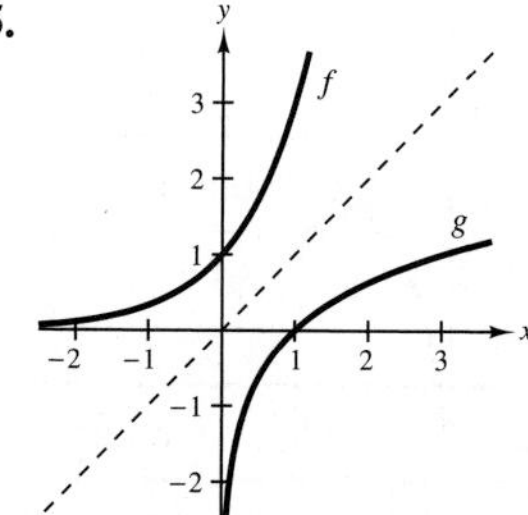

26.

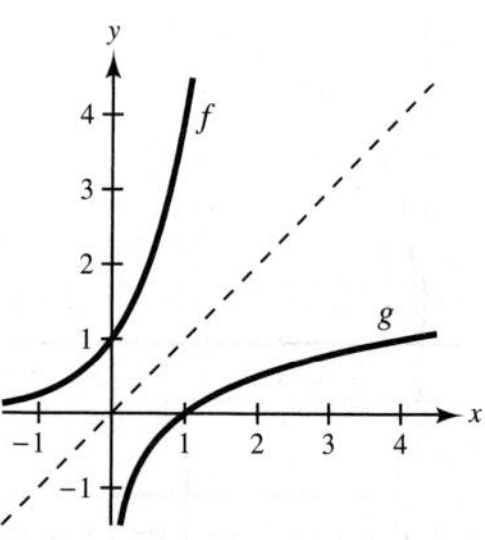

27.

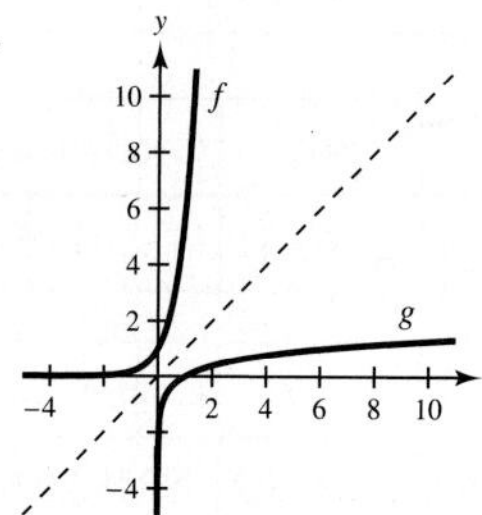

28.

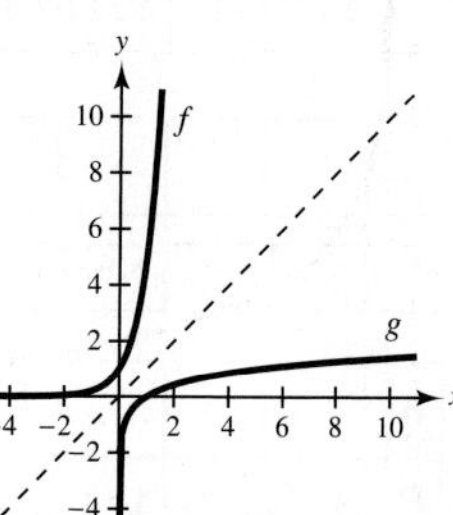

29.

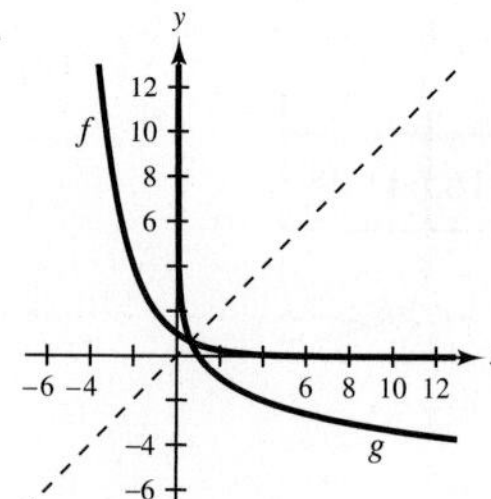

30.

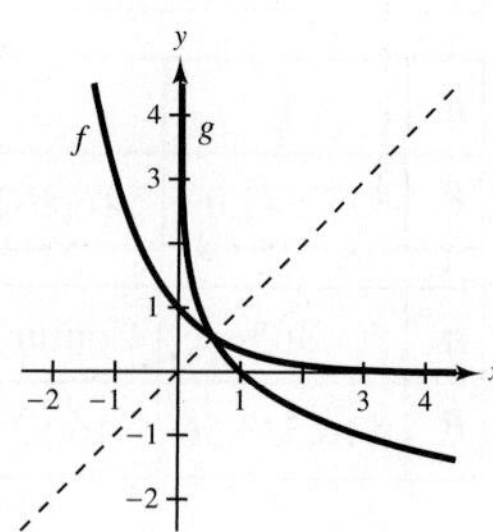

35.

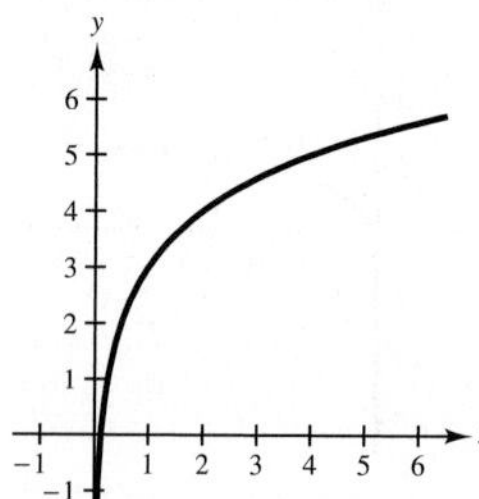

36.

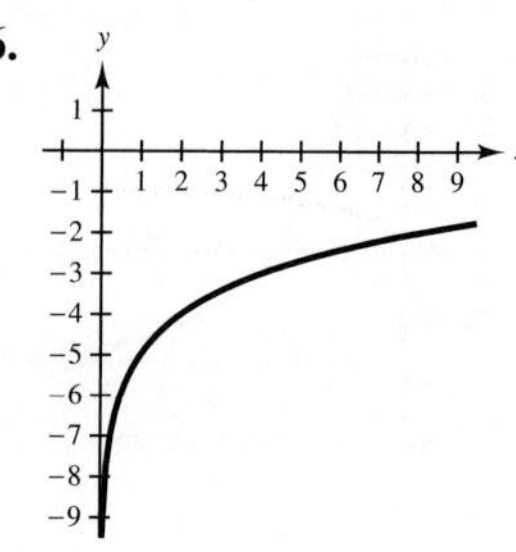

37.

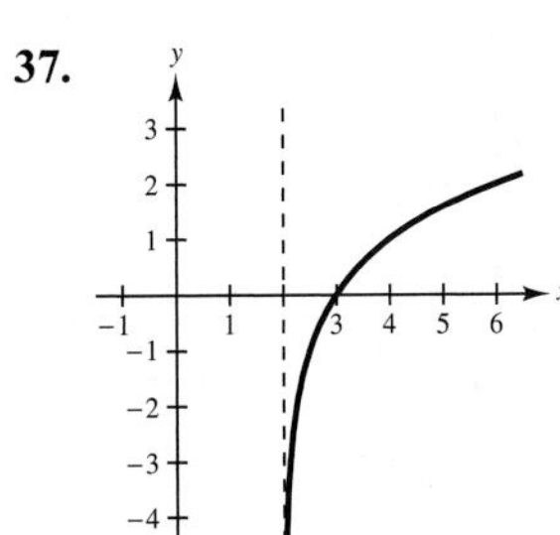

38.

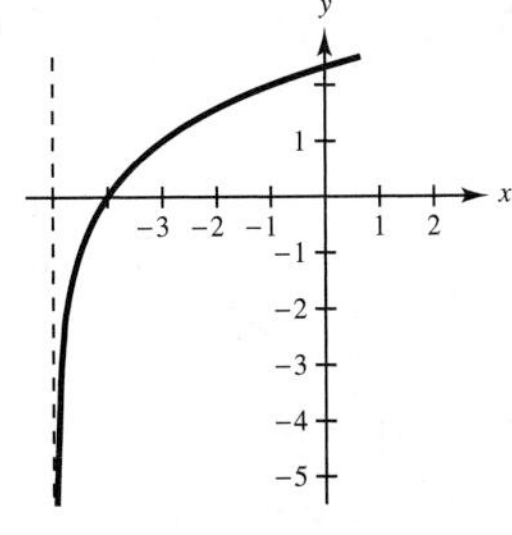

39.

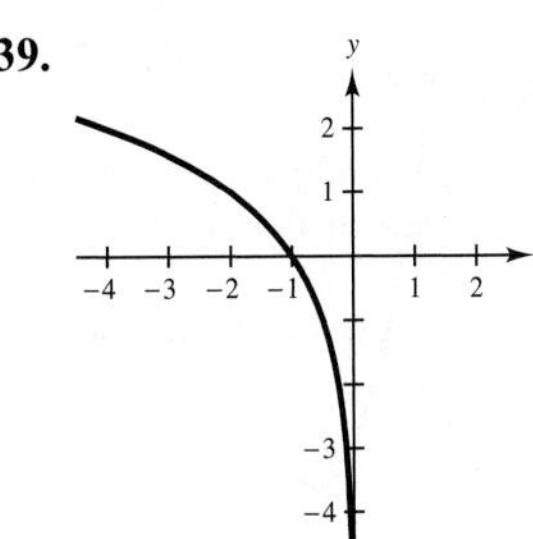

40.

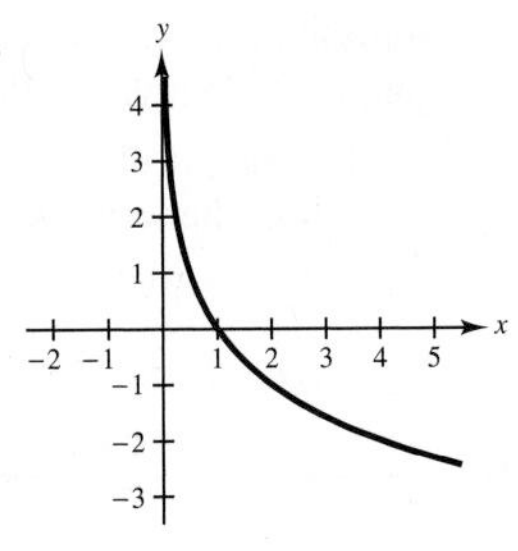

45.

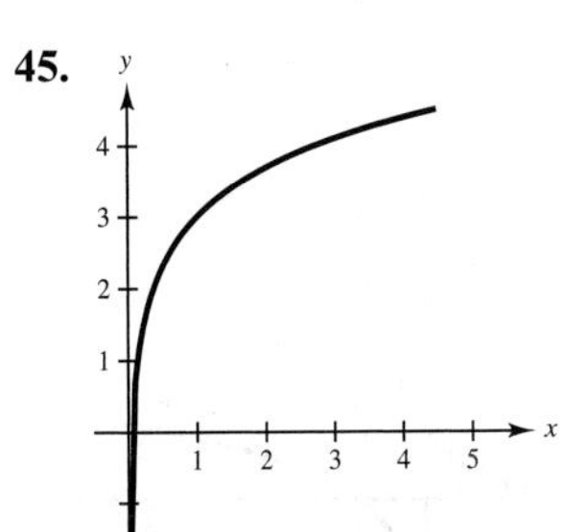

46.

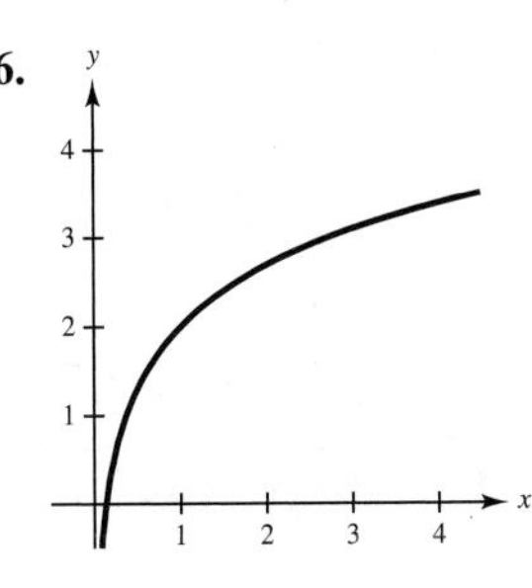

47.

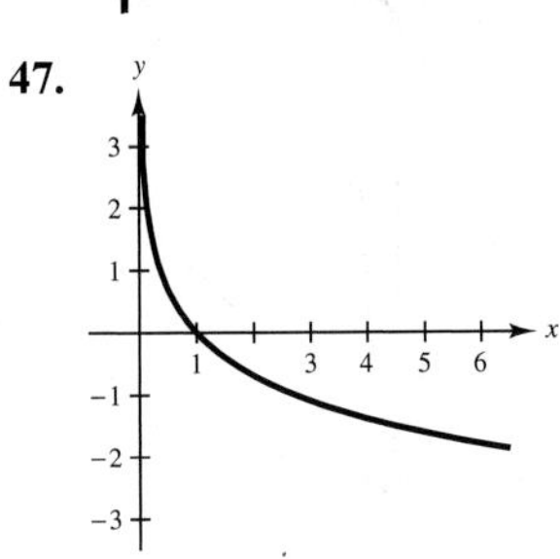

48.

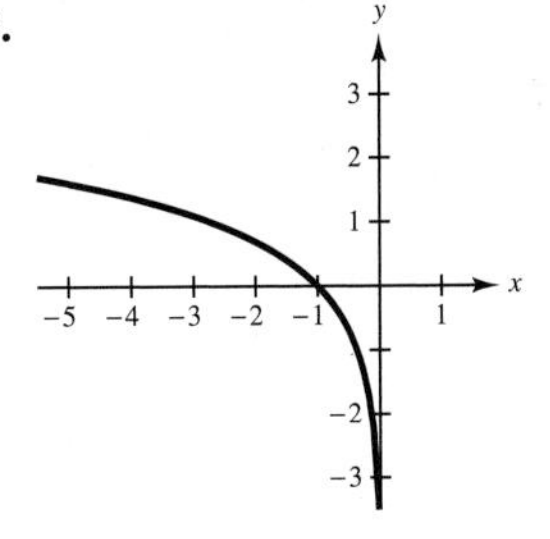

79.

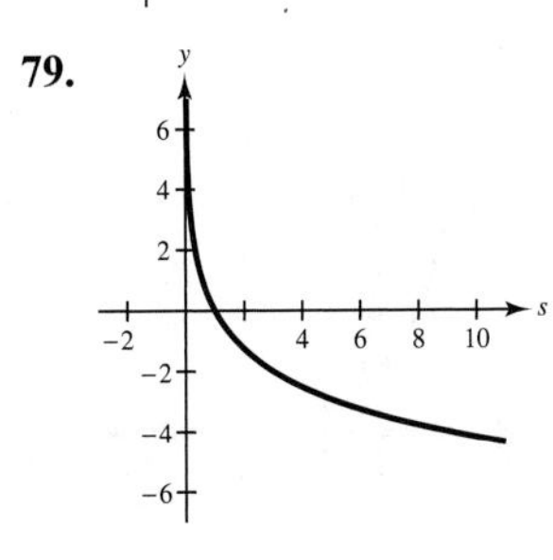

80.

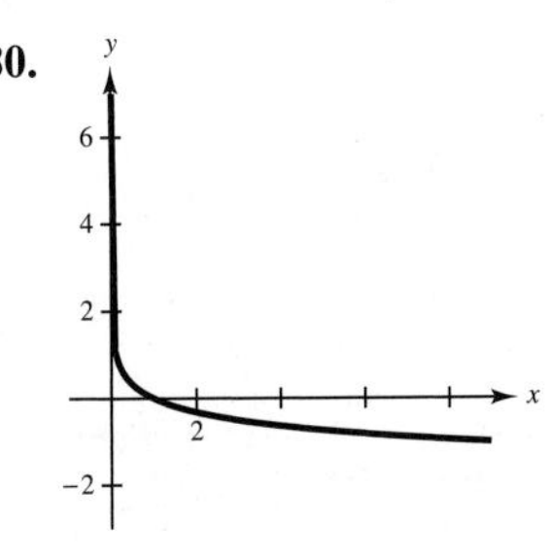

81.

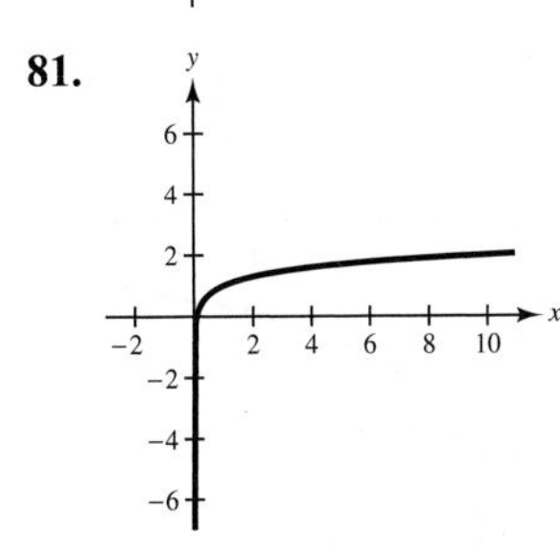

82.

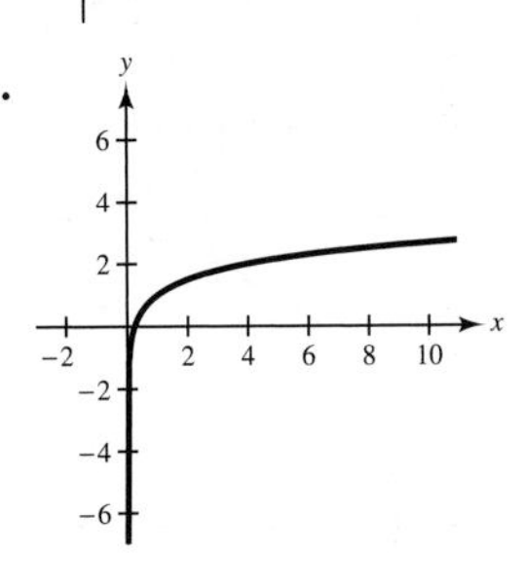

83.

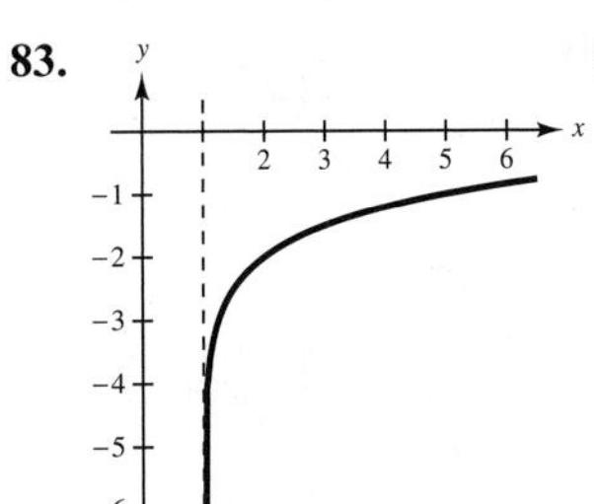

84.

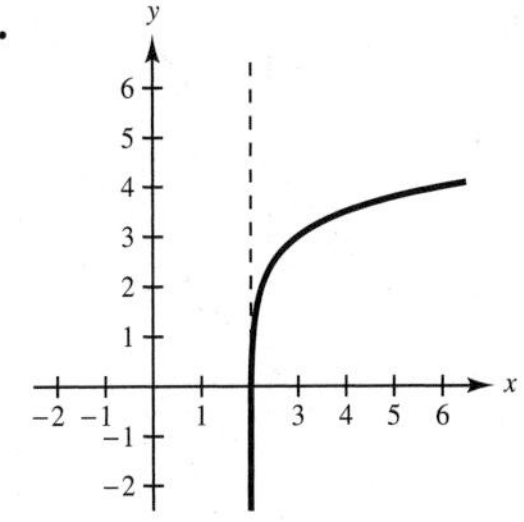

85.

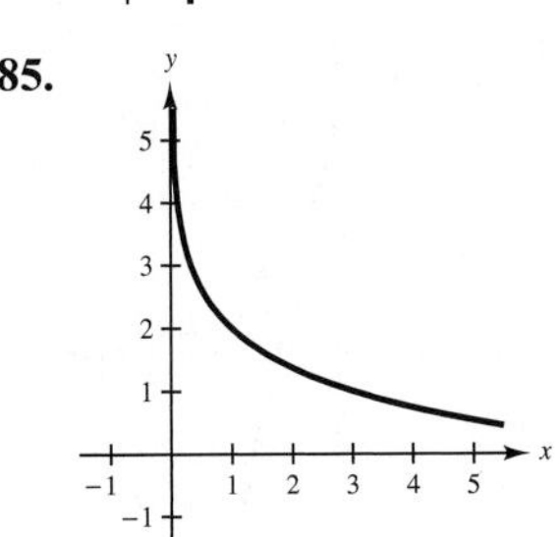

86.

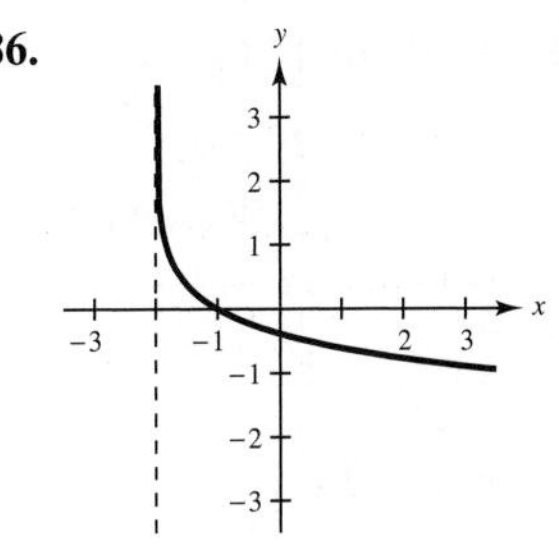

89. (a)

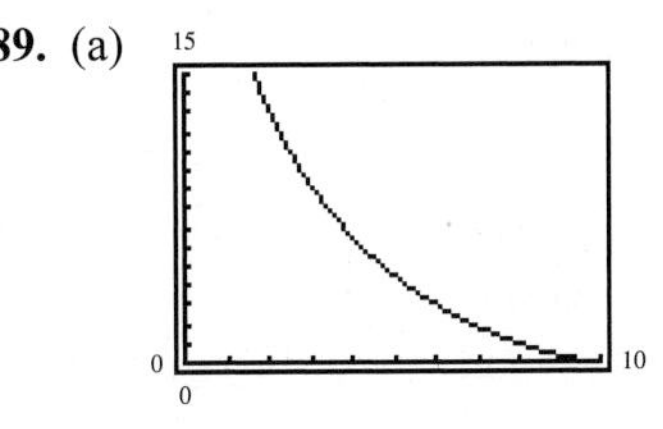

90. (a)

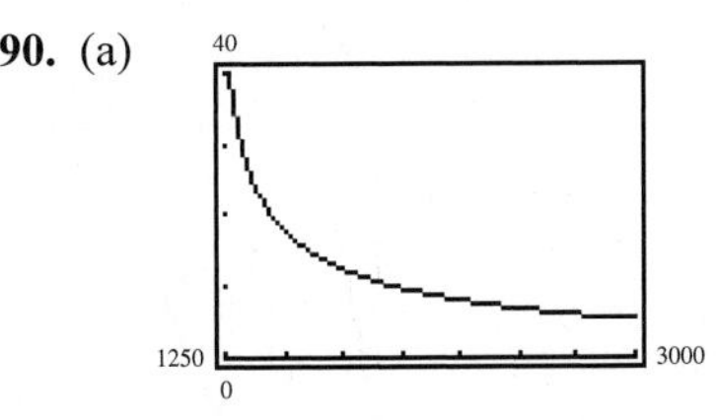

99. Logarithmic functions with base 10 are common logarithms. Logarithmic functions with base e are natural logarithms.

100. The domain of f and the range of g are both $(0, \infty)$

101. A vertical shift or reflection in the x-axis of a logarithmic graph does not affect the domain or range. A horizontal shift or reflection in the y-axis of a logarithmic graph affects the domain, but the range stays the same.

102. By definition, $y = \log_a x$ if and only if $a^y = x$, and because $a^y = x$ is defined when $0 < a < 1$ and $a > 1$, the same must be true for $\log_a x$.

Mid-Chapter Quiz *(p. 498)*

8.

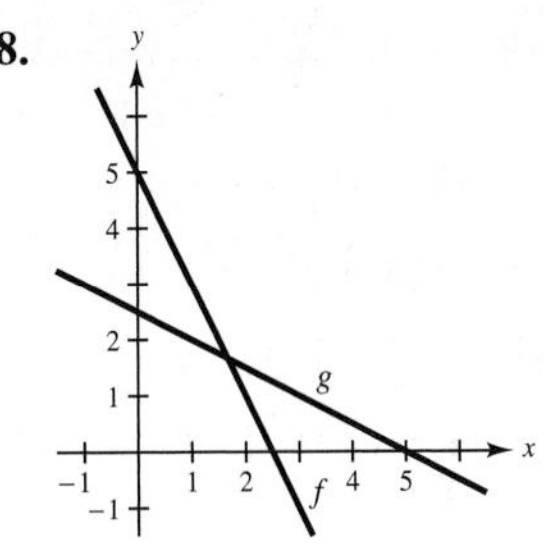

14. 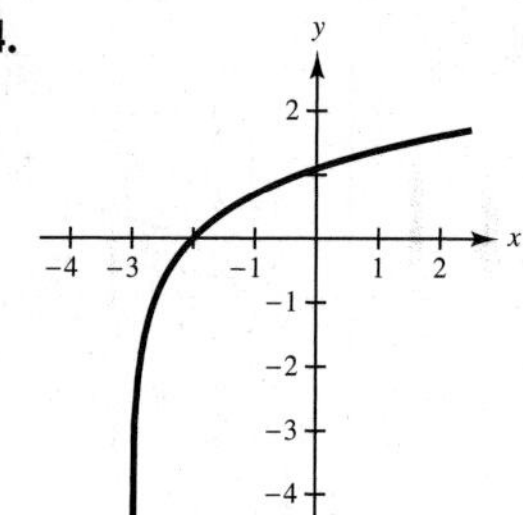

15. 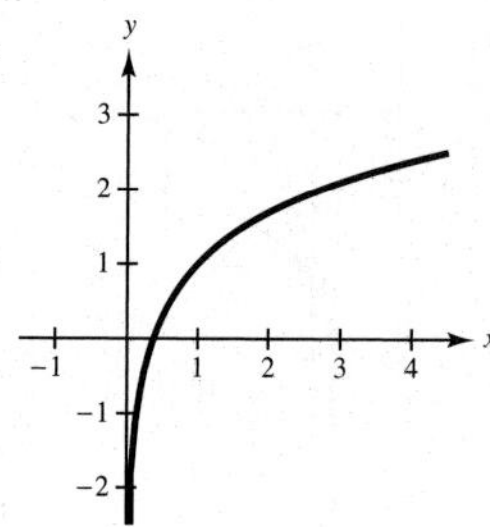

Section 9.4 (pp. 500–507)

99. True. $f(ax) = \log_a ax$
$= \log_a a + \log_a x$
$= 1 + \log_a x$
$= 1 + f(x)$

106.

$\ln 1 = 0$	$\ln 9 \approx 2.1972$
$\ln 2 \approx 0.6931$	$\ln 10 \approx 2.3025$
$\ln 3 \approx 1.0986$	$\ln 12 \approx 2.4848$
$\ln 4 \approx 1.3862$	$\ln 14 \approx 2.6390$
$\ln 5 \approx 1.6094$	$\ln 15 \approx 2.7080$
$\ln 6 \approx 1.7917$	$\ln 16 \approx 2.7724$
$\ln 7 \approx 1.9459$	$\ln 18 \approx 2.8903$
$\ln 8 \approx 2.0793$	$\ln 20 \approx 2.9956$

Explanations will vary. Any differences are due to rounding.

113. 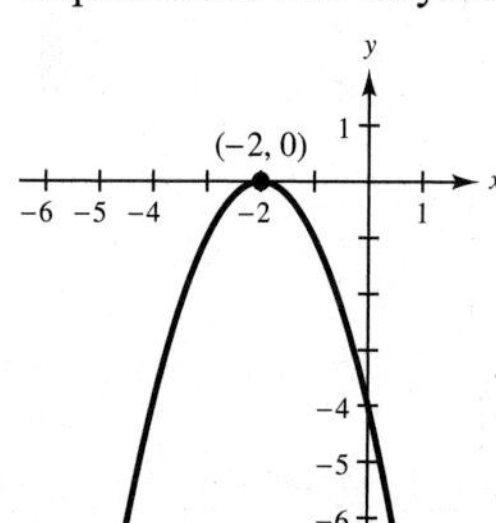

114. 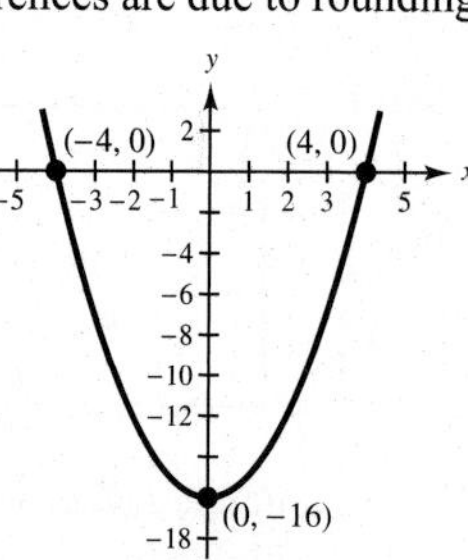

115. 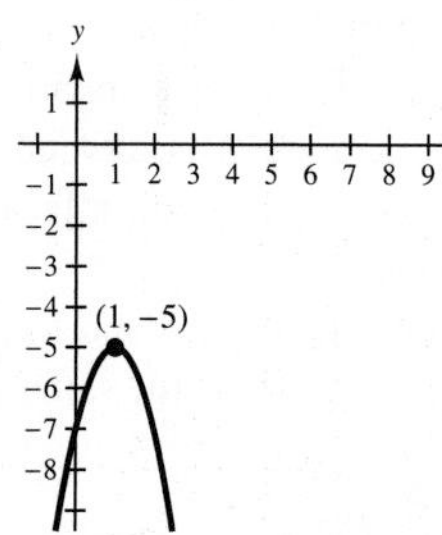

116. 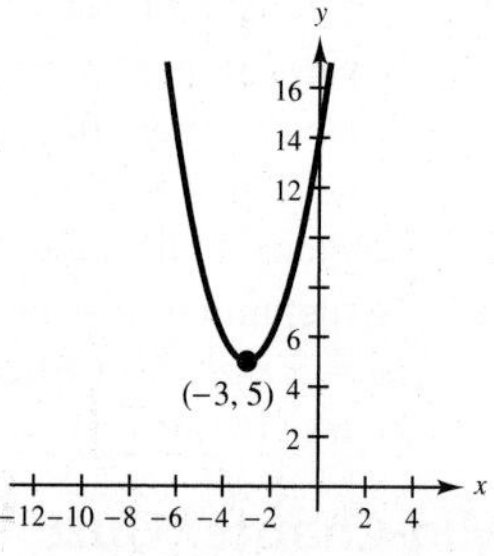

117. (a) $(f \circ g)(x) = 4x - 11$
Domain: $-\infty < x < \infty$
(b) $(g \circ f)(x) = 4x + 4$
Domain: $-\infty < x < \infty$

118. (a) $(f \circ g)(x) = \sqrt{x - 3}$
Domain: $x \geq 3$
(b) $(g \circ f)(x) = \sqrt{x} - 3$
Domain: $x \geq 0$

119. (a) $(f \circ g)(x) = \dfrac{1}{x + 2}$
Domain: $x < -2$ or $x > -2$
(b) $(g \circ f)(x) = \dfrac{1}{x} + 2 = \dfrac{2x + 1}{x}$
Domain: $x < 0$ or $x > 0$

120. (a) $(f \circ g)(x) = \dfrac{5}{x^2 + 2x - 3}$
Domain: $x < -3$ or $-3 < x < 1$ or $x > 1$
(b) $(g \circ f)(x) = \dfrac{5}{x^2 - 4} + 1 = \dfrac{x^2 + 1}{x^2 - 4}$
Domain: $x < -2$ or $-2 < x < 2$ or $x > 2$

Section 9.5 (pp. 508–515)

89. To solve an exponential equation, first isolate the exponential expression, then take the logarithm of each side of the equation and solve for the variable.
To solve a logarithmic equation, first isolate the logarithmic expression, then exponentiate each side of the equation and solve for the variable.

Section 9.6 (pp. 516–523)

55. 56. 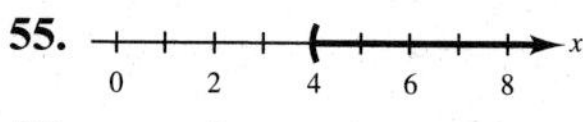

57. 58. 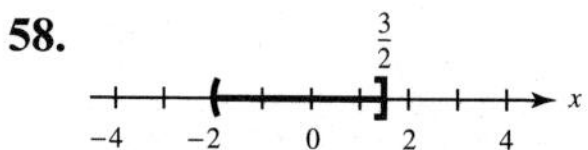

Review Exercises (pp. 526–529)

5. 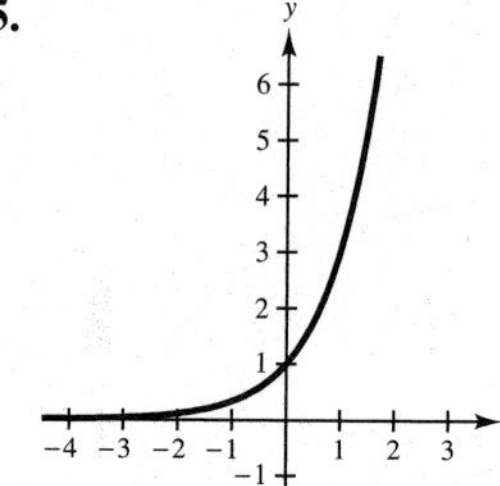

6. 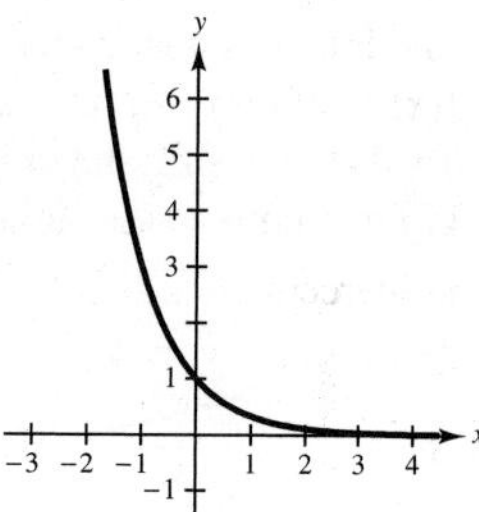

7. 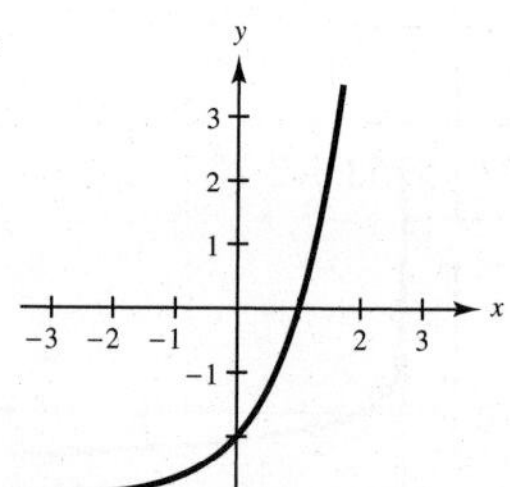

8. 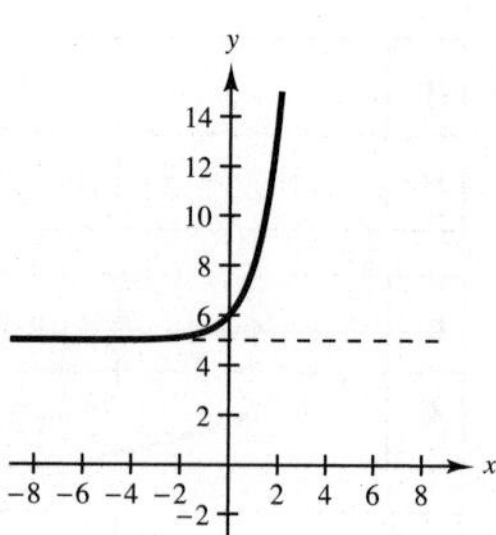

9. 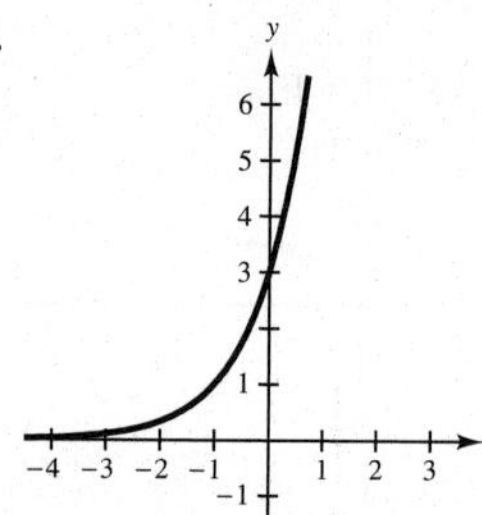

10. 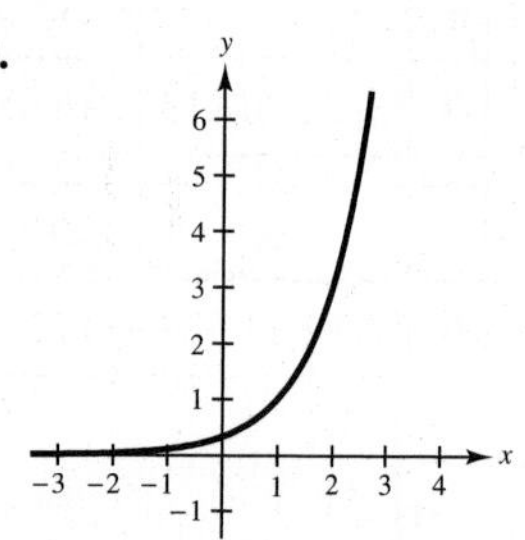

13. The function y is related to $f(x) = e^x$ by $y = f(-x) + 1$. To sketch y, reflect the graph of f in the y-axis, and shift the graph one unit upward.
y-intercept: $(0, 2)$
Asymptote: $y = 1$

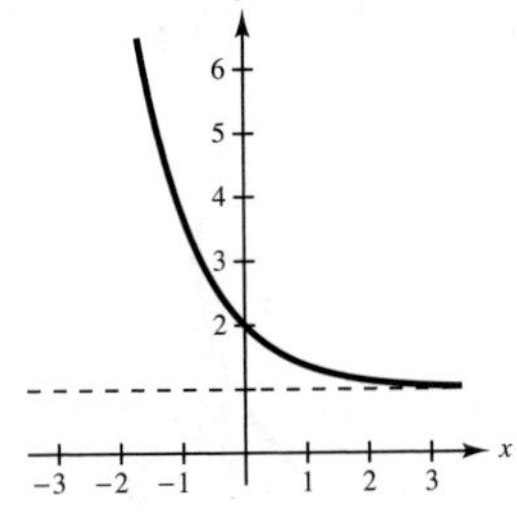

14. The function y is related to $f(x) = e^x$ by $y = -f(x) + 1$. To sketch y, reflect the graph of f in the x-axis, and shift the graph one unit upward.
y-intercept: $(0, 0)$
Asymptote: $y = 1$

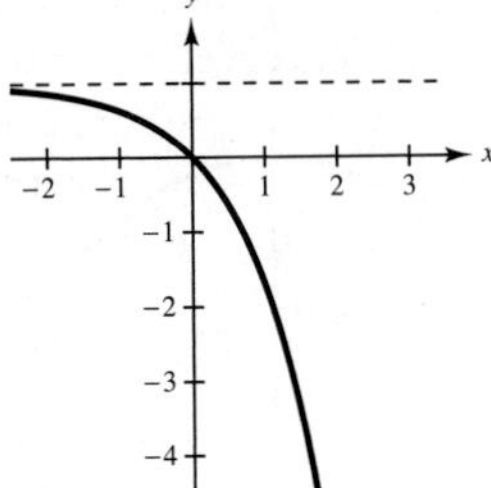

15. The function g is related to $f(x) = e^x$ by $y = f(x + 2)$. To sketch g, shift the graph of f two units to the left.
y-intercept: $(0, e^2)$
Asymptote: $y = 0$

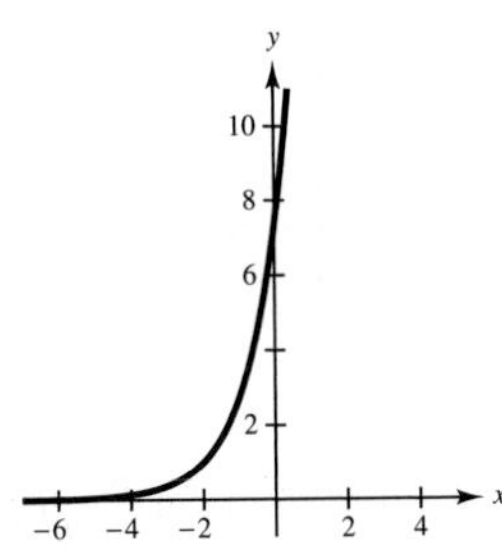

16. The function h is related to $f(x) = e^x$ by $y = f(x - 2)$. To sketch h, shift the graph of f two units to the right.
y-intercept: $(0, e^{-2})$
Asymptote: $y = 0$

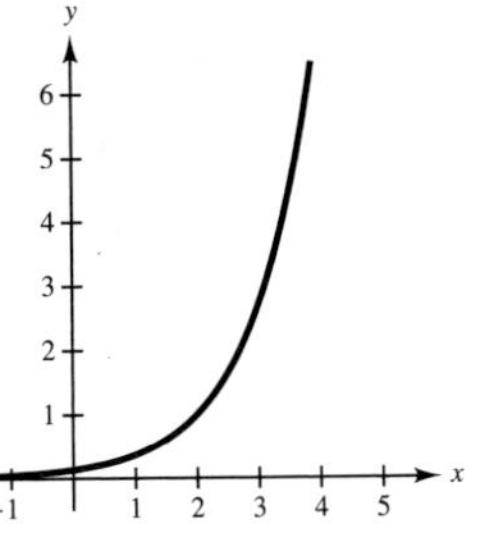

17.

n	1	4	12
A	$226,296.28	$259,889.34	$268,503.32

n	365	Continuous
A	$272,841.23	$272,990.75

18.

n	1	4	12
A	$152,203.13	$167,212.90	$170,948.62

n	365	Continuous
A	$172,813.72	$172,877.82

20.

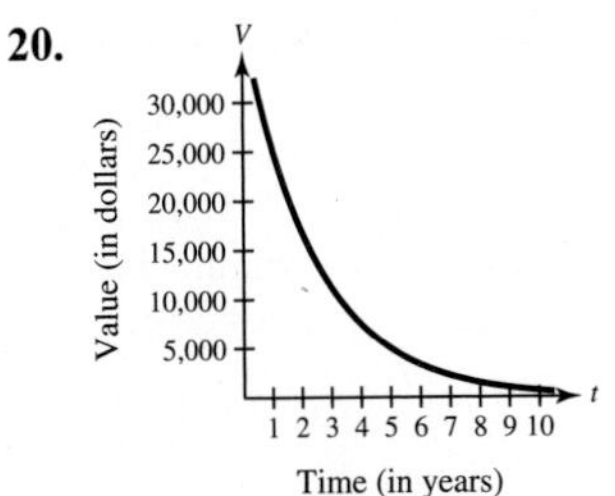

35.

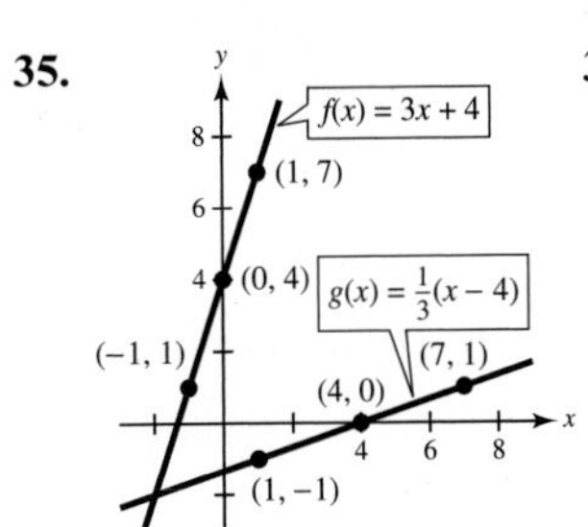

36.

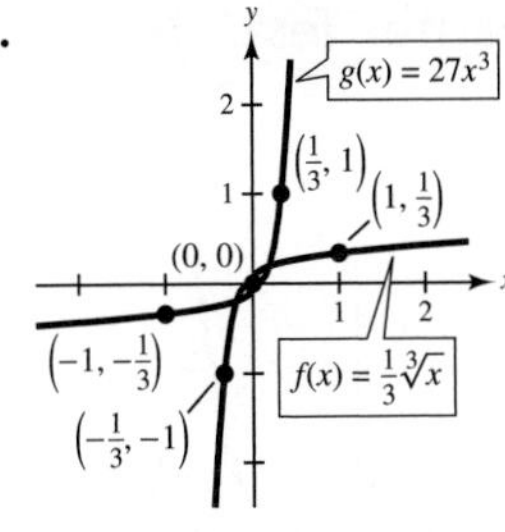

45.

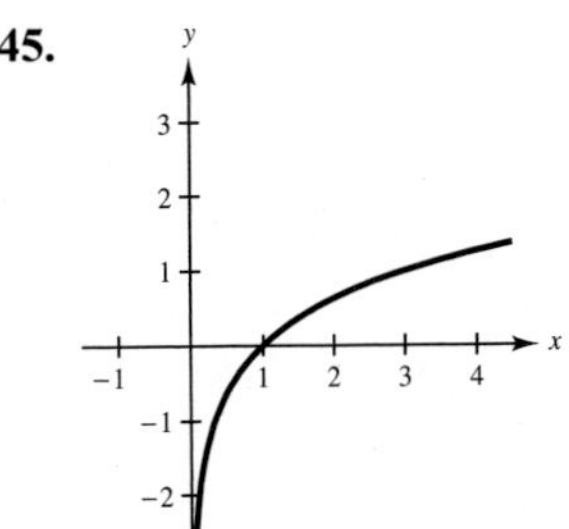

46.

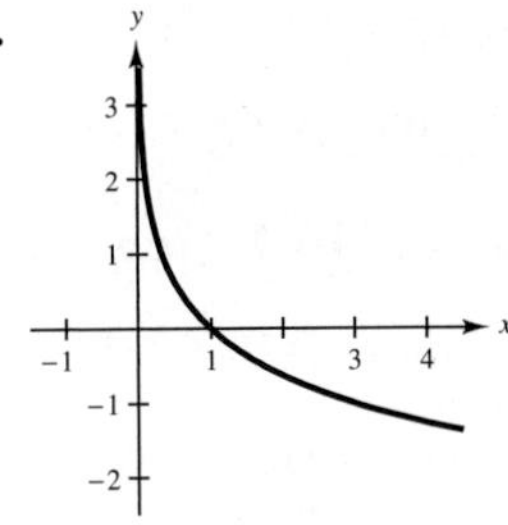

47.

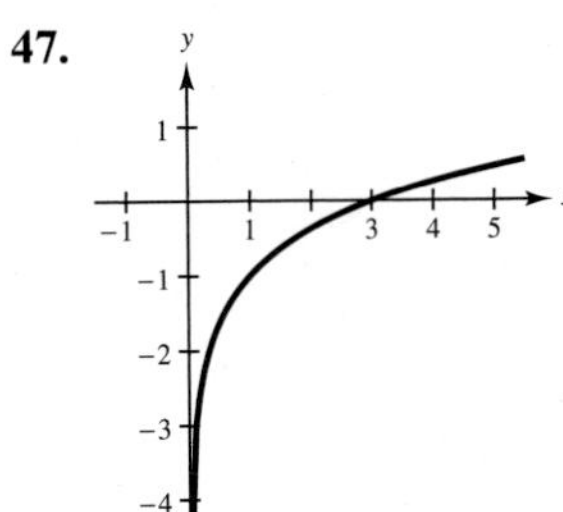

48.

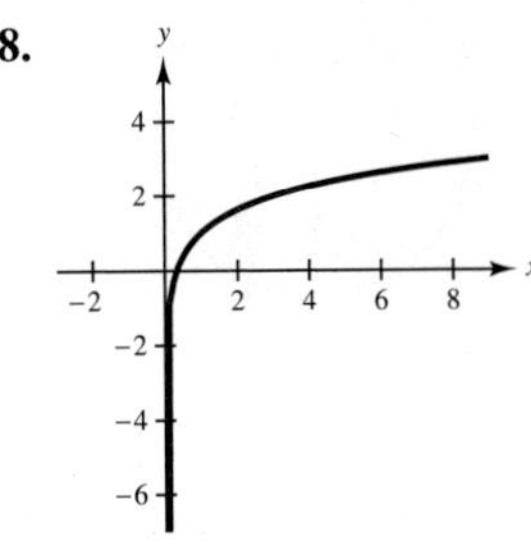

49.

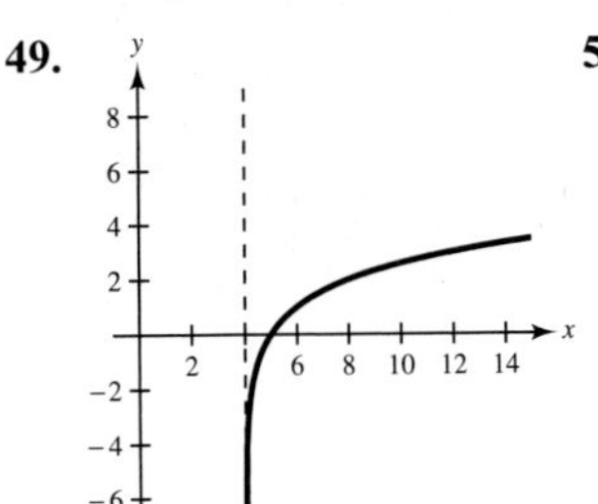

50.

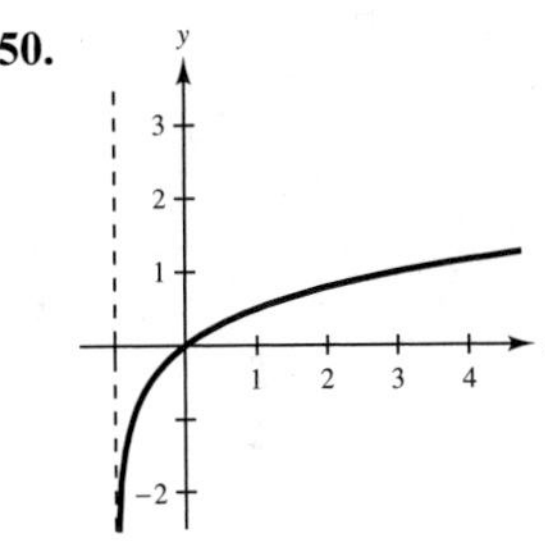

53.

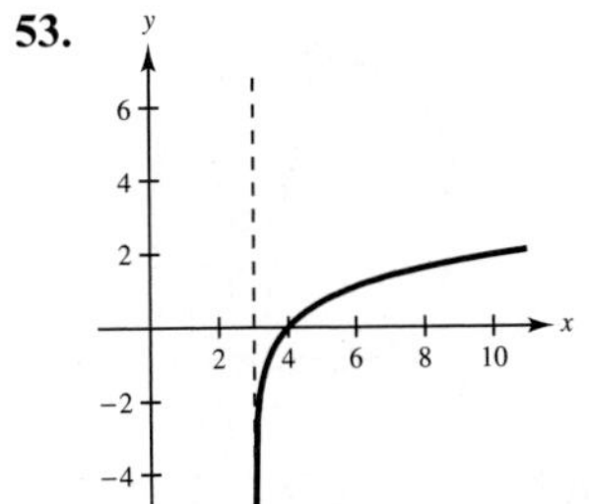

54.

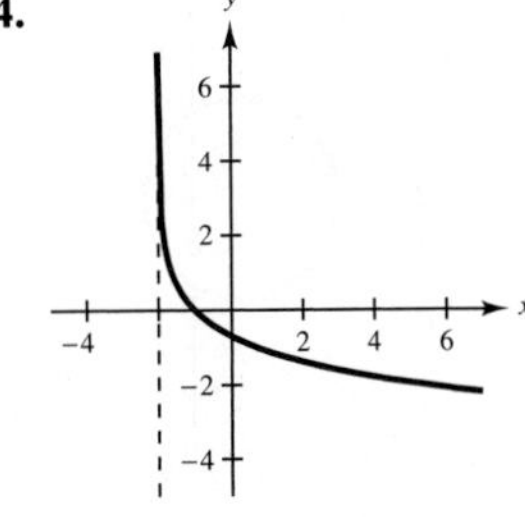

55.

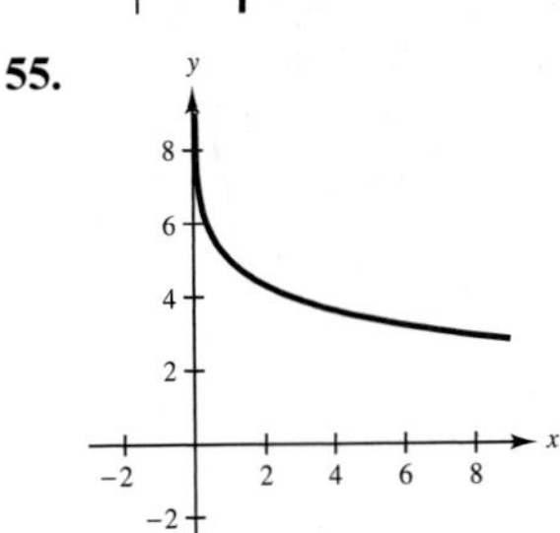

56.

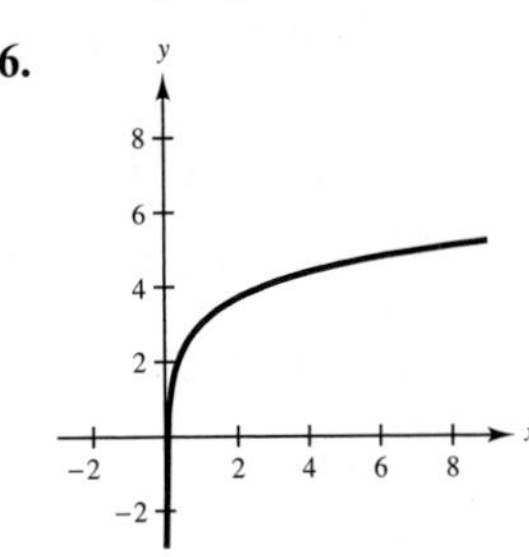

127. The earthquake of 1906 was about 19,953 times as intense.
128. The earthquake in Virginia was about 3.2 times as intense.

ADDITIONAL ANSWERS

Chapter Test *(p. 530)*

2.

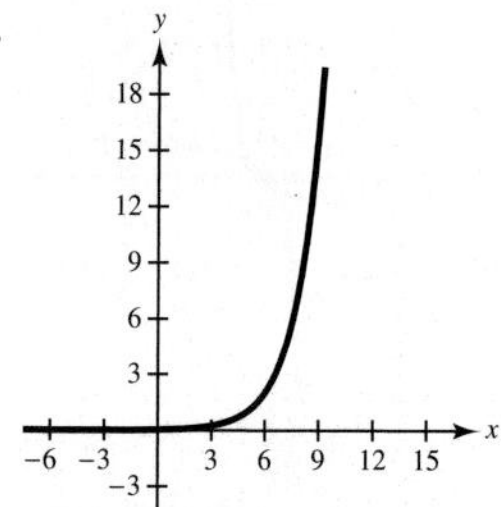

CHAPTER 10

Section 10.1 *(pp. 532–539)*

9. Center: $(0, 0)$, $r = 4$

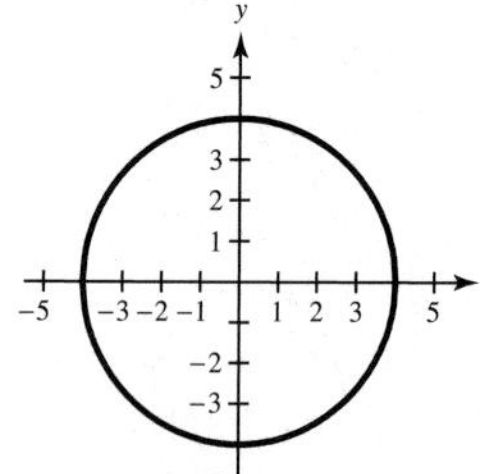

10. Center: $(0, 0)$, $r = 1$

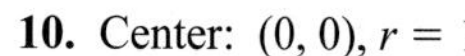

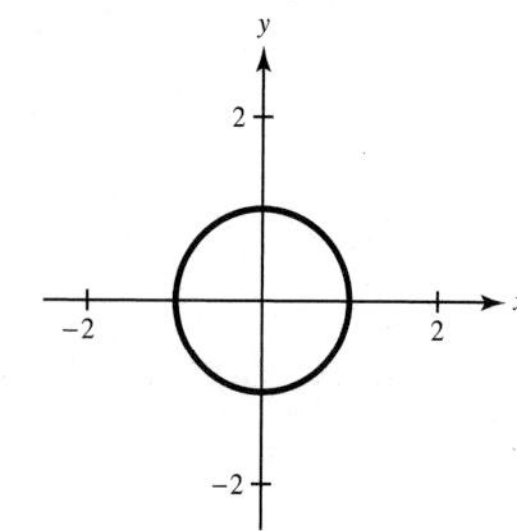

11. Center: $(0, 0)$, $r = 6$

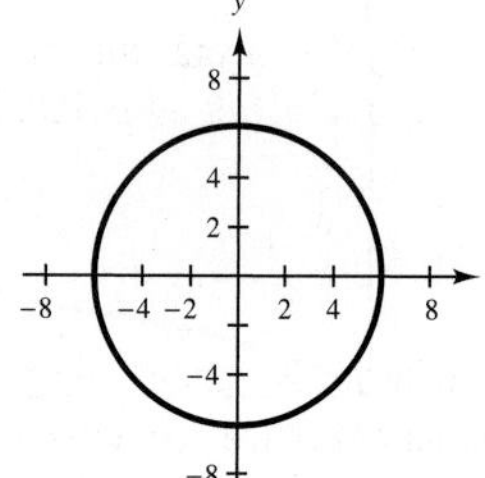

12. Center: $(0, 0)$, $r = \sqrt{15}$

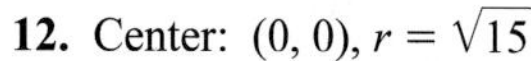

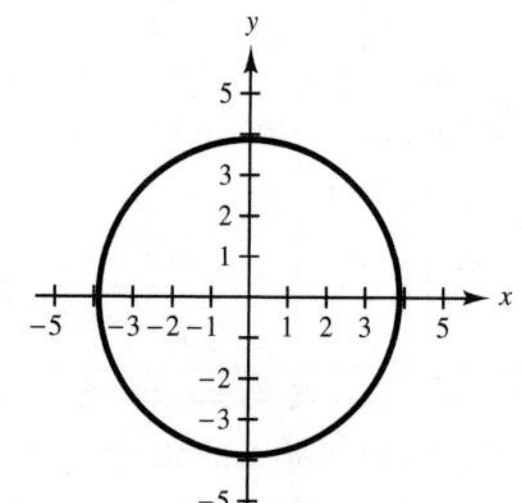

19.

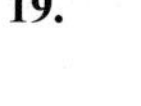

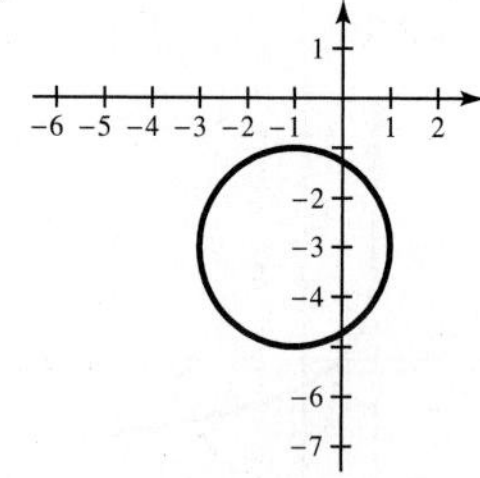

20.

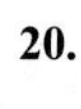

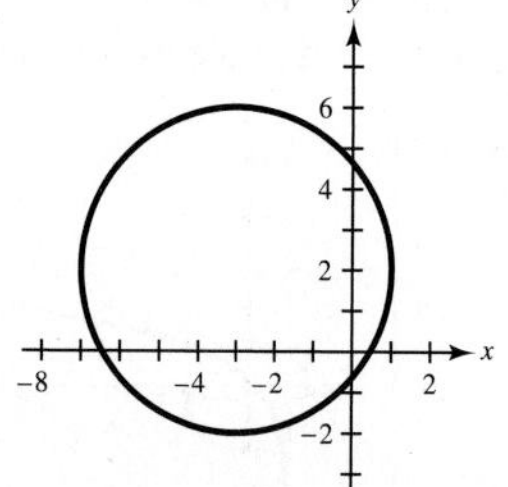

29.

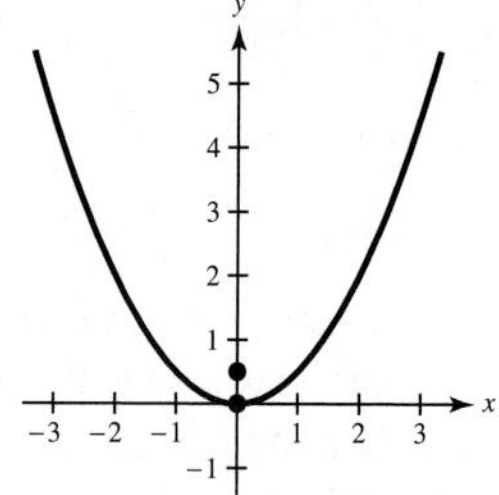

30.

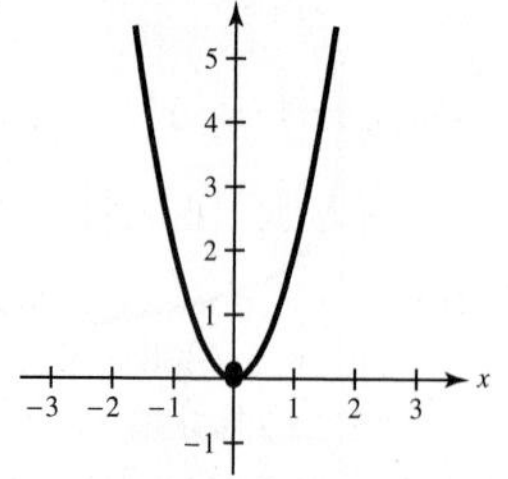

31.

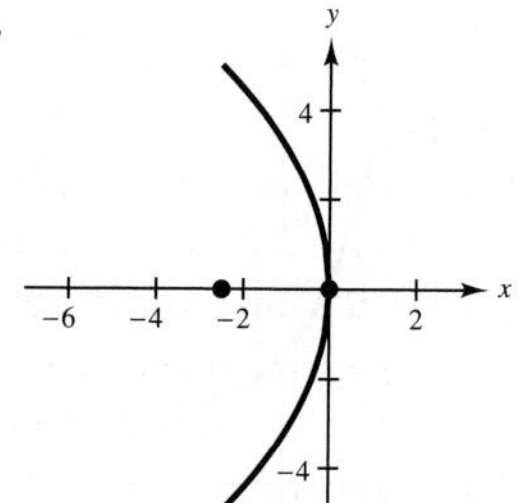

32.

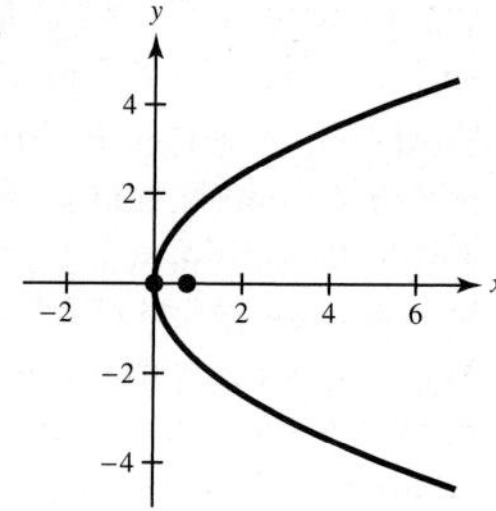

33.

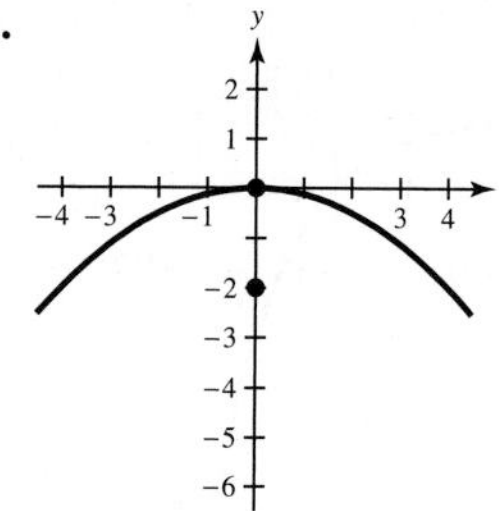

34.

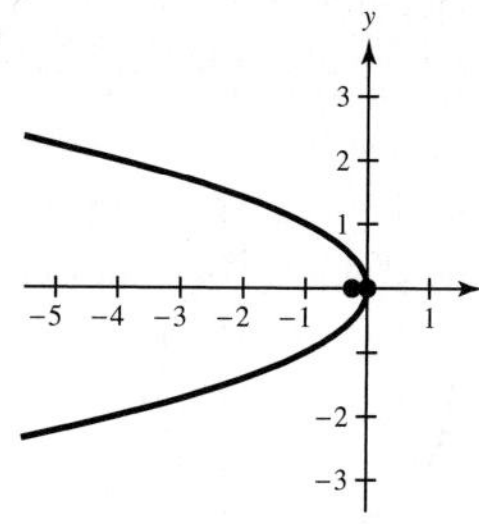

35.

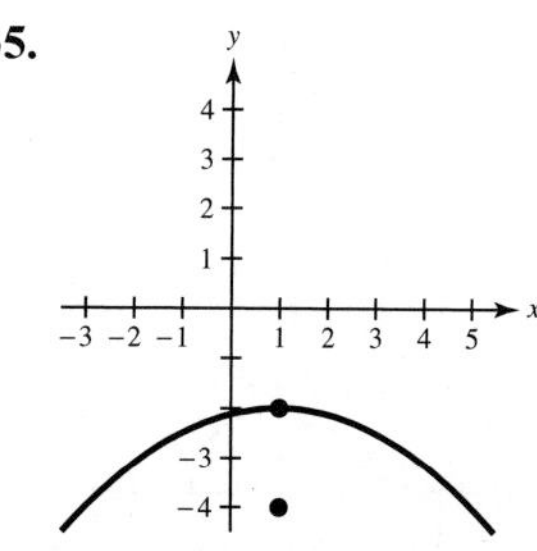

36.

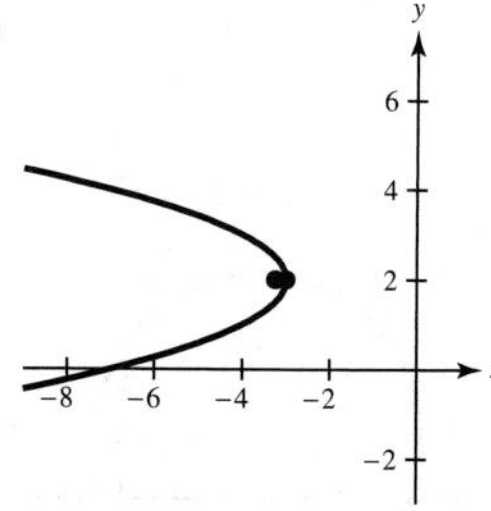

37.

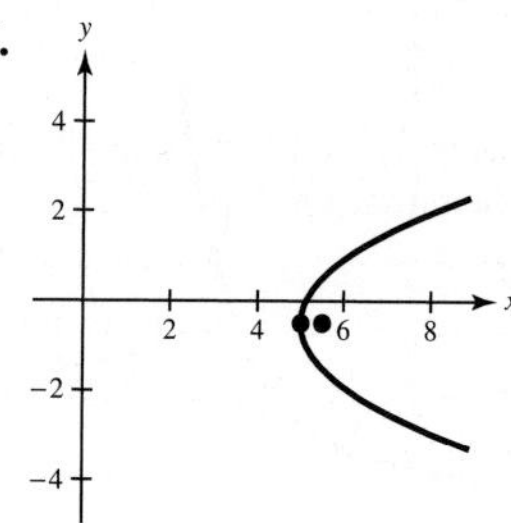

38.

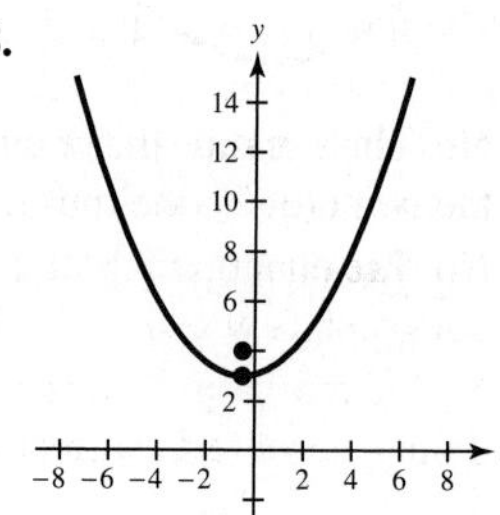

39.

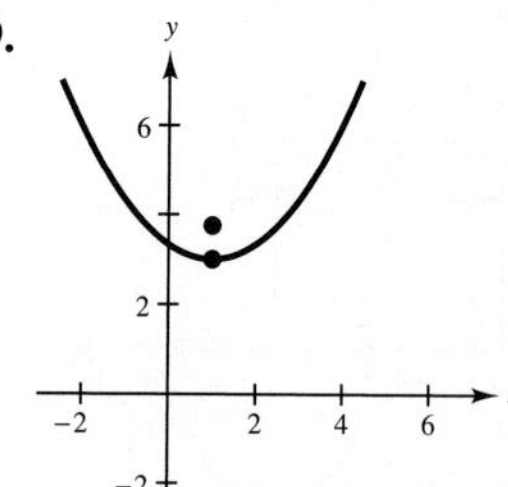

40.

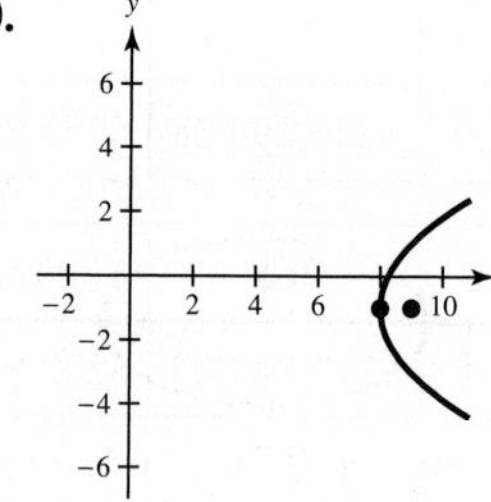

41.

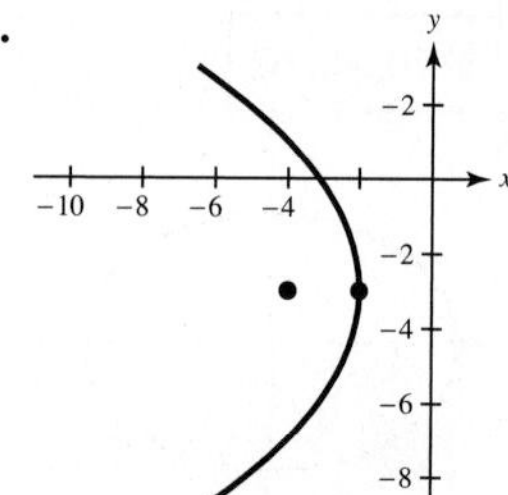

42.

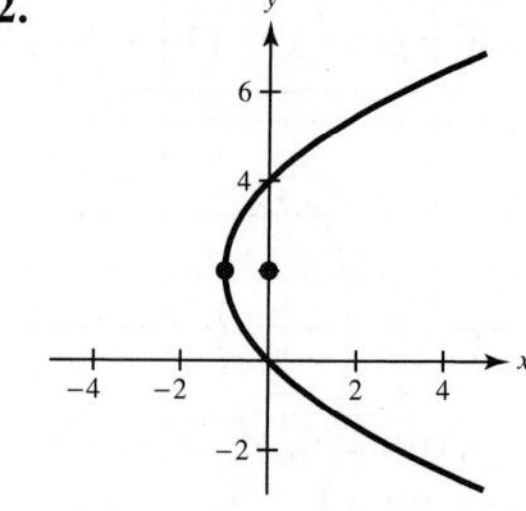

43. First write the equation in standard form: $x^2 + y^2 = 6^2$. Because $h = 0$ and $k = 0$, the center is the origin. Because $r^2 = 6^2$, the radius is 6 units.

44. No. The equation can be rewritten as
$[x - (-2)]^2 + [y - (-4)]^2 = 20$
to show that $h = -2$ and $k = -4$. So, the center is $(-2, -4)$ and is shifted two units to the *left* and four units *downward*.

45. The equation $(x - h)^2 = 4p(y - k)$, $p \neq 0$, should be used because the vertex and the focus lie on the same vertical axis.

46. When x is the variable squared in the standard equation of a parabola, the parabola opens upward when $p > 0$ and downward when $p < 0$. When y is the variable squared in the standard equation of a parabola, the parabola opens to the right when $p > 0$ and to the left when $p < 0$.

47.

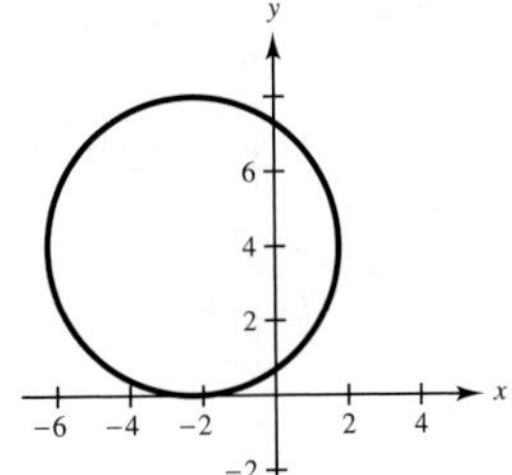

48.

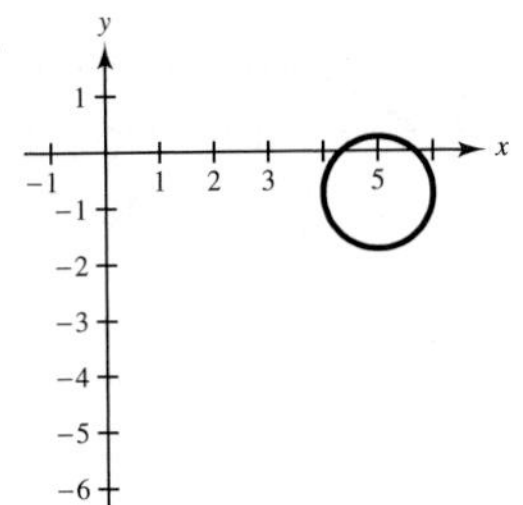

49.

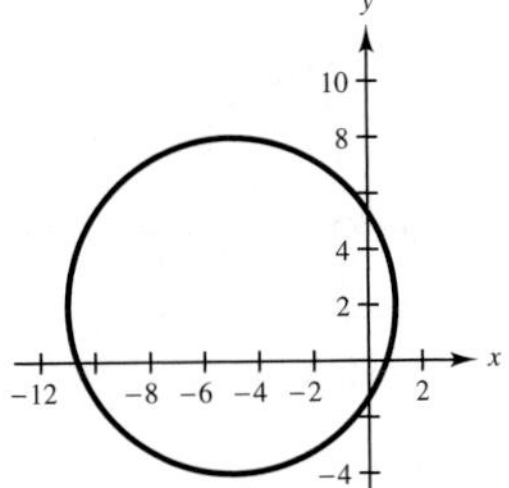

50.

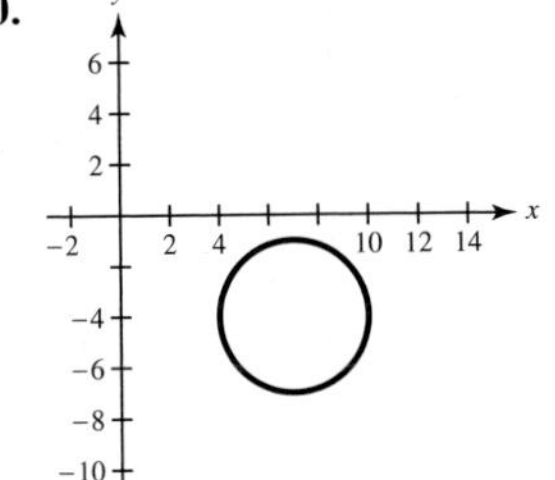

55. No. The equation of the circle is $(x + 1)^2 + (y - 1)^2 = 13$, and the point $(3, 2)$ does not satisfy the equation.

56. No. The equation in standard form is $x^2 + (y - 3)^2 = 4$, which does represent a circle with center $(0, 3)$ and a radius of 2.

58. No. If the graph intersected the directrix, there would exist points closer to the directrix than the focus.

Section 10.2 *(pp. 540–547)*

9.

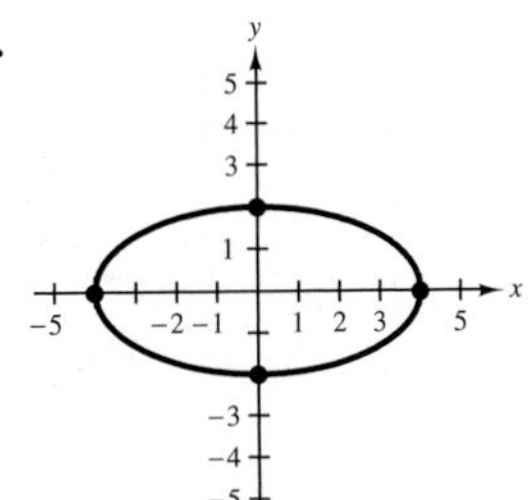

10.

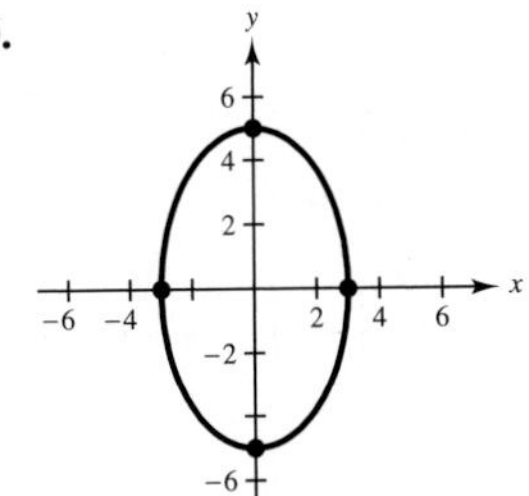

11.

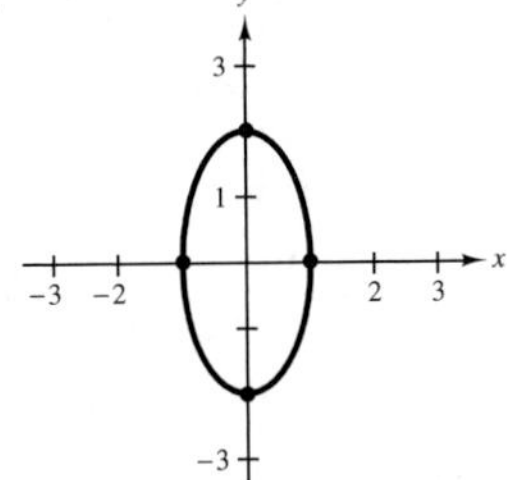

12.

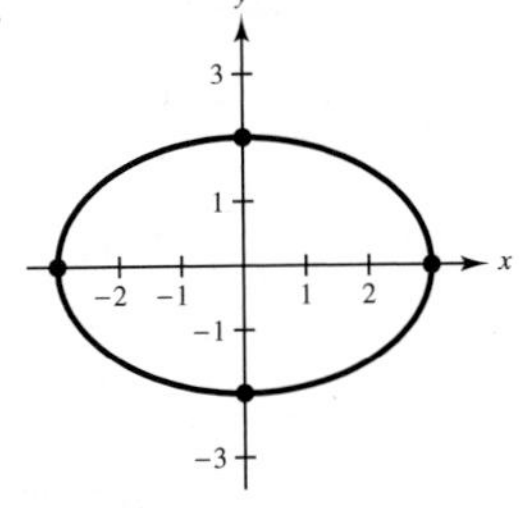

21.

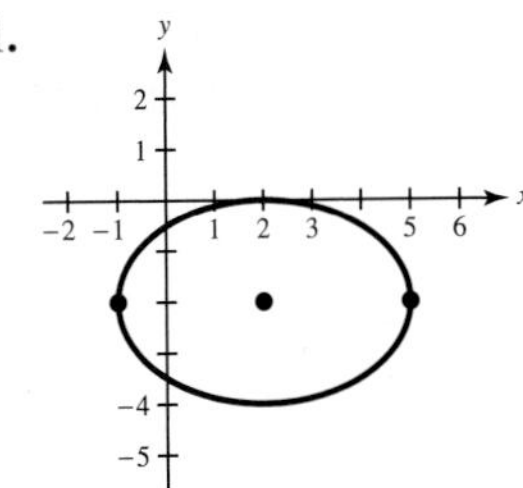

22.

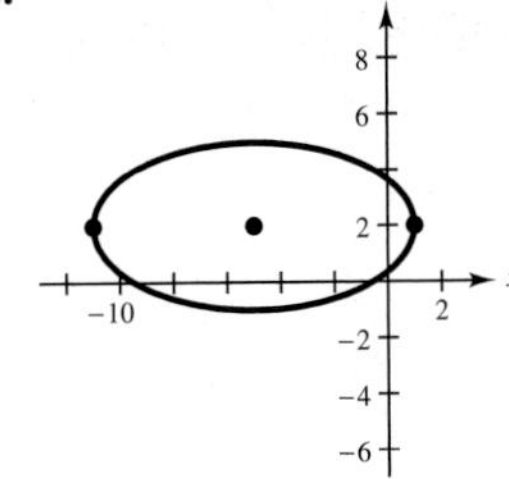

23.

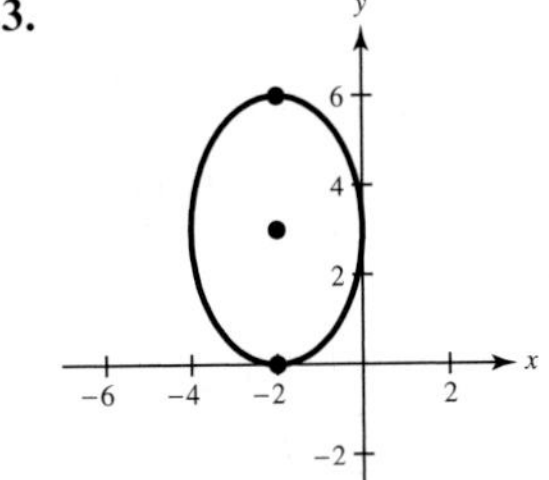

24.

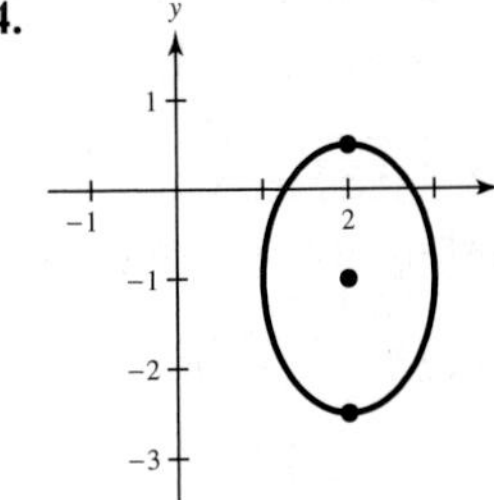

25.

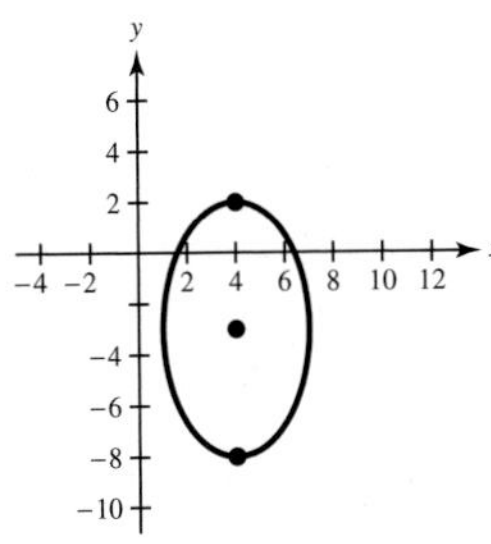

26.

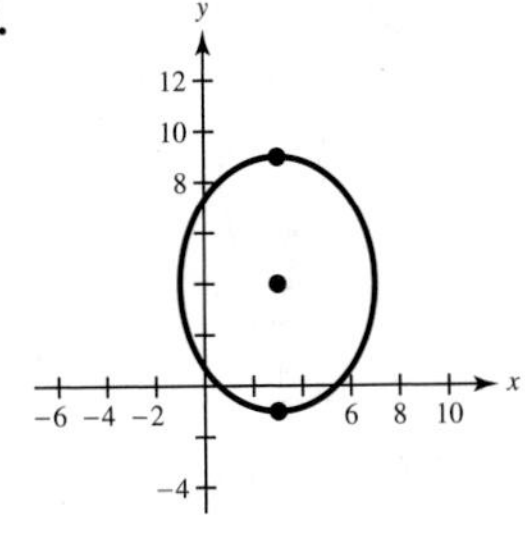

29. An ellipse is the set of all points (x, y) such that the sum of the distances between (x, y) and two distinct fixed points is a constant.

$$\frac{x^2}{a^2} + \frac{y^2}{b^2} = 1 \text{ or } \frac{x^2}{b^2} + \frac{y^2}{a^2} = 1$$

32. When the denominator of the x^2-term is greater than the denominator of the y^2-term, the major axis is horizontal and the minor axis is vertical. When the denominator of the y^2-term is greater than the denominator of the x^2-term, the major axis is vertical and the minor axis is horizontal.

35.

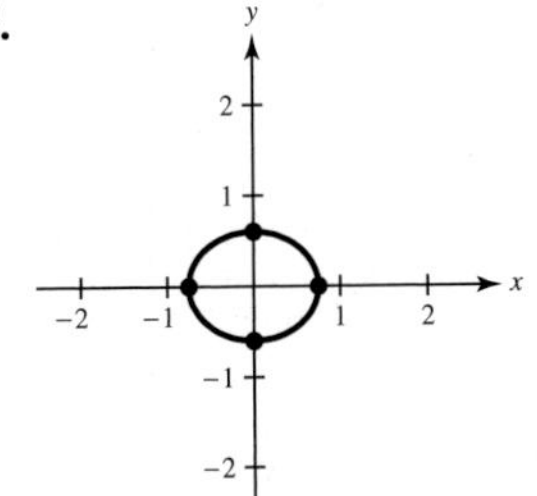

Vertices: $\left(\pm\frac{3}{4}, 0\right)$

Co-vertices: $\left(0, \pm\frac{3}{5}\right)$

36.

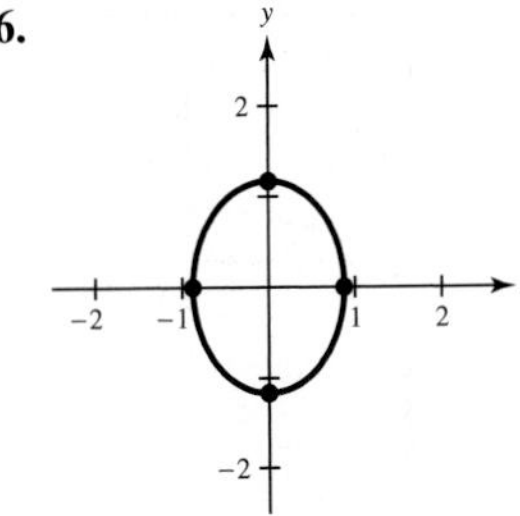

Vertices: $\left(0, \pm\frac{7}{6}\right)$

Co-vertices: $\left(\pm\frac{7}{8}, 0\right)$

39. (a) Every point on the ellipse represents the maximum distance (800 miles) that the plane can safely fly with enough fuel to get from airport A to airport B.

(b) Airport A: $(-250, 0)$; Airport B: $(250, 0)$

(c) 800 miles; Vertices: $(\pm 400, 0)$

(d) $\dfrac{x^2}{400^2} + \dfrac{y^2}{\left(50\sqrt{39}\right)^2} = 1$

(e) $20{,}000\sqrt{39}\pi \approx 392{,}385$ square miles

40. (a) Every point on the ellipse represents the maximum distance (800 miles) that the plane can safely fly with enough fuel to get from airport A to airport B.
(b) Airport A: $(-325, 0)$; Airport B: $(325, 0)$
(c) 800 miles; Vertices: $(\pm 400, 0)$
(d) $\dfrac{x^2}{400^2} + \dfrac{y^2}{(25\sqrt{87})^2} = 1$
(e) $10{,}000\sqrt{87}\pi \approx 293{,}028$ square miles

45. The graph of an ellipse written in the standard form
$$\frac{(x-h)^2}{a^2} + \frac{(y-k)^2}{b^2} = 1$$
intersects the y-axis when $|h| > a$ and intersects the x-axis when $|k| > b$. Similarly, the graph of
$$\frac{(x-h)^2}{b^2} + \frac{(y-k)^2}{a^2} = 1$$
intersects the y-axis when $|h| > b$ and intersects the x-axis when $|k| > a$.

46.
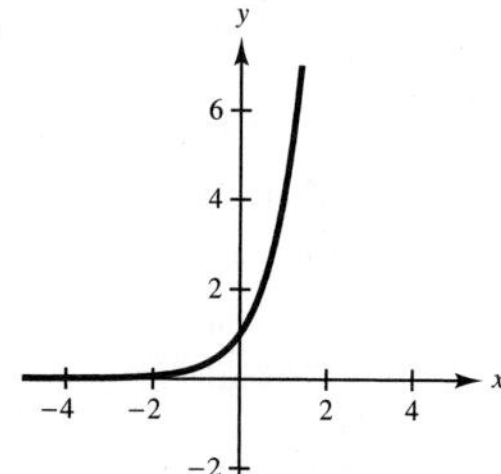

47.
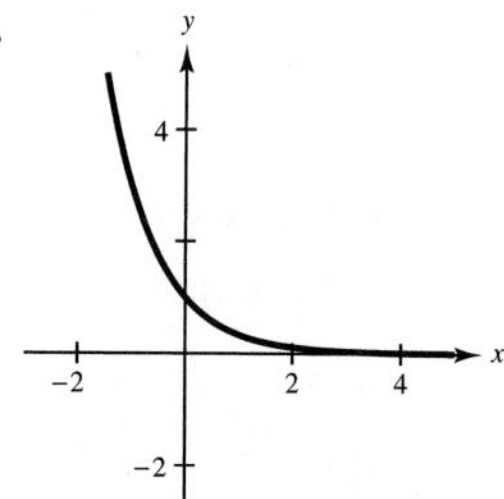

48.
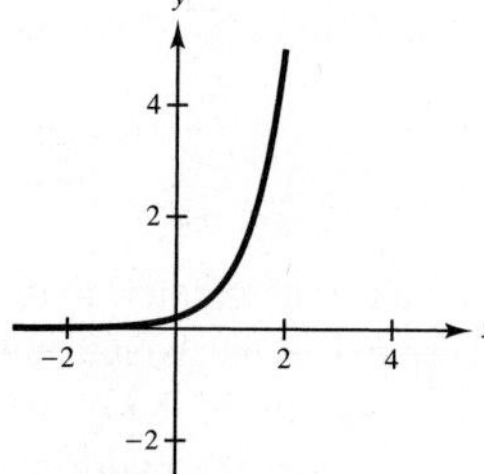

49.
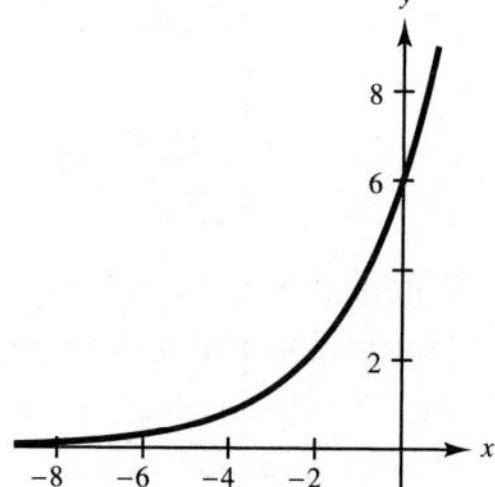

50.
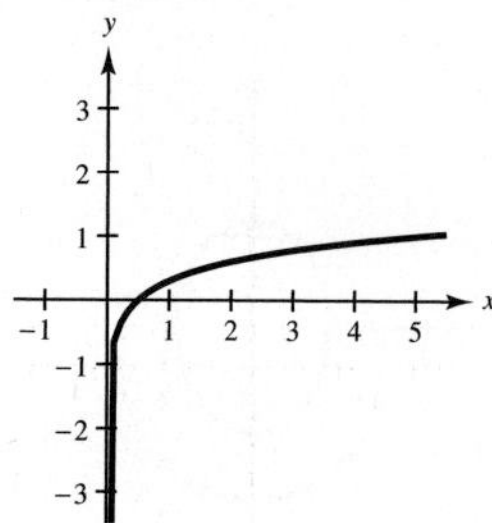

51.
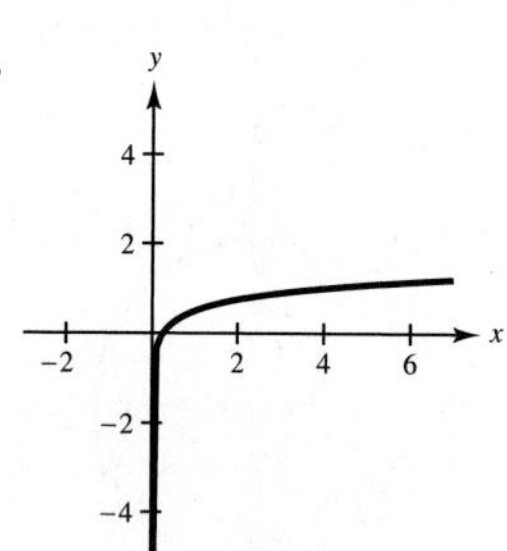

52.
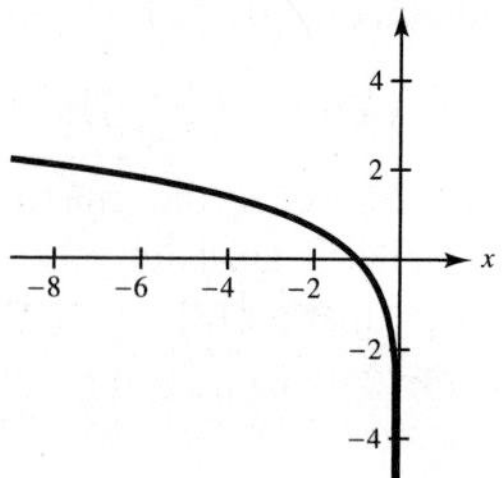

53.
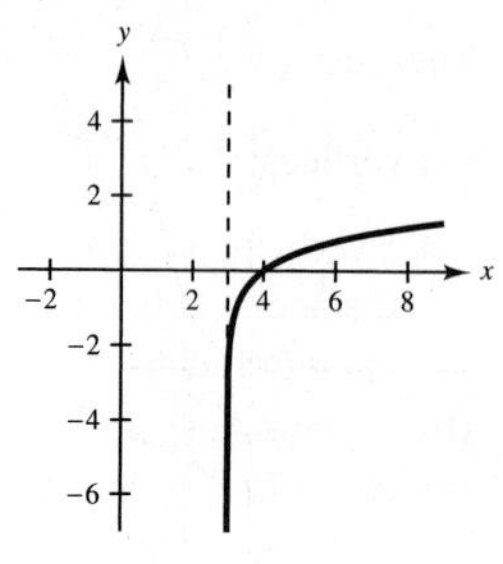

Mid-Chapter Quiz *(p. 548)*

13.
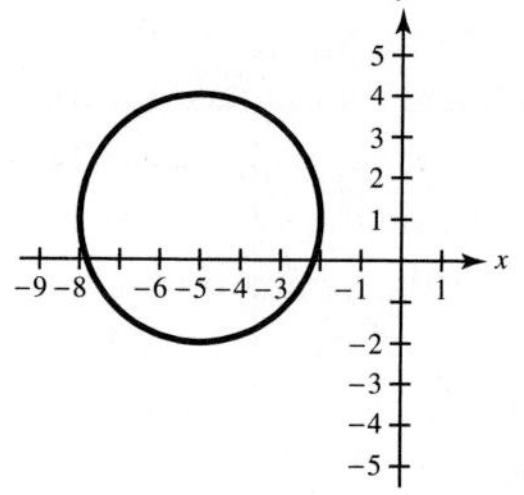

14.
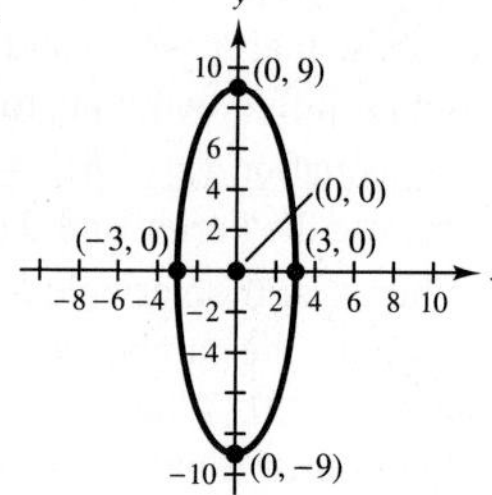

15.
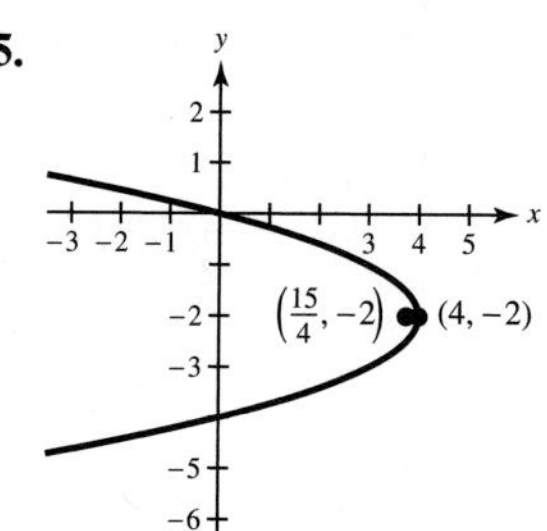

16.
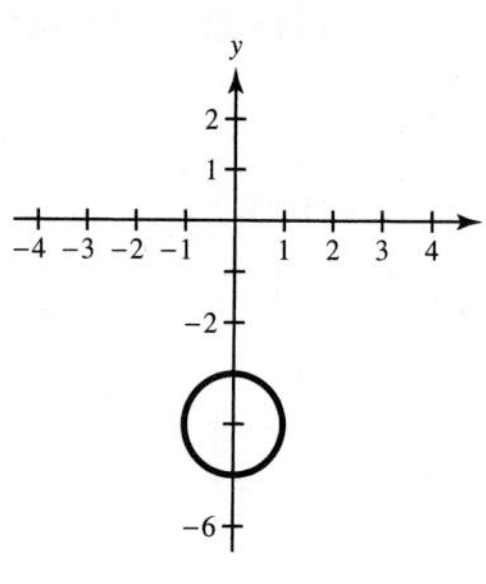

17.
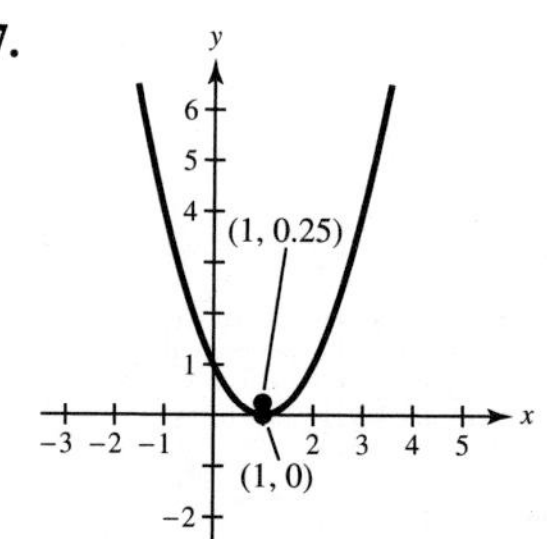

18.
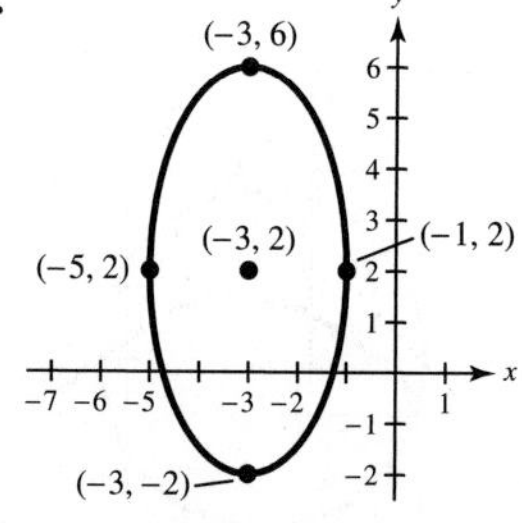

Section 10.3 *(pp. 550–557)*

1. $\dfrac{y^2}{a^2} - \dfrac{x^2}{b^2} = 1$ **2.** $\dfrac{x^2}{a^2} - \dfrac{y^2}{b^2} = 1$

3. $\dfrac{x^2}{a^2} - \dfrac{y^2}{b^2} = 1$ **4.** $\dfrac{y^2}{a^2} - \dfrac{x^2}{b^2} = 1$

5.
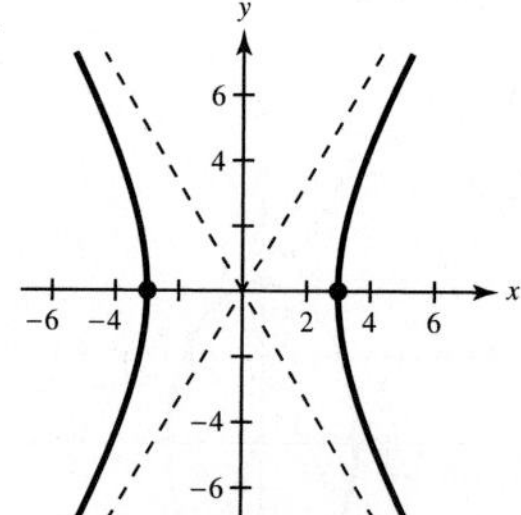

6.
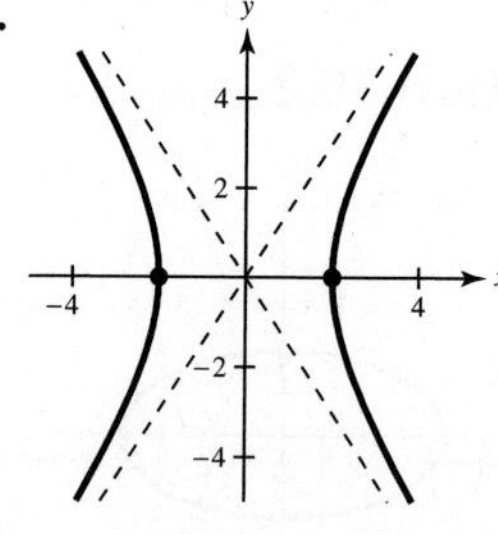

7.
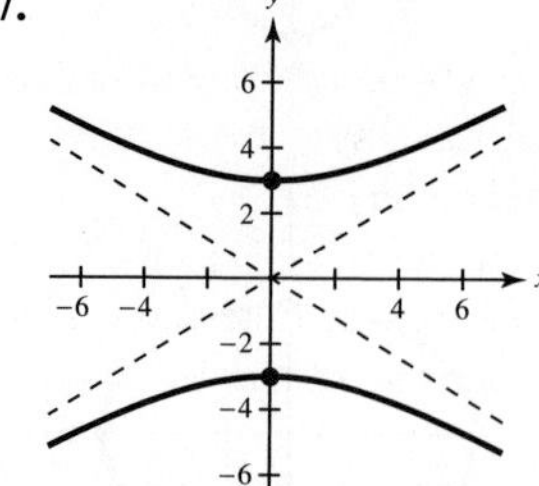

8.
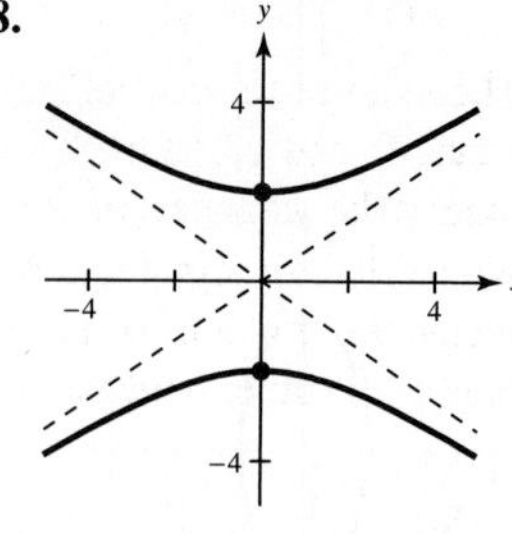

9.
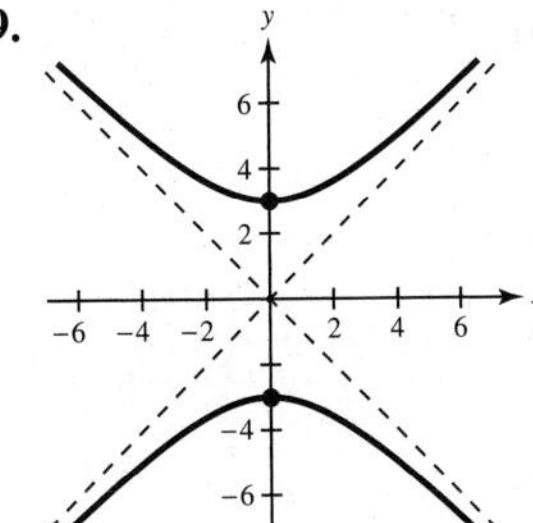

10.
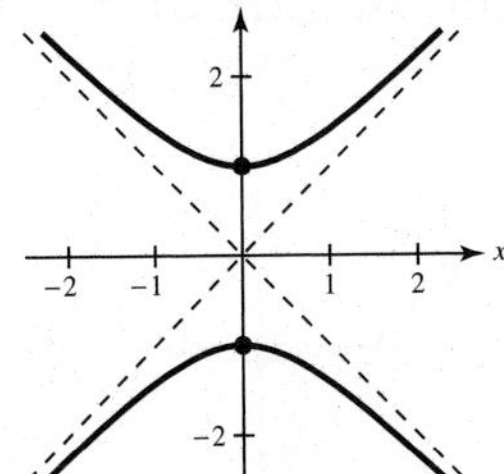

17.
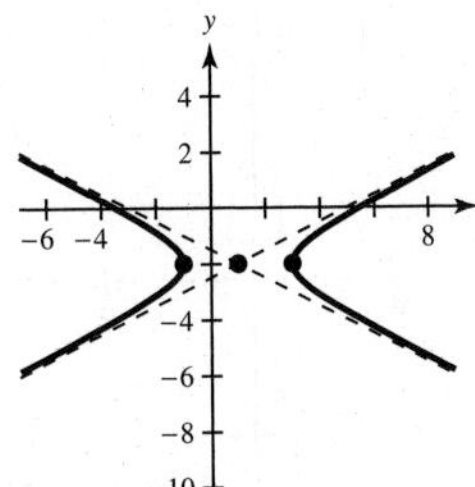

18.
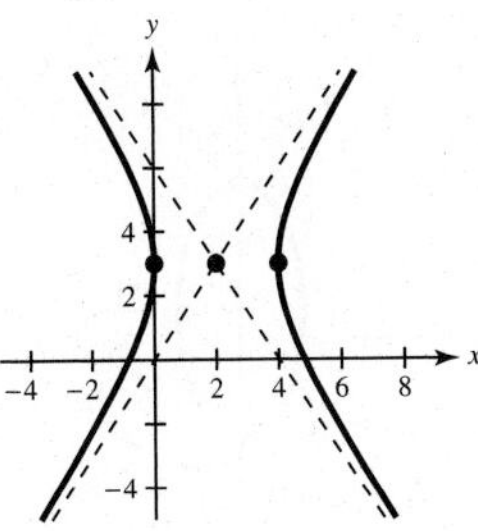

19.
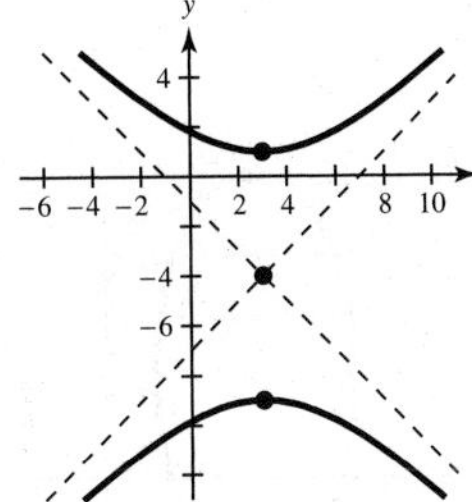

20.
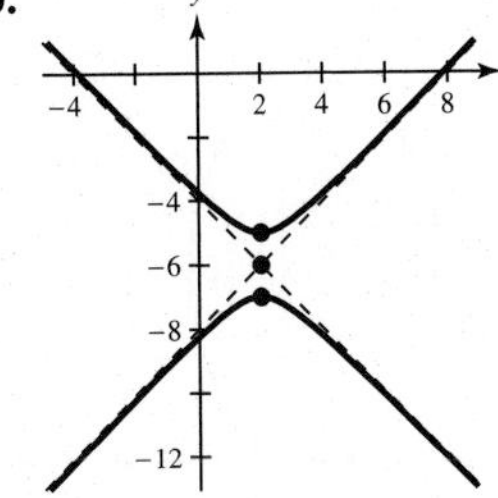

21.
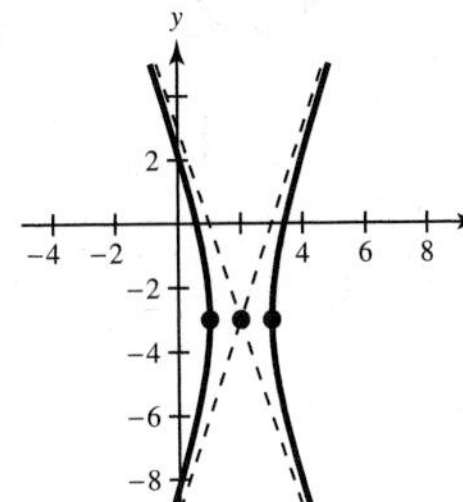

22.
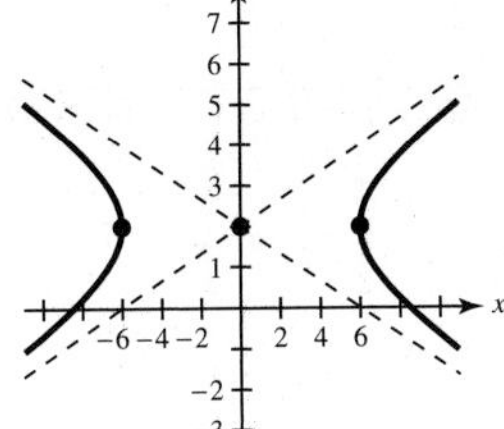

23.
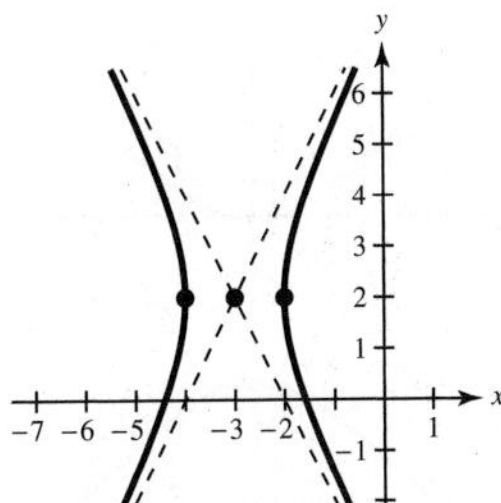

24.
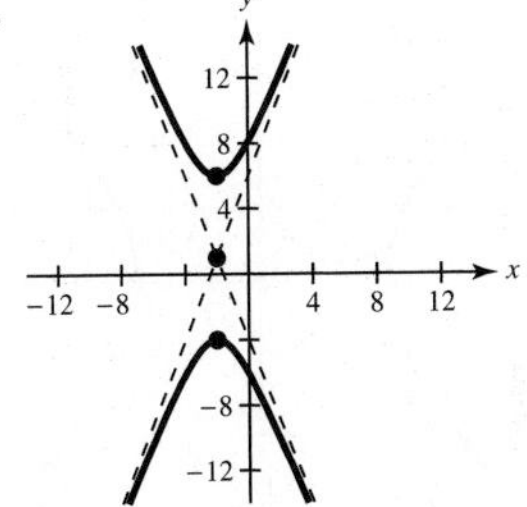

27. The sides of the central rectangle will pass through the points $(\pm a, 0)$ and $(0, \pm b)$. To sketch the asymptotes, draw and extend the diagonals of the central rectangle.

28. Sketch the asymptotes and plot the vertices. Draw the central rectangle. a is half of the distance between the vertices. b is half of the other dimension of the central rectangle.

35.
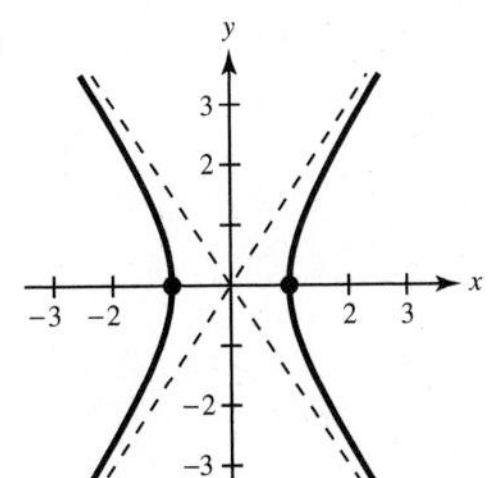

36.
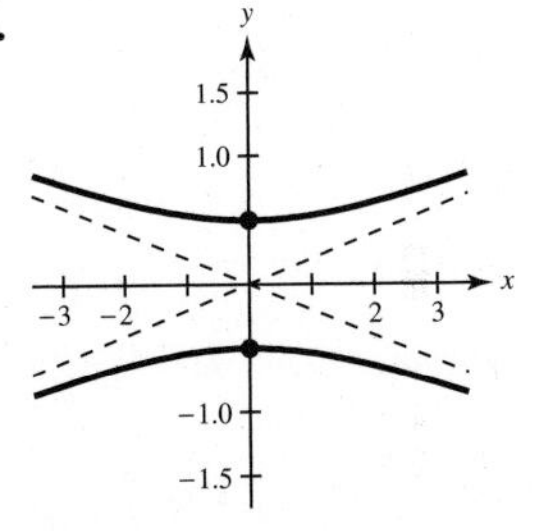

37.
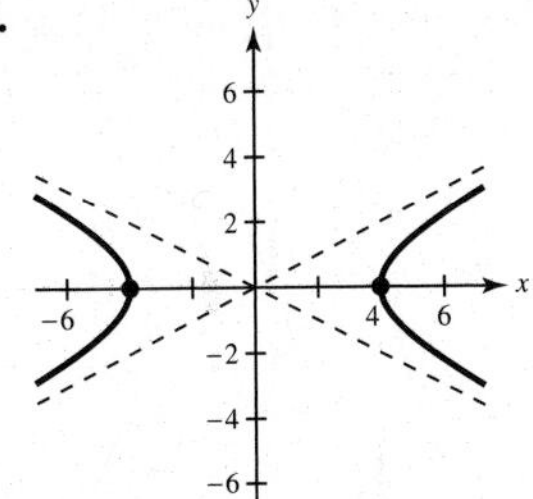

38.
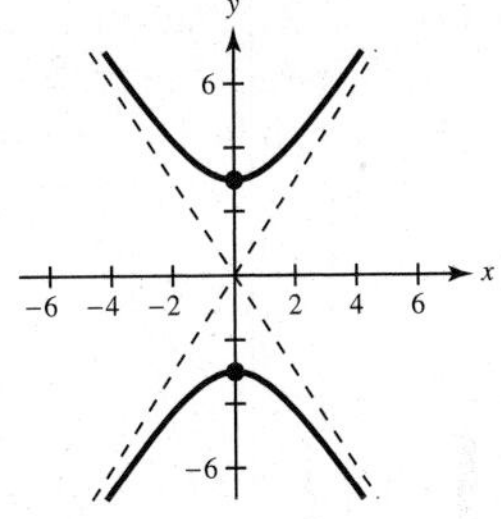

Section 10.4 *(pp. 558–565)*

1.
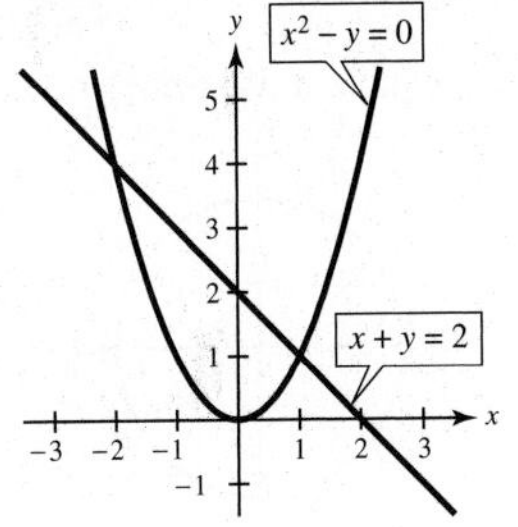

2.
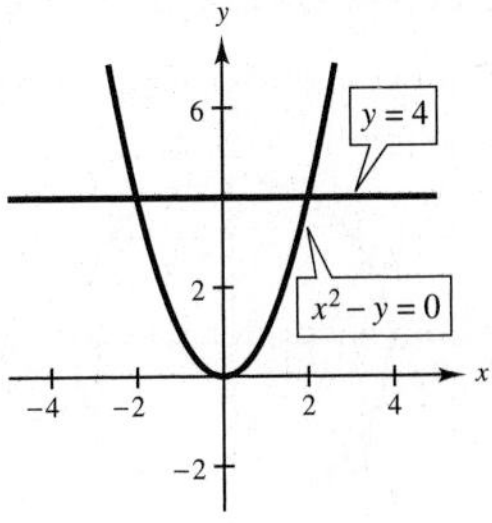

3.
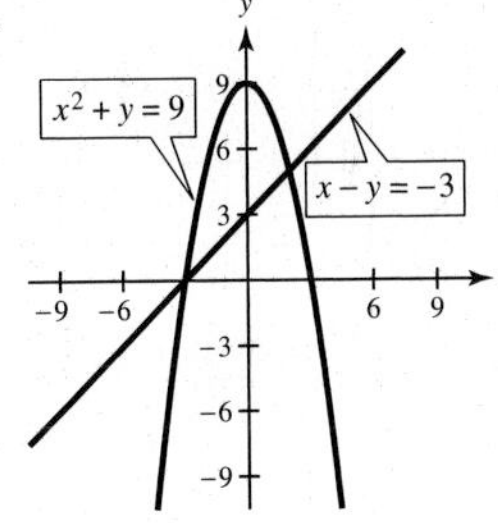

4.
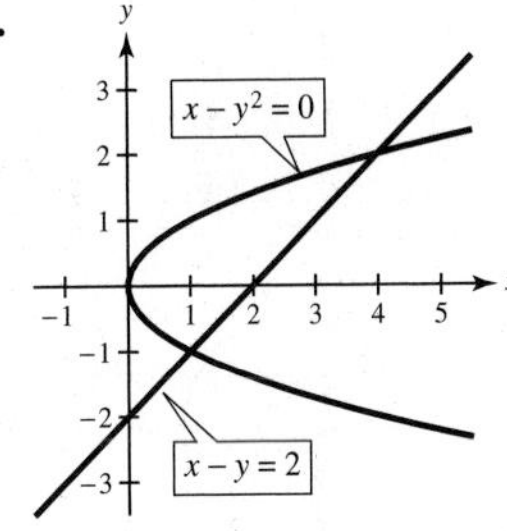

5.
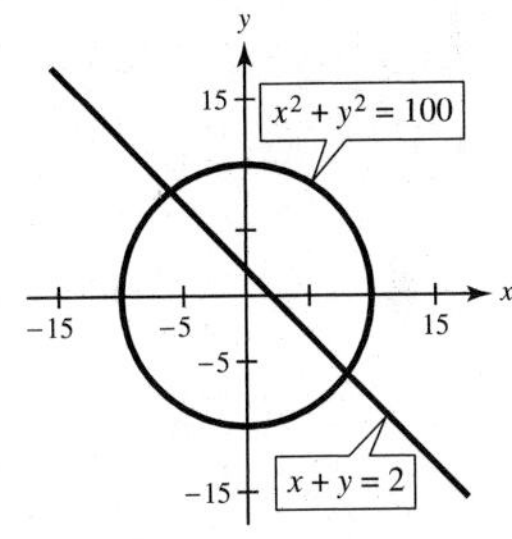

6.
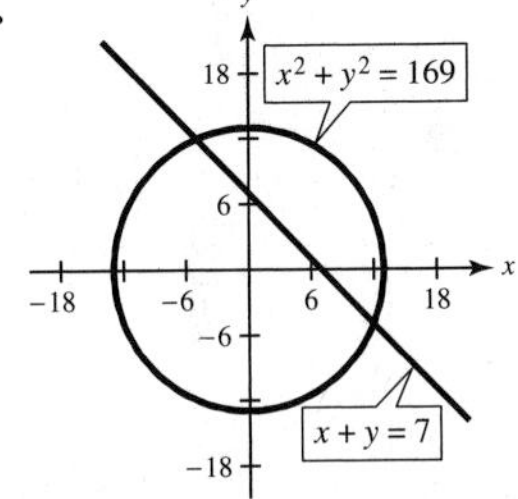

7.
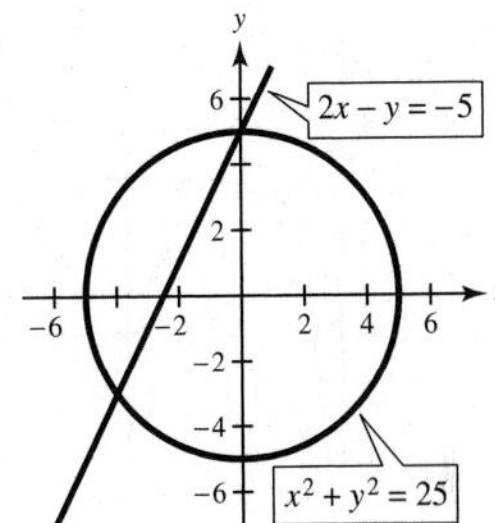

8.
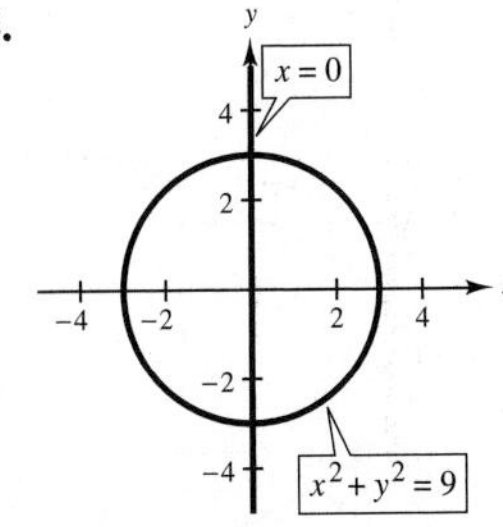

9.
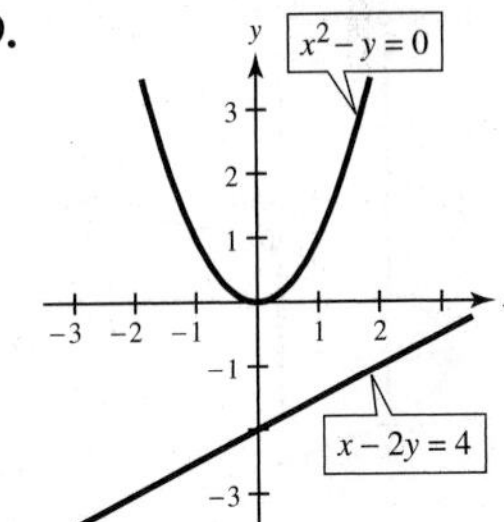

10.
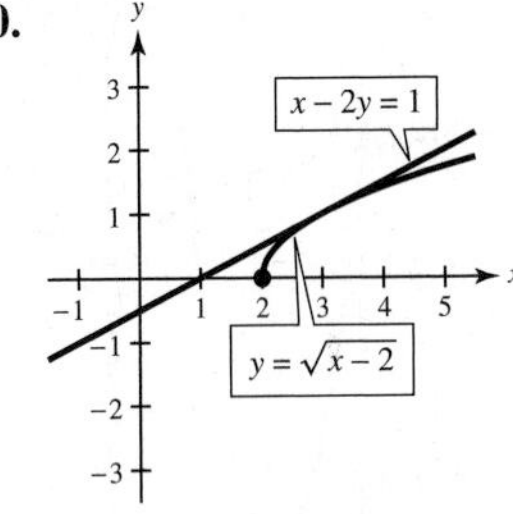

11.

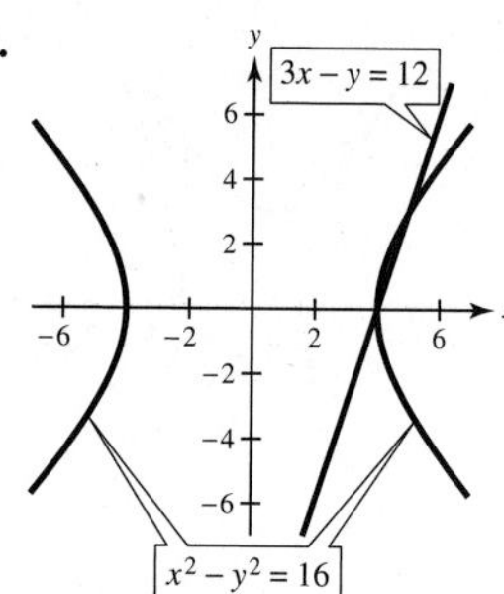

12.

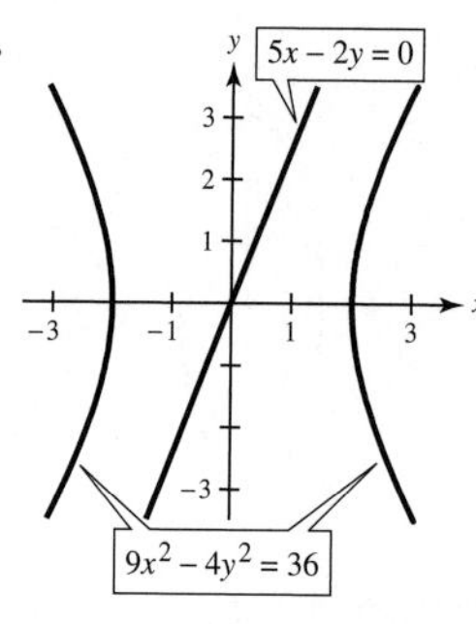

Review Exercises *(pp. 568–571)*

13.

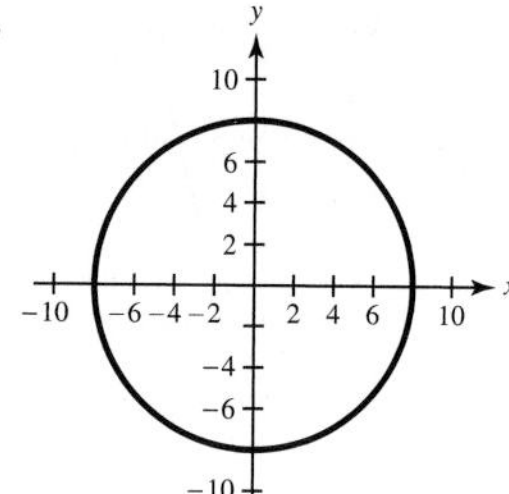

14.

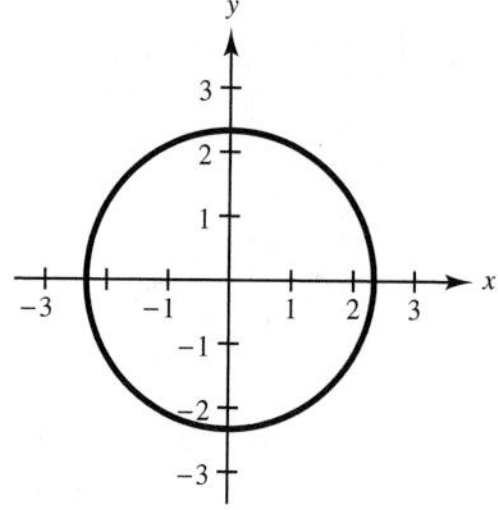

17.

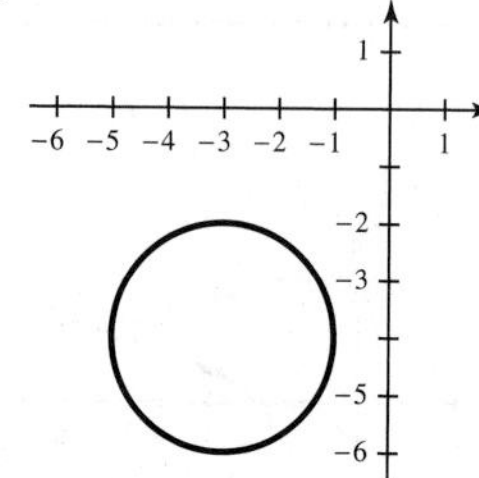

18.

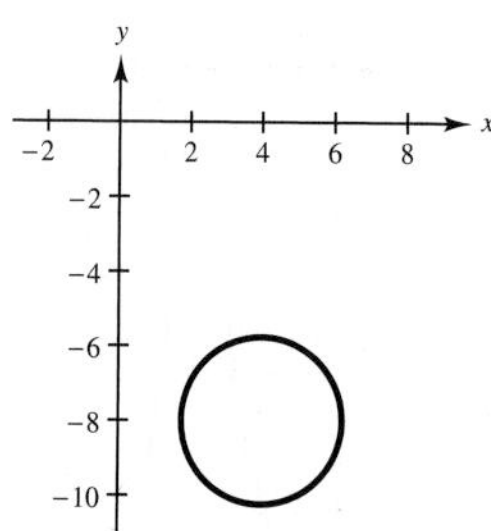

19.

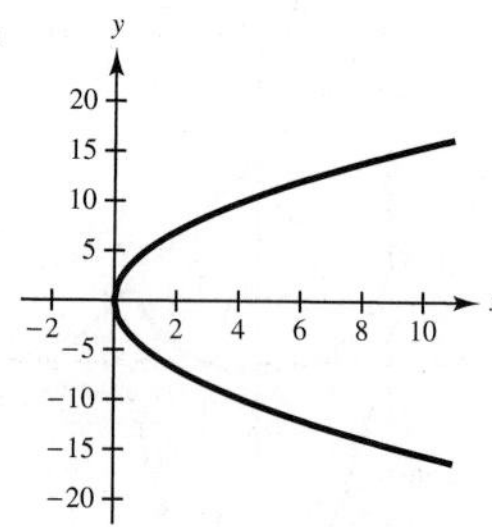

20.

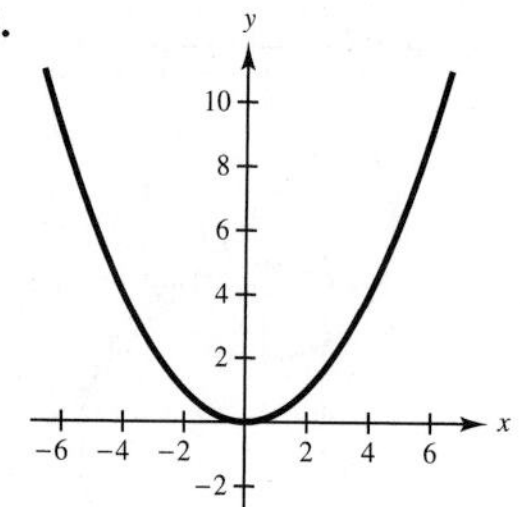

21.

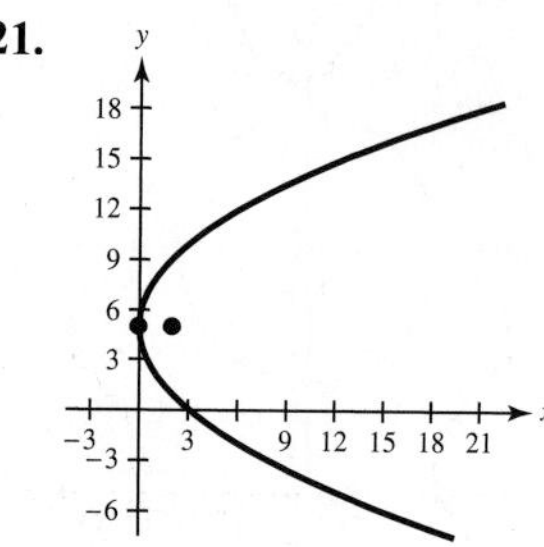

22.

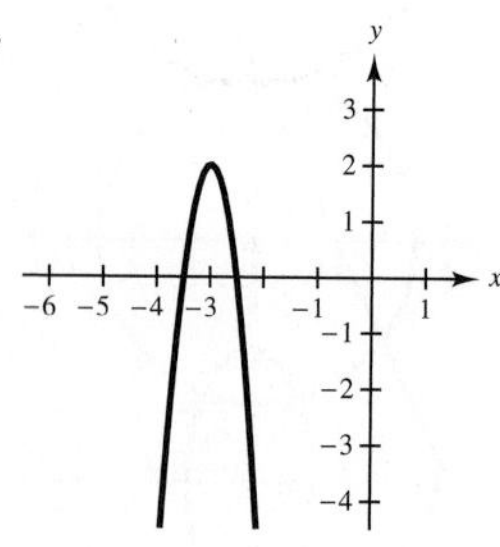

23.

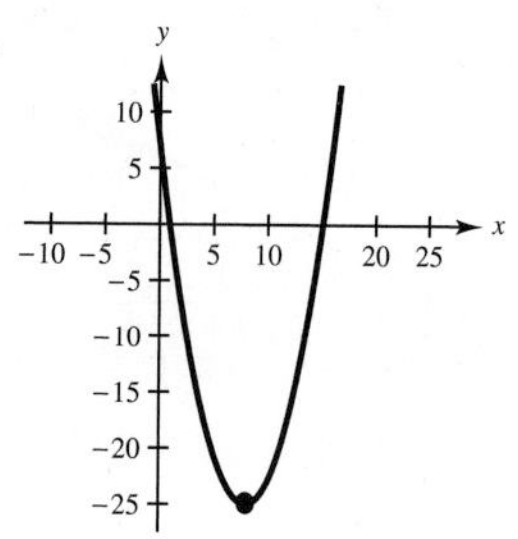

24.

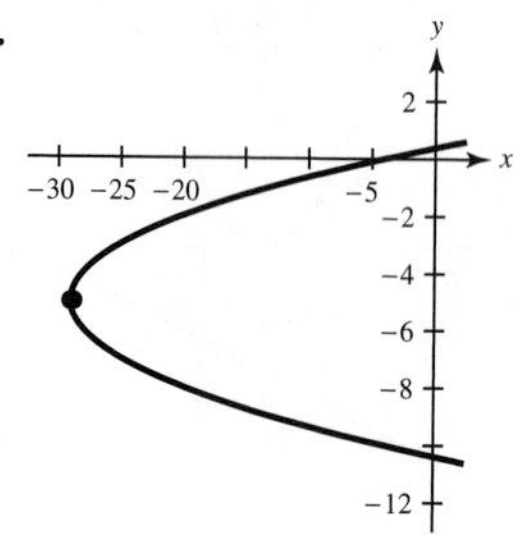

29.

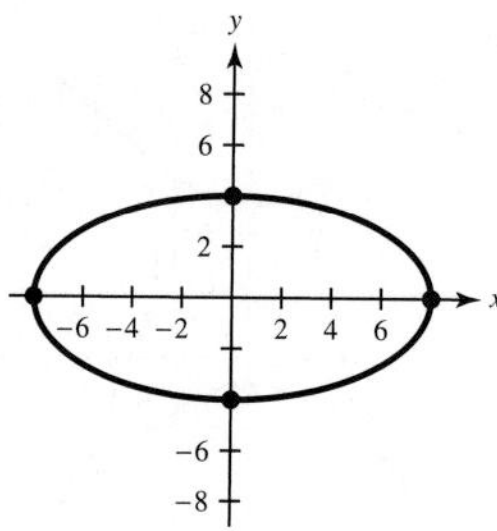

30.

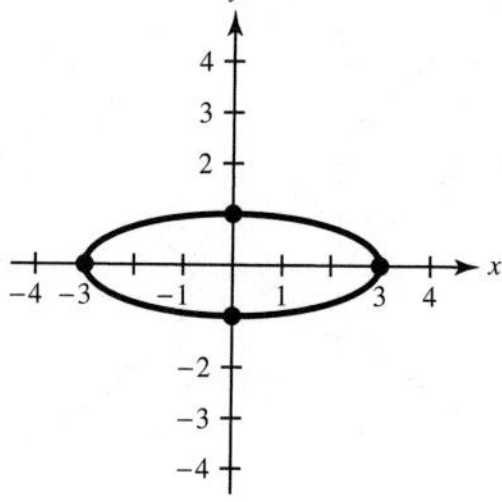

31.

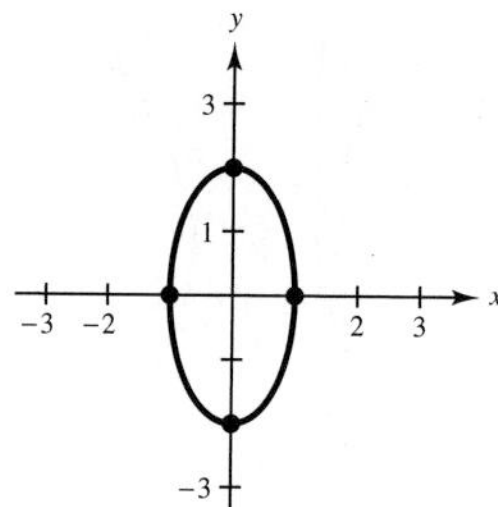

32.

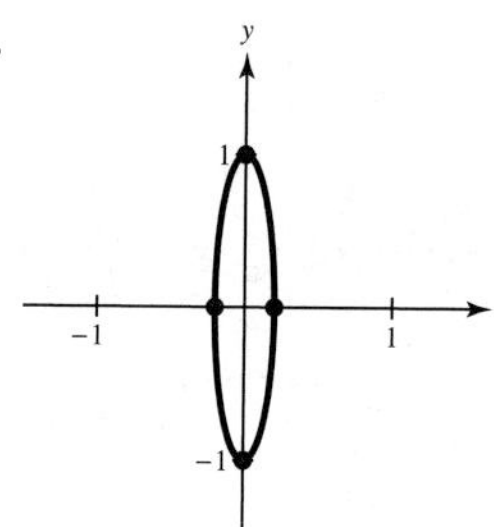

37.

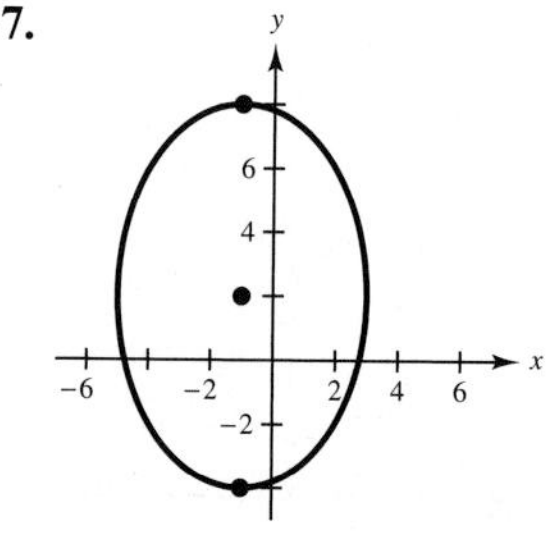

38.

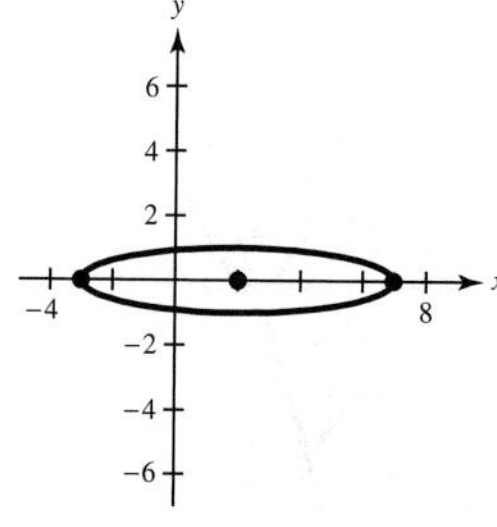

39.

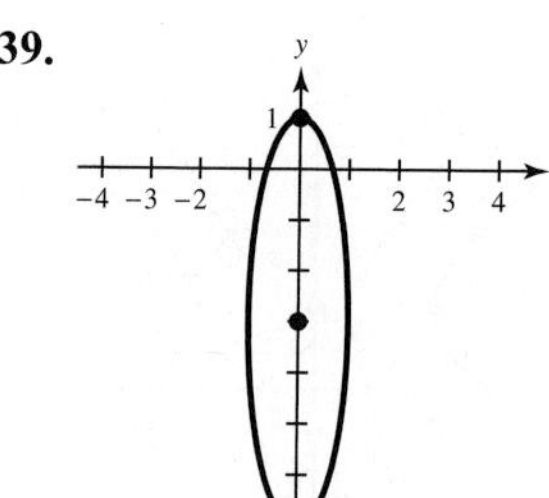

40.

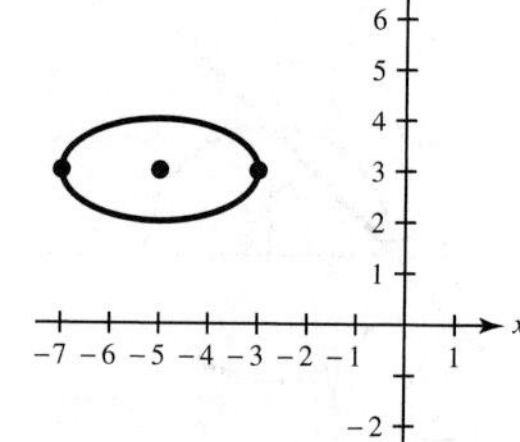

41.

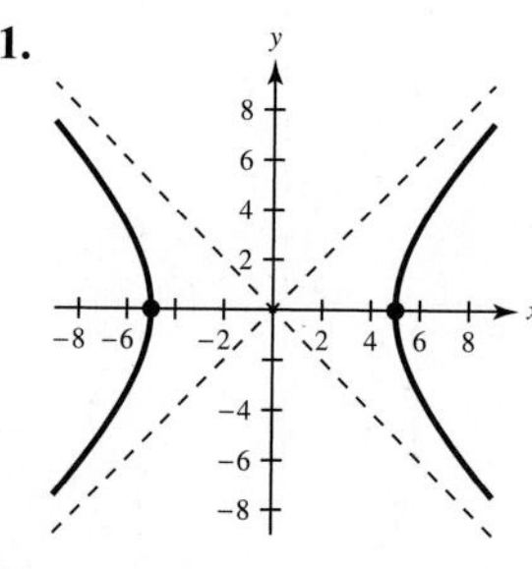

42.

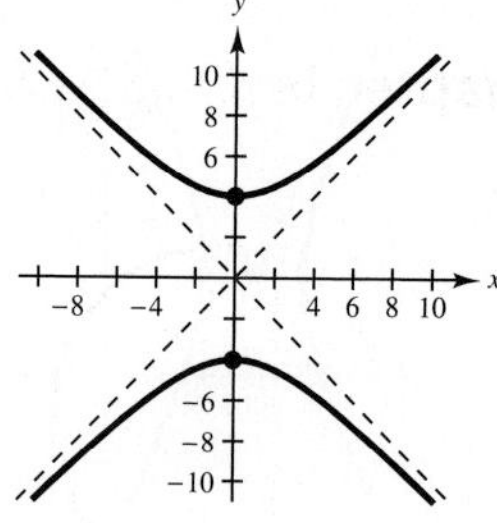

43.

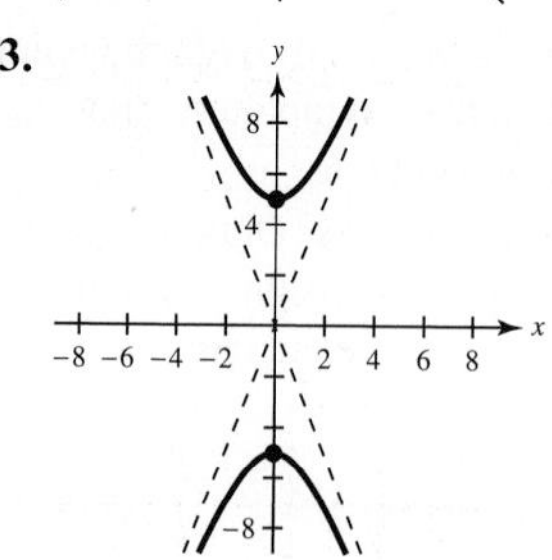

44.

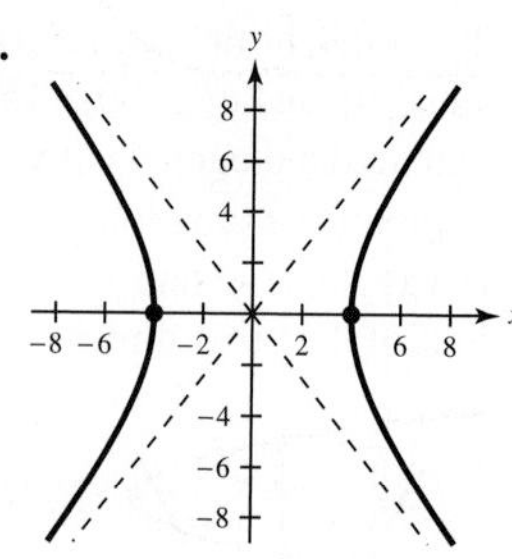

49.

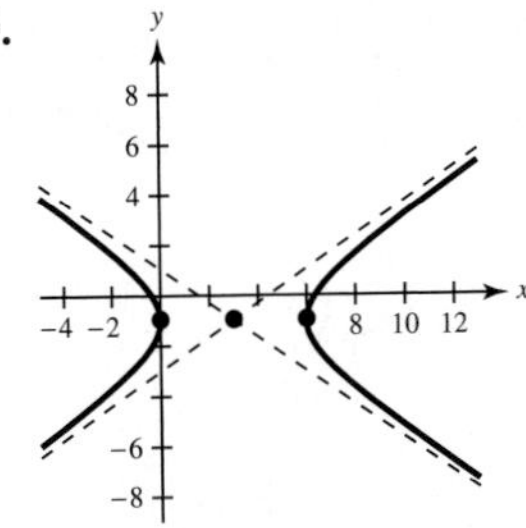

50.

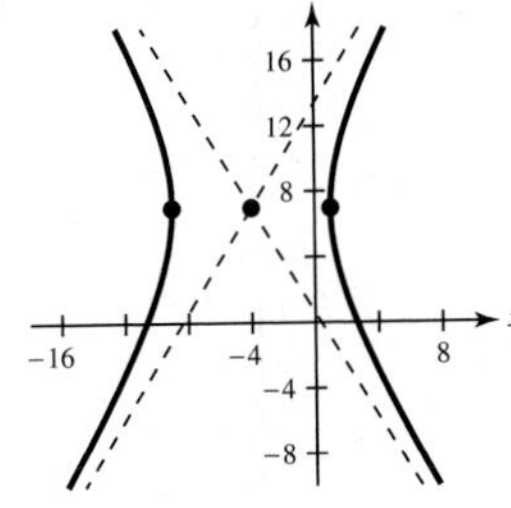

51.

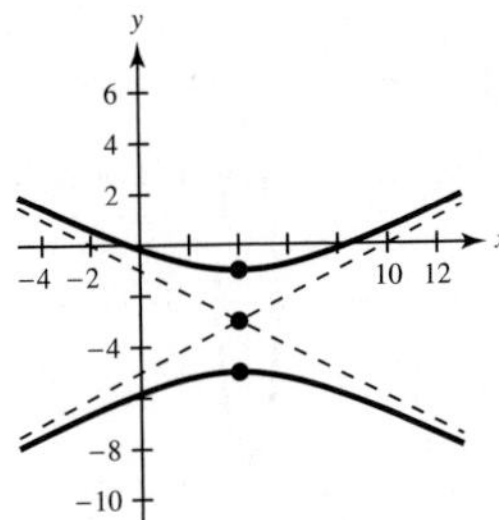

52.

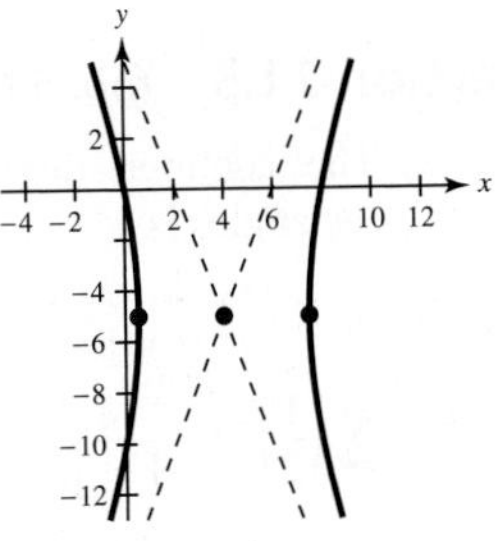

55.

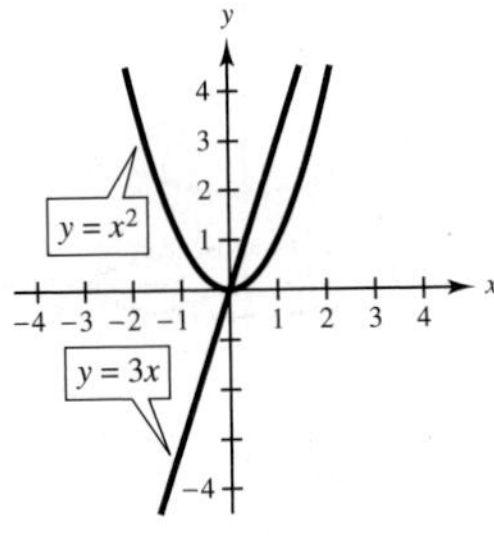

56.

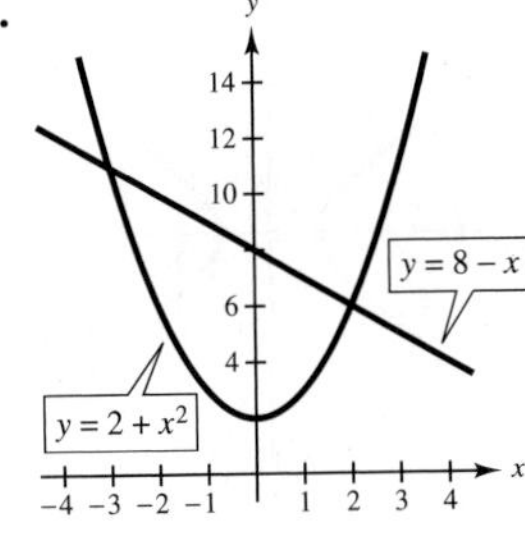

57.

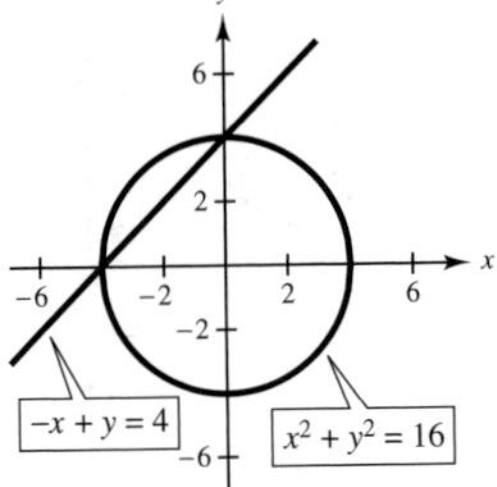

58.

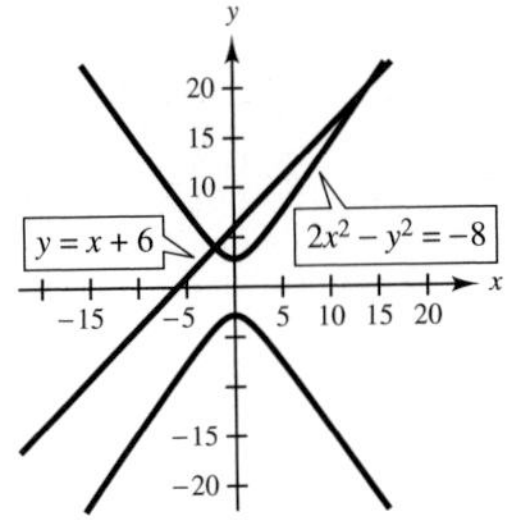

Chapter Test *(p. 572)*

2.

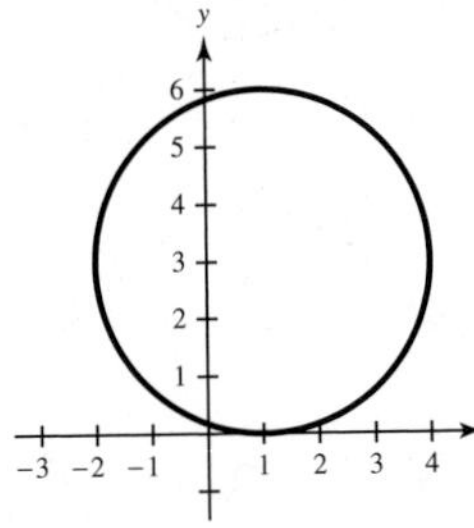

3.

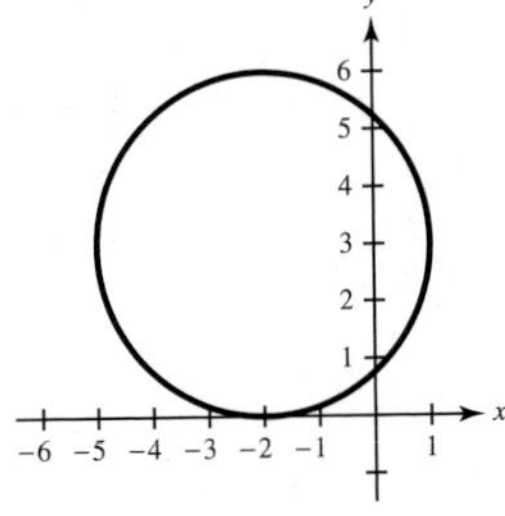

4.

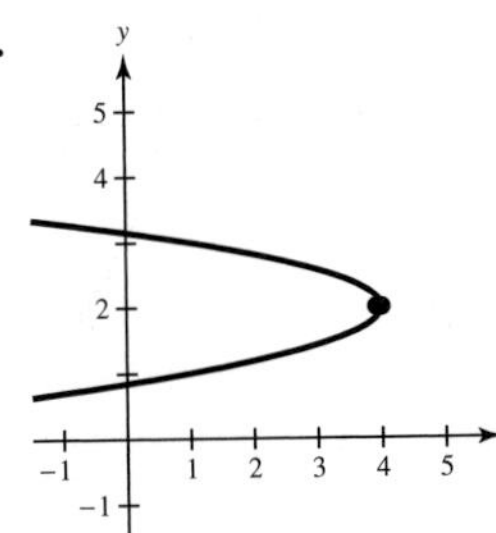

7.

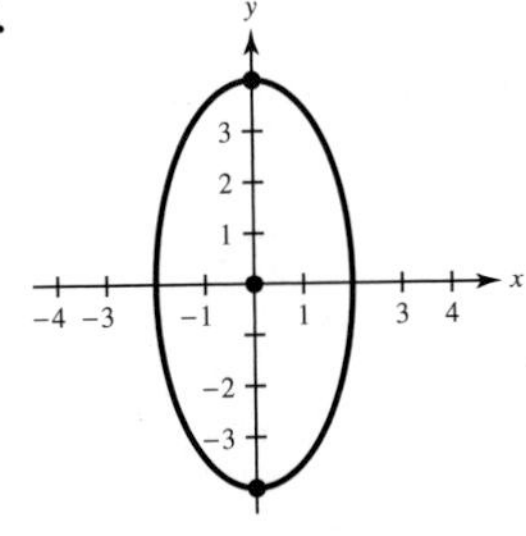

8.

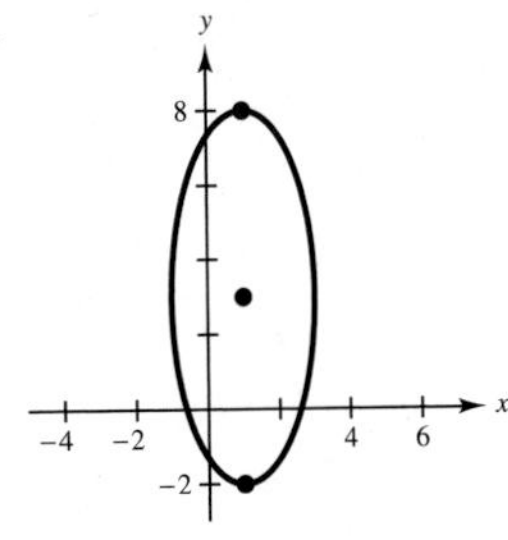

11.

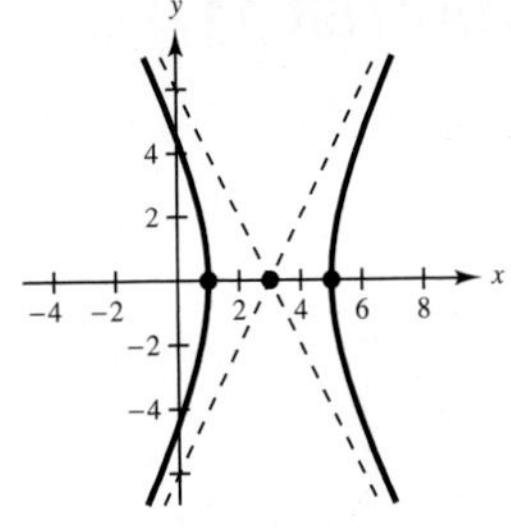

12.

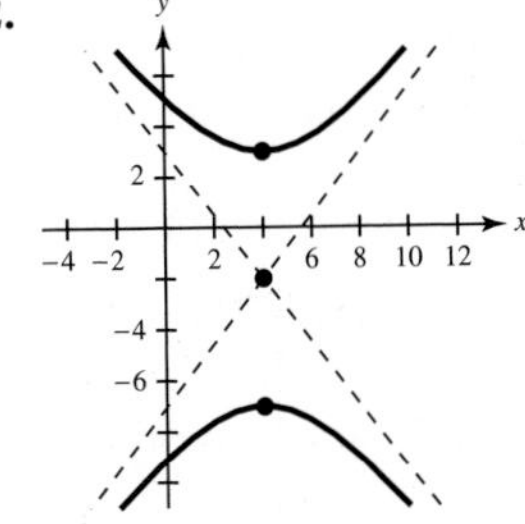

Cumulative Test *(pp. 573–574)*

6.

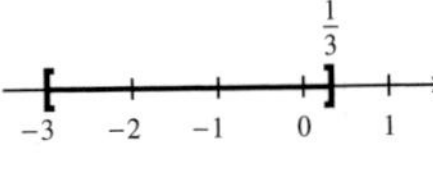

7.

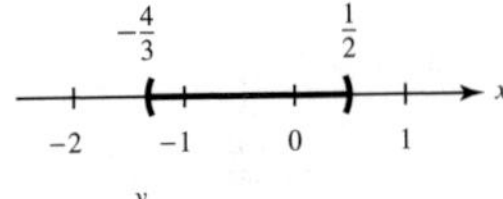

12.

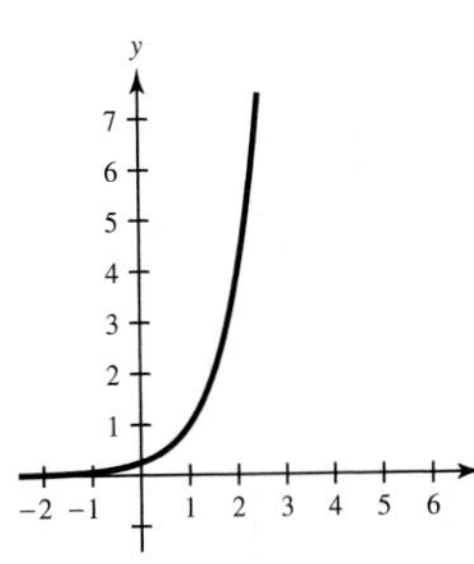

14.

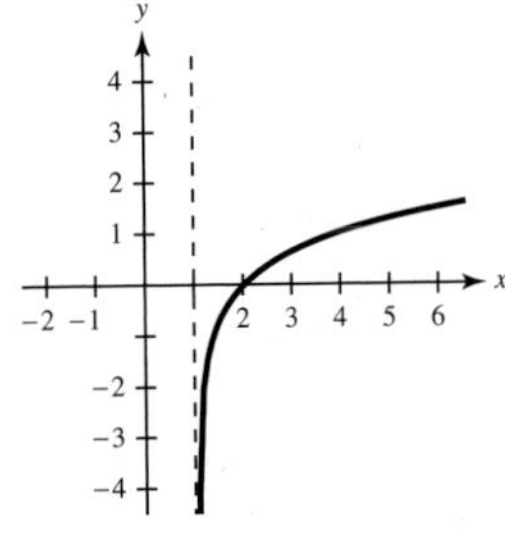

25.

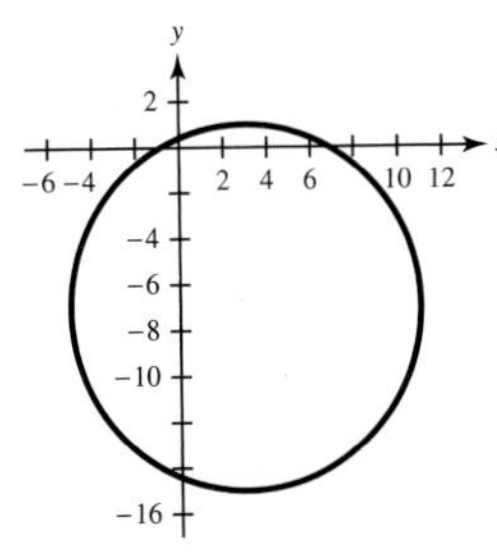

26.

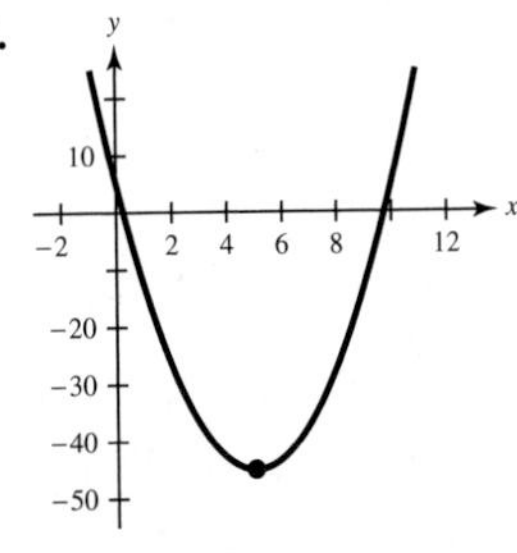

28.

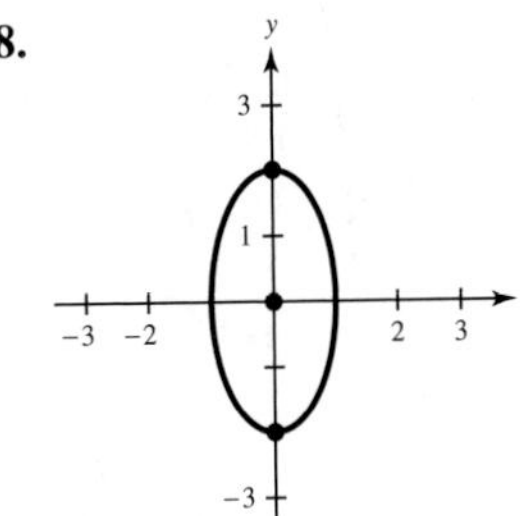

30.

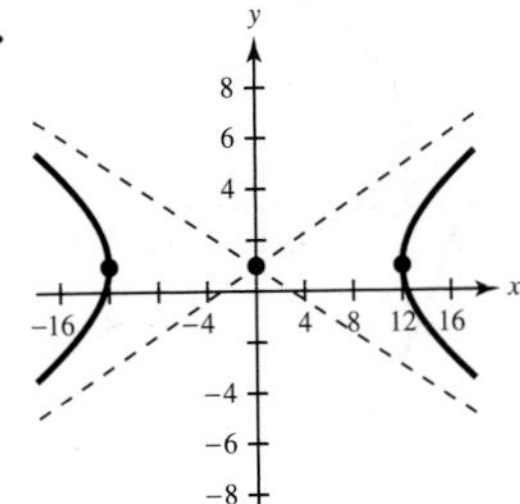

CHAPTER 11

Section 11.1 *(pp. 576–583)*

63. (c)

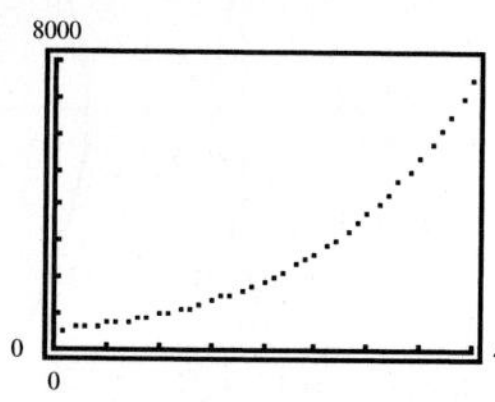

69. $a_n = 4n! = 4[1 \cdot 2 \cdot 3 \cdot 4 \cdot \cdots \cdot (n-1) \cdot n]$

$a_n = (4n)! = 1 \cdot 2 \cdot 3 \cdot 4 \cdot \cdots \cdot (4n-1) \cdot (4n)$

70. $\sum_{i=1}^{4} (i^2 + 2i) = (1 + 2) + (4 + 4) + (9 + 6) + (16 + 8)$

$= (1 + 4 + 9 + 16) + (2 + 4 + 6 + 8)$

$= \sum_{i=1}^{4} i^2 + \sum_{i=1}^{4} 2i$

71. $\sum_{k=1}^{4} 3k = 3 + 6 + 9 + 12$

$= 3(1 + 2 + 3 + 4) = 3\sum_{k=1}^{4} k$

72. $\sum_{j=1}^{4} 2^j = 2^1 + 2^2 + 2^3 + 2^4$

$= 2^{3-2} + 2^{4-2} + 2^{5-2} + 2^{6-2}$

$= \sum_{j=3}^{6} 2^{j-2}$

77.

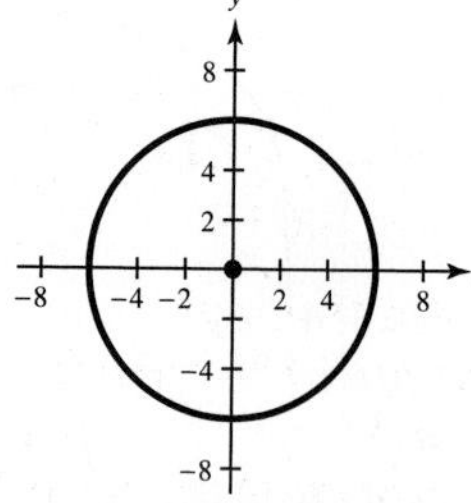

78.

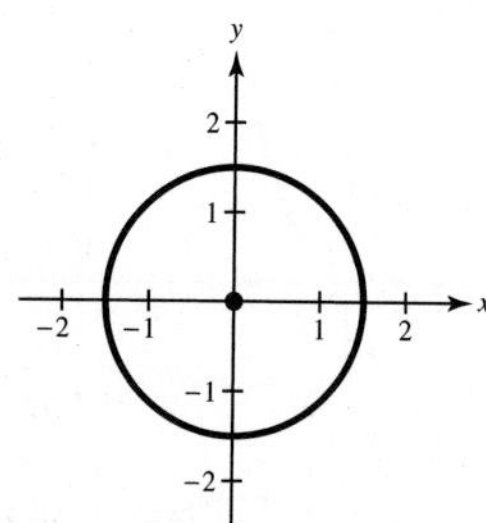

79.

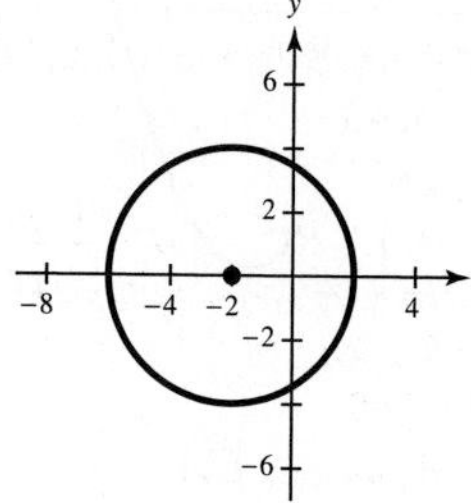

80.

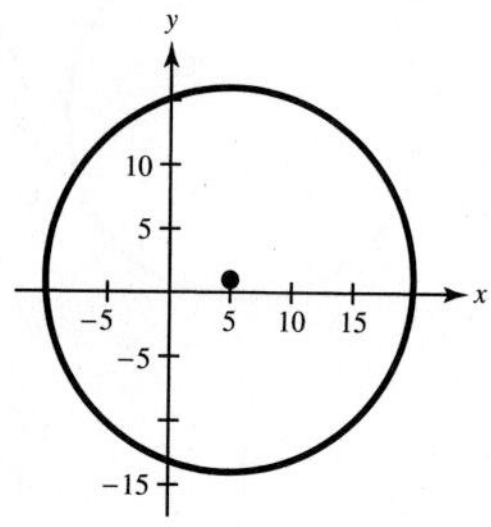

81.

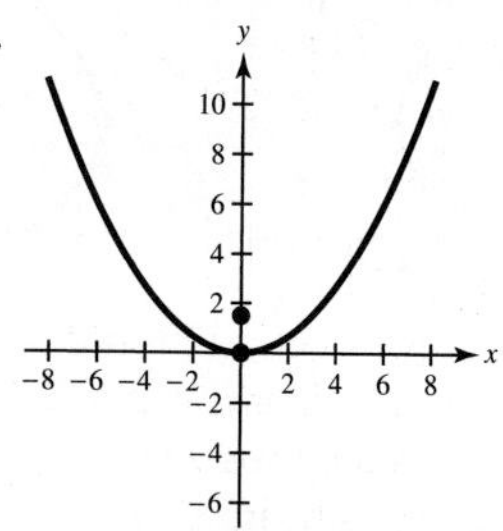

82.

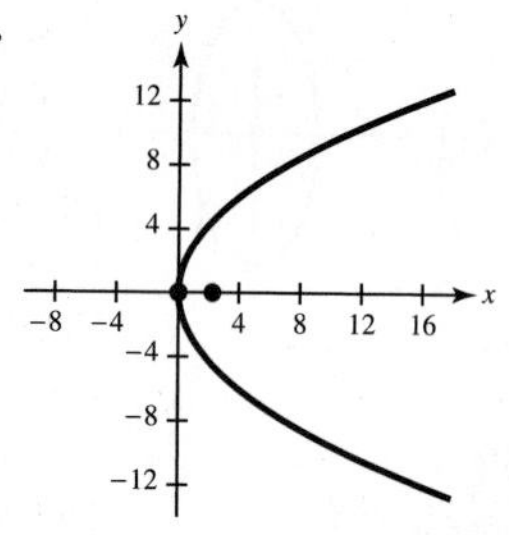

83. **84.**

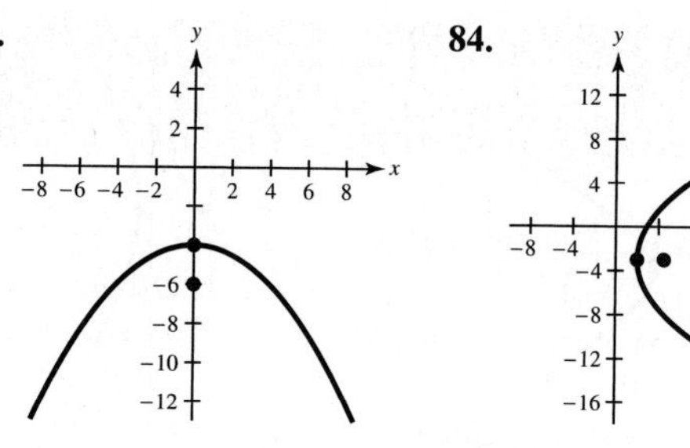

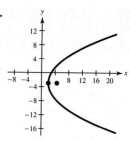

Section 11.3 *(pp. 594–601)*

72. (a) The common ratio of the areas between consecutive rectangles is $\frac{1}{2}$.

(b) $a_n = \left(\frac{1}{2}\right)^n$

(c) $\sum_{i=1}^{\infty} \left(\frac{1}{2}\right)^n = \frac{1/2}{1 - (1/2)} = \frac{1/2}{1/2} = 1$

79.

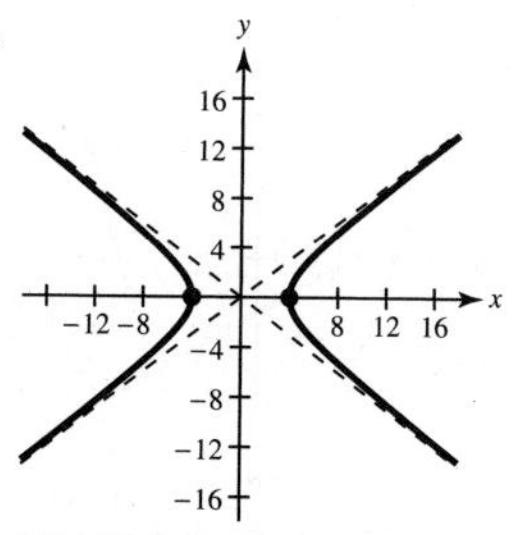

Vertices: $(\pm 4, 0)$

Asymptotes: $\pm\frac{3}{4}x$

80.

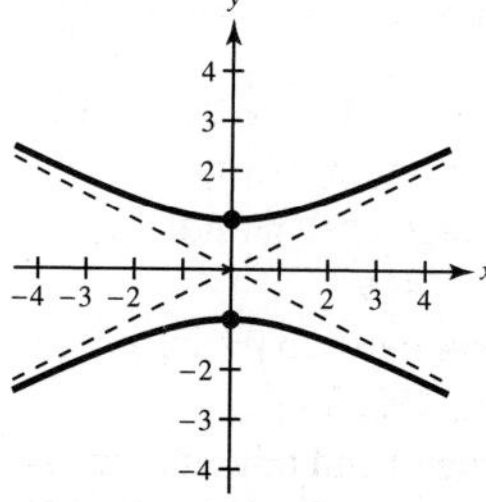

Vertices: $(0, \pm 1)$

Asymptotes: $y = \pm\frac{1}{2}x$

Index of Applications

Biology and Life Sciences

Business

Chemistry and Physics

Construction

Consumer

Geometry

Interest Rates

Miscellaneous

Time and Distance

U.S. Demographics

Index

* Terms that appear in the Math Help feature at AlgebraWithinReach.com

* Terms that appear in the Math Help feature at AlgebraWithinReach.com

* Terms that appear in the Math Help feature at AlgebraWithinReach.com

* Terms that appear in the Math Help feature at AlgebraWithinReach.com

* Terms that appear in the Math Help feature at AlgebraWithinReach.com

Q

R

* Terms that appear in the Math Help feature at AlgebraWithinReach.com

S

* Terms that appear in the Math Help feature at AlgebraWithinReach.com

* Terms that appear in the Math Help feature at AlgebraWithinReach.com

Common Formulas

Distance

$d = rt$

d = distance traveled
t = time
r = rate

Temperature

$F = \frac{9}{5}C + 32$

F = degrees Fahrenheit
C = degrees Celsius

Simple Interest

$I = Prt$

I = interest
P = principal
r = annual interest rate
t = time in years

Compound Interest

$A = P\left(1 + \frac{r}{n}\right)^{nt}$

A = balance
P = principal
r = annual interest rate
n = compoundings per year
t = time in years

Coordinate Plane: Midpoint Formula

Midpoint of line segment joining (x_1, y_1) and (x_2, y_2)

$$\left(\frac{x_1 + x_2}{2}, \frac{y_1 + y_2}{2}\right)$$

Coordinate Plane: Distance Formula

d = distance between points (x_1, y_1) and (x_2, y_2)

$$d = \sqrt{(x_2 - x_1)^2 + (y_2 - y_1)^2}$$

Quadratic Formula

Solutions of $ax^2 + bx + c = 0$

$$x = \frac{-b \pm \sqrt{b^2 - 4ac}}{2a}$$

Rules of Exponents

(Assume $a \neq 0$ and $b \neq 0$.)

$a^0 = 1$ $\qquad$ $a^m \cdot a^n = a^{m+n}$

$(ab)^m = a^m \cdot b^m$ $\qquad$ $(a^m)^n = a^{mn}$

$\frac{a^m}{a^n} = a^{m-n}$ $\qquad$ $\left(\frac{a}{b}\right)^m = \frac{a^m}{b^m}$

$a^{-n} = \frac{1}{a^n}$ $\qquad$ $\left(\frac{a}{b}\right)^{-n} = \frac{b^n}{a^n}$

Basic Rules of Algebra

Commutative Property of Addition

$a + b = b + a$

Commutative Property of Multiplication

$ab = ba$

Associative Property of Addition

$(a + b) + c = a + (b + c)$

Associative Property of Multiplication

$(ab)c = a(bc)$

Left Distributive Property

$a(b + c) = ab + ac$

Right Distributive Property

$(a + b)c = ac + bc$

Additive Identity Property

$a + 0 = 0 + a = a$

Multiplicative Identity Property

$a \cdot 1 = 1 \cdot a = a$

Additive Inverse Property

$a + (-a) = 0$

Multiplicative Inverse Property

$a \cdot \frac{1}{a} = 1, \quad a \neq 0$

Properties of Equality

Addition Property of Equality

If $a = b$, then $a + c = b + c$.

Multiplication Property of Equality

If $a = b$, then $ac = bc$

Cancellation Property of Addition

If $a + c = b + c$, then $a = b$.

Cancellation Property of Multiplication

If $ac = bc$, and $c \neq 0$, then $a = b$.

Zero Factor Property

If $ab = 0$, then $a = 0$ or $b = 0$.